Magnetohydrodynamics of the Sun

Magnetohydrodynamics of the Sun describes the subtle and complex interaction between the Sun's plasma atmosphere and its magnetic field, which is responsible for many fascinating dynamic phenomena. Chapters cover the generation of the Sun's magnetic field by dynamo action, magnetoconvection and the nature of photospheric flux tubes such as sunspots, the heating of the outer atmosphere by waves or reconnection, the structure of prominences, the nature of eruptive instability and magnetic reconnection in solar flares and coronal mass ejections, and the acceleration of the solar wind by reconnection or wave-turbulence. Developed for a graduate course at St Andrews University, this advanced textbook provides a detailed account of our progress towards answering the key unsolved puzzles in solar physics. It is essential reading for graduate students and researchers in solar physics and related fields of astronomy, plasma physics and fluid dynamics. Problem sets and other resources are available at www.cambridge.org/9780521854719.

Eric Priest was elected Fellow of the Royal Society of Edinburgh in 1985, of the Norwegian Academy of Sciences and Letters in 1994, of the Royal Society in 2002 and of the European Academy of Sciences in 2005. He has delivered many named lectures, including the James Arthur Prize Lecture at Harvard and the Lindsay Memorial Lecture at the Goddard Space Flight Center. He was awarded the Hale Prize of the American Astronomical Society, only the second time it has been awarded to a British scientist, and the Gold Medal of the Royal Astronomical Society. Priest created and led an extremely active and successful group at St Andrews, served three times on UK Research Assessment Panels and, as Co-Chair of the PPARC Science Committee, he played an important role when the UK joined the European Southern Observatory.

MAGNETOHYDRODYNAMICS OF THE SUN

Eric Priest
University of St Andrews

CAMBRIDGE
UNIVERSITY PRESS

CAMBRIDGE
UNIVERSITY PRESS

Shaftesbury Road, Cambridge CB2 8EA, United Kingdom

One Liberty Plaza, 20th Floor, New York, NY 10006, USA

477 Williamstown Road, Port Melbourne, VIC 3207, Australia

314–321, 3rd Floor, Plot 3, Splendor Forum, Jasola District Centre, New Delhi – 110025, India

103 Penang Road, #05–06/07, Visioncrest Commercial, Singapore 238467

Cambridge University Press is part of Cambridge University Press & Assessment, a department of the University of Cambridge.

We share the University's mission to contribute to society through the pursuit of education, learning and research at the highest international levels of excellence.

www.cambridge.org
Information on this title: www.cambridge.org/9780521854719

First published 2014
3rd printing 2021

A catalogue record for this publication is available from the British Library

Library of Congress Cataloging in Publication data
Priest, E. R. (Eric Ronald), 1943–
Magnetohydrodynamics of the Sun / Eric Priest.
pages cm
Includes bibliographical references and index.
ISBN 978-0-521-85471-9 (hardback)
1. Sun. 2. Solar activity. 3. Magnetohydrodynamics. 4. Solar magnetic fields.
5. Astrophysics. I. Title. II. Title: Magnetohydrodynamics of the Sun.
QB524.P75 2014
523.7′2–dc23 2013029372

ISBN 978-0-521-85471-9 Hardback

Additional resources for this publication at www.cambridge.org/9780521854719

To Clare,

our wonderful children

(Andrew, Matthew, David and Naomi),

my ever-helpful brother (Gerry) and sister (June)

and amazing 97-year-old mum (Olive)

Contents

Preface *page* xvii

1 A Description of the Sun 1
1.1 Brief History 1
1.2 Overall Properties 4
1.2.1 The Structure of the Solar Interior 5
1.2.2 The Structure of the Solar Atmosphere 6
1.3 The Solar Interior 10
1.3.1 The Solar Core 10
1.3.2 A Model for the Solar Interior 10
1.3.3 Convection Zone 12
1.3.4 Helioseismology and 5-Minute Oscillations 15
1.4 The Photosphere 21
1.4.1 Granulation, Meso- and Super-Granulation 21
1.4.2 Magnetic Field 25
1.4.3 The VAL Model 30
1.5 The Chromosphere and Transition Region 31
1.5.1 Quiet (Internetwork Non-magnetic) Chromosphere 31
1.5.2 Magnetic (Network or Plage) Chromosphere 31
1.6 The Corona 34
1.6.1 The Structure of the Corona from Eclipses and X-ray Images 36
1.6.2 Coronal Loops, X-ray Bright Points and X-ray Jets 37
1.6.3 Coronal Holes 38
1.6.4 Solar Wind 40
1.7 Active Regions, Sunspots and the Solar Cycle 44
1.7.1 Active Regions 45
1.7.2 Sunspots 47
1.7.3 Solar Cycle 53
1.8 Prominences 57
1.8.1 Introduction 57
1.8.2 Plasma and Magnetic Properties 57
1.8.3 Structure 59
1.8.4 Development 62
1.8.5 Eruption 62

1.9 Coronal Mass Ejections (CMEs) and Solar Flares 63
1.9.1 Coronal Mass Ejections 63
1.9.2 Solar Flares 65

2 The Basic Equations of Magnetohydrodynamics (MHD) 74
2.1 Electromagnetic Equations 74
2.1.1 Maxwell's Equations 74
2.1.2 Ohm's Law 76
2.1.3 Generalised Ohm's Law 76
2.1.4 Induction Equation 78
2.1.5 Values for Electrical Conductivity and Magnetic Diffusivity 79
2.2 Plasma Equations 80
2.2.1 Mass Continuity 80
2.2.2 Equation of Motion 80
2.2.3 Perfect Gas Law 81
2.3 Energy Equations 82
2.3.1 Different Forms of the Heat Equation 82
2.3.2 Thermal (or Heat) Conduction 83
2.3.3 Radiation 84
2.3.4 Heating 85
2.3.5 Energetics 85
2.4 Summary of Equations 86
2.4.1 Dimensionless Equations 87
2.4.2 Assumptions 87
2.4.3 Reduced Forms of the Equations 88
2.5 Dimensionless Parameters 89
2.6 Consequences of the Induction Equation 91
2.6.1 Diffusive Limit ($R_m \ll 1$) 91
2.6.2 Ideal Limit ($R_m \gg 1$) 92
2.6.3 Non-ideal Flow 94
2.7 The Lorentz Force 96
2.8 Some Theorems 98
2.8.1 Alfvén's Frozen Flux Theorem 98
2.8.2 The Minimum-Energy Theorem for Potential Fields 98
2.8.3 The Minimum-Energy Theorem for Force-Free Fields 98
2.8.4 Woltjer's Minimum-Energy Theorem 98
2.8.5 Cowling's Anti-dynamo Theorem 99
2.8.6 Taylor-Proudman Theorem 99
2.8.7 Ferraro's Law of Isorotation 99
2.8.8 Virial Theorem 99
2.9 Summary of Magnetic Flux Tube Behaviour 100
2.9.1 Definitions 100
2.9.2 General Properties 101
2.9.3 Flux Surfaces for Two-Dimensional or Axisymmetric Fields 103
2.9.4 Flux Tubes in the Solar Atmosphere 104
2.10 Summary of Current Sheet Behaviour 104
2.10.1 Processes of Formation 106
2.10.2 Properties 106

3 Magnetohydrostatics 107
3.1 Introduction 107
3.1.1 Representation of the Magnetic Field 108
3.1.2 Pressure Scale-Height 108
3.1.3 Plasma Structure in a Given Magnetic Field 110
3.2 Structure of Magnetic Flux Tubes (Cylindrically Symmetric) 111
3.2.1 Purely Axial Field 112
3.2.2 Purely Azimuthal Field 112
3.2.3 One-Dimensional Force-Free Flux Tubes 113
3.2.4 One-Dimensional Magnetostatic Flux Tubes 116
3.3 Current-Free (or Potential) Magnetic Fields 117
3.3.1 Potential Minimum-Energy Theorem 117
3.3.2 General Current-Free Solutions 118
3.3.3 Coronal Magnetic Fields – Roughly Potential? 119
3.4 Force-Free Magnetic Fields 120
3.4.1 General Theorems for Force-Free Fields 121
3.4.2 Simple Linear (Constant-α) Solutions 124
3.4.3 General Linear (Constant-α) Solutions 126
3.4.4 Nonlinear (i.e., Non-constant-α) Solutions 129
3.4.5 Numerical Methods for Force-Free Fields 132
3.4.6 Diffusion of Force-Free Fields 134
3.5 Magnetohydrostatic Fields 135
3.5.1 Magnetostatic Fields 135
3.5.2 Effect of Gravity 138
3.5.3 Transforming to Incompressible Field-Aligned Flows 143

4 Waves 144
4.1 Introduction 144
4.1.1 Fundamental Modes 144
4.1.2 Basic Equations 145
4.2 Sound Waves 147
4.3 Magnetically Driven Waves 147
4.3.1 Alfvén Waves (i.e., Shear or Torsional Alfvén Waves) 149
4.3.2 Compressional Alfvén (or Fast-Mode) Waves 151
4.4 Internal Gravity Waves 152
4.5 Inertial Waves 153
4.6 Magnetoacoustic Waves 155
4.7 Acoustic Gravity Waves 157
4.8 Summary of Magnetoacoustic Gravity Waves 158
4.9 Waves in a Non-uniform Medium 159
4.9.1 Waves in a Continuous Planar Magnetic Field 160
4.9.2 Surface Waves on a Magnetic Interface 161
4.9.3 Surface and Body Waves on a Magnetic Slab 162
4.9.4 Waves in a Continuous Twisted Flux Tube 164
4.9.5 Modes of a Uniform Untwisted Flux Tube 165
4.10 MHD Continua 167
4.10.1 Continuous Spectra 167

4.10.2 Initial-Value Problem 168
4.10.3 Quasi-Modes 170
4.11 Waves on the Sun 172
4.11.1 Waves in the Low Atmosphere 172
4.11.2 Coronal Seismology 172

5 Shock Waves 177
5.1 Introduction 177
5.1.1 Formation of a Hydrodynamic Shock 177
5.1.2 Effects of a Magnetic Field 179
5.2 Hydrodynamic Shocks 181
5.3 Perpendicular Shocks 182
5.4 Oblique Shocks 184
5.4.1 Jump Relations 184
5.4.2 Slow-Mode and Fast-Mode Shocks 186
5.4.3 Switch-Off and Switch-On Shocks 187
5.4.4 Intermediate Wave 188

6 Magnetic Reconnection 189
6.1 Introduction 189
6.1.1 Brief History 190
6.1.2 Overview of Reconnection Concepts 191
6.2 Two-Dimensional Null Points 193
6.2.1 2D Null Point Structure 193
6.2.2 Collapse of 2D Null Points 193
6.3 Current Sheet Formation 196
6.3.1 Planar Motion in 2D Potential Fields 197
6.3.2 Current Sheets in 3D and Non-potential Fields 199
6.3.3 Creation of Sheets at Separatrices by Shearing 202
6.3.4 Braiding by Random Footpoint Motions 203
6.4 Magnetic Annihilation 204
6.4.1 Equations for Steady Annihilation and 2D Reconnection 204
6.4.2 Diffusion and Advection of a 1D Current Sheet 204
6.4.3 Stagnation-Point Flow Model (Sonnerup-Priest 1975) 205
6.4.4 Reconnective Annihilation 207
6.5 Slow 2D Reconnection: The Sweet-Parker Mechanism 208
6.5.1 The Basic Sweet-Parker Model (1958) 208
6.5.2 Effect of Pressure Gradients and Compressibility 210
6.5.3 Energetics 211
6.6 Fast 2D Reconnection: Petschek's Mechanism 211
6.6.1 Petschek's Model (1964) 212
6.6.2 Non-steady Petschek Reconnection 214
6.7 Fast 2D Reconnection: Other Families 214
6.7.1 Almost-Uniform Non-potential Reconnection 215
6.7.2 Early MHD Numerical Experiments 218
6.7.3 Non-uniform Reconnection Theory 219
6.8 Unsteady 2D Reconnection by Resistive Instability 219
6.8.1 Tearing-Mode Instability (Furth et al. 1963) 221

6.8.2 Extensions to the Basic Tearing-Mode Analysis 223
6.8.3 Nonlinear Development of Tearing 223
6.9 3D Reconnection: Geometrical Structures 224
6.9.1 3D Null Points 225
6.9.2 Separatrices and Separators 226
6.9.3 Bifurcations 229
6.9.4 Topological Skeletons in the Solar Corona 229
6.9.5 Quasi-Skeletons 232
6.10 3D Reconnection: Magnetic Helicity 233
6.10.1 Equations for Magnetic Helicity 235
6.10.2 Time-Variation of Magnetic Helicity 235
6.10.3 Magnetic Helicity of Simple Structures 236
6.11 3D Reconnection: Flux and Field-Line Conservation 237
6.11.1 An Ideal Plasma 237
6.11.2 A Non-ideal Plasma 238
6.11.3 Topological Conservation Laws 240
6.12 3D Reconnection Concepts 240
6.12.1 Conditions for Reconnection 241
6.12.2 Reconnection in Two Dimensions ($\mathbf{E} \cdot \mathbf{B} = 0$) 241
6.12.3 Failure of Concept of Flux Velocity in 3D ($\mathbf{E} \cdot \mathbf{B} \neq 0$) 241
6.12.4 Differences Between 2D and 3D Reconnection 242
6.12.5 Definition and Classification of Reconnection 243
6.12.6 Magnetic Helicity and 3D Reconnection 247
6.13 3D Reconnection Regimes 247
6.13.1 Ways to Model Reconnection at a Null Point 247
6.13.2 3D Null Regimes (Priest and Pontin, 2009) 250
6.13.3 Separator Reconnection 252
6.13.4 Quasi-Separator Reconnection (Priest-Démoulin 1995) 254

7 Instability 256
7.1 Introduction 256
7.2 Linearised Equations 258
7.3 Normal-Mode Method 260
7.3.1 Example: Rayleigh-Taylor Instability 260
7.4 Energy (or Variational) Method 264
7.4.1 Example: Helical Kink Instability with Line-Tying 266
7.4.2 Use of the Energy (or Variational) Method 269
7.5 Summary of Instabilities 270
7.5.1 Interchange Instability 270
7.5.2 Rayleigh-Taylor Instability 271
7.5.3 Instability of a Cylindrical Tube 272
7.5.4 Hydrodynamic Instability 274
7.5.5 Resistive Instability 275
7.5.6 Convective Instability 278
7.5.7 Radiatively Driven Thermal Instability 279
7.5.8 Other Instabilities 280

8 Dynamo Theory 281
8.1 Introduction 281
8.1.1 Observed Solar Magnetic Field Patterns 281
8.1.2 Preliminary Comments and Terminology 282
8.1.3 History of Dynamo Ideas 284
8.2 Cowling's Anti-dynamo Theorem 287
8.3 Early Turbulent Dynamos 289
8.3.1 Parker's 1955 Model 289
8.3.2 Mean-Field MHD Theory 291
8.4 Flux-Transport Dynamos 294
8.4.1 Babcock-Leighton Ideas 294
8.4.2 Flux-Transport Dynamo Model 294
8.5 Tachocline Dynamos 298
8.5.1 Overshoot Dynamo 298
8.5.2 Interface Dynamo (Parker 1993) 298
8.5.3 The Solar Tachocline 300
8.6 Other Approaches 301
8.6.1 Global Computations 301
8.6.2 Low-Order Models 302
8.6.3 Stellar Dynamos 303
8.7 Future Directions 304

9 Magnetoconvection and Sunspots 306
9.1 Magnetoconvection 306
9.1.1 Physical Effects 306
9.1.2 Linear Stability Analysis 308
9.1.3 Magnetic Flux Concentration and Expulsion 310
9.1.4 Cooling of Sunspots 312
9.2 Intense Flux Tubes 313
9.2.1 Equilibrium of a Thin Flux Tube 314
9.2.2 Dynamics of a Thin Flux Tube 315
9.2.3 Intensification of a Thin Flux Tube by Convective Collapse 315
9.2.4 Spicule Generation 316
9.3 Magnetic Buoyancy 317
9.3.1 Qualitative Effect 318
9.3.2 Magnetic Buoyancy Instability 318
9.3.3 The Emergence of Magnetic Flux 320
9.4 Overall Equilibrium Structure of Sunspots and Pores 322
9.4.1 Magnetohydrostatic Equilibrium 322
9.4.2 Sunspot Stability 324
9.5 Fine Structure of a Sunspot 325
9.5.1 Umbral Fine Structure 325
9.5.2 The Sunspot Penumbra 326
9.5.3 Oscillations 328
9.6 Evolution of a Sunspot 329
9.6.1 Formation 329
9.6.2 Decay 329
9.7 Unified Model 330

10 Heating of the Upper Atmosphere 334
10.1 Introduction 334
10.1.1 Complex Nature of Chromospheric and Coronal Magnetic Field 337
10.2 Chromospheric and Coronal Loops 339
10.2.1 Observations from Space 340
10.2.2 Physical Properties of Coronal-Loop Models 343
10.2.3 Static Energy-Balance Models for Coronal Loops 344
10.2.4 Dynamic Models for Chromospheric and Coronal Loops 349
10.2.5 Deducing the Form of Heating in Coronal Loops 352
10.2.6 Numerical Experiments on Active-Region Heating 355
10.3 Heating by MHD Waves 356
10.3.1 Propagation of Magnetic Waves 357
10.3.2 Chromospheric and Coronal Heating by Resonant Absorption 358
10.3.3 Chromospheric and Coronal Heating by Phase Mixing 361
10.3.4 Wave Heating of Chromospheric and Coronal Loops 363
10.4 Heating by Magnetic Reconnection 364
10.4.1 Formation of Current Sheets 366
10.4.2 Evidence from Space for Reconnection 367
10.4.3 X-ray Bright Points: The Converging Flux Model 369
10.4.4 Current-Sheet Formation by Parker Braiding 370
10.4.5 Flux-Tube Tectonics: Separator, Separatrix and QSL Heating 374
10.5 Heating by Turbulence 386
10.5.1 Relaxation by MHD Turbulence 386
10.5.2 Avalanches: A Nonlinear Driven Dissipative Process 389
10.6 Conclusion 390

11 Prominences 391
11.1 Summary of Quiescent Prominence Properties 391
11.2 Basic Magnetic Structure (Current Sheet in a Flux Rope) 393
11.2.1 Kippenhahn-Schlüter Model for Prominence Sheet 393
11.2.2 2.5D Magnetic Flux-Rope Models of Inverse Polarity 396
11.2.3 Coronal Cavities 399
11.3 Global Nature of a Filament Channel and Chirality 399
11.4 Three-Dimensional Structure: Barbs (or Feet) 402
11.4.1 3D Aulanier-Démoulin (1998) Flux-Rope Model 402
11.4.2 Numerical Flux-Rope Modelling 404
11.5 Threads 407
11.5.1 Local Structure 407
11.5.2 van Ballegooijen–Cranmer Model for Tangled Fields 408
11.6 Formation of Thermal Structure 409
11.6.1 Injection from Below (Surges) 409
11.6.2 Magnetic Levitation of Cool Plasma 411
11.6.3 Radiative Instability or Non-equilibrium 411
11.6.4 Magnetic Rayleigh-Taylor Instability 415
11.7 Conclusion 415

12 Solar Flares and Coronal Mass Ejections 416
12.1 Introduction 416
12.1.1 Overview 416
12.2 Eruption by Non-equilibrium, Instability or Breakout 418
12.2.1 Non-equilibrium or Catastrophe in 2D 420
12.2.2 The Hoop Force of a Slender Toroidal Flux Rope 424
12.2.3 Non-equilibrium and Torus Instability in 3D 427
12.2.4 Titov-Démoulin Model: An Active-Region Toroidal Flux Rope 431
12.2.5 Helical Kink Instability 433
12.2.6 Resistive Kink Instability 434
12.2.7 Breakout 435
12.2.8 Emerging (or Evolving) Flux Model 437
12.2.9 Other Numerical Experiments on Initiation of Eruption 438
12.3 Reconnection and the Creation of Flare Loops 439
12.3.1 Standard (CSHKP) 2D Reconnection Model 439
12.3.2 3D Modifications to the Standard Model 442
12.4 Concluding Comment 450

13 The Solar Wind 451
13.1 Introduction 451
13.2 Chapman's (1957) Static Corona 452
13.3 Parker's Isothermal Solution 453
13.4 Breeze, Wind or Accretion? 455
13.4.1 Stability of Breeze Solutions 455
13.4.2 Dependence of Solutions on p_∞ and History 456
13.5 More General Models for a Spherical Expansion 457
13.5.1 Polytropic Solar Wind 457
13.5.2 Energy Equation 458
13.6 Effect of Rotation and a Magnetic Field 460
13.6.1 Rotating Wind 460
13.6.2 Magnetic Field 461
13.6.3 Early 1D Coronal-Hole Models 464
13.7 Fast Solar Wind: Self-Consistent Wave-Turbulence Models 466
13.8 Slow Solar Wind: Reconnection Models 468
13.9 2D and 3D Models: Streamers and Coronal Holes 469
13.9.1 2D Helmet-Streamer Model (Pneuman-Kopp) 470
13.9.2 Two-Dimensional Separable Solution for a Rotating Wind 471
13.9.3 3D Topology of Pseudo Coronal Streamers and Coronal Holes 471
13.9.4 Global Three-Dimensional Modelling 474
13.10 Large- and Small-Scale Time-Dependence 475
13.11 Going Beyond a Single-Fluid Picture 475
13.11.1 Two-Fluid Model 476
13.11.2 Kinetic Effects 478
13.12 Conclusion 480

Appendix 1 *Units* 483

Appendix 2 *Useful Values and Expressions* 487

References 493
Index 543

light of new research, such as on waves in non-uniform media. Next comes a substantial new chapter on the fundamental topic of magnetic reconnection, which is dear to my heart. The subsequent chapters on the applications of MHD to solar phenomena are completely different from the original book in view of the enormous changes in understanding. Chapter 10 deals with the theory of chromospheric and coronal loops and mechanisms for heating them by waves, reconnection or turbulence. Chapter 11 discusses flux-rope models for prominences and their fine-scale structure, namely, barbs and threads. Chapter 12 describes mechanisms for the eruption of a coronal flux rope to give an eruptive flare or coronal mass ejection, together with the latest models for energy release by reconnection. The final chapter summarises classical theories for the solar wind. It also describes wave turbulence models for the fast wind and reconnection models for the slow wind. Appendices describe units, useful values and expressions. Problem examples are referred to at various points in each chapter. They, together with their solutions and an appendix on ground- and space-based instruments, can be found on the Web page www.cambridge.org/9780521854719. The book ends with a comprehensive reference list and a detailed index.

The aim of mathematical modelling is not to account for all observational details of a particular phenomenon, or simply to produce an image that resembles an observation, but rather to focus on the key physical mechanisms at work. The full nonlinear equations of MHD are so complex that they often need to be approximated drastically by focusing on a particular mechanism. One begins with a simple model, and then extra effects are added in an attempt to make the model more realistic. Thus, simple analytical models and complex numerical computations can play complementary roles in enhancing understanding. Computational experiments are in many respects similar to laboratory experiments and can be so sophisticated that the underlying processes are far from transparent. When modelling is undertaken successfully, it is possible to predict behaviour with different parameter values and to give deep physical understanding of a phenomenon. However, in order to decide what are the important observations and physical processes, it is invariably crucial to listen carefully to insights from observers.

Single-fluid MHD is a remarkably successful model for describing solar plasma at large scales, even when it is collisionless, for reasons described in Chapter 2. However, it will be important to develop MHD in future to include multi-fluid effects, as well as kinetic models for the internal structure of non-ideal regions in reconnection, shock waves and wave dissipation structures.

Trying to summarise the advances of the past thirty years and the current state of solar MHD has been a formidable task. Although I have done my best, there are bound to be omissions and mistakes in understanding, for which I apologise.

The notations adopted for cylindrical and spherical polars are (R, ϕ, z) and (r, θ, ϕ), respectively. All quantities are measured in rationalised mks units, with the magnetic field in tesla (T) in most formulae. In the text, however, magnetic field strengths are commonly quoted in gauss (G), such that 1 G $= 10^{-4}$ T, while energy is sometimes given in erg instead of joules (J), such that 1 erg $= 10^{-7}$ J, since these are much better known in the community. Lengths in formulae are usually measured in metres (m), although in the text they are often quoted in megametres (Mm) such that 1 Mm $= 10^6$ m, since typical small observed structures in the photosphere and corona are about 1 Mm across.

Figures 1.8, 1.13, 1.18, 1.41, 3.10, 3.11, 3.14–3.16, 9.4, 9.12, 9.16, 9.17, 10.8b, 10.14–10.16, 10.23, 10.24, 10.27, 10.28, 10.30–10.32 and 13.8 appeared in the *Astrophysical Journal* and are reproduced by permission of the American Astronomical Society. Figure 9.11b appeared in *Monthly Notices of the Royal Astronomical Society* and is reproduced by permission of Oxford University Press on behalf of the Royal Astronomical Society.

Figures 1.16, 1.20cd, 1.29a, 1.35b and 1.36 were produced by instruments on Hinode, which is a Japanese space mission developed and launched by ISAS/JAXA, with NAOJ as domestic partner and NASA and STFC (UK) as international partners. It is operated by these agencies in co-operation with ESA and NSC (Norway). Figures 1.3, 1.15, and 1.37b come from SDO and are courtesy of NASA/SDO and the AIA or HMI science teams. Figures 1.12, 1.27 and 1.38 are courtesy of the MDI, LASCO and EIT consortia on SOHO, which is a project of international cooperation between ESA and NASA. Figures 1.13, 1.14a, 1.20ab, 1.28, 1.29b, 1.35a, 9.7 and 11.14c are from the Swedish 1-m Solar Telescope (SST) on La Palma (in the Spanish Observatorio del Roque de los Muchachos of the Instituto de Astrofísica de Canarias); this is operated by the Institute for Solar Physics, which was managed by the Royal Swedish Academy of Sciences until 1 January 2013 and since then by Stockholm University. Figures 1.22, 1.41 and 10.25 are from TRACE, which was a NASA Small Explorer (SMEX) mission. Figures 1.21c and 10.14a are from Yohkoh, which was a Japanese solar mission, developed and launched by ISAS/JAXA, Japan, with NASA and SERC/PPARC (UK) as international partners. Credits to the other figures appear in the captions.

It has been a privilege to be James Gregory Professor at St Andrews University. Many of James Gregory's discoveries are invaluable in modern solar physics. He was one of the inventors of calculus (with Newton and Leibniz), he invented the gregorian telescope and he discovered the diffraction grating by shining a light through a seagull's feather. He was appointed the first Regius Professor of Mathematics in the University in 1668 at the age of 30 and sadly died at the age of 37.

Many people have generously helped with advice and suggestions, for which I am extremely grateful, but they cannot be blamed for any inadequacy in treatment.

St Andrews and Dundee colleagues have given more support than they know and have provided a kindly and highly stimulating environment, including: Vasilis Archontis, Peter Cargill, Ineke De Moortel, Andrew Haynes, Alan Hood, Gunnar Hornig, Duncan Mackay, Julie McCormick, Karen Meyer, Thomas Neukirch, Paolo Pagano, Clare Parnell, David Pascoe, Sarah Platten, David Pontin, Laurel Rachmeler, Bernie Roberts, James Threlfall and Antonia Wilmot-Smith.

Many friends from over the world have also helped in countless ways, such as: Ernest Amouzou, Guillaume Aulanier, Tony Arber, Hubert Baty, Mitch Berger, Tom Berger, Véronique Bommier, Sean Brannon, Philippa Browning, Nic Brummell, Sacha Brun, Paul Cally, Robert Cameron, Dick Canfield, Mats Carlsson, Arnab Choudhuri, Steve Cranmer, Len Culhane, Pascal Démoulin, Bart De Pontieu, Mausumi Dikpati, George Doschek, Alec Engell, Oddbjørn Engvold, Robert Erdelyi, Lyndsay Fletcher, Terry Forbes, Laurent Gizon, Nat Gopalswamy, Mandy Hagenaar, Joanna Haigh, Louise Harra, Richard Harrison, Jack Harvey, David Hathaway, Jean Heyvaerts, Kiyoshi Ichimoto, Elena Khomenko, Bernhard Kliem, Sam Krucker, Jun Lin, Dana Longcope, Jean-Marie Malherbe, Eckart Marsch, Piet Martens, Marian Martinez Gonzalez, Valentin Martinez Pillet, Sarah Matthews, Scott McIntosh, David McKenzie, Nicole Meyer-Vernet, Fernando Moreno Insertis, Valera Nakariakov, Matthew Owens, Susi Parenti, Alex Pevtsov, Joseph Plowman, Jiong Qiu, Matthias Rempel, Alex Russell, Göran Scharmer, Rolf Schlichenmaier, Brigitte Schmieder, Karel Schrijver, Manfred Schüssler, Roger Scott, Kazunari Shibata, Lara Silvers, Sami Solanki, Henk Spruit, Bob Stein, Jack Thomas, Slava Titov, Steve Tobias, Juri Toomre, Javier Trujillo Bueno, Saku Tsuneta, Aad Van Ballegooijen, Luc Rouppe Van Der Voort, Lidia van Driel-Gesztelyi, Marco Velli, Harry Warren, Nigel Weiss, Thomas Wiegelmann, Anthony Yeates and Takaaki Yokoyama.

Chapter 1

A Description of the Sun

The Sun is an object of great beauty and fascination that has been studied with interest for thousands of years. It was born from a contracting, rotating, interstellar cloud that spun up during the collapse. The protostar would have settled down into a state where gravity and a pressure gradient balance one another and where a continued slow contraction heats up the plasma and provides the luminosity. Eventually, the core temperature became high enough for fusion of hydrogen to helium to provide all the luminosity, so that the contraction ceased. The Sun then entered the main ten-billion year (10^{10} yr) phase of its life on the main sequence, during which essentially all of the hydrogen in the core is turned into helium. So far, the Sun is half-way through this stage. In about five billion years, when the core hydrogen is exhausted, fusion will continue to take place in a shell around the helium core, while the Sun will expand greatly into a red giant (see PROBLEM 1.1). (For PROBLEM 1.1 and all the other problems in this book, together with their solutions, see the web page at www.cambridge.org/9780521854719.) Eventually, the red giant will collapse to a white dwarf, containing most of its original mass in a size similar to the Earth.

During the twentieth century, it gradually became clear that much of the Sun's present observed structure and dynamic behaviour owe their existence to the magnetic field. This book is principally concerned with developments during the past few decades in our present understanding of the effect of the magnetic field on the solar atmosphere. To put them into perspective, however, we begin with a list of earlier developments, many of which took place before the solar magnetic field was discovered. Section 1.2 continues by describing the overall characteristics of the solar interior and atmosphere, and is followed by a more detailed account of them (Secs. 1.3–1.6) and of solar activity (Secs. 1.7–1.9).

1.1 Brief History

2000 BC	Eclipses are recorded by the Chinese and from 600 BC by the Greeks.
325 BC	Theophrastus of Athens refers to 'black spots' on the Sun.
280 BC	Aristarchus of Samos suggests that the Earth is travelling around the Sun and estimates the distance to the Sun as five million miles.
190–125 BC	Hipparchus explains the motion of the Moon and Sun in terms of epicycles, with the Earth as the centre of the universe.
165 BC	The Chinese make records of observations of sunspots with the naked eye.
23 BC	Sunspots begin to be observed systematically by the Chinese.
AD 140	Ptolemy maintains that the Sun moves around the Earth, a belief held for the next 1,400 years.
1530	Copernicus suggests that the six known planets revolve about the Sun in concentric circles.

1609 Kepler uses Tycho Brahe's observations to formulate his laws of planetary motion and gives the Sun–Earth distance as fourteen million miles (PROBLEM 1.2). He also suggests that the Sun might have a magnetic field (to keep planets moving in their orbits).

1610 In the West, sunspots have been forgotten until they are recorded by Thomas Harriot in England and later by Johannes Fabricius, Christoph Scheiner and Galileo Galilei through recently invented telescopes. Some who see sunspots regard them as planets, others as the slag of the burning Sun or opaque clouds of smoke!

1633 Galileo is sentenced to house arrest for heretical heliocentric views.

1645–1715 Very few sunspots are present (the *Maunder Minimum*).

1666 Newton formulates the law of universal gravitation and applies it to the motion of planets around the Sun.

1770 Euler gives the Sun–Earth distance correctly as 93 million miles.

1814 Fraunhofer discovers most of the first 547 lines in the solar spectrum, which show up when sunlight is passed through a prism. At this time it is commonly thought that the Sun is inhabited!

1842 At a solar eclipse, *prominences* are rediscovered, after having been mentioned previously in medieval Russian chronicles and observed by Vassenius in 1733. Also, the outer layers of the solar atmosphere (the *chromosphere* and *corona*) are clearly seen.

1843 Schwabe notices that the number of sunspots varies with an *eleven-year cycle.*

1851 The corona is photographed for the first time as a faint halo, visible around the Sun during an eclipse (PROBLEM 1.3).

1852 Sabine, Wolf and Gautier find that the sunspot cycle is related to geomagnetic storms.

1858 Carrington discovers that sunspots appear at lower and lower latitudes as the sunspot cycle progresses.

1859 Carrington and Hodgson observe a *solar flare*, possibly for the first time, although Stephen Gray may have seen one on 27 December 1705 (Hoyt and Schatten 1996).

1861 Spörer discovers his law for the distribution of sunspots.

1868 At an eclipse, Secchi detects the emission line of a new element, which is given the name helium (after Helios, the Greek Sun god).

1869 Another new emission line, in the corona, is ascribed (wrongly) to a hypothetical element called coronium.

1874 Langley gives a detailed description of the fine structure, called *granulation*, of the Sun's visible surface (the *photosphere*).

1875 Secchi describes the change in coronal form during a solar cycle.

1877 Secchi describes *spicules* as a burning prairie.

1889 Hale invents the spectroheliograph.

1908 Hale discovers that sunspots possess a strong magnetic field.

1909 The outward flow in sunspot penumbrae is observed by Evershed.

1919 Hale and Joy discover that pairs of sunspots tend to have opposite polarity and to be tilted with respect to latitude, and that leading sunspots have opposite polarity in the north and south hemispheres.

1920's	The dominance of hydrogen and helium in the atmosphere is realised by Celia Payne (1925) and Russell (1929) and in the interior by Eddington and Strömgren (1932).
1930	Lyot invents his coronagraph to view the corona without an eclipse.
1934	Cowling proposes a theory for sunspots and an anti-dynamo theorem.
1935	Hartmann conducts experiments on mercury flow in a magnetic field.
1938	The carbon-nitrogen and proton-proton chains are proposed by Bethe as an explanation for the source of solar energy.
1939	Grotrian (1939) and Edlén (1943) show that coronal emission lines arise from normal elements ionised by a very hot corona ($> 10^6$ K).
1941	Biermann realises sunspots are cool due to inhibition of convection.
1942	Alfvén sets up a theory for magnetic waves. Also, radio emission from the Sun is detected by radar.
1945	W. Roberts names and describes *spicules* in detail.
1948	Biermann and Schwarzschild propose that the outer atmosphere is heated by sound waves propagating up from the convection zone.
1951	Biermann suggests that the Sun is continuously emitting 'solar corpuscles' that make comet tails point away from it.
1951	Laboratory experiments begin on magnetic containment of a plasma; the ultimate goal is to contain plasma for only about a second at 10^6 K, so that light atoms can fuse together and release energy.
1952	Babcock and Babcock (1952, 1955) invent the magnetograph and discover properties of photospheric magnetic fields.
1955	Parker produces major works on the dynamo and magnetic buoyancy.
1956	The basic theory of magnetohydrodynamics (MHD) is summarised by Cowling in his book on the subject.
1957	The first satellite observations of interplanetary plasma are made. Kippenhahn and Schlüter propose a model for prominence support.
1958	Babcock and Livingston observe polar fields reverse at spot maximum.
1958	Parker predicts the existence of the *solar wind* and proposes a model.
1960	Leighton discovers *five-minute oscillations* in the photosphere.
1962	Leighton, Noyes and Simon find that the *network* (first seen by Hale and Deslandres in the 1890s) outlines *supergranule cells.*
1970	Ulrich suggests that global sound waves can be trapped below the surface of the Sun and proposes a relation between their frequency (ω) and horizontal wavenumber (k_h).
1972	Tousey and Koomen observe a coronal mass ejection from OSO 5.
1973	After early rocket flights, Skylab (1973–4) explores in detail the corona in soft X-rays with its holes, loops and X-ray bright points.
1973	Stenflo discovers kG magnetic fields in the network.
1975	Deubner first observes power in the ω-k_h ridges predicted by Ulrich.
1980	Hickey et al. (1980) discover that the solar irradiance is varying, that is, the *solar constant* is not constant.
1980s	Important advances occur in the MHD theory of equilibria, waves, instabilities and reconnection.
1990s	Yohkoh reveals the dynamic nature of the corona and the presence of magnetic reconnection in solar flares.

	SoHO produces a major increase in understanding, especially the rotation structure of the interior from the MDI instrument and the properties of coronal mass ejections from LASCO.
1998	TRACE revolutionises our view at high resolution of the corona.
2002	RHESSI transforms our knowledge of solar flares.
2000s	STEREO views coronal mass ejections in three dimensions.
	Hinode studies the link between photosphere and corona and changes our paradigm for photospheric magnetic fields.
2010	SDO begins a new revolution in understanding with super-TRACE and super-MDI instruments

In the past fifty years, our understanding of the Sun has been revolutionised by a combination of theoretical advances (both analytical and computational), ground-based observations (notably from helioseismology networks such as GONG and BiSON and from telescopes in La Palma, Tenerife, Sacramento Peak, Big Bear and Kitt Peak) and especially space observations. The satellites include (with their launch dates): Skylab (1973), Yohkoh (1991), Ulysses (1992), SoHO (1995), TRACE (1998), RHESSI (2002), STEREO (2006), Hinode (2006) and SDO (2010) (see APPENDIX 3), which is on the web site www.cambridge.org/9780521854719.

1.2 Overall Properties

Traditionally, solar phenomena have been divided into two classes: quiet and active. In this paradigm, the *quiet Sun* is viewed as a static, spherically symmetric ball of plasma, for which properties depend to a first approximation on radial distance from the centre and for which the magnetic field is negligible. The *active Sun* consists of transient phenomena, such as sunspots (Sec. 1.7), prominences (Sec. 1.8), flares and coronal mass ejections (Sec. 1.9), which are superimposed on the quiet atmosphere and owe their existence to the magnetic field. This division is hinted at here, by first describing aspects of the quiet Sun in Sections1.3–1.6 and then transient phenomena in Sections 1.7–1.9, even though the paradigm needs to be replaced by a much more dynamic, multi-scale view. For instance, the quiet atmosphere is influenced markedly by the magnetic field; it is structured by the magnetic network above and around evolving granule and supergranule cells (Sec. 1.4) and the normal heating of the outer atmosphere is due to the magnetic field (Chapter 10). Furthermore, in the atmosphere, the notion of a static, spherically symmetric structure is not even a first approximation to reality.

General descriptions of the Sun can be found in many books, but those I have found particularly readable and informative are Noyes (1982), Golub and Pasachoff (1997), Schrijver and Zwaan (2000), Stix (2002), Zirker (2002) and Lang (2009).

The Sun is a fairly ordinary star of spectral type G2 V and absolute stellar magnitude 4.8, but of course its proximity to the Earth makes it unique. It has profound effects on the Earth's climate and on space weather, and its study is of central importance for understanding the behaviour of stars and of cosmic plasma in general. Its overall physical properties are as follows:

Age:	$= 4.6\times10^9$ yr
Mass ($M_\odot$):	$= 1.99\times10^{30}$ kg
Radius ($R_\odot$):	$= 695.5$ Mm ($= 695{,}500$ km $= 6.955\times10^8$ m)
Mean density:	1.4×10^3 kg m^{-3}
Mean distance from Earth:	1 AU $= 1.496\times10^{11}$ m ($= 215$ R$_\odot$)
Surface gravity ($g_\odot$):	$= 274$ m s^{-2}
Escape velocity at surface:	$= 618$ km s^{-1}

Radiation (luminosity, $L_\odot$):	$= 3.86\times10^{26}$ W ($= 3.86\times10^{33}$ erg s^{-1})
Equatorial (synodic) rotation period:	= 26.24 days
Angular momentum:	$= 1.7\times10^{41}$ kg m^2 s^{-1}
Mass-loss rate:	$= 10^9$ kg s^{-1}
Effective temperature:	= 5785 K
1 arcsec ($\equiv 1''$):	$\approx$ 726 km

Many of these numbers may be rather meaningless to the reader, so let us put them into perspective. The Earth has a mass of 6×10^{24} kg and a radius of 6 Mm (6×10^6 m); so the Sun is 330,000 times more massive than the Earth and is 109 times larger in radius, while its surface gravity is 27 times greater. The mean density of the Earth is 5.5×10^3 kg m^{-3}, roughly equal to that of the Sun, while the atmospheric density at the Earth's surface is 1 kg m^{-3}, which makes the Sun's surface pressure 0.2 times that of the Earth's sea-level atmosphere. The mean distance from the Earth to the Sun is 149.6 million km (or 93×10^6 miles), which takes light 8 minutes to travel. The radiation emitted by the Sun amounts to about 1 kW m^{-2} at the surface of the Earth. Furthermore, the Sun's equator is inclined at about 7° to the plane of the Earth's orbit and the solar rotation gives the Sun an equatorial velocity of 2 km s^{-1}. An *arcsec* is a unit of angular measurement equal to 1/3,600 of a degree or π/648,000 radians. The 726 km refers to the corresponding distance at the Sun measured by an Earth-bound observer, but that does vary by ±12 km during the Earth's journey around the Sun in an eccentric orbit (PROBLEM 1.4) and could be quite different when measured from a spacecraft out of Earth's orbit.

1.2.1 The Structure of the Solar Interior

The Sun is such a massive plasma ball that it is held together and compressed by its own gravitational attraction. It comprises mainly H (92 per cent) and He (8 per cent) atoms by number, mostly ionised due to the high temperature; the remaining elements, such as C, N and O, total about 0.1 per cent and are present in roughly the same proportions as on Earth, suggesting a common origin such as the interiors of older stars.

The *interior* of the Sun is shielded from our view; only its surface layers can be seen directly. However, the field of *helioseismology* (Sec. 1.3.4) is now being used to infer many properties of the interior, which is divided into three regions, as sketched in Figure 1.1, namely the *core*, *radiative zone* and *convection zone*, where different physical processes are dominant.

In the nineteenth century, it was shown that, if the Sun's energy arose purely from gravitational contraction, it would last only 10^{15} sec $= 3\times10^7$ yr, the so-called *Kelvin-Helmholtz time.* This is calculated simply by dividing the Sun's gravitational potential energy ($M_\odot^2 G/R_\odot = 4\times10^{41}$ J, where G is the gravitational constant) by its present luminosity $L_\odot \approx 4\times10^{26}$ W. Eddington, however, concluded in 1925 that it would last much longer if the core of the Sun were a gigantic reactor converting nuclear energy. This energy (generated in the core) leaks continuously outwards in a very gentle manner across the *radiative zone* by radiative diffusion, as the photons are absorbed and emitted many times, taking many years to cross it.

By contrast, in the *convection zone*, convection is the dominant means of outward energy transport. Convection transports energy because an individual blob of plasma carries heat as it rises and then gives up some of it before falling and picking up more. At the lower boundary of the convection zone, there exists a strong shear layer [called the *tachocline* by Spiegel and Zahn (1992)], where much of the Sun's large-scale magnetic field is probably generated by a *dynamo* (Chapter 8).

The solar interior is so incredibly opaque that, whereas an unimpeded photon would take 2 sec (at the speed of light) to reach the surface from the centre, there are so many collisions (absorptions and

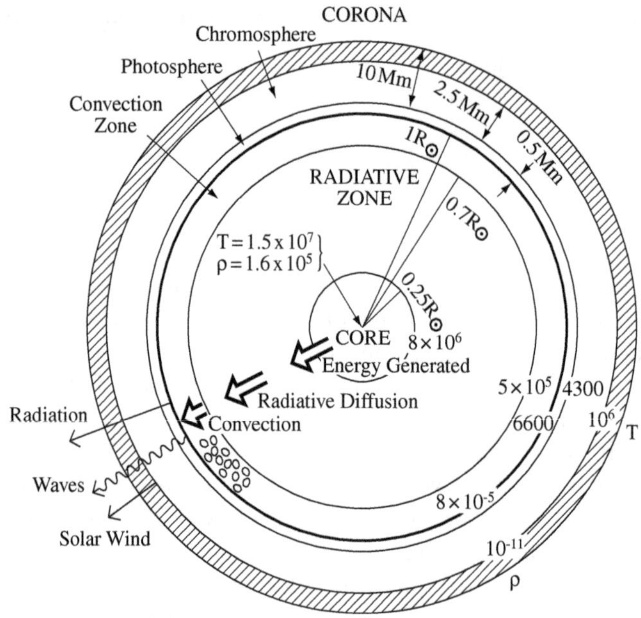

Figure 1.1. The overall structure of the Sun, indicating the sizes of the various regions and their temperatures (in K) and densities (in kg m^{-3}). The thicknesses of the various regions are not drawn to scale, and the boundary between chromosphere and corona is highly variable between 2.5 and 15 Mm, as indicated by the shaded region.

re-emissions) that photons in practice take 170,000 years for the journey (Mitalas and Sills 1992). The effect of these collisions is to increase the typical wavelength from that of high-energy gamma rays in the core to visible light at the solar surface, where most of the energy generated in the core is radiated into space.

The magnetic diffusion-time (Sec. 2.6.1) for the original magnetic flux that threaded the plasma cloud from which the Sun contracted is 10^{11} yr and so is somewhat longer than the Sun's age (4.6×10^9 yr). Since there are mechanisms for enhancing diffusion, it is unknown how much of this primordial magnetic field is still present in the solar interior. The other main global time-scales for the interior are: the viscous diffusion-time (10^{12} yr); the spin-down time (10^{10} yr) due to angular momentum loss in the solar wind, which is a little longer than the solar age; the thermal diffusion-time (1.2×10^7 yr) through the radiative envelope; the thermal relaxation-time of the convection zone (10^5 yr); the Alfvén travel-time (20 yr) across the radiative interior for a magnetic field of 2 kG; and the thermal or rotational equilibration-time (1 yr) of the convection zone. By comparison, the acoustic travel-time from the centre to the solar surface is only about one hour, which is why for most purposes the radiative envelope and core may be regarded as being in hydrostatic equilibrium.

1.2.2 The Structure of the Solar Atmosphere

The atmosphere is defined as the part of the Sun from which photons can escape directly into space. It consists of three regions with different physical properties, which are often conveniently but incorrectly pictured as a series of spherical shells. The lowest is an extremely thin layer of plasma only several hundred kilometers thick, called the *photosphere* (Sec. 1.4), which is relatively dense and opaque and emits most of the solar radiation. It has an optical thickness $\tau \lesssim 1$ in the near-ultraviolet, visible and near-infrared continua, but is optically thick in all except the weakest spectral lines. The *optical depth* or *thickness* (τ), is a measure of the transparency of a medium. It is defined by the equation

$$I = I_0 e^{-\tau},$$

where I_0 is the intensity of radiation at the source and I is the observed intensity after a given path, so that I/I_0 represents the fraction of radiation that remains (i.e., is not scattered or absorbed). Thus, $\tau = 1$ represents the location where the radiation has fallen by a factor e.

Above the photosphere lies the rarer and more transparent *chromosphere* (Sec. 1.5), which is optically thin in the near-ultraviolet, visible and near-infrared continua, but is optically thick in strong spectral lines. It can be glimpsed for a few seconds at the start and end of a solar eclipse as the red colour of the Balmer spectrum.

The *corona* (Sec. 1.6) extends from the top of a narrow *transition region* (Sec. 1.5) and out into the *solar wind,* which fills the *heliosphere.* The outer boundary of the corona may be defined as the Alfvén radius, where the solar-wind speed equals the Alfvén speed, so that Alfvén-wave communication with the Sun is no longer possible (Sec. 13.6.2). The corona is optically thin over the whole electromagnetic spectrum, except for radio waves and a few spectral lines.

The pressure of the solar transition region can be inferred from density-sensitive extreme ultra-violet (EUV) emission lines. Recent solar observations yield 4×10^{-3} N m^{-2} in the quiet Sun and 0.3 N m^{-2} in active region moss (Warren 2005; Warren et al. 2010), where 1 N m^{-2} = 1 0 dyne cm^{-2}. Hydrogen is almost wholly ionised in the upper chromosphere, but neutrals are important in the lower chromosphere and photosphere.

The density (n) decreases rather rapidly with height: typical values are 10^{23} m^{-3}, 10^{19} m^{-3} and 10^{15} m^{-3} in the photosphere, chromosphere and transition region, 10^{12} m^{-3} at a height of $1R_\odot$, 10^{7} m^{-3} at 1 AU, and 10^{6} m^{-3} in the interstellar medium. By comparison, the gas density at the Earth's surface is 10^{25} m^{-3}.

Before 1940, it was thought that the temperature decreases with height above the solar surface. But since then it has been realised that, for a steady, spherically symmetry atmosphere, after falling from about 6,600 K (at the bottom of the photosphere) to a minimum value of about 4,400 K at a height of 500 km, the temperature rises slowly through the lower chromosphere and then dramatically through the transition region (less than 100 km thick) to a few million degrees in the corona (Figure 1.2). Thereafter, it falls slowly in the outer corona, which is expanding outwards as the solar wind, to a value of 10^5 K at 1 AU. The photospheric temperature of a few thousand degrees may be compared with the temperature of red-hot iron (1,400 K) and that of the white-hot filament of an electric bulb (3,900 K).

Table 1.1 gives energy losses (after Withbroe and Noyes 1977) from different parts of the corona as conductive, radiative and solar-wind fluxes in W m^{-2}, where 1 W m^{-2} = 10^3 erg cm^{-2} s^{-1}. Lower estimates from coronal-hole models by Hansteen and Leer (1995) are shown in parentheses.

The *VAL model* of the solar atmosphere is a semi-empirical one-dimensional model that successfully fits a number of spectral lines from different regions and has been extremely useful. However, a representation such as Figure 1.2 of the atmospheric structure is a gross oversimplification, which indicates only the mean properties. In reality, the atmosphere does not consist of static plane-parallel layers but is a constantly seething mass of plasma with a far-from-uniform structure. The temperature and density at any location are continually changing as plasma heats and cools dynamically over tiny as-yet-unresolved length-scales and moves around in response to a variety of different physical processes. Indeed, along any line of sight there is an enormous range of temperatures. Thus, for instance, transition-region emission comes not in reality from a 100 km thick layer, but from plasma at many different heights that is heating up or cooling down through transition-region temperatures.

The bulk of the solar radiation comes from the photosphere (Figure 1.3a), which emits a continuous spectrum with superimposed dark *absorption lines.* Light of all wavelengths is emitted by the photosphere and most of it goes straight through the overlying atmosphere into space. However, at certain specific

Table 1.1. *Order-of-magnitude energy-loss fluxes in W m^{-2}*

	Coronal hole	Quiet Sun	Active region
Corona			
Conduction	60 (15)	200	10^3–10^4
Radiation	10 (15)	100	5000
Solar wind	700 (100)	< 50	< 100
TOTAL	800	300	10,000
Chromosphere			
Low	2000	2000	10,000
Middle	2000	2000	10,000
Upper	300	300	2000
TOTAL	4000	4000	20,000

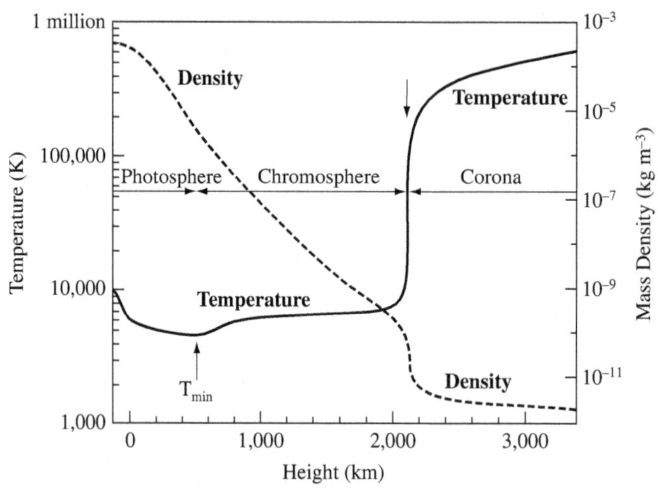

Figure 1.2. A schematic of the mean variation of temperature and density with height in the solar atmosphere according to the VAL (Vernazza-Avrett-Loeset) model (courtesy Eugene Avrett, see Sec. 1.4.3), although in practice the atmosphere is highly inhomogeneous, dynamic and time-varying.

wavelengths it is absorbed by particles in the overlying atmosphere, due to an increased opacity, which gives rise to the absorption lines. For example, the H Balmer line (Hα) is due to absorption of a photon making an H atom jump from its second to its third quantum level. Such lines give us much information on temperature and density (from intensity), magnetic field strength (from Zeeman splitting or Hanle effect) and local line-of-sight plasma motion (from Doppler shifts).

Most spectral lines are formed in the lower photosphere, but some (such as Hα) come from the chromosphere, and most lines in the transition region and corona are emission lines. The transition region emits mainly in UV wavelengths below 2,000 Å, which are strongly absorbed by the Earth's atmosphere. Due to its high temperature, the corona has increased UV, EUV and X-ray emission, but it also emits a pair of visible continua (the K and F coronae), as well as lines such as the green line (5,303 Å) and the red line (6,374 Å), which are due to forbidden transitions in highly ionised iron (Fe XIV and Fe XV, respectively).

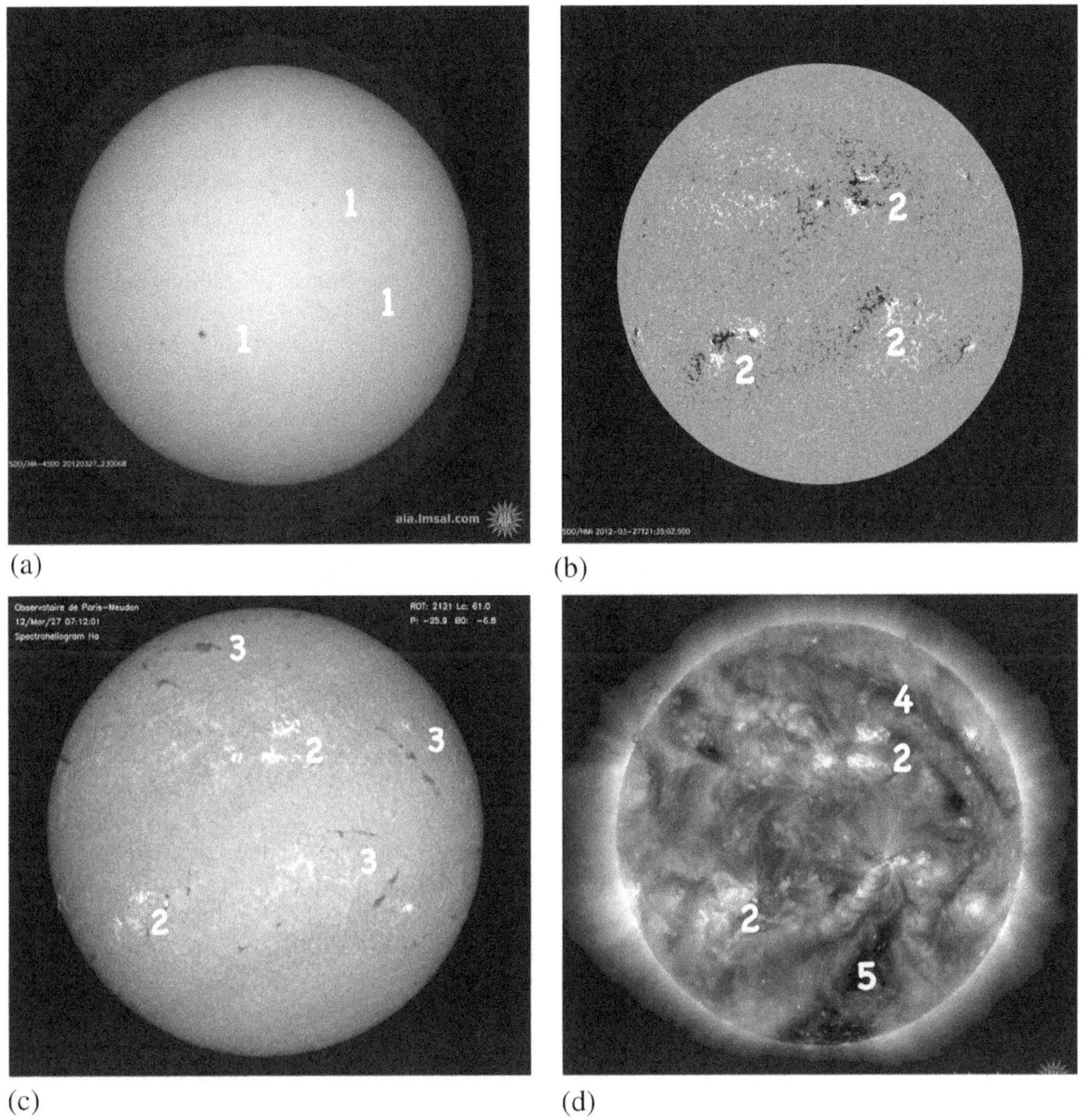

Figure 1.3. (a) The photosphere (SDO/AIA), (b) the line-of-sight photospheric magnetic field with positive polarity in white and negative in black (SDO/HMI), (c) the chromosphere in Hα (courtesy Jean-Marie Malherbe, Meudon Observatory) and (d) the corona (SDO/AIA) in 195 Å at 1.5 MK, all on 27 March 2012. Numbers indicate the presence of a (1) sunspot, (2) active region, (3) prominence or filament, (4) filament channel and (5) coronal hole.

When the Sun is observed through filters of different wavelengths, pictures can be obtained of the Sun's structure at a variety of levels (Figures 1.3a–d). For example, the *chromosphere* is revealed by using an Hα filter (Figure 1.3c), which is most important for following the evolution of *active regions* (Sec. 1.7.1) and *prominences* (Sec. 1.8) and for observing the low-temperature part of a *solar flare* (Sec. 1.9.2). Just at the start of an eclipse, you can see light that has originally come up from the photosphere and is then scattered towards you at the chromospheric level as well as the intrinsic chromospheric emission. This can be a very colourful effect, with reds from Hα and blue from Hβ, and it led Lockyer (1868) to give the chromosphere its name (from the Greek word for 'colour') (PROBLEM 1.5).

1.3 The Solar Interior

Models of the interior give a central temperature (15 million K) and density (1.6×10^5 kg m^{-3}) that are high enough for thermonuclear reactions to take place. The central temperature is so high that the material there remains in a gaseous (plasma) state under a pressure 230 billion times that of the Earth's sea-level air pressure; the central density is 13 times that of solid lead! Moving outwards, the temperature and density fall by 3.5 and 8.5 orders of magnitude to 6,000 K and 2×10^{-4} kg m^{-3} at the visible surface; thus, the temperature falls on average by only 2 degrees per 100 m.

The *core* has a radius of 150 Mm and contains half the mass of the Sun in only about one-fiftieth of its volume, but generates 99 per cent of the energy. The *convection zone* extends from a radius of 500 to 700 Mm and contains more than 60 per cent of the solar interior's volume but less than 2 per cent of its mass.

1.3.1 The Solar Core

For every kilogram of H that is fused to form He, 0.007 kg is converted into energy, so that the great furnace in the solar core uses up 5×10^6 tonne of H per second. In this process, He nuclei are built up from H nuclei mainly by the proton-proton (PP) chain but partly by the CNO cycle. At the end of these cycles, groups of four protons (^{1}H) have been fused into one helium nucleus (^{4}He) according to the reaction

$$4\,{}^1\mathrm{H} \rightarrow {}^4\mathrm{He} + 2e^+ + 2\nu + 26.7\ \mathrm{MeV},$$

and other nuclei have just acted as catalysts. However, each He nucleus is smaller in mass by 3 per cent than the original protons, and this mass deficit appears as energy. According to Albert Einstein's equivalence of mass and energy ($E = mc^2$), each kilogram of mass is equivalent to 9×10^{16} J. In the above reaction, energy is released in the form of two high-frequency γ-rays (26.2 MeV) and two electron neutrinos (0.5 MeV), denoted by ν. The neutrinos are tiny subatomic particles with no electric charge that travel at nearly the speed of light and interact poorly with matter, escaping unimpeded from the core through the rest of the solar interior.

Neutrinos are our only direct diagnostic of core conditions and rain down on the Earth at a rate of about 70 billion per cm^2 per second. Davis, Jr and Evans (1978) observed the flux of solar electron neutrinos a mile underground in the Homestake Gold Mine in South Dakota, but it was lower by a factor of three than predicted from the standard solar model, a discrepancy known as the *solar neutrino problem.* Was it the solar model or the particle physics or the experiment that was wrong? After years of debate, it was established that, because of their tiny but non-zero mass, the electron neutrinos emitted by the Sun change to other forms (muon and tau neutrinos) during their journey to the Earth; this was discovered in 2002 at the Sudbury Neutrino Observatory, Canada, which detected all three types of neutrino and showed that one-third of the arriving solar neutrinos are electron neutrinos.

1.3.2 A Model for the Solar Interior

The *standard model* for the interior of the Sun assumes the pressure $[p(r)]$, density $[\rho(r)]$ and temperature $[T(r)]$ are functions only of radial distance (r) from the centre and that the interior consists of a set of spherical shells in hydrostatic and thermal equilibrium. The basic equations are: a *perfect gas law* (Eq. 2.24)

$$p = \frac{k_B}{m}\,\rho\,T, \tag{1.1}$$

where k_B is the Boltzmann constant and m the mean particle mass; a *hydrostatic force balance* (Sec. 3.1.2) between an outward pressure gradient and an inward force of gravity due to the mass $[M(r)]$ inside a sphere of radius r

$$\frac{dp}{dr} = -\rho(r)\ g(r), \tag{1.2}$$

where the gravitational acceleration is $g(r) = M(r)G/r^2$ (Eq. 2.21), $M(r)$ is given by $dM/dr = 4\pi r^2 \rho$ and $G = 6.67\times10^{-11}$ N m^{2}kg^{-2}; and a steady-state *energy balance*

$$\frac{dL(r)}{dr} = 4\pi r^2 \rho\ \epsilon, \tag{1.3}$$

where $L(r)$ is the outward flow of thermal energy through a sphere of radius r and ϵ is the nuclear energy-generation rate per unit mass. Outside the core, $L(r)$ is constant and is equal to the solar luminosity ($L_\odot$). Below the convection zone, the energy flux is carried by radiation alone, and so in *radiative equilibrium* the diffusion approximation for the radiative flux (F) is $F \equiv L/(4\pi r^2) = -[16\ \sigma_{SB}\ T^3/(3\ \kappa_R\ \rho)]\ dT/dr$. It gives a temperature gradient of

$$\frac{dT}{dr} = -\frac{3\ \kappa_R\ \rho}{16\ \sigma_{SB}\ T^3}\frac{L}{4\pi r^2}, \tag{1.4}$$

where κ_R is the *Rosseland mean opacity* and $\sigma_{SB} = 5.6705\times10^{-8}$ W m^{-2} K^{-4} the Stefan-Boltzmann constant. In the convection zone, the opacity is so high that most of the energy flux is transmitted by convection rather than radiation, and, except in a narrow region near the surface, it is a good approximation to replace Equation (1.4) by the condition that the temperature gradient equals the adiabatic value (Eq. 1.9) for the onset of convective instability, namely (with γ the specific heats ratio),

$$\frac{dT}{dr} = -\frac{\gamma - 1}{\gamma}\frac{gm}{k_B}. \tag{1.5}$$

The above equations are solved subject to the boundary conditions $M = 0$, $dT/dr = 0$ at the solar centre ($r = 0$) and $M = M_\odot = 1.989\times10^{30}$ kg, $L = L_\odot = 3.846\times10^{26}$ W at the solar surface ($r = R_\odot = 6.955\times10^8$ m). (The Stefan-Boltzmann law determines the *effective temperature* of the Sun as $T_{eff} = [L_\odot/(4\pi R_\odot^2 \sigma_{SB})]^{1/4} = 5785$ K. This is the temperature it would have if radiating as a black body.)

In a standard model of the Sun of mass $M_\odot$, radius $R_\odot$ and luminosity $L_\odot$, the resulting entropy at the surface depends on the details of the turbulence in the convective zone. Qualitatively, the turbulence is a function of the ratio of the near-surface *mixing-length* (l) (the mean-free-path of a turbulent eddy) to the local scale-height (Sec. 3.1.2), which is adjusted until the radius of the model Sun equals the observed value (1 $R_\odot$).

A standard model should also have the correct abundances, such as the observed mass-fractions of H (X) and of heavy elements (Z), which are known (with the important exception of He) from spectroscopic observations. As the Sun aged, the main change in composition came from the fusion of H into He, but gravitational settling also played a role in determining the present structure. An initial set of uniform abundances is therefore adjusted to give the present composition (X and Z) after integrating over 4.6 billion years (the time since the onset of core hydrogen burning on the main sequence). The correct $L_\odot$ is obtained by adjusting the initial He abundance (Y).

The results for a standard solar model (Model S, Christensen-Dalsgaard et al. 1996) are shown in Figure 1.4. The core extends to a radius of 0.25 $R_\odot$ and the base of the convection zone is at 0.713 $R_\odot$. Within the convection zone, stratification is essentially adiabatic and abundances are uniform. Below the convection zone, the stratification rapidly becomes substantially sub-adiabatic and the H composition decreases with depth while the heavy-element composition increases. Settling of He and heavy elements produces sharp gradients in abundances just beneath the convection zone. Model S is fine for many purposes, although it has been suggested that the abundances of C, N and O may be substantially lower (by almost a factor of two) than thought before (Asplund et al. 2005), which would give an abundance ratio (Z/X) of 0.0165 rather than the 0.0245 used in Model S. This in turn would lower the opacity and sound speed below

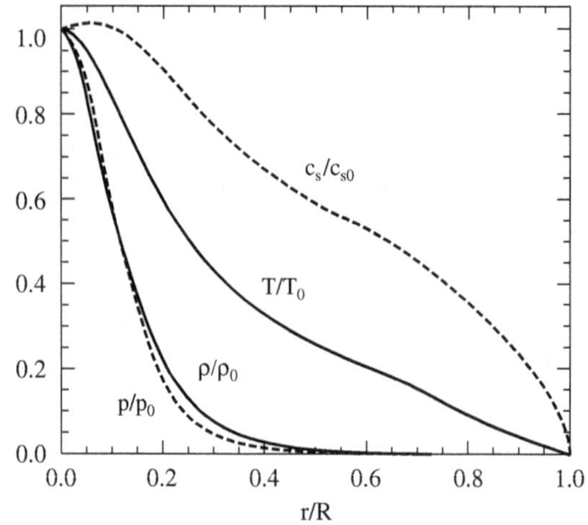

Figure 1.4. A standard model (Model S) for the solar interior, which uses OPAL opacities and equation of state and includes settling of He and heavy elements. Shown are the sound speed (c_s), temperature (T), density (ρ) and pressure (p), plotted as functions of radius ($r/R_\odot$) in terms of their values at the solar centre ($c_{s0} = 5.05\times10^5$ m s^{-1}, $T_0 = 1.57\times10^7$ K, $\rho_0 = 1.54\times10^5$ kg m^{-3} and $p_0 = 2.35\times10^{16}$ N m^{-2}). The known global properties of the present Sun are its mass ($M_\odot$), surface radius ($R_\odot$), surface luminosity ($L_\odot$), age and surface heavy-element abundance (Z) [except for helium (Y)] relative to hydrogen (X). The model is calibrated by fixing $M_\odot$ and the age (4.6 Gyr), and adjusting the initial abundances (Z_0 and Y_0) and mixing-length (l) to produce the observed Z (= 0.0245 X), $L_\odot$ and $R_\odot$ (courtesy Jörgen Christensen-Dalsgaard).

the convection zone and change the base of the convection zone to 0.726 $R_\odot$. However, it is not yet clear whether this is correct. Other uncertainties include effects of: macroscopic interior motions, which could change the composition profile (e.g., lithium is observed to be depleted by a factor of 150 from the initial abundance estimated from meteorites); mass loss or accretion; and magnetoconvection near the surface (Fabbian et al. 2010).

1.3.3 Convection Zone

Every Scot who heats his or her pan of porridge oats knows that, when the bottom of the pan is hot enough, the porridge starts to bubble away and exhibit a cellular circulation. Such turbulent convective motions also take place in the *convection zone*, where the temperature gradient is too great for the material to remain in stable equilibrium (Secs. 7.5.6 and 9.1). In practice, the convective velocities are much slower than sonic and the mean stratification is close to adiabatic, with the relative departure from adiabaticity probably not exceeding 10^{-4}.

As one moves outwards through the radiative zone, the opacity (κ_R) starts to rise rapidly, which increases the magnitude of the temperature gradient (Eq. 1.4). It is at about 0.7 $R_\odot$, where $T \approx 10^6$ K, that *convective instability* sets in. When one reaches the low photosphere, some radiation can escape directly from the Sun, and so the opacity (κ_R) decreases and the material returns to convective stability again. Hot plasma rising from the solar interior in turbulent cells reaches a position where the opacity is no longer sufficient to prevent the escape of radiation. The plasma expands, radiates, and cools and in so doing loses its buoyancy and descends.

The onset of instability when the vertical temperature gradient ($|dT/dr|$) is too large may be explained by the following qualitative argument (Figure 1.5). Consider a vertically stratified plasma in hydrostatic equilibrium with pressure $p(r)$, temperature $T(r)$ and density $\rho(r)$. Suppose an elementary blob of plasma moves upwards so slowly that it remains in *horizontal pressure equilibrium* with its surroundings. Then the

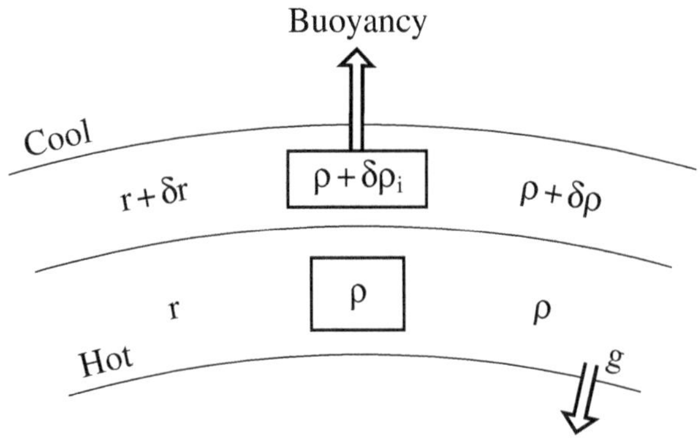

Figure 1.5. The displacement of an element of plasma from r to $r + \delta r$, during which the density decreases by $-\delta\rho_i$ inside the blob and $-\delta\rho$ outside it.

blob will feel a buoyancy force and so continue to rise if

$$\delta\rho_i < \delta\rho, \tag{1.6}$$

where $\delta\rho_i$ and $\delta\rho$ are the density changes inside the blob and in the ambient medium, respectively. By differentiating the perfect gas law (Eq. 1.1), the changes in pressure, density and temperature inside and outside the blob are related by

$$\frac{\delta p_i}{p} = \frac{\delta\rho_i}{\rho} + \frac{\delta T_i}{T} \quad \text{and} \quad \frac{\delta p}{p} = \frac{\delta\rho}{\rho} + \frac{\delta T}{T}. \tag{1.7}$$

But horizontal pressure equilibrium means $\delta p = \delta p_i$, and so Eqs. (1.6) and (1.7) imply $-\delta T > -\delta T_i$. In other words, the blob is unstable and will continue rising if

$$\left|\frac{dT}{dr}\right| > \left|\frac{dT_i}{dr}\right|, \tag{1.8}$$

so that the ambient temperature (T) is falling with height faster than the temperature (T_i) within the blob. (If on the other hand $|dT/dr| < |dT_i/dr|$, then there is a restoring force, the blob is stable and it will undergo oscillations by *gravity waves.*)

Now, the blob properties are governed by

$$p_i = \frac{k_B}{m}\rho_i T_i, \qquad \frac{dp_i}{dr} = -\rho_i g, \qquad \frac{p_i}{\rho_i^\gamma} = constant,$$

under the assumptions that: the gas is perfect; the motion is so slow that it is in hydrostatic equilibrium vertically; and it is so rapid (i.e., adiabatic) that there is no heat exchange with the surroundings. These may be combined to give

$$-\frac{dT_i}{dr} = \frac{\gamma - 1}{\gamma}\frac{gm}{k_B}, \tag{1.9}$$

which is known as the *adiabatic temperature gradient.* The criterion (1.8) for convective instability therefore finally becomes

$$\left|\frac{dT}{dr}\right| > \frac{\gamma - 1}{\gamma}\frac{gm}{k_B}, \tag{1.10}$$

and is known as the *Schwarzschild criterion.* In a fully ionised monatomic gas, $\gamma = 5/3$ and so the factor $(\gamma - 1)/\gamma$ is 2/5.

Partial ionisation has important physical effects (Sec. 2.1.5). Just below the solar surface it influences energy transport, since hydrogen is less than 90 per cent ionised in the top 8 Mm of the convection zone and the factor $(\gamma-1)/\gamma$ in the adiabatic temperature gradient (Eq. 1.9) reaches a minimum value of about 0.1 at a depth of 1 Mm (e.g., Vögler et al. 2005; Perez et al. 2009). In the photosphere and chromosphere, partial ionisation also changes the form of Ohm's law (Secs. 1.4.3 and 2.1.3 and 2.1.5).

Modelling turbulent convection in the (compressible) convective zone is a formidable task that has taken great strides in the past twenty years (e.g., Cattaneo et al. 2003; Miesch and Toomre 2009; Nordlund et al. 2009). In the early days, an order-of-magnitude treatment of turbulent convection (e.g., Böhm-Vitense 1958; Spruit 1974; Canuto and Mazzitelli 1991), was used to describe collisions between turbulent eddies in terms of a *mixing-length*, which is the distance a convecting blob travels before disintegrating and exchanging its excess heat and which was assumed to be proportional to the local pressure scale-height. More recently, numerical modelling has followed two complementary philosophies, namely, either using simplified physics to explore fully the roles of different effects and different parameter regimes (e.g., Hurlburt and Toomre 1988; Weiss et al. 1996; Brun et al. 2004) or trying to adopt as realistic physics as possible (e.g., Nordlund 1985; Carlsson et al. 2004; Stein and Nordlund 2006). There have been impressive advances in understanding and comparisons with observations. It is, however, not possible to run the numerical experiments at realistic values of the Reynolds number (namely, $Re \approx 10^{12}$) or Prandtl number (defined in Sec. 2.5), and so new discoveries are expected as observational and computational resolution improves, so that the computations become less diffusive.

At the photosphere, convection is dominated by cells of two main types (Sec. 1.4.1), namely, *granulation* on scales of 1 Mm and *supergranulation* (20–70 Mm). An intermediate scale of convection, called *mesogranulation*, probably exists, but it possesses a continuum of scales between granulation and supergranulation (Nordlund et al. 2009; Yelles Chaouche et al. 2011). A much weaker convection (*giant cells*) may also be present on a scale of 200 Mm (comparable with the depth of the convection zone). It has not yet been detected but is seen in global convection experiments (e.g., Miesch 2005; Jouve et al. 2013) and is expected on theoretical grounds, since the largest scale of convection in a simulation is invariably the size of the numerical box. Giant cells may well have a surface velocity of only 0.03 km s^{-1} and a lifetime of a year, and they may possibly create large-scale weak unipolar regions and polarity-inversion lines at the solar surface along which prominences are located.

A stronger global motion is solar rotation, but the Sun does not rotate like a solid body, since its surface exhibits *differential rotation*. As viewed from the Earth, equatorial regions rotate once every 26.24 days (the *synodic period*), at a speed of 1.9 km s^{-1}, while at a latitude of 26° typical of sunspots the synodic period is 27.28 days (the *Carrington period*) and at latitudes of 60° it is about 31 days. Few accurate measurements exist at higher latitudes, but polar regions take 36 or 37 days. In contrast, the so-called *sidereal period* (i.e., relative to the stars) is 24.47 days at the equator and 25.38 days at 26°.

Sunspots rotate somewhat more rapidly than the photosphere and, since they are presumably anchored below the surface, this implies that the angular velocity increases slightly with depth. However, coronal holes (Sec. 1.6.3) have very little differential rotation, implying that they continually reconnect as they evolve.

A much weaker *meridional flow* is associated with differential rotation and directed towards the poles at the solar surface at about 20 m s^{-1} (Duvall 1979). It may be faster at sunspot minimum (Komm et al. 1993; Hathaway and Rightmire 2010). The details of its continuation into the convection zone (important for dynamo theory) are unknown: presumably, there is a return flow (possibly multi-cellular) of a few m s^{-1} or less in the convection zone: at 1 m s^{-1}, it would take 25 years to traverse the base of the convection zone in one hemisphere. Hathaway (2012b) has found a hint using supergranules of a return flow at a depth of 50 Mm.

Differential rotation, meridional flow and the rotational structure of the convection zone are caused by the subtle and complex effects of rotation on inhomogeneous compressible turbulent convection: a

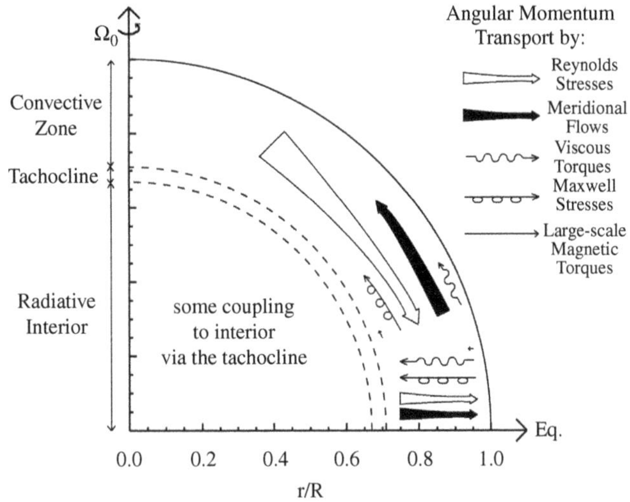

Figure 1.6. Physical effects creating differential rotation in the convection zone. Angular momentum is carried towards the equator by Reynolds stresses, which are opposed by meridional circulation as well as viscous and Maxwell stresses (after Brun 2004).

combination of anisotropy of Reynolds stresses, asphericity of heat flux and meridional flow is involved. Enhanced differential rotation at the equator is associated with the transport of angular momentum from pole to equator in the convection zone by several effects (Figure 1.6). Meridional circulation is thought to be due to a small departure from magnetogeostrophic balance, in which the relatively large effects of buoyancy, Reynolds stresses, pressure gradients and Coriolis forces almost balance one another (Brun et al. 2011). Also, a latitudinal heat flux (originating in either the convection zone or the tachocline) may play a significant role if the poles are, say, 10 K hotter than the equator.

The main effect that drives differential rotation is the Λ-*effect* (Rüdiger 1989). It is the nondiffusive part of the *Reynolds stress tensor* [– that is, nonlinear correlations $\langle v_r v_\phi \rangle$ and $\langle v_\theta v_\phi \rangle$ between radial (v_r), latitudinal (v_θ) and toroidal (v_ϕ) flows, where the angular brackets represent horizontal averages over many small turbulent cells]. These produce mean radial and latitudinal angular momentum fluxes ($F_r = \rho r \sin\theta \langle v_r v_\phi \rangle$ and $F_\theta = \rho r \sin\theta \langle v_\theta v_\phi \rangle$). The correlations arise from convection cells that are elongated in a north-south direction and are tilted preferentially by Coriolis forces towards the rotation axis, thus departing from the local radial direction and from meridional planes. Reynolds stresses appear to be particularly strong in concentrated downflowing plumes that separate broad weak upflowing regions. At lower latitudes, the tilts are bigger and the correlations stronger. The effect is present in local (Brummell et al. 1998) and global (Glatzmaier and Gilman 1982; Brun et al. 2004; Miesch 2005) computational experiments on compressible convection and also in mean-field models that use weak turbulence theory (Kitchatinov and Rüdiger 1993, 2005; Rempel 2005). So far, the best global experiments (Brun et al. 2011) can model from the global scale down to supergranule scales, where differential rotation is still important, so it will be interesting to see what happens when they can be extended down to much smaller scales.

1.3.4 Helioseismology and 5-Minute Oscillations

The ability to infer the properties of the solar interior from observations of solar oscillations has produced many surprises and has, quite simply, revolutionised our knowledge of the solar interior (e.g., Gough and Scherrer 2001; Thompson et al. 2003; Chaplin 2006). Before 1975, our understanding of the interior was based on modelling, in which the global observable quantities (the solar mass, radius and luminosity) were used to calibrate spherically symmetric models (Sec. 1.3.2), assuming a solar age estimated from the age of the Earth and an initial He abundance (Y) of 0.25 deduced from the amount of He in interstellar gas clouds. The main uncertainties were the neutrino flux (Sec. 1.3.1), the efficiency of solar spin-down (and hence the angular velocity of the core), the oblateness of the Sun, the nature of the solar dynamo, and the variation if any of the solar constant or total irradiance.

After the first observation of 5-minute (3 mHz) photospheric oscillations (Leighton et al. 1962), their true nature came to light only with the pioneering observations of Frazier (1968a,b), who suggested they are standing resonant acoustic waves formed within the convection zone. The theory was then developed (Ulrich 1970; Leibacher and Stein 1971), but the idea was only fully accepted five years later when the predicted rich spectrum of overtones was observed as a series of power ridges (Deubner 1975). These are due to high-degree modes (with small horizontal wavelength) that penetrate a shallow distance into the convection zone. Later, low-degree modes extending to the deep interior were also observed from the average line-of-sight velocity of the whole Sun (Claverie et al. 1979). The diagnostic potential was soon recognised (Christensen-Dalsgaard and Gough 1976) and helioseismology was born. For example, Gough (1977) was able to deduce a better value for the depth of the convection zone (i.e., 200 Mm) than the previous one (150 Mm).

The oscillations so far detected are either globally coherent acoustic waves (called *p-modes*), for which the restoring force is pressure, or *f-modes*, which at high degree have the character of surface gravity waves. The p-modes have periods of about 5 minutes (ranging typically from 3 to 10 minutes). They are formed by waves that live long enough to make many complete circuits of the Sun and interfere with themselves to form normal modes, thus providing a measure of internal properties averaged in solar longitude. The waves come from broadband acoustic noise generated by photospheric granulation with a power peaking near 5 minutes, which is roughly the lifetime of granules. Thus, the Sun is ringing like a bell in a sandstorm with many different modes. A typical location on the surface is moving up and down with an amplitude of a few hundred m s^{-1} corresponding to 10 cm s^{-1} per mode.

The sound travel-time through the Sun is about 2 hours. Sound waves with periods above about 3 minutes (frequencies below 5 mHz) cannot propagate in the solar atmosphere and so are reflected back into the interior. The sound speed (c_s) is about 7 km s^{-1} at the surface and increases into the interior to about 500 km s^{-1} at the centre. As a wave propagates inwards, the increase in sound speed causes it to be refracted away from the radial direction and eventually to return to the surface. The turning point for a wave of given frequency (ω) and horizontal wavenumber (k_h) occurs where the horizontal phase speed equals the local sound speed [$\omega/k_h = c_s$ or $\omega r/(l+\frac{1}{2}) = c_s$]. Thus, the larger the wavelength, the smaller the k_h (or the l), the larger the value of c_s, and so the deeper the reflection point.

The terminology (l,m,n) of spherical harmonics in spherical polars (r, θ, ϕ),

$$Y_l^m(\theta,\phi) = c_{lm} P_l^m(\cos\theta)\exp(im\phi),$$

is used to distinguish the modes, where P_l^m is a *Legendre function.* The dimensionless *degree* (l) measures the horizontal wavenumber. Its east-west component is the *azimuthal order* (m), which gives the number of meridional nodal planes (half the number of nodes around the equator). There are $l - m$ half-cones intersecting the solar surface in lines of latitude. The *radial order* (n) is the number of spherical nodes in radius as one goes into the Sun, and the frequencies of the modes increase with n at fixed l and m (see Figure 1.7 for an example). It is traditional to describe moderate- or low-degree modes ($l <$ few hundred) by their cyclic frequency $\nu = \omega/(2\pi)$ (Hz), but high-degree modes (when l is more than a few hundred) by their *angular frequency* ω (radians s^{-1}) and their horizontal wavenumber k (radians m^{-1}).

The low-degree modes (Claverie et al. 1979) are almost radial in nature with 15 or 25 wavelengths in the radial direction. The power spectrum shows a sequence of rather uniformly spaced peaks due to $l = 0$ and $l = 2$ modes (having m even) together with $l = 1$ and $l = 3$ modes (having m odd). The (large) spacing between the two groups is a measure of the acoustic travel time through the Sun. Later, from a 5-day run at

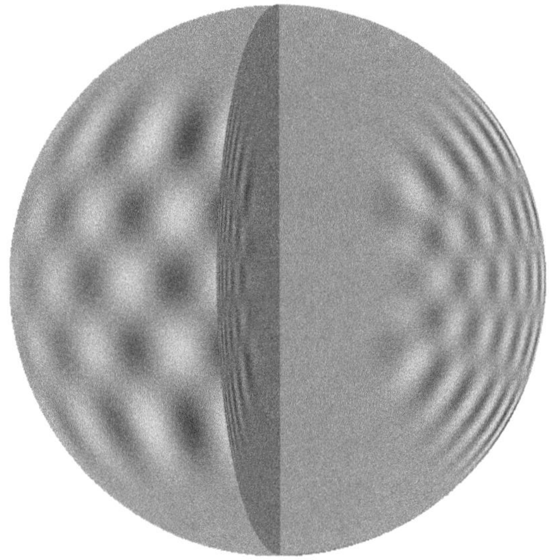

Figure 1.7. Example of a global acoustic oscillation with $l = 20$, $m = 16$, $n = 14$ and having neighbouring cells moving in opposite directions along the line of sight.

the south pole, Grec et al. (1980) found that the modes of even parity produce two peaks of similar power separated by a (small) frequency spacing of about 10 μHz, while the odd peaks have a smaller spacing with most of the power in the $l = 1$ peak. Next, by passing the light through a cylindrical lens and so contracting the Sun's image to a line along the rotation axis, Duvall and Harvey (1983) observed and identified the nature of zonal modes with $m = 0$ and degrees from 1 to 139. An important advance was the discovery by Duvall (1982) that the frequencies obey a waveguide law ($[n + \alpha]/\omega = f(\omega^2/[l(l+1)])$), where $d = R - r_t$ is the penetration depth and r_t is the turning point satisfying $\omega r_t = Lc_s(r_t)$. An inversion of the function f then led to a determination of radial variation $c_s(r)$ of the sound speed, almost to the centre of the Sun (Christensen-Dalsgaard et al. 1985). This favoured models with $Y = 0.25$, which in turn implied that the solar neutrino problem could not be solved by changing the solar models.

Then Duvall and Harvey (1984) and Duvall et al. (1984) had the brilliant idea of rotating their cylindrical lens through 90 degrees and so observed sectoral modes with $l = m$. This enabled them to measure the splitting by rotation with respect to azimuthal order m and to deduce for the first time the radial variation of the Sun's internal rotation. They found several unexpected results: the rotation (Ω) increases slightly with depth just below the photosphere and is almost uniform along radial lines through the convection zone; and the radiative interior rotates almost uniformly at a slower rate than the surface equatorial value (see also Brown et al. 1989).

The next ten years saw measurements at Big Bear Observatory of modes with $l = 4$ up to $l = 1300$ and whole-disc observations from the BiSON (Elsworth et al. 1990) and IRIS (Fossat 1991) networks of those with $l = 0, 1, 2, 3$. These were used to deduce the internal sound speed, which was found to agree with the standard model to within 0.5 per cent. Revisions to nuclear cross-section rates and opacity were made and also the effect of He settling was included. Furthermore, a calculation of the second derivative of the sound speed led Christensen-Dalsgaard et al. (1991) to deduce the location of the base of the convection zone (where the stratification ceases to be adiabatic) at a radius $0.713 \pm 0.003\ R_\odot$.

What was then needed to obtain more precise frequencies was a set of long, continuous observations in order to give cleaner spectra with minimal side-lobes. In 1986, after a meeting in Darmstadt at which I presented the scientific case, ESA and NASA agreed to cosponsor the SoHO mission (Solar and Heliospheric Observatory). SoHO was launched in 1995 with twelve experiments to study the Sun and its environment,

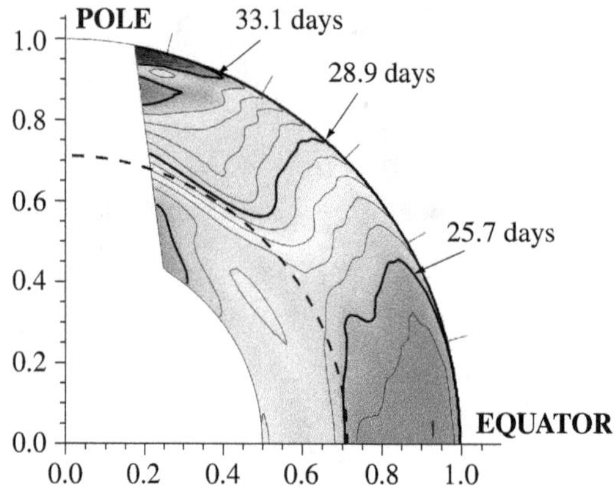

Figure 1.8. Rotation rate on a meridional cut through the solar interior, inferred from rotational splitting of normal modes, with rotation times indicated on three thick curves. Note the shear layers at the solar surface and at the tachocline (after Schou et al. 1998).

including three for the solar interior, namely, GOLF, VIRGO and SOI/MDI. The helioseismology measurements from space have been complemented by more from the ground for both low-degree modes (BiSON and IRIS, Chaplin et al. 1996; Gelly et al. 1997) and medium-degree modes (GONG deployed in 1995).

Observations from SoHO and the ground have given a major advance in our understanding of the solar interior, with several million different normal modes of vibration being observed with much higher precision than before. First of all, an improved measure of the mean internal structure (averaged over latitude and longitude) as a function of radius (r) gives an impressive agreement with the standard solar model (Figure 1.4): the maximum discrepancies are just below the base of the convection zone, where the observed sound speed is higher by only 0.2 per cent (due mainly to a deficit of He decreasing the mean molecular weight) and near the core where it is lower by 0.1 per cent (Kosovichev et al. 1997). Secondly, the small frequency splittings of the $2l+1$ modes of different m and the same degree (l) and order (n) due to waves travelling in different directions give information on the departures from sphericity (from the even splittings) and on the angular velocity (from the odd splittings) (Gough and Thompson 1991).

Some of the most dramatic discoveries from helioseismology concern the internal angular velocity. Prior to the helioseismology age, remembering the *surface* differential rotation with its faster rotation at the equator, the *internal* rotation was widely expected to be constant on cylinders parallel to axis of rotation in view of the Taylor-Proudman theorem. It was also expected that the the dynamo would be acting throughout the convection zone.

However, SoHO (Figure 1.8) confirmed earlier suggestions by Duvall et al. (1984):

(i) At mid-latitudes the angular velocity is almost constant along radial lines – that is, on cones (Schou et al. 1998). The way in which the Taylor-Proudman constraint is broken involves deflection towards the poles by rotation of the convective energy flux (making the poles 5 to 10 K hotter than the equator) as well as the introduction by the tachocline of an entropy variation at the base of the convection zone (Brun et al. 2004; Rempel 2005).

(ii) The radiative interior below the tachocline at the convection-zone base rotates roughly like a solid body at a rate equal to the surface value at a latitude of about 40 degrees. Thus, at the tachocline itself, a very strong shear layer exists, with a radial width at the equator of at most 0.03 $R_\odot$ that increases with latitude (see Sec. 8.5.3). This is now regarded as a prime location for the dynamo generation of magnetic fields, which subsequently rise to produce active regions and sunspots. Indeed, understanding in detail the tachocline is now a major challenge (Hughes et al. 2007).

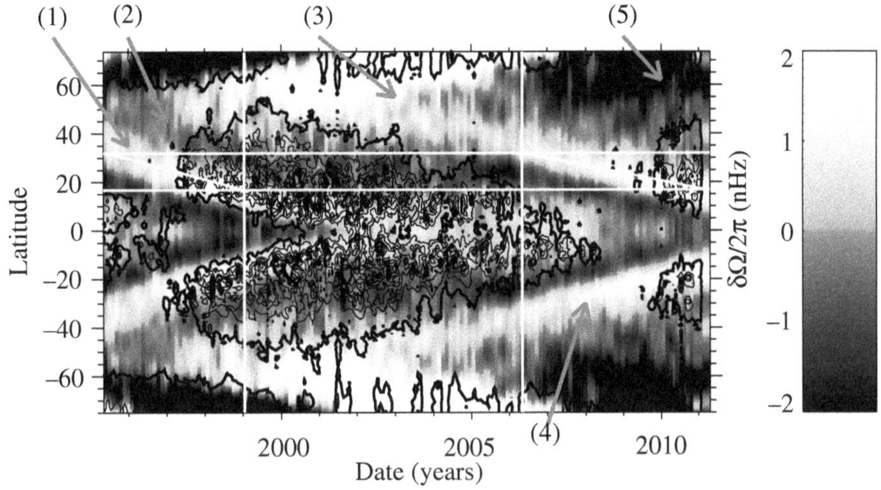

Figure 1.9. Torsional oscillations, showing alternating bands (1) and (2) rotating faster and slower than normal. Low-latitude bands (4) migrate from mid-latitudes towards the equator, while high-latitude bands (3) migrate towards the poles but (5) were late starting in 2010 (courtesy Rachel Howe).

(iii) The internal rotation increases slightly with depth just below the photosphere in a surface shear layer to a maximum at a depth of 35 Mm (at a radius of 0.95 $R_\odot$).

(iv) The poles rotate slowly, although helioseismology inversions near the poles and below the convection zone are difficult, and so these regions have been left blank in Figure 1.8. BiSON data from low-degree modes suggest that the rotation is uniform between 0.2 and 0.65 $R_\odot$ with the suggestion of a slight decrease just below 0.2 $R_\odot$ (Chaplin et al. 1999), but the rotation of the energy-generating core is still uncertain.

Differential rotation (Ω_s) at the solar surface is well approximated by the form

$$\Omega_s(\theta) = \Omega_e(1 - \alpha_2 \cos^2\theta - \alpha_4 \cos^4\theta)$$

in terms of the colatitude (θ), and this remains roughly constant on cones through the convection zone. The values of the constants at a radius $r = 0.995\ R_\odot$ were obtained by Schou and colleagues to be $\Omega_e/2\pi = 455$ nHz, $\alpha_2 = 0.12$, $\alpha_4 = 0.17$, and the corresponding rotation rate of the radiative envelope is $\Omega_0/2\pi = 430$ nHz. The rotation increases by about 20 nHz in the surface shear layer.

Torsional oscillations had been discovered by Howard and Labonte (1980) at the solar surface as weak alternating bands of rotation that are faster and slower than normal. They are centred on the active-region belts and drift equatorwards with the sunspot zones, causing the angular velocity at a given latitude to oscillate with a period of eleven years (Figure 1.9). At higher latitudes, the bands migrate towards the poles from about 60 degrees. Schou (1999) discovered from SoHO that the oscillations extend into the convection zone by at least 0.15 $R_\odot$ as bands about 15-degrees wide.

Additional results from SoHO are: the oblateness of the Sun is just what is expected from the rotation; from asymmetries in lines of the p-mode spectrum, the confirmation of a previous result (Duvall et al. 1993) that the source of the oscillations is only 50 to 100 km below the photosphere; the angular velocities above and below the tachocline vary in antiphase with a 1.3-year period and an amplitude of about 6 nHz (Howe et al. 2000).

The global oscillation frequencies vary with the solar cycle (Elsworth et al. 1990) and have a strong correlation with surface magnetic activity. The main cause of this variability is thought to lie in layers close to the solar surface. In the tachocline or deep interior there is as yet no direct seismological detection

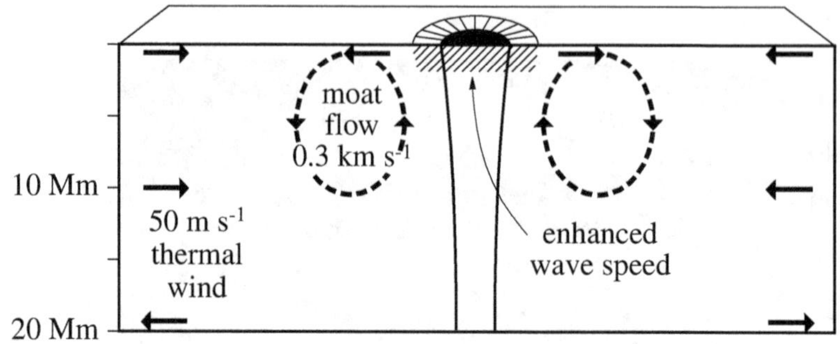

Figure 1.10. The tentative structure below a sunspot, as deduced from local helioseismology. The only region showing a significant change of wave speed is in the first 2 Mm below the spot, where it is enhanced. There is a strong outward moat flow of 0.25–0.3 km s^{-1} at the surface and down to a depth of 5 Mm, which is probably completed below the surface (dashed) like a supergranule. Far from the spot there is a much weaker background inflow of 50 m s^{-1} (a thermal wind) into the active region as a whole that reverses at a depth of 20–50 Mm (courtesy Laurent Gizon).

of magnetic fields, but there is a greater possibility to detect such fields in the upper convection zone, where their dynamical importance is greater.

As a complement to global methods, *local helioseismology* has also been used to probe just below the solar surface, by regarding the disturbances as propagating waves rather than normal modes. For example, in his *ring diagram analysis*, Hill (1989) finds a two-dimensional power spectrum with respect to wavenumber (k) at fixed frequency (ω) in the form of a series of almost circular rings; the displacement of the centre of each ring measures the mean horizontal velocity that advects the waves. An inversion yields the horizontal flow velocity as a function of depth. Instead, in *time-distance seismology*, Duvall et al. (1993, 1997) determine the wave travel-time between pairs of points on the surface from the time-shift that gives the best correlation between wave forms observed at each point.

Local helioseismology has produced several interesting results. One is that poleward meridional flows (Sec. 1.3.3) extend downwards from the solar surface into the convection zone; however, an expected equatorward return flow near the tachocline (of a few m s^{-1} from mass continuity, either as a single cell or multiple cells) has not yet been measured. Another result is that superposed on the meridional flows one finds large-scale flows towards active regions down to a depth of 20 Mm, where they reverse (Gizon et al. 2010). Furthermore, below a sunspot, the only change in wave travel speed is an enhancement in the first 2 Mm (Gizon et al. 2010) (Figure 1.10).

Again, Lindsey and Braun (2000) have been able to image active regions on the far side of the Sun and so detect changes in activity a week or more before a growing active region is visible from Earth. Also, acoustic helioseismic waves generated by a strong solar flare have been observed like ripples propagating from a stone thrown into a pond (Kosovichev and Zharkova 1998). Another intriguing discovery is that supergranule cells move at different speeds according to their size, with the larger cells rotating faster (Hathaway et al. 2000; Beck and Schou 2000).

In conclusion, the Sun's internal dynamics are highly complex and involve subtle interactions between turbulent convection, rotation, shear, instabilities and magnetic fields. The theory of the solar interior, which is largely outside the scope of this book, has many future challenges, including: detecting the g-modes that are expected to have amplitudes of 0.1 mm s^{-1} or less at the surface and can probe the solar core; deducing more reliable information on the deep interior and the region near the poles; understanding the observed angular velocity; modelling the tachocline and understanding the dynamos that are likely to be operating both there and possibly throughout the convection zone and in the shear layer near the solar surface.

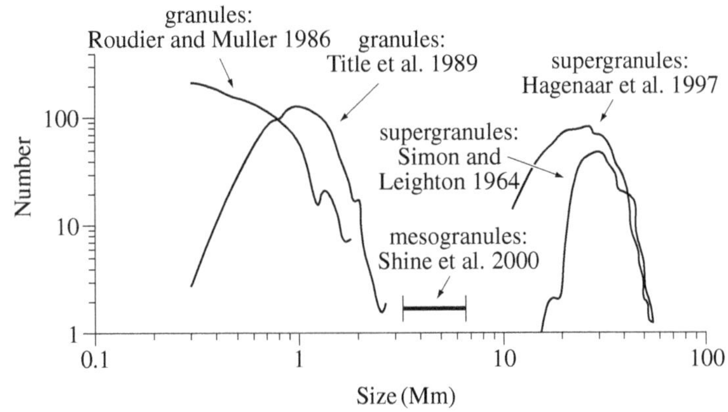

Figure 1.11. The distributions of sizes of: granules according to Roudier and Muller (1986) from Pic du Midi at 0.25 arcsec resolution and Title et al. (1989) from Spacelab2 at 0.13 arcsec; mesogranules (Shine et al. 2000) from SoHO/MDI; and supergranules according to Simon and Leighton (1964) and Hagenaar et al. (1997).

1.4 The Photosphere

The photosphere is the Sun's thin visible surface layer, named after the Greek word for 'light'. It may be defined as the region (a few 100-km thick) from which most of the Sun's visible light at 5,000 Å (green) is emitted (Sec. 1.2.2). However, the photosphere is neither uniformly bright nor perfectly still (PROBLEM 1.6). Several types of convective motion (Sec. 1.3.3) are present on a range of different scales (Figure 1.11), namely, *granulation* and *supergranulation* (Sec. 1.4.1), as well as possibly *mesogranulation* and large-scale *giant cells* (Sec. 1.3.3). In addition, two other types of large-scale flow are seen, namely, *differential rotation* and *meridional flow* (Sec. 1.3.3), and superimposed on these flows one finds *5-min oscillations* (Sec. 1.3.4).

1.4.1 Granulation, Meso- and Super-Granulation

Granulation covers virtually the whole Sun at the photosphere, so that, at any one time, there are several million *granules* present with irregular shapes. They represent the tops of small convective cells, continually forming and disappearing in a turbulent manner. The centre of each granule is bright, because it consists of hot, rising (0.5 to 1.5 km s^{-1}) and horizontally outflowing plasma, whereas the 0.3 Mm-wide boundaries (or *intergranular lanes*) are dark and represent cool, falling material.

A typical granule diameter is 1 Mm, but there is a wide range from 0.3 to 2 Mm. Granules of size 1 Mm with a speed of 1 km s^{-1} have a turnover-time of order 1,000 sec $\approx$ 20 min. They have typical intensity contrasts of 5 to 15 per cent in white light (depending on the resolution) or up to 32 per cent in the near UV (Hirzberger et al. 2010). Parts of granules appear brighter in magnetic regions near the limb, where one looks at deeper layers of a granule through magnetic features. However, in general they are less vigorous in regions with more magnetic flux (e.g., Kobel et al. 2012).

Granules have a mean lifetime of about 5 to 10 min, with a range from 1 to 20 mins: those with the longest lifetimes are the largest and have the smallest random horizontal velocities. The evolution of a granule is strongly influenced by its environment, namely, the nearby magnetic field and its location within meso- and super-granules. Indeed, they move with both meso- and super-granules (Title et al. 1989). So-called *exploding granules* are brighter than normal and form rings that expand at 1.5 to 2.0 km s^{-1} and fragment over about 10 min. Granules are born either by the merging of two granules or the splitting of a large one, such as an exploding granule.

Mesogranulation as a separate mode of convection between granulation and supergranulation was controversial, since it is difficult to observe (November et al. 1981; Shine et al. 2000). An example in Figure 1.12 shows the divergence of horizontal velocity (top right), with mesogranules being bright patches indicated by arrows, as well as velocity vectors of a large-scale, long-lived supergranule (bottom right). Mesogranules tend to form near the centres of supergranules and to be advected to their boundaries, with lifetimes of 3 to 6 hr. Exploding granules often occur inside mesogranules in regions of positive divergence of the horizontal flow. Simulations and observations imply a continuum of scales between granulation and supergranulation in the velocity power spectrum, with no peak for mesogranulation (Nordlund et al. 2009; Rieutord and Rincon 2010). Sunrise observations (see Figure 1.17) agree in finding no preferred scale for mesogranulation, but they do confirm its reality from the way it concentrates magnetic elements at its boundaries (Bonet et al. 2010; Yelles Chaouche et al. 2011), and from the presence of calm areas in mesogranule interiors; here the magnetic field is low and there are fewer granular magnetic loops (Ishikawa and Tsuneta 2010; Martínez González et al. 2012).

Supergranulation (Rieutord and Rincon 2010; Hathaway 2012a) shows up most clearly not in white-light intensity but as a pattern of horizontal motions, either from Doppler measurements near the limb or from local correlation tracking of granules near the Sun's centre (Figure 1.12). It was discovered by Hart (1954) and studied by Leighton et al. (1962) in the first Doppler images of the Sun. Typical sizes are 30 Mm, but they range from 20 to 70 Mm (Rieutord and Rincon 2010) and the mean size depends on the measurement method: 32 Mm with an autocorrelation method (Simon and Leighton 1964), 13–18 Mm with a "basin-finding" algorithm (Hagenaar et al. 1997), 36 Mm from granule tracking (Rieutord et al. 2008) and 27 Mm from Dopplergrams (Hirzberger et al. 2008).

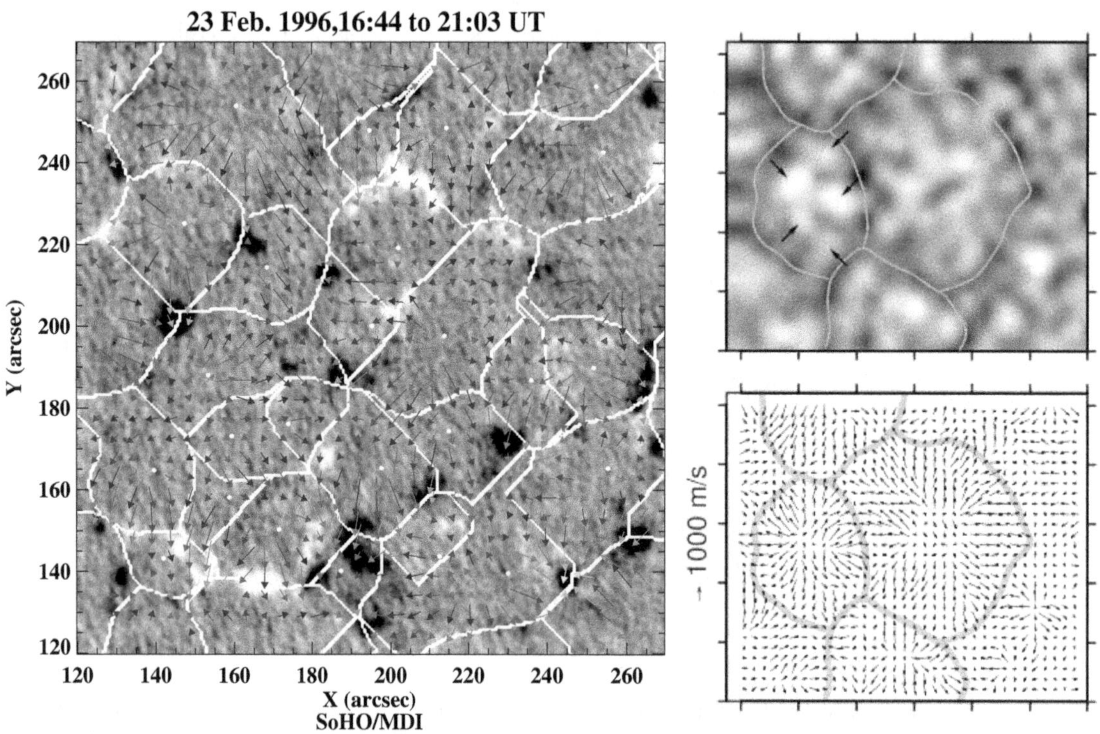

Figure 1.12. Supergranular network determined from the long-term divergence of the horizontal photospheric flow, superimposed on: (left) a magnetogram from SoHO/MDI, with the magnetic field (black and white patches) clustering in downflow lanes and especially network junctions (from Schrijver and Zwaan 2000); (right) mesogranules (top) and supergranular flow (bottom) (from Shine et al. 2000).

Plasma rises at the centre of each supergranule cell at 30 m s^{-1}, moves horizontally outwards at typically 350 m s^{-1} and then descends at the boundaries (Hathaway et al. 2002; Rieutord et al. 2010). The cells are irregular in shape and last for 1 to 2 days with a mean of 1.6 days, approximately the turnover-time (Hirzberger et al. 2008). Some supergranules are born between existing cells, others by fragmentation or splitting off from an existing supergranule. They develop as strongly diverging areas expanding at about 100 m s^{-1} for over a day. Coriolis effects make them into anticyclonic structures, with a vertical vorticity that is negative in the northern hemisphere and positive in the southern. The temperature is slightly higher in the cell centre by 1 to 2 K (Meunier et al. 2008). In the chromosphere, their boundaries outline a bright network where magnetic flux is concentrated and spicules are generated.

Photospheric vortex tubes are often present in convection simulations at downdraft junctions of cells (Nordlund et al. 2009). They have been discovered in SST and Sunrise movies of the photosphere near magnetic elements (Bonet et al. 2008, 2010; Steiner et al. 2010), with lifetimes of 5 to 10 min and rotation periods of about 30 min. They occur preferentially at the edges of mesogranules and supergranules.

One expects a large range of sizes and lifetimes for vortex tubes in the solar atmosphere, just as there are for magnetic flux tubes, and so it is important to distinguish among different types: thus, a twisting erupting prominence is an example of a large vortex tube, which we could call a *prominence tornado*, say; a rotating macrospicule (or *macrospicule tornado*, Sections 1.5.2, 10.4.2) has intermediate size, and so does a *barb tornado* below a prominence (Sec. 1.8.3); if spicules rotate, they are examples of very small vortex tubes and could be called *spicule tornadoes.*

The relationship between photospheric vortex tubes and the different kinds of tornado, and in particular whether they drive them, is not yet clear. The continuation of such a tube up into the chromosphere and corona to give a macrospicule tornado has been detected with SST and SDO/AIA (Wedemeyer-Böhm and Rouppe van der Voort 2009; Wedemeyer-Böhm et al. 2012), but it is not known whether that is a general feature and whether it applies to other kinds of tornado.

Photospheric faculae ("little torches") are bright patches in white light, especially visible near active regions at the limb (Figure 1.13a). They are the edges of granules seen obliquely through kilogauss magnetic elements that reduce the opacity (Keller et al. 2004; Berger et al. 2007). At high resolution, they consist of numerous *facular points* of diameter 100 km or less from Sunrise (Solanki et al. 2011), situated in intergranular lanes, coincident with intense magnetic flux and varying over times of a few minutes. They often have dark lanes or stripes, and in the G-band they are also clearly visible at disc centre. Near active

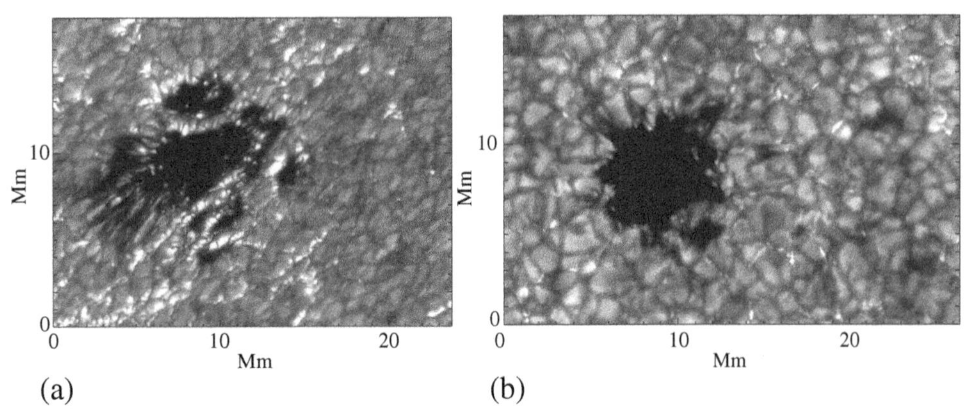

Figure 1.13. High-resolution G-band images from the SST on La Palma of photospheric granulation in a plage near a pore, indicating: (left) faculae near the limb and (right) magnetic bright points near disc centre. Both are located at the edges of granules and form a network around mesogranule and supergranule borders (from Berger et al. 2007, courtesy Luc Rouppe van der Voort).

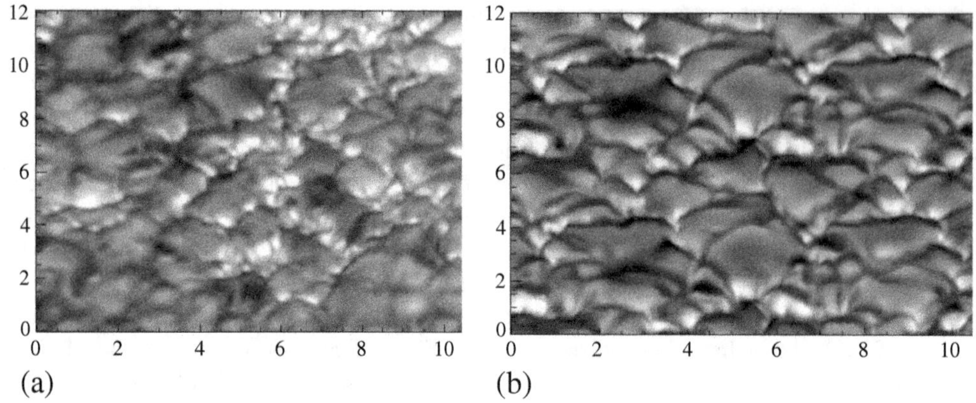

Figure 1.14. Granules viewed obliquely in (a) a G-band image from the SST on La Palma (courtesy Göran Scharmer) and (b) a synthetic G-band image from a numerical simulation (courtesy Mats Carlsson). Scales are in Mm. Locations with strong magnetic fields give bright faculae as one looks through a partially evacuated transparent region towards the bright inclined side of a granule.

regions, they tend to cluster more densely as a crinkle pattern, known as *filigree* (Dunn and Zirker 1973), with a width of 100 km or less and running along intergranular lanes. Individual crinkles are up to 2.5 arcsec (1800 km) long and appear to be jostled around by granules.

Faculae are hotter than the normal photosphere by a small amount at the solar surface, but this increases to 1000 K at the temperature minimum. Observations at high resolution (100 km) and numerical experiments (Lites et al. 2004; Carlsson et al. 2004; Keller et al. 2004) have revealed the subtle three-dimensional structure of granules and faculae, with the $\tau = 1$ surface (Sec. 1.2.2) being raised in the centres of granules and lowered in intergranular lanes (Figure 1.14). Faculae are brighter than normal because the plasma inside magnetic flux concentrations is less dense and more transparent, and so one views deeper and hotter parts of a granular wall.

Outside active regions, faculae exist more sparsely along supergranule boundaries at all latitudes as a *photospheric network*. At high latitudes they are referred to as *polar faculae*. Within the network, downflows are concentrated in small regions, especially at supergranule junctions, with speeds of 3 to 4 km s^{-1} in the upper chromosphere and transition region, falling to 1 to 2 km s^{-1} in the low chromosphere and 0.1 km s^{-1} in the upper photosphere.

Numerous *(photospheric) bright points* (Dunn and Zirker 1973), observed near disc centre, are much smaller than faculae (Figure 1.13b) and are common in supergranule boundaries and interiors (Sánchez Almeida et al. 2004). They have widths of 100 km or less (Solanki et al. 2010) and represent the deep cores of magnetic elements (kilogauss magnetic flux tubes) heated by lateral radiative transfer from the bright walls and scattering from the relatively evacuated interiors. Strings of bright points form *filigree* or sheets. At locations intermediate between disc centre and limb, one sees a mixture of faculae and bright points. When a magnetic element is larger than 300 km, there is not enough lateral radiation to heat the tube interior and it becomes a dark *micropore*.

It had been suggested that granulation and supergranulation are driven by the ionisation of H and He, respectively, but this effect is now thought to be too weak. Instead, there is evidence of a continuous distribution of convection at different scales, with the dominant scale increasing with depth and being proportional to the local scale-height (Stein et al. 2006). Granular scales are likely, therefore, to represent several times the scale-height at the surface (150 km). The cause of supergranular scales is at present unknown, although the fact that supergranules are about 10 per cent smaller near active regions or at solar maximum suggests a magnetic influence on their size (De Rosa and Toomre 2004; Meunier et al. 2008).

1.4.2 Magnetic Field

The photospheric magnetic field possesses a wide range of scales and strengths (e.g., Solanki et al. 2006). The strong-field kG-part is vertical, covers less than 5 per cent of the solar surface and is located mainly in a network at supergranule boundaries (Figure 1.15) (Lagg et al. 2010). Weaker, more-horizontal internetwork fields occupy much of the interior of a supergranule cell, with strengths of 100 to 300 G (Khomenko et al. 2003; Martínez González et al. 2008) and possibly an ambient field of, say, 5 G. It is not yet clear how much flux is present in strong-field and weak-field forms, although the majority of the flux in the quiet Sun is likely to be weak (de Wijn et al. 2009).

The fundamental flux units into which the strong field is concentrated are tiny *magnetic elements* or *intense flux tubes* with typical field strengths of 1 kG, fluxes of 3×10^{9} Wb (3×10^{17} Mx) and diameters of 100 km (Stenflo 1973; Solanki et al. 2006). The field strength decreases with height as the flux tubes spread out from 1,500 to 1,700 G in the deep photosphere to 1,000 to 1,200 G in the middle photosphere and 200 to 500 G at the temperature minimum. The ratio (β) of plasma to magnetic pressure is 0.2 to 0.4 in the tubes and so the magnetic field dominates the plasma. In the neighbouring photosphere $\beta \gtrsim 1$, whereas in the chromosphere and corona (into which the magnetic flux spreads) $\beta \ll 1$ in active regions and $\beta \lesssim 1$ elsewhere.

The above size of 100 km for magnetic elements is near the current spatial resolution, so are the elements likely to be even smaller? A theoretical limit of order 10 km is estimated in Section 9.2, assuming flux concentration by a stagnation flow (at 1 km s^{-1} on a scale of 1 Mm) that balances diffusion, such that below this thickness diffusion dominates. If the flow speed were higher or the flow length-scale smaller, the diameters of the intense flux tubes would be even smaller. With THEMIS, Bommier and Stein (2013) have estimated a mean flux tube diameter in the quiet Sun of 30 km, a separation distance of 230 km and a magnetic field of 1.3 kg.

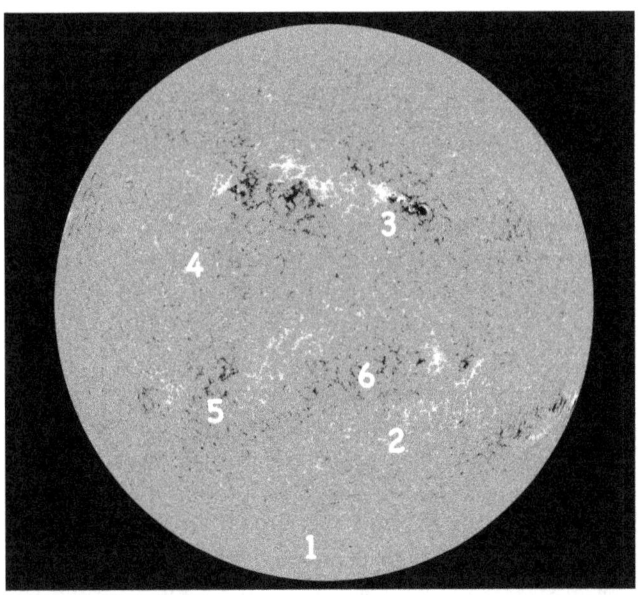

Figure 1.15. A longitudinal magnetic field map of the solar disc from SDO/HMI on 4 April, 2012. Positive and negative polarities are bright and dark, respectively, and (1) indicates polar field, (2) a large-scale unipolar field, (3) an active region, (4) an ephemeral region and normal network, (5) a remnant active region and (6) enhanced network field. The two sunspot bands north and south of the equator can clearly be seen (from SDO/HMI).

The *emergence of magnetic flux* through the photosphere occurs on a wide range of scales. The smallest are tiny *granular magnetic loops* (with a flux as small as 10^{16} Mx), a phenomenon on a granular scale. Next come *ephemeral regions* with fluxes of 10^{19} Mx, which are a supergranule phenomenon that appear in the interior of supergranules, and finally *active regions* with fluxes up to 10^{23} Mx are independent of supergranules and modify the convection pattern as they appear. The basic picture is of a carpet of magnetic loops of many different sizes, closing down at a variety of heights from the photosphere for the smallest, to the chromosphere for intermediate scales (see Figure 1.20a) and the corona at largest scales (see Figures 1.21 and 1.22).

The mean photospheric magnetic field depends on location. In active regions, the area-averaged flux density is about 100 to 500 G. In the quiet Sun, the flux from coronal holes suggests 5 to 10 G for the mean flux that reaches the corona (Sec. 1.6.3), whereas values in the photosphere itself are higher, since most of the small-scale internetwork flux closes down before reaching the corona. Measurements with the longitudinal Zeeman effect depend on spatial resolution, since unresolved mixed-polarity flux cancels; thus, SoHO/MDI gives a mean flux of 3 G, whereas Hinode/SOT implies 10 G. In contrast, the Hanle effect measures the true flux, but is not sensitive to the line-of-sight field (Stenflo 1982; Faurobert-Scholl 1991); it gives 160 G in the low photosphere and 100 to 140 G in the upper photosphere (Trujillo Bueno et al. 2004; Trujillo Bueno et al. 2011), which if the field is isotropic implies a mean vertical field of 80 G in the low photosphere.

Like filigree elements (Sec. 1.4.1), magnetic elements tend to be located in intergranular lanes and to be continually shuffled around by granules with random motions of 1 km s^{-1}. However, these small-scale magnetic structures are organised into larger-scale patterns of several types, as follows.

(1) **Supergranulation Network**. The boundary of a supergranule cell consists of a mixture of weak and strong field elements, but it has a much greater concentration of intense kG flux tubes than in the cell interior. Each network element (of typical flux 3×10^{18} Mx) usually consists of many *magnetic elements*. They coincide with the bright facular and filigree network and are associated with strong downflows of 1 to 2 km s^{-1} in the photosphere and 3 to 4 km s^{-1} in the chromosphere, although the speeds inside the flux tubes are uncertain, since these values would not satisfy mass continuity. In old remnant active regions, the network pattern shows up particularly well, whereas in the quiet Sun (where the flux is lower) the network is much more fragmented. Network fields are concentrated partly by supergranular and granular flow to equipartition strengths of $B = v\sqrt{\mu\rho} =$ 100 to 400 G and even further by convective collapse to kG values (Sec. 9.2.3).

Above a supergranule cell (at about 750 km for active regions and 1,600 km for quiet regions), the magnetic field spreads out to fill the chromosphere and form a horizontal canopy or partial canopy, depending on the ratio of network to internetwork flux. It used to be thought that 90 per cent of the photospheric flux resides in the network, but it is now thought to be smaller than 50 per cent (Schrijver et al. 2006; Solanki et al. 2011).

(2) **Ephemeral Regions**. These small regions of flux emergence are tiny bipolar regions, visible on photospheric magnetograms (Harvey and Martin 1973) and sometimes showing up in the corona as *X-ray bright points* (Sec. 1.6.2). They are much less common than short-lived internetwork flux loops (see below), but they each carry 1 to 3 orders of magnitude more flux. Their emergence rate from SoHO/MDI is two ephemeral regions [each with a typical flux of 10^{11} Wb (10^{19} Mx)] emerging in each 10 Mm square per day (Hagenaar et al. 2008). Fifty to ninety per cent of the magnetic flux of the network comes from ephemeral regions. The typical speed of emergence is 1.8 to 2.6 km s^{-1}, and they last for only about half a day, with the longer-lived features containing more magnetic flux. They vary weakly in antiphase with the solar cycle, such that the emergence rate is a factor of 1.5 larger during solar minimum than solar maximum and a factor of 3 larger in a flux-balanced region than in a unipolar region. It is unclear whether

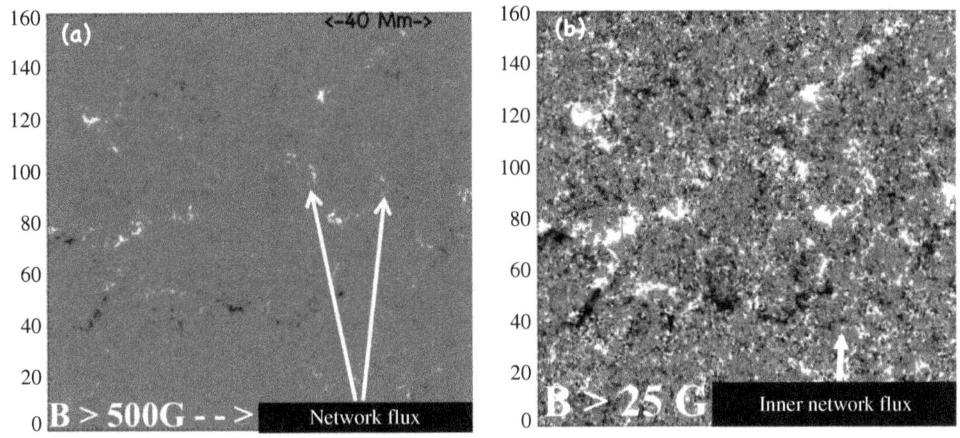

Figure 1.16. Local magnetograms of part of the quiet Sun, when the threshold for magnetic flux is (a) 500 gauss (showing network flux at supergranule boundaries) and (b) 25 gauss (showing the supergranule interiors packed with internetwork flux). Distances are measured in Mm (from Hinode/SOT, courtesy Bruce Lites).

they are produced by: a near-surface small-scale dynamo; or as the decay products of active regions; or as magnetic strands from deeper levels picked up and brought to the surface by supergranules (Moreno Insertis 2012).

(3) **Internetwork (or Inner Network) Field.** This is mixed-polarity, 100 to 300-gauss magnetic flux in supergranule interiors (Keller 1995; Khomenko et al. 2003; Berger et al. 2004; Rouppe van der Voort et al. 2005; Martínez González et al. 2008; de Wijn et al. 2009), originally detected as a weak Stokes V signal (Livingston and Harvey 1971). It carries far more flux than the network and ephemeral regions, but most of it closes down low in the atmosphere. Much of the flux emerges as tiny *granular magnetic loops*, discovered by De Pontieu (2002) with SST and seen by Martínez González et al. (2007) with VTT and later Centeno et al. (2007) with Hinode/SOT.

In the 1980s–90s, the general picture outside active regions was of the photospheric magnetic field being vertical and mainly located in supergranulation boundaries (Figure 1.15). However, this paradigm has now been changed in two ways, earlier glimpsed by Lites et al. (1996). As the threshold magnetic flux with Hinode was reduced, more and more line-of-sight flux came into view, with a huge amount in the interiors of supergranules (Figure 1.16) (Tsuneta et al. 2008). Also, tiny bright points and lines were found between granules (Figures 1.13b, 1.14).

An even greater surprise from Hinode came from studies of the detailed properties of the tiny emerging *granular magnetic loops* (Lites et al. 2008; Martínez González and Bellot Rubio 2009; Martínez González et al. 2010; Ishikawa and Tsuneta 2009, 2010, 2011; Gömöry et al. 2010). They show up as small regions with transient transverse (horizontal) magnetic flux of 2×10^{16} to 2×10^{17} Mx and lifetimes of 1 to 120 min. They are located at the edges of 10 per cent of granules and are situated between vertical flux patches that separate in time, bringing 1.1×10^{24} Mx of internetwork flux to the surface per day. Lifetimes are 1 to 20 mins, but they rarely cancel. Twenty-five per cent of them reach the chromosphere.

Internetwork elements move around between granules until they disappear, fragment, merge or occasionally cancel. They may be produced by the granular processing of flux produced in the convection zone or by a local turbulent dynamo (Moreno Insertis 2012), both of which are consistent with observations that granular magnetic loops are randomly oriented (Martínez González and Bellot Rubio 2009) and the quiet-Sun internetwork flux is independent of the solar cycle (Bühler et al. 2013). In addition, there is

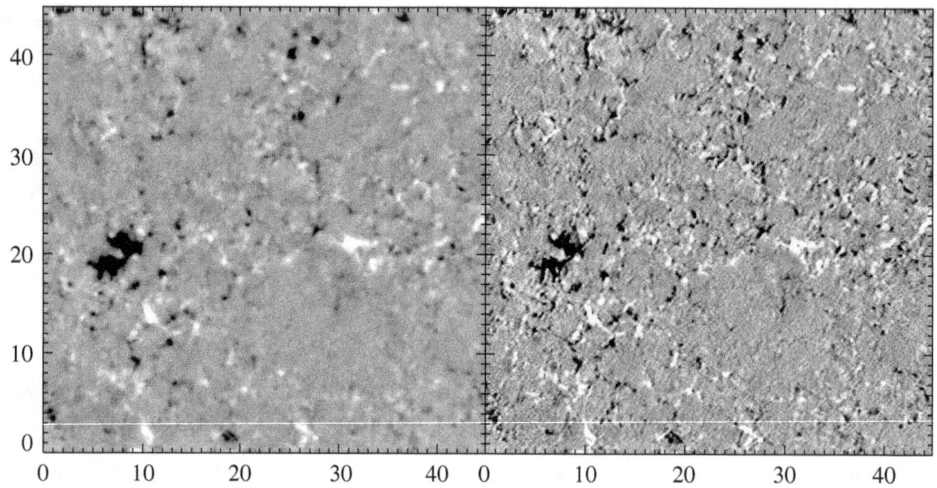

Figure 1.17. Line-of-sight magnetogram from (right) the IMAX instrument on the Sunrise balloon and (left) the same data at the resolution of Hinode. Bipolar fragments (black and white next to one another) represent emerging flux if they are separating and cancelling flux if they are approaching. Distances are measured in arcsec (courtesy Sami Solanki).

perhaps a turbulent magnetic component with fields of a few tens of gauss or higher (Lites 2002; López Ariste and Sainz Dalda 2012; Stenflo and Kosovichev 2012).

The Sunrise balloon mission has detected quiet-Sun magnetic fields with a detection threshold for the flux that is 10 times lower than Hinode (i.e., down to 2×10^{15} Mx) (Danilovic et al. 2010; Solanki et al. 2010). Consequently, it sees ten times as many features (Figure 1.17), but at least half of the flux closes down in the photosphere (Wiegelmann et al. 2010). The lifetimes of features are shorter, having an exponential distribution with a width of 90 sec (Solanki et al. 2011). Interesting results from Sunrise are that tiny magnetic elements often reveal vortex patches (Bonet et al. 2010) and both of them map out mesogranules (Yelles Chaouche et al. 2011). Also, long-lived magnetic patches have oscillations that appear to be driven by granulation (Martínez González et al. 2011).

In coronal holes, most of the kG network component fans out with height and opens into the solar wind. In contrast, most of the internetwork component there closes down at low heights, while some of it interacts with the network and some becomes open. In the quiet Sun outside coronal holes, the internetwork component probably behaves the same, but the network flux closes down and creates a hotter denser corona than in coronal holes. Tsuneta et al. (2008) and Ito et al. (2010) have given values from Hinode/SOT for the area, maximum and average field strengths in a polar coronal hole of 23 Mm^2, 1.6 kG and 520 G. By comparison, corresponding values for the quiet Sun are 7 Mm^2, 1.2 kG and 440 G. In future, it would be interesting to compare polar coronal holes, non-polar holes, and non-coronal-hole parts of the quiet Sun, as well as coronal holes of different strength and lifetime.

(4) **Magnetic Carpet**. The surface of the Sun is covered with a magnetic carpet, in which magnetic fragments are continually emerging, fragmenting, merging and cancelling. Indeed, this is such an efficient process that the magnetic flux at the photosphere is being replaced every 10 to 40 hours (Schrijver et al. 1998). Estimates from the whole Sun (Hagenaar 2001) including smaller ephemeral regions give a flux replacement-time of only 14 hours. This in turn produces highly complex chromospheric (see Figures 1.20 and 9.7) and coronal (see Figures 1.21 to 1.23) structures. Models for the way the quiet-Sun magnetic field distribution is created and maintained by ephemeral flux emergence, fragmentation and cancellation have been constructed (Schrijver et al. 1997; Parnell 2001; Simon et al. 2001).

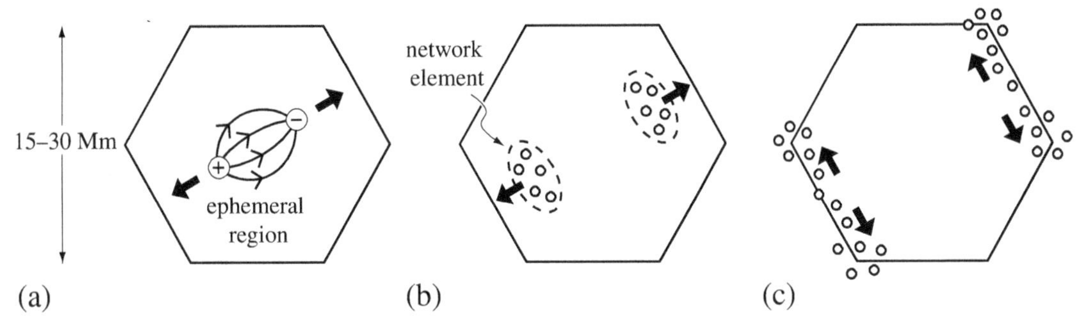

Figure 1.18. Scenario for the evolution of much of the strong flux in the magnetic carpet, each network element consisting of about 10 intense flux tubes (from Priest et al. 2002).

There is a power law for the distribution of flux (F_m) normal to the photosphere of $F_m^{-1.9}$ for fluxes between 10^{17} and 10^{23} Mx from SoHO/MDI, Hinode/SOT and SDO/HMI (Parnell et al. 2009), suggesting that the mechanism for creating or processing the flux is scale-free. This is independent of the solar cycle, with a fall-off above 10^{20} Mx at solar minimum.

The number of emerging elements as a function of magnetic flux (F_m) scales as $F_m^{-2.7}$, and so the small-scale flux dominates the flux emergence process (Thornton and Parnell 2010; Otsuji et al. 2011). The global emergence rate is 1.5×10^{25} Mx/day between 10^{16} and 10^{23} Mx, independently of the solar cycle, with an extra 3×10^{22} Mx/day from large-scale features at solar maximum. The instantaneous global flux is 1.9×10^{24} Mx at solar maximum and 7×10^{23} Mx at solar minimum.

(5) **Magnetic Carpet Evolution**. The typical life story of network photospheric flux has three phases (Figure 1.18). (i) First of all, every 8 hours a bipolar ephemeral region is born over about 30 minutes in a supergranule cell of typical diameter 15 Mm and flux of 3×10^{19} Mx. The two poles of the ephemeral region separate rapidly, initially at a speed of 4 km s^{-1}. (ii) Then each polarity of the ephemeral region moves towards the boundary of the cell at a speed of 0.5 km s^{-1} over about 4 hours and fragments into typically 10 network elements (each having flux 3×10^{18} Mx and so consisting of 10 intense flux tubes). (iii) Over the remaining period, the network elements move along the boundary at typically 0.1 km s^{-1} and eventually cancel. During this period they may also fragment into smaller elements or merge with like-polarity elements. For Hagenaar's (2001) estimate of the flux replacement time, the motion along the boundary takes place for 9 hours and the mean distance travelled is 3 Mm. X-ray bright points occur either during the emergence or more commonly during the subsequent cancellation with network flux. Meyer et al. (2011) have produced a realistic model for the photospheric evolution of the magnetic carpet in response to flux emergence, cancellation, coalescence and fragmentation in a supergranular flow. The importance of the magnetic carpet is described in Section 10.2.6 for global coronal evolution and in Section 10.4.5 for coronal heating.

Given that tiny internetwork magnetic elements contribute at least as much flux as ephemeral regions to the quiet-Sun photosphere, they could have potentially a large, but as yet unknown, impact on the above picture. The life history of internetwork photospheric loops is a mystery, since they usually just disappear from view before their evolution can be determined. However, it would not be unreasonable to expect them to undergo similar processes of fragmentation, coalescence and cancellation in a statistically steady state.

(6) **Active Regions**. At their peak of activity, these large-scale regions of flux emergence contain in their cores groups of sunspots, as described in Section 1.7.2. The mean magnetic field of active regions is a few hundred gauss but is concentrated into the magnetic elements described previously. As a function

of size or lifetime, active regions appear at first sight to form a broad spectrum of flux emergence features which is continuous with ephemeral regions. These range from the largest active regions [with fluxes of nearly 10^{15} Wb (10^{23} Mx) and lifetimes of several months] down to the smallest ephemeral regions [10^{11} Wb (10^{19} Mx)], but they differ in three ways: active regions form two bands north and south of the equator below 45 degrees and their coronal parts are significantly hotter, whereas ephemeral regions are distributed uniformly over the disc; the emergence rate of active regions varies by about a factor of 8 during the solar cycle, whereas ephemeral regions vary very little and may even show a slight variation in antiphase (Hagenaar et al. 2003); active regions are generally orientated in an east-west direction (with a slight tilt such that the leading patch is at a lower latitude), whereas ephemeral regions are orientated in essentially any direction but with a slight excess consistent with Hale's polarity law.

(7) **Large-Scale Unipolar Areas and Coronal Holes**. These extend over several 100 Mm in latitude and longitude. One polarity dominates but they contain elements of both polarities and can be remarkably long-lived, with lifetimes of a year or more. Unipolar regions evolve under the combined influence of differential rotation, meridional flow and magnetic flux dispersal by supergranular motion.

Coronal holes exist above some large unipolar regions (Sec. 1.6.3), especially near the poles. At the north and south poles, the net polarities are opposite most of the time; they reverse 1 to 2 years after sunspot maximum but often not at the same time.

The evolution of the large-scale photospheric field may be described remarkably well by two-dimensional *flux-transport models* (Wang and Sheeley 1994; Mackay and van Ballegooijen 2006), in which the radial magnetic field is assumed to change in response to "turbulent" diffusion as well as the emergence of new flux and its subsequent transport by differential rotation and meridional flow. Thus, as an active region declines, most of the flux disappears in situ (by flux cancellation and submergence), while a small amount may spread into its environment (Dalda and Martínez Pillet 2008) to create unipolar regions. In particular, the net polar field forms from the trailing parts of active regions whose remnant flux migrates polewards.

1.4.3 The VAL Model

A useful mean model of the photosphere and chromosphere, the VAL model assumes thermodynamic properties vary only with height and are static (Vernazza et al. 1981; Avrett and Loeser 2008) (see Figure 1.2). It was built on the earlier *Harvard-Smithsonian Reference Atmosphere (HRSA)* (Gingerich et al. 1971) and was later modified by Fontenla et al. (2009). Its zero level is taken as the point where the optical depth ($\tau_{5,000}$) at a wavelength 5,000 Å in the photosphere is equal to unity.

In the model the photosphere is defined as the region of thickness 550 km from $\tau_{5,000} = 1$ to the temperature minimum, which is 4,300 K (although the HRSA model gives 4,170 K). The density and pressure fall off by two-and-a-half orders of magnitude across the photosphere and a further four orders of magnitude across the next 2 Mm. Between a depth of about 8 Mm and a height of 2 Mm, H is only partially ionised. Indeed, in the photosphere proper, the inclusion of metals such as Mg, Fe and Si increases the proportion of total material that is ionised to 10^{-4} (see Figure 2.1a). Thus, just below and in the photosphere, the adiabatic gradient and energy transport is affected (Sec. 1.3.3), while in the lower chromosphere *ambipolar diffusion* is important due to slippage between the neutrals and ionised fluids (Secs. 2.1.3 and 2.1.5).

The model is important as a standard. However, the solar atmosphere is in reality a highly inhomogeneous, time-dependent non-equilibrium plasma, and so the model is being replaced by more sophisticated dynamic three-dimensional computational models using, for example, the Bifrost code (Gudiksen et al. 2011).

1.5 The Chromosphere and Transition Region

1.5.1 Quiet (Internetwork Non-magnetic) Chromosphere

A quiet, essentially non-magnetic, chromosphere with bright K2V grains may be produced in weak internetwork field between magnetic flux concentrations by dissipation of shock waves (Carlsson and Stein 1995, 1997; Wedemeyer et al. 2004) from high-frequency sound waves (Fossum and Carlsson 2006). In such models, the mean temperature does not vary much in height, which is quite different from the magnetic chromosphere that exists in network and plage regions.

1.5.2 Magnetic (Network or Plage) Chromosphere

According to the VAL model, above the temperature minimum (4,300 K), the average temperature gradually rises through the chromosphere in network or plage regions, while the density decreases rapidly. At a height of about 2,000 km (Figure 1.2), it increases dramatically from 25,000 K to a coronal value of 10^6 K through a narrow *transition region* only a few 100 km thick. However, in reality the chromosphere and transition region are highly non-uniform and dynamic and far from a static plane-parallel "average" state (Figure 1.3c). They are filled with a wide range of fine-scale flows, inhomogeneities and wave motions. The chromosphere was earlier described rather picturesquely as a burning prairie or as the spray of the photosphere thrown up from below, but recent observations and numerical simulations (Figure 1.19b) have revealed the filamentary and time-dependent response of the overlying atmosphere to photospheric convection. The heating requirement for the chromosphere is at least an order of magnitude greater than for the corona.

Middle chromospheric emission (Ca II K line) reveals a *network* of supergranulation boundaries very clearly as an irregular bright pattern. As one goes higher in the solar atmosphere, the magnetic field above supergranule boundaries continues to spread out, causing the network to thicken with increasing temperature, and eventually at coronal heights it ceases to exist. However, much of the chromospheric magnetic field comes from fine-scale internetwork flux rather than network flux.

At the limb, the Hα chromosphere shows up as a forest of plasma jets called *spicules* (Secchi 1877b) (Figure 1.20a). There are at least 100,000 spicules present at a given time, providing a mass flux that is 100 times that of the solar wind. (For theories of their origin, see Section 9.2.4.) *Type I spicules* are seen mostly in active regions and occasionally in quiet regions and coronal holes. It is unclear at what height they originate, but in the upper chromosphere they dominate the scene. They are ejected from supergranule boundaries along the magnetic field, reaching speeds of 10 to 50 km s^{-1} and heights of 3 to 4 Mm in active regions and 5 Mm in the quiet Sun, before falling back (e.g., Beckers 1972; De Pontieu et al. 2007). They often trace out parabolic paths with a deceleration (50 to 400 m s^{-2}) that is not equal to solar gravity. Such spicules have typical lifetimes of 3 to 10 min, diameters of 120 to 700 km, temperatures 10,000 to 15,000 K and electron densities 10^{17} m^{-3}.

Type II spicules were discovered with Hinode/SOT in coronal holes and the quiet Sun (De Pontieu et al. 2007) (Figure 1.20d). They are more dynamic (reaching speeds of 30 to 150 km s^{-1}), longer (2 to 10 Mm) and much shorter-lived (10 to 180 s) than type I spicules. After rising, they fade rapidly over their whole length without falling. Sometimes they accelerate during the rise. They are shorter in active regions (2 to 5 Mm) and are taller in coronal holes (5 to 10 Mm). In coronal holes and the quiet Sun they dominate, whereas in active regions there is a mixture of type Is and type IIs.

Type II spicule movies resemble prairie grass swaying in the wind, with an energy flux of 100 erg cm^{-2} s^{-1} (De Pontieu et al. 2007). As well as swaying motions of 15 to 20 km s^{-1}, they also possess torsional motions of 15 to 20 km s^{-1} (De Pontieu et al. 2012), as *spicule tornadoes* (Sec. 1.4.1). Ca II and C IV spicules (Wilhelm 2000) are well correlated, suggesting that type II spicules are heated to transition-region

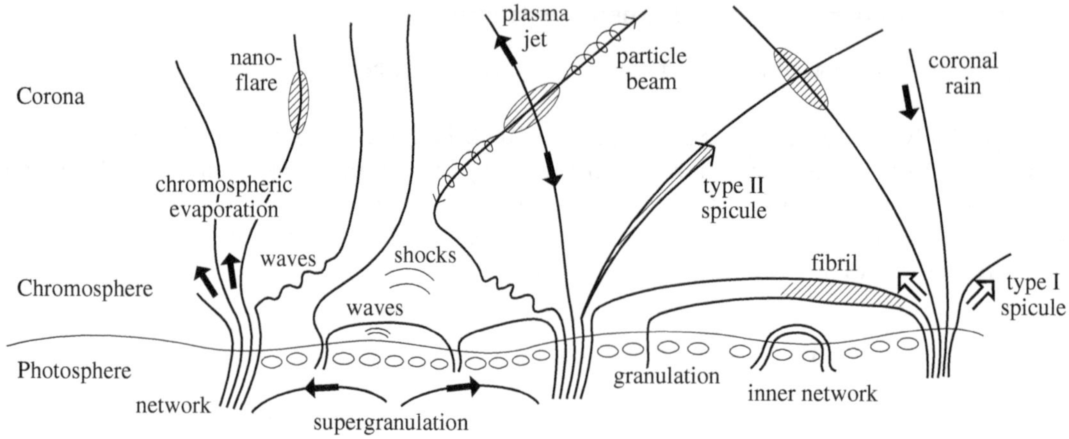

(a)

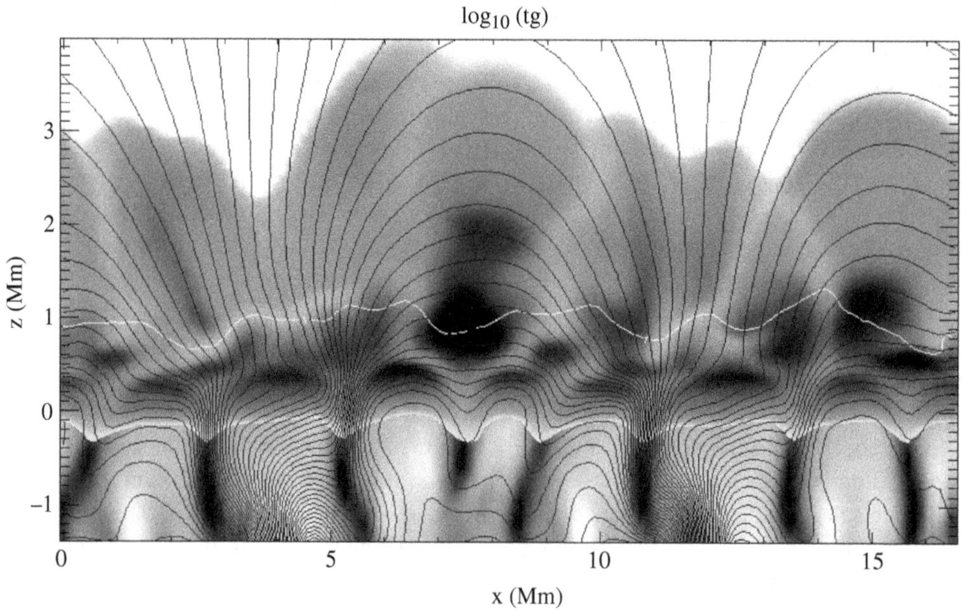

(b)

Figure 1.19. (a) Schematic of the dynamic nature of the "quiet-Sun" chromosphere and corona, with plasma continually being accelerated, heated and cooled due to many physical processes. In the photosphere, granulation and supergranulation concentrates magnetic flux in the network and internetwork. Chromospheric plasma is accelerated as evaporation and draining (in coronal rain), as plasma jets at nanoflares by reconnection, and as type I and type II spicules in the network by reconnection and granular buffeting. In addition, the atmosphere is filled with wave modes and particle beams, generated by the flows and nanoflares. (b) Numerical simulation of the atmosphere (with magnetic field lines shown in black). The shading indicates the temperature above the lower white line (the top of the photosphere) and the vertical velocity below it (with downflows being dark). Below the upper white line $\beta > 1$ and above it $\beta < 1$ (from Carlsson and Hansteen 2005).

and coronal temperatures as they fade (De Pontieu et al. 2011). Also, at temperatures from 65,000 K to 2 MK, one sees a faint upflow (50 to 150 km s^{-1}) in active regions and in the quiet-Sun network at coronal loop footpoints (Hara et al. 2008; De Pontieu et al. 2009). These spicules may play an important role in heating the corona and accelerating the solar wind (McIntosh et al. 2011; McIntosh 2012).

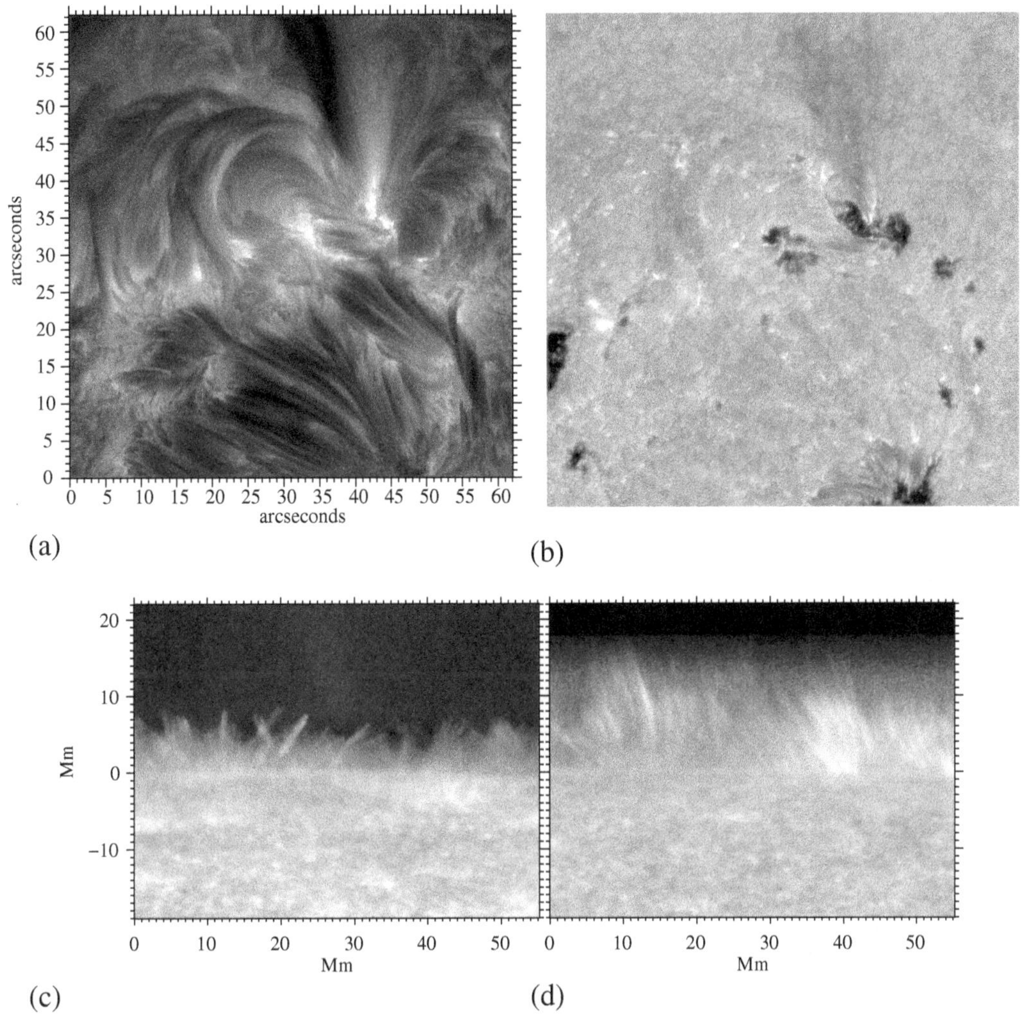

Figure 1.20. (a) Dynamic fine-scale structure of the chromosphere from above (from the SST on La Palma, courtesy Luc Rouppe van der Voort) with (b) a simultaneous photospheric image and from the side (bottom). The bottom images show (c) spicules of type I (in an active region) and (d) type II (in a coronal hole), as seen at the limb in Ca II H from Hinode/SOT (courtesy Bart De Pontieu and Mats Carlsson).

On the disc, the Hα chromosphere is dominated by two types of *fibril* (see Figure 9.7). Long fibrils are more horizontal and longer-lived. They arch over supergranule cells (Figure 1.20, top) or emanate from sunspots as superpenumbral fibrils (see Figure 1.29b). Shorter *dynamic fibrils* are much shorter-lived and are the on-disc counterpart of type I spicules, showing up in the quiet Sun as dark *mottles* at supergranule junctions. The disc counterparts of type II spicules are faint, bright, waving *straws* in Dutch Open Telescope movies (Rutten 2006); they are seen with the Swedish Solar Telescope (SST) as *rapid blue-shift events*, namely, fine upflowing jets, heated during their short lifetime (Rouppe van der Voort et al. 2009; Sekse et al. 2012).

Macrospicules are long jets of chromospheric plasma, resembling small surges and are often seen in polar coronal holes. Ejected along open magnetic field with speeds of 10 to 150 km s^{-1}, they reach heights of 4 to 40 Mm and then fall back or fade away. They last for 8 to 45 min and have diameters of 4 to 11 Mm and densities of 10^{16} m^{-3}. Sometimes they rotate as *macrospicule tornadoes* (Sec. 1.4.1).

The *transition region* is observed mainly in EUV emission lines. It is not a narrow static horizontal layer, but represents all the plasma that happens to be at transition-region temperatures while heating up (in response to coronal heating) or cooling down (either by thermal instability or when the heating is switched off). Some of it may be a thin layer near the base of a coronal loop, while some may be spicular jets that are heating up or cooling down and some represents *explosive events* or *blinkers* (see Sec. 10.4.2). Typical transition-region models give a thickness of only 30 km for the temperature rise from 3×10^4 K to 3×10^5 K, which simply means that there is very little plasma at these temperatures. The average height at 10^5 K is 1700 ± 800 km for quiet regions and is much higher for coronal holes (spanning 1,500 to 15,000 km).

Thus, the chromosphere is full of plasma that is continually accelerated, heated and cooled in a dynamic way. Its mass balance is likely to be dominated by spicular upflow at 10^4 K, with some of the plasma heated to transition-region temperatures and most flowing back down. Only a small fraction is heated to coronal temperatures, and some flows out as the solar wind in open-field regions.

In summary, the main MHD goals here are to understand the complex behaviour of the dynamic, highly inhomogeneous chromosphere. This includes the generation and behaviour of spicules, models for the structure of the magnetic field and plasma above supergranule cells, and an understanding of the nature of the heating, which is an order of magnitude stronger than the corona. Indeed, it is coupled in complex ways to the overlying corona and mediates the energy flux that heats the corona and drives the solar wind. What makes the chromosphere especially tough to model is that it is a non-LTE region of transitions – from high to low beta, from optically thick to optically thin and from partially ionised to fully ionised.

1.6 The Corona

The corona is dominated by the magnetic field and provides MHD theory with a wealth of problems, such as to understand: how to model the different kinds of coronal loop (Sec. 10.2); how the various parts of the corona are heated (Secs. 10.3 and 10.4); how the different types of solar wind are accelerated (Secs. 13.7 and 13.8); and how the corona is structured and evolves globally.

The corona (Latin for 'crown') is observed at eclipses as a faint halo of low density and high temperature (a few$\times10^6$ K), about as bright as the full Moon (Figures 1.21a,b). Its existence has been known for centuries. The Egyptians used to worship the Sun, and on the back of the golden chair of Tutankhamun is an emblem reminiscent of the shape of the corona. The Chinese thought an eclipse is caused by a dragon swallowing up the Sun, and so they employed an astronomer to shoot the beast!

Coronal images seen at an eclipse are created by the scattering of photospheric light coming up from below, both off electrons (the K-corona) and dust (the F-corona). Within 2.3 solar radii, the K-corona is dominant. Its intensity is proportional to electron density, so it appears brighter where there is more plasma. Figure 1.21a shows an example of an eclipse image near sunspot maximum using a radially graded filter that is 10,000 times more sensitive near its edge than its centre. Low down, much of the structure is due to *closed magnetic field lines*, whereas farther out most of it is due to *open field lines*, stretched radially by the outwards-flowing *solar wind*.

It was realised only in the 1930s that many lines in coronal spectra are due to high ionisation states of known elements, whose existence implies that the coronal temperature is several 10^6 K. This is so high that H nuclei (protons) completely lose their planetary electrons and heavier nuclei may lose up to 15 of theirs. The corona is therefore emitting thermally in X-rays and can be imaged directly in soft X-rays or EUV from space (Figures 1.21c and 1.22).

The mean electron density in the quiet inner corona is a few times 10^{14} m^{-3} (10^8 cm^{-3}), but it can be enhanced by factors of 5 to 20 in places. The density rapidly falls off with distance from the solar surface: it is a few times 10^{12} m^{-3} at 1 $R_\odot$ above the surface and less than 10^{10} m^{-3} at 10 $R_\odot$.

The heating of the chromosphere and corona remains a major problem in solar physics and two main MHD mechanisms have been proposed: the first is magnetic waves that are generated in the noisy

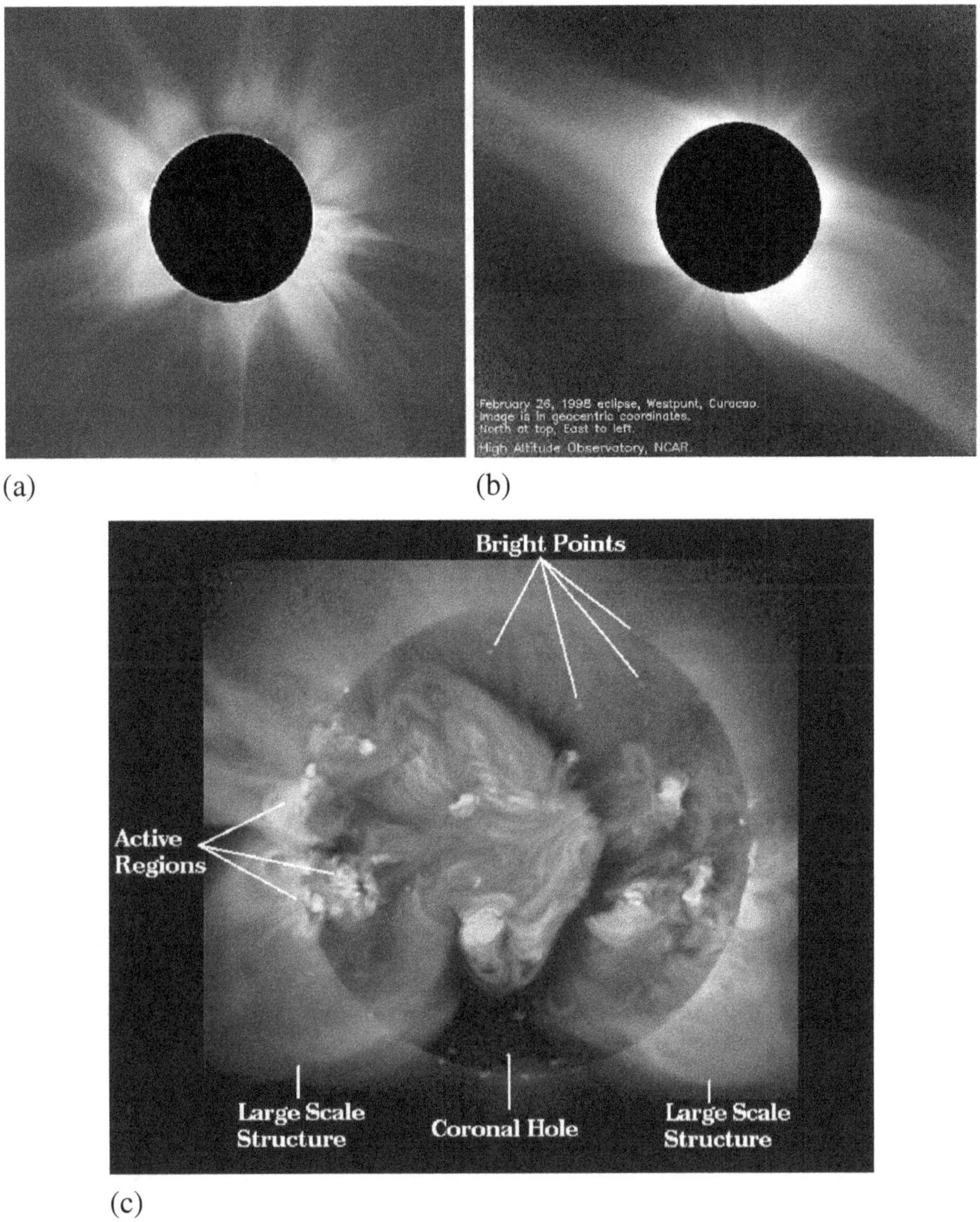

Figure 1.21. Two views of the solar corona: in white light during an eclipse near (a) solar maximum and (b) solar minimum [courtesy Rhodes College, Memphis, Tennessee, and High Altitude Observatory (HAO), University Corporation for Atmospheric Research (UCAR), Boulder, Colorado. UCAR is sponsored by the National Science Foundation]; (c) in soft X-rays from the Yohkoh satellite (courtesy Saku Tsuneta).

convection zone, propagate outwards and then dump their energy (Sec. 10.3); the second general class of mechanism is *magnetic reconnection* in myriads of current sheets (Sec. 10.4). There have been many observations of different types of wave in the corona (Sec. 4.11.2), including *Alfvénic waves* (Sec. 4.9.4), and many detections of the probable consequences of reconnection (Sec. 10.4.2). The corona suffers huge energy losses due to thermal conduction, radiation and mass outflow, making a total of typically 10^{21} W, which is only 0.001 per cent of the solar luminosity (3.86×10^{26} W).

Figure 1.22. Coronal loops at the limb from the NASA TRACE mission (courtesy Alan Title).

Coronal magnetic fields vary typically between a few G and many hundreds of G. They may be measured directly from radio emission or from the Zeeman splitting of infra-red coronal emission lines and compared with potential field extrapolations (e.g., Lin et al. 2000; Liu and Lin 2008).

1.6.1 The Structure of the Corona from Eclipses and X-ray Images

In *white light*, the corona was originally observed only at solar eclipses, since it cannot normally be seen against the dazzling light of the photosphere (a million times brighter than the corona). However, on average, there is only one total eclipse per year lasting about 3 minutes and the area covered by the Moon's shadow is only 100 km in diameter (see PROBLEM 1.7). Such a paucity of opportunities to observe the corona stimulated Lyot to create artificial eclipses by means of a *coronagraph* (1930), which is a telescope containing an occulting disc to eliminate the photospheric glare. The apparent size of the disc needs to be somewhat greater than that of the Moon, but much has been learned from coronagraphs on board satellites such as Skylab, SMM, SoHO and STEREO, where the stray light is much reduced.

In soft X-rays, the corona may be viewed directly, since the contribution from the lower atmosphere is negligible. The disadvantage of observing in soft X-rays (and EUV) is that they are absorbed by the Earth's atmosphere. But images from satellites such as Skylab, Yohkoh, SoHO, TRACE, Hinode and SDO reveal the corona in a completely different guise from solar eclipses (Figures 1.21c and 1.22).

The corona has a three-part structure comprising: relatively dark *coronal holes*, where plasma is escaping outwards along open magnetic fields to give the *fast solar wind*; brighter *coronal loops*, which are magnetically closed and connect photospheric regions of opposite polarity; and small intense features called *X-ray bright points*, scattered over the whole disc and consisting of tiny loops.

A *coronal* or *helmet streamer* is a large bright structure extending above the limb with a mean density enhancement of 3 to 10. It lies above a prominence or an active region and consists of a base (or arcade) of closed field lines surmounted by a blade or fan of open field lines, reminiscent of a Roman brush-topped helmet. It forms when the upper part of a large-scale coronal arcade reaches high enough to be stretched open by the solar wind. Two kinds of magnetic topology can exist under helmet streamers (Hundhausen 1972; Zhao and Webb 2003). A *bipolar streamer* is situated above a single bipolar arcade and between coronal holes having opposite polarities, so that the overlying blade is a current sheet with oppositely

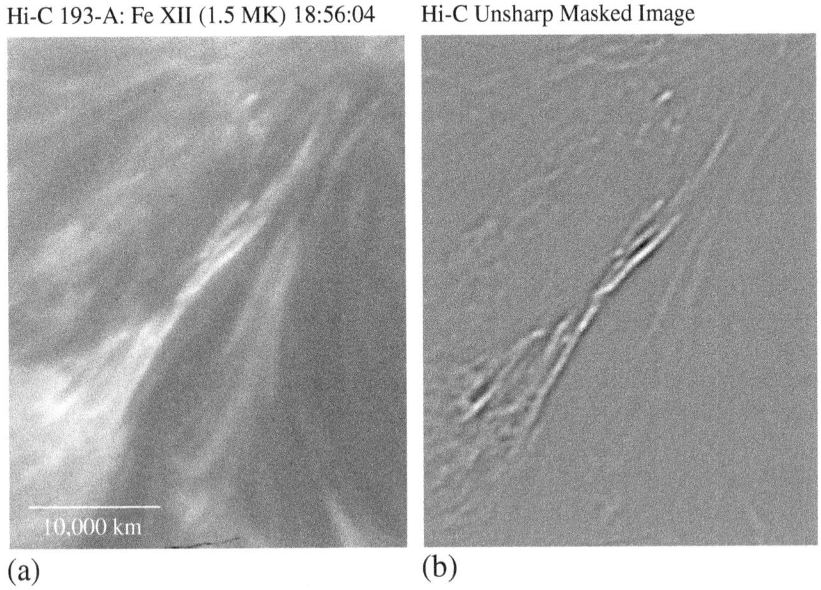

Figure 1.23. Images of active-region coronal loops from the NASA Hi-C rocket flight. (a) Original Hi-C image in the 193 Å passband that images plasma around 1.5 MK. (b) Unsharp mask of the same image, in which the raw data are smoothed and subtracted from the original image (courtesy Jonathan Cirtain and Amy Winebarger).

directed magnetic fields on both sides. A *unipolar streamer* or *pseudo-streamer* (Hundhausen 1972; Wang et al. 2007) separates coronal holes of the same polarity and lies above a quadrupolar field; that is, two oppositely directed bipolar fields side by side (Sec. 13.9.3, Figure 13.10).

The overall shape of the corona varies with the solar cycle: near sunspot maximum, coronal streamers and holes extend out in all directions at any latitude (Figure 1.21a); whereas, near sunspot minimum, streamers are present only in the equatorial region and coronal holes occupy the polar regions (Figure 1.21b).

1.6.2 Coronal Loops, X-ray Bright Points and X-ray Jets

Coronal loops (Sec. 10.2) comprise several morphological types, viewed in white light, soft X-rays, EUV lines and the red and green coronal lines. *Interconnecting loops* join different active regions and form when two loops from separate active regions reconnect or when one loop reconnects with newly emerging flux. They are up to 700 Mm long and in soft X-rays have typical temperature and density of 2 to 3 MK and 10^{14} to 10^{15} m^{-3}. Their ends are rooted in islands of strong magnetic field near the edges of active regions. A single loop lasts about a day, but a whole loop system may endure for many rotations. Loops that connect fully developed active regions have an intermittent visibility, but those that join an active region to an old remnant change little in shape and brightness for up to 12 days. *Quiet-region loops* do not connect active regions, and in soft X-rays they are somewhat cooler, with a temperature of 1.5 to 2.1 MK; their densities range between 2×10^{14} m^{-3} and 10^{15} m^{-3}. *Active-region loops* (Figures 1.22, 10.5) and many more details of coronal loops in general are described in Section 10.2, including the fact that loops are likely to consist of many thin *strands*. These may be braided and twisted, as glimpsed in Hi-C (High-resolution Coronal Imager) 0.25 arcsec resolution images of strands in an active-region loop, obtained on a 5-minute rocket flight (Figure 1.23).

X-ray bright points have typical diameters of 22 Mm and bright cores of 4 to 7 Mm (Golub et al. 1974, 1979) (Sec. 10.4.2). About 1,500 of them appear per day. Their mean lifetime is 8 hours; less than 15 per cent live longer than one day and only a very few survive longer than two days. They appear to consist of two types: one has a lifetime of two days or less and is distributed uniformly over the solar disc, while the other is longer-lived and has a distribution similar to active regions, occurring within $\pm$ 30 degrees of the equator. With much better spatial resolution, bright points are seen not as points but as the coronal manifestation of tiny bipolar areas (*ephemeral regions*, Section 1.4.2), consisting of many small coronal loops. Some of these are caused by emerging flux and others by chance encounters of opposite polarities (*cancelling magnetic flux*). Over the course of a solar cycle, more magnetic flux emerges through the solar surface as bright points than as active regions, and there appear to be more bright points during sunspot minimum than sunspot maximum, although the number density is constant (Hara, 2003).

Coronal X-ray jets and *EUV jets* were first extensively studied by Shibata et al. (1992) with Yokhoh. They are common in coronal holes (Sec. 1.6.3), where they comprise 70 per cent of the coronal brightenings (Subramanian et al. 2010) and can extend for more than half a solar radius with a flow speed in excess of 200 km s^{-1}. Coronal jets lie above magnetic concentrations (Shimojo et al. 2007) and are a clear signature of magnetic reconnection. Indeed, they are a natural consequence of coronal reconnection in association with emerging flux (Secs. 9.3.3, 11.6.1), coronal heating events (Sec. 10.4.2) and solar flares (Sec. 1.9.2.3).

X-ray jets may be accelerated in the corona itself or may be associated with the cooler jets, such as spicules, macrospicules or surges (Sec. 1.5). They sometimes represent the continuation or heating of cooler jets and sometimes they are adjacent to them (Canfield et al. 1996). A brightening may be seen near a jet's point of origin in the chromosphere (Shimojo et al. 1996). Coronal jets are often also associated with subflares or microflares and with type III radio bursts. Half the jets show helical structure (Patsourakos et al. 2008).

Blowout jets are a subclass of coronal hole jets and are the coronal counterparts of erupting Hα macrospicules (Moore et al. 2010). They consist of a normal X-ray jet with a violent eruption at their base, possibly of a flux rope, like a mini-CME (see also Madjarska 2011; Liu et al. 2011).

Since early Yohkoh studies, knowledge of the properties of X-ray jets has increased enormously with observations from TRACE (Alexander and Fletcher 1999), Hinode/XRT (Culhane et al. 2007; Nishizuka et al. 2008) and STEREO (Patsourakos et al. 2008). The properties of X-ray jets are as follows. From Yohkoh, Shimojo and Shibata (2000) find temperatures of 3 to 8 MK and densities of 7×10^{14} to 4×10^{16} m^{-3}. From Hinode/XRT (e.g., Savcheva et al. 2007), temperatures are 1 to 30 MK, lifetimes are 5 to 30 min (peaking at 10 min), jet velocity is 70 to 500 km s^{-1} (although some authors find up to 1,000 km s^{-1}), occurrence rate is 60 per day, jet lengths are 10 to 120 Mm (although some can be up to 400 Mm) and widths are 6 to 10 Mm.

The emerging or interacting flux scenario (Heyvaerts et al. 1977; Priest 1992) is a likely explanation for the reconnection that may produce X-ray jets. Detailed computation models have been developed in two-dimensions (Yokoyama and Shibata 1996) and three-dimensions (Moreno-Insertis and Galsgaard 2013), as described in Sections 9.3.3 and 11.6.1.

1.6.3 Coronal Holes

Coronal holes are extended regions of open magnetic field that are cooler and less dense than the surrounding corona (Zirker 1976; Cranmer 2009). They were named and studied by Waldmeier (1956) with a coronagraph in the green line (5,303 Å). Then, after early rocket flights, images of them were studied in detail from Skylab (1973–74) and other space missions, followed by quantitative measurements of temperature and flow speed from UV coronagraph spectrometers (SoHO/UVCS).

Near solar minimum, large coronal holes cover the north and south polar caps with opposite magnetic polarity. They last for 7 to 8 years and so are among the longest-lived of solar features. For 2 years before solar maximum, polar holes decrease in size and for 1 to 2 years around solar maximum they are not present at all. However, near solar minimum, coronal holes can exist at all other solar latitudes, but they may persist for only a few months. Low-latitude holes are often situated near complex active regions and can join with the polar hole of the same polarity.

Coronal hole boundaries continually change in response to global magnetic evolution by flux emergence, footpoint motion and large-scale reconnection, which can make some holes rotate more like a solid body than differentially. *Transient coronal holes* show up as temporary rapid reductions in UV and X-ray intensity in blue-shifted patches known as *coronal dimmings* that last for a few hours or days after a coronal mass ejection (Sec. 1.9) (Thompson et al. 2000; Harra et al. 2007).

Coronal holes possess an open diverging magnetic structure and lie above extensive unipolar regions in the photosphere that have a larger flux imbalance than elsewhere. For polar holes, such flux imbalance is due to transport and diffusion from the trailing polarity of active regions (see Figure 1.31). The flux in a *polar coronal hole* is typically 10^{15} Wb (10^{23} Mx), with an average field strength of about 5 to 10 G. The conditions for formation of a coronal hole above a unipolar region are unclear, but they are likely to depend on its magnetic flux and the nature of its environment.

Coronal holes are largely indistinguishable from their surroundings in the photosphere and chromosphere (although they do show up in the UV lines of He or in the near-infrared line He 10,830 Å, which is influenced by coronal radiation). But in near-coronal transition-region lines (such as Ne VIII) they are more visible, since the pressure is two or three times smaller and the temperature gradient five times smaller than in the quiet Sun. This makes the transition region five times thicker and the conductive flux (60 W m^{-2}) an order of magnitude smaller than elsewhere in the Sun. In cooler transition-region lines, coronal holes are hardly visible because of the dominance of cool loops (e.g., Stucki et al. 2000).

In coronal holes, the density is ten times lower than in coronal streamers, falling typically from 10^{13} m^{-3} at a height of 0.3 $R_\odot$ to 10^{12} m^{-3} at 1 $R_\odot$. The temperature is lower too (1.4 to 1.8 MK at 2 $R_\odot$). The energy input is similar to the quiet-Sun corona, but, since some goes into flow energy, there is less available for radiative loss, which makes the temperature and pressure lower. The effect of the lower rate of flux emergence in unipolar regions is not yet clear (Hagenaar et al. 2008).

In polar coronal holes, ray-like structures (called *polar plumes* or *coronal rays*) are especially noticeable at sunspot minimum. They arise from network flux concentrations, and typically 40 are present in a coronal hole. Individual plumes last for about 1 day and presumably outline the local magnetic field. Low down in the corona they are seen with EUV imagers such as STEREO/EUVI or SoHO/EIT and higher up their extensions become visible in SoHO/LASCO images out to 30 to 40 $R_\odot$ (DeForest et al. 2001). They expand laterally and their contrast reduces with distance to give a plume/interplume density ratio of 2 but 3 times more interplume plasma at a height of 2 $R_\odot$.

Shorter, more impulsive bursts of heating produce *polar jets* of two types: cool jets with lower temperatures, higher densities and faster outflows than the surrounding coronal hole; and hot X-ray jets (Sec. 1.6.2)

Elemental abundances in the upper chromosphere and low corona of coronal holes are close to photospheric values, whereas active regions possess enhancements in low-FIP (first ionisation potential) elements. Such selective fractionisation occurs in the upper chromosphere, where low-FIP elements become ionised and high-FIP elements remain neutral.

The collisionless nature of the outer corona (especially coronal holes) means that each ion species has its own temperature and outflow speed, as revealed by SUMER and UVCS on SoHO (Figure 1.24). UVCS line widths give a *kinetic temperature*, which is a sum of microscopic motions and unresolved motions due to waves or turbulence. The electron temperature is only 0.8 MK, but heavy ions such as O^{+5} are much hotter normal to the magnetic field (up to 200 MK), with temperature anisotropies possibly up to $T_{i\perp}/T_{i\parallel} \approx 10$, although the effects of density stratification and line-of-sight integration reduce this value considerably (Raouafi and Solanki 2006). UVCS also measures large proton kinetic temperatures ($T_{p\perp} \approx 3$ MK) normal to the field and an anisotropy ($T_\perp > T_\parallel$) above heights of 2 to 3 $R_\odot$ (Kohl et al. 2006). Also, O^{+5} has twice the outflow speed of protons at a height of 2 $R_\odot$. Out in the solar wind from 0.3 to 1 AU, the proton

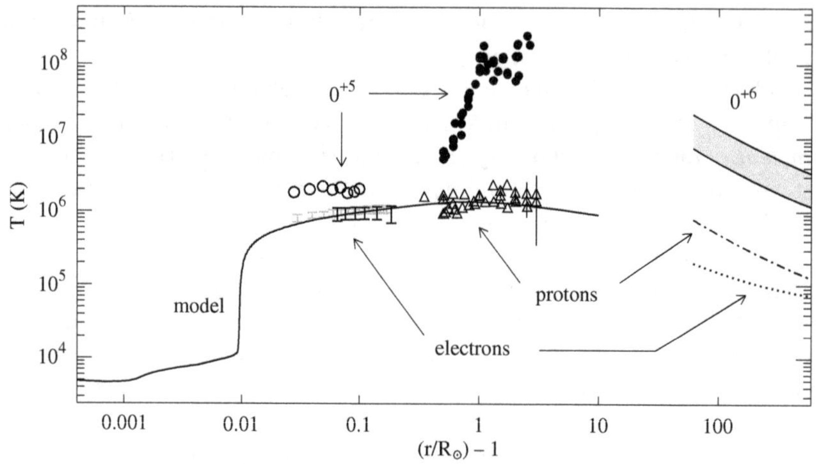

Figure 1.24. Radial dependence of observed temperatures of electrons, protons and ionised oxygen in a polar coronal hole and fast solar wind and their comparison with a model (solid curve). In the hole, T_e is from SoHO/SUMER (Wilhelm 2006), T_p from SoHO/UVCS and perpendicular O^{+5} from UVCS (Landi and Cranmer 2009) (courtesy Steve Cranmer).

anisotropies continue ($T_{p\perp} > T_{p\|}$) and many heavy ions continue to flow faster than protons (Marsch 2006). Plumes show lower kinetic temperatures and flow speeds than interplume regions.

1.6.4 Solar Wind

The outer solar corona is not in hydrostatic equilibrium, but is continuously expanding outwards as the *solar wind*, models of which are described in Chapter 13. Many spacecraft have observed its properties in situ, beginning with Luna 1 (1959) and Mariner 2 (1962), and ranging from 0.3 AU (Helios 1 and 2 launched in 1974 and 1976) to the edge of the solar system (Pioneer 10 and 11 and Voyager 1 and 2). The Earth's orbital velocity is 30 km s^{-1}, greater than current technology allows rockets to escape unassisted, and so most spacecraft have remained in the ecliptic plane. An exception was the European Ulysses mission, launched in 1990, which used an encounter with Jupiter to swing it into an orbit inclined at 80 degrees to the ecliptic that passed over the Sun's poles in 1994–95 (near solar minimum), 2000–01 (near solar maximum) and again in 2007 (near solar minimum).

The dominant energy loss from coronal holes is advective transport by the solar wind (600 W m^{-2}) rather than downward conductive flux. Much of the solar-wind plasma originates in thin intense flux tubes at supergranule and granule boundaries (G-band and network bright points). Such outflow shows up as 3 km s^{-1} blueshifts in the supergranule network (Hassler et al. 1999; Marsch et al. 2008). In polar holes, the fast solar wind accelerates rapidly, reaching half its terminal speed by heights of 2 to 4 $R_\odot$ (Kohl et al. 2006). From the cusps of helmet streamers, a continual release of faint plasma blobs can be seen with SoHO/LASCO, accelerated to 300 to 400 km s^{-1} by 30 $R_\odot$ and comprising a small fraction of the slow solar wind (Sheeley et al. 1997; Chen et al. 2009).

The main characteristics of the solar wind at 1 AU are as follows (Table 1.2):

* A variable supersonic outflow comprising two components, a slow wind (300 to 400 km s^{-1}) and a fast wind (700 to 750 km s^{-1}) of lower density and higher temperature.

* *Slow solar wind* is highly variable, with an unclear origin. It comes from high-latitude coronal-hole/streamer boundaries at solar minimum, but from small coronal holes and active regions at solar maximum. Elements with low FIP (first ionisation potential) have enhanced abundance (relative to the photosphere) and so the wind originates in the corona rather than the photosphere. Its oxygen freeze-in

Table 1.2. *Properties of the solar wind at 1 AU*

	Minimum	Average	Maximum
Velocity	200 km s^{-1}	400 km s^{-1}	900 km s^{-1}
Density	4×10^5 m^{-3}	6.5×10^6 m^{-3}	10^8 m^{-3}
Electron temperature	5×10^3 K	2×10^5 K	10^6 K
Proton temperature	3×10^3 K	5×10^4 K	10^6 K
Magnetic field	2×10^{-6} G	6×10^{-5} G	8×10^{-4} G
Alfvén speed	30 km s^{-1}	60 km s^{-1}	150 km s^{-1}

Table 1.3. *Comparison of fast and slow winds at 1 AU*

	Speed	Electron Density	Mass Loss	Ram Pressure
Fast	750 km s^{-1}	2.5×10^6 m^{-3}	10^9 kg s^{-1}	2.6×10^{-9} Pa
Slow	400 km s^{-1}	7×10^6 m^{-3}	1.5×10^9 kg s^{-1}	2.1×10^{-9} Pa

temperature (which is conserved outside a few solar radii and so reflects the inner coronal temperature) is 1.1 MK. At solar minimum, it is located only near the equator, but at solar maximum it is at all latitudes alternating with fast wind (Table 1.3).

* *Fast solar wind* is relatively steady and originates in magnetically open regions (*coronal holes*) as high-speed streams, which tend to recur every 27 days at the equatorial rotation rate and are correlated with the recurrence of geomagnetic storms. Elements with low FIP have a factor of 2 lower abundance than the slow wind, suggesting a photospheric origin. The fast wind has an oxygen freeze-in temperature of 1.6 MK. At solar minimum, it always occupies the polar regions, but at solar maximum it occurs as narrow, somewhat slower streams at all latitudes.

* The proton temperature is higher in the fast wind than the slow wind, while the electron temperature is the same. The mass flux is, surprisingly, the same (10^{12} m^{-2} s^{-1}) to within a factor of two for all wind conditions.

* The main constituents are electrons, protons and α-particles (at 3 to 4 per cent). In the fast wind, there is preferential heating of heavy ions ($T_{ion} \gg T_p > T_e$) and of ions perpendicular to the magnetic field ($T_\perp > T_\parallel$).

* The magnetic field forms on average a spiral pattern.

* The solar wind is highly time-dependent. A transient wind is associated with coronal mass ejections (CMEs), originates in coronal streamers, varies with the solar cycle and contributes about 16 per cent of the mass flux at solar maximum. Magnetic fluctuations are continually present and consist mainly of outwardly propagating Alfvén waves, possibly generated by supergranulation. *Tangential* and *rotational discontinuities* are common and *fast magnetoacoustic shocks* are often found, generated by CMEs or by complex interactions between streams of different speeds.

At sunspot minimum, oppositely directed magnetic fields from the two polar coronal holes are separated at the equator by a *helmet streamer* (Figure 1.25a). This consists of a closed magnetic arcade lying below a *neutral (heliospheric) current sheet* (HCS) that stretches outwards from a height of one or two solar radii. In an idealised axisymmetric solar atmosphere, the current sheet would lie along the equator, but a lack of symmetry causes it to be *warped*, like the brim of a sombrero (Figure 1.25b). See Section 13.9.3 for a discussion of its topology in more complex situations. The sheet is typically inclined by about 7 degrees to

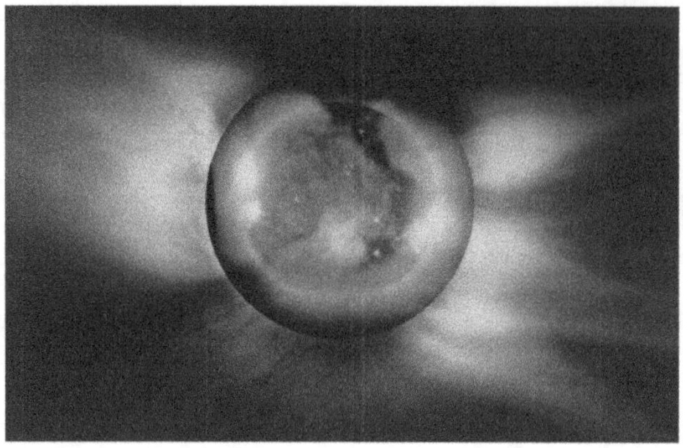

(a)

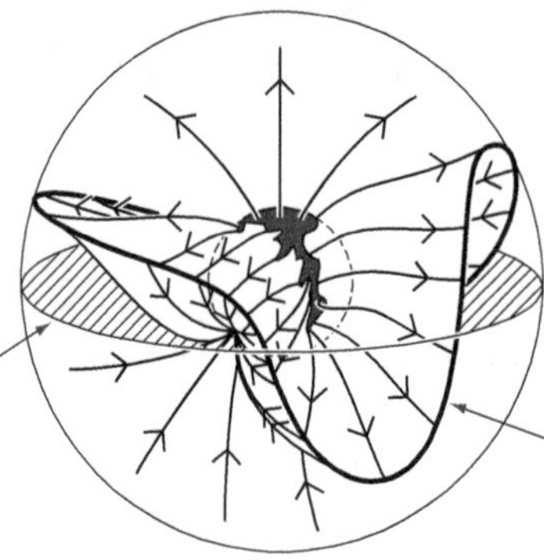

(b)

Figure 1.25. (a) The outer corona in white light during a total eclipse (30 June 1973) (courtesy Joan Burkepile, High Altitude Observatory, Boulder) and the inner corona in soft X-rays (courtesy Leon Golub). (b) A sketch of the warped heliospheric current sheet that was present, crossing the solar equatorial plane four times (after Hundhausen 1977).

the Earth's orbit at solar minimum. As the Sun spins, an observer at 1 AU in the ecliptic lies alternately in the fast winds from one magnetic hemisphere and from the other, with a disturbed period of slow solar wind in between. The current sheet itself is called a *sector boundary* as it moves past the observer and the whole pattern is known as *sector structure.* The most common pattern is a set of four streams, two of high speed and two of low speed, with a unipolar magnetic field in each. Streams last from 1 to 18 rotations and are especially persistent during the declining phase of the solar cycle.

An example of solar-wind measurements at 1 AU near solar minimum in Figure 1.26 shows small fluctuations over small timescales, but also a regular alternation at weekly intervals of high- and low-speed streams, in which the speed varies by a factor of two and the density by a factor of ten. Abrupt speed increases are followed by slow decreases, with the density peaking during the rapid speed increases. The radial magnetic field changes sign when a new fast stream is encountered and remains constant inside it. This demonstrates the existence of a fast solar wind that is tenuous and relatively uniform and a slow solar wind that is much denser and more structured.

Results from Ulysses are summarised in Figure 1.27, showing how the speed changes with latitude and with the solar cycle. The speed is given on a polar plot, such that the distance from the centre is proportional to the speed at each latitude. Time starts on the left and progresses in an anti-clockwise direction, as shown

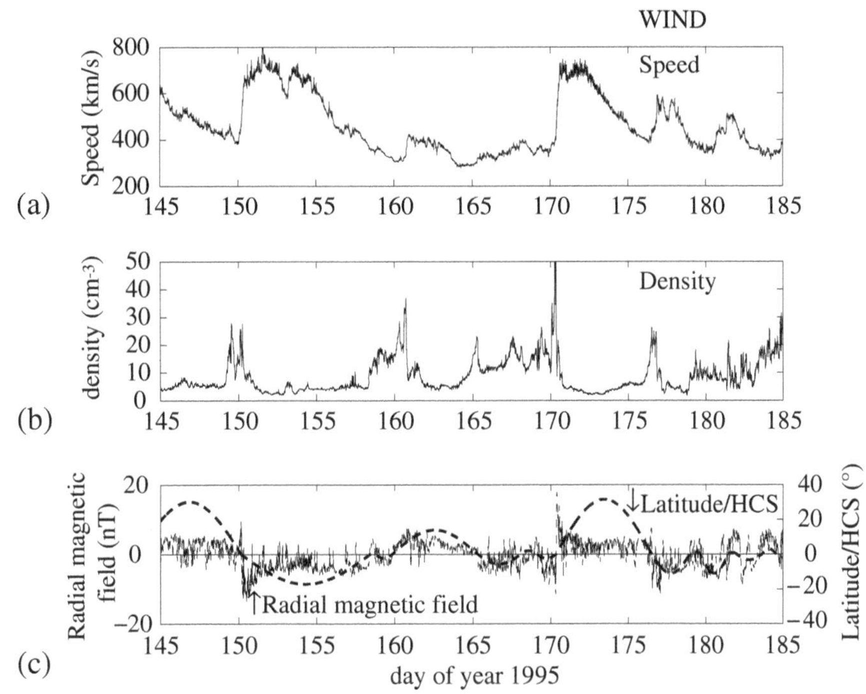

Figure 1.26. Measurements on the NASA Wind spacecraft (June 1995) at 1 AU in the ecliptic of: (a) proton speed, (b) electron density and (c) radial magnetic field and spacecraft latitude relative to the heliospheric current sheet (HCS) (dashed).

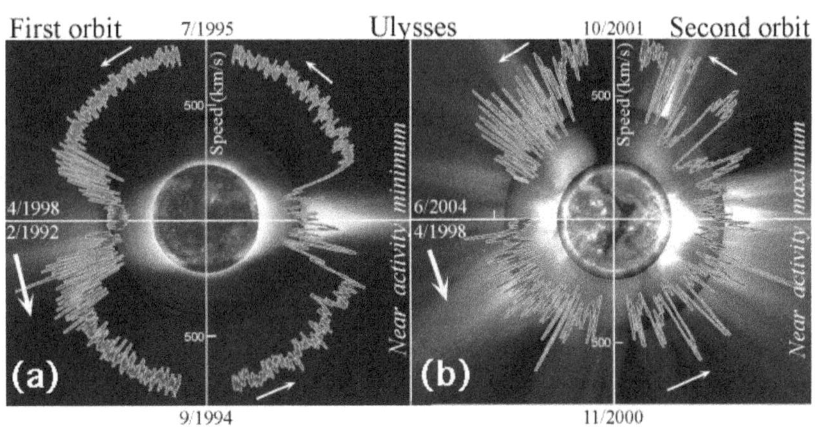

Figure 1.27. Solar-wind speed as a function of latitude in polar coordinates from the Ulysses spacecraft during (a) solar minimum and (b) solar maximum. Superimposed are coronal images from the Mauna Loa K coronameter, plus SoHO/EIT and LASCO (from Meyer-Vernet 2007).

by the white arrows. Near solar minimum, Figure 1.27a shows a simple coronal structure of two large polar holes and an equatorial helmet streamer, while the solar-wind speed is nearly constant and fast at all latitudes, except for a narrow band of $\pm$ 20 degrees of highly variable slow solar wind about the equator. At the same time, the magnetic field is mostly outward in the northern hemisphere and inward in the southern hemisphere.

Near solar maximum, the heliospheric current sheet tends to break up and be highly distorted. Figure 1.27b shows a much more complex and irregular coronal image with bright streamers extending out from all latitudes. The solar-wind structure from Ulysses is equally complex, with a pattern of alternating fast and slow streams at all latitudes that are slower and smaller than at solar minimum.

The flow speed increases monotonically from very low values in the inner corona (where the plasma is close to hydrostatic), and eventually it becomes super-Alfvénic beyond the Alfvén radius (r_A) (Sec. 13.6.2). Within this radius, the magnetic field dominates the plasma dynamics. Beyond it, the magnetic field is carried by the plasma, and the rotation of the Sun makes the magnetic field take on a spiral structure, while the plasma flows roughly radially, like water from a rotating garden hose.

The solar wind does not quite flow radially from the Sun; its velocity is inclined at about 1.5 degrees to the radius vector. This means that angular momentum is being transferred from the Sun to the solar wind, and the Sun is being braked in the process; the effect is sufficient to slow the Sun down significantly over its lifetime.

The solar-wind speed and density, as measured near the Earth, are about 30 per cent lower at sunspot maximum than sunspot minimum, while the magnetic field strength remains constant. Whereas sunlight takes only 8 minutes to reach the Earth's orbit, normal fast solar-wind plasma travels the distance in about 5 days. Beyond the Earth, the solar wind extends out to the heliospheric boundary with the interstellar medium at 50 to 100 AU, where a shock slows the flow from supersonic to subsonic.

Other key advances in understanding include:

* The development of coronal-hole models; the realisation of a likely role for Alfvén waves in solar-wind acceleration and of the need to dump energy beyond the critical point (Leer et al. 1982);

* The observations from SoHO/SUMER of the temperature and velocity in polar plumes and interplume regions (Wilhelm et al. 1998);

* The discovery of 3 to 10 km s^{-1} outflows in coronal-hole network in He I and Ne VIII from SoHO/SUMER formed at 2×10^4 K and 6.3×10^5 K, representing the source of fast solar wind (Hassler et al. 1999; Wilhelm et al. 2000), although McIntosh (2012) also found upflow throughout supergranule interiors;

* The measurement from SoHO/UVCS of large perpendicular temperatures and velocities for minor ions, suggesting a role for a turbulent cascade from large scales to scales where an ion-cyclotron resonance can set in (Cranmer et al. 2007); however, the origin of such high-frequency waves is so far unclear, as is their relationship to the low-frequency Alfvénic waves observed with CoMP and Hinode/SOT/XRT (Secs. 4.9.4 and 4.11.2) (e.g., Tomczyk et al. 2007);

* The discovery from Hinode/EIS of high-temperature (>1 MK) outflows on open or long field lines from the edges of most active regions at typically 50 to 100 km s^{-1} for several days; they have low-FIP composition and are the source of perhaps 10 to 25 per cent of the solar wind (Sakao et al. 2007; Del Zanna 2008; Brooks and Warren 2011; Doschek 2012); and

* The heating of spicules to coronal temperatures (De Pontieu et al. 2011).

1.7 Active Regions, Sunspots and the Solar Cycle

Viewed at low resolution in white light, the Sun appears rather dull, but a closer look reveals a tantalising world, changing dynamically in a rich variety of ways. For example, an Hα photograph (Figure 1.3c) shows up many features superimposed on the quiet atmosphere. *Active regions* appear as bright *plages* of emission in the equatorial belt within $\pm$ 30 degrees of the equator and represent moderate concentrations of magnetic flux with mean fields of 100 G or so. The plages are formed of *intense flux tubes* and also harbour darker pores that sometimes appear bright in Hα. Within a mature active region one finds dark regions of more intense magnetic field called *sunspots*.

All aspects of solar activity owe their existence to the magnetic field and represent different ways in which solar plasma is responding to the underlying magnetic field evolution. An active region has such a large scale and such systematic behaviour that it must arise from a large coherent flux bundle that is formed deep in the convective zone. It rises by magnetic buoyancy as an Ω-shaped loop, breaking through the surface in a globally organised form that is locally fragmented and concentrated by convection to form intense flux tubes, pores and sunspots.

1.7.1 Active Regions

1.7.1.1 Development

When new magnetic flux pushes up from below the photosphere, as an *ephemeral active region* of typical size 10 Mm, it produces in the corona an X-ray bright point. Most bright points fade away in a few hours, but sometimes one swells over a few days into an *active region* of typical size 100 Mm. In the chromosphere, growing bipolar *emerging flux regions* (Sec. 9.3.3) show up as *arch-filament systems*, which are sets of dark low-lying loops (fibrils) expanding outwards into the corona with lengths up to 30 Mm, heights less than 5 Mm and individual lifetimes half an hour. Their summits rise at about 10 km s^{-1}, while plasma falls down near both ends with speeds up to 50 km s^{-1} and their feet separate at 1–3 km s^{-1}.

It takes about 3 or 4 days for a well-developed active region to form. At the photosphere it shows up as a sunspot group surrounded by photospheric *faculae* (Sec. 1.4.1), while in the chromosphere there is a bright *Hα plage*, and in the corona a set of EUV and X-ray loops. The region continues to grow and reaches its maximum activity in 10 to 15 days. Sunspots form during the emergence of flux, but most of the sunspots (except for a dominant leading spot) have usually disappeared after one rotation. However, the region may remain active with an enhanced magnetic field for weeks or months after the sunspots have disappeared.

Since the Sun rotates from left to right (viewed from the Earth), the right part of an active region is known as the *preceding* part, which usually contains a large, regular sunspot (the *p-spot* or *leader spot*). The main spot in the left part is the *f-spot* or *follower spot* (Sec. 1.7.2.6).

Around, within and far away from active regions, there are thin, cool, dark ribbons called *filaments* (when viewed on the disc) or *prominences* (when viewed on the limb). A prominence may erupt sometime during its life and be associated with a CME and, in an active region, with an eruptive *two-ribbon solar flare*, a region of intense emission that represents a violent energy release. The eruption is basically a process of MHD instability or non-equilibrium in a highly sheared magnetic field. Flare-like brightenings can be seen in X-ray bright points or young active regions, but major *solar flares* usually occur near the peak of an active region's development, when sunspots are present and when the magnetic structure is at its most complex.

The decay of an active region is much slower than its rise. It is marked by a continued increase in the size of the plage and a decrease in flare incidence. It occurs by the slow cancellation and dispersal of magnetic flux, by submergence of Ω-loops and by the emergence of U-loops, although their relative contributions are not well known. Typically, the Hα plage decreases in brightness after two rotations and disappears altogether after about four rotations.

In X-rays, the compact active-region core decays away to leave a simple, diffuse loop structure joining opposite polarities; such a *remnant active region* is often marked by a prominence, embedded in an arcade of hot loops, occupying a sheared magnetic *polarity-inversion line.* Subsequently, it may become a huge quiescent prominence, which slowly migrates towards one of the poles.

1.7.1.2 Structure

Most active regions are bipolar with the flux well-ordered into two islands of opposite polarity, but occasionally a more complex region appears either fully formed or by emergence of new flux with a different orientation within an existing region. New active regions have a tendency to develop near existing or remnant active regions and so can form *complexes of activity* or *activity nests.* Most nests last less than 6 months and occupy preferred locations in latitude and longitude. The total flux in a medium-size active region is 10^{14} Wb (10^{22} Mx).

In the photosphere, especially near the limb, the presence of an active region is indicated by faculae, which cluster as filigree (Sec. 1.4.1). In the chromosphere above faculae, there is bright Hα plage and long, thin, dark streaks called *fibrils.* They follow magnetic field that either connects opposite polarities or, for example, in a unipolar supergranule canopy, turns up into the transition region. Individual fibrils have widths 0.7 to 2.2 Mm, an average length of 11 Mm and lifetimes of 10 to 20 min. In the quiet Sun they tend to arch over supergranule cells. Around a mature sunspot they form a sometimes spiral pattern of *superpenumbral fibrils.*

The corona above an active region possesses an enhanced density and temperature, which makes it prominent both at the limb (Figure 1.21b) and also in EUV, X-ray and radio wavelengths against the disc. It consists of a jungle of *active-region loops* (Figure 1.22) with a wide range of lengths (10 to 100 Mm) and temperatures. The active-region corona has a hot core with densities up to 10^{16} m^{-3} and temperatures in excess of 3 MK. It is surrounded by a longer-lived halo with densities of a few times 10^{15} m^{-3} and temperatures 1.5 to 2.5 MK.

1.7.1.3 Internal Motions

The large-scale coronal magnetic field of an active region evolves slowly through a series of mostly force-free states (Sec. 3.4). As far as movement normal to the magnetic field is concerned, the plasma is completely dominated by the field, since coronal plasma has $\beta << 1$ (Sec. 2.5). But, along the field, the chromospheric and coronal plasma is in continual motion with a wide range of flows, including network downflows (Sec. 1.4.2), spicules (Sec. 1.5) and explosive events (Secs. 1.5 and 10.4.2).

Cool Hα *surges* often occur at the leading edge of an active region, ejected upwards along curved paths at 20 to 60 km s^{-1} (occasionally 100 to 200 km s^{-1}), reaching heights 20 to 35 Mm (occasionally 100 Mm) and lasting 10 to 20 min (Shibata et al. 1982; Schmieder et al. 1995). A surge either fades from view or returns back along the same path. At the feet of small surges near tiny satellite sunspots, one finds bright dots or jets in the wings of Hα (*Ellerman bombs*) lasting 10 to 20 min, which occur above emerging or cancelling flux (Park and Chae 2012). Also, adjacent to surges, one often observes much hotter hot X-ray jets (Sec. 1.6.2). Both are likely to be produced by reconnection in emerging flux (Heyvaerts et al. 1977; Yokoyama and Shibata 1995; Moreno Insertis et al. 2008).

Coronal rain is cool plasma in lines such as Hα that is flowing down active-region loops along curved paths at sub-free-fall speeds (50 to 100 km s^{-1}). This may well be part of the natural process of heating, evaporation, cooling and draining of plasma in the dynamic life of a coronal loop (Antiochos et al. 1999; Antolin et al. 2010). Antolin et al. (2012) reports that, for a decaying active region, it consists of small dense cool (< 7000 K) cores, with mean widths of 310 km, lengths of 710 km and downflow speeds of 70 km s^{-1}. Typically 7 to 10 per cent of the coronal volume is filled with coronal rain, with an occurrence time per strand of 5 to 20 hr and a mass drain-rate of 5×10^6 kg s^{-1} for an area 4×10^{14} m^2 [i.e., one-third of the classical cool upflow mass-rate due to spicules, (Beckers 1972)].

Active-region transient brightenings are localised impulsive brightenings in X-rays or EUV by 30 to 100 per cent that last for typically 1 to 15 mins over areas 10 to 100 Mm^2 and have energies 10^{17} to 10^{22} J in the range of nanoflares to microflares. Some are point-like and others are loop-like. They heat the plasma to 5 MK and are associated with emerging flux, especially around sunspots (Shimizu et al. 1992; Berghmans and Clette 1999; Warren et al. 2007; Kano et al. 2010). Observations from SDO/AIA and NST at Big Bear Observatory show upflow from the ends of fine loops that are being filled with hot plasma in heating events (Ji et al. 2012).

1.7.2 Sunspots

Sunspots are cooler and darker than their surroundings and represent exceptionally large concentrations of magnetic flux, in which the intricate and diverse effects of convection in a magnetic field are revealed (see Solanki 2003; Thomas and Weiss 2008; Rempel 2011).

1.7.2.1 History

At sunrise or sunset or when the Sun was partly obscured by thin cloud, sunspots may well have been seen with the naked eye in ancient Egyptian or Chinese societies, but the first recorded mention is in 325 BC by Theophrastus of Athens. He was a student of Aristotle, who had taught that the Sun is perfect, without blemish. The oldest known drawing is in the chronicle of John of Worcester (1128).

The invention of telescopes (Netherlands, 1608) led to the first recorded telescopic observation of sunspots by Thomas Harriot in England (December 1610) and a 1612 letter by Galileo saying that he had been observing them since November 1610. In the first published account, Johann Fabricius (June 1611) noticed spot foreshortening at the limb and deduced that they lie on the surface of a rotating Sun. More systematic sunspot observations by Christoph Scheiner in Bavaria (late 1611) and by Galileo (summer 1612) showed that they always lie near the equator.

In 1769, Alexander Wilson (Glasgow) discovered that the side of a penumbra that is farthest from the limb is narrower. He deduced that a sunspot is a saucer-like depression (of 500 to 700 km) in the visible surface (the *Wilson effect*). Later, it was realised that this is due to the cooler, less-dense umbra being more transparent, so that the umbral light comes from a deeper level than the normal photosphere.

In 1843, Heinrich Schwabe discovered that the number of sunspots varies with an 11-year sunspot cycle. This led Richard Carrington in 1858 (and independently Spörer) to discover that the latitude of sunspot-appearance drifts equatorwards from about 40 degrees to about 5 degrees during the cycle, and also to discover *differential rotation*, namely, that spots closer to the equator rotate more rapidly. In 1859, Carrington (and independently Hodgson) made the first recorded observation of a *solar flare*.

Photography and spectroscopy were developed in the 1860s and 70s, and then George Ellery Hale in 1908 used Zeeman splitting to discover magnetic fields of 3,000 G in sunspots, the first detection of magnetic fields outside Earth. The following year, Evershed (1909) discovered a radial photospheric flow in the penumbra called *Evershed outflow* (Sec. 1.7.2.5). More recently, exquisitely fine structure has been observed from ground- and space-based telescopes.

1.7.2.2 Overall Properties

A sunspot is dark because of partial inhibition by its magnetic field of normal energy transport. It has a typical radius of 10 to 20 Mm and consists of a central dark *umbra* with an almost vertical magnetic field, surrounded by a less-dark and highly filamentary *penumbra*, with a characteristic pattern of bright and dark radial *filaments* and a weaker more-horizontal magnetic field.

The umbral radius is about 40 per cent of the overall spot radius. Its brightness depends on wavelength but the wavelength-integrated value is typically 30 per cent of the surrounding quiet-Sun photosphere (ranging between 20 and 50 per cent). It is cooler by 1,000 to 1,900 K, which implies that a substantial convective flux is still required to maintain this brightness (see PROBLEM 1.8). By comparison, the mean penumbra radiates at about 75 to 85 per cent of the photospheric rate and is cooler by 250 to 400 K, while the normal photospheric radiative energy flux is 6×10^7 W m^{-2} (6×10^{10} erg cm^{-2} s^{-2}).

Sunspots come in a range of shapes and sizes with diameters 3.5 to 60 Mm (Figure 1.28a). Their lifetimes are proportional to their areas, so that a spot of diameter 10 Mm lasts typically 2 to 3 days while a 60 Mm spot lasts 80 to 90 days. Sunspots are possibly darker at the beginning and brighter at the end of the cycle, when the field strength is lower, although some authors find no change (Mathew et al. 2007). Around a

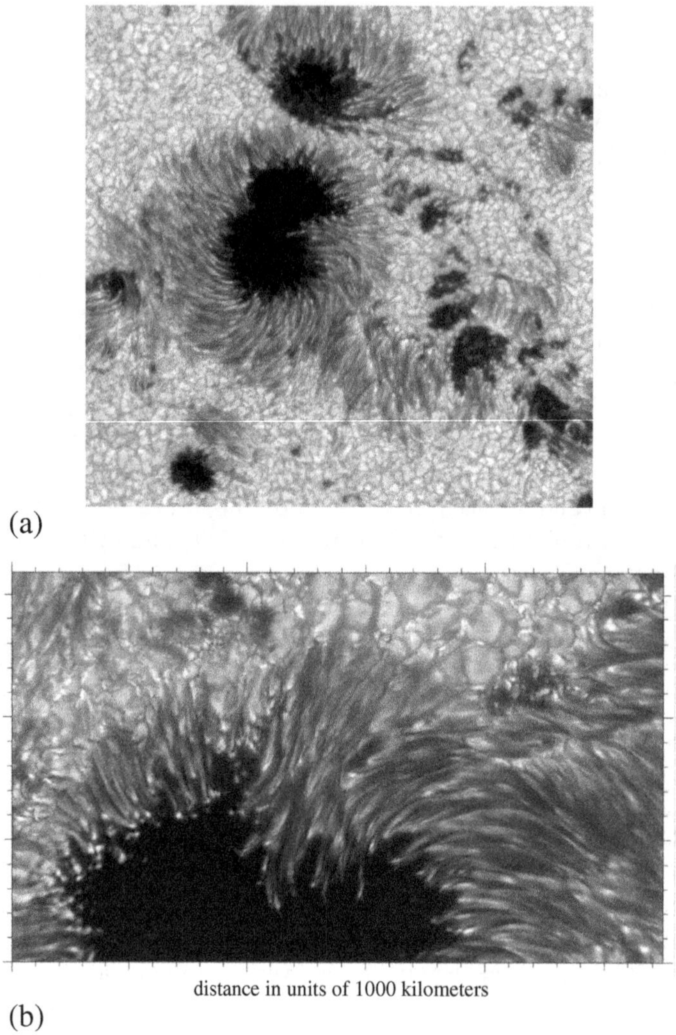

Figure 1.28. (a) An image of an active region with sunspots and pores from the SST on La Palma. (b) A close-up of penumbral structure (courtesy Göran Scharmer).

sunspot is a very faint (0.1 to 0.3 per cent) bright ring; this is too small to balance the intensity deficit in the sunspot, which must spread much more widely.

The magnetic field in the centre of an umbra is vertical and increases with size, while its brightness decreases. It has a typical strength of about 2.8 kG [though it may be as small as 1.5 kG or reach as much as 6 kG, (Livingston et al. 2006)]. The field strength decreases gradually with radius, giving an average field across the whole spot of 1.2 to 1.7 kG. The fluxes of a large spot and a large sunspot group are typically 10^{13} Wb (10^{21} Mx) and 2×10^{14} Wb. In the penumbra, the inclination of the field to the vertical increases with radius to a mean value of 70 to 80 degrees at the edge of the spot, where the field strength has fallen to less than 700–900 G.

Forty-six per cent of sunspot groups are unipolar (α-type); 53 per cent of sunspot groups are bipolar (β-type), with the spots located at the preceding and following sides of a group and having opposite polarity. Only 1 per cent are complex in their polarity (γ-type). A δ-spot has umbrae of opposite polarity inside the same penumbra. Large solar flares are especially likely in such a δ-configuration, when the magnetic gradient is exceptionally high and the field highly complex.

The complexity of an active region may be enhanced by the presence of isolated areas of one polarity, completely surrounded by the opposite polarity; they are called *parasites* or *inclusions* and are preferred locations for flares, surges and Ellerman bombs. Parasites often surround a large spot of opposite polarity, where they sometimes show up as *satellite sunspots.*

Sunspots move faster than the surrounding plasma, so they are probably anchored below the surface where the rotation rate is faster. They tend to drift apart by 0.1 degree per day, since the leader spot is at lower latitude and therefore rotating faster.

Pores are also darker than the surrounding photosphere, but they have no penumbra. They represent intermediate flux concentrations between normal intense flux tubes and sunspots. Pore diameters are 0.7 to 7 Mm, the largest being larger than the smallest sunspots. They have about 50 per cent of the photospheric brightness and central field strengths between 1.8 kG and 2.3 kG, dropping to 1 kG at the edge of the pore.

Superpenumbral fibrils are 5 Mm-high Hα loops around a mature sunspot that form a nearly radial pattern, often spiralling, stretching out far beyond the penumbra (Figure 1.29b). The sense of twist tends to follow a hemispheric rule, with most having anti-clockwise twist in the north and clockwise in the south. They usually begin at the outer edge of the penumbra, but sometimes inside it. Individual fibrils change over a few minutes, but their overall pattern evolves slowly over hours or a day. In transition-region lines between 10,000 K and 100,000 K, one observes bright *sunspot plumes* or *sunspot loops*, extending to a height of 15 Mm and containing downflowing plasma at 20 to 40 km s^{-1}, possibly driven by a *siphon flow* mechanism.

1.7.2.3 Fine Structure of the Umbra

The umbra is far from uniform. *Dark nuclei* are darker than normal, with typical intensity 0.15 of the photospheric brightness (I_0) and diameters of 1 Mm. *Umbral dots* (Thiessen 1950; Danielson 1964) are small bright features (found in all pores and sunspot umbrae) with diameters between 100 and 450 km, upflows of up to 1 km s^{-1} and lifetimes 10 min to 2 hr. At lower resolution, they were formerly called *umbral granulation.* They cover 3 to 10 per cent of the umbral area but account for 10 to 20 per cent of the brightness. They are 1,000 K hotter than the surrounding umbra and 500 to 1,000 K cooler than the normal photosphere. *Peripheral umbral dots* lie close to the penumbra and are associated with bright grains that move into the umbra from the penumbra. They are larger, live longer, have weaker and more inclined magnetic fields, move faster and are brighter than *central umbral dots* (intensity 0.38 to 0.64 I_0). Their motion is inward in the direction of the horizontal magnetic field.

With the Swedish Solar Telescope (SST), spectra of peripheral umbral dots indicate regions of slightly reduced magnetic field strength (1,400 to 2,400 G), localised slow upflow (0 to 210 m s^{-1}), higher temperature (5,300 to 6,600 K) and more horizontal magnetic field (by 10°) (Socas-Navarro et al. 2004). They show up only in lower photospheric lines.

Hinode observations by Riethmüller et al. (2008) of peripheral umbral dots reveal similar properties, namely, temperature enhancement of 570 K, field weakening by 510 G and upflows of 800 m s^{-1}, whereas central umbral dots have weaker excess temperature (550 K) and field weakening (480 G) and no significant upflows. Also, Watanabe et al. (2009) give mean sizes of 180 km, lifetimes of only 7 mins, magnetic fields of strength 2,200 G, magnetic inclination 27 degrees to the vertical, transverse motions of 400 m s^{-1}, and brightness a factor 1.7 above background. Few umbral dots are present where the magnetic field is strongest ($>$ 2,900 G).

Light bridges are narrow low-lying, slightly raised bright features (of penumbral brightness) crossing over an umbra. The magnetic field is reduced and more horizontal. A long narrow dark lane runs along an elevated ridge along their length, and they are segmented by short dark lanes running across them.

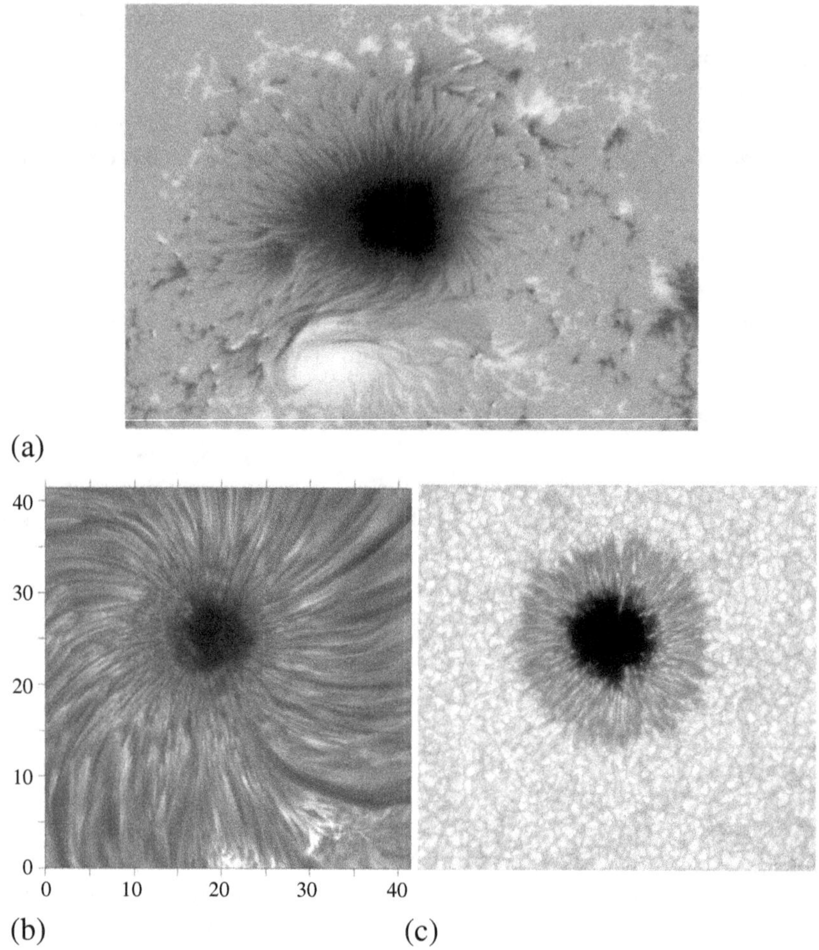

Figure 1.29. (a) A magnetogram of the line-of-sight field, showing the fine interlocking comb structure of the penumbra from Hinode/SOT (courtesy K. Ichimoto). (b) An Hα image of a superpenumbra with tick marks every 5 Mm (left) and the underlying sunspot (right) from the SST (courtesy Luc Rouppe van der Voort, Mats Carlsson and Viggo Hansteen).

Basically, a light bridge is a fissure separating two different components of a spot and the dark lane is produced by an enhancement of density raising the $\tau = 1$ surface. The brightening of a light bridge (or the appearance of a new one) can indicate the imminent breakup and fragmentation of a sunspot, especially if it grows in width and intensity (to normal photospheric brightness) and displays normal granulation.

At the photosphere, there is a broad spectrum of *umbral oscillations*, with periods around 5 min (300 sec) and 3 min (Beckers and Schultz 1972), which may be understood in terms of absorption and partial mode-conversion of *p*-modes (Sec. 9.5.3). The "5-min" oscillations have an angular frequency $\nu = 1/\text{period} = 3$ mHz, but lie between 2 mHz and 5 mHz and are coherent over the whole umbra, often extending into the penumbra. With velocities of 40 to 100 m s^{-1}, their power is a factor 2 or 3 smaller than that of 5-min oscillations in the surrounding photosphere, but their horizontal wavenumbers are smaller due to the higher phase speed in the umbra.

The "3-min" oscillations lie between 4.5 mHz and 7 mHz and are more localised and intermittent. In the chromosphere, there is a broad range of power between 2 and 8 mHz, but most of it is concentrated in the 3-min band, with much higher velocities (up to several km s^{-1}) than in the photosphere. They are nonlinear and compressive and at their peak intensities create sudden brightenings called *umbral flashes* in chromospheric lines such as Ca K (Beckers and Tallant 1969). The flashes have a rapid increase and slow decrease in brightness. They repeat with periods of 140 to 190 s, have diameters of 2 Mm and move towards the penumbra at 40 km s^{-1}.

1.7.2.4 Fine Structure of the Penumbra

The SST and Hinode/SOT have revealed many features of sunspot fine structure (e.g., Rouppe van der Voort et al. 2004; Ichimoto et al. 2007). *Penumbral filaments* are typically 3.5 to 7 Mm long. Individual penumbral filaments endure 0.5 to 6 hr, by comparison with a lifetime of days or months for the sunspot as a whole. The intensity of a bright filament is typically 95 per cent of the surrounding photosphere, while that of a dark filament is only 60 per cent. Bright filaments protrude into the umbra and become less distinct in the outer penumbra, with some of them even becoming dark. Many originate within the penumbra at bright upflowing grains and fan out with distance from the umbra. As well as a predominantly radial flow, there is a weak transverse flow that converges on dark filaments.

At high resolution, the penumbral filament widths are as small as 150 km and possess unresolved features below the current resolution (80 km). Bright filaments consist of elongated *penumbral grains* aligned on a dark background. A grain is typically 300 km (or less) wide and 300 to 2,000 km long with a lifetime 40 min to 3 hr and intensity 0.85 to 1.1 I_0. They move slowly inwards in the inner 60 per cent of the penumbra and outwards in the outer penumbra, in both cases at 0.5 km s^{-1}. Grains often cross into the umbra, where they become peripheral umbral dots and carry on moving for a while. Most outwards-moving grains disappear, but a third cross the boundary and continue moving. Perched on top of many bright filaments there is a slender *dark core* (Scharmer et al. 2002) that extends across the penumbra and often splits into a Y-shape at the umbra.

The magnetic field has an *interlocking comb* structure (see Figure 9.13) that is much more horizontal in dark than bright filaments (Figure 1.29a). The inclination to the vertical in bright filaments increases with radius across the penumbra from 40 degrees to 60 degrees. In dark filaments, the field is weaker, and its inclination increases from 40 degrees to become horizontal at 80 per cent of the spot radius (where it is only a few 100 km above the photosphere) and thereafter to point back towards the surface and become 100 degrees at the edge (Degenhardt and Wiehr 1991; Title et al. 1993; Langhans et al. 2005).

The penumbral magnetic field in this picture consists of three components: bright filaments are more vertical and form loops extending for large distances from the spot; some of the dark filaments make up a shallow canopy about 300 km high and reaching to twice the spot radius; other dark filaments are lower, bending over and dipping below the surface within the penumbra or just beyond. The structure continually evolves and occasionally neighbouring fields that are inclined to one another will reconnect and produce fine-scale jets (Katsukawa et al. 2007).

Tiwari et al. (2013), however, have argued against separate bright and dark filaments. They applied a new inversion technique to Hinode/SOT data and discovered that all penumbral filaments are remarkably similar, irrespective of position. Their lengths are 5 $\pm$1.6 arcsec and their mean width 0.8 arcsec. Their inward end (or head) is bright and hot ($T = 6,500$ K) and has an upward-pointing field (1.5 to 2 kG), inclined at 40 degrees to the vertical with a strong upflow (typically 5 km s^{-1}, but up to 11 km s^{-1}). Their outward end (or tail) is dark and cool ($T = 6,000$ K) and has a downward-pointing field (2 to 3.5 kG), inclined at 40 degrees to the vertical with a strong downflow (typically 7 km s^{-1} but up to 19 km s^{-1}). In the body of a filament, there is an almost-horizontal field of 1 kG coaligned with the Evershed flow (going from bright to dark towards the tail) and with lateral downflow of 0.5 km s^{-1}. The dark cores of some filaments have a temperature reduced by 200 to 400 K. In this picture, filaments are flanked by dark lanes or *spines* of more vertical field (Lites et al. 1993), which provide most of the field lines that go far from a spot.

Running penumbral waves (Zirin and Stein 1972; Giovanelli 1972) are a chromospheric phenomenon that shows up in Hα movies as outwards-propagating concentric circular ripples over a large angle in azimuth. They start at the umbral boundary with a speed of 15 to 20 km s^{-1} and slow to 5 to 7 km s^{-1} at the outer edge of the penumbra. Their velocity amplitude is 1 km s^{-1} and the period is typically 200 to 300 s, so they are weaker and slower than 3-min umbral oscillations. They are probably slow magnetoacoustic gravity waves (Sec. 9.5.3).

1.7.2.5 Flows and Motion

Evershed flow is a radial penumbral outflow in the photosphere parallel to the magnetic field of average value 2.5 km s^{-1} at moderate resolution (Tritschler et al. 2004), but, if the flowing component is isolated, much larger values are found of 6 km s^{-1} (Bellot Rubio et al. 2003). The flow is almost horizontal with upward and downward inclinations of 15 degrees and 5 degrees in the inner and outer penumbra, respectively. It begins as soon as a sunspot's penumbra forms and is confined to narrow channels in parts of both light and dark filaments, although faster in the dark ones and focussed in the dark ones in the outer penumbra. (Since more filaments end in the outer penumbra, there is a preponderance of dark parts of filaments there.) The flow waxes and wanes over 10 to 20 mins and increases with radius, reaching a maximum in the outer penumbra. The strongest upflows (2 km s^{-1}) are associated with bright penumbral grains near the umbral boundary, while strong downflows (up to 9 km s^{-1}) are observed in the tiny patches of return flux just inside the outer penumbral boundary (Westendorp Plaza et al. 1997; Franz and Schlichenmaier 2009).

Higher up in the atmosphere, Evershed outflow becomes slower and eventually reverses its direction at chromospheric levels. This *Evershed inflow* (or *reverse Evershed effect*) extends over a fairly large region surrounding the sunspot. It is observed along *superpenumbral fibrils*, which are Hα loops about 5 Mm high: the inflow speed is typically 20 km s^{-1}, but it may reach as much as 50 km s^{-1}.

A *moat* is a steady annular convection cell around a stable sunspot, similar to a long-lived supergranule. It has a diameter between 15 Mm and 100 Mm and is swept clear of magnetic flux except for tiny *moving magnetic features* (Sheeley 1969; Harvey and Harvey 1973). The moat flow decreases with radius and is focussed in spoke-like channels. The net outward flux-transport is 6×10^9 to 8×10^{11} Wb hr^{-1} and of the same polarity as the sunspot. The moat flow has a speed of 0.3 to 1 km s^{-1} (twice as fast as a normal supergranule) and persists throughout a spot's life. It begins with the formation of a penumbra and exists only outside parts of a spot that possess a penumbra. Vargas Dominguez et al. (2008) found with the SST that moat flows point in the same direction as penumbral filaments and are absent when such filaments are parallel to the penumbral boundary.

Moving magnetic features (MMFs) are of both polarities, with a slight preference for the polarity of the sunspot. They move outwards at typical speeds of 1.5 to 1.8 km s^{-1} (faster than the moat). Type I MMFs are bipolar pairs that appear to be extensions of horizontal penumbral fields that dip down into the moat and back up again (Zhang et al. 2003). Type II MMFs are single magnetic elements of the same polarity as the spot and appear to be eroded away from the spot by granular convection (Kubo et al. 2007; Kubo et al. 2008). Type III MMFs are also single polarities, but of polarity opposite to the spot and with higher speeds (2 to 3 km s^{-1}).

Sunspots can exhibit *proper motion* relative to the surrounding photosphere in several ways. During the first few days of its life, a *p*-spot moves westwards and then, as the region decays, it slowly returns to roughly its original longitude. Occasionally, spots may divide or merge and a single spot or a bipolar pair may rotate.

1.7.2.6 Growth and Decay

Sunspots form by the accumulation of magnetic flux and coalescence of pores (Sec. 9.6.1). Flux first collects at the boundary of a supergranule cell, especially at a junction of three cells. A *pore* can appear when a downflow of 0.5 km s^{-1} with cooling signifies the onset of *convective collapse* (Sec. 9.2.3). Often a pore lasts only hours or days, but sometimes it continues to grow and develops into a small sunspot when the diameter and flux are large enough (3.5 to 7 Mm and 1×10^{20} Mx) that a penumbra quickly appears (over 20 mins) (Schlichenmaier et al. 2010).

During the growth of a sunspot, say between 3 and 10 days, more and more magnetic flux is added at speeds of 0.25 to 1.0 km s^{-1}, while the maximum field strength increases and reaches a peak. Typically, spots are formed in pairs and move apart to a maximum separation of about 150 Mm.

Most sunspots disappear within a few days (the very common smaller ones), but sometimes they last a few weeks and occasionally a few months. Long-lived spots tend to be leader spots with a dark circular umbra and surrounded by a moat and a plage of the same polarity. They are connected to trailing plage of the opposite polarity. Larger leader spots tend to live longer, whereas follower spots tend to be more irregular in shape and to last only a few days.

Short-lived sunspots tend to fragment in less than a day after a *light-bridge* of normal photospheric intensity has formed across the umbra. Light bridges often develop along the boundaries between the pores that originally coalesced to form the spot. Long-lived spots decay gradually (Sec. 9.6.2): as soon as they are fully formed, the area shrinks in size at a constant rate as flux is carried across a moat as moving magnetic features, while the relative size, brightness and field strength of umbra and penumbra remain constant. Most of the magnetic flux in the sunspot is transported to the outer boundary of the moat region (as moving magnetic features) and then removed from the photosphere by flux cancellation around the outer boundary of the moat region (Kubo et al. 2008).

1.7.3 Solar Cycle

The *solar* or *sunspot cycle* refers to the variation in many characteristics of solar activity that follow an 11-year oscillation in sunspot number. Traditionally, the state of the cycle is measured by counting the number of sunspots (f) and sunspot groups (g) visible on the disc and combining them as the *Wolf number*,

$$R = K\ (10g + f), \tag{1.11}$$

where K is an observer correction factor (typically 0.6). The resulting sunspot number has a cyclic behaviour, as indicated in Figure 1.30.

Although the periodicity was discovered only in 1843 (by Schwabe), the activity level has now been extended back before the start (1610) of detailed sunspot records. ^{14}C proxy data in tree rings and ^{10}Be

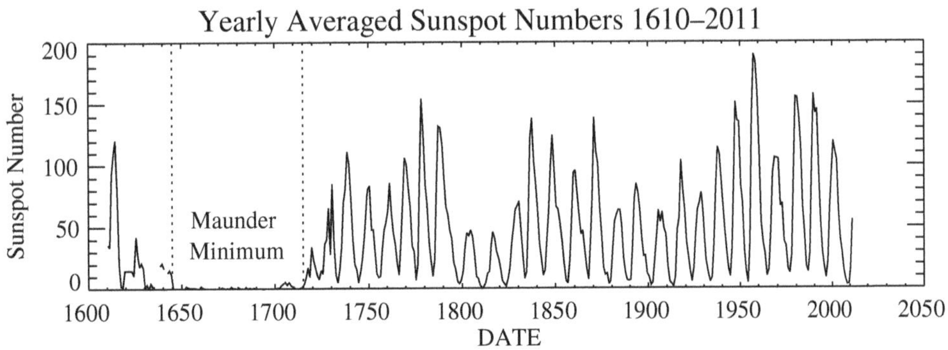

Figure 1.30. The yearly averaged sunspot number from 1610 to 2011, showing the 11-year cycle with a long-term modulation. The Maunder minimum of 1645–1715 had virtually no sunspots and the Dalton minimum of 1800–1820 had very few (courtesy David Hathaway).

in ice cores extend it back 26,000 yr and 50,000 yr, respectively, since the incidence of galactic cosmic rays on the Earth varies in antiphase with solar activity and creates radioactive isotopes such as ^{14}C and ^{10}Be, which decay with half-lives of 5,730 years and 1.5×10^6 years, respectively. Looking back say 10,000 yr, sunspot maxima as high as that in 1990 and minima as low as the one in 2008 are quite common, and in fact 9000 years ago the maxima were considerably larger. However, records going back before the Holocene (12,000 years ago) cannot be used for reliable estimates of solar activity, since the Be records are affected by local climate, which was quite different during the last ice age.

The *sunspot cycle* has a considerable variation in both period, maximum and minimum. Since 1750, the average period between maxima is 10.9 years (ranging from 7.3 to 17.1 years), whereas between minima it is 11.1 years (ranging from 9.0 to 13.6 years). The average maximum in sunspot number is 108 (varying between 49 and 201), while the average minimum is 5.1 (varying between 0 and 11). Another feature of the cycle is that the rise from minimum to maximum (4.5 years on average) is shorter than the fall from maximum to minimum (6.5 years). When the rise phase is faster, the cycle tends to be larger. The length of a cycle is related to the strength of the following cycle: small cycles start late and leave behind a long cycle and low minimum.

The longer-term variation in sunspot number is considerable and appears largely random. A tendency can be seen in Figure 1.30 for very low or high peaks in the solar cycle to be present every seven or eight cycles, which suggests that a longer cycle may be operating on a 70 to 100-year period (the Gleissberg cycle). Around 1920, 1810 and 2008, the minima were deeper and longer than average, while the neighbouring maxima were smaller. Even longer periods, such as the 210-year Seuss (or de Vries) cycle, have also been suggested.

In a piece of historical detective work, Eddy (1976) revived an interest in the fact that for a period of 70 years from 1645, which he dubbed the *Maunder minimum*, the Sun was largely without sunspots at all. This period was also known as the *Little Ice Age*, since the climate in Europe was considerably cooler than normal, with the river Thames occasionally freezing over. Indeed, this is now known to be only the latest of many *grand minima* that show up in proxy data (Beer 2000). It is interesting to note that the ^{10}Be oscillation continues through the Maunder minimum; the magnetic cycle did not switch off but just reduced in strength so that sunspots could not form readily (Beer et al. 1998). Furthermore, the decline into the Maunder minimum was gradual rather than sudden (Vaquero et al. 2011).

Some intriguing *laws of sunspot polarity* were put forward by Hale and Nicholson in 1925 and are found to be obeyed by about 97 per cent of sunspot groups (Figure 1.31):

(i) the polarity of all leading spots in each hemisphere is the same for 11 years (freaks with inverted polarity are often associated with high flare activity);

(ii) the polarity reverses its sense every 11 years at the start of a new activity cycle, two of which make a magnetic (or *Hale*) cycle with a period of 22 years;

(iii) leading spots in the north and south hemisphere have opposite polarity (*Hale's law*);

(iv) the leading polarity is closer to the equator (*Joy's law*), with the inclination angle ranging from 0 degree near the equator to 15 degrees at latitude $\pm$ 35 degrees.

These suggest that active regions are born from a strong subphotospheric toroidal flux tube, with a slight inclination from Coriolis forces as it rises to the surface.

In the rise phase of a cycle, polar flux has the same polarity as leader spots in that hemisphere (Figure 1.31). It reverses sign 1 to 2 years after solar maximum due to migration of trailing-polarity flux from sunspot latitudes. The two polar reversals may occur at different times. At sunspot minimum, the polar fields are strong and extensive, but near sunspot maximum the polar holes contract and disappear.

Most sunspots are confined to two belts between latitudes $\pm$ 5 degrees and $\pm$ 35 degrees. At any one time there may be a considerable spread of latitude, but the average latitude drifts towards the equator as the cycle proceeds (Spörer's Law), as indicated in Maunder's *butterfly diagram* (Figure 1.32). At the beginning of a new solar cycle, the average latitudes are $\pm$ 28 degrees from the equator; after 6 years they

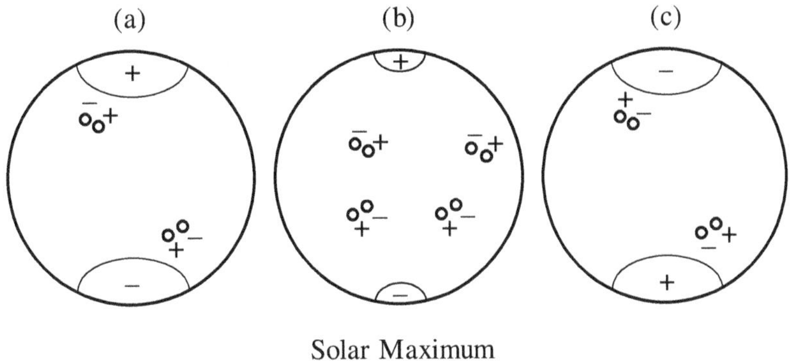

Figure 1.31. The polarity of sunspots and polar regions for (a) the start of one cycle, (b) the maximum of that cycle and (c) the start of the next cycle.

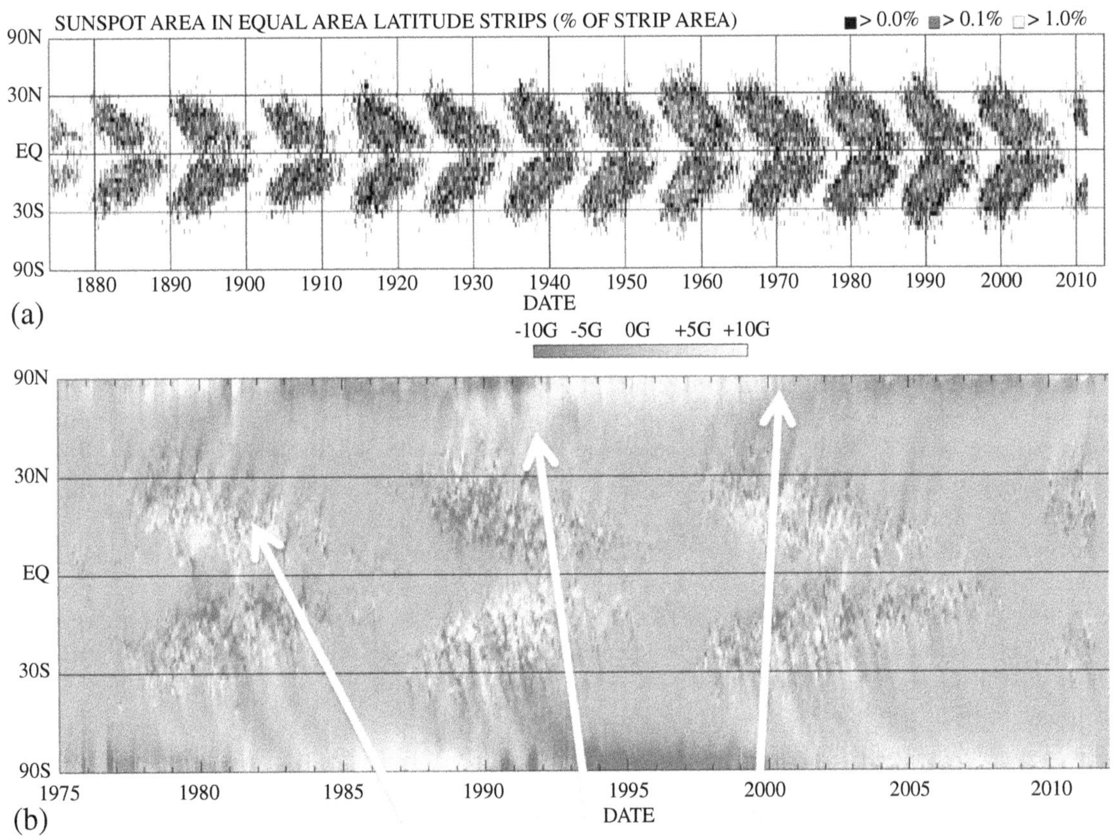

Figure 1.32. (a) Butterfly diagram, showing a latitude drift in time of sunspot occurrence towards the equator. (b) Magnetic butterfly diagram of the longitudinally averaged photospheric magnetic field, showing the variation in positive (light)- and negative (dark)-polarity sunspots as a function of time. Arrows (from left to right) indicate: migration of sunspots towards the equator; poleward migration of trailing flux by a meridional flow; and reversal of the polar field about 2 years after sunspot maximum (courtesy David Hathaway).

are $\pm$ 12 degrees and after 11 years they have fallen to $\pm$ 7 degrees. Consecutive cycles may overlap by typically 2 years, since the last spots of one cycle can be present at low latitudes, while new spots of the next cycle are appearing at higher latitudes and with the opposite polarity.

Magnetic butterfly diagrams such as Figure 1.32b are even more revealing. They show how sunspot locations migrate equatorwards during the solar cycle, and indicate the preferential polewards migration of trailing flux, especially near sunspot maximum. This leads to a reversal of the polar field about a couple of years after maximum.

Many solar features owe their existence to the magnetic field. It is not surprising, therefore, that it is not just the sunspot number that varies with the solar cycle, but also the whole solar atmosphere and features of solar activity such as prominences and flares. For example, the chromosphere has quite a different appearance at solar minimum and solar maximum, as does the global magnetic structure revealed in white-light eclipse images. Thus, at solar minimum the corona often has a dipole shape with prominent open plumes at the poles and helmet streamers at the equator, whereas at solar maximum it is much more isotropic with streamers stretching out from all latitudes. Furthermore, the coronal intensity in soft X-rays increases by a factor of 100 from minimum to maximum, and the corona is much more highly structured and varied at solar maximum. Naturally, variations in the solar wind accompany the solar-cycle variations in coronal structure (see Figure 1.27, Section 1.6.4).

Even differential rotation shows solar-cycle variations, since rotating bands that are slower and faster than normal by 1 per cent migrate from mid-latitudes towards the equator, tracking the sunspots zones (Howard and Labonte 1980) and possibly driven by a $\mathbf{j} \times \mathbf{B}$ force. These 11-year *torsional oscillations* (Figure 1.9) are also detected below the surface by helioseismology techniques as *zonal shear flows* (e.g., Howe et al. 2005).

The total solar irradiance varies by 0.1 per cent with the solar cycle, being brighter at solar maximum due to a competition between sunspots and the smaller but more numerous bright faculae (Sec. 1.4.1). When an active region is present, the excess radiation from faculae more than balances the reduction in radiation from sunspots (Solanki and Fligge 2002). Radiation at UV, EUV and radio wavelengths varies by factors of 0.01 to 10 between solar minimum and maximum, while the soft X-ray flux changes by a factor of 100. Variations in total solar irradiance may slightly affect terrestrial climate, but the much stronger variations in UV radiation are probably more important through their effect on the stratosphere by photochemical reactions involving ozone, which then affect the jet stream (Haigh 2007). The effect of cosmic rays in triggering cloud formation is probably much smaller.

The solar minimum of 2008 between cycles 23 and 24 was unusual in several ways. The decline in sunspot number into the minimum was more gradual and the minimum itself was deeper and longer than in the previous few cycles. The duration of cycle 23 (12.6 years) was the longest for 100 years. From the magnetic butterfly diagram, sunspots approached the equator much more slowly in cycle 23 and the polar fields were much weaker. Torsional oscillations showed the same slow migration to the equator and a delay in the start of the polar branch [numbers (4) and (5) in Figure 1.9].

The butterfly diagram shows that usually the cycles overlap, with the new sunspots appearing at high latitudes at the start of a new cycle at the same time as spots from the old cycle are still appearing at low latitudes. However, this was not the case in 2008, since a clear gap between the two cycles was present (just as in 1900). Meridional flow (Sec. 1.3.3) was faster in cycle 23 than cycle 22. Nandy et al. (2011) found that in a flux-transport dynamo a fast meridional flow produces no overlap in cycles, whereas a slow flow makes the cycles overlap. In their scenario, a fast meridional flow sweeps the poloidal field more rapidly along the base of the tachocline, so that a weaker toroidal field is built up, leading to a smaller sunspot cycle.

Sunspots provide many challenges to MHD theory and have a great fascination. For example, when will the next great flare-producing active region appear? How has a large active region evolved as it has rotated round the far side of the Sun? Also, at the end of one cycle, there is an air of expectancy as one awaits the birth of the first sunspots at high latitudes, heralding the dawn of a new cycle. Even though there is, like weather patterns on Earth, an inherent unpredictability about sunspots, much progress has been made on understanding their structure (Chapter 9). Furthermore, the systematic large-scale properties of solar magnetic fields indicate the existence of a deep-seated oscillatory dynamo, such that many processes observed at the solar surface are controlled by a mechanism operating in the convection zone (Chapter 8).

1.8 Prominences

1.8.1 Introduction

Prominences are amazing objects (see, e.g., Priest 1989; Tandberg-Hanssen 1995; Parenti and Vial 2007; Labrosse et al. 2010; Mackay et al. 2010). They are suspended in the corona but are a hundred times cooler and denser than the corona. In eclipse or coronagraph pictures, they appear bright in emission at the limb (Figure 1.33a), but in Hα photographs of the disc they show up in absorption as thin, dark, meandering ribbons called *filaments* (Figure 1.33c). They are located in highly sheared, mainly horizontal, magnetic field above polarity-inversion lines, where the radial photospheric magnetic field component reverses (Figure 1.33d).

The main type is a *quiescent prominence*, which occurs over a wide range of latitudes outside active-region cores, where *active-region prominences* are found. Quiescent prominences are sometimes split into those that occur in the quiet Sun and those that are located at the boundary of an active region (called instead *intermediate prominences*), but physically they have the same structure and formation mechanism. Outside active regions, prominences may be much thicker and longer, reaching up to 1000 Mm in length as huge *hedgerow prominences*, which migrate slowly towards the polar regions where they form a band of *polar-crown prominences.*

Quiescent prominences have captivated the solar observer for centuries. One was observed at an eclipse in the Middle Ages (1239) and described as a 'burning hole in the Moon' (Secchi 1877a); others in 1733 were called 'red flames' by Vassenius in Russian chronicles. However, by the beginning of the nineteenth century, their existence had been forgotten. At the 1842 eclipse they were rediscovered, but the observers were so surprised that they did not give a reliable description and were unclear whether they were solar or lunar features (PROBLEM 1.3). In 1860, they were photographed, and in 1868 spectroscopic techniques were introduced which led to the discovery of He. Bernard Lyot's invention (1930) of the *coronagraph* enabled him to observe prominences at the limb without waiting for an eclipse; then Kippenhahn and Schlüter (1957) put forward their classic model for the support of dense prominence material against gravity by a magnetic field. Observational and theoretical advances have recently transformed dramatically our understanding of the nature and formation of the overall plasma and magnetic structure of prominences, but the fine-scale structure and internal dynamics remain a mystery (Chapter 11).

1.8.2 Plasma and Magnetic Properties

A *quiescent prominence* has a complex three-dimensional structure (Figure 1.33). It is a huge, vertical sheet of dense, cool plasma, made up of dynamic threads ($<$ 100 km in width), and surrounded by a hotter and rarer coronal environment. Its core density is 10^{15} to 10^{17} m^{-3}, with a ratio of protons to neutral H atoms 0.2–0.9. The core temperature is 7,500 to 9,000 K. The dimensions (typically $200\times50\times5$ Mm) range as follows: length 60 to 600 Mm, height 10 to 100 Mm, and thickness 3 to 10 Mm.

Prominence plasma is highly dynamic (Schmieder et al 1984), with horizontal and/or vertical flows of typically 5 km s^{-1}, but up to 10 to 70 km s^{-1}, while turbulent (i.e., unresolved) velocities are of order 3 to 20 km s^{-1}. The prominence-corona transition region is hotter (10^4 to 10^6 K) and less dense (10^{12} to 10^{14} m^{-3}), but with stronger flows (10 km s^{-1}) and turbulent velocities (30 km s^{-1}).

The *magnetic field* in quiescent prominences is mainly horizontal (Rust 1967; Leroy 1989; Bommier and Leroy 1998). At the limb, the average line-of-sight component (from the Zeeman or Hanle effect) is typically 8 to 15 G, but it may be 20 to 40 G in some prominences and reach 80 G in patches (Casini

(a) (b) (c) (d)

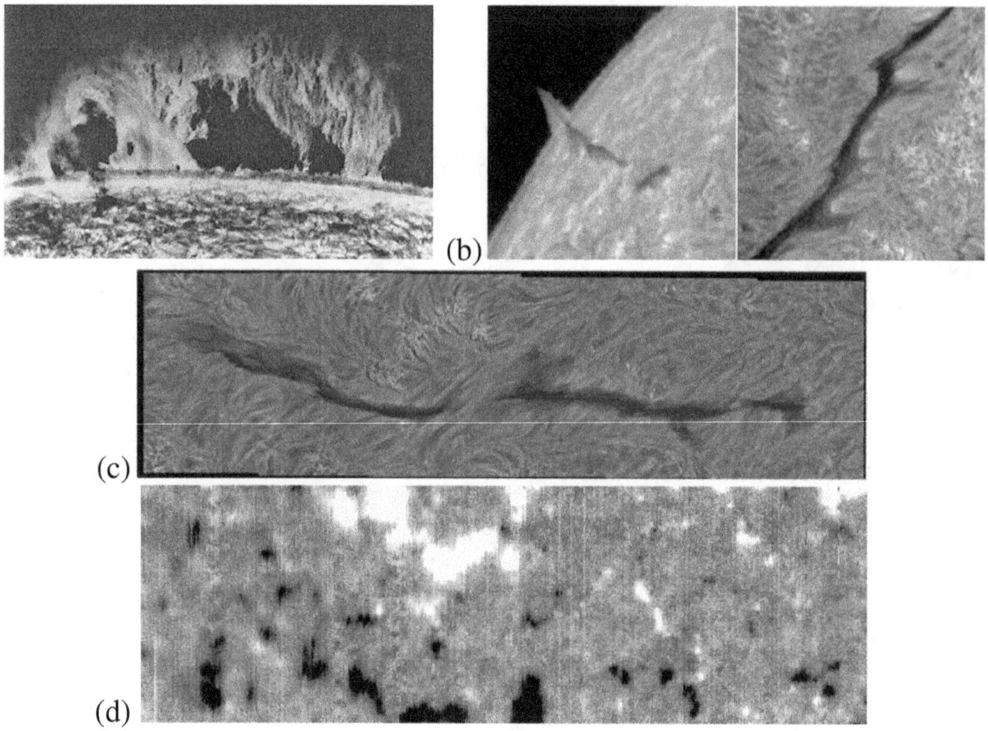

Figure 1.33. Prominences viewed from different directions: (a) at the limb from the side (Big Bear Solar Observatory, operated by the New Jersey Institute of Technology), (b) near the limb from the end (Big Bear Solar Observatory), and at an oblique angle (National Solar Observatory/AURA/NSF, right), (c) on the disc from above (Dutch Open Telescope/La Palma) with (d) the corresponding line-of-sight magnetic field (THEMIS/MTR) from Dudík et al. (2008).

et al. 2005). It increases over the height of the prominence, which is consistent with the presence of a dip. The average angle between the direction of the magnetic field and the long axis of a prominence is only about 20 to 30 degrees, so prominences are located in highly sheared magnetic field above a photospheric polarity-inversion line (PIL), which, in the quiet Sun, is actually a broad zone of mixed polarity (Figure 1.33d, Schmieder et al 1996b). Most prominences have *inverse polarity* in the sense that the horizontal component perpendicular to the prominence axis is oppositely directed to that of a potential field based on the normal photospheric field. Furthermore, the magnetic field below a prominence at the photosphere also has inverse polarity (López Ariste et al. 2006), and so it has the structure of a so-called *bald patch* (Titov et al. 1993).

Mackay et al. (2008) classified the parts of a PIL at which prominences form into four types: Internal Bipolar-Region Filaments (IBR) lie in the interior of a bipolar region; External Bipolar-Region Filaments (EBR) lie between two magnetic bipoles; Internal/External Bipolar-Region Filaments (I/EBR) lie above both the internal PIL of a bipole and the external PIL surrounding the bipole; Diffuse Bipolar-Region Filaments (DBR) are located in a broad diffuse region of bipolar flux. They studied 600 quiescent and intermediate filaments and found that only 7 per cent form in a single bipole, whereas 92 per cent require multiple bipole-interactions: 62 per cent are EBR, 17 per cent DBR and 13 per cent I/EBR. This is a key result that had been long known qualitatively (Kiepenheuer 1953). The EBR filaments are more numerous at sunspot maximum, while none of the other filaments depend on the solar cycle.

Filaments are located in *filament channels*, which are regions surrounding a PIL where chromospheric fibrils (and therefore the magnetic field) are aligned with the PIL. A filament channel or prominence is

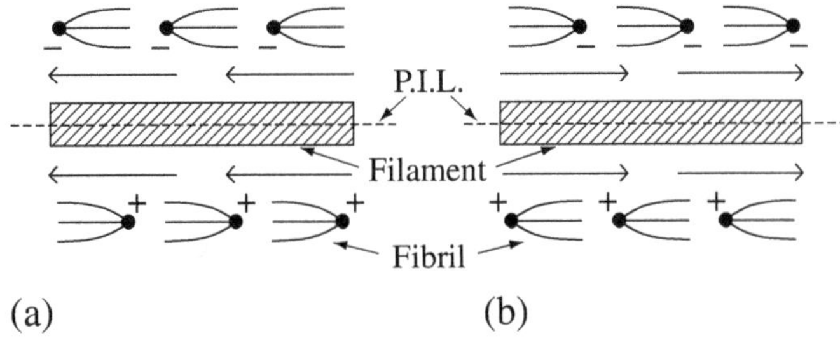

Figure 1.34. Schematic of a filament and filament channel viewed from above. The magnetic field is predominantly horizontal (long arrows) and is directed along the channel. Its direction determines the chirality, namely, either (a) sinistral or (b) dextral. The chromospheric fibrils are nearly parallel to the polarity-inversion line (PIL); they originate in positive magnetic fragments (+) and end in negative ones (-).

said to be *dextral* if its axial magnetic component (along the PIL) points to the right when viewed from the positive polarity side and *sinistral* if it points to the left (Figure 1.34). Martin's law, a global chirality law, says that filaments and channels in the northern hemisphere are mainly dextral, while those in the south are mainly sinistral (Martin et al. 1994). Surprisingly, this is the opposite of what differential rotation acting on a coronal arcade with an east-west PIL would produce, but a good explanation is given in Section 11.3. Furthermore, in EUV lines, a filament appears as a dark corridor that is typically five times thicker and longer-lived than the Hα filament (Heinzel et al. 2001; Aulanier and Schmieder 2002; Dudík et al. 2008) and contains more mass.

An active-region prominence is typically much smaller than a quiescent prominence, with a length of 10 Mm and a height below 2.5 Mm. Its temperature is similar, but its density is rather larger ($\gtrsim 10^{17}$ m^{-3}). The magnetic field is stronger, usually 20 to 70 G, though sometimes much higher: measurements in Tenerife gave longitudinal fields in filaments of 100 to 200 G and transverse fields of 500–600 G, implying a field strength of 600 to 700 G (Kuckein et al. 2009; Guo et al. 2010). The structure is that of a twisted flux-rope with complex flows (Kuckein et al. 2012a,b).

1.8.3 Structure

A filament possesses three structural components, namely: a *spine*, running along the top of its axis; *barbs*, protruding from its sides (Figure 1.33b,c), and which, when viewed at the limb, extend down to the chromosphere as a series of regularly spaced feet resembling great tree trunks (Figure 1.33a); and two *legs* or ends to the filament.

The barbs extend towards patches of parasitic polarity in the photosphere and have vertical extents much greater than the local scale-height. They protrude at an acute angle with respect to the spine, like ramps off an elevated motorway, and so filaments are either *left-bearing* or *right-bearing*, depending on the directions of the barbs seen from above (Figure 1.33b,c). Martin et al. (1992) discovered that dextral filaments have right-bearing barbs, while sinistral filaments have left-bearing ones.

Both the spine and barbs can possess counter-streaming plasma at 10 to 15 km s^{-1} (Schmieder et al. 1991; Zirker et al. 1998; Lin et al. 2005). The nature of the photospheric magnetic field near barbs has been determined by López Ariste et al. (2006), who found that barbs are located in magnetic dips above bald patches next to parasitic polarity. In order to do so, they resolved the 180-degree ambiguity in determining vector magnetic fields by invoking chirality laws. For a theory of barbs, see Section 11.4.

Intriguingly, tiny *barb tornadoes* have been discovered with SDO/AIA and the VTT on Tenerife in the cores of barbs (Su et al. 2012; Li et al. 2012; Orozco Suárez et al. 2012; Wedemeyer-Böhm et al. 2013). They are potentially important for barb support or supply of prominence mass and magnetic helicity. Lifetimes are typically a few days (10 min to 1 week). It is therefore as yet unknown whether they are driven by photospheric vortex tubes, which are currently observed to be much shorter-lived (Sec. 1.4.1).

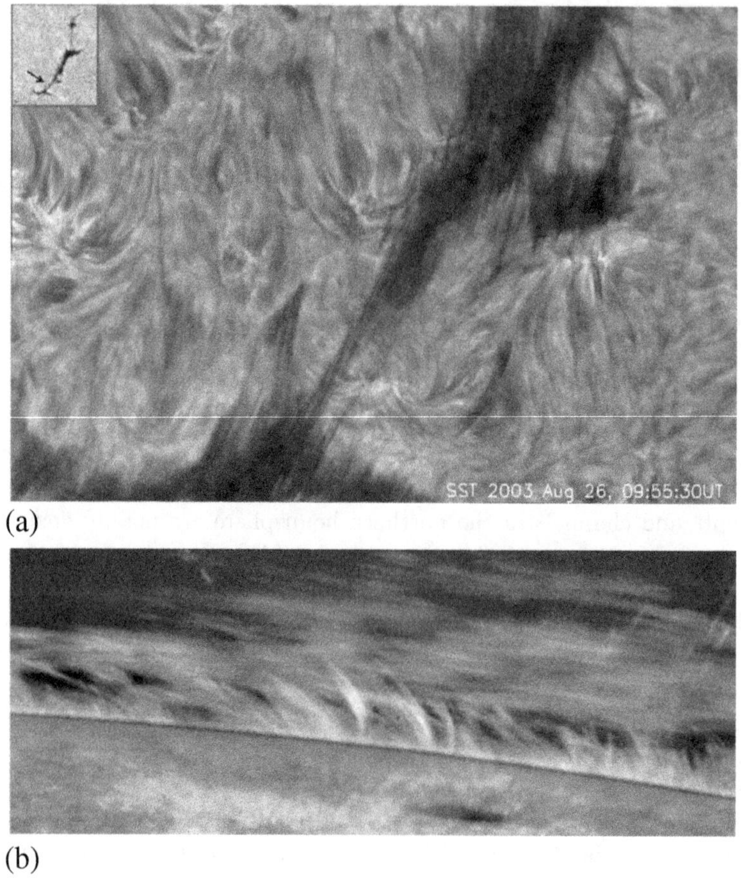

Figure 1.35. Fine-scale prominence threads viewed (a) from above in the spine and barbs of a filament with the SST (courtesy Oddbjørn Engvold) and (b) at the limb with Hinode/SOT in Ca II H (courtesy Saku Tsuneta). Image (a) was was studied by Lin et al. (2008) and (b) by Okamoto et al. (2007).

They represent rotating funnel-like magnetic structures with typical heights of 10 to 20 Mm. The largest are 45 Mm high, 1.5 Mm wide at its foot and 7.5 Mm at its top. They form in groups, appearing and growing upwards before the appearance of a prominence spine, and probably acting as sources or sinks of plasma for a prominence. Rotation periods are 10 s of minutes with rotation speeds of 5 to 10 km s^{-1}. Tornadoes show Doppler shifts of the same order as counter-streaming barb flows. They are somewhat longer-lived and thicker than barbs and have a cool component with a temperature of 7,000 K. Their relation to normal macrospicule tornadoes (Secs. 1.5.2 and 10.4.2) is unclear.

Each of these filament components consists of many thin dark *threads* of width <100 km and lengths 3 to 20 Mm (Figure 1.35a) and having lifetimes of only a few minutes (Lin et al. 2008). They are probably aligned with the magnetic field, and the field lines seem to continue beyond the ends of the threads before reaching the solar surface. Active-region filaments consist of longer threads than quiescent filaments.

Viewed at the limb, quiescent prominences often consist of many bright threads (Figures 1.35b, 1.36) (Okamoto et al. 2007), sometimes nearly horizontal but sometimes vertical. There are likely to be many present along a line of sight. At high resolution, the horizontal threads often give the appearance of being coronal loops and so of tracing field lines, whereas the vertical ones do not. Plasma continually streams slowly both up and down within a prominence at speeds of only 1 to 10 km s^{-1}, much less than the free-fall speed, so that the plasma must be supported somehow. Cork movies that follow these apparent motions

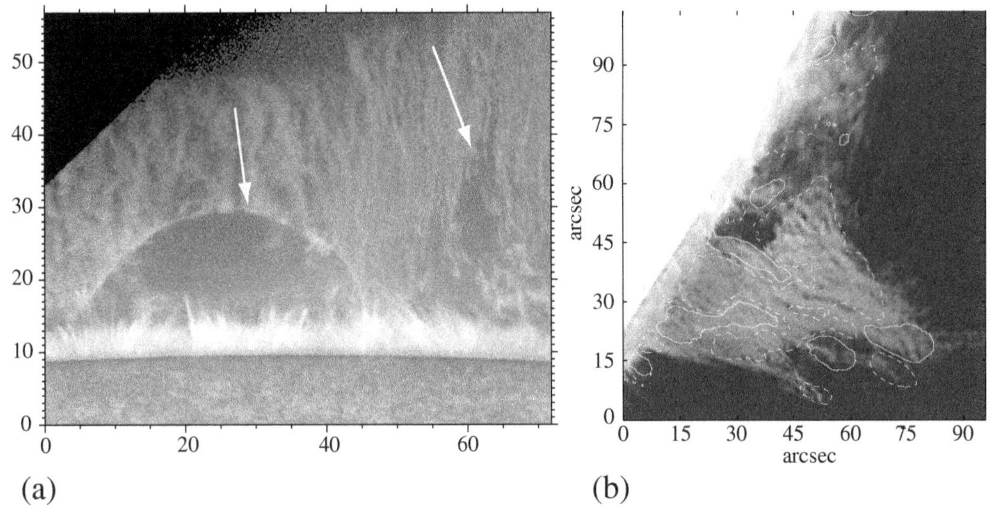

Figure 1.36. Prominences viewed at the limb (Hinode/SOT, courtesy Saku Tsuneta): (a) with arrows indicating rising arches (bubbles) and turbulent upflows (plumes), (b) with a superimposed Doppler-shift map from Meudon MSDP (solid for blue shifts and dashed for red) (Schmieder et al. 2010).

do not show a preferential global up- or down-flow; rather, there are temporary accumulation regions, some being close to a prominence top (Berger et al. 2008). The resulting immense circulation of mass is a mystery and, if sustained as a downflow, it could drain a prominence in a day or so. One possibility for a vertical thread is that the plasma is supported in a series of magnetic dips, stacked one above the other (Sec. 11.4).

Prominences contain large-scale (20 to 50 Mm) dark arches or *bubbles* that inflate from below, as well as smaller (2 to 6 Mm) dark turbulent upflows or *plumes* (Figure 1.36a), which rise from the base of a prominence up to 10 to 15 Mm, with initial speeds of 20 to 30 km s^{-1} and lifetimes of 5 to 15 mins (Berger et al. 2010). Simultaneous observations of bubbles and plumes with SDO/AIA show that they are much hotter (2.5×10^5 to 1.2×10^6 K) than the surrounding prominence plasma, which explains why they are dark and buoyant in Hinode/SOT (Berger et al. 2011). The role of these buoyant plumes in mass supply and their cause are unclear. SDO/AIA has, however, revealed a clear example of condensation of part of a prominence from coronal plasma (Berger et al. 2012).

Velocity vectors have been measured in prominences, combining transverse motions from Hinode/SOT with line-of-sight Doppler motions from the Meudon MSDP instrument (Schmieder et al. 2010): maps of the Doppler motions have a dappled character with nearby counterstreaming patches and flow speeds of 5 km s^{-1} in the prominence body and 15 km s^{-1} at its edges; transverse motions are similar, so that none of the flows are purely vertical. Sometimes there is swirling or twisting flow suggesting a vortex tube or prominence tornado (Sec. 1.4.1).

A filament channel is part of a much larger structure. In many images, one sees a *coronal cavity* (Sec. 11.2.3), a region of reduced density surrounding a prominence and reaching twice as high, but its mass defect is much too small to account for the prominence mass. An arcade of coronal loops is present above and around the cavity and is skewed in a direction along the PIL (forming in some cases a *sigmoid*, Section 1.9.2.3). Dextral (sinistral) channels have left (right)-skewed arcades, opposite to the structure of barbs (e.g., McAllister et al. 2002). Above the coronal loops there lies a *helmet streamer*, which may live for several months and has a broad base (up to a solar radius in diameter) with a pointed top at 1 or 2 $R_\odot$ above the limb.

1.8.4 Development

The formation of a prominence at a polarity-inversion line takes typically a few minutes or hours within an active region or a day or two in the quiet Sun (though sometimes up to several weeks). Active-region prominences tend to last only a few hours or a day, while quiescent prominences change slowly in overall shape, with a lifetime between a few and 300 days and a mean value of about 2 rotations (50 days) at low latitudes or 5 rotations (140 days) at high latitudes. Before the birth of a prominence, the Hα fibrils first align themselves end-to-end to form a *filament channel.* Once the prominence has formed, the fibrils on either side are aligned roughly parallel to the prominence, suggesting that the magnetic field is directed approximately along the prominence. Eventually, a quiescent prominence disappears by either slowly dispersing and breaking up, or erupting.

As a prominence migrates towards a pole, it is stretched more and more by differential rotation into an east-west direction, while its width and height remain relatively constant. Prominences tend to occupy two *zones* about 10 degrees polewards of the sunspot latitudes, as well as *polar zones*, where they sometimes form a *polar crown* around a polar cap at about latitude 70 degrees. The polar zones form about 3 years after spot maximum and migrate towards the poles; at about the next maximum they reach the poles, accompanied by the reversal in polarity of the polar magnetic fields.

A few examples of filament-channel or filament formation have been studied deeply (e.g., Gaizauskas et al. 1997; Martin 1998; Wang and Muglach 2007; Gaizauskas 2008); they imply that most long-lived prominences (i.e., external or diffuse bipolar region ones of quiescent type) form by the reconfiguration (reconnection) of pre-existing coronal fields by surface motions. Two key processes are involved (Sec. 11.3):

(i) creation of a filament channel with coaligned chromospheric fibrils when two separate flux systems interact with one another to form a highly sheared, largely horizontal magnetic field on a polarity-inversion line (PIL); and

(ii) convergence and cancellation of flux at the PIL (Martin et al. 1985) to create a twisted horizontal flux-rope, within which prominence plasma may be supported.

Stage (i) occurs after new flux has emerged and begins to interact with the flux of an old remnant active region or a nearby normal active region. It also follows the interaction of one unipolar region with another due to large-scale motions or supergranular convection, such as at the polar crown. In stage (ii) the filament passes over the location of the flux cancellation. The magnetic helicity may arise from newly emerging non-potential flux, or it may be slowly built up by flux cancellation as one flux system spreads towards the other. When two filament segments reconnect, they merge to form a larger filament if their chiralities are the same, or they create a solar flare if they are opposite (Schmieder et al. 2004).

1.8.5 Eruption

Active-region and quiescent prominences are sometimes *activated*, becoming darker (when viewed on the disc) or brighter (at the limb). At the same time, there may be an increase in turbulent (or helical) motion or flow along the prominence. Such activation may fade away after an hour or so, or sometimes it leads to an *eruption.* In other cases, prominence material may drain away from the summit along curved arcs at speeds of 100 km s^{-1}. During another type of activation, called a *winking prominence*, the prominence performs a damped oscillation (for 2 to 5 oscillations with a period of 6 to 40 min) after the passage of a shock wave from a distant large flare.

At some stage in its life, an active-region or quiescent prominence may become completely unstable and erupt, especially once it exceeds about 50 Mm in height. It ascends as an *erupting prominence* (Figure 1.37) and often shows substantial twist or twisting motions as a prominence tornado (Sec. 1.4.1). It eventually disappears; some of the material escapes from the Sun altogether, while some descends to the chromosphere along helical arches. Often, the prominence reforms in the same place and with much the same shape over the course of 1 to 7 days.

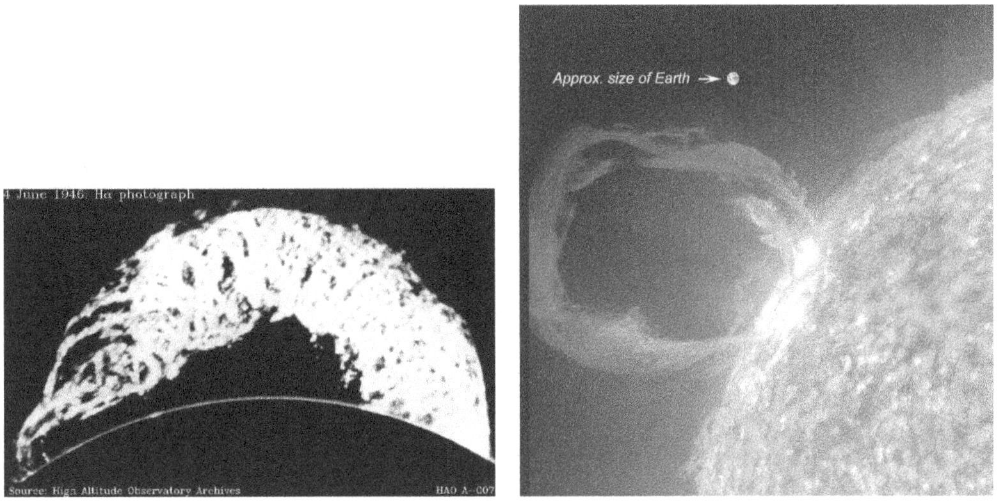

(a) (b)

Figure 1.37. Erupting prominences: (a) in Hα on 4 June 1946 (courtesy Joan Burkepile, High Altitude Observatory); (b) in He II 304 on 10 March 2010, with the size of the Earth indicated (from SDO/AIA).

The eruption of an old *quiescent prominence* is referred to as a *disparition brusque* (sudden disappearance); it starts with a slow rising motion at a few km s^{-1} and may take several hours. It is accompanied by an X-ray brightening and occasionally by the appearance of Hα flare ribbons. The eruption of an *active-region prominence* is much more rapid and takes about 30 min or less; it is accompanied by a two-ribbon solar flare (Sec. 1.9.2) and the prominence generally reforms after only a few hours.

From a theoretical viewpoint there are many questions that need answering. How does a prominence form (Sec. 11.6)? What is the magnetic configuration and how is the plasma supported against gravity (Sec. 11.2)? What causes the fine structure and the barbs or feet (Secs. 11.4 and 11.5)? Why does a prominence erupt (Sec. 12.2)?

1.9 Coronal Mass Ejections (CMEs) and Solar Flares

Most CMEs are associated with erupting prominences, and those which erupt from active regions are also associated with two-ribbon solar flares. When both occur, it is not the case that one causes the other, but rather that they are both aspects of the same eruptive process that drives or is associated with energy release by reconnection [see Secs. 12.2 and 12.3 and Harrison (1996), Webb and Howard (2012)]. We first describe the observational signatures of CMEs and then those of large eruptive two-ribbon flares. As we shall see, many such signatures are consistent with the occurrence of magnetic reconnection.

1.9.1 Coronal Mass Ejections

CMEs are large-scale magnetically driven plasma bubbles expanding out from the Sun and observed in interplanetary space as *magnetic clouds*. They are responsible for many space-weather effects at Earth, such as geomagnetic substorms, aurora and interplanetary shocks with solar energetic particle events (>1 MeV) (Webb et al. 1997; Yashiro et al. 2004; Schwenn 2006; Chen 2011). In extreme events, they may permanently damage sophisticated electronics on space satellites, endanger astronauts, cause aeroplanes to be routed away from high latitudes because of potential radiation damage, cause problems with radio communications and create power cuts.

CMEs represent the outward motion of a coronal streamer, in whose core invariably lies a prominence (Hundhausen 1999). This gives about 30 per cent of them a three-part structure of a bright outer loop (or

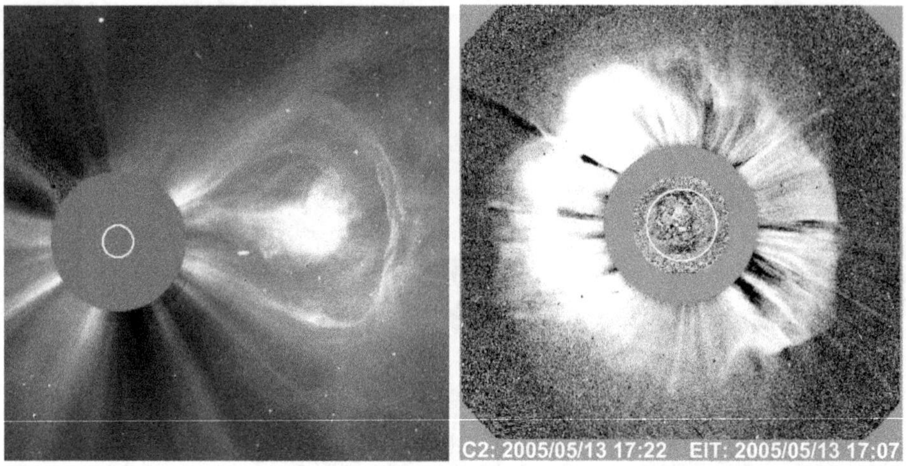

(a) (b)

Figure 1.38. Coronal mass ejections observed by SoHO/LASCO: (a) moving to the right and (b) a halo event on July 5, 2004 (pointed out by Nat Gopalswamy).

shell), a dark cavity and a bright core (due to the prominence) (Figure 1.38a). Sometimes the streamer disappears and sometimes it reforms afterwards. Originally glimpsed during earlier eclipses, they were discovered with OSO-7 (Tousey 1973) and studied in detail by coronagraphs on board other satellites (Skylab, 1973–74; P78-1, 1979–85; SMM, 1980–89; SoHO, 1995-; STEREO, 2006–). They are normally viewed in white light by Thomson scattering off free electrons in the plasma (depending on density along the line of sight), but they may also be seen in soft X-rays or EUV.

Often they go out on one side of the Sun (Figure 1.38a), but occasionally they expand in all directions from it and so move directly towards or away from the Earth as a *halo CME*; these are usually associated with flares within 30 degrees of disc centre (Figure 1.38b). An Earthward-directed CME would take 2 to 5 days to reach Earth, but large flare-related CMEs can accelerate particles to, say, a tenth of the speed of light, which take only 80 mins to reach Earth; these produce a "snow-shower" effect as they hit a CCD detector in LASCO and cause it to scintillate (Figure 1.38).

The occurrence rate of CMEs tracks the solar cycle with a peak delay of 6 to 12 months. From LASCO, it varies from 0.5 per day at solar minimum to 6 per day at solar maximum when detected by eye (increasing to 2 and 8 per day with automated software, Robbrecht et al. 2009). Their spatial distribution is clustered about the equator at solar minimum but is much broader at solar maximum. Often the start time of a CME associated with a flare precedes flare onset by tens of minutes and is associated with pre-flare soft X-ray heating (Harrison 1986). CMEs cause major modifications in the large-scale structure of the corona, such as a depletion of coronal material (known as a *coronal dimming*) and the creation of a transient coronal hole.

The magnetic flux carried out by a CME is usually 10^{20} to 10^{22} Mx for the axial flux and 10^{21} to 10^{22} Mx for the poloidal flux (Owens 2008). Its mass ranges between 10^{11} and 4×10^{13} kg (average 3×10^{12} kg), which is about ten times the prominence mass, so that CMEs account for roughly 5 to 10 per cent of the solar-wind mass-loss, depending on the phase in the solar cycle. The angular width of normal CMEs ranges from about 20 to 360 degrees, though some are much narrower. From SoHO, their average width rose from 47 degrees at solar minimum to 61 degrees at solar maximum. Their kinetic energies are typically 10^{22} to 10^{25} J, which is similar to solar flares.

Most of the acceleration of a CME occurs below 2 $R_\odot$, so that between 2 and 6 $R_\odot$ its speed is either constant or slightly increasing (for slow CMEs) or decreasing (for fast ones). The projected speeds have a wide range, from less than 20 km s^{-1} up to 2,000 km s^{-1} (occasionally 3,500 km s^{-1}), with a mean speed from SoHO of 300 km s^{-1} at solar minimum and 500 km s^{-1} at solar maximum. The mean speed of halo CMEs (960 km s^{-1}) is twice that of normal CMEs (420 km s^{-1}). CMEs associated with large flares escape much more rapidly (775 km s^{-1}) than those associated with quiescent prominences (330 km s^{-1}).

Most of the events moving faster than 400 km s^{-1} are also associated with type II radio bursts generated by shock waves. Assuming synchrotron emission, the radio data imply a rather large magnetic field of 4 to 5 G at 3 to 5 $R_\odot$, which in turn means a value for β of only 0.1, so that the driving force must be magnetic.

Several pieces of evidence indicate the presence of magnetic reconnection, which is thought to be crucial in the initiation and development of CMEs. Disconnected concave-upward structures are observed in one-third of CMEs observed with LASCO (St. Cyr et al. 2000). In half of these, a bright ray (suggestive of a current sheet) appears in the wake of the CME and lasts afterwards for hours or days (Webb et al. 2003). Its typical length is 200 Mm and width 5 Mm, and it often lines up with a plasma sheet lying above flare loops (Savage et al. 2010; Liu et al. 2010). Along the ray, bright blobs often flow away from the Sun, suggestive of reconnection outflows. Reconnection rates of typically 0.002 to 0.02 have been estimated from observed inflow and outflow velocities or current-sheet dimensions (Lin et al. 2005; Savage et al. 2010). UV spectra at 1.5 to 1.7 $R_\odot$ indicate an unusually high temperature (3 to 4 MK) for the corona at this height, while the plasma density in the ray is two to three times the coronal background (Raymond et al. 2003; Ciaravella and Raymond 2008). Quasi-periodic, type III-like radio pulses, called drifting pulsating structures, have been observed slowly drifting towards higher or lower frequencies in association with moving X-ray sources, which is suggestive of magnetic islands formed by impulsive bursty reconnection (Kliem et al. 2000; Bárta et al. 2008).

In interplanetary coronal mass ejections, counterstreaming suprathermal electrons suggest the magnetic field lines are closed – that is, still connected to the Sun at both ends – whereas single electron streams suggest the magnetic field has reconnected and become open. The data show that, by the time CMEs reach 5 AU, 50 per cent of them have become disconnected from the Sun (Owens et al. 2009). In addition, the open solar magnetic flux and its solar-cycle variation are caused by a balance between flux injection in CMEs and its subsequent reconnection (Owens et al. 2007).

The STEREO mission has revealed the three-dimensional structure of CMEs (Harrison et al. 2009). In August 2010 there was a set of four CMEs in quick succession from a complex active region, and the HI instrument was able to track each, to assess their onsets at the Sun and their arrivals at Earth (Harrison et al. 2012). STEREO also sees a plethora of outward-propagating structures from blobs to CMEs and has imaged corotating interaction regions.

As stressed by Forbes (2000), the cause of a CME must be the release of energy stored in the magnetic field, as the following argument demonstrates. The energy requirement for a moderately large CME is: kinetic energy 10^{25} J (10^{32} erg); heating and radiation 10^{25} J; work done against gravity 10^{24} J. Over a volume of 10^{24} m^3, this represents an energy density of 20 J m^{-3}. However, the available energy densities in J m^{-3} in the corona are kinetic ($\frac{1}{2}\rho v^2$) $= 10^{-6}$ for a density of 10^{15} m^{-3} and a speed of 1 km s^{-1}; thermal (nk_BT) $= 0.01$ for a temperature of 1 MK; gravitational ($\rho g h$) $= 0.05$ for a height (h) of 100 Mm; and magnetic $[B^2/(2\mu)] = 40$ for a field of 100 G. Thus, only the magnetic energy source is large enough to drive a CME, and the energy is likely to be stored in a force-free magnetic field.

1.9.2 Solar Flares

The solar flare is a truly remarkable and beautiful phenomenon. It varies from being a simple, localised brightening to a bewilderingly complex event – the most violent in the solar system. Before the space age, it was defined as a *rapid transient brightening in Hα*, but that has now been extended to include X-rays and other parts of the electromagnetic spectrum. It often accelerates high-energy particles and ejects plasma and magnetic flux into the solar wind. The Hα brightening is probably only a secondary chromospheric response to the conversion of magnetic energy into heat and particle energy much higher up in the coronal part of an active region.

Flares are classified as A, B, C, M or X depending on their peak 1 to 8 Å X-ray flux near Earth, as measured on the GOES spacecraft. Each class has a peak flux ten times greater than the preceding one, with X-class flares having a peak flux of order 10^{-4} W m^{-2}, namely, 0.01 per cent of the solar luminosity. Within each class there is a scale from 1 to 9, such that the flux in an X-3 flare is 50 per cent greater than an X-2 flare. There are typically 1 to 10 X-class (and 20 to 300 M-class) events per year. The largest flare in modern times was a colossal X-28 flare on 4 November 2003, although the *Carrington flare*, the first ever observed (by Richard Carrington in 1859), was probably larger.

There are many types of flare, which have been categorised in several different ways in addition to their X-ray flux. One is between *impulsive flares* (with strong impulsive behaviour) and *gradual flares* (with few high-energy particles). Another is between *compact* or *confined flares* and large *two-ribbon* or *eruptive flares* (Pallavacini et al. 1977; Priest 1981). The former are confined by a strong overlying magnetic field to such an extent that there is no associated eruption, whereas the latter (most large flares) are accompanied by an erupting prominence in the core of a CME. Usually, confined flares have half the duration of eruptive flares and have no pre-flare heating or long decay. There are many more flares than CMEs, but the correlation between the two increases with flare importance (Yashiro et al. 2006). Thus, eruptive flares comprise 30 per cent of C-class, 56 per cent of M-class and 90 per cent of X-class flares.

The main type of solar flare which we shall focus on here is the large eruptive *two-ribbon* flare. It takes place near an active-region prominence which often snakes its way through a complex active region. During the rise phase, two ribbons of Hα emission form, one on each side of the prominence (or prominence channel), and throughout the main phase the ribbons move apart. They are connected by a rising arcade of Hα, UV and soft X-ray *flare loops*. Occasionally, the prominence remains intact, though slightly disturbed, but usually it rises and disappears completely. Such an eruption of the prominence and CME begins slowly in the pre-flare phase, typically 10 minutes (but up to an hour) before flare onset, and continues at the rise phase with a much more rapid acceleration than before.

The rest of this section describes the basic features of a large flare, and then Chapter 12 outlines the MHD theory, including the cause of an eruption and the nature of reconnection (see also Tandberg-Hanssen and Emslie 1988; Priest and Forbes 2002; Forbes 2010; Hudson 2010; Moore et al. 2011; Shibata and Magara 2011).

1.9.2.1 Flare Phases

A flare is an intermingling of signatures across the spectrum of high-energy particles (the impulsive part) and heated plasma (the gradual part). Soft X-ray emission results from a thermal distribution of plasma and so contributes to the gradual part, whereas the non-thermal tail of the X-ray spectrum ($>$ 10 keV) is due to bremsstrahlung when high-energy particles are slowed down. Four basic stages are often present (Figure 1.39).

* During the *preflare phase* for tens of minutes or half an hour, the soft X-ray intensity ($<$ 10 keV) increases, while an active-region prominence becomes activated and starts to rise slowly.

* At the start of the flare proper, an *impulsive phase* may occur for 100 to 1000 secs, indicated by the appearance of a microwave burst and a hard X-ray burst ($>$ 30 keV), with internal 1-sec (or shorter) structure caused by highly accelerated electrons.

* The *rise* (or *flash) phase* lasts typically 5 minutes (but sometimes an hour). It is marked by a rapid increase in both intensity and area of soft X-ray and Hα emission, while the prominence rapidly erupts. Acceleration of fast particles continues and coronal plasma is heated to tens of millions of degrees.

* Then, in the *main phase*, the intensity slowly declines over about an hour (though sometimes as much as a day). In many flares the pre-flare phase is not seen. Some (i.e., *impulsive flares*) are highly impulsive

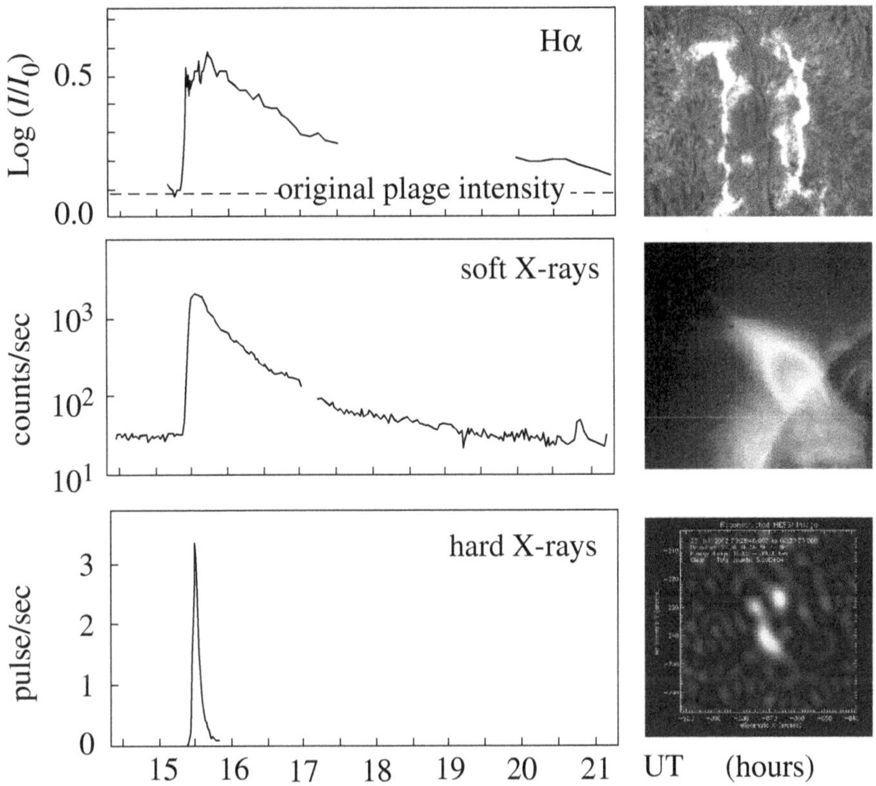

Figure 1.39. The time evolution of flare intensity in Hα, soft X-rays and hard X-rays during a particular flare, together with typical images from other flares (courtesy T. Forbes).

and produce copious amounts of high-energy particles and hard X-rays, while others (*gradual flares*) are much slower and produce much less (if any) high-energy emission.

In the impulsive phase, the hard X-rays usually come from two footpoints, with a thermal loop joining them (the classic thick-target picture, Figure 1.40). However, there may also be a hard X-ray *loop-top source*, as in the famous *Masuda flare* (Masuda et al. 1994), which had a source 7 Mm above the thermal loop for 2 minutes with a similar time evolution to the feet; after the impulsive phase, all the coronal emission became thermal at about 40 MK. A reconnecting current sheet can sometimes produce double coronal hard X-ray sources, with the peak temperature and double X-ray jets (Sec. 1.6.2) lying between the loop and the highest source (Sui et al. 2004; Wang et al. 2007); one example has 100 keV hard X-rays at a height of over 150 Mm (Krucker et al. 2007).

Usually, both the footpoint and coronal hard X-ray sources have a *soft-hard-soft* spectral evolution, probably intrinsic to the acceleration mechanism itself, in which the slope of the X-ray spectrum starts out steep (i.e., soft), flattens at the flare peak (i.e., hard) and steepens again in the decay phase. Some long-duration hard X-ray events, however, show a soft-hard-harder evolution due to trapping of particles in the corona and gradual erosion of the low-energy region by collisional losses. Coronal hard X-rays may be present without footpoint emission, when the loop density is so great ($> 10^{16}\text{m}^{-3}$) that the corona itself acts as a thick target to the injected electron beam (Veronig and Brown 2004).

Great flares begin at several points and expand over a large area, giving the impression that the initial flare energy does not simply spread out but stimulates extra energy release over a large region. In impulsive flares, *knots* or *kernels* of Hα emission appear during the impulsive phase almost simultaneously with the

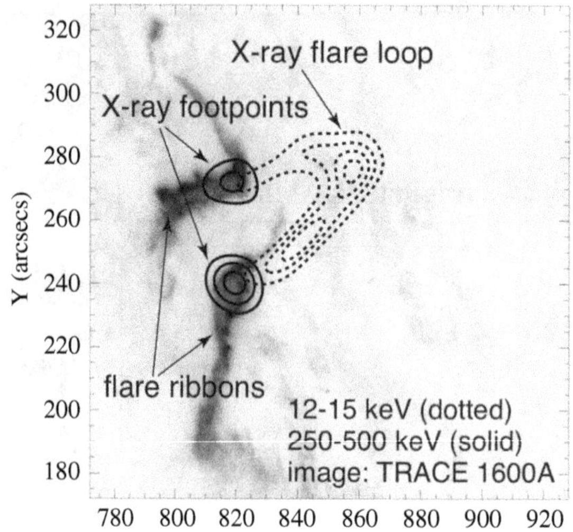

Figure 1.40. X-ray emission from RHESSI at the peak of a typical solar flare, showing a soft X-ray loop linking two hard X-ray footpoints, superimposed on a 1600 Å image from TRACE, with distances in arcsec (after Krucker et al. 2008).

hard X-ray and microwave bursts. The Hα is usually delayed by 1 to 2 secs (but occasionally by 10 to 20 secs), consistent with energy transfer by electron beams along a loop leg (or in the latter case by conduction fronts) (Radziszewski et al. 2011). Sometimes, the initial brightenings occur simultaneously at opposite ends of a magnetic field line. In the rise phase, the chromospheric knots merge to form a pair of bright *Hα ribbons*, which move apart initially very fast (100 km s^{-1}), but slowing quickly to only 4 km s^{-1} by flare maximum and about 1 km s^{-1} after 4 hours (van Driel-Gesztelyi et al. 1997).

The ribbons are joined by a rising arcade of Hα, UV and soft X-ray *flare loops* in the corona at temperatures between 10^4 K and 30 MK, with cooler loops nested below hotter loops. These flare loops were often wrongly referred to as *"post-flare"* loops, since it was thought that all the energy was released in the impulsive phase. But, in fact, the emission in these loops may continue for as much as 2 days in a really large event, much longer than the natural decay time of hot plasma by radiation or thermal conduction, and so the energy release continues through the main phase.

Chromospheric evaporation is the upward expansion of the chromosphere after it has been heated to coronal temperatures by high-energy particles or conduction fronts that have propagated down from the coronal energy-release site during the impulsive, rise and main phases. The "evaporation" fills the flare loops with hot dense plasma. The gradual part of the flare consists of this increase in temperature followed first by an increase in coronal density (and so emission measure) and then by a decline in both as the coronal plasma cools and drains back down to the chromosphere.

When prominences and CMEs erupt outside active regions, where the magnetic field is much weaker, little significant hard X-ray emission is produced. However, huge rising soft X-ray loops and separating ribbons are still produced. The ribbons can be seen in He 10830 Å, but are usually too faint to be observed in Hα.

The energy released in a flare varies from, say, 10^{22} J in a sub-flare to 6×10^{25} J (6×10^{32} erg) or more in the largest of events. The division among various types of energy varies from one flare to another, since some are much more non-thermal than others and some have coronal mass ejections. For a large eruptive flare, the radiation from the flare and the kinetic energy of the accompanying CME are similar and a typical estimate is (Emslie et al. 2005):

Total radiant energy	2×10^{25} J (2×10^{32} erg)
[Fast electrons that produce hard X-rays	2×10^{24} J]
[Fast ions that produce γ-rays	4×10^{24} J]
Coronal mass ejection	2×10^{25} J
Total energy output	4×10^{25} J

Most of the total radiation appears during the main phase, with a minority (3×10^{24} J) in the impulsive phase. Most of it is in the visible and UV continuum, with Hα being particularly noticeable; there is typically only 2×10^{24} J in soft X-rays. The energy in the fast electrons and ions is here put in brackets, since it does not contribute separately to the total energy but is converted (in the impulsive phase) into radiant energy by Coulomb collisions. The fast ion estimate is particularly uncertain, since the lower limit to the ion spectrum is unknown (it has been assumed to be 1 MeV here). Extra fast-particle energy and heat conduction contribute to the optical and UV radiation. Solar energetic particles are also accelerated efficiently at a CME-driven shock wave and may possess 15 per cent of the CME energy (3×10^{24} J).

The radiant energy is in turn split up into the following proportions according to Canfield et al. (1980) for a well-studied small flare:

Soft X-rays (1–20 Å)	6×10^{18} J s^{-1} (6×10^{25} erg s^{-1})
XUV lines	2×10^{18} J s^{-1}
EUV continuum	1×10^{19} J s^{-1}
Lyman α	2×10^{18} J s^{-1}
Other EUV lines	2×10^{18} J s^{-1}
Hα	4×10^{18} J s^{-1}
Other visible lines	3×10^{19} J s^{-1}
Radio continuum, cm–m wavelengths	6×10^{12} J s^{-1}

The total radiative power output is itself 5.4×10^{19} J s^{-1}, which for a duration of 800 sec gives a total radiant energy for this small flare of 4×10^{22} J. Later, Kretzschmar et al. (2010) used SoHO/VIRGO to measure the total solar irradiance (TSI) in M-class and X-class flares over a solar cycle. They found that the emission in near-UV and visible dominates and that the TSI is typically a factor of a 100 greater than the output in soft X-rays.

Since other sources of energy are inadequate, it is clearly the magnetic field which supplies the energy for a flare. The energy is stored in nonlinear force-free fields low down (below 15 Mm) in the corona, often around an active-region prominence (Regnier and Priest 2007; Schrijver et al. 2008). 3×10^{25} J of energy would be released if, for instance, the whole of a 500 G magnetic field in a cube of side 30 Mm (or of a 100 G field in a cube of side 90 Mm) were annihilated; it also corresponds to the energy released if the field in a cube of side 40 Mm falls from 500 to 400 G.

Several properties of a flare or CME are of importance when it comes to constructing a theoretical model. For example, apart from very large events, there are generally no large-scale changes in the photosphere beneath a flare or CME; variations in the magnetic flux are no different from normal evolutionary changes, with slow movements of sunspots and other magnetic features being unaffected by the occurrence of a flare or CME. This is entirely reasonable since the photospheric plasma is almost 10^9 times denser than the coronal plasma where the flare originates, which *inertially line-ties* the footpoints of coronal field lines and so keeps them stationary during the eruption. Furthermore, the flare energy is only a small fraction of the energy (10^{27} J) of a large sunspot group.

Another interesting property of flares is that active regions which are *complex* and *rapidly evolving* are the most likely flare producers; by contrast, simple bipolar regions show little activity. Also, some flares exhibit: *homology*, when a flare recurs in the same place and with similar characteristics; or *sympathy*, when one flare is triggered by the occurrence of another at a different location.

1.9.2.2 Effects of Flare Energy Release

The disturbance of the solar atmosphere by a large flare may be manifested in many ways. In the primary energy-release sites of a large flare, magnetic energy is converted into fast-particle energy, heat and plasma motions, which in turn have secondary consequences, such as radio bursts, waves and shock waves. Also, the energy in the primary flare products may be converted into other forms. For example, fast particles may be slowed down and heat the plasma and drive mass motions, so it is not always obvious whether a given plasma jet or ejection is a primary or secondary consequence of the energy conversion.

The standard *collisional thick-target model* (Brown 1971) suggests that, during the impulsive and rise phases, large numbers (10^{35} to 10^{37} per sec) of electrons are accelerated in the corona to high energies (10 to 100 keV) and travel as beams down the legs of a coronal loop to its two footpoints, producing microwaves by gyrosynchrotron emission from spiralling electrons. At the chromospheric footpoints, they produce simultaneous (to within 0.1 sec) hard X-ray emission and γ-ray continuum by thick-target bremsstrahlung radiation from fast electrons (Krucker et al. 2008), as well as UV and optical emission.

Hard X-rays have been imaged by Yohkoh (Kosugi et al. 1992) and RHESSI (Lin et al. 2002) and imply an electron energy of 10^{19} to 10^{25} J (10^{26} to 10^{32} erg). In addition, low-frequency radio emission arises from plasma radiation, soft X-rays arise from thermal bremsstrahlung continuum and atomic lines, and EUV emission from hot thermal plasma. Type III radio bursts indicate the presence of beams of high-energy electrons in the corona, accelerated to about a third the speed of light. Also, long-lasting type IV bursts show the presence of an energetic plasma cloud above the flare site.

Occasionally, one also finds γ-ray lines (due to interaction with ambient nuclei of ions having 10s of MeV per nucleon) and γ-ray continuum greater than 10 MeV (due to pion decay). The presence of fast neutrons is revealed by the neutron-capture line at 2.223 MeV, and of positrons by the 511 keV annihilation line.

There is a good correlation between electrons above 300 keV and ions above 30 MeV/nucleon. Sometimes, in γ-ray flares, electrons are accelerated to as much as 10 MeV in a few seconds and non-thermal ions to 10 GeV in about a second. However, there are some *electron-rich* flares with bremsstrahlung from electrons above 1 MeV but little γ-ray emission.

Timing measurements of hard X-ray spikes, when interpreted as time-of-flight effects, imply that the electron-acceleration site is located above the soft X-ray loops. To produce the hard X-rays non-thermally requires typically 3×10^{35} to 3×10^{36} electrons per second with an energy of 3×10^{24}J (3×10^{31} erg). The number of ions accelerated above 1 MeV and their energy are similar (Holman et al. 2003).

The low chromosphere and photosphere can be affected in the impulsive phase, with white-light emission produced by free-bound emission and radiation back-warming; indeed, tiny spots of white-light emission may be a common feature of even small flares (Jess et al. 2008). Often, the photospheric magnetic field may change (Sudol and Harvey 2005; Petrie and Sudol 2010).

Many different types of wave phenomenon (not all distinct) are naturally initiated by flares (and CMEs) and propagate outwards from the flare site:

(i) *Moreton waves*, originally discovered in Hα (Athay and Moreton 1961) and now detected in several wavelengths, are fast waves in the chromosphere at 1000 km s^{-1}. The standard interpretation (Uchida 1970) is that they are the skirts of global fast-mode shock waves propagating in the corona.

(ii) *Type II radio bursts* are evidence for such outwards-propagating coronal shock waves. In the interplanetary medium, type IIs can be observed in situ as bow waves driven by CMEs.

(iii) *EIT waves* were discovered with SoHO/EIT (Thompson et al. 1998). The faster ones are likely to be fast-mode shocks generated by a flare or driven by a CME (Warmuth 2010; Liu et al. 2012), but many are only a few 10s or 100s km s^{-1} and so may instead represent a slow-mode shock or the outer boundary of the CME structure (Chen et al. 2002; Delannée et al. 2008). For example, using SDO/AIA and STEREO, Schrijver et al. (2011) observed loop structures expanding ahead of an erupting active-region flux-rope at 500 km s^{-1}; a coronal propagating front (an EIT wave) developed at the outer boundary of the propagating loops (the leading edge of a CME); then the front compressed and heated adiabatically and propagated into an opening angle equal to that of the pre-eruptive helmet streamer.

(iv) *Helioseismic waves* occur occasionally when intense flares, such as white-light events with a strong photospheric disturbance, generate a wave that propagates down into the solar interior, where it is refracted back up to the surface and is seen at the photosphere as an outwards-moving ripple (Kosovichev and Zharkova 1998).

(v) *Coronal loop oscillations* in active regions in response to flares (Nakariakov et al. 1999) are also analysed by coronal seismology techniques (Sec. 4.11.2).

1.9.2.3 Evidence for Magnetic Reconnection

Conversion of magnetic energy by reconnection in a flare is a widely accepted paradigm [see Sec. 12.3, Forbes (2010), Shibata and Magara (2011)]. However, it is difficult to prove, since, although individual coronal flux tubes or loops can be defined and labelled at one time, it is not easy to follow them in time as loops change their temperature and structure. Nevertheless, many observed aspects of flares are fully consistent with reconnection, and so it is highly likely to be occurring, although the detailed way in which it does so in three-dimensions is not yet well understood. Furthermore, reconnection is a natural way of accelerating particles, by either shocks, direct electric fields or turbulence, although the details about how it works in a flare are also not yet well established (e.g., Miller et al. 1997).

Examples of observations suggestive of reconnection are: connectivity changes, flux emergence, X-ray jets, separating ribbons, rising flare loops, current sheets, supra-arcade downflows, slow and fast-mode shocks, as described in the following.

Connectivity changes during a flare are sometimes inferred from changes in fibrils or in coronal-loop structures (Nishio et al. 1997). Also, the reported location of flare ribbons, where coronal separatrix or quasi-separatrix surfaces (QSLs) intersect the chromosphere, is good evidence for reconnection in a coronal separator or quasi-separator (Mandrini et al. 1991; Démoulin et al. 1993; Schmieder et al. 1997; Masson et al. 2009; Longcope et al. 2007). In addition, the presence of coronal null points and separators in flaring configurations has been reported (Démoulin et al. 1994; Aulanier et al. 2000; Fletcher et al. 2001; Masson et al. 2009).

Many flares occur after the emergence of new magnetic flux from below the photosphere, as sometimes evidenced by the appearance of small satellite sunspots close to large sunspots of opposite polarity. Flares are especially likely when the satellite is of *following* polarity and appears just in front of the main spot with *preceding* polarity. The most flare-productive active regions have steep field gradients and *polarity-inversion lines* at which the longitudinal magnetic field component changes sign and the field is largely horizontal, resulting from the arrival of satellite spots. Often, it is the local emergence of new flux which seems to initiate the activation or eruption of a prominence, and the brightest flare knots appear near the points of emerging flux. During the course of the flare, changes in fibril and coronal-loop direction may sometimes be seen, which suggest reconnection and relaxation of the field to a lower energy state (van Driel-Gesztelyi et al. 1996).

Soft X-ray jets or *coronal jets* have properties described in Section 1.6.2. They are often associated with subflares or microflares, and are a natural consequence of reconnection in emerging flux (Sec. 9.2.3), cancelling flux (Sec. 1.4.2), coronal heating events (Sec. 10.4.2) and solar flares (Sec. 12.3). Jets are also often associated with type III radio bursts and electron events in interplanetary space.

A flare arcade comprises a set of nested loops about 100 Mm high, whose heights and footpoint separations increase in time. The loops have many different temperatures, with hot soft X-ray loops above cooler UV and Hα loops. The outer edge of the hot loops maps to the outer edge of the Hα ribbons and are strongly blue-shifted (upflows), while the inner edge of the cool loops maps to the inner edge of the ribbons and exhibits downflows (Schmieder et al. 1995; Czaykowska et al. 1999).

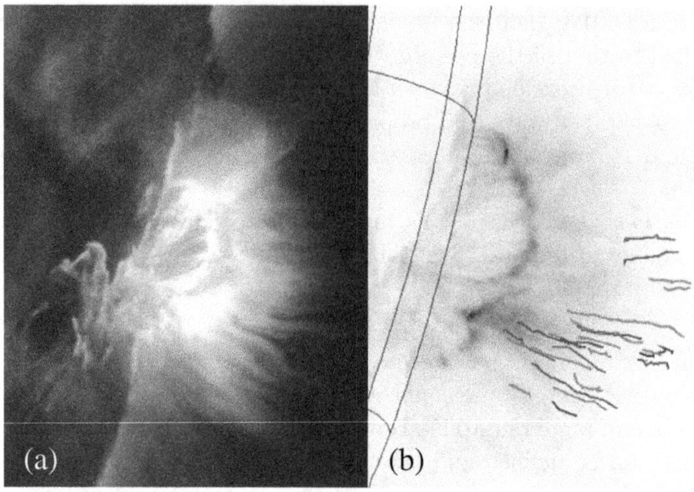

Figure 1.41. (a) TRACE image in 195 Å (emitting at both 1.5 MK and 30 MK) during the main phase of a flare and (b) locations of supra-arcade downflows (SADs). In (a) the spiney appearance of the image above the flare loops is caused by many downflowing dark voids (SADs) (from McKenzie and Savage 2009).

An immense amount of plasma drains down the legs of the cooler loops – far exceeding the entire coronal content, which used to be a major puzzle. The explanation lies in the response of plasma to reconnection and the way the loops rise in time (Sec. 12.3). The rise-speed is initially more than 20 km s^{-1}, falling to typically 1.5 km s^{-1} after 5 h and 0.5 km s^{-1} after 10 h. Their long duration is clear evidence for continued heating (by reconnection), often for a day or more.

Observations from Yohkoh in e.g. the "Tsuneta flare" revealed the candlelight or cusp-shaped flaring loops that had been predicted by reconnection theory (Cargill and Priest 1982), as well as hot ridges that are suggestive of slow-mode shock waves (Tsuneta 1996). Large CMEs outside active regions, sometimes associated with a polar-crown filament eruption, produce much larger giant arcades with cusp-shaped loops. A standing fast-mode shock in the outflow of a reconnecting region may be indicated by a super-hot loop-top source (Tsuneta et al. 1997) and a stationary type II radio burst (Aurass et al. 2002). In addition, observation of type III radio bursts with both forward and reverse slopes suggests particles moving both outwards and inwards from a reconnection site (Aschwanden and Benz 1997).

Sigmoids are S-shaped (or reverse S-shaped) structures in soft X-rays, sometimes seen in active regions and suggestive of highly sheared non-potential fields (Rust and Kumar 1996) with large magnetic helicity (Pevtsov et al. 2008). Canfield et al. (1999) found sigmoidal active regions more likely to erupt (i.e., 85 per cent) than nonsigmoidal regions (i.e., 50 per cent). Also, 50 per cent of CMEs possess sigmoids as precursors. Sigmoids may have complex substructure and are not necessarily just one twisted flux tube. They tend to form several hours or days before eruption in situations that are building up the magnetic helicity due to convergence and cancellation of flux (Green and Kliem 2009; Green et al. 2011), such as a decaying active region or a rotating sunspot or emerging flux. McKenzie and Canfield (2008) described a long-lived two-day sigmoid that consisted of two separate J-shaped bundles of loops. The straight sections of the Js lay anti-parallel to each other on opposite sides of a polarity-inversion line. At the start of the eruption, a diffuse linear structure (possibly the pre-flare brightening around a prominence) lifted off from the middle of the S with a slight clockwise rotation. The flare began with the appearance of a sheared arcade of short loops between the two Js.

SADs (supra-arcade downflows) are small dark plasma voids with a quarter of the surrounding density that are evidence of fragmented reconnection (McKenzie and Hudson 1999). They are located within high-temperature (10 to 20 MK) emission that exists during the main phase above an arcade of flare loops and that was discovered with Yohkoh/SXT by Švestka et al. (1998) as a likely location of a reconnecting current sheet. An example from TRACE is shown in Figure 1.41. SADs flow downwards at 50 to 500 km s^{-1} and decelerate as they approach the flare loops (Asai et al. 2004; Savage and McKenzie 2011). At the same time, thin downflowing loops (sometimes called called *Supra-Arcade Downflowing Loops* or SADLs) are observed to shrink as they relax after reconnecting to a more dipolar shape (Forbes and Acton 1996). Both SADs and SADLs are probably caused by the fragmentation or patchy nature of reconnection in the current sheet, with the SADs representing wakes behind the shrinking loops (SADLs) (Savage and McKenzie 2011; Savage et al. 2012; Warren et al. 2011). The loop cross-sections (the reconnection patches) are <400 km in size, while the wakes are typically 1 to 4 Mm. The downflows are slower than the Alfvén speed (500 to 3,000 km s^{-1}), perhaps because of drag forces or the build-up of downstream magnetic pressure.

Other suggestive indications of reconnection in flares (van Driel-Gesztelyi 2009; Isobe and Shibata 2009) include upwards-moving plasmoids above flare loops (Shibata et al. 1995; Tsuneta et al. 1997); EUV inflows above cusp-shaped flare loops at 1 to 5 km s^{-1} implying reconnection inflows with an inflow Alfvén Mach number typical of fast reconnection (Yokoyama et al. 2001; Narukage and Shibata 2006); and magnetic disconnections in CMEs with upwards-curving structures (Dere et al. 1999; Savage et al. 2010). Reconnection outflows in flares have been seen with sizes of 800 to 3,500 km s^{-1} from SoHO/SUMER (Innes et al. 2003; Wang et al. 2007). Also, an example of a 12-MK loop-top source with a 1.2-MK reconnection inflow at 20 km s^{-1} together with a 9-MK outflow jet at 200 km s^{-1} was found with Hinode/EIS (Hara et al. 2011). Indications of reconnection from space in association with coronal heating are summarised in Section 10.4.2.

Chapter 2

The Basic Equations of Magnetohydrodynamics (MHD)

In MHD, the behaviour of a continuous plasma is governed by a simplified form of Maxwell's equations (Sec. 2.1.1), together with Ohm's Law (Sec. 2.1.2), a gas law (Sec. 2.2.3) and equations of mass continuity (Sec. 2.2.1), motion (Sec. 2.2.2) and energy (Sec. 2.3). For simplicity, we shall use these equations as a basis for our modelling, although they can be derived by taking moments of the Boltzmann equations for electrons and protons and making a closure hypothesis (see, e.g., Choudhuri 1998; Boyd and Sanderson 2003; Goedbloed et al. 2004). We shall here eliminate the electromagnetic fields between Maxwell's equations (and Ohm's Law) and so reduce them to the core equation of MHD, known as the *induction equation* (Sec. 2.1.4), which describes the fundamental behaviour of the magnetic field ($\mathbf{B}$) in terms of the *plasma velocity* ($\mathbf{v}$). (Formally, $\mathbf{B}$ is the *magnetic induction* but in a solar or astrophysical context it is normally referred to as the magnetic field.)

Most of this book deals with single-fluid MHD, which has been a remarkably successful model for plasma behaviour, even when the plasma is collisionless. This success arises for several reasons: ideal MHD embodies conservation of mass, momentum and energy, which are universal in collisionless and collisional plasmas; gyro-motion prevents particles travelling perpendicular to the magnetic field, whose net effect is described by MHD-like equations; often wave-particle interactions impede particle motion along the magnetic field; a collisionless plasma with the $\mathbf{E} \times \mathbf{B}$-drift dominant leads to $\mathbf{E}_{\perp} + \mathbf{v} \times \mathbf{B} = \mathbf{0}$; adding the Lorentz force on individual particles gives a net force of $\mathbf{j} \times \mathbf{B}$ when the drift speed and Debye length are small enough.

However, in future it will be increasingly important to go beyond a single-fluid picture, especially in the high corona or in very narrow structures where particle collisions no longer dominate. These include shock waves, reconnecting diffusion regions (Birn and Priest 2007) and resonant absorption layers. The simplest extension is to adopt a two-fluid or three-fluid model (e.g., Sec. 13.11.1), which replaces the normal form of Ohm's law by a generalised Ohm's law (Sec. 2.1.3). Beyond that, a kinetic treatment is needed that calculates the distribution function f (e.g., Sec. 13.11.2). For example, a promising approach for some applications is *gyrokinetic theory*, which averages over the gyro-phase (when the length-scale for magnetic field variations is much larger than the ion gyro-radius) to give $f(x, y, z, t, v_{\parallel}, v_{\perp})$, where the last two variables are the particle speed parallel and perpendicular to the magnetic field (e.g., Taylor and Hastie 1968; Littlejohn 1979; Brizard and Hahm 2007; Schekochihin et al. 2009).

2.1 Electromagnetic Equations

2.1.1 Maxwell's Equations

We begin with Maxwell's equations in *mks* units,

$$\nabla \times \mathbf{B} = \mu \mathbf{j} + \frac{1}{c^2}\frac{\partial \mathbf{E}}{\partial t}, \tag{2.1}$$

$$\nabla \cdot \mathbf{B} = 0, \tag{2.2}$$

$$\nabla \times \mathbf{E} = -\frac{\partial \mathbf{B}}{\partial t}, \tag{2.3}$$

$$\nabla \cdot \mathbf{E} = \frac{\rho^*}{\epsilon}, \tag{2.4}$$

where the constitutive relations ($\mathbf{H} = \mathbf{B}/\mu$, $\mathbf{D} = \epsilon\mathbf{E}$) have been used to eliminate the magnetic field ($\mathbf{H}$) and electric displacement ($\mathbf{D}$). (Throughout this book for solar plasma, μ and ϵ are approximated by their vacuum values, μ_0 and ϵ_0, respectively, but the 0 subscripts are dropped for ease of writing.) In these equations, $\mathbf{E}$ is the *electric field*, ρ^* the *charge density*, $\mathbf{j}$ the *current density*, μ_0 the *magnetic permeability* and ϵ_0 the *permittivity of free space*, such that the *speed of light* in a vacuum is

$$c = (\mu_0 \epsilon_0)^{-1/2} \approx 2.998 \times 10^8 \text{ m s}^{-1},$$

$$\text{where} \quad \mu_0 = 4\pi \times 10^{-7} \text{ henry m}^{-1}, \quad \text{and} \quad \epsilon_0 \approx 8.854 \times 10^{-12} \text{ farad m}^{-1}.$$

$\mathbf{E}$ is measured in volts per metre (V m^{-1}), $\mathbf{B}$ in tesla (T) or webers m^{-2} (Wb m^{-2}), j in amps m^{-2} (A m^{-2}) and v in m s^{-1}. Definitions of henry and farad and the relationship between mks and gaussian units are given in Appendix 1, but note in particular that 1 G $= 10^{-4}$ T. In mks, magnetic induction is measured in T, but in this book its values are usually quoted in gauss (and flux in maxwells), since they are much more familiar to solar physicists and it is an acceptable option in the SI system (see Appendix 1). Furthermore, it is sometimes convenient to quote lengths in Mm, which is a useful measure for many solar features (1 Mm $= 10^6$ m); for example, it is the typical diameter of a granule and is somewhat larger than an arc second (Sec. 1.2).

The first Maxwell equation (2.1) shows that either currents or time-varying electric fields may produce magnetic fields, whereas the third and fourth equations (2.3 and 2.4) imply that either electric charges or time-varying magnetic fields may give rise to electric fields. The second equation assumes that there are no magnetic monopoles and implies that a magnetic flux tube has a constant strength along its length (Sec. 2.9.2).

A fundamental supposition of normal MHD is that the electromagnetic variations are non-relativistic or 'quasi-steady'. In other words,

$$V_0 \ll c, \tag{2.5}$$

where $V_0 = l_0/t_0$ is a characteristic electromagnetic (or plasma) speed, while l_0 and t_0 are a typical length and time. In addition, assume that

$$\frac{E_0}{l_0} \approx \frac{B_0}{t_0}, \tag{2.6}$$

where E_0 and B_0 are typical values of E and B, so that the two sides of Eq. (2.3) are the same in order of magnitude. In Eq. (2.1), the second term on the right-hand side (the displacement current) then has magnitude

$$\frac{E_0}{c^2 t_0} \approx \frac{B_0 l_0}{c^2 t_0^2} = \frac{V_0^2}{c^2}\frac{B_0}{l_0} \approx \frac{V_0^2}{c^2}|\nabla \times \mathbf{B}|,$$

which by Eq. (2.5) is much smaller than the left-hand side of Eq. (2.1).

Thus, one consequence of Eq. (2.5) is that the term $c^{-2}\partial \mathbf{E}/\partial t$ may be neglected in Eq. (2.1). Another is that the equation of charge continuity, which is obtained from the divergence of Eq. (2.1), becomes $\nabla \cdot \mathbf{j} = 0$; this implies physically that local accumulations in time of charge are negligible and electric currents flow in closed circuits. A further consequence [upon using Eq. (2.6)] is that the ratio of electrostatic to magnetic energy density, namely,

$$\frac{\epsilon_0 E_0^2}{B_0^2/\mu} \approx \frac{l_0^2}{c^2 t_0^2} = \frac{V_0^2}{c^2},$$

is much less than unity.

A characteristic of many processes in the solar atmosphere is that the plasma is, to a high degree of approximation, *electrically neutral* (otherwise, the resulting electric fields would be enormous); in other words,

$$n_+ - n_- \ll n, \tag{2.7}$$

where n_+ and n_- are the *number densities* of positive and negative ions per unit volume and n is the total number density. The magnitude of the charge imbalance [$\rho^* = (n_+ - n_-)e$, where $e \approx 1.6\times10^{-19}$ C] is determined by Eq. (2.4). After substituting for E_0, the above condition for charge neutrality becomes

$$6 \times 10^7 \frac{V_0 B_0}{l_0} \ll n,$$

with B_0 in tesla.

This condition and Eq. (2.5) are well satisfied for many solar applications. For example, in the photosphere near a sunspot, one may observe motions with speeds $V_0 \approx 10^4$ m s^{-1}, which is a factor of 3×10^4 smaller than c. Typical values for magnetic field strength, length-scale and number density there are $B_0 \approx 0.1$ T (10^3 G), $l_0 \approx 10^5$ m, $n \approx 10^{20}$ m^{-3}, so that $6\times10^7\ V_0 B_0/l_0 \approx 6\times10^5 \ll n$, as required.

In practise, a local charge imbalance produces an electrical field with spatial range the *Debye length* (e.g., Boyd and Sanderson 2003; Gurnett and Bhattacharjee 2005)

$$\lambda_D = \left(\frac{\epsilon_0 k T}{n e^2}\right)^{\frac{1}{2}} \approx 69 \left(\frac{T}{n}\right)^{\frac{1}{2}} \text{ m}, \tag{2.8}$$

which is a measure of the distance over which n_- can deviate appreciably from n_+ (with T in K and n in m^{-3}). In fact, a *plasma* may be defined as an ionised gas for which the number of particles in a sphere of radius λ_D is very large.

2.1.2 Ohm's Law

Plasma moving at a non-relativistic velocity $\mathbf{v}$ in the presence of a magnetic field is subject to an electric field ($\mathbf{v} \times \mathbf{B}$) in addition to the electric field ($\mathbf{E}$) which would act on material at rest. Ohm's Law asserts that the current density is proportional to the *total* electric field (in a frame of reference moving with the plasma), and so it may be written

$$\mathbf{j} = \sigma(\mathbf{E} + \mathbf{v} \times \mathbf{B}), \tag{2.9}$$

where σ is the *electrical conductivity*, measured in mho m^{-1}.

2.1.3 Generalised Ohm's Law

A generalisation of Ohm's Law may be more appropriate in some regions of the photosphere and corona and, indeed, determining its consequences in detail is an important task. It arises from a 'three-fluid' model for electrons, protons and neutral atoms (with number densities n_e, $n_i = n_e$ and n_a,

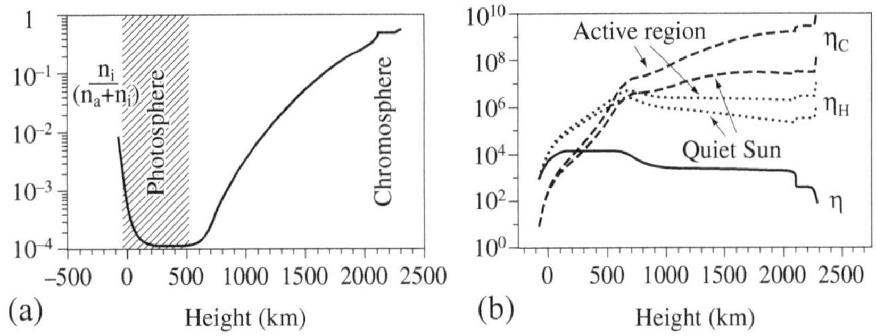

Figure 2.1. (a) The degree of ionisation $n_i/(n_i+n_a)$ in the low solar atmosphere as a function of height for the VAL-C model. (b) The corresponding Cowling (ambipolar) (η_C, dashed), Hall (η_H, dotted) and normal Coulomb (η, solid) magnetic diffusivities in magnetic flux tubes that are typical of the quiet Sun and of an active region (courtesy Elena Khomenko).

respectively) and may be written

$$n_e e\left(\mathbf{E}_0+\frac{\boldsymbol{\nabla} p_e}{n_e e}\right)=\frac{m_e}{e}\left(\frac{\partial \mathbf{j}}{\partial t}+\boldsymbol{\nabla}\cdot(\mathbf{vj}+\mathbf{jv})\right)+\left(\frac{1}{\Omega_e\tau_{ei}}+\frac{1}{\Omega_e\tau_{en}}\right)B\,\mathbf{j}$$
$$+\,\mathbf{j}\times\mathbf{B}+\frac{f^2\Omega_i\tau_{in}}{B}[\boldsymbol{\nabla} p_e\times\mathbf{B}-(\mathbf{j}\times\mathbf{B})\times\mathbf{B}], \tag{2.10}$$

where $\mathbf{v}$ is the centre of mass velocity, $\mathbf{E}_0 \equiv \mathbf{E}+\mathbf{v}\times\mathbf{B}$ is the *total electric field*, $\Omega_e = eB/m_e$ and $\Omega_i = eB/m_p$ are the electron and ion (proton) gyration frequencies, $f = n_a(n_a+n_e)^{-1}$ is the fraction of ions not ionised, τ_{ei} is the electron-ion collision time, while τ_{en} and τ_{in} are collision times for neutrals with electrons and ions (Cowling, 1976). In the derivation of this version of the generalised Ohm's Law, terms of order m_e/m_p by comparison with unity have been neglected.

The *degree of ionisation* or *ionisation fraction* is the proportion $[n_i/(n_i+n_a)]$ of neutrals and ions that are ionised (Figure 2.1a). When it is small, the plasma is *partially ionised* and, when it is close to unity, the plasma is *fully ionised.* Usually, $n_e = n_i$ and so the degree of ionisation is $1-f$.

In many applications, both the electron inertia and electron pressure gradient terms are negligible. The generalised Ohm's Law (2.10) then reduces to

$$\mathbf{E}_0=\frac{\mathbf{j}}{\sigma}+\frac{1}{n_e e}\mathbf{j}\times\mathbf{B}-\frac{f^2\Omega_i\tau_{in}}{n_e eB}(\mathbf{j}\times\mathbf{B})\times\mathbf{B}, \tag{2.11a}$$

where

$$\sigma=\frac{n_e e^2 m_e^{-1}}{\tau_{ei}^{-1}+\tau_{en}^{-1}} \tag{2.11b}$$

is the electrical conductivity in the absence of a magnetic field.

When the current is parallel to $\mathbf{B}$, Eq. (2.11a) reduces to $\mathbf{E}_0 = \mathbf{j}/\sigma$ with conductivity σ. On the other hand, when the current is normal to the magnetic field,

$$\mathbf{E}_0=\frac{\mathbf{j}}{\sigma_C}+\frac{1}{n_e e}\mathbf{j}\times\mathbf{B},\quad (a)\quad \text{where}\quad \frac{1}{\sigma_C}=\frac{1}{\sigma}+\frac{f^2B\Omega_i\tau_{in}}{n_e e}\quad (b) \tag{2.12}$$

is the *Cowling conductivity.* It may be interpreted as the conductivity normal to the magnetic field (defined as the ratio of j to the electric field component $(\mathbf{E}_0{\cdot}\mathbf{j}/j)$ in the direction of $\mathbf{j}$). Equation (2.12a) may be solved for $\mathbf{j}$ to give

$$\mathbf{j} = \sigma_1 \mathbf{E}_0 + \sigma_H \mathbf{B} \times \mathbf{E}_0 / B,$$

$$\text{where} \quad \sigma_1 = \frac{\sigma_C}{1 + [\sigma_C B/(n_e e)]^2} \quad \text{and} \quad \sigma_H = \frac{\sigma_C B}{n_e e} \sigma_1$$

are the *direct* (or *Pedersen*) *conductivity* and *Hall conductivity*, respectively.

Also, from Eq. (2.12a) the rate of energy dissipation may be written $\mathbf{E}_0 \cdot \mathbf{j} = j^2/\sigma_C$, which is larger than the dissipation (j^2/σ) in the absence of neutrals, because collisions of neutrals with charged particles make σ_C smaller than σ. In the photosphere, for example, σ_C may be much smaller than σ because of a significant diffusion of charged particles relative to neutrals (known as *ambipolar diffusion*).

The Cowling conductivity is evaluated in Section 2.1.5, where it is shown to make ambipolar diffusion up to 10^5 times stronger than normal magnetic diffusion in the chromosphere. This important effect has not yet been fully studied, but it has far-reaching implications for wave excitation and propagation (e.g., Kumar and Roberts 2003; Vranjes et al. 2008; Soler et al. 2010; Zaqarashvili et al. 2011), for heating the chromosphere (Khomenko and Collados Vera 2012a), for reconnection (Sakai and Smith 2009; Zweibel and Yamada 2009), flux emergence (Leake and Arber 2006; Arber et al. 2007) and the rapid dissipation of perpendicular currents to create force-free fields (Arber et al. 2009).

When the gas is fully ionised so that $n_a = 0$, the conductivities reduce to

$$\sigma_1 = \frac{\sigma}{1 + \Omega_e^2 \tau_{ei}^2}, \qquad \sigma_H = \Omega_e \tau_{ei} \sigma_1, \qquad \sigma_C = \sigma = \frac{n_e e^2 \tau_{ei}}{m_e}.$$

By comparison with an unmagnetised medium (where $j = \sigma E$), the effect of a magnetic field is to reduce the current parallel to $\mathbf{E}_0$ and to produce an extra (Hall) current normal to $\mathbf{E}_0$ and $\mathbf{B}$, the total dissipation (j^2/σ) being unaltered. The Hall term is produced by the drifting of charged particles across the magnetic field. It dominates when $\Omega_e \tau_{ei} \gg 1$, so that electrons spiral freely between collisions, and it is much less important than the direct conduction term when $\Omega_e \tau_{ei} \ll 1$. (However, it is only in extremely thin regions of high current density that Ohm's Law departs at all from $\mathbf{E} + \mathbf{v} \times \mathbf{B} = \mathbf{0}$, as stressed in Secs. 2.1.4 and 2.6.) Some of the consequences of Hall MHD can be found in Huba and Rudakov (2002) and Huba (2003).

2.1.4 Induction Equation

It is convenient to eliminate $\mathbf{E}$ and $\mathbf{j}$ among (2.1), (2.3) and (2.9) to give

$$\frac{\partial \mathbf{B}}{\partial t} = -\nabla \times (-\mathbf{v} \times \mathbf{B} + \mathbf{j}/\sigma) = \nabla \times (\mathbf{v} \times \mathbf{B}) - \nabla \times (\eta \nabla \times \mathbf{B}),$$

where $\eta = 1/(\mu\sigma)$ is the *magnetic diffusivity* (which must not be confused with the *electrical resistivity* σ^{-1}). Corresponding *Hall* and *Cowling diffusivities* are defined by $\eta_H = 1/(\mu\sigma_H)$ and $\eta_C = 1/(\mu\sigma_C)$, respectively. After using Eq. (2.2) and the identity $\nabla \times (\nabla \times \mathbf{B}) = \nabla(\nabla \cdot \mathbf{B}) - \nabla^2 \mathbf{B}$, one finds (for uniform η) the *induction equation*

$$\frac{\partial \mathbf{B}}{\partial t} = \nabla \times (\mathbf{v} \times \mathbf{B}) + \eta \nabla^2 \mathbf{B}. \tag{2.13}$$

When $\mathbf{v}$ is known, the induction equation can be used to determine $\mathbf{B}$, subject to

$$\nabla \cdot \mathbf{B} = 0. \tag{2.14}$$

The resulting current density and electric field follow from Ampère's Law,

$$\mathbf{j} = \nabla \times \mathbf{B}/\mu, \tag{2.15}$$

Table 2.1. *Variation of ln Λ with T(K) and n(m^{-3})*

T \ n	10^{12}	10^{15}	10^{18}	10^{21}	10^{24}	10^{27}
10^4 K	16.3	12.8	9.43	5.97		
10^5 K	19.7	16.3	12.8	9.43	5.97	
10^6 K	22.8	19.3	15.9	12.4	8.96	5.54
10^7 K	25.1	21.6	18.1	14.7	11.2	7.85

and Ohm's Law

$$\mathbf{E} = -\mathbf{v} \times \mathbf{B} + \mathbf{j}/\sigma, \tag{2.16}$$

while the charge density can be determined, if need be, from Eq. (2.4). Physical consequences of the induction equation are discussed further in Section 2.6.

It is important to note that in MHD the magnetic field and plasma velocity are the primary variables, while the electric current and electric field are secondary. In particular, it is Eq. (2.15) rather than Eq. (2.16) that determines the current. Indeed, in many solar applications, the length-scales are so large that the current term in Eq. (2.16) is completely negligible; thus, save in regions of high current concentration such as *current sheets*, the electric field is given by $\mathbf{E} \approx -\mathbf{v} \times \mathbf{B}$.

When there are no motions, the magnetic field is determined from *magnetohydrostatic* considerations (Chapter 3), and the resulting electric current and electric field are given by Eqs. (2.15) and (2.16); such electric fields are, in practice, extremely small due to the large length-scale (l_0) for magnetic variations. For example, a typical active-region magnetic field (B_0) of 100 G and plasma speed (V_0) of 10^3 m s^{-1} give $E_0 \approx V_0 B_0 \approx 10$ V m^{-1}. But, if there are no motions, the current density and electric field become only $j_0 \approx B_0/(\mu l_0) \approx 8\times10^{-4}$ A m^{-2} and $E_0 \approx j_0/\sigma \approx 8\times10^{-7}$ V m^{-2} when $l_0 = 10^7$ m and $\sigma = 10^3$ mho m^{-1}.

2.1.5 Values for Electrical Conductivity and Magnetic Diffusivity

For a *fully ionised*, collision-dominated plasma Eq. (2.11b) gives $\sigma = n_e e^2 \tau_{ei}/m_e$, where $\tau_{ei} = 0.543\times10^6 T^{3/2}/[n_e \ln\Lambda]$ sec is the *effective electron collision-time* for a hydrogen plasma (Spitzer 1962), with n_e measured in m^{-3}. The *Coulomb logarithm* (ln Λ) lies between 5 and 20 in value and has a weak dependence on temperature and density (Table 2.1). The electrical conductivity therefore becomes

$$\sigma = 1.53 \times 10^{-2} \frac{T^{3/2}}{\ln\Lambda} \text{ mho m}^{-1}, \tag{2.17}$$

while the corresponding expression for magnetic diffusivity is

$$\eta = \frac{m_e}{\mu n_e e^2 \tau_{ei}} = 5.2 \times 10^7 \ln\Lambda\, T^{-3/2} \text{ m}^2\text{s}^{-1}. \tag{2.18}$$

Typical values for η in the solar chromosphere and corona are $8\times10^8\ T^{-3/2}$ and $10^9\ T^{-3/2}$ m^2 s^{-1}, respectively, for a fully ionised plasma. In particular, in the photosphere, chromosphere (at 10^4 K) and corona (at 10^6 K), we find η roughly equal to 10^4, 10^3 and 1 m^2 s^{-1}, respectively.

Partial ionisation influences energy transport in the top 8 Mm of the convection zone (Sec. 1.3.3) and (ambipolar) magnetic diffusion in the photosphere and chromosphere up to a height of 2,000 km. The degree of ionisation [$n_i/(n_i + n_a)$, Sec. 2.1.3] is plotted as a function of height in Figure 2.1a for the VAL-C model (Sec. 1.4.3). The resulting Cowling (ambipolar) (η_C) and Hall (η_H) diffusivities depend on the magnetic and plasma properties. They are plotted in Figure 2.1b for examples of quiet-Sun or active-region flux tubes, starting from a field of 1500 G in the photosphere and decreasing to 10 G or 100 G, respectively, in the chromosphere. $\eta_C < \eta$ in the photosphere, but η_C rapidly increases through the chromosphere to a value that is 10^4 to 10^6 times larger than η at a height of 2,000 km. η_H equals η at the base of the photosphere, but is a factor of 100 larger throughout the chromosphere. The larger magnetic field in the active-region chromosphere increases both η_C and η_H above the quiet-Sun values. Both η_C and η_H have strong effects on MHD simulations of the photosphere and chromosphere (e.g., Martínez-Sykora et al. 2012).

For a turbulent plasma, the collision time and corresponding electrical conductivity can often be much smaller than the Spitzer values. For example, low-frequency ion-sound turbulence has an anomalous collision-time $\tau^* = \omega_{pe}^{-1}(nk_BT/W)$ and corresponding *anomalous conductivity* $\sigma^* = n_e e^2 \tau^*/m_e$, where W is the turbulent energy density and ω_{pe} is the electron plasma frequency. After putting, for instance, $W \approx 0.01\ nk_BT$, this gives numerically $\sigma^* \approx 5.0 \times 10^{-8} n_e^{1/2}$ mho m^{-1}. Such turbulence, however, occurs only for extremely high current densities (j_0) and correspondingly small magnetic field scale-lengths (l_0) (typically less than 2 m).

If the plasma is known to be in a fluid turbulent state, such as, for instance, in the convection zone, it may be useful to introduce eddy transport coefficients. Thus, if the small-scale eddies have a speed v and a scale-length d, an *eddy magnetic diffusivity* may be defined as $\tilde{\eta} = vd$. For granulation (where $v \lesssim 1$ km s^{-1} and $d \approx 10^3$ km) it has a typical value of at most 10^9 m^2 s^{-1}, whereas for supergranules (where $v \approx 0.2$ km s^{-1} and $d \approx 3\times10^4$ km) it is somewhat larger (6×10^9 m^{2}s^{-1}). By comparison, it is possible to infer values of 2×10^7 m^{2}s^{-1} and 10^9 m^{2}s^{-1} from the observed decay of sunspots (Sec. 9.6.2) and the dispersal of active regions.

2.2 Plasma Equations

2.2.1 Mass Continuity

The behaviour of a magnetic field, described by the induction equation, is coupled to that of the plasma by the presence of the velocity term in Eq. (2.13). The plasma motion is in turn governed by equations of continuity, motion and energy. Consider first the equation of mass conservation, which may be written

$$\frac{d\rho}{dt} + \rho \boldsymbol{\nabla}\cdot\mathbf{v} = 0 \quad \text{(a)} \qquad \text{or} \qquad \frac{\partial\rho}{\partial t} + \boldsymbol{\nabla}\cdot(\rho\mathbf{v}) = 0, \quad \text{(b)}, \tag{2.19}$$

where ρ is the mass density, $d/dt \equiv \partial/\partial t + \mathbf{v}\cdot\boldsymbol{\nabla}$ is the *material, total* or *convective derivative* for time variations following the motion (sometimes written as D/Dt). d/dt used in this sense is rarely confused with an ordinary derivative of a function of t alone. Equation (2.19b) expresses the fact that the density at a point *increases* ($\partial\rho/\partial t > 0$) if mass flows *into* the surrounding region $[\boldsymbol{\nabla}\cdot(\rho\mathbf{v}) < 0]$ and decreases if it flows out.

2.2.2 Equation of Motion

Under conditions of electrical neutrality, the equation of motion may be written

$$\rho\frac{d\mathbf{v}}{dt} = -\boldsymbol{\nabla}p + \mathbf{j}\times\mathbf{B} + \mathbf{F}, \tag{2.20}$$

where p is the plasma pressure (assumed scalar) and the material is assumed to be subject to a plasma pressure gradient ∇p, a *Lorentz force* $\mathbf{j}\times\mathbf{B}$ per unit volume (see Sec. 2.7), and a force $\mathbf{F} = \mathbf{F}_g + \mathbf{F}_v$, which represents the effects of *gravity* ($\mathbf{F}_g$) and *viscosity* ($\mathbf{F}_v$) (see PROBLEM 2.1: for all problems and solutions, see web page www.cambridge.org/9780521854719).

The gravitational force is

$$\mathbf{F}_g = -\rho g(r)\,\hat{\mathbf{r}}, \tag{2.21}$$

where the unit vector ($\hat{\mathbf{r}}$) acts radially outwards from the centre of the Sun and the local gravitational acceleration may be written $g(r) = M(r)G/r^2$, in terms of the mass $[M(r)]$ of the Sun inside a radius r and the gravitational constant (G). At the solar surface $r = R_\odot$ ($= 6.96\times10^8$ m), $M = M_\odot$ ($= 1.991\times10^{30}$ kg) and the gravitational acceleration is $g_\odot = 274$ m s^{-2}.

The viscous force is

$$\mathbf{F}_v = \rho\nu[\nabla^2\mathbf{v} + \tfrac{1}{3}\boldsymbol{\nabla}(\boldsymbol{\nabla}\cdot\mathbf{v})], \tag{2.22}$$

where ν is the *coefficient of kinematic viscosity* (assumed uniform). It simplifies to $\mathbf{F}_v = \rho\nu\nabla^2\mathbf{v}$ when the flow is incompressible. For a fully ionised hydrogen plasma, Spitzer (1962) gives

$$\rho\nu = 2.21\times10^{-16}\frac{T^{5/2}}{\ln\Lambda}\ \text{kg m}^{-1}\ \text{s}^{-1}.$$

For a discussion of the viscous stress tensor in the corona and of the different viscous coefficients parallel and perpendicular to the magnetic field, see Hollweg (1986).

In a frame of reference that is rotating with instantaneous *angular velocity* $\boldsymbol{\Omega}$ relative to an inertial frame, the equation of motion (2.20) at a distance $\mathbf{r}$ from the rotation axis is modified to

$$\rho\left(\frac{d\mathbf{v}}{dt} + 2\boldsymbol{\Omega}\times\mathbf{v}\right) = -\boldsymbol{\nabla}p + \mathbf{j}\times\mathbf{B} + \mathbf{F} + \rho\mathbf{r}\times\frac{d\boldsymbol{\Omega}}{dt} + \tfrac{1}{2}\rho\boldsymbol{\nabla}|\boldsymbol{\Omega}\times\mathbf{r}|^2,$$

under the assumption that $|\boldsymbol{\Omega}\times\mathbf{r}+\mathbf{v}| \ll c$. In particular, when $\boldsymbol{\Omega}$ and ρ are both constant and $\boldsymbol{\omega} = \boldsymbol{\nabla}\times\mathbf{v}$ is the *relative vorticity*, the curl of the above equation gives

$$\rho\frac{d\boldsymbol{\omega}}{dt} = \rho[(\boldsymbol{\omega} + 2\boldsymbol{\Omega})\cdot\boldsymbol{\nabla}]\mathbf{v} + \boldsymbol{\nabla}\times(\mathbf{j}\times\mathbf{B}) + \boldsymbol{\nabla}\times\mathbf{F}.$$

2.2.3 Perfect Gas Law

The gas pressure is determined by an equation of state, which is taken for simplicity as the perfect gas law

$$p = \frac{\tilde{R}}{\tilde{\mu}}\rho T, \tag{2.23}$$

where $\tilde{R}$ is the *gas constant*, $\tilde{\mu}$ is the *mean atomic weight* (a dimensionless number which represents the average mass per particle measured in units of m_p). The reader should beware that $\tilde{\mu}$ is sometimes incorporated into $\tilde{R}$ in Eq. (2.23). Also, the *mean particle mass* (m) may be used in place of $\tilde{\mu} = m/m_p$, and the *Boltzmann constant* (k_B=1.381$\times10^{-23}$ J deg^{-1}) can be introduced, so that $\tilde{R}$ is replaced by $\tilde{R} = k_B/m_p$. Equation (2.23) then becomes

$$p = \frac{k_B}{m}\rho T. \tag{2.24}$$

In some cases it is more convenient to write the equations in terms of the *total number* (n) of particles per unit volume rather than ρ. Then the perfect gas law becomes

$$p = nk_BT, \tag{2.25}$$

and ρ is replaced by $\rho = \tilde{\mu}m_pn = mn$ in the equation of motion.

For a *fully ionised hydrogen plasma*, there are two particles (a proton and an electron, denoted by subscripts p and e, respectively) for every proton; this implies that $\tilde{\mu} = 0.5$ and, in terms of the electron number density (n_e), we have

$$n \equiv n_p + n_e = 2n_e, \qquad \rho \equiv n_p m_p + n_e m_e \approx n_e m_p,$$

assuming electrical neutrality. In the solar atmosphere the presence of extra elements makes $\tilde{\mu} \approx 0.6$ (and $n \approx 1.9\, n_e$), except close to the photosphere (where H and He are not fully ionised) and in the core (where the composition is different).

2.3 Energy Equations

2.3.1 Different Forms of the Heat Equation

The last fundamental equation is the heat equation

$$\rho T \frac{ds}{dt} = -\mathcal{L}, \tag{2.26}$$

where the *energy loss function* $\mathcal{L}$ is the net effect of all the sinks and sources of energy and s is the *entropy* per unit mass of the plasma. The heat equation simply states that the rate of increase of heat for a unit volume as it moves in space is due to the net effect of the energy sinks and sources, which are described in the following. When the energy losses and gains balance, so that $\mathcal{L} \equiv 0$, the entropy is conserved.

The heat equation may be written in many alternative ways. For example, in terms of the *internal energy* (e) per unit mass, it becomes

$$\rho \frac{de}{dt} - \frac{p}{\rho} \frac{d\rho}{dt} = -\mathcal{L}. \tag{2.27}$$

For an ideal polytropic gas the internal energy is $e = c_v T$, where c_v is the *specific heat at constant volume.* Furthermore, the *specific heat at constant pressure* (c_p) and the *ratio of specific heats* (γ) are defined by $c_p = c_v + k_B/m$ and $\gamma = c_p/c_v$, which together imply that

$$c_p = \frac{\gamma}{\gamma - 1} \frac{k_B}{m}, \quad c_v = \frac{1}{\gamma - 1} \frac{k_B}{m} \quad \text{and} \quad e = \frac{p}{(\gamma - 1)\rho}. \tag{2.28}$$

The ratio of specific heats may also be written as $\gamma = (\tilde{N} + 2)/\tilde{N}$, in terms of $\tilde{N}$, the number of degrees of freedom in the plasma. For fully ionised hydrogen $\tilde{N} = 3$ and so $\gamma = 5/3$, but in general γ lies between 1 and 5/3. In a model solar atmosphere, the effect of partial ionisation is to make γ fall typically from a maximum of 5/3 at $\tau_{5,000} = 1$ to a minimum of 1.1 at an altitude of a few thousand km and then to rise rapidly to 5/3 again (Secs. 1.4.3 and 2.1.5).

The energy equation (2.26) may be rewritten in several other ways as follows. After substituting for e in (2.27) and using the perfect gas law (2.23) for T, the terms on the left-hand side may be combined to give the concise form

$$\frac{\rho^\gamma}{\gamma - 1} \frac{d}{dt} \left(\frac{p}{\rho^\gamma} \right) = -\mathcal{L}, \qquad \text{or} \qquad \rho c_v T \frac{d}{dt} \left(\log \frac{p}{\rho^\gamma} \right) = -\mathcal{L}, \tag{2.29a}$$

$$\text{or} \qquad \frac{dp}{dt} - \frac{\gamma p}{\rho} \frac{d\rho}{dt} = -(\gamma - 1)\mathcal{L}, \qquad \text{or} \qquad \rho \frac{d}{dt}(c_p T) - \frac{dp}{dt} = -\mathcal{L}, \tag{2.29b}$$

where $c_p T = \gamma p/[(\gamma - 1)\rho]$ is the *enthalpy* per unit mass. Alternatively, Eq. (2.19) may be used to substitute for $d\rho/dt$ in Eq. (2.27) to give

$$\rho \frac{de}{dt} + p\, \nabla \cdot \mathbf{v} = -\mathcal{L}, \tag{2.30}$$

where the internal energy e is given by Eq. (2.28).

When the *pressure remains constant*, the second of Eqs. (2.29b) reduces to

$$\rho c_p \frac{dT}{dt} = -\mathcal{L}. \tag{2.31}$$

When the plasma is *thermally isolated* from its surroundings, in the sense that there is no exchange of heat ($\mathcal{L} = 0$), the change of state is said to be *adiabatic* and the second of Eqs. (2.29a) shows that $p/\rho^\gamma =$ *constant* for each plasma element following the motion. This in turn means that the entropy of each element remains constant, since a comparison of Eqs. (2.26) and the second of (2.29a) implies in general

$$s = c_v \log (p/\rho^\gamma) + \textit{constant}.$$

Formally, $\mathcal{L}$ is negligible in Eq. (2.29a) when the time-scale for changes in p, ρ and T is much smaller than the time-scales for radiation, conduction and heating, which is often valid for rapid changes associated with wave motions or instabilities.

2.3.2 Thermal (or Heat) Conduction

Now consider in turn the various terms combined in the energy loss function ($\mathcal{L}$), which, in general, may be written as the rate of energy loss minus the rate of energy gain, namely,

$$\mathcal{L} = \nabla \cdot \mathbf{q} + L_r - j^2/\sigma - F_H, \tag{2.32}$$

where $\mathbf{q}$ is the *heat flux* due to particle conduction, L_r is the net radiation, j^2/σ is the *Ohmic dissipation*, and F_H represents the sum of all the other heating sources.

The heat flux vector may be written

$$\mathbf{q} = -\boldsymbol{\kappa}\nabla T, \tag{2.33}$$

where $\boldsymbol{\kappa}$ is the *thermal conduction tensor*, and the divergence of the heat flux may be split into two parts,

$$\nabla_\parallel \cdot (\kappa_\parallel \ \nabla_\parallel T) + \nabla_\perp \cdot (\kappa_\perp \ \nabla_\perp T),$$

where subscripts $\parallel$ and $\perp$ refer to values along and across the magnetic field. Conduction along the magnetic field is primarily by electrons, and, for a fully ionised hydrogen plasma, Spitzer (1962) gives

$$\kappa_\parallel = \kappa_0 T^{5/2} \equiv 1.8 \times 10^{-10} \frac{T^{5/2}}{\ln \Lambda} \text{ W m}^{-1}\text{K}^{-1}. \tag{2.34}$$

Typical values are $4\times10^{-11}\ T^{5/2}$, $10^{-11}\ T^{5/2}$ and $9\times10^{-12}\ T^{5/2}$ for photospheric, chromospheric and coronal regions, respectively. Conduction across the magnetic field is mainly by the protons in a fully ionised hydrogen plasma, depending on the product $\Omega_i \tau_{ii} = 1.63\times10^{15}\ B\ T^{3/2}/(n \ln \Lambda)$ of the *ion gyro-frequency* Ω_i and the *ion-ion collision-time* τ_{ii} [which is $(m_i/m_e)^{1/2}$ times the electron collision time τ_{ei}]. In most solar applications, the ions spiral many times between collisions, so that $\Omega_i \tau_{ii} \gg 1$ and Spitzer gives $\kappa_\perp/\kappa_\parallel = 2\times10^{-31}\ n^2/(T^3 B^2)$, with B in tesla. More general expressions for transport coefficients are given by Braginsky (1965b).

In a magnetic field that is strong enough to make $\kappa_\perp \ll \kappa_\parallel$, conduction is mainly along the field and the heat conduction term may be approximated by $\nabla_\parallel \cdot (\kappa_\parallel \ \nabla_\parallel T)$, or, in terms of the distance s along a particular magnetic field line, $(d/ds)(\kappa_\parallel\ dT/ds) - (\kappa_\parallel/B)\ dB/ds\ dT/ds$; the second term arises because the magnetic field, in general, varies with s. An alternative expression is simply

$$\frac{1}{A}\frac{d}{ds}\left(\kappa_\parallel \frac{dT}{ds} A\right),$$

where $A(s)$ is the cross-sectional area of a flux tube, related to the magnetic field strength by $(d/ds)(BA)=0$, as shown in Section 2.9.2.

For a discussion of other expressions for heat conduction when the classical treatment breaks down (e.g., in the high corona), see Marsch (2006).

2.3.3 Radiation

In the *solar interior*, energy transport by radiation (or convection) dominates particle conduction, and the net radiation may be written as the divergence of a radiative flux ($\boldsymbol{q}_r$), namely, $L_r = \boldsymbol{\nabla} \cdot \boldsymbol{q}_r$, where $\boldsymbol{q}_r = -\kappa_r \boldsymbol{\nabla} T$ and $\kappa_r = 16\sigma_s T^3/(3\tilde{\kappa}\rho)$ is the *coefficient of radiative conductivity*; here σ_s is the Stefan-Boltzmann constant, $\tilde{\kappa}$ the *opacity* or *mass absorption coefficient* and $\tilde{\kappa}\rho$ is the *absorption coefficient*. When κ_r is locally uniform, the radiative loss reduces to the simple form $L_r = -\kappa_r \nabla^2 T$. Furthermore, it is sometimes convenient to introduce the *thermal diffusivity* (κ) defined by $\kappa = \kappa_r/(\rho c_p)$, so that at constant pressure the energy Eq. (2.31) reduces to

$$\frac{dT}{dt} = \kappa \nabla^2 T.$$

When waves or instabilities are being modelled (Chapters 4 and 7) on a time-scale τ, say, it is usually assumed that the perturbations are adiabatic. However, in the photosphere and chromosphere the effects of *radiative damping* must often be included on the *radiative relaxation time-scale* (τ_R). When $\tau_R > \tau$, the plasma variations are approximately adiabatic; but when $\tau_R < \tau$ a displaced plasma element comes into thermal equilibrium with its surroundings before the restoring forces have time to act. For long wavelengths $[\lambda \gg (\tilde{\kappa}\rho)^{-1}]$, the plasma is optically thick to the disturbance and the above expressions may be used for L_r. For $\lambda \ll (\tilde{\kappa}\rho)^{-1}$, however, the plasma is effectively optically thin, and $\tau_R = c_v/(16\sigma_s\tilde{\kappa}T^3)$. In this case, if equilibrium values and small disturbances are denoted by subscripts 0 and 1, respectively, Eq. (2.27) becomes (with $\mathcal{L} = L_r$)

$$\frac{\partial T_1}{\partial t} - (\gamma - 1)\frac{T_0}{\rho_0}\frac{\partial \rho_1}{\partial t} = -\frac{T_1}{\tau_R}.$$

Thus, at constant density, $\partial T_1/\partial t$ becomes proportional to T_1, and so the temperature evolution is governed by *Newton's Law of Cooling*.

For the *optically thin* part of the atmosphere ($T \gtrsim 2\times10^4$ K in the chromosphere and corona), the *radiative loss* (L_r) is no longer coupled to the radiation field by the radiation transfer equations and it takes the form

$$L_r = n_e n_H Q(T), \tag{2.35}$$

where n_e is the electron density and n_H the density of hydrogen atoms or protons. (When the plasma is fully ionised, $n_H = n_e$.) The temperature variation $[Q(T)]$ has been evaluated by a number of authors (Cox and Tucker 1969; Rosner et al. 1978; Martens et al. 2000), the main differences arising from different assumptions about the elemental abundances. Figure 2.2 shows two functions calculated for coronal and photospheric abundances. The coronal iron abundance is a factor of 3 higher than the photospheric value, which leads to a radiative loss rate that is a factor of 4 higher at 10^6 K and in turn a radiative cooling time that is a factor of 4 smaller than previously thought. The most important features of $Q(T)$ are the presence of a maximum around 2×10^5 K and a minimum around 10^7 K. A reasonable approximation that is useful for analytical modelling over the whole temperature range 2×10^4 K$< T < 10^7$ K for the coronal abundance case is (with $\chi = 10^{-32}$)

$$Q(T) = \chi T^{-1/2} \text{ W m}^3. \tag{2.36}$$

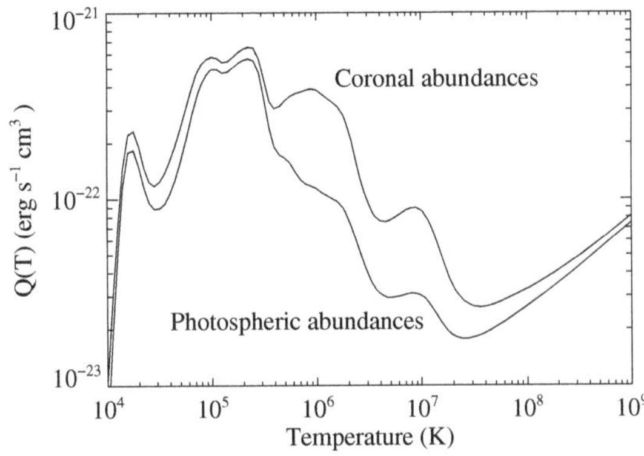

Figure 2.2. The optically thin radiative loss function $[Q(T)]$ for coronal (upper curve) and photospheric (lower curve) abundances, derived by Dere et al. (2009) from the CHIANTI atomic database (1 erg sec^{-1} cm$^3 = 10^{-13}$ W m^3).

A more accurate piecewise linear approximation (Klimchuk et al. 2008) is

$$Q(T) = \begin{cases} 1.09\times10^{-31}T^2 & \text{for} \quad T \leqslant 10^{4.97}, \\ 8.87\times10^{-17}T^{-1} & \text{for} \quad 10^{4.97} < T \leqslant 10^{5.67}, \\ 1.90\times10^{-22} & \text{for} \quad 10^{5.67} < T \leqslant 10^{6.18}, \\ 3.53\times10^{-13}T^{-3/2} & \text{for} \quad 10^{6.18} < T \leqslant 10^{6.55}, \\ 3.46\times10^{-25}T^{1/3} & \text{for} \quad 10^{6.55} < T \leqslant 10^{6.90}, \\ 5.49\times10^{-16}T^{-1} & \text{for} \quad 10^{6.90} < T \leqslant 10^{7.63}, \\ 1.96\times10^{-27}T^{1/2} & \text{for} \quad 10^{7.63} < T. \end{cases}$$

2.3.4 Heating

The heating term in Eq. (2.32) may be written

$$F_H = \rho\epsilon + H_\nu + H_c,$$

where ϵ is the *nuclear energy* generation rate (per unit mass) in the interior, H_ν is the *viscous dissipation rate* (important for strong flows) and H_c is the *coronal (or chromospheric) heating* term for the outer atmosphere. The viscous heating is

$$H_\nu = \rho\nu[\tfrac{1}{2}e_{ij}e_{ij} - \tfrac{2}{3}(\boldsymbol{\nabla}\cdot\mathbf{v})^2],$$

where $e_{ij} = \partial v_i/\partial x_j + \partial v_j/\partial x_i$ is the rate of strain tensor. The form of the coronal heating term is not well known, but is often assumed to be either uniform or proportional to density $H_w = constant \times n$. For heating by waves or reconnection, see Sections 10.3 and 10.4. For many applications, where one is modelling a phenomenon on a particular scale, other sources and sinks of energy can be included on the right-hand side of Eq. (2.29a) in order to take account of smaller-scale heating processes.

2.3.5 Energetics

The various equations we have set up in this chapter imply several relationships between the different types of energy such as heat, electrical energy and mechanical energy. First of all, the heat equation (2.29a) states that an '*increase in entropy* is due to *heat flow* − *radiation* + *heat sources*'.

Also, the divergence of the *Poynting flux* may be written $\nabla\cdot(\mathbf{E}\times\mathbf{H}) = -\mathbf{E}\cdot\nabla\times\mathbf{H}+\mathbf{H}\cdot\nabla\times\mathbf{E}$ and transformed, using Eqs. (2.15) for $\nabla\times\mathbf{H}$ and (2.3) for $\nabla\times\mathbf{E}$, to

$$-\nabla\cdot(\mathbf{E}\times\mathbf{H}) = \mathbf{E}\cdot\mathbf{j}+\frac{\partial}{\partial t}\left(\frac{B^2}{2\mu}\right). \tag{2.37}$$

Its physical interpretation is that an '*inflow of electromagnetic energy* $\mathbf{E}\times\mathbf{H}$ produces *electrical energy* $\mathbf{E}\cdot\mathbf{j}$ *for the plasma + a rise in magnetic energy* $B^2/(2\mu)$.

In turn, the electrical energy given to the plasma by the electromagnetic field may be rewritten, after substituting for $\mathbf{E}$ from Eq. (2.16), as

$$\mathbf{E}\cdot\mathbf{j} = j^2/\sigma+\mathbf{v}\cdot\mathbf{j}\times\mathbf{B}, \tag{2.38}$$

so that the '*electrical energy* appears as *heat by Ohmic dissipation + work done by the* $\mathbf{j}\times\mathbf{B}$ *force*'.

Furthermore, the scalar product of $\mathbf{v}$ with the equation of motion (2.20) gives the so-called *mechanical energy equation*

$$\rho\frac{d}{dt}\left(\tfrac{1}{2}v^2\right) = -\mathbf{v}.\nabla p+\mathbf{v}\cdot\mathbf{j}\times\mathbf{B}+\mathbf{v}\cdot\mathbf{F}, \tag{2.39}$$

implying that '*plasma speed is changed* by the *work done by* $-\nabla p$, $\mathbf{j}\times\mathbf{B}$ *and* $\mathbf{F}$'.

Finally, Eqs. (2.19, 2.30, 2.38, 2.39) may be combined together to give another alternative to Eqs. (2.29a, 2.29b), namely,

$$\frac{\partial}{\partial t}(\rho e+\tfrac{1}{2}\rho v^2)+\nabla\cdot[(\rho e+\tfrac{1}{2}\rho v^2)\mathbf{v}] = -(\mathcal{L}+j^2/\sigma)+\mathbf{E}\cdot\mathbf{j}-\nabla\cdot(p\mathbf{v})+\mathbf{v}\cdot\mathbf{F}, \tag{2.40}$$

where $-(\mathcal{L}+j^2/\sigma) = -\nabla\cdot\mathbf{q}-L_r+F_H$. Eq. (2.40) expresses the fact that '*the gain in material energy* (internal plus kinetic) is due to *heat flow, radiation, viscous dissipation, heat sources, electrical energy* and the *work done* by the *pressure* (and other forces $\mathbf{F}$) in propulsion and compression'. In a steady state, it reduces to

$$\nabla\cdot\{[\gamma p/(\gamma-1)+\tfrac{1}{2}\rho v^2]\mathbf{v}\} = -(\mathcal{L}+j^2/\sigma)+\mathbf{E}\cdot\mathbf{j}+\mathbf{v}\cdot\mathbf{F},$$

where $\gamma p/[(\gamma-1)\rho]$ is the enthalpy per unit mass.

2.4 Summary of Equations

The fundamental MHD equations employed throughout this book are

$$\frac{\partial\mathbf{B}}{\partial t} = \nabla\times(\mathbf{v}\times\mathbf{B})+\eta\nabla^2\mathbf{B}, \tag{2.13}$$

$$\frac{d\rho}{dt}+\rho\nabla\cdot\mathbf{v} = 0, \tag{2.19}$$

$$\rho\frac{d\mathbf{v}}{dt} = -\nabla p+\mathbf{j}\times\mathbf{B}+\mathbf{F}, \tag{2.20}$$

$$p = \frac{k_B}{m}\rho T\ \left(=\frac{\tilde{R}}{\tilde{\mu}}\rho T\right), \tag{2.24}$$

$$\frac{\rho^\gamma}{\gamma-1}\frac{d}{dt}\left(\frac{p}{\rho^\gamma}\right) = -\nabla\cdot\mathbf{q}-L_r+j^2/\sigma+F_H, \tag{2.29a}$$

where $\mathbf{F}=\mathbf{F}_g+\mathbf{F}_v$ is given by Eqs. (2.21) and (2.22), $\mathbf{q}$ is defined by Eq. (2.33), L_r follows from Eq. (2.35) and $F_H=\rho\epsilon+H_v+H_c$ is given in Section 2.3.4. These equations are, in general, coupled and serve to determine the primary variables $\mathbf{v}$, $\mathbf{B}$, p, ρ and T. In addition, the secondary variables $\mathbf{j}$ and $\mathbf{E}$ are given explicitly by

$$\mathbf{j} = \boldsymbol{\nabla} \times \mathbf{B}/\mu, \tag{2.15}$$

$$\mathbf{E} = -\mathbf{v} \times \mathbf{B} + \mathbf{j}/\sigma, \tag{2.16}$$

while $\mathbf{B}$ is subject to the condition

$$\boldsymbol{\nabla} \cdot \mathbf{B} = 0. \tag{2.2}$$

This last equation plays the role of an initial condition for the time-dependent equations (2.13, 2.19, 2.20, 2.24, 2.29a), which form a complete set, representing nine equations for nine variables, while the divergence of (2.13) shows that, if $\boldsymbol{\nabla} \cdot \mathbf{B}$ vanishes initially, then it continues to vanish for all time (see PROBLEMS 2.2 and 2.3).

2.4.1 Dimensionless Equations

Writing the equations in dimensionless form is extremely useful, both when seeking numerical solutions and also when looking for analytical solutions in powers of some dimensionless parameter that is much smaller than unity. For example, if the magnetic field, plasma velocity, density, temperature and pressure have typical magnitudes B_0, v_0, ρ_0, T_0 and p_0, respectively, and if they vary over length and time-scales of L_0 and t_0, then one can change the variables to $\mathbf{B}=B_0\mathbf{B}'$, $\mathbf{v}=v_0\mathbf{v}'$, $\rho=\rho_0\rho'$, $T=T_0T'$, $p=p_0p'$, $\mathbf{r}=L_0\mathbf{r}'$ and $t=t_0t'$. If we assume that $v_0=L_0/t_0$ and $p_0=(k_B/m)\rho_0T_0$, then the basic equations become

$$\frac{\partial \mathbf{B}'}{\partial t'} = \boldsymbol{\nabla}' \times (\mathbf{v}' \times \mathbf{B}') + (R_m^{-1})\nabla'^2\mathbf{B}', \tag{2.13$'$}$$

$$\frac{d\rho'}{dt'} + \rho'\boldsymbol{\nabla}' \cdot \mathbf{v}' = 0, \tag{2.19$'$}$$

$$M_{A0}^2\ \rho'\frac{d\mathbf{v}'}{dt'} = -\tfrac{1}{2}\beta_0\boldsymbol{\nabla}'p' + \mathbf{j}' \times \mathbf{B}' + \epsilon_F\mathbf{F}', \tag{2.20$'$}$$

$$p' = \rho'T', \tag{2.24$'$}$$

$$\rho'^{\gamma}\frac{d}{dt'}\left(\frac{p'}{\rho'^{\gamma}}\right) = -\epsilon_q\boldsymbol{\nabla}' \cdot \mathbf{q}' - \epsilon_r L_r' + \epsilon_J\, j'^2 + \epsilon_H F_H', \tag{2.29$'$}$$

where each of the dashed variables is of order unity and the dimensionless parameters appearing in the dimensionless equations (which may be large, small or of order unity, depending on the application) are the magnetic Reynolds number ($R_m=L_0v_0/\eta$), the Alfvén Mach number ($M_{A0}=v_0\sqrt{\mu\rho_0}/B_0$), the plasma beta ($\beta_0=2\mu p_0/B_0^2$), the ratio $[\epsilon_F=L_0F_0/(\rho_0v_0^2)]$ of the magnitude (F_0) of the force $\mathbf{F}$ to that of the inertial term, and the ratios $[\epsilon_q,\ \epsilon_r,\ \epsilon_J=2(\gamma-1)/(\beta_0R_m),\ \epsilon_H]$ of conduction, radiation, Joule heating and heating to adiabatic variations.

2.4.2 Assumptions

The assumptions made in deriving the above set of basic equations are as follows:

(1) A collection of particles is a *plasma* if the number of charged particles of density n inside a sphere of radius λ_D [the *Debye length*, (Eq. 2.8)] is large ($4\pi n\lambda_D^3/3 \gg 1$). The plasma is approximately charge-neutral over distances larger than λ_D.

(2) The plasma is treated as a *continuum*, which is valid provided the length-scale for variations greatly exceeds internal plasma lengths such as the ion gyroradius.

(3) The plasma is assumed to be in thermodynamic equilibrium with distribution functions close to Maxwellian. This holds for time-scales much larger than the collision times and length-scales much longer than the mean free paths.

(4) The coefficients η and μ have been supposed uniform, and most of the plasma properties assumed *isotropic*. A more comprehensive theory using tensor transport coefficients can be developed.

(5) The equations are written for an *inertial frame*. The extra terms that arise for a frame rotating with the Sun are important for large-scale processes.

(6) *Relativistic* effects are neglected, since the flow speed, sound speed and Alfvén speed are all assumed to be much smaller than the speed of light.

(7) The simple form of *Ohm's Law* (2.16) is adopted for many applications, rather than its more general version (2.10).

(8) The plasma is treated as a *single fluid*, although two- or three-fluid models may be more relevant for the coolest or rarest parts of the solar atmosphere (Sec. 13.11).

2.4.3 Reduced Forms of the Equations

Often, simplified forms of the basic equations are considered. For instance, approximations to the induction equation are discussed in Section 2.6. Also, if the density changes of a moving plasma element are negligible ($d\rho/dt=0$), Eq. (2.19) reduces to $\nabla \cdot \mathbf{v}=0$, so that one needs to solve only the two coupled equations (2.13) and (2.20) for $\mathbf{v}$ and $\mathbf{B}$, subject to the conditions that $\nabla \cdot \mathbf{B}$ and $\nabla \cdot \mathbf{v}$ vanish. This assumption of *incompressibility* is often made for simplicity in magnetohydrodynamics when one is more interested in other effects than compressibility and wishes to make analytical progress. But such a neglect of density variations is valid only when the *sound speed* $c_s=(\gamma p/\rho)^{1/2}$ is much larger than typical plasma speeds. When plasma speeds are of the order or less than the *Alfvén speed* $v_A=B/(\mu\rho)^{1/2}$, the condition for incompressibility becomes $c_s \gg v_A$, so that $\beta \equiv (2\mu p/B^2) \gg 1$ (see PROBLEM 2.4). When performing a stability analysis, it is often found that the most unstable mode has an incompressible velocity perturbation, so that the density and pressure remain constant during a perturbation and no work is wasted in compressing the plasma unnecessarily.

In the incompressible case with viscosity, Eqs. (2.13) and (2.20) become

$$\frac{\partial \mathbf{B}}{\partial t} = (\mathbf{B}\cdot\nabla)\mathbf{v} - (\mathbf{v}\cdot\nabla)\mathbf{B} + \eta\nabla^2\mathbf{B},$$
$$\frac{\partial \mathbf{v}}{\partial t} + (\mathbf{v}\cdot\nabla)\mathbf{v} = -\nabla\left(\frac{P}{\rho}\right) + (\mathbf{B}\cdot\nabla)\left(\frac{\mathbf{B}}{\mu\rho}\right) + \nu\nabla^2\mathbf{v},$$

where $P=p+B^2/(2\mu)$ is the total (plasma plus magnetic) pressure. Also, the magnetic field may be written $\mathbf{B}=\mathbf{B}_0+\mathbf{B}_1$, where $\mathbf{B}_0$ is uniform and $\mathbf{B}_1$ is not necessarily small. Then, by adding and subtracting these equations, they may be written elegantly in terms of Elsässer (1950) variables

$$\boldsymbol{z}^+ = \mathbf{v} + \frac{\mathbf{B}_1}{\sqrt{\mu\rho}}, \qquad \boldsymbol{z}^- = \mathbf{v} - \frac{\mathbf{B}_1}{\sqrt{\mu\rho}}$$

as

$$\frac{\partial \boldsymbol{z}^\pm}{\partial t} \mp (\mathbf{v}_A\cdot\nabla)\boldsymbol{z}^\pm + (\boldsymbol{z}^\mp\cdot\nabla)\boldsymbol{z}^\pm = -\nabla\left(\frac{P}{\rho}\right) + \nu^\pm\nabla^2\boldsymbol{z}^\pm + \nu^\mp\nabla^2\boldsymbol{z}^\mp,$$

where $\mathbf{v}_A=\mathbf{B}_0/\sqrt{\mu\rho}$, $\nu^\pm=\frac{1}{2}(\nu\pm\eta)$ and $\nabla\cdot\boldsymbol{z}^\pm=0$ (from $\nabla\cdot\mathbf{B}=\nabla\cdot\mathbf{v}=0$). These equations are useful for incompressible turbulence or wave calculations; for example, when they are linearised about $\mathbf{B}_0$ and diffusion is neglected, they become

$$\frac{\partial \boldsymbol{z}^\pm}{\partial t} \mp (\mathbf{v}_A\cdot\nabla)\boldsymbol{z}^\pm = 0,$$

so that $z^-(x - v_A t)$ and $z^+(x + v_A t)$ describe Alfvén waves propagating along $\mathbf{B}_0$ in the positive and negative directions (Sec. 4.3).

The MHD equations can also be simplified by prescribing $\mathbf{v}$ in Eq. (2.13) at the outset and neglecting Eq. (2.20), so that only $\mathbf{B}$ remains to be determined. With this so-called *kinematic approach*, frequently used in dynamo theory, only the effect of motions on the magnetic field is considered, while the influence of the magnetic field on the motions is neglected. This is valid provided the magnetic field is so small that $\beta \gg 1$ and pressure gradients dominate the Lorentz force.

Other reduced forms are obtained when $v \ll v_A$ and $v \ll c_s$, so that the inertial term in Eq. (2.20) is negligible, and then magnetohydrostatic force balance results (Chapter 3). In particular, if also $\beta \ll 1$, the magnetic field is *force-free* (Sec. 3.4) and is determined by $\mathbf{j} \times \mathbf{B} = \mathbf{0}$, a particular case of which is a *potential field* with $\mathbf{j} = \mathbf{0}$ (Sec. 3.3). In these cases, the plasma motion along each field line is governed by the component of

$$\rho \frac{d\mathbf{v}}{dt} = -\nabla p + \mathbf{F}$$

along the magnetic field; in particular, the equations for flow along a prescribed moving field line have been derived by Pneuman (1981). When the flow speed remains much smaller than the sound speed, this implies an evolution of the plasma through a series of equilibria parallel to the field. If the slow evolution is due to magnetic diffusion, $\mathbf{v}$ is determined by Eq. (2.13), as in Section 3.4.6.

For convection, a common simplification is the *Boussinesq approximation* for a thin layer (Sec. 9.1.2), which filters out sound waves and allows density variations only in the buoyancy force. A generalisation for thicker layers is the *anelastic approximation*, which also excludes sound waves and approximates the continuity equation by $\nabla \cdot (\rho_0 \mathbf{v}) = 0$ (Ogura and Phillips 1962; Gough 1969; Mizerski and Tobias 2011).

Often, when the energetics of a process are not the prime consideration, the energy equation is approximated either by $T = constant$ or, more generally, by the *polytropic approximation* $(p/\rho^\alpha = constant)$, where α is a constant (distinct from γ). The latter approximation is simply meant to model temperature variations in a rough manner, but it may be derived from the full energy equation (2.29a) when the only contribution to $\mathcal{L}$ is the conduction term and the conductive flux $\kappa \nabla T$ is proportional to the work done ($\mathbf{v}p$) by the pressure.

2.5 Dimensionless Parameters

In terms of a typical plasma speed (V_0) and length-scale (l_0), the magnitude of the convective term in Eq. (2.13) divided by the diffusive term is a dimensionless parameter,

$$R_m = \frac{l_0 V_0}{\eta}, \tag{2.41}$$

known as the *magnetic Reynolds number.* (When V_0 equals the Alfvén speed this becomes the Lundquist number.) It is a measure of the strength of the coupling between flow and magnetic field. In the laboratory, usually $R_m \ll 1$ and the coupling is weak, whereas in the solar atmosphere $R_m \gg 1$ and the coupling is strong.

The *Reynolds number,*

$$Re = \frac{l_0 V_0}{\nu}, \tag{2.42}$$

gives the ratio of the sizes of the inertial and viscous terms in the equation of motion.

The *Mach number*,

$$M = \frac{V_0}{c_s}, \tag{2.43}$$

measures the flow speed (V_0) relative to the sound speed $c_s = (\gamma p_0/\rho_0)^{1/2}$.

The *Alfvén Mach number*,

$$M_A = \frac{V_0}{v_A}, \tag{2.44}$$

gives the size of the flow speed in terms of the Alfvén speed $v_A = B_0/(\mu\rho_0)^{1/2}$, where B_0 and ρ_0 are a typical magnetic field strength and plasma density.

The *plasma beta*,

$$\beta = \frac{2\mu p_0}{B_0^2}, \tag{2.45}$$

is the plasma pressure (p_0) divided by the magnetic pressure, so that when $\beta \ll 1$ we have a so-called "low-β" plasma, for which the magnetic force dominates.

The *Rossby number* [$Ro = V_0/(l_0\Omega)$] is the ratio of inertial to Coriolis terms in the equation of motion.

Certain other dimensionless numbers, not independent of those above, are sometimes met. For instance, the *magnetic Prandtl number* ($P_m = R_m/Re = \nu/\eta$) compares viscous and magnetic diffusion (sometimes the term is used for the *diffusivity ratio* $\zeta = \eta/\kappa$), while the *Ekman number* [$E = Ro/Re = \nu/(l_0^2\Omega)$] is the ratio of the viscous force to the Coriolis force; furthermore, the *Taylor number* ($\mathcal{T} = E^{-2}$) measures the strength of the rotation.

For magnetoconvection the following numbers are also important (Sec. 9.1): the *Rayleigh number* [$Ra = \alpha g \Delta T\ d^3/(\kappa\nu)$] measures the importance of the buoyancy force relative to the stabilizing effects of non-magnetic diffusion and is much larger than unity in the convection zone; the *Chandrasekhar number* is $Q = Ha^2$, where $Ha = B_0 d/(\mu\rho\eta\nu)^{1/2}$ is the *Hartmann number*, representing the ratio of magnetic to viscous diffusion forces; the *Prandtl number* ($Pr = \nu/\kappa$) compares viscous and thermal diffusion and is generally $\ll 1$ (or of order unity) for the Sun, depending on whether molecular (or eddy) values are adopted; the *diffusivity ratio* ($\zeta = \eta/\kappa$) is usually $\ll 1$ in the convection zone but exceeds unity between 2 and 20 Mm below the photosphere due to the effect of ionisation on κ; finally, the *Nusselt number* [$Nu = 1 + F_c/(\kappa\Delta T/d)$], where F_c and $\kappa\Delta T/d$ are the convected and conducted heat fluxes, is a measure of the vigour of convection.

In the solar atmosphere (for $\tilde{\mu} = 0.6$ and $\gamma = 5/3$),

$$R_m = 1.9 \times 10^{-8} l_0\ V_0\ T_0^{3/2}/\ln\Lambda, \tag{2.46}$$

$$c_s = 152\ T_0^{1/2}\ \text{m s}^{-1}, \tag{2.47}$$

$$v_A = 2.8 \times 10^{12}\ B_0\ n_0^{-1/2}\ \text{m s}^{-1}, \tag{2.48}$$

$$\beta = 3.5 \times 10^{-21}\ n_0\ T_0\ B_0^{-2}, \tag{2.49}$$

where l_0, V_0, n_0 are in mks units, B_0 in gauss and Λ is given in Sec. 2.1.5. For example, above a sunspot, where typically $l_0 \approx 10^7$ m, $V_0 \approx 10^3$ m s^{-1}, $T_0 \approx 10^4$ K, $n_0 \approx 10^{20}$ m^{-3}, $B_0 \approx 10^3$ G, the relevant values are $R_m \approx 3{\times}10^7$, $c_s \approx 2{\times}10^4$ m s^{-1}, ($M \approx 0.05$), $v_A \approx 3{\times}10^5$ m s^{-1}, ($M_A \approx 4{\times}10^{-3}$), $\beta \approx 3{\times}10^{-3}$; also, with $v \approx 10^{-2}$ m^{2}s^{-1} and $\Omega \approx 10^{-6}$ s^{-1}, we have $Re \approx 10^{12}$ and $Ro \approx 100$. The plasma beta rapidly decreases with altitude above the solar surface, while the Alfvén speed increases. Whereas the plasma mostly dominates magnetic fields in the photosphere ($\beta > 1$), the opposite is true in the corona ($\beta < 1$).

2.6 Consequences of the Induction Equation

The induction equation,

$$\frac{\partial \mathbf{B}}{\partial t} = \nabla \times (\mathbf{v} \times \mathbf{B}) + \eta \nabla^2 \mathbf{B}, \tag{2.13}$$

has been derived in Section 2.1.4, and the magnetic Reynolds number,

$$R_m = \frac{l_0 V_0}{\eta}, \tag{2.41}$$

has been defined (Sec. 2.5) as the ratio of the convective to diffusive terms on the right of Eq. (2.13). Once $\mathbf{v}$ is known, the induction equation determines the behaviour of the magnetic field, which depends crucially on whether $R_m \ll 1$ or $R_m \gg 1$. When R_m is small, the magnetic field diffuses through the plasma (Sec. 2.6.1), whereas, when it is large, the magnetic field is frozen to the plasma (Sec. 2.6.2).

For some very small-scale phenomena, such as current sheets a kilometre or less thick, R_m may be of order unity, but usually it is much larger. For instance, consider typical sunspot motions at a speed $V_0 \approx 10^3$ m s^{-1} with a length-scale $l_0 \approx 10^7$ m. They make $R_m \approx 10^6$ for a magnetic diffusivity $\eta \approx 10^4$ m^2 s^{-1}, and so the magnetic field is tied very closely indeed to the plasma. According to Eq. (2.51) below, the field would diffuse away over 300 years, whereas sunspot fields are observed, in practice, to decay over only 100 days. Clearly, this cannot be caused by the above classical Ohmic dissipation and would require an eddy diffusivity (due to, for instance, small-scale instability) that is a factor of 10^3 larger than the classical value.

Interestingly, Eq. (2.13) may be rewritten (for incompressible flow) as

$$\frac{d\mathbf{B}}{dt} = (\mathbf{B} \cdot \nabla)\mathbf{v} + \eta \nabla^2 \mathbf{B},$$

which is similar in form to the *vorticity equation* $[d\boldsymbol{\omega}/dt = (\boldsymbol{\omega} \cdot \nabla)\mathbf{v} + \nu \nabla^2 \boldsymbol{\omega}]$ for an incompressible fluid with no magnetic field (in an inertial frame). This is the basis of the so-called *vorticity-magnetic field analogy*, which implies that magnetic field lines respond in a similar way to the classical behaviour of vortex lines. For instance, field lines are, in general, partly transported with the flow and partly diffuse through it. Also, they may be stretched and increase in strength due to motions along their length. However, the analogy holds only for incompressible flow, and it is in any case not exact since the extra relation $\boldsymbol{\omega} = \nabla \times \mathbf{v}$ does not hold for the magnetic field.

2.6.1 Diffusive Limit ($R_m \ll 1$)

Here the induction equation (2.13) reduces to a simple diffusion equation

$$\frac{\partial \mathbf{B}}{\partial t} = \eta \nabla^2 \mathbf{B}, \tag{2.50}$$

implying that field variations on a length-scale l_0 are destroyed over a *diffusion time-scale* $\tau_d = l_0^2/\eta$. With a fully ionised value [Eq. (2.18)] for η, this becomes

$$\tau_d = 1.9 \times 10^{-8}\, l_0^2\, T^{3/2}/\ln \Lambda \text{ sec.} \tag{2.51}$$

Thus, for instance, with $T = 10^6$ K and a length-scale $l_0 = 10^7$ m (10 Mm), τ_d is approximately 10^{14} s, whereas a length-scale of 1 m gives $\tau_d \approx 1$ s. Since *solar flares* represent a release of magnetic energy over a time-scale of 100 or 1000 s, this suggests that a length-scale as small as 100 or 1000 m is required (Chapter 12).

The smaller the length-scale, the faster the field diffuses away. As an example, the diffusion of a unidirectional magnetic field is described in Section 6.4.2. The diffusion of magnetic fields in one, two and three dimensions has been studied by Wilmot-Smith et al. (2005), who show how magnetic fields can diffuse through the plasma and disappear at null sheets in one dimension or at O-lines in two dimensions (Sec. 6.2); in three dimensions the diffusion is much more subtle (Secs. 6.11 and 6.12).

We return to flow in a non-ideal plasma in Section 2.6.3 after dealing with ideal flow.

2.6.2 Ideal Limit ($R_m \gg 1$)

This is sometimes referred to as the perfectly conducting or infinite conductivity limit, but it would be better called the 'large length-scale' limit, since what makes the diffusion term so small in astrophysical plasmas (compared with most terrestrial plasmas) is the relatively large length-scale rather than the value of the electrical conductivity (which varies relatively little). Here Eq. (2.13) becomes approximately

$$\frac{\partial \mathbf{B}}{\partial t} = \boldsymbol{\nabla} \times (\mathbf{v} \times \mathbf{B}), \tag{2.52}$$

while Ohm's Law (2.16) reduces to

$$\mathbf{E} + \mathbf{v} \times \mathbf{B} = \mathbf{0}. \tag{2.53}$$

Equation (2.52) follows from taking $\boldsymbol{\nabla}\times$ Eq. (2.53) and using Eq. (2.3). Although the total electric field vanishes [Eq. (2.53)], currents are still present and are given by Eq. (2.15).

In this large magnetic Reynolds number limit, Alfvén's frozen-flux theorem holds and implies that for ideal plasma flows:

(i) magnetic flux is conserved;
(ii) magnetic field lines are conserved.

Results (i) and (ii) are discussed in the following sections, with field line conservation being a consequence of flux conservation.

Furthermore, *magnetic topology* includes any property of a magnetic field that is preserved by an ideal displacement, such as the linkage and knottedness of magnetic field lines. The magnetic structure may be stretched and deformed by such a flow, but its topology is not changed. Thus, a third property follows from field line conservation, namely,

(iii) magnetic topology is conserved.

Alfvén's result implies that magnetic field changes at a particular point are the same as if magnetic field lines move with the plasma. One refers to field lines being *frozen into the plasma*; plasma can move freely *along* field lines, but, in motion perpendicular to them, either the field lines are dragged with the plasma or the field lines push the plasma.

Another result (see PROBLEM 2.5) is that

(iv)

$$\mathbf{w}_\perp = \mathbf{v}_\perp = \frac{\mathbf{E} \times \mathbf{B}}{B^2}, \tag{2.54}$$

where $\mathbf{w}_\perp$ and $\mathbf{v}_\perp$ are the components of the magnetic flux velocity and plasma velocity perpendicular to the magnetic field. The fact that these two are identical (and also equal to the field line velocity) follows from the result that the magnetic flux is frozen to the plasma. Their common value follows from taking the vector product of Eq. (2.53) with $\mathbf{B}$.

Now what happens to these properties in a non-ideal flow, for which Eq. (2.53) does not hold? As we shall see in Section 2.6.3, flux conservation and field line conservation may sometimes still hold but they are in general no longer equivalent and the flux velocity is no longer unique. We here present the basic results and develop them further in Section 6.11, applying them to reconnection in Section 6.12.

2.6.2.1 Magnetic Flux Conservation

When the magnetic Reynolds number is large, Alfvén's theorem states that magnetic flux is conserved, so that magnetic field behaves as if it moves with the plasma. (By the vorticity-magnetic field analogy this is similar to the classical vorticity theorem of Helmholtz and Kelvin.) An intuitive proof is as follows and a more exact proof is given in Section 2.6.3. Consider a curve C (bounding a surface S) which is moving with the plasma, such that in a time dt an element $\mathbf{dl}$ of C sweeps out an element of area $\mathbf{v} \times \mathbf{dl}\, dt$. The rate of change of magnetic flux through C is

$$\frac{d}{dt}\iint_S \mathbf{B}\cdot d\mathbf{S} = \iint_S \frac{\partial \mathbf{B}}{\partial t}\cdot d\mathbf{S} + \oint_C \mathbf{B}\cdot \mathbf{v}\times d\mathbf{l}. \tag{2.55}$$

As C moves, so the flux changes, both because the magnetic field changes with time (the first term on the right of Eq. 2.55) and also because the boundary moves in space (the second term on the right). By setting $\mathbf{B}\cdot\mathbf{v}\times d\mathbf{l} = -\mathbf{v}\times\mathbf{B}\cdot d\mathbf{l}$ and applying Stokes' theorem to the second term on the right, we obtain

$$\frac{d}{dt}\iint_S \mathbf{B}\cdot d\mathbf{S} = \iint_S \left(\frac{\partial \mathbf{B}}{\partial t} - \boldsymbol{\nabla}\times(\mathbf{v}\times\mathbf{B})\right)\cdot d\mathbf{S},$$

which vanishes in the ideal limit. Thus, the total magnetic flux through S bounded by C remains constant as it moves with the plasma. In other words, we have proved *magnetic flux conservation*, namely the plasma elements that initially form a flux tube continue to do so at later times (Figure 2.3).

2.6.2.2 Magnetic Field Line Conservation

There is also *magnetic field line conservation*, namely, if two plasma elements lie on a field line initially, they will always do so (Figure 2.4). Suppose that at $t = t_1$ the elements P_1 and P_2 lie on a field line,

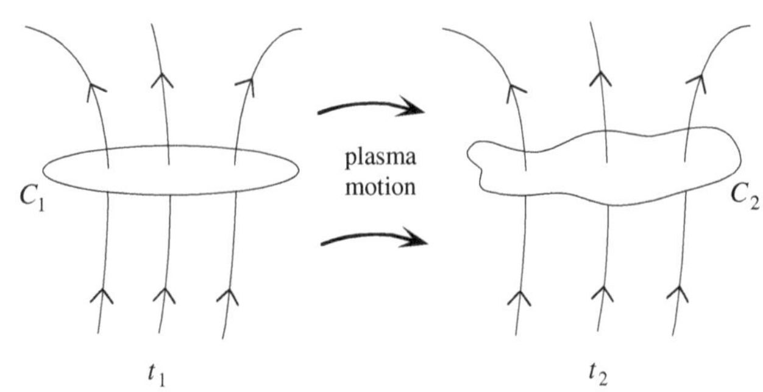

Figure 2.3. Magnetic flux conservation: if a curve C_1 is distorted into C_2 by plasma motion, the flux through C_1 at t_1 equals the flux through C_2 at t_2.

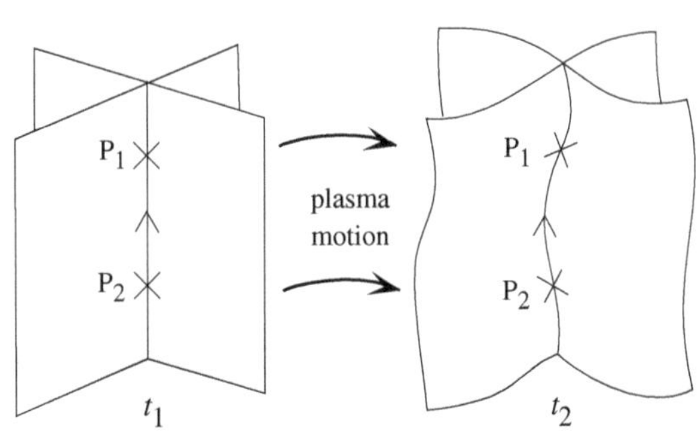

Figure 2.4. Magnetic field line conservation: if plasma elements P_1 and P_2 lie on a field line at time t_1, then they will lie on the same line at a later time t_2.

which may be defined as the intersection of two flux surfaces. Then, at some later time ($t = t_2$), according to magnetic flux conservation, P_1 and P_2 will still lie on both surfaces, and so they will lie on the field line defined by their intersection. We have therefore established that, if Eq. (2.53) is true, then flux conservation holds and therefore line conservation holds.

Line conservation also may be proved directly from Eqs. (2.53) and (2.19) as follows. Applying the appropriate vector identity to the ideal induction equation gives

$$\frac{\partial \mathbf{B}}{\partial t} + (\mathbf{v} \cdot \boldsymbol{\nabla})\mathbf{B} = (\mathbf{B} \cdot \boldsymbol{\nabla})\mathbf{v} - \mathbf{B}(\boldsymbol{\nabla} \cdot \mathbf{v}). \tag{2.56}$$

This implies that changes in the magnetic field following the motion (i.e., the left-hand side, which is actually $d\mathbf{B}/dt$) are produced when a flux tube is stretched, sheared or expanded: according to the first term on the right-hand side, an accelerating motion along the field causes the field strength to increase, whereas a shearing motion normal to the field makes the field change direction by increasing the field component along the flow direction; the second term on the right-hand side implies that an expansion ($\boldsymbol{\nabla} \cdot \mathbf{v} > 0$) causes the field strength to decrease, whereas a compression ($\boldsymbol{\nabla} \cdot \mathbf{v} < 0$) makes it increase.

Using the mass continuity equation (2.19) to eliminate $\boldsymbol{\nabla} \cdot \mathbf{v}$, we then obtain

$$\frac{d}{dt}\left(\frac{\mathbf{B}}{\rho}\right) = \left(\frac{\mathbf{B}}{\rho} \cdot \boldsymbol{\nabla}\right)\mathbf{v}, \tag{2.57}$$

where $d/dt\,(= \partial/\partial t + \mathbf{v} \cdot \boldsymbol{\nabla})$ is the material derivative (Sec. 2.2.1). To see how this result leads to the conclusion that the field lines are 'frozen' to the plasma, consider an elemental segment $\delta\mathbf{l}$ along a line moving with the plasma. If $\mathbf{v}$ is the plasma velocity at one end of the element and $\mathbf{v} + \delta\mathbf{v}$ is the velocity at the other end, then the differential velocity between the two ends is $\delta\mathbf{v} = (\delta\mathbf{l} \cdot \boldsymbol{\nabla})\mathbf{v}$. During the time interval dt, the segment $\delta\mathbf{l}$ changes at the rate

$$\frac{d\delta\mathbf{l}}{dt} = \delta\mathbf{v} = (\delta\mathbf{l} \cdot \boldsymbol{\nabla})\mathbf{v}.$$

Since this has exactly the same form as Eq. (2.57) for the vector $\mathbf{B}/\rho$, it follows that, if $\delta\mathbf{l}$ and $\mathbf{B}/\rho$ are initially parallel, then they will remain parallel for all time. Therefore, any two neighbouring plasma elements initially on a field line will stay on the same field line, with the distance between them proportional to $\mathbf{B}/\rho$.

2.6.3 Non-ideal Flow

Let us suppose that Ohm's Law has the form

$$\mathbf{E} + \mathbf{v} \times \mathbf{B} = \mathbf{N}, \tag{2.58}$$

where $\mathbf{N}$ is any general non-ideal term due to for instance collisions, fluctuations, particle inertia or classical resistivity; in the latter case it is written $\mathbf{N} = \eta \boldsymbol{\nabla} \times \mathbf{B}$.

2.6.3.1 *Flux and Field Line Conservation*

It may be shown (e.g., Hornig and Schindler 1996) that flux conservation (i) and line conservation (ii) are no longer equivalent and that the field-line velocity is no longer unique. Specifically, it is shown in Section 6.11.2 that the condition

$$\mathbf{B} \times (\boldsymbol{\nabla} \times \mathbf{N}) = \mathbf{0} \tag{2.59}$$

implies field line conservation (ii). When $\mathbf{B} \neq 0$, Eq. (2.59) is equivalent to

$$\nabla \times \mathbf{N} = \lambda_L \mathbf{B},$$

where λ_L is a scalar function.

On the other hand [when $\nabla \times (\mathbf{u} \times \mathbf{B}) = \mathbf{0}$], the condition

$$\nabla \times \mathbf{N} = \mathbf{0}, \tag{2.60}$$

implies flux conservation (i) and is equivalent to

$$\mathbf{N} = \nabla G,$$

where G is a scalar. For the case with $\nabla \times (\mathbf{u} \times \mathbf{B}) \neq \mathbf{0}$, see Eq. (2.63).

Thus, the classical result (Newcomb 1958) for an ideal plasma that flux conservation (i) implies line conservation (ii) also holds in a non-ideal plasma. However, in the non-ideal case line conservation does not necessarily imply flux conservation.

2.6.3.2 *Magnetic Flux Velocity*

Although the concept of moving magnetic flux first appeared in connection with ideal plasmas, it can sometimes be extended to include non-ideal plasmas as well, although the definition is not unique. All that is required in order to define a flux velocity ($\mathbf{w}$) in a resistive plasma is that we ensure that it has the same flux-preserving property as in the ideal MHD case. That is, $\mathbf{w}$ should satisfy

$$\frac{\partial \mathbf{B}}{\partial t} = \nabla \times (\mathbf{w} \times \mathbf{B}). \tag{2.61}$$

However, for an Ohm's Law of the form (2.58), Faraday's Law (2.3) implies that

$$\frac{\partial \mathbf{B}}{\partial t} = \nabla \times (\mathbf{v} \times \mathbf{B} - \mathbf{N}). \tag{2.62}$$

Comparing Eq. (2.61) with Eq. (2.62), we see that a flux velocity exists provided the nonideal term in Eq. (2.58) can be written in the form

$$\mathbf{N} = \mathbf{u} \times \mathbf{B} + \nabla \Phi, \tag{2.63}$$

in terms of a slippage velocity $\mathbf{u} = \mathbf{v} - \mathbf{w}$ and a potential (Φ).

From Eqs. (2.58) and (2.63), we find that

$$\mathbf{E} + \mathbf{w} \times \mathbf{B} = \nabla \Phi, \tag{2.64}$$

and so, after subtracting the vector products of Eqs. (2.58) and (2.64) with $\mathbf{B}$ and assuming $(\mathbf{w} - \mathbf{v}) \cdot \mathbf{B} = 0$, we obtain a general expression for the flux velocity, namely,

$$\mathbf{w} = \mathbf{v} + \frac{(\mathbf{N} - \nabla \Phi) \times \mathbf{B}}{B^2}. \tag{2.65}$$

The case when $E_\parallel = 0$ is described in PROBLEM 2.6, and comments on the nature of Φ are given in Section 6.11.2.2.

When there is an isolated three-dimensional diffusion region, a flux velocity no longer exists (Sec. 6.12.3), but it may be replaced by a dual flux velocity (Sec. 6.12.4).

2.7 The Lorentz Force

Consider the physical effect of the magnetic force ($\mathbf{j} \times \mathbf{B}$). A key point to notice is that it is directed *across* the magnetic field, so that any motion or density variation *along* field lines must be produced by other forces, such as gravity or pressure gradients. Another point is that the Lorentz force may be decomposed into two parts with important physical consequences. Thus, after substituting for $\mathbf{j}$ from Eq. (2.15), it becomes $\mathbf{j} \times \mathbf{B} = (\nabla \times \mathbf{B}) \times \mathbf{B}/\mu$, which reduces, by means of a vector identity, to

$$\mathbf{j} \times \mathbf{B} = \frac{(\mathbf{B} \cdot \nabla)\mathbf{B}}{\mu} - \nabla\left(\frac{B^2}{2\mu}\right). \tag{2.66}$$

The first term on the right-hand side is the *magnetic tension force*, which is non-zero if $\mathbf{B}$ varies along the direction of $\mathbf{B}$ and may be regarded as being produced by the effect of a *tension* along $\mathbf{B}$ of magnitude B^2/μ per unit area. Mathematically, we may put $\mathbf{B} = B\,\hat{\mathbf{s}}$, in terms of the unit vector ($\hat{\mathbf{s}}$) along the field, so that the tension force can be decomposed into

$$\frac{B}{\mu}\frac{d}{ds}(B\,\hat{\mathbf{s}}) = \frac{B}{\mu}\frac{dB}{ds}\hat{\mathbf{s}} + \frac{B^2}{\mu}\frac{d\hat{\mathbf{s}}}{ds} = \frac{d}{ds}\left(\frac{B^2}{2\mu}\right)\hat{\mathbf{s}} + \frac{B^2}{\mu}\frac{\hat{\mathbf{n}}}{R_c}, \tag{2.67}$$

where $\hat{\mathbf{n}}$ is the *principal normal* to a magnetic field line and R_c is its *radius of curvature*. Note that we are distinguishing between a tension and a tension force and are interested only in the component of the tension force normal to the magnetic field. The tension only has a resultant effect when the magnetic field is curved – indeed, the smaller the radius of curvature, the larger the tension force becomes.

The second term in Eq. (2.66) represents a *magnetic pressure force* due to a scalar *magnetic pressure* of magnitude $B^2/(2\mu)$ per unit area, the same in all directions. Again, we are interested only in the magnetic pressure force normal to $\mathbf{B}$, since its component parallel to $\mathbf{B}$ cancels with the corresponding tension force component in Eq. (2.66), as it must, since the $\mathbf{j} \times \mathbf{B}$ force is normal to $\mathbf{B}$.

The Lorentz force therefore has two effects. It acts both to shorten magnetic field lines through the tension force and also to compress plasma through the magnetic pressure force. The tension produces a resultant force (normal to $\mathbf{B}$) when the field is curved, while the magnetic pressure gives a resultant effect when the magnitude of $\mathbf{B}$ is changing in a direction normal to $\mathbf{B}$.

A few simple examples may provide some physical insight. In constructing these, several forms for $\mathbf{B}$ that satisfy $\nabla \cdot \mathbf{B} = 0$ are considered, and, after calculating the resulting current (from $\mathbf{j} = \nabla \times \mathbf{B}/\mu$), it is shown how the Lorentz force may be interpreted in terms of magnetic pressure and tension forces (see PROBLEM 2.7).

Consider first the forces exerted by a *uniform field* $B_0\,\hat{\mathbf{y}}$ on an element of plasma shown in Figure 2.5a. Since the current density vanishes, there is no resultant Lorentz force. Equal magnetic pressures (P) act

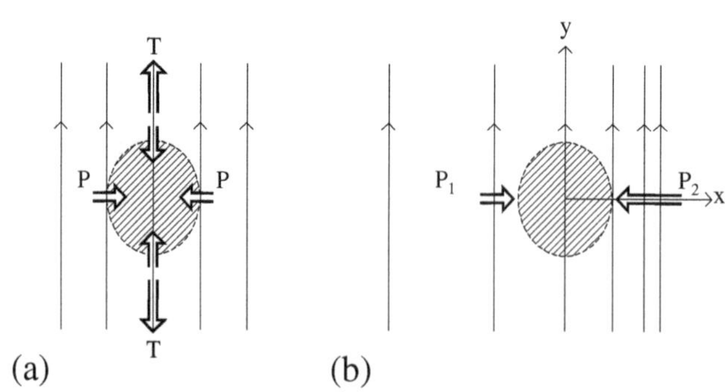

Figure 2.5. (a) In a uniform field, the magnetic pressures (P) and tensions (T) balance. (b) In a field $B(x)\hat{\mathbf{y}}$ with $dB/dx > 0$, there is an imbalance ($P_2 > P_1$).

from each direction, while equal and opposite tensions (T) act along the field lines, so that the net effect is that the plasma element remains in equilibrium; the presence of neighbouring elements means that it is not stretched out along the field.

Next, suppose there is a field

$$\mathbf{B} = B_0 \ \exp(x) \ \hat{\mathbf{y}}, \tag{2.68}$$

with corresponding current density $\mathbf{j} = (dB_y/dx) \ \hat{\mathbf{z}}/\mu = B_0 \exp(x) \ \hat{\mathbf{z}}/\mu$ and Lorentz force $\mathbf{j} \times \mathbf{B} = -B_0^2 \exp(2x) \ \hat{\mathbf{x}}/\mu$, which is directed along the negative x-axis. In the y-direction there is equilibrium, with the tension and pressure being just the same as before (Figure 2.5b). But, in the x-direction, the magnetic pressure (P_1) on the left of the element is less than the magnetic pressure (P_2) on the right, since the field strength increases from left to right. The resultant magnetic pressure force acts from right to left. As expected, the Lorentz force due to the magnetic field (Eq. 2.68) comes entirely from the magnetic pressure force $-\nabla[B^2/(2\mu)]$, since the (resultant) magnetic tension force $(\mathbf{B} \cdot \nabla)\mathbf{B}/\mu$ vanishes.

As another example, consider a (potential) field

$$\mathbf{B}_p = y \ \hat{\mathbf{x}} \ + \ x \ \hat{\mathbf{y}}, \tag{2.69}$$

for which the current density and Lorentz force vanish. The origin is a so-called X-*type magnetic neutral point*, and the field lines are shown in Figure 2.6. They are the solutions of $dy/dx = B_y/B_x = x/y$, namely, the hyperbolae $y^2 - x^2 =$ *constant*. In sketching the field lines, account has been taken of the fact that the field strength increases with distance from the origin, so that the hyperbolae must be situated successively closer to one another. Any element of plasma, such as the one shown near the x-axis, experiences a resultant magnetic tension force (T) due to the outwardly curving field lines. It acts outwards from the origin and is exactly balanced by a magnetic pressure force (P), which acts inwards because the magnetic field strength weakens as the origin is approached.

Finally, consider a (non-potential) field of the form

$$\mathbf{B}_{np} = y \ \hat{\mathbf{x}} \ + \ \alpha^2 x \ \hat{\mathbf{y}}, \tag{2.70}$$

where $\alpha^2 > 1$. The corresponding field lines are given by $y^2 - \alpha^2 x^2 = constant$ and are sketched in Figure 2.6b. Again, this represents an X-type magnetic neutral point, but the field lines ($y = \pm \alpha x$)

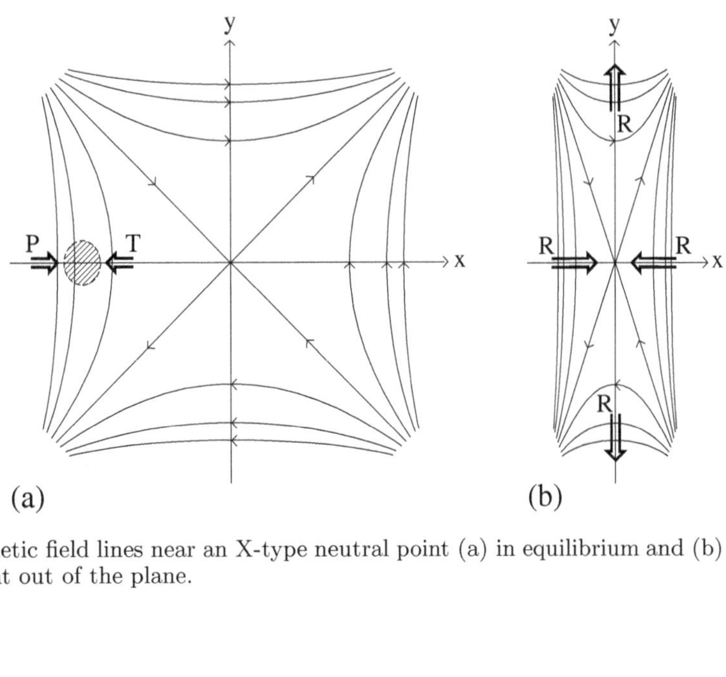

Figure 2.6. The magnetic field lines near an X-type neutral point (a) in equilibrium and (b) away from equilibrium with a uniform current out of the plane.

that pass through it are now no longer inclined at $\frac{1}{2}\pi$ to one another. On the x-axis, the field lines are more closely spaced than those in Figure 2.6a, so that the magnetic pressure is stronger; but their curvature is smaller than before, so that the magnetic tension force has not increased as much as the pressure force. The dominance of the magnetic pressure force therefore produces a resultant force (R) acting inwards. On the y-axis, the field lines have the same spacing as in Figure 2.6a, but they are more sharply curved, so that the pressure force remains the same while the tension force increases. The resultant force (R) therefore acts outwards as shown. These comments are borne out by evaluating the current density $j_{npz} = (\partial B_{npy}/\partial x - \partial B_{npx}/\partial y)/\mu = (\alpha^2 - 1)/\mu$ and the resulting Lorentz force $\mathbf{j}_{np} \times \mathbf{B}_{np} = -(\alpha^2 - 1)\,\alpha^2 x\,\hat{\mathbf{x}}/\mu + (\alpha^2 - 1)\,y\,\hat{\mathbf{y}}/\mu$.

Dungey (1953) pointed out an interesting property of these fields, namely that, when $\mathbf{B}_p$ is perturbed to $\mathbf{B}_{np}$, the magnetic force is such as to *increase* the perturbation. Thus, the equilibrium X-type point of Figure 2.6a is unstable (provided that conditions at distant boundaries allow such a displacement). Furthermore, as the instability proceeds, α increases and the limiting field lines through the origin close up, so the current density and Ohmic heating (j^2/σ) also increase. This idea has been included in a solar flare model (Syrovatsky 1966), and a similarity solution for the collapse has been presented (Sec. 6.2.2.2; Imshennik and Syrovatsky 1967).

Much qualitative information can be gained from magnetic field configurations such as Figure 2.6. As well as seeing at a glance the magnetic field direction everywhere, some impression of relative field strengths at different points may be gained by comparing the spacing of field lines. Also, the directions of the magnetic pressure gradient and tension force can be estimated from spatial variations in field strength and field-line curvature.

2.8 Some Theorems

Several theorems are proved in other sections about equilibria or steady motions which follow very simply from the equation of motion or the induction equation.

2.8.1 Alfvén's Frozen Flux Theorem

When the global magnetic Reynolds number is much larger than unity, the magnetic flux through a curve moving with the plasma is conserved (see Sec. 2.6.2.1).

2.8.2 The Minimum-Energy Theorem for Potential Fields

When the normal component of the magnetic field on a closed surface is given, the potential field in the volume enclosed by the surface has the minimum energy (see Sec. 3.3.1).

2.8.3 The Minimum-Energy Theorem for Force-Free Fields

When the magnetic flux and topological connections are given on a closed surface and the field within the surface possesses a minimum energy, then it is force-free (Sec. 3.4.1.1).

2.8.4 Woltjer's Minimum-Energy Theorem

When the normal component of the magnetic field on a closed surface and the total magnetic helicity within the surface are given, then the field with minimum energy is a linear force-free field (Sec. 3.4.1.7).

2.8.5 Cowling's Anti-dynamo Theorem

Steady plasma motions cannot maintain an axisymmetric magnetic field that is confined to a finite region of space (Sec. 8.2).

2.8.6 Taylor-Proudman Theorem

Steady, slow motions of a perfectly conducting plasma permeated by a uniform magnetic field ($\mathbf{B}_0$) *must be two-dimensional, with no variation along* $\mathbf{B}_0$.

Proof. For steady flow with $\eta = 0$, the induction equation (2.13) reduces to

$$\nabla \times (\mathbf{v} \times \mathbf{B}) = \mathbf{0}, \tag{2.71}$$

where $\nabla \cdot \mathbf{B} = 0$ and, from Eq. (2.19b),

$$\nabla \cdot (\rho \mathbf{v}) = 0. \tag{2.72}$$

Consider small departures from a uniform plasma (of density ρ_0) at rest in a uniform field ($\mathbf{B}_0$). After putting $\rho = \rho_0 + \rho'$, $\quad \mathbf{v} = \mathbf{v}'$, $\quad \mathbf{B} = \mathbf{B}_0 + \mathbf{B}'$, and linearising, Eq. (2.72) becomes

$$\nabla \cdot (\rho_0 \mathbf{v}') = 0. \tag{2.73}$$

Also, Eq. (2.71) gives $\nabla \times (\mathbf{v}' \times \mathbf{B}_0) = \mathbf{0}$, which, after using Eq. (2.73) and the uniformity of $\mathbf{B}_0$, reduces to $(\mathbf{B}_0 \cdot \nabla)\mathbf{v}' = \mathbf{0}$. This implies that $\mathbf{v}'$ has no variation in the direction of $\mathbf{B}_0$, as required. (The original, *non-magnetic* version of this theorem applies to slow, steady inviscid flow in a uniformly rotating fluid; it states that such motions must be two-dimensional, with no variation along the axis of rotation.)

2.8.7 Ferraro's Law of Isorotation

For a steady, axisymmetric flow and magnetic field, the angular speed (v_ϕ / R in cylindrical polars) is constant along field lines.

In the special case when the plasma is perfectly conducting ($\eta = 0$) and the motion is around the z-axis with $\mathbf{v} = v_\phi(R, z)\hat{\boldsymbol{\phi}}$, the proof is particularly simple. Then the axisymmetric magnetic field $\mathbf{B} = B_R(R, z)\ \hat{\mathbf{R}} + B_z(R, Z)\ \hat{\mathbf{z}}$ reduces the steady induction equation $\nabla \times (\mathbf{v} \times \mathbf{B}) = \mathbf{0}$ to

$$(\mathbf{B} \cdot \nabla)\left(\frac{v_\phi}{R}\right) = 0.$$

In other words, v_ϕ / R remains constant along magnetic field lines; this is a reasonable consequence of flux freezing, since, otherwise, the resulting differential rotation would generate a toroidal field component (B_ϕ).

2.8.8 Virial Theorem

Consider the equation of motion $\rho d\mathbf{v}/dt = -\nabla p + \mathbf{j} \times \mathbf{B} + \rho\ \mathbf{g}$, where the gravitational acceleration may be written in terms of a *gravitational potential* (Φ) such that $\mathbf{g} = -\nabla\Phi$ and $\nabla^2\Phi = 4\pi G\rho$ for a self-gravitating plasma. This equation of motion holds locally everywhere in the plasma, and, after taking the scalar product with $\mathbf{r}$, it may be integrated over the volume occupied by the plasma to yield a relationship among the global mechanical energy contributions. The result is known as the *scalar virial theorem* and takes the form

$$\frac{1}{2}\frac{d^2\mathcal{I}}{dt^2} = 2\mathcal{K} + 3(\gamma - 1)\mathcal{E} + \mathcal{M} + \mathcal{G} + \mathcal{S}, \tag{2.74}$$

where $\mathcal{I} = \int \rho r^2 dV$ is the polar moment of inertia, $\mathcal{K} = \int \frac{1}{2}\rho v^2 dV$ is the kinetic energy, $\mathcal{E} = \int p/(\gamma - 1) dV$ is the internal energy, $\mathcal{M} = \int B^2/(2\mu) dV$ is the magnetic energy, $\mathcal{G} = -\frac{1}{2} G \int\int \rho(\mathbf{r})\rho(\mathbf{r}') \mid \mathbf{r} - \mathbf{r}' \mid^{-1} dV\, dV'$ is the gravitational potential energy, $\mathcal{S} = \int_s (\mathbf{r} \cdot \mathbf{B})\mathbf{B}/\mu \cdot d\mathbf{S} - \int_s [p + B^2/(2\mu)]\mathbf{r} \cdot d\mathbf{S}$ is the surface contribution.

In Eq. (2.74) all the terms on the right are positive except for $\mathcal{G}$ (which is always negative) and $\mathcal{S}$ (which may sometimes be negative). Thus, there can be no equilibrium (i.e., $d^2\mathcal{I}/dt^2 = 0$) or deceleration ($d^2\mathcal{I}/dt^2 < 0$) without the effect of $\mathcal{G}$ and/or $\mathcal{S}$. A more general *tensor virial theorem* is proved in Chandrasekhar (1961).

2.9 Summary of Magnetic Flux Tube Behaviour

The two types of building block found in magnetic configurations are a *magnetic flux tube* and a *current sheet*. When discussing their characteristics, it is helpful to regard them simply as isolated entities, although it should be remembered that they are not really isolated at all and interact in an intimate way with the surrounding magnetic field. Their basic properties are described in detail at various points throughout the book and are summarised in this and the next sections.

Examples of a flux tube in the photosphere include a *sunspot* (Secs. 9.3 to 9.7), where a particularly large tube breaks through the solar surface, and *intense magnetic tubes* (Sec. 9.2), which are much tinier tubes that are concentrated along granule and supergranule boundaries. Examples in the corona include the field around a *solar prominence* (Chapter 11) and the myriads of coronal loops (Sec. 10.2) that pervade much of the Sun's outer atmosphere.

2.9.1 Definitions

A *magnetic field* line is such that the tangent at any point is in the direction of $\mathbf{B}$. Thus, in rectangular Cartesians, it is the solution of $dy/dx = B_y/B_x$ in two dimensions or $dx/B_x = dy/B_y = dz/B_z$ in three dimensions. For cylindrical or spherical polar coordinates, the equations of field lines are instead given by solving $dR/B_R = R\, d\phi/B_\phi = dz/B_z$ or $dr/B_r = r\, d\theta/B_\theta = r \sin\theta\, d\phi/B_\phi$, respectively.

A *magnetic flux tube* is the volume enclosed by the set of field lines that intersect a simple closed curve (Figure 2.7). However, if you are considering a particular magnetic configuration, it is entirely up to you which particular closed curve you choose to take and so which of an infinite set of possible flux tubes you select. The *strength* (F_M) of such a flux tube is defined as the magnetic flux crossing a section (S),

$$F_M = \int_S \mathbf{B} \cdot d\mathbf{S}, \tag{2.75}$$

where $d\mathbf{S}$ is taken in the same sense as $\mathbf{B}$, so that F_M is always positive.

A *magnetic flux rope* is a twisted magnetic flux tube. An *isolated flux tube* is one outside of which the magnetic field vanishes.

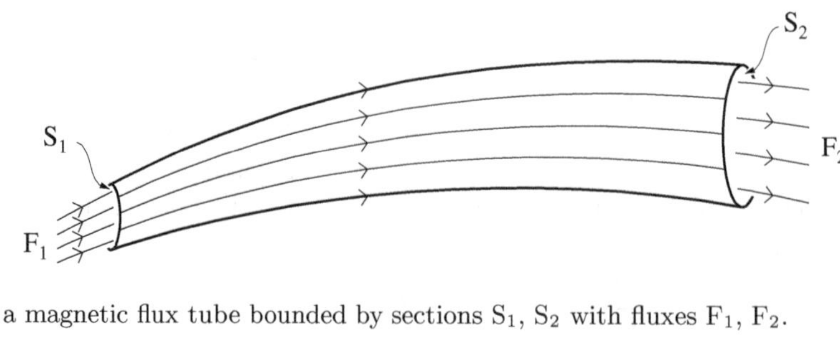

Figure 2.7. Part of a magnetic flux tube bounded by sections S_1, S_2 with fluxes F_1, F_2.

2.9.2 General Properties

(1) *The strength of a flux tube remains constant along its length.*
This is a consequence of the basic equation

$$\nabla \cdot \mathbf{B} = 0. \tag{2.76}$$

Integrate the flux [Eq. (2.75)] over the surface (S) enclosing the volume (V) between the two sections (S_1 and S_2). Then, since the contribution from the curved surface vanishes, $\int_S \mathbf{B} \cdot d\mathbf{S} = \int_{S_1} \mathbf{B} \cdot d\mathbf{S} + \int_{S_2} \mathbf{B} \cdot d\mathbf{S}$, where $d\mathbf{S}$ is along the outward normal from V. But, by the divergence theorem, $\int_S \mathbf{B} \cdot d\mathbf{S} = \int \nabla \cdot \mathbf{B} \, dV$, and this vanishes by Eq. (2.76), so that $F_{M(1)} \equiv \int_{S_1} \mathbf{B} \cdot d\mathbf{S} = -\int_{S_2} \mathbf{B} \cdot d\mathbf{S} \equiv F_{M(2)}$. In other words, the *strength* F_M [Eq. (2.75)] remains constant along the length of a flux tube.

(2) *The mean field strength of a flux tube increases when it narrows and decreases when it widens.* Equation (2.75) may be written in terms of the mean field ($\bar{B}$) across the tube as $F_M = \bar{B}A$, where A is the cross-sectional area. Thus, if the flux tube narrows as one moves along it, so A decreases and the mean field strength ($\bar{B}$) increases, and vice versa. Strong-field regions have the field lines close together, while weak-field regions have them farther apart. This is a direct consequence of Eq. (2.76).

(3) *Flux tube compression in ideal MHD increases B and ρ in the same proportion.*
Consider a cylindrical flux tube whose dimensions change from l_0 and L_0 to λl_0 and $\lambda^* L_0$, as shown in Figure 2.8. If the original uniform density and field strength are ρ_0 and B_0, and the field is frozen to the plasma, the conservation of matter gives $\rho\pi(\lambda l_0)^2(\lambda^* L_0) = \rho_0 \pi l_0^2 L_0$, so that the final density is

$$\rho = \frac{\rho_0}{\lambda^2 \lambda^*}. \tag{2.77}$$

Also, conservation of magnetic flux, namely, $B\pi(\lambda l_0)^2 = B_0 \pi l_0^2$, implies a final magnetic field strength of

$$B = \frac{B_0}{\lambda^2}. \tag{2.78}$$

Thus, if, for example, the length of the tube remains constant ($\lambda^* = 1$), we see that $B/\rho =$ *constant* [see also Eq. (2.57)]; in other words, a transverse compression ($\lambda < 1$) increases B and ρ by the same proportion, while a transverse expansion decreases them. This holds only if a constant length of plasma is being considered. For example, it would not apply to a *coronal loop* whose magnetic field strength was observed to increase, since plasma could flow in or out of the loop footpoints.

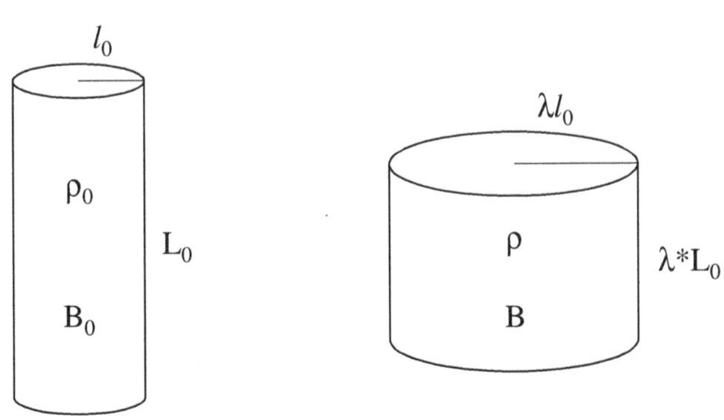

Figure 2.8. A magnetic flux tube whose field strength changes from B_0 to B while the plasma density changes from ρ_0 to ρ and the dimensions alter by fractions λ and λ^*.

(4) *An extension of a flux tube without compression increases the field strength.*

If the plasma is not compressed, so that the density is unchanged, Eq. (2.77) implies that $\lambda^2\lambda^* = 1$. Then Eq. (2.78) gives $B = \lambda^* B_0$, and so an extension of the tube ($\lambda^* > 1$) produces a rise in field strength, while a shortening causes the field to weaken. An increase in the length of a flux tube may be produced by, for example, the shearing motions that arise in subphotospheric *convection* or *differential rotation.*

(5) *For a cylindrical flux tube in magnetohydrostatic equilibrium,* the plasma pressure $[p(R)]$ and magnetic field components $[B_\phi(R), B_z(R)]$ are related by

$$0 = \frac{dp}{dR} + \frac{d}{dR}\left(\frac{B_\phi^2 + B_z^2}{2\mu}\right) + \frac{B_\phi^2}{\mu R},$$

where R is the radial distance from the axis. The second term represents the magnetic pressure force and acts outwards if the magnetic pressure $[B^2/(2\mu)]$ decreases with R. The third term gives the effect of a magnetic tension force and acts inwards. The corresponding *twist* $[\Phi(R)]$ of a field line about the axis in going from one end of a tube (of length L) to the other is given by

$$\Phi(R) = \frac{LB_\phi(R)}{RB_z(R)}. \tag{2.79}$$

The twist on the surface ($R = a$) of the flux tube is $\Phi(a) = [LB_\phi(a)]/[aB_z(a)]$, whereas the mean twist is $\bar{\Phi} = a^{-1}\int_0^a \Phi(R)dR$.

Related measures which are often met in laboratory literature are: the *pitch* $L_p \equiv 2\pi L/\Phi$, which is the length travelled along the axis that makes a field line encircle the axis once (so that $\Phi = 2\pi$); and the *reciprocal pitch*

$$\tilde{\mu} = \frac{\Phi(R)}{2\pi L}. \tag{2.80}$$

Force-free and magnetostatic flux tubes are described in Sections 3.2.3 and 3.2.4.

(6) *Cylindrical flux tubes in force-free equilibrium with free ends* have several interesting properties (Sec. 3.2.3), many of which have been demonstrated in masterly fashion by Parker (1979a).

Suppose the magnetic field $[B(a)]$, at some variable distance $R = a$, is held fixed in value. Then the mean-square value of B_z across the tube can be shown to possess the value $B^2(a)$, and so it is *invariant with respect to twisting.* Also, during the twisting up of a tube, the mean value of B_z across the tube *decreases* [and is always less than $B(a)$], since an increasing part of the total field $[B(a)]$ goes into B_ϕ.

Suppose a flux tube expands, with its axial and azimuthal magnetic flux held fixed. Then the *tube becomes increasingly more and more twisted.* (Ultimately, when the magnetic pressure of B_ϕ exceeds the magnetic tension of B_z, the tube will no longer be under tension and will *buckle*; that is, provided an instability does not set in first.)

Suppose just one section of a flux tube expands, with its axial flux and azimuthal torque held fixed. Then *coils of flux are transferred into the expanded portion,* which becomes more twisted than the remainder of the tube.

(7) *A twisted flux tube with free ends is unstable to the helical kink instability.*

According to the Kruskal-Shafranov-Tayler criterion, all helical kink perturbations whose axial wavelength ($-2\pi/k$) is so long that $-kL \leqslant \Phi$ are unstable. The effect of line-tying at the ends of a flux tube is stabilising: for example, a uniform-twist force-free tube requires a twist (Φ) larger than 2.6π before it becomes kink unstable (Sec. 7.5.3). A plasma pressure that increases away from the axis is also a stabilising influence.

(8) *The fundamental wave modes in a uniform plasma are modified by the geometry when they propagate along a flux tube.*

Torsional Alfvén waves propagate at the Alfvén speed (Sec. 4.3.1). *Magnetoacoustic tube waves* possess an amplitude that is radially dependent. A *slow (sausage) surface wave* propagates at a speed slower than the *tube speed* $c_T = c_s v_A/\sqrt{c_s^2 + v_A^2}$ (Sec. 4.9.5), which is smaller than both the sound speed (c_s) and Alfvén speed (v_A). There are also slow (kink) surface waves with speed $v_A/\sqrt{2}$, slow (kink and sausage) body waves at c_T and fast (body or surface) waves at the external sound speed. At critical radii, where there are Alfvén or cusp resonances, waves may be absorbed and so heat the plasma (Secs. 4.9, 4.10, and 10.3.2).

(9) *An isolated magnetic flux rope has vanishing axial current.*

This follows by integrating the field along a closed loop encircling the flux rope in the surrounding medium where the field vanishes. Then, by Stokes' theorem,

$$0 = \int \mathbf{B} \cdot d\mathbf{s} = \int \boldsymbol{\nabla} \times \mathbf{B} \cdot d\mathbf{S} = \int \mu \mathbf{j} \cdot d\mathbf{S}.$$

Thus, the right-hand side, which is the total current along the flux rope, vanishes. In particular, if the current is directed in one direction along the interior of the rope, then there must be an equal and opposite current along the surface of the flux rope.

2.9.3 Flux Surfaces for Two-Dimensional or Axisymmetric Fields

Section 3.1.1 describes how to represent a general three-dimensional magnetic field in terms of essentially two scalar functions, namely, either a magnetic vector potential or Euler (Clebsch) potentials or toroidal-poloidal potentials. However, for two-dimensional or axisymmetric fields, only one scalar function is needed, namely, a *flux function* F, such that $F = constant$ represents a series of flux surfaces as the constant is varied.

In order to find the relation between $\mathbf{B}$ and F, note that they need to satisfy two equations, namely,

$$\boldsymbol{\nabla} \cdot \mathbf{B} = 0 \quad \text{(a)} \qquad \text{and} \qquad \mathbf{B} \cdot \boldsymbol{\nabla} F = 0. \quad \text{(b)} \tag{2.81}$$

Equation (2.81b) expresses the fact that, if magnetic field lines lie in surfaces $F = constant$, then the vector $\mathbf{B}$ will lie in those surfaces and $\boldsymbol{\nabla} F$ will be perpendicular to them.

Thus, in two-dimensioanl Cartesians (x, y), Eq. (2.81b) becomes $B_x \partial F/\partial x + B_y \partial F/\partial y = 0$, so that both this and Eq. (2.81a) are satisfied identically if (as in Sec. 3.1.1) we write

$$\mathbf{B} = \boldsymbol{\nabla} \times [F(x, y)\hat{\mathbf{z}}] = \frac{\partial F}{\partial y}\hat{\mathbf{x}} - \frac{\partial F}{\partial x}\hat{\mathbf{y}}.$$

In two-dimensional polars (r, θ), Eq. (2.81b) becomes $B_r \partial F/\partial r + (B_\theta/r)\partial F/\partial\theta = 0$, so that, if we write

$$\mathbf{B} = \boldsymbol{\nabla} \times [F(r, \theta)\hat{\mathbf{z}}] = \frac{1}{r}\frac{\partial F}{\partial \theta}\hat{\mathbf{r}} - \frac{\partial F}{\partial r}\hat{\boldsymbol{\theta}},$$

both parts of Eq. (2.81) are satisfied identically.

In axisymmetric polars (r, θ, ϕ), we can satisfy Eq. (2.81a) by putting

$$\mathbf{B} = \boldsymbol{\nabla} \times [F(r, \theta)G(r, \theta)\hat{\boldsymbol{\phi}}] = \frac{1}{r \sin\theta}\frac{\partial}{\partial\theta}(FG\sin\theta)\,\hat{\mathbf{r}} - \frac{1}{r}\frac{\partial}{\partial r}(FGr)\,\hat{\boldsymbol{\theta}}.$$

Then Eq. (2.81b) becomes $B_r \partial F/\partial r + (B_\theta/r)\partial F/\partial\theta = 0$ or

$$\frac{1}{r\sin\theta}\frac{\partial}{\partial\theta}(FG\sin\theta)\,\frac{\partial F}{\partial r} = \frac{1}{r^2}\frac{\partial}{\partial r}(FGr)\,\frac{\partial F}{\partial\theta},$$

which is satisfied by putting $G = 1/(r\sin\theta)$, so that

$$(B_r, B_\theta) = \left(\frac{1}{r^2\sin\theta}\frac{\partial F}{\partial\theta}, -\frac{1}{r\sin\theta}\frac{\partial F}{\partial r}\right).$$

The axisymmetric case in cylindrical polars is described in PROBLEM 2.8.

2.9.4 Flux Tubes in the Solar Atmosphere

(1) *Convection* can expel magnetic flux from a convecting eddy and concentrate it to form a vertical flux tube with a field strength that exceeds the photospheric equipartition value of typically a few hundred gauss (Sec. 9.1.3).

(2) A horizontal flux tube embedded in a gravitationally stratified medium is subject to a *magnetic buoyancy* force, which tends to make it rise. It can remain in equilibrium as an arch if the feet are anchored at points that are separated by less than a few scale-heights (Sec. 9.3.1). Also, magnetic buoyancy may destabilise an equilibrium magnetic field whose strength declines too rapidly with height (Sec. 9.3.2); the resulting rate of rise is discussed in Section 9.3.3.

(3) Sunspots are cool because of the inhibition of convection (Sec. 9.1.4).

(4) A sunspot probably consists of a single large flux tube in equilibrium, whose diameter decreases monotonically with depth (Sec. 9.4). It is magnetohydrodynamically stable when its flux exceeds about 10^{11} Wb (10^{19} Mx). Slender flux tubes with less flux than this are unstable to *fluting* (Sec. 9.4.2).

(5) Intense flux tubes may be formed when a static photospheric tube of a few hundred gauss is subject to an instability which causes a downflow of plasma and a *compression* of the field to a new equilibrium at typically 1 to 2 kG (Sec. 9.2.3).

(6) A slender subphotospheric tube in thermal and hydrostatic equilibrium widens with height (Sec. 9.2.1). If it is cooler than its surroundings it becomes evacuated over a few scale-heights.

(7) The static equilibrium of a coronal loop is in general governed by $\mathbf{j}\times\mathbf{B} - \nabla p + \rho\,\mathbf{g} = \mathbf{0}$. In active regions, the magnetic field is approximately force-free, so that $\mathbf{B}$ satisfies $\mathbf{j}\times\mathbf{B} = \mathbf{0}$, while the plasma structure along each field line follows from $-\nabla p + \rho\,\mathbf{g} = \mathbf{0}$ together with an energy equation (Sec. 10.2).

(8) *Siphon flows* may be driven along photospheric and coronal loops by an imposed pressure difference, while downflows and upflows may arise in a variety of ways (Sec. 10.2.4).

2.10 Summary of Current Sheet Behaviour

In most of the Sun's atmosphere, the length (L) over which the magnetic field varies is large, typically 1 to 10 Mm, and the corresponding current density,

$$j \approx \frac{B}{\mu L}, \tag{2.82}$$

is extremely small. However, *current sheets* may exist with thicknesses (l) much smaller than L and corresponding current densities very much larger than Eq. (2.82). They are likely to be rather transient in nature, and the energy that is liberated within them may well heat the corona (Sec. 10.4). Current sheets may play a key role in the *solar flare* process (Chapter 12) and on a large scale they show up in the corona as *solar prominences* (Sec. 11.2). Also, they may be present on the surface of a flux tube such as a *sunspot* (Sec. 9.4) and on the edges of *helmet streamers* (Sec. 13.9) at the boundary between closed and open fields.

A simple current sheet may be defined as a non-propagating boundary between two plasmas, with the magnetic field tangential to the boundary. In other words, it is a *tangential discontinuity* (Sec. 5.4.2) with no flow. The tangential field components are arbitrary in magnitude and direction, subject only to the condition that the total pressure be continuous (when the sheet is in transverse equilibrium):

$$p_2 + \frac{B_2^2}{2\mu} = p_1 + \frac{B_1^2}{2\mu}, \tag{2.83}$$

where subscripts 1 and 2 denote conditions on the two sides of the current sheet. This holds even when the sheet is curved, since the magnetic tension has no effect.

In many applications such as active regions the magnetic field is so strong that the plasma pressures (p_1 and p_2) outside the current sheet may be neglected. Then Eq. (2.83) implies that the magnetic field strength is the same on both sides of the sheet, while the field directions may differ (Figure 2.9a). Suppose the x-axis is taken normal to the plane of the sheet and the y-axis bisects the angle between the two field directions. Then, as one passes through the current sheet, B_y may remain constant, while B_z reverses its direction; at the centre of the sheet, the plasma pressure is enhanced by an amount

$$p_0 = \frac{B_{z1}^2}{2\mu} = \frac{B_{z2}^2}{2\mu}.$$

The particular case when B_y vanishes gives a *neutral current sheet* (or *neutral sheet*); the magnetic field vanishes completely in the centre of such a sheet and the field reverses its direction from one side of the sheet to the other (Figure 2.9b). It is the neutral current sheet that is, for simplicity, most often analysed.

A current sheet is rather like a *shock wave* in the sense that it may be regarded as a discontinuity separating two regions where the equations of ideal MHD hold; also, its width and the details of its interior are determined by diffusive processes. However, the similarity ends there. Unlike shocks, simple current

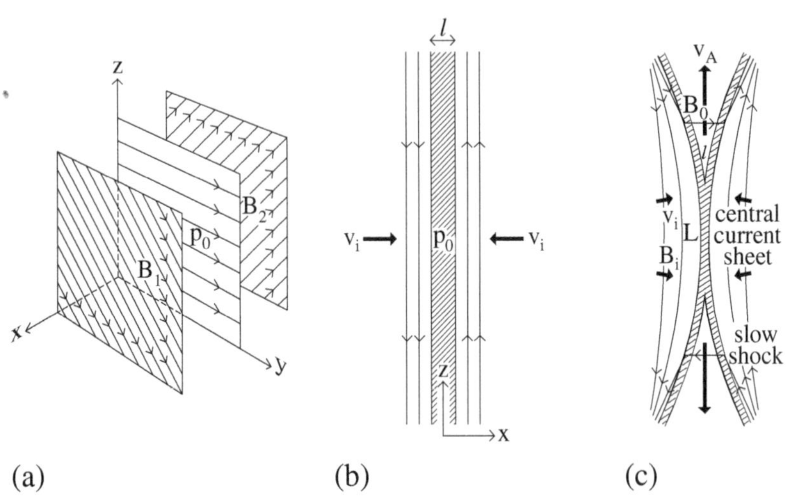

Figure 2.9. (a) A current sheet in the yz-plane across which the magnetic field rotates from $\mathbf{B}_1$ to $\mathbf{B}_2$. (b) A section across a neutral current sheet, in the centre of which the magnetic field vanishes and the plasma pressure is p_0. (c) The reconnection of magnetic field lines by their passage through a current sheet, which bifurcates into two pairs of slow shocks.

sheets do not propagate; they tend to diffuse away in time and jets of plasma are squirted from their ends at Alfvénic speeds. The ways in which such sheets may form and some elements of their behaviour are summarised in the following and detailed in other chapters.

2.10.1 Processes of Formation

There are three ways in which a current sheet may be formed (Sec. 6.3):

(1) The region near an X-type neutral point can collapse (Sec. 6.2.2);

(2) When topologically separate parts of a magnetic configuration are pushed together (Sec. 6.3.1) or sheared relative to each other (Sec. 6.3.3), a current sheet can appear at the boundary (or *separatrix surface*) between them;

(3) When a magnetohydrostatic equilibrium becomes unstable or even ceases to exist (i.e., non-equilibrium), current sheets may develop. In particular, as the photospheric footpoints of a complex coronal force-free field move, the coronal field cannot always adjust to a new stable force-free equilibrium.

2.10.2 Properties

(1) In the absence of flow, a current sheet *diffuses away* at a speed η/l, where η is the magnetic diffusivity. The magnetic field is annihilated and magnetic energy converted into heat by Ohmic dissipation (Sec. 6.4.2).

(2) The region outside a current sheet is effectively frozen to the plasma. Plasma and magnetic flux may be brought towards the sheet from the sides at speed v_i, say. If $v_i < \eta/l$ the sheet expands, whereas if $v_i > \eta/l$ it contracts. When $v_i = \eta/l$ a steady state is maintained.

(3) The enhanced plasma pressure in the centre of the sheet expels material from the ends of the sheet at the Alfvén speed (v_A) based on the external magnetic field and internal density. Magnetic flux is ejected with the material and so one effect of the sheet is to *reconnect the field lines* (Figure 2.9c). The centre of the sheet is an X-type neutral point. In a steady flow, the rate at which magnetic flux is transported remains constant (Sec. 6.5.1); in other words, the rate ($v_i B_i$) at which flux enters the sheet equals the rate ($v_A B_0$) at which it leaves, where subscripts i and 0 denote input and output values, respectively. Thus, for sub-Alfvénic inflow ($v_i < v_A$), the outflow field strength $[B_0 = (v_i/v_A)B_i]$ is smaller than the inflow field strength (B_i). An important effect of a current sheet is therefore to convert magnetic energy into heat and flow energy.

(4) Pairs of *slow-mode shock waves* propagate from the ends of the current sheet and remain as standing waves in a steady flow. The dimensions of the central current sheet are given by $l = \eta/v_i$ and $L = \rho_0 v_A/(\rho_i v_i)l$; the second equation is an expression of mass conservation. Magnetic reconnection may occur for a wide range of inflow speeds (v_i); as v_i is varied, so the dimensions of the sheet respond. There is a maximum inflow speed (or reconnection rate), which lies typically between 0.01 v_{Ae} and 0.1 v_{Ae} and is weakly dependent on the magnetic Reynolds number (Sec. 6.6), where v_{Ae} is the external inflow Alfvén speed at large distances.

(5) Consider an equilibrium current sheet, which has a unidirectional field $[B(x)\hat{\mathbf{y}}]$ reversing at $x = 0$ and a uniform total pressure $p(x) + B(x)^2/(2\mu) = constant$. Such a sheet is subject to the tearing-mode instability on a time-scale of typically the geometric mean $(\tau_d \tau_A)^{1/2}$ of the diffusion time (τ_d) and the Alfvén travel time (τ_A) (Sec. 6.8). In its nonlinear development, this can sometimes (depending on the boundary conditions) develop into a state of fast quasi-steady reconnection.

Chapter 3

Magnetohydrostatics

3.1 Introduction

A comparison of the sizes of terms in the equation of motion (2.20),

$$\rho \frac{d\mathbf{v}}{dt} = -\nabla p + \mathbf{j} \times \mathbf{B} - \rho g \,\hat{\mathbf{z}},$$

shows that the inertial terms on the left-hand side may be neglected when the flow speed is much smaller than both the sound speed $(\gamma p_0/\rho_0)^{1/2}$, Alfvén speed $B_0/(\mu\rho_0)^{1/2}$ and gravitational free-fall speed $(2gL_0)^{1/2}$ for a vertical scale-length L_0. The result is a magnetohydrostatic balance,

$$\mathbf{0} = -\nabla p + \mathbf{j} \times \mathbf{B} - \rho g \,\hat{\mathbf{z}}, \tag{3.1}$$

between a *pressure gradient*, *Lorentz force* and *gravitational force*. The object here is to investigate the implications of Eq. (3.1) together with the subsidiary equations

$$\mathbf{j} = \nabla \times \mathbf{B}/\mu, \tag{3.2}$$

$$\nabla \cdot \mathbf{B} = 0, \tag{3.3}$$

$$\rho = \frac{mp}{k_B T}, \tag{3.4}$$

and (in general) an energy equation for the temperature.

In many applications not all the terms in Eq. (3.1) are equally important. For example, the force of gravity may be neglected by comparison with the pressure gradient when the height of a structure is much less than the pressure scale-height (Sec. 3.1.2). When, in addition, the ratio ($\beta \equiv 2\mu p_0/B_0^2$) of plasma to magnetic pressure is much smaller than unity [see Eq. (2.49) for its numerical value], any pressure gradient is dominated by the Lorentz force and Eq. (3.1) reduces to

$$\mathbf{j} \times \mathbf{B} = \mathbf{0}. \tag{3.5}$$

Magnetic fields satisfying this condition are called *force-free* (or *Beltrami*). A significant Lorentz force is not allowed in this approximation because a pressure gradient or gravity would not be strong enough to balance it. The particular case when $\mathbf{j} = \mathbf{0}$ is called *current-free* or *potential*.

Magnetohydrostatics is relevant to a variety of solar structures that appear to remain motionless or move slowly for long periods of time. It has been applied, for example, to active-region loops or the global structure of the coronal magnetic field, which often appear stationary for times that are long compared with the Alfvén travel time (e.g., Sakurai 1989). This section discusses the different ways of representing magnetic fields, the notion of a scale-height and plasma structure in a prescribed magnetic field. The chapter then continues with straight flux tubes (Sec. 3.2) and general theories for current-free, force-free and magnetohydrostatic fields (Secs. 3.3 to 3.5). The properties of a slender flux tube are discussed in Section 9.2 and a summary of qualitative flux tube behaviour is given in Section 2.9.

At present, several aspects of the theory are well understood, such as potential fields, linear force-free fields, and the equilibria of simple flux tubes or simple two-dimensional fields. In future, however, it will be interesting to see how our understanding of nonlinear force-free fields develops, as well as the general theory of magnetohydrostatic fields: in particular, it is important to understand how coronal potential or force-free fields link to the lower non-force-free atmosphere and also how magnetic structures are affected by subtle effects in the energy equation.

3.1.1 Representation of the Magnetic Field

There are several ways of writing the magnetic field so that it automatically satisfies Eq. (3.3). The first is to put

$$\mathbf{B} = \boldsymbol{\nabla} \times \mathbf{A}, \tag{3.6}$$

where $\mathbf{A}$ is the *magnetic vector potential*, but it is not unique since the gradient of a scalar (a *gauge*) may be added to $\mathbf{A}$ without changing $\mathbf{B}$ (PROBLEM 3.1: for all problems and solutions, see web page www.cambridge.org/9780521854719).

In the particular case of a two-dimensional field (B_x, B_y), say, the vector potential may be written $\mathbf{A} = A\,\hat{\mathbf{z}}$, where $A(x, y)$ is the *magnetic flux function* and

$$(B_x, B_y) = \left(\frac{\partial A}{\partial y}, -\frac{\partial A}{\partial x}\right).$$

For a two-dimensional field, functions $A =$ *constant* represent field lines in two-dimensions or flux surfaces in three-dimensions. The extension to flux surfaces of axisymmetric fields is described in Section 2.9.3.

Another way, which is especially useful in a three-dimensional spherical geometry, is to adopt a toroidal-poloidal decomposition (Morse and Feshbach 1953)

$$\mathbf{B} = \boldsymbol{\nabla} \times [T\,\mathbf{a} + \boldsymbol{\nabla} \times (P\,\mathbf{a})], \tag{3.7}$$

in terms of toroidal (T) and poloidal (P) scalars, where either $\mathbf{a} = \hat{\mathbf{z}}$ or $\mathbf{a} = \mathbf{r}$.

A further representation, in terms of *Euler* (or *Clebsch*) *potentials* (f and g), is

$$\mathbf{B} = \boldsymbol{\nabla} f \times \boldsymbol{\nabla} g. \tag{3.8}$$

It follows that $\mathbf{B} \cdot \boldsymbol{\nabla} f = \mathbf{B} \cdot \boldsymbol{\nabla} g = 0$, and so f and g are constant along magnetic field lines. A pair of values (f, g) therefore labels a field line, while surfaces $f =$ *constant* and $g =$ *constant* represent magnetic flux surfaces that intersect in the field line (see PROBLEMS 3.2 and 3.3). Although this seems an elegant way of representing magnetic field lines and is always possible locally, it is not always possible globally, since the Euler potentials do not always exist as regular single-valued functions (e.g., in a torus or near null points). Furthermore, the magnetohydrostatic equations become highly complex and nonlinear when written in terms of f and g.

The final representation when the field is current-free (Sec. 3.3) is to write

$$\mathbf{B} = \boldsymbol{\nabla} \Psi, \tag{3.9}$$

in terms of the *magnetic* (*scalar*) *potential* (Ψ). (Sometimes this definition is used with a minus sign.)

3.1.2 Pressure Scale-Height

If we consider the effect of stratification above a locally plane part of the solar surface, such that gravity acts along the negative z-axis and s measures the distance along magnetic field lines inclined at

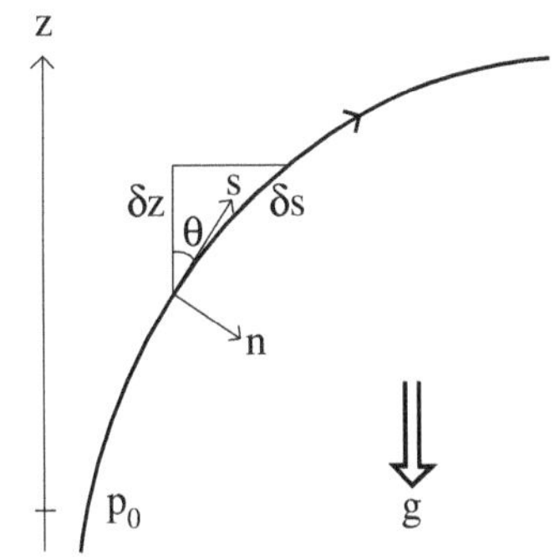

Figure 3.1. A magnetic field line inclined at θ to the vertical z-axis. Distance s is measured along the field line and p_0 is the base pressure at the reference height $z = 0$.

an angle θ to the vertical, the component of Eq. (3.1) parallel to $\mathbf{B}$ is

$$0 = -\frac{\partial p}{\partial s} - \rho g \cos\theta,$$

with no effect from the Lorentz force. Since $\delta s \cos\theta = \delta z$ (Figure 3.1), this becomes

$$0 = -\frac{\partial p}{\partial z} - \rho g, \tag{3.10}$$

where p and ρ are functions of z along a particular field line. After substituting for ρ from Eq. (3.4) into Eq. (3.10) and integrating, we find

$$p = p_0 \exp\left(-\int_0^z \frac{dz}{H(z)}\right), \tag{3.11}$$

where p_0 is the base pressure (at $z = 0$), which [like $H(z)$] may vary from one field line to another; also,

$$H(z) = \frac{k_B T(z)}{mg} = \frac{p}{\rho g} = \frac{\tilde{R} T(z)}{\tilde{\mu} g} \tag{3.12}$$

is the (*pressure*) *scale-height*, which represents the vertical distance over which the pressure falls by a factor e. In terms of density, Eq. (3.11) becomes

$$\frac{\rho}{\rho_0} = \frac{T_0}{T(z)} \exp\left(-\int_0^z \frac{dz}{H(z)}\right). \tag{3.13}$$

Equation (3.11) shows that the pressure along a given magnetic field line decreases exponentially with height. The rate of decrease depends on the temperature structure as determined by the energy equation. It is here that the magnetic field enters implicitly, since the length of the field line depends on the magnetic structure and may influence both the conductive and heating terms in the energy balance. The corresponding density variation follows from Eq. (3.13). When the temperature increases with height, the density decreases faster than the pressure; but, when the temperature falls with height, the density may either increase or decrease locally, depending on whether the factor T^{-1} or the exponential term dominates in Eq. (3.13).

For the particular case of uniform temperature along a field line (due to, e.g., the dominance of thermal conduction), H is constant and Eq. (3.11) reduces to

$$p = p_0\, e^{-z/H}. \tag{3.14}$$

More generally, we may include the effect of the curvature of the solar surface by remembering that the gravitational acceleration (g) varies with distance (r) from the solar centre like $g_\odot r_\odot^2/r^2$, where $r_\odot$ is the solar radius and $g_\odot$ is the value of g at the photosphere. Thus, assuming $\tilde{\mu} = 0.6$, the scale-height (3.12) can be written numerically in terms of the temperature (T) as

$$H = 50T\ (r/r_\odot)^2 \text{ m.} \tag{3.15}$$

For example, temperatures of 10^4 K and 10^6 K give scale-heights of 500 km (for $r \approx r_\odot$) and 50 $(r/r_\odot)^2$ Mm, respectively.

3.1.3 Plasma Structure in a Given Magnetic Field

When $\beta \ll 1$, the magnetic structure is determined by the appropriate force-free solution to Eq. (3.5), as described in Section 3.4. The plasma structure follows in general by solving along each field line the time-dependent equations of motion, mass continuity and energy for the plasma velocity, density and temperature. When the plasma is static, p is given by Eq. (3.11), together with an appropriate energy equation for T. For example, a *coronal arcade* is modelled in Section 3.4.2.

Consider the case when g is uniform and β is no longer small, but the magnetic field is either vertical or horizontal. A *purely vertical field* (independent of y) of the form $\mathbf{B} = B(x)\ \hat{\mathbf{z}}$ gives Eq. (3.10) as the z-component of Eq. (3.1). This has solution (3.11), with the base pressure $[p_0(x)]$ and temperature varying from one field line to another. The horizontal x-component is

$$0 = -\frac{\partial}{\partial x}\left(p + \frac{B^2}{2\mu}\right),$$

which has the solution

$$p + \frac{B^2}{2\mu} = f(z), \tag{3.16}$$

say, where $f(z)$ describes the vertical variation of total pressure at any value of x. Then the z-derivative of Eq. (3.16) gives $\partial p/\partial z = df/dz$, which is independent of x. This in turn shows from Eq. (3.10) that the density is unaffected by the presence of the magnetic field. In addition, Eq. (3.16) implies that, for a given z, the pressure (and therefore temperature) is strongest where the magnetic field is weakest (Figure 3.2).

When the magnetic field is *purely horizontal* (independent of y) and of the form $\mathbf{B} = B(z)\hat{\mathbf{x}}$, the horizontal component of Eq. (3.1) implies that the pressure is a function of z alone, while its vertical component gives

$$0 = -\frac{\partial}{\partial z}\left(p + \frac{B^2}{2\mu}\right) - \rho g.$$

Since the field lines are straight, the only contribution from the Lorentz force is the magnetic pressure force. In general, ρ and T can vary along the field in such a way as to keep the pressure at that level constant. If $B(z)$ and T are prescribed, the above equation with $\rho = pm/(k_B T)$ can be integrated to give p. In particular, when T is uniform we find

$$p = \left[p_0 - \int_0^z e^{z/H}\ \frac{d}{dz}\left(\frac{B^2}{2\mu}\right)\ dz\right]\ e^{-z/H},$$

which shows how a magnetic field that decreases (or increases) with height raises (or lowers) the pressure over its hydrostatic value. When both the sound speed and Alfvén speed are uniform, the equilibrium solution is

$$p = p_0\ e^{-z/H_B}, \qquad \rho = \rho_0\ e^{-z/H_B}, \qquad B = B_0\ e^{-z/(2H_B)},$$

with a modified scale-height $H_B = [p_0 + B_0^2/(2\mu)]/(\rho_0 g)$. Magnetic buoyancy instability in such a configuration is discussed in Section 9.3.2.

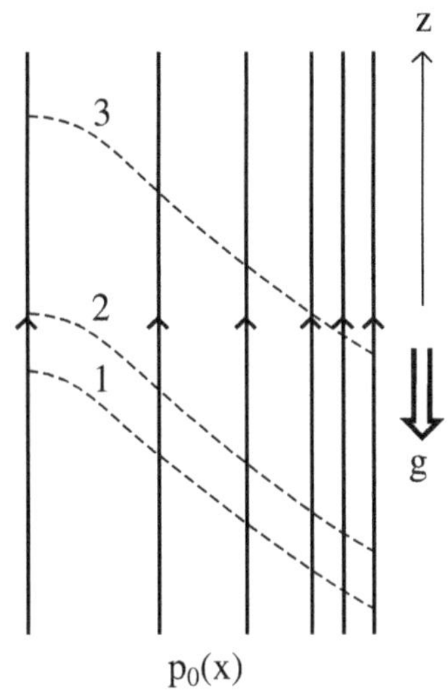

Figure 3.2. For plasma situated in a vertical magnetic field, the isodensity contours are horizontal, while the isobars (constant-pressure contours, dashed) are inclined as shown, with labels 1, 2, 3 indicating successively lower pressure values.

3.2 Structure of Magnetic Flux Tubes (Cylindrically Symmetric)

Consider a cylindrically symmetric flux tube whose magnetic field components

$$[0,\ B_\phi(R),\ B_z(R)] \tag{3.17}$$

in cylindrical polar coordinates are functions of R alone. Its field lines are helical and lie on cylindrical surfaces, as indicated in Figure 3.3, as do the electric current components, which follow from Eq. (3.2) as

$$[0,\ j_\phi(R),\ j_z(R)] = \left(0,\ -\frac{1}{\mu}\frac{dB_z}{dR},\ \frac{1}{\mu R}\frac{d}{dR}(RB_\phi)\right). \tag{3.18}$$

For a structure whose height is much lower than a scale-height, gravity is unimportant and force-balance reduces to

$$\frac{dp}{dR} + \frac{d}{dR}\left(\frac{B_\phi^2 + B_z^2}{2\mu}\right) + \frac{B_\phi^2}{\mu R} = 0. \tag{3.19}$$

The second term represents the magnetic pressure force and the third term the magnetic tension force due to the azimuthal component (B_ϕ) that encircles the axis. On each cylindrical surface, the field lines have a constant inclination, but this may vary from one radius to another. The field lines are given by $R\,d\phi/dz = B_\phi/B_z$, and the amount by which a given line is twisted in going from one end of the tube (of length $2L$) to the other is $\Phi = \int d\phi = \int_0^{2L} B_\phi/(RB_z)dz$, or

$$\Phi(R) = \frac{2LB_\phi(R)}{RB_z(R)}. \tag{3.20}$$

($2\pi L/\Phi$ is sometimes called the *pitch* of the field and gives the axial length of a field line that encircles the axis once.)

Equation (3.19) contains three dependent functions $p(R)$, $B_\phi(R)$, $B_z(R)$, so that any two can be prescribed and the third deduced. The most natural choice for many applications is to impose the longitudinal component (B_z) and the twist Φ (or equivalently B_ϕ) and deduce p. But sometimes one may want to fix B_z and p from observations and determine B_ϕ. Several special cases arise as follows.

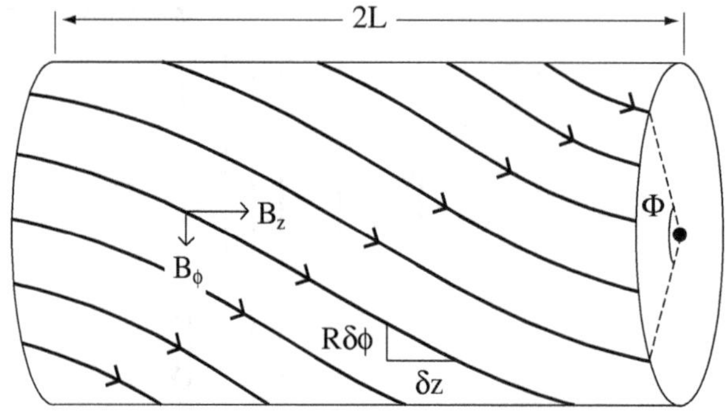

Figure 3.3. The notation for a cylindrically symmetric flux tube of length $2L$.

3.2.1 Purely Axial Field

When no azimuthal component (B_ϕ) is present, Eq. (3.19) reduces to

$$\frac{d}{dR}\left(p + \frac{B^2}{2\mu}\right) = 0,$$

with solution $p + B^2/(2\mu) = constant = P_T$, say, so that the *total pressure* (P_T, plasma plus magnetic) is conserved.

3.2.2 Purely Azimuthal Field

When the axial component (B_z) vanishes, Eq. (3.19) becomes

$$\frac{dp}{dR} + \frac{d}{dR}\left(\frac{B_\phi^2}{2\mu}\right) + \frac{B_\phi^2}{\mu R} = 0, \tag{3.21}$$

where, according to Eq. (3.2), B_ϕ is related to the current by

$$j_z = \frac{1}{\mu R}\frac{d}{dR}(RB_\phi). \tag{3.22}$$

If, in particular, the current flows with uniform total value I within a cylinder of radius a, an integration of Eq. (3.22) yields

$$B_\phi = \begin{cases} (\mu I R)/(2\pi a^2), & R < a, \\ (\mu I)/(2\pi R), & R > a, \end{cases}$$

assuming B_ϕ to be finite and continuous (the magnetic field lines are shown in Figure 3.4). The corresponding plasma pressure results from integrating Eq. (3.21). Assuming that it takes the value p_∞ outside the current column, we find

$$p = \begin{cases} p_\infty + \frac{1}{4}\mu[I/(\pi a^2)]^2(a^2 - R^2), & R < a, \\ p_\infty, & R > a. \end{cases}$$

Within the cylinder of radius a, B_ϕ increases linearly with R, while the plasma pressure decreases, so that the outward plasma pressure force is balanced by inward magnetic pressure and tension forces. Outside the cylinder, the plasma pressure is uniform and the magnetic field is potential, so that the outward magnetic pressure and inward tension forces balance one another.

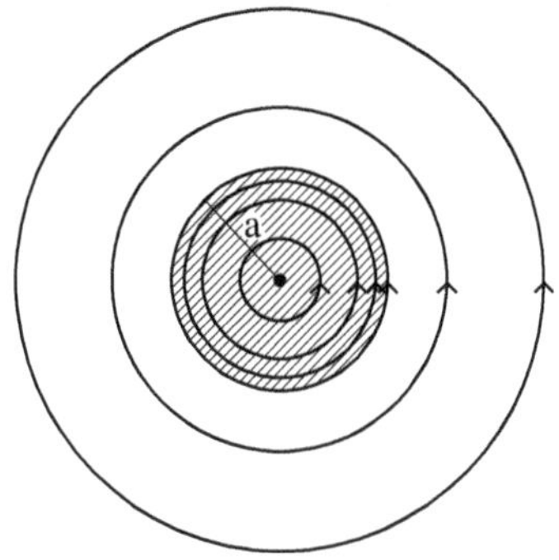

Figure 3.4. Purely azimuthal magnetic field lines in a section across a column of uniform current and radius a.

In the laboratory, a plasma configuration with an axial current and azimuthal magnetic field is known as a *linear pinch.* A simple relation may be derived, as follows, in this case between the current ($I \equiv \int_0^{R_0} j_z \; 2\pi R \; dR$) flowing through a plasma column (of radius R_0) and the number ($N \equiv \int_0^{R_0} n \; 2\pi R \; dR$) of particles per unit length of the column. Equation (3.21) may be multiplied by R^2 and integrated to give

$$\int_0^{R_0} R^2 dp = -\int_0^{R_0} (RB_\phi/\mu) \; d(RB_\phi).$$

Then, assuming that the plasma pressure vanishes at R_0 and the temperature $[T = p/(nk_B)]$ is uniform across the column, an integration by parts of the left-hand side together with the use of Eq. (3.22) on the right-hand side yields *Bennett's relation*:

$$I^2 = (8\pi/\mu) \; k_B T N.$$

3.2.3 One-Dimensional Force-Free Flux Tubes

3.2.3.1 Linear (Constant-α) Fields

In the absence of plasma pressure, Eq. (3.19) reduces to

$$\frac{d}{dR}\left(\frac{B_\phi^2 + B_z^2}{2\mu}\right) + \frac{B_\phi^2}{\mu R} = 0. \tag{3.23}$$

Here, either $B_\phi(R)$ or $B_z(R)$ may be prescribed and the other deduced. For a so-called '*constant-α*' *field* (Sec. 3.4), the ϕ-component of $\nabla \times \mathbf{B} = \alpha \mathbf{B}$ (where α is assumed uniform) becomes

$$-\frac{dB_z}{dR} = \alpha B_\phi.$$

Then, an elimination of B_ϕ between the above two equations yields Bessel's equation, whose solution subject to $B_z = B_0$ and $dB_z/dR = 0$ at $R = 0$ is

$$B_\phi = B_0 J_1(\alpha R), \qquad B_z = B_0 J_0(\alpha R), \tag{3.24}$$

where J_0, J_1 are Bessel functions. This solution is due to Lundquist (1951), but it has the rather undesirable feature (for solar applications) that the axial component possesses field reversals at the zeros of $J_0(\alpha R)$.

3.2.3.2 Nonlinear (Non-Constant-α) Fields

An easy way to generate solutions to Eq. (3.23) is to choose

$$B^2 = f(R), \tag{3.25}$$

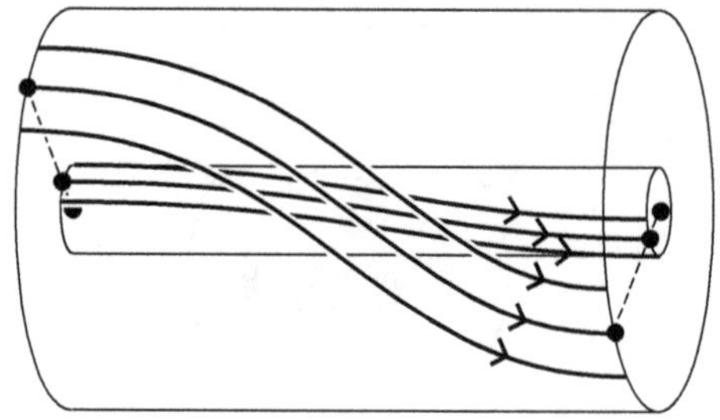

Figure 3.5. Magnetic field lines at two radii for the uniform-twist field.

and then, following Lüst and Schlüter (1954), it gives

$$B_\phi^2 = -\tfrac{1}{2}R\frac{df}{dR} \qquad \text{and} \qquad B_z^2 = B^2 - B_\phi^2. \tag{3.26}$$

The restrictions that B_ϕ^2 and B_z^2 be positive imply that df/dR is negative and that f approaches zero slower than R^{-2} as R tends to ∞. The limiting case $f = R^2$ gives the purely azimuthal field $R^{-1}\hat{\phi}$.

Another simple example of a force-free field is the *'uniform-twist force-free' field*, for which the twist Φ [given by Eq. (3.20)] is constant and the field components are

$$B_\phi = \frac{B_0\Phi R/(2L)}{1+\Phi^2R^2/(2L)^2}, \qquad B_z = \frac{B_0}{1+\Phi^2R^2/(2L)^2}, \tag{3.27}$$

$$\text{while} \qquad \alpha = \frac{\Phi/L}{1+\Phi^2R^2/(2L)^2}.$$

This has the property that field lines at different radii are twisted through the same angle, as shown in Figure 3.5, so that the whole tube is twisted like a rigid body.

3.2.3.3 Effect of Twisting a Tube

Parker (1977) explored the effect of twisting a force-free flux tube of finite radius (a), confined by a fixed plasma pressure $[B^2(a)/(2\mu)]$. While the tube is being twisted up, the field is assumed to remain cylindrically symmetric, but the radius (a) is allowed to vary from an initial value $[a^{(0)}]$. Then the mean-square axial field $< B_z^2 >$ is unaffected by twisting, since

$$\begin{aligned} < B_z^2 > &\equiv \frac{2}{a^2}\int_0^a RB_z^2\,dR, \\ &= \frac{2}{a^2}\int_0^a Rf + \tfrac{1}{2}R^2\frac{df}{dR}\,dR \qquad \text{from Eq. (3.26)}, \\ &= \frac{1}{a^2}\int_0^a \frac{d}{dR}(R^2f)\,dR = f(a), \end{aligned}$$

or, by Eq. (3.25), $< B_z^2 > = B^2(a)$.

In particular, Parker considered the uniform-twist field (3.27) with the value of B_z on the axis $R = 0$ set equal to $B_0 = B^{(0)}(1+\Phi^2a^2/(2L)^2)^{1/2}$. This varies with twist (Φ) in such a way as to keep the total field equal to a constant ($B^{(0)}$) at the edge of the tube, where $B^{(0)}$ is the initial uniform axial field strength when the tube is untwisted. The variation of the tube radius with twist is determined by the condition

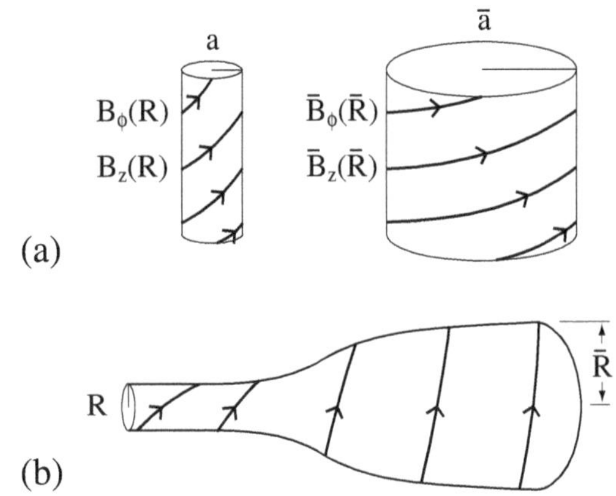

Figure 3.6. (a) The radial expansion of a twisted flux tube from radius a to radius $\bar{a}$. (b) The concentration of azimuthal flux in the widest part of a flux tube.

that the longitudinal magnetic flux through the tube be conserved, namely,

$$2\pi \int_0^a RB_z \, dR = \pi a^{(0)2} B^{(0)},$$

or

$$\left(1 + \frac{\Phi^2 a^2}{(2L)^2}\right) \log_e \left(1 + \frac{\Phi^2 a^2}{(2L)^2}\right) = \frac{\Phi^2 a^{(0)2}}{(2L)^2}.$$

It transpires from this equation that, as Φ increases, so B_ϕ and a increase. At the same time $B_z(0)$ increases (since an outward magnetic pressure is required to balance the inward tension force produced by twisting), while $B_z(a)$ decreases (since an increasing proportion of $B(a)$ goes into the azimuthal component).

3.2.3.4 *Effect of Expanding a Tube*

Parker (1974) showed how the radial expansion of a tube makes B_ϕ/B_z increase, since the twist $[2LB_\phi/(RB_z)]$ remains constant while the mean value of R increases. To demonstrate the effect, suppose the confining pressure decreases on the surface of a cylindrical tube with radius a and field $[B_\phi(R), B_z(R)]$, and so causes it to expand to a radius $\bar{a}$, in such a way that the radius R becomes $\bar{R}$ and the field becomes $[\bar{B}_\phi(\bar{R}), \bar{B}_z(\bar{R})]$, as indicated in Figure 3.6a.

Suppose the plasma pressure is negligible (or uniform), with the initial and final fields being given by generating functions f and $\bar{f}$ according to Eq. (3.25). Then conservation of longitudinal and azimuthal flux initially through the annulus $(R, R+dR)$ and finally through $(\bar{R}, \bar{R}+d\bar{R})$ gives $B_z(R)R\, dR = \bar{B}_z(\bar{R})\bar{R}\, d\bar{R}$ and $B_\phi(R)\, dR = \bar{B}_\phi(\bar{R})\, d\bar{R}$. After using Eqs. (3.25) and (3.26) to rewrite these in terms of f and $\bar{f}$, $\bar{f}$ may be eliminated to yield an equation for the mapping $R(\bar{R})$, which may be solved once the initial generating function (f) is prescribed.

For example, with $a^2 = \frac{1}{2}$ and $f(R) = 1 - R^2$ for $0 \leqslant R^2 \leqslant \frac{1}{2}$, so that $B_z^2 = 1 - 2R^2$ and $B_\phi^2 = R^2$, Parker found for a large expansion ($\bar{a} \gg \frac{1}{2}$) that

$$R^2 = \frac{\log_e(1+\bar{R}^2)}{4\ \log_e \bar{a}} \qquad \text{and} \qquad \frac{\bar{B}_\phi^2}{\bar{R}^2} = \bar{B}_z^2 = \frac{1}{16(1+\bar{R}^2)^2\ \log_e^2 \bar{a}}.$$

Thus, the initial field has a B_ϕ that increases with R and a B_z that is uniform near the axis and falls to zero at the surface ($R = a$), while the final field is mainly azimuthal ($\bar{B}_\phi \gg \bar{B}_z$) over most of the radius ($1 < \bar{R} < \bar{a}$). A similar increase in B_ϕ/B_z occurs if the flux tube is compressed in the axial direction, while the opposite effect may be produced by compressing the tube radially or stretching it axially. It may be noted that three types of field are invariant with respect to expansion, namely, a purely axial field, a purely azimuthal field and a uniform-twist field [Eq. (3.27)].

3.2.3.5 A Tube of Non-Uniform Radius

Parker (1974) also considered a flux tube whose radius increases with axial distance (z) due to a fall in the confining pressure (Figure 3.6b). He supposed that the radius (a) is uniform for $z < -h$ with field $[B_\phi(R),\ B_z(R)]$ and generating function f, and that the radius ($\bar{a}$) is uniform for $z > h$ with field $[\bar{B}_\phi(\bar{R}), \bar{B}_z(\bar{R})]$ and generating function $\bar{f}$. Then again the problem is to determine the relationship between the two fields. Conservation of longitudinal flux gives $B_z(R)R\, dR = \bar{B}_z(\bar{R})\bar{R}\, d\bar{R}$, as before, but conservation of azimuthal flux is no longer appropriate, since it may be concentrated at certain locations along the tube. However, the azimuthal Maxwell stress is $B_\phi B_z/\mu$, and so conservation of its torque gives

$$R(B_\phi B_z)\ RdR = \bar{R}(\bar{B}_\phi \bar{B}_z)\ \bar{R}d\bar{R}.$$

The field components may be written in terms of f and $\bar{f}$, and then $\bar{f}$ eliminated between these two equations. The result for an initial generating function $f = 1 - R^2$ $(0 \leqslant R^2 \leqslant \frac{1}{2})$ and tube radius $a = 1/\sqrt{2}$ is a mapping $R(\bar{R})$ that is singular at the tube surface where B_z vanishes. If the tube is expanded greatly, so that the new radius is $\bar{a} \gg 1/\sqrt{2}$, the resulting field components at large z are

$$B_z^2 = \frac{J_0^2(2\bar{R})}{2\bar{a}^2 J_1^2(2\bar{a})} \qquad \text{and} \qquad \bar{B}_\phi^2 = \frac{J_1^2(2\bar{R})}{2\bar{a}^2 J_1^2(2\bar{a})},$$

which are valid only as far out as $\bar{R} \approx 1.2$ where B_z^2 vanishes. Beyond this radius the mapping $\bar{R} = \bar{R}(R)$ is no longer single-valued, and so Parker suggested that the field becomes purely azimuthal. Browning and Priest (1982) have included B_R in the calculation, in which case the surface becomes cusp-like in shape; when the internal gas pressure exceeds the external pressure, it is then possible that the flux tube bursts, with its outer layers being stripped off.

3.2.4 One-Dimensional Magnetostatic Flux Tubes

One simple solution to the magnetostatic equation (3.19) has field components

$$B_z = \frac{B_0}{1 + R^2/a^2}, \qquad B_\phi = \frac{\Phi_0 R B_0}{2L(1 + R^2/a^2)}. \tag{3.28}$$

The axial component (B_z) decreases from a maximum on the axis over a length-scale a, while the azimuthal component (B_ϕ) is chosen to make the twist ($\Phi = \Phi_0$) uniform. The plasma pressure for this *uniform-twist magnetostatic* field is

$$p(R) = p_\infty + \frac{[\Phi_0^2 a^2/(2L)^2 - 1]B_0^2}{(1 + R^2/a^2)2\mu}.$$

For an untwisted tube, this has a minimum on the axis, but, when Φ_0 exceeds $2L/a$, it turns into a maximum. As the twist increases, so $p(0)$ increases in value.

Another simple solution occurs with a uniform axial field ($B_z = B_0$) and a twist,

$$\Phi(R) \equiv \frac{2LB_\phi}{RB_0} = \frac{\Phi_0}{1 + R^2/a^2},$$

that decreases from a maximum Φ_0 on the axis. This *variable-twist field* has a pressure [from Eq. (3.19)]

$$p(R) = p_\infty + \frac{\Phi(R)^2 B_0^2 a^2}{8\mu L^2},$$

whose maximum is located on the axis and increases in value with increasing Φ_0. In future, there is a need to study in detail the effect of flux tube curvature by considering solutions in toroidal geometry (Sec. 12.2.2) with a varying cross-section.

3.3 Current-Free (or Potential) Magnetic Fields

When the current density vanishes everywhere, Eq. (3.2) implies that $\nabla \times \mathbf{B} = \mathbf{0}$, and the curl of this, together with Eq. (3.3), yields the vector Laplace equation

$$\nabla^2 \mathbf{B} = \mathbf{0}.$$

Thus, the magnetic field is *potential* and the whole body of potential theory may be applied. It is often convenient to write

$$\mathbf{B} = \nabla \Psi, \tag{3.9}$$

where Ψ is the *scalar magnetic potential* satisfying the scalar Laplace equation

$$\nabla^2 \Psi = 0. \tag{3.29}$$

The beauty of this is that, once Ψ is found, the field components follow immediately from Eq. (3.9), and both $\nabla \times \mathbf{B} = \mathbf{0}$ and $\nabla \cdot \mathbf{B} = 0$ are automatically satisfied.

Many computer codes now exist as a standard part of a solar theorist's armoury to construct a potential field model for the corona using the measured line-of-sight photospheric magnetic field as a boundary condition. The usefulness of such a model for the corona should not be underestimated. It is often an extremely good approximation which can be used to estimate the shapes of coronal loops, the strength of the coronal magnetic field or the global topology of the field. Moreover, with modern computer power it can be obtained very easily. Thus, a potential field is often a useful first approximation, since finding a realistic non-potential field can be extremely difficult and time-consuming.

3.3.1 Potential Minimum-Energy Theorem

Many results of potential theory are of interest to solar physics. For instance, the solution within a closed volume is unique when the value of Ψ or the normal component of the magnetic field ($\partial\Psi/\partial n \equiv B_n$) on the boundary is imposed. In the latter case, the corresponding potential field contains the smallest possible amount of magnetic energy $W = \int B^2/(2\mu)\, dV$. Thus, magnetic fields with non-zero currents but the same B_n on the boundary must contain more energy than the potential field. This result is true for a semi-infinite region such as the solar atmosphere above a (plane or spherical) photosphere when the normal field at the photosphere is given, provided there are no sources at infinity, so that the magnetic field at large distances R falls off faster than R^{-2}. It also holds if, on part of the boundary, B_n is not prescribed but instead the tangential field component vanishes. The proof of the basic *Potential Minimum-Energy Theorem* is as follows.

Suppose the potential magnetic field is $\mathbf{B}_0$ and write any other magnetic field with the same normal component on the boundary (S) as $\mathbf{B} = \mathbf{B}_0 + \mathbf{B}_1$, where $\mathbf{B}_1$ is not necessarily small and the normal components of $\mathbf{B}$ and $\mathbf{B}_0$ are the same on S, so that the normal component (B_{1n}) of $\mathbf{B}_1$ vanishes there. Then the magnetic energy of any magnetic field is

$$W = \int \frac{B^2}{2\mu} dV = \int \frac{(\mathbf{B}_0 + \mathbf{B}_1) \cdot (\mathbf{B}_0 + \mathbf{B}_1)}{2\mu} dV = \int \frac{B_0^2 + 2\mathbf{B}_0 \cdot \mathbf{B}_1 + B_1^2}{2\mu} dV,$$

where, since $\mathbf{B}_0$ is potential, it may be written $\mathbf{B}_0 = \nabla \Psi_0$, and the middle term is

$$\int \mathbf{B}_0 \cdot \mathbf{B}_1 dV = \int (\nabla \Psi_0) \cdot \mathbf{B}_1 dV = \int \nabla \cdot (\Psi_0 \mathbf{B}_1) - \Psi_0 (\nabla \cdot \mathbf{B}_1) dV.$$

However, $\nabla \cdot \mathbf{B}_1 = 0$ and so applying the divergence theorem gives

$$\int \mathbf{B}_0 \cdot \mathbf{B}_1 dV = \int \hat{\mathbf{n}} \cdot (\Psi_0 \mathbf{B}_1) dS = \int \Psi_0 B_{1n} dS,$$

which vanishes since $B_{1n} = 0$ on S. Thus, the magnetic energy reduces to

$$W = \int \frac{B^2}{2\mu} dV = \int \frac{B_0^2 + B_1^2}{2\mu} dV,$$

which in general exceeds the energy of the potential field, as required, since $B_1^2 > 0$.

3.3.2 General Current-Free Solutions

Standard solutions of Eq. (3.29) may be derived by using the method of separation of variables. Thus, for example, in Cartesians the solution above an xy-plane that decays to zero as z tends to infinity and is periodic in x and y has the form

$$\Psi = a \ \exp\,(ik_x x + ik_y y - kz), \tag{3.30}$$

where k_x, k_y, $k > 0$ are real and $k^2 = k_x^2 + k_y^2$. If the normal magnetic field also vanishes on side planes at $x = 0,\ y = 0,\ x = a,$ and $y = b$, the general solution is

$$\Psi = \sum_{n=0}^{\infty} \sum_{m=0}^{\infty} a_{nm} \ \sin\left(\frac{2n\pi x}{a}\right) \ \sin\left(\frac{2m\pi y}{b}\right) e^{-k_{nm} z}, \tag{3.31}$$

where $k_{nm}^2 = (2n\pi/a)^2 + (2m\pi/b)^2$ and the coefficients a_{nm} can be found in terms of a prescribed field B_z on the base $z = 0$. If the field instead decays to a constant value $(B_0\hat{\mathbf{z}})$ at large z, the above solution is just modified by adding $B_0 z$.

Again, in *spherical polar coordinates* (r,θ,ϕ) the general solution is

$$\Psi = \sum_{l=0}^{\infty} \sum_{m=-l}^{l} [a_{lm} r^l + b_{lm} r^{-(l+1)}] \ P_l^m(\cos\ \theta)\ e^{im\phi}, \tag{3.32}$$

in terms of associated Legendre polynomials (P_l^m); the special case that is independent of ϕ is

$$\Psi = \sum_{l=0}^{\infty} [a_l r^l + b_l r^{-(l+1)}] \ P_l(\cos\ \theta),$$

where P_l is the Legendre polynomial.

In *cylindrical polars* (R,ϕ,z), the general solution may be written

$$\Psi = \sum_{n=-\infty}^{\infty} [c_n J_n(kR) + d_n Y_n(kR)] \ e^{in\phi \pm kz}, \tag{3.33}$$

where J_n and Y_n are Bessel functions, or, when there is no z-dependence,

$$\Psi = C \log_e R + \sum_{n=-\infty}^{\infty} [C_n R^n + D_n R^{-n}] \ e^{in\phi}.$$

In each of the above cases, the constants can be determined by applying appropriate boundary conditions (e.g., from line-of-sight magnetograms).

3.3.3 Coronal Magnetic Fields – Roughly Potential?

The actual solar magnetic field above the coronal base is expected to be potential when three conditions hold: the magnetic force dominates; the region contains no magnetic helicity; and the photospheric sources have remained stationary long enough that the field has reduced to its minimum energy configuration. Furthermore, a region that evolves with no magnetic helicity at a speed much slower than the Alfvén speed will pass through a series of potential states [unless there is an extra constraint such as the topological degree (Sec. 10.4.4)].

Locations in the solar atmosphere, however, where potential fields may be a poor approximation include the following:

(i) *active regions* that are highly sheared or twisted, for which force-free models are better;

(ii) the *photosphere*, *chromosphere* and *transition region* below the coronal base, where pressure gradients and gravity come into play, so that in a stationary case a full magnetohydrostatic model is required;

(iii) the high corona, where pressure gradients and inertia become important;

(iv) times, such as early in a solar flare, when the magnetic field is evolving rapidly (i.e., at speeds that are a significant fraction of the Alfvén speed).

Several techniques have been used to find potential fields, either for a local part of the corona, for which the coronal base may be treated as a plane, or for the global corona, in which case it is a sphere. Treating the coronal base as a plane, there are two main techniques. The first (*Green's function*) method (Schmidt 1964) is to place a series of monopoles on the lower boundary at points $(x', y', 0)$, each having a magnetic flux $B_n dx'dy'$. Since the magnetic potential at distance r from a single source of half-flux F_m is simply $-F_m/(\pi r)$, the resulting potential at (x, y, z) due to a normal magnetic field $B_n(x', y')$ on $z = 0$ is

$$\Psi(x, y, z) = \int B_n(x', y')\, G_n(x, y, z, x', y')\, dx'dy', \tag{3.34}$$

where $G_n(x, y, z, x', y') = -1/(\pi|\mathbf{r}-\mathbf{r}'|)$ is the Green's function and $|\mathbf{r}-\mathbf{r}'|$ is the distance between (x, y, z) and $(x', y', 0)$.

The second (*Fourier expansion*) method is to expand the imposed boundary value in its Fourier components over appropriate values of k_x and k_y as

$$B_n(x, y) = B_0 + \sum_{\mathbf{k}} B_{\mathbf{k}} e^{ik_x x + ik_y y},$$

and then using Eq. (3.30) the solution follows as

$$\Psi(x, y, z) = B_0 z - \sum_{\mathbf{k}} \frac{B_{\mathbf{k}}}{k} e^{ik_x x + ik_y y - kz}. \tag{3.35}$$

Treating the coronal base as a sphere, Sakurai (1982) adapted the first technique by calculating the Green's function for a spherical surface S (i.e., the field of a monopole on S, modified to make the component along the line-of-sight vanish everywhere else on S). The second technique approximates the global magnetic field by a series [Eq. (3.32)], the coefficients of which are determined to obtain a fit at the photosphere (Altschuler and Newkirk 1969). The effect of the solar wind in dragging out field lines is roughly simulated by forcing the field to become radial at a distance of typically 2.5 $R_\odot$. The resulting *potential-field source-surface* magnetic configuration is sketched as a so-called 'hairy ball' (e.g., Figure 3.7), which has been widely used to compare with coronal images. For more details and more realistic global MHD models, see Section 13.9.4, Mikić et al. (2007), Linker et al. (2011), Riley et al. (2013).

Alternatively, one may employ a finite-difference method, including the effect of a non-spherical source surface. Limitations on global calculations include: the poor quality of data near the poles; the fact that

Figure 3.7. A 'hairy ball', indicating potential magnetic field lines in the solar corona for 23 April 2001, with a source surface at 2.5 $R_\odot$ (courtesy Thomas Wiegelmann).

variations on a time shorter than the solar rotation period cannot be studied; and the failure of the potential approximation as one approaches the photosphere, where the boundary field is usually measured.

3.4 Force-Free Magnetic Fields

The Lorentz force dominates the pressure gradient and gravitational force in Eq. (3.1) when the plasma has a low beta ($\beta \ll 1$) and its vertical extent (L_0) is much smaller than H/β, where H is the scale-height. Such conditions exist above an active region and, to the lowest order in β and $\beta L_0/H$, the force balance reduces to

$$\mathbf{j} \times \mathbf{B} = \mathbf{0}. \tag{3.5}$$

Thus, the electric current flows along magnetic field lines and, since the current is given by Ampère's Law (3.2), we may write Eq. (3.5) as

$$\boldsymbol{\nabla} \times \mathbf{B} = \alpha \mathbf{B}, \tag{3.36}$$

where α is some function of position. The only restriction on α follows by taking the divergence of Eq. (3.36), so that the left-hand side vanishes identically, while the right-hand side reduces [after using Eq. (3.3)] to

$$(\mathbf{B} \cdot \boldsymbol{\nabla})\alpha = 0. \tag{3.37}$$

This implies that α is constant along magnetic field lines (and also along electric current lines). If such magnetic field lines lie on a closed surface S, it cannot in general be simply connected. For, otherwise, integrating along a magnetic field line C gives

$$\int_C \mathbf{B}\cdot d\mathbf{s} = \iint_S \nabla \times \mathbf{B}\cdot d\mathbf{S}, \qquad \text{by Stokes' Theorem,}$$

$$= \mu \iint \mathbf{j}\cdot d\mathbf{S}, \qquad \text{by Eq. (3.2),}$$

which in turn vanishes since $\mathbf{j}$ lies in the surface S. Thus, the simplest form for a constant-α closed surface is a torus on which the line of force spirals. The field lines may either close on themselves after a finite number of circuits (when we have a *rational surface*) or they may cover the entire surface *ergodically*. If, on the other hand, closed flux surfaces do not exist, then the field lines fill a volume in which α is constant and they wind around *stochastically* (Bateman 1978; Friedberg 1987).

When α takes the same value on each field line, we have a *linear* or *constant-α* field, for which the curl of Eq. (3.36) reduces to a vector Helmholtz equation,

$$(\nabla^2 + \alpha^2)\,\mathbf{B} = \mathbf{0}, \tag{3.38}$$

so that the governing equation is linear. Much attention has been paid to constant-α fields because of the difficulty that exists of finding more general nonlinear solutions to $\mathbf{j}\times\mathbf{B} = \mathbf{0}$, despite its disarmingly simple form! (See PROBLEMS 3.4, 3.5.)

This section first presents some rather surprising results that hold for general force-free fields. Constant-α fields are then described, both simple analytical solutions and general solutions. Next, nonlinear force-free solutions and numerical methods are presented, and, finally, the diffusion of force-free fields is discussed briefly.

3.4.1 General Theorems for Force-Free Fields

3.4.1.1 Force-Free Minimum-Energy Theorem

Because of their inherent nonlinearity, the nature of force-free fields is not well understood at all. Indeed, several general results lend an air of mystery to their structure (Roberts 1967; Cowling 1976). The generalisation of the *Minimum-Energy Theorem* for potential fields (Sec. 3.3.1) is as follows: suppose the normal magnetic field component is prescribed on the bounding surface S of a volume V, together with the correspondence between points of entrance and exit (i.e., the footpoint positions are fixed and therefore also the value of the vector potential $\mathbf{A}$ on S); then, if the field energy in V is stationary (i.e., an extremum), it must be force-free. A corollary of this is that, *if the flux and topological connections on S are given and the field possesses a minimum energy, then it is force-free.* However, the converse is not true, so that a force-free field does not necessarily produce a minimum energy.

3.4.1.2 Magnetic Energy

From the *Virial Theorem* (2.74), the magnetic energy (W) within a volume V may be written simply in terms of the magnetic field on the surface S alone as

$$W \equiv \int_V B^2/(2\mu)\;dV = \int_S [(\mathbf{r}\cdot\mathbf{B})\mathbf{B} - \tfrac{1}{2}B^2\mathbf{r}]\cdot d\mathbf{S}/\mu. \tag{3.39}$$

When, in particular, V is the region above the plane $z = 0$, this becomes

$$W = \int_S (xB_x + yB_y)\frac{B_z}{\mu}\;dxdy.$$

For example, suppose a force-free ($\mathbf{B}^{ff}$) and a potential field ($\mathbf{B}^{pot}$) have the same normal component (B_z). Then the difference ($W^{ff} - W^{pot}$) in stored energy is

$$W^{ff} - W^{pot} = \int_S [x(B_x^{ff} - B_x^{pot}) + y(B_y^{ff} - B_y^{pot})] \frac{B_z}{\mu} \, dxdy.$$

3.4.1.3 Necessity for Stress

Another interesting theorem is that, if $\mathbf{j} \times \mathbf{B}$ vanishes everywhere within a volume (V) and on its surface (S), then the magnetic field is identically zero. (This follows from the Virial Theorem above by taking a new surface at large distances so that the right-hand side vanishes.) Thus, a *non-trivial magnetic field* ($\mathbf{B} \neq \mathbf{0}$) *that is force-free within V must be stressed somewhere on S.* In other words, force-free fields are possible, but they must be anchored down somewhere on the boundary. Therefore, attempts to build force-free fields from currents enclosed entirely within a simple volume are doomed to failure.

3.4.1.4 Force-Free Everywhere?

A similar theorem is that *no magnetic field having a finite energy can be force-free everywhere.* For, if a magnetic field falls off like r^{-2} (or faster) at large distances from the origin, the energy [$W = \int B^2/(2\mu) \, dV$] can be transformed to

$$W = \int \frac{\mathbf{r} \cdot \mathbf{j} \times \mathbf{B}}{\mu} \, dV,$$

which vanishes when the field is force-free everywhere. Thus, a non-trivial field that is force-free (or potential) everywhere must possess a singularity.

3.4.1.5 Axisymmetric Fields

The following result warns against trying to construct fields in polar coordinates that are too simple: *an axisymmetric, force-free, poloidal magnetic field must be current-free.* A poloidal field possesses no azimuthal component and so has the form $\mathbf{B} = B_R\hat{\mathbf{R}} + B_z\hat{\mathbf{z}}$, where axisymmetry implies that $\mathbf{B}$ is independent of azimuth (ϕ). The current from Ampère's law is then $\mathbf{j} = (\partial B_R/\partial z - \partial B_z/\partial R)\hat{\boldsymbol{\phi}}/\mu$, so that

$$\mathbf{j} \times \mathbf{B} = j \, (B_z\hat{\mathbf{R}} - B_R\hat{\mathbf{z}}).$$

If this is to vanish, then so must j and the theorem is established.

3.4.1.6 Linear Force-Free Fields

For a constant-α force-free field, Molodensky (1974) obtained the lower limit

$$\delta W \geqslant \tfrac{1}{2}\mu^{-1} \left[\int B_1^2 \, dV - \left(\int \alpha^2 B_1^2 \, dV \right)^{\frac{1}{2}} \left(\int r^2 B_1^2 \, dV \right)^{\frac{1}{2}} \right],$$

for the change in energy (δW) produced by a perturbation ($\mathbf{B}_1$) from a force-free field. If d is the maximum dimension of the finite volume, so that $r \leqslant d$, we have

$$\delta W > \tfrac{1}{2}\mu^{-1} \int B_1^2 \, dV (1 - |\alpha| d),$$

which is positive provided $|\alpha| < d^{-1}$. Thus, the field is stable if the electric current scale-length exceeds the size of the region. Later, the result was strengthened by showing that a non-constant-α field is stable if the maximum value of α is small enough (Molodensky 1975, 1976). However, general conditions for *instability* are difficult to obtain.

Boström (1973) established the result that, in a plasma of finite electrical conductivity, a force-free solution of constant α within a *finite* region cannot be matched to a potential field outside the region, which limits the applicability of such solutions.

3.4.1.7 Woltjer's Minimum-Energy Theorem

Woltjer (1958) has proved two important results: (i) *for a perfectly conducting plasma in a closed volume* (V_0), *the total magnetic helicity*

$$H_0 = \int_{V_0} \mathbf{A} \cdot \mathbf{B}\, dV \tag{3.40}$$

is invariant; and (ii) the *state of minimum magnetic energy for a given total magnetic helicity is a linear (constant-α) force-free field.*

The proof is as follows. For a perfectly conducting plasma, $\partial \mathbf{B}/\partial t = \boldsymbol{\nabla} \times (\mathbf{v} \times \mathbf{B})$, or, in terms of the magnetic vector potential ($\mathbf{A}$),

$$\frac{\partial \mathbf{A}}{\partial t} = \mathbf{v} \times \mathbf{B}, \quad \text{(a)} \qquad \text{where} \qquad \mathbf{B} = \boldsymbol{\nabla} \times \mathbf{A}, \quad \text{(b)} \tag{3.41}$$

and the gauge has been chosen to make the scalar potential vanish. Thus,

$$\begin{aligned} \frac{\partial}{\partial t} \int_{V_0} \mathbf{A} \cdot \mathbf{B}\, dV &= \int_{V_0} \mathbf{A} \cdot \frac{\partial \mathbf{B}}{\partial t}\, dV + \int_{V_0} \frac{\partial \mathbf{A}}{\partial t} \cdot \mathbf{B}\, dV, \\ &= \int_{V_0} -\boldsymbol{\nabla} \cdot \left[\mathbf{A} \times \frac{\partial \mathbf{A}}{\partial t} \right] + 2 \frac{\partial \mathbf{A}}{\partial t} \cdot \boldsymbol{\nabla} \times \mathbf{A}\, dV, \end{aligned}$$

after putting $\mathbf{B} = \boldsymbol{\nabla} \times \mathbf{A}$ and rewriting a triple scalar product. Now, the second term vanishes by Eq. (3.41a) and the first may be transformed by Gauss' Theorem to

$$\frac{\partial}{\partial t} \int_{V_0} \mathbf{A} \cdot \mathbf{B}\, dV = - \int_{S_0} \mathbf{A} \times \frac{\partial \mathbf{A}}{\partial t} \cdot d\mathbf{S}.$$

This vanishes for a closed volume with $\mathbf{A} =$ *constant* on S_0 and so part (i) follows.

For part (ii) of the proof, consider the magnetic energy $[W = \int B^2/(2\mu)\, dV]$ contained within the volume (V_0) and the effect on it of a small arbitrary change of $\mathbf{A}$ and $\mathbf{B}$ to $\mathbf{A} + \delta\mathbf{A}$ and $\mathbf{B} + \delta\mathbf{B}$, such that $\delta\mathbf{A} = \mathbf{0}$ on S and $\delta\mathbf{B} = \boldsymbol{\nabla} \times \delta\mathbf{A}$. Then, by linearising and subtracting α_0 times (3.40), one finds

$$2\mu\, \delta W = \int_{V_0} 2\mathbf{B} \cdot \delta\mathbf{B} - \alpha_0 (\delta\mathbf{A} \cdot \mathbf{B} + \mathbf{A} \cdot \delta\mathbf{B})\, dV,$$

or, after substituting for $\delta\mathbf{B}$ and rewriting the triple scalar products,

$$2\mu\, \delta W = \int_{V_0} \boldsymbol{\nabla} \cdot (-2\mathbf{B} \times \delta\mathbf{A} + \alpha_0 \mathbf{A} \times \delta\mathbf{A})\, dV + 2 \int_{V_0} (\boldsymbol{\nabla} \times \mathbf{B} - \alpha_0 \mathbf{B}) \cdot \delta\mathbf{A}\, dV.$$

The first integral reduces to a surface integral that vanishes since $\delta\mathbf{A} = \mathbf{0}$ on S, and the second then implies that $\delta W = 0$ for all perturbations ($\delta\mathbf{A}$) if and only if

$$\boldsymbol{\nabla} \times \mathbf{B} = \alpha_0 \mathbf{B}. \tag{3.42}$$

Thus, if the energy is a minimum, the field must satisfy Eq. (3.42) for some α_0, and, if a field satisfies Eq. (3.42), the energy must be an extremum (but not necessarily a minimum). The possible values of α_0 may be found in terms of the total magnetic helicity (H_0) and the amount of flux that is present. Note too that the theorem also holds if the magnetic field component normal to the boundary (S_0) is prescribed (but not necessarily zero), so that again $\partial \mathbf{A}/\partial t$ and $\delta\mathbf{A}$ vanish on S_0.

3.4.1.8 Taylor Hypothesis: Minimum Energy at Large R_m

Taylor (1974, 1976) has developed Woltjer's Theorem for the case of force-free fields in a laboratory torus, whose walls are rigid and perfectly conducting, so that the normal magnetic field component vanishes and $\mathbf{A}$ is constant on the surface. In such experiments, an initial toroidal magnetic field ($\mathbf{B}_0$) is produced by external coils, and then a toroidal current (I) is induced. It pinches the plasma, and, after a dynamic dissipative phase, the magnetic field relaxes to a stable state which depends on the value of the *pinch ratio* $\theta = (\mu I)/(2\pi a B_0)$, where a is the minor radius of the torus and R the major radius. When $\theta \ll a/R$, one has the *tokamak regime*, whereas when $\theta \gtrsim 1.2$ the configuration is known as a *reversed-field pinch* and contains a reversed field in the outer regions. The stable field must possess a minimum energy subject to whatever constraints are imposed on the possible motion, but what are the relevant constraints? Taylor points out that in a *perfectly conducting plasma*, for every infinitesimal flux tube, the magnetic helicity (Sec. 6.10)

$$H_0 = \int_V \mathbf{A} \cdot \mathbf{B}\, dV \tag{3.43}$$

is invariant; the minimum-energy field subject to the constraint that all the H_0's be invariant is then the nonlinear force-free field $\nabla \times \mathbf{B} = \alpha(\mathbf{r})\mathbf{B}$, where α is constant on each field line. In other words, in an ideal plasma the final state can be any force-free equilibrium. On the other hand, most plasmas, such as a reversed-field pinch, are non-ideal and have a large magnetic Reynolds number. Taylor suggests that for them the resulting changes in topology are accompanied by small changes in total magnetic helicity (H_0), in such a way that the helicity density ($\mathbf{A} \cdot \mathbf{B}$) is redistributed among the field lines, but its integral over all field lines is almost unchanged. In other words, *Taylor's hypothesis* is that: *the total magnetic helicity* $H_0 = \int_{V_0} \mathbf{A} \cdot \mathbf{B}\, dV$ *integrated over the total volume is approximately invariant when* $R_m \gg 1$, so that by Woltjer's Theorem the *minimum-energy configuration is a linear force-free field.* (Other topological invariants are discussed in Section 10.4.4.)

3.4.2 Simple Linear (Constant-α) Solutions

The simplest solution has the form $\mathbf{B} = [0, B_y(x), B_z(x)]$, in which case Eq. (3.5) implies $d/dx(B_y^2+B_z^2) = 0$, and $\nabla \cdot \mathbf{B} = 0$ is satisfied identically. An integration with respect to x gives $B_y^2 + B_z^2 = B_0^2$, say, and the solution becomes

$$(B_x, B_y, B_z) = (0, B_y, (B_0^2 - B_y^2)^{1/2}). \tag{3.44}$$

As one moves along the x-axis the field rotates, keeping a constant magnitude (B_0).

For the particular case of constant-α, the z-component of Eq. (3.36) is

$$\frac{dB_y}{dx} = \alpha(B_0^2 - B_y^2)^{1/2},$$

so that, with the origin chosen to be a zero of B_y, the solution becomes

$$B_y = B_0 \sin \alpha x, \qquad B_z = B_0 \cos \alpha x. \tag{3.45}$$

Another useful one-dimensional force-free field with constant field magnitude (but without a constant α) is (0, tanh x, sech x). The cylindrical analogue to (3.45) is the *Lundquist solution* (3.24), but in spherical polars the only solution to $\nabla \cdot \mathbf{B} = 0$ and $\mathbf{j} \times \mathbf{B} = \mathbf{0}$ having the form $\mathbf{B}(r)$ is singular (being proportional to $r^{-2}\hat{\mathbf{r}}$).

There are several simple *two-dimensional constant-α* solutions of separable form. For example, in rectangular Cartesians with $\alpha^2 = k^2 - l^2$ the field

$$(B_x, B_y, B_z) = B_0\,[-(l/k)\cos kx,\ -(1 - l^2/k^2)^{1/2} \cos kx,\ \sin kx]\ e^{-lz} \tag{3.46}$$

is periodic in the x-direction and provides a useful model for a coronal arcade of lateral extent π/k (Priest and Smith 1979), as shown in Figure 3.8. The inclination of the field lines to the x-direction is

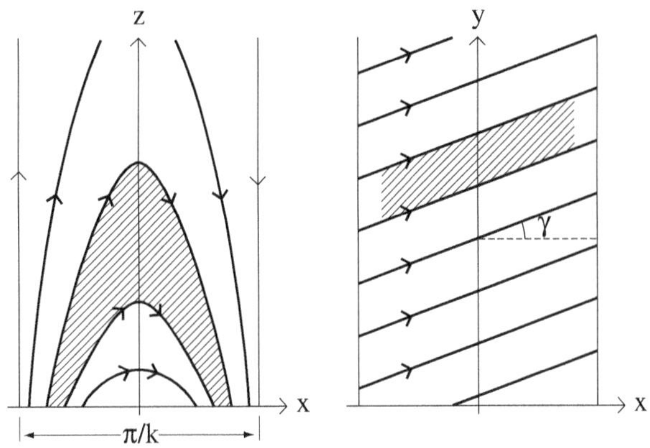

Figure 3.8. Vertical and horizontal sections through a model coronal arcade described by Eq. (3.46) with $B_0 < 0$. The shaded loop possesses a pressure that is enhanced at the base and therefore also at all heights (Sec. 3.1.3).

$\gamma = \tan^{-1}(k^2/l^2 - 1)^{1/2}$ at their summits. As l decreases from k to 0, α increases from 0 to k and the shear angle γ increases from 0 to $\frac{1}{2}\pi$. Thus, when $l = k$, α, γ and B_y all vanish and we have the potential field

$$B_x = -B_0 \cos\ kx\ e^{-kz}, \qquad B_z = B_0 \sin\ kx\ e^{-kz}.$$

Since the governing equation is linear, more complex solutions may be set up by summing solutions with different values of k and the same value of α.

The special case when the atmosphere is isothermal leads to a pressure that decreases exponentially with height along each field line according to Eq. (3.14). If the base pressure (p_0) is uniform, we have a simple plane-parallel atmosphere, with the magnetic field having no influence on the plasma structure at all. But, if the base of a particular flux loop has an enhanced pressure, it will possess a higher pressure than its surroundings throughout its length (see Figure 3.8).

The analogous axisymmetric solution in cylindrical polars (R,ϕ,z) is

$$(B_R, B_\phi, B_z) = B_0\ [(l/k)J_1(kR),\ (1 - l^2/k^2)^{1/2} J_1(kR),\ J_0(kR)]\ e^{-lz}, \tag{3.47}$$

in terms of Bessel functions J_0 and J_1, with the z-axis vertical (Schatzman 1965). As one moves outwards in such a structure in the R-direction, so the horizontal field component reverses in sign at each zero of $J_1(kR)$. However, if it is limited at the first zero, the solution may be used as a simple model for a twisted sunspot field above the photosphere (Figure 3.9). The spiral angle $\gamma = \tan^{-1}(B_\phi/B_R) = \tan^{-1}(k^2/l^2-1)^{1/2}$ is uniform throughout the structure. It vanishes for the potential field ($l = k$)

$$B_R = B_0 J_1(kR) e^{-lz}, \quad B_z = B_0 J_0(kR) e^{-lz},$$

and, as l decreases from k to 0, so γ increases from 0 to $\frac{1}{2}\pi$ and the field winds up.

Again, in spherical polar coordinates (r, θ, ϕ) the simplest axisymmetric solution represents a spherical plasmoid of radius $4.49/\alpha$ and has the form

$$B_r = \frac{1}{r^2 \sin\theta} \frac{\partial \tilde{A}}{\partial \theta}, \quad B_\theta = -\frac{1}{r \sin\theta} \frac{\partial \tilde{A}}{\partial r}, \quad B_\phi = \frac{\alpha \tilde{A}}{r \sin\theta}, \tag{3.48}$$

with $\tilde{A} = B_0 r^{1/2} J_{3/2}(\alpha r) \sin^2\theta$ and $J_{3/2}(4.49) = 0$ (e.g., Wilmot-Smith et al. 2005).

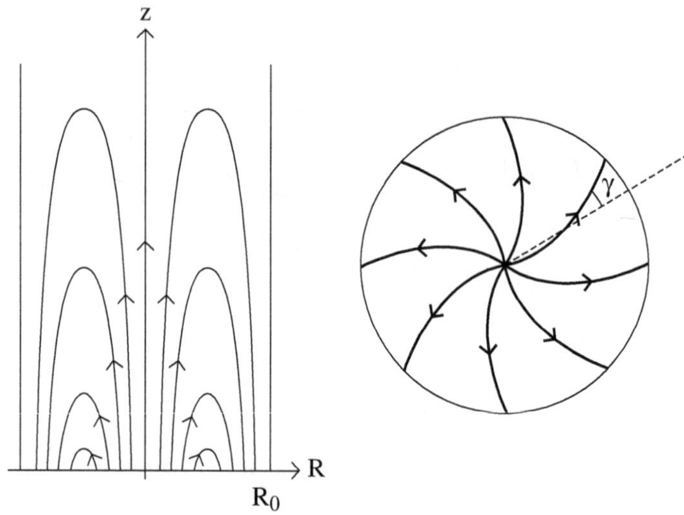

Figure 3.9. Vertical and horizontal sections through a magnetic structure described by Eq. (3.47), which may model the field above a twisted sunspot.

3.4.3 General Linear (Constant-α) Solutions

Here we seek general solutions to

$$(\nabla^2 + \alpha^2)\,\mathbf{B} = \mathbf{0} \tag{3.38}$$

that satisfy $\boldsymbol{\nabla} \cdot \mathbf{B} = 0$. Using the toroidal-poloidal decomposition [Eq. (3.7)], namely, $\mathbf{B} = \boldsymbol{\nabla} \times [T\,\hat{\mathbf{z}} + \boldsymbol{\nabla} \times (P\,\hat{\mathbf{z}})]$ in *Cartesian coordinates* with α uniform, Eq. (3.36) implies that $T = \alpha P$ and that P satisfies the *Helmholtz equation*:

$$(\nabla^2 + \alpha^2)P = 0. \tag{3.49}$$

As for current-free magnetic fields, solutions may be sought using either a Green's function or a Fourier method.

Several difficulties and inadequacies, however, should be borne in mind when attempting to use linear force-free fields to model the corona. The first is that they cannot continue in a physically realistic manner to infinity. Potential fields are fine: they either grow or decay at infinity and so we can apply a boundary condition there that selects only the decaying solution; but all the solutions to Eq. (3.49) oscillate at large distances and decay like $\exp(\pm i\alpha r)/r$, so that we have no effective boundary condition at infinity. Furthermore, there is no physically realistic linear force-free solution extending to large distances, since they all have unphysical reversals of sign.

The second difficulty is that the magnetic field decays so slowly at large distances [again like $\exp(\pm i\alpha r)/r$] that the magnetic energy diverges and again the solution is physically unacceptable. The third problem concerns the uniqueness of the solutions: for a given boundary, Eq. (3.49) possesses eigensolutions that make $B_n \equiv 0$ for particular values of α; thus, whenever α coincides with one of these eigenvalues, the solution is not unique, since an eigenfunction of any magnitude may be added. The fourth problem is that the quiet-Sun photospheric magnetic field consists of a series of small magnetic fragments in each of which, in practice, there are often α's of both signs. This implies that any resulting force-free magnetic fields are highly nonlinear. However, the analysis of nonlinear force-free fields, which we proceed to describe after this section, is still in its infancy.

In the Fourier method (Nakagawa and Raadu 1972), the simple solution of Eq. (3.49) that decays away to zero as z tends to ∞ may be used to construct solutions that decay with height in the domain $0 \leqslant x \leqslant L_x$, $0 \leqslant y \leqslant L_y$, $0 \leqslant z \leqslant L_z$ of the form

$$P(x, y, z) = \sum_{n,m} P_{n,m}^{(0)} \sin k_{xn}x \sin k_{ym}y \; e^{-l_{nm}z}, \tag{3.50}$$

where Eq. (3.49) implies that $l_{nm}^2 = k_{xn}^2 + k_{ym}^2 - \alpha^2$, and the condition that P vanish on the side boundaries of the domain implies that $k_{xn} = 2n\pi/L_x$ and $k_{ym} = 2m\pi/L_y$. Furthermore, if B_z is imposed on the base and has Fourier components $B_{n,m}^{(0)}$, then the fact that $B_z = -(\partial^2 P/\partial x^2 + \partial^2 P/\partial y^2)$ implies that $P_{n,m}^{(0)} = (k_{xn}^2 + k_{ym}^2) B_{n,m}^{(0)}$. Then the magnetic field components become

$$(B_x, B_y, B_z) = \left(\alpha \frac{\partial P}{\partial y} + \frac{\partial^2 P}{\partial x \partial z}, -\alpha \frac{\partial P}{\partial x} + \frac{\partial^2 P}{\partial y \partial z}, -\frac{\partial^2 P}{\partial x^2} - \frac{\partial^2 P}{\partial y^2} \right). \tag{3.51}$$

However, this boundary-value problem also possesses eigenfunctions (which vanish on all the boundaries) of the form

$$P_{nmN}^*(x, y, z) = \sin k_{xn}x \sin k_{ym}y \sin K_N z,$$

where $K_N = 2N\pi/L_z$ and Eq. (3.49) implies that the eigenvalue is given by

$$\alpha^2 = \alpha_{nmN}^2 \equiv k_{xn}^2 + k_{ym}^2 + K_N^2,$$

where n, m and N are integers (at least one of them being non-zero). Thus, if L_z is finite, then the solution (3.50) is unique provided α does not equal any of the discrete eigenvalues (α_{nmN}). If, on the other hand, $L_z = \infty$, the values of k_N become dense, and so the spectrum of eigenvalues is continuous in the region $|\alpha| \geqslant \alpha_{min} = \min(2\pi/L_x, 2\pi/L_y)$. The solution is then unique and is of the above form only when $|\alpha| \leqslant \alpha_{min}$. When $|\alpha| \geqslant \alpha_{min}$, eigenfunctions of arbitrary amplitude may be added to P of the form P^* above but with K_N replaced by $K = \sqrt{(\alpha^2 - \alpha_{min}^2)}$. As the dimensions of the region increase, the allowable range of α with unique solutions decreases. The method is therefore useful only for relatively small regions and assumes particular boundary conditions on the side and top boundaries. Similar comments apply to linear force-free fields in other coordinate systems.

In the case when $\mathbf{a} = \hat{\mathbf{r}}$, Eq. (3.7) also satisfies Eq. (3.36), and the general solution of Eq. (3.49) in *spherical polars* (r,θ,ϕ) for the poloidal scalar is

$$P = \sum_{n=0}^{\infty} \sum_{m=0}^{\infty} A_{n,m} \, r^{-\frac{1}{2}} J_{n+1/2}(\alpha r) \, P_n^m(\cos\,\theta) \, e^{im\phi}, \tag{3.52}$$

in terms of Bessel functions (J_n) and associated Legendre functions $[P_n^m(\cos\,\theta)]$. It may be used to find the solution outside a spherical surface (such as the photosphere) and leads to the following magnetic field components with complex constants C_n^m

$$(B_r, B_\theta, B_\phi) = \sum_{n=0}^{\infty} \sum_{m=0}^{\infty} C_{n,m} \, (b_r, b_\theta, b_\phi) \, e^{im\phi},$$

where $\quad b_r = (n+1) \dfrac{J_{n+1/2}(\alpha r)}{r^{3/2}} \, P_n^m(\cos\,\theta),$

$$b_\theta = \left[-n \frac{J_{n+1/2}(\alpha r)}{r^{3/2}} + \alpha \frac{J_{n-1/2}(\alpha r)}{r^{1/2}} \right] \frac{dP_n^m(\cos\,\theta)}{d\theta} + \alpha \frac{J_{n+1/2}(\alpha r)}{r^{1/2}} im \frac{P_n^m(\cos\,\theta)}{\sin\,\theta},$$

$$b_\phi = \left[-n \frac{J_{n+1/2}(\alpha r)}{r^{3/2}} + \alpha \frac{J_{n-1/2}(\alpha r)}{r^{1/2}} \right] im \frac{P_n^m(\cos\,\theta)}{\sin\,\theta} - \alpha \frac{J_{n+1/2}(\alpha r)}{r^{1/2}} \frac{dP_n^m(\cos\,\theta)}{d\theta}.$$

In *cylindrical polars* (R,ϕ,z) the general solution in a vertical cylinder of radius a and height L_z, say, for which each term decays to zero as z tends to ∞ and remains finite on the axis is

$$P(R,\phi,z) = \sum_{n=0}^{\infty} e^{in\phi} \sum_m C_{n,m}\, J_n(k_{nm}R)\, e^{-l_{nm}z}, \tag{3.53}$$

where the condition that P vanish on $R=a$ implies that $l_{nm} = (k_{nm}^2 - \alpha^2)^{1/2}$ with $k_{nm} = j_{nm}/a$ in terms of the mth zero (j_{nm}) of the Bessel function J_n. Eigenfunctions that vanish on the boundary have the form

$$P_{nmN}^*(R,\phi,z) = e^{in\phi}\, J_n(k_{nm}R)\, \sin K_N z,$$

where $K_N = N\pi/L_z$; the corresponding eigenvalue is determined by Eq. (3.49) as $\alpha_{nmN}^2 = k_{nm}^2 + K_N^2$.

Following a similar path to the discussion about the Cartesian solution, if the height (L_z) of the cylinder is finite and α is not equal to an eigenvalue, then the solution is unique. If, on the other hand, L_z is infinite, then the solution is unique if α is smaller in magnitude than the smallest eigenvalue $(\alpha_{min} = k_{01} = j_{01}/a)$; if it is larger, then eigenfunctions of arbitrary magnitude may be added of the above form, but with K_N^2 replaced by $K^2 = \alpha^2 - k_{nm}^2$. If the radius (a) of the cylinder is also infinite, then $\alpha_{min} = 0$ and the sum over m in Eq. (3.53) becomes an integral over k, so that (Chiu and Hilton 1977)

$$P(R,\phi,z) = \sum_{n=0}^{\infty} e^{in\phi} \left\{ \int_\alpha^\infty J_n(kR) I_1(z)\, dk + \int_0^\alpha J_n(kR)[I_2(z) + I_3(z)]\, dk \right\}, \tag{3.54}$$

where $\quad I_1(z) = C_n(k)\exp[-(k^2-\alpha^2)^{1/2}z], \quad I_2(z) = D_n(k)\cos[(\alpha^2-k^2)^{1/2}z]$

and $\quad I_3(z) = E_n(k)\sin[(\alpha^2-k^2)^{1/2}z]$.

This has the surprising effect that, if only α and the value of B_z on the plane $z=0$ are prescribed, the solution is not unique, by contrast with the special case of a potential field $(\alpha = 0)$. This can be seen most easily by remembering that $B_z = (\partial^2/\partial z^2 + \alpha^2)P$ and deducing from Eq. (3.54) that

$$B_z(R,\phi,0) = \sum_{n=0}^{\infty} e^{in\phi} \left[\int_\alpha^\infty k^2 J_n(kR) C_n(k)\, dk + \int_0^\alpha k^2 J_n(kR) D_n(k)\, dk \right],$$

so that only $C_n(k)$ and $D_n(k)$ are determined by $B_z(R,\phi,0)$, while $E_n(k)$ remains arbitrary. In order to obtain a unique boundary-value problem, one needs to impose additional information about, for example, the tangential field component on the boundary. The extra freedom may used to improve linear force-free fitting (Wheatland 1999).

When the magnetic field is *axisymmetric* (i.e., independent of ϕ) the poloidal scalar becomes

$$P = \sum_{n=0}^{\alpha} \int_0^\alpha A_n(k)\, \frac{\partial^n}{\partial l^n}\, [J_0(kR)e^{-lz}]\, dk,$$

with again $k = (l^2 + \alpha^2)^{1/2}$, and the resulting field components are

$$[B_R, B_\phi, B_z] = \left[-\frac{\partial}{\partial z}\left(R\frac{\partial P}{\partial z} - z\frac{\partial P}{\partial R}\right), \alpha\left(R\frac{\partial P}{\partial z} - z\frac{\partial P}{\partial R}\right), \frac{1}{R}\frac{\partial}{\partial R}\left(R^2\frac{\partial P}{\partial z} - Rz\frac{\partial P}{\partial R}\right)\right].$$

An alternative prescription to Eq. (3.7) for axisymmetric solutions is obtained by writing (Lüst and Schlüter 1954) $\mathbf{B} = \boldsymbol{\nabla}\times(\tilde{\psi}\,\hat{\boldsymbol{\phi}}) + \alpha\tilde{\psi}\,\hat{\boldsymbol{\phi}}$, where

$$\left(\nabla^2 + \frac{2}{R}\frac{\partial}{\partial R} + \alpha^2 \right)\tilde{\psi} = 0.$$

Solutions behaving like e^{-lz} that are finite on the z-axis take the form $B_R = -\partial\tilde{\psi}/\partial z$, $B_\phi = \alpha\tilde{\psi}$, $B_z = R^{-1}\partial/\partial R(R\tilde{\psi})$, where $\tilde{\psi} = \sum_0^\infty \int_0^\infty A_n(k)\ \partial^n/\partial l^n(J_1(kR)e^{-lz})dk$. The value $n = 0$ gives our previous solution [Eq. (3.47)].

The solution to Eq. (3.38) may also be set up in terms of a *Green's function* ($\mathbf{G}_\alpha$), which is often the easiest way to compute the solution to a linear partial differential equation. It is expressed as an integral over the (known) Green's function times an imposed boundary value. Chiu and Hilton (1977) have shown that the most general solution for the linear force-free field above the (photospheric) plane $z = 0$ when the normal field component $[B_z(x, y, 0)]$ is prescribed on $z = 0$ may be written

$$\mathbf{B}(x, y, z) = \frac{1}{2\pi}\iint_{-\infty}^{\infty} [\mathbf{G}_\alpha(x, y, z, x', y')B_z(x', y', 0) + \bar{\mathbf{G}}_\alpha(x, y, z, x', y')C(x', y')]\ dx'dy', \tag{3.55}$$

where $C(x', y')$ is an arbitrary integrable function and $\mathbf{G}_\alpha$ is the normal Green's function with components

$$(G_{\alpha,x}, G_{\alpha,y}, G_{\alpha,z}) = \left(\frac{x - x'}{\tilde{R}}\frac{\partial\Gamma}{\partial z} + \alpha\Gamma\frac{y - y'}{\tilde{R}}, \frac{y - y'}{\tilde{R}}\frac{\partial\Gamma}{\partial z} - \alpha\Gamma\frac{x - x'}{\tilde{R}}, -\frac{\partial\Gamma}{\partial\tilde{R}} - \frac{\Gamma}{\tilde{R}}\right),$$

$$\text{where} \quad \Gamma = \frac{z}{\tilde{R}\tilde{r}}\cos(\alpha\tilde{r}) - \frac{1}{\tilde{R}}\cos(\alpha z),$$

while $\tilde{R}$ and $\tilde{r}$ are the distances from the point $(x', y', 0)$ to $(x, y, 0)$ and (x, y, z), respectively (i.e., $\tilde{R}^2 = (x - x')^2 + (y - y')^2$ and $\tilde{r}^2 = \tilde{R}^2 + z^2$). Also, the eigenfunction part of the solution gives the extra term containing a complementary Green's function ($\bar{\mathbf{G}}_\alpha$), which is of the same form as $\mathbf{G}_\alpha$ except that the cosine function in Γ is replaced by a sine function. Since $C(x', y')$ is arbitrary, the extra term represents the nonuniqueness of the normal boundary-value problem when only $B_z(x, y, 0)$ and α are prescribed. The solution reduces to the corresponding solution [Eq. (3.34)] of Laplace's equation as α tends to zero.

3.4.4 Nonlinear (i.e., Non-constant-$\boldsymbol{\alpha}$) Solutions

3.4.4.1 Grad-Shafranov Equation for 2.5 Dimensions

When the assumption of constant α is dropped, it becomes much more difficult to solve $\mathbf{j} \times \mathbf{B} = \mathbf{0}$, where $\mathbf{j} = \boldsymbol{\nabla} \times \mathbf{B}/\mu$ and $\boldsymbol{\nabla} \cdot \mathbf{B} = 0$, even when one seeks magnetic fields that are independent of one of the coordinates.

In Cartesians, a so-called '2.5 D field' $[\mathbf{B}(x, z)]$ has three components that depend on two variables (x, z). Its components written in terms of the flux function (A) are

$$B_x = \frac{\partial A}{\partial z}, \quad B_y(x, z), \quad B_z = -\frac{\partial A}{\partial x}, \tag{3.56}$$

so as to satisfy $\boldsymbol{\nabla} \cdot \mathbf{B} = 0$ automatically. The components of $\mathbf{j} \times \mathbf{B} = \mathbf{0}$ become

$$\nabla^2 A\ \frac{\partial A}{\partial x} + B_y\ \frac{\partial B_y}{\partial x} = 0, \tag{3.57}$$

$$\frac{\partial B_y}{\partial z}\frac{\partial A}{\partial x} - \frac{\partial B_y}{\partial x}\frac{\partial A}{\partial z} = 0, \tag{3.58}$$

$$\nabla^2 A\ \frac{\partial A}{\partial z} + B_y\ \frac{\partial B_y}{\partial z} = 0. \tag{3.59}$$

Equation (3.58) implies that B_y is a function of A alone, and so it remains constant along any field line in the flux surface $A =$ *constant*. Either Eq. (3.57) or (3.59) then yields

$$\nabla^2 A = -\frac{d}{dA}(\tfrac{1}{2}B_y^2) = -\mu j_y(A), \tag{3.60}$$

known as the *Grad-Shafranov equation*, where $j_y(A)$ is the y-component of electric current. It determines A and so (B_x, B_z) once $B_y(A)$ or $j_y(A)$ and the boundary conditions are prescribed. The simple particular cases $j_y = 0$, $j_y = c$ and $j_y = cA$, where c is constant, give potential, constant-current and constant-α fields, respectively. However, analytical solutions have also been discovered for other functional forms, namely, $j_y = cA^n$ with $n = -3$ (Priest and Milne 1979), $j_y = c \sin A$ [which gives the *sine-Poisson equation* (Zagrodziński 1976; Roumeliotis 1993)], $j_y = ce^{-2A}$ (Low 1977; Birn et al. 1978) or, by replacing A with $-A$, $\mu j_y = ce^{2A}$. In the latter case, Eq. (3.60) becomes *Liouville's equation*, which has solutions of the form

$$A = -\log_e \left(\frac{1 + \frac{1}{4}c|f(u)|^2}{|df/du|} \right)$$

when $f(u)$ is any analytic function of $u = x + iz$.

The difficulty in dealing with Eq. (3.60) lies in the fact that it is in general a *nonlinear* equation, so that, for instance, the possibility of multiple solutions arises (see Secs. 12.2.1 and 12.2.3.3). One general result is that, *if* the solution of Eq. (3.60) is unique, *then* $d^2/dA^2(\frac{1}{2}B_y^2) \leqslant 0$, for all A (Courant and Hilbert 1963).

In cylindrical polars (R,ϕ,z), axisymmetric field (independent of ϕ) have the form

$$(B_R, B_\phi, B_z) = \left(-\frac{1}{R}\frac{\partial \tilde{A}}{\partial z}, \frac{b_\phi}{R}, \frac{1}{R}\frac{\partial \tilde{A}}{\partial R} \right), \tag{3.61}$$

where b_ϕ is a function of $\tilde{A}$ alone, $\alpha = db_\phi/d\tilde{A}$ and the notation $\tilde{A}$ is used to distinguish from components of the vector potential $\mathbf{A}$. It reduces Eq. (3.5) to

$$\frac{\partial^2 \tilde{A}}{\partial R^2} - \frac{1}{R}\frac{\partial \tilde{A}}{\partial R} + \frac{\partial^2 \tilde{A}}{\partial z^2} = -\frac{d}{d\tilde{A}}(\tfrac{1}{2}b_\phi^2) = -\mu R j_\phi. \tag{3.62}$$

Note that here the first three terms are $R^2\boldsymbol{\nabla} \cdot (R^{-2}\boldsymbol{\nabla}\tilde{A})$ rather than simply $\nabla^2\tilde{A}$. Analytical solutions were first found by Solov'ev (1968) with $Rj_\phi = c_1\tilde{A}$ and later generalised [e.g., see Atanasiu et al. (2004) for $Rj_\phi = c_1\tilde{A} + c_2R^2\tilde{A} + c_3 + c_4R^2$].

Again, in spherical polars (r,θ,ϕ), axisymmetric fields exist in the form

$$(B_r, B_\theta, B_\phi) = \frac{1}{r \sin\,\theta}\left(\frac{1}{r}\frac{\partial \tilde{A}}{\partial \theta}, -\frac{\partial \tilde{A}}{\partial r}, b_\phi(\tilde{A}) \right), \tag{3.63}$$

where b_ϕ is a function of $\tilde{A}$ alone, $\alpha = db_\phi/d\tilde{A}$ and Eq. (3.5) implies that

$$\frac{\partial^2 \tilde{A}}{\partial r^2} + \frac{\sin\theta}{r^2}\frac{\partial}{\partial \theta}\left(\frac{1}{\sin\theta}\frac{\partial \tilde{A}}{\partial \theta} \right) + \frac{d}{d\tilde{A}}(\tfrac{1}{2}b_\phi^2) = 0. \tag{3.64}$$

Analytical solutions confined between two cones with $rj_\phi = c\tilde{A}^{1+2/l}$ and decreasing with distance as r^{-l} have been discovered and applied to astrophysical jets and coronal arcades (Lynden-Bell and Boily 1994; Gourgouliatos 2008).

These solutions may be generalised in a natural way to include the effects of pressure gradients and gravity (Sec. 3.5). Furthermore, as well as *translational symmetry* (Eq. 3.56) and *axisymmetry* or *rotational symmetry* (Eq. 3.61), it may be shown (Solov'ev 1967; Edenstrasser 1980) that there is only one other spatial symmetry that reduces the magnetohydrostatic equations to a single elliptic second-order partial differential equation, namely, *helical symmetry* with $\mathbf{B} = \mathbf{B}(R, n\phi - kz)$.

3.4.4.2 Low's Models of Active Regions

Low (1982a) first constructed one-dimensional models of an active region of the form Eq. (3.63), in which A is a simple function of r alone, such that

$$(B_r, B_\theta, B_\phi) = \frac{B_0 a}{r \sin\theta}\,[0, \cos f(r), \sin f(r)].$$

The field lines, therefore, lie on spherical surfaces; they have an inclination angle $[f(r)]$ on them, which varies linearly with r out to a radius r_0 and then remains constant.

Later, Low considered an interesting class of so-called *laminated fields* (Low 1988),

$$(B_x, B_y, B_z) = \left(0, \frac{\partial \Psi}{\partial y}, \frac{\partial \Psi}{\partial z}\right), \tag{3.65}$$

where $\Psi(x, y, z)$ satisfies a two-dimensional Laplace equation $(\partial^2\Psi/\partial y^2 + \partial^2\Psi/\partial z^2 = 0)$. The magnetic field and electric current lie in vertical planes of constant x and so there is no magnetic tension force in the x-direction, which implies that there is also no magnetic pressure force in that direction. In other words, the magnetic pressure (and so the magnetic field magnitude) may vary within each plane but not from one plane to another $[\partial/\partial x(B^2) = 0]$. The general solution has the form

$$B_y - iB_z = F(\omega) \exp iP(x),$$

where $F(\omega)$ is an arbitrary analytic function of the complex variable $\omega = y + iz$ and $P(x)$ is an arbitrary phase function such that $\alpha = dP/dx$.

Laminated force-free solutions having field lines lying on spherical surfaces with $\alpha(r) = -d\omega(r)/dr$ also exist in the form (Low 2005)

$$(B_r, B_\theta, B_\phi) = \frac{1}{r \sin\theta}\,[0, P(\theta, \phi), Q(\theta, \phi)],$$

where $P + iQ = We^{iw(r)}$ and W is a meromorphic function of $[\log_e \tan(\frac{1}{2}\theta) + i\phi]$. A simple example is to put $W \equiv 1$ with $w = c(r - r_0)^2 + \frac{1}{2}\pi$ when $r < r_0$ and $w = \frac{1}{2}\pi$ when $r > r_0$, as shown in Figure 3.10.

Low and Lou (1990) found a set of separable solutions to Eq. (3.64) describing axisymmetric nonlinear force-free fields in spherical polars of the form $A = F(\cos\theta)r^{-n}$ associated with a point source at the origin. The azimuthal field is assumed to have the form $b_\phi(A) = aA^{1+1/n}$, where n is a positive odd integer and a is constant. Eq. (3.64) then reduces to an ordinary differential equation

$$(1 - \bar{\mu}^2)\frac{d^2F}{d\bar{\mu}^2} + n(n+1)F + a^2(1 + 1/n)F^{1+2/n} = 0$$

that needs to be solved numerically for $F(\bar{\mu})$, where $\bar{\mu} = \cos\theta$, and the condition that $F(\pm 1) = 0$ determines the eigenvalues (a_m) of a. In particular, they focused on a quadrupolar source with $n = 1$ and $m = 1$ for which $F(0) = 0$. In order to obtain a realistic-looking magnetic field, they placed the point source below

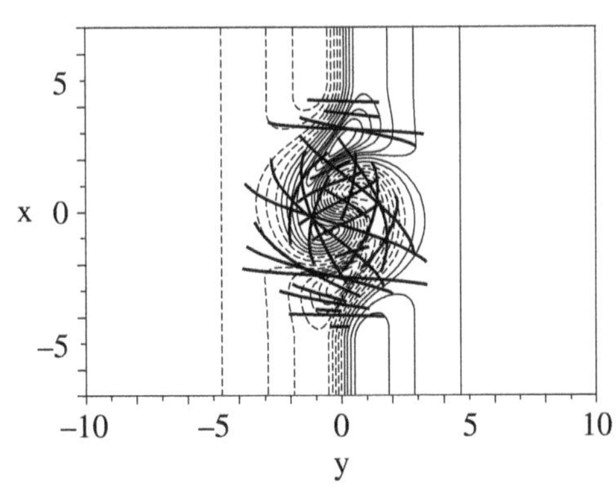

Figure 3.10. A *laminated magnetic field* modelling a pair of sunspots near a polarity inversion line, with field lines lying on spherical surfaces. Shown are projections of field lines viewed from above, together with contours of constant B_z at the photosphere (after Low, 2005).

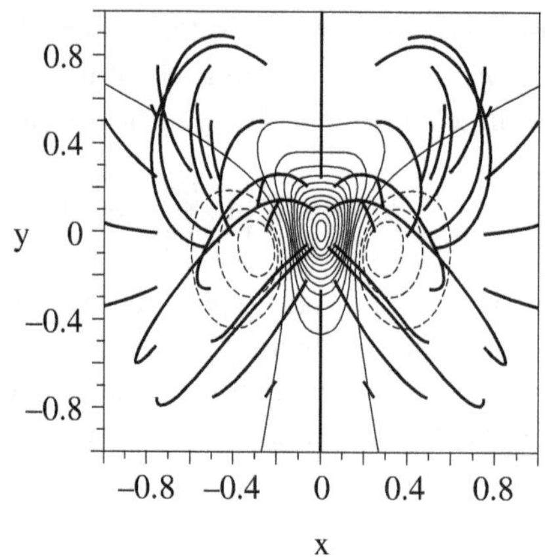

Figure 3.11. Example of a Low–Lou active-region magnetic field with a point source situated a distance 0.3 below the surface and its magnetic symmetry axis horizontal. Magnetic field lines projected onto $z = 0$ are shown (dark curves) as well as the normal magnetic flux distribution on $z = 0$ (light continuous and dashed curves) (after Low and Lou 1990).

the photosphere ($z = 0$) and took the magnetic axis of symmetry to lie in the xz-plane making an angle with the (vertical) z-axis. An example is shown in Figure 3.11.

Finally, Chou and Low (1994) presented another axisymmetric solution to Eq. (3.64) but with $b_\phi(A) = 2cA^{1/2}$ inside a sphere of radius r_0. The magnetic flux function is written as the sum of two parts ($A = A_p + A_{np}$). The potential part is due to a magnetic monopole at the origin together with its image in the sphere. The nonpotential part has $A_{np} = -c^2(r - r_0)^2$ for $r \leqslant r_0$ and $A_{np} = 0$ for $r \geqslant r_0$ and is a simple circular poloidal field $[B_\theta = (r - r_0)/(r \ \sin\theta)]$, with an azimuthal current $[j_\phi = -2c^2/(r \ \sin\theta)]$ that is singular along the whole $\theta = 0$ axis. However, in modelling an active region they avoid the singularity by placing this axis below the photosphere. They explore the effect of photospheric shearing and flux emergence.

3.4.5 Numerical Methods for Force-Free Fields

Several numerical techniques have been proposed to compute nonlinear force-free fields in the active-region corona using measurements of the photospheric or chromospheric magnetic field as a boundary condition, each with their own pros and cons. For reviews and comparisons, see, for example, Sakurai (1989), Amari et al. (1997), Démoulin et al. (1997), Neukirch (2005), Schrijver et al. (2006) and Wiegelmann (2008).

Potential extrapolation is a quick and useful way to have a first overview of the magnetic field structure and has been described in Section 3.3. It can been used for local studies of a small region in Cartesian coordinates or for global studies of the corona, treating the source surface as a sphere rather than a plane. The two main techniques for finding current-free fields are by Green's functions or a Fourier series expansion and these have been extended in natural ways to find linear force-free fields. Linear force-free solutions (Sec. 3.4.3) can have problems of non-uniqueness, and the value of $|\alpha|$ has to be smaller than a maximum value (α_{max}), which is of order $2\pi/L$, where L is the smallest dimension of the computational box. One reaction to the non-uniqueness is to propose a so-called α_{best}-method, in which the deviation between the theoretical and measured transverse magnetic field at the photosphere is minimised (Pevtsov et al. 1995). Another is to optimise the fit between the coronal magnetic field and the observed shapes of coronal loops (Wiegelmann and Neukirch 2002; Carcedo et al. 2003). Nonlinear fields, however, in which one can no longer superpose solutions, demand quite different methods, summarised in the following.

3.4.5.1 Formulation of the Problem

The aim is simply to solve the equations

$$\nabla \times \mathbf{B} = \alpha \mathbf{B}, \tag{3.36}$$

$$\text{and} \quad (\mathbf{B} \cdot \nabla)\alpha = 0, \tag{3.37}$$

where $\nabla \cdot \mathbf{B} = 0$. This is a mixed elliptic-hyperbolic system of equations which makes formulation of a well-defined boundary-value problem non-trivial. For a given α, the equation for $\mathbf{B}$ is elliptic and so a natural boundary condition is to impose the normal field (B_n) on the boundary. For a given $\mathbf{B}$, on the other hand, the equation for α is hyperbolic, and so α should be prescribed at only one end of each field line, for exmaple, on those parts of the boundary where $B_n > 0$, which is equivalent to imposing the normal current ($j_n = \alpha B_n$) there. Note that imposing all three components of $\mathbf{B}$ (provided by, for instance, vector magnetograms) would be too much and would produce an ill-posed problem; thus, instead, one may want to seek a solution that comes as close as possible to the measured values.

Even if the problem is formulated correctly, it is not clear whether nonlinear force-free solutions exist or whether they are unique. Existence and uniqueness have so far been established only for small values of α, both in simple (Bineau 1972) and complex (Boulmezaoud and Amari 2000) topologies.

3.4.5.2 Different Methods of Solution

(i) The Vertical Integration Method

In this simple method (Nakagawa 1974; Démoulin et al. 1992; Amari et al. 1998), Eqs. (3.36) and (3.37) are cast as four equations for $\partial B_x/\partial z$, $\partial B_y/\partial z$, $\partial B_z/\partial z$ and $\partial \alpha/\partial z$ and are integrated upwards as a Cauchy problem from the lower boundary ($z = 0$), using $\mathbf{B}(x, y, 0)$ and $\alpha(x, y, 0)$ as starting values. However, it is ill-posed and produces unphysical growing modes, making the method of limited use.

(ii) The Grad-Rubin Method

By contrast, this iterative method (Grad and Rubin 1958; Sakurai 1981; Wheatland 2004) is the only one with a firm mathematical foundation. It has been applied with great success to extrapolate vector magnetograph data. In the form used by Amari et al. (1997, 1999, 2006), Régnier et al. (2002), Inhester and Wiegelmann (2006), the iterative scheme is written as

$$\mathbf{B}^n \cdot \nabla \alpha^n = 0, \qquad \nabla \times \mathbf{B}^{n+1} = \alpha^n \mathbf{B}^n, \qquad \nabla \cdot \mathbf{B}^{n+1} = 0.$$

(iii) MHD Relaxation Methods

Here time-dependent MHD codes are used to relax an initial field towards a state where the required boundary conditions are satisfied. Mikić and McClymont (1994) start from a potential equilibrium and then drive the system on the boundaries, using viscosity to dissipate the flow energy, while resistivity allows a change of topology if needed. In the *stress-and-relax* method (Roumeliotis 1996), the system is driven from the boundary and relaxed in alternating steps. Others use a simpler *magneto-frictional method*, in which the true viscous term is replaced by a term of the form $k\mathbf{v}$, where k is a positive constant (e.g., Wiegelmann and Neukirch 2003).

(iv) Optimization Methods

Based on an earlier variational principle of Sakurai (1979) (Figure 3.12), Wheatland et al. (2000) proposed an ingenious optimization method, later developed by Wiegelmann (2004) and Wiegelmann et al. (2006). They set up a functional

$$L = \int [B^{-2}|(\nabla \times \mathbf{B}) \times \mathbf{B}|^2 + |\nabla \cdot \mathbf{B}|^2] \, dV,$$

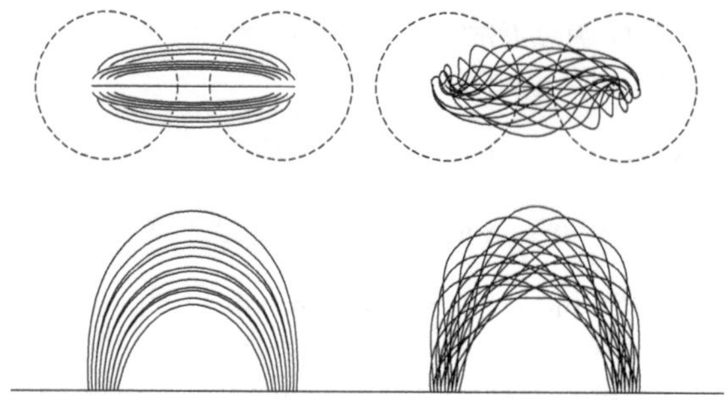

Figure 3.12. (a) Current-free and (b) force-free coronal loop models calculated by a variational method (after Sakurai 1979).

and minimised it in an evolutionary manner. Taking the derivative of L with respect to t gives a result of the form $dL/dt = -\int \partial\mathbf{B}/\partial t \cdot \mathbf{F}_V \, dV - \int \partial\mathbf{B}/\partial t \cdot \mathbf{F}_S \, dS$. If $\mathbf{B}$ is evolved according to an equation of the form $\partial\mathbf{B}/\partial t = f\mathbf{F}$ with f being an arbitrary positive function, together with $\partial\mathbf{B}/\partial t = 0$ on the surface (S), then $dL/dt = -\int fF^2 dV \leqslant 0$ and so L decreases in time. Clearly, if L does reduce to zero, then a force-free equilibrium has been reached.

(v) Boundary-Element Methods

Here Yan (1995, 2003) and Yan and Sakurai (2000) use a tensor Green's function to rewrite the problem as three nonlinear integral equations over the volume, together with a surface integral which is discretised using the boundary-element method.

All of these methods face challenges to make them more effective [see Schrijver et al. (2006), Metcalf et al. (2008) for a comparison]. One problem is to make them as fast and accurate as possible and another is to cope with noise in the measurements (Wiegelmann and Inhester 2010). Furthermore, photospheric magnetic fields are measured in a non-force-free region: one way around this is to make chromospheric magnetograms more accurate; another is to develop magnetohydrostatic codes which can model accurately up from the photosphere; and a third is to use *preprocessing* to derive boundary conditions that are force-free consistent from the photospheric measurements (Wiegelmann et al. 2006, 2008). Finally, there is the question of lateral and upper boundaries for local models, where the boundary conditions affect greatly the solutions: one method here is to use a less sensitive global model to provide the boundary conditions for a more accurate local analysis.

3.4.6 Diffusion of Force-Free Fields

The slow resistive diffusion of a magnetic field through a series of force-free equilibria is governed by the equations

$$\frac{\partial\mathbf{B}}{\partial t} = \nabla \times (\mathbf{v} \times \mathbf{B}) + \eta\nabla^2\mathbf{B}, \tag{3.66}$$

$$\nabla \times \mathbf{B} = \alpha\mathbf{B}, \tag{3.36}$$

where
$$\nabla \cdot \mathbf{B} = 0 \quad \text{and} \quad (\mathbf{B} \cdot \nabla)\alpha = 0. \tag{3.67}$$

If the medium is stationary ($\mathbf{v} \equiv \mathbf{0}$) and $\alpha =$*constant*, an initially force-free field diffuses while remaining force-free (Chandrasekhar and Kendall 1957). This is because Eqs. (3.36) and (3.67) imply $\nabla^2\mathbf{B} = -\alpha^2\mathbf{B}$, and so Eq. (3.66) reduces to

$$\frac{\partial \mathbf{B}}{\partial t} = -\eta\, \alpha^2\, \mathbf{B}, \tag{3.68}$$

with solution $$\mathbf{B} = \mathbf{B}_0(x,y,z)\, e^{-\eta\alpha^2 t}, \quad \mathbf{j} = \mathbf{j}_0(x,y,z)\, e^{-\eta\alpha^2 t}. \tag{3.69}$$

Thus, given an initial magnetic field ($\mathbf{B}_0$) and current ($\mathbf{j}_0$) that are parallel, they subsequently remain parallel for all time (e.g., Wilmot-Smith et al. 2005).

Low (1973, 1974) sought non-constant-α solutions for a so-called '*passive*' medium whose plasma velocity is determined by Eq. (3.66). He considered one-dimensional solutions similar to Eq. (3.45) having the form

$$B_y = B_0\, \cos f(x,t), \quad B_z = B_0\, \sin f(x,t),$$

where $\alpha = -\partial f/\partial x$. The functions $f(x,t)$ and $v_x(x,t)$ are determined by the y- and z-components of Eq. (3.66), which may be written

$$\frac{\partial f}{\partial t} - \eta \frac{\partial^2 f}{\partial x^2} + v_x \frac{\partial f}{\partial x} = 0, \qquad \eta \left(\frac{\partial f}{\partial x}\right)^2 + \frac{\partial v_x}{\partial x} = 0.$$

This pair of nonlinear equations possesses steady solutions that are linearly unstable and have a singularity at the origin. Self-similar solutions based on the variable $xt^{-1/2}$ also exist for both relaxing and steepening forms. One interesting class of solutions evolves slowly at first and then rapidly develops an electric current singularity (Reid and Laing 1979). In solar applications, the diffusion would need to be greatly enhanced over classical values to give relevant time-scales.

3.5 Magnetohydrostatic Fields

In general, we saw in Section 3.1.2 that the component of the force-balance

$$\mathbf{0} = -\nabla p + \mathbf{j} \times \mathbf{B} - \rho g\, \hat{\mathbf{z}} \tag{3.1}$$

along the magnetic field has the solution

$$p = p_0\, \exp\left(-\int_0^z \frac{dz}{H(z)}\right) \tag{3.11}$$

along each field line, with both p_0 and $H(z)$ varying from one field line to another.

3.5.1 Magnetostatic Fields

When the vertical extent of the region being considered is much less than a scale-height, the exponential in Eq. (3.11) approximates to unity, so that the pressure becomes uniform along the particular magnetic field line and Eq. (3.1) reduces to

$$\mathbf{0} = -\nabla p + \mathbf{j} \times \mathbf{B}. \tag{3.70}$$

Taking the scalar product with $\mathbf{B}$ and $\mathbf{j}$ in turn, we have

$$\mathbf{B} \cdot \nabla p \;=\; \mathbf{j} \cdot \nabla p \;= 0, \tag{3.71}$$

so that p is constant along both magnetic field lines and electric current lines. If the magnetic lines lie on a set of closed nested magnetic flux surfaces (as in Figure 3.13), then the current lines will lie on them too and they will also be surfaces of constant p (PROBLEM 3.6). Three distinct classes of magnetic field line are possible: rational, ergodic and stochastic (Bateman 1978; Friedberg 1987). A constant-p surface in which all the field lines close on themselves after a finite number of toroidal circuits is called a *rational*

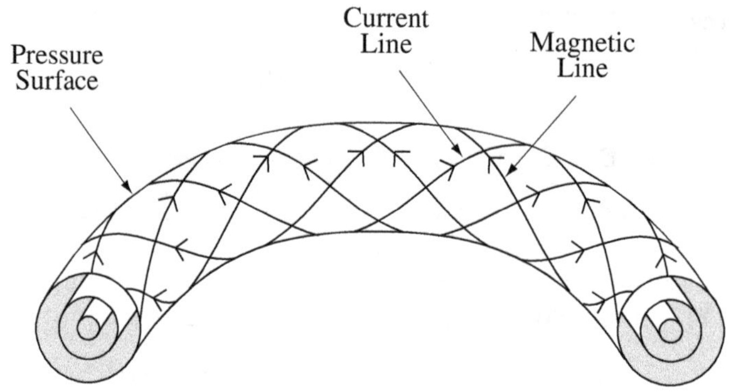

Figure 3.13. A set of nested toroidal flux surfaces of constant pressure, on which lie magnetic field lines and electric current lines when there is magnetostatic equilibrium.

surface. If instead the field lines cover the whole surface and never close, it is an *ergodic surface.* On the other hand, if distinct surfaces of constant p do not exist, the field line fills a volume known as a *stochastic region.* Such regions often occur near separatrices and are clearly unattractive for plasma confinement in a laboratory machine such as a tokamak, since they imply that distinct surfaces of different constant pressure have broken down, making the pressure uniform.

The particular cases of magnetic fields that are uniformly directed or have cylindrical symmetry have already been dealt with in Sections 3.1.3 to 3.2. Moreover, many of the general methods for treating force-free fields (Sec. 3.4) can be extended easily to include the additional pressure gradient term in Eq. (3.70) (PROBLEM 3.7).

Consider an *axisymmetric magnetic field* in cylindrical polar components

$$(B_R, B_\phi,\ B_z) \equiv \frac{1}{R}\left(-\frac{\partial \tilde{A}}{\partial z},\ b_\phi,\ \frac{\partial \tilde{A}}{\partial R}\right), \tag{3.72}$$

so as to satisfy $\nabla \cdot \mathbf{B} = 0$ identically, with the notation $\tilde{A}$ being used to distinguish from the vector potential $\mathbf{A}$. Then the electric current components are

$$(j_R, j_\phi, j_z) = \frac{1}{R}\left(-\frac{\partial b_\phi}{\partial z},\ -\frac{\partial^2 \tilde{A}}{\partial R^2} - \frac{1}{R}\frac{\partial \tilde{A}}{\partial R} + \frac{\partial^2 \tilde{A}}{\partial z^2},\ \frac{\partial b_\phi}{\partial R}\right).$$

For an axisymmetric pressure, $\partial p/\partial \phi = 0$ and the ϕ-component of Eq. (3.70) becomes

$$\frac{\partial b_\phi}{\partial R}\frac{\partial \tilde{A}}{\partial z} - \frac{\partial b_\phi}{\partial z}\frac{\partial \tilde{A}}{\partial R} = 0,$$

which implies that $b_\phi = b_\phi(\tilde{A})$. Also, Eq. (3.71) gives

$$\frac{\partial p}{\partial R}\frac{\partial \tilde{A}}{\partial z} - \frac{\partial p}{\partial z}\frac{\partial \tilde{A}}{\partial R} = 0,$$

so that $p = p(\tilde{A})$ is also a function of $\tilde{A}$ alone, and the R-component of Eq. (3.70) is

$$\frac{\partial^2 \tilde{A}}{\partial R^2} - \frac{1}{R}\frac{\partial \tilde{A}}{\partial R} + \frac{\partial^2 \tilde{A}}{\partial z^2} + \frac{d}{d\tilde{A}}(\tfrac{1}{2}b_\phi^2) = -\mu R^2\,\frac{dp}{d\tilde{A}} \tag{3.73}$$

as the generalisation of Eq. (3.62). It determines $\tilde{A}(R, z)$ after $b_\phi(\tilde{A})$ and $p(\tilde{A})$ have been prescribed.

Again consider the *Cartesian case* when the variables depend on x and z alone so that the magnetic field components may be written

$$(B_x, B_y, B_z) \equiv \left(\frac{\partial A}{\partial z},\ B_y(A),\ -\frac{\partial A}{\partial x}\right), \tag{3.74}$$

in terms of the flux function $A(x, z)$, which defines a field line in the xz-plane or a flux surface in three dimensions. Then Eq. (3.11) becomes

$$p(A, z) = p_0(A)\ \exp\left(-\int_0^z \frac{mg}{k_B T(A, z)}\ dz\right), \tag{3.75}$$

and the component of Eq. (3.1) normal to the magnetic field reduces to

$$\nabla^2 A + \frac{d}{dA}\left(\tfrac{1}{2}B_y^2(A)\right) = -\mu\frac{\partial}{\partial A}\left(p(A, z)\right) \tag{3.76}$$

as the generalisation of Eq. (3.60). Equation (3.75) is in general coupled to an energy equation through the presence of $T(A, z)$, but in special cases the coupling is absent. When the temperature is uniform along each field line, Eq. (3.75) simplifies to

$$p(A, z) = p_0(A) \exp\left[-z/H(A)\right],$$

where the scale-height $H(A) = k_B T(A)/(mg)$ varies from one field line to another. Moreover, if the scale for z-variations is much less than the scale-height, this reduces still further to $p(A) = p_0(A)$, so that Eq. (3.76) may in principle be solved once the total pressure $[p_0(A) + B_y^2(A)/(2\mu)]$ is prescribed on each field line.

It is clearly possible to use standard techniques to seek both linear and nonlinear solutions to Eq. (3.76) that will represent the structure of a *magnetic arcade*. For example, a simple case (Zweibel and Hundhausen 1982) is a linear solution for an isothermal plasma [$H(A) =$*constant*] in a shear-free arcade [$B_y(A) \equiv 0$] with

$$\mu p(A, z) = \tfrac{1}{2}c^2 A^2\ e^{-z/H} + constant,$$

so that Eq. (3.76) reduces to

$$\nabla^2 A = -c^2 A\ e^{-z/H}.$$

The solution subject to $B_z = B_0 \sin(\pi x/L)$, say, at the base is just

$$A = A_0\ \cos(\pi x/L)\ J_n(2cH\ e^{-z/(2H)}), \tag{3.77}$$

where $A_0 = LB_0/[\pi\ J_n(2cH)]$ and $n = 2\pi H/L$.

When $c = 0$ the magnetic field is potential and the isobars are horizontal. When $2cH$ is smaller than the first maximum of the Bessel function J_n, both the field lines and isobars are inflated slightly. When $2cH$ lies between the first maximum and the first zero of J_n, the magnetic field includes a magnetic island, below which there lies a pressure maximum. For an extension in spherical geometry to model the global solar atmosphere, see Hundhausen et al. (1981) and Uchida and Low (1981).

Other nonlinear solutions to Eq. (3.76) with $H(A) = constant$ and with $p_0(A)$ equal to either e^{-2A} or $\sin A$ have been discovered by Khater et al. (1997, 1998, 2002).

3.5.2 Effect of Gravity

For many years, BC Low has been a master at discovering analytical solutions for the force-free or magnetohydrostatic equations, some of which we describe in the following. In general, the object is to solve as the basic equation

$$\mathbf{0} = -\nabla p + \mathbf{j} \times \mathbf{B} - \rho g\,\hat{\mathbf{z}}, \tag{3.1}$$

where $\nabla \cdot \mathbf{B} = 0$, $p = (k_B/m)\rho T$ and the temperature is determined by an energy equation. The general set is daunting indeed, but one way forward for Low has been essentially to ignore the energy equation and replace it by the imposition of some prescribed functional form for p, say, which allows analytical progress and gives insight into some of the properties to be expected of the full system. The basic theory can easily be extended to the radial or more general versions of Eq. (3.1), in which $-\rho g\,\hat{\mathbf{z}}$ is replaced by $-\rho G M_\odot \hat{\mathbf{r}}/r^2$ or by $-\rho \nabla \Phi$, respectively, where Φ is any gravitational potential or it can include inertial forces such as a centrifugal force.

3.5.2.1 Laminated Magnetic Fields

One of Low's most important ideas (Low 1982b) was to seek *laminated solutions* for a magnetic field whose components lie purely in yz-planes, so that

$$(B_x, B_y, B_z) = \left(0, \frac{\partial A(x,y,z)}{\partial z}, -\frac{\partial A(x,y,z)}{\partial y}\right). \tag{3.78}$$

These are a natural generalisation of his force-free laminated solutions [Eq. (3.65)] and are relevant for the structure of prominences, coronal arcades and coronal loops. Equation (3.78) implies that in the x-direction, there is no tension force, as well as no gravitational force, and so the total pressure force $(\partial P_T/\partial x)$ in that direction must vanish, too. In other words, the total pressure distribution,

$$p + \frac{B^2}{2\mu} \equiv P_T(y,z), \tag{3.79}$$

must be the same in each yz-plane, even though p and B^2 can vary with x. The two horizontal components of Eq. (3.1) become

$$\nabla_\perp^2 A \frac{\partial A}{\partial y} + \mu \frac{\partial p}{\partial y} = 0, \qquad \nabla_\perp^2 A \frac{\partial A}{\partial z} + \mu \frac{\partial p}{\partial z} + \mu \rho g = 0, \tag{3.80}$$

where $\nabla_\perp^2 \equiv \partial^2/\partial y^2 + \partial^2/\partial z^2$. After substituting for p from Eq. (3.79), these imply that, for a given $P_T(y,z)$, the following nonlinear condition on A must hold:

$$\frac{\partial^2 A}{\partial y \partial z}\frac{\partial A}{\partial z} - \frac{\partial^2 A}{\partial z^2}\frac{\partial A}{\partial y} = \mu \frac{\partial P_T}{\partial y}. \tag{3.81}$$

In particular, if we suppose that $P_T = P_T(z)$ is a function of z alone, then the right-hand side of Eq. (3.81) vanishes, which implies that $\partial A/\partial z$ is an arbitrary function of A (and x). This result may be in turn integrated to show that in general A is an arbitrary function of x and ζ

$$A = A(x, \zeta), \qquad \text{where} \qquad \zeta \equiv z + \xi(x,y)$$

and $\xi(x,y)$ is an arbitrary function that fixes the geometry. (In fact, x and ζ are Euler potentials.) Once A and therefore $\mathbf{B}$ have been found, Eqs. (3.79) and (3.80b) determine p and ρ, respectively.

Low constructed four examples: the first is a model having no variation with height for a prominence by assuming $A(x,y,z) = B_0 \cos f(x)[z + \tan f(x)\xi(y)]$, for which $B_y = B_0 \cos f(x)$, $B_z = -B_0 \sin f(x)(d\xi/dy)$. The function $f(x)$ allows the downwards-bowing of the field lines, and therefore the mass supported in an

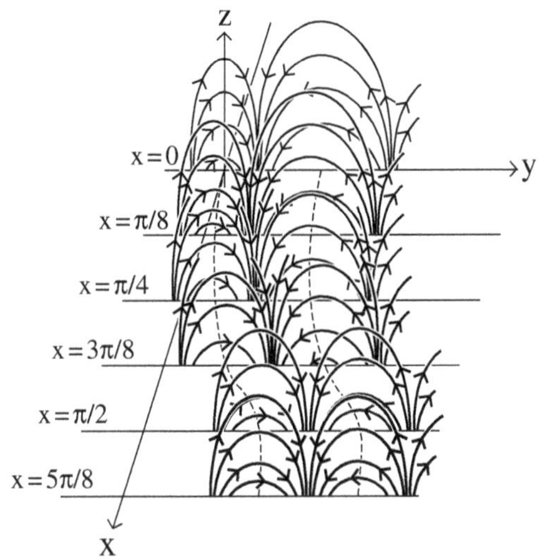

Figure 3.14. A sketch of the magnetic field lines of a laminated coronal arcade with $L(x) = \pi \cos x$ (after Low 1982b).

infinite vertical prominence sheet can vary along the prominence axis. Setting $f(x) \equiv$ *constant* reproduces the *Kippenhahn-Schlüter solution* (Sec. 11.2.1).

The second example has $A = -B_0 e^{-\zeta}$, where $\zeta \equiv z - \log\{2 + \cos[y + L(x)]\}$. The resulting magnetic field,

$$(B_x, B_y, B_z) = B_0 e^{-z}[0, 2 + \cos(y + L), -\sin(y + L)],$$

represents another generalisation of the Kippenhahn-Schlüter model whose horizontal field varies both with z and also [through the arbitrary function $L(x)$] with x.

The third example models a coronal arcade (Figure 3.14) and has $A = -B_0 e^{-\zeta}$, where $\zeta \equiv z - \log\{a_0 \exp(-x^2/l^2) + \cos[y + L(x)]\}$. The resulting magnetic field is

$$(B_x, B_y, B_z) = B_0 e^{-z}[0, a_0 \exp(-x^2/l^2) + \cos(y + L), -\sin(y + L)].$$

A final example with $A = -B_0 y_0 \exp(-z) \cos(y/y_0)$ gives a magnetic field

$$(B_x, B_y, B_z) = B_0 e^{-z}[0, y_0 \cos(y/y_0), -\sin(y/y_0)],$$

to model a curved coronal loop whose field lines lie in vertical planes (Figure 3.15).

Later, Low (1993b) went on to prove that any force-free laminated solution (b_y, b_z) of the general form $b_y - ib_z = F(y+iz) \exp if(x)$ may be transformed to a magnetohydrostatic field by means of the transformation $B_y(x, y, z) = b_y(x, Y, z)$, $B_z(x, y, z) = dY/dy\ b_z(x, Y, z)$, where $Y = Y(x, y)$. For example,

$$(B_x, B_y, B_z) = B_0 e^{-z}[0, \cos(Y/H), -dY/dy\ \sin(Y/H)],$$

where $Y^2 = y^2 + y_0^2$, gives prominence-like condensations.

3.5.2.2 *Laminated Electric Current*

In general, eliminating the horizontal pressure gradients between the two horizontal components of the force-balance equation (3.1) yields a compatibility condition for the magnetic fields, namely,

$$\frac{\partial}{\partial y}(\mathbf{B} \cdot \nabla B_x) = \frac{\partial}{\partial x}(\mathbf{B} \cdot \nabla B_y). \tag{3.82}$$

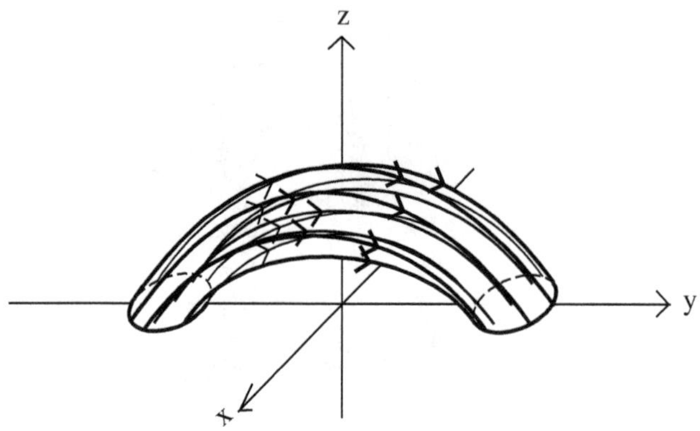

Figure 3.15. The magnetic field lines of a laminated coronal loop (after Low 1982b).

If the magnetic tension force is purely vertical, $\mathbf{B}\cdot\nabla B_x = \mathbf{B}\cdot\nabla B_y = 0$ and so both sides of Eq. (3.82) vanish identically. Also, the projections of field lines on the $z = 0$ plane are straight lines. Low (1984) investigated a range of nonlaminar prominence solutions of this form with $(B_x, B_y, B_z) = [k\sin\alpha, -k\cos\alpha, -a\tanh(ay)\cos\alpha]$, where α is a function of $\zeta = z + f(x, y)$. When $\alpha \equiv 0$, these reduce to the standard Kippenhahn-Schlüter solution $(B_x, B_y, B_z) = (0, -k, -a\tanh ay)$.

Next, Low (1985) explored a class of solutions in which the electric current is perpendicular to gravity. In a Cartesian geometry, this means that $j_z = 0$, and so the magnetic field may be written automatically in the form

$$(B_x, B_y, B_z) = \left(\frac{\partial\Psi}{\partial x}, \frac{\partial\Psi}{\partial y}, B_z\right). \tag{3.83}$$

Equation (3.82) then implies that $B_z = B_z(z, \partial\Psi/\partial z)$ is a function of z and $\partial\Psi/\partial z$ alone, while the remaining equation $\nabla\cdot\mathbf{B} = 0$ becomes

$$\frac{\partial^2\Psi}{\partial x^2} + \frac{\partial^2\Psi}{\partial y^2} + \frac{\partial B_z}{\partial z} = 0, \tag{3.84}$$

which determines Ψ for a given B_z. The special case of a linear field, for which $B_z = a\partial\Psi/\partial z$, reduces Eq. (3.84) to

$$\frac{\partial^2\Psi}{\partial x^2} + \frac{\partial^2\Psi}{\partial y^2} + a\frac{\partial^2\Psi}{\partial z^2} = 0.$$

For example, suppose we have a given potential equilibrium $w(x, y, z)$; then $\Psi(x, y, z) = a^{-1/2}w(x, y, a^{-1/2}z)$ represents a uniform expansion or contraction in the z-direction, and solutions starting from a set of dipoles show the effect of gravity in locally raising or lowering field lines.

The corresponding solutions in a spherical geometry (Bogdan and Low 1986) with a gravitational term $-\rho(GM_\odot/r^2)\hat{\mathbf{r}}$ and $j_r = 0$ have a magnetic field

$$(B_r, B_\theta, B_\Psi) = \left(B_r, \frac{1}{r}\frac{\partial\Psi}{\partial\theta}, \frac{1}{r\,\sin\theta}\frac{\partial\Psi}{\partial\Psi}\right), \tag{3.85}$$

and the special case of a linear field with $B_r = a(r)\partial\Psi/\partial r$ reduces $\nabla\cdot\mathbf{B} = 0$ to

$$\frac{1}{\sin\theta}\left[\frac{\partial}{\partial\theta}\left(\sin\theta\frac{\partial\Psi}{\partial\theta}\right) + \frac{1}{\sin\theta}\frac{\partial^2\Psi}{\partial\phi^2}\right] + \frac{\partial}{\partial r}\left(r^2 a\frac{\partial\Psi}{\partial r}\right) = 0.$$

This can be solved analytically when, for instance, $a(r) = (1 + r_0/r)^2$ by separating the variables, and the resulting density and pressure structures can be deduced both for axisymmetric and nonaxisymmetric

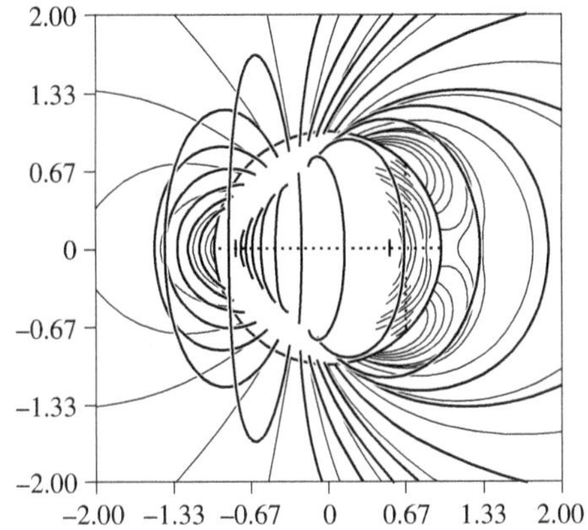

Figure 3.16. Magnetic field lines for a global magnetohydrostatic model of the corona (after Bogdan and Low 1986).

magnetohydrostatic atmospheres (Figure 3.16). In particular, the solutions have been used to compare with eclipse and coronagraph observations of the white-light corona (Bagenal and Gibson 1991; Zhao and Hoeksema 1994; Gibson and Bagenal 1995).

Low (1991) introduced a more general formalism by writing $\mathbf{j} = \nabla\alpha_J \times \nabla\beta_J$ in terms of current Euler potentials (α_J and β_J). Then the magnetic field can be written as $\mathbf{B} = \alpha_J\nabla\beta_J + \nabla f$, where f is an arbitrary function, which by $\nabla \cdot \mathbf{B} = 0$ satisfies

$$\nabla^2 f + \nabla \cdot (\alpha_J \nabla \beta_J) = 0.$$

In particular, if $\mathbf{j}$ is perpendicular to gravity ($g\ \hat{\mathbf{z}}$), then we can develop the analysis by writing $\beta_J = z$, and so $\mathbf{j} \times \mathbf{B}$ has components only in the $\nabla\alpha_J$ and z directions.

3.5.2.3 Effect of Pressure and Gravity on a Linear Force-Free Field

Low (1991, 1992) has explored the effect of gravity and pressure gradients on a linear force-free field by writing $\mu\mathbf{j} = \nabla \times \mathbf{B}$ as the sum of a force-free part and a non-force-free part in the xy-direction in the form

$$\nabla \times \mathbf{B} = \alpha\mathbf{B} + \nabla \times (F\ \hat{\mathbf{z}}), \tag{3.86}$$

where α is constant and $F(x, y, z)$ is an arbitrary function, such that $\mathbf{j} \times \mathbf{B}$ has components only in the ∇F and z directions. The most elegant way forward (Neukirch and Rastätter 1999; Petrie and Neukirch 2000) is to guarantee $\nabla \cdot \mathbf{B} = 0$ by writing $\mathbf{B} = \nabla \times [\nabla \times (P\hat{\mathbf{z}}) + T\hat{\mathbf{z}}]$ in poloidal-toroidal form, for then $T = \alpha P$ and the Helmholtz equation (3.49) (for the linear force-free case) simply generalises to

$$\nabla^2 P + \alpha^2 P + F = 0. \tag{3.87}$$

The aim is to choose forms for F in such a way that Eq. (3.87) can be solved analytically, and the components of the force-balance [Eq. (3.1)] along ∇F and $\hat{\mathbf{z}}$ can be easily integrated to obtain ρ and p. For example, if we let $F = f(z)B_z$, with $f(z)$ arbitrary, Eq. (3.87) becomes (with $\nabla_\perp^2 \equiv \partial^2/\partial x^2 + \partial^2/\partial y^2$)

$$\nabla^2 P + \alpha^2 P + f(z)\nabla_\perp^2 P = 0,$$

which may be Fourier transformed in x and y, and so $B_z = [1 - f(z)]^{-1}(\partial^2 P/\partial z^2 + \alpha^2 P)$ may be deduced as well as B_x and B_y. The resulting pressure and density follow from the force-balance as $p = p_0(z) - gf(z)B_z^2/(2\mu)$ and $\rho = -g^{-1}dp_0/dz + \frac{1}{2}df/dz B_z^2/\mu + f(\mathbf{B} \cdot \nabla)B_z/\mu$.

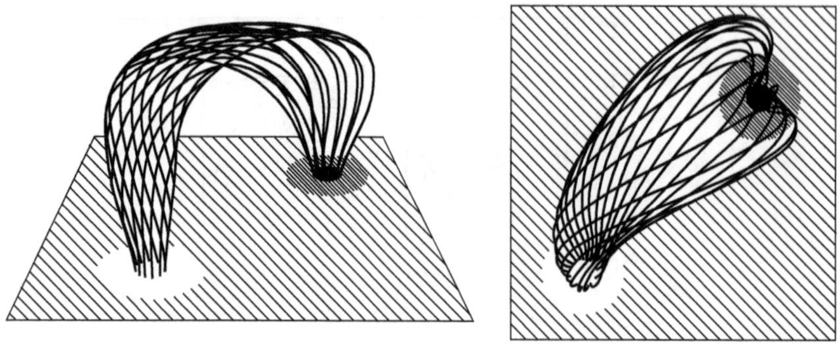

(a) (b)

Figure 3.17. Magnetic field lines viewed from (a) the side and (b) above for a magnetohydrostatic coronal loop generated by two submerged sources of equal but opposite strength (from Petrie and Neukirch 2000).

Petrie and Neukirch (2000) were able to generalise the Chiu-Hilton Green's function method (Sec. 3.4.3) to find the general solution of this class for any boundary conditions. As an example, they calculated for $f(z) \equiv constant = -0.1$ and $\alpha = 0.5$ the field of a coronal loop (Figure 3.17) and showed how the pressure and density contours are raised inside the loop. To make analytical progress, Low (1992) instead assumed that $f(z) = ae^{-kz}$ and found solutions which could be used to model an active region. For instance, the simplest solution is

$$(B_x, B_y, B_z) = B_0[-(\alpha/k)\sin ky\ J_s(\zeta), (\tfrac{1}{2}\kappa/k)\sin ky\ \zeta dJ_s/d\zeta, \cos ky J_s(\zeta)],$$
$$p = B_0^2 a/(2\mu)e^{-\kappa z}[J_s(\zeta)]^2 \sin^2 ky, \qquad \rho = p\kappa/g,$$

where J_s is a Bessel function, $\zeta = 2\sqrt{a}k/\kappa$, $s^2 = 4(k^2 - \alpha^2)/\kappa^2$ and the case $\alpha = 0$ reduces to the Zweibel-Hundhausen solution [Eq. (3.77)].

3.5.2.4 General Magnetic Field

Low (1993a, 2005) considered the generalisation,

$$\mathbf{0} = -\nabla p + \mathbf{j} \times \mathbf{B} - \rho \nabla \Phi, \tag{3.88}$$

of Eq. (3.1) to include a potential Φ, which includes the Cartesian ($\Phi = gz$) and spherical cases ($\Phi = -GM_\odot/r$) as well as more general inertial forces such as a centrifugal force. By taking $\nabla\Phi\times$(3.88), we may write $\mu\mathbf{j}$ as

$$\nabla \times \mathbf{B} = \alpha \mathbf{B} + \frac{\mu}{B_\Phi} \nabla p \times \nabla \Phi, \tag{3.89}$$

in terms of one current component parallel to $\mathbf{B}$ and another normal to both ∇p and $\nabla\Phi$ (i.e., in the intersection of surfaces $p = constant$ and $\Phi = constant$), where $B_\Phi = \mathbf{B}\cdot\nabla\Phi$ and $\alpha = (\nabla\times\mathbf{B}\cdot\nabla\Phi)/B_\Phi$. Now, if p is a function of only B_Φ and Φ, α is constant along field lines and the electric currents along and normal to $\mathbf{B}$ become decoupled; also, the second term on the right of Eq. (3.89) becomes $(\mu/B_\Phi)(\partial p/\partial B_\Phi)\nabla B_\Phi \times \nabla\Phi$, which reduces to a form of the second term on the right of Eq. (3.86) for $\Phi = gz$. However, in general, p will also be a function of a third variable such as α, so that the currents are coupled. For a given $p = p(\alpha, B_\Phi, \Phi)$, Eq. (3.89) together with $\nabla \cdot \mathbf{B} = 0$ then represent four scalar equations in four unknowns (B_x, B_y, B_z, α). The density follows by taking the scalar product of Eq. (3.88) with $\mathbf{B}$ as $\rho = -(\mathbf{B}\cdot\nabla p)/B_\Phi$. Continuing the formulation by writing the magnetic field in terms of Euler potentials highlights the mathematical structure of the equations and shows how the particular cases that we have discussed in this section arise, but it has not yet led to any new classes of analytical solution.

3.5.3 Transforming to Incompressible Field-Aligned Flows

Consider the three-dimensional magnetohydrostatic equations

$$\nabla p_s = \mathbf{j}_s \times \mathbf{B}_s, \quad \mathbf{j}_s = \frac{1}{\mu}\nabla \times \mathbf{B}_s, \quad \nabla \cdot \mathbf{B}_s = 0, \tag{3.90}$$

for a pressure p_s and magnetic field $\mathbf{B}_s$, or the two-dimensional Grad-Shafranov equation

$$\nabla^2 A = -\mu \frac{dp_s}{dA}, \qquad \mathbf{B}_s = \nabla \times (A\hat{\mathbf{z}}). \tag{3.91}$$

Compare these with the three-dimensional ideal, incompressible, steady MHD equations

$$\rho(\mathbf{v}\cdot\nabla)\mathbf{v} = \mathbf{j}\times\mathbf{B} - \nabla p, \quad \nabla\cdot(\rho\mathbf{v}) = \nabla\cdot\mathbf{B} = 0, \quad \nabla\times(\mathbf{v}\times\mathbf{B}) = \mathbf{0}. \tag{3.92}$$

Gebhardt and Kiessling (1992) discovered that magnetohydrostatic solutions can be transformed to give new MHD solutions in the form of steady, incompressible, field-aligned flows. Thus, if p_s and $\mathbf{B}_s$ satisfy Eq. (3.90), and if $\mathbf{B}_s \cdot \nabla\rho = \mathbf{B}_s \cdot \nabla M_A = 0$ (so that ρ and M_A are constant on field lines), then Eq. (3.92) are satisfied by solutions of the form

$$\mathbf{v} = \pm M_A \frac{\mathbf{B}}{\sqrt{\mu\rho}}, \quad \mathbf{B} = \frac{\mathbf{B}_s}{\sqrt{1-M_A^2}}, \quad p = p_s - \frac{M_A^2}{1-M_A^2}\frac{B_s^2}{2\mu}.$$

In particular, two-dimensional solutions may be found by choosing a function $\Pi(A)$, solving the Grad-Shafranov equation

$$\nabla^2 A = -\mu \frac{d\Pi}{dA},$$

choosing a function $M_A(A)$, evaluating $\alpha(A) = \pm \int dA/\sqrt{1-M_A^2}$, and finally deducing the magnetic field $\mathbf{B} = \nabla\alpha \times \hat{\mathbf{z}}$ and pressure $p = \Pi - \frac{1}{2}\rho v^2$. This can easily be performed for $\partial\Pi/dA = c^2 A$ (Helmholtz equation) or $d\Pi/dA = \exp(-2A)$ (Liouville's equation). It has been applied to Evershed flow, two- and three-dimensional coronal arcades with flows (Petrie and Neukirch 1999; Nickeler and Wiegelmann 2010) and to vortex current sheets (Nickeler and Wiegelmann 2012).

Chapter 4

Waves

4.1 Introduction

4.1.1 Fundamental Modes

In a gas such as the air one breathes, one is aware of the continual presence of *sound (or acoustic) waves* – especially when there are children in the vicinity, willing to act as a source! Sound waves exist because of the presence of a pressure *restoring force*, so that a local compression or rarefaction of the gas sets up a pressure gradient in opposition to the motion, which tries to restore the original equilibrium. If the gas is uniform, the waves propagate equally in all directions at the sound speed c_s. The main role of the waves is to carry energy away from the source, but, for the most part, they possess such a small amplitude that the ambient gas is disturbed only slightly. When its amplitude is large enough, however, a wave may sometimes steepen into a *shock wave* (Chapter 5), as with the sonic boom from a supersonic aircraft.

In a plasma such as that found in the solar atmosphere, there are typically four wave modes, driven by different restoring forces. The magnetic tension and Coriolis forces can drive so-called *Alfvén waves* and *inertial waves*, respectively. The magnetic pressure, plasma pressure and gravity can act separately and generate *compressional Alfvén waves*, *sound waves* and (*internal*) *gravity waves*, respectively; however, when acting together, these three forces produce only two *magnetoacoustic gravity* modes. In the absence of gravity, the two modes are referred to as *magnetoacoustic waves*, and when the magnetic field vanishes they are called *acoustic gravity waves.*

Over the past thirty years there has been a huge growth in the theory of waves on the Sun, notably in helioseismology (Sec. 1.3.4) and coronal seismology (Sec. 4.11.2), which diagnose the physical properties of the interior or atmosphere, respectively. The result is that here I can only give a brief introduction to the field, by describing the properties of each of the pure modes and touching on the applications. For a much more in-depth treatment, the reader is referred to the excellent books by Goossens (2003) and Goedbloed et al. (2004) and the review by Nakariakov and Verwichte (2005).

Waves are always present on the Sun, because it is such a restless body, containing features that are continually in motion and are initiating waves over a wide range of periods. In the convection zone, turbulence is continually generating global acoustic modes that are trapped in the solar interior and are viewed directly at the photosphere as five-minute oscillations. In the solar atmosphere, a wide range of small-scale dynamic phenomena (such as granulation, spicules and microflares) are also continually generating a variety of wave modes that may well contribute to its heating. Haunting evidence of such waves has been obtained of spicules waving to and fro in Hinode movies. Again, movies of sunspots reveal *running penumbral waves* propagating outwards from the umbra, and a large flare often emits a *Moreton wave.*

All mathematical discussion of wave motion traces a standard pattern, which is followed here. First, One sets up an equilibrium and then perturbs it slightly to see whether the resulting disturbance propagates as a wave. The basic equations are linearised, and, if the equilibrium is uniform, the perturbation quantities are assumed to vary like $\exp i(\mathbf{k} \cdot \mathbf{r} - \omega t)$, the object being to find the *dispersion relation*, which expresses ω in terms of $\mathbf{k}$ and the parameters of the medium.

In this chapter, our aim is to describe the basic properties of the main types of wave in a uniform medium (Secs. 4.1 to 4.8) and then to touch briefly on the effects of an inhomogeneous medium (Secs. 4.9 and 4.10) and some examples of waves on the Sun (Sec. 4.11). Wave dissipation is treated in Section 10.3. More complex situations when the perturbation is nonlinear are outside the scope of this chapter.

4.1.2 Basic Equations

The basic equations for our discussion of waves are the continuity of mass, momentum and energy, together with the induction equation, in the form

$$\frac{d\rho}{dt} + \rho \boldsymbol{\nabla} \cdot \mathbf{v} = 0, \tag{4.1}$$

$$\rho \frac{d\mathbf{v}}{dt} = -\boldsymbol{\nabla} p + (\boldsymbol{\nabla} \times \mathbf{B}) \times \mathbf{B}/\mu - \rho g\, \hat{\mathbf{z}} - 2\rho\, \boldsymbol{\Omega} \times \mathbf{v}, \tag{4.2}$$

$$\frac{d}{dt}\left(\frac{p}{\rho^{\gamma}}\right) = 0, \tag{4.3}$$

$$\frac{\partial \mathbf{B}}{\partial t} = \boldsymbol{\nabla} \times (\mathbf{v} \times \mathbf{B}), \tag{4.4}$$

$$\boldsymbol{\nabla} \cdot \mathbf{B} = 0. \tag{4.5}$$

The resulting electric current and temperature follow from

$$\mathbf{j} = \boldsymbol{\nabla} \times \mathbf{B}/\mu, \qquad T = \frac{mp}{k_B \rho}.$$

The equations are written in a reference frame rotating with the Sun at a (constant) angular velocity ($\boldsymbol{\Omega}$) relative to an inertial frame. The rotation has a negligible effect on Maxwell's equations provided the speed $|\boldsymbol{\Omega} \times \mathbf{r} + \mathbf{v}|$ is much less than the speed of light. It gives rise to the *Coriolis force* ($-2\rho\, \boldsymbol{\Omega} \times \mathbf{v}$) in the equation of motion together with a *centrifugal force* ($\frac{1}{2}\rho \boldsymbol{\nabla} |\Omega \times \mathbf{r}|^2$), which has here been omitted since it may be combined with the gravitational term. The gravitational force is locally $-\rho g\, \hat{\mathbf{z}}$, with g assumed constant and the z-axis directed along the outward normal to the solar surface. For simplicity, the plasma has been assumed frozen to the magnetic field and *adiabatic*: that is, thermally isolated from its surroundings, so that p/ρ^{γ} remains constant following the motion. From Eq. (2.29a) it can be seen that this is only valid provided $\tau \ll p/\mathcal{L}$, where τ is the wave-period and the energy loss $\mathcal{L}$ [given by Eq. (2.32)] may include thermal conduction, Ohmic heating, radiative cooling (L_r) and (smaller-scale) wave heating. For example, p/L_r takes a value of about 1 s in the low photosphere and 1 h in the upper chromosphere, whereas $p/\mathcal{L}$ due to thermal conduction is about 500 s for a scale-length of 1 Mm (10^6 m).

Consider a vertically stratified, stationary plasma, with a uniform magnetic field ($\mathbf{B}_0$), temperature (T_0), and a density and pressure which behave like

$$\rho_0(z) = constant \times e^{-z/H}, \quad p_0(z) = constant \times e^{-z/H}, \tag{4.6}$$

and satisfy $0 = -dp_0/dz - \rho_0 g$. Here

$$H = \frac{p_0}{\rho_0 g} \tag{4.7}$$

is the scale-height, typically 150 km in the photosphere and 100 Mm (10^8 m) in the corona (see Sec. 3.1.2). Consider small departures from the equilibrium

$$\rho = \rho_0 + \rho_1, \quad \mathbf{v} = \mathbf{v}_1, \quad p = p_0 + p_1, \quad \mathbf{B} = \mathbf{B}_0 + \mathbf{B}_1,$$

and linearise the basic Eqs. (4.1) to (4.5) by neglecting squares and products of the small quantities (denoted by subscript 1). The result is

$$\frac{\partial \rho_1}{\partial t} + (\mathbf{v}_1 \cdot \boldsymbol{\nabla})\rho_0 + \rho_0(\boldsymbol{\nabla} \cdot \mathbf{v}_1) = 0, \tag{4.8}$$

$$\rho_0 \frac{\partial \mathbf{v}_1}{\partial t} = -\boldsymbol{\nabla} p_1 + (\boldsymbol{\nabla} \times \mathbf{B}_1) \times \mathbf{B}_0/\mu - \rho_1 g\, \hat{\mathbf{z}} - 2\rho_0\, \boldsymbol{\Omega} \times \mathbf{v}_1, \tag{4.9}$$

$$\frac{\partial p_1}{\partial t} + (\mathbf{v}_1 \cdot \boldsymbol{\nabla})p_0 - c_s^2 \left(\frac{\partial \rho_1}{\partial t} + (\mathbf{v}_1 \cdot \boldsymbol{\nabla})\rho_0 \right) = 0, \tag{4.10}$$

$$\frac{\partial \mathbf{B}_1}{\partial t} = \boldsymbol{\nabla} \times (\mathbf{v}_1 \times \mathbf{B}_0), \tag{4.11}$$

$$\boldsymbol{\nabla} \cdot \mathbf{B}_1 = 0, \tag{4.12}$$

where
$$c_s^2 = \frac{\gamma p_0}{\rho_0} = \frac{\gamma k_B T_0}{m} \tag{4.13}$$

is the sound speed. Equations (4.8) to (4.12) may be reduced to a single equation by differentiating Eq. (4.9) with respect to time and substituting for $\partial\rho_1/\partial t$, $\partial p_1/\partial t$ and $\partial \mathbf{B}_1/\partial t$ from Eqs. (4.8), (4.10) and (4.11), respectively. After some manipulation, a generalised wave equation may be derived for the disturbance velocity ($\mathbf{v}_1$):

$$\frac{\partial^2 \mathbf{v}_1}{\partial t^2} = c_s^2 \boldsymbol{\nabla}(\boldsymbol{\nabla} \cdot \mathbf{v}_1) - (\gamma - 1) g \hat{\mathbf{z}}(\boldsymbol{\nabla} \cdot \mathbf{v}_1) - g \boldsymbol{\nabla} v_{1z} - 2\boldsymbol{\Omega} \times \frac{\partial \mathbf{v}_1}{\partial t} + \{\boldsymbol{\nabla} \times [\boldsymbol{\nabla} \times (\mathbf{v}_1 \times \mathbf{B}_0)]\} \times \frac{\mathbf{B}_0}{\mu \rho_0}. \tag{4.14}$$

The next step is to seek *plane-wave* solutions of the form

$$\mathbf{v}_1(\mathbf{r}, t) = \mathbf{v}_1 e^{i(\mathbf{k}\cdot\mathbf{r} - \omega t)},$$

in terms of the *wavenumber vector* ($\mathbf{k}$) and *frequency* (ω). The period of the wave is then just $2\pi/\omega$, while its *wavelength* (λ) is $2\pi/k$ and its *direction of propagation* is $\hat{\mathbf{k}}$ ($\equiv \mathbf{k}/k$). The effect of the plane-wave assumption is simply to replace $\partial/\partial t$ by $-i\omega$ and $\boldsymbol{\nabla}$ by $i\mathbf{k}$ in Eq. (4.14). For a vanishing magnetic field, Eq. (4.14) reduces to

$$\omega^2 \mathbf{v}_1 = c_s^2 \mathbf{k}(\mathbf{k} \cdot \mathbf{v}_1) + i(\gamma - 1) g \hat{\mathbf{z}}(\mathbf{k} \cdot \mathbf{v}_1) + i g \mathbf{k} v_{1z} - 2i\omega \boldsymbol{\Omega} \times \mathbf{v}_1. \tag{4.15}$$

However, when a magnetic field is present, the last term in Eq. (4.14) is proportional to $e^{z/H}$ through the presence of ρ_0. Nevertheless, provided the wavelength ($2\pi/k$) of the perturbations is much smaller than the scale-height $H[\equiv c_s^2/(\gamma g)]$, ρ_0 may be regarded as locally constant, which can be justified formally by the WKB approximation (Bender and Orszag 1999). The equation then becomes

$$\omega^2 \mathbf{v}_1 = c_s^2 \mathbf{k}(\mathbf{k} \cdot \mathbf{v}_1) + i(\gamma - 1) g \hat{\mathbf{z}}(\mathbf{k} \cdot \mathbf{v}_1) + i g \mathbf{k} v_{1z} - 2i\omega \boldsymbol{\Omega} \times \mathbf{v}_1 + \{\mathbf{k} \times [\mathbf{k} \times (\mathbf{v}_1 \times \mathbf{B}_0)]\} \times \frac{\mathbf{B}_0}{\mu \rho_0}. \tag{4.16}$$

Equations (4.15) and (4.16) will be employed in this chapter as a basis for discussing the fundamental wave modes in a uniform medium. The object is to determine the dispersion relation $\omega = \omega(\mathbf{k})$ for the frequency as a function of the magnitude of $\mathbf{k}$ and its inclination to gravity and the magnetic field. Since

Eqs. (4.15) and (4.16) each represent a set of three homogeneous equations for the three velocity components of $\mathbf{v}_1$, the relation between its coefficients (and so between ω and $\mathbf{k}$) can be found in principle by setting the determinant of coefficients equal to zero. But, in practice, this is often carried out by various vector manipulations.

The velocity $\mathbf{v}_p = (\omega/k)\ \hat{\mathbf{k}}$ is known as the *phase velocity* of the wave. Its magnitude (ω/k) gives the speed of propagation in the direction $\hat{\mathbf{k}}$ for a wave specified by a single wavenumber. By contrast, a *packet* (or *group*) of *waves* possesses a range of wave-numbers and travels at the *group velocity* ($\mathbf{v}_g$) with Cartesian components

$$v_{gx} = \frac{\partial \omega}{\partial k_x}, \quad v_{gy} = \frac{\partial \omega}{\partial k_y}, \quad v_{gz} = \frac{\partial \omega}{\partial k_z}. \tag{4.17}$$

It is this which gives the velocity at which energy is transmitted, and in general it differs both in magnitude and direction from the phase velocity. When the phase speed varies with wavelength, the wave is said to be *dispersive*. However, in the special case when ω is linearly proportional to k, the wave is said to be *non-dispersive*, and the phase and group velocities are identical. In general, it transpires that the propagation is *anisotropic*, since the phase speed varies with the direction of propagation. There are three preferred directions, due to the magnetic field, gravity and rotation. This causes great complexity in general, and so in the next few sections each of the modes will be isolated in turn to derive their basic characteristics.

4.2 Sound Waves

When $g = B_0 = \Omega = 0$, so that the only restoring force is the pressure gradient, Eq. (4.15) for the disturbance velocity ($\mathbf{v}_1$) reduces to

$$\omega^2\ \mathbf{v}_1 = c_s^2\ \mathbf{k}\ (\mathbf{k} \cdot \mathbf{v}_1). \tag{4.18}$$

After taking the scalar product with $\mathbf{k}$ and assuming $\mathbf{k} \cdot \mathbf{v}_1 \neq 0$, we find

$$\omega^2 = k^2 c_s^2.$$

This *dispersion relation* for *sound waves* becomes

$$\omega = \pm k c_s. \tag{4.19}$$

Sound (or *acoustic*) waves therefore propagate equally in all directions at a phase speed v_p $(\equiv \omega/k) = c_s$ and a group velocity v_g $(\equiv d\omega/dk) = c_s$ in the directions $\pm\mathbf{k}$. From the definition (4.13) for the sound speed, we find numerically, when $\gamma = 5/3$ and $m = 0.5\ m_p$ $(\tilde{\mu} = 0.5)$, that $c_s \approx 166\ T_0^{1/2}$ m s^{-1}, which varies typically from about 10 km s^{-1} in the solar photosphere to about 200 km s^{-1} in the corona. The requirement that $\mathbf{k} \cdot \mathbf{v}_1$ (and so $\nabla \cdot \mathbf{v}_1$) be non-vanishing implies that sound waves owe their existence to the compressibility of the plasma. Also, it is noteworthy that the waves are *longitudinal* in the sense that, according to Eq. (4.18), the velocity perturbation ($\mathbf{v}_1$) is in the direction of propagation ($\mathbf{k}$). The effect of a uniform flow is to replace ω by a Doppler-shifted value (PROBLEM 4.1: for all problems and solutions, see web page www.cambridge.org/9780521854719).

4.3 Magnetically Driven Waves

In Section 2.7 it was pointed out that the Lorentz force may be interpreted as the sum of a magnetic tension of amount B_0^2/μ and a magnetic pressure of amount $B_0^2/(2\mu)$ per unit area. Now, one of the effects of a tension ($\mathcal{T}$) in an elastic string (of mass density ρ_0 per unit length) is to permit transverse waves to propagate along the string with speed $(\mathcal{T}/\rho_0)^{1/2}$. So, by analogy, it is reasonable to expect the

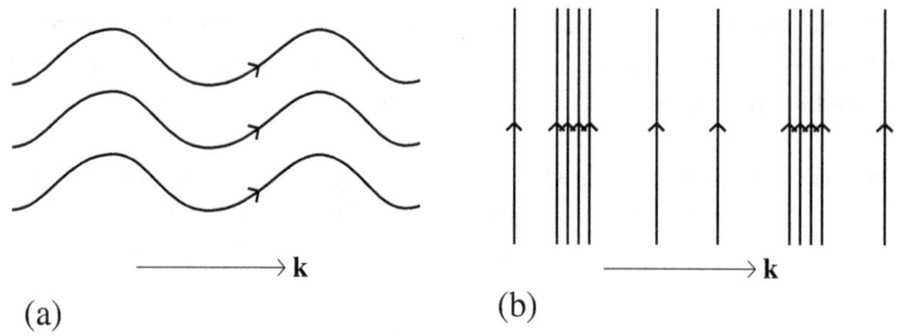

Figure 4.1. (a) The ripples of magnetic field lines caused by an Alfvén wave propagating along the field in the direction **k**. (b) The compression and rarefaction of magnetic field lines due to a compressional Alfvén wave propagating across the field in the direction **k**.

magnetic tension to produce transverse waves that propagate along the magnetic field $\mathbf{B}_0$ (Figure 4.1a) with speed $[(B_0^2/\mu)/\rho_0)]^{1/2}$. This is known as the *Alfvén speed*

$$v_A = \frac{B_0}{(\mu\rho_0})^{1/2},$$

and, according to Eq. (2.48), it may be written numerically as $v_A = 2.8{\times}10^{16}\ (B_0/n_0^{1/2})$ m s^{-1}, with B_0 in tesla. Thus, the values $B_0 \approx 100$ G and $n_0 \approx 10^{16}$ m^{-3} = 10^{10} cm^{-3} (characteristic of the corona above an active region) give typically 3,000 km s^{-1}, whereas $B_0 \approx 10^3$ G and $n_0 \approx 10^{23}$ m^{-3} (characteristic of the photospheric network) give an Alfvén speed of only 10 km s^{-1}.

In Section 4.2 it was shown that the pressure of a gas obeying the adiabatic law, $p/\rho^\gamma = constant$, produces (longitudinal) sound waves with phase speed $(\gamma p_0/\rho_0)^{1/2}$. Thus, by analogy, we might expect the magnetic pressure $p_m = B_0^2/(2\mu)$ to generate longitudinal magnetic waves propagating *across* the magnetic field (Figure 4.1b). If the magnetic field is frozen to the plasma (Sec. 2.6.2), the field strength and plasma density vary such that $B/\rho = constant$, which means that, in terms of the magnetic pressure, $p_m/\rho^2 = constant$. By comparing with the adiabatic law, this suggests that the effective value of γ for these magnetic waves is 2. Their wave speed must then be $(2p_m/\rho_0)^{1/2} = B_0/(\mu\rho_0)^{1/2}$, which is again the Alfvén speed.

The above physical intuition suggests that a purely magnetic wave should exist, driven by the $\mathbf{j} \times \mathbf{B}$ force along or across the field. The mathematical analysis below supports this; it also demonstrates that the two types of magnetic wave are distinct and finds their speeds of propagation oblique to the field, not just along or across it.

When the magnetic field dominates the equilibrium, so that p_0 (and therefore c_s^2), Ω and g may all be set equal to zero, Eq. (4.16) becomes

$$\omega^2\ \mathbf{v}_1 = \{\mathbf{k} \times [\mathbf{k} \times (\mathbf{v}_1 \times \hat{\mathbf{B}}_0)]\} \times \hat{\mathbf{B}}_0\ v_A^2, \tag{4.20}$$

where $\hat{\mathbf{B}}_0$ is a unit vector in the direction of the magnetic field ($\mathbf{B}_0$). The vector products on the right-hand side may be expanded to give

$$\omega^2\ \mathbf{v}_1/v_A^2 = (\mathbf{k}\cdot\hat{\mathbf{B}}_0)^2\mathbf{v}_1 - (\mathbf{k}\cdot\mathbf{v}_1)(\mathbf{k}\cdot\hat{\mathbf{B}}_0)\hat{\mathbf{B}}_0 + [(\mathbf{k}\cdot\mathbf{v}_1) - (\mathbf{k}\cdot\hat{\mathbf{B}}_0)(\hat{\mathbf{B}}_0\cdot\mathbf{v}_1)]\mathbf{k},$$

or, in terms of the angle (θ_B) between the direction of propagation ($\hat{\mathbf{k}}$) and $\mathbf{B}_0$,

$$\begin{aligned}\omega^2\ \mathbf{v}_1/v_A^2 = k^2\cos^2\theta_B\ \mathbf{v}_1 &- (\mathbf{k}\cdot\mathbf{v}_1)k\ \cos\ \theta_B\ \hat{\mathbf{B}}_0 \\ &+[(\mathbf{k}\cdot\mathbf{v}_1) - k\ \cos\ \theta_B(\hat{\mathbf{B}}_0\cdot\mathbf{v}_1)]\ \mathbf{k}.\end{aligned} \tag{4.21}$$

Magnetic waves have several interesting properties. First, note from Eq. (4.12) that $\mathbf{k}\cdot\mathbf{B}_1 = 0$, so that the magnetic field perturbation is normal to the direction of propagation. Next, the scalar product of

Eq. (4.21) with $\hat{\mathbf{B}}_0$ yields

$$\hat{\mathbf{B}}_0 \cdot \mathbf{v}_1 = 0, \tag{4.22}$$

so that the perturbed velocity is normal to $\mathbf{B}_0$. This is not surprising, since only the Lorentz force ($\mathbf{j} \times \mathbf{B}$) is driving the motions and it is, of course, normal to $\mathbf{B}_0$ in the linear approximation. Finally, the scalar product of Eq. (4.21) with $\mathbf{k}$ gives

$$(\omega^2 - k^2 v_A^2)\,(\mathbf{k} \cdot \mathbf{v}_1) = 0, \tag{4.23}$$

which yields two distinct solutions that we discuss in the following.

4.3.1 Alfvén Waves (i.e., Shear or Torsional Alfvén Waves)

If the perturbation is *incompressible* ($\boldsymbol{\nabla}.\mathbf{v}_1 = 0$), so that

$$\mathbf{k} \cdot \mathbf{v}_1 = 0, \tag{4.24}$$

Equation (4.21) gives, after taking the positive square root,

$$\omega = k\ v_A\ \cos\,\theta_B \tag{4.25}$$

for *Alfvén waves,* sometimes known as *shear Alfvén waves.* (The positive or negative root gives a wave propagating in the same or the opposite direction to the magnetic field, respectively.) These waves have a phase speed $v_A \cos\,\theta_B$, which, for propagation along the magnetic field ($\theta_B = 0$), is just the Alfvén speed, in agreement with our earlier intuition. The variation of the phase speed with θ_B is most conveniently exhibited in a *polar diagram* (Figure 4.2), which takes the form of two circles of diameter v_A. In particular, it may be noted that the waves propagate fastest along the field but do not propagate at all in a direction normal ($\theta_B = \frac{1}{2}\pi$) to the field.

Taking the Z-axis, say, along the magnetic field ($\mathbf{B}_0 \equiv B_0\hat{\mathbf{B}}_0$), Eq. (4.25) may be written alternatively as $\omega = k_Z v_A$, and so differentiating with respect to k_Z gives the group velocity [according to Eq. (4.17)] as $\mathbf{v}_g = v_A\ \hat{\mathbf{B}}_0$. Energy is therefore carried at the Alfvén speed along the magnetic field, in spite of the fact that individual waves can travel at any inclination to the field (save $\frac{1}{2}\pi$).

The property (4.24) means that Alfvén waves are *transverse* in the sense that the velocity perturbation is normal to the propagation direction. Furthermore, with ρ_0 and p_0 uniform, Eqs. (4.8) and (4.10) imply that

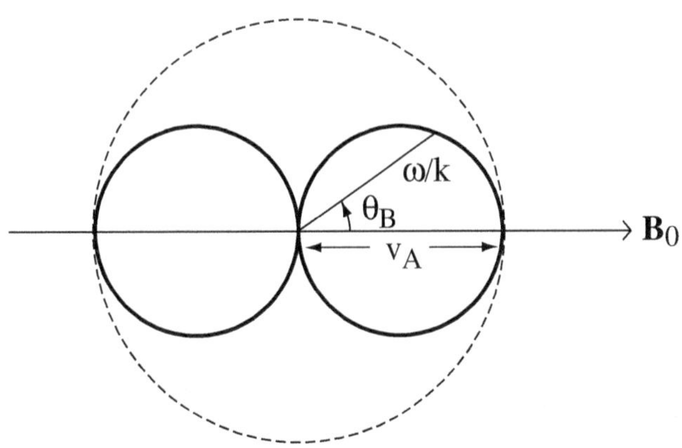

Figure 4.2. A polar diagram for Alfvén waves (solid curve) and compressional Alfvén waves (dashed curve). The length of the radius vector at an angle of inclination θ_B to the equilibrium magnetic field ($\mathbf{B}_0$) is equal to the phase speed (ω/k) for waves propagating in that direction.

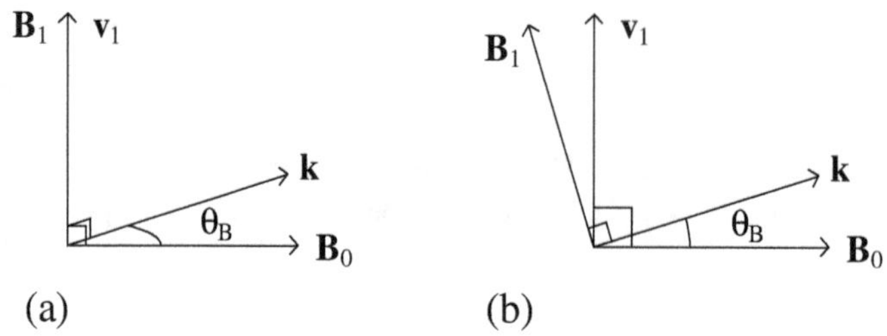

Figure 4.3. The directions of the perturbed velocity ($\mathbf{v}_1$) and magnetic field ($\mathbf{B}_1$) relative to the equilibrium magnetic field ($\mathbf{B}_0$) and the wave propagation direction ($\mathbf{k}$). For (a) the vectors $\mathbf{v}_1$ and $\mathbf{B}_1$ are both normal to the plane of $\mathbf{k}$ and $\mathbf{B}_0$, whereas for (b) $\mathbf{v}_1$ and $\mathbf{B}_1$ lie in the same plane as $\mathbf{k}$ and $\mathbf{B}_0$.

there are no accompanying density or pressure changes. Another characteristic can be shown by considering Eq. (4.11), which for the plane-wave solution becomes $-\omega\ \mathbf{B}_1 = \mathbf{k} \times (\mathbf{v}_1 \times \mathbf{B}_0)$ or

$$-\omega\ \mathbf{B}_1 = (\mathbf{k} \cdot \mathbf{B}_0)\ \mathbf{v}_1 - (\mathbf{k} \cdot \mathbf{v}_1)\ \mathbf{B}_0. \tag{4.26}$$

Since, by Eq. (4.24), $\mathbf{k} \cdot \mathbf{v}_1$ vanishes, Eqs. (4.25) and (4.26) give

$$\mathbf{v}_1 = -\frac{\mathbf{B}_1}{(\mu\rho_0)^{1/2}}, \tag{4.27}$$

which implies that $\mathbf{B}_1$ and $\mathbf{v}_1$ are in the same direction, both lying in a plane parallel to the wave-front. [For propagation in the opposite direction to the magnetic field one would obtain $\mathbf{v}_1 = \mathbf{B}_1/(\mu\rho_0)^{1/2}$.] By Eq. (4.22), this in turn means that

$$\mathbf{B}_0 \cdot \mathbf{B}_1 = 0, \tag{4.28}$$

so that the magnetic field perturbation is normal to $\mathbf{B}_0$. The directions of the perturbed quantities are indicated in Figure 4.3.

It is interesting to rewrite the Lorentz force as $\mathbf{j}_1 \times \mathbf{B}_0 = (\mathbf{k} \times \mathbf{B}_1) \times \mathbf{B}_0/\mu$, or

$$\mathbf{j}_1 \times \mathbf{B}_0 = (\mathbf{k} \cdot \mathbf{B}_0)\ \mathbf{B}_1/\mu - (\mathbf{B}_0 \cdot \mathbf{B}_1)\ \mathbf{k}/\mu. \tag{4.29}$$

The first term on the right comes from magnetic tension and second from magnetic pressure. Thus, in view of Eq. (4.28), the driving force for Alfvén waves is magnetic tension alone. Also, the ratio of magnetic to kinetic energy is

$$\frac{B_1^2/(2\mu)}{\frac{1}{2}\rho_0 v_1^2},$$

which, by virtue of Eq. (4.27) is unity, so that Alfvén waves involve an equipartition between magnetic and kinetic energy.

In a cylindrical geometry with an axial field $B_0\ \hat{\mathbf{z}}$, the shear Alfvén wave is called a *torsional Alfvén wave* and possesses only an azimuthal perturbation $\mathbf{B}_1 \sim \cos\ k(v_A t \pm z)\ \hat{\boldsymbol{\phi}}$. Superposing two such travelling waves propagating in opposite directions produces a standing wave in the form of a *torsional oscillation* of a flux tube with $\mathbf{B}_1 \sim 2\cos(kv_A t)\ \cos(kz)\ \hat{\boldsymbol{\phi}}$, as indicated in Figure 4.4.

When their amplitudes are no longer small, most disturbances cease to propagate with a constant profile, but instead distort in time. For instance, nonlinear shear Alfvén and torsional Alfvén waves possess compressive motions that lead to wave self-interaction and steepening (Vasheghani Farahani et al. 2012). However, *circularly polarised Alfvén waves* are most unusual in that they can continue to propagate without distortion when their amplitudes are large (see PROBLEMS 4.2 and 4.3). It can easily be verified that an

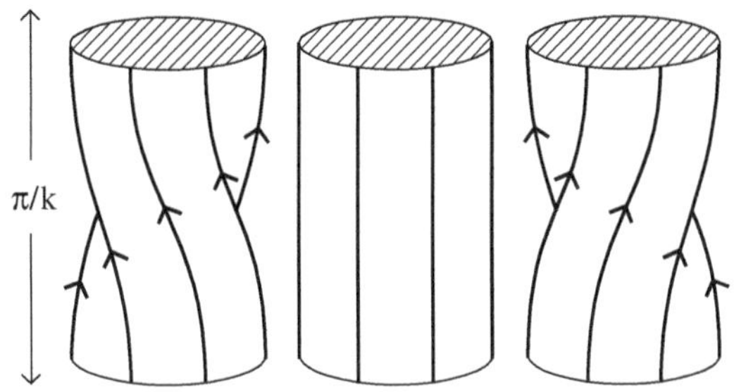

Figure 4.4. Oscillations of a magnetic flux tube of length π/k due to torsional Alfvén waves.

Alfvén-wave disturbance with the properties

$$\mathbf{v}_1 = -\frac{\mathbf{B}_1}{(\mu\rho_0)^{1/2}}, \quad |\mathbf{B}_0 + \mathbf{B}_1| = constant,$$

satisfies the full equations, without the need to linearise. The latter property reduces to Eq. (4.28) in the linear limit and it means that, as the wave passes by, the magnetic field vector just rotates, preserving a constant magnitude. An important consequence of the existence of finite-amplitude circularly polarised Alfvén waves is that they do not steepen and so tend to dissipate much less readily than other wave modes.

The time taken for a linear Alfvén wave (of wavelength λ) to decay away due to Ohmic diffusion is simply the diffusion time λ^2/η (Sec. 2.6.1) (see PROBLEMS 4.4 and 4.5). By contrast, other finite-amplitude waves will decay much more rapidly, since they steepen to form length-scales shorter than λ. Moreover, nonlinear interactions can transfer energy from Alfvén waves to magnetoacoustic waves, which then dissipate rapidly. All of the above plasma properties are unaltered by the presence of a significant plasma pressure, provided the pressure variations (p_1) are adiabatic, so that Eq. (4.24) implies the vanishing of ρ_1 (and hence of p_1).

4.3.2 Compressional Alfvén (or Fast-Mode) Waves

The second solution to Eq. (4.23) is

$$\omega = k\ v_A \tag{4.30}$$

for *compressional Alfvén waves.* The phase speed is v_A, regardless of the angle of propagation, as indicated in Figure 4.2, and the group velocity is $\mathbf{v}_g = v_A\ \mathbf{k}$, so that the energy is propagated isotropically.

Equations (4.21) and (4.22) imply that the velocity perturbation ($\mathbf{v}_1$) lies in the ($\mathbf{k}$, $\mathbf{B}_0$) plane in a direction normal to $\mathbf{B}_0$. It therefore possesses components both along and transverse to $\mathbf{k}$ in general, and it gives rise to both density and pressure changes – hence the name *compressional*, which is allied with *Alfvén wave* since its propagation speed is the Alfvén speed. When pressure gradients are included, the compressional Alfvén wave becomes a fast magnetoacoustic or fast-mode wave (Sec. 4.6), which for some authors is the preferred name even when pressure gradients are unimportant, so as to avoid confusion with shear Alfvén waves.

From Eq. (4.26), the vector $\mathbf{B}_1$ lies in the plane of $\mathbf{v}_1$ and $\mathbf{B}_0$ but is normal to $\mathbf{k}$ (Figure 4.3b). Furthermore, the Lorentz force [Eq. (4.29)] is in the direction of $\mathbf{v}_1$, and in general it contains a contribution from both the magnetic tension and the magnetic pressure. In the particular case of propagation directly across the magnetic field ($\theta_B = \frac{1}{2}\pi$), Eq. (4.21) shows that $\mathbf{v}_1$ is parallel to $\mathbf{k}$; the wave is thus longitudinal, and from Eq. (4.29) only the magnetic pressure is playing a part, as anticipated in our previous intuitive discussion. By contrast, propagation along the field ($\theta_B = 0$) makes the compressional Alfvén wave transverse and degenerate into an ordinary (i.e., shear) Alfvén wave: it is now driven wholly by the magnetic tension and produces no compression in spite of its name!

4.4 Internal Gravity Waves

Consider a blob of plasma that is displaced vertically a distance δz from equilibrium in the absence of a magnetic field (Figure 4.5), and make the assumptions that

(i) it remains in pressure equilibrium with its surroundings,

(ii) the density changes inside the blob are adiabatic.

These are valid for a motion that is slow enough that sound waves rapidly traverse the system and fast enough that entropy is preserved.

At the original height (z), the pressure and density inside the blob are the same as those outside, namely p_0 and ρ_0. These satisfy hydrostatic equilibrium ($dp_0/dz = -\rho_0 g$). The pressure and density outside the blob at height $z + \delta z$ are then $p_0 + \delta p_0$ and $\rho_0 + \delta\rho_0$, where

$$\delta p_0 = -\rho_0 g\, \delta z, \quad \delta\rho_0 = -\frac{d\rho_0}{dz}\, \delta z. \tag{4.31}$$

Inside the blob at height $z+\delta z$, the pressure and density are $p_0+\delta p$ and $\rho_0+\delta\rho$, say, where, by assumption (i), $\delta p = \delta p_0 = -\rho_0 g\, \delta z$. Now, assumption (ii) means that, as the blob rises, its pressure and density obey $p/\rho^\gamma = constant$, so that $\delta p = c_s^2 \delta\rho$, where the sound speed (c_s) is given by Eq. (4.13). Substituting for δp therefore gives the internal density change as

$$\delta\rho = -\frac{\rho_0 g\, \delta z}{c_s^2}. \tag{4.32}$$

Since the new density inside the blob differs from the ambient density at its new height, the blob experiences a buoyancy force of amount

$$g(\delta\rho_0 - \delta\rho) = -N^2 \rho_0\, \delta z, \tag{4.33}$$

say, where by Eqs. (4.31) and (4.32)

$$N^2 = -g\left(\frac{1}{\rho_0}\frac{d\rho_0}{dz} + \frac{g}{c_s^2}\right). \tag{4.34}$$

When it is real, N is known as the *Brunt-Väisälä* (or *Brunt*) *frequency.*

An alternative expression using equations of state and hydrostatic equilibrium is

$$N^2 = \frac{g}{T_0}\left[\frac{dT_0}{dz} - \left(\frac{dT}{dz}\right)_{ad}\right], \tag{4.35}$$

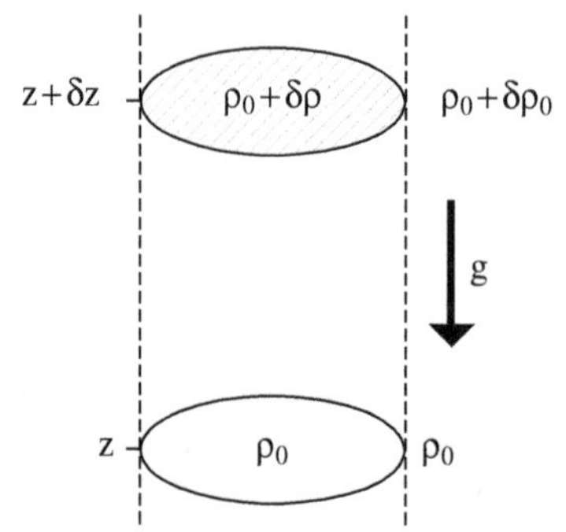

Figure 4.5. An element of plasma moves vertically against gravity from a height z, where the external density is ρ_0, to a height $z + \delta z$, where the external density is $\rho_0 + \delta\rho_0$.

where $(dT/dz)_{ad} = -(\gamma-1)T_0 g/c_s^2$ is the temperature gradient for an adiabatic equilibrium. (It is obtained by eliminating p_0 and ρ_0 between the equations $p_0 = k_B \rho_0 T_0/m$, $p_0/\rho_0^\gamma = constant$ and $dp_0/dz = -\rho_0 g$.) In general, N varies with height z, but, when the equilibrium temperature (T_0) is uniform, Eq. (4.35) becomes

$$N^2 = \frac{(\gamma-1)\, g^2}{c_s^2}. \tag{4.36}$$

Furthermore, in the presence of a horizontal magnetic field with Alfvén speed v_A, the Brunt-Väisälä frequency [Eq. (4.34)] is increased to (Chen and Lykoudis 1972)

$$N^2 = -g\left(\frac{1}{\rho_0}\frac{d\rho_0}{dz} + \frac{g}{c_s^2 + v_A^2}\right),$$

or, in the case of a uniform temperature,

$$N^2 = \frac{g^2}{c_s^2}\left(\gamma - \frac{c_s^2}{c_s^2 + v_A^2}\right).$$

If the only resultant force acting on the plasma element is due to buoyancy [Eq. (4.33)], the equation of motion becomes

$$\rho_0\,\frac{d^2(\delta z)}{dt^2} = -N^2 \rho_0\, \delta z. \tag{4.37}$$

Thus, the element executes simple harmonic motion with frequency $\omega = N$ provided

$$N^2 > 0, \tag{4.38}$$

so that T decreases with height slower than adiabatic $[-dT_0/dz < -(dT/dz)_{ad}]$.

Condition (4.38) is the *Schwarzschild criterion for convective stability.* If the temperature decreases with height faster than adiabatic, so that Eq. (4.38) is violated, the solution to Eq. (4.37) is exponentially growing and we have *convective instability* (Sec. 7.5.6). This is the case in the convection zone (Sec. 1.3.3).

The preceding intuitive discussion leads us to expect the existence of *internal gravity waves* when $N^2 > 0$ due to the tendency for plasma to oscillate slowly with frequency N. When $g/c_s \ll kc_s$, their dispersion relation is

$$\omega = N \,\sin\, \theta_g \tag{4.39}$$

in terms of N and the inclination $[\theta_g = \cos^{-1}(k_z/k)]$ between $\mathbf{k}$ and the z-axis, and they have interesting properties (PROBLEM 4.6). The word 'internal' is added to distinguish them from the *surface gravity waves* that propagate along an interface.

4.5 Inertial Waves

Consider now the effect of the Coriolis force alone, so that the linearised equation of motion [Eq. (4.9)] becomes

$$\frac{\partial \mathbf{v}_1}{\partial t} = -2\,\mathbf{\Omega} \times \mathbf{v}_1.$$

With the Z-axis along the axis of rotation, its components are $\partial v_{1X}/\partial t = 2\Omega v_{1Y}$ and $\partial v_{1Y}/\partial t = -2\Omega v_{1X}$, which possess a wave-like solution

$$v_{1X} = A\,\cos\,(kZ - 2\Omega t), \quad v_{1Y} = A\,\sin\,(kZ - 2\Omega t).$$

This *inertial wave* propagates along the axis of rotation with frequency 2Ω. The Coriolis force acts perpendicular to the direction of motion and causes each plasma element to execute a circular orbit in the XY-plane.

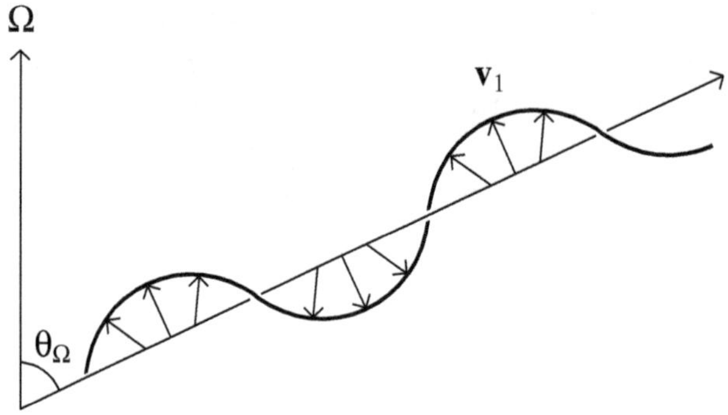

Figure 4.6. The directions of the velocity vector ($\mathbf{v}_1$) in an inertial wave propagating along the direction $\mathbf{k}$ with frequency $2\Omega \cos \theta_\Omega$. $\mathbf{v}_1$ is normal to $\mathbf{k}$ at each point. Relative to an inertial frame, the $\mathbf{k}$-vector itself rotates with angular speed Ω about the rotation axis.

Waves driven by the Coriolis force may also propagate away from the rotation axis. With $g = B_0 = 0$, the basic equation [Eq. (4.15)] becomes

$$\omega^2 \, \mathbf{v}_1 \;=\; c_s^2 \, \mathbf{k}(\mathbf{k}\cdot\mathbf{v}_1) - 2i\, \mathbf{\Omega} \times \mathbf{v}_1. \tag{4.40}$$

Incompressible solutions may be sought such that $\mathbf{k}\cdot\mathbf{v}_1 = 0$, and so the vector product of Eq. (4.40) with $\mathbf{k}$ gives $\omega \, \mathbf{k}\times\mathbf{v}_1 = 2i(\mathbf{k}\cdot\mathbf{\Omega})\mathbf{v}_1$. Equating the magnitude of both sides then yields

$$\omega = \pm\frac{2(\mathbf{k}\cdot\mathbf{\Omega})}{k} \tag{4.41}$$

as the dispersion relation for *inertial waves.* In terms of the angle θ_Ω between the rotation axis and the propagation direction, it may be written $\omega = \pm 2\Omega \cos \theta_\Omega$. As the wave propagates, the velocity vector rotates about the direction of propagation (Figure 4.6). At a fixed location, $\mathbf{v}_1$ performs a circular motion in a plane parallel to the wavefronts. In other words, the waves are *transverse* and *circularly polarised.* They also possess the unusual property that the vorticity ($\nabla \times \mathbf{v}_1$) is either parallel or antiparallel to the velocity $\mathbf{v}_1$. The value $\mathbf{v}_1 \cdot \nabla \times \mathbf{v}_1 = \mp k v_1^2$ is the *helicity.*

The group velocity from Eqs.(4.17) and (4.41) is

$$\mathbf{v}_g = \pm\frac{\mathbf{k}\times(2\mathbf{\Omega}\times\mathbf{k})}{k^3}.$$

It has a magnitude $v_g = 2\Omega \sin \theta_\Omega / k$, such that $(v_g^2 + v_p^2)^{1/2} = 2\Omega/k$ is independent of θ_Ω, where v_p is the phase speed (ω/k). Also, note that $\mathbf{v}_g \cdot \mathbf{k} = 0$, so that energy is transported in a direction normal to the phase velocity.

The effect of the Coriolis force on Alfvén waves can be seen from the basic equation [Eq. (4.16)] with $g = 0$ and $\mathbf{k}\cdot\mathbf{v}_1 = 0$, namely,

$$\omega^2 \, \mathbf{v}_1 = -2i\omega \, \mathbf{\Omega}\times\mathbf{v}_1 + \{\mathbf{k}\times[\mathbf{k}\times(\mathbf{v}_1\times\hat{\mathbf{B}}_0)]\}\times\hat{\mathbf{B}}_0 \, v_A^2.$$

After taking the vector product of this equation with $\mathbf{k}$, it reduces to

$$\omega^2 \mp \omega_1\omega \; - \; \omega_A^2 = 0, \tag{4.42}$$

where $\omega_1 = 2(\mathbf{k}\cdot\mathbf{\Omega})/k$ and $\omega_A = \mathbf{k}\cdot\hat{\mathbf{B}}_0 \, v_A$ are pure inertial and Alfvén frequencies.

When $\mathbf{k}$, $\mathbf{\Omega}$ and $\mathbf{B}_0$ are parallel, the nature of the solutions depends on the ratio $\omega_A/\omega_1 = (kv_A)/(2\Omega)$. On the Sun, this ratio is generally large (typically 10 to 10^3) and the solutions of Eq. (4.42) approximate to $\omega^2 = \omega_A^2(1 \pm \omega_I/\omega_A)$, so that the Coriolis force produces a small *frequency-splitting* of the *Alfvén wave.* In the opposite extreme when $\omega_A/\omega_I \ll 1$, the solutions become

$$\omega^2 \approx \omega_1^2, \quad \omega^2 \approx \frac{\omega_A^4}{\omega_1^2}, \tag{4.43}$$

the former representing inertial waves and the latter so-called *hydromagnetic inertial waves*, which propagate much slower than Alfvén waves. General accounts of flow in a rotating frame of reference may be found in the books by Batchelor (1967), Egbers and Pfister (2000) and Chemin et al. (2006).

4.6 Magnetoacoustic Waves

When both the magnetic force and the pressure gradient are important (but g and Ω vanish), Eq. (4.16) gives a generalisation of Eq. (4.21), namely,

$$\omega^2\ \mathbf{v}_1/v_A^2 = k^2\ \cos^2\theta_B\ \mathbf{v}_1\ -\ (\mathbf{k}\cdot\mathbf{v}_1)\ k\ \cos\ \theta_B\ \hat{\mathbf{B}}_0 \\ +\ [(1+c_s^2 v_A^2)(\mathbf{k}\cdot\mathbf{v}_1)\ - k\ \cos\ \theta_B\ (\hat{\mathbf{B}}_0\cdot\mathbf{v}_1)]\ \mathbf{k}.$$

Since $\mathbf{v}_1$ appears here in the combinations $\mathbf{k}\cdot\mathbf{v}_1$ and $\hat{\mathbf{B}}_0\cdot\mathbf{v}_1$, we may proceed easily by taking the scalar product with $\mathbf{k}$ and $\hat{\mathbf{B}}_0$ in turn to give

$$(-\omega^2 + k^2 c_s^2 + k^2 v_A^2)\ (\mathbf{k}\cdot\mathbf{v}_1) = k^3 v_A^2\ \cos\ \theta_B\ (\hat{\mathbf{B}}_0\cdot\mathbf{v}_1)$$

and

$$k\ \cos\ \theta_B\ c_s^2\ (\mathbf{k}\cdot\mathbf{v}_1) = \omega^2\ (\hat{\mathbf{B}}_0\cdot\mathbf{v}_1). \tag{4.44}$$

If $\mathbf{k}\cdot\mathbf{v}_1$ vanishes, one finds the Alfvén wave solution [Eq. (4.25)] as before, but otherwise $(\mathbf{k}\cdot\mathbf{v}_1)/(\hat{\mathbf{B}}_0\cdot\mathbf{v}_1)$ may be eliminated between the two equations above to yield the dispersion relation

$$\omega^4\ -\ \omega^2 k^2(c_s^2+v_A^2)\ +\ c_s^2 v_A^2 k^4\ \cos^2\theta_B = 0 \tag{4.45}$$

for *magnetoacoustic* (or *magnetosonic*) *waves.* For outwards-propagating disturbances ($\omega/k > 0$) there are two distinct solutions

$$\omega/k = \left[\tfrac{1}{2}(c_s^2+v_A^2)\ \pm\ \tfrac{1}{2}\sqrt{(c_s^4+v_A^4-2c_s^2v_A^2\ \cos\ 2\theta_B)}\right]^{1/2}. \tag{4.46}$$

The higher-frequency mode is known as a *fast magnetoacoustic wave* (or *fast-mode wave*) and the other one a *slow magnetoacoustic wave* or *slow-mode wave.* The Alfvén wave phase speed lies between that of the slow and fast waves, and so the Alfvén wave is sometimes referred to as the *intermediate mode*, a nomenclature that carries over into shock-wave terminology (Sec. 5.4.4).

The variation of the phase speeds with propagation angle (θ_B) is shown in Figure 4.7. For propagation along the field ($\theta_B = 0$), ω/k is either c_s or v_A, whereas across the field ($\theta_B = \frac{1}{2}\pi$), ω/k is $(c_s^2+v_A^2)^{1/2}$ or 0. In an oblique slow-mode wave, the phase and group velocities are surprisingly directed in opposite directions (e.g., Nakariakov and Zimovets 2011). For the slow mode, as θ_B tends to $\frac{1}{2}\pi$, both ω/k and the wavenumber component ($k\ \cos\ \theta_B$) along the field approach zero, but the phase velocity component along the field ($\omega/[k\ \cos\ \theta_B]$) approaches a value

$$\frac{\omega}{k\ \cos\ \theta_B} = \frac{v_A c_s}{(v_A^2+c_s^2)^{1/2}} \equiv c_T,$$

known as the *slow speed* or *cusp speed* or *tube speed.* It represents the group velocity of the slow-mode wave, as can be seen at the cusp in Figure 4.7b and is also the propagation speed of slow-mode surface or body waves in a tube (Secs. 4.9.3 to 4.9.5). It would be preferable and more fundamental to denote this speed by c_S, say, for "slow", but that would risk confusion with the sound speed, so we retain here the notation c_T.

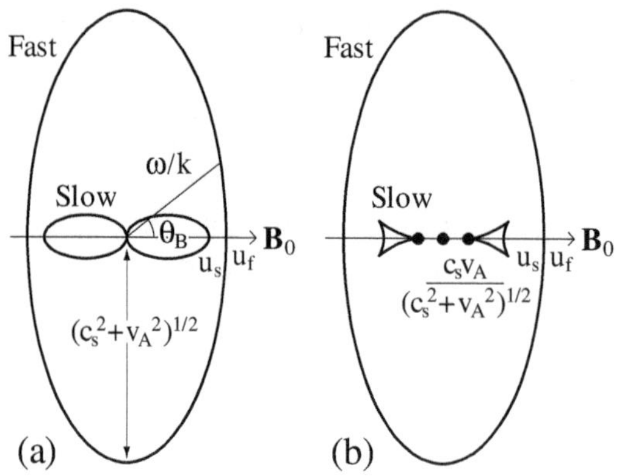

Figure 4.7. Polar diagrams for fast and slow magnetoacoustic waves propagating at an angle θ_B to the equilibrium magnetic field. The speeds u_s and u_f are the slower and faster, respectively, of the Alfvén speed (v_A) and sound speed (c_s). (a) shows the phase velocities and (b) the group velocities.

The two magnetoacoustic waves may be regarded as a sound wave modified by the magnetic field, and a compressional Alfvén wave modified by the gas pressure, the modification being most marked for propagation away from the field direction. In the case of a vanishing magnetic field ($v_A = 0$), the slow mode disappears and the fast mode becomes a sound wave. On the other hand, if the pressure vanishes ($c_s = 0$), the slow mode disappears and the fast mode becomes a compressional Alfvén wave.

When β ($\equiv 2\mu p_0/B_0^2$, the ratio of gas to magnetic pressure) is much larger than unity so that $c_s^2/v_A^2 \gg 1$, the fast- and slow-mode dispersion relations reduce to

$$\frac{\omega}{k} \approx c_s, \quad \frac{\omega}{k} \approx v_A \cos\theta_B,$$

respectively. For the latter case, Eq. (4.44) becomes $\hat{\mathbf{k}} \cdot \hat{\mathbf{v}}_1 \approx (v_A^2/c_s^2) \cos\theta_B\, (\hat{\mathbf{B}}_0 \cdot \hat{\mathbf{v}}_1)$, which is much less than unity, so the disturbance is nearly incompressible. This is a particular case of the more general result that, when $\beta \gg 1$, the plasma may be regarded as incompressible as far as magnetic effects are concerned, since sound waves propagate almost instantaneously.

Polar diagrams for the group velocity are sketched in Figure 4.7. The slow-mode energy is seen to propagate in a narrow cone about the magnetic field direction, but the fast-mode energy propagates more isotropically. When $c_s \gg v_A$, the group velocities become $v_A\, \hat{\mathbf{B}}_0$ and $c_s\, \hat{\mathbf{k}}$ for the slow and fast waves, respectively.

An important limitation on magnetohydrodynamic theory is that $\omega < \Omega_i$, where $\Omega_i = eB/m_i$ is the *ion-gyration* (or *cyclotron*) frequency with which ions gyrate about the magnetic field. When this is violated, the slow-mode wave ceases to exist, while the fast-mode and Alfvén waves are modified in a way described by, for instance, Boyd and Sanderson (2003). Numerically, we have for a hydrogen plasma $\Omega_i = 9.6 \times 10^7 B$, with B in tesla. Thus, for example, with a magnetic field (B) of 1 G, our analysis is valid provided the wave period ($2\pi/\omega$) exceeds 7×10^{-4} s.

It may be noted that there is a marginal solution ($\omega = 0$), which we have overlooked in the preceding analysis (Goedbloed et al. 2004). It represents a transition from stability to instability and is an *entropy wave* (PROBLEM 4.7). The wave is genuine, but is quite degenerate in the sense that it does not propagate or involve perturbations in flow, magnetic field or pressure. It just represents a perturbation of density (or entropy) which could be carried with a background flow and which suggests the possibility of new waves when extra physics is included. Its relevance has been suggested for cool moving blobs in flare loops or coronal rain (Murawski et al. 2011).

4.7 Acoustic Gravity Waves

When compressibility and buoyancy are present, the acoustic and gravity waves remain as distinct modes, but both are modified somewhat. Taking the scalar product of the basic equation [Eq. (4.15)] with $\mathbf{k}$ and $\hat{\mathbf{z}}$ in turn yields a pair of equations for v_{1z} and $\mathbf{k}\cdot\mathbf{v}_1$, which give the dispersion relation

$$\omega^2(\omega^2 - N_s^2) = (\omega^2 - N^2 \sin^2 \theta_g')\, k'^2 c_s^2, \tag{4.47}$$

where

$$N_s = \frac{\gamma g}{2c_s}\left(\equiv \frac{c_s}{2H}\right), \quad N = \frac{(\gamma-1)^{1/2} g}{c_s}, \quad \sin^2\theta_g' = 1 - \frac{k_z'^2}{k'^2}, \quad \mathbf{k}' = \mathbf{k} + \frac{i}{2H}\hat{\mathbf{z}}.$$

Here N is the Brunt-Väisälä frequency, H is the scale-height and θ_g' is the angle between the wave vector $\mathbf{k}'$ and the vertical. N is always greater than or equal to N_s, equality occurring when $\gamma = 2$, but usually the difference between the two is rather small. When $\gamma = 5/3$, for instance, $N_s \approx 1.02\, N$. The change of wavenumber from k to k' means that the disturbances take the form

$$\mathbf{v}_1(\mathbf{r}, t) = \mathbf{v}_1 e^{z/(2H)} e^{i(\mathbf{k}'\cdot\mathbf{r} - \omega t)},$$

and so their amplitude increases exponentially over a vertical distance $2H$.

In the limit $\omega \ll k'c_s$, Eq. (4.47) reduces to $\omega = N \sin\,\theta_g$ for gravity waves, whereas when $\omega \gg N$ it becomes $\omega = k'c_s$ for pure acoustic waves. Vertical propagation ($\theta_g = 0$) gives $\omega^2 = N_s^2 + k'^2 c_s^2$, for which the wave exists ($\omega^2 > 0$) only if $\omega > N_s$.

Wave-like solutions propagating away from the vertical ($k'^2 > 0$, $\omega^2 > 0$) are allowed in two frequency domains, $\omega < N \sin\,\theta_g$ and $\omega > N_s$. The higher-frequency mode is basically acoustic in nature, with group speed $v_g < c_s$ and phase speed $\omega/k > c_s$; the former approaches zero, while the latter tends to infinity as the *lower cut-off* (N_s) is approached. The lower-frequency wave is basically a gravity mode, with a phase speed $\omega/k < c_s$ that approaches zero at the *upper cut-off* ($N \sin\,\theta_g$). For frequencies between $N \sin\,\theta_g$ and N_s, the disturbance is *non-propagating* (or *evanescent*). Here k' is purely imaginary and so the disturbance grows or decays exponentially in space. In this band of frequencies, therefore, only standing waves can exist (Figure 4.8a) and no energy is propagated (since p_1 and v_1 are out of phase).

It is sometimes of interest to know whether disturbances of a given horizontal wavenumber (k_x) and frequency (ω) are vertically propagating or not. Since $k'^2 = k_z'^2 + k_x^2$, Eq. (4.47) may be recast in the form

$$k_z'^2 \omega^2 c_s^2 = \omega^2(\omega^2 - N_s^2) - (\omega^2 - N^2)\, k_x^2 c_s^2.$$

Thus, for $\omega^2 > 0$ and $k_x^2 > 0$, vertically propagating waves ($k_z'^2 > 0$) exist provided

$$\omega^2(\omega^2 - N_s^2) > (\omega^2 - N^2)\, k_x^2 c_s^2.$$

This condition divides the ωk_x-plane into three domains, as indicated in Figure 4.8b, sometimes referred to as a *diagnostic diagram.* For very small horizontal wave-numbers (k_x), it reduces to $\omega > N_s$ or $\omega < (N/N_s)k_x c_s$, whereas, for large values of k_x, it becomes $\omega < N$ or $\omega > k_x c_s$. For small wavenumbers, it can be seen that the acoustic waves are inhibited by gravity at frequencies less than N_s, and the internal gravity waves are inhibited by compressibility at frequencies larger than kc_s.

The inclusion of the magnetic field further complicates matters by introducing an extra restoring force and an extra preferred direction in addition to that of gravity. The Alfvén wave remains unaffected by gravity, but the magnetic field modifies the acoustic gravity waves (or gravity modifies the magnetoacoustic waves) to give two *magnetoacoustic gravity* (or '*magneto-atmospheric*') *modes.* For example, an isothermal plasma in a strong uniform vertical magnetic field with $v_A \gg c_s$ supports a fast mode for which $k_x^2 + k_y^2 + k_z^2 = 0$; thus, if k_x and k_y are real, k_z will be imaginary, and so it represents a wave that is propagating horizontally and is evanescent vertically. The slow mode is similar to an acoustic wave; it has $2ik_z = 1 - (1 - \omega^2/N_s^2)^{1/2}$ and so is evanescent below the cut-off frequency (N_s).

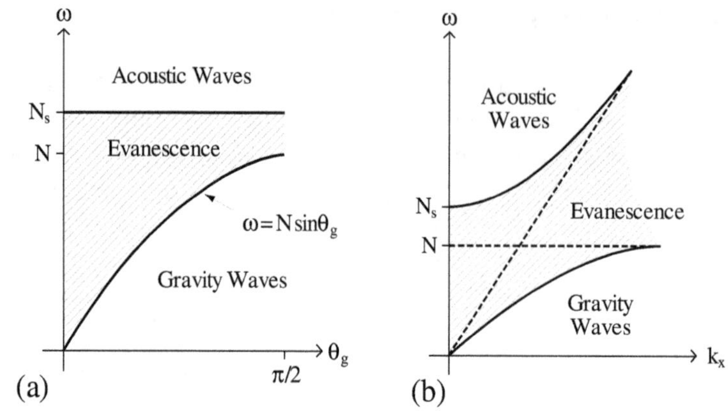

Figure 4.8. Diagnostic diagrams indicating the allowable domains for the propagation of acoustic gravity waves of frequency ω. Disturbances in the shaded region are non-propagating (evanescent). (a) propagation at an angle θ_g to the vertical. (b) vertical propagation of waves of horizontal wavenumber k_x, in which the asymptotes $\omega = N$ and $\omega = k_x c_s$ are indicated by dashed lines.

4.8 Summary of Magnetoacoustic Gravity Waves

In the absence of a Coriolis force, the basic equations for wave propagation in a plasma with a uniform magnetic field, a uniform temperature and a density proportional to exp $(-z/H)$ allow the solution

$$\omega^2 = k^2 v_A^2 \cos^2\theta_B \qquad (\textit{Alfvén wave}),$$

together with two magnetoacoustic gravity modes.

When there is no magnetic field, Eq. (4.47) gives their dispersion relation as

$$\omega^4 - \omega^2\,(N_s^2 + k'^2 c_s^2) + N^2 \sin^2\theta'_g\, k'^2 c_s^2 = 0 \ \ (\textit{acoustic gravity waves}), \tag{4.48}$$

$$\text{where} \quad N_s = \frac{\gamma N}{2(\gamma-1)^{1/2}}, \quad N = (\gamma-1)^{1/2}\frac{g}{c_s}, \quad \sin^2\theta'_g = 1 - \frac{k_z'^2}{k'^2},$$

$$k'_z = k_z + \frac{i}{2H}, \quad k'^2 = k^2 + k_z'^2 - k_z^2, \quad H = \frac{c_s}{2N_s}\left(= \frac{c_s^2}{\gamma g} = \frac{p_0}{\rho_0 g}\right).$$

In the limit $\omega \gg N$, the effect of gravity can be neglected and Eq. (4.48) reduces to

$$\omega^2 = k^2 c_s^2 \qquad (\textit{acoustic wave}).$$

The alternative limit, namely, $\omega \ll kc_s$ and $N_s \ll kc_s$ (i.e., $kH \gg 1$) makes buoyancy dominate and gives rise to

$$\omega^2 = N^2 \sin^2\theta_g \qquad [(\textit{internal})\ \textit{gravity wave}].$$

The presence of a magnetic field means that plane-wave solutions arise locally if $kH \gg 1$, or, from the definition of H, if

$$N_s \ll k\, c_s. \tag{4.49}$$

This means that the wavelength is much shorter than a scale-height, which is a severe restriction near the photosphere where H may be as small as a few hundred kilometres [Eq. (3.12)]. The magnetoacoustic gravity dispersion relation can be obtained by eliminating $\mathbf{v}_1$ from Eq. (4.16) and invoking (4.49). This yields

$$\omega^4 - \omega^2 k^2 (c_s^2 + v_A^2) + k^2 c_s^2 N^2 \sin^2\theta_g + k^4 c_s^2 v_A^2 \cos^2\theta_B = 0, \tag{4.50}$$

where θ_B is the angle between $\mathbf{k}$ and the magnetic field. There are several limiting cases. When $N = v_A = 0$, acoustic waves are recovered, whereas, when $v_A = 0$ and $\omega \ll kc_s$, the result is internal gravity waves. But, in the case of vanishing v_A, the full acoustic gravity wave dispersion relation (4.48) is not recovered, since the coupling between acoustic and gravity waves is excluded by the condition (4.49).

When $N = c_s = 0$, Eq. (4.50) reduces to

$$\omega^2 = k^2 v_A^2 \qquad \textit{(compressional Alfvén wave)},$$

while the condition $N = 0$ alone gives

$$\omega^4 - \omega^2 k^2 (c_s^2 + v_A^2) + k^4 c_s^2 \cos^2\theta_B = 0 \quad \textit{(magnetoacoustic waves)}. \tag{4.51}$$

It should be noted that, if $c_s^2 \lesssim v_A^2$, the validity condition [Eq. (4.49)] reduces Eq. (4.50) to (4.51), so that the interaction with gravity is excluded and the limit $c_s^2 \to 0$ produces just compressional Alfvén waves. However, if $c_s^2 \gg v_A^2$, the two magnetoacoustic gravity solutions to Eq. (4.50) become

$$\omega^2 = k^2 c_s^2$$

$$\text{and} \qquad \omega^2 = N^2 \sin^2\theta_g + k^2 v_A^2 \cos^2\theta_B.$$

While the former expression describes an acoustic wave, the latter reduces to an internal gravity wave at low frequencies (when $kv_A \ll N$) and a slow magnetoacoustic wave at high frequencies (when $kv_A \gg N$); it is known as a *magneto-gravity wave.*

In view of the condition (4.49), a coupling between the magnetic field and gravity is not allowed by the magnetoacoustic gravity wave dispersion relation (4.50), save in the limit $c_s^2 \gg v_A^2$. However, there are alternative approaches which do allow a coupling. A simple example is to consider a non-uniform horizontal magnetic field that behaves like $\exp(-\frac{1}{2}z/H)$, so that the Alfvén speed is uniform. Again, one can seek solutions in the form $v_1(z)\exp i(k_x x + k_y y - \omega t)$, with a variation that is sinusoidal in the horizontal but not the vertical direction. Equation (4.14) then reduces in general to a fourth-order ordinary differential equation for $v_1(z)$.

4.9 Waves in a Non-uniform Medium

Up to this section, the analysis has rested on the assumption that the ambient medium in which the waves are propagating is *uniform*, which is valid if the wavelength (λ) is much smaller than the length-scale (l_0) for variations in the medium. As such waves propagate, their properties change slowly, which may be analysed with a WKB approximation (Bender and Orszag 1999); for example, acoustic waves may steepen into shocks when they travel into a region of decreasing density (Sec. 5.1.1).

The mathematical advantage of considering a *uniform* medium is that the partial differential equations (4.8) to (4.12) for the perturbed quantities reduce to algebraic equations (4.15) or (4.16), from which the dispersion relation $\omega = \omega(k_x, k_y, k_z)$ may readily be deduced. But when $\lambda \gtrsim l_0$ the inhomogeneous nature of the medium determines the structure of the disturbance, which can no longer be assumed to be sinusoidal. If, for example, the medium is structured in the z-direction, the perturbation equations reduce to ordinary differential equations in z; their solution subject to certain boundary conditions determines the z-structure of the disturbance as well as the dispersion relation $\omega = \omega(k_x, k_y)$ for sinusoidal variation in the other two directions. A complication is that when the differential equation is singular there may exist a *continuous spectrum* of modes in addition to a *discrete spectrum* (Sec. 4.10). If the medium is structured in two directions, the result is a set of linear partial differential equations in those coordinates and, if the disturbance is no longer small, they become nonlinear – a continuing topic of current research.

The main agents for creating inhomogeneity on the Sun are gravity and the magnetic field. Gravity causes the pressure to increase inwards towards the solar centre, while the magnetic field and its associated Lorentz force cause the plasma pressure to increase in a direction normal to the magnetic field and away from regions of magnetic flux concentration. The inhomogeneity introduces several new effects:

(i) *amplification* – the wave amplitude may increase (or decrease) as it propagates;

(ii) *evanescence* – a time-oscillating disturbance may have a wave-like character in one region where it oscillates spatially, but then it may become evanescent and decay exponentially in another region;

(iii) *surface waves* – a discontinuity in the basic state may give rise to extra waves that decay away from the interface in addition to the 'body' waves that are present in a uniform medium;

(iv) *dissipation* – an increase in spatial gradients in the wave may lead eventually to dissipation; and

(v) *mode coupling* – different wave modes may be linearly or nonlinearly coupled.

In this section, we focus on a medium that is structured in one dimension (x or R). We set up the equations for perturbations of a continuous planar field [$B_0(x)\ \hat{\mathbf{z}}$] and use them to deduce wave properties on a single interface and a magnetic slab (bounded by two interfaces separating three uniform regions). Next, the comparable problem in a cylindrical geometry is studied by setting up the equations for perturbations of a continuous twisted cylindrical flux tube [$B_{0\phi}(R), B_{0z}(R)$] (Sec. 4.9.4), and these in turn are applied to deduce wave modes of a simple flux tube with uniform field and no twist: the results are similar to a slab, but there are some differences. Then, in Section 4.10, we discuss important aspects of waves in a continuous medium that are not present in a uniform medium (continuous spectra and quasi-modes). Damping by resonant absorption and phase mixing are treated in Section 10.3.

4.9.1 Waves in a Continuous Planar Magnetic Field

Consider a unidirectional magnetic field [$\mathbf{B} = B_0(x)\ \hat{\mathbf{z}}$] containing plasma at rest with equilibrium properties $p_0(x), \rho_0(x), T_0(x)$, such that $p_0(x) + B_0(x)^2/(2\mu) = constant$. A small perturbation velocity

$$v_{1x} = v_{1x}(x)e^{i(k_y y + k_z z - \omega t)},$$

with wave numbers k_y and k_z in the y- and z-directions, can be shown to satisfy

$$\frac{d}{dx}\left[\frac{\varepsilon(x)}{k_y^2 + m_0(x)^2}\frac{dv_{1x}}{dx}\right] = \varepsilon(x) v_{1x}, \tag{4.52}$$

where, in terms of $v_A^2 = B_0^2/(\mu\rho_0)$ and $c_s^2 = \gamma p_0/\rho_0$,

$$\varepsilon = \rho_0(\omega^2 - \omega_A^2), \qquad m_0^2 = \frac{(k_z^2 c_s^2 - \omega^2)(\omega_A^2 - \omega^2)}{(c_s^2 + v_A^2)(\omega_T^2 - \omega^2)}, \qquad c_T^2 = \frac{c_s^2 v_A^2}{c_s^2 + v_A^2},$$

and $\omega_A = k_z v_A$ and $\omega_T = k_z c_T$ are the Alfvén and slow (or cusp) frequencies. Here B_0, ρ_0 and p_0, and therefore ε, m_0, v_A, c_s and c_T, are all functions of x.

An alternative way of writing Eq. (4.52) is

$$\frac{d}{dx}\left[\frac{(\omega^2 - \omega_A^2)(\omega^2 - \omega_T^2)}{(\omega^2 - \omega_+^2)(\omega^2 - \omega_-^2)}(c_s^2 + v_A^2)\rho_0 \frac{dv_{1x}}{dx}\right] = -(\omega^2 - \omega_A^2)\rho_0 v_{1x}, \tag{4.53}$$

where ω_+ and ω_- are the slow- and fast-mode turning-point frequencies given by Eq. (4.45) when $k_x = 0$, namely, $\omega^4 - \omega^2 k^2 (c_s^2 + v_A^2) + k^2 k_z^2 c_s^2 v_A^2 = 0$. Thus, the variation of magnetic field (B_0) with x implies that, in general, the perturbation velocity (v_{1x}) is not sinusoidal in x but satisfies Eq. (4.53). The function m_0^2 may be positive or negative, since it changes sign at the places where the phase speed ω/k_z (assumed real) is equal to the sound speed (c_s), the Alfvén speed (v_A) and the 'slow speed' (c_T), whose significance was discussed in Section 4.6.

In two cases, the fundamental equation (4.52) or (4.53) takes on a simpler form. When the basic state is uniform, ε and m_0 are constants, and we may recover the previous results of Sections 4.3.1 and 4.6. One solution has $\varepsilon = 0$ for Alfvén waves, while the other has v_{1x} proportional to $e^{ik_x x}$ with $k_x^2 + k_y^2 + m_0^2 = 0$, which is just the dispersion relation (4.45) for magnetoacoustic waves.

The second simple case is that of transverse propagation (with ω^2 not close to ω_T^2), so that $k_y^2 \gg m_0^2(x)$ and Eq. (4.52) simplifies to

$$\frac{d}{dx}\left[\frac{\varepsilon(x)}{k_y^2}\frac{dv_{1x}}{dx}\right] = \varepsilon(x)v_{1x} \tag{4.54}$$

for Alfvén ($\varepsilon = 0$) and compressional Alfvén (i.e., fast-mode) disturbances.

The solutions of Eq. (4.52) over some range of x (subject to relevant boundary conditions) are quite complicated and, in general, need to be found numerically. If the profile is made up of discrete sections, the solution may be written as a sum of discrete normal modes with different values of ω, including both surface and body waves. However, if the profile is continuous and $v_A(x)$ or $c_T(x)$ equals ω/k_z at some location, the differential equation (4.53) possesses a singularity there. This leads to a *continuous spectrum* of frequencies (Sec. 4.10) and the general solution consists of an integral over some range of ω in addition to a possible sum of discrete modes.

4.9.2 Surface Waves on a Magnetic Interface

Consider next the case of an interface at $x = 0$ between uniform media with

$$B_0(x) = B_- \quad \text{for } x < 0 \qquad \text{and} \qquad B_0(x) = B_+ \quad \text{for } x > 0,$$

where B_+ and B_- are constants and subscripts + and − denote values on the two sides. Supposing further that $k_y = 0$, Eq. (4.52) reduces to either $\varepsilon(x) \equiv 0$ or

$$\frac{d^2v_{1x}}{dx^2} = m_-^2\, v_{1x} \quad \text{for } x < 0 \qquad \text{and} \qquad \frac{d^2v_{1x}}{dx^2} = m_+^2\, v_{1x} \quad \text{for } x > 0, \tag{4.55}$$

where m_+ and m_- are constants defined in a similar way to m_0 in Section 4.9.1. There are two basic classes of solution to Eq. (4.55) in the region $x < 0$:

(i) *Unbound states* represent waves that are propagating from or towards infinity and are being reflected and transmitted at the interface. They occur for frequencies such that $m_-^2 < 0$ and have $v_{1x} \sim \exp\, i(-m_-^2)^{1/2}x$. (An example would be a *leaky mode*, which propagates away from the interface.) In the absence of the interface, they become the normal waves of a uniform medium.

(ii) *Bound states* decay away spatially to zero at minus infinity like $e^{m_- x}$ and so they occur for $m_-^2 > 0$. These *surface* waves (Figure 4.9a) owe their existence to the interface and cannot exist when it is absent (i.e., when $B_- = B_+$ and $\rho_- = \rho_+$).

When $m_-^2 > 0$ and $m_+^2 > 0$, there are discrete *magnetoacoustic surface mode* solutions such that

$$v_{1x} \sim \begin{cases} e^{m_- x}, & x < 0, \\ e^{-m_+ x}, & x > 0. \end{cases}$$

The boundary conditions at the interface (continuity of v_{1x} and total pressure) yield their dispersion relation as

$$\frac{\omega^2}{k^2} = v_{A-}^2 - \frac{\rho_+ m_-}{\rho_+ m_- + \rho_- m_+}(v_{A-}^2 - v_{A+}^2). \tag{4.56}$$

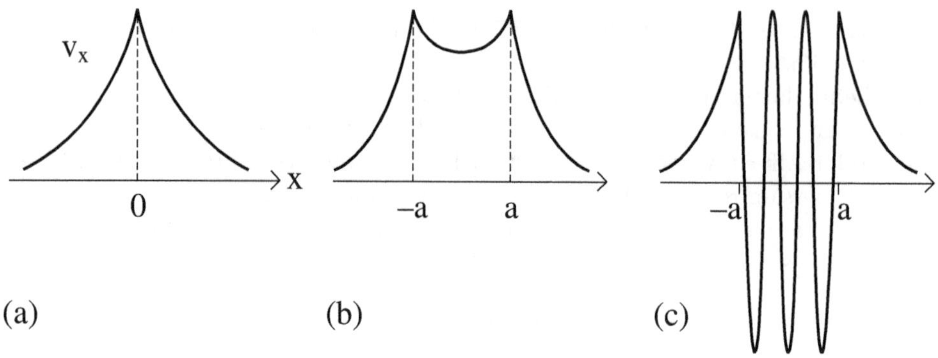

Figure 4.9. A surface wave on (a) an interface and (b) a slab or a flux tube. (c) A body wave on a slab or flux tube.

It can be seen that their phase speed lies between the Alfvén speeds v_{A-} and v_{A+} in the two media, but the fact that m_- and m_+ are functions of ω means that there can be several such solutions.

In the particular case when the second medium has a negligible field strength ($B_+ = 0$), Eq. (4.56) reduces to

$$\frac{\omega^2}{k^2} = \frac{\rho_- m_+}{\rho_- m_+ + \rho_+ m_-} v_{A-}^2, \tag{4.57}$$

where m_+^2 becomes $m_+^2 = k^2 - \omega^2/c_{se}^2$. There is then always a slow surface wave solution to Eq. (4.57) with ω/k smaller than both c_{T-} and c_{s+}. However, when $c_{s+} > c_{s-}$ and $v_{A-} > c_{s-}$, a fast surface wave also exists with ω/k larger than c_{s-} but less than both c_{s+} and v_{A-}.

4.9.3 Surface and Body Waves on a Magnetic Slab

In order to explore further the effect of magnetic structuring on wave propagation, Edwin and Roberts (1982) considered a uniform slab of width $2a$ and magnetic field B_i surrounded by a uniform medium with field B_e, with equilibrium values inside and outside the slab denoted by subscripts i and e, respectively (Figure 4.10a). They distinguished between *sausage modes*, for which the two interfaces bounding the slab oscillate exactly out of phase so that v_x is an odd function of x (Figure 4.10b), and *kink modes*, for which they are in phase and v_x is even in x (Figure 4.10c). It is assumed that $m_e^2 > 0$, so that the waves are evanescent outside the slab and decay away to zero as $x \to \pm\infty$ like $\exp(\mp m_e x)$. The term *surface wave* (Figure 4.9b) refers here to a wave that is also evanescent inside the slab ($m_i^2 > 0$) with a behaviour like $\exp i(-\omega t + kz) \pm m_i x$, whereas a *body wave* (Figure 4.9c) is oscillatory there ($m_i^2 = -n_i^2 < 0$) and has the form $\exp i(-\omega t + kz \pm n_i x)$.

In a uniform medium, a fast-mode wave propagates faster than both v_A and c_s whereas a slow-mode wave is slower than v_A or c_s. In a slab or a flux tube, a *fast wave* propagates faster than both the minimum of c_{si} and c_{se} and the minimum of v_{Ai} and v_{Ae}, whereas a *slow wave* is slower than one minimum or the other. Thus, the division between fast and slow is marked by the greater of the two minima.

The dispersion relations are derived in a similar manner to Eq. (4.56) for a single interface and take the forms

$$\rho_i\,(k^2 v_{Ai}^2 - \omega^2)\, m_e \;+\; \rho_e\,(k^2 v_{Ae}^2 - \omega^2)\, m_i \tanh\,(m_i a) \;=\; 0 \tag{4.58}$$

for the sausage mode and

$$\rho_i\,(k^2 v_{Ai}^2 - \omega^2)\, m_e \;+\; \rho_e\,(k^2 v_{Ae}^2 - \omega^2)\, m_i \coth\,(m_i a) \;=\; 0 \tag{4.59}$$

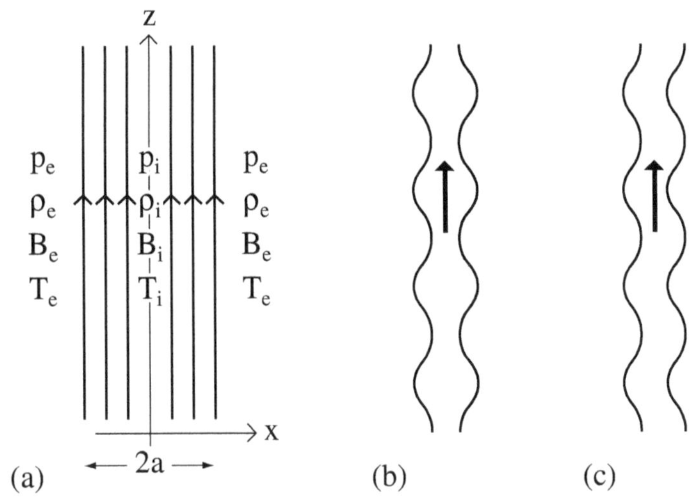

Figure 4.10. (a) A uniform magnetic slab surrounded by a uniform medium. (b) A sausage mode disturbance travelling along the slab. (c) A kink mode.

for the kink mode, where

$$m_i^2 = \frac{(k^2c_{si}^2 - \omega^2)(k^2v_{Ai}^2 - \omega^2)}{(c_{si}^2 + v_{Ai}^2)(k^2c_{Ti}^2 - \omega^2)}, \qquad m_e^2 = \frac{(k^2c_{se}^2 - \omega^2)(k^2v_{Ae}^2 - \omega^2)}{(c_{se}^2 + v_{Ae}^2)(k^2c_{Te}^2 - \omega^2)},$$

in terms of the usual sound speed (c_s), Alfvén speed (v_A) and slow (or tube) speed $[c_T = c_s v_A/(c_s^2+v_A^2)^{1/2}]$. In view of the constancy of total pressure, the sound and Alfvén speeds are related by $\rho_i(c_{si}^2 + \frac{1}{2}\gamma v_{Ai}^2) = \rho_e(c_{si}^2 + \frac{1}{2}\gamma v_{Ai}^2)$.

The fact that m_i and m_e are themselves functions of ω and k makes Eqs. (4.58) and (4.59) into a pair of complicated transcendental equations for $\omega = \omega(k)$. A rich variety of waves is possible (fast or slow, kink or sausage, surface or body), depending on the relative Alfvén and sound speeds inside and outside the slab.

The dispersion relations for one set of conditions appropriate to the photosphere and one to the corona are shown in Figure 4.11. In the slender slab limit ($ka \ll 1$) of the photospheric case, there is a *slow surface sausage wave* (sometimes called a *tube wave*) with $\omega = kc_T$, a slow kink surface wave with $\omega = kv_{Ae}$ and a fast wave with $\omega = kc_{se}$, which may be of surface or body type depending on the sign of m_i^2. Also, there are *slow body waves* (both sausage and kink) with $\omega = kc_T$ that cause rapid oscillations across the slab.

Under coronal conditions (Figure 4.11b), when the Alfvén speed exceeds the sound speed, there are no surface waves, but harmonics of the fast and slow body waves arise, and the fast body waves are cut off at v_{Ae} rather than c_{se}. In the slender slab limit ($ka \ll 1$), the slow body modes propagate at c_{Ti}, while the fundamental kink fast mode travels much faster at v_{Ae}.

The two relations $\omega = kv_{Ai}$ and $\omega = kv_{Ae}$ appear at first sight to be solutions of Eqs. (4.58) and (4.59), but they violate the conditions $m_e^2 > 0$ and $m_e^2 > 0$ for validity of the analysis. Indeed, as we shall see in Section 4.10, there also exist continuum Alfvén and slow modes, which propagate at the local Alfvén and cusp speeds. In a flux tube, the equivalent would be a torsional wave. The other modes are magnetoacoustic in nature, so that there are no Alfvén surface waves in a compressible plasma: they arise only in the incompressible limit from slow modes.

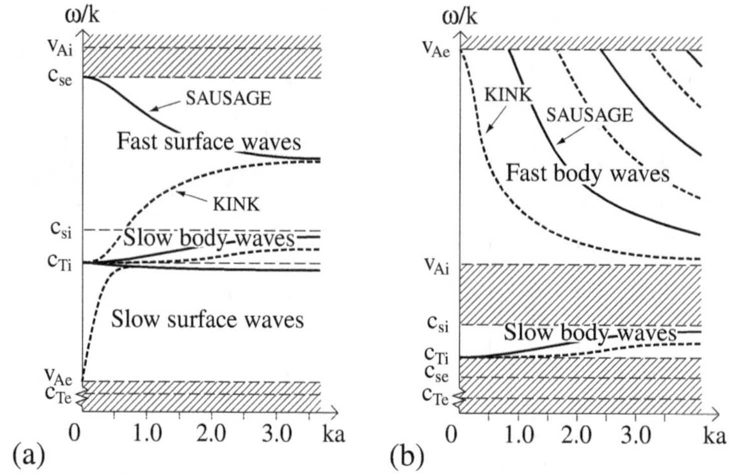

Figure 4.11. Waves in a slab of width $2a$, showing the phase speed (ω/k) as a function of wavenumber (k) under conditions typical of (a) the photosphere (i.e., $v_{Ai} > c_{se} > c_{si} > v_{Ae}$) and (b) the corona (i.e., $v_{Ae}, v_{Ai} > c_{si}, c_{se}$). Sausage modes are shown by solid curves and kink modes by dashed ones. In the shaded region, there are no modes evanescent outside the slab (after Edwin and Roberts 1982).

In the incompressible limit (i.e., $c_{si}/v_{Ai} \to \infty$), $c_{Ti} = v_{Ai}$, $m_i = m_e = k$ and we have *Alfvén surface waves* with dispersion relation

$$\frac{\omega^2}{k^2} = \frac{\rho_i v_{Ai}^2 + \rho_e v_{Ae}^2 \tanh\,(ka)}{\rho_i + \rho_e \tanh\,(ka)}$$

for the sausage mode and the same expression with tanh ka replaced by coth ka for the kink mode. For a slender slab ($ka \ll 1$), this gives slow surface sausage waves with $\omega/k = v_{Ai}$ and slow surface kink waves with $\omega/k = v_{Ae}$. For a wide slab ($ka \gg 1$), both modes have phase speed given by $\omega^2/k^2 = (\rho_i v_{Ai}^2 + \rho_e v_{Ae}^2)/(\rho_i + \rho_e)$.

There are many ways of generalising this analysis. Propagating modes outside the slab (so-called *leaky modes*) can be considered by allowing $m_e^2 < 0$, so that the waves approach (or recede from) the slab and the disturbance in the slab grows (or decays) in time as the energy accumulates (or declines) there. The analysis can be performed in a cylindrical rather than a Cartesian geometry for a flux tube, which leads to a few quantitative changes (see Sec. 4.9.5). The effect of gravity can also be included in the basic state, so that, for instance, the slow sausage surface wave becomes evanescent when the wave period is too large.

4.9.4 Waves in a Continuous Twisted Flux Tube

Consider a cylindrically symmetric flux tube, whose pressure varies with radial distance (R) from some axis and whose magnetic field lines lie on cylindrical surfaces. In other words, the equilibrium state is specified by

$$\mathbf{B}_0 = [0, B_{0\phi}(R), B_{0z}(R)], \quad p_0 = p_0(R), \quad \rho_0 = \rho_0(R).$$

The gravitational and Coriolis forces are neglected, and the equilibrium variables satisfy the magnetostatic condition

$$0 = \frac{dp_0}{dR} + \frac{d}{dR}\left(\frac{B_0^2}{2\mu}\right) + \frac{B_{0\phi}^2}{\mu R}.$$

The basic Eqs. (4.1) to (4.5) are linearised for small disturbances from equilibrium, and each of the perturbation variables is written in the form

$$f_1(R, \phi, z, t) = f_1(R)e^{i(kz+m\phi-\omega t)},$$

since the azimuthal and axial variations can be Fourier analysed whereas the radial variations cannot. Then the equations may be reduced to the *generalised Hain-Lüst equation* (Goedbloed et al. 2004) for the radial velocity $v_{1R}(R)$, namely,

$$\frac{d}{dR}\left(\rho_0 \, F \frac{d}{dR}(R \, v_{1R})\right) + \rho_0 \, G \, v_{1R} = 0, \tag{4.60}$$

where

$$F = (\omega^2 - \bar{\omega}_A^2) \, [(c_s^2 + v_A^2)\omega^2 - c_s^2\bar{\omega}_A^2]/D, \qquad \bar{\omega}_A^2 = \frac{(\mathbf{k} \cdot \mathbf{B}_0)^2}{\mu\rho_0} = \frac{1}{\mu\rho_0}\left(\frac{mB_{0\phi}}{R} + kB_{0z}\right)^2,$$

$$G = (\omega^2 - \bar{\omega}_A^2) - \frac{R}{\mu\rho_0}\left(\frac{B_{0\phi}^2}{R^2}\right)' - \frac{4k^2 \, v_{A\phi}^2}{R^2 \;\; D}(\omega^2 v_A^2 - \bar{\omega}_A^2 c_s^2)$$
$$+\frac{R}{\mu\rho_0}\left\{\frac{2kB_{0\phi}(mB_{0z}/R - kB_{0\phi})}{R^2 D}[(c_s^2 + v_A^2)\omega^2 - \bar{\omega}_A^2 c_s^2]\right\}',$$

$$D = \omega^4 - (m^2/R^2 + k^2)(c_s^2 + v_A^2)\omega^2 + (m^2/R^2 + k^2)c_s^2\bar{\omega}_A^2.$$

In the particular case when there is no twist in the field, $B_{0\phi}$ vanishes, $\bar{\omega}_A^2 = \omega_A^2 \equiv k^2 v_A^2$ and $G = \omega^2 - \omega_A^2$, so that Eq. (4.60) reduces to

$$\frac{d}{dR}\left[\frac{(\omega^2 - \omega_A^2)(\omega^2 - \omega_T^2)}{(\omega^2 - \omega_+^2)(\omega^2 - \omega_-^2)}(c_s^2 + v_A^2)\frac{\rho_0}{R}\frac{d}{dR}(R \, v_{1R})\right] + (\omega^2 - \omega_A^2)\rho_0 v_{1R} = 0, \tag{4.61}$$

which is similar to Eq. (4.53) except for extra factors of R.

In a flux tube with a discrete Alfvén profile, one finds pure slow-mode, fast-mode and Alfvén waves (Sec. 4.9.5), but, when the profile is instead continuous, MHD waves have a mixture of the properties of the pure modes, which vary as they propagate through a non-uniform medium (Goossens 2008; Pascoe et al. 2011; McIntosh et al. 2011). The labelling of waves is then based on the restoring force and the location of their frequency in the spectrum.

For example, the term *Alfvénic wave* is used by some authors to denote a largely transverse mode in a continuous Alfvén profile, with magnetic tension as the restoring force (e.g., Cargill and De Moortel 2011; McIntosh et al. 2011). It propagates at the kink speed (see Eq. 4.64), which is a mean Alfvén speed, and so some authors prefer to refer to it as a kink mode (Nakariakov and Verwichte 2005). It changes its characteristics as it propagates, so that at some times and locations it looks like a fast-mode kink wave and at others a torsional Alfvén wave. In the case of a steeply changing continuous Alfvén profile, it is almost completely fast-mode kink in its behaviour. For the usual case of wavelengths much longer than the tube width, it is almost incompressible, but otherwise it is compressible.

4.9.5 Modes of a Uniform Untwisted Flux Tube

Consider now a cylinder of radius a containing uniform plasma of density ρ_i, pressure p_i and magnetic field $B_i\hat{\mathbf{z}}$, embedded in a uniform plasma of density ρ_e, pressure p_e and magnetic field $B_e\hat{\mathbf{z}}$ (Zaitsev and Stepanov 1975; Edwin and Roberts 1983). This is a generalisation to a cylindrical geometry of the slab that we considered in Section 4.9.3. We here neglect twist and plasma flows in the basic state, so that lateral total pressure balance gives $p_i + B_i^2/(2\mu) = p_e + B_e^2/(2\mu)$. In the internal medium, the sound speed, Alfvén speed and tube speed are, respectively, c_{si}, v_{Ai}, $c_{Ti} = c_{si}v_{Ai}/(c_{si}^2 + v_{Ai}^2)^{1/2}$, while in the external medium outside the tube they become c_{se}, v_{Ae} and $c_{Te} = c_{se}v_{Ae}/(c_{se}^2 + v_{Ae}^2)^{1/2}$.

Linear perturbations of the total pressure (P_1) and of the other variables are taken in the form $P_1(r)\exp[i(kz + m\phi - \omega t)]$, so that $m = 0$ waves are called *sausage modes*, $m = 1$ are *kink modes* and $m > 1$ are *flute* or *ballooning modes.* The linearised MHD equations then give

$$D\frac{d}{dr}(r\xi_r) = (v_A^2 + c_s^2)(\omega^2 - k_z^2c_T^2)\left(\kappa^2 + \frac{m^2}{r^2}\right) rP_1,$$

$$\frac{dP_1}{dr} = \rho_0(\omega^2 - k_z^2v_A^2)\ \xi_r,$$

$$\rho_0(\omega^2 - k_z^2v_A^2)\ \xi_\phi = -\frac{im}{r}P_1,$$

where ξ_r and ξ_ϕ are the radial and azimuthal displacement perturbations, while D and κ are defined by

$$D = \rho_0(v_A^2 + c_s^2)(\omega^2 - k_z^2v_A^2)(\omega^2 - k_z^2c_T^2)$$

$$\text{and} \qquad \kappa^2(\omega) = -\frac{(\omega^2 - k_z^2c_s^2)(\omega^2 - k_z^2v_A^2)}{(c_s^2 + v_A^2)(\omega^2 - k_z^2c_T^2)}.$$

Inside the flux tube, the above equations reduce to

$$(\omega^2 - k_z^2v_{Ai}^2)\left[\frac{d^2}{dr^2} + \frac{1}{r}\frac{d}{dr} - \left(\kappa_i^2 + \frac{m^2}{r^2}\right)\right] P_1, \tag{4.62}$$

and similarly for the external region where the subscript i is replaced by e. The first term here represents torsional Alfvén waves with $\omega = v_{Ai}k_z$, whereas the second term describes magnetoacoustic waves and the function κ_i is a transverse wavenumber. Modes that are evanescent in the external medium have $\kappa_e^2 > 0$, for which the following dispersion relation may be derived by using continuity of total pressure and normal velocity at the tube boundary:

$$\rho_e(\omega^2 - k_z^2v_{Ae}^2)\kappa_i\frac{I_m'(\kappa_i a)}{I_m(\kappa_i a)} + \rho_i(k_z^2v_{Ai}^2 - \omega^2)\kappa_e\frac{K_m'(\kappa_e a)}{K_m(\kappa_e a)} = 0, \tag{4.63}$$

where I_m and K_m are modified Bessel functions of order m and the prime denotes a derivative with respect to the argument of the function.

Solutions to Eq. (4.63) are shown in Figure 4.12 for photospheric and coronal conditions. *Torsional Alfvén waves* have phase speed $\omega/k_z = v_{Ai}$ and are incompressible, whereas in a twisted flux tube they perturb the density (Zhugzhda and Nakariakov 1999). Just as for the slab, *body modes* are oscillatory inside the tube, whereas *surface modes* are evanescent (Figure 4.9). The modes guided by the tube with phase speeds larger than the minima of the sound and Alfvén speeds (c_{si}, c_{se}, v_{Ai}, v_{Ae}) are called *fast,* whereas those smaller than one minimum or the other are called *slow.* The modes in a flux tube are very similar to those in a slab (Figure 4.11), except that in the slender flux tube limit (i.e., long wavelength $k_za \ll 1$), the phase speed of the fundamental fast kink mode tends to the *kink speed* (instead of v_{Ae} or c_{se}) given by

$$c_K = \left(\frac{B_i^2 + B_e^2}{\mu(\rho_i + \rho_e)}\right)^{1/2} = \left(\frac{\rho_i v_{Ai}^2 + \rho_e v_{Ae}^2}{\rho_i + \rho_e}\right)^{1/2}, \tag{4.64}$$

which is a mean Alfvén speed.

In the photosphere, the fast sausage mode is a surface wave and has a cut-off at the external sound speed (c_{se}), beyond which it becomes *leaky* and propagates outwards in the external medium with a complex eigenfrequency obtained by replacing the modified Bessel functions (K_m) in Eq. (4.63) by Hankel functions. In the corona, it is a body wave that is cut off at the external Alfvén speed (v_{Ae}). The slow body waves are

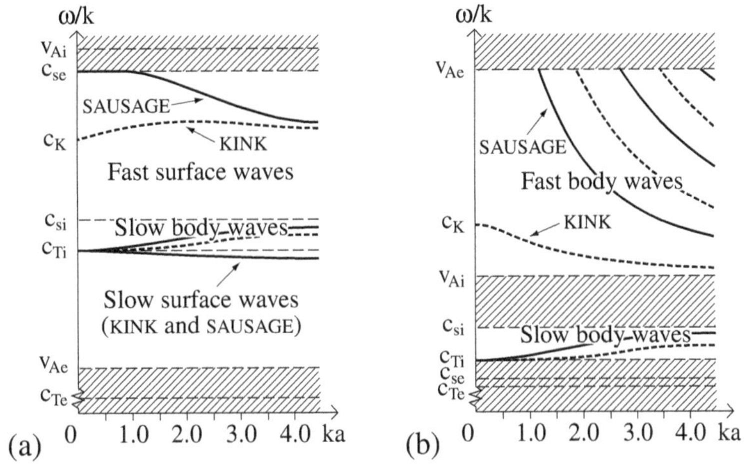

Figure 4.12. Waves in a flux tube of radius a and magnetic field B_i surrounded by an external field B_e, showing the phase speed (ω/k) as a function of wavenumber (k_z), for (a) photospheric conditions (i.e., $v_{Ai} > c_{se} > c_{si} > v_{Ae}$) and (b) coronal conditions (i.e., $v_{Ae} > v_{Ai} > c_{si} > c_{se}$). Sausage modes are shown by solid curves and kink modes by dashed ones (after Edwin and Roberts 1983).

always present with a phase speed equal to the slow (or tube) speed (c_T) in the slender flux tube limit, but the slow surface waves are present only in the photospheric conditions of Figure 4.12, and the fast body modes in the corona disappear if $v_{Ae} < v_{Ai}$. For an asymptotic theory based on a small aspect ratio that allows density variations along a tube, see Dymova and Ruderman (2005), and for an analysis of modes in a twisted tube, see Verth et al. (2010).

4.10 MHD Continua

In our treatment of waves in a non-uniform media, a striking feature of the differential equations (4.53) or (4.61) for v_{1x} or v_{1R} in a plane or cylindrical geometry is the appearance of factors $(\omega^2 - \omega_A^2)$ and $(\omega^2 - \omega_T^2)$ in front of the terms having highest derivatives. These factors vanish at frequencies equal to the Alfvén frequency ($\omega_A = k_z v_A$) or slow (or cusp) frequency ($\omega_T = k_z c_T$), and so make the differential equations singular as $\omega \to \omega_A$ and $\omega \to \omega_T$. They give rise to a continuous range of frequencies known as *continuous spectra*, which we describe in Section 4.10.1, largely following Goedbloed et al. (2004). These Alfvén and slow continua are in turn intimately related to the presence of two key enhanced dissipation processes in a non-uniform medium, called resonant absorption and phase mixing (Sec. 10.3).

4.10.1 Continuous Spectra

Shear Alfvén waves in a one-dimensional Cartesian geometry (or torsional Alfvén waves in a straight cylinder) have an unusual property, namely, that the group velocity is directed along the magnetic field, so that magnetic field lines may oscillate locally with their own local Alfvén frequency without affecting the rest of the medium. Thus, in a non-uniform magnetic field $B_0(x)\ \hat{\mathbf{z}}$, there is a continuous range of eigenfrequencies (or continuous spectrum in frequency space) of the form (Uberoi 1972)

$$\omega = \omega_A(x) \equiv k_z\ v_A(x),$$

which satisfy Eq. (4.53) identically. Each separate mode in this *Alfvén continuum* oscillates with a particular frequency ω and wavenumber k_z and is localised around a particular value of x given by $v_A(x) = \omega/k_z$.

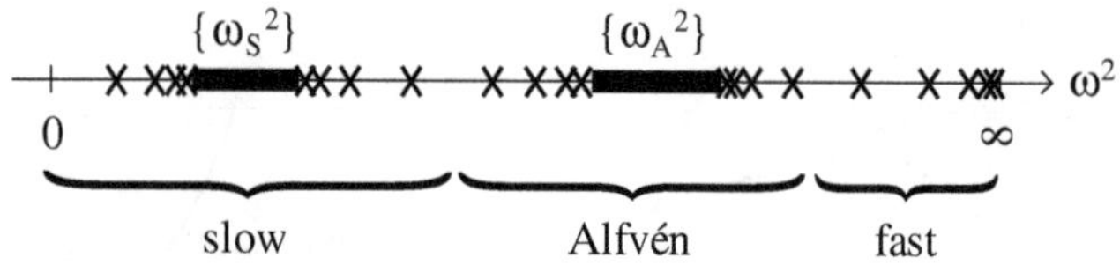

Figure 4.13. Schematic of the frequencies of discrete (X) and continuous ($\{\omega^2\}$) spectra in a structure that varies with one dimension (x or R, say). The discrete slow and Alfvén spectra cluster at the edges of the corresponding continua, while the discrete fast spectra clusters at infinity (after Goedbloed et al. 2004).

In ideal MHD, in addition to the Alfvén continuum, there is a separate range of frequencies called the *slow* (or *cusp*) *continuum* with $\omega = \omega_T(x) \equiv k_z c_T(x)$. Solutions to the differential equation (4.53) show that the continuum modes are characterised by non-square integrable tangential components that are perpendicular to the magnetic field for the Alfvén modes and parallel to the field for the slow modes (Goedbloed et al. 2004). The relation between the discrete and continuous spectra in general is shown in Figure 4.13.

The Alfvén and slow continua also exist in a cylindrical flux tube with field $B_0(r)\,\hat{\mathbf{z}}$. This time the Alfvén continuum consists of torsional Alfvén waves, in which each cylindrical shell oscillates independently at a frequency $\omega = \omega_A(r) \equiv k_z v_A(r)$, while the slow continua have frequencies $\omega = \omega_T(r) \equiv k_z c_T(r)$.

The processes of phase mixing and resonant absorption (Sec. 10.3) have a deep connection with the presence of the continuous spectra that exist in a non-uniform plasma in addition to the usual discrete wave modes. Studying the initial-value problem allows us to gain insight into some of the new features that a non-uniform medium brings. Furthermore, when one replaces a discontinuity by a continuous transition, a surprising object – the *quasi-mode* – makes an appearance.

4.10.2 Initial-Value Problem

If a discontinuous interface is replaced by an interface of finite width with a continuous magnetic profile, a surface wave (with a single frequency) no longer exists! This is because the presence of the singularity of Eq. (4.52) at $\varepsilon(x) = 0$ [i.e., $v_A(x)^2 = \omega^2/k_z^2$] implies there is no continuous nontrivial solution that decays away exponentially as x approaches $\pm\infty$. However, although the normal-mode technique fails, the initial-value problem may be tackled as follows (Sedlacek 1971; Tataronis and Grossmann 1973; Rae and Roberts 1981; Lee and Roberts 1986). The linearised equation of motion for $\boldsymbol{v}_1(\mathbf{r}, t)$ is

$$\rho_0 \frac{\partial \boldsymbol{v}_1}{\partial t} = \mathbf{F}_1 = -\boldsymbol{\nabla} p_1 - \mathbf{j}_1 \times \mathbf{B}_0 - \mathbf{j}_0 \times \mathbf{B}_1,$$

subject to given initial velocity and acceleration [$\boldsymbol{v}_i = \boldsymbol{v}_1(\mathbf{r}, 0)$ and $\dot{\boldsymbol{v}}_i = \partial \boldsymbol{v}_1/\partial t(\mathbf{r}, 0)$]. One first takes Laplace transforms $\hat{\boldsymbol{v}}(\mathbf{r};\omega) = \int_0^\infty \boldsymbol{v}_1(\mathbf{r}, t) e^{i\omega t}\, dt$ to give an inhomogeneous equation for $\hat{\boldsymbol{v}}(\mathbf{r};\omega)$, which is then solved in terms of a Green's function. Finally, an inverse Laplace transform is taken to give the required solution as

$$\boldsymbol{v}_1(\mathbf{r}, t) = \frac{1}{2\pi} \oint_C \hat{\boldsymbol{v}}(\mathbf{r};\omega) e^{-i\omega t} d\omega, \tag{4.65}$$

where C is a contour taken above the largest eigenvalue of $\mathbf{F}_1$ in the complex ω-plane.

Consider a unidirectional magnetic field [$B_{0z}(x)$] in the range $x_1 < x < x_2$ with a linear profile for the square of the Alfvén frequency in the form (Figure 4.14)

$$\omega_A^2(x) = \omega_0^2 + (\omega_A')_0^2 (x - x_0),$$

where $(\omega_A')_0^2$ is the gradient and $\omega_0^2 = \frac{1}{2}(\omega_{A1}^2 + \omega_{A2}^2)$. We focus on the Alfvén continuum by assuming for simplicity that: ρ_0 is constant; $\beta \ll 1$; and the propagation is nearly perpendicular ($k_z \ll k_y \approx k$). Then

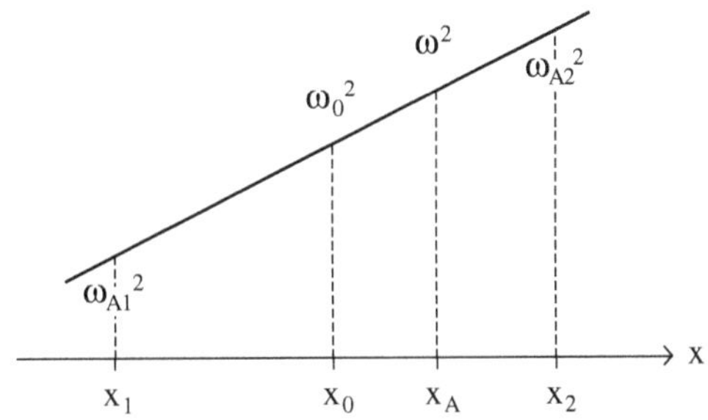

Figure 4.14. Notation for a linear profile of the Alfvén frequency squared $[\omega_A^2(x)]$.

the x-component of $\hat{\boldsymbol{v}}(\mathbf{r};\omega)$ satisfies

$$\frac{d}{dx}\left(P(x)\frac{d\hat{v}_x}{dx}\right) - Q(x)\ \hat{v}_x = R(x), \tag{4.66}$$

where $P(x) = -\rho_0[\omega^2 - \omega_A^2(x)]/k^2$, $Q(x) = \rho_0[\omega^2 - \omega_A^2(x)]$ and $R(x) = \rho_0 i\omega(v_{xi} + v'_{yi}/k) - \rho_0(\dot{v}_{xi} + \dot{v}'_{yi}/k)$. The solution that vanishes at x_1 and x_2 is

$$\hat{v}_x(x;\omega) = \int_{x_1}^{x_2} G(x,x';\omega)R(x')\ dx'$$

in terms of the Green's function $G(x,x';\omega)$, which also satisfies Eq. (4.66) but with $R(x)$ replaced by the delta function $\delta(x-x')$.

The Green's function is constructed from solutions of the homogeneous version of (4.66), in which $R(x)$ is replaced by zero. For a given ω, this equation is singular where $P(x) = 0$, that is, at $x = x_A \equiv x_0 + (\omega^2 - \omega_0^2)/(\omega'_A)_0^2$. It may be transformed by putting $X = k(x - x_A) \equiv k(\omega_A^2 - \omega^2)/(\omega'_A)_0^2$ into the modified Bessel equation

$$X\frac{d^2\hat{v}_x}{dX^2} + \frac{d\hat{v}_x}{dX} - X\hat{v}_x = 0, \tag{4.67}$$

whose solutions are the zeroth-order modified Bessel functions $I_0(X) \approx 1 + X^2/4 + ...$ and $K_0(X) = -u(X)\log_e X - v(X) \approx (-\gamma_0 - \log_e X/2) + (-\gamma_0 + 1 - \log_e X/2)X^2/4 + ...$, where γ_0 is Euler's constant and $v(X)$ is the nonsingular part of $K_0(X)$.

The resulting Green's function in terms of $u(X) = I_0(X)$ and $v(X)$ has the form

$$G(x,x';\omega) = \left[\left(\log_e\frac{\omega^2-\omega_A^2(x_S)}{\omega^2-\omega_{A1}^2} - \frac{v_1}{u_1}\right)u(x_S) + v(x_S)\right]$$
$$\times\left[\left(\log_e\frac{\omega^2-\omega_A^2(x_L)}{\omega^2-\omega_{A2}^2} - \frac{v_2}{u_2}\right)u(x_L) + v(x_L)\right] \Big/ \left(\log_e\frac{\omega^2-\omega_{A2}^2}{\omega^2-\omega_{A1}^2} - \frac{v_1}{u_1} + \frac{v_2}{u_2}\right),$$

where $u(X)$, $v(X)$, $u_1 = u(x_1)$, $v_1 = v(x_1)$, $u_2 = u(x_2)$ and $v_2 = v(x_2)$ are all functions of ω^2, and x_S and x_L are, respectively, the smaller and larger of x and x'.

Since ω^2 is complex, terms such as $\log_e(\omega^2 - \omega_A^2)$ in $G(x,x';\omega)$ are multi-valued, and so they may be made single-valued by inserting branch cuts in the complex ω-plane corresponding to the continuous spectrum $\pm\{\omega_A\}$, whereas zeros of G constitute a discrete spectrum. Then the asymptotic solution of Eq. (4.65) for $\boldsymbol{v}_1(\mathbf{r},t)$ may be obtained by deforming the contour C around the branch cuts and poles, so that the contribution of a branch cut is due to the jump in the logarithm at the cut. The resulting behaviour at a particular frequency (Tataronis and Grossmann 1973) is

$$v_{1x} \sim \frac{\omega}{t}\mathrm{e}^{-i\omega t}, \qquad v_{1y} \sim -i\frac{\omega\omega'}{k}\mathrm{e}^{-i\omega t}. \tag{4.68}$$

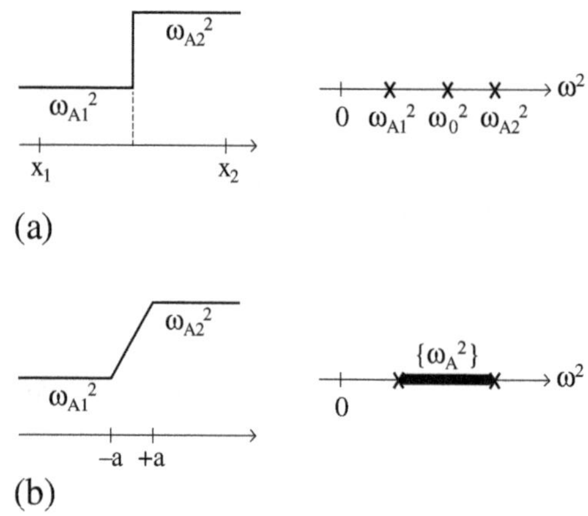

Figure 4.15. (a) A step profile for the square Alfvén frequency $\omega_A^2(x)$ producing a surface mode (ω_0) and a discrete spectrum of Alfvén modes (ω_{A1} and ω_{A2}) localized to the left and right of the step. (b) The effect of smoothing the inhomogeneity is to produce a continuous spectrum $\{\omega_A^2\}$.

Thus, the normal component (v_{1x}) damps like $1/t$, whereas the tangential component (v_{1y}) is undamped. As t increases, there is an ever more fluctuating spatial structure (*phase mixing*), with each point oscillating at its own local Alfvén frequency (Sec. 10.3.3).

The result is that a surface wave decays away in time and feeds energy into the interface, where tangential motions having a continuum of frequencies grow with an oscillating amplitude. The accompanying build-up of large gradients can eventually produce substantial viscous or Ohmic dissipation, which in turn dissipates the energy of the continuum modes and heats the plasma (Sec. 10.3).

4.10.3 Quasi-Modes

There is, however, another kind of motion that possesses a coherent oscillation. For example, consider first a region $x_1 \leq x \leq x_2$ with an Alfvén profile $[\omega_A^2(x)]$ that has a step discontinuity at $x = x_0 \equiv \frac{1}{2}(x_1 + x_2)$ from one uniform value (ω_{A1}^2) to another (ω_{A2}^2) (Figure 4.15a). The singularities of the continuous spectrum are then concentrated at the point $x = x_0$, which gives rise to a *surface mode* (Sec. 4.9.2). In the particular case when the density is homogeneous ($\rho_0 =$ *constant*), $\beta \ll 1$ and $k_z \ll k_y$, Eq. (4.53) or the homogeneous version of Eq. (4.66) becomes

$$[(\omega^2 - \omega_A^2)v_{1x}']' - k^2[\omega^2 - \omega_A^2(x)]v_{1x} = 0, \tag{4.69}$$

where a prime denotes d/dx. Either side of the step at $x = x_0$, $\omega_A(x)$ is uniform and so Eq. (4.69) reduces to $v_{1x}'' - k^2 v_{1x} = 0$. The solution that vanishes at $x = x_1$ and $x = x_2$ and is continuous at $x = x_0$ has a cusped shape at $x = x_0$ and is

$$v_{1x} = \frac{\sinh k(x - x_1)}{\sinh k(x_0 - x_1)} \quad \text{for } x < x_0, \qquad v_{1x} = \frac{\sinh k(x_2 - x)}{\sinh k(x_2 - x_0)} \quad \text{for } x > x_0,$$

(Figure 4.9a). Integrating Eq. (4.69) across $x = x_0$ implies that the jump in $(\omega^2 - \omega_A^2)v_{1x}'$ be continuous, so that

$$\omega^2 = \omega_0^2 \equiv \tfrac{1}{2}(\omega_{A1}^2 + \omega_{A2}^2).$$

This is simply the eigenfrequency of the surface wave and agrees with the earlier result [Eq. (4.56)] when ρ is uniform.

Next, for simplicity set $x_0 = 0$, $x_1 \to -\infty$, $x_2 \to +\infty$ and smooth out the step profile of $\omega_A^2(x)$ by replacing it with a linear profile between $x = -a$ and $x = a$. The spectrum of the system changes drastically, with discrete modes ($\omega^2 = \omega_{A1}^2, \omega_{A2}^2$) localised at the ends of a continuous range of Alfvén frequencies (Figure 4.15b). The fact that there are infinitely many discrete modes may be seen from the fact that in the homogeneous intervals Eq. (4.69) reduces to

$$(\omega^2 - \omega_A^2)(v_{1x}'' - k^2 v_{1x}) = 0,$$

so that for $x < -a$ a proper eigenfunction may be obtained by putting $\omega^2 = \omega_{A1}^2$ and choosing v_x arbitrarily, and similarly for $x > a$.

So, what has been the effect on the surface wave of introducing the linear profile? Has the coherent oscillation of the surface wave been replaced by a chaotic continuous spectrum of frequencies behaving like Eq. (4.68)? The answer can be found by using the initial-value approach of the previous section to determine what has happened to the Green's function pole that gave rise to $\omega^2 = \omega_0^2$.

The Green's function is constructed as before from the homogeneous version of Eq. (4.66), whose fundamental solutions $U_1(x)$ and $U_2(x)$ satisfy the left- and right-hand boundary conditions $U_1(x_1) = 0$ and $U_2(x_2) = 0$, respectively. It has the form

$$G(x, x') = \frac{U_1(x_S)U_2(x_L)}{\Delta}, \qquad \text{where } \Delta = P(x)[U_1(x)U_2'(x) - U_1'(x)U_2(x)].$$

Thus, for $-\infty < x < -a$ [where $\omega_A(x) = \omega_{A1}$ is constant], $U_1 = e^{kx}$ and $U_2 = C_2 e^{kx} + D_2 e^{-kx}$, whereas for $a < x < +\infty$ [where $\omega_A(x) = \omega_{A2}$], $U_1 = C_1 e^{kx} + D_1 e^{-kx}$ and $U_2 = e^{-kx}$. In the inhomogeneity ($-a < x < a$), U_1 and U_2 satisfy Eq. (4.67) with $X = 2ka(\omega_A^2 - \omega^2)/(\omega_{A2}^2 - \omega_{A1}^2)$, and so they are linear combinations of $I_0(X)$ and $K_0(X)$. Applying jump conditions at $x = \pm a$ determines the constants C_1, C_2, D_1, D_2 and implies

$$\begin{aligned}\Delta = CX_1X_2 \{&[I_0(X_1) - I_1(X_1)]\,[K_0(X_2) - K_1(X_2)] \\ &-[K_0(X_1) + K_1(X_1)]\,[I_0(X_2) + I_1(X_2)]\},\end{aligned}$$

whose zeros give the poles of $G(x, x')$, where $X_1 = 2ka(\omega_{A1}^2 - \omega^2)/(\omega_{A2}^2 - \omega_{A1}^2)$ and $X_2 = 2ka(\omega_{A2}^2 - \omega^2)/(\omega_{A2}^2 - \omega_{A1}^2)$.

Now put $\Delta = 0$ and expand the Bessel functions for $X \ll 1$ (a thin inhomogeneity with $ka \ll 1$), to obtain the dispersion equation $\log_e(X_2/X_1) + X_1^{-1} + X_2^{-1} = 0$ or

$$\log_e \frac{\omega^2 - \omega_{A2}^2}{\omega^2 - \omega_{A1}^2} - \frac{\omega_{A2}^2 - \omega_{A1}^2}{2ka}\left[\frac{1}{\omega^2 - \omega_{A1}^2} + \frac{1}{\omega^2 - \omega_{A2}^2}\right] = 0.$$

This has complex solutions for $\omega = \omega_0 + i\nu_0$, due to the movement of the surface-wave pole ($\omega = \omega_0$) off the real axis to give

$$\omega_0 = \sqrt{\tfrac{1}{2}(\omega_{A1}^2 + \omega_{A2}^2)}, \qquad \nu_0 = \tfrac{1}{8}\pi ka(\omega_{A2}^2 - \omega_{A1}^2)/\omega_0.$$

The asymptotic contribution of this pole for large t is to make

$$v_{1x} \sim v_{1y} \sim e^{-\nu_0 t} e^{-i\omega_0 t}.$$

Thus, the effect of the smooth inhomogeneity is surprisingly to make the previous surface mode become exponentially damped in time. It is a so-called *quasi-mode* or *collective mode*, which arises when a discrete mode exists in the range of a continuous spectrum, so that it couples to the continuum modes: it is this coupling which transfers energy from the quasi-mode to the continuum and causes it to damp. The quasi-mode is not a normal eigenmode of the plasma, since there is a general proof that complex eigenvalues do not exist for a self-adjoint operator such as arises in ideal MHD. However, it does represent a coherent orderly oscillation of the inhomogeneous plasma at a definite frequency. (When performing the integral along the Laplace contour in the complex-ω plane in order to take the inverse Laplace transform, the quasi-mode pole does not exist on the principal Riemann sheet: its contribution is picked up by deforming the contour across the branch cuts onto a neighbouring Riemann sheet.) If the density is no longer uniform, the real part of the quasi-mode frequency becomes $\sqrt{[(\rho_1\omega_{A1}^2 + \rho_2\omega_{A2}^2)/(\rho_1 + \rho_2)]}$.

4.11 Waves on the Sun

4.11.1 Waves in the Low Atmosphere

Leighton (1960) discovered that the quiet solar surface (photosphere and low chromosphere) is covered with regions oscillating up and down with a range of periods that peaks at about 5 minutes. These *five-minute oscillations* (Sec. 1.3.4) represent global acoustic modes that are trapped in a resonant cavity below the solar surface within which they can propagate. The velocity amplitude is typically 0.4 km s^{-1} in the photosphere and increases slowly with height. Upward vertical phase speeds of 30–100 km s^{-1} are observed, as are horizontal phase speeds of the same magnitude. In the photosphere, there is no phase difference between velocity oscillations at different heights, implying that the waves are evanescent rather than propagating.

The chromosphere is also seen to oscillate with a wide range of periods, but with the power concentrated near 300 sec. The lower periods increase in importance with height. In contrast to the photosphere, the chromospheric 5-minute oscillations appear to be propagating upwards at roughly the sound speed. In the transition region, there are fluctuations of 2 to 3 km s^{-1} but no well-defined period, since they are destroyed during their propagation through the inhomogeneous chromosphere.

Local diagnostic diagrams such as Figure 4.8b at different temperatures can be used to show whether a wave of a given frequency (ω) and horizontal wavenumber (k_x) is evanescent or propagating in a particular region of the atmosphere. For example, acoustic waves with a horizontal wavelength of, say, 5 Mm and a period of 5 minutes are propagating in the upper convection zone (i.e., they lie above the acoustic wave-evanescence boundary in Figure 4.8b), but, by the time they reach the temperature-minimum region of the photosphere, they have become evanescent (since they lie below the boundary for that temperature). In other words, the acoustic waves are trapped in a resonant cavity below the temperature minimum.

In the convection zone, gravity waves do not propagate, since gravity drives turbulent motions (by the buoyancy force). These in turn produce pressure perturbations that propagate away as sound waves. Sound is generated, according to the *Lighthill mechanism* (Lighthill 1953), mainly by quadrupole emission and most efficiently at a wavelength comparable to the turbulence length-scale (L). For a sound speed (c_s) of 20 km s^{-1} characteristic of the low photosphere, one would need $L \approx 6000$ km to obtain a period (L/c_s) of 300 s.

Oscillations are observed in many strongly magnetic regions, such as sunspot umbrae, plages and supergranulation boundaries. In particular, properties of umbral oscillations and running penumbral waves are described in Section 1.7.2, and explanations in terms of absorption and mode conversion of p-modes are given in Section 9.5.3. Oscillations are also found in prominences (Oliver 2009; Tripathi et al. 2009).

4.11.2 Coronal Seismology

Many different wave modes have now been detected in the corona (standing and propagating, slow- and fast-mode) and this has stimulated the development of a new field which is still in its infancy – *coronal seismology*. It uses the properties of the waves to infer physical properties of the corona that are hard to measure directly, such as the magnetic field, transport coefficients, fine structuring and heating rate. The original idea was suggested by Uchida (1970) and developed by Roberts et al. (1984). An invaluable living review has been written by Nakariakov and Verwichte (2005), which complements other reviews by

Roberts (2000), Banerjee et al. (2007), Goossens (2008), Ruderman and Erdélyi (2009) and De Moortel and Nakariakov (2012).

In future, it is expected that this field will develop greatly, but so far the analysis of observed waves has been fairly unsophisticated, mainly to determine qualitatively what mode is being observed. Several basic questions arise:

(i) **Geometry**. What is the underlying structure: can it be adequately modelled as a slab or a tube or is a more complex geometry necessary?

(ii) **Mode**. Which mode is it? As well as the observed phase speed, other properties of the wave will be diagnostic, such as whether it is propagating or standing, longitudinal or transverse, compressible or incompressible, isotropic or anisotropic and what are the phase differences in the wave.

(iii) **Damping**. What is the main cause of observed damping? Is it leakage by radiating modes or resonant absorption or phase mixing (Sec. 10.3)?

(iv) **Corona**. What can be deduced about the physical properties of the corona by comparing the observations with wave theory?

There are essentially two main wave speeds in the corona, namely, the sound speed, which depends on the temperature ($c_s \approx 166\ T_0^{1/2}$ m s^{-1}) and is typically 200 km s^{-1}, and the Alfvén speed, which depends on density and magnetic field [$v_A = 2.18\times10^{12}\ (B_0/n_0^{1/2})$ m s^{-1}, with B_0 in gauss] and is typically 3000 km s^{-1} (for $B_0 \approx 100$ G and $n_0 \approx 10^{16}$ m^{-3}).

When interpreting observations of coronal oscillations, there are three levels of sophistication:

(A) Since c_s and v_A are widely separated in the low-β corona, the first step is to distinguish from its phase speed between a mode that is essentially pressure-driven or magnetically driven: in other words, is the mode basically acoustic (or slow magnetoacoustic) in nature, or is it strongly magnetic, in the sense of being a wave that has properties of shear Alfvén, torsional Alfvén or fast magnetoacoustic waves?

(B) Next, if the geometry can be well modelled as varying predominantly in one-dimension (i.e., a slab or a tube), then key results of Section 4.9, as summarised in Figures 4.11 and 4.12, may be used to identify the mode, from the observed phase speed and the other wave properties. Here, however, the distinctions between sound and Alfvén speeds in different parts of a structure or between the cusp (c_T) and sound speeds or the kink (c_K) and Alfvén speeds are much more subtle.

(C) Finally, the modes that occur in much more realistic inhomogeneous three-dimensional structures may be determined and compared with the observations.

So far, step (A) and a partial use of step (B) have been the norm, but I expect step (C) to come to the fore in future, in which the understanding from step (B) will be essential. For step (B), Figures 4.11 and 4.12 have rich diagnostic potential which, as yet, remains relatively untapped. In a simple discrete slender slab or tube (with $k_z a < 1$ and uniform interior), they imply that several wave modes are possible (the first three being magnetoacoustic body modes), as follows.

(1) ***Fast kink waves*** have a phase speed of v_{Ae} for a slab; for a tube it is the kink speed $\{c_K = [B_i^2 + B_e^2]/[\mu(\rho_i + \rho_e)]^{1/2}\}$, which is a mean Alfvén speed. They are almost incompressible, transverse waves ($m = 1$) relative to the magnetic field, so that the slab or tube is moved laterally, producing little variation of the cross-section, density or intensity. For thick slabs or tubes ($k_z a > 1$), higher-order fast kinks also exist with higher speeds.

(2) ***Fast sausage waves*** propagate at a speed between the internal (v_{Ai}) and external (v_{Ae}) Alfvén speeds. Such waves possess a long wavelength cut-off where the phase speed becomes v_{Ae}, beyond which they exist only in a leaky form. Thus, at a given wavelength they are present in their pure form only in inhomogeneities that are sufficiently thick. They are basically modes with $m = 0$ that throb or

pulsate. In the process, they perturb the loop cross-section and produce intensity variations. The mode is mainly transverse, so that velocities normal to the magnetic field are stronger than those along the field. Higher-order fast sausage modes also exist in thicker inhomogeneities.

(3) ***Slow kink and sausage waves*** propagate at the internal cusp (or tube) speed $[c_{Ti} = v_{Ai}c_{si}/(v_{Ai}^2 + c_{si}^2)^{1/2}]$, which reduces to the internal sound speed (c_{si}) when $\beta \ll 1$. These modes are basically longitudinal, with velocities directed along the magnetic field. They are compressible and so produce density and intensity variations. They should not be confused with or misnamed *acoustic waves*, which exist in a medium with low or non-existent magnetic field and are isotropic.

(4) ***Alfvén waves*** give rise to oscillations at the internal Alfvén speed (v_{Ai}). They are incompressible and so produce no intensity variations in the linear regime. The resulting motions are normal to the magnetic field but are not transverse in the sense of being normal to the slab or tube – rather, they lie within magnetic surfaces, and so for a tube the Alfvén waves are torsional.

When the Alfvén speed profile is continuous (rather than discontinuous), MHD waves have a mixture of properties of the above pure modes. For example, the term

(5) ***Alfvénic waves*** denotes largely transverse modes, which propagate at essentially the Alfvén speed (Sec. 4.9.4). Almost incompressible for thin flux tubes, they have a mixture of fast-mode kink and torsional Alfvén wave characteristics.

A comprehensive summary of current coronal seismology observations is given by Nakariakov and Verwichte (2005) and De Moortel and Nakariakov (2012), the main features of which are as follows. A major discovery by Aschwanden et al. (1999) and Nakariakov et al. (1999) from the TRACE satellite is the presence of decaying transverse coronal loop oscillations generated by a flare, with more examples given by Schrijver et al. (2002). The example shown in Figure 4.16 is typical, with a period of 250 sec, amplitude of 90 km s^{-1} and phase speed of 1020 km s^{-1} in a loop of length 260 Mm and cross-section 1 Mm. These oscillations are interpreted as standing *fast kink waves* and decay rapidly over typically 3 periods.

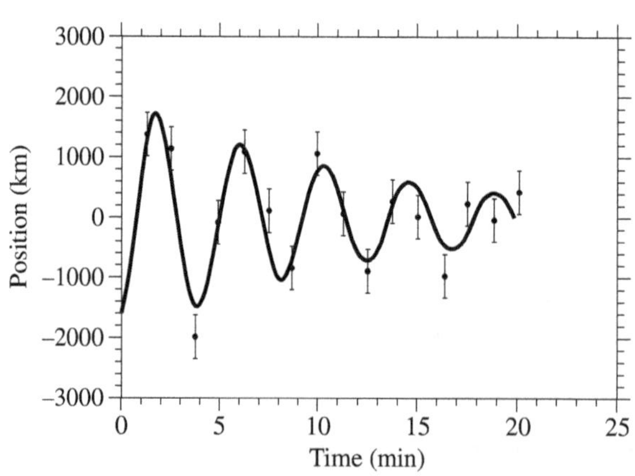

Figure 4.16. The temporal evolution of the displacement of a coronal loop observed with TRACE. The solid curve is a best fit to a decaying sinusoid with amplitude 2,030 km, period 4.3 min and decay time 14.5 min (courtesy Valery Nakariakov).

In the low-β limit, where $B_i \approx B_e$, the kink speed [Eq. (4.64)] becomes

$$c_K \approx \frac{B_i}{\sqrt{(\mu \rho_i)}} \left(\frac{2}{1 + \rho_e/\rho_i} \right)^{1/2},$$

which illustrates the current difficulty of making confident deductions; for an observed value of c_K, neither B_i, ρ_i nor ρ_e is well-known. If reasonable values of ρ_i and ρ_e are adopted, however, then the magnetic field is determined to be about 13 G. One way to reduce the uncertainty in future would be to use accurate magnetic extrapolations for off-limb observations (to deduce the coronal magnetic field) and then the above formula to give a relation between the densities.

Fast kink modes have also been observed in spatially resolved microwave pulsations of short flaring loops (only 16 Mm long) by Asai et al. (2001) with a period of 6.6 sec. In this case, the same argument as above implies a magnetic field of 400 G. The reason for the decay of the fast kink modes has not yet been definitively identified. Leakage of wave energy from the loop or through the footpoints has been ruled out and the two main candidates are currently resonant absorption and phase mixing (Sec. 10.3). However, so far the geometry has been oversimplified as a single isolated loop rather than a complex three-dimensional structure. The damping time of standing global kink modes due to resonant absorption has been calculated as a function of density ratio and aspect ratio by Ruderman and Roberts (2002). There is also some evidence for damping by resonant absorption in propagating kink modes (Verth et al. 2010).

Transverse loop oscillations (fast kink modes) can in practice produce intensity variations by changes in line-of-sight column depth (Verwichte et al. 2009), by density variations from coupling to slow modes (Terradas et al. 2011) or by variations in loop length and cross-section in a vertically polarised mode (Wang and Solanki 2004). Detailed analyses of loop oscillations observed with SDO/AIA show in one case periods 1.7 to 10 min, damping times 3 to 13 min and intensity changes due to variations in line-of-sight column depth (White and Verwichte 2012) and in other cases coronal Alfvén speeds of 2,100 or 2,500 km s^{-1} (Verwichte et al. 2013).

Fast sausage waves may explain microwave oscillations of flares at 10 to 20 sec in short loops of length 25 Mm (Nakariakov et al. 2003), as well as X-ray and radio oscillations at 0.5 to 10 sec (Zaitsev and Stepanov 2002). Furthermore, rapidly propagating kink or sausage fast-mode waves of period 6 sec and speed 2,100 km s^{-1} have been observed in an eclipse with the SECIS instrument (Williams et al. 2002).

Slow magnetoacoustic waves have been inferred from intensity and Doppler oscillations. SoHO/SUMER and Yohkoh/BCS reveal fundamental standing modes in coronal loops at high temperatures (6 to 10 MK), with periods 5 to 30 min, in response to flaring (Wang 2003; Ruderman and Erdélyi 2009; Wang 2011). Many examples of upwards-propagating slow modes along plumes and loops at active region edges have been seen with SoHO (UVCS, EIT) and TRACE at speeds of 100 km s^{-1}, variations of $\lesssim 5\%$ in density, and periods of 2 to 10 min (Ofman et al. 1997; De Moortel et al. 2000, 2002). They are thought to be caused by leakage of p-modes and damped by conduction. Some authors have used Hinode/EIS observations to suggest they are upflows triggered by periodic heating events (De Pontieu and McIntosh 2010), while others interpret the observations as waves (Verwichte et al. 2010). The trapping of slow modes in a chromospheric resonator may explain the presence of 3-min oscillations above sunspots (Botha et al. 2011).

Alfvén waves (i.e., shear Alfvén waves or torsional Alfvén waves) are transverse and incompressible, with perturbations lying in magnetic flux surfaces. They do not perturb the density and so remain unseen in intensity but could be observed by Doppler shifts. So far there have been no convincing reports in coronal observations, although of course non-thermal broadening of emission lines may include an Alfvén wave component; an increase in broadening observed with Hinode/EIS from 26 km s^{-1} at a height of

8 Mm above the limb to 42 km s^{-1} at 110 Mm may be caused by a growth of Alfvén wave amplitude with height (Banerjee et al. 2009). However, torsional Alfvén waves have been observed in the chromosphere (Jess et al. 2009) above photospheric bright points with vortex motions between granules (Bonet et al. 2008, 2010; Wedemeyer-Böhm and Rouppe van der Voort 2009).

Ubiquitous propagating Alfvénic waves in the corona have been discovered by Tomczyk et al. (2007), Tomczyk and McIntosh (2009) by using 'CoMP' (the Coronal Multi-Channel Polarimeter) to measure the four Stokes parameters in the corona between 1.05 and 1.35 solar radii above the limb every 29 sec for many hours. In a discrete flux tube with uniform interior, they would be called fast kink modes, but in a flux tube with continuous Alfvén profile they have both fast-mode and Alfvén characteristics and are sometimes termed *Alfvénic waves* (Sec. 4.9.4). The same waves are seen lower down in the atmosphere as type II spicules (Secs. 1.5 and 9.2.4) swaying to and fro like prairie grass (De Pontieu et al. 2007). In CoMP, the waves have an amplitude of 0.3 km s^{-1} (compared with 30 km s^{-1} for coronal line widths) and have also been observed with Hinode/SOT/XRT. The power spectrum of coronal velocity fluctuations shows a broad peak at 3.5 mHz (5-min period), which is below the cut-off frequency, suggesting that the waves have tunnelled through the lower atmosphere or have been able to propagate along intense flux tubes at the boundaries of supergranules or granules. There is no evidence of significant intensity perturbations, implying the disturbances are incompressible. In the plane of the sky, the waves propagate with amplitudes of 300 m s^{-1} and phase speeds of 550 km s^{-1} in the same direction as the magnetic field. There is a predominance of upwards over downwards propagating waves, suggesting that they dissipate before reaching the other ends of coronal loops. An energy flux of 10–100 erg cm^{-2} s^{-1} (0.01–0.1 W m^{-2}) is estimated, which is 3 or 4 orders of magnitude smaller than required (3×10^5 erg cm^{-2} s^{-1}) to heat the quiet corona. Whether or not they can heat at least part of the corona is not yet clear, but they will in any case prove invaluable to coronal seismology through the inference of physical properties of the corona.

Quasi-periodic pulsations (QPP) during solar flares in radio and hard X-rays are evidence of MHD oscillations (Nakariakov and Melnikov 2009). They reflect modulation of flare emission due to either an oscillation of the magnetic arcade in which high-energy particles are propagating or an oscillation of the reconnection process itself. They often exhibit multiple periodicities as harmonics of a fundamental mode. Nakariakov and Zimovets (2011) invoke repeated bouncing to and fro of a slow-mode wave in a reconnecting arcade in order to explain pulsations in a two-ribbon flare and the changing location of energy release along the arcade.

In future, what is needed are carefully designed three-dimensional numerical experiments that can give theoretical and physical insight into the subtle nature of MHD waves on the Sun. One aim is to develop coronal seismology much more so that it can be used to infer accurate physical properties of the corona, and another is to determine the quantitative role of MHD waves in heating the corona and driving the solar wind.

Chapter 5

Shock Waves

5.1 Introduction

The Sun is in a far from static state, and often the motions that are universally present produce shock waves. For example, magnetic reconnection is likely to be taking place continually when distinct magnetic flux systems interact with one another or new flux emerges from below the photosphere, and this reconnection invariably generates magnetic shock waves (Sec. 6.6). Also, the occurrence of a coronal mass ejection containing a huge erupting solar prominence is likely to generate a wave that propagates ahead of the ejection and whose presence is sometimes revealed by a Type II radio burst. On a smaller scale, shock waves are probably generated by surges, spicules and the continual granular motion of the solar surface. Applications are discussed elsewhere, but the present chapter develops the basic theory for both hydrodynamic and magnetic shocks.

5.1.1 Formation of a Hydrodynamic Shock

Consider wave propagation in a non-conducting gas. When the amplitude is so small that linear theory applies, we saw in Section 4.2 that a disturbance propagates as a sound wave. If the gas has a uniform pressure (p_1) and a density (ρ_1), the speed of propagation is the sound speed

$$c_{s1} = \left(\frac{\gamma p_1}{\rho_1}\right)^{\frac{1}{2}},$$

and the wave profile maintains a fixed shape, since each part of the wave moves with the same speed. However, when the wave possesses a large enough amplitude that nonlinear effects become important, the crest of the sound wave moves faster than its leading or trailing edge. This causes a progressive steepening of the front portion of the wave as the crest catches up. Ultimately, the gradients of pressure, density, temperature and velocity become so large that dissipative processes, such as viscosity or thermal conduction, are no longer negligible (Figure 5.1). Then a steady wave-shape can be attained, called a *shock wave*, with a balance between the steepening effect of the nonlinear convective terms and the broadening effect of dissipation. The shock wave moves at a speed in excess of the sound speed (c_{s1}), so that information cannot be propagated ahead to signal its imminent arrival, since such information would travel at only c_{s1} relative to the undisturbed medium ahead of the shock. When the shock speed greatly exceeds c_{s1} we have a *strong shock*, but when it is only slightly larger than c_{s1} the shock is said to be *weak*. The dissipation inside the shock front leads to a gradual conversion of the energy being carried by the wave into heat. Thus, the effects of the passage of a shock wave are to convert ordered (flow) energy into random (thermal) energy through particle collisions and also to compress and heat the gas.

In water, for example, surface gravity waves approaching a sloping beach exhibit the increase in wave-amplitude and steepening of the wave-front that we have just described. These processes are expected to

Figure 5.1. The steepening of a finite-amplitude wave profile to form a shock wave.

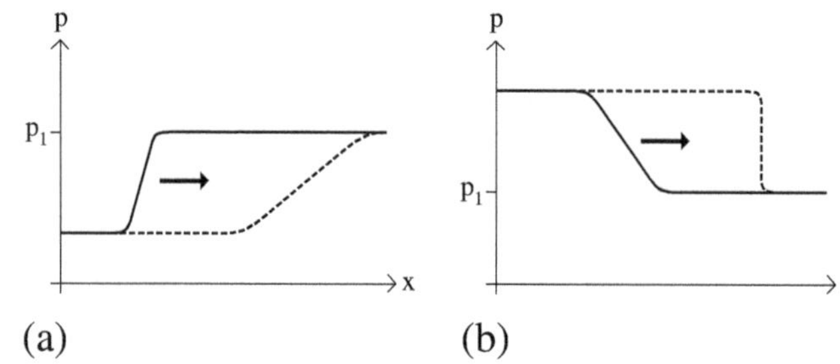

Figure 5.2. The pressure profiles at two times caused by a piston which is accelerated from $x = 0$ in the directions (a) $x < 0$ and (b) $x > 0$. The profile at the later time is shown dashed, and p_1 is the initial uniform gas pressure.

occur also in the solar atmosphere, where magnetoacoustic waves increase in amplitude as they propagate upwards from the photosphere. They eventually form into shock waves, which probably dissipate their wave energy in the low chromosphere.

One way of producing sound waves is to move the walls of a gas container. For example, suppose a long tube contains gas initially at rest, and a piston at one end of the tube is accelerated into uniform motion. If the piston is being withdrawn from the tube, an *expansion wave* travels from the initial location of the piston into the gas. A fall in pressure occurs as the expansion wave passes by. However, nonlinear effects make the wave flatten out as it propagates (Figure 5.2a) and so no shock wave develops. By contrast, if the piston is pushed into the tube, a *compression wave* is generated, across which the pressure is increased, and this eventually steepens into a *shock wave* (Figure 5.2b).

The shock front itself is in reality a very thin transition layer from one quasi-uniform region to another. Its width is typically only a few mean-free paths, with particle collisions establishing the new uniform state behind the shock. Normally, one models a shock front mathematically by a plane discontinuity, and the two states that it separates are taken to be uniform (and hence dissipationless). They are denoted here by subscripts (1) for the undisturbed gas (ahead of the the shock) and (2) for the shocked gas (behind the shock). In the frame of reference of the unshocked gas (i.e., at rest), the shock speed is U, while the speed of the shocked gas is, say, U_2 $(< U)$ (Figure 5.3a). However, it is more convenient to use a frame of reference moving with the shock (Figure 5.3b), so that the undisturbed gas enters the front of the shock with speed

$$v_1 = U, \tag{5.1}$$

while the shocked gas leaves the back of the shock with speed

$$v_2 = U - U_2. \tag{5.2}$$

Since U_2 is positive, we see from Eqs. (5.1) and (5.2) that

$$v_2 \leqslant v_1, \tag{5.3}$$

equality occurring when there is no shock.

In Section 5.2, a set of *conservation relations* or *jump conditions* is derived. They relate properties on both sides of the shock front, regardless of the detailed structure within it. To obtain physically relevant solutions, these relations are supplemented by an *entropy condition*, namely that the entropy increases following the flow.

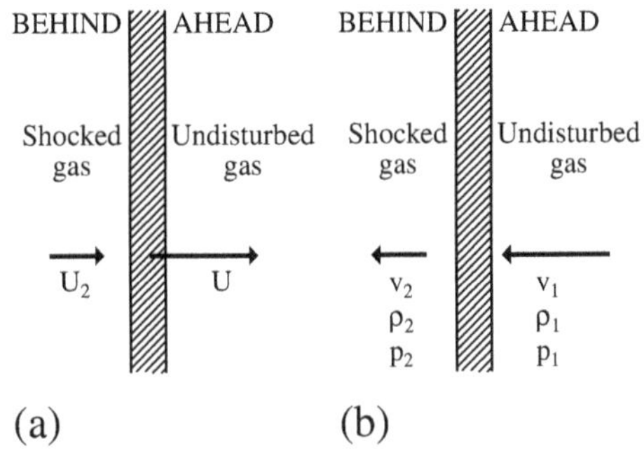

Figure 5.3. Notation for a plane hydrodynamic shock wave moving to the right with speed U into a gas at rest. Properties ahead of and behind the shock are denoted by 1 and 2.

A detailed determination of the thickness of the shock and its internal structure is in practice very complicated. For example, in addition to the effects of viscosity and thermal conduction, it often transpires that the electrons heat up first and then share their energy with the ions, a process whose description needs at least a *two-fluid model.* Also, the effect of *ionisation* in the shock may need to be included if the unshocked gas is not completely ionised, for instance near the *temperature-minimum region* of the solar atmosphere. However, if the dominant dissipation mechanism is known, an order-of-magnitude estimate of the *shock-width* (δx) may be obtained. For instance, in the case of viscous dissipation, the amount of energy (δW) dissipated during a small time (δt) is given by

$$\frac{\delta W}{\delta t} \approx \rho\nu \left(\frac{\delta v}{\delta x}\right)^2, \tag{5.4}$$

where ν is the kinematic viscosity. After taking

$$\delta t \approx \frac{\delta x}{v_1} \tag{5.5}$$

as the time for the shock to move distance δx and putting $\delta v \approx v_1 - v_2$, Eq. (5.4) gives

$$\delta x \approx \frac{\rho\nu(v_1 - v_2)^2}{v_1 \delta W}, \tag{5.6}$$

where δW can be found by comparing the energies on both sides of the shock front. In particular, it transpires that for a strong shock (Sec. 5.2) $\delta W \approx \frac{1}{2}\rho_1 v_1^2$, so that

$$\delta x \approx \frac{\nu}{v_1}. \tag{5.7}$$

In other words, the *Reynolds number* ($v_1 \delta x/\nu$) is of order unity (see Sec. 2.5).

5.1.2 Effects of a Magnetic Field

In a conducting gas, a magnetic field can interact strongly with the flow. The analysis of shock waves therefore becomes more complex, but the basic principles remain the same as for a non-conducting gas, where only sound waves are present. We have seen (Sec. 4.3) that the magnetic field introduces two extra waves, and so, in the absence of rotation and gravity, three wave modes become possible. They are classified according to their phase speeds as *slow, intermediate* and *fast.* When the wave amplitude is large, the intermediate mode (i.e, an Alfvén wave) can propagate without steepening, whereas the slow and fast magnetoacoustic modes steepen to form *slow* and *fast magnetoacoustic shock waves*, respectively. Furthermore, in the motion of a piston, the three degrees of freedom (two transverse and one longitudinal) give rise to the three distinct modes of wave propagation, two of which can steepen as in Figure 5.2b. A discussion of MHD shock waves can be found in Anderson (1963) and Kantrowitz and Petschek (1966), as well as in many of the standard text books on magnetohydrodynamics such as Ferraro and Plumpton (1961), Jeffrey and Taniuti (1966) and Goedbloed et al. (2004).

A set of jump conditions may again be derived (Sec. 5.4.1), but they are considerably more complicated than in the purely hydrodynamic case. The extra complexity arises both from the presence of an extra variable, namely the magnetic field strength, and also from the fact that the magnetic field and plasma velocity may be inclined away from the shock normal. Furthermore, the entropy condition that supplemented the jump conditions for the hydrodynamic shock is now insufficient. The system is no longer isolated, since the magnetic field may do work from the outside on the plasma, with the result that the entropy may, in some cases, decrease. A (stronger) so-called *evolutionary condition* must be employed instead (Jeffrey and Taniuti 1964), which ensures that the perturbation caused by small disturbances incident on the shock front is both small and unique. For simplicity, just the results of its application will be quoted here. In practice, it implies that the tangential magnetic field component cannot reverse its sign across a shock.

For wave propagation normal to the magnetic field, the slow magnetoacoustic mode ceases to exist, while the fast mode propagates with speed $(c_{s1}^2+v_{A1}^2)^{\frac{1}{2}}$ in a medium with sound speed c_{s1} and Alfvén speed v_{A1}. Thus, only one type of shock wave can propagate directly across the magnetic field, namely the fast shock, and its speed exceeds $(c_{s1}^2+v_{A1}^2)^{\frac{1}{2}}$. It is sometimes called a *perpendicular shock.* In any other direction, both the slow and fast shocks can propagate. These *oblique shocks* can travel with speeds in excess of the slow and fast wave-speeds, respectively. The slow shock has the effect of decreasing the magnetic field strength as it passes and making the magnetic field rotate towards the shock normal, whereas the fast shock has the opposite effect (see Figure 5.6). In the particular case when the magnetic field *behind* the shock is aligned with the shock normal, so that the tangential field component has been 'switched off', the slow shock is called a *switch-off shock.* In the case when the magnetic field *ahead* of the shock is directed along the normal (and the incident speed is Alfvénic), so that a tangential component is 'switched on' by the passage of the shock, the fast shock is termed a *switch-on shock* (see Figure 5.7).

Another effect in a conducting (or ionised) gas is the presence of an extra dissipative mechanism, namely *Ohmic heating,* due to a finite electrical conductivity. This complicates the shock structure but has no influence on the jump relations, provided Ohmic dissipation and the slipping of field lines can be neglected on both sides of the shock front. If Ohmic heating dominates the energetics, the shock thickness (δx) may be estimated as follows. The rate of dissipation of energy is

$$\frac{\delta W}{\delta t} \approx \frac{j^2}{\sigma}, \tag{5.8}$$

where, according to Ampère's law, the shock current is $j \approx (B_{1y} - B_{2y})/(\mu\delta x)$, in terms of the transverse magnetic field components (B_{1y} and B_{2y}) ahead of and behind the shock (Figure 5.5). Thus, with δt given by Eq. (5.5), Eq. (5.8) implies

$$\delta x \approx \frac{(B_{1y} - B_{2y})^2}{\mu^2 \sigma v_1\, \delta W},$$

where δW can be determined from the jump conditions. For example, a strong shock propagating perpendicular to the magnetic field (Sec. 5.3) has $\delta W \approx \frac{1}{2}\rho_1 v_1^2$ and $B_{2y} \approx 4B_{1y} = 4B_1$, so that its width is

$$\delta x \approx \frac{18B_1^2}{(\mu\sigma v_1)(\mu\rho_1 v_1^2)}. \tag{5.9}$$

In terms of the *magnetic Reynolds number* R_{m1} ($= \mu\sigma v_1\, \delta x$) based on δx and the *Alfvén Mach number* M_{A1} $[= v_1(\mu\rho_1)^{\frac{1}{2}}/B_1]$, this may be written $R_{m1} \approx 18/M_{A1}^2$.

Finally, it must be added that an ionised gas may support shock waves with a thickness much smaller than a collision mean-free path. They are known as *collisionless shocks*, and a magnetohydrodynamic description is inadequate to describe their structure. The conversion of ordered energy into random motion occurs not by particle collisions but by either plasma oscillations (that are subsequently damped) or plasma microinstabilities that generate a turbulent state. In the former case, the shocked plasma is in an oscillatory state rather than a uniform one. The thickness of a collisionless shock is typically of the order of one of the plasma length-scales, such as an ion gyro-radius or the geometric mean of the ion and electron gyro-radii. For details, see Burgess (1995) and Boyd and Sanderson (2003).

5.2 Hydrodynamic Shocks

Consider a steady plane shock wave propagating with constant speed into an ideal, stationary gas, with density ρ_1 and pressure p_1. Suppose that, in a frame of reference moving with the shock, the speed of the shocked gas is v_2 and its density and pressure are ρ_2 and p_2 (Figure 5.3). Then ρ_2, v_2, p_2 are determined in terms of ρ_1, v_1, p_1 by the equations of *conservation of mass, momentum and energy*, namely,

$$\rho_2 v_2 = \rho_1 v_1, \tag{5.10}$$

$$p_2 + \rho_2 v_2^2 = p_1 + \rho_1 v_1^2, \tag{5.11}$$

$$p_2 v_2 + (\rho_2 \epsilon_2 + \tfrac{1}{2}\rho_2 v_2^2) v_2 = p_1 v_1 + (\rho_1 \epsilon_1 + \tfrac{1}{2}\rho_1 v_1^2) v_1, \tag{5.12}$$

where (for a perfect gas) the internal energy per unit mass (2.28) is $\epsilon = p/[(\gamma-1)\rho]$, so that Eq. (5.12) reduces to

$$\frac{\gamma p_2}{(\gamma-1)\rho_2} + \tfrac{1}{2}v_2^2 = \frac{\gamma p_1}{(\gamma-1)\rho_1} + \tfrac{1}{2}v_1^2. \tag{5.13}$$

These equations are referred to as the *jump* or *Rankine-Hugoniot* relations.

Equation (5.10) arises because ρv is the mass that crosses a unit area in unit time. The significance of the terms in Eq. (5.11) is that $(\rho v)v$ gives the rate at which momentum is transported across a unit surface area, while p is the force which acts on that area. In Eq. (5.12), $(\rho\epsilon + \frac{1}{2}\rho v^2)v$ represents the rate of transport of internal and kinetic energy, whereas pv is the rate at which the gas pressure does work. The jump relations may be derived from the fundamental Eqs. (2.19), (2.20) and (2.40) of mass, momentum and energy (with losses $\mathcal{L} \equiv 0$ and $\mathbf{B} = \mathbf{F} = \mathbf{0}$). The procedure is to assume that each variable is purely a function of distance through the shock and then to integrate from one side of the shock front to the other.

The non-trivial solution of Eqs. (5.10), (5.11) and (5.13) can be written

$$\frac{\rho_2}{\rho_1} = \frac{(\gamma+1)M_1^2}{2+(\gamma-1)M_1^2}, \tag{5.14}$$

$$\frac{v_2}{v_1} = \frac{2+(\gamma-1)M_1^2}{(\gamma+1)M_1^2}, \tag{5.15}$$

$$\frac{p_2}{p_1} = \frac{2\gamma M_1^2-(\gamma-1)}{\gamma+1}, \tag{5.16}$$

in terms of the shock *Mach number* $M_1 \equiv v_1/c_{s1}$, where $c_{s1} \equiv (\gamma p_1/\rho_1)^{\frac{1}{2}}$. Furthermore, the second law of thermodynamics states that the *entropy* (s) of an isolated system must increase, where, for a perfect gas, $s \equiv c_v \log(p/\rho^\gamma) + constant$. Thus, Eqs. (5.14) to (5.16) need to be supplemented for our shock wave by

$$s_2 \geqslant s_1; \tag{5.17}$$

equality arises only when conditions are the same on both sides of the shock front, so that the shock ceases to exist.

Several interesting properties arise from Eqs. (5.14) to (5.17). For instance, the *shock speed must exceed the sound speed ahead of the shock* (see PROBLEM 5.1), i.e.,

$$v_1 \geqslant c_{s1} \qquad \text{or} \qquad M_1 \geqslant 1. \tag{5.18}$$

Furthermore, we find that

$$v_2 \leqslant c_{s2}, \tag{5.19}$$

and so, in the shock frame of reference, the flow is supersonic in front of the shock but subsonic behind it. This means that information cannot be transmitted ahead of the shock front but can catch it up from behind. Another property is that *the shock must be compressive with*

$$p_2 \geqslant p_1, \tag{5.20}$$

$$\rho_2 \geqslant \rho_1. \tag{5.21}$$

Equations (5.21) and (5.10) imply in turn that $v_2 \leqslant v_1$, while Eqs. (5.21) and (5.17) lead to $T_2 \geqslant T_1$, so that the *shock wave slows the gas down but heats it up*, converting flow energy into thermal energy in the process.

The above results may be established as follows. From the expression for s, the entropy jump across the shock is

$$s_2 - s_1 = c_v \log\frac{p_2}{p_1} - \gamma c_v \log\frac{\rho_2}{\rho_1},$$

where s_1, p_1, ρ_1, may be regarded as prescribed constants, while p_2 and ρ_2 vary with the Mach number (M_1) according to Eqs. (5.16) and (5.14). Differentiating this with respect to M_1 gives $ds_2/dM_1 = (c_v/p_2)dp_2/dM_1 - (\gamma c_v/\rho_2)d\rho_2/dM_1$, or, after using Eqs. (5.16) and (5.14),

$$\frac{ds_2}{dM_1} = \frac{4\gamma(\gamma-1)(M_1^2-1)^2 c_v}{M_1[2\gamma M_1^2-(\gamma-1)][2+(\gamma-1)M_1^2]}. \tag{5.22}$$

But $\gamma > 1$ and $p_2 > 0$, so Eq. (5.16) gives $2\gamma M_1^2 - (\gamma-1) > 0$ and Eq. (5.22) implies

$$\frac{ds_2}{dM_1} \geqslant 0. \tag{5.23}$$

Now, when $M_1 = 1$, we have $p_2 = p_1, \rho_2 = \rho_1$ and $s_2 = s_1$, so Eqs. (5.17) and (5.23) lead to the conclusion that $M_1 \geqslant 1$, as required. The results (5.20) and (5.21) then follow immediately from Eqs. (5.16) and (5.14), respectively, while Eq. (5.19) is a consequence of Eqs. (5.14) to (5.16), which give

$$\frac{v_2^2}{c_{s2}^2} = \frac{2+(\gamma-1)M_1^2}{2\gamma M_1^2-(\gamma-1)} = 1 - \frac{(\gamma+1)(M_1^2-1)}{2\gamma M_1^2-(\gamma-1)} \leqslant 1.$$

As the Mach number (M_1) increases from unity to infinity, so the pressure ratio p_2/p_1 [from Eq. (5.16)] increases without bound like $2\gamma M_1^2/(\gamma+1)$, but the density jump (5.14) varies only in the range

$$1 \leqslant \frac{\rho_2}{\rho_1} < \frac{\gamma+1}{\gamma-1}.$$

(Thus, e.g., a monatomic gas with $\gamma = 5/3$ has a maximum density jump of only 4.)

5.3 Perpendicular Shocks

The simplest form of magnetic shock wave is the perpendicular shock. In this case, the velocities of both the shock and the plasma are perpendicular to the magnetic field, which itself is unidirectional and parallel to the shock front. In a frame of reference moving with the shock front (Figure 5.4), the properties

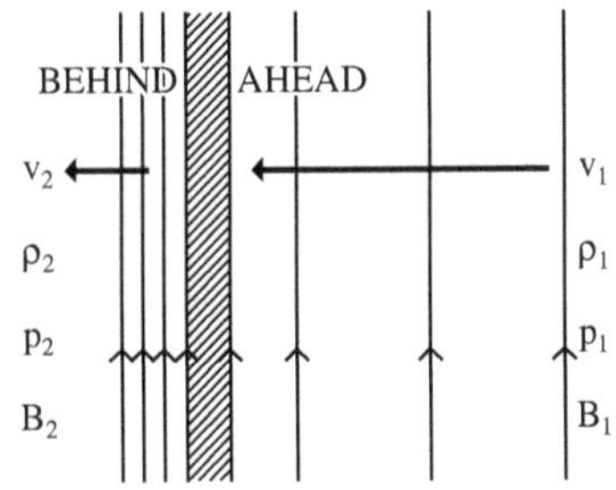

Figure 5.4. The notation for a plane perpendicular shock wave in the shock frame of reference. The magnetic field is parallel to the wavefront and perpendicular to the flow velocity.

of the shocked plasma (v_2, ρ_2, p_2, B_2) are related to those of the unshocked plasma (v_1, ρ_1, p_1, B_1) by the equations for conservation of mass, momentum, energy and magnetic flux, namely,

$$\rho_2 v_2 = \rho_1 v_1, \tag{5.24}$$

$$p_2 + B_2^2/(2\mu) + \rho_2 v_2^2 = p_1 + B_1^2/(2\mu) + \rho_1 v_1^2, \tag{5.25}$$

$$\begin{aligned}[p^2 + B_2^2/(2\mu)]v_2 &+ [\rho_2 e_2 + \tfrac{1}{2}\rho_2 v_2^2 + B_2^2/(2\mu)]v_2 \\ &= [p_1 + B_1^2/(2\mu)]v_1 + [\rho_1 e_1 + \tfrac{1}{2}\rho_1 v_1^2 + B_1^2/(2\mu)]v_1,\end{aligned} \tag{5.26}$$

$$B_2 v_2 = B_1 v_1, \tag{5.27}$$

where the internal energy (e) is given by Eq. (2.28). The first three conservation equations are natural extensions of Eqs. (5.10) to (5.12) for the hydrodynamic shock, with the gas pressure replaced by the total pressure (plasma plus magnetic) and extra terms $B^2/(2\mu)v$ in Eq. (5.26) representing the transport of magnetic energy.

In Eq. (5.27) the quantity Bv gives the rate at which magnetic flux is transported across a unit surface area, and this is assumed to be conserved. An alternative derivation of Eq. (5.27) may be obtained as follows. For our steady flow we have $\nabla \times \mathbf{E} = \mathbf{0}$, which implies that the tangential electric field component is continuous across the shock, $E_2 = E_1$. But, if the magnetic field is frozen to the plasma on both sides of the shock, the total electric field ($\mathbf{E} + \mathbf{v} \times \mathbf{B}$) vanishes, so that

$$E_2 = -v_2 B_2, \qquad E_1 = -v_1 B_1, \tag{5.28}$$

and Eq. (5.27) follows. It is also of interest to note that, on each side of the energy equation (5.26), the two terms involving the magnetic field may be combined to give $(B^2/\mu)v$, which by virtue of Eqs. (5.28) is just the magnitude of the Poynting vector $\mathbf{E} \times \mathbf{H}$. After some manipulation, the solution to the set (5.24) to (5.26) may be written in terms of the density ratio, shock Mach number and plasma beta, namely,

$$\frac{\rho_2}{\rho_1} \equiv X, \tag{5.29}$$

$$M_1 \equiv v_1/c_{s1}, \tag{5.30}$$

$$\beta_1 \equiv \frac{2\mu p_1}{B_1^2} \equiv \frac{2c_{s1}^2}{\gamma v_{A1}^2}, \tag{5.31}$$

where c_{s1} is the sound speed and $v_{A1} = B_1/(\mu\rho_1)^{1/2}$ is the Alfvén speed for the unshocked plasma. The result is

$$v_2/v_1 = X^{-1}, \tag{5.32}$$

$$B_2/B_1 = X, \tag{5.33}$$

$$p_2/p_1 = 1 + \gamma M_1^2(1 - X^{-1}) + \beta_1^{-1}(1 - X^2), \tag{5.34}$$

where X is the positive solution of

$$f(X) \equiv 2(2-\gamma)X^2 + (2\beta_1 + (\gamma-1)\beta_1 M_1^2 + 2)\gamma X - \gamma(\gamma+1)\beta_1 M_1^2 = 0. \tag{5.35}$$

The fact that $1 < \gamma < 2$ implies that Eq. (5.35) has just one positive root. Furthermore, we note that the solution reduces to the hydrodynamic value (5.14) in the limit of large β_1. The effect of the magnetic field is to reduce X below its hydrodynamic value, since the flow energy can be converted into magnetic energy as well as heat.

Comparable results to those for the hydrodynamic shock can be established. For example, *the shock is compressive* with $X \geqslant 1$, which, since $f(X)$ is a quadratic function with a single minimum, implies that $f(1) \leqslant 0$, or $M_1^2 \geqslant 1 + [2/(\gamma\beta_1)]$.

In terms of the sound and Alfvén speeds this may be written $v_1^2 \geqslant c_{s1}^2 + v_{A1}^2$, so that the *shock speed* (v_1) *must exceed the fast magnetoacoustic speed* $(c_{s1}^2 + v_{A1}^2)^{1/2}$ *ahead of the shock.* The latter speed therefore plays the same role as the sound speed does for the hydrodynamic shock. Furthermore, when M_1 increases without bound, the compression ratio (X) increases to the limiting value $(\gamma+1)/(\gamma-1)$ and the magnetic compression is therefore restricted to the range

$$1 < \frac{B_2}{B_1} < \frac{\gamma+1}{\gamma-1},$$

or, for $\gamma = \frac{5}{3}$, $1 < B_2/B_1 < 4$.

5.4 Oblique Shocks

5.4.1 Jump Relations

Apart from the special case of a perpendicular shock, the magnetic field and plasma velocity contain components both parallel and normal to the shock front; here we assume they lie in the xy-plane. Variables ahead of and behind the front are as usual denoted by subscripts 1 and 2, respectively, as indicated in Figure 5.5. Then the equations for conservation of mass, x-momentum, y-momentum, energy and magnetic flux are

$$\rho_2 v_{2x} = \rho_1 v_{1x}, \tag{5.36}$$

$$p^2 + B_2^2/(2\mu) - B_{2x}^2/\mu + \rho_2 v_{2x}^2 = p_1 + B_1^2/(2\mu) - B_{1x}^2/\mu + \rho_1 v_{1x}^2, \tag{5.37}$$

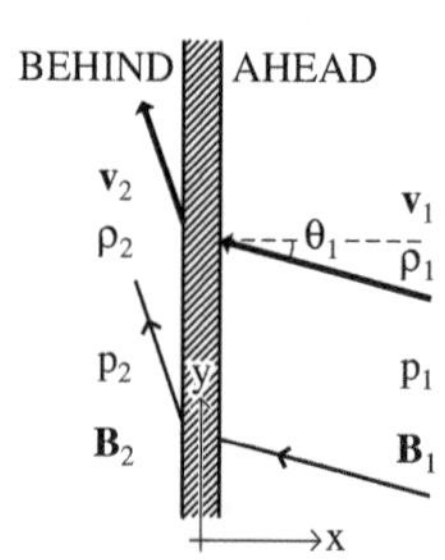

Figure 5.5. The notation for an oblique shock wave in a frame of reference such that the velocity component *normal* to the shock front is the same as that of the shock front, whereas the velocity component *along* the shock front is chosen to make the plasma velocity (**v**) everywhere parallel to the magnetic field (**B**).

$$\rho_2 v_{2x} v_{2y} - B_{2x} B_{2y}/\mu = \rho_1 v_{1x} v_{1y} - B_{1x} B_{1y}/\mu, \tag{5.38}$$

$$[p_2 + B_2^2/(2\mu)]v_{2x} - B_{2x}(\mathbf{B}_2 \cdot \mathbf{v}_2)/\mu + [\rho_2 e_2 + \tfrac{1}{2}\rho_2 v_2^2 + B_2^2/(2\mu)]v_{2x}$$
$$= [p_1 + B_1^2/(2\mu)]v_{1x} - B_{1x}(\mathbf{B}_1 \cdot \mathbf{v}_1)/\mu + [\rho_1 e_1 + \tfrac{1}{2}\rho_1 v_1^2 + B_1^2/(2\mu)]v_{1x}, \tag{5.39}$$

$$B_{2x} = B_{1x}, \tag{5.40}$$

$$v_{2x} B_{2y} - v_{2y} B_{2x} = v_{1x} B_{1y} - v_{1y} B_{1x}, \tag{5.41}$$

where the integral energy (e) is given by Eq. (2.28).

There are some extra terms by comparison with the jump relations (5.24) to (5.27) for a perpendicular shock, in which the components v_y and B_x were absent. In Eqs. (5.37) and (5.38), B_x^2/μ and $B_x B_y/\mu$ represent the x- and y-components, respectively, of the tension force $(B_x \mathbf{B}/\mu)$ acting across a plane $x =$ *constant*, whereas $\rho v_x v_y$ is the rate of transport of y-momentum (ρv_y) across a unit surface area of the shock front. The new term $B_x(\mathbf{B}\cdot\mathbf{v})/\mu$ in Eq. (5.39) arises from the rate of working of the magnetic tension, while $[p+B^2/(2\mu)]v_x$ gives the rate of working of the total pressure and $[\rho e+\frac{1}{2}\rho v^2+B^2+B^2/(2\mu)]v_x$ gives the rate of transfer of energy (internal, kinetic and magnetic) across the shock front. The conservation of the normal component of magnetic flux B_x [Eq. (5.40)] is a consequence of $\nabla\cdot\mathbf{B} = 0$. Finally, the continuity of $|\mathbf{v}\times\mathbf{B}|$ (Eq. 5.41) follows from the results that the tangential electric field component is continuous and that the total electric field $(\mathbf{E}+\mathbf{v}\times\mathbf{B})$ vanishes outside the shock front since dissipation is negligible there.

An analysis of the jump relations can be considerably simplified by choosing axes moving along the y-axis parallel to the shock front at such a speed that

$$v_{1y} = v_{1x}\frac{B_{1y}}{B_{1x}}. \tag{5.42}$$

[This choice is not possible in the case of the perpendicular shock (Sec. 5.3), for which B_{1x} vanishes.] The result is that in this frame of reference both sides of Eq. (5.41) vanish, and the plasma velocity becomes parallel to the magnetic field on both sides of the shock front. This in turn implies the vanishing of both the electric field (E_z) and the Poynting vector $(\mathbf{E}\times\mathbf{H})$, which represents the total flux of magnetic energy across a surface. But on either side of Eq. (5.39) all the terms involving the magnetic field may be combined to give $\mathbf{E}\times\mathbf{H}$, so that their sum vanishes and Eq. (5.39) reduces to the purely hydrodynamic form

$$\frac{\gamma p_2}{(\gamma-1)\rho_2} + \frac{v_2^2}{2} = \frac{\gamma p_1}{(\gamma-1)\rho_1} + \frac{v_1^2}{2}.$$

Once the solutions to the jump relations have been derived, a whole family of extra solutions may be obtained by adding the same constant value to both v_{2y} and v_{1y}. Such solutions are discussed by, for instance, Cargill and Priest (1982).

In terms of the compression ratio $(X \equiv \rho_2/\rho_1)$, sound speed $[c_{s1} \equiv (\gamma p_1/\rho_1)^{1/2}]$ and Alfvén speed $[v_{A1} \equiv B_1/(\mu\rho_1)^{1/2}]$, Eqs. (5.36) to (5.42) may be combined to give

$$\frac{v_{2x}}{v_{1x}} = X^{-1}, \tag{5.43}$$

$$\frac{v_{2y}}{v_{1y}} = \frac{v_1^2 - v_{A1}^2}{v_1^2 - X v_{A1}^2}, \tag{5.44}$$

$$\frac{B_{2x}}{B_{1x}} = 1, \tag{5.45}$$

$$\frac{B_{2y}}{B_{1y}} = \frac{(v_1^2 - v_{A1}^2)X}{v_1^2 - X v_{A1}^2}, \tag{5.46}$$

$$\frac{p_2}{p_1} = X + \frac{(\gamma-1)X v_1^2}{2c_{s1}^2}\left(1 - \frac{v_2^2}{v_1^2}\right). \tag{5.47}$$

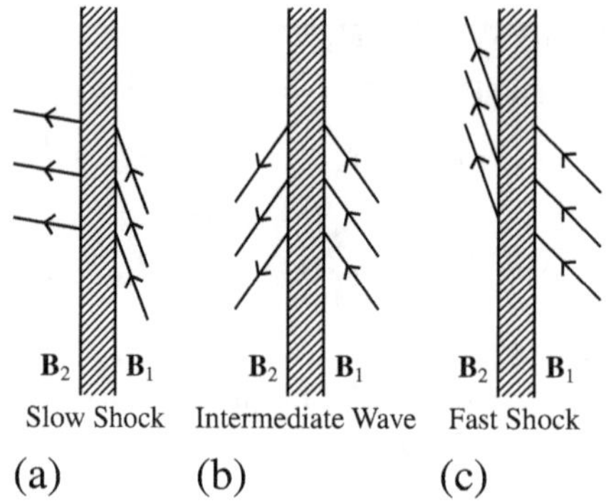

Figure 5.6. The changes in magnetic field direction that are caused by the three types of oblique wave.

Here X is a solution of

$$(v_1^2 - Xv_{A1}^2)^2\{Xc_{s1}^2 + \tfrac{1}{2}v_1^2\cos^2\theta[X(\gamma-1) - (\gamma+1)]\} + \tfrac{1}{2}v_{A1}^2 v_1^2\sin^2\theta\, X$$
$$\times\{[\gamma + X(2-\gamma)]v_1^2 - Xv_{A1}^2[(\gamma+1) - X(\gamma-1)]\} = 0, \tag{5.48}$$

and $\theta \equiv \theta_1$ is the inclination of the upstream magnetic field to the shock normal (Figure 5.5) such that $v_{1x} = v_1 \cos\, \theta$.

Corresponding to the three solutions of Eq. (5.48), there are three waves, namely a *slow shock*, an *intermediate (or Alfvén) wave* and a *fast shock* (Figure 5.6). Indeed, in the limit as $X \to 1$, they reduce to the three infinitesimal waves, since Eq. (5.48) becomes $v_1^2 = v_{A1}^2$, for an Alfvén wave, together with $v_{1x}^4 - (c_{s1}^2 + v_{A1}^2)v_{1x}^2 + c_{s1}^2 v_{A1}^2\cos^2\theta = 0$, for the propagation speeds of slow and fast magnetoacoustic waves (see Eq. 4.45).

5.4.2 Slow-Mode and Fast-Mode Shocks

A full derivation of the properties of oblique shock waves can be found in, for instance, Bazer and Ericson (1959) or Jeffrey and Taniuti (1964). Consider first the slow-mode and fast-mode shocks (sometimes called slow and fast shocks, for short). They are *compressive*, with

$$X > 1, \tag{5.49}$$

which from Eq. (5.47) implies that $p_2 > p_1$. They also conserve the sign of the tangential magnetic field component, so that B_{2y}/B_{1y} is positive. This makes both the numerator and denominator on the right-hand side of Eq. (5.47) either negative or positive. In the first case we have

$$v_1^2 \leqslant v_{A1}^2 \quad (< Xv_{A1}^2), \tag{5.50}$$

and so Eq. (5.46) implies that $B_2 < B_1$. This is the basic property of a *slow-mode shock*, for which the magnetic field is refracted *towards* the shock normal and its strength *decreases* as the shock front passes by (Figure 5.6a). In the second case,

$$v_1^2 \geqslant Xv_{A1}^2 \quad (> v_{A1}^2), \tag{5.51}$$

and we find from Eq. (5.46) that $B_2 > B_1$ for the *fast-mode shock*. Here the shock front has the effect of refracting the magnetic field away from the normal and increasing its strength. It can be shown from the evolutionary condition that the *shock speed* (v_{1x}) *relative to the unshocked plasma (ahead) must exceed the characteristic wave speed* for that plasma, namely the slow magnetoacoustic wave speed for a slow shock and the fast magnetoacoustic speed for a fast shock. Also, *behind* the shock front, the slow speed (v_{2x}) must be *less* than the relevant characteristic wave speed. Furthermore, we see from Eqs. (5.43) and (5.49) that the effect of the shock is to *slow down the flow in the x-direction* ($v_{2x} < v_{1x}$). The flow in the y-direction is slowed down for a slow shock ($v_{2y} < v_{1y}$) but speeded up for a fast shock ($v_{2y} > v_{1y}$).

In the limit as the normal magnetic field component (B_x) approaches zero, so that the field becomes purely tangential, the fast shock becomes a perpendicular shock. On the other hand, the slow shock reduces to a *tangential discontinuity*, for which both the flow velocity and magnetic field are tangential to the plane of discontinuity, since $v_{2x} = v_{1x} = B_{2x} = B_{1x} = 0$. It is therefore simply a boundary between two distinct plasmas, at which the jumps in the tangential components of velocity (v_y) and magnetic field (B_y) are arbitrary, subject only to the condition

$$p_2 + \frac{B_2^2}{2\mu} = p_1 + \frac{B_1^2}{2\mu},$$

that the total pressure be continuous.

Consider lastly the case

$$B_{2x} = B_{1x} \neq 0, \qquad v_{2x} = v_{1x} = 0,$$

so that magnetic field lines cross the boundary but there is no flow across it. One trivial solution has the velocity, magnetic field and pressure continuous. But also the density and temperature may be discontinuous in such a way that $p_2 = p_1$. This boundary between plasmas at different densities and temperatures is known as a *contact* (or *entropy*) *discontinuity.*

5.4.3 Switch-Off and Switch-On Shocks

Two special cases of slow and fast shocks are of particular interest, namely, so-called *switch-off* and *switch-on* shocks (Figure 5.7). They occur in the limit when equality holds in Eq. (5.50) or (5.51), respectively. When

$$v_1 = v_{A1} \tag{5.52}$$

and $X \neq 1$, Eq. (5.46) implies that the tangential magnetic field component behind the shock (B_{2y}) must vanish, and so, if B_{1y} is non-zero, we have a *switch-off shock.* Since $\mathbf{v}_1$ and $\mathbf{B}_1$ are parallel by Eq. (5.42), the condition (5.52) is equivalent to $v_{1x} = B_{1x}/(\mu\rho_1)^{1/2}$. In other words, a switch-off shock propagates at a speed (v_{1x}) equal to the Alfvén speed based on the normal magnetic field component. Furthermore, Eq. (5.48) for the compression ratio X ($\neq 1$) reduces to

$$(2c_{s1}^2/v_{A1}^2 + \gamma - 1)X^2 - [2c_{s1}^2/v_{A1}^2 + \gamma(1+\cos^2\theta)]X + (\gamma+1)\cos^2\theta = 0,$$

with exactly one solution greater than unity. When $c_{s1}^2/v_{A1}^2 > \frac{1}{2}$, the solution for X increases from 1 to $1+(2c_{s1}^2/v_{A1}^2+\gamma-1)^{-1}$ as the angle of incidence θ increases from 0 to $\frac{1}{2}\pi$. When $2c_{s1}^2/v_{A1}^2 < \frac{1}{2}$, X decreases from $(\gamma+1)/(2c_{s1}^2/v_{A1}^2 + (\gamma-1)$ to $1 + (2c_{s1}^2/v_{A1}^2 + \gamma - 1)^{-1}$ as θ increases from 0 to $\frac{1}{2}\pi$.

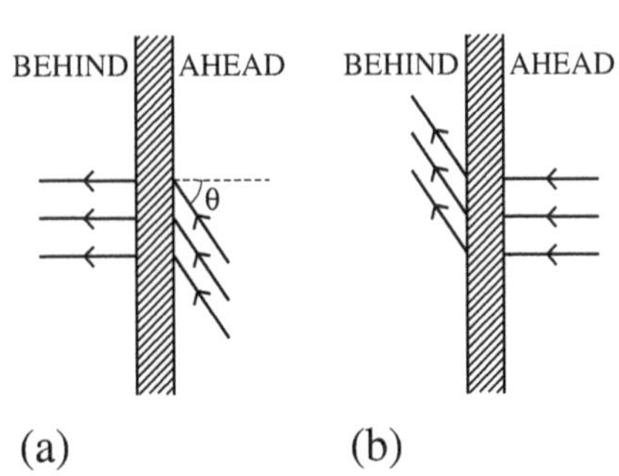

Figure 5.7. The magnetic field changes for switch-off and switch-on waves.

For a shock propagating along the magnetic field (so that B_{1y} and θ vanish), Eq. (5.48) reduces to

$$(v_1^2 - Xv_{A1}^2)^2 Xc_{s1}^2 + \tfrac{1}{2}v_1^2[X(\gamma-1)-(\gamma+1)] = 0.$$

The slow-mode shock solution for the compression ratio (X) is here given by the vanishing of the expression in curly brackets, and so it represents a purely hydrodynamic shock (Sec. 5.2). The fast-mode shock solution, on the other hand, is

$$X = \frac{v_1^2}{v_{A1}^2}, \tag{5.53}$$

corresponding to a *switch-on shock*. Since $X > 1$, *this occurs only when the shock speed exceeds the Alfvén speed* ($v_1 > v_{A1}$) by Eq. (5.53). Furthermore, $B_{2x} = B_1$, and an elimination of p_2 between Eqs. (5.37) and (5.39) yields

$$B_{2y}^2/B_{2x}^2 = (X-1)(\gamma+1) - (\gamma-1)X - 2\mu\gamma p_1/B_1^2.$$

Since the right-hand side must be positive, the density ratio has to lie in the range

$$1 < X \leqslant \frac{\gamma + 1 - 2c_{s1}^2/v_{A1}^2}{\gamma - 1}. \tag{5.54}$$

The upper limit is at its largest, namely $(\gamma+1)(\gamma-1)$, when $v_{A1} \gg c_{s1}$. Moreover, Eq. (5.54) implies $v_{A1} > c_{s1}$, which means that a switch-on shock can exist *only when the Alfvén speed exceeds the sound speed* in the unshocked plasma. As X increases from unity, so the deflection of the field lines, as measured by B_{2y}^2/B_{2x}^2, increases from zero to a maximum value of $4(1-c_{s1}^2/v_{A1}^2)^2/(\gamma-1)^2$ at $X = (\gamma - c_{s1}^2/v_{A1}^2)/(\gamma-1)$, and thereafter it decreases to zero at

$$X = \frac{\gamma + 1 = 2c_{s1}^2/v_{A1}^2}{\gamma - 1}.$$

5.4.4 Intermediate Wave

When the wave front propagates at the Alfvén speed in the unshocked plasma, so that $v_1 = v_{A1}$, one solution to Eq. (5.48) is simply $X = 1$. In this case, Eqs. (5.44) and (5.46) become meaningless, but the more fundamental Eqs. (5.41) and (5.42) imply

$$\frac{v_{2y}}{v_{1y}} = \frac{B_{2y}}{B_{1y}},$$

while Eqs. (5.37) and (5.39) give

$$p_2 = p_1, \qquad B_{2y}^2 = B_{1y}^2.$$

Thus, in addition to the trivial solution $\mathbf{B}_2 = \mathbf{B}_1$, we have

$$B_{2y} = -B_{1y}, \qquad B_{2x} = B_{1x},$$

$$v_{2y} = -v_{1y}, \qquad v_{2x} = v_{1x},$$

for an *intermediate* (or *transverse*) *wave* (or *rotational discontinuity*). The tangential magnetic field component is reversed by the wave, and within the wave front the magnetic field simply rotates out of the plane of Figure 5.6b while maintaining a constant magnitude. This is therefore just a finite-amplitude Alfvén wave (Secs. 4.3.1), and, since no change in pressure or density is produced, it is not a shock at all. Also, by contrast with a shock wave, it broadens in time due to Ohmic dissipation, and so it may be regarded as possessing a constant width only over times small by comparison with the Ohmic decay-time.

Chapter 6

Magnetic Reconnection

6.1 Introduction

Magnetic reconnection is a fundamental process in an almost-ideal plasma, whose magnetic Reynolds number based on typical scales (L_e) external to current sheets is much larger than unity. Here we give a brief overview and refer the reader for further details to Priest and Forbes (2000) or Birn and Priest (2007). A more extended version, which is regularly updated, can be found in Priest and Parnell (2014).

In an ideal medium, plasma elements preserve their magnetic connections. However, the presence of a *localised region* (of length $L \ll L_e$), where non-ideal effects are important, can lead to a *change of connectivity of plasma elements* (Figure 6.1). This magnetic reconnection may be fast or slow (in a sense defined in Section 6.1.1), although in many dynamic phenomena (such as solar flares) it is fast.

The main effects of magnetic reconnection are often:

(i) to convert some of the magnetic energy into heat by Ohmic dissipation;
(ii) to accelerate plasma by converting magnetic energy into bulk kinetic energy;
(iii) to create shock waves, current filamentation and turbulence, all associated with strong electric fields that may accelerate fast particles; and
(iv) to change the global connections of the field lines and so affect the fluxes of fast particles and heat, which are directed mainly along the magnetic field.

In the solar interior and low solar atmosphere, reconnection is well modelled by resistive MHD with classical Ohmic dissipation. However, in the outer corona, Hall MHD with a two-fluid approach or a kinetic model are more appropriate (see Birn and Priest 2007) (Sec. 13.11). Nevertheless, even in the latter case, an MHD

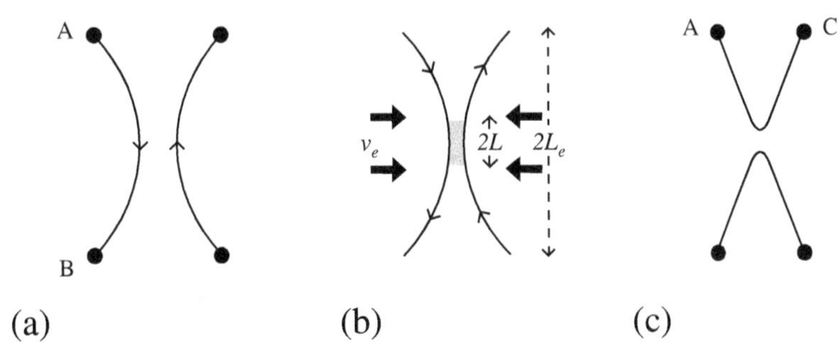

Figure 6.1. Breaking and reconnection of magnetic field lines when a localised diffusion region (shaded) leads to a change of connectivity of plasma elements (AB to AC).

approach can capture much of the essence of the process and provide an overall macroscopic picture or mould within which the detailed micro-plasma physics operates.

This chapter describes in detail two-dimensional MHD reconnection and summarises the main features of three-dimensional MHD reconnection that have been so far discovered, although this is very much a matter of current research.

6.1.1 Brief History

Dungey (1953) was the first to suggest that *"lines of force can be broken and rejoined"*. Sweet (1958a) then presented a model for the flattening of the magnetic field to form a current sheet when two bipolar regions come together at an X-type neutral point (see Figure 6.6). The magnetic field squeezes out the plasma between them in a process of steady-state reconnection. Parker (1957) developed scaling laws for the model and coined the phrase *"reconnection of field lines"*.

The *Sweet-Parker model* has a reconnection rate (or inflow plasma speed) of $v_i = v_{Ai}/\sqrt{R_m}$, where v_{Ai} is the inflow Alfvén speed and $R_m = Lv_{Ai}/\eta$ is the magnetic Reynolds number based on the length L of the sheet. This rate is a small fraction of the Alfvén speed (if $R_m \gg 1$), much too slow for solar flares, and so it is referred to as *slow reconnection*.

Next, Furth et al. (1963) discovered the *tearing-mode instability* for time-dependent reconnection (Sec. 6.8). Furthermore, Petschek (1964) realised that slow-mode shock waves also convert magnetic energy into heat and kinetic energy and are naturally generated by a tiny diffusion region. His (steady) mechanism possesses four standing slow-mode shock waves extending from a tiny central Sweet-Parker current sheet. It is the first of many regimes of *fast reconnection*, and is indeed rapid enough (at typically 0.01 to $0.1v_A$) for a flare.

Petschek's mechanism was widely accepted as the answer to fast flare energy release, especially when self-similar solutions for the external region were discovered (Soward and Priest 1977). A state of calm ensued until the watershed year of 1986 when new resistive MHD computational and theoretical models led to a state of ferment. Numerical experiments (Biskamp 1986) revealed solutions very different from Petschek's and so, at first, they seemed to cast doubt on the validity of the Petschek mechanism. However, Priest and Forbes (1986) realised that the reason for the difference was the different boundary conditions being imposed by Biskamp. They discovered a family of *Almost-Uniform* solutions for *fast reconnection*, including the solutions of both Petschek and Biskamp as special cases.

It is now well established that, when the magnetic diffusivity is enhanced at the X-point, Petschek's mechanism and other Almost-Uniform reconnection regimes can indeed occur, and an enhancement of diffusivity is a common effect in practice. However, what happens when the magnetic diffusivity is spatially uniform has only just been clarified. High-resolution numerical experiments (Baty et al. 2009a,b) and an analytical model (Forbes et al. 2013) have demonstrated that fast reconnection is stable when the diffusion region diffusivity is enhanced, but it is structurally unstable when the diffusivity is uniform or reduced. Thus, in the case of uniform diffusivity, reconnection can occur only at the slow Sweet-Parker rate.

Fast collisionless reconnection may be assisted by the Hall effect (Shay and Drake 1998; Huba 2003), where the resistive diffusion region is replaced by an ion diffusion region (with the size of an ion inertial length) and a smaller electron diffusion region. Indeed, the *GEM Challenge* has shown that full-particle, hybrid and Hall MHD codes all tend to give the same fast rate of reconnection (Birn et al. 2001).

Most of the attention is now focussed on three-dimensional reconnection, which is completely different from two-dimensional reconnection (see Priest et al. 2003). A landmark paper by Schindler et al. (1988) proposed a concept of *General Magnetic Reconnection*, in which reconnection can occur either at null points or in the absence of null points whenever a parallel electric field ($E_{\|}$) is produced by any region where the plasma is locally nonideal. The condition for reconnection to occur is simply that

$$\int E_{\parallel}\, ds \neq 0,$$

evaluated along a magnetic field line passing through the region of local nonidealness: indeed, the maximum value of this integral gives the rate of reconnection.

Later, Priest and Pontin (2009) updated an earlier classification (Priest and Titov 1996) by proposing several types of 3D reconnection, namely: *torsional spine or fan reconnection* when rotational motions concentrate the current along the spine or fan of a null point; *spine-fan reconnection* when shearing motions concentrate it along both (Sec. 6.13.2); and *separator reconnection* when it focusses along a separator field line (Sec. 6.13.3). A separator joins two null points and represents the intersection of two separatrix surfaces. Furthermore, in the absence of null points, Priest and Démoulin (1995) proposed a concept of reconnection at a *quasi-separator*, which is the intersection of two *quasi-separatrix layers (QSLs)* (Secs. 6.9.5 and 6.13.4). In a QSL, the mapping of magnetic field lines changes continuously but very rapidly, whereas across a separatrix surface it changes discontinuously.

6.1.2 Overview of Reconnection Concepts

In this chapter, we focus first (Secs. 6.2 to 6.8) on classical 2D reconnection. In the final five sections (Secs. 6.9 to 6.13), we describe deeper concepts and more recent ideas on 3D reconnection. To start with (Sec. 6.2), we discuss the structure of null points where the magnetic field vanishes, both in 2D and 3D, as well as the ways in which such nulls collapse. Then, we move on in Section 6.3, to show how current sheets may be formed by different types of motion, namely planar, shearing and braiding.

Once a current sheet has formed, it tends to diffuse away. If magnetic field and plasma are brought in at the same rate as the outward diffusion, however, a steady state may be set up. For straight field lines and a stagnation-point flow, there exists an exact solution of the steady nonlinear MHD equations, the *Stagnation-Point Flow Model* (Sec. 6.4.3), for both a 2D and a 3D stagnation flow. A generalisation of this solution was later discovered, referred to as *reconnective annihilation*, which has an X-point field in place of a one-dimensional field (Sec. 6.4.4).

Steady 2D reconnection is described in Sections 6.5 to 6.7, including the *Sweet-Parker model*, *Petschek's mechanism* and the *Almost-Uniform* family of fast solutions. Unsteady solutions due to tearing-mode instability are also presented (Sec. 6.8).

New concepts are involved in moving from 2D to 3D (Sec. 6.12). When nonideal plasma effects are important in a localised region, there are several classes of evolution of a magnetic field that satisfy Faraday's law and $\nabla \cdot \mathbf{B} = 0$ (see Figure 6.2 and Sec. 6.11.3). The largest subclass conserves electromagnetic flux

$$\int_{S(t)} \mathbf{B} \cdot d\mathbf{S} + \int_{S(t)} \mathbf{E} \cdot d\mathbf{l}\; dt = const.$$

One subclass of solutions conserves magnetic flux by itself ($\int_{S(t)} \mathbf{B} \cdot d\mathbf{S} = const$), while another represents 3D reconnection. The subclass of 3D reconnection that preserves magnetic flux represents 2D reconnection. Furthermore, magnetic flux conservation implies field line conservation (Sec. 6.11), but the reverse is not true since there are solutions that conserve field lines but not flux.

When $\mathbf{E} + \mathbf{v} \times \mathbf{B} = \mathbf{0}$, the magnetic flux and field line connections are both conserved and there is no reconnection, so that magnetic topology is conserved. The term *magnetic topology* refers here to any property that is preserved by an ideal displacement, such as the linkage and knottedness of the field.

When the plasma is instead non-ideal with $\mathbf{E} + \mathbf{v} \times \mathbf{B} = \mathbf{N}$, where $\mathbf{N}$ represents any non-ideal term such as $\mathbf{N} = \eta \mathbf{j}$, the condition $\mathbf{B} \times (\nabla \times \mathbf{N}) = \mathbf{0}$ implies field-line conservation, whereas $\nabla \times \mathbf{N} = \mathbf{0}$ implies

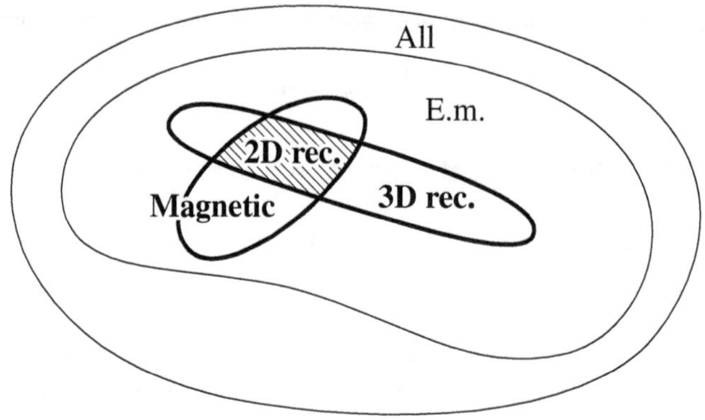

Figure 6.2. The different classes of evolution of a magnetic field. Within all possible types of evolution (indicated as 'All', there is a large class that conserves electromagnetic flux (called 'E.m.'). Entirely within that class is a family of three-dimensionally reconnecting situations ('3D rec') and a second family that conserves magnetic flux ('Magnetic'). Two-dimensional reconnection ('2D rec') is a special case of 3D reconnection that also conserves magnetic flux.

flux conservation. Thus, line conservation and flux conservation are no longer equivalent; although flux conservation implies field-line conservation, the reverse is not true (Sec. 6.11.2).

It is important to distinguish between *diffusion* and *reconnection* (Sec. 6.12.1): reconnection is a global process that always involves diffusion in a localised region surrounded by an ideal region, but there are examples of diffusion without reconnection. This occurs when there is no change of magnetic connectivity of plasma elements, or when diffusion is taking place throughout the region of consideration. The presence or absence of diffusion or reconnection depends on the nature of $\mathbf{N}$. If $\mathbf{N}$ can be written as $\mathbf{N} = \mathbf{u} \times \mathbf{B} + \nabla\Phi$, then it transpires that:

(a) if $\mathbf{u}$ is smooth, there is slippage of magnetic field but no reconnection;

(b) if $\mathbf{u}$ is singular, there is 2D reconnection;

(c) if $\mathbf{N} \neq \mathbf{u} \times \mathbf{B} + \nabla\Phi$, there is reconnection in 2.5D or 3D.

For 2D MHD (Sec. 6.12.2), $\mathbf{E} \cdot \mathbf{B} = 0$ and a flux-preserving flow $\mathbf{w}$ may always be found, such that $\mathbf{E} + \mathbf{w} \times \mathbf{B} = \mathbf{0}$. In this case, three types of behaviour are possible:

(a) if $\mathbf{B} \neq \mathbf{0}$, then $\mathbf{w}$ is smooth everywhere and there is magnetic field slippage;

(b) if $\mathbf{B} = \mathbf{0}$ at a point and the neighbouring magnetic field is elliptic, then there is magnetic flux destruction or generation as flux disappears or appears at the null; and

(c) if $\mathbf{B} = \mathbf{0}$ at some point and the neighbouring magnetic field is hyperbolic, then there is magnetic flux reconnection.

Several important differences arise between 2D and 3D reconnection (Sec. 6.12.4).

In 2D: reconnection can occur only at an X-point; a flux velocity ($\mathbf{w}$) always exists and is singular at the X-point; in the diffusion region, field lines slip through the plasma and change their connections only at the X-point.

In 3D, none of the above properties hold: reconnection can occur at a null or in the absence of a null; a single flux velocity does not exist, but can be replaced by a *dual flux velocity* [i.e., a pair $(\mathbf{w}_{in}, \mathbf{w}_{out})$ of flux velocities that describe separately the parts of a field line that point into or out of a diffusion region]; while passing through the diffusion region, field lines continually change their connections.

Fast outflow jets are often accelerated by 2D reconnection when the diffusion region is long and thin, but they are not always present in 3D reconnection when all the diffusion region's dimensions are similar.

Several different regimes of reconnection in 3D have been proposed (Sec. 6.13). These include: *quasi-separator reconnection* in the absence of a null point at the intersection of two *quasi-separatrix layers (QSLs)*; and *separator reconnection* at the intersection of two *separatrices*. Also, three kinds of reconnection may occur at a null point, namely, *torsional spine reconnection*, *torsional fan reconnection* and *spine-fan reconnection.*

6.2 Two-Dimensional Null Points

Null points are locations where the magnetic field vanishes. In particular, X-type (i.e., hyperbolic) null points are potential weak spots in a 2D magnetic configuration, in the sense that current sheets tend to be created at them in response to external motions. This section describes their structure and the way they tend to collapse.

6.2.1 2D Null Point Structure

It can be shown (PROBLEM 6.1: for all problems and solutions, see web page www.cambridge.org/9780521854719) that any linear null with field components $B_X = bX + 2cY$, $B_Y = -2aX - bY$, (such that $\nabla \cdot \mathbf{B} = 0$) can be transformed to

$$B_x = B_0 \frac{y}{r_0}, \qquad B_y = B_0 \bar{\alpha}^2 \frac{x}{r_0}, \tag{6.1}$$

where
$$\frac{B_0}{r_0} = (a+c) - \sqrt{b^2 + (a-c)^2}, \qquad \bar{\alpha}^2 = \frac{\sqrt{b^2+(a-c)^2}+(a+c)}{\sqrt{b^2+(a-c)^2}-(a+c)}.$$

When $\bar{\alpha}^2 < 0$, the field lines are elliptical and the origin is an *O-type neutral point* or *O-point.* The particular case when $\bar{\alpha}^2 = -1$ produces circular field lines.

When $\bar{\alpha}^2 > 0$, the field lines are hyperbolic around an *X-type neutral point* or *X-point.* The limiting field lines ($y = \pm \bar{\alpha} x$) through the origin are known as *separatrices* and are inclined at the angles $\pm \tan^{-1} \bar{\alpha}$ to the x-axis. The separatrices form an 'X', from which the term 'X-type' null point is derived.

The value of $\bar{\alpha}$ is related to the current density. Taking the curl of $\mathbf{B}$ gives $j_z = B_0(\bar{\alpha}^2 - 1)/(\mu r_0)$. Thus, when $\bar{\alpha} = 1$, $j_z = 0$ and the separatrix angle is 90 degrees, whereas an imaginary $\bar{\alpha}$ produces an O-point and $j_z \neq 0$.

6.2.2 Collapse of 2D Null Points

6.2.2.1 Qualitative Analysis

Provided the sources of a magnetic field containing an *X-type neutral point* are free to move, the X-point tends to be locally unstable (Dungey 1953). This may be demonstrated qualitatively by considering the equilibrium current-free field

$$B_x = B_0 \frac{y}{r_0}, \quad B_y = B_0 \frac{x}{r_0}, \tag{6.2}$$

where B_0 and r_0 are constant. The field lines are the rectangular hyperbolae $y^2 - x^2 = constant$, as shown in Figure 6.3a. Any element of plasma, such as the one shown on the negative x-axis, experiences a magnetic tension force (T) that acts outwards from the origin due to the outwardly curving field lines. It is exactly

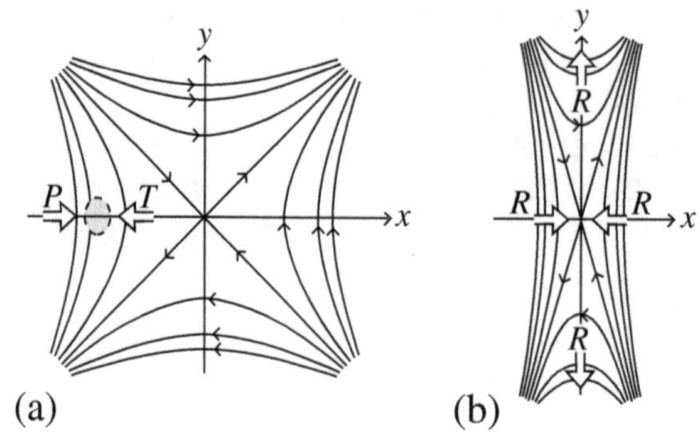

Figure 6.3. Magnetic field lines near an X-type neutral point: (a) in equilibrium with no current, showing a plasma element (shaded) acted on by magnetic pressure (P) and tension (T) forces; (b) away from equilibrium with a uniform current and a resultant force R.

balanced by a magnetic pressure force (P), which acts inwards because the magnetic field strength weakens as one approaches the origin.

Now suppose the magnetic field Eq. (6.2) is distorted to the form Eq. (6.1), where $\bar{\alpha}^2$ (>1) is constant. The field lines are given by $y^2 - \bar{\alpha}^2 x^2 = $ *constant* and are sketched in Figure 6.3b. The limiting field lines ($y = \pm\bar{\alpha}x$) through the origin are no longer inclined at $\frac{1}{2}\pi$, but have closed up a little, like a pair of scissors.

On the x-axis, the field lines are more closely spaced than in Figure 6.3a, so the magnetic pressure force has increased. They also have smaller curvature, so the magnetic tension force has increased less than the pressure. The dominance of the magnetic pressure produces a resultant force (R) acting inwards. On the y-axis, the field lines have the same spacing as in Figure 6.3a, but they are more sharply curved, so the magnetic pressure force remains the same, while the tension force increases; the resultant force (R) therefore acts outwards as shown. These comments may be borne out by evaluating the current density and Lorentz force.

The magnetic force is therefore such as to increase the original perturbation by closing up the 'scissors' even more, which means that the equilibrium (6.2) is unstable. As the instability proceeds, $\bar{\alpha}$ increases and the limiting field lines through the origin close up, so that the current density and Ohmic heating (j^2/σ) also increase.

The magnetic field (6.2) increases indefinitely with x and y and so cannot be valid in an infinite domain; rather, it should be regarded as the local field near an X-point. The collapse process takes place only if conditions at distant boundaries permit. It cannot occur in, for instance, a potential field with line-tied boundaries (since that would be a minimum-energy state and therefore completely stable), but, if a more complex magnetic configuration with an X-point contains energy stored in excess of potential, collapse is likely to be an efficient means of dissipating that energy.

Later, Imshennik and Syrovatsky (1967) made Dungey's qualitative analysis more quantitative by discovering an exact nonlinear self-similar solution with an explosive growth of the current density as it becomes infinite in a finite time. The next subsections describe it (Sec. 6.2.2.2) and a linear stability analysis (Sec. 6.2.2.3).

6.2.2.2 Self-Similar Imshennik-Syrovatsky Solution

Self-similar solutions are particular solutions that are valid when there is no natural length-scale, so the time-behaviour at each location is similar and the governing partial differential equations reduce to much simpler ordinary differential equations. The general self-similar form for the solution of a partial differential equation in x and t, say, is $f(x,t) = F(x)G(x^n t)$, where n and the functions F and G are to be determined. The particular case $n=0$ gives separable solutions.

The full system of equations describing compressible X-type collapse is:

$$\frac{\partial \rho}{\partial t} = -\left(\frac{\partial(\rho v_x)}{\partial x} + \frac{\partial(\rho v_y)}{\partial y}\right), \qquad \left(\frac{\partial}{\partial t} + \mathbf{v}\cdot\boldsymbol{\nabla}\right)\frac{p}{\rho^\gamma} = (\gamma-1)\frac{j^2}{\sigma\rho^\gamma}, \tag{6.3}$$

$$\frac{\partial B_x}{\partial t} = -\frac{\partial(v_y B_x - v_x B_y)}{\partial y} + \eta\nabla^2 B_x, \qquad \frac{\partial B_y}{\partial t} = -\frac{\partial(v_x B_y - v_y B_x)}{\partial x} + \eta\nabla^2 B_y, \tag{6.4}$$

$$\rho\left(\frac{\partial v_x}{\partial t} + \mathbf{v}\cdot\boldsymbol{\nabla} v_x\right) = -\frac{\partial p}{\partial x} - jB_y, \qquad \rho\left(\frac{\partial v_y}{\partial t} + \mathbf{v}\cdot\boldsymbol{\nabla} v_y\right) = -\frac{\partial p}{\partial y} + jB_x, \tag{6.5}$$

where $\mu j = \partial B_y/\partial x - \partial B_x/\partial y$, $\mathbf{v}\cdot\boldsymbol{\nabla} = v_x\partial/\partial x + v_y\partial/\partial y$ and $\nabla^2 = \partial^2/\partial x^2 + \partial^2/\partial y^2$.

The initial conditions are: $B_x = B_0\, y/y_0$, $B_y = (1+\epsilon)\, B_0\, x/y_0$, $v_x = 0$, $v_y = 0$, $\rho = \rho_0$, $p = p_0$, with $1+\epsilon = y_0/x_0$ and $\beta_0 = 2\mu p_0/B_0^2$. Here the single scale r_0 in Eq. (6.2) has been replaced by two separate scales x_0 and y_0. The initial flux function and current density are $A = [B_0/(2y_0)][y^2 - (1+\epsilon)x^2]$ and $j = -\epsilon B_0/(\mu y_0)$.

Pressure gradients are neglected and a solution is sought in the form

$$B_x = \frac{B_0\, y}{a(\bar{t})^2\, y_0}, \quad B_y = \frac{B_0\, x}{b(\bar{t})^2\, x_0}, \quad v_x = \chi_x(\bar{t})\frac{v_{A0}}{x_0}x, \quad v_y = \chi_y(\bar{t})\frac{v_{A0}}{y_0}y,$$

where $\bar{t} = t\, v_{A0}/y_0$, $v_{A0} = B_0/(\mu\rho_0)^{1/2}$, and the functions a and b satisfy

$$\frac{da}{d\bar{t}} = \left(\frac{(1+\epsilon)}{b} - \frac{b}{a^2}\right), \qquad \frac{db}{d\bar{t}} = \left(\frac{(1+\epsilon)}{a} - \frac{(1+\epsilon)^2 a}{b^2}\right), \tag{6.6}$$

with $a(0) = b(0) = 1$ and $da/dt(0) = db/dt(0) = 0$ as initial conditions.

The density $[\rho = \rho_0/(ab)]$, current density $\{j = [B_0/(\mu\, y_0)][(1+\epsilon)/b^2 - 1/a^2]\}$ and pressure $(p = s\rho^\gamma)$ are uniform, as are the velocity gradients (χ_x and χ_y) and entropy (s), which are determined by $\chi_x = b^{-1}(1-\epsilon)^{-1}db/d\bar{t}$, $\chi_y = a^{-1}da/d\bar{t}$ and $ds/dt = (\gamma-1)j^2/(\sigma\rho^\gamma)$. Since j is uniform, the diffusion terms in Eq. (6.4) vanish.

Figure 6.4a shows the behaviour of the system after numerically solving Eq. (6.6) for initial perturbations $\epsilon = 0.9$ and 10^{-5}. When a reaches zero, the current density becomes infinite. As ϵ is reduced, the time for the singularity to appear takes longer.

In this ideal low-β model, if collapse is initiated by a disturbance at a finite radius, then the singularity occurs at the same time as an inwards-propagating fast-mode wave reaches the X-point. In practice,

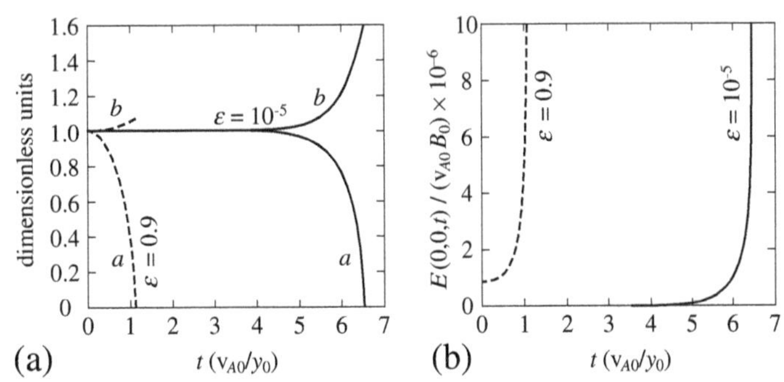

Figure 6.4. (a) The growth of the magnetic field parameters a and b as functions of time in the self-similar solution of Imshennik and Syrovatsky (1967) for two values of the initial current parameter (ϵ). The corresponding reconnection rate (E) as a function of time (t).

magnetic diffusion or pressure gradients will speed up the wave and stop self-similar collapse before the singularity.

6.2.2.3 Linear Solution for Collapse

Surprisingly, it was not until the 1990s that the first thorough linear analyses of the collapse process were carried out (e.g., Bulanov et al. 1990; Craig and McClymont 1991, 1993; Titov and Priest 1993). An important result is that reconnection in the linear regime is fast, scaling as $1/(\ln\,\eta)$.

First, linearise the MHD equations (6.3) to (6.5) by expressing the flux function as $A = A_0 + A_1$, where A_1 is the linear perturbation and A_0 is the current-free state $\frac{1}{2}B_0 r_0(\bar{y}^2 - \bar{x}^2)$, with $\bar{y} = y/r_0$ and $\bar{x} = x/r_0$. Neglecting the pressure then leads to

$$\frac{\partial^2 A_1}{\partial \bar{t}^2} = (\bar{x}^2 + \bar{y}^2)\nabla^2 A_1 + \bar{\eta}\nabla^2 \frac{\partial A_1}{\partial \bar{t}},$$

where $\bar{t} = t v_{A0}/r_0$ and $\bar{\eta} = \eta/(v_{A0} r_0)$. Then seek separable solutions of the form

$$A_1 = \tfrac{1}{2} B_0\, r_0\, \epsilon\, Re\left[f(\bar{r}) e^{im\theta + \omega t}\right], \tag{6.7}$$

where $\omega = \omega_R + i\omega_I$, $\bar{r} = r/r_0$, $\bar{\omega} = \omega r_0/v_{A0}$, ϵ is the dimensionless magnitude of the perturbation, and $f(\bar{r})$ is a complex function satisfying

$$\bar{r}\frac{d}{d\bar{r}}\left(\bar{r}\frac{df}{d\bar{r}}\right) = \left(\frac{\bar{r}^2\,\bar{\omega}^2}{\bar{r}^2 + \bar{\eta}\bar{\omega}} + m^2\right) f. \tag{6.8}$$

Frozen-flux conditions are imposed at the surface $r = r_0$ by setting $f = 0$ at $\bar{r} = 1$. Only the $m = 0$ mode changes A at the origin and so corresponds to reconnection.

The solution of Eq. (6.8) for $m = 0$ determines the eigenvalue ($\bar{\omega} = \bar{\omega}_R + i\bar{\omega}_I$) as

$$\bar{\omega}_R = \frac{-(2n+1)^2\pi^2}{2(n+1)\ln^2\bar{\eta}}, \quad \bar{\omega}_I = -\frac{(2n+1)\pi}{\ln\bar{\eta}},$$

where n is the number of radial nodes. It describes radial oscillations propagating between the boundary and the origin. The travel-time depends on diffusivity (η) because the wave speed vanishes as the origin is approached, and so diffusion allows a perturbation to reach the origin and reflect.

The total flux function (A) for the lowest-order mode ($m = 0, n = 0$) is

$$A(\bar{r},\theta,\bar{t}) = \tfrac{1}{2}B_0 r_0\{-\bar{r}^2\cos(2\theta) + \epsilon\exp(\bar{\omega}_R\bar{t})[f_R(\bar{r})\cos(\bar{\omega}_I\bar{t}) - f_I(\bar{r})\sin(\bar{\omega}_I\bar{t})]\},$$

where $f = f_R + i f_I$. This describes oscillations which periodically flatten the field configuration along the x- and y-axes, with the amplitude of the oscillations decaying at a rate $\bar{\omega}_R = -\pi^2/(2\ln^2\bar{\eta})$.

At the X-line, the flux function becomes $A(0,0,\bar{t}) = \frac{1}{2}\epsilon\, B_0\, r_0\, e^{\bar{\omega}_R\bar{t}}\cos(\bar{\omega}_I\bar{t})$, since $f_R = 1$ and $f_I = 0$ at $\bar{r} = 0$, where $\bar{\omega}_I = -\pi/\ln\bar{\eta}$. Thus, the collapse-time is $\bar{t}_f \equiv \pi/(2\bar{\omega}_I) = -\frac{1}{2}\ln\bar{\eta}$, and for the first collapse cycle ($t = \bar{t}_f$) the flux reconnected (A_f) and average electric field (E_{ave}) are $A_f \approx \frac{1}{2}\epsilon B_0 r_0$ and $E_{ave} = A_f/t_f \approx -\epsilon v_{A0} B_0/(\ln\bar{\eta})$. The reconnection rate (E) depends only weakly on $\bar{\eta}$ and so is fast.

Nonlinear collapse beyond t_f has been studied numerically by, for instance, McClymont and Craig (1996), who established that collapse is likely whenever an X-line configuration is perturbed in a low-resistivity, low-β plasma. If the perturbations are small ($\epsilon < \bar{\eta}$), the reconnection rate scales as $1/\ln\,\bar{\eta}$, while if they are large ($\epsilon > \bar{\eta}$) the average reconnection rate appears to be nearly independent of $\bar{\eta}$.

6.3 Current Sheet Formation

If photospheric footpoints of a coronal field move rapidly for a short time, the response is often the collapse of an X-point field towards a configuration containing a current sheet. On the other hand, if the footpoint motions are much slower than the wave travel-time across the system, it passes through a series of equilibria which may also contain a current sheet. This section supposes that no reconnection takes place and determines the configurations resulting from motions that are planar (Secs. 6.3.1 and 6.3.2), shearing (Sec. 6.3.3) or braiding (Sec. 6.3.4).

Once a current sheet has formed, or indeed is in the process of forming, it will tend to diffuse away and it will often tend to become unstable to a reconnection instability known as the tearing mode (Sec. 6.8). Furthermore, in some cases a quasi-steadily reconnecting state will be established (Secs. 6.5 to 6.7).

6.3.1 Planar Motion in 2D Potential Fields

6.3.1.1 Modelling a Simple Current Sheet

Let us start with a magnetic field $B_x = y, B_y = x$ (Figure 6.5a), and suppose that a series of equilibria containing a current sheet forms, driven by the slow motion of the sources of the field.

The question is how best to describe mathematically the resulting equilibrium that has the appearance of Figure 6.5b. Outside the sheet, if the current vanishes, the magnetic field satisfies $\nabla \times \mathbf{B} = \mathbf{0}$ and $\nabla \cdot \mathbf{B} = 0$ or, in 2D,

$$\frac{\partial B_y}{\partial x} - \frac{\partial B_x}{\partial y} = 0 \quad \text{and} \quad \frac{\partial B_x}{\partial x} + \frac{\partial B_y}{\partial y} = 0. \tag{6.9}$$

A powerful result is that, if the combination $B_y + iB_x = f(z)$ is any analytic (i.e., differentiable) function of the complex variable $z = x + iy$, then Eqs. (6.9) are automatically satisfied. Thus, the current sheet can be treated as a cut in the complex plane and a function $f(z)$ may be sought which has such a cut.

In the initial state, $B_y + iB_x$ has the elegant form

$$B_y + iB_x = x + iy \equiv z. \tag{6.10}$$

In a stroke of inspiration, Green (1965) realised that, when a sheet stretches from $z = -iL$ to $z = iL$, the surrounding field may be represented by

$$B_y + iB_x = (z^2 + L^2)^{1/2}, \tag{6.11}$$

which behaves like z when $z \gg L$ and reduces to z (as required) when $L = 0$. Thus, the evolution through a series of equilibria with a slowly growing sheet may simply be modelled by letting L increase in value in Eq. (6.11).

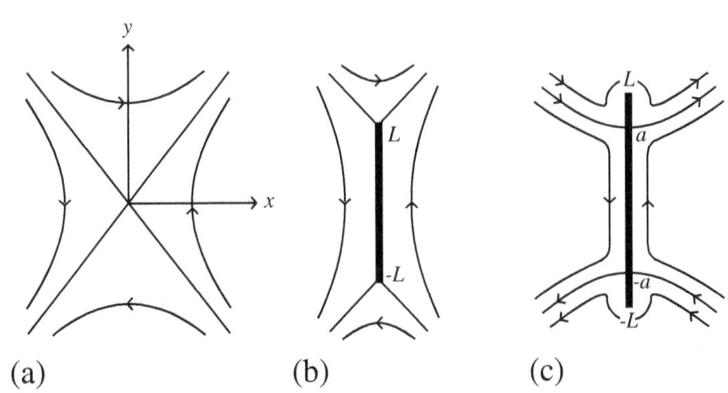

Figure 6.5. (a) Magnetic field near an X-type neutral point, which evolves to a field with a current sheet having at its ends either (b) Y-points or (c) reversed currents and singularities.

6.3.1.2 Current Sheet with Singular End-Points

Instead of evolving into the field (6.11), the initial field (6.10) may evolve into the more general field (Somov and Syrovatsky 1976)

$$B_y + iB_x = \frac{z^2 + a^2}{(z^2 + L^2)^{1/2}}, \tag{6.12}$$

where $a^2 < L^2$ (Figure 6.5c). The neutral points at the ends ($z = \pm iL$) of the sheet have now in general been replaced by singularities, where the field behaves like $(z \mp iL)^{-1}$ as z tends to $\pm iL$. The special case when $a = L$ reduces to Eq. (6.11).

The magnitude of the field at the right of the current sheet ($x = 0+, y^2 < L^2$) is $B_x(0+, y) = (a^2 - y^2)/(L^2 - y^2)^{1/2}$, which vanishes at $y = \pm a$. The current per unit length in the sheet is then simply

$$J \equiv \int j dy = \int \frac{1}{\mu}\frac{\partial B_x}{\partial y} dy = \frac{1}{\mu}[B_x(0+, y) - B_x(0-, y)] = \frac{2(a^2 - y^2)}{\mu(L^2 - y^2)^{1/2}},$$

giving reversed currents between the neutral points and the ends of the sheet.

Near an end ($Z = d$) of such a current sheet, the field behaves like $B \sim K/(Z - d)^{1/2}$, but the solution (6.12) fails at these singularities, where diffusive effects become important and slow-mode shocks may be generated.

6.3.1.3 Current Sheets in Other Configurations

The formation of a current sheet in the solar corona between two approaching bipolar regions (Figure 6.6) was first proposed by Sweet (1958a) as an explanation for solar flares. It can be modelled in terms of complex variables (Priest and Raadu 1975).

Consider an initial potential field (B_{0x}, B_{0y}) in $y > 0$ given by

$$B_{0y} + iB_{0x} = \frac{iD}{(z + a_0)^2} + \frac{iD}{(z - a_0)^2},$$

due to two bipoles of moment $(2\pi D/\mu)$ initially situated at $(a_0, 0)$ and $(-a_0, 0)$.

Suppose that significant reconnection does not take place, and that the field passes through a series of equilibria with a vertical current sheet stretching from $P(0, p)$ to $Q(0, q)$ along the y-axis (when the dipole separation is $2a$). Outside the current sheet where $j = 0$, the magnetic field $(B_y + iB_x)$ forms an analytic function of z, namely

$$B_y + iB_x = \frac{\bar{D}(z^2 + p^2)^{1/2}(z^2 + q^2)^{1/2}}{(z^2 - a^2)^2},$$

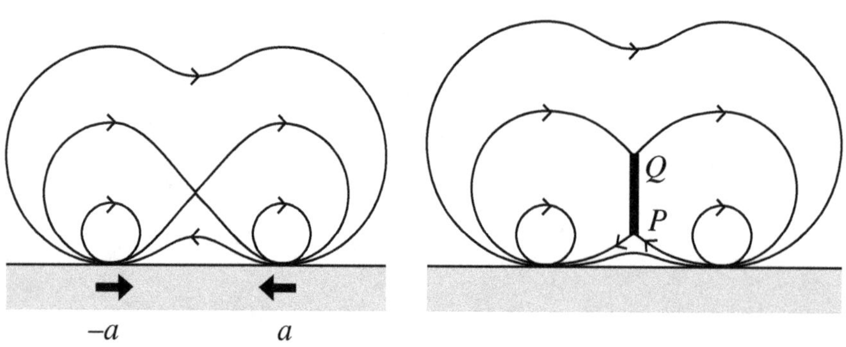

Figure 6.6. The interactions of the coronal fields of two aligned bipolar photospheric magnetic regions.

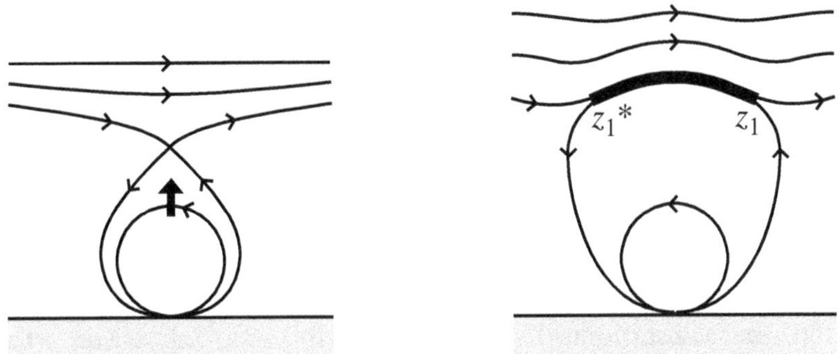

Figure 6.7. The formation of a curved current sheet by the emergence of new magnetic flux through the photosphere (shaded).

where $\bar{D} = 4iDa^2(a^2 + p^2)^{-1/2}(a^2 + q^2)^{-1/2}$ is chosen to give the correct behaviour near the sources. The values of p and q are found as functions of a by assuming frozen-in flux and equating the fluxes crossing the positive y-axis above and below the original neutral point to those crossing above P and below Q, respectively.

The analysis was extended by Tur and Priest (1976) to include curved current sheets forming between unequal dipoles.

They also consider the emergence of new flux into a uniform horizontal field (Figure 6.7). The initial potential field is $B_{0y} + iB_{0x} = ib + iD_0/z^2$ and later it becomes

$$B_y + iB_x = \frac{\bar{K}\left[(z^2 - z_1^2)(z^2 - z_1^{*2})\right]^{1/2}}{z^2},$$

where $\bar{K}$ is constant. The shape and end-points (z_1 and $z_2 = -z_1^*$) of the current sheet are determined by flux conservation. Dissipation of such a curved current sheet was part of a solar flare model (Heyvaerts et al. 1977) and may be invoked to explain the one-third of X-ray bright points that are associated with emerging flux.

General properties have been established by Amari and Aly (1990) and Titov (1992), including a method for computing arbitrary 2D potential fields with multiple curved current sheets and a given normal field component at the boundary.

6.3.2 Current Sheets in 3D and Non-potential Fields

The technical advantage of considering 2D potential fields is that complex variable theory can be employed, with current sheets treated as cuts in the complex plane. When modelling 3D potential fields or magnetic fields that are force-free or magnetohydrostatic, however, other approaches have to be developed.

6.3.2.1 *Axisymmetric Current Sheet in 3D*

Tur (1977) developed a technique for computing the annular 3D current sheet that forms between two approaching 3D symmetric dipoles of moment $2\pi D/\mu$ (the axisymmetric version of Figure 6.6). The magnetic field is $\mathbf{B} = -\nabla\Phi_m$, where Φ_m is the sum of parts (Φ_{ms}) arising from the current sheet and (Φ_{mD}) from the dipole sources. The solution for Φ_{ms} may be built up as an integral sum of potentials $[\phi_m(R)]$ of the individual circular current loops of radius R that make up the current sheet. For further properties of 3D current sheet formation, see Longcope and Cowley (1996), Longcope (1996), Longcope and Silva (1998) and Longcope et al. (1998).

6.3.2.2 Current Sheets in Non-Potential Fields

Consider a 2.5D force-free field $(B_x, B_y, B_z) = (\partial A/\partial y, -\partial A/\partial x, B_z)$ with $B_z^2(A) = 2\mu J_0\,(A + A_0)$, where J_0 and A_0 are constant (Bungey and Priest 1995). This represents a *constant-current force-free field*, for which A satisfies

$$\nabla^2 A + \mu J_0 = 0.$$

The conformal mapping $z = \frac{1}{2}(\tilde{z} + a^2/\tilde{z})$ maps the region outside a cut (a current sheet) stretching from $z = -a$ to $z = +a$ onto the region outside a circle of radius $\tilde{r} = a$. As the current (J_0) varies, the resulting configurations vary greatly in structure, with the ambient field being either closed and elliptical or X-type.

Consider next a 2D magnetostatic field $(B_x, B_y) = (\partial A/\partial y, -\partial A/\partial x)$, arising when a potential field with an X-point experiences 2D changes in source positions and pressure. Suppose it relaxes to a new equilibrium, but the topology is preserved (i.e., no reconnection). In general, the X-point bifurcates, and current sheets form near the X-point and along the separatrices (Figure 6.8). The force balance is

$$\nabla^2 A + \mu \frac{dp}{dA} = 0,$$

in terms of the flux function (A), where $p = p(A)$ and where there is a balance in total pressure $[p + B^2/(2\mu)]$ from one side to the other of each separatrix.

If the sources to left and right of the X-point separate or approach, a current sheet forms between two Y-points (Figures 6.8b and 6.8c). If instead the pressure is enhanced below and above the X-point, a *cusp-point* forms (Figure 6.8d), with current sheets extending along the separatrices. The magnetic field tends to zero above and below the cusp, whereas to the left and right it is non-zero, such that the decrease in plasma pressure crossing from a shaded to an unshaded region is balanced by an increase in magnetic pressure. The same feature occurs in Figure 6.8f due to an increase in plasma pressure above and below the Y-points of Figure 6.8c.

If, on the other hand, the plasma pressure is enhanced above and below a sheet (Figure 6.8b), then *T-points* form (Figure 6.8e). As the T-points are approached from the high-pressure regions, the magnetic field tends to zero, whereas in the low-pressure regions it is non-zero, such that again total pressure balance

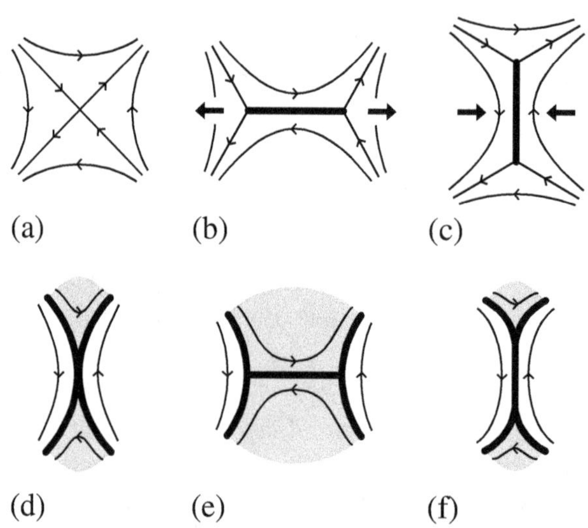

Figure 6.8. The effect of stretching or squashing motions on (a) an X-point is to form simple current sheets with Y-points at their ends [(b) or (c)]. Increasing the pressure in the shaded regions forms curved separatrix sheets: the X-point in (a) becomes a cusp-point in (d), while the Y-points in (b) and (c) become T-points in (e) and cusp-points in (f).

(plasma plus magnetic) is maintained. If the pressure enhancements are asymmetric, the current sheets joining the T-points or cusp-points become curved.

6.3.2.3 Squashing a Sheared Magnetic Field

Bobrova and Syrovatsky (1979) realised that one may form current sheets by moving the boundary of a force-free field $B_{0x} = \cos\alpha_0 z$, $B_{0y} = \sin\alpha_0 z$. They move the top ($z = L$) of the region up or down by a small distance $\xi_L(t)$, while the bottom ($z = 0$) remains at rest. The resulting magnetic field has a singularity where $k_{||}(z) = 0$, which is just the condition ($\mathbf{k}\cdot\mathbf{B} = 0$) for a resonant surface.

More generally, consider a diffuse current sheet with field $B_0(y)\ \hat{\mathbf{x}}$, pressure $p_0(y)$ and flux function $A_0(y)\ \hat{\mathbf{z}}$. Suppose $B_0(y)$ is an odd function and apply a periodic perturbation ($A_1 = \epsilon\cos kx$) at the boundaries $y = \pm a$. Then allow the field to relax to an equilibrium satisfying $\nabla^2 A + \mu\, dp/dA = 0$, whose linearized form is $\nabla^2 A_1 + \mu(d^2 p_0/dA_0^2)A_1 = 0$. The general smooth solution consists of the sum of an odd function of y, which does not satisfy the boundary conditions, and an even function, which violates the frozen-flux condition. Thus, no smooth ideal solution exists, and the only solution is the odd function for $y > 0$ together with $A_1(x,y) = A_1(x,-y)$ for $y < 0$, which possesses a current sheet at $y = 0$. Such a sheet, whose formation is driven by boundary motions, may reconnect in a process called *forced magnetic reconnection* (Kulsrud and Hahm 1982; Vekstein and Bian 2005).

6.3.2.4 Magnetic Relaxation

Magnetic relaxation is a technique for producing magnetic equilibria (e.g., Sturrock and Woodbury 1967; Klimchuk et al. 1988), which has been developed to include current sheets by Arnol'd (1974) and Moffatt (1985, 1990). They consider a magnetic field threading an incompressible fluid ($\boldsymbol{\nabla}\cdot\mathbf{v} = 0$, $\boldsymbol{\nabla}\cdot\mathbf{B} = 0$) contained in a domain $\mathcal{D}$ with boundary $\partial\mathcal{D}$ on which $\mathbf{v}\cdot\mathbf{n} = 0$ and $\mathbf{B}\cdot\mathbf{n} = 0$. The magnetic field is frozen to the plasma and a simple model equation of motion is adopted, namely,

$$\rho\frac{\partial \mathbf{v}}{\partial t} = -\boldsymbol{\nabla}p + \mathbf{j}\times\mathbf{B} - K\mathbf{v}, \tag{6.13}$$

with ρ and K constant. Starting from a smooth static state, force imbalance accelerates the plasma, after which dissipation slows it as it relaxes towards equilibrium.

It is the final state that is of interest here rather than the evolution towards it, and so (6.13) is easier to deal with numerically than the full equation of motion. Furthermore, it can be proved that the total energy ($W_m + W_k$) has a rate of change

$$\frac{d}{dt}(W_m + W_k) = -2KW_k, \tag{6.14}$$

and so it decreases monotonically, since the magnetic and kinetic energies [$W_m(t) = \int_{\mathcal{D}} B^2/(2\mu)dV$, $W_k(t) = \frac{1}{2}\rho\int_{\mathcal{D}} v^2 dV$] are both positive (see PROBLEM 6.2). The total energy tends to a finite limit, which is nonzero if the initial topology is nontrivial. The asymptotic state is magnetostatic and may possess current sheets, across which $\mathbf{n}\cdot\mathbf{B} = 0$ and the total pressure $[p + B^2/(2\mu)]$ is continuous.

As examples of this technique, Linardatos (1993) considered 2D fields with and without X-points, while Moffatt (1985) described qualitatively a 3D configuration consisting of two linked flux tubes.

6.3.2.5 DYNAMIC TIME-DEPENDENT SHEET FORMATION

The Green and Syrovatsky solutions represent two solutions for the slow, passive evolution through a series of equilibria from an X-point field. Priest et al. (1994) have discovered an alternative nonlinear, compressible pathway for the dynamic formation of a current sheet. They assume $c_s \ll v \ll v_A$ and solve the low-beta equation of motion and the ideal induction equation,

$$\rho\frac{d\mathbf{v}}{dt} = \mathbf{j}\times\mathbf{B}, \qquad \frac{\partial\mathbf{B}}{\partial t} = \boldsymbol{\nabla}\times(\mathbf{v}\times\mathbf{B}).$$

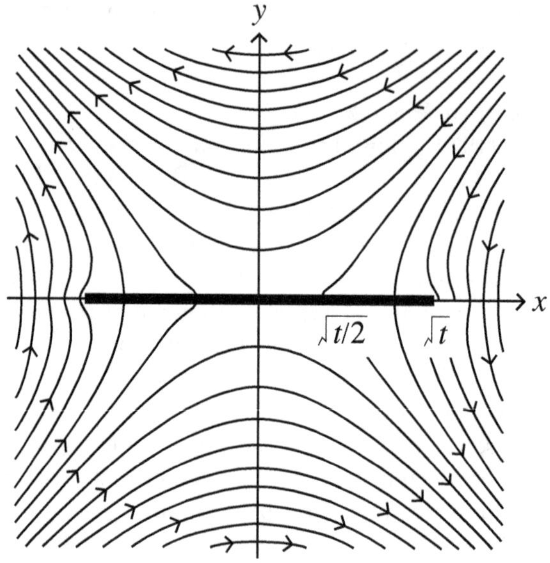

Figure 6.9. Magnetic field lines during the dynamic self-consistent collapse of a current sheet.

They found a self-similar, quasi-static collapse through current-free states surrounding a growing current sheet (Figure 6.9). The current sheet stretches from $-\sqrt{t}$ to $+\sqrt{t}$ along the x-axis. As the sheet grows in length the magnetic dissipation increases, and it swallows up half of the magnetic flux ahead of it, so creating a transverse y-component of field threading the sheet. The other half piles up ahead of the sheet and creates regions of reversed current near the ends.

In terms of $z = x + iy$, the solution for the magnetic field may be written as

$$B_y + iB_x = -\frac{[z + \sqrt{(z^2 - t)}]^2}{4\sqrt{(z^2 - t)}},$$

while the velocity is $v_x + iv_y = \partial z/\partial t = 1/[z + 2(z^2 - t)^{\frac{1}{2}}]$.

6.3.3 Creation of Sheets at Separatrices by Shearing

Current sheets may also be created without an X-point (Low and Wolfson 1988; Amari and Aly 1990; Vekstein et al. 1990). Consider a quadrupolar field with a separatrix that meets the photosphere at a so-called *bald patch* (Titov et al. 1993) (Figure 6.10a). Converging motions will produce an X-point on the boundary and later a current sheet extending upwards so as to preserve the topology (Figure 6.10b).

Shearing motions have an even more dramatic effect, since a much longer curved current sheet is created along the whole length of the separatrix (Figure 6.10c). This may be demonstrated as follows. In a 2.5D field the magnetic components are $(B_x, B_y, B_z) = [\partial A/\partial y, -\partial A/\partial x, B_z(A)]$ in terms of the flux function (A), while the force-free equation $\mathbf{j} \times \mathbf{B} = \mathbf{0}$ reduces to the *Grad-Shafranov equation*

$$\nabla^2 A + B_z \frac{dB_z}{dA} = 0. \tag{6.15}$$

Starting with a potential field (Figure 6.10a), impose a smooth perturbation $[\xi_z(x)]$ of the footpoints while keeping B_y at the base $y = 0$ as a smooth fixed function. The equation for a field line is $dz/B_z = ds/B_p$, where $\mathbf{B}_p = (\partial A/\partial y, -\partial A/\partial x)$ is the poloidal field in the xy-plane. Therefore integrating the field-line

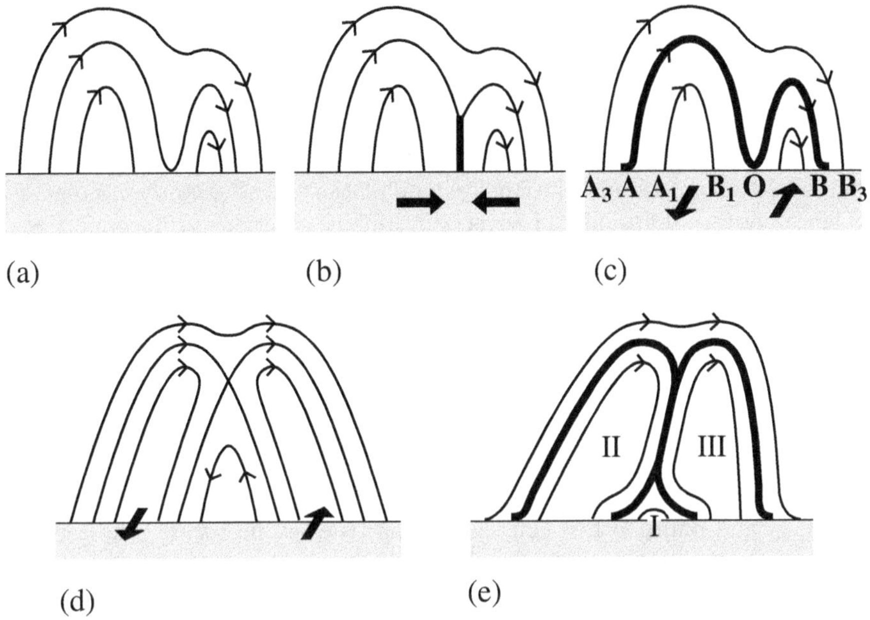

Figure 6.10. A quadrupolar field with (a) no X-points and (d) an X-point. The effect of (b) converging motions and (c),(e) shearing motions to create current sheets (thick curves).

equation gives

$$B_z(A) = \frac{d(A)}{\bar{V}(A)},$$

where $d(A)$ is the difference in footpoint displacements in the z-direction and $\bar{V}(A) = \int ds/B_p$ is the *differential flux volume* of the poloidal field.

The component $B_z(A)$ is constant all along a given field line, but it may have quite different values on the two field lines A_3B_3 and A_1B_1 that lie just above and below the separatrix AOB, since their footpoint displacements are in general quite different. The whole separatrix therefore becomes a current sheet (Figure 6.10c).

Consider next a 2D potential field containing an X-point (Figure 6.10d). Such a field is *structurally unstable*, since ideal converging motions will make an X-point split into a pair of Y-points joined by a current sheet (Figure 6.8c). Shearing motions transform the X-point into a pair of cusp-points joined by a current sheet, together with current sheets all along the separatrices (Vekstein and Priest 1992, 1993) (Figure 6.10e). A cusp-point has the useful property that B_p tends to zero from one side and to constant values from the other two sides, so there is a jump in B_p^2 across the separatrix which can balance the jump in B_z^2 (see PROBLEM 6.3).

6.3.4 Braiding by Random Footpoint Motions

As we have seen, for a magnetic field containing topologically distinct flux systems, a smooth transition from one equilibrium to another is not always possible, and current sheets form on separatrices and near null points. However, Parker (1972, 1994) went a step further by arguing qualitatively that magnetic fields with simple topology (i.e., no separatrices) but complex geometry (such as braiding) may also form current sheets. For further details and its importance for coronal heating, see Section 10.4.3.

6.4 Magnetic Annihilation

The term *magnetic annihilation* is an important ingredient of magnetic reconnection; it refers to the carrying in and cancelling of oppositely directed straight field lines in a one-dimensional current sheet. After presenting the equations (Sec. 6.4.1), diffusion and advection in such a sheet are discussed (Sec. 6.4.2), followed by the classical solution for annihilation by a stagnation-point flow (Sec. 6.4.3), with a steady balance between diffusion and advection. Finally, solutions for reconnective annihilation are described, which are closer to annihilation than reconnection (Sec. 6.4.4).

6.4.1 Equations for Steady Annihilation and 2D Reconnection

The induction equation

$$\frac{\partial \mathbf{B}}{\partial t} = \boldsymbol{\nabla} \times (\mathbf{v} \times \mathbf{B}) + \eta \nabla^2 \mathbf{B}, \tag{6.16}$$

describes how the magnetic field ($\mathbf{B}$) changes in time due to advection of the field with the plasma (the first term on the right) and diffusion through the plasma (the second term). The ratio of these terms is the magnetic Reynolds number ($R_m = l_0 V_0/\eta$).

Consider a 2D steady state with flow and magnetic field lying in the xy-plane. In this case, Faraday's Law ($\boldsymbol{\nabla} \times \mathbf{E} = \mathbf{0}$) implies that $\partial E/\partial y = \partial E/\partial x = 0$, so that $\mathbf{E} = E\hat{\mathbf{z}}$ is uniform. Also, the current is $\mathbf{j} = \boldsymbol{\nabla} \times \mathbf{B}/\mu$, so that Ohm's law ($\mathbf{E} + \mathbf{v} \times \mathbf{B} = \mathbf{j}/\sigma$), which is the integral of Eq. (6.16), becomes

$$\mathbf{E} + \mathbf{v} \times \mathbf{B} = \eta \boldsymbol{\nabla} \times \mathbf{B}. \tag{6.17}$$

In most parts of the universe, the magnetic field is frozen very effectively indeed to the plasma and the diffusion term is negligible, so that Eq. (6.17) reduces to

$$\mathbf{E} + \mathbf{v} \times \mathbf{B} = \mathbf{0}, \tag{6.18}$$

to a very high accuracy. In particular, this holds in the ideal external region around a diffusion region and outside any shock waves.

The other main equation for steady-state MHD reconnection (when the dominant forces are a pressure gradient and a magnetic force) is the equation of motion

$$\rho(\mathbf{v} \cdot \boldsymbol{\nabla})\mathbf{v} = -\boldsymbol{\nabla} p + \mathbf{j} \times \mathbf{B}, \tag{6.19}$$

$$= -\boldsymbol{\nabla} p + (\boldsymbol{\nabla} \times \mathbf{B}) \times \mathbf{B}/\mu,$$

$$= -\boldsymbol{\nabla}[p + B^2/(2\mu)] + (\mathbf{B} \cdot \boldsymbol{\nabla})\mathbf{B}/\mu. \tag{6.20}$$

These are supplemented by an energy equation and the continuity equation $[\boldsymbol{\nabla} \cdot (\rho \mathbf{v}) = 0]$, which for an incompressible plasma reduces to $\boldsymbol{\nabla} \cdot \mathbf{v} = 0$.

6.4.2 Diffusion and Advection of a 1D Current Sheet

When $R_m \ll 1$, the advection term in the induction equation is negligible and it becomes a diffusion equation, so that magnetic variations on a scale l_0 are smoothed out on a time $\tau_d = l_0^2/\eta$ and at a speed $v_d = \eta/l_0$. Consider, for example, a 1D field $[B(x,t)\ \hat{\mathbf{y}}]$ satisfying

$$\frac{\partial B}{\partial t} = \eta \frac{\partial^2 B}{\partial x^2}, \tag{6.21}$$

whose solution may be written in general as

$$B(x,t) = \int G(x - x', t)\ B(x', 0)\ dx',$$

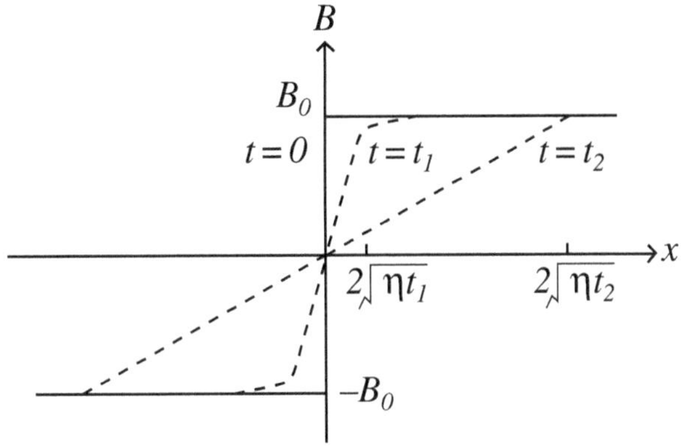

Figure 6.11. The magnetic field (B) as a function of distance (x) in a 1D sheet that is diffusing from zero thickness initially, for times $t=0$, t_1, t_2, where $0<t_1<t_2$.

in terms of any initial magnetic profile $[B(x',0)]$ and the *Green's function* $G(x-x',t)=(4\pi\eta t)^{-1/2}\exp[-(x-x')^2/(4\eta t)]$.

Consider, as an example, initially an infinitesimally thin current sheet with a piecewise-constant magnetic field ($B=B_0$ for $x>0$ and $B=-B_0$ for $x<0$), as shown in Figure 6.11. Physically, the steep magnetic gradient at $x=0$ is expected to spread out, and that is indeed what the mathematical solution of Eq. (6.21) reveals: it is given in terms of the *error function* $[\mathrm{erf}(\zeta)]$ as

$$B(x,t)=B_0\,\mathrm{erf}\left(\frac{x}{\sqrt{(4\eta t)}}\right)\equiv\frac{2B_0}{\sqrt{\pi}}\int_0^{x/\sqrt{(4\eta t)}}e^{-u^2}du.$$

Thus, the half-width of the sheet is of order $(4\eta t)^{1/2}$ and increases in time. The resulting magnetic field strength at a fixed position decreases with time, so the field is said, in a rather dramatic fashion, to be *annihilated.* The field lines diffuse inwards through the plasma and cancel at $x=0$. The total magnetic flux ($\int_{-\infty}^{\infty}B\,dx$) remains constant (namely zero) and the total current $J=\int_{-\infty}^{\infty}j\,dx=\mu^{-1}\int_{-\infty}^{\infty}(\partial B/\partial x)dx=2B_0/\mu$ is conserved, since it simply spreads out in space. However, the magnetic energy $[\int_{-\infty}^{\infty}B^2/(2\mu)dx]$ is converted entirely into heat by Ohmic dissipation (j^2/σ per unit volume). It decreases in time at a rate (see PROBLEM 6.4)

$$\frac{\partial}{\partial t}\int_{-\infty}^{\infty}\frac{B^2}{2\mu}dx=-\int\frac{j^2}{\sigma}dx. \tag{6.22}$$

If instead $R_m\gg 1$, the induction equation reduces to $\partial\mathbf{B}/\partial t=\boldsymbol{\nabla}\times(\mathbf{v}\times\mathbf{B})$, and so the magnetic field lines are frozen into the plasma. For instance, PROBLEM 6.5 shows the effect on a field $B_0(x)\hat{\mathbf{y}}$ of a flow $v_x=-V_0x/a$, $v_y=V_0y/a$.

6.4.3 Stagnation-Point Flow Model (Sonnerup-Priest 1975)

A current sheet naturally tends to diffuse outwards as the field lines diffuse inwards and cancel. Thus, a steady state may be set up and maintained if magnetic flux is carried in at the same rate as it is trying to diffuse (Figure 6.12). Although the magnetic field may be annihilated as it comes in, the plasma itself cannot be destroyed and needs to flow out sideways. This is illustrated in the following 2D steady-state incompressible model (Parker 1973; Sonnerup and Priest 1975), using Eqs. (6.17) and (6.20), where $\boldsymbol{\nabla}\cdot\mathbf{B}=0$, $\boldsymbol{\nabla}\cdot\mathbf{v}=0$, and v_x, v_y, B_x, B_y depend just on x and y.

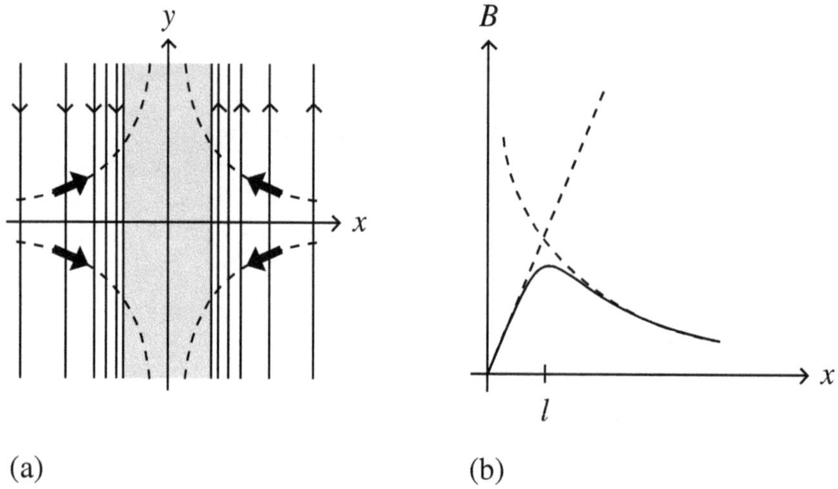

Figure 6.12. (a) Stagnation-point flow creating a steady current sheet of width $2l$ (shaded). (b) Magnetic field profile, with small-x and large-x approximations shown dashed.

Consider the incompressible, *stagnation-point flow*

$$v_x = -\frac{V_0\ x}{a}, \quad v_y = \frac{V_0\ y}{a} \tag{6.23}$$

(with V_0/a constant), which vanishes at the origin and for which $\nabla \cdot \mathbf{v} = 0$. The steady-state continuity equation $[(\mathbf{v}\cdot\nabla)\rho + \rho(\nabla\cdot\mathbf{v}) = 0]$ then reduces to $(\mathbf{v}\cdot\nabla)\rho = 0$, which implies that the density (ρ) is uniform if it is constant at the inflow.

The effect of such a flow on a unidirectional magnetic field $\mathbf{B} = B(x)\hat{\mathbf{y}}$ that reverses sign at $x = 0$ is determined by Ohm's Law (6.17), namely,

$$E - \frac{V_0 x}{a} B = \eta \frac{dB}{dx}. \tag{6.24}$$

When x is large, the field is frozen to the plasma and the right side of Eq. (6.24) is negligible, so that $B \approx (Ea)/(V_0 x)$. When x is small, the field diffuses through the plasma and the second term is negligible, so that $B \approx Ex/\eta$. The division between these two extremes (at the half-width of the current sheet) occurs at $x = (a\eta/V_0)^{1/2}$ (see Figure 6.12b). The full solution follows (by using an integrating factor of Eq. 6.24) in terms of the *Dawson integral function (Daw)* with $l^2 = V_0/(2\eta a)$ as

$$B = \frac{2E_0 a}{V_0 l} \mathrm{Daw}\left(\frac{x}{l}\right) \equiv \frac{2E_0 a}{V_0 l} \exp\left(-\frac{x^2}{l^2}\right) \int_0^{x/l} \exp(X^2)\, dX.$$

For $x/l \ll 1$, *Daw* $(x/l) \approx x/l$, while, for $x/l \gg 1$, *Daw* $(x/l) \approx l/(2x)$.

This represents one of the few exact solutions of the nonlinear MHD equations (see PROBLEM 6.6). It may be generalised to a 3D stagnation-point flow by taking

$$v_x = -\frac{V_1}{a}x, \quad v_y = \frac{V_2}{a}y, \quad v_z = \frac{V_1 - V_2}{a}z, \quad B_x = 0, \quad B_y = B_y(x), \quad B_z = B_z(x).$$

In this case, the y- and z- components of Eq. (6.16) reduce for a steady state to

$$a\eta \frac{d^2 B_y}{dx^2} + V_1 x \frac{dB_y}{dx} + V_2 B_y = 0, \qquad a\eta \frac{d^2 B_z}{dx^2} + V_1 x \frac{dB_z}{dx} + (V_1 - V_2) B_z = 0.$$

Solutions exist joining two arbitrary magnetic fields at large positive and negative x. As the fields approach, their directions rotate and their magnitudes change, annihilating part of the magnetic flux in the process.

One limitation is the assumption of incompressibility and another is that the flow into the current sheet be much slower than the Alfvén speed, so that the sheet length (L) far exceeds its width (l). Furthermore, the solution fails near the ends of the sheet, where there are significant transverse components of magnetic field (in the x-direction). Relaxing these conditions is a far from trivial undertaking.

6.4.4 Reconnective Annihilation

The stagnation-point solution (Sec. 6.4.3), namely $v_x = -x, v_y = y, \mathbf{B} = B_y(x)\hat{\mathbf{y}}$, works as an MHD solution for several reasons: $\boldsymbol{\nabla}\cdot\mathbf{B}=0$ and $\boldsymbol{\nabla}\cdot\mathbf{v}=0$ are satisfied identically, while $\mathbf{v}\times\mathbf{B} = -xB_y(x)\,\hat{\mathbf{z}}$ and $\eta\boldsymbol{\nabla}\times\mathbf{B} = \eta dB_y/dx\,\hat{\mathbf{z}}$ are functions of x alone; also, both the magnetic tension $[(\mathbf{B}\cdot\boldsymbol{\nabla})\mathbf{B}/\mu]$ and curl of $\rho(\mathbf{v}\cdot\boldsymbol{\nabla})\mathbf{v}$ vanish.

Another exact solution has been discovered (Craig and Henton 1995) that generalises the stagnation-point solution in a cunning way (Figure 6.13). Consider a 2D velocity $[v_x(x,y), v_y(x,y)]$ and a magnetic field of the form

$$\mathbf{B} = \lambda\mathbf{v} + G(x)\hat{\mathbf{y}}, \tag{6.25}$$

which is the superposition of a 1D term $[G(x)\hat{\mathbf{y}}]$, as before, and a term $(\lambda\mathbf{v})$ parallel to the flow (where λ is a constant). Suppose again that the terms in Eq. (6.17) depend only on x, so that both $v_x = v_x(x)$ and the vorticity $(\omega = \partial v_y/\partial x)$ depend only on x. Then the only form that is also consistent with $\boldsymbol{\nabla}\cdot\mathbf{v}=0$ is proportional to

$$v_x = -x, \quad v_y = y - F(x), \tag{6.26}$$

where $F(x)$ is any function of x. In other words, both the vorticity $(\omega = -F')$ and electric current $[j = (G' - \beta F')/\mu]$ depend only on x, and the stagnation-point flow is distorted by the addition of a shear flow $[v_y = -F(x)]$. In Eq. (6.17), $\mathbf{v}\times\mathbf{B}$ becomes $v_x B_y\,\hat{\mathbf{z}}$, so that only the $G(x)$ part of B_y is advected; furthermore, $\eta\boldsymbol{\nabla}\times\mathbf{B}$ becomes $(\eta\partial B_y/\partial x\,\hat{\mathbf{z}})$, so only B_y diffuses and diffusion is purely in the x-direction.

The forms (6.25) and (6.26) give non-vanishing x- and y-components of the inertial and magnetic tension terms in the equation of motion (6.20), but the key point is that the curls of these terms are functions of x alone. This means that the curl-free part of Eq. (6.20) determines the pressure and the curl of Eq. (6.20) determines a relation between the two free functions, $F(x)$ and $G(x)$, namely $F(x) = -\lambda/(1-\lambda^2)G(x)$.

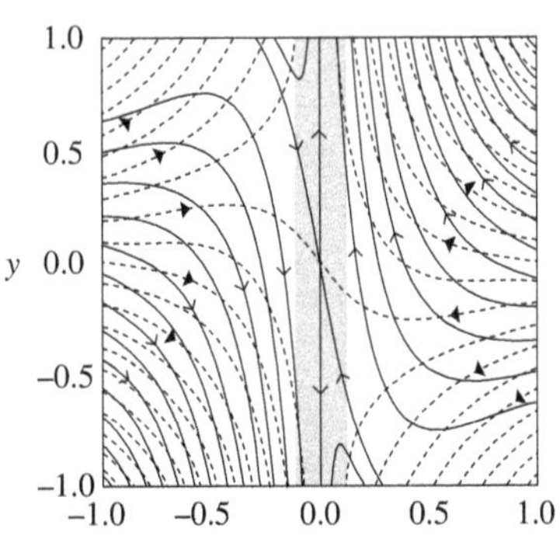

Figure 6.13. Streamlines (dashed) and magnetic field lines (solid) for Craig-Henton reconnective annihilation with $\lambda = 0.9$. The 1D diffusion region is shaded.

Ohm's Law (6.17) then reduces to

$$E - xG = \frac{\eta}{1-\lambda^2}\frac{dG}{dx},$$

which determines $G(x)$. The parameter λ measures the departure of the solution from simple magnetic annihilation. It can be seen that the magnetic field [Eq. (6.25)] consists of a passive part parallel to the flow and a part $[G(x)\hat{\mathbf{y}}]$ that is the same as the stagnation-point flow solution but with a scaling factor $(1-\lambda^2)$ for the x-variable. The plasma velocity [Eq. (6.26)] now consists of a stagnation-point flow plus a shear flow proportional to $G\hat{\mathbf{y}}$.

The solution is closer in spirit to standard magnetic annihilation than to classical reconnection, since the current sheet is 1D and extends to infinity along the y-axis, while the advection and diffusion are inherently 1D. For these reasons we refer to the process as 'reconnective annihilation'.

Craig et al. (1995) extended the solution to a 3D sheared stagnation-point flow $v_x = -x,\ \ v_y = Ky - F_y(x),\ \ v_z = (1-K)z - F_z(x),\ \ \mathbf{B} = \lambda\ \mathbf{v} + G_y(x)\ \hat{\mathbf{y}} + G_z(x)\ \hat{\mathbf{z}}$, for *fan reconnection* (Sec. 6.13.1.1). Craig and Fabling (1996) have also discovered *spine reconnection* solutions (Sec. 6.13.1.1) in cylindrical polars (R, ϕ, z). Flow across the fan produces a cylindrical diffusion region of radius $\sqrt{\eta}$ with a linearly increasing field surrounding the spine, outside which the field falls off as R^{-2}.

6.5 Slow 2D Reconnection: The Sweet-Parker Mechanism

The early researchers tried to determine the rate at which reconnection occurs – that is, the speed with which field lines are carried into the reconnection site. The focus was on the steady-state process, since it is simpler to analyse and the main phase of a large solar flare is a quasi-steady-state or slowly-varying energy release process for many hundreds of Alfvén travel-times (with a superimposed time-modulation).

Sweet (1958a,b) and Parker (1957, 1963b) considered a steady diffusion layer occupying the whole boundary between two opposing magnetic fields. Its length is therefore the global external length-scale (L_e), and the reconnection rate is just the speed with which field lines enter the diffusion region. The basic equations are Ohm's Law (6.17) and the equation of motion (6.20) with mass continuity $[\boldsymbol{\nabla}\cdot(\rho\mathbf{v}) = 0]$.

6.5.1 The Basic Sweet-Parker Model (1958)

Sweet and Parker modelled a simple *diffusion region* of length 2L and width $2l$, say, lying between oppositely directed magnetic fields (Figure 6.14), for which an order-of-magnitude analysis may be conducted as follows. For a steady state, magnetic field of strength B_i is carried into the diffusion layer at the same speed (v_i) as it is trying to diffuse outwards, so that

$$\boxed{v_i = \frac{\eta}{l}.} \tag{6.27}$$

If mass is conserved, the rate ($4\rho L v_i$) at which mass is entering both sides of the sheet must equal the rate ($4\rho l v_o$) at which it is leaving both ends at speed v_o, so that, if the density is uniform,

$$\boxed{L v_i = l v_o.} \tag{6.28}$$

Eliminating the width (l) between Eqs. (6.27) and (6.28) gives an inflow speed

$$v_i = \left(\frac{\eta\, v_o}{L}\right)^{1/2}. \tag{6.29}$$

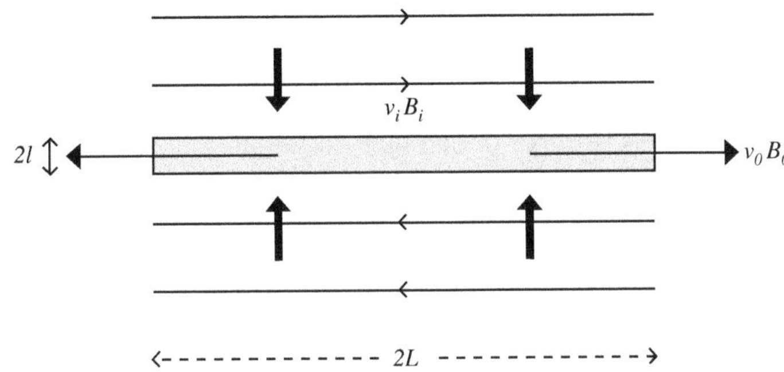

Figure 6.14. Sweet-Parker reconnection. The diffusion region is shaded. The plasma velocity is indicated by solid-headed arrows and the magnetic field lines by light-headed arrows.

In dimensionless variables this becomes $M_i = \sqrt{v_o/v_{Ai}}/\sqrt{R_{mi}}$, where $M_i \equiv v_i/v_{Ai}$ is the *inflow Alfvén Mach number* (or dimensionless *reconnection rate*) and $R_{mi} \equiv L\, v_{Ai}/\eta$ is the *magnetic Reynolds number* based on the inflow Alfvén speed.

Once the outflow speed v_o [and therefore v_i from (6.29)] are known for a given L, (6.28) in turn determines the sheet width as $l = L\, v_i/v_o$, and the outflow magnetic field strength (B_o) is determined from flux conservation ($v_i\, B_i = v_o\, B_o$) as $B_o = B_i\, v_i/v_o$. However, a key question is: what is the outflow speed v_o? This is determined as follows by the equation of motion (or equivalently the *mechanical* energy equation).

The order-of-magnitude electric current is $j \approx B_i/(\mu l)$, and so the Lorentz force along the sheet is $(\mathbf{j} \times \mathbf{B})_x \approx j B_o = B_i B_o/(\mu l)$. This accelerates plasma from rest at the neutral point to v_o over a distance L, and so equating it to the magnitude of the inertial term $\rho(\mathbf{v} \cdot \boldsymbol{\nabla}) v_x$ and neglecting the plasma pressure gradient gives

$$\rho \frac{v_o^2}{L} \approx \frac{B_i\, B_o}{\mu l}. \tag{6.30}$$

However, from $\boldsymbol{\nabla} \cdot \mathbf{B} = 0$, $B_o/l \approx B_i/L$, and so the right-hand side of Eq. (6.30) may be rewritten as $B_i^2/(\mu L)$, with the result from Eq. (6.30) that

$$\boxed{v_o = \frac{B_i}{\sqrt{\mu \rho}} \equiv v_{Ai},} \tag{6.31}$$

where v_{Ai} is the Alfvén speed at the inflow. Not surprisingly, the implication is that the magnetic force accelerates plasma to the Alfvén speed.

In this basic model, the fields reconnect at a speed given by Eq. (6.29) as

$$v_i = \frac{v_{Ai}}{{R_{mi}}^{1/2}}, \tag{6.32}$$

in terms of the (inflow) Alfvén speed (v_{Ai}) and magnetic Reynolds number ($R_{mi} = L v_{Ai}/\eta$). The plasma is ejected from the sheet of width $l = L/{R_{mi}}^{1/2}$ at a speed $v_o = v_{Ai}$ and with a magnetic field strength $B_o = B_i/{R_{mi}}^{1/2}$. Since $R_{mi} \gg 1$, these imply $v_i \ll v_{Ai}, B_o \ll B_i$ and $l \ll L$.

The basic equations (6.27), (6.28) and (6.31) for Sweet-Parker reconnection lead to a reconnection rate (6.32) or in dimensionless form $M_i = R_{mi}^{-1/2}$. In the Sweet-Parker mechanism, the sheet length (L) is identified with the global external length-scale (L_e) and R_{mi} therefore with the *global magnetic Reynolds number* $R_{me} = L_e\, v_{Ae}/\eta$. Since in practice $R_{me} \gg 1$, the reconnection rate is very small: in the solar corona, where $10^6 \lesssim R_{me} \lesssim 10^{12}$, the fields reconnect at between 10^{-3} and 10^{-6} of the Alfvén speed – much too slow for a flare.

6.5.2 Effect of Pressure Gradients and Compressibility

The above simplest version of the Sweet-Parker model has an outflow speed (v_o) equal to the Alfvén speed (v_{Ai}). It assumes that the plasma pressures at the neutral point (p_N) and outflow (p_o) are the same, so that pressure gradients along the sheet play no role in accelerating the plasma from rest at the neutral point to v_o at the outflow. However, in principle the outflow pressure (p_o) may be imposed as any other (positive) value and so lead to a different outflow speed and reconnection rate.

Suppose the inflow speed is much slower than the Alfvén speed, that the sheet is long and thin, and that x- and y-axes are set up along and normal to the sheet. Then the inertial and magnetic tension forces may be neglected in the y-component of (6.20), which reduces to $0 = -\partial/\partial y[p + B^2/(2\mu)]$. By integrating from the inflow point $(0,l)$ to the neutral point $(0,0)$, the neutral point pressure is found to be $p_N = p_i + B_i^2/(2\mu)$. Furthermore, the x-component of Eq. (6.20) is $\rho v_x \partial v_x/\partial x = jB_y - \partial p/\partial x$, which may be evaluated at $(\frac{1}{2}L, 0)$ to give

$$\rho \frac{v_o}{2} \frac{v_o}{L} = \frac{B_i}{\mu l} \frac{B_o}{2} - \frac{p_o - p_N}{L}.$$

Here $j \approx \partial B_x/\partial y \approx B_i/l$, $\partial v_x/\partial x \approx v_o/L$, $v_x \approx \frac{1}{2}v_o$ and $B_y \approx \frac{1}{2}B_o$. After substituting for $B_o/l \approx B_i/L$ and p_N, it becomes

$$v_o^2 = 2v_{Ai}^2 + \frac{2(p_i - p_o)}{\rho}. \qquad (6.33)$$

This is the required generalisation of the usual Sweet-Parker result [Eq. (6.31)]. The outflow speed now depends on the imposed values of p_i and p_o, so the reconnection rate [Eq. (6.29)] may be written, after substituting for v_o/v_{Ai} from Eq. (6.33), as

$$M_i = \frac{2^{1/4}\left(1 + \frac{1}{2}\beta_i(1 - p_o/p_i)\right)^{1/4}}{\sqrt{R_{mi}}}.$$

Thus, it is the values of beta ($\beta_i = 2\mu p_i/B_i^2$) and the pressure ratio (p_o/p_i) which determine the departure from the Sweet-Parker rate ($M_i = 1/\sqrt{R_{mi}}$). If $p_o = p_N$, we recover the Sweet-Parker result ($v_o = v_{Ai}$), whereas if $p_o = p_i$, then $v_o = \sqrt{2}\, v_{Ai}$.

The above equations are only order of magnitude estimates. Surprisingly, a more accurate treatment has not yet been accomplished properly. It would need a solution throughout the diffusion layer, a determination of the shape of its boundary and an integration along the boundary of the conservation equations, taking full account of the variation of the magnitude and inclinations of the velocity and magnetic field.

The original Sweet-Parker model did not include compressibility, and so the density was regarded as uniform. Consider, for example, a simple current sheet with input and output densities ρ_i and ρ_o, respectively, and acceleration by the magnetic force. The equation of mass continuity (6.28) is then modified to

$$\rho_i L v_i = \rho_o l v_o,$$

while the momentum balance (6.30) is altered to $\rho_o v_o^2/L = B_i B_o/(\mu l)$, so that the inflow (6.29) and outflow (6.31) speeds become

$$v_i = \left(\frac{\rho_o}{\rho_i}\right)^{\frac{1}{2}} \left(\frac{\eta v_o}{L}\right)^{\frac{1}{2}} \quad \text{and} \quad v_o = \frac{B_i}{\sqrt{\mu \rho_o}} = v_{Ai}\left(\frac{\rho_i}{\rho_o}\right)^{\frac{1}{2}}.$$

Compressibility therefore slows down the outflow by a factor $(\rho_i/\rho_o)^{\frac{1}{2}}$ when $\rho_o > \rho_i$.

The resulting reconnection rate $[v_i = (\rho_o/\rho_i)^{\frac{1}{4}} (\eta v_{Ai}/L)^{\frac{1}{2}}]$ is enhanced by a factor $(\rho_o/\rho_i)^{\frac{1}{4}}$ when $\rho_o > \rho_i$, since the width of the sheet is decreased by the compression. The density ratio $\rho_o/\rho_i = (p_o/p_i)(T_i/T_o)$ is given in terms of the pressure ratio and the temperature ratio, which is in turn determined by the heat energy equation.

6.5.3 Energetics

Consider a Sweet-Parker diffusion layer which is long and thin ($l \ll L$) and so has $v_i \ll v_{Ai}$. The rate of inflow of electromagnetic energy is the *Poynting flux* ($\mathbf{E} \times \mathbf{H}$ per unit area), which, after writing $E = v_i B_i$, becomes $E B_i L/\mu = v_i B_i^2 L/\mu$. Therefore, the ratio of the inflows of kinetic to electromagnetic energy is

$$\frac{\text{inflow}\, K.E.}{\text{inflow}\, E.M.} = \frac{\frac{1}{2}\rho v_i^2}{B_i^2/\mu} = \frac{v_i^2}{2v_{Ai}^2}.$$

By Eq. (6.32) this is much smaller than unity, so most of the inflow energy is magnetic.

Conservation of flux gives $v_o B_o = v_i B_i$, and so $B_o \ll B_i$. Also, the outflow of electromagnetic energy is $EB_o l/\mu$, which is much less than the inflow of electromagnetic energy ($EB_i L/\mu$), since $B_o \ll B_i$ and $l \ll L$. However, the ratio of outflowing kinetic to inflowing magnetic energy is

$$\frac{\text{outflow}\, K.E.}{\text{inflow}\, E.M.} = \frac{\frac{1}{2}\rho v_o^2\,(v_o l)}{v_i B_i^2 L/\mu} = \frac{\frac{1}{2}v_o^2}{v_{Ai}^2} = \frac{1}{2}.$$

Thus, half of the inflowing magnetic energy is converted to kinetic energy and the rest to thermal energy. In other words, the effect of the reconnection is to create hot fast streams of plasma with a rough *equipartition* between flow and thermal energy.

The equations of electromagnetic energy and mechanical energy are secondary, since they may be derived from the equations of MHD and Maxwell. They provide no extra information about the primary variables, but they do determine the energy partition. Thus, for example, the *electromagnetic energy equation* is

$$-\boldsymbol{\nabla}\cdot(\mathbf{E} \times \mathbf{H}) = j^2/\sigma + \mathbf{v}\cdot\mathbf{j} \times \mathbf{B},$$

which implies that the inflow of electromagnetic energy into a volume produces Ohmic heat and work done by the Lorentz force. By integrating over the diffusion region, it can be shown (PROBLEM 6.7) that half of the inflowing electromagnetic energy goes into Ohmic heat and half into work done by the magnetic force (which in turn goes into kinetic energy when the pressure gradient along the sheet is negligible).

6.6 Fast 2D Reconnection: Petschek's Mechanism

In a pioneering advance, Petschek suggested that the Sweet-Parker diffusion region is limited to a small segment (of length $L \ll L_e$) of the boundary between opposing fields. The diffusion region is thinner, so reconnection can take place faster. Petschek also considered the nature of the external flow outside the diffusion region. He suggested that slow-mode MHD shock waves propagate from the diffusion region, which acts as a kind of obstacle in the flow, since the flow is supersonic relative to the slow-mode wave speed across the magnetic field. Petschek found that the maximum rate of reconnection is typically a tenth or a hundredth of the Alfvén speed.

To distinguish the outer region from the Sweet-Parker region, the flow speed and magnetic field at large distances (L_e) upstream of the X-point are denoted by v_e and B_e (Figure 6.15). The properties of fast reconnection models depend on two dimensionless parameters, namely the *external* reconnection rate ($M_e = v_e/v_{Ae}$) and the *external,* or global, magnetic Reynolds number ($R_{me} = L_e v_{Ae}/\eta$).

The term *fast reconnection* refers to a reconnection rate (M_e) much greater than the Sweet-Parker rate (6.32). Properties at the inflow to the diffusion region (denoted by subscript i) are now related to the 'external' values at large distances (denoted by e). Suppose that the values v_e, B_e at large distances are imposed as typical ambient or external values far from the reconnection site. Consider how the flow speed and magnetic field vary as the reconnection region is approached; that is, what are the unknown values

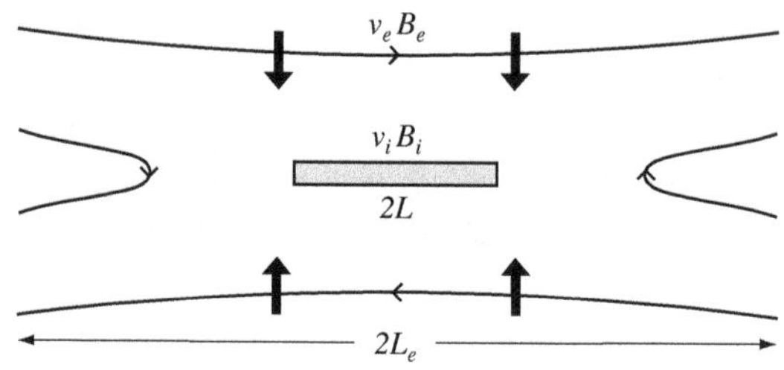

Figure 6.15. The notation for fast reconnection regimes, with the diffusion region shaded.

v_i and B_i close to the diffusion region? In dimensionless terms, what is the value of M_i for a given value of M_e?

Conservation of flux for a steady state ($v_i B_i = v_e B_e$) may be used to relate the inner and external regions. In dimensionless form, it becomes

$$\frac{M_i}{M_e} = \frac{B_e^2}{B_i^2}. \tag{6.34}$$

Furthermore, the *Sweet-Parker relations* (6.27) and (6.28) for the dimensions of the central diffusion region may be rewritten in dimensionless form as

$$\frac{L}{L_e} = \frac{1}{R_{me}} \frac{1}{M_e{}^{1/2}} \frac{1}{M_i{}^{3/2}}, \qquad \frac{l}{L_e} = \frac{1}{R_{me}} \frac{1}{M_e{}^{1/2}} \frac{1}{M_i{}^{1/2}}. \tag{6.35}$$

Once B_i/B_e has been determined from a model of the external region, Eqs. (6.34 and 6.35) determine M_i/M_e and the diffusion region dimensions in terms of M_e and R_{me} alone. The following sections consider different models for the external region, starting with Petschek's mechanism (Sec. 6.6.1), and then moving on to a larger family of solutions, called *Almost-Uniform* reconnection (Sec. 6.7.1).

6.6.1 Petschek's Model (1964) *(Almost-Uniform, Potential Reconnection)*

Petschek (1964) realised that a *slow-mode shock* provides another way (as well as a diffusion region) of converting magnetic energy into heat and kinetic energy. He suggested that four such shocks would stand in the flow when a steady state is reached. Indeed, most of the energy conversion takes place at the shocks (Figure 6.16a), where the magnetic field turns towards the normal and decreases its magnitude.

In the *switch-off* limit, the shock propagates at a speed $v_s = B_N/\sqrt{(\mu\rho)}$ into a medium at rest (where B_N is the normal field component), and it accelerates plasma to the Alfvén speed (v_A) parallel to the shock front. For a specific heat ratio (γ) of $\frac{5}{3}$, $\frac{2}{5}$ of the inflowing magnetic energy is converted to heat and $\frac{3}{5}$ to kinetic energy.

The terms 'potential', 'non-potential', 'uniform' and 'non-uniform' are used to refer to the nature of the magnetic field in the inflow region upstream of the diffusion region and slow-mode shocks. Petschek's regime is *almost-uniform* in the sense that the field in the inflow region is a small perturbation to a uniform field (B_e). It is also potential in the sense that there is no current in the inflow region.

The Petschek analysis is disarmingly simple. The magnetic field decreases substantially from a uniform value (B_e) at large distances to a value B_i at the entrance to the diffusion region, while the flow speed increases from v_e to v_i. The object is to determine for a given B_e the maximum value of v_e (in dimensionless form $M_e = v_e/v_{Ae}$). The shocks provide a normal field component (B_N), which is associated with a small

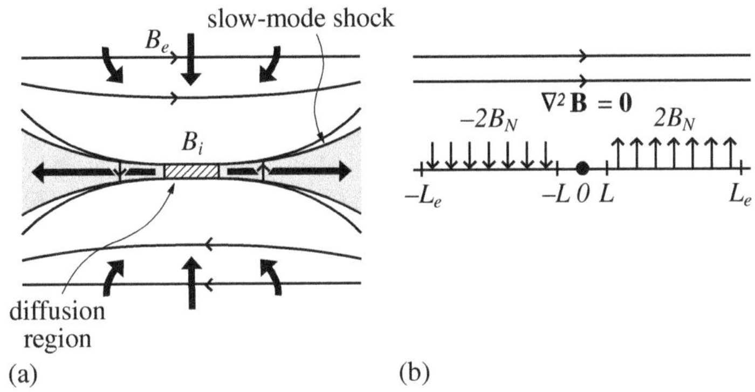

Figure 6.16. (a) Petschek's model, in which the central shaded region is the diffusion region and the other two shaded regions represent plasma that is heated and accelerated by the shocks. (b) Notation for the analysis of the upper inflow region.

distortion in the inflow field from the uniform value (B_e) at large distances, so that the inflow regions consist of slightly curved field lines.

The magnetic field in the upper inflow region is a uniform horizontal field ($B_e\hat{\mathbf{x}}$) plus a potential field, whose normal component equals B_N along the shock waves and vanishes at the diffusion region. To lowest order, the inclination of the shocks may be neglected, apart from imposing B_y on the x-axis equal to $2B_N$ between L and L_e, and $-2B_N$ between $-L_e$ and $-L$.

The normal component on the x-axis may be regarded as being due to a continuous series of poles. If each pole produces a field m/r at distance r, then the flux produced in the upper half-plane by that pole will be πm. If the pole occupies a length dx of the x-axis, the flux is also $2B_N dx$, so that $m = 2B_N dx/\pi$. Integrating along the x-axis gives the field at the diffusion-region inflow as the sum of (B_e) and

$$\frac{1}{\pi}\int_{-L_e}^{-L}\frac{2B_N}{x}dx - \frac{1}{\pi}\int_{L}^{L_e}\frac{2B_N}{x}dx \qquad \text{or} \qquad B_i = B_e - \frac{4B_N}{\pi}\log\frac{L_e}{L}. \tag{6.36}$$

Remembering that, in the switch-off limit, slow-mode shocks travel at the Alfvén speed based on the normal field, $B_N/\sqrt{\mu\rho} = v_e$, Eq. (6.36) can be rewritten as

$$B_i = B_e\left(1 - \frac{4M_e}{\pi}\log\frac{L_e}{L}\right), \tag{6.37}$$

which is the required expression for B_i.

Using $M_e \ll 1$ and $B_i \approx B_e$, the scalings (6.35) reduce to $L/L_e \approx 1/(R_{me}M_e^2)$ and $l/L_e \approx 1/(R_{me}M_e)$, implying that the dimensions of the central region decrease as the magnetic Reynolds number (R_{me}) or reconnection rate (M_e) increase. Petschek suggested that the mechanism chokes itself off when B_i becomes too small. By putting $B_i = \frac{1}{2}B_e$ in Eq. (6.37), he estimated a *maximum* reconnection rate (M_e^*) of

$$\boxed{M_e^* \approx \frac{\pi}{8\log R_{me}}}. \tag{6.38}$$

This lies in practice between 0.01 and 0.1, and so is much faster than Sweet-Parker.

Whereas in the Sweet-Parker model $L = L_e$ and so Eq. (6.28) determines v_i, a key to Petschek's mechanism is that Eq. (6.28) merely determines L once v_i has been deduced from external conditions. Thus, the diffusion region responds to the driving, and the Sweet-Parker relations (6.27) and (6.28) simply determine the dimensions (l and L) of the diffusion region, which shrinks as η becomes smaller. These relations imply that, as v_i varies between $v_A/\sqrt{R_{me}}$ and v_A, the length of the diffusion region decreases from a maximum value of L_e to a minimum of l.

Although in 1D annihilation (Sec. 6.4) it is possible to have a formal linking of the external (ideal) and internal (diffusive) regions, this is not feasible for Petschek's much more complex 2D magnetic field. Instead, the matching is done in an integral or approximate sense, using the order-of-magnitude Sweet-Parker relations. This is entirely reasonable and is shown by numerical experiments to work well when the resistivity is enhanced in the diffusion region, as expected in many applications.

For twenty years, it was thought that Petschek had completely solved the problem of fast reconnection, until in the 1980s a new generation of reconnection solutions was discovered (Sec. 6.7.1), with Petschek's mechanism as a special case.

6.6.2 Non-steady Petschek Reconnection

Time-dependent Petschek-type solutions arise during temporal evolution triggered by a localised resistivity increase in a pre-existing current sheet (Figure 6.17a) (Semenov et al. 1983; Biernat 1987; Rijinbeek and Semenov 1993; Heyn 1996). Reconnection onset launches fast and slow magnetoacoustic waves. In the incompressible case, the fast-mode waves propagate outwards instantaneously and set up an inflow towards the X-point. The inflow decreases with distance and vanishes at infinity. After reconnection ceases, there is a switch-off phase, whose effect propagates outwards.

The inflow near the X-line is super-slow-magnetosonic, and so forms slow-mode shock pairs. The shocks are curved, enclosing the rear portion of a tear-drop-shaped outflow region (Figure 6.17b). In the incompressible limit, the speed at which the outflow disturbance propagates along the x-axis is just v_{A0}, the ambient Alfvén speed, so the external scale-length (L_e) of the system continuously increases with time as $L_e = v_{A0}\, t$. Close to the origin, a steady-state Petschek solution is found.

The inflow in the incompressible case is current-free to first order in $\epsilon(t) = E^*(t)/(v_{A0}B_0)$, where $E^*(t)$ is the electric field at the X-line. In steady-state theory, Petschek's is the only solution which is free of slow-mode effects in the inflow. Petschek's solution is therefore associated with undriven reconnection, while other types, such as flux-pile-up, may be generally associated with driven reconnection.

In the outflow region, the zeroth-order solution is $B_{x0} = 0, B_{y0} = B_0 g(x - v_{A0}\, t), v_{x0} = v_{A0}, v_{y0} = 0$, and the location of the shock transition separating outflow from inflow is $f(x,t) = x\, g(x - v_{A0}\, t)$, where $f(x,t)$ is the height of the shock.

6.7 Fast 2D Reconnection: Other Families

Petschek's mechanism has been generalised in two distinct ways by adopting different boundary conditions to give regimes of *Almost-Uniform Reconnection* (Sec. 6.7.1) and *Non-Uniform Reconnection* (Sec. 6.7.3). Petschek's mechanism is almost-uniform and potential, but these others are non-uniform or non-potential.

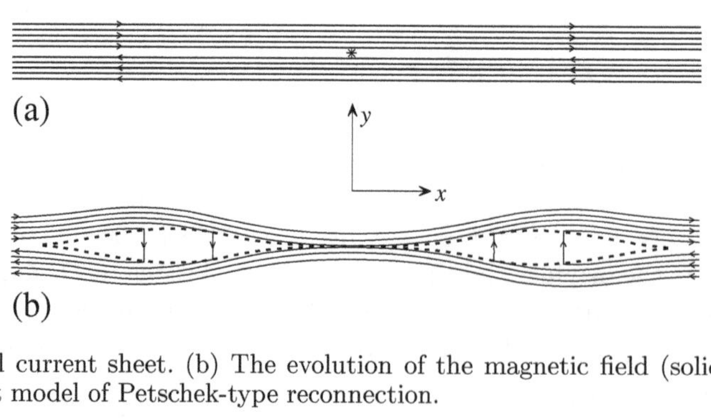

Figure 6.17. (a) An initial current sheet. (b) The evolution of the magnetic field (solid) and shocks (dashed) in Semenov's time-dependent model of Petschek-type reconnection.

The "correct" boundary conditions depend on the particular application; *spontaneous reconnection*, due to some localised instability and unaffected by distant magnetic fields, would require free boundary conditions and would tend to produce potential reconnection; *driven reconnection*, on the other hand, depends on the details of the driving but in general tends to give rise to non-potential reconnection.

Numerical experiments confirm that fast reconnection does indeed exist (Sec. 6.7.2), provided there are appropriate boundary conditions and sources of locally enhanced magnetic diffusivity in the diffusion region, such as current-induced micro-instabilities, which are highly likely in solar coronal, space and astrophysical current sheets. Therefore, fast reconnection, either Almost-Uniform or Non-Uniform, is a prime candidate for rapid energy conversion in solar, space and astrophysical plasmas.

6.7.1 Almost-Uniform Non-potential Reconnection

The inflow region in Petschek's mechanism is a diffuse *fast-mode expansion*, in which the pressure and field strength decrease and the flow converges as the magnetic field is carried in. A *fast-mode* disturbance has plasma and magnetic pressures increasing or decreasing together, while a *slow-mode* disturbance has them changing in the opposite sense. An *expansion* makes the pressure decrease while a *compression* makes it increase, even in the incompressible limit.

Priest and Forbes (1986) wanted to explore different types of inflow and were puzzled at many strange features of numerical reconnection experiments. We sought fast, steady, almost-uniform reconnection solutions to Eqs. (6.18) and (6.19) for 2D, ideal, incompressible flow, where $\boldsymbol{\nabla}\cdot\mathbf{v}=0$, $\boldsymbol{\nabla}\cdot\mathbf{B}=0$, $\mathbf{j}=\boldsymbol{\nabla}\times\mathbf{B}/\mu$ and $\mathbf{E}$ is constant. The object was to determine a relation between the external and inflow Alfvén Mach numbers ($M_e=v_e/v_{Ae}$ and $M_i=v_i/v_{Ai}$) at large distances and at the inflow to the diffusion region (i.e., the top and bottom of the box in Figure 6.18a). The velocity in the inflow region is uniform to lowest order, while the magnetic field strengths (B_e and B_i) are calculated by seeking solutions in powers of the *reconnection rate* ($M_e \ll 1$) in the form $\mathbf{B}=\mathbf{B}_e+M_e\,\mathbf{B}_1+\ldots$, $\mathbf{v}=M_e\,\mathbf{v}_1+\ldots$, representing a small perturbation to a uniform field ($\mathbf{B}_e=B_e\,\hat{\mathbf{x}}$).

If the pressure gradient is negligible at first order, Eq. (6.19) implies $j_1 B_e=0$ or

$$\nabla^2 A_1=0, \tag{6.39}$$

where $(B_{1x},B_{1y})=(\partial A_1/\partial y,-\partial A_1/\partial x)$ and A_1 is a flux function. As boundary conditions (Figure 6.18a), we impose $B_{1x}=0$ on the top boundary and $\partial B_{1y}/\partial x=0$ on the sides ($y=\pm L_e$) (a rough free-floating

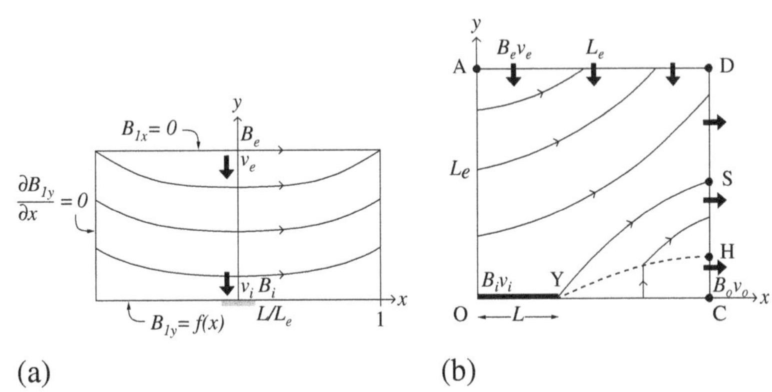

Figure 6.18. Notation for: (a) *Almost-Uniform Reconnection*, where subscripts e and i refer to values at $(0,L_e)$ and at the inflow to the diffusion region (shaded), respectively; and (b) *Non-Uniform Reconnection* with separatrix YS and shock YH. The field is approximately straight with weak curvature in (a) and is X-type with strong curvature in (b).

condition). On the lower boundary ($y=0$), the effect of the diffusion region and shocks are modelled by putting

$$B_{1y}=f(x)=\begin{cases} 2B_N & \text{for} \quad L<x<L_e, \\ 2B_N x/L & \text{for} \quad -L<x<L, \\ -2B_N & \text{for} \quad -L_e<x<-L. \end{cases}$$

Here $B_N \approx M_e B_e$ is the normal field component at the slow-mode shocks.

The required separable solution of Laplace's equation (6.39) for A_1 gives

$$B_{1x}=-\sum_0^\infty a_n \cos\left[\left(n+\tfrac{1}{2}\right)\pi\frac{x}{L_e}\right]\sinh\left[\left(n+\tfrac{1}{2}\right)\pi\left(1-\frac{y}{L_e}\right)\right],$$

$$B_{1y}=\sum_0^\infty a_n \sin\left[\left(n+\tfrac{1}{2}\right)\pi\frac{x}{L_e}\right]\cosh\left[\left(n+\tfrac{1}{2}\right)\pi\left(1-\frac{y}{L_e}\right)\right],$$

where $$A_1=\sum_0^\infty \frac{a_n L_e}{(n+\frac{1}{2})\pi}\cos\left[\left(n+\tfrac{1}{2}\right)\pi\frac{x}{L_e}\right]\cosh\left[\left(n+\tfrac{1}{2}\right)\pi\left(1-\frac{y}{L_e}\right)\right]$$

and $$a_n=\frac{4B_N\sin\left[\left(n+\frac{1}{2}\right)\pi L/L_e\right]}{L/L_e\left(n+\frac{1}{2}\right)^2\pi^2\cosh\left[\left(n+\frac{1}{2}\right)\pi\right]}.$$

This represents a *Petschek-type* solution with a *weak fast-mode expansion*. The first-order flow $[\mathbf{v}_1 = (E/B_0)\hat{\mathbf{y}}]$ is uniform, but the second-order flow is converging. A calculation of B_i/B_e and its substitution into Eq. (6.34) in place of Petschek's original expression (6.37) produces graphs of M_e against M_i for given $R_{me}(=L_e v_{Ae}/\eta)$. These confirm that, for a given R_{me}, there is indeed a maximum reconnection rate ($M_e{}^*$), as Petschek had surmised.

Deriving other people's results in your own way is often a good exercise, so as to understand their motivation and assumptions. While carrying out the above analysis, we realised how to generalise it by adding a pressure gradient to the right-hand side of Eq. (6.39), which becomes a Poisson equation $[\nabla^2 A_1 = -(\mu/B_e)(dp_1/dy)]$. The effect is simply to add a constant (b) to the x-dependent part of each term, so that

$$A_1=-\sum_0^\infty \frac{a_n}{(n+\frac{1}{2})\pi}\left\{b-\cos\left[\left(n+\tfrac{1}{2}\right)\pi\frac{x}{L}\right]\right\}\cosh\left[\left(n+\tfrac{1}{2}\right)\pi\left(1-\frac{y}{L}\right)\right]$$

and $$\left(\frac{M_e}{M_i}\right)^2\approx\frac{4M_e(1-b)}{\pi}\left\{0.834-\log_e\tan\left(\frac{4R_{me}M_e^{1/2}M_i^{1/2}}{\pi}\right)^{-1}\right\}.$$

Introducing the parameter b produces a new family of regimes with a rich diversity of properties. When $b=0$, Petschek's regime (a weak fast-mode expansion) is recovered, whereas, when $b=1$, the inflow field on the y-axis is uniform. The solutions are determined by the nature of the flow on the inflow boundary, since the horizontal flow speed at the corner $(x,y)=(L_e,L_e)$ is proportional to $(b-2/\pi)$. The way that the reconnection rate (M_e) varies with M_i and b for a given R_{me} is shown in Figure 6.19b. When $b>0$, the reconnection rate is faster than the Petschek rate.

When $b<0$, the streamlines (Figure 6.20) near the y-axis converge and tend to compress the plasma, producing a *slow-mode compression*. When $b>1$, the streamlines diverge and tend to expand the plasma, producing a *slow-mode expansion*. We refer to this type as the *flux pile–up regime*, since the magnetic field lines come closer as they approach the diffusion region. The intermediate range ($0<b<1$) gives a hybrid

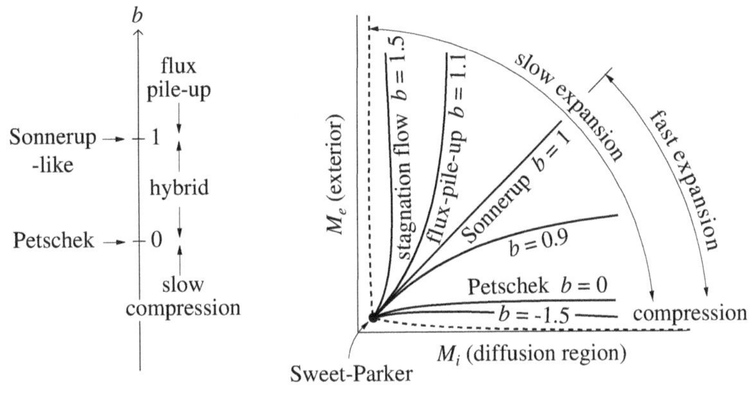

(a) (b)

Figure 6.19. (a) The different regimes of fast reconnection for different values of b. (b) Reconnection rate ($M_e = v_e/v_{Ae}$) as a function of $M_i = v_i/v_{Ai}$ for different b.

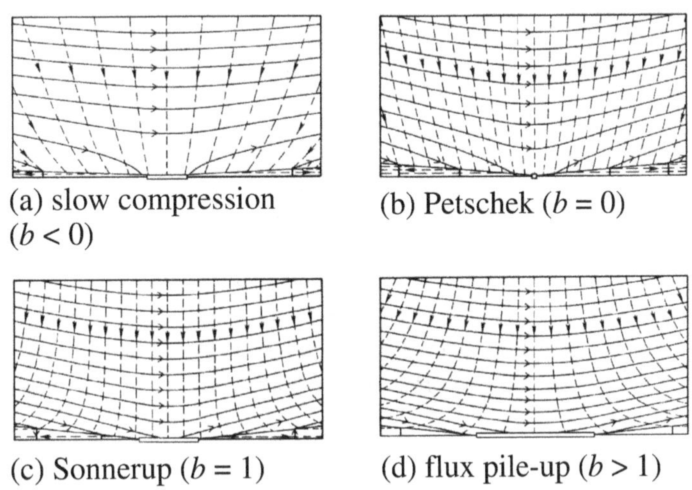

Figure 6.20. Magnetic field lines (solid) and streamlines (dashed) in the upper half-plane for several different regimes of *Almost-Uniform* reconnection (Priest and Forbes 1986).

family of slow- and fast-mode expansions. Another feature is that the central diffusion regions are much larger for flux pile-up than for the Petschek regime.

An important conclusion is that the type and rate of reconnection depend sensitively on the parameter b, characterising inflow conditions. Petschek ($b = 0$) reconnection is just one particular member of a much wider class. When the diffusion region becomes too long, it may become unstable to secondary tearing (Sec. 6.8), and a new regime of *impulsive bursty reconnection* results (Priest 1986; Lee and Fu 1986b); indeed, Sweet-Parker current sheets are unstable to tearing when $R_m > 3\times10^4$ (Loureiro et al. 2007; Bhattacharjee et al. 2009). Almost-Uniform theory has been compared with several numerical experiments (Forbes and Priest 1987).

6.7.2 Early MHD Numerical Experiments

The first numerical experiments to achieve a steady state (Ugai and Tsuda 1977; Sato 1979) helped confirm the reality of the standing slow-mode shocks proposed by Petschek (1964). But it was not until much later that a closer comparison with Petschek's solution was made (Biskamp 1986; Scholer 1989; Yan et al. 1992, 1993).

Biskamp's (1986) reconnection is slow and possesses a Sweet-Parker scaling. Both the diffusion-region length (L) and thickness (l) increase with M_e or R_{me}, which completely contradicts Petschek's result. Thus, despite the presence of slow shocks, Biskamp's numerical experiments had not found the Petschek scaling, and he concluded that Petschek's mechanism, and therefore fast reconnection, does not exist.

However, his boundary conditions are quite different from those in Petschek's solution (Priest and Forbes 1992a), and Petschek's scaling may be produced by adopting the appropriate boundary conditions (i.e., the Strachan-Priest solution, Sec. 6.7.3). Biskamp's runs are shown in Figure 6.21 for $M_e = 0.042$ with $R_{me} = 1746$, 3492, 6984, respectively, in the first, second and third columns. They reveal an increase in diffusion-region length with R_{me} and several other interesting features:

(i) an inflow whose streamlines vary from converging to diverging;
(ii) highly curved inflow field lines with a large shock angle;
(iii) strong jets of plasma flowing out along the separatrices;
(iv) spikes of reversed current at the ends of the diffusion region.

Most of these properties are explained by the models of Sections 6.7.1 and 6.7.3. The family of Almost-Uniform models explains feature (i), while the family of Non-Uniform models reproduces (ii) to (iv) and the appropriate reconnection rate.

Yan et al. (1992) carried out numerical experiments with Petschek's boundary conditions, and they obtained fast reconnection with Petschek's scaling, provided that a non-uniform resistivity is used. By proper choice of boundary conditions, they recovered the entire family of Almost-Uniform solutions.

One aspect of the simulations remained puzzling. When η is uniform, the steady-state solutions are usually no longer sustained. Possible causes were that the slow-mode shocks were not properly resolved numerically or that the boundary conditions were not prescribed well enough. However, Baty et al. (2009a,b) clarified numerically the conditions under which fast reconnection is stable with an enhanced resistivity, and later Forbes et al. (2013) demonstrated that Petschek reconnection with a uniform or reduced resistivity is structurally unstable.

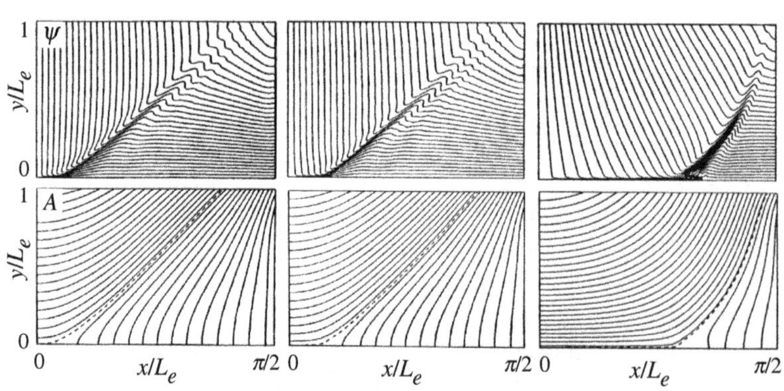

(a) (b) (c)

Figure 6.21. Numerical experiment showing streamlines (top) and magnetic field lines (bottom) in the first quadrant (after Biskamp 1986).

6.7.3 Non-uniform Reconnection Theory

When the inflow magnetic field possesses highly curved field lines, the reconnection is said to be 'Non-Uniform'. *Non-uniform Potential Reconnection* may be solved approximately in three steps, as follows (Priest and Forbes 1990), when the flow (v_e) and field strength (B_e) on the inflow boundary are imposed (Figure 6.18b).

In the upstream region ahead of the slow shock (YH), suppose first that both the plasma speed and sound speed are much smaller than the Alfvén speed. Then the magnetic field is potential and may be written $B_y + iB_x = B_i(z^2/L^2 - 1)^{1/2}$, where $z = x + iy$ and there is a cut (a current sheet) from $z = -L$ to $z = L$. Furthermore, Eq. (6.18) implies that the stream function for the flow velocity is an integral $(\psi = v_e B_e \int ds/B)$ taken along a field line.

As a second step, the position of the shock (YH) that passes through the end-point (Y) of the diffusion region is calculated. The shock relations are applied to deduce the conditions just downstream of the shock (YH). Finally, the MHD equations in the downstream region are solved, subject to the appropriate boundary conditions.

This exhibits many of the properties of Biskamp's experiment. The field lines are highly curved because of the form of the inflow boundary conditions. Reversed current spikes at the ends of the diffusion region are revealed as regions of reversed magnetic curvature, which slow down the inflowing plasma and divert it along the separatrices. Also, separatrix jets occur for long diffusion regions because the inflow to the diffusion region is diverted along the separatrices (YS).

The above potential solution has been generalised to include pressure gradients and produce *Nonpotential Non-Uniform* solutions (Strachan and Priest 1994). In the inflow region, the equation of magnetostatic balance $(\mathbf{j} \times \mathbf{B} = -\boldsymbol{\nabla} p)$ reduces to $\nabla^2 A = -\mu j(A)$. For a uniform current, the magnetic field becomes $B_y + iB_x = B_i(z^2/L^2 - 1)^{\frac{1}{2}} - i\bar{c}(B_e/L)y$, which contains a parameter $\bar{c}$ that is analogous to b in Almost-Uniform theory. If its value is determined by imposing the same value of B_y as Biskamp at the corner $(x, y) = (L_e, L_e)$, then the relation between M_i and M_e reduces to $M_i/M_e \approx 1/(M_e^2 R_{me})$ when $L/L_e \ll 1$, quite different from Almost-Uniform theory. The length L also behaves very differently, since an X-type background field implies that $B_i \to 0$ as $M_e \to 0$, while the background field in the Uniform theory gives $B_i \to B_e$ instead.

When $L = L_e$, the maximum Alfvén Mach number becomes $M_e^* = [2/(\sqrt{5} - 1)]^{3/4}/R_{me}^{1/2} \approx 1.43/R_{me}^{1/2}$. Thus, when the normal magnetic field component is fixed at the inflow boundary, the Strachan-Priest reconnection scales like the slow Sweet-Parker rate, as in Biskamp's experiment.

6.8 Unsteady 2D Reconnection by Resistive Instability

Theory has often focused on steady-state reconnection, since it is far simpler to analyse than non-steady reconnection and many examples of reconnection are either steady or have a quasi-steady component lasting for many Alfvén travel-times. For example, a solar flare has two intermingled components, namely, a "gradual" phase for which the emission slowly rises and falls, and an "impulsive" phase with extremely rapid rises and falls over a second or less (Sec. 1.9.2).

One aspect of time-varying reconnection is *impulsive bursty reconnection* (Priest 1986), in which the reconnection either rapidly switches on and off or rapidly changes between slow and fast reconnection. Such a process can occur in a manner that is either quasi-periodic or random. Its observational consequences on the Sun are impulsive energy release, particle acceleration and generation of waves.

Impulsive bursty reconnection occurs in several ways:

(i) in response to the switching on and off of an enhanced turbulent magnetic diffusivity (η_{turb}) when the current density exceeds a threshold (j_{crit}) for microinstabilities;

(ii) or as part of the nonlinear development of resistive instability when magnetic islands form, fragment and coalesce;
(iii) or due to fragmentation of magnetic reconnection when a current sheet splits up into many interacting reconnection locations (e.g., *SADs*, Sec. 1.9.2.3);
(iv) or when fast reconnection switches on and off near a metastable state and slow-mode shocks form and are shed (Baty et al. 2009b);
(v) or when the diffusion region of a steadily reconnecting field becomes so long that it goes unstable to resistive instability (Priest 1986; Lee and Fu 1986a; Biskamp 1986; Forbes and Priest 1987).

So far, impulsive bursty reconnection has only been described qualitatively. Still needed are its physical properties such as peak current, time between energy-release spikes, net energy release, peak electric field, and their dependency on parameters such as β, j_{crit} and η_{turb}.

A current sheet of width $2l$ tends to diffuse outwards on a time-scale $\tau_d = l^2/\eta$ (Sec. 6.4.2). Such a time is often in practice much longer than time-scales of interest, but Furth et al. (1963) realised that diffusion can couple to magnetic forces and drive three kinds of much faster *resistive instability* that cause current sheets or sheared magnetic fields to reconnect in a time-dependent way.

All three resistive instabilities occur when $\tau_d \gg \tau_A$, where $\tau_A = l/v_A$ is the Alfvén travel-time across the sheet. They grow on time-scales $\tau_d^{(1-\lambda)}\tau_A^{\lambda}$, where $0 < \lambda < 1$, and create many small-scale magnetic loops (or flux ropes in 3D) or current filaments, which subsequently diffuse away. These instabilities are often overlooked in solar physics, when numerical computations have insufficient resolution to allow them, but they may play an important role in coronal filamentation, diffusion and heating or in magnetoconvection.

Gravitational and *rippling modes* are driven by gradients in density $[\rho_0(x)]$ or resistivity $[\eta(x)]$ in a direction (x) across a sheet. They have short wavelengths, of the order of the sheet width $(kl \simeq 1)$. Their main role is to create fine-scale filamentary structure in coronal loops, prominences and sunspots. This may be responsible for turbulent diffusion of plasma across the magnetic field, with accompanying heating. Their growth-rates (ω_g and ω_r, respectively) are

$$\omega_g = \left(\frac{(kl)^2\,\tau_A^2}{\tau_d\,\tau_G^4}\right)^{1/3}, \qquad \omega_r = \left[\left(\frac{d\eta_0}{dx}\frac{l}{\eta_0}\right)^4 \frac{(kl)^2}{\tau_d^3\tau_A^2}\right]^{1/3},$$

where $\tau_G = (-g/\rho_0\, d\rho_0/dx)^{-1/2}$ is the gravitational time-scale. A non-uniform diffusivity $[\eta_0(x)]$ may arise from a temperature variation from one field line to another.

The *tearing-mode instability* has a long wavelength, much greater than the width of the sheet $(kl < 1)$, and so it may have more global consequences. Its growth-rate is $\omega = [\tau_d^3\,\tau_A^2\,(kl)^2]^{-1/5}$ for wave-numbers (k) in the range $(\tau_A/\tau_d)^{1/4} < kl < 1$. The smallest allowable wavelengths (l) grow in a time $\tau_d^{3/5}\tau_A^{2/5}$, whereas the longest wavelength has the fastest growth-rate, namely,

$$\omega = \left(\frac{1}{\tau_d\,\tau_A}\right)^{1/2}.$$

Diffusion is important only in a narrow layer of width $\epsilon l = (kl)^{-3/5}(\tau_A/\tau_d)^{-2/5} l$.

Starting with a 1D sheet with straight field lines and then making a perturbation such as in Figure 6.22, the resulting forces make the perturbation grow. The magnetic tension tends to pull the new loops of field to left and right away from the X-points, while the magnetic pressure gradient tends to push plasma towards them from above and below. The large-scale field line curvature does produce a restoring magnetic tension force, but this is minimised for long wavelengths.

Resistive modes also occur in a sheared magnetic field, since the stability analysis is unaffected by the addition of a uniform field normal to the plane of Figure 6.22. Sheared fields are, in general, resistively unstable at many thin sheaths throughout a structure. At any specified location, the instability has a vector

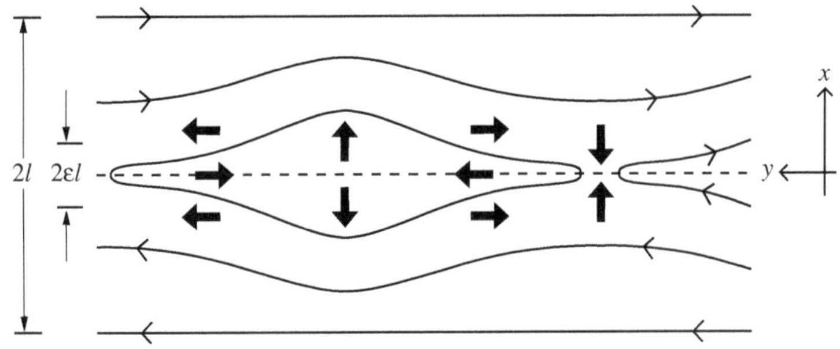

Figure 6.22. Magnetic field lines (light arrows) and plasma velocity (solid arrows) for the tearing-mode instability when a current sheet or sheared field of width $2l$ is perturbed from an initial state consisting of straight field lines at rest.

wavenumber ($\mathbf{k}$) in a direction normal to the equilibrium field ($\mathbf{B}_0$), that is,

$$\mathbf{k}\cdot\mathbf{B}_0 = 0.$$

This section first presents the linear stability analysis of the tearing mode (Sec. 6.8.1) and then various modifications (Sec. 6.8.2), such as effects of a flux tube geometry and nonlinear development (Sec. 6.8.3).

6.8.1 Tearing-Mode Instability (Furth et al. 1963)

Begin with a static plasma in an equilibrium sheared magnetic field $\mathbf{B}_0 = B_{0y}(x)\,\hat{\mathbf{y}} + B_{0z}(x)\,\hat{\mathbf{z}}$, whose field lines are confined to the yz-planes and rotate with x. Make small perturbations to this initial state in the form $\mathbf{v}_1(x)\exp[i(k_y y + k_z z) + \omega t]$, $\mathbf{B}_1(x)\exp[i(k_y y + k_z z) + \omega t]$, satisfying the equations of induction and vorticity for an incompressible plasma with a uniform diffusivity (η), namely:

$$\frac{\partial \mathbf{B}}{\partial t} = \nabla\times(\mathbf{v}\times\mathbf{B}) + \eta\nabla^2\mathbf{B}, \tag{6.40}$$

$$\mu\rho\frac{d}{dt}(\nabla\times\mathbf{v}) = \nabla\times[(\nabla\times\mathbf{B})\times\mathbf{B}]. \tag{6.41}$$

Define dimensionless variables $\bar{\mathbf{B}} = \mathbf{B}/B_0$, $\bar{\mathbf{v}}_1 = -\mathbf{v}_1 ikl^2/\eta$, $\bar{k} = kl$, $\bar{\omega} = \omega l^2/\eta$, $\bar{x} = x/l$, in terms of a typical field strength (B_0) and scale-length (l), and use $\nabla\cdot\mathbf{B} = 0$ and $\nabla\cdot\mathbf{v} = 0$ to eliminate v_{1y} and B_{1y}. The linearised x-component of Eq. (6.40) and the z-component of Eq. (6.41) are then

$$\bar{\omega}\bar{B}_{1x} = -\bar{v}_{1x} f + (\bar{B}''_{1x} - \bar{k}^2\bar{B}_{1x}), \tag{6.42}$$

$$\omega(\bar{v}''_{1x} - \bar{k}^2\bar{v}_{1x}) = R_m^2\,\bar{k}^2 f[-\bar{B}_{1x} f''/f + (\bar{B}''_{1x} - \bar{k}^2\bar{B}_{1x})], \tag{6.43}$$

where a prime denotes a derivative with respect to $\bar{x}$ and $f = \mathbf{k}\cdot\bar{\mathbf{B}}_0/k$.

The magnetic Reynolds number ($R_m = lv_A/\eta = \tau_d/\tau_A \gg 1$) is assumed to be so large that plasma is frozen to the magnetic field almost everywhere and the diffusion term ($\eta\,\nabla^2\bar{B}_{1x}$) in Eq. (6.40) is negligible. The exception is in thin sheets, where $\nabla\times(\mathbf{v}\times\mathbf{B})$ vanishes (i.e., where f vanishes or $\mathbf{k}\cdot\mathbf{B}_0 = 0$). In such a sheet (of width $\sim 2\epsilon l$), the field lines diffuse through the plasma and reconnect. If the centre of the sheet is located at $x = 0$ and $k_z = 0$, the condition $\mathbf{k}\cdot\mathbf{B}_0 = 0$ reduces to $B_{0y} = 0$. Solutions to Eqs. (6.42) and (6.43) are obtained in an outer region ($|x| > \epsilon$) and in an inner region ($|x| < \epsilon$), and are then patched together (or, in a more precise analysis, mathematically matched) at their boundary to give the growth-rate.

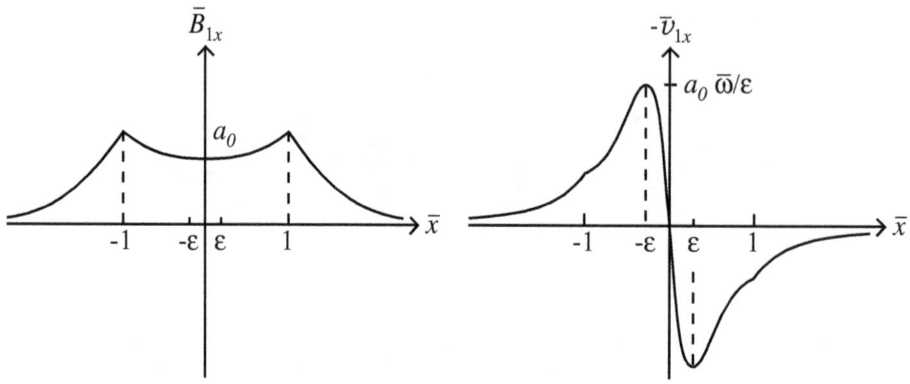

Figure 6.23. A sketch of the perturbed magnetic field ($\bar{B}_{1x}$) and velocity ($\bar{v}_{1x}$) for the tearing mode, as functions of distance ($\bar{x}=x/l$) across a current sheet. The half-widths of the sheet and the inner diffusive layer are l and ϵl, respectively.

In the *outer region*, terms of order R_m^{-1} are negligible compared with unity, and Eqs. (6.42) and (6.43) become

$$\begin{aligned}\bar{\omega}\bar{B}_{1x} &= -\bar{v}_{1x}\bar{B}_{0y}+(\bar{B}''_{1x}-\bar{k}^2\bar{B}_{1x}),\\ 0 &= -\bar{B}_{1x}\bar{B}''_{0y}/\bar{B}_{0y}+(\bar{B}''_{1x}-\bar{k}^2\bar{B}_{1x}).\end{aligned} \tag{6.44}$$

Given an equilibrium profile for $\bar{B}_{0y}(\bar{x})$, Eq. (6.44) can be solved for $\bar{B}_{1x}$. For instance, with a simple step-profile ($\bar{B}_{0y}=1$ for $\bar{x}>1$, $\bar{x}$ for $|\bar{x}|<1$, and -1 for $\bar{x}<-1$), the solution for $\bar{x}>0$ that vanishes at large distances is

$$\bar{B}_{1x}=\begin{cases} a_1\sinh\bar{k}\bar{x}+b_1\cosh\bar{k}\bar{x} & \bar{x}<1,\\ a_0\exp(-\bar{k}\bar{x}) & \bar{x}>1,\end{cases} \tag{6.45}$$

where continuity of $\bar{B}_{1x}$ at $\bar{x}=1$, and an integral of Eq. (6.44) across $\bar{x}=1$, give $a_1=a_0[\exp(-\bar{k})(\cosh\bar{k}/\bar{k})-1]$ and $b_1=a_0[1-\exp(-\bar{k})(\sinh\bar{k}/\bar{k})]$ for $\bar{x}>0$.

These conditions at $\bar{x}=-1$ give the same b_1 for $\bar{x}<0$ but the sign of a_1 is reversed. Thus, although $\bar{B}_{1x}$ is continuous at $\bar{x}=0$, its slope is not, so resistivity becomes important in an *inner region* or *singular layer*. The jump in $\bar{B}'_{1x}/\bar{B}_{1x}$ across the singular layer, denoted by Δ' and called *delta prime*, is given from Eq. (6.45) as

$$\Delta'=\left[\frac{\bar{B}'_{1x}}{\bar{B}_{1x}}\right]_{0-}^{0+}=\frac{2a_1\,\bar{k}}{b_1}.$$

In the *inner region*, $\bar{B}_{0y}$ is so small that diffusion and inertia are important in Eqs. (6.42) and (6.43). Its width is of order ϵl, where $\epsilon^4=\bar{\omega}/(4\bar{k}^2R_m^2)$. In terms of new variables $X=\bar{x}/\epsilon$, $V_{1x}=\bar{v}_{1x}(4\epsilon/\bar{\omega})$, Eqs. (6.42) and (6.43) become

$$\ddot{\bar{B}}_{1x}=\epsilon^2\bar{k}^2\bar{B}_{1x}+\epsilon^2\bar{\omega}(\bar{B}_{1x}+\tfrac{1}{4}V_{1x}X), \tag{6.46}$$

$$\ddot{V}_{1x}=V_{1x}(\bar{k}^2\epsilon^2+\tfrac{1}{4}X^2)+\bar{B}_{1x}X, \tag{6.47}$$

where dots denote X-derivatives. In order of magnitude, Eq. (6.46) implies $\ddot{\bar{B}}_{1x}\sim\epsilon^2\bar{\omega}\bar{B}_{1x}$ for $\bar{k}\ll 1$, and so $\Delta'=2(\bar{B}'_{1x}/\bar{B}_{1x})_{x=\epsilon}=2[\dot{\bar{B}}_{1x}/(\epsilon\bar{B}_{1x})]_{X=1}\sim\epsilon\bar{\omega}$.

A key assumption is the so-called *constant-psi approximation*, namely, that $\bar{B}_{1x}$ be constant in the inner region. It enables Eq. (6.46) to be solved to give $\Delta'=3\epsilon\bar{\omega}$ when $\bar{k}\epsilon\ll 1$. The outer and inner regions may be patched (Figure 6.23) by equating the above two expressions for Δ', to give $\bar{\omega}=[(8R_m)/(9\bar{k})]^{2/5}$.

Two conditions restrict the size of $\bar{k}$. Firstly, $\bar{B}'_{1x}/\bar{B}_{1x}$ must be positive at the edge of the diffusion region, which implies that $\bar{k} < 0.64$ from Eq. (6.45). Secondly, the condition ($\epsilon|\bar{B}'_{1x}/\bar{B}_{1x}| < 1$) that $\bar{B}_{1x}$ does not change significantly in the internal region implies $\bar{k} > 2[8/(3R_m^2)]^{1/8}$. These give an instability growth-rate (ω) between the diffusion ($\omega_d = \eta/l^2$) and Alfvén ($\omega_A = v_A/l$) rates.

Eqs. (6.46) and (6.47) have also been solved when $\bar{B}_{1x}$ is not constant in $|x| < \epsilon$. The fastest-growing mode has a growth-rate $\bar{\omega}_{max} \simeq 0.6\,R_m^{1/2}$ and a wave-number of $\bar{k}_{max} \simeq 1.4\,R_m^{-1/4}$. Thus, the growth-time $[\tau_m = (\tau_d\,\tau_A)^{1/2}]$ is the geometric mean of the diffusion and Alfvén times, so that long narrow islands are formed.

6.8.2 Extensions to the Basic Tearing-Mode Analysis

The basic analysis has been developed to include effects of diffusion, compressibility, steady flow (Bulanov and Sasorov 1978; Dobrowolny et al. 1983; Ofman et al. 1993), a transverse field component (Bulanov et al. 1979; Somov and Vernata 1994; DeBenedetti and Pegoraro 1995), a formal asymptotic matching (Baldwin and Roberts 1972) and triggering by fast-mode waves (Sakai 1983). When two singular surfaces are close together, the linear growth is greatly enhanced in so-called *double tearing*, which scales like $R_m^{1/3}$ (Rechester and Stix 1976; Pritchett et al. 1980).

The theory has also been extended to a cylindrical flux tube (e.g., Furth et al. 1973; Bateman 1978; Wesson 1997), where ideal modes grow fastest, but resistive modes have lower instability thresholds (see Sec. 12.2.6). Suppose the tube has *major radius* R and *minor radius* $a \ll R$, with poloidal and toroidal field components $B_p(r)$ and $B_\phi(r)$ that depend on distance (r) from the magnetic axis. Key parameters are the normal and poloidal plasma betas ($\beta = 2\mu p/B^2$ and $\beta_p = 2\mu p/B_p^2$) and the amount $[\Phi_T(r) = 2\pi\,R\,B_p/(r\,B_\phi)]$ by which a field line is twisted around the axis in going from one end of the tube to the other. (For a loop of length L, replace R by $L/(2\pi)$.) A related quantity is the *safety factor* $q(r) = 2\pi/\Phi_T = r\,B_\phi/(R\,B_p)$. Thus, $q = 1$ means the field line twists once around the major axis of a torus.

A radial perturbation of the form $\boldsymbol{\xi} = \boldsymbol{\xi}(r)\exp[i(m\theta - n\phi)]$ produces a single helix if $m = 1$ or a double helix if $m = 2$. The *resonant surface* occurs at a radius (r_s) where $\mathbf{k}\cdot\mathbf{B} = 0$, namely $kB_\phi(r_s) + (m/r_s)B_p(r_s) = 0$. Since $k = n/R$, this implies that $q(r_s) = m/n$. On this *rational surface*, the orientation of the perturbation matches that of the field, so that crests and troughs of the helix follow the field lines.

The ideal modes for such a flux tube include *kink modes* when $q(a) < 1$ or $\Phi_T(a) > 2\pi$ and *interchange modes* (Secs. 7.5.1 to 7.5.3). The resistive form of the kink mode is a *tearing mode*, driven by the current gradient but now occurring when $q_a > m$, so that the resonant surface ($r = r_s$) lies inside the tube. The mode is unstable when $\Delta' = [(1/B_{r1})dB_{r1}/dr]_{r_s-\epsilon}^{r_s+\epsilon} > 0$; its growth-time behaves like $\tau_d^{3/5}\tau_A^{2/5}$.

The resistive form of an internal mode with $m > 1$ is called a *resistive interchange* (or resistive-g) mode with a growth-time of $\tau_d^{1/3}\tau_A^{2/3}$. Pressure gradients may also drive *resistive ballooning modes* (e.g., Strauss 1981). The $m = 1$ *internal resistive kink mode* becomes unstable when $q_0 < 1$. As the twist is increased, the flux tube first becomes tearing-mode unstable and later it becomes unstable to the ideal mode.

For solar flares, an important extra effect is *line-tying* of the ends of coronal magnetic fields in the dense photosphere. In order of magnitude, an ideal kink mode needs $2\pi/k \le L$, so that a wavelength ($2\pi/k$) can fit into a loop of length L. But k satisfies $\mathbf{k}\cdot\mathbf{B}_0 = 0$, and so this condition reduces to $\Phi_T \ge 2\pi$ in terms of the twist $[\Phi_T = LB_p/(rB_\phi)]$. A precise analysis gives instead a threshold for ideal instability of $\Phi_T = 2.5\pi$ for a field of uniform twist (Hood and Priest 1981). During its nonlinear development, a flux tube may become highly kinked, with many reconnecting current sheets (Browning et al. 2008; Hood et al. 2009).

6.8.3 Nonlinear Development of Tearing

In its nonlinear development, tearing can follow several pathways, depending on the value of R_m, the equilibrium and the boundary conditions. A full treatment is still needed. Tokamak physicists have emphasised *nonlinear saturation*, while solar flare theorists have stressed a rapid nonlinear phase, which may develop from linear tearing (with free boundary conditions) or may be driven by an ideal instability.

For standard tearing, the constant-psi approximation holds, and the instability quickly nonlinearly saturates (Rutherford 1973). When the island width exceeds the resistive layer width (ϵl) and nonlinear terms are included in the singular layer, the linear exponential growth is replaced by a slower algebraic growth ($\sim t^2$) of the perturbed magnetic field and of the island width ($w = 4(-A_1/B'_{0y})^{1/2}$) like t. The growth eventually saturates at an island width of order $\Delta'_1(0)l^2$ (White et al. 1977).

The Rutherford scaling has been confirmed numerically for $m = 2$ tearing (Park et al. 1984), but nonlinear growth of $m = 1$ tearing is quite different from higher m-modes. Since ψ is no longer constant in the island, Rutherford's analysis fails. Kadomtsev (1975) gave instead a heuristic argument that reconnection continues to grow at the linear rate until the current density becomes flat inside the $q = 1$ surface.

Nonlinear coupling of tearing modes of different pitch in cylindrical geometry has been followed numerically (Waddell et al. 1976; Diamond et al. 1984). The interaction of modes can enhance the growth of some modes, drive others unstable, and extend the region of island activity across a significant fraction of the minor radius, thereby increasing greatly the heat transport across the toroidal field. Overlapping of modes can produce high-m turbulence via an energy cascade.

A chain of magnetic islands created by tearing can also be subject to an ideal *coalescence instability* (Finn and Kaw 1977; Longcope and Strauss 1993). Neighbouring islands coalesce because parallel currents at the O-points attract one another. Numerical computations find that nonlinear coalescence depends on the island amplitude (ϵ) and R_m (Prichett and Wu 1979; Biskamp 1982).

The sequence of events expected in a solar current sheet with a large R_m is as follows. If the sheet is shorter than the fastest-growing mode in an infinite medium, tearing favours the longest-wavelength mode that fits. If it is longer, the sheet first tears linearly at the wavenumber ($1.4\ R_m^{-1/4}/l$) of the fastest growing mode, and then primary coalescence combines neighbouring islands. This may lead to a fast reconnection regime (such as Petschek or flux pile-up), often disrupted by impulsive bursty reconnection due to secondary tearing and coalescence of the diffusion region. However, in a current sheet that is line-tied to the photosphere at one end, dominant tearing may develop at the neutral point closest to the surface. Rather than coalesce with its neighbour, this neutral point develops fast nonlinear reconnection, which undergoes secondary tearing and coalescence.

6.9 3D Reconnection: Geometrical Structures

Null points and separatrix surfaces are key features of 3D topology. Separatrices originate at null points or bald patches, and in a complex field they form a web of surfaces called a *topological skeleton* (Priest et al. 1996). This can be calculated in a robust way (Haynes and Parnell 2010) and is crucial for understanding the structure of the field and the nature of many types of reconnection (Longcope 2005).

Complex magnetic configurations have many components (Figure 6.24). Laminar field lines lie on flux surfaces, but chaotic field lines are space-filling. Laminar regions have a skeleton of separatrix surfaces (Sec. 6.9.2) and a *quasi-skeleton* of quasi-separatrix layers (QSLs) (Sec. 6.9.5). Pairs of separatrices intersect in *separator curves* (Lau and Finn 1990) and QSLs intersect in *quasi-separators* (sometimes called *hyperbolic flux tubes*). The term *structural skeleton* (Titov 2007; Titov et al. 2009) refers to the sum of the topological and quasi-skeleton (i.e., both the separatrices and QSLs) and it is best identified by Titov's Q-factors (Titov 2007; Titov et al. 2009).

Reconnection tends to occur wherever strong currents concentrate as sheets and so rapidly dissipate. This occurs especially near null points, and along separators and quasi-separators (Secs. 6.13.3 and 6.13.4), but also on separatrices and QSLs.

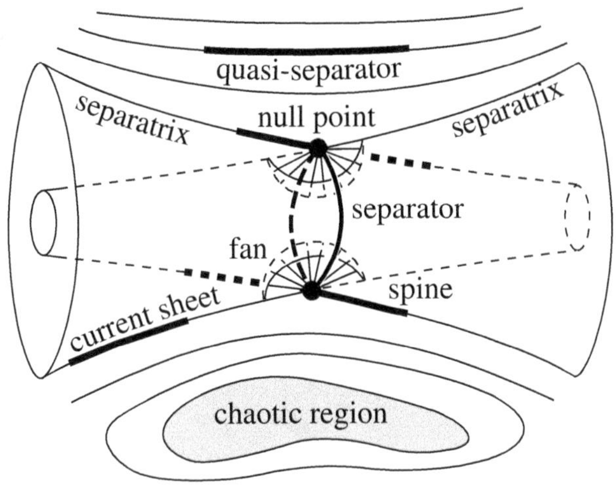

Figure 6.24. The components of a complex magnetic field, containing both laminar and chaotic (shaded) regions. Separatrix surfaces intersect in separator curves, which link null points. The null point regions contain spine curves (which lie on the separatrices) and fan surfaces (which spread out from a null to become separatrices).

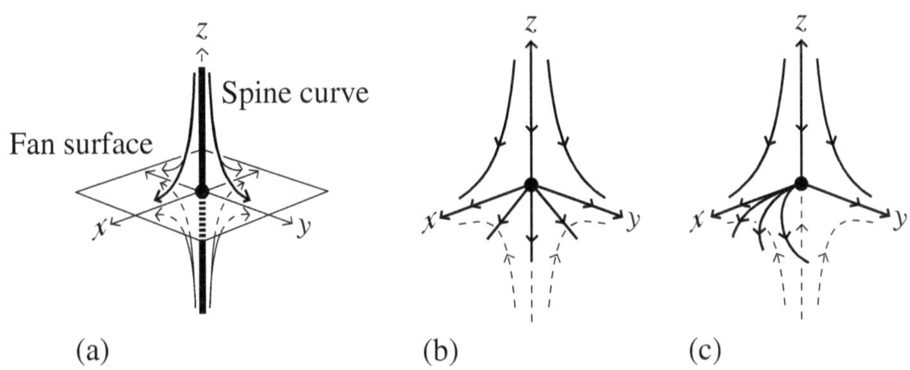

Figure 6.25. (a) Structure of the field lines near a 3D null point with a spine and fan, (b) a proper radial null and (c) an improper radial null with $a > 1$.

6.9.1 3D Null Points

The magnetic field vanishes at a null point, and the field near a *linear null point* increases linearly with distance from it. The simplest example has field components

$$(B_x, B_y, B_z) = (x, y, -2z),$$

or in cylindrical polars $(B_r, B_\theta, B_z) = (r, 0, -2z)$, so that $\nabla \cdot \mathbf{B} = 0$ is satisfied identically. The resulting field lines satisfy $dx/B_x = dy/B_y = dz/B_z$ and have a completely different structure from nulls in 2D. The field lines are given by intersections of the two sets of surfaces $y = Cx$ and $z = K/x^2$.

Two families of field lines, the *spine* and *fan* (Priest and Titov 1996), link to a linear null (Lau and Finn 1990). A spine curve is an isolated field line in Figure 6.25a approaching or receding from the null along the z-axis. Its neighbouring field lines form two bundles that spread out as they approach the fan surface (the xy-plane).

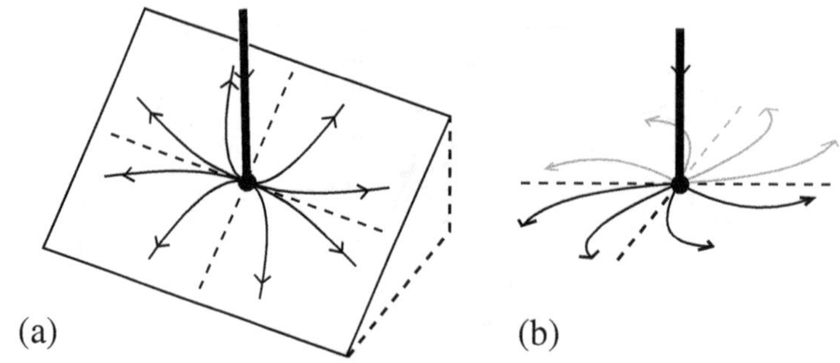

Figure 6.26. (a) Skeleton of an oblique null ($j_\perp \neq 0$) and (b) a spiral null, for which $j_\parallel$ exceeds a critical value. (Nearby field lines are similar to those in Figure 6.25.)

This null is known as a *proper radial null*, since its fan field lines are straight (Figure 6.25b). It is the $a = 1$ case from a wider class of null point fields

$$(B_x, B_y, B_z) = (x, ay, -(a+1)z).$$

$a \neq 1$ gives the generic case of an *improper radial null* with curved fan field lines, which touch the y-axis when $0 < a < 1$ and the x-axis when $a > 1$ (Figure 6.25c).

The magnetic field of a general linear null contains nine constants, three for each field component. By using $\nabla \cdot \mathbf{B} = 0$, normalising and rotating axes, they may be reduced to four constants, namely, $a, b, j_\parallel, j_\perp$ such that (Parnell et al. 1996)

$$\begin{pmatrix} B_x \\ B_y \\ B_z \end{pmatrix} = \begin{pmatrix} 1 & \frac{1}{2}(b - j_\parallel) & 0 \\ \frac{1}{2}(b + j_\parallel) & a & 0 \\ 0 & j_\perp & -a-1 \end{pmatrix} \begin{pmatrix} x \\ y \\ z \end{pmatrix}. \tag{6.48}$$

If the current ($j_\perp$) normal to the spine is non-zero, the fan surface is inclined to the spine at an angle not equal to $\frac{1}{2}\pi$, which we call an *oblique null* (Figure 6.26a). If the current ($j_\parallel$) along the spine exceeds a critical value, the eigenvalues of the matrix in Eq. (6.48) are no longer all real and we term it a *spiral null*, with the field lines in the fan spiralling into or out of the null (Figure 6.26b).

6.9.2 Separatrices and Separators

Most coronal magnetic fields enter or leave the corona through the photosphere in locations that may be considered sources or sinks of magnetic flux as far as the corona is concerned. When the corona is modelled as a 2D region, it contains *separatrix curves*, field lines that separate the plane into topologically distinct regions, in the sense that all the field lines in one region start at a particular source and end at a particular sink (Figure 6.27a,b).

Two types of 2D separatrix are possible. One originates at an *X-point*, where the field vanishes and is locally hyperbolic, while the other simply touches the boundary in what is called a *bald patch* (Titov et al. 1993). Reconnection occurs by the breaking and rejoining of field lines at an X-point and the transferring of flux across the separatrices from one topological region to another.

Complex configurations in 3D have similar properties, with *separatrices* being surfaces across which the mapping of field lines jumps discontinuously. The separatrices are of two types: the first is a *separatrix fan surface*, which is an extension of the fan surface of a null point (examples of which are shown in Figure 6.32); and the second is a *bald-patch separatrix* (or *separatrix touching surface*), which touches a boundary in a curve known as a *bald patch* (or *touch curve*), as described in Section 6.9.4 and shown in Figure 6.33.

Separatrices intersect in a *separator*, a special field line that ends at null points or on the boundary (Figure 6.27c) (Longcope and Cowley 1996; Priest and Titov 1996; Longcope 2005; Parnell et al. 2010). Isolated null points (far from other nulls) can be destroyed or created in pairs (N_1 and N_2, say) by a local separator bifurcation (Sec. 6.9.3) in the generic case (Figure 6.28a). The spines (S_1 and S_2) of each null lie in the fans (F_2 and F_1, respectively) of the other null. The fans form sheets which intersect in a *single separator* field line, going from one null to the other, and the fans are finite in the transverse direction, being bounded by the spines of the opposite nulls. An example of a sequence of fields (satisfying $\nabla \cdot \mathbf{B} = 0$)

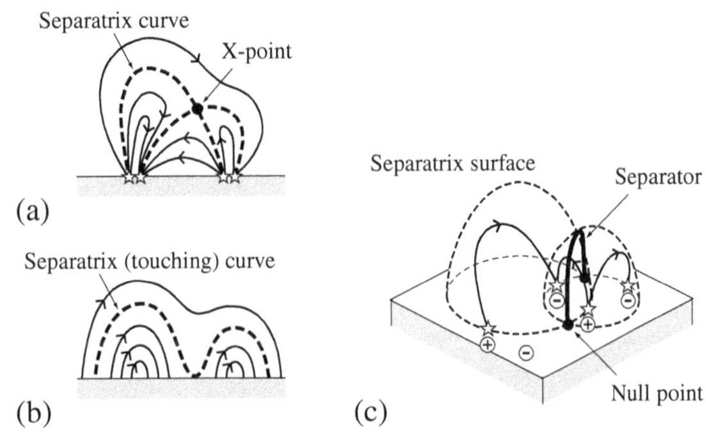

Figure 6.27. 2D separatrix curves that (a) intersect at an X-point or (b) touch the boundary in a bald patch. (c) 3D separatrix surfaces (due to four starred sources) that intersect in a separator (and that may instead touch the boundary).

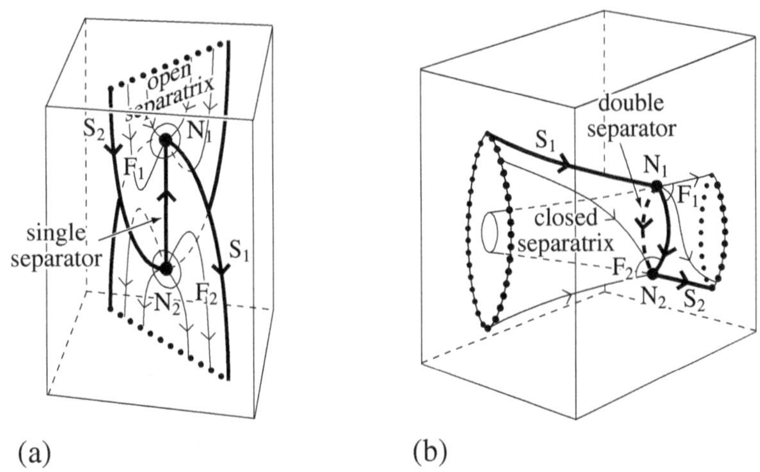

Figure 6.28. The topological structure of the field near a pair of nulls (N_1 and N_2, with spines S_1 and S_2 and fans F_1 and F_2). In case (a) the fans of the two null points are *open separatrices* that intersect the boundary in non-closed curves (indicated by dotted lines) and intersect each other in a *single separator* line; they are bounded by the spines of the other nulls. This is the generic case for the creation or destruction of a separator by a local bifurcation. In case (b) the fans are *closed separatrices* that intersect the boundary in closed curves and each other in a *double separator* forming a closed curve; the spines of the other nulls lie on them but do not bound them: they just represent the only field lines that connect to the other nulls.

in which such a null pair is created has the form (Gunnar Hornig, private correspondence)

$$(B_x, B_y, B_z) = [-x(1+z), y(1-z), z^2 - t + 1 - e^{-4(x^2+y^2)}].$$

A pair of nulls can, however, be joined by extra separators that are added by a global bifurcation. For instance, a *double separator* (Figure 6.28b) consists of a pair of separators, both going from one null to the other. An example of a sequence of fields in which a null pair and a double separator are created by a global separator bifurcation that brings them in from infinity is

$$(B_R, B_\phi, B_z) = [Rz + R\cos\phi, -2R\sin\phi, (tR^2 - z^2 - 2)].$$

Figure 6.28 illustrates another distinction, namely, between *open separatrices* and *closed separatrices.* Closed separatrices form closed flux surfaces or surfaces that intersect the boundary of a domain in closed curves (e.g., Figure 6.28b), and so they separate the volume into topologically different regions, in the sense that all the field lines in one region join one domain on the boundary to another domain. Open separatrices, on the other hand, are open surfaces that are bounded by spines and intersect the boundary in non-closed curves, as exemplified in Figure 6.28a. The field lines on both sides of such a separatrix occupy the same region, but their mapping has a discontinuity as the separatrix is crossed. For closed separatrices, reconnection (which by definition changes the magnetic connectivity of plasma elements) can transfer flux across a separator from one region to another. For open separatrices, there is also flux transfer across the separator, but it goes from one part of a region to another part of the same region.

Null-point pairs can also be created near pre-existing nulls in several ways. One example is given by the field

$$(B_x, B_y, B_z) = [-x(3z^2 + t + 1), (t+2)y, z(z^2 - 1)],$$

for which there is a single null at the origin when $t < 0$ (Brown and Priest 2001) (Figure 6.29a). When t increases through $t = 0$, the original null moves to $z = +\sqrt{t}$ and a new pair of nulls is created by a local double-separator bifurcation at $z = 0, -\sqrt{t}$ (Sec. 6.9.3). Tiny separators join the lateral nulls to the central null. The fan of the central null is a narrow open separatrix, bounded by the spines of the other two nulls (Figure 6.29b). Such a situation occurs commonly in the corona when the open separatrix is bounded by spines from two nulls whose fans form a separatrix dome (Figure 6.29c, Sec. 6.9.4, Sarah Platten, private correspondence).

Null points are always created in pairs. In many numerical experiments, such as those on coronal heating or flux emergence, null points can be highly numerous and form like beads on a string or chain, joined in multiple ways by short separators (Haynes and Parnell 2010; Parnell et al. 2010). Thus, separators tend

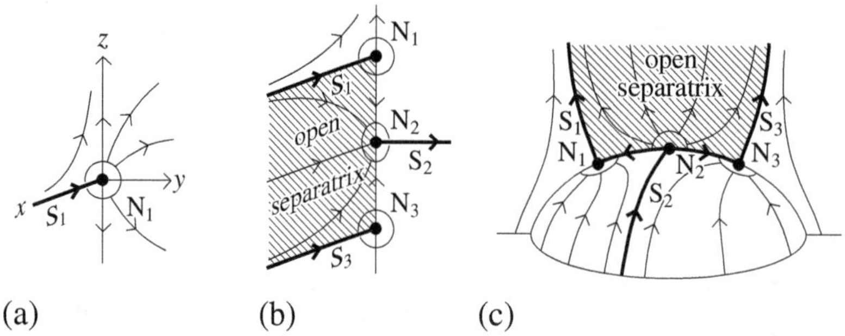

Figure 6.29. The topological structure of the field (a) before and (b) after the creation of a new pair of nulls (N_2 and N_3) near a pre-existing null (N_1), with corresponding spines (S_i) and and fans (F_i). In (b) a narrow open separatrix is created from the fan F_2 bounded by spines S_1 and S_3. (c) A common situation in the corona where an open separatrix is created when the fans F_1 and F_3 form a dome that arches down to the solar surface.

to occur as either tiny *intracluster separators* joining nulls within a cluster or as much longer *intercluster separators* linking to distant nulls or joining separate clusters.

Several complementary tools are useful in finding topology or quasi-topology. Topology can be revealed from changes of footpoint connectivity (e.g., Parnell and Galsgaard 2004; Haynes et al. 2007; Longcope et al. 2007) and they may be displayed in a *connectivity map*, where regions with different connectivity are coloured differently (Parnell et al. 2010). This reveals the presence of closed separatrices but not open separatrices or QSLs. On the other hand, a map of the discontinuities in footpoint connectivity by calculating the skeleton from nulls and bald patches shows up both open and closed separatrices (Titov et al. 2011). Instead, the squashing factor (Q, Sec. 6.9.5) reveals both separatrices and QSLs but does not distinguish precisely between them, apart from the separatrices appearing generally as thinner structures in practice (e.g., Titov 2007; Masson et al. 2012; Titov et al. 2012).

6.9.3 Bifurcations

If the essential features of the topology of a magnetic field (such as its skeleton) are unaffected by any small change in the field, the field is said to be *structurally* or *topologically stable*. Conversely, if an arbitrary change in a magnetic field causes a change in topology, the field is *structurally unstable*. Such a change in topology is known as a *bifurcation* of the magnetic field.

A *local bifurcation* involves a change in the number or nature of null points. In 2D or 3D, isolated linear nulls are structurally stable, but *null lines* or *null sheets* (consisting of curves or surfaces where the field vanishes) are structurally unstable, since a small perturbation may break them up into a series of nulls. More general null points are themselves structurally unstable if they are *degenerate* [i.e., when the Jacobian matrix ($D\mathbf{B}$) in Eq. (6.48) is singular at the null] or if the null points are of second or higher order (i.e., when $D\mathbf{B}$ vanishes at the null).

A *global bifurcation* in 2D involves a change in connectivity of the separatrix field lines. At the moment of bifurcation, there exists either a *homoclinic* separatrix (starting and ending at the same null) or a *heteroclinic* separatrix (linking one null to another). However, homoclinic and heteroclinic field lines are structurally unstable in 2D and so do not survive the bifurcation.

In 3D, on the other hand, a *global bifurcation* can involve the creation and continued existence of a separator, or the destruction of such a separator. Separators that are (heteroclinic) field lines linking two nulls and represent the intersection of two fan surfaces are structurally stable.

Linear null points may coalesce at a second-order null (for which the field increases quadratically from zero) or a second-order null may split and give birth to linear nulls. Such local bifurcations may produce global changes of magnetic topology.

In 2D, the field components may be written as $dx/ds = B_x = \partial A/\partial y, \quad dy/ds = B_y = -\partial A/\partial x$. The flux function (A) is a Hamiltonian and so the nature of the bifurcations is well understood (e.g., Priest et al. 1996). In 3D, the system ($dx/ds = B_x$, $dy/ds = B_y$, $dz/ds = B_z$) is no longer Hamiltonian (but it is conservative since $\nabla \cdot \mathbf{B} = 0$). The null points and their bifurcations are therefore much more complex. A first step has been to consider fields with cylindrical symmetry, for which a general categorisation exists (Priest et al. 1996).

6.9.4 Topological Skeletons in the Solar Corona

The solar corona is incredibly complex, with myriads of magnetic flux sources that poke through the photosphere into the overlying atmosphere. The photospheric magnetic field is concentrated by convection in many intense flux tubes, and each such photospheric source is itself joined through the corona to many other sources. Similarly, whenever a numerical MHD experiment gives rise to many null points, the fans of the nulls spread out to form a complex web of separatrices.

As mentioned earlier, a powerful way to understand the topology of complex fields is to construct the *skeleton* of the field, namely, the set of separatrix surfaces that originate both in the fans of null points and in bald patches (Priest et al. 1996). This is particularly useful when analysing numerical MHD experiments – for example, Haynes et al. (2007) were only able to understand how 3D reconnection was occurring after they had constructed the evolution of the skeleton (see Sec. 10.4.5.4).

As an example, the coronal skeleton due to two unbalanced photospheric sources consists of a separatrix surface in the shape of a dome encircling the weaker source (Figure 6.30). From this may be built up the topological properties of the building blocks of complex fields created by a finite number of sources. For example, topologies due to three sources are shown in Figure 6.31.

This *magnetic charge topology* approach, which approximates groups of magnetic flux as point sources, is commonly adopted for modelling coronal magnetic field arising from photospheric flux distributions. Some of the resulting photospheric null points (and separators joining them) are real or approximate weak-field regions: for instance, distinct clumps of flux in the original flux distribution separated by weak-field regions. However, when the field is modelled more accurately, some of the resulting nulls and separators become quasi-nulls and quasi-separators (Sec. 6.9.5), while others are spurious and have no remnants at all.

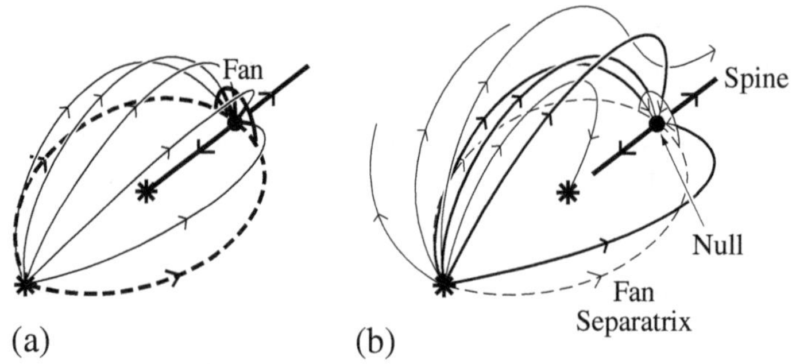

Figure 6.30. (a) The separatrix dome skeleton of the coronal field due to two unbalanced sources. (b) The field lines above and below the separatrix dome.

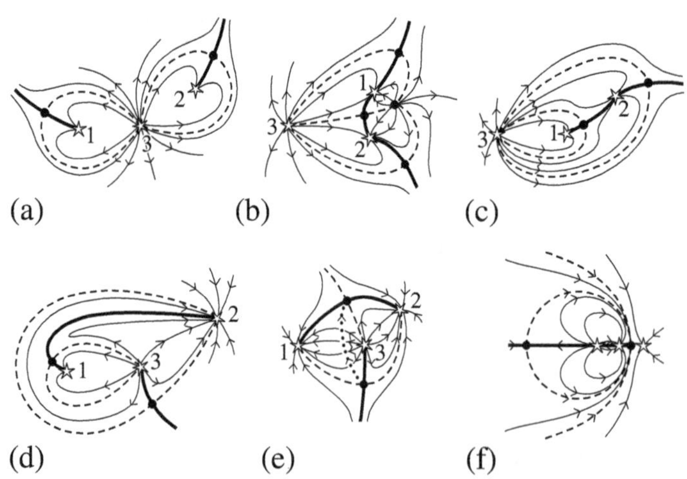

Figure 6.31. Coronal topologies from two negative (1 and 2) and one positive (3) source. The sources are indicated by stars, the nulls by dots, the spines by thick solid curves, the fans by dashed curves and a separator by a dotted curve. Local bifurcations along the top line change from separate to touching to enclosed states. Global bifurcations along the bottom line change from nested to intersecting to detached states (from Priest et al. 1997).

In the corona, the skeleton includes several important elements, which make up, for example, *pseudo-streamers* (Secs 1.6.1 and 13.9.3; Figure 13.11; Titov et al. 2011):

- *separatrix domes* spread out from the fans of coronal nulls and close down to the photosphere, enclosing a region of parasitic polarity (Figures 6.32a);
- *separatrix curtains* also originate as the fans of coronal nulls, but they are more vertical and form open sheets that project out into the solar wind, either as closed separatrices or open separatrices (Figures 6.32b);
- *bald-patch separatrices* touch the solar surface at bald patches (Figures 6.33);
- *streamer separatrices* separate closed field of bipolar regions from open field;
- *closed separators* are the intersections of separatrix curtains and either separatrix domes or bald-patch separatrices; and
- *open separators* are the intersections of separatrix curtains and streamer separatrices.

A bald patch is a segment of a polarity-inversion line (PIL) where the direction of the horizontal field is abnormal in the sense that it is directed from the negative polarity side. Bald patches are rather numerous in practice (Slava Titov, private correspondence). They tend to occur when the polarity-inversion line of a highly sheared field snakes through an active region, or below a large-scale magnetic flux rope (e.g., around

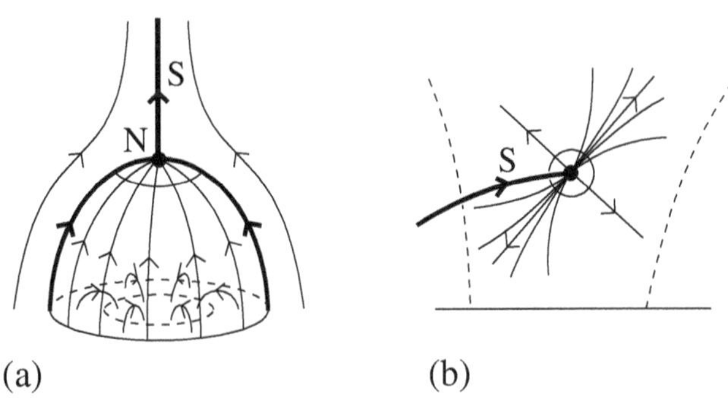

Figure 6.32. Two of the three building blocks of coronal topology arising from coronal nulls with spines (S) and fans (indicated by a circle): (a) a separatrix dome formed when the null-point fan closes down to the solar surface and (b) a separatrix curtain, when the null-point fan is open into the high corona.

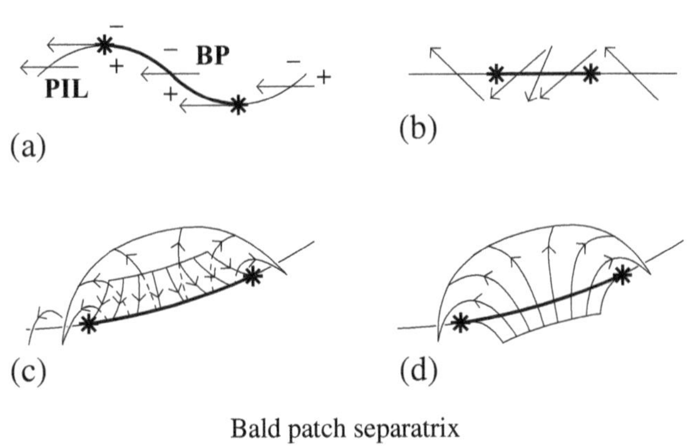

Figure 6.33. The third building block of coronal topology, namely a bald-patch separatrix surface, showing (a) the field line directions across a curved polarity-inversion line (PIL), a segment of which is a bald patch (BP), with a strong overlying field having a uniform direction (b) the field line directions when the PIL is straightened out, and the field lines in the bald-patch separatrix surface viewed (c) from the side and (d) from above.

a prominence), or when the large-scale background field is oppositely directed to and stronger than the small-scale field that is trying to close down across a polarity-inversion line.

6.9.5 Quasi-Skeletons

Skeletons are important since the magnetic connectivity of plasma elements jumps discontinuously across separatrix surfaces, and reconnection can occur at the intersections of separatrices, namely separators. However, quasi-skeletons are equally important: they consist of *quasi-separatrix layers* (QSLs), across which the connectivity has a steep but finite gradient. QSLs, which are remnant separatrices, intersect in *quasi-separators*, at which reconnection may take place just as readily as at separators. The concept of *quasi-separator reconnection*, first proposed by Priest and Démoulin (1995), built on earlier ideas of a *singular field line* (Priest and Forbes 1989) and of *magnetic flipping* (Priest and Forbes 1992b). Magnetic flipping is the core process inside a QSL whereby magnetic field lines flip past one another in opposite directions. Application to solar flares was soon developed (Sec. 12.3.2.2), where magnetic flipping in quasi-separator reconnection was instead called *slip-running reconnection* by Aulanier et al. (2006). Furthermore, Aulanier et al. (2005) found that current sheets tend to form spontaneously along quasi-separators in response to any smooth and large-scale footpoint motion.

2D reconnection at a null point is associated with a discontinuity in the field-line mapping from one footpoint to another. For the simple X-point field

$$B_x = x, \quad B_y = -y, \tag{6.49}$$

the point (x_0, y_0) on one part of a boundary will map to (x_1, y_1), say, on another part in such a way that, when (x_0, y_0) crosses a separatrix, the point (x_1, y_1) suddenly jumps in location (Figure 6.34a).

In 3D, such a mapping discontinuity is also present at the separatrix surfaces that spread out from nulls or bald patches. When nulls or bald patches are absent, there are no separatrices and the mapping of one footpoint to another is continuous (Schindler et al. 1988). However, Priest and Démoulin (1995) realised that, even in the absence of separatrices, there often exists a remnant structure known as quasi-topology consisting of a web of quasi-separatrix layers.

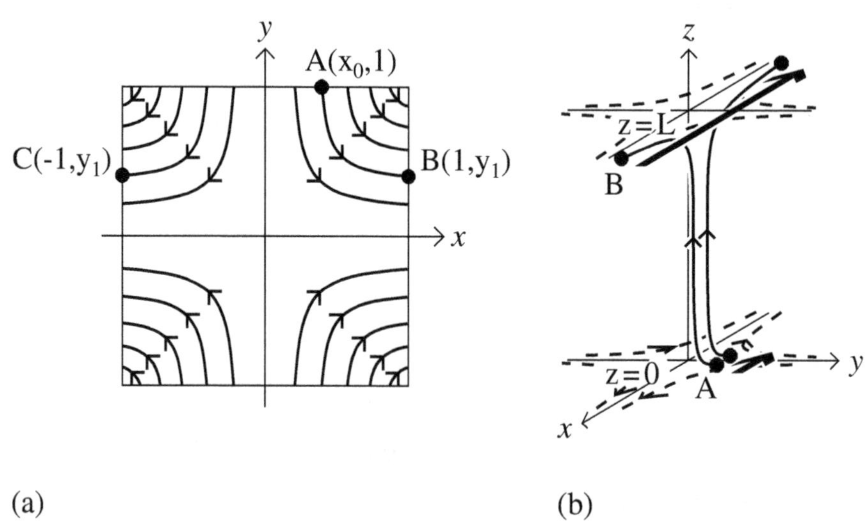

Figure 6.34. Mapping of footpoints for (a) a 2D X-field from the top or bottom boundary of a square to the side boundary and (b) a 3D sheared X-field from $z = 0$ to $z = L$.

When investigating the nature of a 3D configuration, the first step is to surround it by a closed surface S. Then the mapping of field-line footpoints from one part of S to another is calculated. For instance, if a small component ($B_z = l \leq 1$) is added to Eq. (6.49) to create a sheared X-field, the mapping becomes continuous, so that, as the point A(x_0, y_0) crosses the y-axis in the plane $z = 0$, the other end B(x_1, y_1) in the plane $z = 1$ moves continuously (Figure 6.34b).

Next, so-called *quasi-separatrix layers* are identified as sheets where the gradients of the mapping are very large. This definition of a QSL involves a mapping to a boundary and therefore refers to global properties of a configuration. The concept of a QSL may be defined formally as follows. Split the surface S into parts S_0 and S_1 where the field lines enter and leave the volume, respectively, and set up orthogonal coordinates (u, v) in S and w normal to S. Then field lines map (u_0, v_0) in S_0 to (u_1, v_1) in S_1. Next, form the *displacement gradient tensor*

$$\mathcal{F} = \begin{pmatrix} s_1 \partial u_1/\partial u_0 & s_2 \partial u_1/\partial v_0 \\ s_3 \partial v_1/\partial u_0 & s_4 \partial v_1/\partial v_0 \end{pmatrix}$$

from the gradients of the mapping functions $u_1(u_0, v_0)$ and $v_1(u_0, v_0)$ and the scaling factors s_i and evaluate the *norm*

$$N = \sqrt{\left(s_1 \frac{\partial u_1}{\partial u_0}\right)^2 + \left(s_2 \frac{\partial u_1}{\partial v_0}\right)^2 + \left(s_3 \frac{\partial v_1}{\partial u_0}\right)^2 + \left(s_4 \frac{\partial v_1}{\partial v_0}\right)^2}. \tag{6.50}$$

Finally, define a quasi-separatrix layer as the region where $N \gg 1$.

Note that a surface element dS_0 transforms to $dS_1 = \mathcal{J}\, dS_0$, where $\mathcal{J} = s_1 s_4\, (\partial u_1/\partial u_0)(\partial v_1/\partial v_0) - s_2 s_3 (\partial u_1/\partial v_0)(\partial v_1/\partial u_0)$ is the *Jacobian*. Thus, flux conservation ($B_1 dS_1 = B_0 dS_0$) implies that $B_1 = B_0/\mathcal{J}$, where $\mathcal{J}$ is finite and non-zero if the field has no nulls or singularities. Also, the displacement gradient tensor may be written as the product $\mathcal{F} = \mathcal{F}_R \mathcal{F}_0$ of one matrix ($\mathcal{F}_R$) representing a rotation through an angle and another ($\mathcal{F}_0$) representing a stretching by λ_+ (the largest eigenvalue) along $\mathbf{e}_+$ (the corresponding eigenvector) together with a compression by λ_- (the other eigenvalue) along $\mathbf{e}_-$ (the other eigenvector). A quasi-separatrix layer (where $N \gg 1$) is therefore associated with a large expansion along one direction and a large compression along the other, such that N is approximately equal to the largest eigenvalue $N \approx \lambda_{max}$. For instance, see PROBLEM 6.8.

Titov (2007) discovered a better way of diagnosing a QSL by normalising N^2 to give a *squashing factor*

$$Q = -\frac{N^2}{B_{z-}^*/B_{z+}}, \tag{6.51}$$

where N is calculated by mapping along a field line from one end where the normal field is B_{z+} to the other end where it is B_{z-}^*. Then Q is independent of the direction of mapping and a QSL is a layer where $Q \gg 1$. If a flux tube has one footpoint rooted in a small circular patch and the other in a distorted ellipse, then Q measures the aspect ratio of the ellipse. Titov et al. (2009) extended the concept further by defining slip-forth and slip-back squashing factors (Q_{sf} and Q_{sb}), which identify the *reconnection fronts* of field lines that have been or are about to be reconnected in an evolving configuration. For the nature of reconnection at QSLs, see Section 6.13.4.

6.10 3D Reconnection: Magnetic Helicity

Magnetic helicity is a topological quantity lying at the core of 3D reconnection. It measures the twisting and kinking of a flux tube (referred to as *self-helicity*) and the linkage between different flux tubes (referred to as *mutual helicity*). It is a global topological invariant that cannot be changed in an ideal medium and decays very slowly (over the global magnetic diffusion time τ_d) in a weakly resistive medium. To a first approximation, magnetic reconnection on times much smaller than τ_d cannot destroy magnetic helicity but only convert it from one form to another. Thus, when a complex configuration undergoes reconnection, conservation of magnetic helicity provides an important constraint on the final state. During 3D reconnection, there is, however, a tiny change in magnetic helicity whose presence is essential to the occurrence of reconnection (Sec. 6.12.6). In addition, other concepts have been developed, such as the *topological degree*, *topological entropy* and *topological flux function* (Yeates et al. 2010; Yeates and Hornig 2013), which are important in measuring topology and their changes during reconnection (Sec. 10.4.4).

Woltjer (1958) suggested that the lowest-energy state that conserves the total magnetic helicity is a linear force-free field. Taylor (1974) considered its role in laboratory machines called reversed-field pinches, which he suggested are so turbulent that magnetic surfaces break down by multiple reconnection and spread the magnetic helicity uniformly throughout the configuration to create a linear force-free state. Moffatt (1978) developed its basic properties in relation to dynamo theory.

Heyvaerts and Priest (1984) were the first to realise the importance of magnetic helicity for the solar corona. They developed Taylor's theory to allow magnetic flux to thread the photosphere and suggested that coronal magnetic fields are in a state of MHD turbulence with energy fed into the corona by photospheric motions, so that the corona is continually heated by turbulent reconnection as it evolves through a series of linear force-free states. They also suggested that, if the magnetic helicity becomes too great, it may be expelled in a coronal mass ejection. Its importance is now widely accepted (Berger 1999; Pevtsov et al. 2001, 2003). Berger and Field (1984) developed a theory for relative helicity and its evolution. Also, the Heyvaerts-Priest theory of coronal relaxation has been generalised to include constraints other than magnetic helicity (Amari and Luciani 2000), such as the topological degree of the mapping from one footpoint to another (Yeates et al. 2010; Yeates and Hornig 2011; Pontin et al. 2011).

The continual motion of photospheric magnetic footpoints in the solar atmosphere tends to build up the magnetic helicity until it is ejected by magnetic eruptions. Thus, magnetic helicity in the atmosphere can change either by the emergence of twisted or linked structures from the solar interior or by twisting (or untwisting) footpoint motions of coronal loops. Magnetic helicity of a coronal loop can be decreased by the detachment and ejection of magnetic structures from the solar surface (as in prominence eruptions) (Figure 6.35). This section gives definitions of magnetic helicity (Sec. 6.10.1), its evolution (Sec. 6.10.2) and simple examples (Sec. 6.10.3).

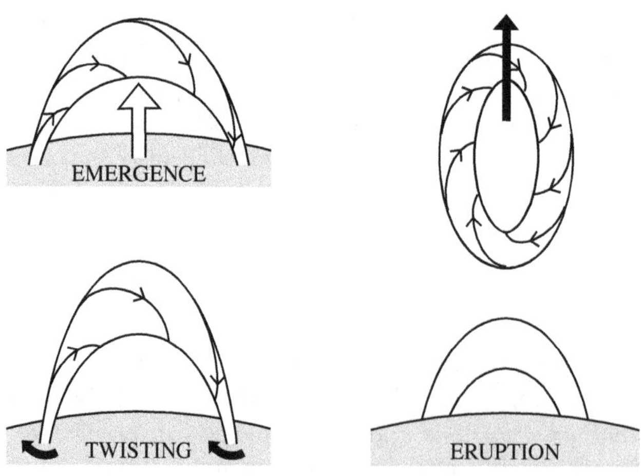

Figure 6.35. Magnetic helicity changes due to flux emergence, twisting or an eruption.

6.10.1 Equations for Magnetic Helicity

Magnetic helicity in a closed volume (V), bounded by a surface S, is defined as

$$H_0 = \int_V \mathbf{A} \cdot \mathbf{B} \, dV, \tag{6.52}$$

where $\mathbf{A}$ is the vector potential. H_0 is gauge invariant if the volume is magnetically closed, but not if magnetic field lines enter or leave it. As a proof, suppose $\mathbf{A}$ is replaced by $\mathbf{A} + \nabla\Phi_A$ (which does not change $\mathbf{B} = \nabla \times \mathbf{A}$), so that H_0 changes by

$$\Delta H_0 = \int_V \nabla\Phi_A \cdot \mathbf{B} \, dV = \int_V \nabla \cdot (\Phi_A \mathbf{B}) \, dV,$$

using $\nabla \cdot \mathbf{B} = 0$. Then the divergence theorem for a simply connected volume implies

$$\Delta H_0 = \int_S \Phi_A \mathbf{B} \cdot \mathbf{n} \, dS,$$

where $\mathbf{n}$ is the unit normal to S, pointing out of V_0. Gauge invariance ($\Delta H_0 = 0$) follows if $\mathbf{B} \cdot \mathbf{n} = 0$ on the surface S, as required.

A gauge-invariant *relative magnetic helicity*,

$$H = \int_{V_\infty} \mathbf{A} \cdot \mathbf{B} - \mathbf{A}_0 \cdot \mathbf{B}_0 \, dV, \tag{6.53}$$

was therefore proposed by Berger and Field (1984) as an alternative to H_0. Here $\mathbf{B}_0 = \nabla \times \mathbf{A}_0$ is a field that is potential inside V, is the same as $\mathbf{B}$ outside V and satisfies $\mathbf{A} \times \mathbf{n} = \mathbf{A}_0 \times \mathbf{n}$ on S. They realised that the difference in helicity integrated over all space (V_∞) of any two fields that differ only inside V is independent of the field outside V, and that a particularly useful reference field inside V is a potential field, since it is completely determined by $\mathbf{B} \cdot \mathbf{n}$ on S.

6.10.2 Time-Variation of Magnetic Helicity

The rate of change of magnetic helicity is

$$\frac{dH}{dt} = -2 \int_V \mathbf{E} \cdot \mathbf{B} \, dV + 2 \int_S \mathbf{A}_p \times \mathbf{E} \cdot \mathbf{n} \, dS,$$

where a gauge ($\mathbf{A}_p$) has been chosen that satisfies $\nabla \cdot \mathbf{A}_p = 0$ and has $\mathbf{A}_p \cdot \mathbf{n} = 0$ on S. If there is no slippage on the boundary and the resistive Ohm's Law ($\mathbf{E} = -\mathbf{v} \times \mathbf{B} + \mathbf{j}/\sigma$) holds, this becomes

$$\frac{dH}{dt} = -2 \int_V \mathbf{j} \cdot \mathbf{B}/\sigma \, dV + 2 \int_S (\mathbf{B} \cdot \mathbf{A}_p)(\mathbf{v} \cdot \mathbf{n}) - (\mathbf{v} \cdot \mathbf{A}_p)(\mathbf{B} \cdot \mathbf{n}) \, dS. \tag{6.54}$$

Here the internal helicity dissipation is represented by the first term and the flow of helicity across the boundary by the surface integral.

Several useful deductions follow. First, if the magnetic field varies on a length-scale (L), equating dH/dt to the first term with $j \sim A/(\mu L^2)$ implies that the time-scale for magnetic helicity dissipation is the global diffusion time ($\tau_d = L^2/\eta$). Second, on time-scales much shorter than τ_d, changes of helicity are given by the surface term. Thus, for example, if the volume is closed, with $\mathbf{B} \cdot \mathbf{n} = \mathbf{v} \cdot \mathbf{n} = 0$, the helicity is conserved. But, if instead footpoint motions along the surface are prescribed, the resulting injection or extraction of helicity may be deduced as follows.

For a plane surface ($z = 0$), Eq. (6.54) gives the magnetic helicity evolution as

$$\frac{dH}{dt} = 2 \int\int (\mathbf{B} \cdot \mathbf{A}_p) v_z - (\mathbf{v} \cdot \mathbf{A}_p) B_z \, dx dy, \tag{6.55}$$

in the ideal limit, where the integration is over the xy-plane. The first term involves only motions in the z-direction and thus represents the effect of the emergence of structures carrying helicity through the surface. The second term involves only motions in the xy-plane and represents injection by footpoint shuffling of helicity into fields already present in the volume.

Consider a plane surface S threaded by the footpoints of N thin flux tubes of magnetic flux $F_{m(i)}$. If the motion of each footpoint consists of a translation plus a uniform rotation at a rate ω_i, the rate of helicity change is

$$\frac{dH}{dt} = -\frac{1}{2\pi}\left[\sum_{i=1}^{N}\omega_i F_{m(i)}^2 + \sum_{i=1}^{N}\sum_{j=1}^{N}\dot{\theta}_{ij}F_{m(i)}F_{m(j)}\right], \tag{6.56}$$

where $\dot{\theta}_{ij}$ is the time derivative of the relative angle (θ_{ij}) between footpoints i and j (Berger 1984). The first term arises because footpoint rotation injects one unit of twist into a flux tube every $2\pi/\omega$ seconds. The second term measures the rate at which footpoints circle each other and braid the flux tubes.

6.10.3 Magnetic Helicity of Simple Structures

The helicity of a magnetic configuration may be deduced from Eq. (6.56) when twisting and linking is built up from an initially potential field that has widely separated, untwisted tubes. The resulting total helicity consists of the self-helicity (H_s) of each tube due to its own internal twist, and the mutual helicity (H_m) due to linking of the tubes. Thus, from Eq. (6.56)

$$H = \sum_{i=1}^{N} H_{si} + \sum_{\substack{i,j=1\\ i<j}}^{N} H_{mij},$$

where the ith flux tube has *self-helicity* $H_{si} = \Phi_{T_i}\ F_{m(i)}^2/(2\pi)$ in terms of its twist Φ_{T_i} (or $2\pi\times$ number of turns) and magnetic flux $F_{m(i)}$, and the *mutual helicity* is

$$H_{mij} = 2L_{ij}\ F_{m(i)}F_{m(j)} \tag{6.57}$$

in terms of the *linking number* L_{ij}, which is described as follows.

The linking number is a topological parameter describing two curves that does not change as the curves are distorted without crossing through each other. Each curve is given a direction, and reversing one of the directions changes L_{ij} by -1. Each crossing has a sign $+1$ or -1, depending on whether the first curve is in front of or behind the other, and the linking number is just half the sum of the signed crossings.

The helicity of a straight flux tube of uniform twist is

$$H = \frac{\Phi_T}{2\pi}F_m^2. \tag{6.58}$$

More generally, if the axis of the tube is itself twisted or kinked, then the self-helicity is increased by the *writhing number* due to the shape of the axis.

Consider a coronal arcade of flux F_m per unit length, uniformly sheared by S per unit length (Figure 6.36b). If it consists of a cylindrical shell of small thickness w, radius r, with field components B_θ and B_z in cylindrical coordinates (r,θ,z), then by Eq. (6.58) the magnetic helicity of the arcade (half a flux tube) is $H = \Phi_T F_{m(0)}^2/(4\pi)$, where $\Phi_T = B_\theta L/(B_z r)$ and $F_{m(0)} = 2\pi r w B_z$ is twice the axial flux along the arcade. By putting $L=1$ and noting that the axial field component is $B_z = B_\theta S/(\pi r)$, the magnetic helicity per unit length of the arcade reduces to

$$H = SF_m^2,$$

where $F_m = wB_\theta$ is the (azimuthal) magnetic flux of the arcade per unit length.

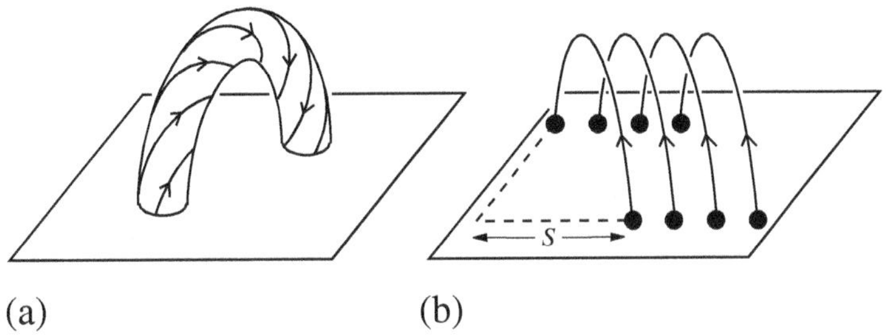

Figure 6.36. The helicity of (a) a twisted flux tube is $H = \Phi_T F_m^2/(2\pi)$ and of (b) a sheared arcade is $H = SF_m^2$ per unit length, where F_m is magnetic flux, Φ_T is twist and S is shear.

6.11 3D Reconnection: Flux and Field-Line Conservation

Before tackling magnetic reconnection in 3D, some subtleties of flux and field-line conservation need to be described, expanding on Section 2.6.2. In ideal MHD, flux and field lines are both conserved and the flux and field-line velocities are equal to the plasma velocity (Sec. 6.11.1). In nonideal MHD, flux and field-line conservation are no longer equivalent (Hornig and Schindler 1996), and neither flux velocity nor field-line velocity is unique (Sec. 6.11.2).

These are part of a deeper concept of electromagnetic flux conservation (Sec. 6.11.3), in which the different kinds of magnetic field evolution may be categorised in a way that distinguishes among conservation of electromagnetic flux, magnetic flux and magnetic field lines (Hornig 2001).

6.11.1 An Ideal Plasma

For an ideal plasma, the induction equation becomes

$$\frac{\partial \mathbf{B}}{\partial t} = \boldsymbol{\nabla} \times (\mathbf{v} \times \mathbf{B}), \tag{6.59}$$

and Ohm's Law reduces to

$$\mathbf{E} + \mathbf{v} \times \mathbf{B} = \mathbf{0}. \tag{6.60}$$

In this large magnetic Reynolds number limit, Alfvén's theorem (Sec. 2.6.2) implies:

(i) magnetic flux is conserved;
(ii) magnetic field lines are conserved;
(iii) magnetic topology is conserved.

Furthermore, the components of plasma velocity ($\mathbf{v}_\perp$), flux velocity ($\mathbf{w}_\perp$) and field-line velocity ($\mathbf{w}_{L\perp}$) perpendicular to the magnetic field are all equal:

(iv) $\mathbf{v}_\perp = \mathbf{w}_\perp = \mathbf{w}_{L\perp} = \dfrac{\mathbf{E} \times \mathbf{B}}{B^2}.$

Magnetic Flux Conservation means that plasma elements initially forming a flux tube will continue to do so (Figure 2.3). By contrast, *Magnetic Field Line Conservation* signifies that two plasma elements initially joined by a field line will always remain joined (see Figure 2.4 and PROBLEM 6.9).

6.11.2 A Non-ideal Plasma

Non-ideal MHD considers in general an Ohm's Law of the form

$$\mathbf{E}+\mathbf{v}\times\mathbf{B}=\mathbf{N}, \tag{6.61}$$

where the non-ideal term on the right is written $\mathbf{N}=\eta\boldsymbol{\nabla}\times\mathbf{B}$ for classical resistivity.

6.11.2.1 *Flux Conservation*

A magnetic field variation is *flux-preserving* if there exists a magnetic flux velocity ($\mathbf{w}$) satisfying

$$\frac{\partial\mathbf{B}}{\partial t}=\boldsymbol{\nabla}\times(\mathbf{w}\times\mathbf{B}). \tag{6.62}$$

However, for an Ohm's Law of the form (6.61), Faraday's equation implies (Sec. 2.6.3.2)

$$\mathbf{N}=\mathbf{u}\times\mathbf{B}+\boldsymbol{\nabla}\Phi, \tag{6.63}$$

in terms of a *slippage velocity* $\mathbf{u}=\mathbf{v}-\mathbf{w}$ and a potential (Φ).

The case when $\boldsymbol{\nabla}\times(\mathbf{u}\times\mathbf{B})=\mathbf{0}$ is particularly simple, since the condition for flux conservation (6.63) reduces (by including any curl-free part of $\mathbf{N}$ in $\boldsymbol{\nabla}\Phi$) to

$$\mathbf{N}=\boldsymbol{\nabla}\Phi \qquad \text{or} \qquad \boldsymbol{\nabla}\times\mathbf{N}=\mathbf{0}.$$

The $\boldsymbol{\nabla}\Phi$ term in Eq. (6.63) is necessary in general because, in a non-ideal plasma, $\mathbf{E}$ may possess a component ($E_{\|}$) parallel to $\mathbf{B}$. Indeed, it is this component which is intimately connected with 3D reconnection (Sec. 6.12.1).

6.11.2.2 *Flux Velocity*

Equations (6.61) and (6.63) imply that

$$\mathbf{E}+\mathbf{w}\times\mathbf{B}=\boldsymbol{\nabla}\Phi, \tag{6.64}$$

in which a general expression for the flux velocity (assuming $\mathbf{u}\cdot\mathbf{B}=0$) is (Sec. 2.6.3.2)

$$\mathbf{w}=\mathbf{v}+\frac{(\mathbf{N}-\boldsymbol{\nabla}\Phi)\times\mathbf{B}}{B^2}. \tag{6.65}$$

This immediately suggests that ($\mathbf{w}$) may become infinite at a null point.

The scalar product of Eq. (6.64) with $\mathbf{B}$ implies that $\mathbf{B}\cdot\boldsymbol{\nabla}\Phi=\mathbf{E}\cdot\mathbf{B}$, which (unless there are problems with nulls or boundary conditions) integrates to

$$\Phi(\mathbf{r},t)=\int_{\mathbf{r}_0}^{\mathbf{r}}\mathbf{E}\cdot d\mathbf{s}+\Phi_0(\mathbf{r}_0,t). \tag{6.66}$$

This integral is evaluated along a field line, and Φ_0 is the value of Φ at some reference surface ($\mathbf{r}_0$). Since the reference potential (Φ_0) is an arbitrary function of $\mathbf{r}_0$ and t, $\Phi(\mathbf{r},t)$ is not unique, and so the flux velocity ($\mathbf{w}$) is also not unique in general.

The assumption $(\mathbf{w}-\mathbf{v})\cdot\mathbf{B}=0$ is a matter of convenience, because the displacement of points along a field line is in general arbitrary. When there is no component of electric field parallel to $\mathbf{B}$ and Ohm's Law is resistive, Eq. (6.65) reduces to

$$\mathbf{w}=\mathbf{v}+\frac{\mathbf{j}\times\mathbf{B}}{\sigma B^2},$$

since Eq. (6.66) implies that $\boldsymbol{\nabla}\Phi=\mathbf{0}$ if $\Phi_0=0$ at $\mathbf{x}_0$.

A necessary condition for the validity of the definition (6.66) is that

$$\Phi = -\oint \mathbf{E}\cdot dl = 0$$

along any closed field line in the plasma. This is necessary to avoid Φ becoming multiple-valued as one integrates around the loop, leading to unphysical jumps in components of $\mathbf{w}$ at the starting point. A steady-state configuration always satisfies this necessary condition, but time-dependent configurations don't necessarily do so.

6.11.2.3 Field-Line Conservation

In non-ideal MHD, the relation between conservation of flux and field lines is as follows. Magnetic field variations that conserve magnetic flux satisfy Eq. (6.62). On the other hand, they conserve magnetic field lines if there exists a magnetic field line velocity ($\mathbf{w}_L$) and a scalar function of position (λ_L) that satisfy

$$\frac{\partial \mathbf{B}}{\partial t} = \nabla \times (\mathbf{w}_L \times \mathbf{B}) + \lambda_L \mathbf{B}, \tag{6.67}$$

or, by expanding $\nabla \times (\mathbf{w}_L \times \mathbf{B})$, putting $\lambda = \lambda_L + \nabla \cdot \mathbf{w}_L$ and using $\nabla \cdot \mathbf{B} = 0$,

$$\frac{\partial \mathbf{B}}{\partial t} + (\mathbf{w}_L \cdot \nabla)\mathbf{B} - (\mathbf{B}\cdot\nabla)\mathbf{w}_L = \lambda \mathbf{B}.$$

From this it may be deduced that, whenever flux is conserved (such as in ideal MHD), field lines are also conserved. This follows simply by picking the particular case where $\lambda_L = 0$ and $\mathbf{w}_L = \mathbf{w}$, so that the equation for field line conservation (6.67) becomes the same as that for flux conservation (6.62).

However, the reverse is not true (i.e., field line conservation does not necessarily imply flux conservation), since there are many solutions for a non-ideal plasma that conserve field lines but don't conserve flux (i.e., all those for which $\lambda_L \neq 0$).

6.11.2.4 Field-Line Velocity

In an ideal plasma, the perpendicular component of magnetic field line or flux velocity ($\mathbf{w}_{L\perp} = \mathbf{w}_\perp = \mathbf{v}_\perp$) satisfying Eq. (6.60) may be defined as $\mathbf{E} \times \mathbf{B}/B^2$. However, in a non-ideal plasma there is no unique definition of field-line velocity. This result is particularly simple in 2D (PROBLEM 6.10), but it also holds in 3D.

The component ($\mathbf{w}_{L\perp}$) of a field-line velocity may be defined if and only if Ohm's Law can be transformed into the form

$$\mathbf{E} + \mathbf{w}_L \times \mathbf{B} = \mathbf{a}, \tag{6.68}$$

where $\nabla \times \mathbf{a} = -\lambda_L \mathbf{B}$, since the curl of Eq. (6.68) and Faraday's law then imply Eq. (6.67). This definition is field-line preserving, but it is only flux preserving when $\lambda_L = 0$, so that $\nabla \times \mathbf{a} = \mathbf{0}$ and $\mathbf{a} = \nabla \Phi$.

The case when $\nabla \times (\mathbf{w}_L \times \mathbf{B}) = \nabla \times (\mathbf{v} \times \mathbf{B})$ is particularly simple, since a comparison of the curls of Eqs. (6.68) and (6.61) then implies that $\nabla \times \mathbf{a} = \nabla \times \mathbf{N}$, and so the condition for field-line conservation becomes

$$\nabla \times \mathbf{N} = -\lambda_L \mathbf{B} \quad \text{or} \quad \mathbf{B} \times (\nabla \times \mathbf{N}) = \mathbf{0}.$$

From Eq. (6.68), the expression $\mathbf{w}_{L\perp} = (\mathbf{E} - \mathbf{a}) \times \mathbf{B}/B^2$ may be deduced for the field-line velocity. However, this is not unique, since, if $\mathbf{a}$ is replaced by $\mathbf{a}' = \mathbf{a} + \nabla \Psi^*$, where $(\mathbf{B}\cdot\nabla)\Psi^* = 0$, then the magnetic field is unaltered and $\mathbf{B}\cdot$(6.68) is unchanged, whereas $\mathbf{B} \times$ (6.68) gives another flow $\mathbf{w}'_{L\perp} = (\mathbf{E} - \mathbf{a}') \times \mathbf{B}/B^2$ in which the magnetic field lines are frozen. The form (6.68) is always possible for ideal flow (when $\mathbf{w}_{L\perp} = \mathbf{E} \times \mathbf{B}/B^2$), but it is also always possible when $\mathbf{E}\cdot\mathbf{B} = 0$.

6.11.3 Topological Conservation Laws

In an ideal medium, Eq. (6.59) holds and Alfvén's theorem implies that magnetic flux is constant through a surface (S) moving with the plasma ($\int_{S(t)} \mathbf{B}\cdot d\mathbf{S} = constant$), as well as conservation of field lines, null points, knots and linkages of field lines.

However, magnetic flux conservation is part of a much deeper theory of topological integral conservation laws (Hornig 2001). For a fluid with a velocity field $\mathbf{v}(\mathbf{x},t)$, the transport of a field (ω^k) by the flow of $\mathbf{v}$ is given by

$$\frac{\partial \omega^k}{\partial t} + L_{\mathbf{v}}\omega^k = 0,$$

where $L_{\mathbf{v}}$ is the Lie-derivative. Then the Lie-derivative theorem implies conservation of an integral over a k-dimensional comoving volume V^k, namely, $\int \omega^k dV^k = const.$ This powerful result may be used to clarify at a basic level the relation between field-line and flux conservation, and to show that magnetic flux conservation is part of a more general concept of electromagnetic flux conservation, as follows.

The notions of magnetic field lines and flux tubes and their transport with a plasma and diffusion through it are fundamental. Most of the Sun can be regarded for many purposes as a magnetically ideal medium, so that the magnetic field is frozen to the plasma. It moves with the plasma and, in a magnetically dominated region such as the solar corona, energy changes directly only between magnetic and kinetic forms (although the kinetic energy may in turn be dissipated by, e.g., viscous effects). However, localised regions of non-ideal behaviour exist, where magnetic fields may diffuse through the plasma and often reconnect. In such *diffusion regions*, the notions of conservation and transport of magnetic flux tubes invariably break down, but can be replaced by *electromagnetic flux conservation.*

When the plasma is nonideal, there are several classes of evolution of a magnetic field $\mathbf{B}(\mathbf{r},t)$ and electric field $\mathbf{E}(\mathbf{r},t)$, as illustrated schematically in Figure 6.2. All types of evolution are assumed to satisfy $\boldsymbol{\nabla}\cdot\mathbf{B} = 0$ and to obey Faraday's law $\partial\mathbf{B}/\partial t = -\boldsymbol{\nabla}\times\mathbf{E}$. The largest subclass within all such solutions is the family that also conserves electromagnetic flux, namely,

$$\int_S \mathbf{B}\cdot d\mathbf{S} + \int_S \mathbf{E}\cdot d\mathbf{l}\; dt = const. \tag{6.69}$$

A smaller family of solutions lying entirely within this consists of those that conserve magnetic flux by itself

$$\int_S \mathbf{B}\cdot d\mathbf{S} = const. \tag{6.70}$$

As far as magnetic reconnection is concerned, the solutions that represent 3D reconnection are a particular class of electromagnetic-flux preserving solutions. Furthermore, as can be seen from Figure 6.2, 2D reconnection represents the particular class of 3D solutions that is also magnetic-flux preserving. Various measures of topology have been developed, such as the *topological degree*, *topological entropy* and *topological flux function*, whose changes are important during reconnection (Sec. 10.4.4).

6.12 3D Reconnection Concepts

In order to understand reconnection in 3D, we need to describe some subtleties about the differences between diffusion and reconnection. In so doing, we will find that in many respects 3D reconnection is very different indeed from the concept of 2D reconnection that has dominated our thinking for decades.

We first determine how the non-ideal term in Ohm's Law determines whether there will be simple slippage of field lines or 2D reconnection or 3D reconnection. In 2D, the concept of a *flux velocity* is useful. In 3D, it fails but can be replaced by a so-called *dual flux velocity* (Sec. 6.12.4). Also, we summarise the differences between 2D and 3D reconnection and comment on the way reconnection can be defined and classified as a prelude to describing the different 3D regimes (Sec. 6.13).

6.12.1 Conditions for Reconnection

A necessary condition for reconnection is that the plasma be non-ideal, with

$$\mathbf{E}+\mathbf{v}\times\mathbf{B}=\mathbf{N}, \tag{6.71}$$

but whether or not reconnection occurs depends on the nature of $\mathbf{N}$.

If $\mathbf{N}$ can be written in the form $\mathbf{N}=\mathbf{u}\times\mathbf{B}+\boldsymbol{\nabla}\Phi$, then, from Eq. (6.63), the curl of Eq. (6.71) becomes $\partial\mathbf{B}/\partial t=\boldsymbol{\nabla}\times(\mathbf{w}\times\mathbf{B})$, where $\mathbf{w}=\mathbf{v}-\mathbf{u}$ is a flux velocity and $\mathbf{u}$ is the slippage velocity. Thus, by analogy with Eq. (6.59), the magnetic field behaves as if it moves with a velocity $\mathbf{w}$ and the following can be deduced:

(a) If $\mathbf{N}=\mathbf{u}\times\mathbf{B}+\boldsymbol{\nabla}\Phi$ and $\mathbf{u}$ is smooth, then there is slippage of magnetic field but no reconnection;

(b) If $\mathbf{N}=\mathbf{u}\times\mathbf{B}+\boldsymbol{\nabla}\Phi$ and $\mathbf{u}$ is singular, then there is 2D reconnection; and

(c) If $\mathbf{N}\neq\mathbf{u}\times\mathbf{B}+\boldsymbol{\nabla}\Phi$, then there is reconnection in 2.5D or 3D.

Furthermore, an implication of Eq. (6.71) is that $\mathbf{E}\cdot\mathbf{B}=\mathbf{N}\cdot\mathbf{B}$, so that the components of $\mathbf{E}$ and $\mathbf{N}$ parallel to the magnetic field are the same.

Thus, reconnection involves diffusion in an isolated region, but diffusion can occur without reconnection, so that reconnection implies diffusion, but diffusion does not necessarily imply reconnection. Furthermore, *fast outflow jets* are sometimes but not always accelerated during 3D reconnection, whereas they are invariably present in slow or fast 2D reconnection.

6.12.2 Reconnection in Two Dimensions ($\mathbf{E}\cdot\mathbf{B}=0$)

Consider the case $\mathbf{E}\cdot\mathbf{B}=0$, for which Eq. (6.71) implies that $\mathbf{N}\cdot\mathbf{B}=0$, so that $\mathbf{E}$ and $\mathbf{N}$ are perpendicular to $\mathbf{B}$. Provided $\mathbf{E}$ and $\mathbf{B}$ are not both zero, $\mathbf{v}$ and $\mathbf{B}$ are confined to a plane and depend only on coordinates in that plane (i.e., they are 2D), while $\mathbf{E}$ and $\mathbf{N}$ are normal to that plane.

Since $\mathbf{N}\cdot\mathbf{B}=0$, it is possible to set $\boldsymbol{\nabla}\Phi\equiv\mathbf{0}$ in Eq. (6.63), so that Eq. (6.64) becomes $\mathbf{E}+\mathbf{w}\times\mathbf{B}=\mathbf{0}$. Thus, the diffusion velocity is $\mathbf{u}=(\mathbf{B}\times\mathbf{N})/B^2$ and a flux-preserving flow $\mathbf{w}_\perp=\mathbf{E}\times\mathbf{B}/B^2$ exists, which is smooth except where $B=0$ (assuming $N\neq0$).

Furthermore, there are three possible types of behaviour:

(a) If $\mathbf{B}\neq\mathbf{0}$, then $\mathbf{w}$ is smooth everywhere and there is slippage of the magnetic field, which is *flux-conserving* in a strong sense;

(b) If $\mathbf{B}=\mathbf{0}$ at some point and the neighbouring magnetic field is elliptic, then $\mathbf{w}$ has a divergent singularity and magnetic flux is destroyed or generated at the null; and

(c) If $\mathbf{B}=\mathbf{0}$ at some point and the neighbouring magnetic field is hyperbolic, then $\mathbf{w}$ has a hyperbolic singularity and there is *magnetic flux reconnection*, with the flux conserved in the weak sense that $\mathbf{w}$ is regular except at one point.

6.12.3 Failure of Concept of Flux Velocity in 3D ($\mathbf{E}\cdot\mathbf{B}\neq0$)

The Flux Velocity Theorem: for an isolated 3D non-ideal region, a flux conservation velocity ($\mathbf{w}$) does not in general exist (Priest et al. 2003).

Proof: If a flux velocity exists, then Eq. (6.62) holds and a function Φ exists such that $\mathbf{E}+\mathbf{w}\times\mathbf{B}=\boldsymbol{\nabla}\Phi$, which implies that $\mathbf{E}\cdot\mathbf{B}=\boldsymbol{\nabla}\Phi\cdot\mathbf{B}$. If this is integrated along a magnetic field line from a point $(\mathbf{r}_1)$ on one side of the diffusion region to a point $(\mathbf{r}_2)$ on the other side, then

$$\int_{\mathbf{r}_1}^{\mathbf{r}_2} E_{\|}ds=\Phi(\mathbf{r}_2)-\Phi(\mathbf{r}_1).$$

Suppose that $E_{\|}$ is positive in the diffusion region, so that $\Phi(\mathbf{r}_2)>\Phi(\mathbf{r}_1)$. Since the diffusion region is isolated and everywhere outside it $\mathbf{w}=\mathbf{v}$ and $\boldsymbol{\nabla}\Phi=\mathbf{0}$, Φ must be uniform outside the diffusion region and so $\Phi(\mathbf{r}_2)=\Phi(\mathbf{r}_1)$. This contradiction implies that a single flux velocity does not exist and the theorem is proved. However, they can be replaced by a dual flux velocity (Sec. 6.12.4).

6.12.4 Differences Between 2D and 3D Reconnection

3D reconnection possesses many new features that are not present in 2D (Priest et al. 2003). These new features, together with the fact that many properties of 2D reconnection do not carry over into 3D, mean that 3D reconnection is completely different from 2D reconnection.

In 2D, reconnection possesses several basic properties, as follows:

(i) Reconnection can occur only at an X-type null point;
(ii) A flux velocity ($\mathbf{w}$) satisfying Eq. (6.62) exists everywhere except at null points; it has a hyperbolic singularity at an X-point; the magnetic flux moves at the velocity ($\mathbf{w}$) and slips through the plasma (which moves at $\mathbf{v}$);
(iii) The mapping of field lines near an X-point from one part of a boundary to another is discontinuous as a footpoint crosses a separatrix (Figure 6.34a);
(iv) While they are in the diffusion region, field lines generally preserve their connections; the exception is the X-point, where the field lines break and their connections are changed;
(v) When two flux tubes are about to reconnect, they approach the diffusion region with velocity $\mathbf{w}=\mathbf{v}$, and then they break and rejoin perfectly to form two new flux tubes that are moving out at $\mathbf{w}=\mathbf{v}$ (Figure 6.37a); and
(vi) When a flux tube is partly in a diffusion region, both parts of the tube that are outside have $\mathbf{w}=\mathbf{v}$, whereas the segment that lies inside has $\mathbf{w}\neq\mathbf{v}$; in other words, the two wings of the tube outside the diffusion region are moving with the plasma, while the central segment is slipping through the plasma (Figure 6.37b).

Surprisingly, none of the above properties carry over into 3D, so the nature of 3D reconnection is profoundly different from the way of thinking that has been developed in two dimensions. Its properties are as follows:

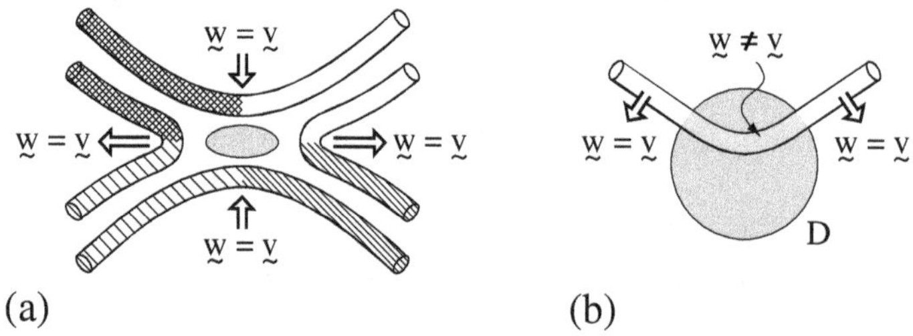

Figure 6.37. In 2D reconnection, (a) two flux tubes break and rejoin perfectly and (b) a flux tube slips through most of the diffusion region while preserving its connection.

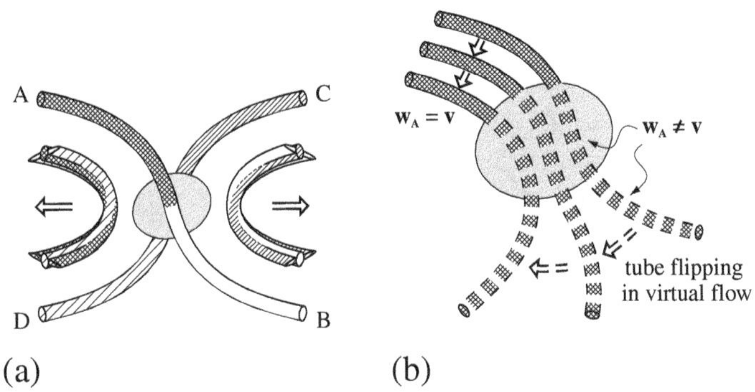

Figure 6.38. In 3D reconnection (a) two flux tubes break and partly rejoin and (b) the projection of a flux tube slips through the diffusion region and flips in a virtual flow.

(i) Reconnection can take place in 3D either at null points or in the absence of nulls (e.g., at *quasi-separators*, Sec. 6.9.5);
(ii) A flux tube velocity (**w**) does not in general exist in 3D (Sec. 6.12.3);
(iii) For regions without a null point, the mapping of field lines from one part of a boundary to another is in general continuous (e.g., Figure 6.34b);
(iv) While in a 3D diffusion region, field lines continually change their connections;
(v) Two flux tubes don't usually break and reform perfectly to give two flux tubes; instead, each section of the original tubes joins to a different part (Figure 6.38a); and
(vi) When two flux tubes are partly in the diffusion region, during the process of reconnecting they split into four parts, each of which flips differently (Figure 6.38b); if field lines forming a flux tube are projected through a diffusion region, beyond the diffusion region it moves with a velocity that is completely different from the plasma velocity: such a *virtual flow* (with which the tube is flipping) is a manifestation of the non-existence of a flux-conserving velocity (Sec. 6.12.3).

In general, the flux velocity $\mathbf{w}$ is non-unique in 2D and 3D. It is usually chosen by assuming $\mathbf{w} = \mathbf{v}$ in the ideal region. The key difference between 2D and 3D is that in 2D a single $\mathbf{w}$ exists (and is singular), while in 3D reconnection no single velocity $\mathbf{w}$ exists satisfying Eq. (6.62) and the constraint that $\mathbf{w} = \mathbf{v}$ in the ideal region.

Thus, if you consider a set of field lines whose directions point in towards a diffusion region and which are attached to moving plasma elements, their instantaneous positions can be calculated and hence their velocities ($\mathbf{w}_{in}$, say). This includes their continuations through the diffusion region (into an ideal region) which are no longer attached to plasma elements but exhibit flipping. Similarly, a set of field lines can be calculated from the other side of the diffusion region whose directions point out from the diffusion region and which are also attached to moving plasma elements. Their velocities ($\mathbf{w}_{out}$, say) can be calculated, and their continuations beyond the diffusion region into the inwards-pointing region also exhibit a flipping behaviour that is not attached to plasma elements. In 2D, $\mathbf{w}_{in}$ and $\mathbf{w}_{out}$ are identical, but in 3D this *dual flux velocity* pair $(\mathbf{w}_{in}, \mathbf{w}_{out})$ is needed to describe the behaviour of field lines, since they continually change their connections while they are passing through the diffusion region.

6.12.5 Definition and Classification of Reconnection

In 2D, the notion of magnetic reconnection is straightforward and has several aspects (Figure 6.39a):

(i) Reconnection occurs at an X-point, where two pairs of separatrices meet, with properties just described in Section 6.12.4;
(ii) The electric field (**E**) is normal to the plane;

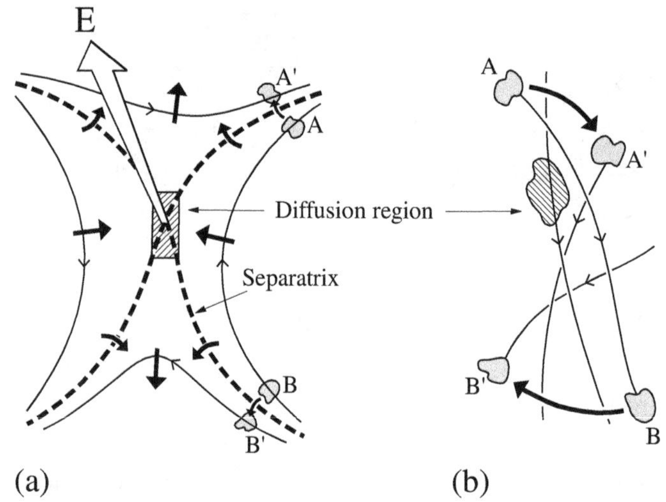

Figure 6.39. General properties of (a) 2D and (b) 3D reconnection, in which plasma elements A and B that are initially joined by a magnetic field line move to locations A' and B', where they are no longer joined magnetically.

(iii) There is a change of magnetic connectivity of plasma elements due to the presence of a localised diffusion region where ideal MHD breaks down; and

(iv) A flow of plasma is present across the separatrices.

So a natural question arises: which of these properties is robust enough to form the basis for a definition of reconnection in 3D? The answer was given in a landmark paper by Schindler et al. (1988), who realised that reconnection can occur in the absence of nulls and separatrices, so that (i) and (iv) cannot provide such a basis (Figure 6.39b). Instead they suggested that (iii) be used as the fundamental definition of *'General Magnetic Reconnection'*. Their concept includes all effects of local non-idealness that produce a component ($E_{||}$) of the electric field along a particular magnetic field line. They therefore pointed out that a 3D generalisation of (ii),

$$\int E_{||}\, ds \neq 0, \tag{6.72}$$

is a necessary and sufficient condition for general magnetic reconnection, where the integral is along that particular magnetic field. An equivalent condition is that the magnetic helicity (Sec. 6.12.6) change in time.

Schindler divides reconnection into *Null Reconnection* and *Non-Null Reconnection*, which produces global effects if Eq. (6.72) holds. This in turn is divided into driven and spontaneous reconnection, depending on whether there is an asymptotic electric field at large distances.

We are led to classify non-ideal processes in the following way (Figure 6.40). Non-ideal processes (Sec. 6.11.2) have $\mathbf{E}+\mathbf{v}\times\mathbf{B}=\mathbf{N}$ and may be divided into those for which $\mathbf{N}=\mathbf{u}\times\mathbf{B}+\boldsymbol{\nabla}\Phi$ (when a smooth finite flux velocity ($\mathbf{w}$) exists and there is slippage) and those for which $\mathbf{N}\neq\mathbf{u}\times\mathbf{B}+\boldsymbol{\nabla}\Phi$ (when no $\mathbf{w}$ exists and there is either a problem at a null point or a mismatch of the boundary conditions). The latter may in turn be divided into two classes.

In the first class, $\mathbf{E}\cdot\mathbf{B}=0$ and there is either a 2D X-point topology (classical 2D reconnection) or a 2D O-point (where magnetic flux may be destroyed or generated). 2D X-points or null sheets may be aligned in 3D to give a 3D X-line or null sheet, where reconnection may occur, but this is topologically unstable.

The second class is generic 3D reconnection. It has a localised diffusion region with $\mathbf{E}\cdot\mathbf{B}\neq 0$, so that $\int E_{||}\, ds\neq 0$, whose maximum value gives the reconnection rate. This class in turn comprises several types,

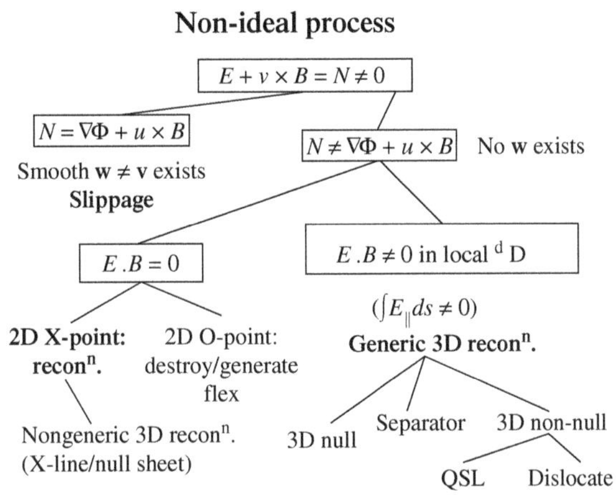

Figure 6.40. Classification of non-ideal processes. Other 3D non-null examples of reconnection include kink instability and braided fields.

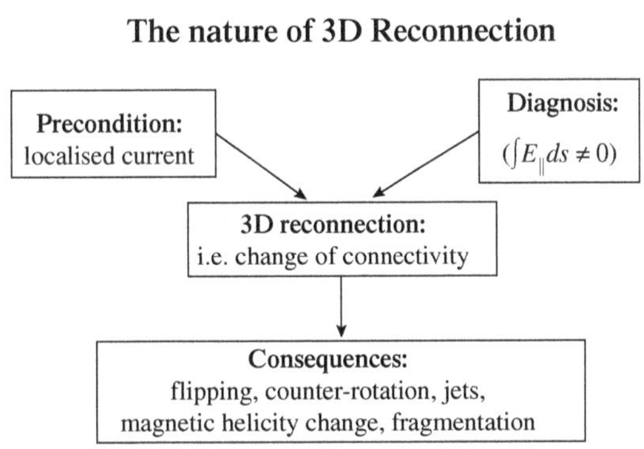

Figure 6.41. The precondition for, diagnosis and consequences of 3D reconnection.

depending on whether a null-point is present (3D null reconnection) or not (3D non-null reconnection). 3D null reconnection includes spine-fan reconnection, torsional reconnection and separator reconnection, depending on whether the current concentrates along the spine, fan or separator (Sec. 6.13.2). Non-null reconnection may be modelled kinematically (Sec. 6.13.1), and includes reconnection at QSLs (Sec. 6.13.4) and flux-tube disconnection.

3D reconnection is defined as happening when there is a change of magnetic connectivity of plasma elements and is diagnosed by the condition $\int E_{||}\, ds \neq 0$ (Figure 6.41). The precondition for such reconnection is the formation of a localised current concentration. Indeed, the key reason why nulls, separators and quasi-separators are natural locations for reconnection is that they are locations where strong currents tend to grow. Furthermore, it is not the nulls, separators and quasi-separators themselves that are the locations for reconnection, but rather reconnection occurs everywhere throughout the finite diffusion regions located around such features.

The possible consequences of reconnection are: magnetic flipping and counter-rotation associated with a small change of magnetic helicity; jets of plasma, naturally accelerated by the Lorentz forces in the current concentrations; and fragmentation of the current that arises either by a resistive or ideal instability or by the action of jets from one reconnection region interacting with the surroundings. Sophisticated diagnostic tools are needed to reveal these effects and to map out the associated skeleton and quasi-skeleton (Haynes et al. 2007; Parnell et al. 2008; Parnell et al. 2010).

Hesse and Schindler (1988) set up a mathematical formalism for *General Magnetic Reconnection* as follows. They express the magnetic field as

$$\mathbf{B} = \nabla\alpha \times \nabla\beta, \tag{6.73}$$

in terms of *Euler potentials* (α and β), which are constant on field lines, since (6.73) implies $\mathbf{B}\cdot\nabla\alpha = \mathbf{B}\cdot\nabla\beta = 0$. Each field line is specified by a particular α and β. [However, care must be taken when using Euler potentials, since normally they may be defined only locally (rather than globally), and not in the presence of null points.]

The time evolution of these potentials may be calculated as follows. First, the vector potential ($\mathbf{A}$) (such that $\mathbf{B} = \nabla \times \mathbf{A}$) can be expressed in a gauge with $\mathbf{A}\cdot\mathbf{B} = 0$ as $\mathbf{A} = \alpha\nabla\beta$. Next, in order to satisfy $\nabla \times \mathbf{E} = -\partial\mathbf{B}/\partial t$ identically, the electric field may be written as

$$\mathbf{E} = -\frac{\partial \mathbf{A}}{\partial t} - \nabla\Phi = -\frac{\partial\alpha}{\partial t}\nabla\beta + \frac{\partial\beta}{\partial t}\nabla\alpha - \nabla\Psi,$$

where $\Psi = \Phi + \alpha\,\partial\beta/\partial t$. Then, in terms of components parallel to $\nabla\alpha, \nabla\beta$ and ∇s, the components of Ohm's Law ($\mathbf{E} + \mathbf{v}\times\mathbf{B} = \mathbf{N}$) become

$$\frac{d\alpha}{dt} = -\frac{\partial\Psi}{\partial\beta} - N^{\beta}, \qquad \frac{d\beta}{dt} = \frac{\partial\Psi}{\partial\alpha} + N^{\alpha}, \qquad \frac{\partial\Psi}{\partial s} = -N^{s}, \tag{6.74}$$

where s measures distance along field lines and d/dt follows the plasma motion.

Outside the diffusion region, the components of $\mathbf{N}$ all vanish, so $\Psi = \Psi(\alpha,\beta,t)$ is independent of s and is a Hamiltonian. *Line Conservation* is an immediate consequence, since the time-evolution of α and β is independent of s: that is, α and β change in the same way for all plasma elements on a given field line. *Flux Conservation* follows from Liouville's theorem, which implies that the area within any closed curve in $\alpha\beta$-space remains constant, and so the magnetic flux through a closed curve moving in physical space with the plasma is constant in time. Inside the diffusion region, where $\mathbf{N} \neq \mathbf{0}$, *Line Conservation* holds if $\mathbf{B}\times(\nabla\times\mathbf{N}) = \mathbf{0}$, whereas *Flux Conservation* holds if $\nabla\times\mathbf{N} = \mathbf{0}$ (PROBLEM 6.11).

It may then be proved that finite-B reconnection with global effects occurs if and only if Eq. (6.72) holds in the diffusion region (D). Inside D, the function Ψ that appears in Eq. (6.74) is a function of $\alpha, \beta, s,$ and t. Outside D, $\partial\Psi/\partial s = 0$, so that Ψ is a function of α, β and t alone. However, for a field line that passes through D, these may be different functions, Ψ_1 and Ψ_2, say, on either side of D. In other words,

$$\Psi_2(\alpha,\beta,t) - \Psi_1(\alpha,\beta,t) = \int_1^2 ds\frac{\partial\Psi}{\partial s},$$

where the integration is along a field line and the difference between Ψ_1 and Ψ_2 depends on the way Ψ varies in D. Thus, if the right-hand side ($\int ds\,E_{\|}$) is non-zero, the diffusion region creates a global effect along field lines that thread it. This effect is to produce different values on both sides of the diffusion region for the α and β of plasma elements that are initially located on the same field line – that is, reconnection. These ideas have been used for several applications (e.g., Birn et al. 1989; Hesse and Birn 1990; Schindler et al. 1991; Van Hoven et al. 1995).

The above definition of general magnetic reconnection is highly attractive, although it could be regarded as too general, since it includes examples of magnetic diffusion that have not been traditionally included in the concept of reconnection. An alternative is therefore to restrict the definition of reconnection to *Singular Field Line Reconnection*, in which the presence of $E_{\|}$ along a field line is supplemented by the condition that the nearby field has a certain topology in a plane perpendicular to the field line (Priest and Forbes 1989). If the transverse topology is X-type, then we would refer to it as *X-type Singular Reconnection*, which has the advantage of being close in spirit to traditional 2D reconnection. If the transverse topology is O-type, then we would refer to it as *O-type Singular Reconnection*, which includes *Flux Tube Disconnection*

(Wilmot-Smith and Priest 2007) and some kinds of separator reconnection discovered in numerical experiments (Parnell et al. 2010).

6.12.6 Magnetic Helicity and 3D Reconnection

During reconnection, the total magnetic helicity is conserved to a high degree of approximation, although it can be changed from mutual to self helicity or vice versa. Reconnection can redistribute magnetic helicity within a configuration and often tends to make it more uniform, so that it approaches a minimum-energy state with the same total helicity, but uniformly distributed in a linear force-free field. However, the small change in magnetic helicity that does occur is intimately related to 3D reconnection (Schindler et al. 1988).

Consider undriven non-null (or finite-B) reconnection that is global and spontaneous (Sec. 6.12.5). The relative magnetic helicity (Sec. 6.10.1) may be written $H = \int_V (\mathbf{A}+\mathbf{A}_0)\cdot(\mathbf{B}-\mathbf{B}_0)\, dV$, where $\mathbf{B}_0$ is the magnetic field at time (t_0) and $\mathbf{A}_0$ is the corresponding vector potential. Using $\mathbf{E} = -\boldsymbol{\nabla}\Phi - \partial\mathbf{A}/\partial t$ then leads to

$$\frac{dH}{dt} = -2\int_V \mathbf{E}\cdot\mathbf{B}\, dV - \int_S \mathbf{n}\cdot[(\mathbf{B}-\mathbf{B}_0)\Phi + \mathbf{E}\times(\mathbf{A}-\mathbf{A}_0)]\, dS.$$

Assuming that $\mathbf{E}=\mathbf{0}$ and $\mathbf{B}=\mathbf{B}_0$ on S and remembering that $\mathbf{E}\cdot\mathbf{B}=0$ outside the diffusion region D_R, this reduces to

$$\frac{dH}{dt} = -2\int_{D_R} \mathbf{E}\cdot\mathbf{B}\, dV = -2\int_{D_R} E_\parallel B\, dV.$$

Thus, if H changes with time, $E_\parallel \neq 0$ and so by Eq. (6.72) there is 3D reconnection. This implies the parallel electric field is roughly $E_\parallel \simeq \delta H/(2V_R B_0\, \delta t)$, in terms of the diffusion region volume (V_R) and rate of change of magnetic helicity ($\delta H/\delta t$).

6.13 3D Reconnection Regimes

This section describes models for reconnection at a null point, at a separator and at non-null regions such as a quasi-separator. Reconnection will tend to occur wherever current sheets form. Null points and separators are natural weak spots in a magnetic field where current can easily build up, as are quasi-separators and other non-null regions, such as braiding locations or resonant surfaces for kinking.

6.13.1 Ways to Model Reconnection at a Null Point

From the kinematics of steady ideal flows in the neighbourhood of a null point or separator (Sec. 6.13.1.1), three distinct types of 3D reconnection were proposed, depending on whether the current focuses along a spine, fan or separator (Priest and Titov 1996). Later, the nature of the flow in the diffusion region was studied (Sec. 6.13.1.2) and computational experiments set up (Sec. 6.13.1.3), leading to a categorisation of null-point reconnection (Sec. 6.13.2) into *spine-fan reconnection*, *torsional spine reconnection* and *torsional fan reconnection* (Priest and Pontin 2009).

6.13.1.1 Kinematic Ideal Models: Spine and Fan Regimes

The effects of steady reconnection in the ideal region around a 3D null may be studied by solving the kinematic equations for $\mathbf{v}$ and $\mathbf{E}$, namely,

$$\mathbf{E}+\mathbf{v}\times\mathbf{B}=\mathbf{0} \quad \text{(a)} \quad \text{and} \quad \boldsymbol{\nabla}\times\mathbf{E}=\mathbf{0}, \quad \text{(b)} \qquad (6.75)$$

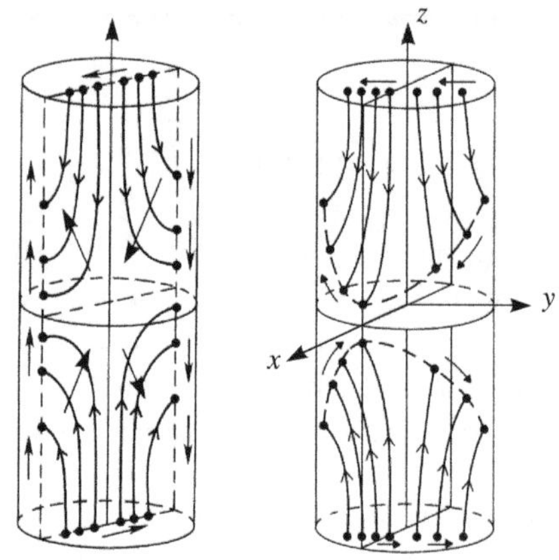

Figure 6.42. (a) Spine reconnection with a strong spine electric field driven by continuous motions across the fan. (b) Fan reconnection with a strong fan electric field and flipping of field lines above and below the fan produced by continuous motions across the spine.

with $(B_x, B_y, B_z) = B_0/L_0(x, y, -2z)$ and various boundary conditions. $\nabla \times \mathbf{E} = \mathbf{0}$ implies that $\mathbf{E} = \nabla\Phi$, so the component of Eq. (6.75a) perpendicular to $\mathbf{B}$ yields

$$\mathbf{B} \cdot \nabla\Phi = 0, \tag{6.76}$$

which, for given values of Φ on a surface, may be integrated along field lines to determine Φ (and therefore $\mathbf{E}$) throughout the volume. Then the component of Eq. (6.75a) perpendicular to $\mathbf{B}$ determines the plasma velocity normal to $\mathbf{B}$ as

$$\mathbf{v}_\perp = \frac{\nabla\Phi \times \mathbf{B}}{B^2}. \tag{6.77}$$

When a continuous flow is imposed across the fan (Figure 6.42a), singularities in $\mathbf{E}$ and $\mathbf{v}$ occur at the spine. It was speculated that this would produce a strong current at the spine in what was dubbed *spine reconnection* (Priest and Titov 1996).

For a continuous flow imposed across the spine (Figure 6.42b), singularities are produced at the fan together with a strong *flipping flow* (previously discovered by Priest and Forbes 1992b). It was suggested that this would produce *fan reconnection*, with a strong current at the fan (Priest and Titov 1996).

6.13.1.2 Kinematic Resistive Hornig-Priest Modelling

Hornig and Priest (2003) modelled the kinematic effect of an isolated 3D diffusion region, where frozen-in flux breaks down. They solved

$$\mathbf{E} + \mathbf{v} \times \mathbf{B} = \eta\, \mathbf{j}, \tag{6.78}$$

where $\nabla \times \mathbf{E} = \mathbf{0}$, $\mathbf{j} = \nabla \times \mathbf{B}/\mu$ and $\nabla \cdot \mathbf{B} = 0$. The idea was to impose a sufficiently simple magnetic field that both the field's mapping and its inverse can be found analytically. Writing $\mathbf{E} = \nabla\Phi$, the integral of the component of Eq. (6.78) parallel to $\mathbf{B}$ determines Φ everywhere as an integral along field lines

$$\Phi = \int \frac{\eta\, \mathbf{j} \cdot \mathbf{B}}{B}\, ds + \Phi_e, \tag{6.79}$$

in terms of values (Φ_e) at one end of the field lines and the distance s along them.

A simple technique to isolate the reconnection region in these kinematic solutions is to choose a form of η that is localised. So-called *pure* solutions have $\Phi_e \equiv 0$ and can produce counter-rotating (or flipping) flows of field lines that link the diffusion region. The rate of reconnection of flux is calculated by evaluating the integral

$$\frac{dF_m}{dt} = \int E_{\|} ds \tag{6.80}$$

along a field line through the diffusion region (Schindler et al. 1991; Hesse et al. 2005). Then the flow normal to the field lines is determined by the component of Eq. (6.78) perpendicular to $\mathbf{B}$ as

$$\mathbf{v}_{\perp} = \frac{(\nabla\Phi - \eta\,\mathbf{j}) \times \mathbf{B}}{B^2}. \tag{6.81}$$

6.13.1.3 Resistive Numerical MHD Experiments

Numerical experiments have been conducted which go beyond the constraints of analytical theory and shed more light on the nature of reconnection at a 3D null. The aim was also to see whether the types of reconnection envisaged qualitatively could indeed take place in practice and to discover other regimes.

Galsgaard et al. (2003) studied the propagation of a helical Alfvén wave towards a fan plane, launched by rotational driving of field lines around a spine. The current concentrated in the fan plane, suggesting the possibility of *torsional fan reconnection* (Sec. 6.13.2.2). In contrast, Pontin and Galsgaard (2007) rotated the field lines near a fan plane and produced strong currents along the spine, symptomatic of *torsional spine reconnection* (Sec. 6.13.2.1).

The formation of current sheets by the compressible collapse of a line-tied 3D null has been modelled with an ideal Lagrangian code (Pontin and Craig 2005). Shearing either the spine axis or fan plane focusses externally generated large-scale stresses near the null. As in linear theory (Rickard and Titov 1996), the fan and spine collapse locally towards each other to form a current sheet singularity.

Pontin et al. (2007a) used a resistive MHD code to investigate the formation and dissipation of the current sheet in response to shearing the spine (Figure 6.43). The results support the idea of *spine-fan reconnection*, in which current concentrates around the null in a sheet spanning the spine and fan (Sec. 6.13.2). In the incompressible limit, the spine-fan current is found to reduce to a purely fan current (Pontin

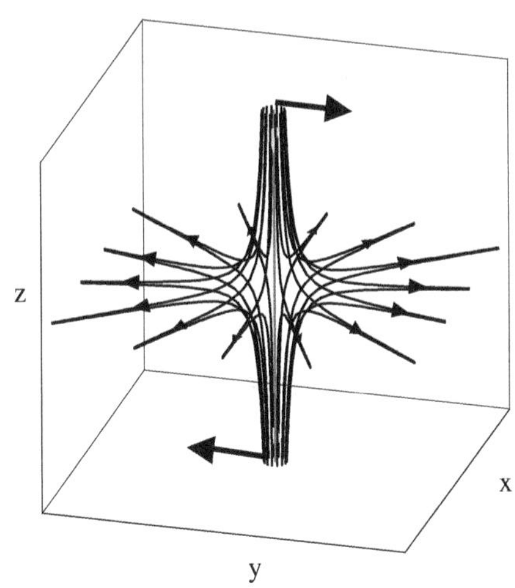

(a)

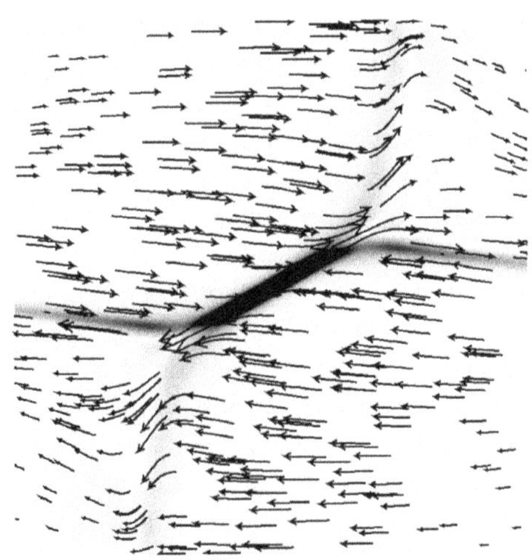

(b)

Figure 6.43. (a) A shearing motion of a spine situated on the z-axis. (b) The resulting collapse of spine and fan to form *spine-fan reconnection*, showing the current-density contours (shaded) and flow velocity arrows in the $x = 0$ plane (after Pontin et al. 2007a).

et al. 2007b) with behaviour closely resembling earlier fan reconnection models. Thus, pure *fan reconnection* can arise as the incompressible limit of *spine-fan reconnection.*

6.13.2 3D Null Regimes (Priest and Pontin 2009)

Which type of reconnection occurs at a 3D null depends crucially on the nature of the flows and boundary conditions. A flow that creates a current along the spine (or fan) gives rise to spine (or fan) reconnection. A kinematic resistive solution for *pure fan reconnection,* with flow across the spine but not the fan, can be found by adopting a field of the form $(B_0/L_0)(x, y - \bar{j}_0 z^3/L_0^2, -2z)$ with a fan x-current $3B_0\bar{j}_0 z^2/(\mu L_0^3)$ (Pontin et al. 2004). *Pure spine reconnection* with flow across the fan but not the spine can be modelled by considering $(B_0/L_0)(x, y, \bar{j}_0 y^3/L_0^2 - 2z)$ with a fan x-current $3B_0\bar{j}_0 y^2/(\mu L_0^3)$. Both of these fields have a vanishing current at the null. However, a key property of a null point is the hyperbolic field structure, which tends to focus disturbances and thus generate non-zero currents at the null for the *primary reconnection modes.* The above pure spine and fan solutions should therefore not be considered as fundamental or primary reconnection modes but as *secondary reconnection modes* in the sense that the current vanishes at the null.

Three primary reconnection modes exist, the first two being *torsional spine reconnection* and *torsional fan reconnection,* which are driven by twisting motions. Surprisingly, the generic null point reconnection mode that is observed in numerical experiments in response to shearing motions possesses a strong fan-aligned current with flow across both spine and fan. It is in some sense a combination of spine and fan reconnection, and so we call it *spine-fan reconnection* (Sec. 6.13.2.3).

6.13.2.1 Torsional Spine Reconnection

The Hornig-Priest formalism (Sec. 6.13.1.2) may be applied to steady kinematic behaviour of an isolated diffusion region containing a spiral null of the form

$$(B_R, B_\phi, B_z) = \frac{B_0}{L_0}\,(R, \tfrac{1}{2}\bar{j}_0 R, -2z) \tag{6.82}$$

in cylindrical polars (Pontin et al. 2004). Here the spine and current are both directed along the z-axis, where $\bar{j}_0$ is a dimensionless current density and the diffusion region is a cylinder of radius a and height $2b$.

A rotation of the fan plane drives a current along the spine and creates twisting flows about the spine (Pontin and Galsgaard 2007). Inside the spine current tube, there is rotational slippage, with the field lines becoming disconnected and rotating around the spine. This is quite different from the spine reconnection that was originally envisaged, and so we called it *torsional spine reconnection* (Figure 6.44a).

The core of such reconnection in the kinematic model is a pure elementary solution, with zero flow outside the volume defined by the 'envelope' (F) of flux that threads the diffusion region. Inside F, the flow and flux velocities are purely rotational (i.e., in the ϕ-direction), so that there is no flow across either spine or fan. The reconnection rate is $\int E_\parallel dl$ along the spine, and measures the rate of rotational mis-matching of the flux velocities of field lines entering and leaving the diffusion region.

Any ideal solution (Φ_{id}) may be added to this, such as a stagnation-point flow ($\Phi_{id} = \phi_0 x_0 y_0$), which brings flux into F and carries it out again. The result is a transition from O-type to X-type flow near the null when ϕ_0 exceeds a critical value.

The (non-ideal) pure elementary solution has $\mathbf{E} = \nabla\Phi_{nid}$, where $\nabla\Phi_{nid} + \mathbf{v}_{nid} \times \mathbf{B} = \eta\nabla \times \mathbf{B}$, and Φ_{nid} is given by Eq. (6.79). For the spiral null with $\Phi_e \equiv 0$

$$\Phi_{nid} \;=\; -\Phi_{nid0} \int \eta/\eta_0 \; e^{-2S} dS,$$

where $\Phi_{nid0} = 2B_0 b\bar{j}_0\eta_0/(\mu L_0)$. Once a form for η has been assumed, this may be integrated to give $\Phi_{nid}(S, R_0, \phi_0)$. Using the inverse mapping, we deduce $\Phi_{nid}(R, \phi, z)$ and therefore $\mathbf{E}$ and $\mathbf{v}_\perp$ everywhere.

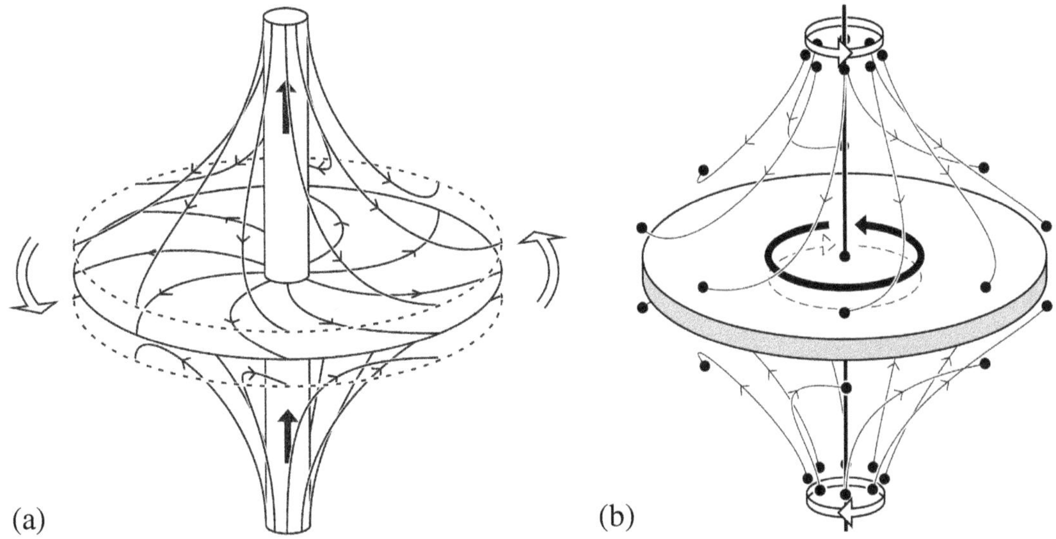

Figure 6.44. Rotational motions (open arrows) of: (a) the fan, driving torsional spine reconnection with a strong spine current (solid arrows); (b) the spine, driving torsional fan reconnection with a strong fan current and slippage of field lines (solid arrow).

The resulting reconnection rate depends on the form of η. As an example, suppose it vanishes outside the diffusion region (D, defined by $R<a$, $|z|<b$), and inside D it has the form $\eta=\eta_0(1-R^4/a^4)(1-z^2/b^2)$, which peaks at the origin and vanishes on the boundary of D. Then we find a reconnection rate of $\int_{-b}^{b} E_z(0,0,z)\, dz \;=\; \frac{4}{3}\, E_0\, b = \frac{2}{3}\Phi_{nid0}$, where the electric field and current at the centre of the diffusion region are $E_0=E_z(0,0,0)=\Phi_{nid0}/(2b)=\eta j_0$ and $j_0=\bar{j}_0 B_0/(\mu L_0)$.

6.13.2.2 Torsional Fan Reconnection

In *torsional fan reconnection*, a rotation of field lines near the spine in opposite directions above and below the fan builds up a fan current sheet. In the sheet, field lines experience rotational slippage in the opposite sense above and below the fan (Figure 6.44b), but there is no flow across either spine or fan. The counter-rotation (above and below the fan) builds up a double-spiral structure near the null point, with a radial current and an axial current that reverses sign at the fan plane.

Such reconnection may be modelled as above, adopting a *double-spiral null*, with

$$(B_R, B_\phi, B_z) = \frac{B_0}{L_0}\left(R,\; 2\bar{j}_0 \frac{z^{2M+1}R^{N-1}}{b^{2M+N-1}},\; -2z\right)$$

and M and N positive integers. The magnetic diffusivity inside the diffusion region (D) is $\eta=\eta_0(1-R^m a^m)(1-z^n b^n)$, which peaks at the null point and vanishes on the boundary of D when m and n are positive and n is even. As before, the pure non-ideal solution may be calculated and a wide range of ideal solutions added.

6.13.2.3 Spine-Fan Reconnection

The Hornig-Priest approach has been applied by Pontin et al. (2005) to a disc-shaped diffusion region (D) containing a null point with a uniform fan-aligned current $[B_0\bar{j}_0/(\mu L_0)]$ in the x-direction and field components

$$(B_x, B_y, B_z) = \frac{B_0}{L_0}(x, y-\bar{j}_0 z, -2z).$$

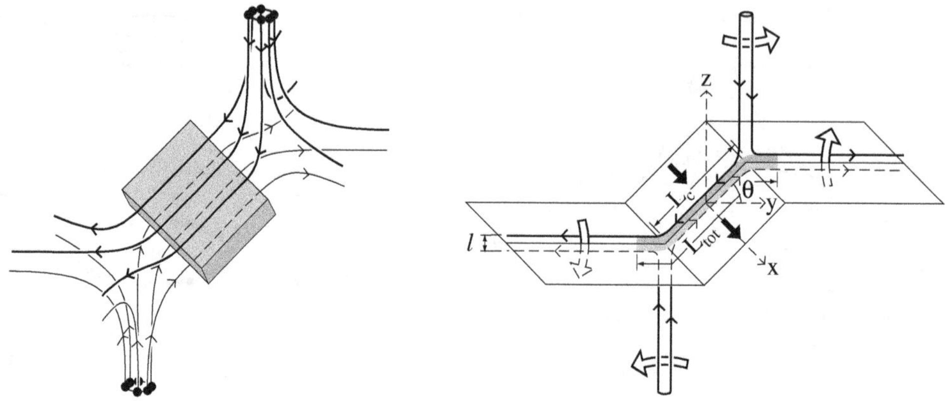

(a) (b)

Figure 6.45. (a) The structure of spine-fan reconnection, showing field lines and the (shaded) diffusion region. (b) Corresponding motion of flux across spine and fan (large light arrows). The sheet current flows in the x-direction (large dark arrows): it has width l, total length L_{tot} (in the yz-plane), and length L_c common to spine and fan.

Inside the diffusion region [$R_1 < a,\ |z| < b$, where $R_1^2 = x^2 + (y - \bar{j}_0 z/3)^2$], the magnetic diffusivity decreases smoothly and monotonically from the null to zero at its boundary, outside which it vanishes.

Surprisingly, the resulting plasma flow is quite different from simple fan reconnection: since it crosses both the spine and fan of the null, we call the resulting reconnection *spine-fan reconnection.* Field lines traced from footpoints anchored in the fan-crossing flow are found to flip up and down the spine, whereas those that are traced from the top and bottom of the domain flip around the spine in the fan plane. The reconnection rate is again given by an integral of the form (6.80), taken along the fan field line parallel to the direction of current flow (here the x-axis): it equals the rate of flux transport across the fan (separatrix) surface (Pontin et al. 2005).

In general, if the driving motions shear a null point rather than rotate it, the result will be *spine-fan reconnection*, with the shear distortion focussing in the weak-field region near the null point, forming a localised current sheet (cf Sec. 6.2.2.1).

It is the flux transfer across both spine and fan that distinguishes spine-fan reconnection. Furthermore, the current concentration is in the form of a localised sheet that is inclined at an intermediate angle between the spine and fan (Figure 6.45). Note that reconnection always occurs at a current sheet of non-zero dimensions rather than at a single point, and so there is a transfer of flux through the current sheet and hence across the fan, although this will generally be smaller than the flux transfer across the more extensive current sheets that can form along separators as follows.

6.13.3 Separator Reconnection

Another type of 3D reconnection can often exist when two or more nearby null points are present. The fan surface of a null consists of field lines linking to a null point, and so, when there are two nulls, their two fans will in the generic (structurally stable) case intersect in a special curve called a *separator* (Figure 6.46). The fundamental reason why reconnection can occur at a separator is that it is a natural location for the formation of a current sheet, which is often a precursor for reconnection.

The formation of such a sheet in *separator reconnection* implies, in the simplest case, that planes across the separator resemble 2D reconnection. A simple example,

$$(B_x, B_y, B_z) = [x(z-3), y(z+3), 1-z^2],$$

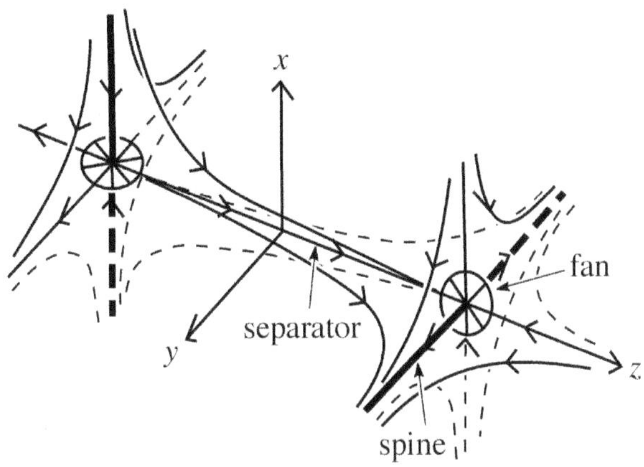

Figure 6.46. Configuration with two nulls including a separator joining one null to the other.

possesses null points at $z=\pm 1$ on the z-axis. Now surround this configuration by a cube with a side whose length is two units long, centred at the origin, and impose a horizontal flow on the sides $y=\pm 1$ with $v_z=0$. This drives fan reconnection at the upper null and spine reconnection at the lower null. The field lines are of the form

$$x(z-1)^{-1}(z+1)^2=C, \qquad y(z-1)^2(z+1)^{-1}=K, \tag{6.83}$$

and so a field line through the footpoint $(x_0,1,z_0)$ has

$$C=\frac{x_0(z_0+1)^2}{z_0-1}, \qquad K=\frac{(z_0-1)^2}{z_0+1}. \tag{6.84}$$

The condition that $v_z=0$ on $y=1$ implies $E_xB_y=E_yB_x$, which in turn implies that the potential has the form $\Phi=f[\frac{1}{2}x_0^2+(z_0+1)^{1/2}(z_0-3)^{9/2}(z_0-1)^{-4}]$, where $x_0(C,K), z_0(C,K)$ are given by Eq. (6.84), and $C(x,z), K(x,z)$ follow from Eq. (6.83).

The nature of separator reconnection was illuminated by a pioneering numerical experiment of Galsgaard and Nordlund (1996b) on the response of a periodic force-free equilibrium containing eight null points to footpoint shearing on two boundaries. They discovered that current sheets develop along the separators joining nearby nulls, with Alfvénic jets of plasma ejected out of the sides of the sheets. Then Parnell and Galsgaard (2004) and Parnell et al. (2008) modelled coronal reconnection driven by the motion of a pair of opposite-polarity photospheric fragments in an overlying horizontal field. They discovered that the main dissipation mechanism is separator reconnection at multiple separator current sheets, with dissipation enhanced by flux recycling in *recursive reconnection.* Furthermore, when the separator current is strong enough, the field in transverse planes changes from X-type to O-type.

Longcope has also developed the idea of the formation and dissipation of separator current sheets (Longcope and Cowley 1996; Longcope 2001) and applied it to coronal heating (Priest et al. 2005). He generalised the 2D theory for current sheet formation in a quadrupolar field (Priest and Raadu 1975) to show how in 3D the current and energy storage are related to the change in magnetic flux (Longcope et al. 1998; Longcope and Magara 2004). The ideas have also been applied to the evolution and brightening of an active region (Longcope et al. 2005) and to flares (Longcope and Silva 1998; Longcope and Beveridge 2007).

Separator reconnection is probably important in coronal heating (Sec. 10.4.5), where it is it at the core of heating by flux-tube tectonics, and it may also play a role in accelerating the slow solar wind according to the S-Web model (Sec. 13.8) and in releasing energy in solar flares (Sec. 12.3.2).

6.13.4 Quasi-Separator Reconnection (Priest-Démoulin 1995)

A Quasi-Separatrix Layer (QSL) may be identified as a region in a magnetic field where the gradient of the footpoint mapping is large (Sec. 6.9.5), the best measure being the *squashing factor, Q* [Eq. (6.51)]. Having located a QSL, kinematic reconnection may now be considered satisfying Eq. (6.75) and producing a potential (Φ) and field-line velocity [Eq. (6.77)]. Suppose the field-line velocity components $v_{\perp 1x}$ and $v_{\perp 1y}$ are imposed at a point $(x_1, y_1, 1)$ on the top side ($z=1$) of a cube and deduce the function $\Phi(x_1, y_1)$ together with $\mathbf{E}$ and $\mathbf{v}_\perp$ throughout the cube. The resulting electric field on the base ($z=0$) of the cube has components

$$E_{x0} = -\frac{\partial \Phi}{\partial x_1}\frac{\partial x_1}{\partial x_0} - \frac{\partial \Phi}{\partial y_1}\frac{\partial y_1}{\partial x_0}, \qquad E_{y0} = -\frac{\partial \Phi}{\partial x_1}\frac{\partial x_1}{\partial y_0} - \frac{\partial \Phi}{\partial y_1}\frac{\partial y_1}{\partial y_0},$$

which depend partly on the electric field components on the top ($E_{x1} = -\partial\Phi/\partial x_1$, $E_{y1} = -\partial\Phi/\partial y_1$) and partly on the gradients of the mapping functions $[x_1(x_0, y_0)$ and $y_1(x_0, y_0)]$. Thus, $\mathbf{E}_0$ is large where the gradients of the mapping are large, namely in a quasi-separatrix layer (Figure 6.47). This is the crux of why QSLs are so important and why reconnection tends to occur in quasi-separators, namely the intersection of two QSLs, which are essentially the remnants of separatrices.

For example, impose on the top ($z = 1$) and side ($x = 12$) of a cube $v_{\perp 1x} = 0$, $v_{\perp 1y} = v_0 x_1$ and $v_{\perp 1x} = 0$, $v_{\perp 1y} = \frac{1}{2} v_0$. Then the base velocity ($z=0$) along the x-axis ($y=0$) peaks at $x_0 = \frac{1}{2}\epsilon$ with a value of $v_0/(2\epsilon)$. If this peak exceeds the Alfvén speed, there will exist two diffusive layers centred on $x_0 = \pm\frac{1}{2}\epsilon$, where the field lines are unfrozen and so flip rapidly through the plasma. In other words, the field lines move quicker than the plasma and become disconnected from it.

Démoulin et al. (1996) showed that any smooth photospheric motion of the footpoints tends to produce strong currents at QSLs. They consider an initially potential field $\mathbf{B}_0 = b(X_0, Y_0)\boldsymbol{\nabla} X_0 \times \boldsymbol{\nabla} Y_0$, where (X_0, Y_0) are the horizontal coordinates for the field line footpoint in the positive flux. They suppose the footpoints are displaced by a small amount and give rise to a force-free field ($\mathbf{B}_0 + \mathbf{B}_1$), where $\mathbf{B}_1$ may be written in the form $\mathbf{B}_1 = U\boldsymbol{\nabla} X_0 + V\boldsymbol{\nabla} Y_0 + W\boldsymbol{\nabla} S_0$ in terms of functions (U, V, W) of (X_0, Y_0, S_0), which are determined by $\mathbf{j}_1 \times \mathbf{B}_0 = \mathbf{0}$. When $W \equiv 0$, the footpoint displacement ($\mathbf{d}_1$) has the form

$$\mathbf{d}_1 = U\int \frac{\boldsymbol{\nabla} X_0}{B_0} dS + V\int \frac{\boldsymbol{\nabla} Y_0}{B_0} dS,$$

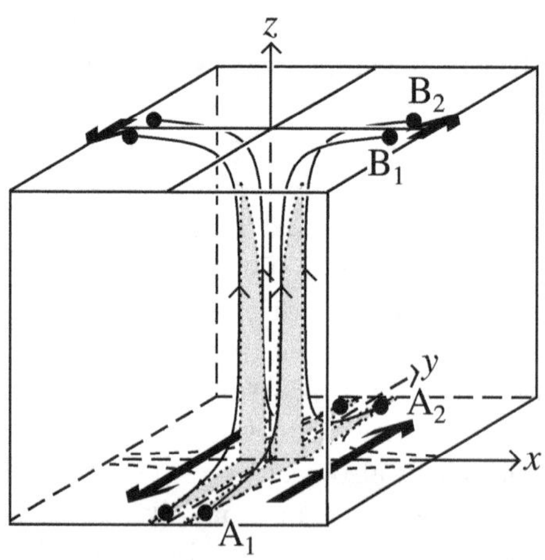

Figure 6.47. Quasi-separatrix layer (shaded) due to large field-line displacement on the bottom of a cube caused by a small displacement on the top. The x, y, z components of the magnetic field are $(x, -y, c)$ with c constant.

where the integrals are taken along field lines. At QSLs, the integrals change their values rapidly with (X_0, Y_0), and so do U and V if $\mathbf{d}_1$ is assumed to vary smoothly. But $\mathbf{j}_1 \times \mathbf{B}_0 = \mathbf{0}$ implies $\mu\mathbf{j}_1 = \alpha(X_0, Y_0)\mathbf{B}_0$, where $\mu\mathbf{j}_1 = \nabla \times \mathbf{B}_1$ in turn implies $\alpha = \partial V/X_0 - \partial U/\partial Y_0$. Thus, the rapid variation of U and V with X_0, Y_0 means that intense currents will be generated along the QSLs.

Later, Titov et al. (2003) showed how boundary motions that create a stagnation point near the quasi-separator make currents build up exponentially along it, while Aulanier et al. (2005) demonstrated the effect in a resistive MHD experiment. Also, the approach has been applied to solar flares, as described in Section 12.3.2.2.

Chapter 7

Instability

7.1 Introduction

Questions of stability and instability are important for many solar phenomena. Sometimes one needs to explain how a structure can remain stable for a long period of time, when preliminary theoretical considerations may suggest it should be unstable. For instance, the Sun seems to maintain the stability of a solar prominence or coronal loop with great ease, in contrast to the immense difficulty of containing plasmas in the laboratory. At other times, one wants to understand why magnetic structures on the Sun suddenly become unstable and produce events of great beauty, such as erupting prominences, coronal mass ejections or solar flares.

The methods employed to investigate the linear stability of a hydromagnetic system are natural generalisations of those for studying a particle in one-dimensional motion. Suppose such a particle has a mass m and moves along the x-axis under the action of a conservative force $F(x)\,\hat{\mathbf{x}} = -(dW/dx)\,\hat{\mathbf{x}}$, where $W(x)$ is the potential energy. Its equation of motion is simply

$$m\ddot{x} = F(x) \equiv -\frac{dW}{dx}.$$

For small displacements from equilibrium at x=0, say, this reduces to the linear form

$$m\ddot{x} \approx F_1(x) \equiv -x\left(\frac{d^2W}{dx^2}\right)_0,$$

where $F_1(x)$ is the first-order approximation to $F(x)$.

One approach is to seek *normal-mode solutions* of the form $x = x_0 e^{i\omega t}$, so that the equation of motion gives

$$\omega^2 = \frac{1}{m}\left(\frac{d^2W}{dx^2}\right)_0.$$

If $W(x)$ has a minimum at the origin (Figure 7.1), then $(d^2W/dx^2)_0 > 0$, which implies that $\omega^2 > 0$, and so the particle oscillates about $x = 0$. The force acting on the particle tends to restore its equilibrium, which is therefore said to be *stable*. If instead $W(x)$ possesses a maximum, $(d^2W/dx^2)_0 < 0$, so that $\omega^2 < 0$, and the displacement ($|x|$) increases in time from the equilibrium position, which is now *unstable*. When $(d^2W/dx^2)_0 = 0$, the particle is said to be *neutrally stable*.

An alternative approach for tackling particle stability is to consider the change (δW) in potential energy due to a displacement (x) from equilibrium. To first order in x, $\delta W = x(dW/dx)_0$, which vanishes by assumption. To second order,

$$\delta W \equiv W(x) - W(0) \simeq \frac{x^2}{2}\left(\frac{d^2W}{dx^2}\right)_0,$$

which may be derived alternatively by noting that the change in potential energy to this order is simply minus the work done by the linear force $[F_1(x)]$, namely,

$$\delta W = -\int_0^x F_1(x)dx = -\tfrac{1}{2}xF_1(x). \tag{7.1}$$

The particle is in stable equilibrium if $\delta W > 0$ for *all* small displacements from $x = 0$, with both $x > 0$ and $x < 0$. It is unstable if $\delta W < 0$ for at least one small displacement, with either $x > 0$ or $x < 0$. The frequency (ω) can be written in terms of δW by eliminating $(d^2W/dx^2)_0$ between the above expressions for ω^2 and δW with the result that

$$\tfrac{1}{2}m\omega^2x^2 = \delta W. \tag{7.2}$$

The stability of an MHD system may be studied in a similar way. First, linearise the equations and then look for either normal modes or the variation of the energy. Each method has its advantage. With the first, a *dispersion relation* can be found, linking the frequency (ω) to the wavenumber (k) of the disturbance, whereas the variational method may be applied to more complex equilibrium states. Occasionally, the normal-mode method may fail, in which case a treatment of the initial-value problem may succeed instead (e.g., Roberts 1967; Goedbloed et al. 2004).

Most of this chapter refers to *linear* stability, but, by considering deviations from equilibrium that are not small, it is possible to investigate the *nonlinear* stability of a system. For instance, as indicated in Figure 7.1, a particle able to move in 1D may be linearly stable but nonlinearly unstable (Figure 7.1e), or linearly unstable but nonlinearly stable (Figure 7.1f). In the former case, the instability is promoted by nonlinear effects and is said to be *explosive*. The term *metastability* refers to a system that is neutrally

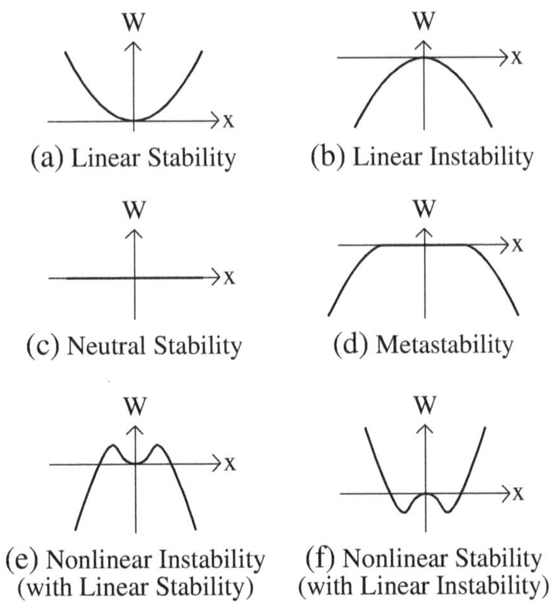

Figure 7.1. Potential energy $W(x)$ for a one-dimensional system in equilibrium at $x = 0$.

stable ($d^2W/dx^2 = 0$) to small-amplitude (linear) perturbations but is unstable to large (finite-amplitude) ones (Figure 7.1d), by having, for instance, $d^3W/dx^3 = 0$ and $d^4W/dx^4 < 0$.

As the parameters of a system are varied, the transfer from stability to instability via a state of *marginal* (or *neutral*) *stability* can occur in two possible ways. If ω^2 is real and it decreases through zero, there is the onset of a monotonic growth in the perturbation (the principle of the exchange of stabilities); the marginal state is then stationary ($\omega = 0$). If, on the other hand, the frequency (ω) in Eq. (7.2) is complex and its imaginary part decreases from positive to negative values, then a state of growing oscillations appears, called *overstability* by Eddington. The marginal state in this case has oscillatory motions at a certain frequency.

The main sources for this chapter are Chandrasekhar (1961) and Bateman (1978). Other good accounts are in Friedberg (1987), Manheimer and Lashmore-Davies (1989) and Wesson (1997) for MHD instabilities in the laboratory and Hood (1985) and Goedbloed et al. (2004) for their role on the Sun. After a derivation of the linearised equations (Sec. 7.2), the use of the normal-mode method (Sec. 7.3) and the energy or variational method (Sec. 7.4) is illustrated, and finally the properties of some of the main instabilities are summarised (Sec. 7.5).

7.2 Linearised Equations

The behaviour of an ideal (dissipationless) MHD system is governed by Eqs. (2.13), (2.19) and (2.20) of Sections 2.1 and 2.2 in the limit as the magnetic diffusivity and viscosity approach zero, namely,

$$\frac{\partial \mathbf{B}}{\partial t} = \boldsymbol{\nabla} \times (\mathbf{v} \times \mathbf{B}), \tag{7.3}$$

$$\rho \frac{d\mathbf{v}}{dt} = -\boldsymbol{\nabla} p + \mathbf{j} \times \mathbf{B} + \rho \mathbf{g}, \tag{7.4}$$

$$\frac{\partial \rho}{\partial t} + \boldsymbol{\nabla} \cdot (\rho \mathbf{v}) = 0, \tag{7.5}$$

where $\mathbf{j} = \boldsymbol{\nabla} \times \mathbf{B}/\mu$ and $\boldsymbol{\nabla} \cdot \mathbf{B} = 0$. Also, the energy equation (2.29a) is approximated here, for simplicity, by

$$\frac{d}{dt}\left(\frac{p}{\rho^\gamma}\right) = 0. \tag{7.6}$$

The condition for the validity of this equation is that the time-scale of interest be much smaller than the time it takes heat to be conducted or radiated away. More general forms can easily be adopted instead (see, for instance, Heyvaerts (1974), who considered thermal instability in the presence of a magnetic field).

Suppose the initial state is in equilibrium with a vanishing plasma velocity ($\mathbf{v}_0 \equiv 0$), together with a magnetic field ($\mathbf{B}_0$), plasma pressure (p_0), density (ρ_0) and electric current ($\mathbf{j}_0$) that are all independent of time; then Eqs. (7.3), (7.5) and (7.6) are satisfied trivially and the remaining equations yield

$$\mathbf{0} = -\boldsymbol{\nabla} p_0 + \mathbf{j}_0 \times \mathbf{B}_0 + \rho_0 \mathbf{g},$$
$$\mathbf{j}_0 = \boldsymbol{\nabla} \times \mathbf{B}_0/\mu,$$
$$\boldsymbol{\nabla} \cdot \mathbf{B}_0 = 0.$$

Next, perturb the equilibrium by setting

$$\rho = \rho_0 + \rho_1, \quad \mathbf{v} = \mathbf{v}_1, \quad p = p_0 + p_1, \quad \mathbf{B} = \mathbf{B}_0 + \mathbf{B}_1, \quad \mathbf{j} = \mathbf{j}_0 + \mathbf{j}_1,$$

and ignore squares and products of the perturbation quantities (denoted by subscript 1); the basic equations then reduce to

$$\frac{\partial \mathbf{B}_1}{\partial t} = \nabla \times (\mathbf{v}_1 \times \mathbf{B}_0), \tag{7.7}$$

$$\rho_0 \frac{\partial \mathbf{v}_1}{\partial t} = -\nabla p_1 + \mathbf{j}_1 \times \mathbf{B}_0 + \mathbf{j}_0 \times \mathbf{B}_1 + \rho_1 \mathbf{g}, \tag{7.8}$$

$$\frac{\partial \rho_1}{\partial t} + \nabla \cdot (\rho_0 \mathbf{v}_1) = 0, \tag{7.9}$$

$$\mathbf{j}_1 = \nabla \times \mathbf{B}_1/\mu, \quad \nabla \cdot \mathbf{B}_1 = 0, \tag{7.10}$$

$$\frac{\partial p_1}{\partial t} = \frac{\gamma p_0}{\rho_0} \frac{\partial \rho_1}{\partial t} - p_0 (\mathbf{v}_1 \cdot \nabla) \log_e \left(\frac{p_0}{\rho_0^\gamma} \right). \tag{7.11}$$

Note in passing that Eq. (7.11) may be rewritten, after using Eq. (7.9), as

$$\frac{\partial p_1}{\partial t} = -(\mathbf{v}_1 \cdot \nabla) p_0 - \gamma p_0 \nabla \cdot \mathbf{v}_1,$$

and so, in the (incompressible) limit as $\gamma \to \infty$, this becomes simply $\nabla \cdot \mathbf{v}_1 = 0$.

The displacement $[\boldsymbol{\xi}(\mathbf{r}_0,t)]$ of a plasma element from equilibrium may be written $\boldsymbol{\xi} = \mathbf{r} - \mathbf{r}_0$, in terms of the resulting position vector ($\mathbf{r}$) and the initial equilibrium position ($\mathbf{r}_0$) (Figure 7.2). By writing Eqs. (7.7) to (7.11) in terms of $\boldsymbol{\xi}$ alone and changing from *Eulerian coordinates* $(\mathbf{r},t)$ to *Lagrangian coordinates* $(\mathbf{r}_0,t)$, it is possible to reduce the whole set of linearised equations to just one equation, as follows.

To lowest order, the vector operator ∇ is identical with $\nabla_0 \equiv \hat{\mathbf{x}}(\partial/\partial x_0) + \hat{\mathbf{y}}(\partial/\partial y_0) + \hat{\mathbf{z}}(\partial/\partial z_0)$. Furthermore, the perturbed velocity vector is $\mathbf{v}_1 = \partial \mathbf{r}/\partial t = \partial(\mathbf{r}_0 + \boldsymbol{\xi})/\partial t$, or

$$\mathbf{v}_1 = \frac{\partial \boldsymbol{\xi}}{\partial t}, \tag{7.12}$$

since $\mathbf{r}_0$ is independent of time. Then the fact that $\mathbf{B}_1$ and $\boldsymbol{\xi}$ vanish initially may be used to integrate Eq. (7.7) with respect to time to give

$$\mathbf{B}_1 = \nabla_0 \times (\boldsymbol{\xi} \times \mathbf{B}_0). \tag{7.13}$$

Similarly, the integral of Eq. (7.9) is

$$\rho_1 = -\nabla \cdot (\rho_0 \boldsymbol{\xi}), \tag{7.14}$$

and so, after substituting for $\mathbf{j}_1, \mathbf{B}_1$ and ρ_1 from Eqs. (7.10), (7.13) and (7.14), respectively, Eq. (7.8) becomes

$$\rho_0 \frac{\partial^2 \boldsymbol{\xi}}{\partial t^2} = \mathbf{F}[\boldsymbol{\xi}(\mathbf{r}_0,t)], \tag{7.15}$$

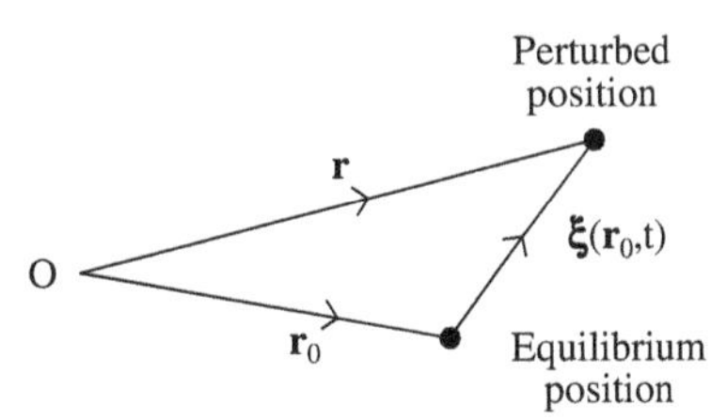

Figure 7.2. Displacement of a plasma element by $\boldsymbol{\xi}$ from an initial equilibrium ($\mathbf{r}_0$) to $\mathbf{r}$.

where

$$\mathbf{F}(\boldsymbol{\xi}) \equiv -\nabla p_1 + \rho_1 \mathbf{g} + \mathbf{j}_1 \times \mathbf{B}_0 + \mathbf{j}_0 \times \mathbf{B}_1,$$
$$= -\nabla p_1 + \nabla \cdot (\rho_0 \boldsymbol{\xi})\, \mathbf{g} + \{\nabla \times [\nabla \times (\boldsymbol{\xi} \times \mathbf{B}_0)]\} \times \mathbf{B}_0/\mu$$
$$+ (\nabla \times \mathbf{B}_0) \times [\nabla \times (\boldsymbol{\xi} \times \mathbf{B}_0)]/\mu. \tag{7.16}$$

To first order, the perturbed total force ($\mathbf{F}$) is a linear functional of $\boldsymbol{\xi}$ and its spatial derivatives, so that, for a displacement of the form

$$\boldsymbol{\xi}(\mathbf{r}_0, t) = \boldsymbol{\xi}(\mathbf{r}_0) e^{i\omega t},$$

the equation of motion (7.15) reduces to

$$-\omega^2 \rho_0 \boldsymbol{\xi}(\mathbf{r}_0) = \mathbf{F}[\boldsymbol{\xi}(\mathbf{r}_0)]. \tag{7.17}$$

Both the normal-mode and energy methods use this equation as a basis. Here the same notation has been adopted as for wave-like perturbations (when ω is real), but this chapter considers instabilities, for which ω is imaginary and $i\omega$ is positive. Note that in some texts our $i\omega$ is replaced by ω.

In general, the time integral of Eq. (7.11) determines p_1 as

$$p_1 = \frac{\gamma p_0}{\rho_0} \rho_1 - p_0 (\boldsymbol{\xi} \cdot \nabla) \log_e \frac{p_0}{\rho_0^\gamma} = -\boldsymbol{\xi} \cdot \nabla p_0 - \gamma p_0 (\nabla \cdot \boldsymbol{\xi}),$$

and two special cases are of interest. Firstly, for a *uniform equilibrium*, ρ_0, p_0, $\mathbf{B}_0$ are all constant and Eq. (7.16) reduces to

$$\mathbf{F} = \rho_0 c_s^2 \nabla (\nabla \cdot \boldsymbol{\xi}) - \nabla \cdot (\rho_0 \boldsymbol{\xi})\, \mathbf{g} + \rho_0 v_A^2 \{\nabla \times [\nabla \times (\boldsymbol{\xi} \times \hat{\mathbf{B}}_0)]\} \times \hat{\mathbf{B}}_0,$$

where $c_s = (\gamma p_0/\rho_0)^{1/2}$ and $v_A = B_0/(\mu \rho_0)^{1/2}$ are the sound and Alfvén speeds. The other special case occurs when density variations are *incompressible* (Sec. 2.4.3). This is obtained by letting the sound speed (c_s) approach infinity. Then the time integral of Eq. (7.11) implies that ρ_1 tends to zero and Eq. (7.11) no longer determines p_1.

7.3 Normal-Mode Method

Once the boundary conditions and the equilibrium configuration have been prescribed, the perturbed variables (ρ_1, $\mathbf{v}_1$, p_1, $\mathbf{j}_1$ and $\mathbf{B}_1$) are determined by the set of linear Eqs. (7.7) to (7.11). Each of these variables may be decomposed into a spectrum of Fourier components, which behave like $e^{i\omega t}$. The resulting normal-mode equations may be solved to determine which values of ω are allowed by the boundary conditions. If *all* the resulting normal modes have real frequencies ($\omega^2 > 0$), the system just oscillates about the equilibrium configuration, which is therefore *stable*. If at least one of the frequencies is imaginary ($\omega^2 < 0$), the system is *unstable*, since the corresponding perturbations grow exponentially.

The method of normal modes is applied below to a particular example using the original set of Eqs. (7.7) to (7.11); but one could just as well solve the normal-mode equation (7.17) for the displacement $\boldsymbol{\xi}(\mathbf{r}_0, t) = \boldsymbol{\xi}(\mathbf{r}_0) e^{i\omega t}$ and deduce the original variables from (7.12) to (7.14), (7.10) and (7.11). Sturm-Liouville theory can be used to show that the eigenvalues (ω^2) of Eq. (7.17) are all real. Thus, overstability cannot occur in a static, ideal MHD system, although the additional presence of shear flow, rotation or dissipation may produce overstability (e.g., Sec. 9.1).

7.3.1 Example: Rayleigh-Taylor Instability

As an example of the use of the normal-mode method, consider a sharp boundary separating two perfectly conducting plasmas with different properties denoted by superscripts + and − (Figure 7.3a).

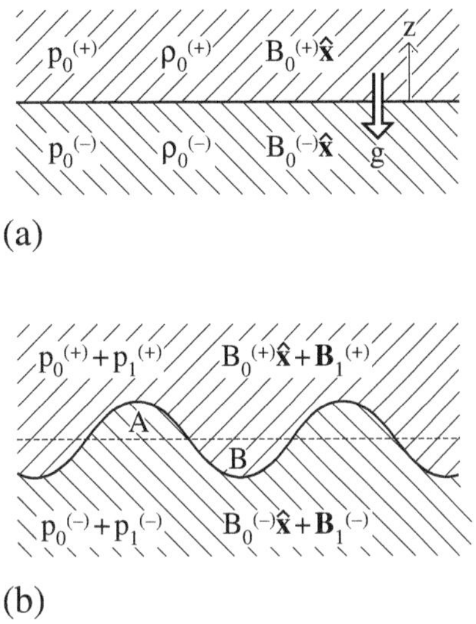

Figure 7.3. (a) The initial and (b) the perturbed interface between two uniform plasmas.

The boundary is situated at $z=0$, and a gravitational force acts normal to it along the negative z-axis; the initial magnetic field is in the x-direction parallel to the boundary.

Consider first the case with no magnetic field, and suppose the interface receives a sinusoidal displacement (Figure 7.3b); plasma of density $\rho_0^{(+)}$ now occupies the region B (previously of density $\rho_0^{(-)}$), while plasma of density $\rho_0^{(-)}$ occupies region A (previously of density $\rho_0^{(+)}$). The net result is that plasma of density $\rho_0^{(+)}$ in A has been exchanged with plasma of density $\rho_0^{(-)}$ in B. If $\rho_0^{(+)} < \rho_0^{(-)}$, this represents a gain of gravitational potential energy, since the net displacement is against gravity. But, if $\rho_0^{(+)} > \rho_0^{(-)}$, so that the denser plasma rests on top of the lighter, a loss of gravitational potential energy results and we have the *Rayleigh-Taylor instability.*

The influence of a horizontal magnetic field on this instability depends on its orientation parallel to the interface. For a magnetic field in the plane of Figure 7.3, the perturbation stretches the field lines and so produces a stabilising force. Field lines normal to the plane of Figure 7.3, however, are just displaced as a whole without being stretched and therefore have no stabilising effect. In order to derive the above results mathematically and find the quantitative effect of the magnetic field, two special cases will be considered of the equilibrium configuration shown in Figure 7.3a. In the first case, plasma above $z=0$ is supported by a magnetic field below $z=0$; in the second, the initial magnetic field is uniform, with a dense plasma resting on top of a lighter one. Incompressible variations alone are considered, so that

$$\nabla \cdot \mathbf{v}_1 = 0, \tag{7.18}$$

which is often the case for the most unstable displacement, since energy is then not wasted in compressing the plasma unnecessarily (see Sec. 7.4.2).

7.3.1.1 Plasma Supported by a Magnetic Field

Consider a plasma of uniform density ($\rho_0^{(+)}$) supported against gravity by a magnetic field ($B_0^{(-)}\hat{\mathbf{x}}$), with the z-axis normal to the plane interface indicated in Figure 7.4a. Assume that the plasma in the region below $z=0$ has a negligible pressure [$p_0^{(-)} \ll B_0^{(-)2}/(2\mu)$] and that the interface itself is in equilibrium under the pressure balance $p_0^{(+)} = B_0^{(-)2}/(2\mu)$. Suppose also that the boundary develops oscillations along the y-direction alone, so that

$$\mathbf{v}_1 = e^{i\omega t}\mathbf{v}_1(z)e^{iky}, \tag{7.19}$$

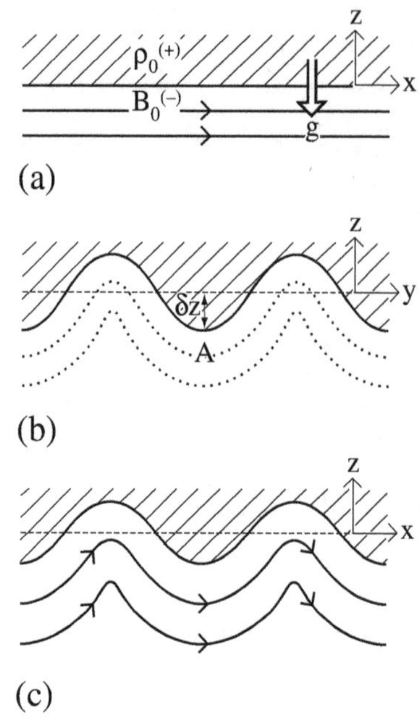

Figure 7.4. Plasma (shaded) supported by a magnetic field ($B_0^{(-)}\hat{\mathbf{x}}$). (a) Equilibrium configuration, plus perturbations that are rippled in (b) the y-direction and (c) the x-direction.

and similarly for all the other variables.

First of all, it is necessary to derive the boundary condition at the interface, which may be regarded as a thin layer where variations in z are much more rapid than those in x or y. The z-component of the time-derivative of Eq. (7.8) is

$$\rho_0 \frac{\partial^2 v_{1z}}{\partial t^2} = -\frac{\partial}{\partial t}\left(\frac{\partial p_1}{\partial z}\right) - \frac{\partial j_{1y}}{\partial t} B_{0x} - j_{0y}\frac{\partial B_{1x}}{\partial t} - \frac{\partial \rho_1}{\partial t} g. \tag{7.20}$$

However, from Eqs. (7.9), (7.10) and (7.7),

$$\frac{\partial \rho_1}{\partial t} = -\frac{d}{dz}(\rho_0 v_{1z}), \quad j_{1y} = \frac{1}{\mu}\frac{\partial B_{1x}}{\partial z} \quad \text{and} \quad \frac{\partial \mathrm{B}_{1x}}{\partial \mathrm{t}} = 0;$$

and so, for the perturbation (7.19), Eq. (7.20) reduces to

$$-\rho_0 \omega^2 v_{1z} = -\omega \frac{dp_1}{dz} + g\frac{d}{dz}(\rho_0 v_{1z}).$$

After being integrated across the interface, this becomes

$$0 = -i\omega[p_1] + g[\rho_0 v_{1z}], \tag{7.21}$$

where square brackets denote the jump in a quantity across the interface, so that, for instance, $[p_1] \equiv p_1^{(+)} - p_1^{(-)}$. Moreover, perturbations of the form (7.19) make Eq. (7.18) and the y-component of Eq. (7.8) reduce to the pair $dv_{1z}/dz + ikv_{1y} = 0$ and $i\rho_0\omega v_{1y} = -ikp_1$, which may be combined to give

$$p_1 = -\frac{i\omega\rho_0}{k^2}\frac{dv_{1z}}{dz}.$$

After using the result from the integral of Eq. (7.18) that v_{1z} is continuous and substituting for p_1, Eq. (7.21) becomes

$$0 = -\frac{\omega^2}{k^2}\left[\rho_0 \frac{dv_{1z}}{dz}\right] + g v_{1z}[\rho_0],$$

which, for our interface, finally reduces to the required boundary condition

$$0 = -\frac{\omega^2}{k^2}\rho_0^{(+)}\left(\frac{dv_{1z}}{dz}\right)^{(+)} + g v_{1z}^{(+)}\rho_0^{(+)}. \tag{7.22}$$

The aim is to find the solution for v_{1z} on both sides of the interface. Assuming that the initial equilibrium values $(\mathbf{B}_0, \rho_0, p_0)$ are uniform and that Eq. (7.18) holds, Eqs. (7.7) and (7.9) give $\partial\mathbf{B}_1/\partial t = \mathbf{0}$ and $\partial\rho_1/\partial t = 0$, so that $\mathbf{j}_1$, $\mathbf{B}_1$ and ρ_1 all vanish and Eq. (7.8) simplifies to $\rho_0 \partial\mathbf{v}_1/\partial t = -\boldsymbol{\nabla} p_1$. The curl of this gives $\boldsymbol{\nabla} \times \mathbf{v}_1 = \mathbf{0}$, which, together with Eq. (7.18), implies that $\nabla^2\mathbf{v}_1 = \mathbf{0}$.

For perturbations (7.19), the z-component of $\nabla^2\mathbf{v}_1 = \mathbf{0}$ becomes

$$\frac{d^2 v_{1z}}{dz^2} - k^2 v_{1z} = 0,$$

and the continuous solution that vanishes as z approaches $\pm\infty$ is

$$v_{1z} = \begin{cases} e^{-kz}, z < 0, \\ e^{+kz}, z > 0. \end{cases} \tag{7.23}$$

From this solution $v_{1z}^{(+)} = 1$ and $dv_{1z}^{(+)}/dz = -k$, so that the boundary condition (7.22) gives $\omega^2 = -gk$. Ripples of the boundary in the y-direction (Figure 7.4b) therefore grow like $\exp(\sqrt{gk}\ t)$, the smallest wavelengths being the fastest.

An alternative explanation for the instability (to the one mentioned above when discussing Figure 7.3) comes from considering the point A in Figure 7.4b. It is depressed by a distance $|\delta z|$, so that a greater mass of plasma exists above it. The plasma pressure at A is therefore larger than $p_0^{(+)}$ by an amount $\rho g|\delta z|$, and this forces the perturbation to grow.

More general perturbations of the form

$$\mathbf{v}_1 = e^{i\omega t}\mathbf{v}_1(z)e^{i(k_x x + k_y y)} \tag{7.24}$$

lead to the boundary condition

$$0 = -\frac{\omega^2}{k^2}\rho_0^{(+)}\left(\frac{dv_{1z}}{dz}\right)^{(+)} - \frac{k_x^2}{k^2}\frac{B^{(-)2}}{\mu}\left(\frac{dv_{1z}}{dz}\right)^{(-)} + g v_{1z}^{(+)}\rho_0^{(+)}$$

and the corresponding dispersion relation

$$\omega^2 = -gk + \frac{k_x^2 B^{(-)2}}{\mu\rho_0^{(+)}},$$

where $k = (k_x^2 + k_y^2)^{1/2}$. Thus, ripples along the direction of the field with $k_y = 0$ (Figure 7.4c) are unstable at long wavelengths ($0 < k < k_{crit}$, where $k_{crit} = g\mu\rho_0^{(+)}/B_0^{(-)2}$), but they are stable at short enough wavelengths ($k > k_{crit}$), since then the magnetic tension dominates. The most rapidly growing mode has wavenumber $\frac{1}{2}k_{crit}$.

This magnetic analogue of the Rayleigh-Taylor instability is known as the *Kruskal-Schwarzschild* (or *hydromagnetic Rayleigh-Taylor) instability.* Both are examples of a wider class of *interchange instabilities,* which includes the *sausage instability* (see Figure 7.10b). Another example is the *flute instability* of plasma contained in a flux tube (Sec. 7.5.1). Other effects that can be included in the above analysis are: dissipation (Furth et al. 1963), non-uniformities in conditions either side of the interface (so that $(dB_0/dz)^{(+)} \neq 0$ in

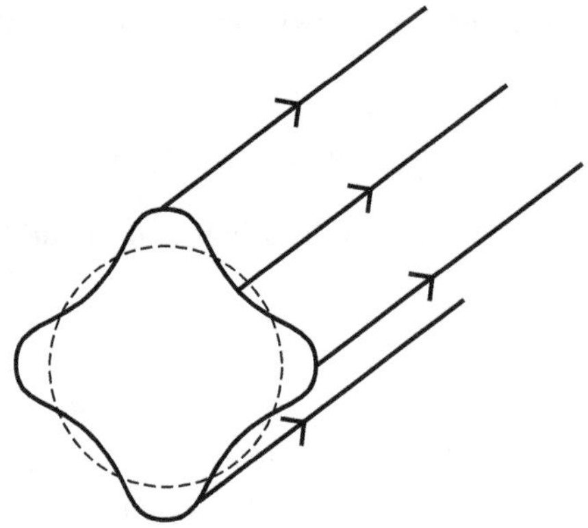

Figure 7.5. The fluted cross-section of a cylindrical column of plasma. Its undisturbed cross-section (shown dashed) is circular, and it is subject to an interchange instability.

deriving the equivalent of Eq. 7.22), and different inclinations of the magnetic field each side of $z=0$. Also, allowance can be made for steady motions in the initial state or for compressible variations.

7.3.1.2 Uniform Magnetic Field ($B_0^{(+)} = B_0^{(-)}$)

Consider instead a uniform horizontal magnetic field ($B_0\hat{\mathbf{x}}$) parallel to the interface that separates uniform plasmas with densities $\rho_0^{(+)}$ and $\rho_0^{(-)}$ (see Figure 7.3a). The velocity perturbation again has the form given by Eq. (7.23), and the boundary condition for perturbations behaving like Eq. (7.24) can be shown to be

$$\left(\frac{\omega^2}{k^2}\rho_0^{(+)} - \frac{k_x^2}{k^2}\frac{B_0^2}{\mu}\right)\left(\frac{dv_{1z}}{dz}\right)^{(+)} + \left(\frac{\omega^2}{k^2}\rho_0^{(-)} - \frac{k_x^2}{k^2}\frac{B_0^2}{\mu}\right)\left(\frac{dv_{1z}}{dz}\right)^{(-)}$$

$$= -g(v_{1z}^{(+)}\rho_0^{(+)} - v_{1z}^{(-)}\rho_0^{(-)}),$$

where $k = (k_x^2 + k_y^2)^{1/2}$. After substituting $(dv_{1z}/dz)^{(+)} = -(dv_{1z}/dz)^{(-)} = -k$ and $v_{1z}^{(+)} = v_{1z}^{(-)} = 1$, it reduces to

$$\omega^2 = -gk\left(\frac{\rho_0^{(+)} - \rho_0^{(-)}}{\rho_0^{(+)} + \rho_0^{(-)}}\right) + \frac{2B_0^2 k_x^2}{\mu(\rho_0^{(+)} + \rho_0^{(-)})}. \tag{7.25}$$

This dispersion relation implies that, when there is no magnetic field present ($B_0 = 0$), the interface is unstable ($\omega^2 < 0$) to the *Rayleigh-Taylor instability* provided the heavy fluid rests on top of the light one ($\rho_0^{(+)} > \rho_0^{(-)}$). Furthermore, disturbances that are uniform along the magnetic field direction ($k_x = 0$) have no effect on stability, but undulations of the boundary purely along the field ($k_y = 0, k_x = k$) make the second term in Eq. (7.25) positive and so allow a stabilising effect. In the latter case, when $\rho_0^{(+)} > \rho_0^{(-)}$ the interface is unstable for wavenumbers $0 < k < k_c$, where $k_c = (\rho_0^{(+)} - \rho_0^{(-)})g\mu/(2B_0^2)$, and the wavenumber for the fastest growing mode is $\frac{1}{2}k_c$. For large wavelengths ($\lambda > 2\pi/k_c$), the magnetic tension is insufficient to counteract gravity; but, for short wavelengths ($\lambda < 2\pi/k_c$), the magnetic tension is strong enough to make the interface stable. This is similar to the effect of surface tension on the Rayleigh-Taylor instability in a liquid.

7.4 Energy (or Variational) Method

By analogy with the method of Section 7.1 for one-dimensional dynamics, the object here is to construct the (second-order) change (δW) in potential energy when a plasma element is displaced by an amount $\boldsymbol{\xi}$ from equilibrium ($\mathbf{r}_0$). This follows the classic treatment of Bernstein et al. (1958), who extended an earlier energy principle of Lundquist (1951). The system is stable if all possible displacements make $\delta W > 0$, and it is unstable if at least one displacement makes $\delta W < 0$.

For an equilibrium plasma, with gravity acting in the negative z-direction, the potential energy is

$$W_0 = \int \left(\frac{B_0^2}{2\mu} + \rho_0 U_0 + \rho_0 g z \right) dV,$$

where the first and last terms represent the magnetic and gravitational energies, U is the internal energy per unit mass, and the integration is performed over the whole volume occupied by the plasma. The first-order perturbations ($\rho_1, p_1, \mathbf{B}_1$) from equilibrium are given in terms of the displacement vector $\boldsymbol{\xi}$ by Eqs. (7.14), (7.11) and (7.13), and so the resulting potential energy may be written in terms of $\boldsymbol{\xi}$ by substituting for $\rho_0 + \rho_1$, $p_0 + p_1$ and $\mathbf{B}_0 + \mathbf{B}_1$ into an integral of the above form. To first order in $\boldsymbol{\xi}$, the change in potential energy vanishes because of the initial equilibrium state. To second order, an expression for δW may be derived most simply in the following way. A displacement of the form $\boldsymbol{\xi}(\mathbf{r}_0,t) = \boldsymbol{\xi}(\mathbf{r}_0)\ e^{i\omega t}$ turns the linearised equation of motion into

$$-\omega^2 \rho_0\ \boldsymbol{\xi}(\mathbf{r}_0) = \mathbf{F}[\boldsymbol{\xi}(\mathbf{r}_0)], \tag{7.26}$$

where $\mathbf{F}$ (given by Eq. 7.16) is the linearised force per unit volume acting on the plasma. The change in potential energy is then just minus the work done by this force during the displacement, namely

$$\delta W = -\tfrac{1}{2} \int \boldsymbol{\xi} \cdot \mathbf{F}\ dV. \tag{7.27}$$

The factor $\frac{1}{2}$ arises because the mean force during the displacement from $\mathbf{0}$ to $\boldsymbol{\xi}$ is $\frac{1}{2}\mathbf{F}(\boldsymbol{\xi})$. Furthermore, the scalar product of Eq. (7.26) with $\boldsymbol{\xi}$ and an integration over the volume give

$$\omega^2 \int \tfrac{1}{2} \rho_0\ \xi^2\ dV = \delta W. \tag{7.28}$$

The left-hand side is minus the plasma kinetic energy, so that, as expected, the kinetic energy gain is positive if the potential energy decreases ($\delta W < 0$) due to a positive amount of work done. Note that Eqs. (7.27) and (7.28) are analogous to Eqs. (7.1) and (7.2) for particle dynamics. For the effect of a steady flow on Eq. (7.26), see PROBLEM 7.1: for all problems and solutions, see web page www.cambridge.org/9780521854719.

The object of the energy method is to investigate stability, subject to a given class of perturbations ($\boldsymbol{\xi}$). This is accomplished by minimising δW and determining its sign. For the particular case when gravity is negligible, it can be shown (Cowling 1976) that, for given components of $\boldsymbol{\xi}$ normal to $\mathbf{B}_0$, the component of $\boldsymbol{\xi}$ along $\mathbf{B}_0$ that minimises δW is given by $\boldsymbol{\nabla} \cdot \boldsymbol{\xi} = 0$. In other words, the most unstable perturbation is incompressible, which is expected since energy would be needed to compress or expand the original plasma. Thus, when searching for instability, one need usually consider only perturbations that make $\boldsymbol{\nabla} \cdot \boldsymbol{\xi} = 0$ and so forget the term arising from plasma pressure in Eq. (7.27).

It proves convenient to rewrite the expression (7.27) for δW, namely

$$\delta W = \tfrac{1}{2} \int \boldsymbol{\xi} \cdot \boldsymbol{\nabla} p_1 - \boldsymbol{\xi} \cdot (\rho_1 \mathbf{g}) - \boldsymbol{\xi} \cdot (\mathbf{j}_1 \times \mathbf{B}_0 + \mathbf{j}_0 \times \mathbf{B}_1)\ dV. \tag{7.29}$$

After substituting for $\mathbf{j}_1$ and $\mathbf{B}_1$ from Eqs. (7.10) and (7.13), the last two terms on the right-hand side may be rearranged, using vector identities as follows:

$$\int \boldsymbol{\xi} \cdot (\mathbf{j}_1 \times \mathbf{B}_0 + \mathbf{j}_0 \times \mathbf{B}_1)\ dV = \int \boldsymbol{\xi} \cdot [(\boldsymbol{\nabla} \times \mathbf{B}_1) \times \mathbf{B}_0/\mu + \mathbf{j}_0 \times \mathbf{B}_1]\ dV,$$

$$= -\mu^{-1} \int \boldsymbol{\xi} \times \mathbf{B}_0 \cdot (\boldsymbol{\nabla} \times \mathbf{B}_1)\ dV + \int \boldsymbol{\xi} \cdot \mathbf{j}_0 \times \mathbf{B}_1\ dV,$$

$$= -\mu^{-1} \int \boldsymbol{\nabla} \cdot [(\boldsymbol{\xi} \times \mathbf{B}_0) \times \mathbf{B}_1] + [\boldsymbol{\nabla} \times (\boldsymbol{\xi} \times \mathbf{B}_0)] \cdot \mathbf{B}_1\ dV + \int \mathbf{j}_0 \cdot \mathbf{B}_1 \times \boldsymbol{\xi}\ dV,$$

or, after using the divergence theorem on the first integral and Eq. (7.13) for $\mathbf{B}_1$,

$$= -\mu^{-1}\int [(\boldsymbol{\xi}\times\mathbf{B}_0)\times\mathbf{B}_1]\cdot d\mathbf{S} - \mu^{-1}\int B_1^2\, dV + \int \mathbf{j}_0\cdot\mathbf{B}_1\times\boldsymbol{\xi}\, dV.$$

If there is no displacement ($\boldsymbol{\xi}$) on the boundary of the volume, the surface integral vanishes and

$$\int \boldsymbol{\xi}\cdot\nabla p_1\, dV = -\int p_1(\nabla\cdot\boldsymbol{\xi})\, dV.$$

Then, after substituting for ρ_1 from Eq. (7.14) and p_1 from Eq. (7.11), the change in potential energy given by Eq. (7.29) becomes

$$\delta W = \tfrac{1}{2}\int [B_1^2/\mu - \mathbf{j}_0\cdot(\mathbf{B}_1\times\boldsymbol{\xi}) + (\boldsymbol{\xi}\cdot\nabla p_0)(\nabla\cdot\boldsymbol{\xi}) + \gamma p_0(\nabla\cdot\boldsymbol{\xi})^2 + (\boldsymbol{\xi}\cdot\mathbf{g})\nabla\cdot(\rho_0\boldsymbol{\xi})]\, dV, \tag{7.30}$$

assuming $p_0/\rho_0^\gamma = constant$, where $\mathbf{B}_1$ is given by Eq. (7.13).

7.4.1 Example: Helical Kink Instability with Line-Tying

Consider a cylindrically symmetric magnetic flux tube, whose equilibrium magnetic field in cylindrical polar coordinates (R, ϕ, z) is

$$\mathbf{B}_0 = B_{0\phi}(R)\ \hat{\boldsymbol{\phi}} + B_{0z}(R)\ \hat{\mathbf{z}},$$

and whose corresponding electric current density is

$$\mathbf{j}_0 = -\frac{dB_{0z}}{dR}\ \hat{\boldsymbol{\phi}} + \frac{1}{R}\frac{d}{dR}(RB_{0\phi})\ \hat{\mathbf{z}}.$$

Suppose now the field is force-free, with $\mathbf{j}_0\times\mathbf{B}_0 = \mathbf{0}$, or equivalently

$$\frac{d}{dR}(B_{0\phi}^2 + B_{0z}^2) = -\frac{2B_{0\phi}^2}{R},$$

which determines $\mathbf{B}_0$ after either $B_{0\phi}(R)$ or $B_{0z}(R)$ has been prescribed.

An intuitive explanation for the instability of the flux tube to a lateral kink-like perturbation (Figure 7.6) is as follows. Suppose the magnetic field lines outside the equilibrium flux tube have only a ϕ-component (Figure 7.6a). Then, after making kink-like displacement (Figure 7.6b), the field lines at A are closer together than those at B. The magnetic pressure is therefore stronger at A than B, and the resulting force from A to B is such as to increase the perturbation further.

This simple argument would be modified if extra features were added, such as an axial magnetic field, a plasma pressure gradient, curvature or line-tying. Bernstein's energy method was first applied to a magnetic

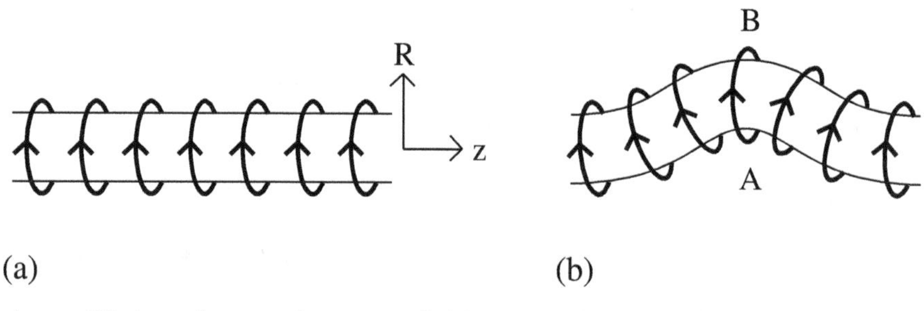

Figure 7.6. (a) An equilibrium plasma tube surrounded by azimuthal magnetic field lines. (b) A lateral kink-like perturbation of the tube.

flux tube by Newcomb (1960) and to a force-free tube by Anzer (1968) and Raadu (1972). Here our object is to illustrate the method by analysing the effect of line-tying at the ends of a force-free coronal tube anchored in the photosphere (Hood and Priest 1979b).

In the force-free approximation, the gravitational and pressure forces are negligible in the equilibrium. Omitting them also in the perturbed state reduces the general expression (7.30) for the second-order potential energy to

$$\delta W = (2\mu)^{-1} \int B_1^2 - \mathbf{B}_1 \cdot \boldsymbol{\xi} \times (\boldsymbol{\nabla} \times \mathbf{B}_0) \, dV, \tag{7.31}$$

where

$$\mathbf{B}_1 = \boldsymbol{\nabla} \times (\boldsymbol{\xi} \times \mathbf{B}_0). \tag{7.32}$$

Here, only a certain class of perturbation ($\boldsymbol{\xi}$) will be considered for simplicity. If the resulting smallest value of δW is negative, the system is certainly unstable; but, if it is positive, the system is stable only to that class of perturbation.

The form adopted for the perturbation is

$$\boldsymbol{\xi} = f(z)\, \boldsymbol{\xi}^* \equiv f(z) \left[\xi^R(R)\hat{\mathbf{R}} - i\frac{B_{0z}}{B_0}\xi^0(R)\hat{\boldsymbol{\phi}} + i\frac{B_{0\phi}}{B_0}\xi^0(R)\hat{\mathbf{z}} \right] e^{i(m\phi+kz)}, \tag{7.33}$$

where $f(0) = f(L) = 0$, so that it vanishes at the ends ($z = 0$ and $z = L$) of the tube. This form also has other useful properties. If the functions $\xi^0(R)$ and $\xi^R(R)$ are purely real, the radial component is out of phase by $\frac{1}{2}\pi$ with the other components. In addition, the perturbation lies in a direction normal to $\mathbf{B}_0$ (i.e., $\boldsymbol{\xi} \cdot \mathbf{B}_0 = 0$), since, by Eqs. (7.31) and (7.32), perturbations parallel to $\mathbf{B}_0$ have no effect on the energy. Furthermore, the presence of the factor $e^{i(m\phi+kz)}$ means that the kinking is helical, screwing out of the plane of Figure 7.6, rather than just being lateral (which could be represented by a factor cos kz cos ϕ). The object now is to choose the arbitrary functions $f(z)$, $\xi^0(R)$ and $\xi^R(R)$ so as to minimise δW.

First, in order to choose $f(z)$, integrate Eq. (7.31) with respect to R and ϕ to give

$$\delta W = \int_0^L A \left(\frac{df}{dz} \right)^2 + Cf^2 \, dz, \tag{7.34}$$

$$\text{where} \qquad A = \int_{R=0}^{\infty} \int_{\phi=0}^{2\pi} [\hat{\mathbf{z}} \times (\boldsymbol{\xi}^* \times \mathbf{B}_0)]^2 \, R \, dR \, d\phi \tag{7.35}$$

and

$$C = \int_{R=0}^{\infty} \int_{\phi=0}^{2\pi} [\boldsymbol{\nabla} \times (\boldsymbol{\xi}^* \times \mathbf{B}_0)] \cdot [\boldsymbol{\nabla} \times (\boldsymbol{\xi}^* \times \mathbf{B}_0) - (\boldsymbol{\nabla} \times \mathbf{B}_0) \times \boldsymbol{\xi}^*] R \, dR \, d\phi \tag{7.36}$$

are both constants. The aim is to minimise the integral in Eq. (7.34) subject to a normalisation constraint

$$\int_0^L f^2(z) \, dz = 1, \tag{7.37}$$

so that, after introducing a Lagrange multiplier (λ), we have

$$\delta W + \lambda = \int_0^L \mathcal{F}(f, f') \, dz,$$

where $\mathcal{F}(f, f') = Af'^2 + (C + \lambda)f^2$ and $f' \equiv df/dz$. Then, according to variational theory, the minimising function satisfies the *Euler-Lagrange equation*, $\partial\mathcal{F}/\partial f = d/dz(\partial\mathcal{F}/\partial f')$, or $(C + \lambda)f = Af''$. The lowest-energy solution that satisfies the boundary conditions $f(0) = f(L) = 0$ and the normalisation (7.37) is

$$f(z) = \left(\frac{2}{L} \right)^{1/2} \sin \frac{\pi z}{L},$$

for which Eq. (7.34) becomes

$$\delta W = \left(\frac{\pi}{L}\right)^2 A + C.$$

Forcing the ends of the flux tube to remain fixed therefore has a stabilising effect, as indicated by the resulting presence of the positive-definite term $(\pi/L)^2 A$.

Next, perform the integrals over ϕ in the expressions (7.35) and (7.36) for A and C, with the result that the energy perturbation (δW) reduces to

$$\delta W = \int_0^\infty F\left(\frac{d\xi^R}{dR}\right)^2 - G(\xi^R)^2 + \frac{(k^2+h^2)R^2+1}{R}(B_0\xi^0 - H)^2 \, dR,$$

where

$$F(R) = \frac{R(B_{0\phi} + kRB_{0z})^2 + h^2R^3B_0^2}{1+(k^2+h^2)R^2}, \tag{7.38}$$

$$G(R) = -\frac{(B_{0\phi} + RkB_{0z})^2}{R} - \frac{(B_{0\phi} - kRB_{0z})^2}{R[1+(k^2+h^2)R^2]}$$

$$-\frac{h^2RB_0^2}{1+(k^2+h^2)R^2} + h^2RB_{0z}^2 - \frac{d}{dR}\left(\frac{B_0^2 + h^2R^2B_{0\phi}^2}{1+(k^2+h^2)R^2}\right),$$

$$H\left(R, \xi^R, \frac{d\xi^R}{dR}\right) = \frac{R}{1+(k^2+h^2)R^2}\left(\frac{d\xi^R}{dR}(kRB_{0\phi} - B_{0z}) - \frac{\xi^R}{R}(kRB_{0\phi} + B_{0z})\right),$$

and $h = \pi/(L)$. Since the integrand contains $\xi^0(R)$ only in quadratic form, it may be minimised with respect to ξ^0 by simply choosing $\xi^0 = H/B_0$, for which δW becomes

$$\delta W = \int_0^\infty F\left(\frac{d\xi^R}{dR}\right)^2 - G\,(\xi^R)^2 dR.$$

Finally, the minimisation of this integral with respect to $\xi^R(R)$ is accomplished by solving the associated Euler-Lagrange equation

$$\frac{d}{dR}\left(F\frac{d\xi^R}{dR}\right) + G\xi^R = 0. \tag{7.39}$$

This is subject to the natural boundary condition for the variational problem, namely $d\xi^R/dR = 0$ at $R = 0$, together with $\xi^R = 1$ at $R = 0$; any other value of $\xi^R(0)$, say C, simply multiplies δW by C^2 without changing its sign. It can be seen from the expression (7.38) for $F(R)$ that the differential equation (7.39) has just one singular point, namely $R = 0$. Newcomb (1960) established that, if $\xi^R(R) > 0$ for all R, then $\delta W > 0$ and so the perturbation is stable, whereas if $\xi^R(R)$ vanishes somewhere we have an instability. Once the components $(B_{0\phi}, B_{0z})$ of the equilibrium field are specified, the technique is therefore to test the solution for the presence of zeros for each wavenumber (k). Typical solutions are shown in Figure 7.7.

As a particular example, consider the *uniform-twist* force-free field.

$$B_{0z} = \frac{B_0}{1+[\Phi R/L]^2}, \qquad B_{0\phi} = \frac{B_0[\Phi R/L]}{1+[\Phi R/L]^2},$$

for which the parameter Φ is the angle of twist of a field line about the axis in going from one end of the tube to the other. The resulting stability diagram (Figure 7.8) indicates that the magnetic flux tube first goes unstable to kinking of the form (7.33) when the twist exceeds about 3.3π. This differs from the 2π of the *Kruskal-Shafranov limit* (Sec. 7.5.3), because account has been taken here of the flux-tube structure. It should be noted that only the form (7.33) was tested for stability. Thus, when $\Phi < 2\pi$ the

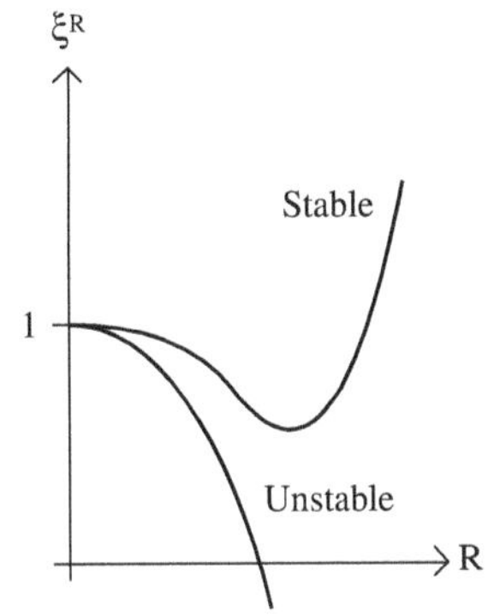

Figure 7.7. Typical solutions for a flux tube to the Euler-Lagrange equation (7.39) for the radial component (ξ^R) of the minimising perturbation as a function of radius (R).

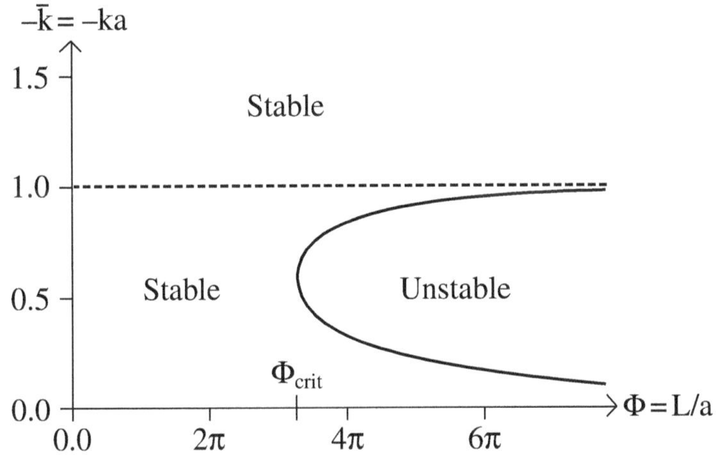

Figure 7.8. The helical kink stability diagram for a line-tied uniform-twist force-free flux tube of length L and effective width a, where k is the wavenumber of the perturbation along the tube and Φ is the twist (from Hood and Priest 1979b).

tube is certainly stable, and when $\Phi > 3.3\pi$ it has certainly become unstable, but there may be some other form for the perturbation that goes unstable at a lower threshold. Indeed, a full solution of the partial differential equations of motion later showed that the real threshold for instability lies at 2.5π (Hood and Priest 1981). Also, line-tying is described in Section 12.2.5.

It is unclear, without modelling stratification properly including pressure gradients, what is the best boundary condition to adopt at the ends of a loop to simulate photospheric line-tying. Possibilities include either $\xi_\perp = 0$, $\boldsymbol{\xi} = \mathbf{0}$, or $\mathbf{B}_1 = \mathbf{0}$.

7.4.2 Use of the Energy (or Variational) Method

There are several points to note about the use of the energy method.

(a) If a displacement ($\boldsymbol{\xi}$) is discovered that makes $\delta W < 0$, the plasma is certainly unstable, although a more unstable mode may exist. It is more difficult to prove stability, since one needs $\delta W > 0$ for *all* possible displacements. In practice, one often considers only a certain class of displacements; some skill is needed to choose the particular class when seeking an instability. Most $\boldsymbol{\xi}$s will not satisfy the equation of motion, but

Laval et al. (1965) established that, if a particular $\boldsymbol{\xi}$ makes δW negative, then a physical perturbation does exist which grows exponentially.

(b) For a given displacement, Eq. (7.28) may used to determine the frequency of the oscillation ($\omega^2 > 0$) or the growth rate of the instability ($\omega^2 < 0$). Note, however, that, if there exists a band of unstable wavenumbers, the most unstable mode (with the largest value of $|\omega^2|$) can be identified as the one with the largest value of $|\delta W|$ only if the normalisation $\int \frac{1}{2}\rho_0 \xi^2 dV = $ *constant* is employed.

(c) The minimisation of δW is often performed with a constraint in order to rule out a trivial solution. For instance, the constraint given by Eq. (7.37) debars the solution $f(z) \equiv 0$, which would make $\delta W = 0$. The choice of such a normalisation is at our disposal, provided it does not change the sign of δW.

(d) The expression (7.30) for δW is derived under the assumption that $\boldsymbol{\xi}$ vanishes on the boundary; so, if some alternative boundary condition is employed, one must return to the more fundamental expression (7.29).

(e) For a uniform equilibrium pressure (p_0), the terms involving p_0 in (7.30) reduce to $\gamma p_0(\boldsymbol{\nabla}\cdot\boldsymbol{\xi})^2$, which is positive. Thus, a uniform plasma pressure has a stabilising influence and the perturbation that minimises this term alone (to zero) is an incompressible one (which is also true when p_0 is non-uniform and gravity is negligible).

(f) If $\delta W = 0$ to second order, we may need to go to third or fourth order.

(g) Bernstein's energy principle has been extended by Frieman and Rotenberg (1960) to include steady flows in the equilibrium state. The result is the addition of extra terms to the expression for δW and the possibility of overstable modes.

(h) Here only *linear stability* has been considered. For nonlinear methods, see, for example, Spies (1974), Drazin and Reid (1981), Biskamp (1993). Also, as well as macroscopic instabilities that are described by magnetohydrodynamics, a plasma may be subject to microscopic instabilities that lead to changes in the particle distribution functions according to kinetic theory (e.g., Mikhailovsky 1974; Boyd and Sanderson 2003).

7.5 Summary of Instabilities

The previous sections have described the normal-mode and energy methods for testing stability. These are used in several other chapters, but here the properties of some of the basic instabilities that may be relevant to the Sun are summarised.

7.5.1 Interchange Instability

Consider an interface between two plasmas with different pressures and magnetic field strengths. A so-called *interchange* (or *exchange*) instability arises when two neighbouring bundles of field lines can be interchanged in such a way that the volumes occupied by the plasma remain constant but the magnetic energy decreases and so feeds the instability. The perturbed interface is rippled, with magnetic field lines parallel to the crests and troughs of the ripples. The instability is sometimes referred to as a *flute instability*, especially when the one plasma occupies a cylinder, so that its perturbed shape resembles a fluted column (Figure 7.5).

A special case occurs when plasma containing no magnetic field is confined by a field whose lines of force are *concave* towards the plasma (Figure 7.9a). (By contrast, plasma contained by *convex* field lines is stable.) Consider the fluted displacement of such an interface (Figure 7.9b). If the volume occupied by the plasma is unaltered, the pressure of the plasma does no work and its internal energy is unchanged. The volume (V) occupied by the magnetic field remains the same, so the curvature of the boundary means that the cross-sectional area of region 2 exceeds that of region 1 (i.e., $A_2 > A_1$). But the net effect of the perturbation is to transfer flux from region 1 to region 2, so that the fluxes in the two regions are the same, namely $B_2A_2 = B_1A_1$.

The final magnetic field strength (B_2) must therefore be smaller than its initial value (B_1). Thus, during the course of the perturbation, the magnetic field lines have shortened and moved farther apart, with

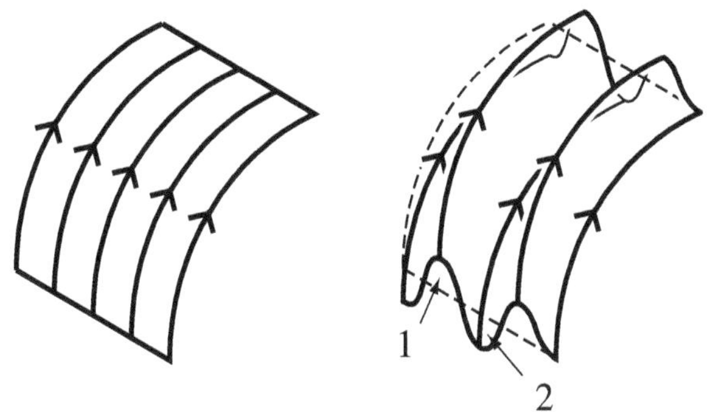

Figure 7.9. (a) Part of the concave surface of a magnetic field that confines plasma in some region. (b) A flute-like displacement of the interface.

an accompanying fall in field strength (cf., Sec. 2.9.2). The change in magnetic energy produced by the displacement is

$$\frac{B_2^2}{2\mu}V - \frac{B_1^2}{2\mu}V = \frac{B_1^2 V}{2\mu}\left(\frac{A_1^2}{A_2^2} - 1\right).$$

This is negative, and so the magnetic energy is reduced and instability ensues. The perturbation grows like $e^{i\omega t}$, where, if the wrinkles have a wavenumber k and the radius of curvature in the interface is R_c, then approximately

$$\omega^2 = -\frac{2pk}{\rho R_c}. \tag{7.40}$$

The instability tends to be inhibited if the lines of force are sheared in such a way that it is difficult for them to be interchanged. Another stabilising effect is the anchoring of the field lines at points a distance L apart, say; this is relevant to coronal magnetic fields that are tied to relatively dense photospheric material. The influence of the line-tying propagates along the field at the Alfvén speed $[v_A \equiv B/(\mu\rho)^{1/2}]$, so that the expression for the growth-rate in Eq. (7.40) becomes modified to

$$\omega^2 = -\frac{2pk}{\rho R_c} + \frac{v_A^2}{L^2},$$

approximately. If the interface is limited in a direction normal to the magnetic field by a distance r_0, so that the minimum wavenumber (k) is roughly r_0^{-1}, it can be seen that line-tying stabilises the longest-wavelength perturbations when $(2\mu p/B^2) > (r_0 R_c/L^2)$; but sufficiently short waves are unstable in this approximation.

There is a wide variety of interchange instabilities, other examples being the *Rayleigh-Taylor instability*, the *Kelvin-Helmholtz instability* and the *sausage instability* treated in the following. Indeed, when an interface is plane, an interchange of magnetic field lines in two regions (with $\mathbf{k}\cdot\mathbf{B} = 0$) is often the most likely to give instability. This is because the field lines are not stretched during such a perturbation, and so no work is done against magnetic tension forces.

7.5.2 Rayleigh-Taylor Instability

Consider two incompressible, inviscid plasmas of uniform densities $\rho_0^{(-)}$ and $\rho_0^{(+)}$, separated by a horizontal boundary, with gravity acting vertically downwards. The plasma of density $\rho_0^{(+)}$ rests on top of the other. If it is rarer ($\rho_0^{(+)} < \rho_0^{(-)}$), the system is stable, but if it is denser ($\rho_0^{(+)} > \rho_0^{(-)}$), the system is unstable to a fluting of the boundary (Figure 7.3). Perturbations like $e^{i\omega t}$ grow at a rate $|\omega|$ given by

$$\omega^2 = -gk\,\frac{\rho_0^{(+)} - \rho_0^{(-)}}{\rho_0^{(+)} + \rho_0^{(-)}}, \tag{7.41}$$

as proved in PROBLEM 7.2.

The effect of a uniform vertical magnetic field ($B_0\,\hat{\mathbf{z}}$) is to modify the growth-rate but not to change the stability. For very long waves ($k \to 0$), the expression (7.41) is unaltered, but, for short waves ($k \to \infty$), the growth-rate is reduced to the value $i\omega \approx (g\sqrt{\mu}/B_0)(\sqrt{\rho_0^{(+)}} - \sqrt{\rho_0^{(-)}})$, which is independent of k.

A uniform horizontal field ($B_0\,\hat{\mathbf{x}}$) gives the dispersion relation

$$\omega^2 = -gk\,\frac{\rho_0^{(+)} - \rho_0^{(-)}}{\rho_0^{(+)} + \rho_0^{(-)}} + \frac{2B_0^2 k_x^2}{\mu(\rho_0^{(+)} + \rho_0^{(-)})},$$

in place of Eq. (7.41). This implies that the magnetic field produces no additional effect when the wavenumber is normal to the field ($k_x = 0$); ripples along the field ($k = k_x$), however, produce a restoring force through the magnetic tension, which allows instability only when $0 < k < k_c$, where $k_c = (\rho_0^{(+)} - \rho_0^{(-)})\, g\mu/(2B_0^2)$. For the effect of a continuously stratified medium, see PROBLEM 7.3.

The *hydromagnetic Rayleigh-Taylor instability* (Kruskal and Schwarzschild 1954) occurs when a plasma of density $\rho_0^{(+)}$ is supported against gravity by a magnetic field ($B_0^{(-)}\hat{\mathbf{x}}$), as shown in Figure 7.4. The dispersion relation is

$$\omega^2 = -gk + \frac{k_x^2 B_0^{(-)2}}{\mu\rho_0^{(+)}}, \tag{7.42}$$

so that the most unstable mode flutes the boundary with $k_x = 0$ and a growth-rate

$$i\omega = (gk)^{1/2}. \tag{7.43}$$

Disturbances with $k_x \neq 0$ distort the field lines, and so the magnetic tension tries to restore equilibrium, as indicated by the second term on the right of Eq. (7.42). If the ends of the field lines are fixed a distance L apart, the lines bend when displaced; an Alfvén wave propagates at speed $v_A = B_0^{(-)}/(\mu\rho_0^{(+)})^{1/2}$ and Eq. (7.43) is modified to $\omega^2 \approx -gk + (v_A^2/L^2)$, so that long waves [with $k < v_A^2/(L^2 g)$] are stabilised. The more general situation of a non-uniform field is dealt with in PROBLEM 7.4.

Other analogues of Rayleigh-Taylor instability occur when: plasma is forced into a vacuum with acceleration f; or its boundary is curved (with radius of curvature R_c), so that the containing field lines have a tension of magnitude $2p/R_c$ (Eq. 7.40); or the plasma is rotating with angular speed Ω, so that the centripetal acceleration at a distance r_0 is $\Omega^2 r_0$. In each of these cases, the growth-rate is given by Eq. (7.43) with g replaced by f, $2p/(\rho R_c)$ or $\Omega^2 r_0$, respectively.

7.5.3 Instability of a Cylindrical Tube

A *linear pinched discharge* is a cylindrical laboratory plasma column (of radius a), confined (or 'pinched') by the azimuthal magnetic field due to a current ($J\,\hat{\mathbf{z}}$) flowing along its surface or through its interior (Figure 7.10a). Since it is similar to a flux tube in the solar interior or atmosphere, we summarise its stability properties here; instability of a curved tube is described in Section 12.2.3. The radially inward $\mathbf{j} \times \mathbf{B}$ force is balanced by an outward pressure gradient. When the plasma column (at pressure p_0 and density ρ_0) contains no magnetic field, it is unstable to an interchange mode, since the confining field is concave to the plasma (see PROBLEM 7.5).

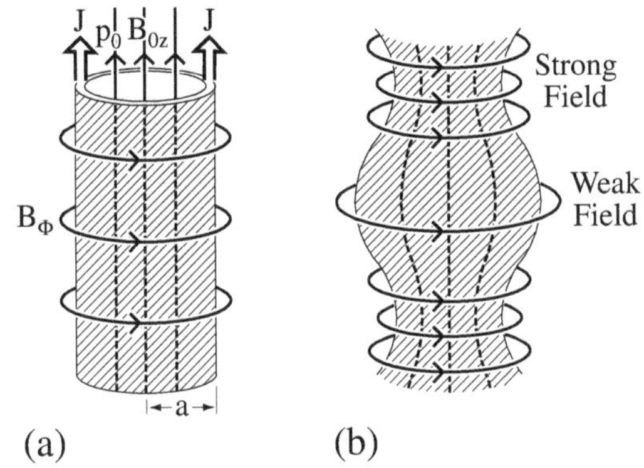

Figure 7.10. (a) A linear pinch containing plasma at pressure p_0 and magnetic field strength $B_{0z}\,\hat{\mathbf{z}}$. A current $(J\,\hat{\mathbf{z}})$ flows on the surface and produces a field $(B_\phi\,\hat{\boldsymbol{\phi}})$ in the surrounding region. (b) A sausage perturbation of the interface.

From Equation (7.40), the growth-rate for this *sausage instability* (Figure 7.10) is $i\omega = [2p_0k/(\rho_0 a)]^{1/2}$. Modes destroying the whole column $(k \approx a^{-1})$ have growth-rate

$$i\omega = \frac{(2p_0/\rho_0)^{1/2}}{a}. \tag{7.44}$$

The linear pinch can be stabilised against the sausage mode by the presence of a large enough axial field (B_{0z}) inside the plasma. If the value of the azimuthal field at the interface is B_ϕ, force balance on the interface gives $p_0 + B_{0z}^2/(2\mu) = B_\phi^2/(2\mu)$, and the effect of an Alfvén wave propagating along the axis with speed $B_{0z}/(\mu\rho_0)^{1/2}$ is to modify the dispersion relation from Eq. (7.44) to approximately

$$\omega^2 = -\frac{2p_0}{\rho_0 a^2} + \frac{B_{0z}^2}{\mu\rho_0 a^2}.$$

Substitution for p_0 from the above force-balance in this expression then gives stability $(\omega^2 > 0)$ when $B_{0z}^2 > \frac{1}{2}B_\phi^2$.

The linear pinch (with a purely azimuthal external field B_ϕ) is also subject to *kink instability* (Lundquist 1951; Kruskal and Schwarzschild 1954). A *helical kink* perturbation $\{\boldsymbol{\xi} = \boldsymbol{\xi}(R)\exp[i(\phi + kz) + i\omega t]\}$ gives instability for all axial wavenumbers. The same is true for a *lateral kink* (Figure 7.6) with a perturbation proportional to $e^{i\phi}\cos kz$, obtained by superposing two oppositely twisted helical perturbations like $e^{i(\phi+kz)}$ and $e^{i(\phi-kz)}$. For such displacements the cross-section of the column remains circular, but other modes have been considered behaving like $e^{i(m\phi+kz)}$ with $m \geq 2$ and leading to a deformation of the section.

Kink instability cannot be stabilised by an axial field outside the plasma column, since the above helical perturbation is unstable provided the wavelength $(-2\pi/k)$ is long enough that $B_\phi/R + kB_z \geq 0$. At equality, the wavenumber vector $(0, R^{-1}, k)$ is perpendicular to the equilibrium field $(\mathbf{k}\cdot\mathbf{B} = 0)$, so that the crests and troughs of the perturbation follow the field as it spirals around the axis. In other words, the perturbation and the initial field have the same twist and pitch (Eqs. 2.79 and 2.80). Another consequence of $\mathbf{k}\cdot\mathbf{B} = 0$ is that it makes the perturbed Lorentz force $(\mathbf{j}_1 \times \mathbf{B}_0)$ vanish. The above criterion may be written in terms of the twist $\Phi = 2LB_\phi/(RB_z)$ of a field line about the tube axis as it goes from one end of the tube to the other, a distance L. Thus, *for any given twist*, the flux tube is locally unstable to a helical kink with wavenumber (the *Kruskal-Shafranov criterion*)

$$k \geq -\frac{\Phi}{L}. \tag{7.45}$$

For the particular case of a laboratory *torus* of major radius R_0, two points located an axial distance $L = 2\pi R_0$ apart refer to the same location on the torus, and so k equals $-2\pi/L$ (or a multiple thereof). Helical kink instability is therefore present in such a torus when $\Phi \geq 2\pi$ (Kruskal and Schwarzschild 1954). An interesting solar example is a coronal loop, for which the Kruskal-Shafranov analysis has been modified to include line-tying (Hood and Priest 1979b) – that is, the stabilising effect of the dense photosphere anchoring the ends of the loop (Sec. 12.2.5).

In general, the electric current may flow through the whole of the pinched plasma and not just through a sheet on its surface. Then the plasma pressure $[p(R)]$ and reciprocal pitch $\tilde{\mu}(R) = B_\phi(R)/[2\pi R B_z(R)]$ (Eq. 2.80) vary with radius. A necessary condition for such an infinitely long 'diffuse' pinch to be stable to kinks is

$$\frac{R}{4}\left(\frac{\tilde{\mu}'}{\tilde{\mu}}\right)^2 + \frac{2\mu p'}{B_z^2} \geq 0 \qquad (7.46)$$

at every point in the plasma, where a prime denotes differentiation with respect to R. This is known as *Suydam's criterion* (Suydam 1959). If the pitch is stable to kinking, then the criterion must be satisfied, but there may be examples satisfying the criterion which are unstable. If Eq. (7.46) is not satisfied, the infinitely long pinch is unstable (without line-tying). A general method for treating the stability of a diffuse pinch has been presented by Newcomb (1960) (see Sec. 7.4) and applied by Hood and Priest (1979b, 1981) to a line-tied force-free coronal loop of uniform twist, which is found to become kink unstable when its twist exceeds 2.5π.

Even if a flux tube were not unstable to sausage or kink modes, too much twist would mean that it is no longer under tension and becomes unstable to *buckling* (Parker 1974). A flux rope with only azimuthal and longitudinal components $[B_\phi(R), B_z(R)]$ buckles when the total stress across a section $(\pi/\mu)\int_0^\infty (B_z^2 - B_\phi^2)R\, dR$ is negative. In other words, the magnetic pressure of B_ϕ exceeds the magnetic tension of B_z, which is true when $B_\phi > B_z$ across enough of the section.

7.5.4 Hydrodynamic Instability

Classical hydrodynamic instability is a huge topic (e.g., Drazin and Reid 1981) with many implications for MHD. Laminar viscous flows in channels, boundary layers, jets and shear layers become unstable and develop turbulence when the *Reynolds number* ($Re \equiv L_0 V_0/\nu$) exceeds a critical value (Re^*), typically 40,000, where L_0 is a typical length (e.g., the half-distance between channel boundaries), V_0 is a typical fluid speed and ν is the kinematic viscosity. A linear viscous flow profile $[v_z(x) = V_0 x/L_0]$ gives so-called *plane Couette flow*, whereas a parabolic profile $[v_z(x) = V_0(1 - x^2/L_0^2)]$ is known as *plane Poiseuille flow* (and the corresponding flow in a circular pipe is *Poiseuille flow*). The linearised hydrodynamic equations lead to a 4th-order equation (the *Orr-Sommerfeld equation*) for the amplitude of the perturbed stream function. The effect of a magnetic field on such flows is to raise the value of Re^*.

The steady circular flow of an incompressible liquid [with angular speed $\Omega(R)$] between two coaxial rotating cylinders is known as *Couette flow*. According to *Rayleigh's criterion*, such a flow is stable if and only if $d/dR(R^4\Omega^2) > 0$ everywhere, and is unstable if the same quantity is negative anywhere. The condition for stability is therefore that the angular momentum per unit mass ($R^2|\Omega|$) increase with distance from the axis, which ensures that an (angular momentum conserving) interchange of (equally massive) elementary fluid rings produces an increase of kinetic energy (Chandrasekhar 1961), which itself requires a source of energy. Both viscosity and a magnetic field have stabilising influences (Chandrasekhar 1961). An arbitrarily small axial field (B_z) can stabilise the flow if $d/dR(|\Omega|) > 0$, while an azimuthal field $[B_\phi(R)]$ makes the motion stable if and only if

$$\frac{d}{dR}(R^4\Omega^2) - \frac{\Omega_2}{\Omega_1}\frac{R^4}{\mu\rho}\frac{d}{dR}\left(\frac{B_\phi^2}{R^2}\right) > 0,$$

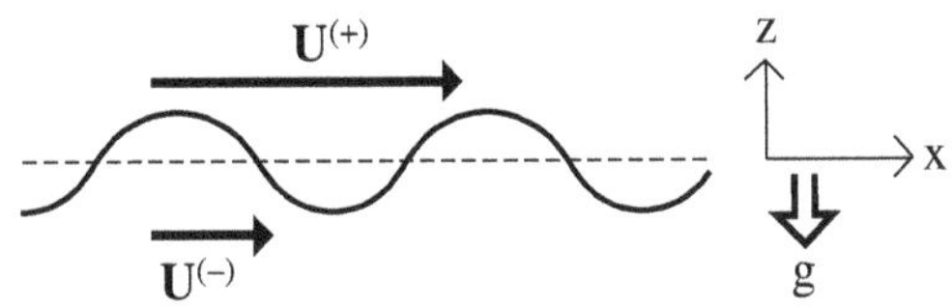

Figure 7.11. Kelvin-Helmholtz instability for superposed fluids in relative motion.

where Ω_1 and Ω_2 are the angular speeds of the inner and outer cylinders.

When one uniform, inviscid fluid (of density $\rho^{(+)}$, say) rests on top of another (of density $\rho^{(-)}$) and the two are in relative horizontal motion, they are subject to the *Kelvin-Helmholtz instability.* Suppose $\rho^{(+)} < \rho^{(-)}$, so that the interface is (Rayleigh-Taylor) stable in the absence of flow; and that their velocities are $U^{(+)}\hat{\mathbf{x}}$ and $U^{(-)}\hat{\mathbf{x}}$ (Figure 7.11). Then interchanges with wavenumber $k\,\hat{\mathbf{x}}$ are unstable if

$$k > \frac{g(\rho^{(-)^2} - \rho^{(+)^2})}{\rho^{(-)}\rho^{(+)}(U^{(-)} - U^{(+)})^2}.$$

Thus, no matter how small the relative speed $(U^{(-)} - U^{(+)})$, instability will occur at a sufficiently small wavelength.

If the density $[\rho(z)]$ and horizontal flow speed $[U(z)]$ are instead continuous functions of z such that the *Richardson number* $Ri \equiv -[g d\rho/dz]/[\rho(dU/dz)^2] > \frac{1}{4}$, then the flow is stable, since the kinetic energy released during a vertical interchange of neighbouring elements is insufficient to do the necessary work against gravity. The role of shear-flow instabilities (when $Ri < \frac{1}{4}$) has not yet been considered much in a solar context (e.g., in the photosphere or at the boundary of a coronal hole).

The effect on Kelvin-Helmholtz instability of uniform horizontal magnetic fields ($\mathbf{B}^{(-)}$ and $\mathbf{B}^{(+)}$) depends on their orientations relative to the flow. If they are both in the y-direction, perturbations with wavenumber $k\hat{\mathbf{x}}$ simply exchange lines of force without stretching them, and the instability is unaffected. On the other hand, if the fields are parallel to the flow, such a perturbation stretches field lines and produces a restoring force: instability is suppressed provided the fields are so strong that

$$\frac{B^{(-)^2} + B^{(+)^2}}{\mu\rho^{(+)}\rho^{(-)}}(\rho^{(-)} + \rho^{(+)}) \geq (U^{(-)} - U^{(+)})^2.$$

[See PROBLEM 7.6 and Chandrasekhar (1961) or Cowling (1976).]

7.5.5 Resistive Instability

The magnetic field in a current sheet of width l diffuses on a time-scale $\tau_d = l^2/\eta$, where $\eta = (\mu\sigma)^{-1}$ is the magnetic diffusivity. During diffusion, magnetic energy is converted ohmically into heat at the same rate, but solar values of τ_d are enormous unless l is tiny (Sec. 2.6.1). However, Furth et al. (1963) showed how the diffusion can drive three distinct instabilities and hence convert magnetic energy into heat and kinetic energy at a much faster rate. These instabilities occur provided the sheet is wide enough that $\tau_d \gg \tau_A$, where $\tau_A = l/v_A$ is the time taken to traverse the sheet at the Alfvén speed $[v_A = B_0/(\mu\rho_0)^{1/2}]$. They occur on time-scales $\tau_d(\tau_A/\tau_d)^\delta$, where $0 < \delta < 1$ (see the following), and have the effect of creating in the sheet many small-scale magnetic loops, which subsequently diffuse away and release magnetic energy.

During the instability, the magnetic field slips through the plasma in a region of width εl about the centre of the sheet, where the magnetic field vanishes. At the edge of this diffusion region, the magnetic field is εB_0, namely, a fraction (ε) of the value (B_0) at the edge of the sheet (Figure 7.12). If the velocity with which plasma enters the diffusion region is $-v_x\hat{\mathbf{x}}$, the resulting current is $\mathbf{j} \approx \sigma(\mathbf{v}\times\mathbf{B}) = \sigma v_x(\varepsilon B_0)\hat{\mathbf{z}}$

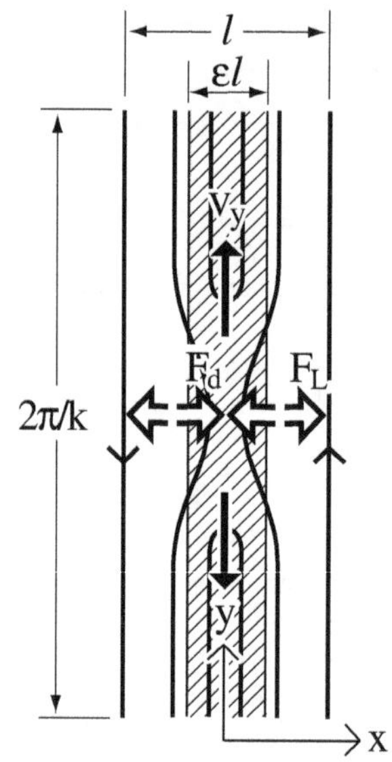

Figure 7.12. Resistive instability in a current sheet for which the driving force (F_d) exceeds the restoring force (F_L). Significant diffusion takes place over a fraction (ϵ) of the width (l) of the sheet. Here just one wavelength ($2\pi/k$) of possibly many is shown.

and the Lorentz force ($\mathbf{F}_L$) which opposes the flow is roughly

$$\mathbf{F}_L \equiv \mathbf{j} \times \mathbf{B} \approx -\sigma v_x (\varepsilon B_0)^2 \hat{\mathbf{x}}, \tag{7.47}$$

where $v_x < 0$ for $x > 0$. Now, if an instability is to take place, this restoring force must be exceeded by a driving force ($\mathbf{F}_d$, say), which has roughly the same magnitude as $\mathbf{F}_L$. The rate of working of $\mathbf{F}_d$ during the instability is therefore

$$-\mathbf{v} \cdot \mathbf{F}_L \approx \sigma v_x^2 (\varepsilon B_0)^2. \tag{7.48}$$

The effect of the driving force is to accelerate plasma at a distance $2\pi/k$ along the sheet to a speed v_y, since the instability has a wavenumber k ($\gg l^{-1}$) in this direction. For an incompressible plasma, the equation $\nabla \cdot \mathbf{v} = 0$ implies that v_y is given in terms of the slow inflow speed ($-v_x$) by $kv_y + (\varepsilon l)^{-1} v_x \approx 0$, or $v_y \approx -v_x/(k\varepsilon l)$. The rate of increase in kinetic energy over a time $(i\omega)^{-1}$ is therefore $i\omega\rho_0 v_y^2 \approx [\omega\rho_0 v_x^2]/(k\varepsilon l)^2$, which, when equated to the work done driving the instability (Eq. 7.48), gives $(\varepsilon l)^4 = [i\omega\rho_0 l^2]/[\sigma k^2 B_0^2]$, or, in terms of the time-scales τ_d and τ_A,

$$\varepsilon^4 = \frac{i\omega}{(kl)^2} \frac{\tau_A^2}{\tau_d}. \tag{7.49}$$

For two of the resistive instabilities, one can equate as follows (in order of magnitude) the particular driving force ($\mathbf{F}_d$) to $-\mathbf{F}_L$ given by Eq. (7.47), and so produce a relationship between ε, ω and k, which, together with Eq. (7.49), determines the dispersion relation for ω as a function of k.

The *gravitational mode* exists when a gravitational (or equivalent) force ($\rho g \hat{\mathbf{x}}$) across the current sheet acts to produce a density stratification $[\rho_0(x)]$. Plasma is transported without compression, so that mass continuity ($\partial\rho_1/\partial t + v_x d\rho_0/dx = 0$) determines the density changes; the driving force ($\rho_1 g \hat{\mathbf{x}}$) then becomes

$$\mathbf{F}_d = -\frac{v_x}{\omega} \frac{d\rho_0}{dx} g\, \hat{\mathbf{x}}, \tag{7.50}$$

and it pushes plasma into the sheet if $d\rho_0/dx > 0$. Equating Eq. (7.50) to Eq. (7.47) and using Eq. (7.49) to eliminate ε gives in terms of a gravitational time-scale (τ_G)

$$i\omega = \left(\frac{(kl)^2\tau_A^2}{\tau_d\tau_G^4}\right)^{1/3}, \qquad \text{where} \qquad \tau_G = \left(-\frac{g}{\rho_0}\frac{d\rho_0}{dx}\right)^{-1/2}.$$

The *rippling mode* instability occurs when there is a spatial variation across the sheet in magnetic diffusivity $[\eta_0(x)]$. This may arise from the temperature structure of the basic state. If the diffusivity at a point varies due to advection with the plasma, we have $\partial\eta/\partial t + \mathbf{v}\cdot\nabla\eta = 0$, and so linear departures (η_1) on a time-scale $(i\omega)^{-1}$ from the original diffusivity (η_0) are given by

$$\eta_1 = -\frac{v_x}{i\omega}\frac{d\eta_0}{dx}. \tag{7.51}$$

This change in diffusivity produces an extra current

$$\mathbf{j}_1 = -\frac{\eta_1}{\eta_0}\,\mathbf{j}_0 \tag{7.52}$$

in Ohm's Law; associated with it is an extra Lorentz force ($\mathbf{j}_1 \times \varepsilon B_0\hat{\mathbf{y}}$) at the ($x > 0$) edge of the diffusion region, where the field is $\varepsilon B_0\hat{\mathbf{y}}$. Since the original current is roughly $\mathbf{j}_0 = B_0/(\mu l)\hat{\mathbf{z}}$, this force may be written [using Eqs. (7.51) and (7.52)] as

$$\mathbf{F}_d = -\frac{v_x}{\omega\eta_0}\frac{d\eta_0}{dx}\frac{\varepsilon B_0^2}{\mu l}\hat{\mathbf{x}}. \tag{7.53}$$

It represents a driving force with the sign shown in Figure 7.12 for $x > 0$ provided $d\eta_0/dx < 0$. Similarly, on the left-hand side of the current sheet ($x < 0$), where the magnetic field is $-\varepsilon B_0\hat{\mathbf{y}}$ and $v_x > 0$, a driving force of the necessary direction is present provided $d\eta_0/dx > 0$ there. Thus, if the diffusivity (η_0) possesses a *maximum* value in the centre of the sheet, the whole of the sheet is subject to the rippling mode. For a classical diffusivity, η_0 varies with temperature (T_0) like $T_0^{-3/2}$, so this would necessitate a temperature minimum at $x = 0$. If, on the other hand, the diffusivity increases (or decreases) monotonically through the sheet, the instability takes place on the left- (or right-) hand side of the sheet. Equating Eq. (7.53) to (7.47) and using Eq. (7.49) for ε determines the growth-rate as

$$i\omega = \left\{\left(\frac{d\eta_0}{dx}\frac{l}{\eta_0}\right)^4\frac{(kl)^2}{\tau_d^3\tau_A^2}\right\}^{1/5}.$$

It should be noted that, if conduction is efficient enough, diffusivity changes are not advected, so that Eq. (7.51) does not hold and the rippling mode is stabilised.

The third type of resistive instability is the *tearing mode*, which, in contrast to the other two, occurs with a wavelength greater than the sheet width ($kl < 1$). Its growth-rate is $i\omega = \{\tau_d^3\tau_A^2(kl)^2\}^{-1/5}$ for wavenumbers in the approximate range $(\tau_A/\tau_d)^{1/4} < kl < 1$. Thus, the mode with longest wavelength has the fastest growth-rate, namely, $i\omega = (\tau_d\tau_A)^{-1/2}$, which is the geometric mean of the diffusion and Alfvén frequencies. The tearing mode is possibly the most important of the three resistive instabilities, since it requires neither a gravitational force nor a resistivity gradient to be excited. In sheared fields, it is likely to produce some coronal heating (Sec. 10.4), since the growth-time for typical coronal conditions is about a week. The tearing mode offers one way in which reconnection can grow (Sec. 6.8.1), and so it is possibly relevant to solar flares (Sec. 12.2.6).

Resistive instabilities can occur whenever the magnetic field is sheared. A neutral current sheet is one example of such a structure; others include fields of the form $B_{0y}(x)\hat{\mathbf{y}} + B_{0z}(x)\hat{\mathbf{z}}$ in plane geometry, or $B_{0\phi}(R)\hat{\boldsymbol{\phi}} + B_{0z}(R)\hat{\mathbf{z}}$ in cylindrical geometry. Sheared fields are in general resistively unstable at many thin sheaths throughout the structure. At any particular location specified by the value of x (or R), the

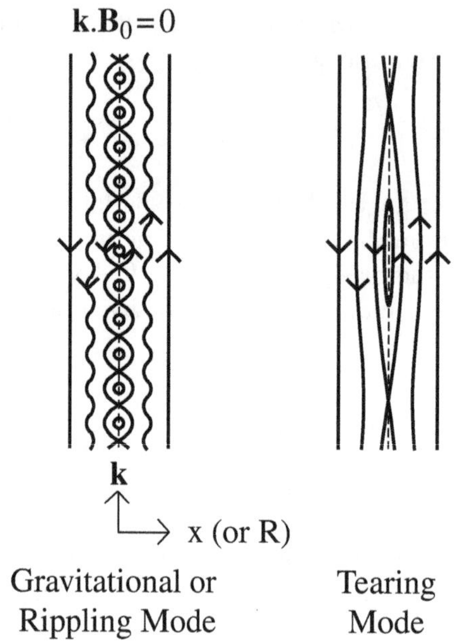

Figure 7.13. Resistive instabilities in a current sheet or a sheared magnetic field.

instability has a wavenumber in a direction normal to the field ($\mathbf{k}\cdot\mathbf{B}_0 = 0$), and in Figure 7.12 the xy-axes refer to the xk- (or Rk-) plane, as indicated in Figure 7.13.

7.5.6 Convective Instability

A horizontal layer of viscous, thermally conducting fluid, heated from below, can become unstable when the temperature difference between the upper and lower surfaces becomes too large (Sec. 1.3.3); this was demonstrated by Bénard (1900), who conducted the first quantitative experiments on the instability. The temperature difference is measured by the *Rayleigh number* $Ra = g\alpha\Delta T d^3/(\kappa\nu)$, where α, κ, ν are the coefficients of volume expansion, thermometric conductivity and kinematic viscosity, while ΔT is the temperature difference maintained across the layer and d is its thickness. Instability sets in when Ra exceeds a critical value Ra^* (Rayleigh 1916). For a Boussinesq fluid (Sec. 9.1.2), $Ra^* = 27\pi^4/4 \approx 658$ when the boundaries are both free. As Ra is increased, instability appears first at a wavenumber $k^* = (\pi d/\sqrt{2}) \approx 2.2d$, and it can develop into convection with a stationary pattern of motions (Figure 7.14). If the cells are infinitely elongated in one direction (x, say), they are called *rolls*, with a vertical velocity component of the form $v_z = V(z)\cos(2\pi y/L)$. Three rolls may be combined to give hexagonal (*Christopherson*) cells with

$$v_z = \tfrac{1}{3}V(z)\left\{\cos\frac{2\pi}{3L}(\sqrt{3}x + y) + \cos\frac{2\pi}{3L}(\sqrt{3}x - y) + \cos\frac{4\pi}{3L}y\right\}.$$

The effect of rotation and magnetic field on the onset of convection is complex (Chandrasekhar 1961; Cowling 1976; Proctor 1994). Suppose first that the fluid is rotating about the vertical (z-axis) with angular speed Ω, measured non-dimensionally by the value of the *Taylor number* ($\mathcal{T} = 4\Omega^2 d^4/\nu^2$). Rotation is found to inhibit convection somewhat, as indicated by the facts that the critical Rayleigh number (Ra^*) for convection onset increases with $\mathcal{T}$ and the size of the cells decreases. For instance, with free boundaries when $\mathcal{T} = 100$, one has $Ra^* \approx 830$ and $k^* \approx 2.6d$, while in the large-$\mathcal{T}$ limit $Ra^* \approx 8.7\ \mathcal{T}$ and $k^* \approx 1.3\ \mathcal{T}$. Usually the instability sets in first as *stationary convection*, but, when the *Prandtl number* ($Pr = \nu/\kappa$)

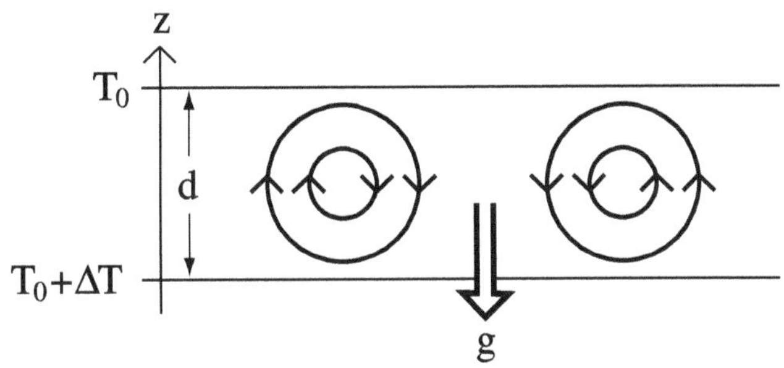

Figure 7.14. Schematic of streamlines for convection that is generated when the temperature difference (ΔT) between the upper and lower surface is large enough.

is less than unity and $\mathcal{T}$ is large enough, instability shows itself first as *overstability*. For instance, when $Pr = 0.1$, overstability occurs first if $\mathcal{T} > 730$.

The effect of a vertical field (parallel to $\mathbf{g}$) is described in Section 9.1.2. It, too, inhibits the onset of convection, because the energy released by buoyancy goes into Joule and viscous dissipation; thus, a higher adverse temperature gradient is needed than in the absence of a magnetic field. The type of instability depends on the ratio of thermal (κ) to magnetic diffusivity (η). If $\kappa < \eta$, the onset occurs as a *leak instability* and stationary convection results. The inhibiting effect of the magnetic field is shown by the fact that the critical Rayleigh number (Ra^*) increases with *Hartmann number* $Ha = B_0 d/(\mu\rho\eta v)^{1/2}$. For example, when $Ha = 10$ one finds $Ra^* \approx 2650$ and $k^* \approx 3.7d$, while as Ha increases so $Ra^* \approx \pi^2 Ha^2$ and $k^* \approx (\pi^2 Ha/\sqrt{2})^{1/3}$. If $\kappa > \eta$, the instability sets in as stationary convection when Ha is less than a critical value (Ha^*), but as overstability when $Ha > Ha^*$. Ha^* depends on Prandtl number (Pr) and magnetic Prandtl number ($P_m = \nu/\eta$).

When the magnetic field is not parallel to gravity, convection sets in first as long horizontal rolls with their axes parallel to the horizontal magnetic field component. When rotation and a magnetic field are both present, their combined effect is to inhibit the onset of instability and to elongate the cells that appear at marginal stability; the critical Rayleigh number depends on Ha and $\mathcal{T}$.

7.5.7 Radiatively Driven Thermal Instability

It was Parker (1953) who first pointed out that, if thermal conduction were ineffective, thermal instabilities would occur in the corona, because of the form of the radiative loss term in the energy equation. To illustrate this effect, suppose that the plasma is initially in equilibrium with temperature T_0 and density ρ_0, under a balance between mechanical heating of amount $h\rho$ per unit volume (where h is constant) and optically thin radiation; the latter takes the form $\chi\rho^2 T^\alpha$, where χ and α are constants (Eqs. 2.36) that depend on the temperature range, so that per unit mass

$$0 = h - \chi\rho_0 T_0^\alpha. \tag{7.54}$$

For a perturbation at constant pressure (p_0), the energy equation is

$$c_p \frac{\partial T}{\partial t} = h - \chi\rho T^\alpha, \quad \text{(a)} \qquad \text{where} \qquad \rho = \frac{m p_0}{k_B T}, \quad \text{(b)} \tag{7.55}$$

or, substituting for h from Eq. (7.54) and for ρ from Eq. (7.55b),

$$c_p \frac{\partial T}{\partial t} = \chi\rho_0 T_0^\alpha \left(1 - \frac{T^{\alpha-1}}{T_0^{\alpha-1}}\right).$$

Thus, if $\alpha < 1$, a small decrease in temperature ($T < T_0$) makes the right-hand side negative, so that $\partial T/\partial t < 0$ and the perturbation continues as a *thermal instability* with time-scale $\tau_{rad} = c_p/(\chi \rho_0 T_0^{\alpha-1})$. Figure 2.2 implies that $\alpha < 1$ for temperatures roughly in excess of 10^5 K.

Usually, thermal instability is prevented from taking place by the efficiency of heat conduction along magnetic field lines; this is represented by an extra term $\rho^{-1}\nabla \cdot (\kappa_{||}\nabla T)$ (per unit mass) on the right-hand side of Eq. (7.55a), where $\kappa_{||} = \kappa_0 T^{5/2}$ is the coefficient of thermal conduction parallel to the magnetic field. Thus, if the length of a field line is L, the conduction time is $\tau_c = L^2 \rho_0 c_p/(\kappa_0 T_0^{5/2})$, in order of magnitude. When L is so small that $\tau_c < \tau_{rad}$, the plasma is thermally stable; but, when L exceeds a value (obtained by equating τ_c and τ_{rad}) of

$$L_{max} = \left(\frac{\kappa_0 T_0^{7/2-\alpha}}{\chi \rho_0^2} \right)^{1/2},$$

the plasma is thermally unstable. This instability may be important in prominence formation (Sec. 11.6.3.1) and coronal loop structure (Sec. 10.2).

The standard work on linear thermal instability theory in the absence of magnetic field and thermal conduction is by Field (1965), who considers a more general energy equation than Eq. (7.55a). The analysis has been generalised by several authors: Hunter (1970) includes both flow and thermal conduction, while Heyvaerts (1974) includes a magnetic field, and Hildner (1974) and Karpen et al. (1989) follow the nonlinear development numerically.

7.5.8 Other Instabilities

In addition to the above instabilities, some others are mentioned elsewhere in the book. Chapter 9 discusses those that are relevant to sunspots and photospheric flux tubes. More details about convective instability are presented in Section 9.1, and the magnetic buoyancy instability is described in Section 9.3. Section 11.6.3 discusses the thermal instability in a coronal loop or flux rope that causes coronal plasma to cool and possibly form a prominence.

Chapter 12 discusses instabilities involved in solar flares. A simple-loop (or compact) flare may be produced by either a thermal non-equilibrium (when the cool core of an active-region loop loses equilibrium) or a (resistive) kink instability (when the loop is twisted too much) or an emerging flux instability (when a current sheet between old and new flux heats up). A two-ribbon flare (Sec. 12.1) starts with the eruption of a magnetic arcade; this may be due to an eruptive MHD instability or the presence of another force-free equilibrium of lower energy when the shear is too great (Sec. 12.2).

Chapter 8

Dynamo Theory

8.1 Introduction

The Sun's magnetic field cannot be due to a simple MHD oscillator, since there is no sign of an accompanying large oscillation in velocity. Understanding how the field can be generated instead by dynamo action in a cyclic way is a key question in solar MHD, which has not yet been fully answered, but on which huge progress has been made over the past 30 years. Here we first summarise the observational clues and describe the terminology and history (Sec. 8.1). Then we describe Cowling's theorem (Sec. 8.2) and the early basis for turbulent dynamo theory (Sec. 8.3), namely, Parker's classic 1955 paper and the more systematic mathematical formalism of *Mean-Field MHD*. This is followed with accounts of *flux-transport dynamos* (Sec. 8.4) and *tachocline dynamos* (Sec. 8.5). The latter includes overshoot and interface dynamos, while the former includes the early heuristic Babcock-Leighton ideas and their more recent development. Finally, two complementary approaches are outlined (Sec. 8.6), namely, complex global computations at one extreme and simple low-order models at the other, and brief comments are given on stellar dynamos.

The present chapter offers no more than a glimpse at a huge topic, but further details may be found in the excellent books by Moffatt (1978), Parker (1979a), Krause and Rädler (1980), Proctor and Gilbert (1994), Hughes et al. (2007), Dormy and Soward (2007) and in reviews by Choudhuri (1999), Mestel (1999), Schüssler (2005), Weiss (2005), Weiss and Thompson (2009) and Charbonneau (2010, 2013). In a mechanical dynamo, which converts mechanical energy into electrical energy, life is simple, since the current flows along given metallic wires. However, in a solar dynamo, the current is a secondary variable whose flow patterns cannot be constrained in this way: they are determined in subtle ways from self-consistent MHD, which, as Cowling showed near the beginning of this dynamo journey, are inherently nonlinear and three-dimensional (3D) – and therein lies the main difficulty.

8.1.1 Observed Solar Magnetic Field Patterns

As well as the fundamental problem of how in principle a magnetic field may be generated in a cyclic fashion and maintained against diffusion, there are many observed features of the Sun's magnetic field that need to be explained. At the photosphere, the magnetic field is concentrated into intense flux tubes and sunspots (Secs. 1.4.2 and 1.7.2 and Chapter 9), whose detailed evolution is highly complex, but whose overall behaviour is remarkably ordered as the solar cycle proceeds (Sec. 1.7.3). The key features of this underlying pattern that need to be taken into account and which offer clues about how the Sun generates its magnetic field are:

(i) The roughly 11-year oscillation in sunspot number (Figure 1.30). The cycle amplitude is modulated (on the Gleissberg cycle of 80 years), but the period varies substantially. Occasionally, the cycle declines to a very low level, so that very few sunspots appear (such as during the Maunder minimum for 70 years in the 17th century). Proxy observations of solar activity are found in terrestrial isotopes produced by cosmic rays, whose flux is modulated by the solar magnetic field. For example, the ^{10}Be stored in ice cores and ^{14}C in tree rings have shown that the activity cycle persists through the Maunder minimum and continues back for at least 50,000 years (Beer 2000), with recurrent grand minima on the de Vries cycle having a period of 205 years (Wagner et al. 2001; Vonmoos et al. 2006; Steinhilber et al. 2008).

(ii) The restriction of sunspots to two belts of latitude between typically $\pm$ 35 degrees, together with a tendency for sunspots to appear in preferred 'active longitudes'.

(iii) The spread and drift towards the equator of the sunspot belts as the solar cycle progresses (Spörer's Law). This is revealed in a 'butterfly diagram' (Figure 1.32a) and suggests a migration in time of dynamo activity from mid-latitudes to equator. The butterfly diagram is normally symmetric on average about the equator, suggesting a dipolar dynamo, but it was asymmetric with sunspots in only one hemisphere when the field was weak as it emerged from the Maunder minimum.

(iv) The inclination or tilt of sunspot groups to lines of latitude (by 4 degrees on average), with the leading sunspot being closer to the equator and the tilt increasing with latitude (Joy's Law).

(v) The Hale Laws of polarity (Figure 1.31), whereby a leading spot has the same polarity as all other leaders in the same hemisphere and opposite to those in the other hemisphere. The polarities reverse at the start of each new eleven-year cycle, implying that there is an underlying 22-year magnetic cycle.

(vi) The dominant polarities of the fields near the two poles, which are opposite to one another and reverse every 11 years near sunspot maximum.

We restrict our attention here to the *solar* dynamo, although rather different dynamos may be operating in the Earth (where the field is relatively steady, with occasional reversals), in Jupiter (which is rapidly rotating), in other stars (Bonnet and Dupree 1981; Tobias and Weiss 2007), and even in galaxies. Future observations of stellar magnetic activity (see Sec. 8.6.3) will be potentially important for inferring additional information about the solar dynamo and how it varies with the important parameters. For example, the Mount Wilson survey of Ca, H and K emission (which is a good indicator of magnetic flux) has revealed that solar-type stars possess a variety of activity that may be cyclic or aperiodic. In particular, as a star spins down, the activity decreases (Noyes et al. 1991), suggesting that the efficiency of a dynamo depends on the rotation rate. Furthermore, rapidly rotating stars may possess huge polar spots, and even fully convective stars can have coronae.

8.1.2 Preliminary Comments and Terminology

Global Diffusion. Since the classical diffusion time $(R_\odot^2/\eta)$ for the decay of a global solar magnetic field is about 10^{10} year, and therefore comparable with the age of the Sun, it might at first be thought that the Sun's magnetic field is primordial. This is, however, unlikely, because the above estimate for the decay time of the field may well be an overestimate. Resistive instabilities may operate much faster (Sec. 6.8), while magnetic buoyancy would tend to expel magnetic flux efficiently (Sec. 9.3), and in the convection zone an eddy magnetic diffusivity $(\tilde{\eta})$ of typically 10^9 m^2 s^{-1} would destroy flux there after only 10 $L^2/\tilde{\eta}$ secs (i.e., about 10 years), where L is the depth of the convection zone). For the radiative interior, both the spindown that has occurred since the Sun's formation and also its present uniform rotation point to the presence of a magnetic field, which may be either primordial or maintained by a dynamo based on Tayler instability (Spruit 2002; Braithwaite 2006). Such a magnetic field in the radiative zone and core would probably be decoupled from the solar cycle (unless it had a subtle effect at the shear layer below the base of the convection zone, called the *tachocline*). Furthermore, the oscillator theories once favoured by some authors (e.g., Piddington 1977) have been ruled out, since they do not explain effectively the solar cycle variations and give the wrong period for angular velocity variations. We are, therefore, impelled to seek some kind of dynamo to explain the maintenance of the solar magnetic field.

Kinematic Dynamo Theory aims to construct a velocity field $[\mathbf{v}(x,y,z,t)]$ such that a magnetic field $[\mathbf{B}(x,y,z,t)]$ satisfying the induction equation

$$\frac{\partial \mathbf{B}}{\partial t} = \nabla \times (\mathbf{v} \times \mathbf{B}) + \eta \nabla^2 \mathbf{B} \tag{8.1}$$

(together with $\nabla \cdot \mathbf{B} = 0$) grows and is maintained by *induction* $[\nabla \times (\mathbf{v} \times \mathbf{B})]$ against diffusion $(\eta \nabla^2 \mathbf{B})$. There is a rich theory of solutions to this equation, which is linear in $\mathbf{B}$ for a given $\mathbf{v}$.

Nonlinear Dynamo Theory, on the other hand, goes one step further by asking whether the magnetic field can be maintained in a self-consistent manner: that is, taking account of the back-reaction of the Lorentz force on the flow. In other words, it seeks a fully MHD dynamo satisfying both the induction equation and the equation of motion. The complete nonlinear coupled problem is so difficult that attention was first concentrated on the kinematic problem (Sec. 8.3), but more recently the focus has turned to the dynamic problem and to understanding the causes of the various observed consequences on the Sun (Sec. 8.1.1).

Slow or Fast Dynamo. In dimensionless form in terms of typical values (v_0, B_0), the induction equation may be written

$$\frac{\partial \mathbf{B}'}{\partial t'} = \nabla' \times (\mathbf{v}' \times \mathbf{B}') + \frac{1}{R_m} \nabla'^2 \mathbf{B}',$$

where $\mathbf{B}' = \mathbf{B}/B_0$, $\mathbf{v}' = \mathbf{v}/v_0$, $\mathbf{r}' = \mathbf{r}/L_0$, $\mathbf{t}' = tv_0/L_0$ and $R_m = L_0 v_0/\eta$ is the magnetic Reynolds number. A kinematic dynamo is said to be *fast* if its growth-rate remains positive as $R_m \to \infty$; otherwise it is said to be *slow*. Since R_m is so large in practice, the aim for a solar dynamo is to seek one that is fast (see PROBLEM 8.1: for all problems and solutions, see web page www.cambridge.org/9780521854719). However, unlike normal fast dynamo theory, one should not hold the velocity fixed as $R_m \to \infty$, since the Sun is at low Prandtl number.

Toroidal (B_ϕ) and Poloidal ($\mathbf{B}_p$) Fields. A key focus in solar dynamo theory concerns the way in which both the *toroidal* (i.e., azimuthal) and *poloidal* (i.e., meridional) field components are generated. A toroidal component can be naturally built up from a poloidal field by the *ω-effect*, namely, differential rotation stretching out field lines (Figure 8.1) (see PROBLEM 8.2). However, the mechanism for completing the cycle by generating poloidal flux from toroidal flux is much more tricky (Sec. 8.3).

Alpha-Omega Dynamo. An *α-ω dynamo* relies on both the *ω-effect* to produce poloidal flux from toroidal flux and also the *α-effect* (due to, e.g., Parker's rising cyclonic eddies, Sec. 8.3.1) to produce toroidal flux

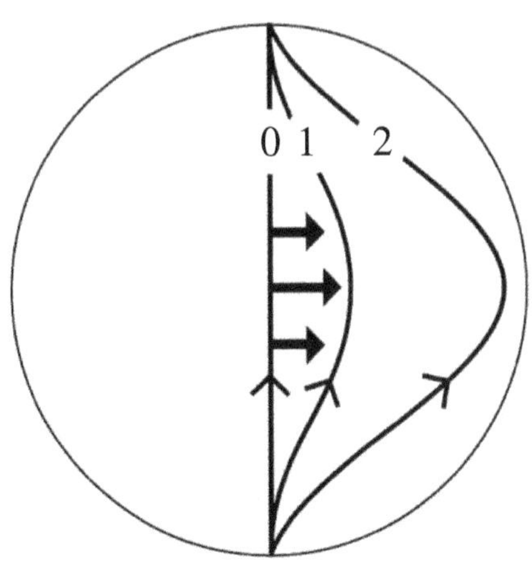

Figure 8.1. The ω-effect (or Ω-effect), whereby poloidal flux is converted into toroidal flux by differential rotation. Light arrows show the magnetic field and dark ones the velocity.

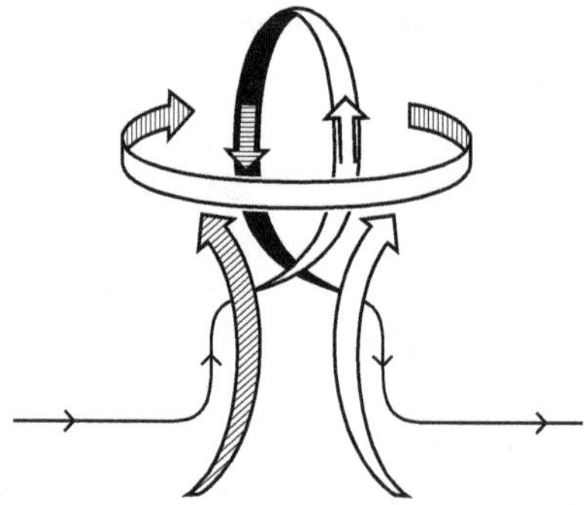

Figure 8.2. The α-effect, whereby toroidal flux is converted into poloidal flux by rising and twisting motions.

from poloidal flux (Figure 8.2). On the other hand, an α^2-*dynamo* has little or no differential rotation but is able to use the α-*effect* to generate both components.

Large-Scale or Small-Scale Dynamos. If the generated magnetic field has a scale much larger than that of the driving flow, we have a *large-scale dynamo*, whereas, if the field is produced on scales comparable to or smaller than that of the flow, then it is a *small-scale dynamo.* A small-scale dynamo generating disordered field seems to be highly likely whenever the fluid is turbulent and the magnetic Reynolds number sufficiently high (Cattaneo 1999; Vögler and Schüssler 2007); large-scale field was traditionally thought to be created from small-scale field in MHD turbulence by an inverse cascade of magnetic helicity, but simulations show this process is inefficient, and so now either boundary effects or magnetic instabilities are thought necessary.

Several Dynamos? Sunspots and active regions are likely to be created by a large-scale dynamo operating (at least partly) near the tachocline at the base of the convection zone, by either a *tachocline dynamo* (Sec. 8.5) or a *flux-transport dynamo* (Sec. 8.4). It is this large-scale dynamo and general dynamo properties that we shall mainly address in this chapter.

However, the magnetic field of the quiet Sun outside active regions (called the *magnetic carpet*) consists of several parts. Some flux spreads from active regions by diffusion and meridional flow: it varies with the solar cycle and is shredded and reprocessed on its way to the poles. Superimposed on it is a background that is largely independent of the solar cycle and arises from two sources: the first is tiny bipolar ephemeral regions (Sec. 1.4.2), whose flux is reprocessed every 10 hours; it is perhaps generated in a second (small-scale) dynamo that operates throughout the convection zone or in a narrow shear layer just under the solar surface on, say, supergranular scales. The second source of the background flux is the inner network field (Sec. 1.4.2) that exists inside supergranules and could perhaps be generated by a third local dynamo operating at even smaller (say, granular) scales just below the surface (Vögler and Schüssler 2007). Indeed, Hinode observations reveal vertical fields between granules and horizontal fields near their edges (Tsuneta et al. 2008).

8.1.3 History of Dynamo Ideas

William Gilbert, physician to Queen Elizabeth I, proposed a bold hypothesis in about 1600 to explain why compass needles point in the way they do, namely, that the Earth is a giant magnet. At first, it was thought that this magnetism was ferromagnetic in origin, but, by the end of the 19th century, the Earth's interior was realised to be hotter than the Curie point (at which a ferromagnet loses magnetism), and so an explanation was sought elsewhere. In 1908 Hale made his momentous discovery of magnetic fields outside the Earth's environment, when he observed strong magnetic fields in sunspots. The puzzle of how a rotating body like the Sun could become a magnet was then posed by Larmor (1919), but Cowling

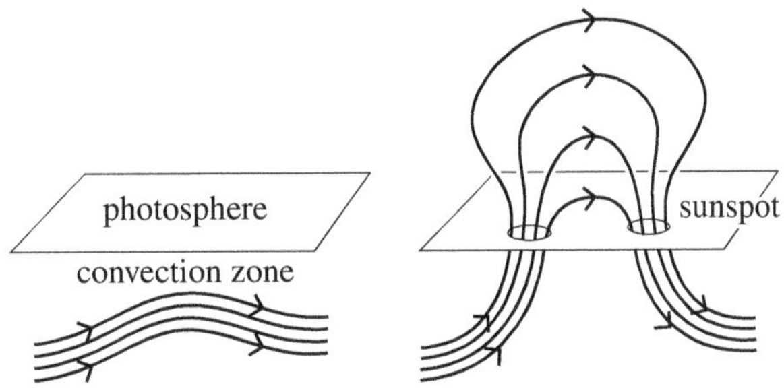

Figure 8.3. Magnetic buoyancy of a magnetic flux tube causes it to rise through the surface and create pairs of sunspots.

(1934) showed in his famous theorem (Sec. 8.2) that a simple solution to the puzzle is impossible: *a steady axisymmetric field cannot be maintained by dynamo action.*

By the 1950s, there was no answer to Cowling's conundrum in sight, even though the basic properties of the Sun's magnetic field had been known for decades (the 11-year sunspot cycle, Hale's polarity laws, the butterfly diagram and the tilts of active regions). It was recognised that toroidal flux can be generated from poloidal flux in a natural way by differential rotation, such that an initially north-south orientated field line will wrap once around the Sun in about 8 months (Figure 8.1). This is called the *ω-effect* [after the notation ($\boldsymbol{\omega}$) for vorticity] or sometimes the *Ω-effect* [after the notation ($\boldsymbol{\Omega}$) for the Sun's angular velocity]. An important part of the dynamo process was solved by Parker (1955a) in his landmark paper on *magnetic buoyancy*, which explains how isolated toroidal magnetic flux tubes rise through the convection zone to give sunspots where they break through the solar surface (Figure 8.3). However, no-one yet knew how to produce poloidal flux from toroidal flux (Cowling 1953).

An answer which has had a major effect on the development of the field was proposed by Parker (1955b), namely, the regeneration of poloidal flux by the collective effect of small-scale helical flows on toroidal flux (Figure 8.2). By physical heuristic arguments, he suggested that these turbulent cyclonic flows could be modelled by a new term [$\boldsymbol{\nabla} \times (\alpha \mathbf{B})$, where α is constant] in the mean induction equation in addition to differential rotation. He found dynamo wave solutions, in which the dynamo-generated magnetic field migrates towards the equator and so explains the main feature of the butterfly diagram. The theory also provided testable predictions:

(a) the Sun's polar fields should reverse during each eleven-year cycle;
(b) and its angular velocity should increase radially inwards in the convection zone.

Next, the Potsdam group (Steenbeck et al. 1966) developed a more systematic mathematical formalism of Parker's dynamo ideas (Moffatt 1978). They set up evolution equations for the mean magnetic field in a turbulent flow (Sec. 8.3.2) and discovered that dynamo action is possible if the small-scale flows are not reflectionally (or mirror-) symmetric: for example, in a rotating stratified plasma sphere, the Coriolis force on radial motions produces a mean helicity $\langle \mathbf{v} \cdot \boldsymbol{\nabla} \times \mathbf{v} \rangle$. The resulting induction equation for the mean poloidal magnetic field $\langle \mathbf{B} \rangle$ takes the form

$$\frac{\partial \langle \mathbf{B} \rangle}{\partial t} = \boldsymbol{\nabla} \times [\langle \mathbf{v} \rangle \times \langle \mathbf{B} \rangle + \alpha \langle \mathbf{B} \rangle] - \boldsymbol{\nabla} \times [(\eta + \beta) \boldsymbol{\nabla} \times \langle \mathbf{B} \rangle],$$

where η is the *magnetic diffusivity*, β is a *turbulent magnetic diffusivity* and $\alpha \langle \mathbf{B} \rangle$ is the *α-effect*, suggested by Parker to regenerate the poloidal field.

In contrast to the mathematical approach of the Potsdam group, Babcock (1961) and Leighton (1969) proposed a totally different qualitative scenario for regenerating a poloidal field. The bipolar sunspots have a definite tilt on average (Joy's law), and so, when pairs of sunspots eventually decay, the magnetic flux diffuses globally in such a way that the flux in the northern hemisphere has one dominant polarity while that in the southern hemisphere has the opposite polarity. In other words, new poloidal flux is created from the toroidal flux that formed sunspots.

The 1970s were highly active years for dynamo theory, due to the development with great optimism of kinematic α-ω models for the Sun and also for other stars, planets, galaxies and accretion discs, culminating in three major textbooks (Moffatt 1978; Parker 1979a; Krause and Rädler 1980). It was thought that almost all sufficiently rapid and complex motions could act as kinematic dynamos. In addition, nonlinear effects were incorporated, such as quenching or saturation of the α-effect, due to feedback by the Lorentz force or the Malkus-Proctor effect (Malkus and Proctor 1975), in which an α^2- or $\alpha\omega$-dynamo is saturated by a Lorentz-induced large-scale differential rotation and meridional flow. During this decade it was assumed that the solar dynamo is working throughout the convection zone.

The 1980s, however, dealt four major blows against classical α-ω dynamo models (with the fourth being the final nail in the coffin), although the ω-effect remains accepted to this day as the mechanism for generating the toroidal field:

(i) It was realised that the properties of emerging fields, such as the latitudes of emergence and the tilts of bipoles, require fields of order 10^4 to 10^5 G (1 to 10 T), but they present two difficulties: if generated in the convection zone they rise too rapidly by magnetic buoyancy (Moreno Insertis 1983; Parker 1984; Choudhuri and Gilman 1987), and also they are too strong for helical turbulence of the expected amplitude to act on them efficiently. It is unclear whether the dynamo region has fields of 10^5 G or mainly 10^4 G with some 10^5 G regions that emerge (Sec. 8.5.2).
(ii) Doubts began to be expressed about validity of mean-field theory and the derivation of the α-effect, and therefore the ability of turbulent diffusion and the α-effect to fulfil the expectations of the models.
(iii) Numerical simulations of thermally driven magnetoconvection in a spherical rotating shell (Gilman 1983; Glatzmaier 1985) did not yield solar-like patterns of migrating solar activity.
(iv) The measurements of rotation (Ω) in the convection zone from helioseismology showed that it is rather constant with radius ($d\Omega/dr = 0$) through the convection zone down to the tachocline, so that it does not have the radial increase with depth ($d\Omega/dr < 0$) required by convection-zone dynamo models.

These new realisations spawned two distinct approaches to building a large-scale dynamo theory responsible for active-region flux (Figure 8.4). The first was a *tachocline dynamo* working near the tachocline at the base of the convection zone (Sec. 8.5). One possibility is an *overshoot dynamo* (Spiegel and Weiss 1980), which locates both the α- and ω-effects in the overshoot region just below the base of the tachocline, while another is an *interface dynamo* (Parker 1993; Charbonneau and MacGregor 1997), which separates these effects spatially and places the ω-effect below the interface with the α-effect above it.

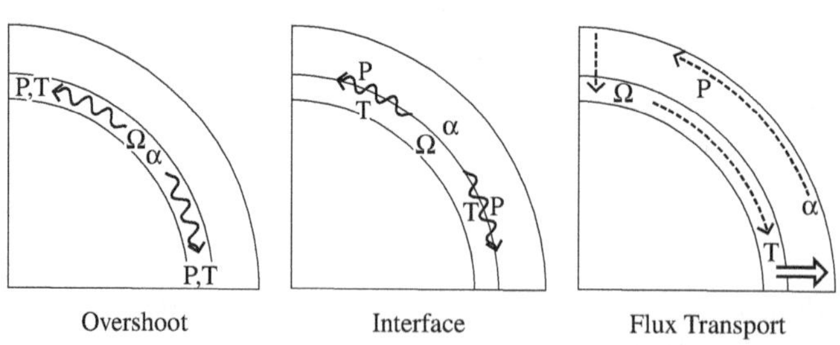

Figure 8.4. Magnetic field generation by overshoot, interface and flux-transport dynamos, indicating where the α- and ω-effects are located and where poloidal (P) and toroidal (T) components are generated. Curly, dashed and double arrows represent transport by a dynamo wave, meridional flow and buoyancy, respectively (courtesy Kristof Petrovay).

The second approach was a *flux-transport dynamo* that develops the earlier Babcock (1961)-Leighton (1969) dynamo by solving the axisymmetric kinematic dynamo equations with an imposed meridional flow and an ω-effect focussed near the tachocline together with an α-effect at the solar surface (Wang et al. 1991; Choudhuri et al. 1995; Dikpati and Choudhuri 1994; Charbonneau and MacGregor 1997; Nandy and Choudhuri 2002). Thus, diffusion of active-region flux at the solar surface and meridional flow through the convection zone play significant roles.

The main difference between these approaches lies in the location of the generation of poloidal flux from toroidal flux: in the former, poloidal flux is generated down in the tachocline, and the surface behaviour of this flux is secondary; in the latter it is near the solar surface, and the surface variation of the polar flux is primary.

Other effects may be important in dynamo theories, such as: shear instabilities, magnetic buoyancy instabilities, flux tube instabilities in the tachocline or overshoot layer; back-reaction of the Lorentz force on the flow and the efficiency of the α-effect; time delays or stochastic forcing to modulate the dynamo; stochastic fluctuations of dynamo coefficients; conservation of magnetic helicity (Charbonneau 2010).

The advances and continual improvements in numerical computation have been impressive, including: a demonstration of dynamo action in turbulent and laminar flows; the discovery of small-scale dynamo action at low magnetic Prandtl numbers; and large-scale dynamo simulations in spherical shells. Full MHD global computations have been conducted by a number of authors (e.g., Brun et al. 2004; Ghizaru et al. 2010), who are now able to resolve supergranulation and generate reasonable behaviour for differential rotation and meridional circulation, as well as a turbulent α-effect and reversals of the magnetic fields.

Predicting the amplitudes and lengths of future solar cycles is, however, a tough endeavour. Many methods have been employed, including dynamo theory, neural networks and geomagnetic indices. The maximum sunspot numbers predicted for cycle 24, for example, ranged between 40 and 170 among the 75 or so attempts. Advocates of flux-transport dynamo theory have been especially active in solar-cycle prediction (e.g., Dikpati and Gilman 2006; Choudhuri et al. 2007), but they adopt different values for the magnetic diffusivity, differential rotation, meridional circulation, poloidal flux source and alpha quenching.

In future, progress is likely to come, as in many branches of solar MHD, from a combination of unexpected new ideas, new observations both of the solar interior and of other stars, and also increasingly complex and realistic computational experiments.

8.2 Cowling's Anti-dynamo Theorem

Whereas for many problems analytical progress can be made by seeking simple solutions, for the dynamo problem Cowling discovered that too much symmetry is a bad thing and that the simplest configuration will not work at all. Interestingly, he made this discovery after having difficulty constructing by hand a numerical solution near a symmetry axis. Cowling's theorem states that: *A steady axisymmetric magnetic field cannot be maintained by dynamo action* (Cowling 1934). A related theorem due to Zel'dovich (1957) is that: *Magnetic fields cannot be maintained by two-dimensional (2D) planar motions or by motions on spherical surfaces* (see PROBLEM 8.3). Cowling's theorem itself may be established as follows.

In cylindrical polars (R, ϕ, z), consider a steady, axisymmetric field $[\mathbf{B}(R, z)]$ that depends on R and z alone, and write it as the sum of a toroidal (i.e., azimuthal) component (B_ϕ) and a poloidal component $(\mathbf{B}_p)$ [which itself represents the sum $(B_R\hat{\mathbf{R}} + B_z\hat{\mathbf{z}})$ of the radial and axial components]

$$\mathbf{B} = B_\phi \mathbf{i}_\phi + \mathbf{B}_p. \tag{8.2}$$

Because of the axisymmetry, the magnetic configuration in all meridional planes (through the axis of symmetry) is the same and the projections of field lines onto such planes must be closed curves. In each

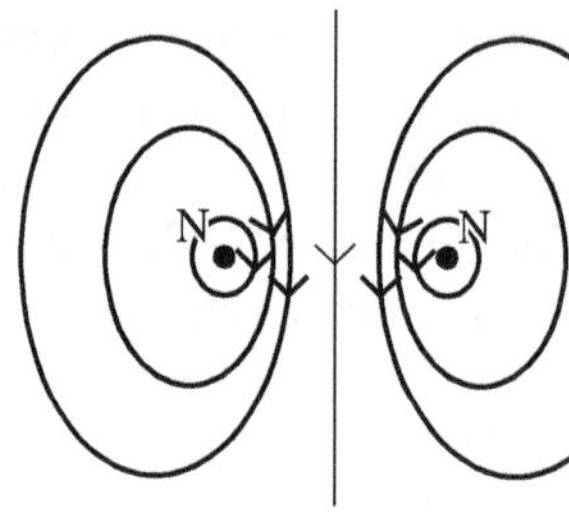

Figure 8.5. The magnetic field lines in a meridional plane for an axisymmetric field.

meridional plane there must therefore exist in general at least one O-type *neutral point* (N), where $\mathbf{B}_p$ vanishes, so that the field is purely azimuthal and the azimuthal current (j_ϕ) is non-zero (Figure 8.5).

Now, Ohm's Law in the form $\mathbf{j}/\sigma = \mathbf{E} + \mathbf{v} \times \mathbf{B}$ may be integrated around the closed azimuthal field line (C) through the neutral points to give

$$\int_C \mathbf{j}/\sigma \cdot d\mathbf{s} = \int_C \mathbf{E} \cdot d\mathbf{s} + \int_C \mathbf{v} \times \mathbf{B} \cdot d\mathbf{s},$$

or, after using Stokes' theorem to transform the first term on the right-hand side,

$$\int_C j_\phi ds/\sigma = \int_S \boldsymbol{\nabla} \times \mathbf{E} \cdot d\mathbf{S} + \int_C \mathbf{v} \times \mathbf{B} \cdot d\mathbf{s}.$$

Since the magnetic field is steady by assumption, $\boldsymbol{\nabla} \times \mathbf{E}$ vanishes due to Faraday's Law. Also, along C, $\mathbf{B}$ is parallel to the path element $d\mathbf{s}$, so that the triple scalar product $\mathbf{v} \times \mathbf{B} \cdot d\mathbf{s}$ vanishes, and the integral of Ohm's Law reduces to $\int_C j_\phi ds = 0$. Since j_ϕ does not vanish along C, this cannot be satisfied, and so a steady magnetic field cannot be axisymmetric. (Even if j_ϕ does vanish locally, the theorem still holds.)

Cowling's argument led to fears that no dynamo could work, but after 25 years the air of pessimism was dispelled when many kinematic dynamo models began to be rigorously established (e.g., Herzenberg 1958). For instance, Roberts (1972a) found dynamo action for a periodic flow in an infinite region, while Braginsky (1965a) and Soward (1972) studied dynamos with a slight departure from axisymmetry. Since then, the emphasis has shifted towards seeking more realistic dynamos.

The difficulty recognised by Cowling with an axisymmetric field is that it is the *poloidal* field component that cannot be maintained. This can be seen by considering a time-dependent axisymmetric field of the form (8.2) acted upon by a flow $\mathbf{v} = v_\phi \mathbf{i}_\phi + \mathbf{v}_p$. Interestingly, the ϕ-component of the induction equation (8.1) becomes

$$\frac{\partial B_\phi}{\partial t} + R\,(\mathbf{v}_p \cdot \boldsymbol{\nabla})\left(\frac{B_\phi}{R}\right) = R\,\mathbf{B}_p \cdot \boldsymbol{\nabla}\left(\frac{v_\phi}{R}\right) + \eta\left(\nabla^2 - \frac{1}{R^2}\right) B_\phi, \tag{8.3}$$

where the first term on the left gives the rate of change of the toroidal field, while the second represents its advection with a flow. On the right-hand side, the first term shows how a shear in angular velocity (v_ϕ/R) (i.e., a *differential rotation*) acting on a poloidal field ($\mathbf{B}_p$) can enhance the toroidal flux. Such a stretching of field lines would continue until balanced by Ohmic diffusion (the second term on the right).

However, the poloidal component of Eq. (8.1) may be integrated [after writing $\mathbf{B}_p = \boldsymbol{\nabla} \times (A_p \mathbf{i}_\phi)$] to give

$$\frac{\partial A_p}{\partial t} + \frac{\mathbf{v}_p}{R} \cdot \boldsymbol{\nabla}(R A_p) = \eta\left(\nabla^2 - \frac{1}{R^2}\right) A_p, \tag{8.4}$$

which does not allow the generation of $\mathbf{B}_p$ from B_ϕ, since it implies that A_p (and therefore $\mathbf{B}_p$) decays away in time. Indeed, for an incompressible flow with $\eta = \eta_0/R$ and appropriate boundary conditions, the proof of decay is immediate (Moffatt 1978), since it may be shown that Eq. (8.4) implies

$$\frac{d}{dt}\int (RA)^2 dV = -2\eta_0 \int |\nabla(RA)|^2 dV \le 0.$$

Turbulent dynamo theory aims to halt such a decay by adding a source term to (8.4).

8.3 Early Turbulent Dynamos

Parker achieved a major breakthrough in 1955 by pointing out that turbulent motions inside the convection zone (which are by nature non-axisymmetric) may be able to sustain the Sun's poloidal field (Sec. 8.3.1). His heuristic ideas were later put on a more systematic basis by the Potsdam group (Sec. 8.3.2).

8.3.1 Parker's 1955 Model

We have already mentioned how differential rotation may stretch out poloidal flux to create a strong toroidal component (Figure 8.6a) and have shown the effect mathematically in Eq. (8.3). Parker (1955b) realised that, since the Sun is rotating, rising blobs of plasma in the convection zone would tend to act like cyclones in the Earth's atmosphere and so would rotate due to the Coriolis effect. These helically upflowing blobs would twist the toroidal magnetic field and produce magnetic loops with a poloidal field component (as shown in Figure 8.6b).

The toroidal field has opposite signs in the two hemispheres, but the helical motions also have opposite senses (clockwise in the northern hemisphere and anti-clockwise in the south), and so Parker realised that the resulting poloidal loops have the same sense in both hemispheres. Provided the loops can reconnect efficiently with one another, they may then coalesce to create a large-scale poloidal field, as shown in Figure 8.6c by the broken curve. He developed a clever heuristic formalism as follows for the way such helical turbulence regenerates poloidal flux from toroidal flux.

The rate of generation of poloidal flux ($\mathbf{B}_p$) is proportional to toroidal flux (B_ϕ), and so Parker modelled the effect of many convection cells by adding an electric field,

$$E_\phi = \alpha B_\phi, \tag{8.5}$$

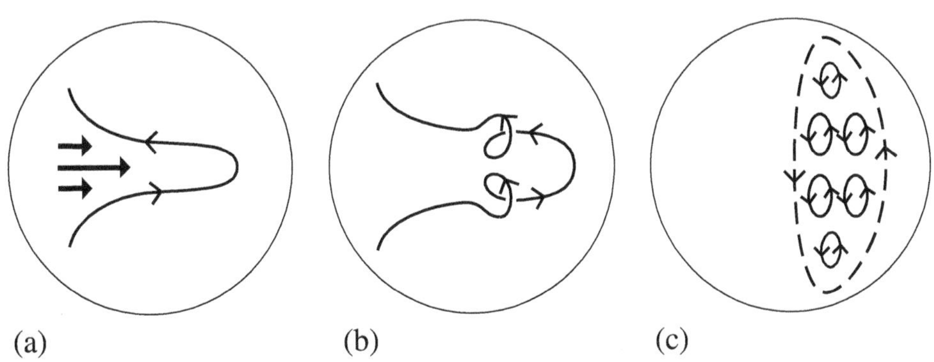

Figure 8.6. Parker's dynamo model in which: (a) toroidal flux is generated from poloidal flux by differential rotation (the ω-effect); (b) two helically rising blobs twist up the toroidal field; (c) the resulting closed loops of many cyclonic eddies merge to give new large-scale (dashed) poloidal flux (the α-effect).

to Equation (8.4), which therefore becomes

$$\frac{\partial A_p}{\partial t} + \frac{\mathbf{v}_p}{R} \cdot \boldsymbol{\nabla}(RA_p) = \alpha B_\phi + \eta \left(\nabla^2 - \frac{1}{R^2} \right) A_p, \tag{8.6}$$

so that dynamo action is now possible.

The constant of proportionality (α) in the mean electric field (E_ϕ) over many eddies gives the *α-effect* its name. It has units of velocity and is a measure of the mean rotational speed of eddies. Its value may be estimated from mean-field MHD (Sec. 8.3.2). Since there are also falling motions, which rotate the field in the opposite direction, there needs to be some asymmetry between up and down motions to create a net effect. The main causes of such asymmetry are stratification (since rising plasma expands while falling plasma contracts) and geometry (since plasma tends to rise at the centre of a cell and fall at its boundary). Other causes are magnetic buoyancy (Sec. 9.3), which aids the rising motions, and hydromagnetic inertial waves (Sec. 4.5), which possess a non-zero helicity ($\mathbf{v} \cdot \boldsymbol{\nabla} \times \mathbf{v}$) (Moffatt 1978).

8.3.1.1 A Simple Solution: Dynamo Waves

The kinematic dynamo equations (8.3) and (8.6) may be solved most simply in a plane geometry: although valid for the Sun only in the limit of a thin spherical shell, Parker (1955b, 1979a) realised (in the spirit of mathematical modelling) that they give interesting insight into some properties of a solar dynamo. He writes

$$\mathbf{B}(x,z) = \left(-\frac{\partial A}{\partial z},\ B_y,\ \frac{\partial A}{\partial x} \right) \tag{8.7}$$

in rectangular Cartesian coordinates, with z normal to the solar surface locally and y in the eastward (toroidal) direction. He considers a 'toroidal' velocity $\mathbf{v} = v_y(z)\hat{\mathbf{y}}$ with a 'radial' (i.e., vertical) shear and supposes $\mathbf{B}$ is 'axisymmetric' (i.e., locally independent of y with $\partial/\partial y = 0$). Equations (8.3) and (8.6) then reduce to

$$\left(\frac{\partial}{\partial t} - \eta \nabla^2 \right) B_y = \frac{dv_y}{dz}\frac{\partial A}{\partial x}, \qquad \left(\frac{\partial}{\partial t} - \eta \nabla^2 \right) A = \alpha B_y. \tag{8.8}$$

When dv_y/dz is constant, these linear dynamo equations possess plane-wave solutions of the form

$$B_y = B_0 \exp[\omega t + i(k_x x + k_z z)], \qquad A = A_0 \exp[\omega t + i(k_x x + k_z z)], \tag{8.9}$$

where

$$\omega = -\eta k^2 \pm (1+i) \left(\frac{\alpha k_x}{2} \frac{dv_y}{dz} \right)^{1/2}.$$

The second term can give a positive contribution to the real part $[\mathcal{R}(\omega)]$ of ω and represents field generation by the α-effect, whereas the first term gives a negative contribution due to simple Ohmic decay (see PROBLEM 8.4). The square of the ratio of these two effects is

$$N_D = \frac{\alpha k_x}{2\eta^2 k^4} \frac{dv_y}{dz}, \tag{8.10}$$

which is known as the *dynamo number*. It can be seen that when $|N_D| > 1$ there are growing solutions $[\mathcal{R}(\omega) > 0]$ in the form of *migrating dynamo waves*, which travel northwards (negative x-direction) when $N_D > 0$ and southwards when $N_D < 0$. For the Sun, the cyclonic motions are such that $\alpha > 0$ in the northern hemisphere and $\alpha < 0$ in the southern hemisphere, and so the waves migrate equatorwards like sunspots on the crest of the wave provided the *toroidal velocity increases with depth* ($dv_y/dz < 0$). Furthermore, the wave period of $\mathcal{R}(\omega)^{-1} \approx R_{\odot}^2/\eta$ agrees with the solar-cycle duration if an eddy value of 10^9 m^2 s^{-1} is adopted for η.

Parker solved the equations in a bounded layer: by varying the value of N_D, he described qualitatively both cyclic and steady ($N_D = 1$) generation of terrestrial, solar and galactic fields. However, the challenge taken up next by the Potsdam group was to try and put these qualitative ideas on a sounder mathematical footing.

8.3.2 Mean-Field MHD Theory

8.3.2.1 The Mean Induction Equation

Parker suggested intuitively that the net effect of averaging many small-scale convective motions is to produce a large-scale electric field (αB_ϕ) in Eq. (8.6) and so allow regeneration of the poloidal field. An attempt to put this idea on a more formal and systematic basis was made by the Potsdam group (Steenbeck et al. 1966; Krause and Rädler 1980). One way to handle MHD turbulence is in terms of an 'ensemble' of many different replicas of the system possessing different values of the fluctuating quantities. The other, which we shall adopt here, is to consider a small-scale fluctuating turbulent motion ($\mathbf{v}$) and magnetic field ($\mathbf{b}$) on a small scale (l), which are statistically steady and homogeneous but not mirror-symmetric. They are superimposed on a field ($\mathbf{B}_0$) and flow ($\mathbf{v}_0$) on a much larger scale (L), so that the total field and flow are $\mathbf{B}_0 + \mathbf{b}$ and $\mathbf{v}_0 + \mathbf{v}$, and the induction equation becomes

$$\frac{\partial}{\partial t}(\mathbf{B}_0 + \mathbf{b}) = \boldsymbol{\nabla} \times [(\mathbf{v}_0 + \mathbf{v}) \times (\mathbf{B}_0 + \mathbf{b})] + \eta \nabla^2 (\mathbf{B}_0 + \mathbf{b}). \tag{8.11}$$

Averaging over a scale intermediate between l and L (denoted by over-bars) makes the means of the fluctuating velocity and magnetic field vanish (i.e., $\bar{\mathbf{v}} = \bar{\mathbf{b}} = \mathbf{0}$). But the average of Eq. (8.11) then produces an induction equation for $\mathbf{B}_0$,

$$\frac{\partial \mathbf{B}_0}{\partial t} = \boldsymbol{\nabla} \times (\mathbf{v}_0 \times \mathbf{B}_0 + \overline{\mathbf{v} \times \mathbf{b}}) + \eta \nabla^2 \mathbf{B}_0, \tag{8.12}$$

which is very similar to the normal induction equation. The only difference is the presence of the mean electric field $\overline{\mathbf{v} \times \mathbf{b}}$ arising from the fluctuating field and flow, which is the secret of a turbulent dynamo, since it has the potential to maintain the large-scale field ($\mathbf{B}_0$) against Ohmic decay. Equation (8.12) may be subtracted from Eq. (8.11) to give an equation for $\mathbf{b}$ in terms of $\mathbf{B}_0$, namely,

$$\frac{\partial \mathbf{b}}{\partial t} = \boldsymbol{\nabla} \times (\mathbf{v} \times \mathbf{B}_0 + \mathbf{v}_0 \times \mathbf{b} + \mathbf{v} \times \mathbf{b} - \overline{\mathbf{v} \times \mathbf{b}}) + \eta \nabla^2 \mathbf{b}. \tag{8.13}$$

In order to close this pair of equations (when $\mathbf{v}$ is given), it is necessary to make an assumption about the form of the mean electric field $\overline{\mathbf{v} \times \mathbf{b}}$. Usually, one considers *pseudo-isotropic* (or non-mirror symmetric) turbulence, such that the flow is not invariant under reflections about the origin. The lack of symmetry may be produced by, for instance, fast rotation or stratification. Assuming that $\mathbf{B}_0$ varies on a scale larger than the turbulence and expanding in terms of $\mathbf{B}_0$ and its derivatives gives

$$\overline{\mathbf{v} \times \mathbf{b}} = \alpha \mathbf{B}_0 - \tilde{\eta} \boldsymbol{\nabla} \times \mathbf{B}_0, \tag{8.14}$$

where $\tilde{\eta}$ is the turbulent diffusivity (usually $\gg \eta$). Equation (8.12) then becomes

$$\frac{\partial \mathbf{B}_0}{\partial t} = \boldsymbol{\nabla} \times (\mathbf{v}_0 \times \mathbf{B}_0) + \boldsymbol{\nabla} \times (\alpha \mathbf{B}_0) + (\eta + \tilde{\eta}) \nabla^2 \mathbf{B}_0,$$

and it can be seen that the effect of the turbulence is to provide an extra electric field ($\alpha \mathbf{B}_0$) and to enhance large-scale diffusion through the term $\tilde{\eta} \nabla^2 \mathbf{B}_0$.

8.3.2.2 Evaluating Alpha

The coefficients α and $\tilde{\eta}$ may be evaluated by transform or Green's-function techniques (e.g., Moffatt 1978). In general, it is the presence of $\nabla \times (\mathbf{v}\times\mathbf{b}-\overline{\mathbf{v}\times\mathbf{b}})$ that makes Eq. (8.13) difficult to solve, but, when the small-scale (turbulent) magnetic Reynolds number (vl/η) is small, these terms and $\partial\mathbf{b}/\partial t$ are negligible. For such a *quasi-linear* (or *first-order smoothing*) *approximation*, when $\mathbf{v}_0 \equiv 0$ and $\nabla\cdot\mathbf{v}=0$, Eq. (8.13) reduces to

$$0=(\mathbf{B}_0\cdot\nabla)\mathbf{v}+\eta\nabla^2\mathbf{b}. \tag{8.15}$$

Since this is linear in $\mathbf{v}$ and $\mathbf{b}$, Fourier transforms may be used as follows to solve for $\mathbf{b}$ and hence obtain $\overline{\mathbf{v}\times\mathbf{b}}$ in terms of $\mathbf{v}$ and $\mathbf{B}_0$. After writing

$$\mathbf{v}(\mathbf{r},t)=\int e^{i\mathbf{k}\cdot\mathbf{r}}\,d\mathbf{Z}(\mathbf{k},t),\qquad \mathbf{b}(\mathbf{r},t)=\int e^{i\mathbf{k}\cdot\mathbf{r}}\,d\mathbf{Y}(\mathbf{k},t),$$

and assuming $\mathbf{B}_0$ is uniform, Eq. (8.15) may be transformed to give $d\mathbf{Y}=(i\mathbf{B}_0\cdot\mathbf{k})/(\eta k^2)\;d\mathbf{Z}$. Substituting this into the expression $\overline{\mathbf{v}\times\mathbf{b}}=\int d\mathbf{Z}^*\times d\mathbf{Y}$ therefore gives $(\overline{\mathbf{v}\times\mathbf{b}})_l=\alpha_{lm}B_{0m}$, where $\alpha_{lm}=(i/\eta)\int(k_m/k^2)(\overline{d\mathbf{Z}^*\times d\mathbf{Z}})_l$. Equivalently, α_{lm} may be written as

$$\alpha_{lm}=\frac{i}{\eta}\int\frac{k_m}{k^2}\,\xi_{lnj}\,\Phi_{nj}\,dk, \tag{8.16}$$

in terms of the *spectrum tensor* $[\Phi_{lm}(\mathbf{k})]$, which is related to the velocity correlation tensor $\overline{v_l\,v_m}=\int\Phi_{lm}\,d\mathbf{k}$.

When the turbulence is *isotropic*, the spectrum tensor is

$$\Phi_{lm}=\Phi^0_{lm}\equiv\frac{E(k)}{4\pi k^4}(k^2\delta_{lm}-k_lk_m),$$

in terms of the *energy spectrum function* $[E(k)]$, which has the property that $\frac{1}{2}\overline{\mathbf{v}\cdot\mathbf{v}}=\int E(k)dk$. According to Eq. (8.16), this makes α_{lm} identically zero, and so dynamo action is not possible. However, when the turbulence is *pseudo-isotropic*,

$$\Phi_{lm}=\Phi^0_{lm}+\frac{iF(k)}{8\pi k^4}\,\xi_{lmn}\,k_n, \tag{8.17}$$

which is invariant with respect to a rotation of axes but not with respect to reflection in the origin. In this case, Eq. (8.16) becomes $\alpha_{lm}=\alpha\delta_{lm}$, so that $\overline{\mathbf{v}\times\mathbf{b}}=\alpha\mathbf{B}$, where

$$\alpha=-\frac{1}{3\eta}\int_0^\infty\frac{F(k)}{k^2}dk. \tag{8.18}$$

The function $F(k)$ introduced in Eq. (8.17) is the *helicity spectrum function*, which is related to the *helicity* $\overline{\mathbf{v}\cdot\nabla\times\mathbf{v}}=[i\int\overline{d\mathbf{Z}^*\cdot\mathbf{k}\times d\mathbf{Z}}=-i\int k_l\;\xi_{lmn}\;\Phi_{mn}\;dk]$, since the form (8.17) implies $\overline{\mathbf{v}\cdot\nabla\times\mathbf{v}}=\int_0^\infty F(k)dk$.

The effect of slow variations in the large-scale field may also be included by writing $\mathbf{B}_0(\mathbf{r})=\mathbf{B}_0+(\mathbf{r}\cdot\nabla)\mathbf{B}_0$. Repeating the above analysis then yields an expression for the *eddy diffusion coefficient* $(\tilde{\eta})$ as

$$\tilde{\eta}=\frac{2}{3\eta}\int_0^\infty\frac{E(k)}{k^2}dk, \tag{8.19}$$

in terms of the energy spectrum function $[E(k)]$.

When the *correlation time* (τ_c) is much smaller than the turnover time (l/v) (Krause and Rädler 1980) or $R_m \ll 1$, Moffatt (1978) finds

$$\alpha=-\tfrac{1}{3}\tau\;\overline{\mathbf{v}\cdot\nabla\times\mathbf{v}}\quad\text{and}\quad\tilde{\eta}=\tfrac{1}{3}\tau\;\overline{\mathbf{v}\cdot\mathbf{v}}. \tag{8.20}$$

It is clear from Eqs. (8.18) and (8.20) that α is a measure of the average helical motion or the lack of symmetry (the right- or left-handedness) of the small-scale flow. It depends on there being a non-zero helicity, corresponding to a twisting motion with a preferred sense of rotation due to, for instance, a Coriolis force. Just such a cyclonic twist was required to generate poloidal flux from toroidal flux in Parker's analysis. Thus, the presence of *gyrotropic* turbulence with a non-vanishing helicity seems highly desirable for a successful dynamo.

8.3.2.3 The α-ω Dynamo: An Early Solar-Cycle Model

Steenbeck and Krause (1969) and Roberts (1972b) solved the dynamo equations in a sphere (of radius $R_\odot$), including an α-effect (with $\alpha = \alpha_0 \cos\theta$ and θ measured from the pole) and a differential rotation due to a shear (Ω_0') in angular velocity (Ω_0). With an axisymmetric flow $\mathbf{v}_0 = R\Omega_0\hat{\boldsymbol{\phi}} + \mathbf{v}_p$ and field $\mathbf{B} = B_\phi\hat{\boldsymbol{\phi}} + \boldsymbol{\nabla}\times(A_p\hat{\boldsymbol{\phi}})$ and a turbulent diffusivity ($\tilde{\eta} \gg \eta$), the dynamo equations become

$$\frac{\partial B_\phi}{\partial t} + R(\mathbf{v}_p\cdot\boldsymbol{\nabla})\left(\frac{B_\phi}{R}\right) = R(\mathbf{B}_p\cdot\boldsymbol{\nabla})\Omega_0 + [\boldsymbol{\nabla}\times(\alpha\mathbf{B}_p)]_\phi + \tilde{\eta}\left(\nabla^2 - \frac{1}{R^2}\right)B_\phi, \tag{8.21}$$

$$\frac{\partial A_p}{\partial t} + \frac{\mathbf{v}_p}{R}\cdot\boldsymbol{\nabla}(RA_p) = \alpha B_\phi + \tilde{\eta}\left(\nabla^2 - \frac{1}{R^2}\right)A_p. \tag{8.22}$$

These are similar to Eqs. (8.8) and (8.4), except that the diffusivity is turbulent and we notice that the α-effect $[\boldsymbol{\nabla}\times(\alpha\mathbf{B}_p)]$ also has a ϕ-component, and so, like the differential rotation term, it is capable of generating B_ϕ from $\mathbf{B}_p$ in Eq. (8.21).

Indeed, when differential rotation is weak ($|L^2\Omega_0'| \ll |\alpha_0|$), the first (i.e., differential rotation) term on the right of Eq. (8.21) is negligible and so the α-effect alone regenerates both the toroidal and poloidal fields. For these so-called 'α^2-dynamos', the fields that are produced tend to be steady rather than oscillatory, and so they are more relevant to the terrestrial dynamo. On the other hand, on the Sun, differential rotation is strong enough ($|L^2\Omega_0'| \gg |\alpha_0|$) that it dominates the α-effect in Eq. (8.21) and we have an 'α-ω dynamo', with the toroidal and poloidal components maintained by differential rotation and the α-effect, respectively.

The structure of the fields and the existence of dynamo action depend on the value of a *dynamo number* $\tilde{N}_D = \alpha_0\Omega_0' R_\odot^3\,\tilde{\eta}^2$, which (like N_D) gives the ratio of field generation to dissipation. Since Eqs. (8.21) and (8.22) are linear in B_ϕ and A_p, they possess solutions behaving like $e^{\omega t}$, so that the object is to determine ω as a function of $\tilde{N}_D$ and find the smallest (i.e., critical) value of $\tilde{N}_D$ that allows the α-effect to overcome diffusion and so produce a growing mode $[\mathcal{R}(\omega) > 0]$.

If $\alpha_0 d\Omega_0/dr < 0$, the most easily excited dynamo is oscillatory [imaginary part $\mathcal{I}(\omega) \neq 0$] and of dipole type ($A_p$ even, B_ϕ odd in z). Its activity progresses from pole to equator (Figure 8.7), just like Parker's

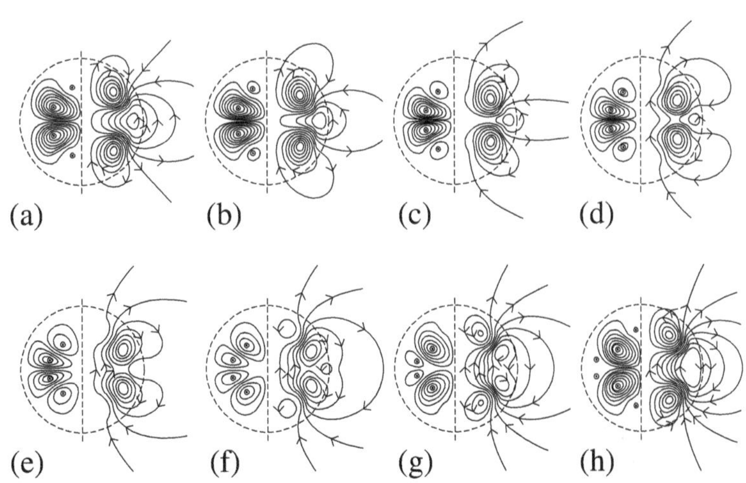

Figure 8.7. A half-cycle for an α-ω dynamo showing poloidal ($\mathbf{B}_p$) field lines in each right-hand hemisphere and lines of constant toroidal field (B_ϕ) on the left, at intervals of one-sixteenth of a cycle period (after Roberts 1972b).

dynamo waves (Roberts 1972b). When the toroidal or poloidal fields from the two hemispheres reach the equator, they disappear by diffusing into one another. If $\alpha_0 d\Omega_0/dr > 0$, the preferred mode (i.e., at the lowest $\tilde{N}_D$-value) is oscillatory and quadrupolar (A_p odd, B_ϕ even), with polar migration. Adding a small meridional circulation ($\mathbf{v}_p$) halves the critical value of $\tilde{N}_D$ and changes the character of the preferred mode to that of a dipole.

The periods of these dynamos is $2\pi R_\odot^2/(100\tilde{\eta})$, which is smaller than the crude estimate of $R_\odot^2/\tilde{\eta}$ since diffusion is taking place on smaller scales than $R_\odot$. Adopting a value for the turbulent eddy diffusivity $\tilde{\eta}$ of 10^9 m^2 s^{-1}, typical of granules or active-region dispersal, gives a period of only 1 year, whereas, in order to obtain a period comparable with that of a solar cycle (22 years), it is necessary to take $\tilde{\eta} \approx 10^8$ m^2 s^{-1}.

During the 1970s many authors continued to focus just on the induction equation and to produce more sophisticated kinematic α-ω dynamos in an attempt to reproduce detailed features of the solar cycle. One approach was to impose the spatial variation of α on physical grounds and another was to model feedback of the Lorentz force on the motions by assuming a variation of α with magnetic field strength (Stix 1972). In addition, various functional forms for α and Ω_0 were adopted to model flux eruption by magnetic buoyancy or time-delay. However, this approach eventually outran its usefulness and the main emphasis shifted away from simple kinematic α-ω dynamos to more fundamental questions and to nonlinear dynamo models that incorporate the equation of motion, and in particular to dynamos that are driven (at least partly) near the tachocline at the base of the convection zone. For a comparison of computations of some standard 2D axisymmetric mean-field dynamo models, namely, an α-ω dynamo with uniform magnetic diffusivity and a flux-transport dynamo (Sec. 8.4.2), see Jouve et al. (2008).

8.4 Flux-Transport Dynamos

8.4.1 Babcock-Leighton Ideas

Babcock (1961) was stimulated by magnetograph observations of the photosphere to develop a qualitative model of the solar cycle. He suggested that the toroidal field is produced by differential rotation in a thin layer just below the surface and rises by magnetic buoyancy, as in the standard turbulent dynamo model (Stages 1 to 3 in Figure 8.8). However, he proposed a rather different scenario for the production of the poloidal field, based on the observation that bipolar sunspots are tilted with respect to lines of latitude, such that the preceding spots are closer to the equator than the following spots (Joy's Law). He suggested that, when the bipolar regions decay, some of the following flux migrates to the poles, cancels the pre-existing flux and so generates a new poloidal field component of opposite sign.

Then Leighton (1969) suggested that the migration of poloidal flux is due to supergranular eddy diffusion and wrote down qualitative kinematic equations for the evolution with time and latitude of B_ϕ and B_r (averaged over r and ϕ) under the effect of shear (creating the toroidal field), magnetic buoyancy and Coriolis forces (creating the poloidal field) and diffusion of the poloidal field. Essentially, he incorporated an α-effect. A numerical integration reproduced the main mean qualitative features of the solar cycle. (Note that the tilt of sunspots reflects the cyclonic twisting of rising Parker loops, which is why it can be described by an α-effect.)

8.4.2 Flux-Transport Dynamo Model

The Babcock-Leighton model was more qualitative than turbulent dynamo theory (the main-stream approach in the 1970s and 1980s). However, four developments paved the way for a flux-transport model. Firstly, the NRL group (e.g., Sheeley et al. 1985; Wang et al. 1989) developed a theory for diffusion of radial magnetic field (B_r) over the solar surface from decaying active regions that reproduces the observed flux distribution. B_r is treated as a *passive scalar*, so that it does not feed back on the flow. Secondly, there was growing evidence for the existence of a meridional flow at the solar surface of 15 to 20 m s^{-1}

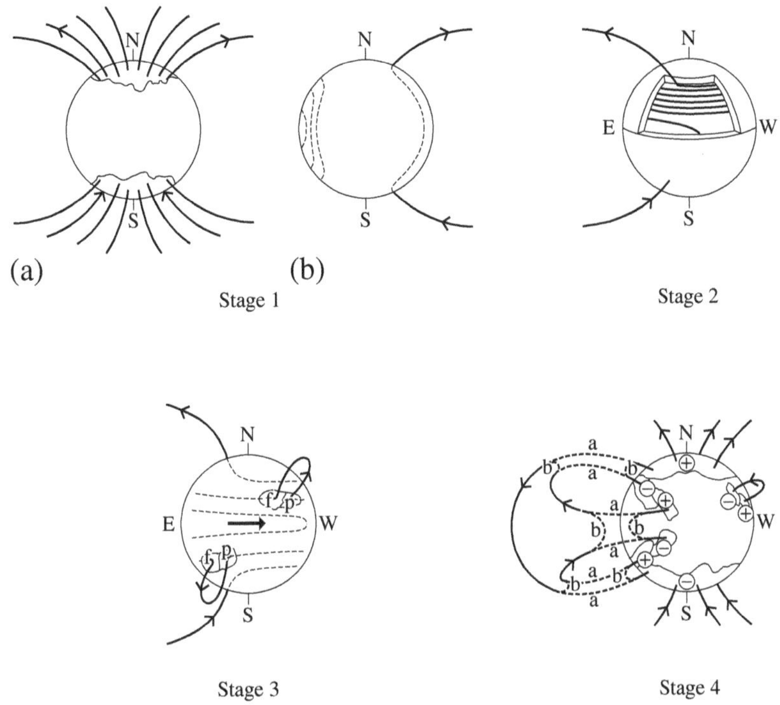

Figure 8.8. Babcock's qualitative scenario for dynamo action. Stage 1: the initial poloidal dipolar field has magnetic field lines closing in the interior at shallow depths. Stage 2: submerged flux is stretched out by differential rotation to produce toroidal flux. Stage 3: magnetic buoyancy causes flux to emerge as sunspot groups. Stage 4: following flux of active regions migrates to the poles and reverses the polar field, creating a new poloidal dipolar field of opposite polarity.

(Hathaway 1996), which carries decaying remnants of active regions polewards more effectively than supergranular diffusion. Presumably, there is a counterflow deep in the convection zone, but it is not yet known whether it reaches down right to the tachocline, whether it varies with time and whether it is in the form of just one cell or multiple cells.

Thirdly, Choudhuri and Gilman (1987) modelled the rise of magnetic flux tubes through the convection zone under the influence of magnetic buoyancy and Coriolis forces, in order to understand the relation between the field generated at the base of the convection zone and sunspots at the solar surface. They found that the field strength of an initial toroidal flux tube at the base of the convection zone needs to be at least as strong as 10^5 G in order for buoyancy to overpower the Coriolis effect and enable the tube to emerge at low latitudes rather than high ones. D'Silva and Choudhuri (1993) extended the calculations to examine the tilts of emerging bipolar regions. Comparing them with Joys Law, they found that the theory fits the observations well only for an initial field of 10^5 G, which is much larger than the equipartition value of 10^4 G between the magnetic field and the fluid kinetic energy $[B^2/(2\mu) = \frac{1}{2}\rho v^2]$ when an expected convective velocity is adopted.

Fourthly, a series of flux tube papers reinforced this conclusion (Spruit and van Ballegooijen 1982; Ferriz-Mas and Schüssler 1994; Schüssler et al. 1994) and showed that thin flux tubes stored in the overshoot layer would be unstable and rise through the convection zone, with a growth-time less than a year provided the field strength is as much as 10^5 G. More realistic numerical models were later developed by Emonet and Moreno-Insertis (1998), Cheung et al. (2006) and Fan (2008).

A *flux-transport dynamo* model was first proposed by Wang et al. (1991), Dikpati and Choudhuri (1994), Durney (1995) and developed by others (Durney 1997; Choudhuri and Dikpati 1999; Nandy and Choudhuri

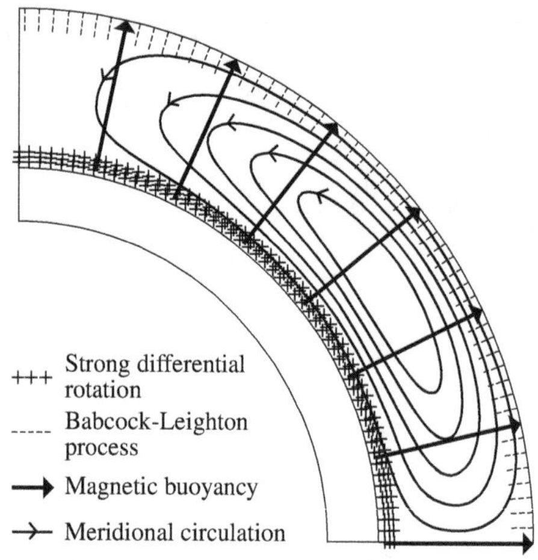

Figure 8.9. The processes involved in a flux-transport dynamo (courtesy Piyali Chatterjee).

2002; Chatterjee et al. 2004; Dikpati and Gilman 2006, 2007; Nandy 2006; Charbonneau et al. 2007). It is motivated to explain the observed surface flux and finds a way around a problem with early dynamo models, namely, that the expected helical turbulence is too weak to generate poloidal flux from a toroidal field of 10^5 G. Two differences from the simple Babcock-Leighton picture are that the toroidal field is produced at the tachocline rather than just below the solar surface and the transport of poloidal flux is by meridional flow as well as diffusion. Transport by diffusion alone gives poleward rather than equatorward propagation (Choudhuri et al. 1995).

In the flux-transport dynamo (Figure 8.9), the α and ω are widely separated in space. Poloidal flux is produced at the solar surface by the Babcock-Leighton decay of tilted active regions. Then it is transported partly by diffusion and partly by meridional flow, acting as a conveyor belt, to the poles and down below the surface to the tachocline, where it is sheared as usual into a toroidal field. This toroidal field rises by buoyancy and rotates, producing new poloidal flux at the surface in addition to the poloidal flux that is being swept up from the tachocline by meridional flow.

Mathematically, equations similar to standard kinematic α-ω dynamo are adopted, except that α is assumed to be concentrated near the solar surface to represent the Babcock-Leighton process. In spherical polar coordinates, an axisymmetric flow $\mathbf{v} = r\sin\theta\ \Omega(r,\theta)\hat{\boldsymbol{\phi}} + \mathbf{v}_p$ is assumed, where $\Omega(r,\theta)$ is the solar angular velocity, chosen to have a strong gradient at the tachocline (consistent with helioseismology), and $\mathbf{v}_p$ is an assumed meridional circulation. The magnetic field has the form $\mathbf{B} = B_\phi(r,\theta)\hat{\boldsymbol{\phi}} + \boldsymbol{\nabla}\times[A_p(r,\theta)\hat{\boldsymbol{\phi}}]$, and the resulting two scalar equations for the toroidal field (B_ϕ) and the poloidal field's vector potential component (A_p) are

$$\frac{\partial B_\phi}{\partial t} + \frac{1}{r}\left[\frac{\partial}{\partial r}(rv_{pr}B) + \frac{\partial}{\partial\theta}(v_{p\theta}B)\right] = r\sin\theta(\mathbf{B}_p\cdot\boldsymbol{\nabla})\Omega$$
$$+\tilde{\eta}\left(\nabla^2 - \frac{1}{r^2\sin^2\theta}\right)B_\phi + \frac{1}{r}\frac{d\tilde{\eta}}{dr}\frac{\partial}{\partial r}(rB_\phi),$$

$$\frac{\partial A_p}{\partial t} + \frac{1}{r\sin\theta}(\mathbf{v}_p\cdot\boldsymbol{\nabla})(r\sin\theta\ A_p) = \alpha B_\phi + \tilde{\eta}\left(\nabla^2 - \frac{1}{r^2\sin^2\theta}\right)A_p + S,$$

where $S = S(r,\theta,B_\phi)$ is a surface poloidal field source term that is taken from observations and represents the emergence of new poloidal flux through the solar surface. In some versions of the model, S and α are 'quenched' above a certain field strength in order to prevent unbounded growth of the field.

Two types of flux-transport dynamo have been proposed. The first (Dikpati and Gilman 2006, 2007) invokes a low value of magnetic diffusivity ($\eta = 5\times10^6$ m^{2}s^{-1}) in the bulk of the convection zone with a diffusion time across the convection zone of 200 years, so the main transport of flux downwards is by meridional flow. An extra α-effect at the base of the convection zone is added in order to favour an antisymmetric dynamo (i.e., one with a dipolar toroidal field satisfying the Hale polarity laws). Also, the strength of the poloidal field is assumed proportional to sunspot area. The time-scale for meridional circulation is about 20 years, and so the strength of the resulting toroidal field (i.e., of the next sunspot cycle) depends on the nature of the two or three previous cycles and how much poloidal flux is brought down to be reprocessed.

The second kind of flux-transport dynamo (Chatterjee et al. 2004; Choudhuri et al. 2007; Jiang et al. 2007) invokes a much higher magnetic diffusivity ($\eta = 10^8$ m^{2}s^{-1}) comparable with the surface value and having a convection-zone diffusion time of order only 10 years. The strong diffusion connects field lines across the equator and produces a dipolar field without needing an α-effect at the base of the convection zone. Furthermore, it suggests (by analogy with terrestrial cyclonic weather patterns) that the poloidal field has a random nature independent of the cycle (in agreement with observations). The resulting strength of the toroidal field in cycle $n+1$ depends on the strength of the polar field at the end of cycle n, again in agreement with observations (Choudhuri et al. 2007).

There are various other observational consequences of the models. The dynamo period is inversely proportional to the speed of the equatorward return meridional flow at the base of the convection zone (where the diffusivity is assumed to be much smaller than the convection-zone value in all models); a value of a few m s^{-1} is needed to give an 11-year cycle. By contrast with interface or classical α-ω dynamos, flux-transport models are rather insensitive to differential rotation, the α-effect or the diffusivity. The timing of the reversal of the polar field (namely, at sunspot maximum) is due to the strength of the polar meridional flow at the solar surface. Furthermore, active regions in the northern hemisphere are observed predominantly to have negative magnetic helicity, while those in the south have the opposite sign (Pevtsov et al. 1995). This may be explained as a natural property of a flux-transport dynamo (Choudhuri 2003; Choudhuri et al. 2004) or alternatively in terms of the effect of helical turbulence on a rising tube (Longcope et al. 1998).

Several possible nonlinear feedback mechanisms have been proposed that limit the amplitude of the magnetic field and control the amplitude of the solar cycle. These include the back-reaction of the magnetic field on: differential rotation; tilt angles of sunspot groups (Cameron et al. 2010); cross-equator transport of magnetic flux and converging flows towards activity belts (Cameron and Schüssler 2012).

Strengths of flux-transport models are: their attempts to reproduce the amplitudes of the observed solar cycle; their lack of reliance on a turbulent α-effect, so that problems associated with α-quenching and its effectiveness are avoided; their polar migration of flux from active regions; their reproduction of the phase relation between polar flux and sunspots; their removal of restrictions of classical dynamo theory without meridional flow (i.e., that the dynamo number must be negative and the shear velocity must increase with depth), since a sufficiently strong meridional flow in a flux-transport dynamo [or an advection-dominated classical dynamo (Rüdiger and Küker 2002)] forces equatorward propagation of the toroidal field even when the dynamo number is positive (Choudhuri et al. 1995). However, like all kinematic models, the flows in flux-transport dynamos are not determined consistently. Moreover, the flux emergence source term $S(r,\theta,B_\phi)$ is imposed from observations rather than being determined by the model, and at present the behaviour throughout the convection zone of the meridional flow (which is prescribed in the model) is unknown. Also, if a turbulent α-effect were indeed at work near the tachocline, it may well be more effective than a distant flux-transport α-effect at the photosphere (Mason et al. 2002). A key question is whether the polar flux at the solar surface is indeed representative of the Sun's poloidal flux (as flux-transport dynamos assume) or whether it is a secondary consequence of poloidal flux generation in the tachocline.

8.5 Tachocline Dynamos

Interest in the concept of a tachocline dynamo began with suggestions of an *overshoot dynamo* (Sec. 8.5.1), stimulated by helioseismology observations (Sec. 1.3.4) and by Parker's *interface dynamo* model (Sec. 8.5.2). A fully developed tachocline model has not yet emerged, but a scenario has been widely discussed and the nature of the tachocline itself has been debated (Sec. 8.5.3). The scenario incorporates several physical processes, as enumerated in Figure 8.10. Magnetic flux in the convection zone is transported by convection (0). It is amplified by small-scale dynamo action (1) and is carried down into the tachocline and held there by either strong overshooting plumes or turbulent pumping (2) (Tobias et al. 1998, 2001; Dorch and Nordlund 2001). In the tachocline, the strong shear (3) amplifies the toroidal field, which then rises into the convection zone by magnetic buoyancy (4) or turbulent pumping. In the convection zone, the α-effect (5) creates poloidal flux and the weak field is recycled (6), while the toroidal flux erupts through the solar surface (7).

8.5.1 Overshoot Dynamo

The idea that the dynamo works throughout the convection zone was questioned when Parker (1975) calculated that magnetic buoyancy would remove magnetic field from the convection zone so quickly that it would not have time to be amplified by dynamo action. In an inspired step, Spiegel and Weiss (1980) and van Ballegooijen (1982) then suggested that a thin *overshoot region* with a thickness of, say, 10 Mm, just below the bottom of the convection zone would be much more favourable for operation of a dynamo, since it is convectively stable and so would suppress magnetic buoyancy and hold down the magnetic field there. Convective plumes from the overlying unstable layers would overshoot and penetrate into this layer, providing turbulent motions that may drive a dynamo. This idea was later given strong support from the discovery of the tachocline in helioseismology observations as a layer of strong differential rotation where a large toroidal field would naturally be generated. From the mid-1980s the idea of an overshoot dynamo became popular (e.g., DeLuca and Gilman 1986; Schüssler 1987; Choudhuri 1990; Stix 1991).

8.5.2 Interface Dynamo (Parker 1993)

If the toroidal field at the base of the convection zone is indeed 10^5 G (D'Silva and Choudhuri 1993; Schüssler et al. 1994), as now seems likely (at least in part), then the turbulent flow there would not be strong enough to twist up the magnetic field lines, and so the α-effect associated with convective

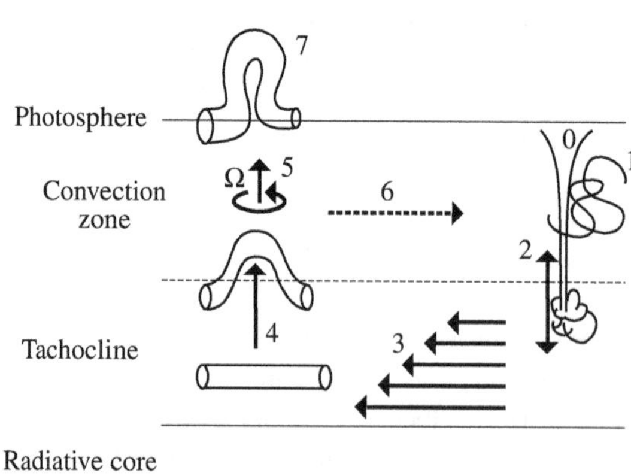

Figure 8.10. A scenario for tachocline dynamo action with several features which are described in the text (courtesy Nic Brummell).

turbulence would be switched off. This, together with the discovery of the tachocline, led Parker (1993), Tobias et al. (1995) and Charbonneau and MacGregor (1997) to propose an 'interface' dynamo with the main shear and α-effect in two distinct regions. The mean toroidal field (that subsequently produces sunspots) is created by radial differential rotation and is stored in the tachocline. The mean poloidal field, on the other hand, is located in the lower part of the convection zone above the tachocline. It is created by an α-effect and then pumped down into the tachocline by convection. If the field strength were 10^4 G, the α-effect there would be produced by convective turbulence, but, if it were 10^5 G, it could be due to some kind of magnetic buoyancy instability (Ferriz-Mas et al. 1994). Perhaps the toroidal tachocline field is mainly 10^4 G but with some 10^5 G-regions, which subsequently rise to produce active regions.

Parker's (1993) analysis showed that the dynamo in this scenario takes on the character of a surface wave tied to the lower surface ($z=0$) of the convection zone. He extended the simple planar dynamo model of Section 8.3.1.1 to a two-layer model, with an α-effect in the region $z>0$ and a uniform shear ($S_0 = dv_y/dz$) in $z<0$. Thus, in the lower part ($z>0$) of the convection zone, where there is no shear, the 'toroidal' field component $B_y(x,z,t)$ and 'poloidal' field ($B_x = -\partial A/\partial z, B_z = \partial A/\partial x$) with flux function $A(x,z,t)$ satisfy

$$\left(\frac{\partial}{\partial t} - \eta_{cz}\nabla^2\right) B_y = 0, \qquad \left(\frac{\partial}{\partial t} - \eta_{cz}\nabla^2\right) A = \alpha B_y. \tag{8.23}$$

In the tachocline $z<0$, on the other hand, where there is a strong shear and the diffusivity (η_t) is much smaller than the turbulent convection-zone value (η_{cz}), the 'toroidal' field $b_y(x,z,t)$ and 'poloidal' field ($b_x = -\partial a/\partial z, b_z = \partial a/\partial x$) with flux function $a(x,z,t)$ satisfy

$$\left(\frac{\partial}{\partial t} - \eta_t\nabla^2\right) b_y = S_0\frac{\partial a}{\partial x}, \qquad \left(\frac{\partial}{\partial t} - \eta_t\nabla^2\right) a = 0. \tag{8.24}$$

Plane-wave solutions are then sought above $z=0$ to Eqs. (8.23) in the form

$$B_y = B_0 \exp[\omega t + i(k_x x + K_z z)], \; A = (A_{0x} + A_{0z} z)\exp[\omega t + i(k_x x + K_z z)],$$

where $\omega = \omega_r + i\omega_i$ and $K_z = K_{zr} + iK_{zi}$ are complex with $K_{zi}>0$, so that they decay to zero as $z\to\infty$. Below $z=0$ the adopted form for solutions to Eqs. (8.24) is

$$b_y = b_0 \exp[\omega t + i(k_x x + k_z z)], \quad a = (a_{0x} + a_{0z} z)\exp[\omega t + i(k_x x + k_z z)],$$

where $k_z = k_{zr} + ik_{zi}$ is complex with $k_{zi}<0$, so that they decay to zero as $z\to-\infty$. Equations (8.23) and (8.24) therefore determine A_{0z}, a_{0z} and K_z and k_z in terms of ω.

To complete the solution, boundary conditions are imposed at the interface ($z=0$), namely, continuity of the magnetic field components ($B_x = b_x$, $B_y = b_y$, $B_z = b_z$) and of the tangential electric field ($E_x = j_x/\sigma$), so that the diffusion of flux is continuous ($\eta_{cz}\partial B_y/\partial z = \eta_t \partial b_y/\partial z$). These conditions determine b_0, a_{0x}, A_{0x} and a dispersion relation for the dimensionless growth-rate and oscillation frequency [$\bar{\omega}_r = \omega_r/(\eta_{cz}k_x^2)$, $\bar{\omega}_i = \omega_i/(\eta_{cz}k_x^2)$] of the surface dynamo wave in terms of the ratio (η_t/η_{cz}) of magnetic diffusivities and a *dynamo number* [$N_d = \alpha S_0/(\eta^2 k_x^3)$] that is similar in form to Eq. (8.10). The result, in the limit when $\eta_t/\eta_{cz} \ll 1$ (so that diffusion is negligible in the shear region), is

$$(\eta_i/\eta_{cz})^2 N_d^2 = 64(\tfrac{1}{2} + \bar{\omega}_r)^2(1+\bar{\omega}_r)\bar{\omega}_r, \qquad \bar{\omega}_i^2 = (1+\bar{\omega}_r)\bar{\omega}_r,$$

in which the first equation determines the growth-rate $\bar{\omega}_r$ implicitly in terms of N_d and the second determines the frequency $\bar{\omega}_i$.

The resulting scale $1/K_{zr}$ over which the magnetic field decays away from the interface in the convection zone is of order the horizontal wavelength $1/k_x$, whereas the corresponding scale for the tachocline field is a factor of $\eta_t/\eta_{cz} \ll 1$ smaller. Thus, the azimuthal field in $z<0$ is confined to a thin layer pressed up against the underside of $z=0$. However, the azimuthal fluxes of b_y and B_y are comparable in magnitude, and so the maximum azimuthal field in the tachocline is much greater than in the convection zone, being of order $B_0\eta_{cz}/\eta_t \gg B_0$.

Strengths of the interface model include the natural ways in which the radial shear generates a large toroidal field and the stable stratification below the interface enables the field to be stored effectively. However, the details of how flux is transported to and from the tachocline and how the α-effect operates are unclear.

8.5.3 The Solar Tachocline

The nature of the tachocline (Sec. 1.3.4) and its role in the dynamo (not yet fully worked out) have been well debated in Tobias et al. (1995) and Hughes et al. (2007). It is a narrow shear layer below the base of the convection zone at $r = 0.713$ R$_\odot$, where the angular velocity rapidly changes structure from being uniform in the radiative interior to being roughly constant on cones in the convection zone (e.g., Kosovichev 1996). The tachocline is slightly prolate, having a radius and thickness that increase with latitude from 0.692±0.002 R$_\odot$ and 0.033±0.007 R$_\odot$ at the equator to 0.710±0.002 R$_\odot$ and 0.076±0.010 R$_\odot$ at 60 degrees, according to Basu and Antia (2003). Thus, it lies essentially below the base of the convection zone at low latitudes and is roughly centred on it at high latitudes.

The tachocline couples the radiative interior (which has 90 per cent of the angular momentum) to the convection zone, which is in turn being spun down by the solar wind. The tachocline also controls the boundary conditions at the base of the convection zone and is probably the seat of the main solar dynamo. It is natural to talk in terms of three separate regions, but in future it would be good to develop a holistic model that treats the interior, tachocline and convection zone as a single system.

One key question is why is the tachocline so thin and another is why does the radiative interior rotate uniformly in the face of the differentially rotating convection zone [whose rotation is due to a combination of Reynolds stresses, asphericity of heat flux and angular momentum transport by meridional flow (Sec. 1.3.3)]. There is likely to be a meridional circulation within the tachocline descending from the convection zone at the poles and equator and rising at mid-latitudes, driven by the stresses associated with differential rotation in the convection zone. Some agent is needed, however, to prevent the meridional circulation from penetrating the interior.

Two answers have been proposed. The first is that the tachocline is dominated by 2D shear-driven turbulence that produces a horizontal Reynolds stress to oppose latitudinal variations in angular velocity (Spiegel and Zahn 1992). The second is that it has a magnetic origin (Kitchatinov and Rüdiger 1996; Rüdiger and Kitchatinov 1997; Gough and McIntyre 1998), later developed by Brun and Zahn (2006), Garaud (2007) and Garaud and Garaud (2008). Here the required rigidity is produced by a relic or fossil magnetic field in the radiative interior, so that the tachocline is a thin magnetic boundary layer without turbulent transport. Indeed, a field there as small as 0.01 G could enforce uniform rotation (Mestel and Weiss 1987).

The overshoot layer just below the convection zone consists of tongues of plasma and magnetic field plunging into the stably stratified interior, generating gravity waves and either mixing with it or returning to the convection zone. They provide some mixing of flux, entropy and chemical species and essentially smooth the boundary between the convection zone and radiative interior. The thickness of the overshoot layer is rather uncertain, but it is likely to be thinner than the tachocline, and so it may well dominate the lower tachocline but not the upper tachocline.

Several kinds of instability could occur in the tachocline, contributing to a dynamo. These include shear flow instability (Giman 2007), magnetic buoyancy instability (Hughes 2007) and magnetorotational instability (Balbus and Hawley 1991; Ogilvie 2007), which occurs when Ω^2 decreases outwards and develops weak 3D turbulence that reduces the shear. Interestingly, the resulting upwelling part of the meridional circulation within the tachocline (which could help flux to rise) is probably located at the latitude where sunspots first emerge at the start of a new cycle.

In the tachocline, an α-effect may well be produced by the Coriolis force acting on magnetic buoyancy instability of flux tubes (Spruit and van Ballegooijen 1982; Ferriz-Mas et al. 1994; Schüssler et al. 1994; Brandenburg and Schmidt 1998; Tobias and Weiss 2007). Furthermore, the observation of zonal shear flows (or *torsional oscillations*), in which bands of slightly faster rotation accompany the activity zones as they drift towards the equator, suggests a suitable nonlinear saturation mechanism for a tachocline dynamo, namely, the *Malkus-Proctor effect* (Malkus and Proctor 1975), in which saturation is by a Lorentz-induced differential rotation.

The possibility of mixing below the convection zone down to a region where the temperature exceeds 2.5×10^6 K (and lithium burns), at least at some time during the Sun's evolution, is suggested by the *lithium problem* (Brun et al. 1999), namely, the fact that lithium in the solar atmosphere has an abundance that is 150 times lower than the meteoritic abundance, which is presumably the same as in the nebula from which the Sun was formed. But, since beryllium is not depleted, such mixing cannot go down to a beryllium-burning region where the temperature is 3.5×10^6 K.

8.6 Other Approaches

8.6.1 Global Computations

Undertaking a global computation of dynamo action in the solar interior is a substantial challenge, since one needs first a realistic model of convection-zone fluid dynamics, including convection over a wide range of scales, differential rotation, meridional circulation and a tachocline (e.g., Miesch 2005; Miesch and Toomre 2009). Then, including the magnetic field in a self-consistent way also involves magnetoconvection (over a range of scales and field strengths), flux pumping, flux expulsion (see PROBLEM 8.5), magnetic buoyancy, turbulent diffusion and an α-effect.

Gilman (1977) made a start by modelling Boussinesq convection in a rotating spherical shell heated uniformly from the base with a Taylor number (T) of 10^5, a Prandtl number (Pr) of 1 and a Rayleigh number (Ra) of 2×10^4. The resulting differential rotation is driven by Reynolds stresses and possesses equatorial acceleration provided the angular velocity *decreases* with depth ($d\omega/dr > 0$). Gilman and Miller (1981) then added a magnetic field, with turbulent values for the diffusion coefficients but no α-effect term in the induction equation. The initial magnetic field is purely toroidal, and the results show that dynamo action is indeed produced, as evidenced by a growing magnetic field when the magnetic Prandtl number (η/κ) exceeds 0.2 (with a magnetic Reynolds number of 150). However, the magnetic field generated in this dynamo differs markedly in character from the solar magnetic field, since there is no evidence for field-reversals or equatorial migration. The main reason is that the helicity (and therefore the effective strength of the α-effect) is much too large. Subsequently, by lowering the diffusivities, Gilman (1983) did find cyclic behaviour, but the period was only a year or so. These pioneering models could not reproduce the butterfly diagram, but they did establish the generic sensitivity of the dynamo equations to the choice of parameters, such that they allow steady or oscillatory behaviour, poleward or equatorward waves, dipole or quadrupole symmetry.

Next, Gilman and Glatzmaier (1981) and Glatzmaier and Gilman (1981, 1982) set up an anelastic model (Sec. 2.4.3) for compressible convection in rotating shells. Equatorial acceleration is found when rotation has a larger effect than viscous diffusion and buoyancy, or when viscous diffusion is smaller than thermal diffusion. In both cases, in the upper part of the convection zone the angular velocity increases with depth, whereas in the lower convection zone it decreases and is nearly constant on cylinders, in contrast to helioseismology observations.

Juri Toomre and his group later built on the Gilman-Glatzmaier code to develop more sophisticated fully compressible models for global convection with the ASH (anelastic spherical harmonic) code, which takes advantage of advances in high-performance computing and aims to understand the mechanisms that are driving the observed differential rotation and meridional flow. Brummell et al. (1996, 1998) and Miesch et al. (2000) discovered that for moderate forcing the resulting laminar convection is dominated by *banana cells* with distinctive tilts of the cells away from the vertical, such that the Reynolds stresses can drive differential rotation. Under more realistic, vigorous forcing the convection is, however, highly turbulent and more complex, with mean flows and coherent vortical downflows that align with the rotation axis, driven by inverse cascades from smaller-scale turbulence. The upper convection zone has an intricate rapidly evolving downflow network, while the lower convection zone and overshoot region have an intermittent plume-dominated structure. The resulting angular velocity is roughly constant with depth, similar to helioseismology observations at mid-latitudes. Miesch et al. (2008) extended the calculation to a spatial resolution of $(257 \times 1,024 \times 2,048)$ grid points in the (r,θ,ϕ) directions and reaching from the base of the convection zone to within 15 Mm of the photosphere; they describe the properties of giant cells and find again that coherent downflows drive a strong differential rotation.

Brun et al. (2004) found that the resulting dynamo-generated magnetic fields have radial fields concentrated in downflow lanes and toroidal fields organised into twisted ribbons that are extended in longitude and reach field strengths of 5,000 G. However, in the absence of a tachocline there is no solar cycle. Browning et al. (2007) then included a simple tachocline with an underlying stable layer and found that a 3,000 G axisymmetric toroidal field is created in the stable layer (in contrast to the fluctuating fields that occupy the convection zone); the toroidal field is antisymmetric, as required, but it is non-reversing. Later, with a global large-eddy anelastic simulation, Ghizaru et al. (2010) were able to produce polarity reversals with decadal periods, while Brun et al. (2011) coupled the radiative and convection zones and Nelson et al. (2013) described the properties of buoyant magnetic loops.

8.6.2 Low-Order Models

A self-consistent treatment of all the main processes thought to be behind solar and stellar dynamos is a formidable task that is beyond current computational capability. A complementary approach, therefore, is to construct low-order ordinary differential equation (ODE) models in order to seek to understand the bifurcation structure that may be present in the full partial differential equations.

The possibility of a simple set of ODEs exhibiting complex behaviour (such as a transition from periodic solutions to chaos) was highlighted by Lorenz (1963), and since then there has been a huge growth in determining the nonlinear dynamics of such equations. One of the aims of low-order theory is to understand the low level of activity during the Maunder Minimum that may have been caused by the dynamo temporarily working less efficiently. Another is to model the different kinds of activity observed in other stars (Sec. 8.6.3).

There are two ways of constructing a low-order dynamo model for the time-dependence of the magnetic field and plasma velocity. The first is to truncate the partial differential dynamo equations (e.g., Weiss et al. 1984). This has the advantage that each term in the ODEs has an obvious physical interpretation. For example, Jennings and Weiss (1991) have included α-quenching in a simple model and found a complex bifurcation structure, with symmetry-breaking to multiple solution branches. The drawback with truncated models is that the dynamics are often fragile and sensitive to the level of truncation. A second approach is to construct low-order models based on '*normal-form equations*' using the theory of nonlinear dynamics, by employing either symmetry arguments or bifurcation analysis. Here the dynamics are generic and so robust, but the physical interpretation of the different terms is less transparent, since they may not have physical analogues.

Examples have been constructed (e.g., Tobias et al. 1995; Wilmot-Smith et al. 2005) in which, as a control parameter is increased, a steady state bifurcates to give first oscillatory solutions (in which trajectories are attracted to a limit cycle in phase space) and later doubly periodic modulated solutions (attracted to a two-torus). Eventually, they in turn break down to chaotic solutions. In this process, symmetry breaking can also occur (Knobloch et al. 1998; Beer et al. 1998). Indeed, low-order models that examine the interaction between dipolar and quadrupolar solutions give much insight into the possible working of dynamos (Knobloch et al. 1998; Tobias 2002). Also, time-delay may be built into low-order models when different effects occur at different spatial locations (e.g., Wilmot-Smith et al. 2006).

8.6.3 Stellar Dynamos

There has been an enormous growth in detailed observations of other late-type stars, which reveal stellar activity taking place when the stellar mass, rotation, convection and magnetic field are quite different from the Sun. There is therefore a two-way flow of science, which is certain to blossom in future, as solar understanding is applied to other stars and stellar observations of the fundamental processes under different parameter regimes offer clues for deeper understanding. For a magnificent in-depth account of the theory of stellar magnetic fields, see Mestel (2012).

A systematic record of activity in late-type stars began in 1966 with the Mt Wilson Ca II H and K project (Wilson 1978; Baliunas et al. 1995; Lockwood et al. 2007). Some stars have been found to have no significant variability, others have long-term changes in emission (on times greater than 20 years), while still others show irregular emission or cyclic variations on periods between 2.5 years and 20 years.

Dynamos depend on various dimensionless parameters, such as a dynamo number (proportional to differential rotation) and a Rossby number (Ro), namely, the ratio of stellar rotation period to convective turnover time. Rapid rotators are more active and possess prominent polar spots. Stars arrive on the main sequence in a highly active state with a variety of angular velocities, but magnetic braking by a stellar wind soon makes them spin down to an angular velocity that depends mainly on their age (Mestel and Spruit 1987). Thereafter, their rotation decreases gently (and their Rossby number increases) until they become middle-aged, slow rotators with cyclic activity similar to the Sun's. It is found that stars with $Ro < 1$ show irregular and strong emission, while those with $Ro > 1$ are cyclic or constant. A possible model for this behaviour is for magnetic activity to be governed by a nonlinear dynamical system whose output changes from constant to periodic to periodically modulated to chaotically modulated as the rotation rate increases (Sec. 8.6.2).

A wide range of different types and strengths of activity has now been observed on other stars, including activity cycles in chromospheric and coronal emission (Güdel 2004), stellar spots (Berdyugina 2005), differential rotation (Cameron 2007), magnetic fields (Donati 1999) and stellar flares (Aschwanden et al. 2008). There is therefore a major incentive to determine which parameters are correlated with the differences in activity and which physical processes are causing them. For example it is likely that rapid rotators have very different internal rotation profiles.

Stellar magnetic fields were first detected by Robinson et al. (1980) by Zeeman broadening and are now found in all late-type stars that are rapidly rotating and bright enough. More exciting are the maps of magnetic fields and their directions on stellar surfaces that have since been obtained by the Zeeman-Doppler imaging technique, based on the rotational modulation of Stokes V parameters (Donati and Semel 1990). For example, Donati (1999) and Donati et al. (2003) found a strong toroidal magnetic field at the surfaces of young K0 dwarf rapid rotators that is probably generated in the convection zone rather than the tachocline. Also, Donati et al. (2006) and Morin et al. (2008) found surprisingly that a rapidly rotating fully convective dwarf star possesses a large poloidal magnetic field and star spots but a weak surface differential rotation (about 1/10th of the solar value).

Once the solar dynamo is well understood, the challenge to develop dynamo theories for other stars is a substantial one, since, when judged against the extremes of stellar activity, solar activity seems rather benign. For example, star spots can be over 100 times larger than on the Sun and cover 30 per cent of the surface, global magnetic fluxes can be 100 times larger, stellar flares can be 1,000 times more energetic and rotation rates can be 50 or more times faster (for which the convection will be quite different from the Sun and dominated by *Taylor convection* aligned with the rotation axis). Furthermore, fully convective stars can be highly active and so need quite a different process from the Sun's main dynamo, since they possess no tachocline.

8.7 Future Directions

Dynamo theory has made huge progress towards providing an explanation for the existence of the Sun's magnetic field and its variation with the solar cycle. Nevertheless, many key problems remain to be overcome.

(1) Developing a better understanding of the *tachocline* is a prime aim. What is the detailed nature of the instabilities at work there and how is the tachocline coupled to the solar interior and convection zone? How is the observed surface meridional flow closed in the convection zone and how does it vary with the solar cycle?

(2) Producing much better models for the *tachocline dynamo* and the *flux-transport dynamo* is also important. Which of these two approaches describes best what is actually responsible in the Sun for sunspots and active regions? Is poloidal flux carried down to the tachocline by flux pumping or by meridional flow? Does the α-effect take place most effectively in the tachocline or near the solar surface? At present, both approaches suffer from a lack of detailed understanding of the tachocline: the tachocline model needs a better treatment for the regeneration of poloidal flux from toroidal flux; and the flux-transport dynamo is qualitative in its transport of poloidal flux down to the tachocline and its destruction of toroidal flux. There is a healthy tension between the focus of tachocline theorists on the fundamentals and the emphasis of flux-transport theorists on explaining observed solar-cycle features. Both approaches are needed.

(3) The details of the rise of magnetic tubes through the convection zone and *flux emergence* through the solar surface need to be modelled more realistically (Hughes 2007). So far, we have simple models for the rise of thin flux tubes across the convection zone and initial attempts to model numerically the emergence through the surface, starting with artificial initial conditions (e.g., Archontis and Hood 2008, 2012), but these approaches need to be joined together and conducted in a turbulent convection zone.

(4) *Mean-field MHD* gives illustrative models, since the α-effect is just a qualitative (but plausible) parametrisation of an MHD turbulence process. It has represented a highly stimulating line of attack, but the fact remains that the α-ω dynamo equations cannot be justified as valid for the Sun or other stars, since the conditions under which they hold do not apply: there is no clear separation of scales in practice, with convection simulations showing a continuous spectrum of scales (Nordlund et al. 2009); and the following parameters are not small – the turbulent magnetic Reynolds number, the ratio of the correlation time to the turnover time, and the ratio of fluctuations in magnetic field to the mean values. In future, hopefully a deeper theory will evolve or a deeper understanding from next-generation computations – and then perhaps some effect resembling the α-effect will indeed be justified, including its variation within the convection zone and dependence on dimensionless parameters, together with a theory for its quenching when the field is too large.

(5) Attempts to compute an *α-effect* from MHD turbulence computations show that α is negligible even when small-scale dynamo action is present (Cattaneo and Hughes 2006, 2009) and α is not necessarily related to helicity (Courvoisier et al. 2006). Other numerical experiments on turbulence suggest a catastrophic quenching of α in the nonlinear regime at very low magnetic field values to give $\alpha = \alpha_0/(1+R_m^{-1}B^2/B_0^2)$, which for $R_m \gg 1$ on the Sun would limit magnetic fields to only 1 G (Cattaneo and Hughes 1996; Ossendriver et al. 2001). There are also difficulties with turbulent diffusion: the value of $\tilde{\eta}$ estimated from observed active-region dispersal is typically 10^9 m^2 s^{-1}, but how can we predict its value and functional form rigorously from dynamo theory and its variation in the convection zone?

(6) Massive *global computations* will be invaluable in studying the compressible hydrodynamics of the convection zone, so that the nature of convection over a wider range of scales and the way that differential rotation and meridional flow are driven can be fully understood. Then the way will be open for a much better computational modelling of MHD turbulence and global dynamo models. Is the magnetic field in a filamentary state and how does it interact with convection? What is the role of fast reconnection at convection cell boundaries? What is the effect of flux pumping and how is it parametrised? Can a mean-field dynamo be constructed that mimics the results, with an α-effect that captures the essential physics?
(7) Developing the above elements of a better theory for the solar cycle should enable a more confident prediction of the strengths and dates of future solar maxima and minima. Also, what is the relation between sunspots and the strength of a magnetic cycle? What is the best way of defining solar minimum. When will the next Maunder minimum take place? What is the precise effect of the Sun on the Earth's climate?
(8) In addition to a tachocline dynamo, the details of other dynamos at work in the Sun should be worked out. Is there a dynamo that gives birth to ephemeral active regions and where is it operating (throughout the convection zone or near the surface)? Is there a separate dynamo creating a smaller-scale inner network field and driven by subsurface shear?
(9) The nature of dynamos on other stars needs to be determined. To what extent are they scaled-up versions of what is happening on the Sun and to what extent are they novel dynamos operating with qualitatively different physics?

In conclusion, bearing in mind these questions, it is clear that dynamo theory remains one of the most highly developed and interesting branches of solar MHD. Certainly, it will remain a lively field of interest in future, with great advances expected from helioseismology, computational modelling and comparison with stellar activity. The current grand maximum in solar activity is likely to terminate within the next few cycles and possibly lead to a grand minimum with a substantial cooling effect on the Earth's climate – but too small to counteract the global warming that we humans are causing.

Chapter 9

Magnetoconvection and Sunspots

The existence of sunspots has been known since ancient times (Sec. 1.1), but it was only in 1908 that they were found to be the sites of very strong magnetic fields, where huge magnetic flux tubes emerge through the solar surface. In the past ten years, there has been a sea change in understanding due to high-resolution observations of their fine-scale structure from the ground and space (Figures 1.28 and 1.29), as well as an initial probing of subsurface structure by local helioseismology and an increase in computational power that has made much more realistic simulations viable.

Aspects related to sunspots that are covered in this chapter include magnetoconvection and sunspot cooling (Sec. 9.1), intense magnetic flux tubes (Sec. 9.2), magnetic buoyancy (Sec. 9.3), the global equilibrium of sunspots (Sec. 9.4), fine-scale structure of umbra and penumbra (Sec. 9.5), sunspot evolution (Sec. 9.6), and, to conclude, a numerical model that unifies many aspects of sunspots (Sec. 9.7).

In preparation, Sections 1.4.2, 1.7, and 2.9 describe the observed properties and behaviour of photospheric magnetic fields, active regions, sunspots and flux tubes. Excellent accounts can be found in the reviews by Solanki (2003), Thomas and Weiss (2008).

9.1 Magnetoconvection

Before tackling sunspots, it is worth summarising magnetoconvection (i.e., thermal convection in a magnetic field). Modelling the convection zone is a formidable task, since the convection is nonlinear, compressible, three-dimensional, unsteady, rotating and threaded by intense flux tubes, while many parameter values are far too extreme to be adopted in computations. But great strides have been taken in numerical simulation and analytical theory. Convection is what concentrates tiny intense flux tubes (Sec. 9.2) at both granulation and supergranulation boundaries (Sec. 1.4.1), and high-resolution observations of sunspots reveal them as amazing natural laboratories for studying the complexities of magnetoconvection. The linear instability theory of photospheric magnetic fields was first studied by Cowling (1953), Cowling (1958) and Chandrasekhar (1961). Since then, two complementary approaches have been adopted: a systematic study of idealised situations pioneered by Nigel Weiss; and more realistic simulations championed by Åke Nordlund. For more details of magnetoconvection theory, including nonlinearity and compressibility, see Schüssler (2001), Proctor (2005), Thomas and Weiss (2008) and Weiss (2011).

9.1.1 Physical Effects

Without a magnetic field, an ideal fluid heated from below is unstable to convective motions when the temperature gradient exceeds its adiabatic value (Sec. 1.3.3). Effects of viscosity and thermal diffusion on convective onset are as follows.

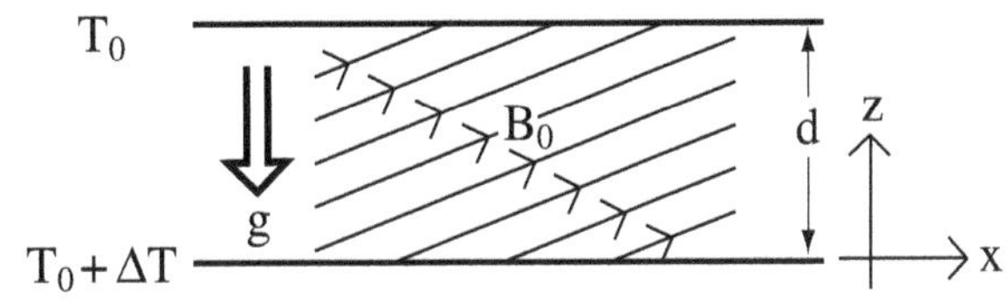

Figure 9.1. The nomenclature for convection in a horizontal layer of plasma, heated from below and containing a uniform magnetic field ($\mathbf{B}_0$.)

As a plasma element rises, buoyancy balances the viscous force ($g\Delta\rho = \pi^2\rho\nu v/d^2$), where v is the vertical speed, d the distance moved and $\Delta\rho$ the density change, which may be written ($\rho\alpha_T\Delta T$) in terms of the *coefficient of thermal expansion* [$\alpha_T = -(1/\rho)(\partial\rho/\partial T)_p$]. Now, suppose (unlike Sec. 1.3.3) that heat leaks into the parcel due to a diffusivity κ. Then the instability that was present in the absence of diffusion will still occur if the thermal diffusion is slower than the time for particle motion, namely, $d^2/(\pi^2\kappa) > d/v$. Eliminating v gives the criterion for convective onset in terms of the *Rayleigh number* (Ra) as

$$Ra \equiv \frac{g\alpha_T\Delta T d^3}{\kappa\nu} > \frac{27\pi^4}{4}.$$

The influence of a magnetic field on convective motions is largely stabilising due to the restoring effect of magnetic tension. Consider a horizontal layer, of thickness d, heated from below (Figure 9.1), with gravity acting in the negative z-direction and a temperature difference of ΔT. The equilibrium is characterised by uniform density (ρ_0), uniform magnetic field ($\mathbf{B}_0$) in the xz-plane, zero velocity, and a temperature stratification given by $T_0(z) = T_0 + \Delta T(1 - z/d)$.

Next, suppose the plasma moves by a small amount (ξ) vertically, such that the magnetic field lines form sine curves of wave-number k. The radius of curvature is $(k^2\xi)^{-1}$, and so the magnetic tension (B_0^2/μ) produces a restoring force $k^2\xi B_0^2/\mu$. At the same time, since the temperature gradient is $\Delta T/d$, the temperature and density decrease by $\xi\Delta T/d$ and $\rho_0\alpha_T(\xi\Delta T/d)$, respectively. The resulting buoyancy force $g(\rho_0\alpha_T\xi\Delta T/d)$ is able to overcome the tension if

$$\frac{\rho_0 g\alpha_T\Delta T}{d} > \frac{k^2B_0^2}{\mu}. \tag{9.1}$$

Thus, *overturning convection* occurs when a horizontal magnetic field is sufficiently weak. Moreover, convection in the form of rolls parallel to the field is unaffected by the field, since the field lines can remain straight as they move.

When the temperature difference is too small for rapid overturning (Eq. 9.1), dissipation can still sometimes allow much gentler convection due to either a leak instability or overstability, since magnetic and thermal diffusivity (η, κ) allow the magnetic field and heat to diffuse through the plasma. Thus, there occurs *leak instability* when

$$\frac{\rho_0 g\alpha_T\Delta T\eta}{d} > \frac{k^2B_0^2\kappa}{\mu} \quad \text{and} \quad \kappa < \eta, \tag{9.2}$$

but *overstability*, with the oscillation amplitude increasing in time (Figure 9.2), when

$$\frac{\rho_0 g\alpha_T\Delta T\kappa}{d} > \frac{k^2B_0^2\eta}{\mu} \quad \text{and} \quad \kappa > \eta. \tag{9.3}$$

In the latter case, for large enough B_0, tension dominates buoyancy and produces oscillations, with conduction and diffusion reducing buoyancy and tension forces by factors proportional to κ and η, respectively. Equation (9.3) represents the condition that the resultant force be larger during the return than during the outgoing motion.

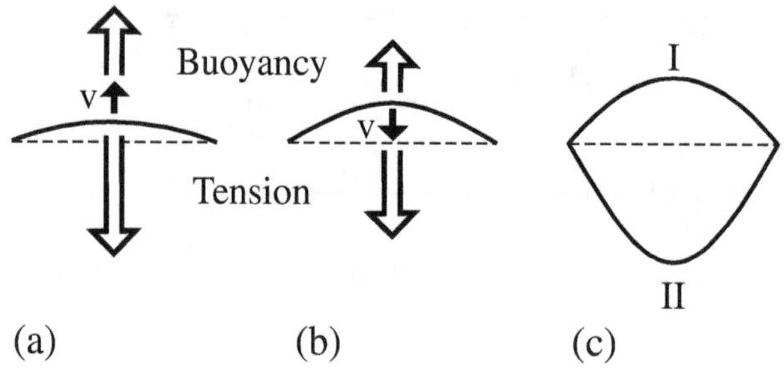

Figure 9.2. An overstable oscillation. (a) As the plasma rises, the restoring tension force exceeds the destabilising buoyancy force. (b) As it moves down, both tension and buoyancy have been decreased by diffusion, but the resultant force has increased. (c) The amplitude of the second half-oscillation (II) exceeds that of the first (I), and the process continues.

9.1.2 Linear Stability Analysis

The incompressible MHD equations, including uniform viscous, magnetic and (radiative) thermal diffusion and gravity, are (Sec. 2.4.3) $\nabla\cdot\mathbf{v}=\nabla\cdot\mathbf{B}=0$ with

$$\rho\frac{d\mathbf{v}}{dt}=-\nabla p+\mathbf{j}\times\mathbf{B}+\rho\nu\nabla^2\mathbf{v}-\rho g\hat{\mathbf{z}},$$

$$\frac{\partial\mathbf{B}}{\partial t}=\nabla\times(\mathbf{v}\times\mathbf{B})+\eta\nabla^2\mathbf{B},\qquad\qquad \frac{dT}{dt}=\kappa\nabla^2 T.$$

Departures from an equilibrium plasma, with a linear temperature profile $[T_0(z)]$, temperature difference ΔT and uniform magnetic field $(\mathbf{B}_0)$, are written $\mathbf{B}=\mathbf{B}_0+\mathbf{B}_1, \mathbf{v}=\mathbf{v}_1, T=T_0(z)+T_1$. The equations are linearised and for simplicity the *Boussinesq approximation* is adopted, which holds when convection is highly subsonic and the depth is much smaller than a scale-height. This filters out sound waves and incorporates density variations only in the buoyancy force, where they are written $\rho_1=-\rho_0\alpha_T T_1$. After eliminating all variables except v_{1z}, the equations reduce to

$$\left(\frac{\partial}{\partial t}-\kappa\nabla^2\right)\left(\frac{\partial}{\partial t}-\eta\nabla^2\right)\left(\frac{\partial}{\partial t}-\nu\nabla^2\right)\nabla^2 v_{1z}$$

$$=\frac{(\mathbf{B}_0\cdot\nabla)^2}{\mu\rho_0}\left(\frac{\partial}{\partial t}-\kappa\nabla^2\right)\nabla^2 v_{1z}+\frac{g\alpha_T\Delta T}{d}\left(\frac{\partial}{\partial t}-\eta\nabla^2\right)\left(\frac{\partial^2 v_{1z}}{\partial x^2}+\frac{\partial^2 v_{1z}}{\partial y^2}\right)$$

(see PROBLEM 9.1: for all problems and solutions, see web page www.cambridge.org/9780521854719). A solution of the form

$$v_{1z}\sim e^{\omega t}e^{i(k_x+k_y y)}\sin k_z z$$

vanishes at the boundaries $(z=0,d)$ if $k_z=\pi/d$ and reduces the above equation to

$$(\omega+\kappa k^2)(\omega+\eta k^2)(\omega+\nu k^2)k^2$$

$$=-\frac{(\mathbf{B}_0\cdot\mathbf{k})^2}{\mu\rho_0}(\omega+\kappa k^2)k^2+\frac{g\alpha_T\Delta T}{d}(\omega+\eta k^2)(k_x^2+k_y^2). \tag{9.4}$$

Stability for this system depends on four dimensionless parameters, namely, the Rayleigh number (Ra), the Chandrasekhar number $[Q=B_0^2d^2/(\mu\rho\eta\nu)]$, the Prandtl number $(Pr=\nu/\kappa)$ and the diffusivity ratio $(\zeta=\eta/\kappa)$.

For the case of no magnetic field with rolls ($k_y = 0$), convection sets in as overturning motion when $Ra > 27\pi^4/4$ at $k_x = \pi d/\sqrt{2}$ (see PROBLEM 9.2).

For a horizontal field ($\mathbf{B}_0 = B_0\hat{\mathbf{x}}$) with no dissipation ($\kappa = \eta = \nu = 0$), rolls parallel to the field (with $k_x = 0$, i.e., having ripples perpendicular to $\mathbf{B}_0$) are unstable ($\omega > 0$) over all wavenumbers $k_y \neq 0$; but, for given $k_x \neq 0$, there are ripples along $\mathbf{B}_0$, and convection is inhibited for all k_y if the field is strong enough that $B_0^2/(\mu\rho_0) > g\alpha_T \Delta T/(dk_x^2)$, which is equivalent to Eq. (9.1) (PROBLEM 9.3).

Next, consider a vertical magnetic field ($\mathbf{B}_0 = B_0\hat{\mathbf{z}}$), and include the dissipative terms, so that in Eq. (9.4) $\mathbf{B}_0 \cdot \mathbf{k}$ becomes $B_0 k_z$. This allows instability (ω real and positive) or overstability ($\omega = \omega_r + i\omega_i$ complex with $\omega_r > 0$). The marginal condition for instability onset is $\omega = 0$, for which Eq. (9.4) yields

$$Ra(d^2k^2 - \pi^2) = \pi^2 Q d^2 k^2 + d^6 k^6.$$

For a sketch of k against Ra, see Figure 9.3a, in which the smallest value of Ra (Ra^*, say) for which there exists a real k^2 is determined by the condition that the three roots of the above cubic in k^2 be real, namely $(Ra^* - \pi^2 Q)^3 = (27/4)\pi^4 Ra^{*2}$.

Thus, when the vertical magnetic field vanishes, Q vanishes and the classical result $Ra^* = 27\pi^4/4$ is recovered (Sec. 7.5.6). On the Sun $Q \gg 1$, so that instead $Ra^* \sim \pi^2 Q$, which is similar in form to Eq. (9.2). As Q increases, so Ra^* and k^* increase, and therefore the effect of raising the magnetic field strength (and so Q) is to make the plasma more stable and any unstable cells smaller in size.

In a vertical field, the marginal condition for the onset of growing oscillations (i.e., overstability) is $\omega_r = 0$, for which Eq. (9.4) gives

$$Ra(d^2k^2 - \pi^2) = \pi^2 Q d^2 k^2 \frac{\eta(\nu + \eta)}{\kappa(\nu + \kappa)} + d^6 k^6 \frac{(\eta + \kappa)(\eta + \nu)}{\kappa\nu}.$$

Again, the k-Ra curve has a form similar to Figure 9.3a, and this time overstability first begins (for $Q \gg 1$) at a Rayleigh number of $Ra^{**} \equiv \pi^2 Q\eta^2/\kappa^2$, which is essentially the same as Eq. (9.3). The overstability is exhibited as a standing Alfvén wave, with the amplitude growing and wave energy being gained from gravity.

In the photosphere $\eta < \kappa$, and so the onset of convective overstability occurs before the onset of convective instability ($Ra^{**} < Ra^*$), as shown on the Ra axis of Figure 9.3b. This figure is a bifurcation diagram, which indicates the nonlinear consequences of linear overstability or instability and how they depend on the value of Ra and the amplitude of the mean square velocity $\langle v^2 \rangle$, as revealed by numerical experiments of two-dimensional Boussinesq convection. Thus, there is a branch of oscillatory convection that originates

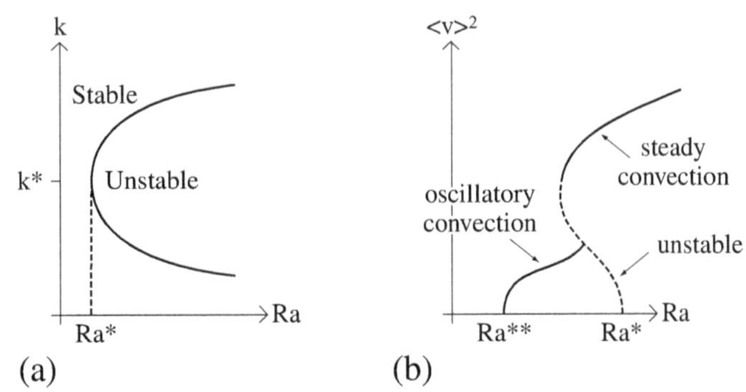

Figure 9.3. (a) Linear magnetoconvection: the form of the marginal stability curve for onset of steady convection, where k is the wavenumber and Ra the Rayleigh number. (b) Nonlinear magnetoconvection: bifurcation diagram for the mean square velocity $\langle v^2 \rangle$ as a function of the Rayleigh number Ra, showing one branch of overstability that leads to oscillatory convection and another of instability that leads to steady convection.

at a Hopf bifurcation at the point Ra^{**} on the Ra-axis, and a branch of steady convection that originates at a pitchfork bifurcation at Ra^*. Part of the steady branch is unstable (dashed), and so does not show up in the experiments, and part is stable. As Ra is increased above Ra^*, at first only oscillatory convection is possible, and then both oscillatory and steady convection are possible (depending on the history), and finally for large enough Ra only steady convection can occur.

9.1.3 Magnetic Flux Concentration and Expulsion

Consider first the effect of a 2D stagnation-point flow ($v_x = -V_0x/L, v_y = V_0y/L$) on a unidirectional field $B(x)\hat{\mathbf{y}}$. Then the induction equation possesses a steady-state solution with no electric field and a balance between advection and diffusion of the form (PROBLEM 9.4a)

$$B(x) = 2B_0\left(\frac{R_m}{\pi}\right)^{1/2} \exp\left(-\frac{R_m x^2}{L^2}\right),$$

where $R_m = LV_0/(2\eta)$. It represents a sheet of flux $2B_0L$, thickness $2L/\sqrt{R_m}$ and peak field strength $2B_0\sqrt{(R_m/\pi)}$ that is concentrated by the flow. (The corresponding solution with a field reversal for magnetic annihilation is given in Sec. 6.4.3.)

The effect of a 3D flow ($v_R = -V_0R/L, v_z = 2V_0z/L$) in cylindrical polar coordinates on a field $B(R)\ \hat{\mathbf{z}}$ is similarly to make

$$B(R) = B_0R_m \exp(-R_mR^2/L^2)$$

(PROBLEM 9.4b). Thus, a flux πL^2B_0 is concentrated to a tube of peak field B_0R_m and radius $a = L/R_m^{1/2}$. In this kinematic approach, the plasma velocity is prescribed, but in practice the build-up of field will be countered by the $\mathbf{j}\times\mathbf{B}$ force, which will slow the flow and exclude it from the flux rope.

Consider a weak magnetic field threading a convective cell. If the magnetic energy density $[B^2/(2\mu)]$ is much less than the kinetic energy density ($\frac{1}{2}\rho v^2$), and the magnetic Reynolds number (R_m) is much larger than unity, the motion is unhindered by the magnetic field, which is simply carried round and wound up inexorably by the flow. This stretches the field lines and increases the magnetic field strength until either the magnetic energy becomes comparable with the kinetic energy (and the flow is slowed down) or the local magnetic Reynolds number becomes of order unity (and the field lines slip through the plasma). In the latter case, flux is expelled from the centre of the cell and accumulates around its boundaries.

For example, the kinematic concentration of an initially uniform vertical field ($B_0\hat{\mathbf{z}}$) by an incompressible convective flow $\mathbf{v} = v_0 \sin kx\ \hat{\mathbf{x}} - v_0kz \cos kx\ \hat{\mathbf{z}}$ is determined by the z-component of the ideal induction equation, namely,

$$\frac{\partial B_z}{\partial t} = -\frac{\partial}{\partial x}(v_0 \sin kx B_z).$$

This may be solved by the method of characteristics to give

$$B_z = \frac{B_0\ e^{-kv_0t}}{\cos^2 \frac{1}{2}kx + \sin^2 \frac{1}{2}kx\ e^{-2kv_0t}}.$$

At $x = 0$ there is an upwelling and $B_z = B_0e^{-kv_0t}$, so the field is dispersed, but at $x = \pi/k$ there is a downdraft, which concentrates the field like $B_z = B_0e^{kv_0t}$.

In a classic paper, Weiss (1966) simulated the kinematic process of flux expulsion numerically by solving the 2D resistive induction equation for the effect on an initially uniform magnetic field ($B_0\hat{\mathbf{z}}$) of an incompressible eddy with stream function $\psi = (UL/\pi)\cos(\pi x/L)\cos(\pi z/L)$, where U and L are constant and $B_x = 0$ on all sides (Figure 9.4). The clockwise motion winds up the field and concentrates the flux at the edges of the cell. Near the cell centre, the field is first amplified, but then it decays by diffusion. After

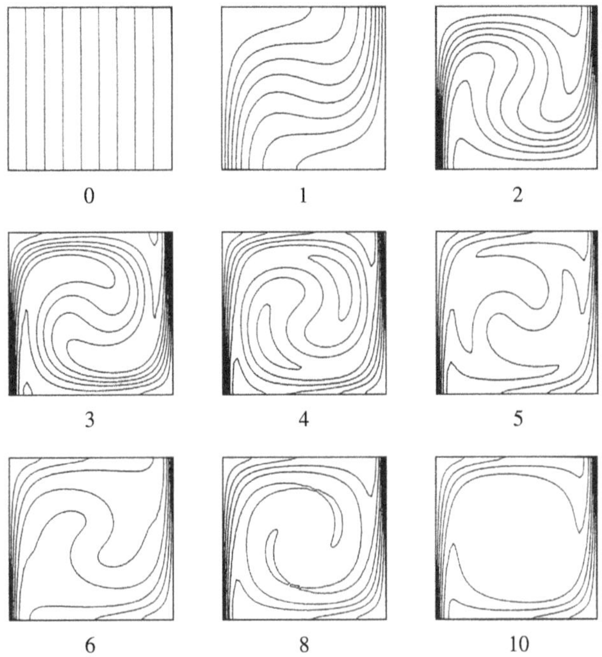

Figure 9.4. Flux expulsion from a single eddy. The magnetic field lines for $R_m \equiv UL/\eta = 250$ after times of $t/\tau = 0-10$, where $\tau = 5L/(8U)$ (after Galloway and Weiss 1981).

a time of $6L/U$, where L/U is the *turnover-time*, nearly all the flux has been expelled from the centre of the eddy and a steady state is established.

An order-of-magnitude estimate for the field (B_m) in the boundary layers (of thickness d, say) at the cell edges may be obtained from a balance between convection and diffusion. In a steady state, the flow speed ($d\,U/L$) into the layer must balance the diffusion speed (η/d), which gives $d \approx R_m^{-1/2} L$. The flux is therefore concentrated to a field strength B_m given by flux conservation ($B_m d \approx B_0 L$) as

$$B_m \approx R_m^{1/2}\, B_0. \tag{9.5}$$

Naively, one might imagine that the field can be concentrated only to the *equipartition field strength* (B_e), given by

$$B_e^2/(2\mu) = \tfrac{1}{2}\rho U^2, \tag{9.6}$$

which would imply fields of 200 G at the photosphere and 10^4 G near the base of the convective zone. However, larger values can be obtained when the Lorentz force is balanced by a pressure gradient instead of the inertial term $\rho(\mathbf{v}\cdot\boldsymbol{\nabla})\mathbf{v}$. Thus, in a static photosphere the maximum field strength is limited by the photospheric pressure (p_e) to be at most the *vacuum field strength* $B_p \approx 1{,}900$ G such that

$$B_p^2/(2\mu) = p_e, \tag{9.7}$$

when the plasma pressure within the flux tube becomes negligible. It may be even larger when dynamic pressure ($\frac{1}{2}\rho v^2$) due to flow is included.

Weiss and collaborators have carried out a comprehensive programme over many years of idealised magnetoconvection computations using the full MHD equations (Weiss 1991; Weiss 2003; Proctor 2005). For 2D cells of Boussinesq convection, nonlinear numerical simulations show that the flux is concentrated equally at both sides. However, in a cylindrical geometry with downflow on the axis, the flux is concentrated preferentially at the centre (Galloway et al. 1978). Also, the field strength is much higher than the 2D value (Eq. 9.5), since flux conservation (in a tube of radius d) now becomes $B_m d^2 \approx B_0 L^2$, so that $B_m \approx R_m\ B_0$. For strong enough fields, the flux is expelled from the cell to form a static flux tube containing no flow. As the Rayleigh number is increased, a cascade of period-doubling bifurcations can occur, leading to chaos (Weiss 1981; Rucklidge 1994).

In 3D, with $\zeta < 1$, convection first sets in as 2D rolls unless the up-down symmetry is broken by, for example, a dependence of thermal conductivity on depth. Then a hexagonal pattern of cells is set up with plasma rising in the centre and falling at the edges: the flux at the top is concentrated at the (converging) vertices of each cell, while at the bottom it is concentrated at the (converging) cell centres (Galloway and Proctor 1983).

In a wide box with $Ra = 5\times10^5$, Cattaneo et al. (2003) studied the effect of increasing the imposed magnetic field (B_0). When $B_0 = 0$, a small seed field grows by dynamo action to a level of 20 per cent of the kinetic energy. The flux has an irregular small-scale structure and forms a mesoscale network, with strong fields of either sign at network junctions enclosing groups of smaller cells. When B_0 is moderately strong (measured by the Chandrasekhar number Q), the system moves from kinematic to dynamic flux expulsion with internally suppressed convection. Cellular overturning convection sets in, with the magnetic flux concentrated at the upper boundary into a narrow tessellated network surrounding broad rising plumes. At higher B_0, this gives way to unsteady oscillatory convection in an irregular pattern with a strong interaction between magnetic field and flow, including periodic flow reversals.

The effect of compressibility in 2D is to break the symmetry between up- and down-flows by producing concentrated rapid downflows and gentle extended upflows, so that most of the flux is concentrated in the downflows with their upper parts partially evacuated (Hurlburt and Toomre 1988). Also, nonlinear oscillatory convection with standing waves can give way to travelling waves (Hurlburt et al. 1989). 3D convection has a greater variety of behaviour that depends on the aspect ratio of the box (Weiss et al. 1996; Rucklidge et al. 2000). Weiss et al. (2002) found (for an $8 \times 8 \times 1$ box and $Ra = 10^5$) regular hexagonal convection when $Q = 3000$ resembling supergranules, but at $Q = 1600$ the plumes vary in t, and at $Q = 1000$ the plumes are more vigorous and field-free, squeezing the field into the gaps between them. At still smaller Q, the magnetic field is in the form of continually evolving thin intense flux tubes nestling between convection cells and confined by a combination of plasma pressure and convective motion (Bushby et al. 2008).

Complementary, more realistic simulations of magnetoconvection by Nordlund and colleagues include more elaborate physics, such as radiative transfer, partial ionisation and spectral line diagnostics. Although realistic values of some parameters (such as R_m) cannot be adopted, it is hoped that parametrisation of the effects at smaller scales than can be included does adequately describe their effect on larger scales. This approach has been used to model convection over a wide range of scales (Stein and Nordlund 2006; Vögler and Schüssler 2007; Nordlund et al. 2009; Stein et al. 2011), to demonstrate convective collapse (Nordlund and Stein 1990) (Sec. 9.2.3), to understand the transition from granulation to 'abnormal' granulation to pores as the field strength is increased, to model the interaction of intense flux tubes with granulation (Steiner et al. 1998; Vögler et al. 2005) and to model umbral dots (Schüssler and Vögler 2006) and even a complete sunspot (Sec. 9.7).

9.1.4 Cooling of Sunspots

Typically, the umbra and penumbra radiate only about 20 to 30 per cent and 75 to 85 per cent, respectively, of the normal photospheric flux, making the umbra 1,000 to 1,900 K cooler and the penumbra 250 to 400 K cooler. The cooling of a sunspot is due to the inhibition of convection by its magnetic field. In the absence of convection, the temperature below the photosphere would increase with depth faster than adiabatic. But convection mixes up the different layers, trying to equalise their temperatures and thus reducing the temperature gradient below the static value. Biermann (1941) suggested that a sunspot's magnetic field inhibits this convection in the top few megameters of the convection zone and so removes the mixing-up process. The temperature, therefore, decreases outwards faster inside the spot than in the surrounding convection region, which leads to a lower temperature in the sunspot. In other words, the magnetic field reduces the heat flow, damming it back and cooling the spot.

Cowling (1953) modified this idea by suggesting that there is *some* convection inside the spot but that it is less vigorous than outside and so transports less heat. He set up a theory for magnetoconvective instability, showing that in an ideal medium buoyancy can be dominated by magnetic tension if the field strength is large enough. This was later developed to include thermal, magnetic and viscous diffusion by Chandrasekhar (1961), who showed how linear instability tends to set in first as growing (overstable) oscillations (Eq. 9.3) rather than purely growing modes. The suggestion of partial inhibition of convection was borne out by observations of weak convective motions inside umbrae including umbral dots (Secs. 1.7.2.3 and 9.5.1).

Another potential way of cooling a sunspot is by a flux of MHD waves, but this is now known to be ineffective, since there is a strong downward reflection of Alfvén waves in the umbra, and so the observed upward flux of waves is not large enough to produce significant cooling. Furthermore, an early objection to inhibition of convection, namely, that the missing heat flux in a sunspot would produce a bright ring around it, is now discounted, since the missing flux is spread widely through the convection zone by turbulent thermal diffusion on a timescale of about a month.

9.2 Intense Flux Tubes

Observations indicate that some of the photospheric flux is concentrated into intense flux tubes, with a field strength of 1 to 2 kG and diameters of only 100 to 300 km (Sec. 1.4.2), representing a flux of typically 3×10^9 Wb (3×10^{17} Mx) (Figures 1.15 and 1.16). They are located between granules and supergranules and are associated with G-band bright points, filigree and small-scale faculae (Figures 1.13 and 1.14).

Their concentration by magnetoconvection to high field strengths in excess of equipartition with the flow is due to a combination of effects, namely, convective compression, flux expulsion, and complete or partial evacuation of plasma from the tube (Sec. 9.1.3). These processes overlap, but an early picture (Figure 9.5) was to suppose a diffuse vertical field is swept into supergranule boundaries and concentrated by convective compression to the equipartition field strength (Eq. 9.6) of several hundred gauss. This is subject to *convective collapse* (Sec. 9.2.3), in which the tube plasma cools radiatively and falls, while the field increases until it is strong enough (1 to 2 kG) to gain convective stability. In practice, instead,

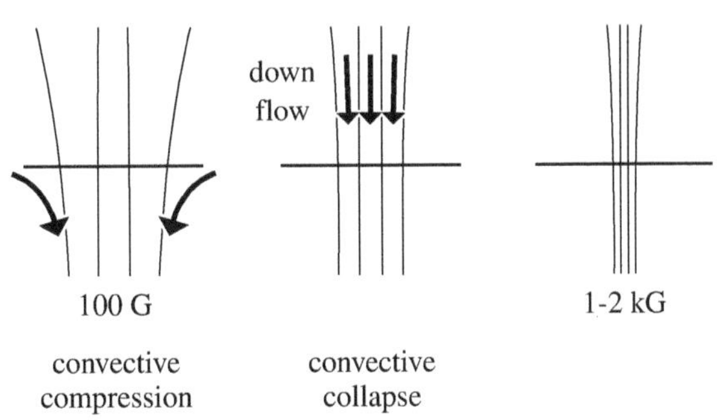

Figure 9.5. The formation of kilogauss flux tubes by convective compression, followed by convective collapse driven by radiative cooling and accompanied by a downflow.

much of the flux emerges within supergranules as internetwork flux and ephemeral active regions and is concentrated at the edges of granules and supergranules.

Once an intense tube is formed, it is then shuffled around as the convection pattern changes, with a tendency for it to remain in the lanes between granules and supergranules. Ultimately, the thin flux tubes will merge or cancel with other tubes or will perhaps fragment or diffuse. The convective collapse process is particular to the solar surface, but it is likely that the whole convection zone is full of convecting cells interspersed with a tangled web of intense flux tubes.

A natural question is "how much smaller than the current observational estimate of 100 km are thin flux tubes?" If one simply equates observed flows of, say, $v_0 = 1$ km s^{-1} to a magnetic diffusion speed based on a photospheric value of, say, $\eta = 10^4$ m^{2}s^{-1} for the Coulomb diffusivity (Sec. 2.1.5), then the tube width would be only $a = \eta/v_0 = 10$ m. But the theory for flux concentration in Section 9.1.3 implies instead a tube diameter of $2a = 2L/\sqrt{R_m} = 2\sqrt{(2L)}\sqrt{(\eta/v_0)}$ with magnetic Reynolds number $R_m = Lv_0/(2\eta)$, which is $2\sqrt{2}$ times the geometric mean of the scale (L) for the stagnation-point flow and the above simple diffusion value. Thus, it depends on the assumed flow scale, but for, say, $L = 1$ Mm typical of granules it gives a thin flux tube width in the photosphere of about 10 km. If the flow had a higher speed or smaller length-scale, the diameters of the intense flux tubes would be even smaller.

9.2.1 Equilibrium of a Thin Flux Tube

A *thin* or *slender* flux tube (Parker 1979a) is one that is so narrow that its magnetic field is approximately uniform across its width; thus, its flux may be written as $F = \pi a^2 B(s)$, in terms of its radius (a) and field strength (B) at a distance s along the tube; also, the field strength at the tube surface does not differ greatly from its value at the tube centre. Consider such a tube in equilibrium with a hydrostatic ambient medium (Figure 9.6). For a vertical tube with $s \equiv z$, the external and internal pressures (p_e,p_i) are given by

$$p_e(z) = p_e(0)e^{-\bar{m}_e(z)}, \qquad p_i(z) = p_i(0)e^{-\bar{m}_i(z)}, \tag{9.8}$$

where $\bar{m}_e(z) = \int_0^z H_e(z)^{-1}dz$ and $\bar{m}_i(z) = \int_0^z H_i(z)^{-1}dz$ are the corresponding numbers of scale-heights (H_e, H_i) (Sec. 3.1.2) above some reference level ($z = 0$).

Across the surface of the flux tube, the total pressure is conserved, and so $p_i + B_i^2/(2\mu) = p_e$. Substitution for p_i and p_e therefore gives the magnetic pressure as

$$\frac{B(z)^2}{2\mu} = p_e(0)e^{-\bar{m}_e(z)} - p_i(0)e^{-\bar{m}_i(z)}; \tag{9.9}$$

if the pressures inside and outside the tube are specified at $z = 0$ and the temperature is specified everywhere, the field strength along the whole tube is determined by Eq. (9.9). The tube radius $[a(z)]$ follows

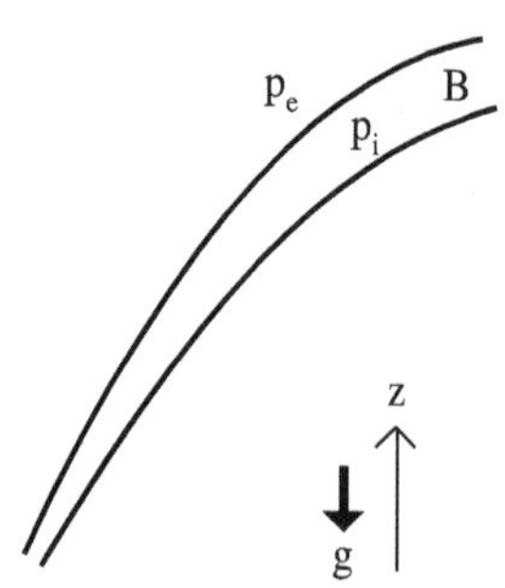

Figure 9.6. An isolated (slender) flux tube confined by an external pressure (p_e).

by flux conservation as

$$a(z)^2 = \frac{a(0)^2 B(0)}{B(z)}. \tag{9.10}$$

For the particular case when the internal and external temperatures are the same, so that $\bar{m}_e(z) = \bar{m}_i(z) \equiv \bar{m}(z)$, Eqs. (9.9) and (9.10) reduce to $B(z) = B(0)e^{-\bar{m}(z)/2}$ and $a(z) = a(0)e^{\bar{m}(z)/4}$. Thus, the tube widens and the field weakens with height.

When the tube interior is cooler than the exterior, $p_i(z)$ declines more rapidly than $p_e(z)$. Over large heights, this has the effect of draining most of the material from the upper parts of the tube. For a tube whose plasma pressure is negligible and whose external medium is isothermal, the magnetic field is given by

$$\frac{B(z)^2}{2\mu} = p_e(z) = p_e(0)e^{-z/H_e},$$

The radius follows from $\pi a(z)^2 B(z) = F_m$, in terms of the magnetic flux (F_m).

9.2.2 Dynamics of a Thin Flux Tube

The waves that may be present in a slender flux tube are described in detail in Section 4.9.5 and Figure 4.12 (in the limit $ka \ll 1$). They include a torsional Alfvén wave propagating at the internal Alfvén speed (v_{Ai}), a fast kink wave at the kink speed (c_K) and a slow wave at the tube speed (c_T).

The equations for the path of a thin tube of magnetic field $\mathbf{B} = B\,\hat{\mathbf{s}}$, density ρ, curvature $\boldsymbol{\kappa} = \partial\hat{\mathbf{s}}/\partial t$ and internal Alfvén speed v_A moving in the convection zone (of density ρ_e) were first written down by Spruit (1981) and subsequently modified and used by many authors in order to study the emergence of flux tubes (Sec. 9.3.3). Spruit writes the equation of motion in the form

$$\frac{d\mathbf{v}}{dt} = -\frac{1}{\rho}\frac{\partial p}{\partial t}\hat{\mathbf{s}} + (\mathbf{g}\cdot\hat{\mathbf{s}})\ \hat{\mathbf{s}} + \frac{\rho}{\rho+\rho_e}v_A^2\boldsymbol{\kappa} + \frac{\rho-\rho_e}{\rho+\rho_e}(\hat{\mathbf{s}}\times\mathbf{g})\times\hat{\mathbf{s}}.$$

Here the terms on the right represent (in order) the pressure gradient and force of gravity along the tube, the tension force and the magnetic buoyancy force. In addition, a drag force normal to the tube can be incorporated, of the form $\rho_e C_D v_\perp^2/(\pi a)$ in terms of a drag coefficient C_D and the tube radius a.

9.2.3 Intensification of a Thin Flux Tube by Convective Collapse

A simple model for convective collapse may be set up as follows. Vertical motions in a slender tube that is slowly diverging with height may be described by the (nonlinear) *slender flux tube equations* (Roberts and Webb 1979). The density $\rho_i(z,t)$, pressure $p_i(z,t)$, vertical speed $v_i(z,t)$ and field $B_i(z,t)$ on the tube axis are given by equations of continuity, vertical and transverse momentum, and isentropic energy:

$$\frac{\partial}{\partial t}\left(\frac{\rho_i}{B_i}\right) + \frac{\partial}{\partial z}\left(\frac{\rho_i v_i}{B_i}\right) = 0, \qquad \rho_i\left(\frac{\partial v_i}{\partial t} + v_i\frac{\partial v_i}{\partial z}\right) = -\frac{\partial p_i}{\partial z} - \rho_i g,$$

$$p_i + \frac{B_i^2}{2\mu} = p_e, \qquad \frac{\partial p_i}{\partial t} + v_i\frac{\partial p_i}{\partial z} = \frac{\gamma p_i}{\rho_i}\left(\frac{\partial \rho_i}{\partial t} + v_i\frac{\partial \rho_i}{\partial z}\right),$$

where p_e is the external pressure. The first equation is obtained by eliminating $\boldsymbol{\nabla}\cdot\mathbf{v}_i$ between the ideal induction and continuity equations. The equations $\boldsymbol{\nabla}\cdot\mathbf{B}_i = 0$ and $\partial\rho_i/\partial t + \boldsymbol{\nabla}\cdot(\rho_i\mathbf{v}_\mathrm{i}) = 0$ then determine $\partial B_{iR}/\partial R$ and $\partial v_{iR}/\partial R$ on the tube axis.

For linear perturbations $v_i(z,t)=\hat{v}_i(z)e^{i\omega t}$ about a basic state given by Eqs. (9.8), with $T_i(z)=T_e(z)$ and no external pressure perturbation, the above equations imply that the velocity amplitude $[\hat{v}_i(z)]$ satisfies

$$\frac{d^2\hat{v}_i}{dz^2}-\frac{1}{2H_i}\frac{d\hat{v}_i}{dz}+\left(\frac{\omega^2-N_i^2}{c_T^2}+(1-\tfrac{1}{2}\gamma)\frac{N_i^2}{c_{si}^2}\right)\hat{v}_i=0. \tag{9.11}$$

Here $c_T(z)=c_{si}v_A/(c_{si}^2+v_A^2)^{1/2}$ is the *tube speed* (Sec. 4.6), $c_{si}(z)=(\gamma p_i/\rho_i)^{1/2}$ and $v_A(z)=B_i/(\mu\rho_i)^{1/2}$ are the sound and Alfvén speeds, and $N_i(z)^2=(g/H_i)[(\gamma-1)/\gamma+dH_i/dz]$ is the *Brunt-Väisälä frequency* squared. Equation (9.11) with two boundary conditions on $\hat{v}_i(z)$ represents an eigenvalue problem for the frequency (ω).

A local approximation to Eq. (9.11) gives $\hat{v}_i(z)\approx e^{z/(4H_i)}\sin\pi z/d$ and

$$\omega^2=\left(\frac{\gamma}{2c_{si}^2}+\frac{1}{v_A^2}\right)c_T^2N_i^2+\left(\frac{\pi^2}{d^2}+\frac{1}{16H_i^2}\right)c_T^2,$$

which implies that the plasma is stable for $N_i^2>0$ but can be unstable for $N_i^2<0$. For conditions typical of an intense tube ($c_{si}=v_A$, $\gamma=1.2$, $H_i=150$ km, $dH_i/dz=-0.25$), the time-scale for instability is typically 100s, and the tube is stable if the field is so large (1 to 2 kG) that $\beta_i\equiv(2\mu p_i/B_i^2)<\frac{1}{2}$.

The above local linear explanation for field intensification during magnetoconvection was subsequently confirmed and modelled more realistically in 3D computational experiments on the nonlinear concentration of magnetic flux in downdrafts driven by radiative cooling (Stein and Nordlund 1989; Nordlund and Stein 1990). Numerical simulations of convective collapse (Danilovic et al. 2010) are consistent with observations of several events of diameter 0.1 to 0.2 arcsec with Hinode/SOT (Nagata et al. 2008; Shimizu et al. 2008); they both show advection of magnetic flux to a granular vertex, development of a strong downflow (5 to 10 km s^{-1}), evacuation of the magnetic feature, increase of field strength and appearance of a bright point.

9.2.4 Spicule Generation

Intense flux tubes provide natural channels for fluid motions between the photosphere and corona. They represent paths along which spicules of types I and II (Sec. 1.5) (Figure 1.20) are ejected upwards in response to photospheric motions and reconnection. Also, they act as a sink for returning spicular material from the corona. The numbers of spicules and intense tubes on the solar disc are comparable ($\approx 10^5$) and both are concentrated in the network above supergranule boundaries.

Spicules are clearly caused by some kind of interaction of plasma with intense fields at supergranule boundaries, and so two main models were originally proposed (see review by Sterling 2000). The first is granular buffeting, which naturally increases the tube pressure and drives an upflow along an intense tube. This was developed as a *shock rebound model* (Hollweg 1982; Sterling and Hollweg 1989), in which a slow magnetoacoustic-gravity wave, excited by a single weak pressure pulse, propagates upwards as a front followed by an oscillating wake. The front and oscillations develop into a set of shock waves that drive chromospheric material upwards, but the model was unable to produce tall enough or fast enough spicules. The second mechanism is acceleration by magnetic reconnection (Uchida 1969), but there was initially thought to be a lack of opposite-polarity magnetic interactions to drive them.

Two developments have, however, implied that both mechanisms may be at work. Firstly, the discovery by De Pontieu et al. (2004) of a 5-minute periodicity in spicule occurrence. Normally, in a vertical magnetic field, *p*-mode oscillations are reflected down by the acoustic cut-off, but along inclined fields they can leak through from the photosphere and grow into slow-mode magnetoacoustic shocks that drive the chromosphere upwards (cf Suematsu 1990). Secondly, the discovery by Hinode of the ubiquitous presence around

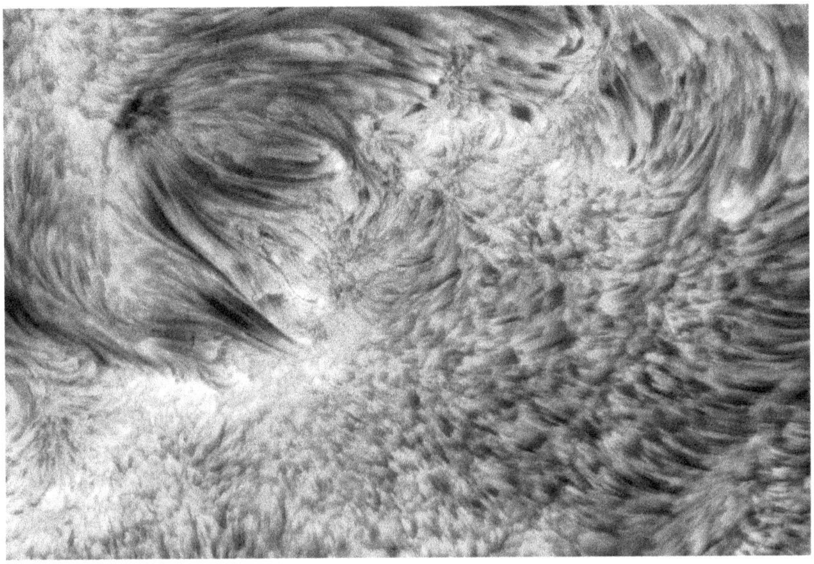

Figure 9.7. Hα image (Swedish Solar Telescope) of an active region near disc centre, showing long dark fibrils (upper left) linking to small sunspots, bright plage and a magnetic carpet (Sec. 1.4.2) of short dynamic fibrils or spicules (lower right) (courtesy Bart De Pontieu).

supergranules and granules of tiny magnetic fragments of opposite polarity, which are continually interacting (Lites et al. 2008) and which may well be driving spicules by reconnection: indeed, reconnection is probably the only way of accelerating tall (i.e., 10 Mm) type II spicules and it can even occur in a unipolar region when nearby fields are differently inclined.

By imposing an observed photospheric driver (including granular flows) with a train of oscillations rather than just a single pulse in a 1D numerical model, De Pontieu et al. (2004) produce type I spicules that agree well with observations (heights of 3 Mm in a vertical field or 4 to 5 Mm for an inclination of 50 degrees). Larger spicules in the quiet Sun may be due to a greater photospheric power and magnetic inclination. Later, more realistic 2D and 3D radiative MHD experiments (Figure 1.19b and Sec. 10.2.6) show how reconnection or buffeting can produce magnetoacoustic shocks and how shock waves can drive type I spicules, while reconnection drives type II spicules (Hansteen et al. 2006; Martínez-Sykora et al. 2009, 2011).

Observed twisting motions in many spicules (De Pontieu et al. 2012) (and chromospheric swirls, Wedemeyer-Böhm et al. 2012) may be evidence of torsional Alfvén waves, produced either by 3D reconnection or by photospheric vorticity (Wedemeyer-Böhm and Rouppe van der Voort 2009). For instance, Matsumoto and Shibata (2010) simulated spicules as the propagation of such waves in response to an observed spectrum of granular motions from Hinode, with the chromosphere becoming an Alfvén resonant cavity at frequencies of 1, 3 and 5 mHz.

9.3 Magnetic Buoyancy

The term 'magnetic buoyancy' refers to three distinct situations (e.g., Hughes 2007). Firstly, a horizontal flux tube that is lighter than its surroundings is not in equilibrium and so rises (Sec. 9.3.1). Secondly, a stratified magnetic field in equilibrium may become unstable to the formation of rising flux tubes (Sec. 9.3.2). Thirdly, an isolated flux tube that is in equilibrium can also become unstable.

9.3.1 Qualitative Effect

The fundamental paper by Parker (1955a) suggest that, once a flux tube has formed in the convection zone, it rises by magnetic buoyancy and produces a pair of sunspots where it breaks through the photospheric surface (Figure 9.8). Suppose the gas pressure and magnetic field strength of the tube are p_i and B_i, and the ambient external pressure is p_e. Lateral total pressure balance implies $p_e = p_i + B_i^2/(2\mu)$. If the temperature (T) is uniform and the densities are ρ_e, ρ_i, this becomes

$$\frac{k_B T \rho_e}{m} = \frac{k_B T \rho_i}{m} + \frac{B_i^2}{2\mu}, \tag{9.12}$$

so that ρ_e must exceed ρ_i. The plasma in the tube therefore feels a buoyancy force of $(\rho_e - \rho_i)g$ per unit volume, which tends to make the tube rise. When it becomes curved, the tube will also experience a restoring force due to magnetic tension, which cannot overcome buoyancy as long as $(\rho_e - \rho_i)g > B_i^2/(\mu L)$, or, using Eq. (9.12),

$$L > \frac{2k_B T}{mg} \equiv 2H. \tag{9.13}$$

In other words, a tube that is longer than twice the local scale-height will rise.

Equation (9.12) implies that the buoyancy force is relatively much stronger in the upper part of the convection zone. Thus, for $B_i \approx 1000$ G, as ρ_e and T change from 0.25 kg m^{-3} and 2.5×10^5 K at 20 Mm below the photosphere to 0.8×10^{-5} kg m^{-3} and 1.5×10^4 K at 1 Mm, so $(\rho_e - \rho_i)/\rho_e$ increases from 10^{-5} to 4×10^{-3}. Furthermore, the shape of a buoyant flux tube in equilibrium can be calculated (PROBLEM 9.5).

9.3.2 Magnetic Buoyancy Instability

Consider a horizontal field $[B_0(z)\hat{\mathbf{x}}]$ in equilibrium, with gravity acting downwards (Figure 9.9a). It is unstable if the field strength decreases fast enough with height, at a rate that depends on what effects are included. For example, if the field remains straight, its strength needs to fall off faster than the density for instability, so that

$$\frac{d}{dz}\left(\frac{B_0}{\rho_0}\right) < 0, \tag{9.14}$$

which may be established as follows (e.g., Moffatt 1978; Acheson 1979). Suppose a flux tube rises by δz, with changes in its internal properties of $\delta B, \delta\rho$ and δp, while the corresponding ambient changes are $\delta B_0, \delta\rho_0, \delta p_o$. Conservation of mass and magnetic flux imply that B/ρ remains constant in the flux tube,

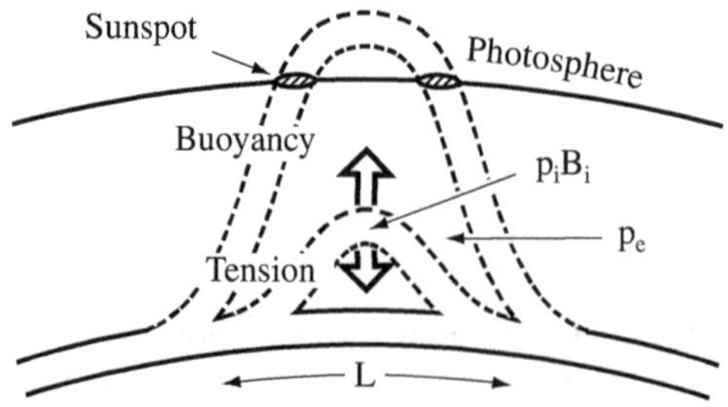

Figure 9.8. A schematic of the way a magnetic flux rope rises through the convection zone by magnetic buoyancy and breaks through the photosphere to create a pair of sunspots.

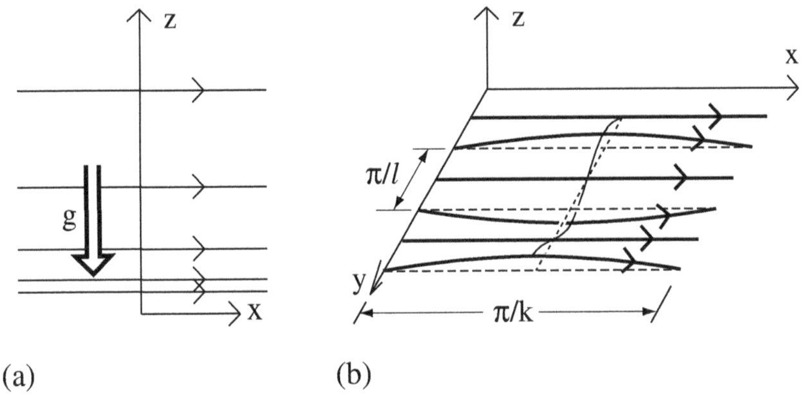

Figure 9.9. Field lines for the magnetic buoyancy effect: (a) equilibrium field in a vertical plane; (b) perturbed field lines at one particular height.

so that

$$\frac{\delta B}{B_0} = \frac{\delta \rho}{\rho_0}. \tag{9.15}$$

Also, horizontal equilibrium means a balance in total pressure $[p + B^2/(2\mu)]$, so that $\delta p + B_0 \delta B/\mu = \delta p_0 + B_0 \delta B_0/\mu$, or, since $\delta p = (k_B/m)T_0\ \delta\rho$ and $\delta p_0 = (k_B/m)T_0\ \delta\rho_0$ for an isothermal medium,

$$\frac{k_B T_0 \delta\rho}{m} + \frac{B_0 \delta B}{\mu} = \frac{k_B T_0 \delta\rho_0}{m} + \frac{B_0 \delta B_0}{\mu}. \tag{9.16}$$

The condition $(\delta\rho < \delta\rho_0)$ for the tube to continue rising then implies that $\delta B_0 < \delta B$, and so $(\delta B_0/B_0) < (\delta\rho_0/\rho_0)$ from Eq. (9.15), which establishes Eq. (9.14).

After adding the stabilising influence of normal buoyancy, the instability criterion is modified [in terms of the *Brunt-Väisälä frequency* N (Secs. 4.4 and 9.2.3)] to

$$\frac{\rho_0}{B_0}\frac{d}{dz}\left(\frac{B_0}{\rho_0}\right) < -\gamma H \frac{N^2}{v_A^2} = -\frac{c_s^2}{g}\frac{N^2}{v_A^2}, \tag{9.17}$$

so that a field that decreases sufficiently rapidly with height can destabilise a convectively stable atmosphere (i.e., one with $N^2 > 0$).

Many other effects can be included. If the field is allowed to bend (Parker 1966, 1979a; Gilman 1970), it is unstable if its strength decreases with height, namely,

$$\frac{dB_0}{dz} < 0.$$

The bending of the field allows plasma to drain from the summits, which enhances the buoyancy effect. This magnetic buoyancy instability is really only a form of hydromagnetic Rayleigh-Taylor instability (Secs. 7.3.1.1 and 7.5.2), since a negative value for dB_0/dz means that the field is holding up extra mass against gravity. A formal stability analysis may be easily carried out for a 1D equilibrium with uniform sound and Alfvén speed (PROBLEM 9.6).

However, magnetic and thermal diffusion reduce the stabilising effect of stratification and modify condition (9.17) to

$$\frac{\rho_0}{B_0}\frac{d}{dz}\left(\frac{B_0}{\rho_0}\right) < -\gamma H \frac{\eta}{\kappa}\frac{N^2}{v_A^2}. \tag{9.18}$$

Instability is aided if magnetic diffusion (η) is small (maintaining the unstable field gradient) and thermal diffusion (κ) is large (eroding the stabilising entropy gradient).

Diffusion can also lead to oscillatory instability or *overstability* when

$$\frac{\rho_0}{B_0}\frac{d}{dz}\left(\frac{B_0}{\rho_0}\right) < -\frac{\gamma H}{v_A^2}\left[\frac{(\kappa+\nu)(\kappa+\eta)(\nu+\eta)k^6/k_x^2+(\kappa+\nu)N^2}{\eta+\nu-\kappa(\gamma-1)}\right]. \tag{9.19}$$

Rapid rotation (Ω) inclined at θ to the vertical tends to inhibit vertical motions and stabilise against magnetic buoyancy if $2\Omega\sin\theta(H/c_{s0})>1$ (Gilman 1970). However, the full picture with diffusion included is highly complicated and a local analysis gives a fifth-order complex dispersion relation (Acheson 1978), which has been solved numerically (Schmitt and Rosner 1983; Hughes 1985).

The important effect of shear flow needs an eigenvalue analysis. Tobias and Hughes (2004) used an energy principle to derive general conditions for stability and also investigated the stabilising effect of shear flow on particular unstable profiles.

The effect of large-scale curvature is to allow magnetic buoyancy instability only outside a critical radius of $2\gamma H$, in the form of growing waves that propagate in an easterly direction (Acheson 1979; Spruit and van Ballegooijen 1982), so that the instability criterion (9.14) is modified to

$$\left(\frac{r}{\gamma H}-2\right)\frac{d}{dr}\log\left(\frac{B_0}{\rho_0 r}\right) < -1. \tag{9.20}$$

Later a comprehensive linear stability treatment including spherical geometry, rotation and stratification was given (Ferriz-Mas and Schüssler 1995).

Nonlinear development of magnetic buoyancy instability has been studied numerically, beginning with 2D instability of a slab of magnetic field (Cattaneo and Hughes 1988), for which rising magnetic mushrooms appear, followed by secondary Kelvin-Helmholtz vortices and pair-wise interaction of neighbouring mushrooms. Later, Matthews et al. (1995) and Fan (2001) considered 3D effects that produce an arching of the magnetic field as it rises through the convection zone.

9.3.3 The Emergence of Magnetic Flux

The emergence of magnetic flux through the solar photosphere over a wide range of scales and its interaction with the overlying chromospheric and coronal magnetic field is a key process in solar physics that is responsible for a wide range of phenomena, including: the presence of sunspots, active regions, ephemeral active regions (in supergranules) and inner network flux (in granules); the eruption of prominences and coronal mass ejections (Sec. 12.2); and small-scale heating phenomena such as explosive events, microflares, nanoflares, spicules and X-ray jets (Secs. 1.5, 1.6, and 10.4.2).

9.3.3.1 Rise through the Convection Zone

Parker (1955a) suggested that toroidal flux tubes rise in sections a few scale-heights (H) long by magnetic buoyancy and create sunspots where they break through the solar surface. The slight inclination (typically about 10 degrees) of a sunspot pair is caused by the action of the Coriolis force as the flux tube rises. If the time-scale for the rise is just the Alfvén time ($\tau_A = H/v_A$), a 100 G field would rise from a depth of 20 Mm in only 2 months, from the middle of the convection zone in a year, and from the base of the convection zone in 10 years.

However, estimating the rise-time of a flux tube is uncertain, since it depends on its field strength, depth of origin and nonlinear instability rate, as well as effects that may slow down its rise, such as aerodynamic drag, expansion and interaction with convection. Nevertheless, the consensus is that the flux tubes that produce sunspots would rise much too quickly, since fields much stronger than 100 G are needed to account for active regions. Magnetic flux is therefore likely to be held down by a downflow or flux pumping and the main dynamo region is probably located in the tachocline at the base of the convection zone (Sec. 8.5, Galloway and Weiss 1981).

The rise of a thin flux tube from the tachocline has been modelled, including the effect of the Coriolis force on the shape, path and orientation of the tube (Moreno Insertis 1983, 1992; Choudhuri and Gilman 1987; Caligari et al. 1995; Fisher et al. 2000; Fan and Gibson 2004). Some twist is needed in order to prevent fragmentation during the rise (Leka et al. 1996; Emonet and Moreno-Insertis 1998). Also, in order to reproduce the observed properties of sunspot groups, the required field strength at the tachocline is estimated to be of order 10^5 G.

9.3.3.2 Emergence across the Photosphere into the Corona

The thin flux tube approximation fails during the emergence of flux, but it has been replaced by sophisticated numerical experiments, beginning in 2D and proceeding to 3D in an increasingly realistic manner, as reviewed by Archontis (2008). The first aim was to investigate the nonlinear evolution of magnetic buoyancy instability in 2D. Then the early 3D experiments studied the effect of various parameters, such as the twist and field strength of the emerging field, in enabling a tube to rise as a coherent structure.

For small values of twist, the tube flattens at the photosphere and comes to a halt, piling up below the surface for a considerable time, before the apex of the tube becomes unstable and starts rising into the corona. For larger twist values, the tube experiences a partial emergence: its main axis remains at 2 to 3 pressure scale-heights above the photosphere, while a new flux rope that forms above it emerges into the outer solar atmosphere. In all cases, the emergence is followed by an outward expansion that leads to a fan-like shape for the outermost field lines, with the innermost field lines forming an undulating or sigmoid shape (Figure 9.10a). The rapid rise and expansion of the field causes strong downflows at the edges of the emergence.

One improvement of the models was to include convection in and below the photosphere (Amari et al. 2005; Cheung et al. 2007, 2010), demonstrating how the granulation is disturbed by the flux emergence and how the flux emergence is fragmented near the photosphere by convection. Another was to include the effect of ambipolar diffusion due to partial ionisation around the photosphere (Arber et al. 2007), which enhances the emergence and expansion and makes the coronal magnetic field force-free by preferentially diffusing the perpendicular current.

Another improvement was to include a coronal magnetic field and to study reconnection between emerging and pre-existing fields, which accelerates a cool surge or hot X-ray jet (Sec. 1.6.2), as in the emerging flux model of flares (Sec. 12.2.8). 2D emergence into a horizontal field was modelled first (Forbes and Priest 1984a; Shibata et al. 1989), followed by the effect of an inclined overlying field in accelerating asymmetric

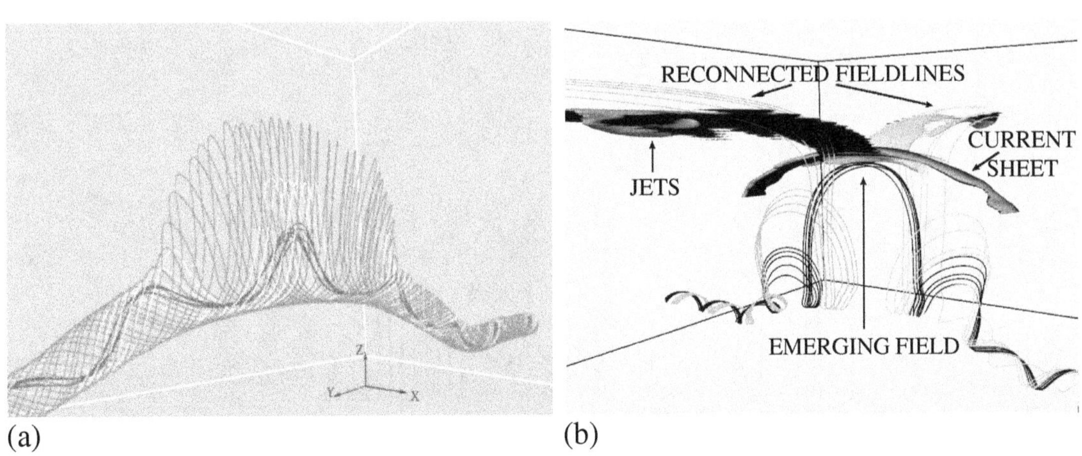

Figure 9.10. An emerging flux tube from a numerical experiment showing (a) the magnetic field lines and (b) reconnection with an overlying field (courtesy Vasilis Archontis).

up- and down-flowing jets (Yokoyama and Shibata 1996; Shibata et al. 2007; Rachmeler et al. 2010) (see Sec. 11.6.1).

In 3D, the process of flux emergence is much more complex and produces a curved 3D current sheet (Figure 9.10b) with strong Joule dissipation (Magara and Longcope 2003; Abbett and Fisher 2003; Fan and Gibson 2004; Hood et al. 2009; Archontis and Hood 2010). New features are filamentation of the current sheet by Rayleigh-Taylor instability (Isobe et al. 2005), the formation of plasmoids, intermittent heating and sigmoid formation (Fan 2010; Archontis and Hood 2012).

In a coronal hole with an open, inclined magnetic field of 10 G, Moreno-Insertis and Galsgaard (2013) modelled 3D formation by flux emergence of a coronal jet, whose properties compare well with observations. They find a complicated, rapidly evolving process with several phases and a topologically complex structure. In the main phase of activity, the jet is accelerated by spine-fan reconnection (Sec. 6.13.2) to 150 to 250 km s^{-1} at the interface between an emerging flux dome and the overlying field. Around the dome there forms a dense cool wall. Later, a series of eruptions occur, with null points and flux ropes that go kink or torus unstable in a way that resembles a blowout jet (Sec. 1.6.2).

Studying the dynamics of flux emergence is a huge challenge in view of the enormous range of scales, parameter values and wave speeds and the transition from a convecting environment with radiative transfer to a low-beta coronal environment with different thermal properties. So far, the focus has been on the basic physical processes, but more realistic physics is beginning to be incorporated.

9.4 Overall Equilibrium Structure of Sunspots and Pores

This section describes models for global magnetohydrostatic equilibrium of a sunspot or pore, treated as a static, axisymmetric field. It provides a background (mean) state within which umbral or penumbral fine-structure may be studied (Sec. 9.5) and naturally leads to questions of convective and interchange stability (Secs. 9.1 and 9.4.2).

9.4.1 Magnetohydrostatic Equilibrium

For simplicity, consider first the effect of a vertical magnetic flux tube $[B(R)\hat{\mathbf{z}}]$ on the equilibrium of a stratified atmosphere (Figure 9.11a). Suppose $B(R) = B_i$ on the axis $(R=0)$ of the tube, and that B vanishes outside the tube, with corresponding pressures $p_i(z)$ and $p_e(z)$. Then vertical equilibrium, $\partial p/\partial z = -\rho g$, becomes

$$\frac{dp_i}{dz} = -\rho_i\, g \qquad \text{and} \qquad \frac{dp_e}{dz} = -\rho_e\, g \tag{9.21}$$

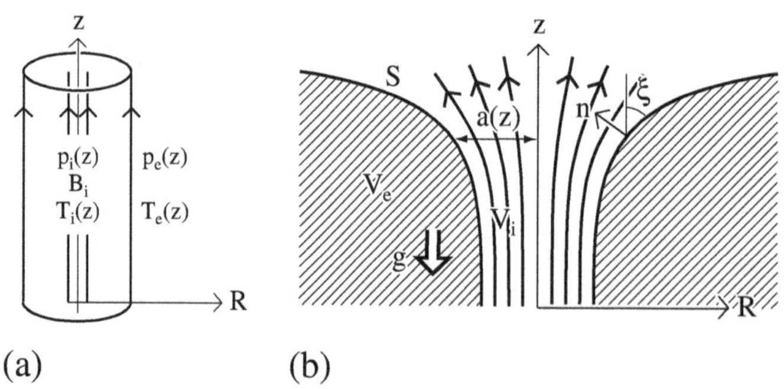

Figure 9.11. (a) A vertical magnetic flux tube in a vertically stratified medium. (b) A magnetic flux rope surrounded by a field-free plasma (after Meyer et al. 1977).

on the axis and outside the flux tube, respectively. Equation (9.21) needs to be supplemented by an energy equation before one can determine the temperature stratification $[T_e(z)]$. Furthermore, horizontal total pressure balance is $p(R,z)+B^2(R)/(2\mu)=p_e(z)$, which on the axis determines p_i when p_e and B_i are known from

$$p_i(z)+\frac{B_i^2}{2\mu}=p_e(z). \tag{9.22}$$

Now, the derivative of Eq. (9.22) with respect to z implies that $dp_e/dz=dp_i/dz$, and so from Eq. (9.21) $\rho_i=\rho_e$. Also, Eq. (9.22) implies that the plasma pressure (p_i) inside the spot is smaller than the ambient pressure (p_e), and the corresponding temperature deficit is $T_i/T_e=1-B_i^2/[2\mu p_e(z)]$. Thus, the presence of the vertical magnetic field has no effect on the density, but it produces a pressure deficit (and hence a temperature deficit) in order to maintain horizontal pressure balance. Furthermore, the result implies that only if the thermal equilibrium of the plasma produces this particular form for the temperature difference will the magnetic lines be straight; otherwise, the tube expands with height.

According to Eq. (9.22), the maximum possible magnetic pressure $[B_i^2/(2\mu)]$ in the vertical flux rope is the external plasma pressure $[p_e(z)]$. However, a puzzle is that the magnetic pressure due to a sunspot field of 3000 G is 2.4×10^4 N m^{-2} (2.4×10^5 dyne cm^{-2}), whereas the photospheric pressure is only 1.4×10^4 N m^{-2} (at $\tau_{5000}=1$). The answer to this conundrum lies in the *Wilson depression*, namely, the fact that the umbral magnetic field is being measured at a depth of 500 to 1500 km below the normal photosphere, where the ambient plasma pressure is higher. Also, the magnetic field lines are not purely vertical but spread out (see Figure 9.11b). The resulting decrease in B_i^2 with height means from Eq. (9.22) that the spot pressure gradient (dp_i/dz) exceeds the ambient value, and so from Eqs. (9.21) $\rho_i<\rho_e$. It is this density deficit within the sunspot that produces the Wilson effect (Sec. 1.7.2.1). Furthermore, since the maximum field that can be balanced by the photospheric pressure at $\tau_{5000}=1$ is 1900 G, the penumbral field cannot be larger than this value.

Below a few megameters the plasma pressure exceeds 10^7 Nm^{-2} and, unless it is twisted, the flux tube may be shredded by turbulent motions. The temperature deficit in a sunspot is probably a rather shallow phenomenon extending to only 10 Mm or less in depth. The flux tube itself, however, may continue to be attached to the base of the convection zone, although Schüssler and Rempel (2005) suggest that it becomes disconnected within 3 days of emergence due to a pressure build-up.

A simple model for magnetostatic equilibrium of a sunspot (Schlüter and Temesváry 1958) is based on the force balance $\mathbf{0}=-\nabla p+\mathbf{j}\times\mathbf{B}-\rho g\hat{\mathbf{z}}$, with gravity acting along the negative z-axis. For a cylindrically symmetric ($\partial/\partial\phi=0$), untwisted ($B_\phi=0$) magnetic field $[B_R(R,z),0,B_z(R,z)]$, the force balance has components

$$0=-\frac{\partial p}{\partial R}+\frac{B_z}{\mu}\left(\frac{\partial B_R}{\partial z}-\frac{\partial B_z}{\partial R}\right),\qquad 0=-\frac{\partial p}{\partial z}-\frac{B_R}{\mu}\left(\frac{\partial B_R}{\partial z}-\frac{\partial B_z}{\partial R}\right)-\rho g. \tag{9.23}$$

A similarity solution automatically satisfying $\nabla\cdot\mathbf{B}=0$ in terms of $\zeta=RB_i^{1/2}(z)$ has the form $B_R=-\frac{1}{2}Rf(\zeta)dB_i/dz$, $B_z=f(\zeta)B_i(z)$. The solution for p depends on the energy balance. Schlüter and Temesváry (1958) demonstrated that a purely radiative model is impossible, but they were able to construct a satisfactory model with partly inhibited convection. The model is developed in PROBLEM 9.7 and was extended by Deinzer (1965) and Osherovich (1982) to include an energy equation, an azimuthal field component (B_ϕ) and return flux.

Meyer et al. (1977) set up magnetohydrostatic *flux tube models* for a sunspot in a volume (V_i) surrounded by a field-free region (V_e) with pressure (p_e) and density (ρ_e) (Figure 9.11b). The equilibrium equations in the two regions are

$$\nabla\left(p_i+\frac{B_i^2}{2\mu}\right)=\rho_i g+(\mathbf{B}_i\cdot\nabla)\frac{\mathbf{B}_i}{\mu}\quad\text{and}\quad\nabla p_e=\rho_e g\hat{\mathbf{z}}, \tag{9.24}$$

subject to total pressure balance $[p_i+B_i^2/(2\mu)=p_e]$ at the interface $[a=a(z)]$.

Near the photosphere, where $p_i \ll B_i^2/(2\mu)$, simpler vacuum models (Meyer et al. 1977) are often useful, for which p_i and $\rho_i g$ are negligible and the field may be written $\mathbf{B}_i = \boldsymbol{\nabla}\Phi_i$ in cylindrical polars, where $\Phi_i(R,z)$ satisfies $\nabla^2\Phi_i = 0$. As $z \to \infty$ the field is assumed to be monopolar, so that on the interface the field is nearly horizontal with $B_R \approx B_i \approx F/(2\pi a^2)$. As $z \to -\infty$, the field becomes vertical with $B_z \approx B_i \approx F/(\pi a^2)$ on S.

The object is to find the radius $[a(z)]$ of the interface and the flux function $[\psi(R,z)]$ for a given pressure $[p_e(z)]$ in the field-free region and a given flux $[F_m = 2\pi(\psi)_s]$. First of all, Meyer et al. (1977) instead solved the simpler inverse problem of choosing analytical solutions to $\nabla^2\Phi_i = 0$ and then deducing $p_e(z)$. They also set up a mean model, by assuming that the mean field $[\bar{B}_i(z)]$ across the flux tube is approximately equal to the field at the boundary. Then $\bar{B}_i(z) \approx [2\mu p_e(z)]^{1/2}$, where $p_e(z)$ is imposed from a standard model of the convection zone based on mixing-length theory; the radius of the tube for a given flux (F_m) follows from $\pi a(z)^2 \bar{B}_i(z) = F_m$. Later, Schmidt (1991) solved the full free boundary-value problem with a prescribed $p_e(z)$. Earlier, Simon and Weiss (1970) had constructed a simple Bessel-function solution $\Phi_i = AJ_0(kr)\exp(-kz)$, which can give pressure balance at only two heights. This was subsequently developed by Simon et al. (1983) for an imposed uniform vertical photospheric field over a disc of given radius.

A further complication is that the observed Wilson depression is so large that a spot field of 5,000 G would be needed to balance the pressure difference if the spot field were potential. Jahn (1989) therefore suggested that strong azimuthal volume currents exist in the penumbra, which are associated with a sharper bend in the field lines and a tension force that balances the extra radial pressure gradient.

Parker (1979b) challenged the conventional picture of a sunspot as a single large coherent flux tube by suggesting that it divides into many independent thin tubes at about 1 Mm beneath the photosphere, like the tentacles of a jellyfish. The tubes are separated by field-free plasma. In this *spaghetti model*, the loose cluster of flux tubes (or strands) is held together at the solar surface by magnetic buoyancy and a downdraft beneath the sunspot. However, although the convection zone is likely to be full of flux strands, it is unclear why they should assemble to form a sunspot. Such models with field-free regions have therefore been replaced in favour of ones with inhomogeneous fields within a sunspot.

9.4.2 Sunspot Stability

Convective instability is important for the structure of sunspots (Sec. 9.1). On linear theory, the onset of steady (overturning) convection is unaltered by the presence of a weak field, whereas moderate fields allow only a weaker form of overstable (oscillatory) convection (when $\kappa > \eta$), and strong fields suppress convection altogether.

The equilibrium structure of a sunspot is quite complex, with its balance between gravity, magnetic field and plasma pressure. For local stability near the photosphere (Meyer et al. 1977) there are two competing effects. The first is destabilising, because the fanning of the field lines makes the boundary of the sunspot curve around the higher-pressure external region, which would be unstable to the interchange or flute instability (Sec. 7.5.1) if gravity were absent. The second effect is stabilising, because the magnetic field of the outer part of the spot lies on top of the denser external plasma, a situation that would be stable to the hydromagnetic Rayleigh-Taylor mode if there were no curvature (Sec. 7.5.2). Meyer et al. (1977) determine the resulting criteria for stability of magnetohydrostatic sunspot models as follows.

Bernstein et al. (1958) energy principle (Sec. 7.4) provides the energy change ($\delta W_i + \delta W_e + \delta W_s$) due to an adiabatic perturbation. The surface contribution is

$$\delta W_s = \frac{1}{2}\int_s (\mathbf{n}\cdot\boldsymbol{\xi})^2 \mathbf{n}\cdot\left[\boldsymbol{\nabla}\left(p_i + \frac{B_i^2}{2\mu}\right) - \boldsymbol{\nabla}p_e\right] dS, \tag{9.25}$$

and the volume terms (δW_i and δW_e) are positive for a surface instability. A necessary and sufficient condition for stability is that the integrand of Eq. (9.25) be positive, so that the restoring force acts in opposition to the displacement. After using Eqs. (9.24) for an asymmetric field (B_R,0,B_z), the stability condition reduces to $dB_R^2/dZ < 0$, taken along a field line in the surface S. Thus, the sunspot's surface is stable if the magnitude of the radial magnetic field component decreases upwards along the tube boundary. In particular, near the top of a sunspot, $B_R^2 \simeq 2\mu p_e(z)$, which decreases with height, and so the surface is stable. However, below the photosphere when the field is nearly vertical, it becomes unstable to interchange modes, although one possibility for stabilising it is the effect of an inflowing collar or moat flow at some depth – indeed, a moat is known to be a property of a long-lived sunspot.

An alternative form for the stability condition at uniform temperature is

$$R_c \sin\chi > 2H_e \tag{9.26}$$

in terms of the scale-height $H_e = p_e/(\rho_e g)$, the radius of curvature (R_c) of a field line in S and the inclination (χ) of S to the vertical. Thus, the boundary is stable if its radius of curvature or inclination are large enough.

Meyer et al. (1977) use Eq. (9.26) to conclude that a vacuum model of a flux tube that converges monotonically with depth is stable when its flux exceeds about 10^{11} Wb (10^{19} Mx). They also suggest that leading spots in a sunspot group tend to outlive following spots because the leading part of a rising flux tube becomes nearly vertical and so is stable, whereas the following part is inclined to the vertical so that its inner edge is subject to fluting, which tends to break up the following spot.

For a thin flux tube, Meyer et al.'s (1977) mean models are appropriate, and the stability condition can be written

$$F_m > 8\pi(2\mu p_e)^{1/2} H_e^2 \left(-4\frac{dH_e}{dz} - 1\right). \tag{9.27}$$

For a given $p_e(z)$ and $H_e(z)$, flux tubes with a small enough flux are therefore locally unstable to fluting when $dH_e/dz < -\frac{1}{4}$, which holds for the standard model atmosphere over the height range -700 km $< z <$ 30 km.

9.5 Fine Structure of a Sunspot

The umbra contains tiny umbral dots and the penumbra is by its nature fine-structured, for which many explanations have been proposed. Furthermore, oscillations occur in both umbra and penumbra. These elements are part of an impressive unified sunspot model that is emerging from computational experiments (Sec. 9.7).

9.5.1 Umbral Fine Structure

Boussinesq convection in a vertical field occurs in narrow vertical cells and depends on the ratio (η/κ): when $\eta < \kappa$ oscillatory convection sets in, but when $\eta > \kappa$ steady overturning convection is preferred. Umbral convection may be modelled most simply as 2D magnetoconvection with $\eta < \kappa$ in the top 2 Mm and $\eta > \kappa$ below 2 Mm. Then convection starts as a steady mode but, as the Rayleigh number (Ra) is increased, this changes to spatially modulated oscillations in the form of slender rising plumes that wax and wane in vigour (Weiss et al. 1990). 3D behaviour is richer, with convection starting as a hexagonal array of steady plumes, but then turning into an irregular time-dependent pattern as Ra is increased (Weiss et al. 2002).

With umbral dots in mind, Schüssler and Vögler (2006) set up a more realistic 3D MHD simulation of magnetoconvection with radiative transfer and partial ionization in a strong background magnetic field (Figure 9.12). They found that convective energy transport is dominated by narrow plumes of rising hot plasma with strongly reduced magnetic field, having a size of 200 to 300 km, lifetime of around 30 mins,

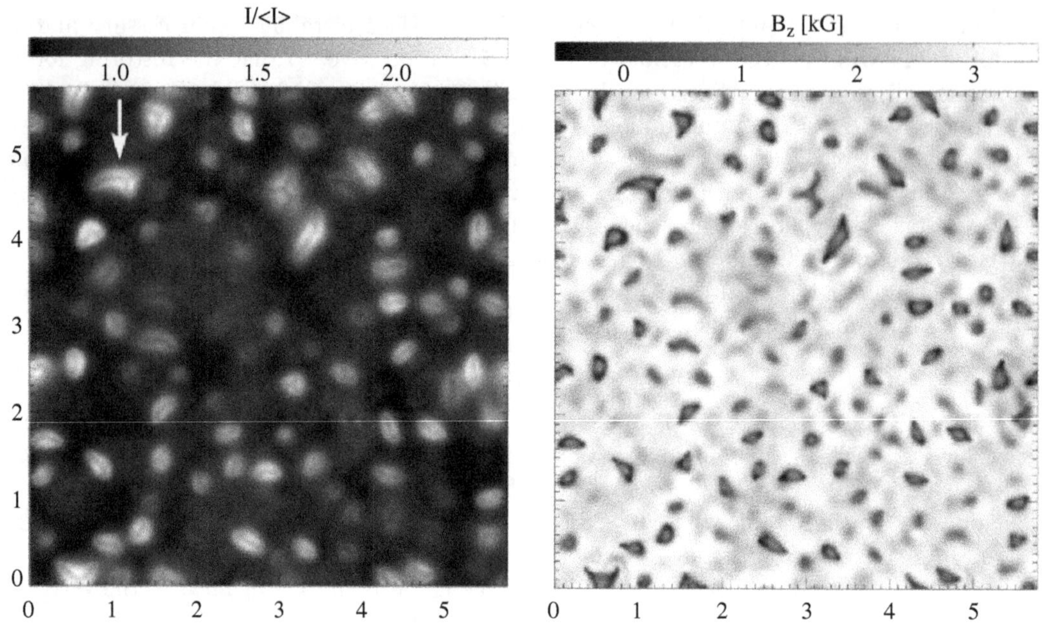

Figure 9.12. Umbral dots as seen in intensity (left) and vertical magnetic field strength (right) in a realistic simulation of umbral magnetoconvection (Schüssler and Vögler 2006).

and a brightness of 2.4 times the rest of the umbra. The difference between umbral dots and the dark umbral background is largest in the lowest layers. These features closely resemble the observed properties of the larger umbral dots at the solar surface.

They used the MURaM code with a box of 5800 $\times$ 5800 km horizontally and 1600 km vertically, extending from 1200 km below to 400 km above optical depth $\tau = 1$, with grid cells of $20\times20\times10$ km. The magnetic field is assumed to be vertical on top and bottom boundaries with a mean value of 2,500 G and the side boundaries are periodic. Instability starts off in the form of narrow columns of rising plasma, but the density drop near the photosphere causes the upflowing plasma to expand and weaken the field so much that it turns into overturning convection with narrow cells. The strong expansion of the rising plasma decreases the magnetic field to only a few 100 G, so that the plasma is almost field-free near the surface.

Near the solar surface ($\tau = 1$) a plume loses it buoyancy due to radiative cooling and is strongly braked: it develops a high-density horizontally elongated shape with a central dark lane, above which is a narrow upflowing jet. The $\tau = 1$ surface bulges upwards in the plume into the dense region above the stagnation point, which creates the dark lane and also hides the extent of the large flow speeds and small magnetic field strength from spectroscopic observations.

In the simulations, larger umbral dots are brighter and live longer, but the short-lived end of the spectrum of observed umbral dots is not reproduced, probably because of insufficient spatial resolution. However, they also ran a simulation with a field of 3,500 G and found very few umbral dots formed, agreeing with the observed dark cores of umbrae, which are virtually devoid of umbral dots.

9.5.2 The Sunspot Penumbra

Several explanations for penumbral structure (Figures 1.28 and 1.29) have been proposed, the classic one being in terms of convective rolls, with light (hot) plasma rising and dark (cool) plasma falling (Danielson 1961). The dispersion relation (9.4) for linear stability in an inclined uniform magnetic field

(B_x, B_z) shows that a horizontal field is unstable to rolls with $k_x = 0$ (Sec. 9.1.2). Also, for an inclined field the most unstable perturbation is a roll aligned with the horizontal field direction, again such that $k_x = 0$. As the inclination of a magnetic field increases, convection sets in when the field is nearly horizontal. Such a picture may well be essentially correct, especially in the inner penumbra, where the inclinations of bright and dark filaments do not differ much and the inwards-moving bright grains may represent travelling waves.

Meyer et al. (1977) suggested instead that convection under the spreading penumbra may drive local interchanges and penetrate through it to give a bright filament. Parker (1979b) regarded penumbral filaments as part of the loose cluster of flux tubes which make up a sunspot in his model. The *embedded flux tube model* (Solanki and Montavon 1993) explained small-scale variations of magnetic inclination in terms of a horizontal magnetic flux tube with outward Evershed flow embedded in an inclined magnetic field. The *gappy penumbra model* (Spruit and Scharmer 2006; Nordlund and Scharmer 2010) interpreted penumbral filaments as due to convection in radial gaps that have weak magnetic field, with the cusp-shaped top of the gap as the location of Evershed flows and dark penumbral cores. However, the magnetic field is greater in bright filaments and the outflow may be localised to the inner penumbra.

The outer penumbra remains enigmatic, since the steeply inclined dark regions carrying Evershed outflow may not be simply continuations of dark lanes in the inner penumbra. Thus, Weiss et al. (2004) proposed a *flux pumping model* for the outer penumbra (Figure 9.13), comprising an interlocking comb structure of fairly horizontal fields (dark radial filaments) interspersed with more vertical fields (bright radial filaments). They suggested that, as a pore grows, the inclination of the field at its outer edge becomes large enough that it goes unstable to a filamentary fluting instability. Within the dark filaments, some magnetic flux tubes extend beyond the penumbra along an elevated magnetic canopy, while others are captured by granules and dive back below the surface, being held down by turbulent pumping (indicated by thick vertical arrows in Figure 9.13). In compressible convection, the dominant effect that gives rise to downward flux pumping into a region of less vigorous convection is the flow pattern asymmetry, namely, the presence of strong concentrated downdrafts and weaker broader updrafts. Thus, in the region surrounding a sunspot,

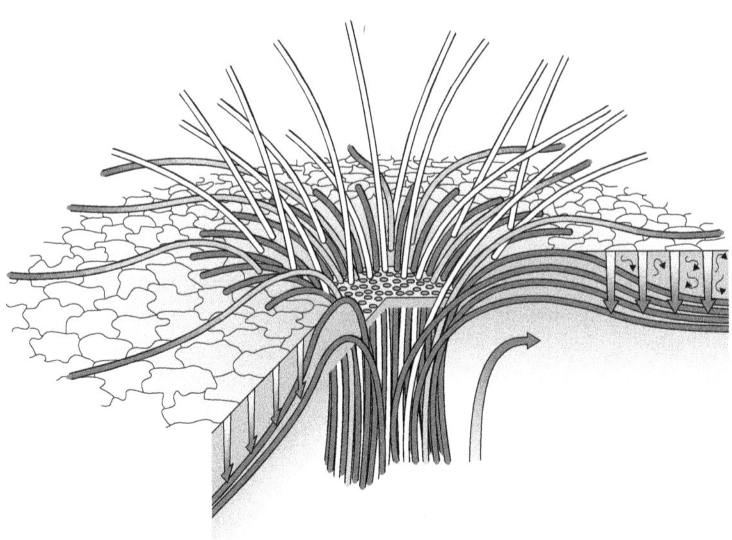

Figure 9.13. Interlocking-comb penumbra in the flux pumping model showing low-lying (dark) and oblique (white) penumbral filaments (after Thomas et al. 2002).

a balance can be set up between downward flux pumping and the combined effect of magnetic buoyancy and the upflow near a spot associated with moat flow.

Evershed outflow is mostly confined to nearly horizontal filaments. One explanation was nonlinear interactions in roll-like convection, while another was siphon flow along field lines from high to low plasma pressure (Meyer and Schmidt 1967; Thomas and Montesinos 1993). These are now elegantly combined in the unified computational model of Rempel (Sec. 9.7), which also explains penumbral filaments.

Another view of penumbral structure from Hinode observations by Tiwari et al. (2013) is described in Section 1.7.2.4, in which all penumbral filaments are similar with a bright head, a horizontal main section and a dark tail.

9.5.3 Oscillations

A broad spectrum of MHD waves modified by stratification and buoyancy is expected to be excited in a sunspot by weak convection and, more especially, by external acoustic waves. More than a few hundred km below the surface of the umbra $c_s > v_A$ and above it $c_s < v_A$, so that the idea of slow and fast modes is potentially confusing, since they will swap their labels as they cross the surface $c_s = v_A$. Furthermore, in a fully 3D gravitationally stratified medium, the sharp distinction between pure slow, fast and Alfvén modes breaks down. Nevertheless, a simple unified explanation for the observed oscillations is emerging, although complications include the breakdown of the slow-fast distinction near $c_s = v_A$, the height of line formation for the observations, the variation of wave properties with height and the fact that a wave can still have substantial amplitude even when it is evanescent.

5-minute oscillations in the umbra are compressive with motions directed along the magnetic field, and in the low-β umbral photosphere (where $c_s < v_A$) they are *slow magneto-acoustic-gravity (MAG) waves.* Locally, in an isothermal uniform magnetic field in the limit $c_s \ll v_A$ (cf. Schunker and Cally 2006) their dispersion relation is

$$k_\parallel^2 = \frac{\omega^2}{c_s^2} - \frac{1}{4H_\parallel^2}, \tag{9.28}$$

where $H_\parallel$ is the local scale-height along the magnetic field. The cause of the reduced amplitude of 5-minute oscillations by comparison with the quiet Sun is likely to be the absorption and partial mode-conversion of *p*-modes that are incident on a sunspot (Spruit and Bogdan 1992; Cally and Bogdan 1997; Cally et al. 2003; Cally and Goossens 2008), as follows.

P-modes are trapped in a sub-surface cavity, and, since a sunspot is a relatively shallow pancake, they are incident at a steep angle from below (Figure 9.14a). They enter the sunspot and propagate as fast-mode waves until they reach the surface where $c_s = v_A$, a few hundred km below the photosphere. Here they are converted to slow MAG waves and fast MAG waves. The slow MAG waves propagate upwards along the magnetic field with velocities along the field and are reflected at the acoustic cut-off in the upper umbral photosphere. The fast MAG waves continue upwards and are reflected from the fast-mode cut-off ($\omega = v_A k_x$) in the umbral photosphere. These downwards-propagating slow and fast MAG waves are later converted to slow and fast modes when they again encounter the $c_s = v_A$ surface. The observed 5-min umbral oscillations along the line of sight are therefore likely to be a mixture of slow and fast modes, but with slow modes dominating since fast modes are reflected lower down and have velocities mainly normal to the magnetic field.

Three-minute oscillations are also observed in the umbral photosphere and dominate in the chromosphere, because they are the high-frequency tail of the photospheric oscillations that are allowed to propagate into the chromosphere (Bogdan and Judge 2006). This arises because Eq. (9.28) implies that the slow MAG waves can propagate provided ω exceeds the acoustic cut-off frequency $[c_s/(2H_\parallel)]$, which holds for 5-minute modes in the photosphere (but not the chromosphere) and for 3-minute modes in both. An alternative

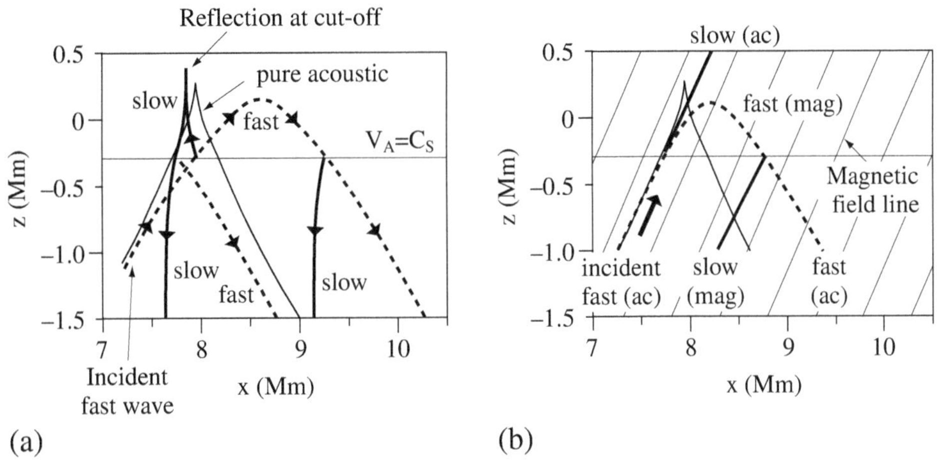

Figure 9.14. Ray diagrams for propagation and mode-conversion of an incident fast-mode MAG wave in a sunspot whose locally uniform magnetic field is (a) vertical and (b) inclined at 30 degrees. Fast-mode waves are dashed and slow-mode waves are thick solid curves. Vertical and horizontal distances from the photosphere are measured by z and x, with the height $z=-0.3$ being the surface where $c_s = v_A$. The ray path when there is no magnetic field is shown as a light solid curve and labelled "pure acoustic" (courtesy Hannah Schunker).

explanation for 3-minute oscillations is in terms of the trapping of slow-modes in a chromospheric resonator (Botha et al. 2011).

Running penumbral waves are a largely chromospheric phenomenon and are something of a puzzle, although an attractive suggestion is that they are another manifestation of slow MAG waves propagating upwards along the inclined magnetic field (Bloomfield et al. 2007; Felipe et al. 2010). Thus, the main difference in ray diagrams in the inclined field of the penumbra is that the scale-height ($H_{\parallel}$) along the field is larger and so the cut-off frequency $[c_s/(2H_{\parallel})]$ for slow 5-min MAG waves is so low that they can propagate outwards along the penumbra (Figure 9.14b).

9.6 Evolution of a Sunspot

9.6.1 Formation

A reasonable scenario for the growth and decay of sunspots is as follows (Meyer et al. 1974). To begin with, a magnetic flux tube is buoyed up to the photosphere as an omega-shaped structure with a field strength of over 1,000 G. Such a flux tube is strong enough to halt overturning convection, and so plasma in the tube cools and falls (Figure 9.15a). The field increases to a value at which the external gas pressure roughly balances the internal magnetic pressure (Eq. 9.22), implying a field strength of 3,500 G for a pressure of $5{\times}10^4$ N m^{-2} at a depth of 200 km. An annular convection cell (or *moat*) is formed around the spot with an upflow at the flux tube and an outflow at the photosphere, having the appearance of a large and persistent supergranulation cell with the spot located at its centre. The moat cell is driven by the leakage of heat below the spot and allows the spot to survive for many weeks while it slowly decays away. Perhaps there is also a counter-rotating collar flow below the surface helping to stabilise the spot (Hurlburt and Rucklidge 2000).

9.6.2 Decay

During the slow decay phase, a sunspot is likely to have the structure shown in Figure 9.15b (Meyer et al. 1974). Outside the spot, small-scale overturning convection in the annular moat flow can be represented

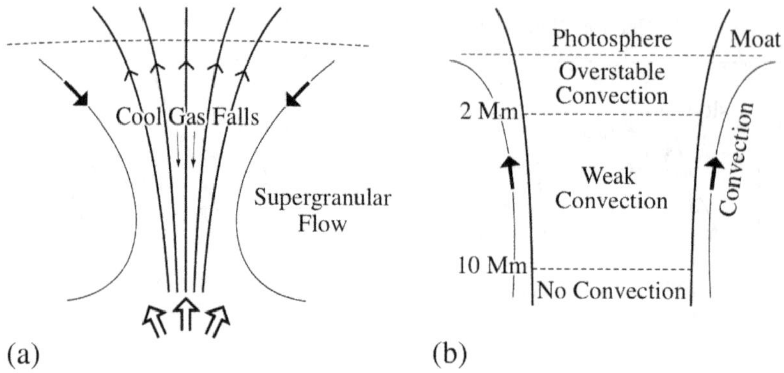

Figure 9.15. (a) The growth of a sunspot. (b) The slow decay of a long-lived sunspot.

by eddy diffusion coefficients. Inside the flux rope, convection is less efficient at transporting heat. In the top 2 Mm of the spot, $\kappa > \eta$ and overstable oscillations produce motions in columns parallel to the magnetic field. At a depth of 2 to 10 Mm, where $\kappa < \eta$, small-scale (leak) convection has a vigour limited by the magnetic field. For example, at a depth of 6 Mm, a cell-size (d) of about 2 Mm, a temperature difference (ΔT) of 50 K and a vertical plasma speed of 20 m s^{-1} give a heat flux (10^7 W m^{-2}) comparable with the observed umbral value.

The slow decay of a sunspot cannot be due to simple Ohmic diffusion, which is far too slow. For example, a scale-length of 3 Mm and a magnetic diffusivity (η) of 300 m^2 s^{-1} would give a diffusion-time $\tau_d = l^2/\eta \approx 1000$ yr. Meyer et al. (1974) suggest instead that magnetic flux leaks slowly from the spot due to an *eddy diffusivity* ($\tilde{\eta}$) and is then carried across the moat by the surrounding annular circulation.

The diffusion of a vertical field $[B(R,t)\hat{\mathbf{z}}]$ may be modelled by the vertical component of the induction equation in cylindrical polars and with no flow, namely,

$$\frac{\partial B}{\partial t} = \frac{\tilde{\eta}}{R}\frac{\partial}{\partial R}\left(R\frac{\partial B}{\partial R}\right).$$

A self-similar solution [in terms of the total flux ($F_{m(0)}$) out to infinity] is

$$B = \frac{F_{m(0)}}{4\pi\tilde{\eta}t}\exp\left(-\frac{R^2}{4\tilde{\eta}t}\right). \tag{9.29}$$

If the edge of the spot $[R = a(t)$, say] is defined such that the field reaches a value B_s, say, the sunspot flux becomes $F_m = \int_0^a 2\pi B R dR = F_{m(0)} - 4\pi\tilde{\eta}B_s t$, whose rate of decrease is $-dF_m/dt = 4\pi\tilde{\eta}B_s$. The fact that this is constant agrees with observations, which give $-dF_m/dt = 1.2\times10^8 B_{max}$ m^2 s^{-1}, in terms of the maximum field strength (B_{max}) in tesla. A value for $\tilde{\eta}$ may be deduced by adopting values of, say, $B_s = 1.5$ kG and $B_{max} = 3$ kG. The result is $\tilde{\eta} \approx 2\times10^7$ m^2 s^{-1}, which is consistent with a rough estimate for the eddy diffusivity due to convection at a depth of 6 Mm. However, a full explanation of sunspot decay needs to address the cause of such an eddy diffusivity, the nature of moving magnetic features and the reason why leader spots are so much more stable than followers (Martínez Pillet 2002), as well as studying further the role of instability and twist.

9.7 Unified Model

In a major advance, Rempel presented remarkable numerical simulations of a sunspot embedded in a realistic convection zone that provide a unified explanation for the global and fine-scale structure of a sunspot solely in terms of magnetoconvection in a non-uniform field. He built on earlier work (Heinemann

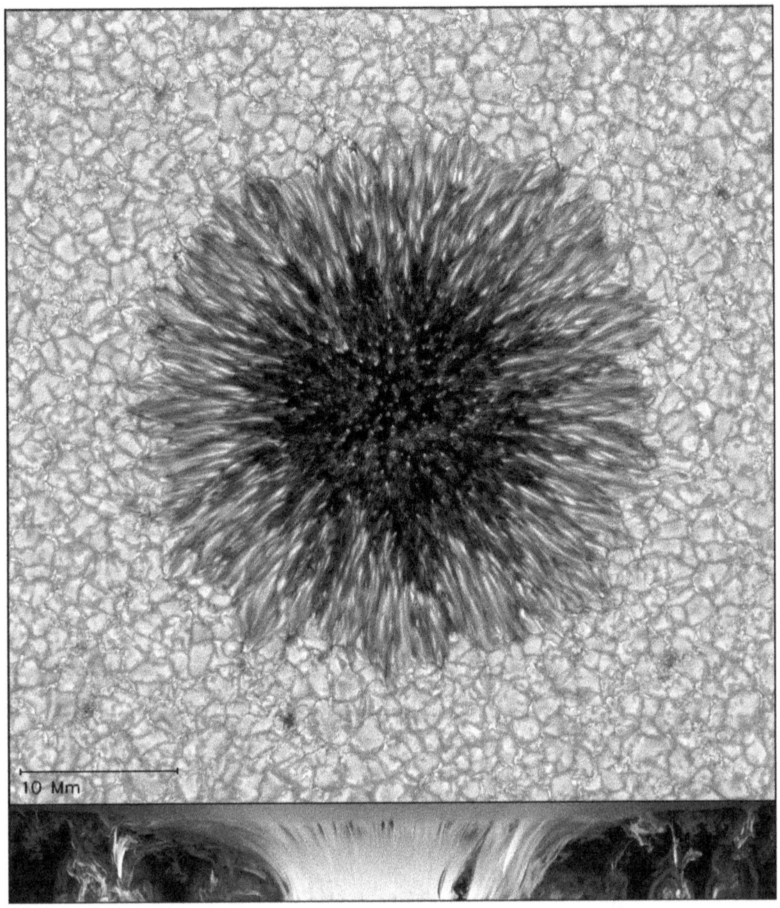

Figure 9.16. Snapshot from a numerical sunspot simulation, showing (above) surface brightness and surrounding granulation and (below) the magnetic field strength (greyscale) in a vertical section with the vertical direction stretched by a factor of 2 (from Rempel 2011).

et al. 2007) and used a 3D compressible MHD code with partial ionization and radiative transfer to model a sunspot having an umbral radius 10 Mm and penumbral width 10 Mm. The side boundaries are periodic, while the top has an imposed magnetic field direction and the bottom an imposed vertical magnetic field. He started with a slab model of a sunspot (Rempel et al. 2009), in which a coherent picture first appeared, and later simulated a pair of opposite-polarity sunspots (Rempel et al. 2009), which produced Evershed flow. This was followed by simulation of a single sunspot with better resolution (3,072×3,072×512 grid points) and a more realistic penumbra, as shown in Figures 9.16 and 9.17 (Rempel 2011, 2012).

While there is room for improving upper boundary conditions and outer penumbral structure, the simulated sunspots are consistent with observations of overall umbral and penumbral structure and average profiles of intensity, velocity and magnetic field, as well as many umbral and penumbral fine-scale properties. The intensity is found to obey the relation $I \sim \sqrt{(v_z^{rms})}$, where v_z^{rms} is the root-mean-square vertical velocity at $\tau = 1$. The umbra is dark and contains umbral dots. It has a brightness of 0.15 to 0.2 I_0 (where I_0 is the surrounding quiet-Sun brightness), a Wilson depression for the visible surface of 550 to 600 km and vertical fields up to 4 kG.

The model penumbra possesses an average brightness of 0.7 I_0, horizontal fields of peak value 2 kG in the inner penumbra and a filamentary structure coupled with Evershed outflow. Its radial extent depends

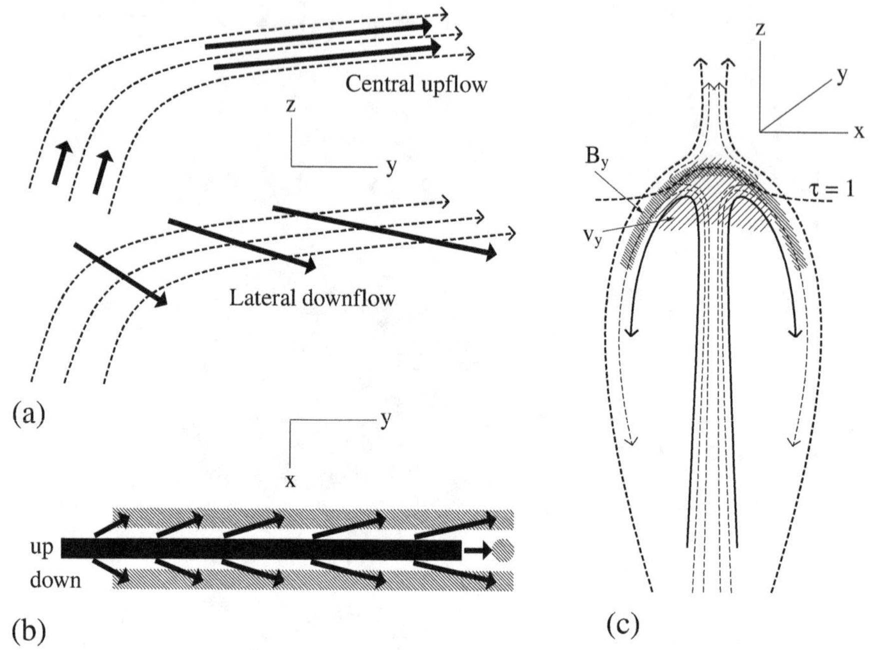

Figure 9.17. A sketch of the structure of a penumbral filament, as deduced from Rempel's simulations with magnetic field (dashed) and velocity (solid). (a) Side view of the central upflow region (where mean flow and magnetic field are aligned) and lateral regions (where a downflow is submerging the magnetic field). (b) Top view of the central upflow region moving outwards (Evershed flow) and mass returning downwards in (shaded) lateral regions plus a small (shaded) region of downwards-pointing field at the outer end. (c) Cross section of the strong overturning convection. A core of vertical (dashed) field lines spreads out, with some of them continuing upwards and some being bent back down. Horizontal field is present everywhere, but Evershed flow is focussed in a shaded region near $\tau = 1$. It is associated with a horizontal field (B_y) that is twice as strong as normal and spreads out in a thin boundary layer (dark shading) by the overturning motions (after Rempel 2012).

on the top boundary condition and increases with magnetic field inclination, a value greater than that of a potential field being needed for a realistic penumbra. A deep magnetic structure lies under the visible penumbra, so that it is not a shallow structure. Indeed, roughly the same amount of vertical magnetic flux is present in the penumbra as in the umbra. Above $\tau = 1$, the penumbral field is rather smooth since $\beta << 1$, and so plasma pressure cannot maintain strong field inhomogeneities. However, for a depth of 2 Mm below $\tau = 1$, the field is highly inhomogeneous due to hot rising magnetoconvective plumes. As the plumes rise, they expand and reduce the field strength before overturning. The rising convective plumes produce penumbral filamentation when the average magnetic field inclination to the vertical exceeds 45 degrees. The plumes are similar to those that produce umbral dots (Schüssler and Vögler 2006), but are modified by a strong horizontal field component. Vertical up- and down-flows occur in narrow regions throughout the penumbra.

Penumbral filaments are thus slender features about 2 Mm deep and a few 100 km wide with significantly reduced field strength. They are neither fully developed convective rolls nor intrusions of field-free plasma from below, nor embedded flux tubes disconnected from their environment. Instead, they develop in the inner penumbra in strong-field regions as a natural consequence of magnetoconvection in an inclined field. Deep-reaching upflow and overturning convection make the model distinct. In the inner penumbra, the magnetic field inclination varies strongly between 40 and 90 degrees over scales of 200 km. In the outer penumbra, the more horizontal regions dominate and downward convective pumping by granules may well play a role in holding down the magnetic field and keeping it fairly horizontal (Weiss et al. 2004).

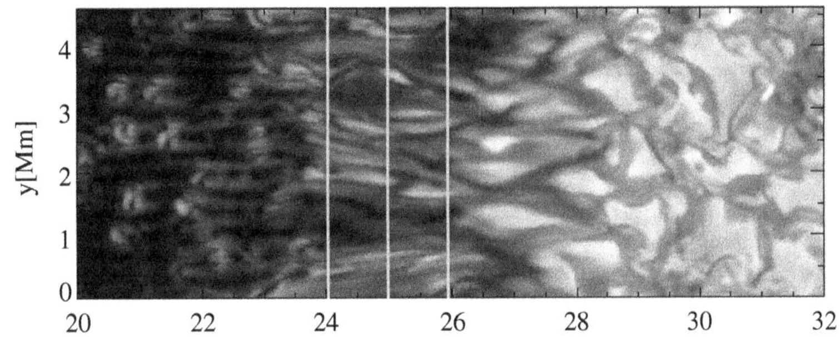

Figure 9.18. Close-up of a snapshot from a numerical simulation of a sunspot, showing umbral dots on the left, penumbral filaments (a few hundred km wide) with dark cores (only 100 km wide) in the centre and granulation on the right (from Rempel et al. 2009).

Rempel also gives an explanation for Evershed flow from his simulations. He finds strong outflows of up to 6 km s^{-1} along the magnetic field in penumbral filaments wherever the field inclination to the vertical exceeds 45 degrees. The outflows are concentrated in the near-surface layers due to the strong vertical pressure and density gradients there. With increasing distance from the umbra, the outflow and magnetic field inclination increase and they become almost horizontal in the outer penumbra. The outflow is therefore concentrated in the more horizontal filaments and becomes stronger and more extended azimuthally in the outer penumbra. In the azimuthal and vertical directions, the dominant terms in the equation of motion are a balance between $\mathbf{j} \times \mathbf{B}$ and $\boldsymbol{\nabla} p$, whereas in the radial direction the acceleration is provided by the $\mathbf{j} \times \mathbf{B}$ force. Thus, Evershed flow is the horizontal component of penumbral convection and a natural consequence of magnetoconvection in an inclined field. However, the usual explanation in terms of siphon flow is also appealing if the pressure difference that drives the flow is provided by the magnetoconvection.

Regarding energy transport, granular and penumbral convection are very similar, apart from the radial elongation of penumbral convection. Rising hot plasma turns over and drives a horizontal flow that is guided out from the spot by the strong inclined magnetic field. In the umbra, the strong vertical field allows convection only in the form of narrow columns, but in the penumbra the inclined field favours upflows in the form of sheets that are extended in radius and narrow in the azimuthal direction. In the inner penumbra, such sheet-like upflows and the resulting weakening of the field by flux expulsion create large azimuthal variations of field inclination. In the outer penumbra, the outflows become strong enough to bend over the field lines and create more extended but localised patches of horizontal field and flow.

Also, a general weak moat-like outflow is present, transporting flux away from the sunspot. It is driven mainly (2/3) by the build up of heat flux below the spot and partly (1/3) as a geometrical effect of placing granules around a circular object.

Many of the penumbral filaments in the simulation possess dark cores or lanes near optical depth unity ($\tau = 1$) with strong outflows and lifetimes of about an hour (Figure 9.18) that were discovered observationally by Scharmer et al. (2002). The dark lanes appear above the strongest upflows with downflows on both sides. They are caused by an upward bulging of the $\tau = 1$ surface and a piling up of plasma in a cusp-shaped region at the top of the filament – above this, the more vertical field outside the filament is more homogeneous. At the umbral end of such a dark core there is a bright head which moves inwards into the umbra, due to the expansion of the sheet-like upflow plumes along the filament.

Chapter 10

Heating of the Upper Atmosphere

10.1 Introduction

Understanding how the chromosphere is heated to 10^4 K and the corona to 10^6 K, by comparison with the photosphere at 6,000 K, is a major challenge in astronomy (Klimchuk 2006; Reale 2010; Parnell and De Moortel 2012). Coronal heating is only a few per cent of chromospheric heating, so a comprehensive model needs to treat the generation, propagation and dissipation of energy through the whole atmosphere. This includes the continual transfer to and fro of mass and energy between chromosphere and corona and the fact that transition-region plasma is material that happens to be heating up and cooling down dynamically through 10^5 K.

Although the present chapter focuses mainly on coronal heating, many of the principles also apply to the chromosphere and some mechanisms work even more effectively there: for example, the tectonics model generates more heat low down in the atmosphere, where the magnetic field is stronger and more complex (Sec. 10.4.5). Until the space age, the main way of viewing the Sun's outer atmosphere was during a solar eclipse (Figures 1.21 and 10.1), when the corona appears with beautiful structures dominated by the magnetic field. Since then, great progress has been made in coronal observations from space, in refining the questions about the nature of the corona, and in proposing viable models.

After an introduction stressing the complexity of the corona (Sec. 10.1.1), this chapter reviews the theory of coronal loops (Sec. 10.2) and of how the corona may be heated by waves (Sec. 10.3) or current sheets (Sec. 10.4). It also describes space observations of loops (Sec. 10.2.1), wave heating (Sec. 10.3.4) and reconnection (Sec. 10.4.2).

The corona is filled with magnetic field. It has a diffuse background together with a threefold structure of *coronal holes, coronal loops* and *X-ray bright points* (Sec. 1.6.2). It is an MHD world, with myriads of coronal loops continually interacting with one another. Thus, an observer of a particular coronal loop should not be misled into regarding it as an isolated structure; rather, it is coupled to its magnetic environment in subtle, complex ways. In magnetically open regions (coronal holes), the plasma is streaming rapidly outwards as the fast solar wind, but in magnetically closed regions the magnetic field is able to contain the plasma at a higher density. Active regions lying above sunspot groups show up as a highly complex collection of loops in soft X-rays or EUV (Figures 10.5, 1.21 and 1.25), which outline the magnetic field because of the efficiency of heat conduction along field lines.

The energy required to heat the corona (see Table 10.1) is typically 300 W m^{-2} (3×10^5 erg cm^{-2} s^{-1}) in the quiet Sun and 5000 W m^{-2} in an active region (Withbroe and Noyes 1977). Corresponding values for the chromosphere are 4,000 W m^{-2} (quiet Sun) and $2{\times}10^4$ W m^{-2} (active region). More recent coronal estimates are 100 W m^{-2} for the quiet Sun and 10^4 W m^{-2} in an active-region core (Katsukawa and Tsuneta 2002). One indication that the energy comes from the magnetic field is that the hotter loops in general possess a stronger magnetic field (Fisher et al. 1998). Another is that the energy flux at the photosphere due to horizontal motions of coronal magnetic footpoints is plentiful.

Figure 10.1. The corona viewed in white light during the eclipse of 1 August, 2008 (courtesy M Druckmüller, P Aniol and V Rusin).

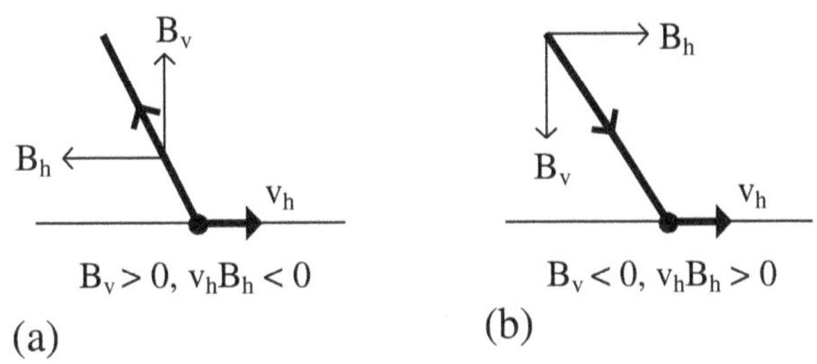

Figure 10.2. The trailing of a magnetic field line with vertical (B_v) and horizontal (B_h) field components when its footpoint (large dot) is moved by a horizontal velocity (v_h), showing that (a) when $B_v > 0$ then $v_h B_h < 0$, whereas (b) when $B_v < 0$ then $v_h B_h > 0$.

Since $\mathbf{E} = -\mathbf{v} \times \mathbf{B}$ in ideal MHD, the Poynting flux ($\mathbf{F}_H$) through the photosphere into the corona is

$$\mathbf{F}_H \equiv \frac{\mathbf{E} \times \mathbf{B}}{\mu} = -\frac{(\mathbf{v} \times \mathbf{B}) \times \mathbf{B}}{\mu} = -\frac{\mathbf{B}(\mathbf{v} \cdot \mathbf{B})}{\mu} + \frac{\mathbf{v}B^2}{\mu}.$$

Thus, if $\mathbf{v}$ has only a horizontal component v_h, the vertical component (F_{Hv}) of the Poynting flux is

$$F_{Hv} = -\frac{v_h B_h B_v}{\mu}, \tag{10.1}$$

where B_h, B_v are the horizontal and vertical components of the magnetic field. Whether there is an inflow of energy to the corona ($F_{Hv} > 0$) or an outflow from it depends on the sign of ($v_h B_h B_v$). However, random motions cause field lines to be wrapped around one another, so that they tend to trail behind the moving footpoints (Figure 10.2). This tilting of field lines backward away from the direction of motion implies that, if $B_v > 0$ then $v_h B_h < 0$, whereas if $B_v < 0$ then $v_h B_h > 0$, so that $F_{Hv} > 0$ and the Poynting flux into the corona is always positive.

Typical mean vertical fields (B_v) are 100 to 500 G in active regions and 5 to 30 G in the quiet Sun, while flux tubes wander about over the solar surface with speeds of 0.1 to 1 km s^{-1}. Thus, adopting a typical v_h of 0.1 km s^{-1}, B_v of 200 G and B_h of 100 G would give a Poynting flux of 10^4 W m^{-2}. However, although B_v is measured well, the value of B_h is highly uncertain and depends on the nature of the coronal currents, and therefore on the heating mechanism itself.

According to Section 2.3.5, the Poynting influx of energy through the photosphere into the chromosphere and corona (which also includes the effect of emerging flux through vertical motions) can be written as

$$\int \frac{\mathbf{E}\times\mathbf{B}}{\mu}\cdot d\mathbf{S} = \int \frac{j^2}{\sigma} + \mathbf{v}\cdot\mathbf{j}\times\mathbf{B} + \frac{\partial}{\partial t}\left(\frac{B^2}{2\mu}\right) dV,$$

and so it can perform several functions: heating the plasma by Ohmic dissipation, accelerating the plasma by the $\mathbf{j}\times\mathbf{B}$ force and raising the magnetic energy content.

In the corona, the plasma velocity is generally much smaller that the Alfvén speed, which is typically 1,000 km s^{-1} [e.g., for a coronal field (B) of 10 G and a density (n) of 10^{15} m^{-3}]. When the magnetic force dominates, this implies that the forces are in approximate equilibrium. The ratio of the pressure gradient to the magnetic force is the plasma beta (with T in K)

$$\beta \approx \frac{3.5\, nT}{10^{15} B^2} \quad (n \text{ in cm}^{-3},\ B \text{ in G}), \qquad \text{or} \qquad \beta \approx \frac{3.5\, nT}{10^{29} B^2} \quad (n \text{ in m}^{-3},\ B \text{ in T}).$$

This is of order unity in the high corona (where $n = 10^{14}$ m^{-3} $= 10^8$ cm^{-3}, $B = 2\times10^{-4}$ T $=$ 2 G, $T = 10^6$ K). However, in the low corona or in active regions (where $n = 10^{15}$ m^{-3}, B = 200 G, $T = 2\times10^6$ K, say), $\beta \ll 1$, and so, since the magnetic force dominates, the equilibrium is force-free with

$$\mathbf{j}\times\mathbf{B} = \mathbf{0}. \tag{10.2}$$

Several different heating mechanisms are probably at work in different parts of the chromosphere and corona, and so the challenge is to refine the current models to estimate quantitatively their relative importance. Two general classes of model have been proposed, namely, MHD waves (Sec. 10.3) and magnetic reconnection (Sec. 10.4). In both cases, this chapter treats Ohmic dissipation, but it must be recognised that the current layers are so thin that resistive MHD breaks down. Dissipation is in reality occurring at kinetic scales, so that in future more work on collisionless models is needed (see, e.g., Birn and Priest 2007; Shay et al. 2007; Drake et al. 2009).

MHD waves may dissipate by phase mixing or resonant absorption. Reconnection can heat the plasma either directly by Ohmic heating or indirectly by generating waves or plasma jets that subsequently dissipate ohmically or viscously (Sec. 10.4.5.6). The complex nature of the coronal heating process (whether waves or reconnection) can also be described in terms of MHD turbulence (Sec. 10.5); indeed, the observed coronal non-thermal line broadening (20 to 30 km s^{-1}) represents unresolved flows that could come from either reconnecting current sheets or superposition of waves (De Moortel and Nakariakov 2012; McIntosh 2012).

One part of the coronal heating problem appears to have been solved, since X-ray bright points are probably heated (according to the Converging Flux Model) by coronal reconnection driven by footpoint motions (Sec. 10.4.3). However, the heating mechanisms of coronal loops and holes are at present unknown. A strong candidate for the outer corona is high-frequency ion-cyclotron or kinetic Alfvén waves, which may also be driving the fast solar wind and explaining the huge line-broadening seen with the UVCS instrument on SoHO (Kohl et al. 2006) (Sec. 13.7).

In most of the low corona (especially active regions), reconnection is widely regarded as the most likely mechanism for coronal heating, since SoHO, Yohkoh, Hinode and SDO observations have given a wide range of evidence of reconnection at work in the corona (Sec. 10.4.2), such as explosive events, jets, spicules, blinkers, microflares, tornadoes and active-region transient brightenings (Sec. 1.7.1.3). Several ways have been proposed to heat the chromosphere and corona by magnetic reconnection at myriads of current sheets, namely: driven reconnection (Sec. 10.4.3), turbulent relaxation (Sec. 10.5.1) and flux-tube tectonics (Sec. 10.4.5), which grew out of the earlier braiding ideas (Sec. 10.4.4) and incorporates binary reconnection, separator heating and separatrix heating. Whereas braiding acts on a simple background magnetic field, tectonics incorporates the complexity of the chromospheric and coronal field (i.e., the *magnetic carpet*, Section 1.4.2, Figures 9.7 and 10.5). However, observations of coronal wave motion and swaying type II spicules suggest they also may play an important role in coronal heating (Sec. 10.3.4).

10.1.1 Complex Nature of Chromospheric and Coronal Magnetic Field

The atmospheric magnetic field is incredibly complex (Figures 1.20 to 1.23), both temporally (Sec. 10.1.1.1) and spatially in closed (Sec. 10.1.1.2) or open regions (Sec. 10.1.1.3). The spatial complexity may be described in terms of the *magnetic skeleton* (Sec. 6.9.4, Priest et al. 1996), which consists of a series of null points, *separatrices* (surfaces of field lines that generally originate from the fans of nulls and separate topologically distinct regions of space) and *separators* (field lines which join one three-dimensional null point to another and represent the intersection of two separatrices). Current sheets tend to form and dissipate most readily at separators, where the magnetic connectivity of coronal footpoints is discontinuous, but they can also do so at null points (Priest and Pontin 2009) and at *quasi-separatrices*, where the magnetic connectivity has steep gradients (Priest and Démoulin 1995; Titov et al. 2002; Aulanier et al. 2006).

10.1.1.1 Chromospheric and Coronal Recycling Time

Photospheric sources of the atmospheric magnetic field are highly fragmentary and concentrated into intense flux tubes threading the solar surface (Figure 1.17). They are also highly dynamic, with magnetic flux emerging continually in the quiet Sun and undergoing fragmentation, merging and cancellation, so that the medium-scale quiet Sun photospheric flux is reprocessed very quickly, in only 14 hours (Hagenaar 2001).

Close et al. (2004) wondered what the corresponding atmospheric reprocessing time is, and used observed quiet-Sun magnetograms from the MDI instrument on SoHO to construct the overlying magnetic field lines, under the simplifying assumption that the field is potential. (In practice, the field may well be close to potential, especially in a time-averaged sense.) They then tracked the motion of individual magnetic fragments in the magnetogram and recalculated the potential atmospheric field lines and their connectivity. In so doing, they discovered the startling fact that the time for all the field lines in the quiet Sun to change their connections is only 1.4 hours. In other words, if their assumption of evolution through potential states is reasonable, an incredible amount of reconnection is continually taking place – indeed, enough to provide the required heating of the chromosphere and corona.

It will be interesting in future to recalculate the recycling time using MHD simulations and higher-resolution magnetograms, which is likely to lead to even more complex structure and shorter recycling times.

10.1.1.2 Loop Structure

Magnetic complexity has been demonstrated in several studies. Close et al. (2004) evaluated the statistical properties of potential field lines that are extrapolated from observed photospheric magnetograms. For the region they considered, 50 per cent of the flux closed down within 2.5 Mm of the photosphere and 90 per cent within 25 Mm, the remaining 10 per cent extending to larger distances or being open (Figure 10.3).

The importance of this result is that it shows, as expected, that low down in the solar atmosphere the field is much more complex than higher up. Thus, in the coronal part of a coronal hole or simple active region, the magnetic field is expected to have a simple structure, but, down in the underlying photosphere, the presence of internetwork granular magnetic loops (Sec. 1.4.2) implies a highly complex structure.

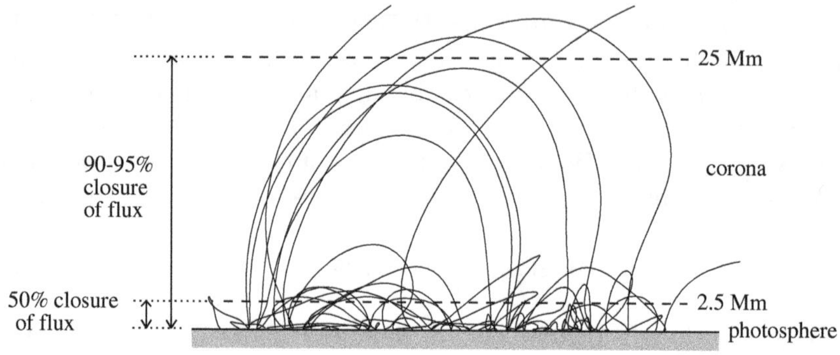

Figure 10.3. Magnetic field lines in the quiet Sun, extrapolated from a SoHO/MDI photospheric magnetogram (Close et al. 2004).

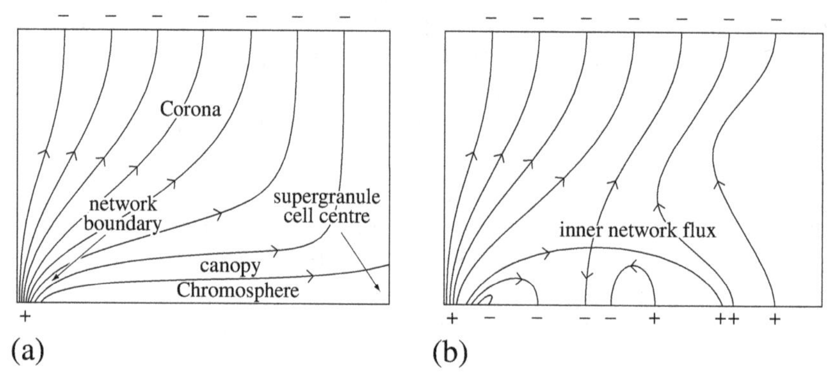

Figure 10.4. Magnetic field lines above half a supergranule cell for (a) the Gabriel model with the photospheric flux concentrated in the network at the left-hand corner and (b) a better model with half the photospheric flux located in the cell interior. The field lines start at positive polarity on the base and end at negative polarity on the top or bottom, as shown.

Quantitatively, it will be interesting to repeat the calculation of Close et al. (2004) with higher-resolution photospheric magnetograms for different regions in order to determine the height-variation through the chromosphere and corona of the complexity in coronal holes, the quiet Sun and active regions.

Global extrapolations of the field in the corona (e.g., Figure 3.7) also reveal a highly complex structure, which can be analysed by calculating the skeleton (e.g., Maclean and Priest 2007; Titov et al. 2011).

10.1.1.3 Structure above a Supergranule Cell

Gabriel (1976) proposed a model for the atmosphere above a (quiet-region) supergranule cell. In the photosphere, magnetic flux is concentrated at supergranule boundaries by the convective flow (Sec. 9.2), but higher up in the atmosphere, the flux expands until it has become relatively uniform in the corona (Figure 10.4a). Thus, images formed in transition-region lines follow the supergranulation pattern, with intensities over supergranule boundaries about a factor of ten higher than those over cell centres, while in the corona the pattern has disappeared.

Gabriel assumes that the magnetic field is potential with all the photospheric flux concentrated at cell boundaries (Figure 10.4a). He next assumes the plasma is in hydrostatic equilibrium and in thermal equilibrium under a balance between conduction and radiation. This enables him to solve the energy-balance equation along each field line to give the temperature and density structure everywhere.

In the Gabriel model, the outermost field line from the network forms a canopy over the field-free cell interior, but the internetwork (inner network) field (Sec. 1.4.2) may provide much of the flux in the quiet Sun and puncture the canopy or restrict it to a collar of closed loops around the network (Schrijver and Title 2003). An example, in which it provides half the flux, is sketched in Figure 10.4b, but the total interior flux may well be larger, thus restricting the collar even further. In three dimensions, with many small sources of flux in the cell interior, the field overlying a supergranule cell is therefore much more complex than in the earlier Gabriel picture. Furthermore, the rapid evolution of the internetwork field and the resulting reconnection could make a substantial contribution to coronal heating and solar wind acceleration.

10.2 Chromospheric and Coronal Loops

The solar atmosphere is highly complex and dynamic. It is structured by gravity and the magnetic field. At a microscopic level, where the magnetic energy is probably being converted in myriads of thin current sheets a few ion gyro-radii thick, a full collisionless plasma treatment is required, but so far only some aspects have been considered (Birn and Priest 2007; Drake et al. 2009). At a macroscopic level, the MHD of the structure can be modelled in two steps. Firstly, since the magnetic field dominates the equation of motion and coronal loops vary slowly by comparison with the Alfvén time, a force-free configuration is determined by solving $\mathbf{j} \times \mathbf{B} = \mathbf{0}$. Then, secondly, the plasma properties are found by solving the hydrodynamic and energy-balance equations along each field line.

Step 1 is highly nontrivial, since there is a multitude of discrete flux sources in the photosphere, which are still not fully resolved in current magnetograms. Vector magnetographs give the transverse field components in strong-field regions, and now with Hinode in the quiet Sun (Centeno et al. 2007; Lites et al. 2008; Zhao et al. 2009). Furthermore, in general one would like nonlinear solutions to $\mathbf{j} \times \mathbf{B} = \mathbf{0}$, although potential field solutions are often a useful first step and are much easier to find. Linear force-free fields are likely to be of little value, since nonlinear solutions suggest that each magnetic source tends to carry both positive and negative currents, especially when currents are produced by localised footpoint twisting. Another difficulty is that $\mathbf{j} \times \mathbf{B} = \mathbf{0}$ will be a poor approximation as the photosphere is approached, where a full magnetohydrostatic solution is preferable.

Step 2, to determine the plasma structure along each field line, is also a tricky business. Even if one assumes a steady state, one problem is to determine the form of the heating term in the energy-balance equation; another is to resolve the very narrow transition region and a third is to link the corona properly to the chromosphere and photosphere, where the optically thin radiative loss assumption fails and the nature of the lower boundary condition is uncertain. Added to these difficulties, the whole atmosphere is highly time-dependent. Thus, for example the transition region is certainly not a thin static layer between chromosphere and corona but is dynamic plasma that is either heating up to coronal temperatures or cooling down towards chromospheric ones.

The magnetic field influence on chromospheric and coronal plasma is threefold.

(i) *It exerts a force that is usually in balance with itself.* The condition $\mathbf{j}\times\mathbf{B} = \mathbf{0}$ determines the shapes of loops, which contain plasma with an enhanced pressure in features such as X-ray bright points, coronal loops and active regions.

(ii) *It stores energy.* The energy $[B^2/(2\mu)$ per unit volume$]$ that is stored in the magnetic field is a source of heating due to Ohmic dissipation (j^2/σ) of electric currents and viscous dissipation of flows.

(iii) *It channels heat.* The coefficient of thermal conduction $(\kappa_\parallel)$ along the field is much larger than the coefficient $(\kappa_\perp)$ across the field, so the magnetic field acts as a "blanket" and thermally insulates the plasma very effectively. Heat is constrained to flow largely along the field, so that in the transition region and corona, where conduction is an important means of energy transport, the temperature and density are strongly affected by the structure of the magnetic field. This explains why fine coronal structures in eclipse, EUV and

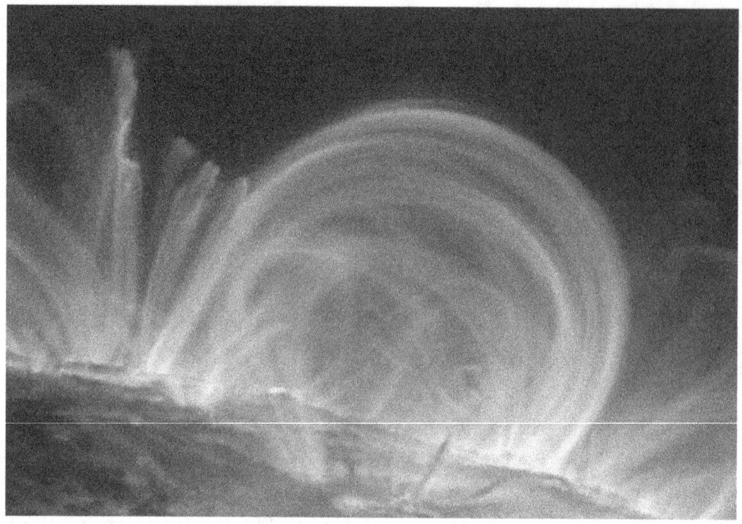

Figure 10.5. Coronal loops imaged by the NASA TRACE instrument at 171 Å. The loops outline plasma that happens to be at the particular temperature (1 MK) to which the 171 Å line is sensitive. The loops are very narrow because the heating events are tiny and heat conduction is essentially along magnetic field lines ($\kappa_{\|} \gg \kappa_{\perp}$) (courtesy Alan Title).

> X-ray images often outline the magnetic field, so that the corona (outside coronal holes) consists largely of multi-thermal loops.

An early classification of coronal loops was by morphological type, namely, *interconnecting loops*, *quiet-region loops*, *active-region loops*, *'post'-flare loops* (an example being shown in Figure 10.5) and *simple (compact) flare loops*. A more recent classification is by temperature regime into *cool loops* (with 0.1 MK $< T <$ 1 MK, detected in UV lines by, e.g., SoHO), *warm loops* (with 1 MK $< T <$ 1.5 MK, well observed by EUV imagers such as SoHO/EIT and TRACE) and *hot loops* (with $T >$ 2 MK, observed in soft X-rays and hot UV lines such as Fe XVI). General observational characteristics are summarised in Sections 1.6.2 and 10.2.1, and Martens (2010) and Reale (2010). Physical properties are described in Section 10.2.2, followed by models for static (Sec. 10.2.3) and dynamic (Sec. 10.2.4) loops. These lead on to a method for deducing the form of heating (Sec. 10.2.5) and 3D computational experiments (Sec. 10.2.6).

10.2.1 Observations from Space

Following earlier rocket flights, the S-054 X-ray telescope on Skylab (Vaiana et al. 1973) imaged the corona in soft X-rays and revealed loop structure at 2 to 5 MK. More details were added at higher spatial resolution and with lower stray light by the Soft X-ray Telescope (SXT) on Yohkoh (Tsuneta et al. 1991) and at a resolution of 2 arcsec by the X-ray Telescope (XRT) on Hinode (Golub et al. 2007) (Figure 1.21). A summary of solar instruments from the ground or space can be found on the web page at www.cambridge.org/9780521854719.

Coronal plasma at 1 to 2 MK has also been imaged in EUV emission lines. Full-disc images were obtained for a full solar cycle by EIT on SoHO (Delaboudinière et al. 1995). There followed local images at 1 arcsec with TRACE (Handy et al. 1999) (Figure 1.25), full-disc images with AIA on SDO (Lemen et al. 2012) and images with a five-fold increase in spatial resolution from the Hi-C (High-resolution Coronal Imager) instrument on a sounding rocket (Figure 1.23). These cooler EUV coronal loops are much more finely structured than the fuzzier soft X-ray loops.

The lengths of loops vary from 1 Mm for X-ray bright points to 10 Mm for small active-region loops, 100 Mm for typical active-region loops, and up to 1,000 Mm for giant arches. Their temperatures may be only 10^5 K for cool loops or a few MK for active-region loops or a few times 10 MK for flaring loops. Typical densities of bright loops in active regions are 10^{15} to 10^{16} m^{-3}, but they may be a factor of 10 higher in flaring loops. The corresponding pressures lie between 10^{-4} and 1 N m^{-2}, corresponding to confining magnetic fields between 0.1 and 10 G.

Hinode/EIS has revealed complex flows in active regions (Del Zanna 2008). Cooler lines at 1 MK show persistent redshifts of 5 to 30 km s^{-1} in most loops. Hotter lines at 2 MK have persistent blueshifts of 5 to 30 km s^{-1} in areas of weak emission. Active regions possess downflows in the core and upflows at the boundary between low-lying 3 MK loops and higher 1 MK loops (Tripathi et al. 2009). Rapid localised outflows at 50 km s^{-1} are seen in Doppler maps (Doschek et al. 2008), and there is a strong correlation between flows and nonthermal line widths (Hara et al. 2008).

At solar maximum, the coronal soft X-ray and EUV emission is dominated by active regions, which account for over 80 per cent of the heating. Typically, active regions contain bright *hot loops* in their cores at 3 to 5 MK and longer *warm loops* on their peripheries at 1 to 2 MK. The hot loops are relatively steady over times longer than the radiative and conductive cooling times, whereas the warm loops are more dynamic, often exhibiting cooling collapse or jets and siphon flows. The footpoints of brighter hot soft X-ray loops contain EUV emission at 1 to 2 MK called *moss* (Berger et al. 1999; Fletcher and De Pontieu 1999).

Active-region cores are thought to be heated by *nanoflares* (Sec. 10.4.2) with an energy of 10^{17} J (10^{24} erg) or less. However, a key question is whether they are high-frequency or low-frequency nanoflares; in the former case, the repeat-time is shorter than the cooling time, so that strands never cool to 10^5 K before being reheated. The picture that is emerging is of high-frequency events early in the life of an active region (Antiochos et al. 2003; Warren et al. 2010) and low-frequency ones later in life, when there are more warm loops (Bradshaw et al. 2012; Tripathi et al. 2012). This is supported by the absence of significant warm (1 to 2 MK) emission in young active-region cores (apart from moss), as well as observations that averaged moss emission from hot-loop feet is constant (Winebarger et al. 2008) and soft X-ray emission from a young active-region core is steady for 6 hours (Warren et al. 2011).

Coronal loops (in EUV or soft X-rays) have aspect ratios of 10 per cent or smaller and radii of several Mm, down to 1 Mm for the thinnest TRACE loops. However, as the flux-tube tectonics model has stressed (Sec. 10.4.5), such loops may in reality consist of many much thinner *strands* (Gomez et al. 1993; Cargill 1994). Due to the small thermal conduction across the magnetic field ($\kappa_\perp$), each strand is thermally isolated from its neighbour, so the energy equation along each strand may be solved independently. When the loops are static and have low plasma beta, the modelling is particularly simple, since the shape of each loop or strand is determined from force-free equilibrium and the hydrodynamics along separate loop is independent.

Currently, coronal loops are not resolved spatially, and so observed soft X-ray and EUV loops are usually not isothermal but multi-thermal (with many temperatures along the line of sight in different strands), as seen with XRT (Reale et al. 2007), EIS (Warren et al. 2008; Schmelz et al. 2011) and SDO/AIA (Schmelz et al. 2011). Furthermore, COMP observations imply the existence of strands with different densities (van Doorsselaere et al. 2008). Indeed, Hi-C images have glimpsed braiding and twisting of such strands in an active-region loop (Figure 1.23). Coronal loops also often possess a fairly constant cross-section along their length (Klimchuk 2000), although that is not necessarily true of the strands within them (DeForest 2007).

Other observations from space that are suggestive of reconnection are described in Section 10.4.2, including X-ray cusps, explosive events, blinkers, microflares, nanoflares, active-region transient brightenings, tornadoes and coronal jets.

Nanoflares (Sec. 10.4.2) are unresolved impulsive energy releases probably due to reconnection, with braiding (Sec. 10.4.4) or tectonics (Sec. 10.4.5) producing impulsive reconnection in fine coronal strands. The width of a strand would be determined by the width of the heating location, and each pulse of heating would heat one long thin strand at a time. Such pulse heating in many strands can be described with an EBTEL model (Sec. 10.2.4.2). If the strands were heated to 10 MK, their detection would be difficult because of the small filling factor, fast cooling and ionisation non-equilibrium. However, the substructuring of coronal loops into many thin strands that are pulse-heated to such temperatures would naturally explain why loops generally live longer than a cooling time and why warm loops (1 MK) are less fuzzy than hot loops (2–3 MK) (Tripathi et al. 2009; Guarrasi et al. 2010), which in turn are fuzzier than hotter loops (6 MK) revealed by SDO (Reale et al. 2011). Furthermore, a brief storm of nanoflares would produce a narrow distribution of temperatures, whereas a long storm would produce a broad distribution (Klimchuk 2009).

The response of an isolated cool loop in equilibrium to a sudden localised transient heating event (such as a *nanoflare*) consists of four physically different phases, which in practise overlap and which can be modelled using the equations of Section 10.2.4:

(I) The temperature increases rapidly, first locally (on the time-scale τ_H for the heating increase) and then spreads through the whole loop by conduction;

(II) The density increases by *evaporation* (i.e., upflow) from the chromosphere and photosphere, since the loop is underdense compared with hydrostatic equilibrium at the new temperature, and so the upward pressure gradient exceeds gravity; if the heating continues long enough, a new hot equilibrium is produced with a balance among conduction, radiation and heating;

(III) After the transient heating ceases, the temperature falls slowly, first by conduction and later by radiation (Cargill et al. 1995) or a downward enthalpy flux (Bradshaw and Cargill 2010); these processes have time-scales $\tau_C \approx L^2 p/(\kappa_0 T^{7/2})$, $\tau_R \sim T^{3/2}/n$ and $\tau_v \approx L/v$, respectively, and so the cooling time depends on which of these three processes dominates (Sec. 10.2.4.2); conduction is important for high temperatures and low densities, radiation dominates at low temperatures and high densities, while enthalpy is important when the downflow speeds are large enough;

(IV) The density decreases by *draining* or *condensation* (i.e., downflow) to the lower atmosphere, since the loop is now overdense relative to hydrostatic equilibrium, and so gravity exceeds the upward pressure gradient.

Phases II and III overlap in practice, since the density continues to rise well after the temperature starts to fall, and most evaporation occurs during Phase III unless the nanoflare is long-lived.

This four-phase response to a nanoflare heating event in an isolated loop needs to be modified to allow reconnection between loops, but it has several observational implications. In Phase I (heating), extremely hot plasma in excess of 4 MK should be produced (Cargill and Klimchuk 1997), but the emission should be weak since the density is low. Such plasma has been detected at 2 MK by Del Zanna (2008) and at 10 MK in active-region cores by Schmelz et al. (2009) and Reale et al. (2009). In Phases II (evaporation) and IV (draining), hot plasma should be moving upwards and cool plasma moving downwards, which is consistent with the observation that hot UV and EUV lines with $T > 0.5$ MK are blue-shifted while cool lines ($T < 0.5$ MK) are red-shifted (Peter and Judge 1999); however, whether a line is red- or blue-shifted does depend on the relative brightness of the evaporating and condensing plasma. It is also consistent with observations from SDO/AIA and NST at Big Bear of upflow from the ends of fine loops that are being filled with hot plasma in heating events (Ji et al. 2012). Another implication is that, since the density evolution lags the temperature evolution in Phases I to III, loops should be underdense (i.e., less dense than hydrostatic) (Cargill and Klimchuk 2004); this is indeed observed for "*hot*" loops (i.e., $T > 2$ MK).

Bradshaw and Cargill (2010) considered the nature of Phases III and IV when enthalpy downflow dominates. For a static loop, the top of the transition region is defined as being the location where the divergence of the heat flux vanishes, so that conduction changes from being a loss term [with heating (H) exceeding radiation (R) in the corona] to a gain term (with $R > H$ in the transition region) (Figure 10.6). However, they suggested that an enthalpy-powered transition region should be defined instead as the location where the divergence of the enthalpy flux vanishes.

So-called '*warm*' loops (i.e., $T \approx 1$ to 2 MK) are observed to be overdense by a factor 10 (Winebarger et al. 2003) and to be multi-thermal with a narrow temperature range (0.7 to 1.3 MK) and a filling factor of only 10 per cent (Warren et al. 2008). They have a temperature profile that is much flatter than hydrostatic and their scale height is too large by a factor of 2, so that their density decreases with height more slowly than expected (Aschwanden et al. 2001). Their lifetime is 1,000 to 5,000 sec, whereas hot loops have a wider range and can persist for many hours (López Fuentes et al. 2007). These warm loops are thought to be highly filamentary with many strands and to be cooling from a hotter state (Warren et al. 2003). Thus, if each strand in a loop undergoes a short-lived heating followed by a longer-lived cooling phase, overall more plasma would be in Phase IV at any time and so the loop would appear overdense (Klimchuk 2006).

10.2.2 Physical Properties of Coronal-Loop Models

In a static model, the temperature increases dramatically from the chromosphere up through a narrow transition region to the corona (Figure 10.6). The temperature gradient increases from small values in the chromosphere to extremely large values in the transition region and then vanishes at the temperature maximum.

The inflexion point in the temperature profile corresponds to the maximum temperature gradient and is located typically near the base of the transition region ($\approx 2 \times 10^4$ K), whereas the inflexion point (at T_i, say) in the profile of $T^{\frac{7}{2}}$ gives the maximum heat flux and occurs typically at about 10^6 K.

Even though the atmosphere is highly time-dependent, a static model helps elucidate important processes at work. Thus, for thermal equilibrium, there is a balance

$$C = H - R,$$

between some kind of heating (H), radiative losses ($-R$) and a conductive loss ($C = \nabla \cdot \mathbf{F}_c$), which is the divergence of the downward conductive flux ($\mathbf{F}_c = -\kappa_0 T^{5/2} \nabla T$). Below T_i, C is negative, so that conduction deposits heat, and radiation (R) exceeds heating (H), whereas above T_i, C is positive and heating dominates radiation.

In thermal equilibrium, the temperature is determined by the relative sizes of R and H and the response of C. In the chromosphere, R and H are both relatively large, while their difference (C) is small, so that the spatial change in ∇T is slow. As T rises, the radiation (R) increases to a maximum between 10^4 and 10^5 K (see Figure 2.2); here $R \gg H$ and is balanced by C. This in turn forces ∇T to increase to large values in the lower transition region (which is therefore being heated by conduction from above and cooled by radiation). Above about 10^5 K, the radiation falls dramatically and eventually reaches equality with heating. Then, through the upper transition region, the conductive flux $\mathbf{F}_c$ stays relatively constant and at a high value. In the lower corona, R and H have fallen to much smaller values and H is balanced mainly

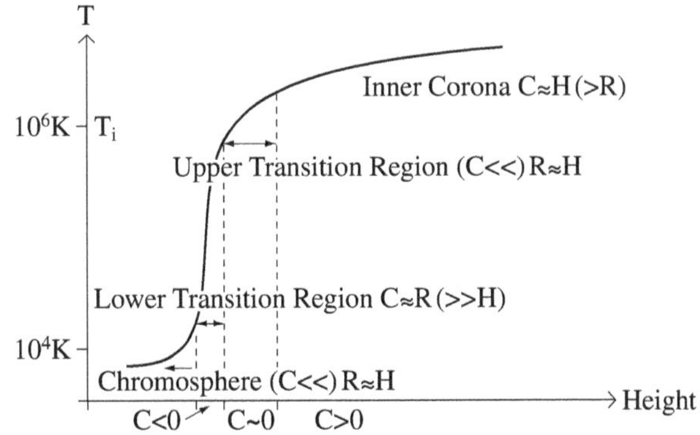

Figure 10.6. Sketch of the temperature structure in a static model of the upper solar atmosphere, indicating the relative roles of conduction (C), radiation (R) and heating (H).

Table 10.1. *Energy losses from the upper atmosphere (after Withbroe and Noyes 1977).* (1 W $m^{-2} \equiv 10^3$ erg $cm^{-2}s^{-1}$)

	Conduction (C) (Wm^{-2})	Radiation (R) (Wm^{-2})	Temperature (MK)	Pressure (Nm^{-2})
Quiet region:				
Low chromosphere		4×10^3		
Upper chromosphere		3×10^2		2×10^{-2}
Corona	2×10^2	10^2	1.1–1.6	
Coronal hole:				
Lower chromosphere	4×10^3			
Upper chromosphere		3×10^2		7×10^{-3}
Corona	6×10	10	1	
Active region:				
Lower chromosphere		2×10^4		
Upper chromosphere		2×10^3		2×10^{-1}
Corona	10^2–10^4	5×10^3	1–5	

by C. For an open magnetic field region, above the temperature maximum, energy transport by the solar wind becomes increasingly important and eventually dominates heat conduction.

Thus, the cause of the extremely steep rise of temperature in the lower transition region is that the radiation around 10^4 to 10^5 K is so large that it cannot be supplied by mechanical heating but must be provided by conduction (or enthalpy) from above. Furthermore, at greater altitudes the deposited heat cannot be radiated away, and so it must be conducted both inwards and outwards from the temperature maximum.

Table 10.1 gives typical values for conductive and radiative losses and plasma pressure. The heating H may be obtained by summing the losses (C and R). Table 10.1 also shows how a coronal hole is cooler and rarer than a normal quiet region, while an active region is hotter and denser.

10.2.3 Static Energy-Balance Models for Coronal Loops

For a loop in hydrostatic equilibrium and in thermal equilibrium between conduction, optically thin radiation and heating, the temperature (T), electron density (n_e) and pressure (p) at a distance s along the loop (Figure 10.7) satisfy

$$\frac{1}{A}\frac{d}{ds}\left(\kappa_0 T^{\frac{5}{2}}\frac{dT}{ds}A\right) = n_e^2 Q(T) - H_c, \tag{10.3}$$

$$\frac{1}{\cos\theta}\frac{dp}{ds} = -m_p n_e g, \tag{10.4}$$

$$p = 2 n_e k_B T, \tag{10.5}$$

where $A(s)$ is the cross-sectional area of the loop, $\theta(s)$ is the inclination of the loop to the vertical, κ_0 is the standard heat conduction coefficient (Eq. 2.34), and $Q(T)$ is the optically thin radiative loss function (Eq. 2.36). The heating (H_c) is usually assumed to be proportional to powers of density and temperature as

$$H_c = H_{c0}\, n_e^{\beta^*} T^{(\alpha^* + \beta^*)},$$

where H_{c0}, α^*, β^* are constants. (Sometimes, H_c also depends on position or magnetic field.) The case of a uniform heating is obtained by setting $\alpha^* = \beta^* = 0$.

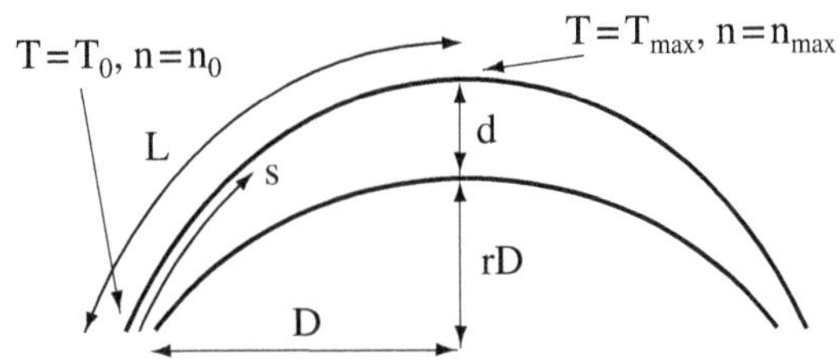

Figure 10.7. The notation for a symmetric coronal loop of length $2L$ with temperature T_0 and density n_0 at the footpoint ($s=0$), and T_{max} and n_{max} at the summit ($s=L$). r is the ratio of loop height to half the base length (D), and d is the ratio of the diameter of the loop cross-section at the top to that at the footpoint.

For given forms of $A(s)$ and $\theta(s)$ in a symmetric loop of half-length L, the third-order system of Eqs. (10.3) to (10.5) is solved subject to three boundary conditions:

$$p=p_0, \quad T=T_0 \quad \text{at the base } (s=0) \text{ of the loop} \tag{10.6}$$

and a symmetry condition

$$\frac{dT}{ds}=0 \qquad \text{at the summit } (s=L). \tag{10.7}$$

The solution determines the summit temperature $T_{max}=T_{max}(L,p_0,H_{c0})$ as a function of three parameters, the loop length ($2L$), base pressure (p_0) and heating rate (H_{c0}). When a value near the temperature minimum or a chromospheric plateau is taken for T_0, we have a so-called *thermally isolated loop* (physically, the most relevant case), for which the above three boundary conditions are supplemented by

$$\frac{dT}{ds}=0 \quad \text{at} \quad s=0, \tag{10.8}$$

which determines the heating rate [$H_{c0} = H_{c0}(L,p_0)$] and therefore the summit temperature [$T_{max} = T_{max}(L,p_0)$] as functions of loop length and pressure.

10.2.3.1 Uniform-Pressure Loops: Scaling Laws

For low-lying loops, whose summits are much below a coronal scale-height of roughly 50 Mm [Eq. (3.12)], the loop pressure is uniform, and so one needs to solve just Eq. (10.3), with n_e given by Eq. (10.5) and $p \equiv p_0$ constant. Suppose also the loops have uniform cross-sectional area and a radiative loss function $Q(T)=\chi\, T^{-\alpha}$. Then $n_e=p_0/(2k_BT)$ and the energy equation (10.3) reduces to

$$\frac{d}{ds}\left(\kappa_0 T^{\frac{5}{2}}\frac{dT}{ds}\right)=\chi_0\, p_0^2\, T^{-2-\alpha}-H_c, \tag{10.9}$$

where $\chi_0=\chi/(4k_B^2)$ and the heating becomes

$$H_c=H_{c0}^*\, p_0^{\beta^*}T^{\alpha^*}, \tag{10.10}$$

with $H_{c0}^*=H_{c0}/(2k_B)^{\beta^*}$. Thus, if T increases from $s=0$ to $s=L$, $\alpha^*<0$ represents footpoint heating and $\alpha^*>0$ represents apex heating.

The *RTV scaling law* (Rosner et al. 1978; Craig et al. 1978; Hood and Priest 1979a) for the summit temperature of a thermally isolated loop with $Q(T)=\chi\, T^{-1/2}$ (i.e., $\alpha=\frac{1}{2}$) is

$$T_{max}\approx c(pL)^{\frac{1}{3}}, \tag{10.11}$$

where $c=(\chi_0/\kappa_0)^{\frac{1}{6}}\approx 10{,}000$ in mks units. It is derived simply by assuming that the orders of magnitude of the conduction and radiation terms are the same in Eq. (10.9). Since $p\sim nT$, it implies

$$T_{max}\sim(n_{max}L)^{\frac{1}{2}}.$$

A second scaling law for the heating may be derived by assuming that it has the same order of magnitude as conduction and radiation and substituting for T_{max} from Eq. (10.11). In the case of uniform heating ($\alpha^* = \beta^* = 0$), it becomes

$$H_c \sim p^{\frac{7}{6}} L^{-\frac{5}{6}}. \tag{10.12}$$

An important question is: which parameters should be regarded as prescribed and which deduced? For a coronal loop, it is most natural to regard the loop heating and length as being given, so that its pressure and temperature just respond to preserve equilibrium. In this case, Eqs. (10.11) and (10.12) may be rearranged to give

$$T_{max} \sim H^{\frac{2}{7}} L^{\frac{4}{7}}, \qquad p \sim H^{\frac{6}{7}} L^{\frac{5}{7}}, \tag{10.13}$$

which imply that T_{max} and p increase when a loop is stretched or heated more.

10.2.3.2 Uniform-Pressure Loops: Analytical Solution

The energy equation (10.9) can be solved analytically to give the *Martens model* for a coronal loop (Martens 2010). First of all, non-dimensionalise by changing the variables to

$$Y = \left(\frac{T}{T_{max}}\right)^{7/2}, \qquad \bar{s} = \frac{s}{L},$$

for which the energy equation becomes

$$\bar{\kappa}\frac{d^2Y}{d\bar{s}^2} = Y^{\bar{\alpha}} - \bar{H}_{c0}\, Y^{\bar{\alpha}^*}, \tag{10.14}$$

where

$$\bar{\kappa} = \frac{2\kappa_0 T_{max}^{11/2+\alpha}}{7\chi_0\, p_0^2 L^2}, \quad \bar{H}_{c0} = \frac{H_{c0} T_{max}^{2+\alpha+\alpha^*}}{\chi_0\, p_0^{2-\beta^*}}, \quad \bar{\alpha}^* = \frac{2\alpha^*}{7}, \quad \bar{\alpha} = \frac{-2(2+\alpha)}{7}.$$

The boundary conditions (10.6) to (10.8) plus $T(L) = T_{max}$ become

$$Y(0) = 0, \quad \frac{dY}{d\bar{s}}(0) = 0, \quad Y(1) = 1, \quad \frac{dY}{d\bar{s}}(1) = 0.$$

The first condition should be $Y(0) = [T(0)/T_{max}]^{7/2} \leqslant 10^{-7}$ for a chromospheric value of $T(0)$, but we have approximated it by zero, since the origin is not a singular point of the differential equation and so the resulting error is tiny. Since there are four boundary conditions, the solution to this second-order equation determines $\bar{\kappa}$ (and therefore T_{max}) and $\bar{H}_{c0}$ as eigenvalues, and so our two scaling laws.

Multiplying Eq. (10.14) by $dY/d\bar{s}$ and applying conditions at $\bar{s} = 0$ gives as a first integral

$$\frac{\bar{\kappa}}{2}\left(\frac{dY}{d\bar{s}}\right)^2 = \frac{Y^{\bar{\alpha}}}{\bar{\alpha}+1} - \bar{H}_{c0}\frac{Y^{\bar{\alpha}^*}}{\bar{\alpha}^*+1}. \tag{10.15}$$

Now, applying the conditions at $\bar{s} = 1$ to this equation implies that $\bar{H}_{c0} = (\bar{\alpha}^*+1)/(\bar{\alpha}+1)$ or, in terms of the initial variables,

$$H_{c0} = \frac{(7/2+\alpha^*)\chi_0}{3/2-\alpha}\frac{p_0^{2-\beta^*}}{T_{max}^{2+\alpha+\alpha^*}}. \tag{10.16}$$

Assuming that the temperature has a monotonic rise from footpoint to apex, Eq. (10.15) may be written as an expression for $d\bar{s}/dY$ as a function of Y, which may be integrated to give

$$\bar{s} = \left[\frac{\bar{\kappa}(\bar{\alpha}+1)}{2}\right]^{1/2} \int_0^Y \frac{dy}{(y^{\bar{\alpha}+1} - y^{\bar{\alpha}^*+1})}.$$

By changing the variable to $x = y^{\bar{\alpha}^*-\bar{\alpha}}$, this may be written in terms of the β-function $\beta(u;\lambda+1,\frac{1}{2}) \equiv \int_0^u x^\lambda/(1-x)^{1/2}dx$ (Spanier and Oldham 1987) as

$$\bar{s} = \left[\frac{\bar{\kappa}(1-2(2+\alpha)/7)}{2}\right]^{1/2} \frac{7\beta(u;\lambda+1,\frac{1}{2})}{2(2+\alpha+\alpha^*)}, \tag{10.17}$$

where

$$u = Y^{\bar{\alpha}^*-\bar{\alpha}} \quad \text{and} \quad \lambda = \frac{3/2-\alpha-2\alpha^*}{2(2+\alpha+\alpha^*)}. \tag{10.18}$$

Now, applying the boundary condition $Y(1) = 1$ at the loop top implies

$$1 = \left[\frac{\bar{\kappa}(1-2(2+\alpha)/7)}{2}\right]^{1/2} \frac{7\beta_C(\lambda+1,\frac{1}{2})}{2(2+\alpha+\alpha^*)}, \tag{10.19}$$

where $\beta_C(\lambda+1,\frac{1}{2}) \equiv \int_0^1 x^\lambda/(1-x)^{1/2}dx$ is the complete β-function. This determines $\bar{\kappa}$, which may be eliminated from Eq. (10.17) to give the solution for the temperature profile in the loop. In terms of the initial variables, the solution is

$$T(s) = T_{max}[\beta_r^{-1}(s/L;\lambda+1,\tfrac{1}{2})]^{1/(2+\alpha+\alpha^*)},$$

where β_r^{-1} is the inverse of the regularised β-function ($\beta_r = \beta/\beta^*$), λ is given by Eq. (10.18) and T_{max} depends on the parameters (p_0, L, α^*). Higher-order eigenvalues also exist that give non-monotonic solutions: for example, the first harmonic gives a single oscillation with a minimum temperature at the summit that may be relevant for prominences. Solutions may also be set up with a non-zero heat flux at the apex to model loop-top heating in a flaring loop (Kano and Tsuneta 1995).

Equation (10.19) determines T_{max} and therefore generalises the RTV scaling law to

$$T_{max} = C(p_0L)^{\frac{4}{11+2\alpha}}, \quad \text{where} \quad C = \left(\frac{\chi_0}{\kappa_0}\right)^{\frac{2}{11+2\alpha}} \left(\frac{4+2\alpha+2\alpha^*}{(3-2\alpha)^{1/2}\beta_C(\lambda+1,\frac{1}{2})}\right)^{\frac{4}{11+2\alpha}}. \tag{10.20}$$

For $\alpha = \frac{1}{2}$, we recover the RTV law [$T_{max} \sim (p_0L)^{1/3}$], but with a precise expression for the constant in terms of the coefficients α^*, β^* in the form of heating Eq. (10.10).

The second scaling law is given in Eq. (10.16) with T_{max} given by Eq. (10.20). It reproduces the order of magnitude result Eq. (10.12) when the heating is uniform ($\alpha^* = \beta^* = 0$).

10.2.3.3 Numerical Solutions for Static Loops

If the conductive flux does not vanish at T_0, H_{c0} is an extra parameter and thermal non-equilibrium becomes possible. Hood and Priest (1979a) studied the case when $T_0 = 10^6$ K and heating is proportional to density ($H_c = H_{c0}n_e$): as p increases (Figure 10.8a), T_{max} rises to a maximum and then decreases to a critical value (T_{crit}) at which dT_{max}/dp becomes infinite and a catastrophe occurs. If $p > p_{crit}$, the loop is therefore in a state of *thermal nonequilibrium.* There is no neighbouring equilibrium, and so the plasma cools down seeking a new equilibrium below 10^5 K. Such a lack of equilibrium is also present when a loop is stretched at constant p and H_{c0} or its heating is decreased at constant p and L. It is a natural explanation for the existence of extremely cool loops and active-region prominences.

Temperature profiles for thermally isolated, uniform-pressure loops with different heating functions are shown in Figure 10.8b. $\alpha^* < 0$ corresponds to footpoint heating and $\alpha^* > 0$ to summit heating, so that, the more the heating is concentrated near the feet, the flatter the profiles become. In the Martens solutions of the previous section, solutions do not exist for $\alpha^* < -2$.

Thermal nonequilibrium has also been demonstrated to occur when heating is highly concentrated at low heights in both legs of a coronal loop (Antiochos and Klimchuk 1991). It produces a dynamic state in which a loop undergoes periodic convulsions in the search for an equilibrium: cold, dense condensations

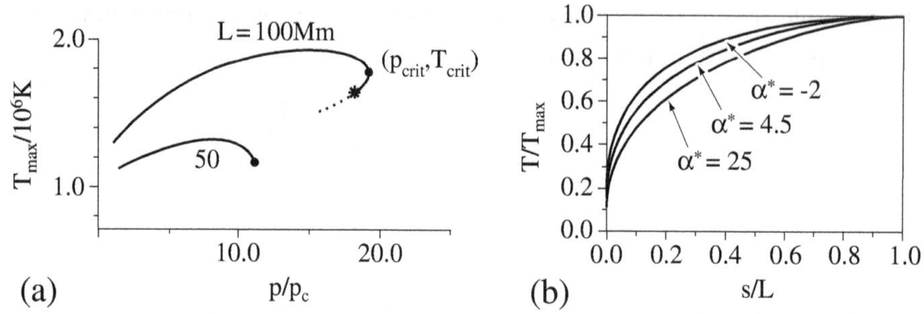

Figure 10.8. (a) Summit temperature (T_{max}) as a function of pressure (p) and half-length (L) for a low-lying static coronal loop with arbitrary base heat flux. p_c is the pressure for a plasma of density $5\times10^{14}\text{m}^{-3}$ and temperature 10^6 K. The curves bend back at critical conditions (p_{crit}, T_{crit}). The star indicates a thermally isolated loop (from Hood and Priest 1979a). (b) The temperature profiles for thermally isolated loops with heating $H_c \sim T^{\alpha^*}$ (after Martens 2010).

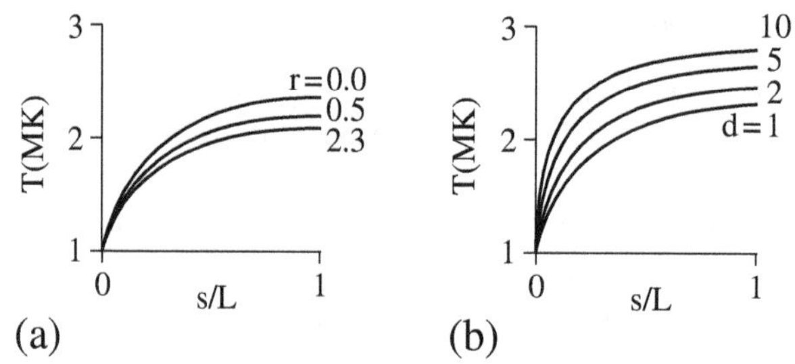

Figure 10.9. The effect on the temperature profile for a hydrostatic loop of (a) its shape and (b) its divergence. The loop shape is an arc of a circle, with $r/2$ the ratio of loop height to footpoint separation ($r=0$ gives uniform pressure). d is the ratio of summit diameter to base diameter. In (a) the loop length is 225 Mm and its heating is $h = 7$. In (b) the loop is semicircular with length 80 Mm and heating $h = 20$ (from Wragg and Priest 1981).

form and slide down the legs of a loop, and then they later reform in a cycle that repeats over tens of minutes or hours (Karpen et al. 2006; Antolin et al. 2012). It may play an important role in the formation of prominences and in producing coronal rain.

The energy balance [Eq. (10.3)] has been solved more generally for a cylindrically symmetric structure (Hood and Priest 1979a) and a force-free arcade of loops (Priest and Smith 1979) rather than just a single field line. Also, the effect of a hydrostatic fall-off with height has been calculated for thermally isolated loops taller than a scale-height (Serio et al. 1981; Wragg and Priest 1982). Increasing the loop height lowers the summit temperature, whereas increasing its divergence raises it (Figure 10.9). The RTV scaling law is accurate for short loops, but, for sufficiently long (or rare) loops, the temperature may be lower by a factor of 2, and too large an increase in loop length may cause the temperature to fall.

Numerical solutions by many authors have investigated the effect of non-uniform heating (e.g., Serio et al. 1981; Aschwanden 2001; Müller et al. 2005) and confirm the result that, for a loop with a given length and heating profile, the temperature maximum and pressure are determined [cf. Eq. (10.13)]. They also show that when heating is concentrated at the footpoints (as would be suggested by, e.g., flux-tube tectonics, Section 10.4.5), the loops tend to have an *over-pressure*, in the sense that the pressure exceeds that of a uniformly heated equilibrium. Strands that are cooling down also tend to have an over-pressure (Winebarger et al. 2003).

10.2.4 Dynamic Models for Chromospheric and Coronal Loops

Many types of flow are present in the solar atmosphere, including Evershed flow (Sec. 1.7.2.5), network downflow (Sec. 1.4.1), surges (Sec. 1.7.1.3), spicules (Sec. 1.5) and coronal rain (Sec. 1.7.1).

There are several ways of generating flow in a loop. *Siphon flow* is driven by a pressure difference between the two footpoints (Figure 10.10a), which could be set up by converging photospheric motion or footpoint heating. Flows are also generated by impulsive coronal heating events such as *nanoflares* (Secs. 10.2.1 and 10.4.2). In order to model such events, it is important to treat the corona, transition region and chromosphere as a coupled system. For example, conduction transports more than half of the coronal heat down to the transition region, where it is radiated. Thus, the transition-region pressure is determined by the heat flux, and the location of the transition region can move dynamically up and down during the evolution of a coronal loop. An increase in coronal heating raises the coronal temperature and therefore increases the downward heat flux: if the transition region cannot radiate this additional energy, heated plasma flows up by *evaporation* (Figure 10.10d). On the other hand a decrease in coronal heating has the opposite effect of *draining* or condensing excess plasma down to the chromosphere (Figure 10.10c).

10.2.4.1 Steady Flows

A simple case is that of steady flow along a loop of uniform cross-section, satisfying conservation of mass, momentum and energy (Cargill and Priest 1980):

$$\frac{d}{ds}(\rho v) = 0; \qquad \rho v \frac{dv}{ds} = -\frac{dp}{ds} - \rho g \cos\theta, \qquad \frac{d}{ds}\left(\frac{p}{\rho^{\gamma}}\right) = 0,$$

where s is the distance measured along a loop of length L and $\theta(s)$ is the local inclination of a section of the loop to the vertical, so that for a semi-circular loop $\theta = \pi s/L$. The adiabatic law has been assumed for simplicity. Eliminating p and ρ between the three equations and writing $c_s(s)$ as the sound speed yields

$$\left(v - \frac{c_s^2}{v}\right)\frac{dv}{ds} = -g \cos\frac{\pi s}{L}, \tag{10.21}$$

which is similar in form to the solar wind equations (Sec. 13.3) and possesses a critical point ($v = c_s$) at the loop summit ($s = \frac{1}{2}L$). An important feature is that for small pressure differences the flow is subsonic, but for larger pressure differences the flow becomes supersonic near the loop summit and is then slowed down by a shock wave in the downflowing leg (Figure 10.11).

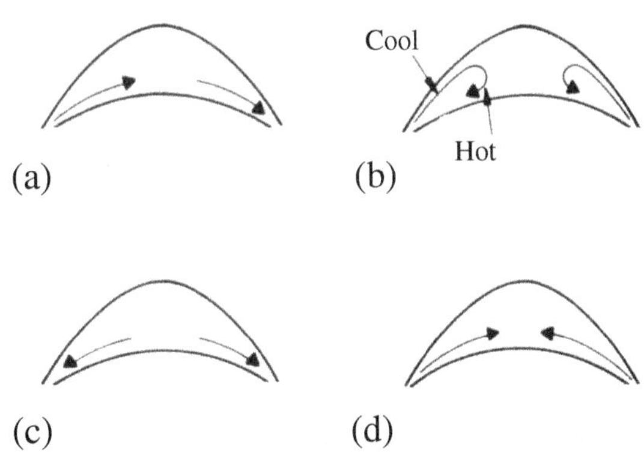

Figure 10.10. The main types of flow in coronal loops are: (a) siphon flow, (b) spicule flow, (c) loop draining and (d) loop filling by evaporation.

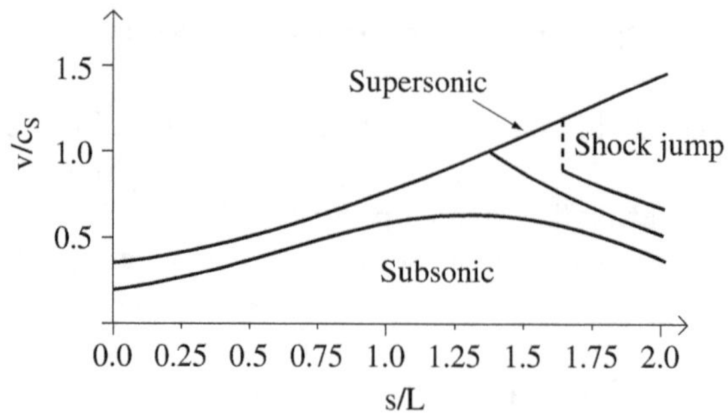

Figure 10.11. The siphon flow speed (v) at a distance along a converging coronal loop of length L= 100 Mm (from Cargill and Priest 1980).

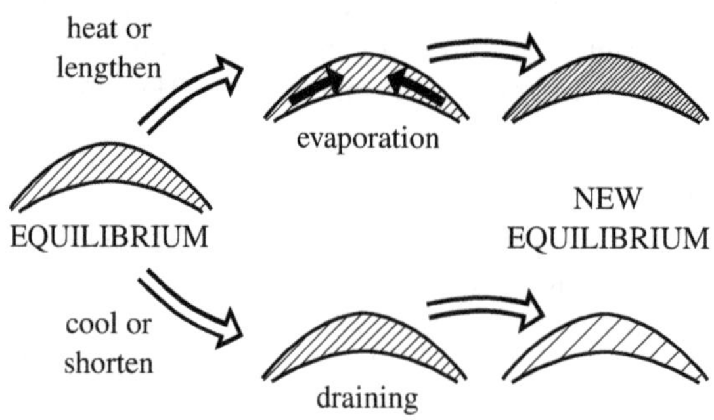

Figure 10.12. The evolution of a coronal loop from one equilibrium to another by means of an evaporation or draining, depending on whether the heating rate (or loop length) increases or decreases in value.

Surges may be caused by reconnection between newly emerging or evolving satellite flux and an ambient sunspot field. *Evershed flow* may be evidence of siphon flow, with a pressure difference caused by magnetoconvection (Sec. 9.7). On the other hand, upflows and downflows in both legs of a loop may be a natural response to heating and cooling of loops, in which heating produces evaporation and cooling leads to draining back down (Figure 10.12).

10.2.4.2 Time-Dependent Flows: 1D and 0D (Ebtel) Models

Ever since the recognition during the 1970s following Skylab that the corona is composed of a multitude of coronal loops, there has been an interest in the temporal evolution of coronal plasma. The 1D time-dependent hydrodynamic equations of conservation of mass, momentum and energy for density (ρ), speed (v) and pressure (p) are, in conservative form from Eqs. (2.19, 2.20 and 2.29) with $\gamma = 5/3$

$$\frac{\partial \rho}{\partial t} + \frac{\partial}{\partial s}(\rho v) = 0,$$

$$\frac{\partial}{\partial t}(\rho v) + \frac{\partial}{\partial s}(\rho v^2) = -\frac{\partial p}{\partial s} + \rho g \cos\theta,$$

$$\frac{3}{2}\frac{\partial p}{\partial t} + \frac{\partial}{\partial t}(\tfrac{1}{2}\rho v^2) = -\frac{5}{2}\frac{\partial}{\partial s}(pv) - \frac{\partial}{\partial s}(\tfrac{1}{2}\rho v^3) + \rho g v \cos\theta - \frac{\partial F_c}{\partial s} - L_r + H_c,$$

where θ is inclination to the vertical, $F_c = -\kappa_0 T^{5/2}\partial T/\partial s$ is conductive flux, $L_r = n_e^2 Q(T)$ is radiative loss (Sec. 2.3.3), $n_e \approx \rho/(1.9m)$ and H_c is atmospheric heating.

They have been solved along a fixed magnetic field line (or coronal loop) by many groups in response to an imposed energy input (e.g., Peres 2000). Applications include modelling a nanoflare (Bradshaw and

Cargill 2006) or a solar flare, including the chromosphere (Fisher et al. 1985a), chromospheric evaporation (Nagai 1980; Fisher et al. 1985b), cool downflows called *coronal rain* (Müller et al. 2005), prominence formation (Antiochos et al. 1999) and loop brightenings (Reale et al. 2000).

Several time-scales arise from equating the time-dependent terms with forcing terms on the right in these equations. Thus, the equation of motion implies that flows will be accelerated on the pressure (τ_p) and gravitational (τ_g) times given by

$$\tau_p = \frac{\rho v L}{p}, \qquad \tau_g = \frac{v}{g},$$

whereas the energy equation gives conduction (τ_C), radiation (τ_R) and enthalpy (τ_v) times of

$$\tau_C = \frac{L^2 p}{\kappa_0 T^{7/2}}, \qquad \tau_R = \frac{p}{n^2 Q} \sim \frac{T^{3/2}}{n}, \qquad \tau_v = \frac{L}{v}.$$

The expression $\tau_R \sim T^{3/2}/n$ arises when $Q(T) \sim T^{-1/2}$.

One difficulty with 1D loop modelling is treating the optically thick chromosphere, which requires a radiative-hydrodynamic treatment, although from a coronal viewpoint the main effect of the chromosphere is to provide a large mass source or sink. Another is the time-step limitation imposed by thermal conduction in the transition region, where spatial gradients of 1 km are implied by static loop models (even shorter in hot flaring loops). These, together with observed coronal loops being perhaps composed of many hundreds or thousands of thin strands (as implied by the flux-tube tectonics model, Section 10.4.5), have stimulated the development of simpler "zero-dimensional" (0D) models, which average over the length of a loop and may be applied to each strand.

Currently, the most commonly used 0D model for dynamic loops or strands is the EBTEL (Enthalpy-based Thermal Evolution of Loops) model (Klimchuk et al. 2008; Cargill et al. 2012a,b). Here we generalise it to make it more complete. The loop is divided into coronal and transition-region parts, with a boundary where thermal conduction changes from a loss to a gain.

Five equations for enthalpy flux ($5v_0p_0/2$) and mean coronal pressure [$\bar{p}(t)$], density [$\bar{\rho}(t)$], temperature [$\bar{T}(t)$] and velocity [$\bar{v}(t)$] are derived as follows, in terms of coronal base (p_0, ρ_0, T_0, v_0) and apex (p_a, n_a, T_a, $v_a = 0$) values, by integrating over coronal and transition-region sections of half a symmetric loop and making a drastic assumption that a mean of products of variables equals the product of the means. Then they may be reduced to the more approximate EBTEL equations.

Firstly, integrating over the transition region and assuming its length ($l \ll L$) is much shorter than the coronal length (L) gives

$$\tfrac{5}{2} v_0 p_0 + \tfrac{1}{2} \rho_0 v_0^3 = -F_0 - R_{tr}, \tag{10.22}$$

where R_{tr} is the total transition-region radiation (assumed much larger than the transition-region heating) and $F_0 = 2\kappa_0 (T_0^{7/2} - T_a^{7/2})/(7L)$. Equation (10.22) implies a key insight, namely, that the enthalpy flux is determined by the ability of the transition region to radiate (e.g., Antiochos and Sturrock 1978; Cargill et al. 2012a). If downward heat flux exceeds transition-region radiation, the excess powers an upward enthalpy flux, which 'evaporates' material into the corona and increases the coronal density. If, on the other hand, the downward heat flux is smaller than the transition-region radiation, a downward enthalpy flux decreases the coronal density.

Then, integrating the equations of continuity and motion in the corona gives

$$\frac{d\bar{\rho}}{dt} = \frac{\rho_0 v_0}{L}, \tag{10.23}$$

$$\frac{d(\bar{\rho}\,\bar{v})}{dt} = \frac{p_0 + \rho_0 v_0^2 - p_a}{L} + \bar{\rho}\bar{g}_{\|}, \tag{10.24}$$

where $g_{\|} = g\cos\theta$. After substituting for F_0 from Eq. (10.22), the mean of the energy equation becomes

$$\frac{3}{2}\frac{d\bar{p}}{dt} + \frac{d}{dt}(\tfrac{1}{2}\bar{\rho}\,\bar{v}^2) = \frac{H_c - R_{tr} - R_c}{L} + \bar{\rho}\bar{g}_{\|}\bar{v}, \tag{10.25}$$

where H_c is the total coronal heating and R_c is the total coronal radiation [assumed to equal $L\bar{n}_e^2 Q(\bar{T})$, where $n_e \approx n/1.9 = \rho/(1.9m)$]. The flow speed and heat flux vanish at the top of the loop, while their values (v_0 and F_0) at the base of the corona are assumed to be much larger than at the base of the transition region. Finally, the mean temperature is determined from the perfect gas law as

$$\frac{1}{\bar{T}}\frac{d\bar{T}}{dt} = \frac{1}{\bar{p}}\frac{d\bar{p}}{dt} - \frac{1}{\bar{n}}\frac{d\bar{n}}{dt}. \tag{10.26}$$

Here, if the form of a variable along the loop is known, its mean may be related to its end values. One possibility is to take quadratic forms, which imply $\rho_a = \frac{1}{2}(3\bar{\rho} - \rho_0)$, $\bar{v} = \frac{1}{3}v_0$ and $p_a = \frac{1}{2}(3\bar{p} - p_0)$. Another possibility is to assume a hydrostatic variation or to take a form from numerical simulations. Then Eqs. (10.22) to (10.26) may be solved for five unknowns provided the appropriate constants are prescribed; however, it is not immediately obvious which ones to prescribe and which to deduce, since values such as v_0 or p_0 at the loop base will change during dynamic evolution.

The EBTEL equations drop the equation of motion (10.24) and gravity and assume $\rho_0 v_0^2 \ll p_0$, so that an expression for v_0 may be substituted from Eq. (10.22) into Eq. (10.23). Then Eqs. (10.23), (10.25) and (10.26) become

$$\frac{d\bar{\rho}}{dt} = -\frac{2m(F_0 + R_{tr})}{5Lk_B T_0}, \qquad \frac{3}{2}\frac{d\bar{p}}{dt} = \frac{H_c - R_c - R_{tr}}{L}, \qquad \frac{1}{\bar{T}}\frac{d\bar{T}}{dt} = \frac{1}{\bar{p}}\frac{d\bar{p}}{dt} - \frac{1}{\bar{\rho}}\frac{d\bar{\rho}}{dt},$$

where $F_0 \approx -2\kappa_0 T_a^{7/2}/(7L)$, $R_c \approx L\bar{n}_e^2 Q(\bar{T})$ and $\frac{5}{2}v_0 p_0 \approx -F_0 - R_{tr}$.

Now, when $H_c(t)$ and L are imposed, these equations will determine $\bar{\rho}(t)$, $\bar{p}(t)$, $\bar{T}(t)$ and $v_0 p_0$ (but not v_0 or p_0 separately or $\bar{v}$), provided R_{tr}, T_0 and T_a are known. In the EBTEL approach, these three parameters are imposed from physical considerations and examination of numerical results as follows. First of all, it assumes that $R_{tr} = c_1 R_c$, where $c_1 = 2$ up to the start of the radiative phase, $c_1 = 0.6$ in the radiative phase and c_1 joins these two values smoothly during the transition from one phase to the other. Secondly, it sets $T_a = \bar{T}/c_2$ and $T_0 = c_3 T_a$, where $c_2 = 0.9$ and $c_3 = 0.6$; the assumption about c_2 is more reliable than that about c_3. In practice, differences of 10 to 20 per cent are found between EBTEL and full 1D simulations, but the advantage of EBTEL is that it is simple and obeys conservation of mass and energy. A useful extension is to allow the length of the loop to change (Cargill et al. 2012b).

Another extension that we suggest here is to incorporate the extra equation (10.24) of mass conservation, which includes important physics regarding the forces that accelerate the plasma and slow it to rest at the summit. This equation allows the value of $\bar{v}$ to be determined and therefore the remaining unknowns, such as v_0 (from, e.g., $v_0 = 3\bar{v}$ for a quadratic variation), p_0 from (10.22), as well as p_a from $\bar{p}$, ρ_a from T_a and ρ_0 from T_0. A further extension, when the evolution is so dynamic that flows are no longer small, is to solve the full equation (10.22) as a cubic for v_0.

10.2.5 Deducing the Form of Heating in Coronal Loops

The form of heating can be deduced from the temperature profile along a loop if it is known accurately enough. Priest et al. (1998, 2000) applied this philosophy in a preliminary way to a large loop system in soft X-ray images from the Yohkoh satellite. Large magnetic loop systems dominate the global corona at solar minimum (e.g., Figures 10.1 and 10.14), and are also present outside active regions at solar maximum when the global X-ray intensity is much higher.

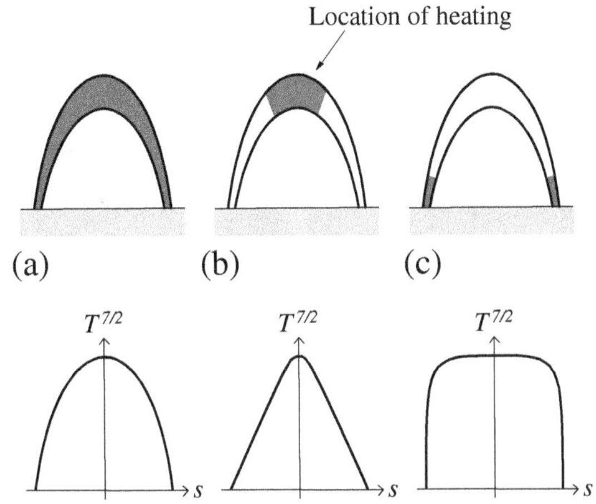

Figure 10.13. In a coronal loop, heating may be: (a) located throughout the loop, (b) focussed near the summit, or (c) focussed near the feet. Each produces a different shape for the temperature profile, as indicated schematically (from Priest and Forbes 2000).

Chromospheric and coronal heating by turbulent reconnection in many small current sheets due to flux-tube tectonics (Sec. 10.4.5) would tend to deposit heat fairly uniformly through the high coronal part of the loop and more intensely in the more complex parts near the coronal base and chromosphere. On the other hand, heating by long-wavelength Alfvén waves standing in a loop with a maximum amplitude at the summit would dump heat near the summit. A steady thermal balance between such heating and conduction would then make $T^{\frac{7}{2}}$ a quadratic function of distance from the footpoint in the first case, or a pointed function in the second case, or having a steep footpoint rise and a flat summit profile for footpoint heating (Figure 10.13).

Realising that the temperature profile is, therefore, highly sensitive to the nature of the heating mechanism, Priest et al. (1998, 2000) used the Yohkoh Soft X-ray Telescope to compare the temperature along a large loop (Figure 10.14) with a series of models.

The observed temperatures rise from about 1.4 MK near the observed feet to about 2.2 MK at the summit, and it can be seen in Figure 10.15 that the model with the heat concentrated near the feet gives a very poor fit, whereas heat focused at the summit produces a better fit. However, uniform heating fits best of all and therefore provides preliminary evidence that the heating mechanism deposits the energy fairly uniformly along the upper coronal part of the loop.

Of the existing models, the one that can most easily explain uniform heating in the high-temperature part of a loop is flux-tube tectonics (Sec. 10.4.5). Moreover, tectonics would suggest that, lower down at the feet of the loop below the measured temperatures, the heating should be much greater due to the carpet dissipation there. However, the importance of this work was not so much in the tentative conclusions for the particular loop system (which may well change when better observations and models are used in future) as in the suggestion that the observed temperature can indeed be used to deduce the form of the heating and therefore to put limitations on the likely heating mechanism, provided the profile can be determined accurately enough. The method is likely to work well for quasi-steady heating (truly steady or high-frequency nanoflares) when a simple form of energy balance can be adopted, but it is harder to apply to more dynamic low-frequency nanoflares.

Martens (2010) suggested, on the basis of his analytical solution, that the heating mechanism may possibly be determined instead from the magnitude of the constant C that multiplies the RTV scaling law and its generalisation in Eq. (10.20). In particular, C depends on the coefficient α^* that appears as the

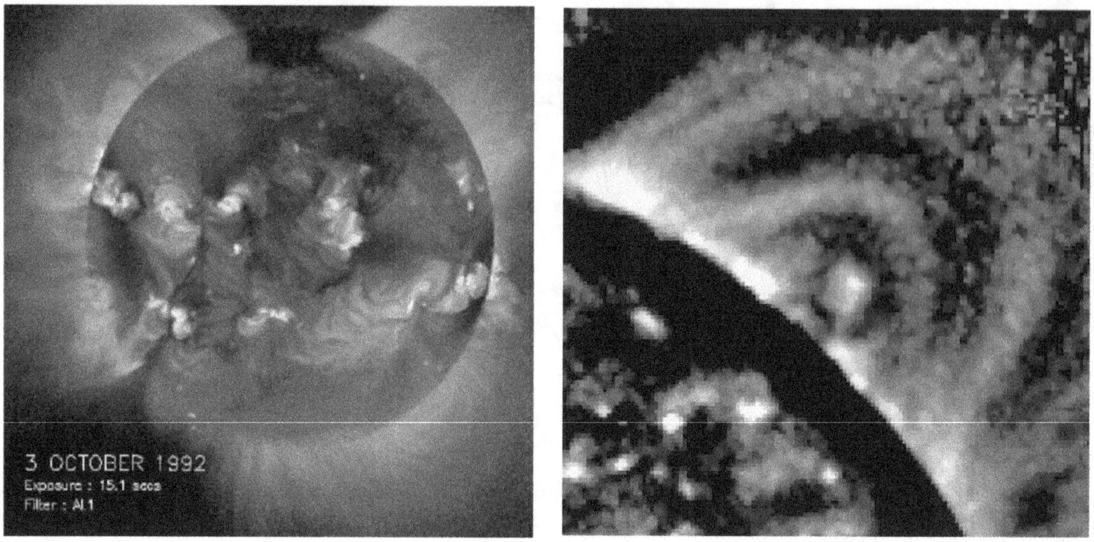

(a) (b)

Figure 10.14. (a) Global image of the Sun in soft X-rays from Yohkoh (courtesy Saku Tsuneta). (b) Close-up of the loops analysed in the top right corner of (a) (from Priest et al. 2000).

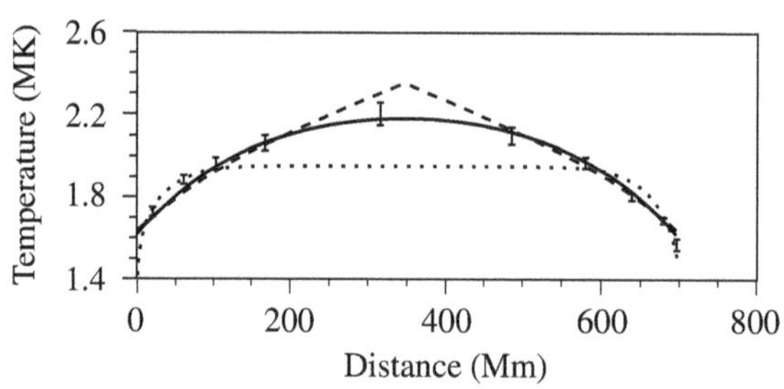

Figure 10.15. A comparison of observed coronal loop temperature profiles (points with error bars) with models in which heat is localised near the feet (dotted), localised at the summit (dashed) or deposited uniformly through the loop (solid) (after Priest et al. 2000)

power of T in the heating function [Eq. (10.10)]. He considers several mechanisms. For Ohmic heating,

$$H_c = \frac{j^2}{\sigma},$$

where $\sigma = \sigma_0 T^{3/2}/\ln \Lambda$ [Eq. (2.17)], so that $\alpha^* = -1.5$. Assuming that a heating flux of, say, 5,000 W m^{-2} is distributed over a height of 10 Mm, the heating power per unit volume in an active region is $5{\times}10^{-4}$ W m^{-3}, and so he finds that equating this to the above Ohmic heating implies a current density of about 20 A m^{-2}. This is more than four orders of magnitude stronger than the typical volume currents in an active region, so that the Ohmic heating must be focused in a set of current sheets.

Other mechanisms considered by Martens (2010) include: viscous dissipation due to phase mixing (Heyvaerts and Priest 1983), for which

$$H_c = \nu\rho\frac{v_{wave}^2}{d^2},$$

where v_{wave} is the wave amplitude, d a local length-scale and viscosity $\nu \sim T^{5/2}$, so that $\alpha^* = 5/2$; and nanoflare heating (uniform in space with $\alpha^* = 0$). Schrijver et al. (2004) set up a potential field model for the corona and populated 50,000 coronal field lines with static plasma. The best match to observed X-ray and EUV images in and around active regions was obtained with a heating flux $F_H \sim B/L$, which suggests coronal heating at many current sheets. This agrees with an earlier comparison of scaling laws due to different heating mechanisms (Mandrini et al. 2000; Klimchuk 2002), for which nanoflare heating fits the observations better than wave heating.

10.2.6 Numerical Experiments on Active-Region Heating

The response of the corona to photospheric footpoint motions is in most models (such as flux-tube tectonics) likely to be a localised heating in many small regions, and to be highly intermittent and impulsive in space and time. Gudiksen and Nordlund (2005) have conducted an interesting 3D MHD computational experiment that demonstrates this using as realistic physics as possible. They started with a stratified atmosphere and a magnetic field that is a potential extrapolation of an MDI magnetogram. At the photosphere they imposed a simulated random granular pattern with a maximum amplitude of 300 m s^{-1}. A grid of 150^3 was used in a computational box of $60 \times 60 \times 37$ Mm, although this is certainly insufficient to resolve the expected physics. Horizontal periodicity was assumed with vanishing vertical velocity and vertical temperature gradient on the upper boundary.

In the numerical experiment they found that the Poynting flux through the lower boundary dissipates ohmically in the atmosphere, maintaining a temperature of 10^6 K. The Ohmic heating decreases with height through the photosphere and chromosphere by a factor of 10^4. Most of the coronal heating is from intermittent short-period reconnection events, representing about 8 per cent of the Poynting flux or about 2×10^3 J m^{-2} (2×10^6 erg cm^{-2}). It is a factor of 4 larger than the radiation and is proportional to the square of the magnetic field. The transition region, where the temperature reaches 10^5 K, is highly intermittent in space and time, ranging between 2.7 Mm and 12.3 Mm in height, with an average of 5 Mm. At a height of 25 Mm, the plasma density and temperature varies continually in space and time, by a factor 100 from 10^{14} to 10^{16} m^{-3} (10^8 to 10^{10} cm^{-3}) and 10^4 to 10^6 K, respectively. The average mean density in the corona is roughly constant with height due to flows as high as 400 km s^{-1} (average 20 km s^{-1}) as plasma is continually heated and cooled. The resulting configuration is approximately a nonlinear force-free field, but close to potential. Simulated images in the TRACE 171 Å and 195 Å bands are qualitatively reasonable as a first attempt (Figure 10.16).

The hope is that, at more realistic solar parameters, where, for example, the magnetic Reynolds number (R_m) is at least a factor of a million higher, the energy would cascade down to appropriate scales and dissipate independently of R_m. In contrast, the braiding experiments of Galsgaard and Nordlund (1996a) give a heating rate that increases by a factor of 1.9 as the spatial resolution is increased by a factor 1.5. Therefore, alternative scenarios and models that consider such dissipation at much higher R_m values (e.g., Sec. 10.4) are of great complementary value.

A later computation on a smaller (16 Mm)^3 region at higher resolution (512^3 grid), including non-LTE radiative transfer and thermal conduction along field lines, shows heating in the chromosphere mostly by Ohmic dissipation at the edges of magnetic flux concentrations (Carlsson et al. 2010), which naturally produce transition-region downflows (Hansteen et al. 2010). Furthermore, 100 structures resembling type I spicules (Sec. 9.2.4) are observed reaching heights of a few Mm with velocities of 10 to 30 km s^{-1} and lifetimes of 3 to 5 min (Martínez-Sykora et al. 2009). They are produced by upwardly propagating magnetoacoustic shock waves, driven by either p-modes, collapsing granules, convective buffeting of intense flux tubes, and magnetic heating in the low atmosphere. Furthermore, when a horizontal flux tube is injected below the photosphere and perpendicular to the overlying magnetic field, it reconnects and drives a type II spicule upwards at 95 km s^{-1} (Martínez-Sykora et al. 2011).

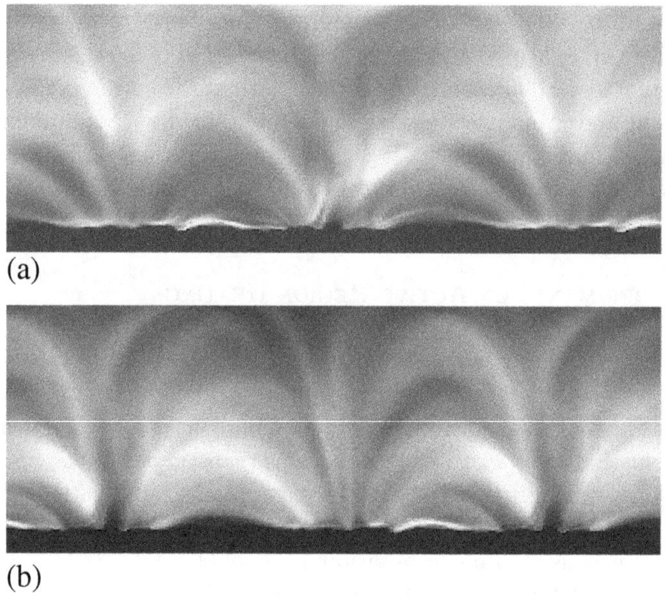

Figure 10.16. Emulated (a) TRACE 171 Å and (b) 195 Å images from the computational experiment on coronal heating by Gudiksen and Nordlund (2005).

A simulation of the same small active region with a 256^3 grid included smaller-scale network patches and magnetic footpoint motions driven by a simulated granulation pattern of horizontal motions (Bingert and Peter 2011). Reconnection in a series of current sheets produces coronal heating events that eject cool plasma into the corona and Doppler shifts in good agreement with observations (Zacharias et al. 2011).

Mackay et al. (2011) have modelled the evolution of an active region through a series of nonlinear force-free fields in response to footpoint motions in the magnetic carpet (Sec. 1.4.2) determined from a series of line-of-sight MDI magnetograms. During the decay of an active region, they find that over 4 days, small-scale motions inject $(2.5 \text{ to } 3)\times10^{18}$ J s^{-1} [$(2.5 \text{ to } 3)\times10^{25}$ erg s^{-1}] of free energy into the coronal field. Most of the energy is stored in the low corona, below 30 Mm. After 4 days, the energy buildup is 10 per cent of the initial potential energy. Such an energy buildup is sufficient to explain the coronal radiation from the active region. Later, Meyer et al. (2012) modelled the interaction of two flux elements in the magnetic carpet and showed how the free energy buildup in nonlinear force-free fields is sufficient to explain X-ray bright points and nanoflares and to provide the quiet-Sun coronal heating.

Complex 3D numerical experiments are a valuable complement to 1D models and detailed theories of coronal heating, since they can drop simplifying assumptions inherent in such models. However, they do not provide a complete answer to coronal heating, since they can never achieve the much higher spatial resolution that is possible in 1D models and that is an essential ingredient of coronal heating. The details of the dissipation mechanism are crucial, both in giving an in-depth physical understanding of coronal heating and in providing the photospheric Poynting flux. This flux depends on the values and spatial structure of the dissipation currents, since they help determine the horizontal photospheric magnetic field. Thus, 3D computations, 1D models and basic theory are all needed to make progress.

10.3 Heating by MHD Waves

Biermann (1946) and Schwarzschild (1948) suggested heating by sound waves that are generated by turbulence in the convection zone and steepen to form shock waves as they propagate upwards. Indeed, the generation of a dynamic atmosphere by photospheric oscillations has been well modelled numerically (Carlsson et al. 2007).

However, it is now known that the corona is dominated by the magnetic field and so the relevant wave modes are MHD in nature. Understanding the propagation and dissipation of such modes in complex magnetic configurations is important in *sunspot seismology* (Bogdan and Knölker 1991) and *coronal seismology* (Nakariakov and Verwichte 2005) (Sec. 4.11.2), with their ability to determine physical plasma parameters. Furthermore, they remain a natural mechanism for heating the outer corona and coronal holes, where reconnection is much less effective. The cause of heating in MHD waves and reconnection is the same, namely Ohmic or viscous dissipation in narrow regions, but the means of producing such regions is different. Sometimes, the current sheets are propagating (i.e., magnetic shocks), and at other times they are non-propagating (resonant absorption regions or reconnecting current sheets).

Photospheric convection generates a copious supply of waves (Narain and Ulmschneider 1996), and so the main issues are how sufficient wave energy reaches the corona and how it then dissipates. Slow magnetoacoustic modes rapidly shock and damp, while fast magnetoacoustic modes reflect and refract as they propagate up, and also mode couple where $c_s \approx v_A$ (e.g., Rosenthal et al. 2002; Bogdan et al. 2003), and so Alfvén waves are the most likely to carry energy up to the corona. However, most Alfvén waves are also reflected, the exception being in narrow frequency bands centred on discrete values where the waves are resonant in a loop (Hollweg 1981). Transmission is, of course, not an issue for waves that are generated in the corona, by, for example, rapid time-dependent reconnection. Furthermore, since Alfvén waves dissipate much too slowly in a uniform medium, a key challenge was to discover efficient dissipation mechanisms in a non-uniform medium.

The details of MHD waves and their behaviour in a non-uniform medium are described in Chapter 4 and in Goossens et al. (2011). Here we complement that treatment by discussing how waves propagate in the solar atmosphere (Sec. 10.3.1) and dissipate by resonant absorption (Sec. 10.3.2) or phase mixing (Sec. 10.3.3). Many different types of wave motion have now been observed in the corona (Sec. 4.11.2), including ubiquitous propagating Alfvénic waves (Tomczyk and McIntosh 2009) and swaying type II spicules (Secs. 1.5 and 10.3.4), although their energy budget may be too small to explain coronal heating in active regions.

10.3.1 Propagation of Magnetic Waves

Granular and supergranular flows continually shuffle around the photospheric footpoints of coronal field lines and generate upwards-propagating waves. In non-magnetic parts of the photosphere, acoustic waves soon become MHD waves when they reach the chromosphere, whereas, in intense photospheric magnetic flux tubes, MHD waves of several types are generated directly. Magnetoacoustic waves steepen into shocks and dissipate in a similar manner to pure acoustic waves, with Ohmic heating providing extra dissipation. Alfvén waves dissipate much less readily. They are not subject to resonant absorption, but they can be the product of it. Thus, kink and other $m > 0$ magnetoacoustic modes are linearly coupled at resonant surfaces to torsional Alfvén waves; then these localised waves with very steep gradients dissipate by classical dissipation or phase mixing.

Fast-mode waves propagate in any direction, but slow modes transmit energy only in directions close to the magnetic field. Alfvén waves possess a group velocity of v_A along the field and produce no change in density or pressure. In weak-field regions ($v_A < c_s$), most of the wave energy generated by convective turbulence is in the form of fast-mode waves, so Osterbrock (1961) proposed dissipation of fast-mode shocks as the dominant heating mechanism in the chromosphere. The acoustic and Alfvén speeds (c_s, v_A) increase with height from typically 10 km s^{-1} at the photosphere to 200 km s^{-1} and 10^3 km s^{-1}, respectively, in the corona. This makes the wave direction rotate away from the vertical as it propagates upwards and into regions of low wave speed. The ray paths $[\mathbf{r}(t)]$ of a wave may be calculated from

$$\frac{d\mathbf{r}}{dt} = \frac{\partial\omega}{\partial\mathbf{k}}, \quad \frac{d\mathbf{k}}{dt} = -\frac{\partial\omega}{\partial\mathbf{r}}.$$

The rate at which fast-mode waves steepen into shocks depends on the wave amplitude $[v_1(z)]$, which varies with height since the wave energy flux $(\frac{1}{2}\rho v_1^2)v_A$ is constant. However, in a stratified isothermal atmosphere, the density falls off as $\rho(z) \sim e^{-z/H}$ and $v_A \sim \rho^{-\frac{1}{2}}$. Thus, the wave amplitude increases as $v_1(z) \sim \rho^{-\frac{1}{4}} \sim e^{z/(4H)}$. Once they have shocked, the waves rapidly dissipate (in the chromosphere).

In a uniform medium, the damping of magnetic waves of wavelength λ by Ohmic dissipation occurs on a time-scale $\tau_d = \lambda^2/\eta$, where η is the magnetic diffusivity. Thus, the distance that a wave can travel at the Alfvén speed before it dissipates (the so-called *damping length*) is

$$L_d = v_A\tau_d = \frac{v_A\,\lambda^2}{\eta} = \frac{v_A^3}{\eta\,\omega^2},$$

where $\omega = v_A/\lambda$ is the wave frequency and $\tau = 2\pi/\omega$ its period. Such dissipation is faster for waves of higher frequency or lower Alfvén speed, but in practice it is extremely slow, since $L_d/\lambda = v_A\lambda/\eta = R_m >> 1$, with R_m based on v_A and λ.

Alfvén waves are generated at supergranulation boundaries, where intense kilogauss fields allow them to penetrate into the photosphere and propagate up into the corona with little attenuation. Indeed, the normal damping of Alfvén waves in a uniform field larger than 10 to 100 G is so small that the problem was to explain how they give up their energy before propagating away, either into the solar wind along open field lines or back down to the photosphere along closed fields. This was the motivation for considering the effect of a non-uniform field in producing damping by phase mixing (Sec. 10.3.2) or resonant absorption (Sec. 10.3.3).

Hollweg (1981) was one of the first to model heating by Alfvén waves. The propagation of axisymmetric twists near the axis of a vertical flux tube obeys

$$\left(v_A^2\frac{\partial^2}{\partial s^2} + \omega^2\right)(B_0^{\frac{1}{2}}\,\delta v) = 0, \qquad i\omega\delta B = B_0^{\frac{1}{2}}\frac{\partial}{\partial s}(B_0^{\frac{1}{2}}\,\delta v),$$

where ω is the frequency and s is the distance along the ambient field (B_0). He solves these equations for the perturbed velocity (δv) and magnetic field (δB) in a model atmosphere, in response to imposed photospheric motions.

In an open region, he suggested that long-period waves ($\tau >$ 10 min) with an energy flux of 10 W m^{-2} may drive the solar wind. Short-period waves (10 sec $< \tau <$ 5 min) with a flux of $10^3 - 10^4$ W m^{-2} may drive spicules and heat the chromosphere and corona. From a nonlinear analysis, Hollweg et al. (1982) suggested that Alfvén waves with photospheric amplitudes of order 1 km s^{-1} can steepen into fast shocks in the chromosphere, provided $\tau <$ few min. Later, Nakariakov et al. (2000) demonstrated the formation of shocks in coronal holes from shear or torsional Alfvén waves.

In a closed loop, resonant frequencies appear at multiples of $v_A/(2L)$, where L is the coronal length of the loop, due to reflections off the transition regions at the ends of the loop. They act like windows, which allow a large energy flux (typically 1.5 $\times 10^4$ W m^{-2}) to pass unimpeded up into the corona, rather than being reflected off the steep Alfvén-speed gradient.

Furthermore, numerical solutions demonstrate that magnetoacoustic waves and torsional Alfvén waves with a wide range of frequencies may leak through the transition region along magnetic flux tubes (Fedun et al. 2011).

10.3.2 Chromospheric and Coronal Heating by Resonant Absorption

The corona is highly non-uniform and is likely to be filled with a broad spectrum of MHD waves propagating in all directions relative to the magnetic field, generated by photospheric motions and

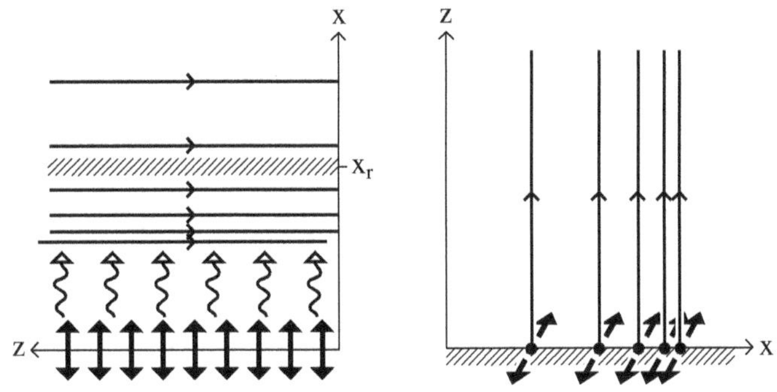

Figure 10.17. Wave dissipation by (a) resonant absorption at $x = x_r$ due to sideways driving (i.e., x-motions) and by (b) phase mixing due to footpoint driving (y-motions).

small-scale dynamic processes in the corona. Sideways driving of a magnetic structure (e.g., in the interior of a supergranule cell or below a coronal loop) would generate waves that propagate across the magnetic field (Figure 10.17a). These would heat the plasma by *resonant absorption* when the waves reach a location where their frequency matches the local Alfvén or cusp frequency. Waves that propagate along the field, on the other hand (Figure 10.17b), dissipate mainly by *phase mixing* as neighbouring waves propagate at different speeds and gradually become out of phase. Both processes are closely related (Ruderman et al. 1997). They involve a non-uniform background, and so the theory here builds on Sections 4.9 and 4.10 for the basic theory of MHD waves in a non-uniform medium.

10.3.2.1 Resonant Absorption by Sideways Driving

The nature of resonant absorption by MHD waves depends on whether a magnetic structure is driven from the side or from the end. Consider first an ideal MHD theory for sideways driving (Chen and Hasegawa 1974). Suppose perturbation velocities in a field $B_0(x)\hat{\mathbf{z}}$ have the form

$$v_{1x} = v_{1x}(x)e^{i(k_y y + k_z z - \omega t)},$$

and focus on Alfvén wave absorption by assuming $\beta \ll 1$, so that the ideal MHD equation (4.52) with $c_s \equiv 0$ reduces to

$$\frac{d}{dx}\left[\frac{(\omega^2 - \omega_A^2)}{(\omega^2 - k^2 v_A^2)} v_A^2 \rho_0 \frac{dv_{1x}}{dx}\right] + (\omega^2 - \omega_A^2)\rho_0 v_{1x} = 0, \tag{10.27}$$

$$\text{where} \qquad \omega_A^2 = k_z^2 v_A^2 \qquad \text{and} \qquad i v_{1y} = \frac{k_y v_A^2}{\omega^2 - k^2 v_A^2}\frac{dv_{1x}}{dx}. \tag{10.28}$$

This describes Alfvén and fast magnetoacoustic modes (compressional Alfvén waves).

When driving from the side (by, e.g., photospheric motions) at a frequency $\omega = \omega_d$ lying in the continuum, the energy propagates across the magnetic field in the fast mode. This couples to continuum Alfvén modes at the resonant layer ($x = x_r$) given by $\omega_A(x_r) = \omega_d$, where Eq. (10.27) possesses a logarithmic singularity (Figure 10.18). Equation (10.28) implies that the Alfvén and fast modes are coupled for $k_y \neq 0$, but not for $k_y = 0$, since the Alfvén waves cannot transport energy across the field.

The singularity at the resonant layer may be smoothed out by putting $\omega = \omega_d - i\delta$ to mimic resistive or viscous dissipation, with δ ($\ll \omega_d$) a small positive constant (Chen and Hasegawa 1974). Note that the energy absorption is independent of the values of δ, resistivity or viscosity. However, it does depend on the driving frequency (ω_d) and in fact maximises (at 100 per cent of the input energy) at the *quasi-mode*

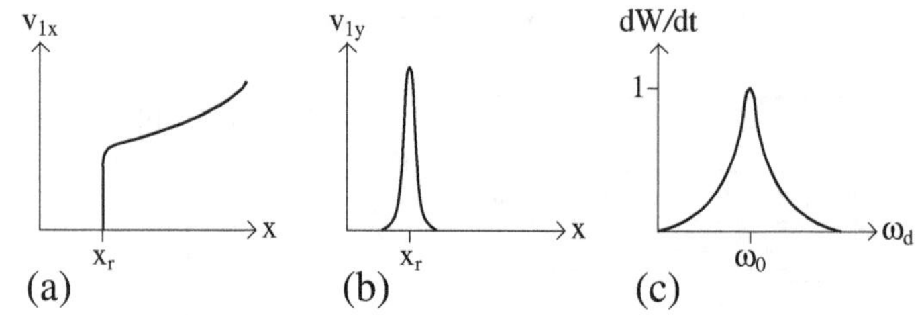

Figure 10.18. (a) A fast mode with velocity v_{1x} propagates from the right and is resonantly absorbed at $x = x_r$, where (b) Alfvén modes with velocity v_{1y} are resonantly excited and dissipate. (c) The dependence of the energy absorption rate (dW/dt, as a fraction of the energy input) on the driving frequency ω_d, where ω_0 is the quasi-mode frequency.

frequency (ω_0) (Figure 10.18). This quasi-mode is the natural oscillation of the system. Its global nature makes it easy to excite, while its coupling to the continuum is efficient and leads to eventual dissipation.

The qualitative features of the analysis were confirmed by Kappraff and Tataronis (1977), who included resistivity in the calculation. Later, Sakurai et al. (1991) and Goossens et al. (1995) repeated the analysis in a cylindrical geometry (relevant to coronal loops) and derived jump conditions that connect external ideal MHD solutions through a resistive resonant absorption layer (PROBLEM 10.1: for all problems and solutions, see web page www.cambridge.org/9780521854719).

At frequencies in the continuum but away from the quasi-mode (ω_0), the efficiency of heating can still be high (often more than 90 per cent of the energy input), and the time it takes to reach a steady-state dissipation is of order $\eta^{-1/3}$ or $\nu^{-1/3}$. At ω_0 itself, the efficiency is 100 per cent, but it takes almost an order of magnitude longer (a time of order $\eta^{-1/5}$ or $\nu^{-1/5}$). For typical coronal loop parameters, this resonant absorption time is a few minutes or hours (Goedbloed et al. 2004).

The response to a broad range of photospheric driving frequencies is that the energy associated with each frequency is dissipated in a different layer. Thus, the input wave energy that reaches the corona (rather than being scattered or refracted), is likely to contribute to heating over a large coronal volume.

In order to demonstrate the practicality of the mechanism in the corona, more detailed models are needed, including line-tying, realistic 3D coronal magnetic and thermal structures, and nonlinearity in the resonant layer. Nonlinearity (Ofman and Davila 1995; Poedts and Goedbloed 1997) is crucial within the resonance layer, since the amplitude of the resonant mode scales as $(\nu + \eta)^{-1/3}$, while line-tying couples the Alfvén and fast magnetoacoustic modes, modifies the continuum and produces multiple resonance layers (Goedbloed and Halberstadt 1994; Beliën et al. 1999; Ofman et al. 1998; De Groof and Goossens 2002).

10.3.2.2 Resonant Absorption by Footpoint Driving

Footpoint driving (Figure 10.17b) is much simpler than sideways driving, since Alfvén modes can be excited directly without needing the fast modes to carry energy across field lines. Consider for simplicity a coronal loop with a uniform magnetic field $B_0\hat{\mathbf{z}}$, a density $\rho_0(x)$ and an Alfvén frequency profile $\omega_A(x) = k_z B_0/\sqrt{\mu\rho_0(x)}$ bounded by $z = 0$ and $z = L$. A typical loop length of $L = 100$ Mm with an Alfvén speed $v_A = 2000$ km s^{-1} has a characteristic time $\tau_A = 50$ s.

Suppose Alfvén waves with motions $v_{1y}(x,z,t)\hat{\mathbf{y}}$ are excited in the loop by a footpoint driver of amplitude $A(x)$ and frequency ω_d in the continuous spectrum. A resonance occurs at the location x_{res} given by $\omega_A(x_{res}) = \omega_d$. It is described by the visco-resistive MHD equations, which reduce to

$$\frac{\partial^2 v_{1y}}{\partial t^2} - (\eta + \nu)\left(\frac{\partial^2}{\partial x^2} + \frac{\partial^2}{\partial z^2}\right)\frac{\partial v_{1y}}{\partial t} + \omega_A^2(x) v_{1y} = A(x)e^{i(k_z z - \omega_d t)}. \tag{10.29}$$

Near the resonance, replacing $A(x)$ by $A(x_{res})$ and $\omega_A(x)$ by $\omega_A(x_{res})$ and seeking a solution of the form $v_{1y} = V_0 \exp[i(k_x x + k_z z - \omega_d t)]$ yields an amplitude

$$V_0 = \frac{A(x_{res})}{[\omega_A^2(x_{res}) - \omega_d^2] - i\omega_d(\eta + \nu)(k_x^2 + k_z^2)}.$$

When $\eta = \nu = 0$, the amplitude of V_0 is real (so that the response is in phase with the driver). It becomes infinite at the resonance when the driving frequency ω_d matches the natural frequency $\omega_A(x_{res})$ of the field line. Including dissipation ($\eta \neq 0$ or $\nu \neq 0$) reduces the magnitude of V_0 to a finite value and introduces a phase shift between driver and response that becomes $\frac{1}{2}\pi$ at resonance. Furthermore, in the steady state, $\partial/\partial z \ll \partial/\partial x$ and a Taylor expansion of $\omega_A(x)$ about x_{res} in [Eq. (10.29)] yields a length-scale and time-scale $[\tau_{res} = l_{res}^2/(\eta + \nu)]$ for resonant absorption of

$$l_{res} \sim \left(\frac{\eta + \nu}{2\omega'_{A0}}\right)^{1/3} \quad \text{and} \quad \tau_{res} \sim \frac{1}{(\nu + \eta)^{1/3} \omega'^{\,2/3}_{A0}},$$

where $\omega'_{A0} \equiv (d\omega_A/dx)_{x=x_0}$.

In the photosphere, there is driving over a broad range of frequencies for each field line, but on each field line it is only the energy associated with its particular resonant frequency that is dissipated. This is much less than the total energy input and so it is unlikely to be a viable coronal heating mechanism.

10.3.3 Chromospheric and Coronal Heating by Phase Mixing

As a response to footpoint driving, a more effective coronal heating mechanism than resonant absorption is phase mixing (Heyvaerts and Priest 1983), since it is spread over the whole volume rather than localised in a narrow resonance layer. Phase mixing may occur either spatially in a propagating wave in an open magnetic region (such as a coronal hole or on a long coronal loop), or in time in a standing wave on a short coronal loop. In both cases, an Alfvén wave is excited on each field line, which oscillates independently of its neighbours with a frequency in the Alfvén continuum. Phase mixing is also part of resonant absorption: it is the process within a resonant layer whereby small scales are produced, but, in the phase-mixing model of coronal heating, resonance is no longer needed, and so the mechanism is effective over the whole range of frequencies, not just the resonant ones.

10.3.3.1 Phase Mixing in Space

On a magnetically open magnetic field (or a long coronal loop), consider a locally planar geometry with an equilibrium magnetic field $B_0(x)\hat{\mathbf{z}}$, density $\rho_0(x)$ and Alfvén speed $v_A(x) = B_0(x)/\sqrt{[\mu\rho_0(x)]}$. Alfvén-wave perturbations in the y-direction and independent of y have the form

$$\mathbf{B} = B_0(x)\ \hat{\mathbf{z}} + B_{1y}(x,z,t)\ \hat{\mathbf{y}}, \qquad \mathbf{v} = v_{1y}(x,z,t)\ \hat{\mathbf{y}},$$

which in ideal MHD satisfy the Alfvén wave equation

$$\frac{\partial^2 v_{1y}}{\partial t^2} = v_A^2(x)\frac{\partial^2 v_{1y}}{\partial z^2}. \tag{10.30}$$

Since the frequency (ω_d) is fixed for propagating waves, each magnetic surface with $x =$ *constant* oscillates independently of its neighbour with a wavenumber $k_z(x) = \omega_d/v_A(x)$, amplitude $V(x)$ and velocity

$$v_{1y}(x,z,t) = V(x)e^{i[k_z(x)z - \omega_d t]}. \tag{10.31}$$

Thus, we have phase mixing in space as the field lines become more and more out of phase (Figure 10.19). At the same time the x-gradients grow, since the variations in phase $k_z(x)z$ in Eq. (10.31) give a contribution

$$\frac{\partial v_{1y}}{\partial x} = i\frac{dk_z}{dx} z\ v_{1y}.$$

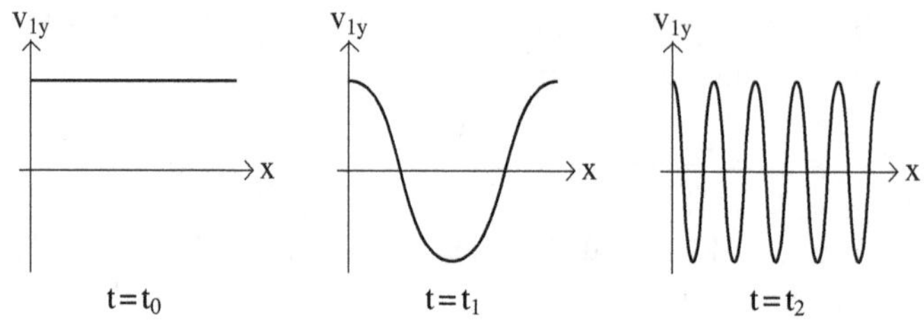

Figure 10.19. The plasma velocity as a function of x for several times $t_0 < t_1 < t_2$, showing the effect of phase mixing in producing a cascade of energy to smaller length-scales.

In other words, the phase mixing produces an effective wavenumber

$$k_x^* = \frac{dk_z}{dx} z,$$

which grows in space with z, so that the effective wavelength becomes smaller and smaller. Such a cascade of energy to smaller length-scales continues until the wave reaches a height where dissipation kicks in and the energy is converted into heat.

This can be seen by including dissipation in the MHD equations. In the case when either viscosity (ν) or magnetic diffusivity (η) or their product ($\nu\eta$) is negligible and phase mixing is well-developed ($\partial/\partial x \gg \partial/\partial z$), Eq. (10.30) is replaced by

$$\frac{\partial^2 v_{1y}}{\partial t^2} = v_A^2(x)\frac{\partial^2 v_{1y}}{\partial z^2} + (\nu + \eta)\frac{\partial^2}{\partial x^2}\frac{\partial v_{1y}}{\partial t}. \tag{10.32}$$

A solution may now be sought in the form

$$v_{1y}(x,z,t) = V(x,z)e^{i[k_z(x)z - \omega_d t]}. \tag{10.33}$$

In the case of weak damping and strong phase mixing (PROBLEM 10.2), the solution has an amplitude that decays with height like

$$V(x,z) = V(x,0)\exp\left\{-\frac{(\nu+\eta)k_z^2}{6v_A^3}\left(\frac{dv_A}{dx}\right)^2 z^3\right\}.$$

The height (typically a few wavelengths) for dissipation by phase mixing is therefore

$$z_{phase} = \frac{\lambda}{2\pi}\left(12\pi Re^*\frac{a}{\lambda}\right)^{1/3},$$

where $\lambda = 2\pi/k_z$ is the longitudinal wavelength, $a = v_A/(dv_A/dx)$ is the length-scale for variations in the Alfvén speed and $Re^* = av_A/(\nu+\eta)$ is a Reynolds number. At this *phase-mixing height*, the x-scale becomes a fraction $(k_x^* a)^{-1} = (12\pi Re^* a/\lambda)^{-1/3}$ of the original scale (a).

See PROBLEM 10.3 for other solutions of Eq. (10.32) by the multiple-scales method.

10.3.3.2 Phase Mixing in Time

On a short coronal loop, standing waves will instead have the wavenumber (k_z) fixed by the geometry, so that each magnetic surface with $x =$ *constant* oscillates independently of its neighbour with a frequency $\omega(x) = k_z v_A(x) \equiv \omega_A(x)$. Its velocity satisfies Eq. (10.30) and has the form

$$v_{1y}(x,z,t) = V(x)e^{i[k_z z - \omega_A(x)t]}. \tag{10.34}$$

This implies phase mixing in time as the field lines become more and more out of phase (Figure 10.19). The time-scale (typically a few periods) for dissipation here is

$$\tau_{phase} = \frac{1}{\omega_A}\left(12\pi Re^* \frac{a}{\lambda}\right)^{1/3},$$

where ω_A^{-1} is the Alfvén time (PROBLEM 10.4).

Dissipation of such standing waves is likely to be enhanced by the growth of tearing-mode instability at the magnetic nodes (Heyvaerts and Priest 1983) and of Kelvin-Helmholtz instability at the velocity antinodes (Browning and Priest 1984). They therefore suggested that such waves are likely to be in a permanent state of MHD turbulence (Sec. 10.5) fed by Poynting flux from the footpoints.

In a coronal hole, phase mixing in space may be important for coronal heating and accelerating the solar wind (Heyvaerts and Priest 1983; Ofman and Davila 1995). In future, more realistic modelling is needed, especially in 3D. A start has been made by several authors, as follows. Ruderman et al. (1998), De Moortel et al. (1999) and Smith et al. (2007) included the effect of a diverging magnetic field with a realistic atmosphere. Hood et al. (2002, 2005) realised that a harmonically generated wave-train decays exponentially, but single pulses decay more slowly (i.e., algebraically) and so transport energy higher into the corona. Mocanu et al. (2008) included an anisotropic viscosity and magnetic diffusivity. McLaughlin et al. (2011) considered nonlinear effects, including generation of flows along the field by localised heating. Finally, Threlfall et al. (2011) found that Hall MHD terms produce wave dispersion and reduce the damping.

10.3.4 Wave Heating of Chromospheric and Coronal Loops

Observations of wave motion (Sec. 4.11.2), such as swaying type II spicules (Secs. 1.5 and 9.2.4) and propagating Alfvénic waves (Tomczyk and McIntosh 2009), suggest they may play a role in chromospheric and coronal mass supply and heating.

Propagating coronal disturbances of low frequency (< 5 mHz) were observed with SoHO and TRACE and interpreted as slow magnetoacoustic waves (e.g., De Moortel et al. 2000; Verwichte et al. 2010), but they contain far too little energy to heat the corona. Type II spicules are swaying chromospheric jets that may provide a substantial part of the coronal mass (De Pontieu et al. 2009, 2011; De Pontieu and McIntosh 2010), but their separate existence and role in heating the corona and driving the solar wind is not yet clear quantitatively (Zhang et al. 2012).

For instance, Klimchuk (2012) has suggested that the heating of type II spicules can supply only a small fraction of coronal plasma (< 2 per cent in active regions and < 5 per cent in the quiet Sun) and the traditional picture of the corona being filled by chromospheric evaporation fits the observations much better. Type II spicules would produce red-blue asymmetries in EUV spectral lines and a ratio of transition region to coronal emission measure that are much larger than observed. Thus, downflows of 10 km s^{-1} in 1 MK lines, which McIntosh et al. (2012) suggest represent the return of heated spicular material, are more likely to be draining of evaporated plasma. Furthermore, heating of spicular material material to $\leqslant$ 0.1 MK and its fall-back to the solar surface may well explain observed transition-region emission.

Transverse propagating kink and Alfvénic waves (Sec. 4.11.2) have been observed in many structures, including spicules (Nikolsky and Platova 1971; Kulidzanishvili and Zhugzhda 1983; De Pontieu et al. 2007), X-ray jets (Sec. 1.6.2) (Cirtain et al. 2007), prominence threads (Harvey 1969; Okamoto et al. 2007; Arregui et al. 2012) and coronal loops (Tomczyk et al. 2007; Tomczyk and McIntosh 2009). In the chromosphere, the energy flux carried by Alfvénic waves is sufficient to heat the quiet-Sun corona and accelerate the solar wind (De Pontieu et al. 2007). In spicules, Alfvénic modes of period 45 sec and amplitude 7 km s^{-1} have a mixture of upwards (59 per cent) and downwards propagating (21 per cent) and standing (20 per cent) waves (Okamoto and De Pontieu 2011). Furthermore, in the chromosphere above magnetic flux concentrations, torsional Alfvén waves have been detected (Jess et al. 2009).

In the corona, COMP observations at the limb have revealed that propagating Alfvénic waves are ubiquitous (Tomczyk and McIntosh 2009), and their observed energy flux of 100 to 200 W m^{-2} (1 to 2 $\times$ 10^5 erg cm^{-2} s^{-1}) is sufficient to heat coronal holes and the quiet Sun (McIntosh et al. 2011). They are too small to heat active regions, but this may be because the superposition of randomly directed oscillating loops or strands along the line of sight reduces the observed Doppler velocity, as demonstrated in numerical experiments (De Moortel and Pascoe 2012).

Although it is now clear that wave modes of many different types are present in the atmosphere and that spicules may perhaps play a role in coronal mass supply, the viability of wave heating and the nature of the heating mechanism needs to be demonstrated (Parnell and De Moortel 2012). For example, the damping of Alfvénic waves in COMP observations may represent coupling to another mode rather than dissipation (De Moortel and Nakariakov 2012).

10.4 Heating by Magnetic Reconnection

In view of the amazing complexity of the chromospheric and coronal magnetic field and the fragmentation and restless motion of its myriads of photospheric footpoints, coronal structures are continually changing and interacting with one another in incredibly intricate ways, so that reconnection in nanoflares (Sec. 10.4.2) is a natural way of heating the corona (Parker 1972; Cargill and Klimchuk 2004). High-resolution observations of the photospheric magnetic field exhibit a highly complex magnetic pattern with frequent changes of polarity (Figure 1.17). The chromospheric and coronal field is equally complex, with many distinct flux tubes showing up in X-ray and EUV pictures (Figures 1.20 to 1.23).

When photospheric motions are sufficiently slow, coronal magnetic configurations evolve passively through a series of equilibria and store energy in excess of potential. However, the complexity of the field implies that these equilibria will inevitably contain a multitude of current sheets and filaments. It is Ohmic dissipation of such current concentrations that is likely to produce much of the necessary coronal heating.

During current-sheet dissipation, magnetic reconnection produces a rapid conversion of magnetic energy into heat, bulk kinetic energy and fast-particle energy (Chapter 6). The suggestion is therefore that, especially in the strong magnetic field of an active region, the corona is in a state of ceaseless activity and is being heated by many tiny micro-flarings (10^{19} J = 10^{26} erg) or nanoflarings (10^{16} J = 10^{23} erg) that are continually generated in tiny current sheets by photospheric motions.

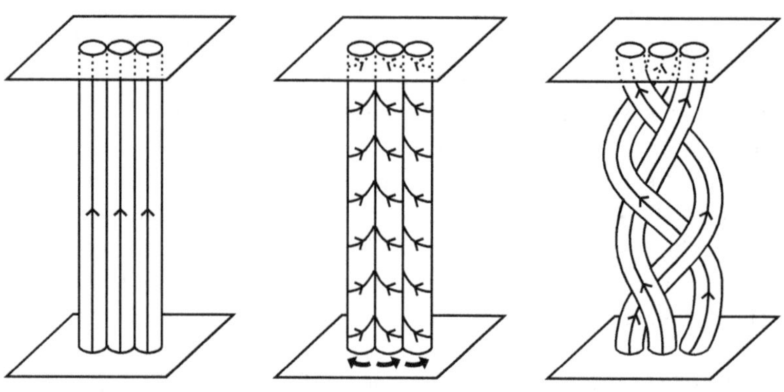

(a) (b) (c)

Figure 10.20. The effect of footpoint motions on (a) a set of straight flux tubes is to (b) twist them up and (c) braid them. Current sheets are likely to form in (b) when neighbouring tubes with the same sense of twist are pressed together and in (c) when braids form.

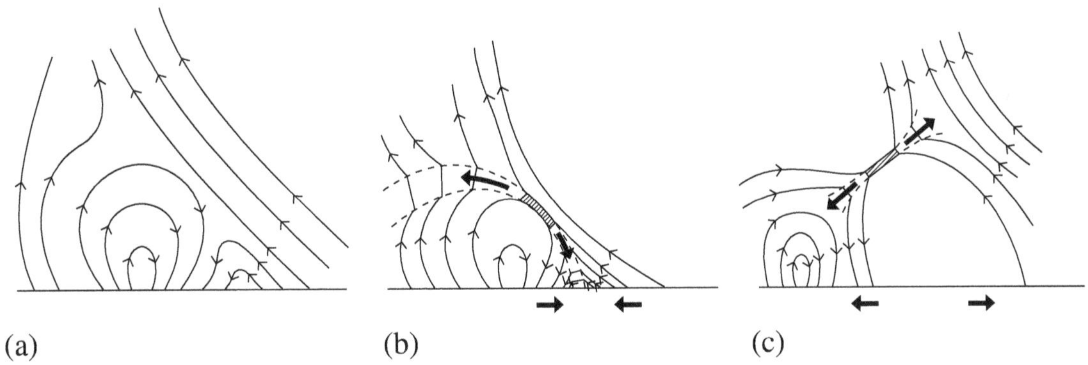

Figure 10.21. Magnetic dissipation due to the relative motion of (a) two neighbouring flux systems when they either (b) approach or (c) separate.

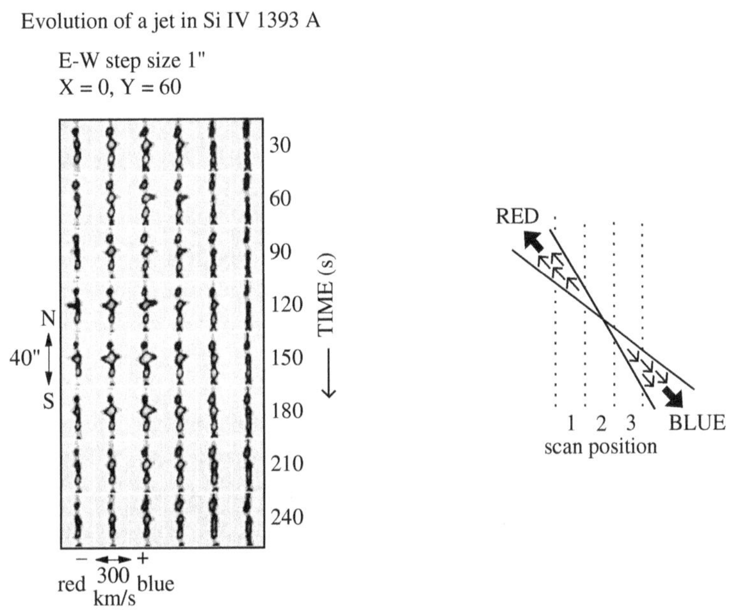

Figure 10.22. SUMER spectra of an explosive event in Si IV with a step-size from left to right in each row of 1 arc sec (courtesy Davina Innes).

The features of heating by *magnetic* (or *current*) *dissipation* that need to be understood are the ways in which current sheets or filaments form, dissipate and decay. For an order-of-magnitude estimate (Tucker 1973; Levine 1974), suppose magnetic energy is being injected with a Poynting flux $\mathbf{E} \times \mathbf{B}/\mu \sim vB^2/\mu$ across an area L^2 at a rate

$$\frac{dW_m}{dt} \sim \frac{vB^2}{\mu} L^2 \tag{10.35}$$

by photospheric motions (v) that twist or shear a magnetic field of strength B. In a steady state, the energy is simultaneously being dissipated ohmically by currents j in a volume L^3 of, say, an active region, at a rate

$$\frac{j^2}{\sigma} L^3. \tag{10.36}$$

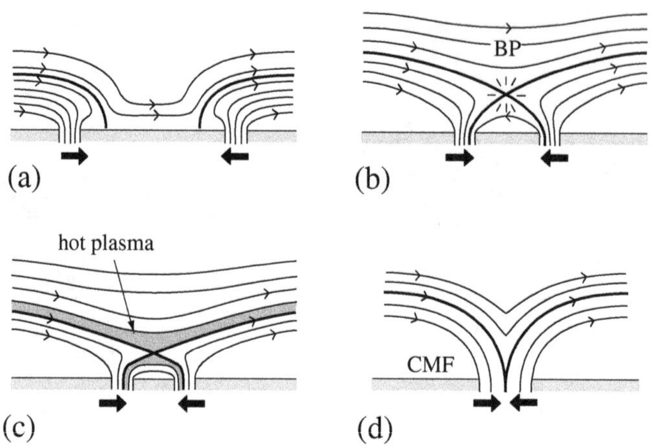

Figure 10.23. The Converging Flux Model for an X-ray bright point. (a) Two equal and opposite magnetic fragments before interacting. (b) When they are closer than the interaction distance, a reconnecting X-point appears at the base and rises upwards into the corona to create (c) an X-ray bright point (BP) with coronal jets. (d) Finally, the fragments meet in the photosphere as a cancelling magnetic feature (CMF) (Priest et al. 1994).

The twisting speed v needed to provide a heat input of, say, 2×10^4 W m^{-2} to the corona is $v \sim 200$ m s^{-1} from Eq. (10.35) for a photospheric field strength of 100 G.

Furthermore, uniform dissipation throughout the active region with a Coulomb conductivity σ ($\approx 10^6$ mho m^{-1}) requires a current density that can be estimated by equating j^2L/σ from Eq. (10.36) to 2×10^4 W m^{-2}, to give $j \sim 15$ A m^{-2} when $L = 100$ Mm. Since this corresponds to an enormous magnetic field gradient (2×10^{-5} tesla m^{-1} or 200 G km^{-1}), it is clear that dissipation is concentrated at many thin current sheets rather than being distributed uniformly. The sheets occupy only a small fraction of the active-region volume and each of them lasts for a short time.

Space satellites have provided much evidence of reconnection at work on the Sun (Sec. 10.4.2), although more details are needed on the way 3D reconnection works and on the observational signatures. One part of the coronal heating problem, namely, X-ray bright points, appears to have been solved (Sec. 10.4.3). There are several ways of forming current sheets to produce nanoflares (Sec. 10.4.1), one of which is Parker's early concept of braiding (Sec. 10.4.4) and another is its more recent development, namely, flux-tube tectonics (Sec. 10.4.5). Once a current sheet has formed, it may rapidly dissipate, so that part of the stored energy is liberated locally as heat and part of it propagates away in a fast-mode wave (Sec. 10.4.5.6).

10.4.1 Formation of Current Sheets

In a simple bipolar magnetic field when the photospheric footpoints move slowly, the low-β corona responds by establishing a series of smooth force-free configurations. In general, however, the coronal field is much more complex, and so current sheets may be formed in a variety of ways (Figure 10.20), as follows, and contribute to chromospheric and coronal heating as they dissipate.

(i) **Interaction of Topologically Distinct Flux Systems** (Sec. 6.3). On the boundaries between separate flux systems (*separatrices*), current sheets easily form, both in 2D and 3D, the magnetic field reconnects, and magnetic energy is therefore released (Priest and Forbes 2000), a process that occurs in particular in flux-tube tectonics (see the following). As the photospheric footpoints of coronal loops move, the neighbouring coronal flux tubes respond by interacting with one another, either moving further apart or coming closer together. When 2D magnetic configurations contain topologically distinct regions with X-type null points, the relative motion of such regions (either pushing together or pulling apart) will create current sheets at the null points

(Figure 10.21, Secs. 6.3.1 and 6.3.2) when reconnection is inhibited. The shearing of regions that contain separatrices will also create current sheets (Sec. 6.3.3).

In 3D configurations, in a similar way, when null points (Sec. 6.9.1), separators (Sec. 6.9.2) or quasi-separators (Sec. 6.9.5) are present, the shearing or relative motion of the separate regions produces current sheets. Such sheets form not only when neighbouring magnetic field lines are oppositely directed, as in Figure 10.21, but also when the field lines are inclined at a non-zero angle. The ways in which reconnection occurs in 3D are modelled in Sections 6.12 and 6.13.

(ii) **Instability or Non-equilibrium**. Current sheets may develop when a magnetostatic equilibrium becomes unstable, such as during the nonlinear development of kink instability of a flux tube (Browning et al. 2008; Hood et al. 2009). They can also be driven when an equilibrium ceases to exist (i.e., *non-equilibrium*).

(iii) **Braiding** (Sec. 10.4.4). Parker (1972, 1994) hypothesised that, for a smooth mapping, such as a uniform magnetic field between two planes, braiding of the footpoints would produce current sheets (Figure 10.20). Infinitesimal braiding has been shown not to work, since it produces a continuous field without current sheets. For finite-amplitude braiding, the jury is still out, since the existence of current sheets remains unproved. However, numerical experiments do indeed demonstrate the formation of current concentrations: even though these are not infinitesimally thin, they can be thin enough to produce dissipation in practice. Thus, although in a purist sense the mechanism has not been proved, in reality it can generate heat.

(iv) **Flux-Tube Tectonics** (Sec. 10.4.5). As a modification of Parker-braiding, the complexity of the coronal field produced by the *magnetic carpet* (Sec. 1.4.2) greatly enhances the ease of current-sheet production and so the effectiveness of heating (Priest et al. 2002). The presence of many small magnetic sources in the photosphere creates a highly complex coronal topology containing myriads of separatrix surfaces, where current sheets form in response to simple photospheric motions. Several aspects of flux-tube tectonics are described, including separator reconnection (Sec. 10.4.5.3), binary reconnection (Sec. 10.4.5.4) and a comparison of separator and separatrix heating (Sec. 10.4.5.2).

(v) **MHD Turbulence**. Alternative and complementary ways to describe the complexity of heating by waves or by reconnection in many small regions are to use ideas of MHD turbulence (Sec. 10.5.1) or avalanches (Sec. 10.5.2).

(vi) **Flux Emergence**. New flux emerges on a variety of scales, from internetwork to whole active regions. In each case, a current sheet will develop where new flux presses up against the pre-existing flux. For example, in a supergranule cell, flux emergence naturally produces an X-ray bright point (Figure 10.24). The same effect is possible when flux is evolving rather than emerging. If photospheric footpoints move relative to one another at a certain speed, the overlying atmosphere will just respond by creating a current sheet and allowing reconnection at that speed. Dissipation is maintained as long as the relative footpoint motion continues, with the dimensions of the current sheet depending on the magnetic field strength and photospheric speed. Conditions inside the sheet will be turbulent if the resulting sheet width is small enough. Also, slow magnetoacoustic shock waves radiate from the ends of the current sheet and fine jets of plasma are emitted between pairs of shocks. As plasma comes in slowly from the sides, the bulk of the heat is released at these shock waves rather than in the central current itself.

Once a current sheet has formed, it may break up into current filaments, for example, by tearing-mode or thermal instability. If the width of such current concentrations is less than (typically) a few metres, the dissipation may be considerably enhanced by plasma turbulence.

10.4.2 Evidence from Space for Reconnection

As well as evidence for its presence in solar flares (Sec. 1.9.2.3), reconnection gives an elegant explanation for many other diverse phenomena described in the following, namely, brightenings in the transition region and corona, such as explosive events, blinkers, microflares, tornadoes and X-ray bright points (Sec. 10.4.3).

You can, of course, never prove a theory with observations – you can only disprove it, which is rather sad for theorists! But there has been a real paradigm shift. Whereas previously reconnection was a fascinating concept which exercised the imagination of theorists like myself, now there are so many observations that fit beautifully into place with the eyes of reconnection that it has become the natural explanation for many coronal phenomena.

Space observations have given important clues about the nature of coronal heating (Sec. 10.2.1) and in particular evidence for reconnection. For example, when large-scale fields close down, Yohkoh discovered that they do so in characteristic cusp-shaped structures. In addition, observations of the temperature in active regions show that all the hottest loops are either cusps or pairs of apparently interacting structures, which is highly suggestive of reconnection (Yoshida and Tsuneta 1996).

Explosive events (Brueckner and Bartoe 1983; Dere et al. 1989) are short-lived bidirectional jets in transition-region lines, with blue and red shifts of 100 km s^{-1} and durations of 1 minute (Sec. 1.5). They have been observed with SUMER, TRACE and SDO (Innes et al. 1997; Innes 2001; Innes et al. 2011). Bidirectional jets in such events are accelerated in opposite directions from reconnection sites (Figure 10.22). The sites are located over regions of complex magnetic field, where magnetic fragments are approaching or interacting with one another and driving reconnection in the overlying atmosphere. They are accompanied by coronal brightenings and may represent swirling jets (Curdt et al. 2012), which are a natural result of 3D reconnection.

Blinkers are much longer-lived transition-region brightenings in the supergranule network with typical lifetimes of 5 to 40 minutes, intensity increases by factors 2–3 and Doppler velocities of 25 to 30 km s^{-1} (Harrison 1997; Bewsher et al. 2005) (Sec. 1.5). They also occur in active regions and above sunspots with lifetimes 16 to 19 min, slightly higher velocities (20 to 40 km s^{-1}), intensity increases of 1.8 to 3.3, and associated coronal brightening (Parnell et al. 2002). Their origin is unclear and could be heating of spicules, containment of plasma in network loops, heating and evaporation or cooling and condensation in response to coronal heating (Priest et al. 2002). Some of them may be macrospicules (Madjarska et al. 2006). In Hinode/XRT, two-thirds correspond to coronal events, such as coronal jets or X-ray bright points, and occur repetitively over 40 minutes, while the remainder are point-like, with a single occurrence over 20 minutes, and have no coronal counterpart (Subramanian et al. 2012).

Tornadoes appear as rotating vortex tubes in a range of different sizes (Sec. 1.4.1), but the relationship between them is unclear. The rotation may be caused by 3D reconnection, since twisted structures are naturally produced by conservation of global magnetic helicity as mutual helicity is converted into self helicity (Sec. 6.10). Alternatively, it may be driven by photospheric vortex tubes. Examples include: macrospicule tornadoes, seen in coronal holes (Pike and Mason 1998; Wedemeyer-Böhm et al. 2012) (Sec. 1.5.2), rotating at typically 20 km s^{-1} (up to 150 km s^{-1}); and barb tornadoes seen below prominences (Secs. 1.8.3).

Coronal X-ray jets are dynamic collimated ejections, which are a clear signature of reconnection (Sec. 1.6.2). At chromospheric temperatures one finds cool jets, such as spicules and surges, which may also be produced by reconnection (Sec. 1.5).

Microflares and *Nanoflares*. Many small hard X-ray spikes were discovered from a rocket flight with energies 10^{17} to 10^{20} J (10^{24} to 10^{27} ergs) (Lin et al. 1984). These led Parker (1988) to speculate that a sea of nanoflares at 10^{24} ergs or less could heat the corona, provided they occur in sufficient numbers. If their energy spectrum is W^{α}, where W is the energy of an event, then a power-law index $\alpha < -2$ is needed for small events to dominate (Hudson 1991). Using SoHO/EIT, Krucker and Benz (1998) observed microflares that contribute 20 per cent of the quiet-Sun heating (Sec. 1.9.2.3), with energies of 10^{25} to 10^{26} ergs and an energy spectrum of $W^{-2.6}$. Smaller-scale events with TRACE have a spectrum that continues down to at least the nanoflare range (10^{24} ergs) at a slope between W^{-2} and $W^{-2.6}$ (Parnell and Jupp 2000). Other authors found a wide range of values for the power-law index (Shimizu 1995; Aschwanden et al. 2000; Benz and Krucker 2002), and so Parnell (2004) made a comprehensive analysis of TRACE data and found $-2.5 < \alpha < -1.6$ with a mean of -1.9, concluding that it was not possible accurately to determine whether $\alpha < -2$ or $\alpha > -2$. Non-thermal events in hard X-rays from RHESSI are seen only in the active-region bands and are too intermittent in space and time to account for coronal heating (Hannah et al. 2011), so if nanoflares are to work they need to be thermal. For other observational consequences of nanoflare heating see Section 10.2.1, and for theories see Sections 10.4.4 and 10.4.5.

Active-region transient brightenings are micro- or nano-flares in active regions (Sec. 1.7.1.3). They last for typically 1 to 15 minutes over areas 10 to 100 Mm^2 and have energies 10^{17} to 10^{22} J (10^{24}–10^{29} ergs). Some are point-like and others loop-like. They heat the plasma to 5 MK and are associated with emerging flux, especially around sunspots (Shimizu et al. 1992; Berghmans and Clette 1999; Warren et al. 2007; Kano et al. 2010). Observations from SDO/AIA and NST (Big Bear Observatory) show upflow from the ends of fine loops being filled with hot plasma (Ji et al. 2012).

10.4.3 X-ray Bright Points: The Converging Flux Model

All over the surface of the Sun, very small regions ($\approx$ 3 arc sec) are bright in X-rays and have a duration of a few hours or less. Because these regions appeared point-like in early X-ray telescopes, they are called X-ray bright points or XBPs for short (Sec. 1.6.2). Some XBPs are located above emerging flux and are explained by the Emerging Flux Model (Heyvaerts et al. 1977), but most are situated in the corona above pairs of opposite polarity magnetic fragments that are approaching one another (Harvey 1996). As they collide, these fragments appear to annihilate one another in a process referred to as *cancellation* (Martin et al. 1985). Because of the overlying coronal brightening, it is likely that reconnection is taking place above the cancellation site (Priest 1987).

The Converging Flux Model (Figures 10.23 and 10.24) explains how cancellation can produce an XBP (Priest et al. 1994; Parnell and Priest 1995). Magnetic flux emerges in a supergranule cell and then moves to the boundary; one polarity accumulates, while the other reconnects with opposite-polarity network and forms a bright point. Because of the overlying field, a null-point (or separator in 3D) forms only when the opposite polarities are sufficiently close. The null point appears at the surface and moves upwards as the polarities approach. Continued motion eventually causes the null point to reverse direction and sink back into the photosphere. Figure 10.25 shows how this structure matches the predictions made by a 3D version of the model.

Consider two oppositely directed photospheric magnetic fragments of flux $\pm F_0$ in a uniform horizontal field of strength B_0. When the fragments are far apart, they are not linked to each other, but, when they are closer than the *interaction distance*, reconnection at a coronal null point (or separator in 3D) between the fragments allows field lines to join them (Priest et al. 1994; Longcope 1998). In a 2D model, the interaction distance is $d = 2F_0/(\pi B_0)$, but in a 3D model it is $d^* = \sqrt{F_0/(\pi B_0)}$ (PROBLEM 10.5). Thus,

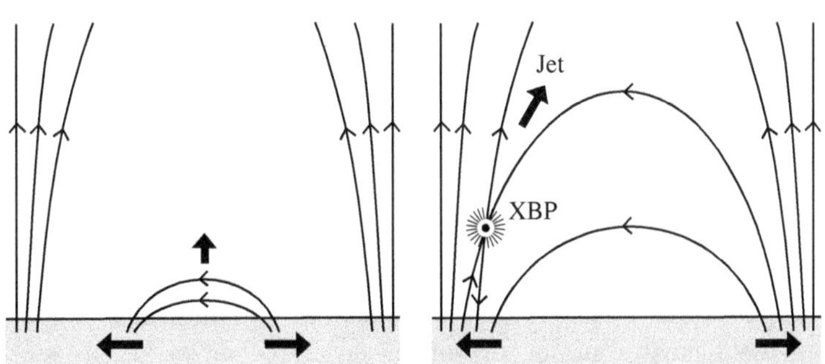

Figure 10.24. The creation of an X-ray bright point at the edge of a supergranule cell, according to the Converging Flux Model (from Priest and Forbes 2000).

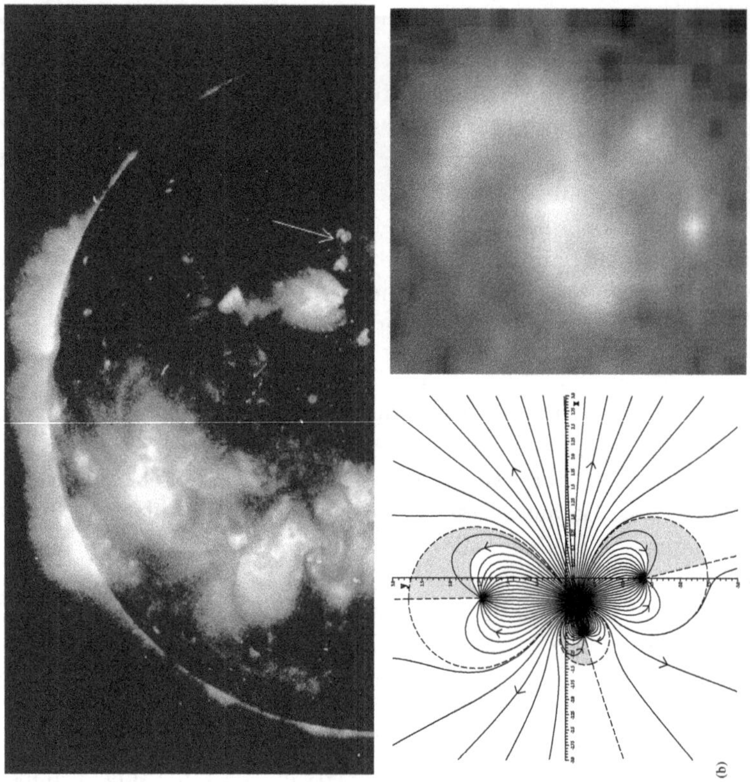

Figure 10.25. Left: X-ray images of the Sun's disc. The arrow shows a faint X-ray bright point. Top right: Close-up of this bright point showing filamentary fine structures. Bottom right; location of the lines of force extrapolated from the observation of the surface magnetic field. The shaded regions show regions of plasma heated by reconnection (Parnell et al. 1994).

for example, fragments of flux 10^{18} Mx in a field of $B_0 = 10$ G have an interaction distance of 2 Mm. For a 3D simulation, see Büchner (2006) and Javadi et al. (2011).

Priest et al. (1994) therefore proposed that coronal reconnection, driven by footpoint motion of neighbouring magnetic fragments that are separated by less than the interaction distance, may be at the root of the nanoflare/microflare process and may be heating normal coronal loops. They suggested that X-ray bright points may be at the large-scale end of a broad spectrum of such coronal heating events. This idea has gained support from a study by Falconer et al. (1999), who found much smaller bright points than previously, called *micro-bright points* in the network over mixed polarity. Comparison with a magnetogram showed that the normal bright points lie over large magnetic fragments of opposite polarity, which are close to each other.

10.4.4 Current-Sheet Formation by Parker Braiding

According to Parker (1972), current sheets may be created by the braiding of magnetic fields around one another, and their dissipation leads to nanoflare heating in many small locations (Figure 10.20). He hypothesises that, if the pattern of small-scale variations is not uniform along a large-scale field, then a force-free equilibrium field must contain current sheets. In other words, a smooth equilibrium exists only if field variations consist of a simple twist extending from one footpoint to the other. More complex mappings, such as braided tubes, must, according to Parker, lead to the formation of current sheets. He first tried (unsuccessfully) to prove this for infinitesimal departures from a uniform field and then later proposed the same hypothesis for a large (finite-amplitude) departure. Many authors have tried to verify it analytically or computationally (e.g., Rosner and Knobloch 1982; Tsinganos et al. 1984; Zweibel and Li 1987; Arendt and Schindler 1988), and evidence in favour of its qualitative viability has grown, but it has not yet been proved formally for finite braiding. A more efficient way of forming current sheets is by tectonics (Sec. 10.4.5).

Parker's first attempt to obtain his result is as follows. Consider a uniform configuration with plasma pressure p_0 and magnetic field $B_0\hat{\mathbf{z}}$, with the footpoints anchored in the planes $z=\pm L$. Suppose that a small displacement of the footpoints by at most $\lambda \ll L$ leads to small deviations (p_1 and $\mathbf{B}_1$) so that

$$p=p_0+p_1(x,y,z), \qquad \mathbf{B}=\mathbf{B}_0+\mathbf{B}_1(x,y,z),$$

where $p_1/p_0 \sim B_1/B_0 \ll 1$. Then the equation $-\nabla p+\mathbf{j}\times\mathbf{B}=\mathbf{0}$ for magnetostatic equilibrium becomes, to first order,

$$-\nabla\left(p_1+\frac{B_0 B_{1z}}{\mu}\right)+\frac{B_0}{\mu}\frac{\partial \mathbf{B}_1}{\partial z}=0. \tag{10.37}$$

But $\nabla\cdot\mathbf{B}=0$ implies $\nabla\cdot\mathbf{B}_1=0$ and so the divergence of Eq. (10.37) implies

$$\nabla^2\left(p_1+\frac{B_0 B_{1z}}{\mu}\right)=0. \tag{10.38}$$

However, variations of $p_1+B_0B_{1z}/\mu$ extend at most a distance of order $\lambda \ll L$ into the region $-L\leqslant z\leqslant L$ from the end planes, and so the solution to (10.38) is

$$p_1+\frac{B_0 B_{1z}}{\mu}=constant, \tag{10.39}$$

except in boundary layers of width λ near $z=\pm L$. Now, the z-component of Eq. (10.37) is $\partial p_1/\partial z=0$ and so the z-derivative of Eq. (10.39) implies $\partial B_{1z}/\partial z=0$: that is, if a field is in smooth magnetostatic equilibrium, its pattern does not change along the general direction of the field. Parker therefore suggested that, if the field is instead braided, it must contain current sheets.

Sakurai and Levine (1981) established that, for a small perturbation about a uniform field, footpoint motions do indeed produce a smooth perturbed field from Eq. (10.37). $\mathbf{B}_1$ is determined uniquely when an arbitrary small footpoint motion is imposed. The perturbation does satisfy the condition $\partial\mathbf{B}_1/\partial z=0$ far away from the boundary, but all the rapid variations occur within boundary layers near $z=\pm L$, with slow variations elsewhere. The equilibrium is, therefore, not of the form shown misleadingly in Figure 10.20c, but instead the braids are located near the boundaries. Interestingly, the force-free fields arise from a difference in rotational motions on the boundaries and are the low-frequency limit of shear Alfvén waves (PROBLEM 10.6).

What small random motions do is to produce a cascade of magnetic energy to small scales: a mean-square current grows exponentially in time, as magnetic and current spectra spread to smaller scales (van Ballegooijen 1985). Finite-amplitude footpoint motions that randomly braid an initially uniform magnetic field have been studied computationally with 3D ideal and resistive MHD numerical experiments (Mikić et al. 1990; Galsgaard and Nordlund 1996a; Craig and Sneyd 2005; Rappazzo et al. 2010). They show evolution through a series of smooth equilibria, with a transfer to small scales and an exponential growth of filamentary currents. The current sheets are highly complex (Figure 10.26), and roughly one turn of braiding is required to initiate reconnection; the resulting heating is rather uniform along the coronal part of a loop, but is much higher in the chromosphere due to the lower Alfvén speed there and reflection off the transition region (Galsgaard et al. 1999; Galsgaard 2002).

A more general statement of Parker's hypothesis is that, if the normal component of the magnetic field is imposed on the boundary and the mapping of the field lines from one boundary to another is imposed,

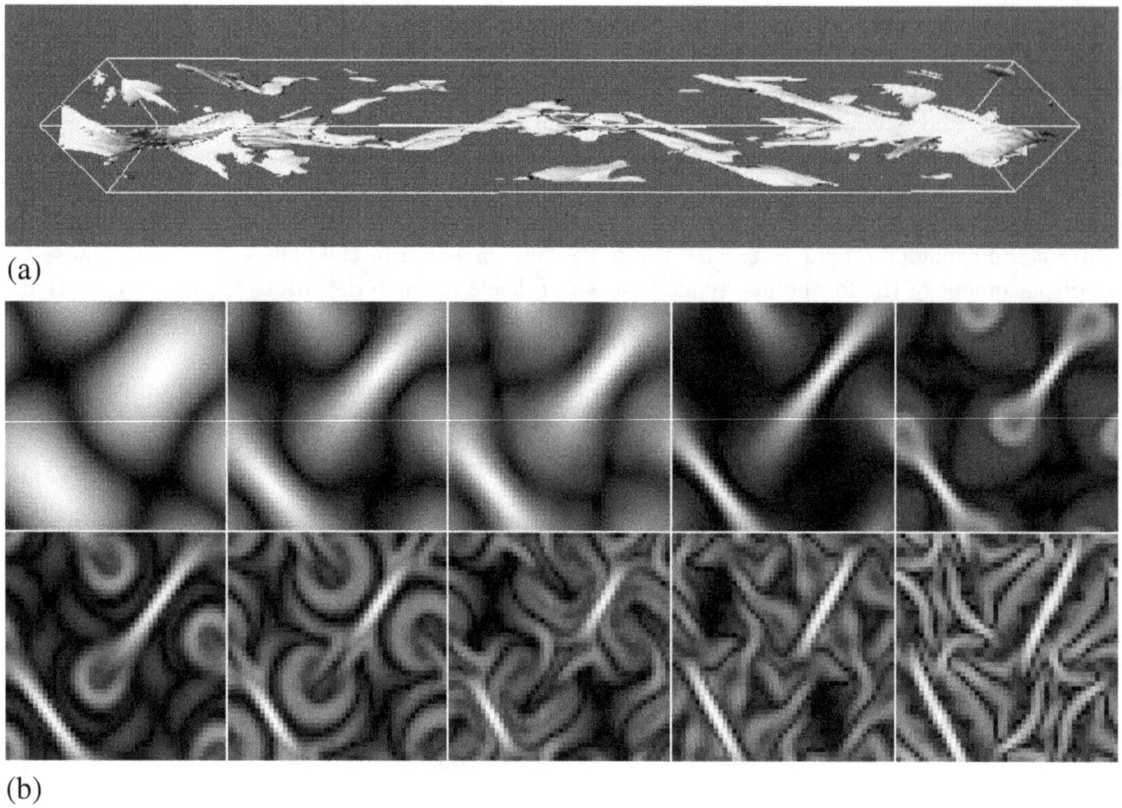

Figure 10.26. The effect of braiding on current filamentation that increases in time as the braiding builds up: (a) in a 3D perspective plot at one time and (b) in a section at several times (courtesy Klaus Galsgaard).

then the resulting force-free field will in general contain current sheets (after Low 2010). For example, Janse and Low (2009) considered an untwisted and unbraided magnetic field in potential equilibrium in a cylinder and suggested that, if the distance between the two ends were decreased and the footpoints held fixed, then there would be no smooth potential field. However, perhaps a smooth force-free field would be possible instead (Pontin and Huang 2012).

Wilmot-Smith et al. (2009a) initiated a different way of analysing braiding. Rather than consider complex braiding motions acting on a simple field, they studied the instability of a complex field. They first set up an analytical field for a pigtail braid that has no net twist and is not in equilibrium (Figure 10.27a). It is the sum of a uniform field and three identical pairs of magnetic flux rings produced by localised currents (intersections of strands) in the form

$$\mathbf{B} = B_0\hat{\boldsymbol{z}} + \sum_{i=1}^{n} \mathbf{B}_{ci},$$

where each of the $\mathbf{B}_{ci}$s is of the form

$$\mathbf{B}_c = 2B_0 k\, \frac{r}{a} \exp\left(-\frac{r^2}{a^2} - \frac{z^2}{l^2}\right) \hat{\boldsymbol{\phi}}$$

in spherical polars, except that it has the parameters $\mathbf{c}_i = (x_i, y_i, z_i, k_i, a_i, l_i)$ for the centre (x_i, y_i, z_i) in place of the origin and (k_i, a_i, l_i) in place of k, a, l.

In particular, they considered $\mathbf{c}_1=(1,0,-20,1,\sqrt{2},2)$, $\mathbf{c}_2=(-1,0,-12,-1,\sqrt{2},2)$, $\mathbf{c}_3=(1,0,-4,1,\sqrt{2},2)$, $\mathbf{c}_4=(-1,0,4,-1,\sqrt{2},2)$, $\mathbf{c}_5=(1,0,12,1,\sqrt{2},2)$, $\mathbf{c}_6=(-1,0,20-1,\sqrt{2},2)$, in a line-tied numerical box with $-4\leqslant x\leqslant 4, -4\leqslant y\leqslant 4, -24\leqslant z\leqslant 24$. The braid has only 3 per cent more energy than the uniform field. They used a 3D Lagrangian ideal relaxation scheme (Craig and Sneyd 1986) to solve

$$\frac{d}{dt}\left(\frac{\mathbf{B}}{\rho}\right)=\left(\frac{\mathbf{B}}{\rho}\cdot\boldsymbol{\nabla}\right)\mathbf{v}, \qquad \nu\,\mathbf{v}=\mathbf{j}\times\mathbf{B},$$

namely, the ideal induction equation and a fictitious equation of motion that guarantees a monotonic decrease of magnetic energy. Magnetic flux and $\boldsymbol{\nabla}\cdot\mathbf{B}$ are both conserved. The current, initially in six isolated closed structures, forms two spirals in the relaxed state, with smooth $\mathbf{j}$ and $\mathbf{B}$ that contains no small-scale structure. The twist in the initial configuration becomes evenly distributed along field lines which are still braided. The initial and final field lines are shown in Figure 10.27b.

A remarkable feature is that the integrated parallel currents ($\mathcal{J}_\parallel\equiv\int j_\parallel ds$) contain extremely fine-scale structure normal to the field (Figure 10.27c), which is conserved during the relaxation. These fine scales are independent of the scales of $\mathbf{j}$ itself and are produced by the complexity of the field-line mapping from one boundary to the other. Since $\int E_\parallel ds=\eta\mathcal{J}_\parallel$, the fine-scale currents are likely to produce complex 3D reconnection and dissipation (Sec. 6.12.5). The width of the $\mathcal{J}_\parallel$ structures decreases exponentially with the number (n) of braids, while the peak value increases linearly.

Wilmot-Smith et al. (2009b) also calculated the squashing factor (Q), which identifies QSLs (Sec. 6.9.5). It possesses a similar highly filamentary structure to $\mathcal{J}_\parallel$; the width of the layers decreases and the maximum value of Q increases exponentially with n, so that the domain becomes filled with a multitude of QSLs. Subsequent reconnection was followed with a resistive MHD code using the force-free near-equilibrium as a starting point (Wilmot-Smith et al. 2010; Pontin et al. 2011). Two thin current sheets form at the locations of peak $\mathcal{J}_\parallel$ and reconnection produces a subtle untwisting of field lines rather than plasma jets.

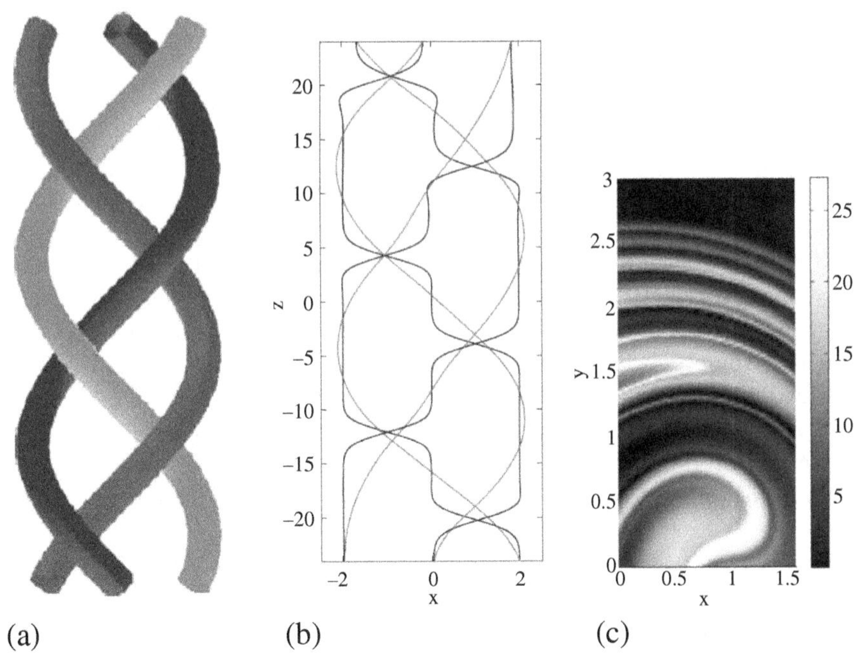

Figure 10.27. (a) A pigtail or Borromean braid. (b) Three braided field lines viewed from the side in the initial and final (less curved) state. (c) Contours of integrated parallel current ($\int j_\parallel ds$) in the initial state as a function of x and y (Wilmot-Smith et al. 2009a).

The dominant $\mathbf{j} \times \mathbf{B}$ forces and flows are in the xy-plane, where the magnetic structure is elliptic rather than hyperbolic (cf Parnell et al. 2010).

The subsequent evolution is by resistive relaxation involving decaying turbulent formation of a complex array of current sheets scattered through the domain. The global reconnection rate summed over all diffusion regions continues to grow due to fragmentation at a rate and duration that increase with magnetic Reynolds number (R_m) and involves *multiple or recursive reconnection* (cf. Parnell et al. 2008), in which the same flux is reconnected 4.4 times for $R_m = 5000$. The reconnection is *fast* in the sense of being independent of R_m. The final effect is to untangle the field lines and leave two weakly twisted flux tubes of opposite twist; the resulting nonlinear force-free field is thought to be caused by an extra constraint (the *topological degree* of the mapping) in addition to the usual conservation of magnetic helicity in Taylor relaxation (Yeates et al. 2010).

Later, Wilmot-Smith et al. (2011) conducted resistive MHD simulations on the relaxation of braided magnetic loops. They showed how the type of braiding produced by footpoint motions affects the resulting loop heating, as measured by the *topological entropy.* This is a measure of the complexity of the footpoint flow, namely, the asymptotic rate of stretching of material lines and so the rate of separation of neighbouring footpoints. For a simple twisting, the topological entropy is low, a small number of current sheets develop, only a little magnetic energy is released, and the result is a patchy heating of the loop. For a complex braid, the relaxation is very efficient with current sheets filling the volume and homogeneous heating of the loop. Furthermore, Yeates and Hornig (2013) have introduced a *topological flux function* to quantify the topology of *magnetic braids* (non-zero, line-tied magnetic fields whose field lines all connect between two boundaries). It is an ideal invariant, written as

$$\mathcal{A} = \int_{S_0}^{S_1} \mathbf{A} \cdot d\mathbf{s},$$

measured along a field line from one boundary (S_0) to the other (S_1). It measures the average poloidal magnetic flux around any given field line and uniquely characterizes the field-line mapping and hence magnetic topology. Physically, it represents the magnetic helicity (Sec. 6.10) in an infinitesimal flux tube (cf. Berger 1988).

10.4.5 Flux-Tube Tectonics: Separator, Separatrix and QSL Heating

The surface of the Sun is punctured by a multitude of flux tubes which act as magnetic sources for the coronal field. They continually move around and produce a highly complex field in the overlying chromosphere and corona, known as the *magnetic carpet* (Schrijver et al. 1997) (Sec. 1.4.2). The idea of the Flux-Tube Tectonics Model (Priest et al. 2002) is that the presence of many small magnetic sources in the photosphere creates a highly complex coronal topology with myriads of separatrix and QSL surfaces. At these surfaces in response to footpoint motions, current sheets form more readily than by Parker braiding and rapidly dissipate (Sec. 10.4.1).

A key question which the model seeks to address is: what is the effect of the relative motions of photospheric sources in driving reconnection and, therefore heating, in the overlying atmosphere? A general possibility is *separator heating* (Sec. 10.4.5.3), but a fundamental mechanism for creating such reconnection is the binary interaction between pairs of sources (Sec. 10.4.5.4).

10.4.5.1 Basic Model

In determining the effect of the magnetic carpet on coronal heating, three important factors are: concentration of flux in the photosphere into discrete intense flux tubes; their continual motion; and the fact that the coronal field consists of a complex collection of topologically separate volumes divided from one another by separatrix surfaces (and by QSLs).

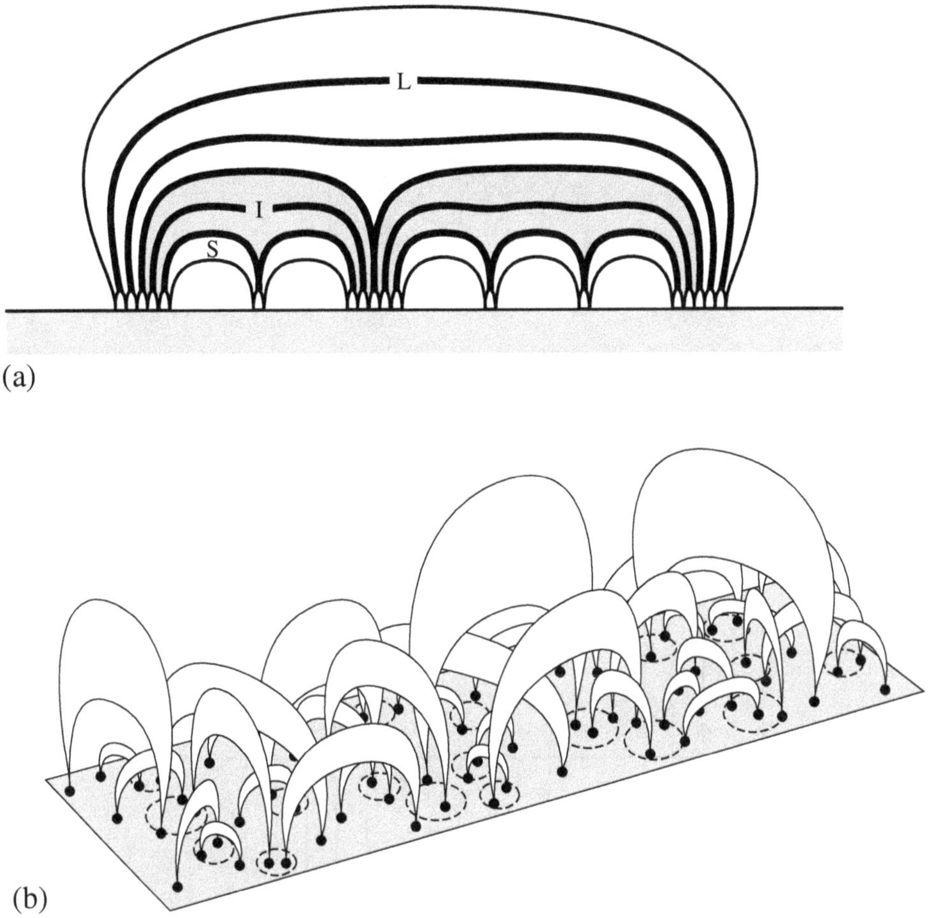

(a)

(b)

Figure 10.28. (a) A schematic of a coronal loop consisting of many sub-volumes, each linked to a separate source and divided from one another by separatrix surfaces. (b) A 3D view within the network indicating that the photospheric flux elements (dashed ovals) are in general connected to a range of neighbours. In practice each coronal loop will link to many more neighbours than indicated here (Priest et al. 2002).

Each coronal loop has a magnetic field that links to the solar surface in many sources. The flux from each source is topologically distinct and is separated from each other by separatrix surfaces (Figure 10.28). As the sources move, the overlying magnetic field slips and forms current sheets along the separatrices, which then reconnect and heat. Thus, in this view, the chromosphere and corona are filled with myriads of separatrix and separator current sheets continually forming and dissipating (Figure 10.29a,b).

The fundamental flux units in the photosphere are likely to be intense flux tubes with fields of 1,200 G, diameters of 100 km (or less) and fluxes of 3×10^{17} Mx (or less). A simple X-ray bright point thus links to a hundred sources and each TRACE loop probably consists of at least 10 finer, as yet unresolved, loops (Figure 10.29c).

Whereas Parker's braiding model assumes complex footpoint motions acting on a uniform field, Priest et al. (2002) consider the effect of simple motions on an array of flux tubes, anchored in small discrete sources. For an array of flux tubes anchored in two parallel planes, they have demonstrated the formation of current sheets by smooth shearing motions and estimated the heating (Figure 10.30).

The simplest model (PROBLEM 10.7) is to schematize the corona as a 2D (x, z) layer between two photospheric surfaces on the planes $z = \pm L$, with an array of line sources of flux F at $z = \pm L$, $x = (2n+1)w$,

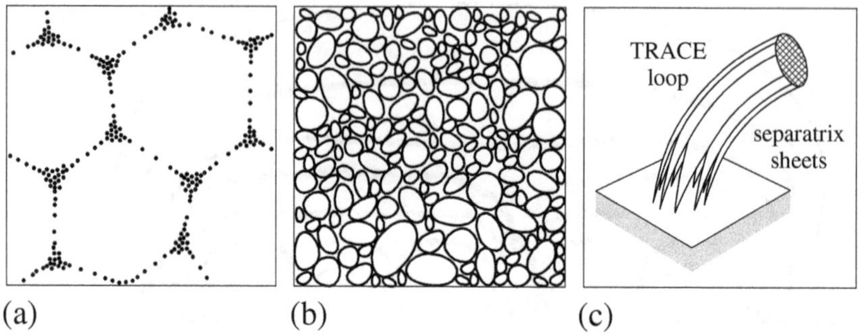

Figure 10.29. (a) A schematic of intense flux-tube sources in the photosphere. (b) Intersections of many current sheets with a horizontal surface in the corona (only a small fraction of the actual sheets are indicated). (c) An observed coronal loop connects to the surface in many sources separated by separatrix current sheets.

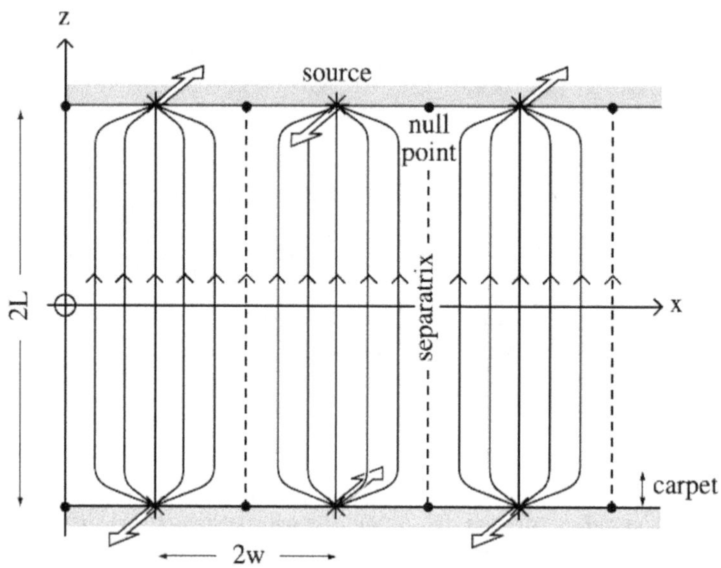

Figure 10.30. 2D idealisation of coronal loops due to magnetic sources (asterisks) in planes $z = \pm L$, showing null points (filled circles) and separatrices (dashed) (Priest et al. 2002).

where $w/(2L) \ll 1$ is the aspect ratio of the loops (Figure 10.30). The initial potential magnetic field is

$$(B_{0x}, B_{0z}) = \frac{F/(2w)}{[\cosh[\pi(L-|z|)/w] + \cos(\pi x/w)]} \left\{ \text{sgn}(z) \sin\left(\frac{\pi x}{w}\right), \sin\left[\frac{\pi(L-|z|)}{w}\right] \right\}, \tag{10.40}$$

which is uniform when z is much further from the boundaries than w. On the boundaries the magnetic sources are separated by an array of null points at $x = 2nw$.

Suppose the footpoints of the nth source at $z = \pm L$ are displaced by $Y_n = \pm(-1)^n Y$, so that the shearing has alternate signs in successive cells, producing opposite field perturbations on each side of any separatrix. The resulting current density can be shown to have the form $\mathbf{j} = (db/dA_0)\mathbf{B}_0$, where A_0 is the flux function associated with the field $\mathbf{B}_0$ (Eq. 10.40) and $b(A_0) = Y/V_0$, where

$$V_0(A_0) = \frac{4w^2}{\pi F} \log_e \frac{\sinh^2(\frac{1}{2}\pi L/w)\cos^2(\pi A_0/F) + \cosh^2(\frac{1}{2}\pi L/w)\sin^2(\pi A_0/F)}{\sin^2(\pi A_0/F)},$$

and the current flows in a narrow layer of width $w/\pi e^{-[\pi L/(2w)]}$ (PROBLEM 10.7).

The model was extended to 3D by considering arrays of sources on a 2D grid. This replaces the 2D null lines on the boundaries by 3D null points, and the current concentrations on the separatrices became singular. A more realistic model would have asymmetrically placed sources (so as to create many more separatrices) or it would place all the sources on one plane and have mixed polarity. The basic principles would, however, be unchanged. For a model with diffusion but no reconnection, the heating is independent of resistivity (Ng and Bhattacharjee 2008).

The results give a uniform heating along each separatrix, so that each (sub-resolution) coronal flux tube would be heated uniformly. But at least 50 per cent of the photospheric flux closes low down in the magnetic carpet (Close et al. 2004), so the remaining flux forms large-scale connections. Thus, the magnetic carpet would be heated more effectively than the large-scale corona. Unresolved observations of coronal loops would give enhanced heating near the loop feet in the carpet, while the upper parts of coronal loops would be heated uniformly but less strongly.

10.4.5.2 Properties of Separator Current Sheets

Separators are important locations for heating. Suppose a current sheet of length L_s, width $2L$ and thickness $2l$ forms along a separator located at the junction of four domains (Figure 6.27c). Its basic properties are as follows (Priest et al. 2005).

(a) Electric Current and Dimensions

In a plane perpendicular to the current sheet (Figure 10.31), suppose the field B_e at a large distance L_e is moved towards the separator by a given distance δ. If there were instantaneous reconnection, the flux that would be reconnected is simply

$$\triangle\psi = B_e\,\delta\,L_s, \tag{10.41}$$

where L_s is the length of the separator out of the plane. However, we suppose that instead a current sheet forms along the separator and there is no reconnection.

The x- and y-components of the local field $[B_x = (B_e/L_e)y,\ B_y = (B_e/L_e)x,\ B_z = B_0]$ may be written in terms of $Z = x + iy$ as $B_y + iB_x = (B_e L_e)Z$. Then, after the current sheet has formed, it may be represented (Sec. 6.3) by

$$B_y + iB_x = B_e(Z^2 - L^2)^{1/2}/L_e,$$

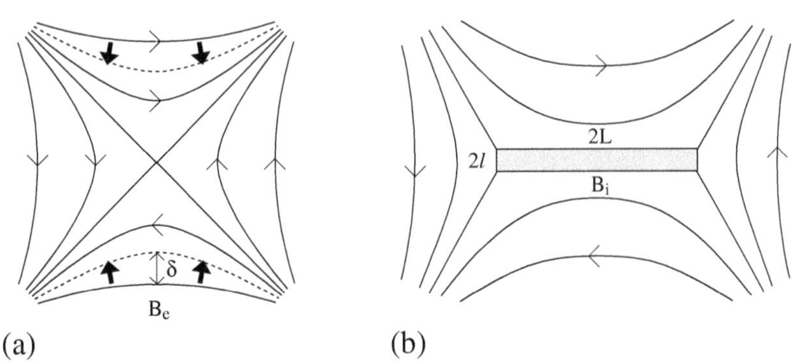

Figure 10.31. Magnetic field in a plane perpendicular to a separator (a) in an initial potential state and (b) after a current sheet has formed without reconnection due to a displacement δ of the initial field B_e at distance L_e from the separator (Priest et al. 2005).

from which several properties may be deduced. By evaluating it at $(x,y)=(0,0+)$ or $(0,0-)$, we find the inflow field (B_i) to be

$$B_i = B_e L/L_e. \tag{10.42}$$

Secondly, using the fields on the top and bottom sides of the sheet at $(x,0+)$ and $(x,0-)$, the current in the sheet becomes

$$J(x)=\frac{B_x(x,0+)-B_x(x,0-)}{\mu}=\frac{2B_e(L^2-x^2)^{1/2}}{\mu L_e}.$$

Integrating along the sheet gives the total current (I) as

$$I=\pi B_i L/\mu. \tag{10.43}$$

Eliminating B_i between (10.42) and (10.43) gives $I(L)$ as

$$\frac{I}{I_e}=\frac{L^2}{L_e^2}, \tag{10.44}$$

where $I_e=\pi L_e B_e/\mu$ is a characteristic current.

Furthermore, the field $B_x(0,y)$ along the y-axis is $B_e y/L$ initially and $B_x(0,y)=(B_e L_e)(L^2+y^2)^{1/2}$ after the sheet has formed. Integrating the difference between these two along the y-axis gives another expression for the extra magnetic flux ($\Delta\psi$) associated with the formation of the current sheet. This may be equated to Eq. (10.41) to give $L=L(\delta)$, in the limit when $\delta/L_e \ll 1$ as

$$\frac{L}{L_e}=\left(\frac{4\delta/L_e}{-\log(\delta/L_e)}\right)^{1/2}. \tag{10.45}$$

The current $[I=I(\delta)]$ follows from Eq. (10.44) as

$$\frac{I}{I_e}=\frac{4\delta/L_e}{-\log(\delta/L_e)}.$$

(b) Energy Release, Voltage and Electric Field during Reconnection

During 3D reconnection, a change in flux is associated with an electric field ($E_\parallel$) and an associated voltage (V) along the separator. This may be estimated by the following thought experiment. Suppose the flux sources are moved through a distance δ and that the field responds by an ideal motion to a new equilibrium containing a current sheet with current I. Then suppose the magnetic field reconnects, while the current declines from I to 0.

First of all, the relation between the flux $\Delta\psi$ and voltage V may be calculated by integrating Faraday's Law over a surface bounded by the separator $\int\int \partial\mathbf{B}/\partial t\cdot d\mathbf{S}=-\int\int\nabla\times\mathbf{E}\cdot d\mathbf{S}$, and so, using Stokes' theorem, $d/dt\int\int\mathbf{B}\cdot d\mathbf{S}=-\int E_\parallel ds$ or, in other words, $\partial\psi/dt=V$. If our flux $\Delta\psi$ takes a time Δt to reconnect, therefore,

$$\Delta\psi = V\Delta t. \tag{10.46}$$

Next, the energy (ΔW) released during the reconnection process can be estimated in terms of the transfer ($\Delta\psi$) of flux. It is simply $\Delta W=\frac{1}{2}\mathcal{P}\Delta t$, where the power ($\mathcal{P}$) is given by $\mathcal{P}=IV$ and the factor $\frac{1}{2}$ arises because the current declines from an initial value of I to zero over the course of the reconnection. Therefore, using Eq. (10.46) for $V\Delta t$, we find

$$\Delta W=\tfrac{1}{2}I\Delta\psi. \tag{10.47}$$

After substituting for I from (10.44) and $\Delta\psi$ from (10.41), this becomes

$$\triangle W = \frac{I_e L^2 B_e \delta L_s}{2L_e^2} = \frac{I_e B_e \delta^2 L_s / L_e}{\log(L_e/L)} = \frac{\pi B_e^2 \delta^2 L_s}{\mu \log(L_e/L)}, \tag{10.48}$$

since $I_e = \pi L_e B_e / \mu$, where L/L_e is determined by (10.45).

Thus, for given values of B_e, L_e, L_s, δ, we have deduced the flux change ($\triangle\psi$) from (10.41), the sheet half-width (L) from (10.45), the current (I) from (10.44), and the heat ($\triangle W$) from (10.47). What is still unknown is the time ($\triangle t$) for flux transfer by reconnection. It could in principal be determined from observations or from the type of reconnection operating if that were known, such as Sweet-Parker (1958) reconnection, fast reconnection (Priest and Forbes 1986) or stick-slip reconnection (Longcope and Cowley 1996). Once $\triangle t$ is indeed known, three other remaining physical quantities can be determined, namely, the heating power or rate of heating $\mathcal{P} = 2\triangle W/\triangle t = I\triangle\psi/\triangle t$, the voltage $V = \triangle\psi/\triangle t$ and the electric field along the separator $E_{\|} = V/L_s$.

(c) Separatrix Current Sheet

The current build-up and energy release at a separatrix surface of length L_l and width L_0 (a typical distance between photospheric sources) may be estimated in a similar way. A footpoint motion (δ) acting on an initially vertical field (B_0) produces a transverse component $B_\delta = (\delta/L_l)B_0$, and so the separatrix current sheet contains a current (I) of $I = 2B_\delta L_0/\mu$, or, after substituting for B_δ,

$$I = \frac{2B_0\, \delta\, L_0}{\mu\, L_l}.$$

The energy release by reconnection [if the transverse component (B_δ) is annihilated] is $\triangle W = \frac{1}{2} I \triangle\psi$, where $\Delta\psi$ is the flux of the annihilated component (B_δ) over an area $L_l L_0$, namely, $\Delta\psi = B_\delta\, L_l\, L_0$. After substituting for B_δ, this becomes $\triangle\psi = B_0\, \delta\, L_0$, so that the energy release becomes

$$\triangle W = \frac{{B_0}^2\, \delta^2\, {L_0}^2}{\mu\, L_l}. \tag{10.49}$$

In order of magnitude, the currents and energy release from separator and separatrix reconnection thus have similar forms and similar size. In order to compare the energies in (10.48) and (10.49), suppose that $L_s = L_l$ and $L_e = L_0$. B_0 is of order the main coronal field, whereas B_e is a field component transverse to the main field. Assume also that $B_e \approx B_0 L_0 / L_l$, since this is the maximum perpendicular field that would be obtained by moving the footpoints of an initially uniform field (B_0) a distance L_0 in a cell of size $L_0 \times L_0$. The resulting ratio of energies is then

$$\frac{\triangle W_{\mathrm{sepr}}}{\triangle W_{\mathrm{sepx}}} = \frac{\pi}{\log(L_e/L)}.$$

For separator heating to be important, the sheet width (L) needs to cover a significant fraction of the separatrix. Thus, the formation of current sheets at separators and separatrices are likely to be equally important in coronal heating by tectonics.

(d) Separator Heating Estimates in the Corona

Consider three cases: very quiet Sun, with mean field B_e of 3 G; quiet Sun at solar maximum, with corresponding field of 20 G; and an active region, with 100 G.

The mean separator length (L_s) is 60 Mm (Close et al. 2004). Then, for a flux displacement (δ) of, say, 300 km, Eq. (10.41) gives a flux transfer ($\triangle\psi$) of 5.3×10^{17} Mx, 3.6×10^{18} Mx or 1.8×10^{19} Mx, for ambient fields (B_e) of 3 G, 20 G or 100 G.

Next, Eq. (10.45) determines the sheet length (L/L_e) as a function of δ/L_e, where L_e, the external scale length, is, say, 30 Mm. Thus, a value $\delta = 300$ km gives $L/L_e = 0.1$, so that the sheet is 3 Mm wide.

The current (I) along the separator follows from Eq. (10.44) as 1.9×10^8 A, 1.3×10^9 A or 6.3×10^9 A for fields (B_e) of 3 G, 20 G or 100 G, respectively. The corresponding magnetic fields ($B_i = B_e L/L_e$) at the sheet are 0.3 G, 2 G and 10 G, respectively.

Energy release ($\triangle W$) in an elementary heating event with flux transfer $\triangle\psi$ follows from Eq. (10.47) as $\triangle W = \frac{1}{2} I \triangle\psi$, which becomes $5{\times}10^{17}$ W, $2.3{\times}10^{19}$ W (i.e., 2.3×10^{26} erg s^{-1}) or $5.6{\times}10^{20}$ W for fields of 3, 20 or 100 G. This corresponds to heating by big events (microflares): smaller flux transfers would produce nanoflares.

For a footpoint motion (v_0) of 0.3 km s^{-1}, say, the time ($\triangle t$) associated with the flux displacement $\delta = 300$ km is 10^3 sec. The corresponding voltage (V) along the separator is then 5.4 GV, 36 GV or 180 GV, respectively, while the electric fields ($E_{\|}$) are 100 V m^{-1}, 720 V m^{-1} or 3600 V m^{-1}.

Finally, the power ($\mathcal{P}$) is found to be $1.0{\times}10^{15}$ W, $4.6{\times}10^{16}$ W ($4.6{\times}10^{23}$ erg s^{-1}) or $1.1{\times}10^{18}$ W for fields of 3 G, 20 G or 100 G, respectively. Dividing by the area (15 Mm)^2 of a typical supergranule gives a power per unit area of 4.8 W m^{-2}, 200 W m^{-2} (i.e., $2{\times}10^5$ erg cm^{-2} s^{-1}) or 5,000 W m^{-2} for the heating in the very quiet Sun, in the quiet Sun at solar maximum or in an active region. These compare well with observed values of 10 W m^{-2}, 300 W m^{-2} (i.e., $3{\times}10^5$ erg s^{-1} cm^{-2}) and 5,000 W m^{-2}, respectively (Withbroe and Noyes 1977; Katsukawa and Tsuneta 2002; Schrijver et al. 1997).

10.4.5.3 Separator Reconnection

Several distinct types of reconnection are associated with null points, namely, spine, fan, and separator reconnection (Priest and Titov, 1996), and, in particular, separator reconnection is a prime candidate for coronal heating. Numerical experiments have been conducted on this possibility (Galsgaard and Nordlund 1996b; Parnell et al. 2010a, 2010b), and the way in which it operates has been studied in detail by Longcope and co-workers.

Having shown how a current sheet may form along a separator (Longcope and Cowley 1996), a stick-slip model for reconnection was developed together with the concept of a *minimum-current corona* (MCC) (Longcope 1996). The assumption is that, after slow motions of the photospheric footpoints, the corona relaxes to a *flux-constrained equilibrium*, in which the magnetic fluxes within each domain are conserved, but the field lines within each domain can slip through the plasma, or move their footpoints (Longcope 2001). Such equilibria have potential magnetic fields in each domain and current sheets along the separators. MCC has been extended to include sunspot rotation (Beveridge and Longcope 2006) and flux emergence (Tarr and Longcope 2012). The theory has been applied to X-ray bright points (Longcope 1998) and to provide a lower bound on pre-flare energy storage for solar flares (Longcope et al. 2007; Kazachenko et al. 2010, 2012).

Priest et al. (2005) stress that, in general, the effect of slow photospheric motions on complex coronal magnetic configurations will be to generate three forms of current, namely, distributed currents throughout the volume, current sheets on separators and current sheets on separatrices. They compare energy storage and heating at separators and separatrices by using reduced MHD to model coronal loops that are long and thin. For example, suppose a flux tube has two sources at one end and two sources at the other end. In the initial potential state, there will be a separator that represents the intersection of the two separatrix surfaces that divide the fluxes from the pairs of sources (Figure 10.32a,b). Then, if the sources are spun or rotated around one another, current sheets will appear along the separatrices and also at the separator (Figure 10.32c), as modelled by De Moortel and Galsgaard (2006). Their dissipation, while conserving magnetic helicity (Sec. 6.10) will in general leave a relaxed state that is a twisted linear force-free field (Figure 10.32d).

10.4.5.4 Binary Reconnection and Fly-by

The skeleton of the field due to two unbalanced sources (stars) in the photosphere is shown in Figure 10.33a, where a null point (filled circle) closer to the smaller source possesses a spine (bold line) that joins the null point to the weaker source and to infinity. It also possesses a fan surface of field lines

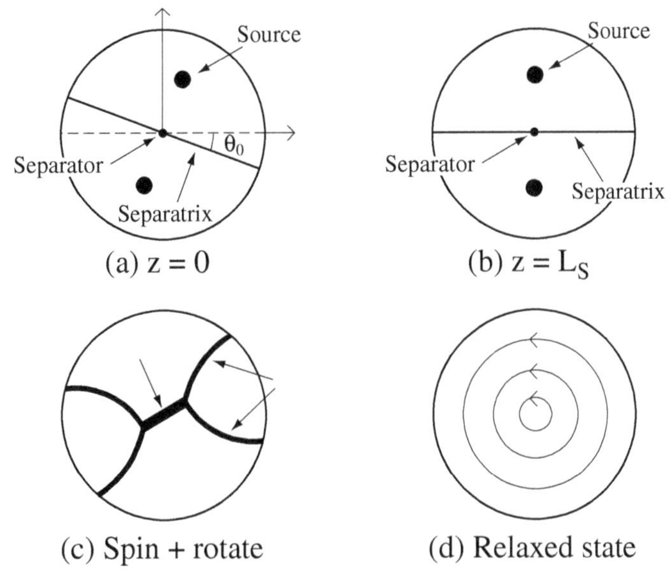

Figure 10.32. Projection of magnetic field lines in a section across the coronal part of a flux tube with two sources at each end for (a) initial state at $z = 0$, (b) initial state at $z = L_s$, (c) configuration after spinning and rotating sources, (d) relaxed state (Priest et al. 2005).

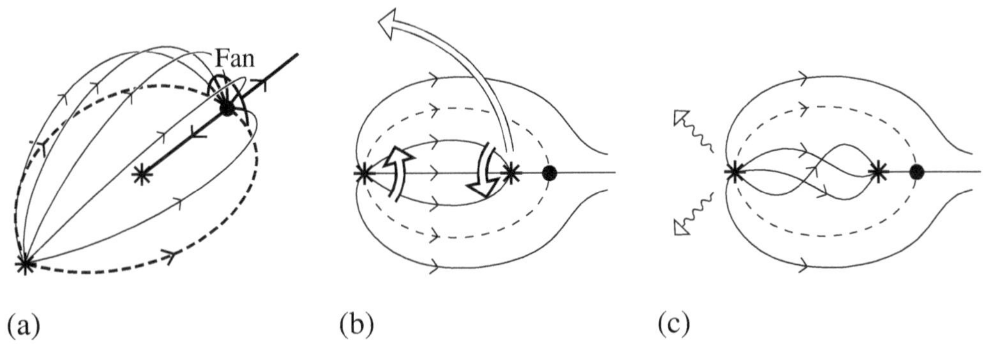

Figure 10.33. (a) Skeleton of the field due to two unbalanced sources (stars), with the fan of a null point arching over to form a separatrix dome. (b) The relative motion of the sources generates (c) twisted force-free fields and waves.

that arch over the weaker source in the form of a dome, which intersects the photosphere in a dashed curve. Part of the magnetic flux from the stronger source lies below the separatrix dome, while the remaining flux lies above it and links out to distant sources.

The fundamental heating mechanism may be one of so-called *binary reconnection* due to the motion of a given magnetic source relative to its nearest neighbour (Priest et al. 2003). The heating comes from several effects: (1) the 3D reconnection of field lines that start up joining the sources and end up joining the largest source to other more distant sources; (2) the viscous or resistive damping of the waves that are emitted by the sources as their relative orientation rotates; and (3) the relaxation of the nonlinear force-free fields that join the two sources and that are built up by the relative motion of the two sources (Figure 10.33).

Priest et al. (2003) modelled kinematically the steady process of laminar 3D reconnection as the position of one magnetic source rotates relative to another, by calculating the resulting electric field and field-line velocity and showing that spine-fan reconnection is driven at the null point (PROBLEM 10.8a). They also evaluated the magnetic helicity injection (Sec. 6.10), showing that part of it propagates away as torsional

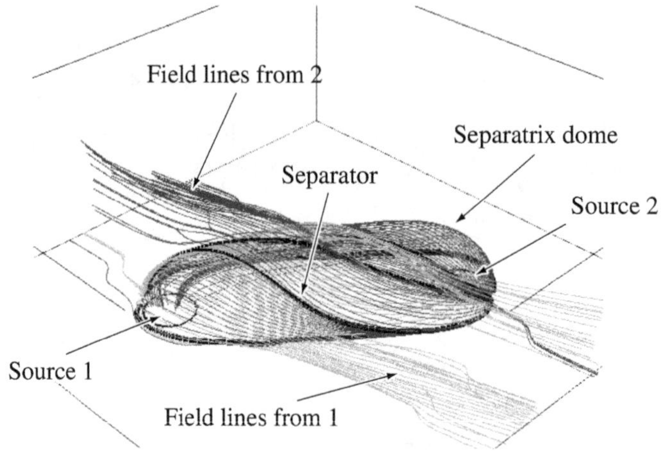

Figure 10.34. A snapshot of a movie of the skeleton during a fly-by numerical experiment, revealing the presence of five separators (thick curves) and the two sets of field lines linking the two sources labelled 1 and 2 (courtesy Andrew Haynes).

Alfvén waves, while the remainder builds up twisted force-free structures. Finally, they showed that, if the mean photospheric flux densities of positive and negative flux are $\bar{B}_+$ and $\bar{B}_-$, then the heat flux due to the source motion (v_0) is (PROBLEM 10.8b)

$$F_{heat} = \frac{\bar{B}_+^2 v_0}{3\pi\mu}.$$

Adopting a relative footpoint motion of $v_0 = 0.3$ km s^{-1} and mean fields of $\bar{B}_+ \approx \bar{B}_- = 20$ G for the quiet Sun and 200 G in active-region cores gives heating rates of 100 and 10^4 W m^{-2}, respectively, in agreement with observations.

An important aspect of binary reconnection is to consider the effect of an overlying large-scale coronal magnetic field on the field produced by the relative motion of two nearby photospheric magnetic fragments. Parnell and co-workers have conducted a series of numerical experiments on this effect, starting with a simple *fly-by* that models a binary interaction between two opposite-polarity photospheric fragments in an overlying uniform field (Figure 10.34) (Parnell and Galsgaard 2004). Initially, the two fragments are not joined, and as they approach at 5 per cent of the Alfvén speed they become joined by reconnection. However, what type of reconnection is occurring? It was only by constructing the magnetic skeleton that Haynes et al. (2007) were able to answer this question as follows.

A vertical section through such a skeleton reveals a much more complex interaction than was expected, with six distinct steps (Figure 10.35). Initially, two nonintersecting separatrix surfaces are present, below which the flux from each source is open, in the sense that it goes to the nearby boundary. Then they touch and intersect in two separators at which reconnection transfers flux from the "open" regions into a "closed" region that represents flux joining one source to the other. In the third step, the lower separator moves down through the lower boundary, while reconnection at the upper separator increases the amount of closed flux. Then the two sides of the closed separatrix move outwards to touch the other separatrices and proceed to intersect them in two new separators each, making five in all. Next, two of those separators descend through the lower boundary, and finally leaving just one separator, with most of the flux reopened.

Later analysis of fly-by experiments (Parnell et al. 2008) showed that individual field lines can reconnect many times in *multiple* or *recursive reconnection*, so increasing the heating rate and spreading it throughout the volume. As the magnetic Reynolds number is increased, reconnection between opposite-polarity sources

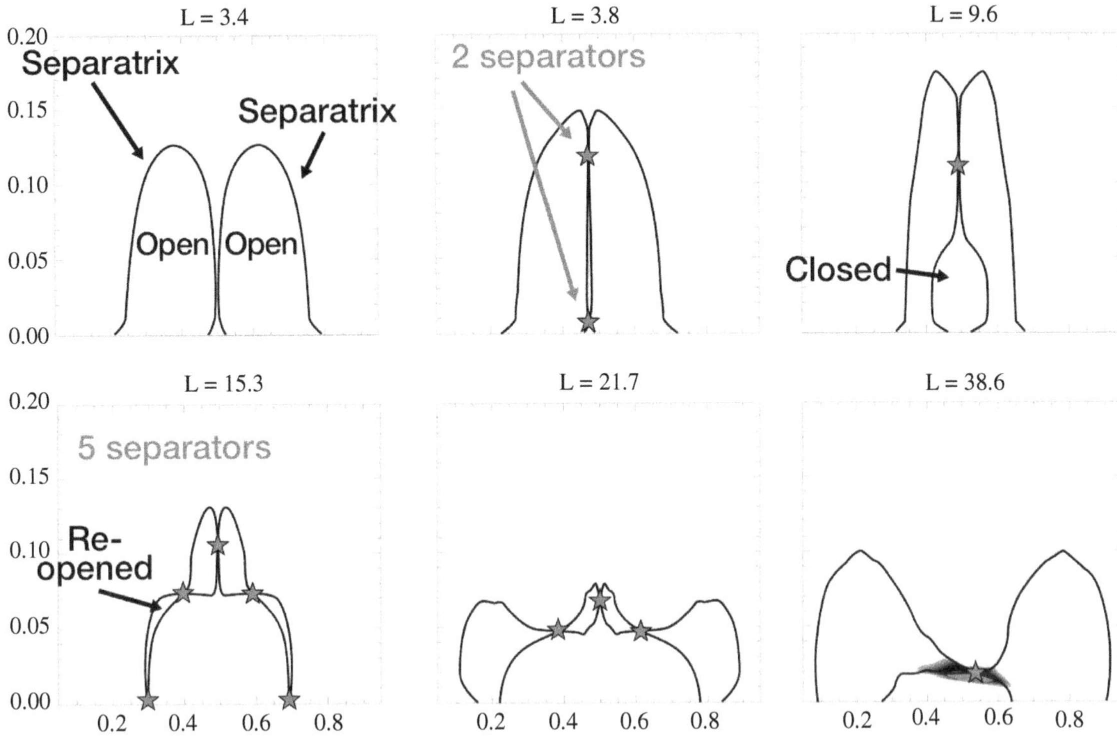

Figure 10.35. Vertical cross-sections at different times through the fly-by skeleton (courtesy Andrew Haynes).

becomes more complex and lasts longer. This in turn increases the horizontal component of the magnetic field at the base and so increases the Poynting flux.

10.4.5.5 Switch-on Property of Dissipation

Equation (10.1) gives the vertical component (F_{Hv}) of the Poynting flux ($\mathbf{F}_H$) as

$$F_{Hv} \approx -\frac{v_h B_h B_v}{\mu}$$

at the photosphere, where v_h is the horizontal velocity and B_h, B_v are the horizontal and vertical components of the magnetic field. Klimchuk (2006) points out that, for typical values of F_{Hv}, v_h and B_v, this implies that the inclination of the field to the vertical $[\tan^{-1}(B_h/B_v)]$ is roughly 20 degrees for the quiet Sun or 10 degrees for an active region. Thus, impulsive reconnection in a nanoflare must remain dormant until the magnetic field is tilted by this critical angle. If the mechanism activates too soon or too late, the corona will be cooler or hotter than observed.

One possibility for the necessity of this critical angle (of 40 degrees between neighbouring flux tubes) is that there is some aspect of the microphysics of fast reconnection that allows it to start only at this critical angle. For example, Dahlburg et al. (2005) have suggested that slow reconnection in the form of the tearing-mode instability saturates at a low level and then undergoes a fast secondary instability that produces a burst of energy release (a nanoflare) when the critical angle is reached.

10.4.5.6 Sudden Dissipation of a Tectonics Current Sheet

If reconnection were heating the corona at many current sheets by coronal tectonics, then several questions would naturally arise. How does the energy spread out – by conduction along the magnetic field,

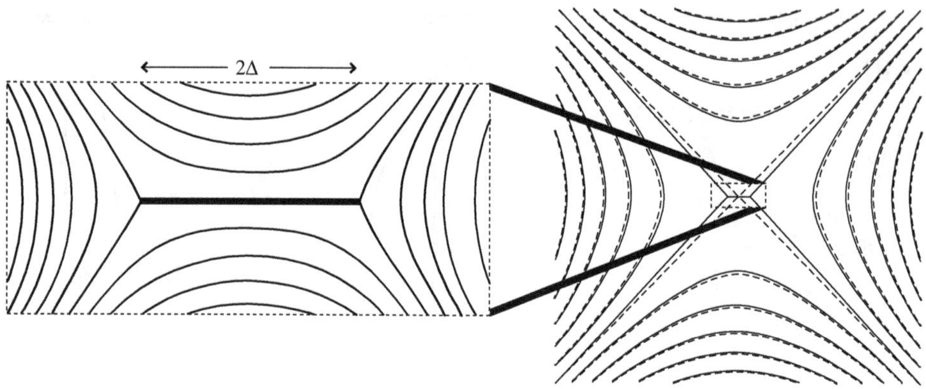

Figure 10.36. A magnetic field which has an X-point geometry at large distances and contains a current sheet near the origin (from Longcope and Priest 2007).

by reconnection jets, by waves across the magnetic field or by fast particles? Furthermore, if reconnection is time-dependent, how much energy is liberated locally and how much globally?

An attempt has been made to start answering these questions by setting up a simple model problem (Longcope and Priest 2007; Fuentes-Fernández et al. 2012). Consider initially the magnetic field of a current sheet in the form

$$B_y + iB_x = B'\sqrt{w^2 - z^2},$$

as shown in Figure 10.36. At large distances (r) we may expand $\mathbf{B} = \mathbf{B}_0 + \mathbf{B}_1$, with

$$\mathbf{B}_0 = -B'(y\,\hat{\mathbf{x}} + x\,\hat{\mathbf{y}}) \qquad \text{and} \qquad B_{1\phi} = \frac{I_0}{2\pi r},$$

where the latter term is the field of a line current I_0. There is thus, surprisingly, lots of potential energy in the magnetic field lying at large distances from the origin, which may be available for dissipation if the current changes.

Suppose for some reason that the current (I) in the sheet dissipates, perhaps due to the sudden onset of an enhanced diffusivity due to current-induced microinstabilities. This is a local process, but it has global consequences. The decrease in current implies that the magnetic field must change at large distances, but how? The effect of reconnection may be modelled by linearising about the X-point field ($\mathbf{B}_0$) in the low-beta limit, so that the induction equation and equation of motion become

$$\frac{\partial \mathbf{B}_1}{\partial t} = \boldsymbol{\nabla} \times (\mathbf{v}_1 \times \mathbf{B}_0) = \eta \nabla^2 \mathbf{B}_1,$$

$$\rho_0 \frac{\partial \mathbf{v}_1}{\partial t} = \mathbf{j}_1 \times \mathbf{B}_0.$$

Assume that $\mathbf{B}_1$ at $t = 0$ is due to the current sheet, and that diffusion is turned on, causing the current to dissipate and the field lines to reconnect.

The natural dependent variable in these equations is $rB_{1\phi} \equiv I(r,t)$, which is twice the current enclosed in a circle of radius r. The equations may then be combined to give in terms of I

$$\frac{\partial^2 I}{\partial t^2} = \omega_A^2 r \frac{\partial}{\partial r}\left(r \frac{\partial I}{\partial r}\right) + \eta r \frac{\partial}{\partial r}\left(\frac{1}{r}\frac{\partial^2 I}{\partial r \partial t}\right), \qquad (10.50)$$

where the first term on the right represents wave behaviour and the second diffusion. Both terms cause an initially uniform I to spread out from the origin in time.

In the large-r advective limit when $R=\ln(r/l_\eta)\gg 1$, where $l_\eta=\sqrt{(\eta/\omega_A)}$ and $\omega_A=B'/\sqrt{4\pi\rho}$, the diffusion term is negligible and Eq. (10.50) becomes a normal wave equation in R and t with solution

$$I(R,t)=I_0-F(t-R).$$

On the other hand, in the small-r diffusive limit, the wave term is negligible and

$$I(r,t)=I_0-I_0\exp\left(-\frac{r^2}{4\eta t}\right),$$

so that the current density $j\equiv(1/r)\partial I/\partial r$ tends to zero at the origin as t increases.

Equation (10.50) has been solved numerically and it does indeed show a transition from diffusive to wave-like behaviour (Figure 10.37a), with a sheath of current propagating outwards (Figure 10.38). However, in the wake of the sheath there is a flow and an associated convective electric field $E_v(r,t)\hat{\mathbf{z}}\equiv-\mathbf{v}_1\times\mathbf{B}_0$ with a surprising behaviour (Figure 10.37b). Instead of the electric field and current at the origin diminishing in time, they slowly increase.

This paradox may be solved by recognising that a third regime exists at large time ($t>1/\omega_A$), when the left-hand side of Eq. (10.50) is negligible and the two terms on the right balance one another to give

$$I(r,t)=\frac{I_0}{2\omega_A t}\left[1-\exp\left(-\frac{\omega_A^2 r^2 t}{\eta}\right)\right].$$

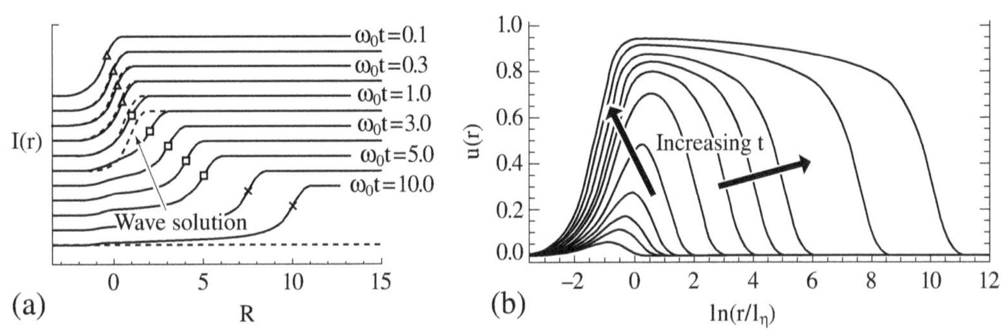

Figure 10.37. (a) A plot of the current I contained within a radius $R=\ln(r/l_\eta)$ as a function of R for several times, and displaced vertically for ease of reading – the dashed curve shows the diffusive solution. (b) The convective electric field (E_v) as a function of dimensionless radius $R=\ln(r/l_\eta)$ at several times (from Longcope and Priest 2007).

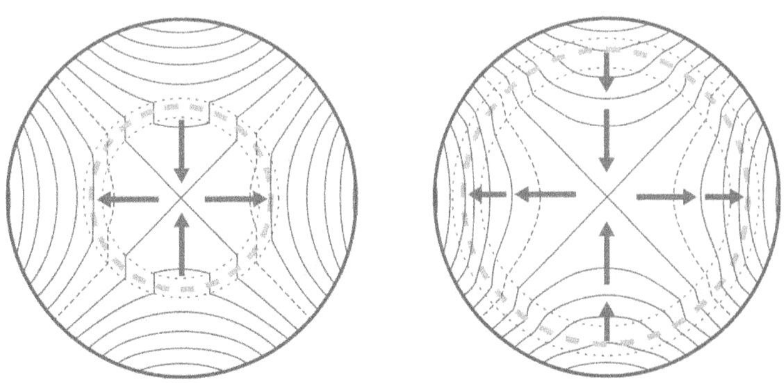

Figure 10.38. Outwards propagation of a sheath of current and reconnection flow in its wake, shown in sketches of the magnetic field at two times (courtesy Dana Longcope).

The resulting current density is

$$j = \frac{1}{r}\frac{\partial I}{\partial r} = \frac{I_0 \omega_A}{\eta} \exp\left(-\frac{\omega_A^2 r^2 t}{\eta}\right),$$

which equals $I_0\omega_A/\eta$ at the origin. Thus, the peak in current remains at the X-point and the outwards-propagating wave continues to drive a steady electric field $(I_0\omega_A)$ there that is independent of η – i.e., fast reconnection.

10.5 Heating by Turbulence

10.5.1 Relaxation by MHD Turbulence

At the same time as coronal structures are trying to evolve through nonlinear force-free equilibria in response to footpoint motions, they also tend to relax by 3D reconnection towards linear force-free states that conserve global magnetic helicity (Sec. 6.10). Turbulent relaxation tends not to destroy magnetic helicity when the magnetic Reynolds number is very large, but it can convert it from one kind to another, such as from mutual to self-helicity. Heyvaerts and Priest (1984) suggested that the corona may be heated by relaxation as it evolves through a series of linear force-free states, satisfying $\nabla \times \mathbf{B} = \alpha \mathbf{B}$, with the footpoint connections not preserved but the force-free constant (α) determined from the evolution of relative magnetic helicity

$$H_m = \int (\mathbf{A} + \mathbf{A}_0)(\mathbf{B} - \mathbf{B}_0)dV, \tag{10.51}$$

where $\mathbf{A}$ is the vector potential, and $\mathbf{A}_0$, $\mathbf{B}_0$ refer to corresponding potential values with the same normal field at the boundaries (Berger and Field 1984).

Boundary motions cause the magnetic helicity to change in time according to

$$\frac{dH_m}{dt} = 2\int_S (\mathbf{B}\cdot\mathbf{A}_0)(\mathbf{v}\cdot\mathbf{n}) - (\mathbf{v}\cdot\mathbf{A}_0)(\mathbf{B}\cdot\mathbf{n})dS. \tag{10.52}$$

Conceptually, photospheric motions tend to build up energy in a nonlinear force-free field and to release it by reconnection. The resulting heating flux is of the form

$$F_H = \frac{B^2 v}{\mu}\frac{\tau_d}{\tau_0},$$

where, as before, τ_d is the dissipation time and τ_0 the time-scale for footpoint motions. This is the same as Parker's (1979a) result when τ_d is replaced by the reconnection time (d/v_R) and τ_0 is replaced by the convection time (L/v_0).

Several extensions of the basic theory have been constructed. Vekstein et al. (1991) suggested *intermediate relaxation* to a state between the nonlinear and linear fields, while Vekstein et al. (1993) suggested the corona is in a state of *partial relaxation*, with the closed fields being relaxed linear force-free states and the open fields being potential. The basic analysis has been applied by Browning et al. (1986) to a set of closely packed flux tubes and by Dixon et al. (1988) to an axisymmetric flux tube.

Important physical quantities are the *global ideal invariants*, which are conserved in the absence of dissipation (Frisch et al. 1975; Montgomery 1983). In 2D MHD, they are energy, correlation (or mean-square vector potential) and cross-helicity,

$$W_2 = \int \tfrac{1}{2}\rho v^2 + \frac{B^2}{2\mu} dS, \quad a = \int \tfrac{1}{2}A^2 dS, \quad H_2 = \int \mathbf{v}\cdot\mathbf{B}\, dS,$$

while, in 3D, the energy, magnetic helicity and cross-helicity are conserved:

$$W_3 = \int \tfrac{1}{2}\rho v^2 + \frac{B^2}{2\mu} dV, \quad H = \int \mathbf{A}\cdot\mathbf{B}\, dV, \quad H_3 = \int \mathbf{v}\cdot\mathbf{B}\, dV,$$

where $\mathbf{B} = \nabla \times \mathbf{A}$.

In Fourier space (and therefore in real space) these global invariants undergo cascades, which are direct if the transfer is from large to small wavelengths and indirect (or inverse) if it is in the other direction. The energy has either a Kolmogorov spectrum ($\sim k^{-\frac{5}{3}}$) or a Kraichnan spectrum ($\sim k^{-\frac{3}{2}}$) and has a direct cascade towards small wavelengths, whereas the correlation A ($\sim k^{-\frac{1}{3}}$) in 2D and magnetic helicity H ($\sim k^{-2}$) in 3D have indirect cascades towards large wavelengths.

Ting et al. (1986) conducted a series of 2D MHD experiments, in which they find *selective decay* when the initial kinetic energy is much smaller than the magnetic energy and the cross-helicity (normalised) is less than the energy. Here the magnetic energy decays faster than the magnetic helicity and so has a direct cascade towards small wavelengths, since nonlinear interactions tend to replenish it. At the same time, the magnetic helicity has an indirect (or inverse) cascade, and the magnetic field tends towards a force-free state. In contrast, when the kinetic and magnetic energies are similar and the cross-helicity is of a similar size to the energy, they find a process of *dynamic alignment* with a tendency towards Alfvénic states having $\mathbf{v} = \pm\mathbf{B}/\sqrt{\mu\rho}$ (see also Biskamp 1994).

Applications of the theory to coronal heating have been made by several authors. Sturrock and Uchida (1981) calculated the rate of increase of stored energy due to random twisting of a flux tube through force-free states as

$$\frac{B^2}{\mu L}\langle v_p^2\rangle \tau_p,$$

where v_p is the photospheric velocity, τ_p is the correlation time of the footpoint motions, and the angular brackets indicate a mean value. They assume that the dissipation time $\tau_d > \tau_p$, so that the free energy can continue to be stored. In contrast, Heyvaerts and Priest (1984) included a dissipation mechanism and assumed $\tau_d < \tau_p$. They obtained a heating flux of

$$F_H = \left(\frac{B^2}{\mu L} v_p^2 \tau_p\right)\left(\frac{L}{L + l_p}\right)^2 \frac{\tau_d}{\tau_p}, \tag{10.53}$$

where L is the loop half-length and l_p is the length-scale for footpoint motions. The last two factors here are less than unity and so limit the efficiency of the process.

van Ballegooijen (1985) discussed the initial stages of energy cascade due to random footpoint motions. He assumed that the states are force-free and the length-scale (l_p) for photospheric motions is much smaller than the loop length. He found that the mean-square current density (j^2) increases exponentially $\sim \exp(t/\tau_p)$. Gomez and Ferro Fontan (1988) applied 2D MHD turbulence theory to twisted coronal loops and suggested the injection of energy at a specific wave number (k_p) produces a cascade of energy like $k^{-\frac{5}{3}}$ to a dissipation wave number k_d, together with an inverse cascade of mean-square potential like $k^{-\frac{1}{3}}$.

Many coronal heating mechanisms, such as braiding and current-sheet formation or resistive instabilities or waves, all lead to a state of MHD turbulence, so how can we analyse such a state? Heyvaerts and Priest (1984) made a start by adapting Taylor's relaxation theory to the coronal environment, in which the field lines thread the boundary rather than being parallel to it. Although many mechanisms produce a turbulent state, they are incomplete in the sense that a free parameter is present, such as a diffusion, correlation or relaxation time. In other words, the mechanisms do not determine the heating flux (F_H) in terms of photospheric motions alone.

Heyvaerts and Priest (1992) therefore began a new approach in two parts, in which photospheric motions inject energy into the corona and maintain it in a turbulent state with a turbulent magnetic diffusivity (η^*) and viscosity (ν^*). First of all, they calculate the global MHD state driven by boundary motions, which gives F_H in terms of ν^* and η^*. Secondly, they invoke cascade theories of MHD turbulence to determine the ν^* and η^* that result from F_H. In other words, the circle is completed and F_H is determined independently of ν^* and η^*. They applied their general philosophy to a simple example of 1D random photospheric motions producing a 2D coronal magnetic field, as follows.

Suppose the dimensionless boundary motions are $\pm V(x)\hat{\mathbf{y}}$ (with Fourier coefficients V_n) at $z = \pm L$ and produce motions $v(x,z)\hat{\mathbf{y}}$ and field $B_0\hat{\mathbf{z}} + B_y(x,z)\hat{\mathbf{y}}$ within the volume between $z = -L$ and $z = L$. Then the steady MHD equations of motion and induction reduce simply to

$$0 = \frac{B_0}{\mu}\frac{\partial B_y}{\partial z} + \rho\nu^*\nabla^2 v_y, \qquad 0 = B_0\frac{\partial v_y}{\partial z} + \eta^*\nabla^2 B_y.$$

From the resulting solutions, the Poynting energy flux through the boundary is

$$F_H = \frac{B_0^2 v_{A0}}{\mu}\sum_0^\infty \frac{V_n^2 H^*}{\eta/(l v_{A0})}\left(1 + \frac{2\lambda_n^2}{\sqrt{1+4\lambda_n^2}}\right) \times \frac{\sinh\left(\sqrt{1+4\lambda_n^2/H^*}\right) + \sinh(1/H^*)}{\cosh\left(\sqrt{1+4\lambda_n^2/H^*}\right) - \cosh(1/H^*)},$$

where $H^* = \sqrt{\eta^*\nu^*}/(L v_{A0})$ is the inverse Hartmann number.

For the second step, invoking Pouquet theory gives $\nu^* = \eta^*$ and

$$F_H = \frac{27[\nu^{*2}/(L v_{A0})]\pi^3}{2a^3/L^3}\frac{B_0^2 v_{A0}}{\mu},$$

where a is the loop half-width. Equating the two above expressions for F_H finally gives a single equation for ν^*. They found typically for a quiet-region loop that a density of 2×10^{16} m^{-3} and a magnetic field of 3 to 5×10^{-3} tesla (30 to 50 G) produces a heating F_H of 2.4 to 5.5 $\times10^2$ W m^{-2} and a turbulent velocity of 24 to 33 km s^{-1}, whereas values of 5×10^{16} m^{-3} and 10^{-2} tesla (100 G) for an active-region loop give 2×10^3 W m^{-2} and 40 km s^{-1}. Given the limitations of the model, these reasonable values are very encouraging. Inverarity and Priest (1995b,a) then went on to apply the theory to a twisted flux tube and to turbulent heating by waves due to more rapid footpoint motions.

A key question is: when a field is stressed to a nonlinear force-free state by footpoint motions, for how long does energy build up before a heating event occurs? Browning et al. (2008) considered a series of discrete nanoflares of varying magnitude, with stored energy built up by shearing and released by sporadic reconnection. They suggested heating events are triggered when a flux rope becomes unstable to an ideal kink instability and that the subsequent energy release is determined by magnetic helicity-conserving relaxation to a minimum-energy state. They conducted 3D numerical experiments on a line-tied cylindrical flux rope with a piecewise-constant force-free α parameter. A kink instability is excited and forms a helical current sheet. When the current density exceeds a critical value, anomalous resistivity and fast reconnection are triggered. In the subsequent relaxation towards a constant-α state, the relative change in magnetic helicity is only one-hundredth of the relative decrease in magnetic energy. For a magnetic loop with zero net current (so that its twist is localised within the loop), the unstable mode is confined in the radial direction; after an initial reconnection phase in a single current sheet, the current sheet fragments into multiple reconnection sites to give almost full relaxation and heating throughout the flux rope (Hood et al. 2009). The idea was developed by Bareford et al. (2011) to model an ensemble of loops that attain instability and to predict the nanoflare energy distribution, together with a total energy flux that is independent of aspect ratio and is sufficient to heat the corona.

Attempts to build self-consistent MHD turbulence models of both coronal heating and solar wind acceleration are important (Cranmer et al. 2007), as summarised in Sec. 13.7, as are models for a turbulent MHD cascade in coronal loops (Rappazzo and Velli 2011) and leakage to the chromosphere (Verdini et al. 2012). Furthermore, van Ballegooijen and Cranmer (2008) have proposed a model for heating a flux rope containing stochastic magnetic fields on a scale of 1 Mm, using the idea of *hyperdiffusion*, which diffuses the field while preserving the magnetic helicity.

A particularly interesting simple model by van Ballegooijen et al. (2011) suggests that the chromosphere and corona can both be heated by Alfvén turbulence, created by interactions between waves propagating up from the photosphere. They extended their solar wind heating model (Sec. 13.7, Cranmer et al. 2007) and used reduced MHD in a single thin flux tube with field $\mathbf{B}_0(z)$ and small incompressible transverse motions with velocity $\mathbf{v}_1 = (\boldsymbol{\nabla}_\perp f) \times \hat{\mathbf{B}}_0$ and magnetic field $\mathbf{B}_1 = \boldsymbol{\nabla} \times (h\mathbf{B}_0)$. The MHD equations then reduce to

$$\frac{\partial \omega_\pm}{\partial t} = \pm v_A \hat{\mathbf{B}}_0 \cdot \boldsymbol{\nabla} \omega_\pm - v_A \frac{dv_A}{dz}\alpha - [\omega_\pm, f_\mp] \pm \left[\frac{\partial f_+}{\partial x}, \frac{\partial f_-}{\partial x}\right] \pm \left[\frac{\partial f_+}{\partial y}, \frac{\partial f_-}{\partial y}\right],$$

where $f_\pm \equiv f \pm v_A h$ and $\omega_\pm \equiv \omega \pm v_A \alpha$ are Elsässer-like variables with vorticity $\omega \equiv -\nabla_\perp^2 f$, $\alpha \equiv -\nabla_\perp^2 h$, $[a,b] \equiv (\partial a \partial x)(\partial b/\partial y) - (\partial a \partial y)(\partial b/\partial x)$ is the bracket operator, and the $\pm$ subscripts describe inward and outward propagating Alfvén waves. In these two equations, the first two terms describe the waves, the term with dv_A/dz describes linear coupling between them and the remaining terms describe nonlinear coupling. They found that part of the wave energy is reflected by the increase in Alfvén speed in the chromosphere and transition region and part is transmitted to the corona. The model produces heating rates observed in active-region coronal loops if the footpoint velocities are 1 to 2 km s^{-1} and the periods are 60–200 Sec. Coronal heating is 10^{-3} to 10^{-2} times smaller than the chromospheric value. It increases with coronal field strength and decreases with loop length.

10.5.2 Avalanches: A Nonlinear Driven Dissipative Process

Photospheric flux elements and atmospheric brightenings are both observed to possess power-law distributions, which in turn implies that the mechanisms creating them are scale-free. Reduced MHD models have been able to produce such power laws (Rappazzo and Velli 2011), but a simple complementary approach to modelling them using complex-systems techniques also offers insights into their origin.

Observations over seven orders of magnitude show that the number of flux elements scales with flux (F) as $F^{-1.85}$ for 10^{16} Mx $< F < 10^{23}$ Mx (Parnell et al. 2009). Furthermore, the number of coronal nanoflare brightenings (in the range 10^{23}–10^{26} ergs) scales with energy (E) as E^{α}, but the power-law index ($-2.5 < \alpha < -1.6$) cannot be accurately determined at present (Sec. 10.4.2).

Having such puzzling observations in mind, Vlahos (2009) proposed an approach with four elements for modelling an active region. First, the photospheric flux elements act as a driver and possess a distribution that may be explained by a *percolation model.* Secondly, the energy is assumed to be stored in self-similar fields associated with current sheets that form at many scales. Thirdly, the observed microflare and nanoflare brightenings are assumed to result from the active region being in a *self-organised critical* (SOC) state. When the current density in the sheets moves above a critical threshold, the sheets are assumed to reconnect and release their energy in an avalanche manner. Finally, the resulting distribution of fractal electric fields is likely to accelerate particles efficiently (Isliker and Vlahos 2003).

The percolation model assumes a competition between emergence of new flux and its disappearance by diffusion, together with random motion of the flux over the solar surface. It gives an explanation for the observed size distribution of flux elements and their fractal properties (Wentzel and Seiden 1992; Vlahos et al. 2002).

The core of the coronal heating idea, however, is the concept of an *avalanche*, put forward by Lu and Hamilton (1991). They set up a cellular automaton, in which the value A_{ijk} of a variable at the points (i,j,k) of a 3D grid is evolved in time subject to certain rules. The variable A_{ijk} is not defined, but it could be the vector potential, magnetic field or current. Starting with a uniform state, the values A_{ijk} are increased at random points, and at each stage the difference (dA_{ijk}) at each point between A_{ijk} and the mean values of its six nearest neighbours is calculated. If dA_{ijk} exceeds a critical value, then an "instability" occurs, which has the effect of reducing dA_{ijk} to zero by adding $dA_{ijk}/7$ to each of its neighbours and subtracting 6 $dA_{ijk}/7$ from A_{ijk}. This in turn can make many of the neighbours go unstable and so produce an avalanche, just like a pile of sand. An estimate of the total energy (E) released from all the reconnections in each avalanche is made. The number of events as a function of energy is found to scale as $E^{-1.5}$.

Subsequently, Vlahos et al. (1995) and Isliker et al. (2000, 2001) developed a more physical model and applied it to the corona, where A_{ijk} is indeed the vector potential, from which $\mathbf{B} = \nabla \times \mathbf{A}$ and $\mathbf{j} = \nabla \times \mathbf{B}/\mu$ are calculated. By adopting a different boundary condition and preferred direction, the magnetic field structure is that of a coronal arcade. The physical picture is of a coronal heating event consisting of many small reconnections in neighbouring locations that trigger one another, so that the size of the event depends on the number of elementary reconnections. The avalanche is triggered when the current density exceeds a threshold, and the Ohmic heating is calculated from the currents: it yields power-law distributions of energy release (with index between 1.6 and 1.8) and of peak flux that are consistent with observations. The resulting current concentrations are highly fragmented in space and time with a fractal dimension of 1.8 (Vlahos et al. 2004).

Later, Parenti et al. (2006), Parenti and Young (2008) produced a distribution of intensities from numerical simulations and concluded that the higher-temperature lines better preserve the signature of the heat input. Furthermore, Rappazzo et al. (2008) produced a reduced MHD model with an artificial forcing term.

In future, it is important to try and discover the physical rule that determines the scale-free power-law distributions, and also to close the gap between MHD and avalanche models, by demonstrating how a fully MHD model can be approximated as a self-organised critical state.

10.6 Conclusion

There have been huge advances in observations of the atmosphere and in theories for the possible ways in which it may be heated. It is now clear that the 'coronal heating problem' is much more complex and multi-faceted than appreciated before. Several aspects need to be further developed before definitive identifications can be made of the mechanisms and their relative contributions to heating different parts of the atmosphere (chromosphere, transition region, corona; X-ray bright points, coronal loops, coronal holes, active regions, diffuse outer corona):

(a) Chromospheric and coronal heating need to be modelled together in the frame of a complex coupled atmosphere, which is highly dynamic and finely structured. Plasma is continually heated and cooled and moves from one height to another, by spicular ejection and fall-back, and by evaporation and draining.

(b) Simplified models are crucial in highlighting fundamental physical processes, but producing detailed observational signatures of different mechanisms demands more realistic and complex approaches, including forward modelling of carefully designed computational experiments.

(c) Reality may be more subtle than the earlier simplistic distinction between reconnection and wave heating, since there are many ways in which both may work together. For example, waves may drive reconnection or reconnection may accelerate spicules or produce waves that subsequently dissipate.

(d) Finally, multi-fluid, microphysics and kinetic aspects of dissipation layers need a careful treatment for both wave damping (due to resonant absorption or phase mixing) and reconnection (due to nanoflares from flux-tube tectonics). These include collisionless reconnection, non-classical particle transport, heat flux saturation and ionization nonequilibrium. For example, ambipolar diffusion is a promising mechanism for heating the chromosphere (Khomenko and Collados Vera 2012a,b).

Chapter 11

Prominences

Observations of these cool, dense sheets of plasma up in the corona have been summarised in Section 1.8 (see Figures 1.33 to 1.37). Here the main aim is to discuss the magnetic properties of large, long-lived *quiescent prominences* (Priest 1989; Tandberg-Hanssen 1995; Mackay et al. 2010). In contrast, the important aspects of radiative transfer and spectroscopy are treated by Labrosse et al. (2010). *Active-region prominences* are much shorter-lived and may perhaps form in a different manner: when they erupt, they normally do so violently, showing up as a spray and giving rise to a two-ribbon solar flare (Chapter 12). *Quiescent prominences* are much larger and have weaker magnetic fields; they erupt more gently and do not usually produce a solar flare, but the basic eruptive magnetic process may well be the same.

This chapter begins with a summary of observational properties and theoretical questions (Sec. 11.1) and then treats the basic magnetic structure of prominences as current sheets within flux tubes (Sec. 11.2). The global nature of filament channel formation is discussed (Sec. 11.3), followed by the three-dimensional (3D) structure of barbs (or feet) (Sec. 11.4) and threads (Sec. 11.5), which are still a mystery, and the formation of thermal structure (Sec. 11.6). Since most observed flow speeds are much slower than the Alfvén speed, the field is roughly magnetohydrostatic (or even force-free when $\beta \ll 1$). As a background, see Chapter 3 for magnetohydrostatic theory and Section 7.5.7 for thermal instability theory. Also, the eruption of prominences and their intimate relation to solar flares and coronal mass ejections is covered in Chapter 12.

New observations and theories have produced a major shift in understanding the overall nature of prominences, with the global structure now well understood, formation less clear and fine-scale structure a mystery. In the previous edition of this text, the main emphasis was on formation by thermal instability or thermal non-equilibrium of a configuration of normal polarity in a coronal loop, arcade or current sheet (Figure 11.1). Now, building on this understanding, the emphasis has shifted to the magnetic structure of a twisted flux rope of inverse polarity, the overall formation in a filament channel, and the fine-scale structure of prominence barbs and threads.

11.1 Summary of Quiescent Prominence Properties

The main observational properties of a large, long-lived prominence that need to be explained are from Section 1.8 as follows.

(i) The density and temperature are a hundred times larger and smaller, respectively, than the surrounding corona. How is such plasma supported and cooled, and what is the origin of the mass? Support is magnetic in some kind of dipped configuration, whereas cooling is probably by radiative thermal non-equilibrium, as indicated in an example from SDO/AIA by Berger et al. (2012). Since the overall mass is so large, it could not come from the surrounding corona but needs to flow up somehow from the chromosphere or photosphere.

(ii) Prominences are located along a polarity-inversion line (PIL) (Figure 1.33). Most have inverse polarity, with the underlying photospheric field also inverse (i.e., a bald patch). The main field component is horizontal

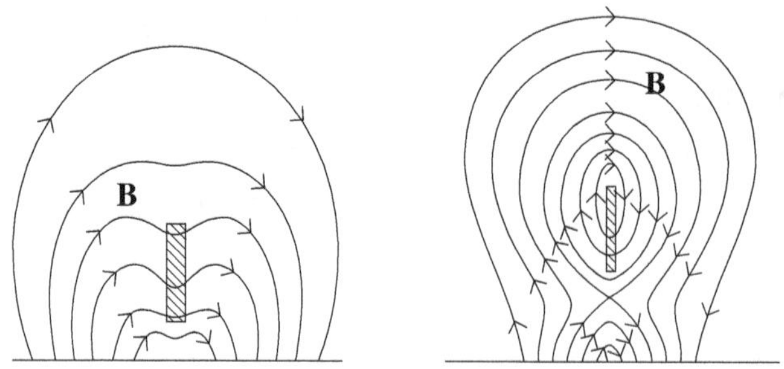

(a) (b)

Figure 11.1. Magnetic configuration normal to a prominence of (a) normal polarity and (b) inverse polarity. The Kippenhahn-Schlüter model for current-sheet support is locally applicable to both types.

and parallel to the PIL. The magnetic structure is thus a largely horizontal (twisted) flux rope with large magnetic helicity. Why does a prominence form only in such a structure above a PIL?

A prominence needs first of all a magnetic dip, and so this is likely to be above a PIL. Secondly, for the thermal condensation to take place against the stabilising feature of thermal conduction, long field lines are needed, and, if such field lines are much lower in the atmosphere than their length, they will form a sheared largely horizontal field. Thirdly, once such a field has built up along a PIL (i.e., a filament channel), reconnection at the PIL by flux cancellation will naturally occur and convert the sheared field into a horizontal flux rope. Indeed, a largely horizontal field that contains magnetic helicity is likely to lie above a PIL.

The alternative possibility of simple emergence by magnetic buoyancy of a flux rope (Rust and Kumar 1994) is much less likely, since emergence tends to be halted by the huge weight of dense plasma trapped in the rope before the tube axis emerges, and the PIL is observed to be present well before a prominence forms.

(iii) The magnetic field in a prominence is observed, surprisingly, to increase with height (z, say) (Rust 1967; Leroy et al. 1983). A natural explanation for this is that it is associated with a field line dip, since the Lorentz force at the summit or dip of a field line (e.g., Figure 11.1a or 11.1b) takes the form

$$\mathbf{j}\times\mathbf{B} = -\frac{1}{2\mu}\frac{\partial B^2}{\partial z}\hat{\mathbf{z}} + \frac{B^2}{\mu R_c}\hat{\mathbf{z}},$$

in terms of the magnetic pressure and tension forces, where R_c is the radius of curvature. Thus, if the field is approximately force-free, these two terms roughly balance, and so an upward curvature (a dip with $R_c > 0$) implies that $\partial B^2/\partial z > 0$, i.e., that the field strength increases with height. It would then be a slight excess of the tension over magnetic pressure force that can provide support against gravity.

(iv) Before a prominence erupts (Figure 1.37), observations of filaments and their flows often imply that the field is weakly twisted, with the axial magnetic component dominating. Highly twisted tubes would be kink unstable, but reconnection during an eruption can increase the twist by transferring magnetic helicity from mutual to self and thereby explaining the twist often seen at the limb during an eruption (Vrsnak et al. 1991); indeed, most of the twist observed in magnetic clouds in interplanetary space is created by the reconnection (e.g., Qiu et al. 2007; Kazachenko et al. 2010).

(v) The overall magnetic field around a prominence is likely to be a weakly twisted flux rope. On the limb this shows up a region of reduced density, called a *coronal cavity* (Sec. 11.2.3) (Figure 1.21a).

(vi) All long-lived (i.e., quiescent) prominences form at a PIL, where two flux systems press up against one another and so probably drive flux cancellation. For example, a new active region may emerge next to a remnant active region or another active region, or unipolar flux may migrate towards a pole with opposite flux and create the polar crown. The reason why such flux cancellation in a filament channel is crucial is that it is a natural way to create a flux rope there by reconnection.

(vii) Filament channels indicate the presence of a sheared horizontal non-potential coronal field. Dextral (sinistral) filaments possess negative (positive) magnetic helicity (Figure 1.34). Why do dextral filaments dominate in the northern hemisphere?

(viii) Prominences possess barbs and consist of fine-scale threads (Figures 1.33 to 1.36). What is their nature? Are threads aligned with the magnetic field? What is the relation of threads seen from above in filaments and from the side in prominences?

(ix) There is a wide range of oscillations of prominences, which are being used to infer their physical properties using techniques of *prominence seismology* (Ballester 2006; Arregui et al. 2012).

One necessary condition for prominences is a filament channel on a polarity-inversion line (containing strong non-potential shear and a horizontal field). Another is likely to be the presence of flux cancellation which builds up the self-magnetic helicity from mutual helicity and creates a flux rope from a horizontal sheared field.

11.2 Basic Magnetic Structure (Current Sheet in a Flux Rope)

Kippenhahn and Schlüter (1957) modelled the local structure of a vertical prominence sheet supported by the magnetic field. They envisaged that the sheet is part of a global arcade or quadrupole structure. We now know that the appropriate surrounding structure is instead probably a flux rope, but their model for the local sheet structure is still valid, so we here describe it (Sec. 11.2.1) before discussing models for a flux rope, both analytical (Secs. 11.2.2 and 11.4.1) and numerical (Sec. 11.4.2).

11.2.1 Kippenhahn-Schlüter Model for Prominence Sheet

Consider a thin isothermal 1D sheet with all variables depending on one coordinate (x) alone. The field lines are bowed down by the dense prominence plasma (Figure 11.2b) and perform two roles: the magnetic tension provides an upward force to balance gravity and support the prominence, while the magnetic pressure increases with $|x|$ and so provides a transverse force to compress the plasma and balance the plasma pressure gradient.

The prominence sheet equilibrium is governed by the force balance (Eq. 3.1)

$$\mathbf{0} = -\nabla p - \rho g \hat{\mathbf{z}} - \nabla\left(\frac{B^2}{2\mu}\right) + (\mathbf{B}\cdot\nabla)\left(\frac{\mathbf{B}}{\mu}\right), \tag{11.1}$$

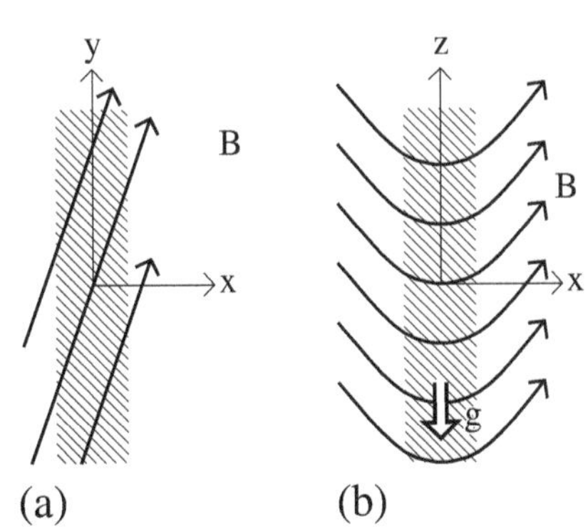

Figure 11.2. Notation for the Kippenhahn-Schlüter model for local support and compression of a vertical prominence sheet (a) from above and (b) from the end. The x-, y- and z-axes are across and along the prominence and normal to the solar surface, respectively.

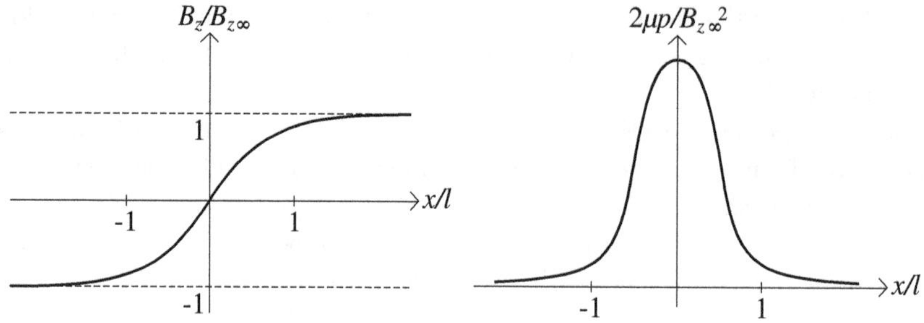

Figure 11.3. The vertical magnetic field (B_z) and plasma pressure (p) for the Kippenhahn-Schlüter model shown schematically as functions of distance (x) across the prominence.

together with the equation of state

$$\rho = \frac{mp}{k_B T}, \tag{11.2}$$

where T and the horizontal components (B_x, B_y) are all assumed uniform. However, the axial field (B_y) plays a purely passive role and doesn't affect the structure of the model at all. The pressure $[p(x)]$, density $[\rho(x)]$ and vertical field $[B_z(x)]$ are all assumed functions of x alone, so that $\nabla \cdot \mathbf{B} = 0$ automatically and the x- and z-components of Eq. (11.1) reduce to

$$0 = -\frac{d}{dx}\left(p + \frac{B^2}{2\mu}\right), \tag{11.3}$$

$$0 = -\rho g + \frac{B_x}{\mu}\frac{dB_z}{dx}. \tag{11.4}$$

The integral of Eq. (11.3), together with the boundary conditions that $p \to 0$ and $B_z \to \pm B_{z\infty}$ as $x \to \pm\infty$, gives

$$p = \frac{B_{z\infty}^2 - B_z^2}{2\mu}. \tag{11.5}$$

Equation (11.4) may be solved [with ρ given by Eq. (11.2) and p by Eq. (11.5)], subject to the boundary condition that $B_z = 0$ at $x = 0$ for a symmetric sheet. If $H = k_B T/(mg)$ is the scale-height (Sec. 3.1.2), the solution is

$$B_z = B_{z\infty} \tanh \frac{B_{z\infty} x}{2B_x H}, \qquad p = \frac{B_{z\infty}^2}{2\mu} \operatorname{sech}^2 \frac{B_{z\infty} x}{2B_x H}, \tag{11.6}$$

as sketched in Figure 11.3. The central plasma pressure is equal to the external magnetic pressure associated with the vertical field. The half-width ($l = 2B_x H/B_{z\infty}$) of the sheet is just the pressure scale-height increased by the factor $2B_x/B_{z\infty}$ and represents the distance over which the pressure falls appreciably from its central value.

The model has been extended and generalised in several ways. One limitation is that the ambient plasma pressure vanishes, since otherwise the ambient plasma density would need a vertical force to support it. However, this small effect has been incorporated by linearising about the present model and dropping the 1D assumption (Ballester and Priest 1987), with the result that the sheet width decreases slightly with height while the field lines become less curved and the horizontal field increases, in agreement with observations (PROBLEM 11.1: for all problems and solutions, see web page www.cambridge.org/9780521854719).

Another extension is to remove the isothermal assumption by coupling an energy equation to the magnetohydrostatics. Heasley and Mihalas (1976) included radiative transfer, while Low (1975) and Milne et al. (1979) used a simple energy balance between conduction, optically thin radiation and heating

(PROBLEM 11.2): they found that, for a large enough B_x, solutions exist only when the magnetic field is sheared by more than about 80 degrees and when the plasma beta is less than a maximum value, so that a prominence cannot exist below a certain height and active-region prominences can form lower in the atmosphere than quiescents.

Generalising the Kippenhahn-Schlüter model to include either steady flow (Low and Petrie 2005) or unsteady cross-field flow due to ambipolar diffusion (Low et al. 2012) are also important advances.

Low et al. (2012) coupled the magnetohydrostatics with an energy equation that includes conduction and forms for radiation ($\sim \rho^2 T$) and heating ($\sim \rho$) that enabled them to solve the equations analytically. They considered what happens when the total mass (M) is imposed. If it equals a critical value ($M = M_c$), then there is a family of continuous solutions with different central temperatures. If $M > M_c$, there are no equilibrium solutions, but, if $M < M_c$ and other conditions hold, there is an equilibrium with a current sheet and a discontinuous jump of vertical field from one side to the other. Simple solutions are set up for the interior of the sheet and for vertical diffusion of the plasma through the magnetic field.

An important question is: how large is the dip created by the plasma? It may be measured by the inclination of the field to the vertical at the edge of a model prominence sheet, which may be estimated as follows. Equation (11.4) implies that

$$\rho g \approx \frac{2B_z B_x}{\mu w},$$

where B_z is the vertical field at the edge of the prominence of width w, and B_x is the horizontal field normal to the prominence. Thus, for typical values ($\rho = 1.7 \times 10^{-10}$ kg m^{-3}, w = 5 Mm) we find B_z/B_x is 0.15 for B_x = 10 G, giving an inclination angle to the horizontal of 9 degrees. An alternative way to make the estimate is to use Eq. (11.5), namely, $B_z^2 \approx 2\mu p$, so that, the larger the ambient field or width, the smaller is the dip. Indeed,

$$\frac{B_z}{B_x} \approx \left(\frac{2\mu p}{B_x^2}\right)^{1/2} = \beta_{hor}^{1/2},$$

where β_{hor} is the plasma beta based on the prominence pressure and the horizontal magnetic field normal to the prominence. For a temperature, density and magnetic field of 8,000 K, 10^{17}m^{-3} and 10 G, we find β_{hor} = 0.028, giving $B_z/B_x = 0.17$. Since $\beta_{hor} \ll 1$, this in turn means that departures from a force-free field are small.

(On the other hand, *active-region prominences* can have fields of 700 G and pressures a factor of 10 higher, which makes $\beta_{hor} = 6 \times 10^{-5}$ and $B_z/B_x = 0.008$, so that the magnetic field structure is almost unaffected by the presence of the plasma. Thus, such prominences are better modelled by dynamic condensation in a low-lying flux tube than by a Kippenhahn-Schlüter model.)

Of course, quiescent prominence plasma certainly creates a larger dip than would otherwise be present, since it distorts the magnetic field, but a more important question is whether the field needs to be dipped before the prominence will form? One argument in favour is that inverse polarity is not possible without a dip, and another is that dense plasma cannot produce a significant dip in a low-beta plasma, so that condensing plasma would quickly drain. This in turn depends on the ratio of the natural draining time to the formation time, which we discuss in Section 11.6.

The magnetic field around a normal-polarity prominence in a coronal arcade has been modelled by Anzer (1972), treating the prominence as a current sheet and using the normal field in the prominence (on the z-axis) and photosphere (on the x-axis) as boundary conditions. The total Lorentz force across the width of the prominence at a given height is then $F_L = J/B_{x0}$, which is positive above 17 Mm and sufficient to support a particular prominence. Here $J = 2B_{zd}/\mu$ is the total prominence current at a given height, where B_{zd} is the vertical field at the boundary ($x = \frac{1}{2}d$, say) of the prominence sheet. Anzer finds $J\ dB_{x0}/dz \geqslant 0$ and $B_{x0}\ dJ/dz \leqslant 0$ as necessary and sufficient conditions for stability. In a major advance,

Hood and Anzer (1990) then developed a 2.5D non-isothermal magnetohydrostatic model for a prominence in a linear force-free coronal arcade.

11.2.2 2.5D Magnetic Flux-Rope Models of Inverse Polarity

Observations (Sec. 1.8) indicate that a largely horizontal twisted flux tube (i.e., a flux rope) with inverse polarity is the likely global configuration for most prominences. The first such model (Kuperus and Raadu 1974) considered a simple circular prominence of radius R, density ρ, mass $m = \pi R^2 \rho$, field B_ϕ and current $I = 2B_\phi R\pi/\mu$ that has formed at a height h in a vertical current sheet. The photospheric line-tying of the current-sheet field is modelled by placing an image current $-I$ a distance h below the photosphere, so that the horizontal field due to the sum of the prominence and image currents vanishes at the photosphere. The prominence is supported against gravity by the repulsive force between the prominence current and its image, namely,

$$\frac{\mu\, I^2}{4\,\pi\, h} = m\, g,$$

or substitution for I and m gives $B_\phi^2/(\mu h) = \rho g$. For a typical density $\rho = 10^{-10}$ kg m^{-3} and height h = 10 Mm, this implies a reasonable field strength of 6 G. The model may be extended to include prominence oscillation (PROBLEM 11.3).

In an inspirational advance, van Ballegooijen and Martens (1989) suggested that a twisted force-free field capable of supporting an inverse-polarity prominence can be built up by a two-stage process (Figure 11.4), namely, the shearing of a coronal arcade followed by reconnection driven by the process of photospheric flux cancellation discovered by Martin et al. (1985), Martin (1986).

This led to the development of a Flux-Rope Model for prominences (Priest et al. 1989; Démoulin and Priest 1989), in which the twist in a large-scale flux tube is built up most likely by the van Ballegooijen-Martens mechanism of shearing plus flux cancellation (but in principle also by photospheric twist). They suggested that, as the twist slowly increases, first a small dip with upward curvature appears and then the dipped region (within which cool prominence plasma can accumulate and be supported) lengthens, and finally the twist is so great that the prominence erupts (Figure 11.5). The critical twist for the appearance of a dip and the variation of prominence length with twist is calculated.

In order to model prominence support in the flux-rope model, consider a section across the tube (of radius a) perpendicular to the prominence axis (Figure 11.6) and suppose the prominence is a vertical

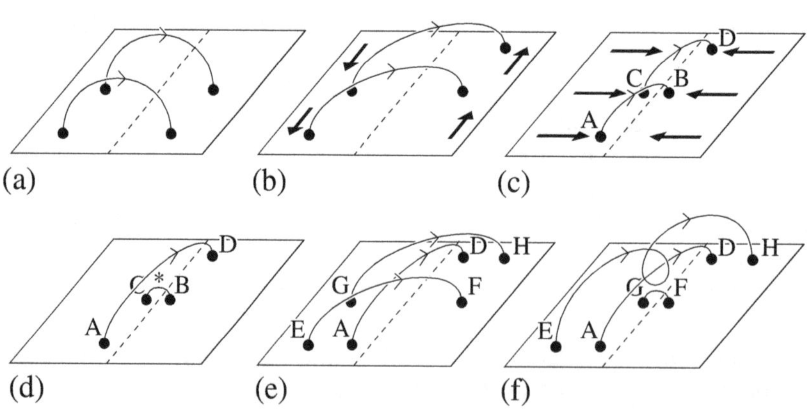

Figure 11.4. The van Ballegooijen-Martens scenario in (a) a coronal arcade for creating a twisted flux tube by (b) shearing and (c) flux cancellation. Reconnection between overlying arcade field lines creates (d) a low-lying sheared field line AD and then (f) a twisted field EH (van Ballegooijen and Martens 1989).

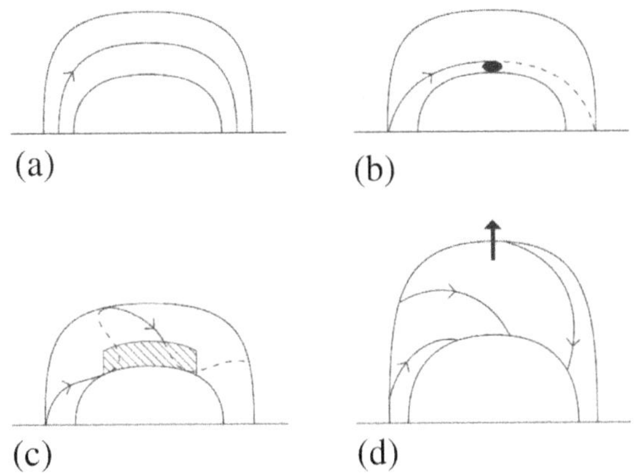

Figure 11.5. Overall scenario in a Flux-Rope Model for prominences (Priest et al. 1989).

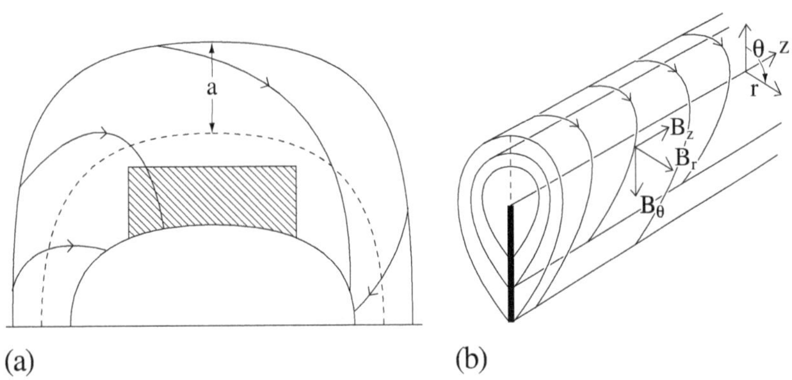

Figure 11.6. Support of a prominence sheet viewed (a) from the side and (b) from the end in a Flux-Rope Model for prominences (Priest et al. 1989).

sheet of mass and current at $\theta = \pi$. A force-free field independent of z may be expressed in terms of a flux function $A(r,\theta)$ as

$$(B_r, B_\theta, B_z) = \left(\frac{1}{r}\frac{\partial A}{\partial \theta}, -\frac{\partial A}{\partial r}, B_z(A)\right),$$

where the force-free equation reduces to the Grad-Shafranov equation,

$$\nabla^2 A + F(A) = 0,$$

with $F(A) = d/dA(\frac{1}{2}B_z^2)$. As a simple example, consider a solution of the form $A(r,\theta) = A_0(r) + A_1(r,\theta)$, where $A_1/A_0 \approx \epsilon \ll 1$, that is close to a cylindrically symmetric field $[A_0(r) = -B_0 r^2/(2\sqrt{2}a)]$ for $r \leqslant a$. The function $A_1(r,\theta)$ satisfies

$$\frac{dA_0}{dr}\left[\frac{1}{r}\frac{\partial}{\partial r}\left(r\frac{\partial A_1}{\partial r}\right) + \frac{1}{r^2}\frac{\partial^2 A_1}{\partial \theta^2}\right] - \frac{d}{dr}\left[\frac{1}{r}\frac{d}{dr}\left(r\frac{dA_0}{dr}\right)\right] A_1 = 0.$$

Separable solutions exist of the form $A_1 = r^K \cos K\theta$ with $K > 0$, so that $B_r = 0$ at $\theta = 0$. The corresponding field components are

$$B_{1r} = -\epsilon K r^{K-1} \sin K\theta, \qquad B_{1\theta} = -\epsilon K r^{K-1} \cos K\theta,$$

for $-\pi < \theta < \pi$. If $1 < K < 1.5$, the field is well behaved at $r = 0$ (the top of the sheet) and provides upward curvature ($B_{1r} > 0$) at the prominence sheet ($\theta = 0$). If there is local force balance ($\rho_1 g = j_{1z} B_{0\theta}$) within the prominence sheet, the prominence mass density (m) integrated across the sheet is given by

$$mg = [B_{1r}] B_{0\theta}/\mu = m_0 g (r/a)^K,$$

where $m_0 = -\sqrt{2}\epsilon K \sin(K\pi) B_0 a^{K-1}/(\mu g)$ and $[B_{1r}]$ is the jump in B_{1r} across it.

Extensions of the analysis include: other flux-rope solutions to $\nabla^2 A + F(A) = 0$ (Ridgway et al. 1991); a force-free solution incorporating boundary conditions at the prominence and photosphere (Ridgway et al. 1991); and a linear force-free tube surrounded by a potential field (Rust and Kumar 1994) (PROBLEMS 11.4, and 11.5).

In addition, conditions for the formation of a dip have been studied as follows for a 2.5D linear force-free field that is independent of y of the form

$$(B_x, B_y, B_z) = \left(-\frac{\partial A}{\partial z}, \alpha A, \frac{\partial A}{\partial x}\right),$$

where the force-free equation reduces to $\nabla^2 A + \alpha^2 A = 0$ (Démoulin and Priest 1989). In particular, the fundamental solution $A = -(1/k)\cos kx \exp(-l_1 z)$ represents a simple coronal arcade of width L with field components

$$(B_x, B_y, B_z) = \{-(l_1/k)\cos kx, -(\alpha/k)\cos kx, \sin kx\} e^{-l_1 z},$$

where $l_1^2 = k^2 - \alpha^2$ and $k = \pi/L$ for $0 \leqslant \alpha \leqslant k$. The shear in the field is given by $\tan\theta = B_y/B_x = \alpha/\sqrt{k^2 - \alpha^2}$, and so it becomes large and $l_1 \to 0$ as $\alpha \to k$.

This field always decreases with height, and so, in order to find a field that increases, subtract from it the nth harmonic of amplitude b_n to give

$$B_x = -(l_1/k)\cos kx\; e^{-l_1 z} + b_n l_n/(nk)\cos nkx\; e^{-l_n z},$$
$$B_y = -(\alpha/k)\cos kx\; e^{-l_1 z} + b_n \alpha/(nk)\cos nkx\; e^{-l_n z},$$
$$B_z = \sin kx\; e^{-l_1 z} - b_n \sin nkx\; e^{-l_n z},$$

where $l_n^2 = n^2 k^2 - \alpha^2 > l_1^2$, so that the nth harmonic is less sheared than the first.

Furthermore, the horizontal field gradient on the z-axis varies with height as

$$\frac{d}{dz}(B_x^2 + B_y^2)(0,0,z) = -2l_1 e^{-2l_1 z} + 2(l_n/n + l_1 n) b_n e^{-(l_1+l_n)z} - 2l_n b_n^2 e^{-2l_n z},$$

which (for each z) vanishes at two positive values of b_n, namely,

$$b_{n1} = (l_1 n/l_n) e^{(l_n - l_1)z} \quad \text{and} \quad b_{n2} = (1/n) e^{(l_n - l_1)z}.$$

The condition that the field increase with height is that b_n lie between b_{n1} and b_{n2}. But $b_{n2} > 1/n$, and the condition that there also be no parasitic flux [i.e., $xB_z(x,y,0) > 0$] is $b_n < 1/n$, so that for both conditions to hold we need $b_{n1} < b_{n2}$. This implies that $l_1/l_n < 1/n^2$ or, after substituting for l_1 and l_n,

$$\frac{\alpha^2}{k^2} > \frac{n^2}{n^2+1},$$

which lies between 0.8 (when $n = 2$) and 1. Moreover, the above conditions ($b_n < 1/n$ and $b_n > b_{n1}$) imply that $B_z(x,y,0)B_x(0,0,z) > 0$, so the field has inverse polarity.

Thus, we have produced a force-free dip of inverse polarity without parasite flux, but only when the shear is very large, which is consistent with observations that prominences form only in highly sheared regions. This is not possible in a potential field, which needs parasitic flux for a dip to form (PROBLEM 11.6).

11.2.3 Coronal Cavities

When a prominence is viewed along its axis on or near the limb in white light, soft X-rays, EUV or in radio, the region around and above it shows up as an elongated, tunnel-like structure of reduced density, called a *coronal cavity* (e.g., Gibson et al. 2006). In the polar crown, it may be long-lived (on the order of months) (Gibson and Fan 2006; Tripathi et al. 2009). A cavity is also observed during a coronal mass ejection, whose classic three-part morphology consists of a bright expanding loop, the dark cavity and a bright core (an erupting prominence) (Hundhausen 1999).

The overall magnetic field in a cavity is likely to be a weakly twisted flux rope. From white-light measurements, the density in a cavity is typically 30 to 50 per cent of the density in a surrounding streamer and twice that of a coronal hole (Fuller and Gibson 2009; Schmit and Gibson 2011).

Gibson et al. (2010) have compared STEREO observations of a cavity with a 3D coronal streamer model, including a cavity of certain dimensions and a density depleted by 50 per cent. The forward-modelled EUV emission tightly constrains the 3D morphology of the cavity. Later, Dove et al. (2011) showed that a flux-rope model for a particular cavity has a robust ring-like signature in linear polarization, which is found in CoMP observations from Mauna Loa Observatory. Then Kucera et al. (2012) used Hinode/EIS observations to reveal multi-thermal structure in the cavity in the range 1.4 to 1.7 MK. Furthermore, Rachmeler et al. (2013) have deduced polarisation features in various cavity models, while Bąk-Stęślicka et al. (2013) have demonstrated from CoMP that many cavities are consistent with flux-rope models.

11.3 Global Nature of a Filament Channel and Chirality

We shall return to the theme of parasitic flux in Section 11.4 when discussing the nature of barbs and chiral structure prominences, but first why do dextral prominences predominate in the northern hemisphere and sinistral ones in the south, and why do only some parts of the global polarity-inversion line (PIL) contain filaments?

Two explanations for the presence of a strong axial field in a filament channel have been proposed. One is that it originates in a surrounding coronal arcade and is transported towards the PIL by convective or diffusion flow (van Ballegooijen and Martens 1989), while the transverse component is reconnected by flux cancellation. However, the problem is that shearing by differential rotation of an arcade arching over an east-west PIL builds up the wrong sign for the axial field. Thus, for the arcade in Figure 11.7a, differential rotation would move the positive footpoints to the right and build up an axial field directed from right to left.

The second explanation is therefore the most likely, namely, that the correct sign for the axial field is built up below the photosphere by differential rotation and Coriolis forces, and that it emerges near the polarity-inversion line (van Ballegooijen and Martens 1990; Rust and Kumar 1994; Martens and Zwaan 2001). Thus, if the footpoints of the arcade in Figure 11.7a are connected below the photosphere by a field that is directed from north to south (i.e., the opposite of the arcade field), differential rotation would build up an axial field directed from left to right. After these flux tubes have emerged, the chiral nature of arcades whose PIL is orientated north-south will be reinforced by another effect, namely, differential rotation: this is appropriate for some active-region filaments and for some young filaments that are located on the boundaries of active regions. Such ideas were incorporated by Priest et al. (1996) in their *Dextral and Sinistral Model* for prominences (Figure 11.7). They showed how it can indeed produce the observed hemispheric pattern, and it was later developed in much greater detail in a superb series of papers by Mackay and van Ballegooijen (see below).

The Dextral and Sinistral Model suggests that: a prominence is a dynamic structure with a continual input of mass and magnetic flux; the barbs are short-lived structures caused by the transient interaction of parasitic flux with the main prominence field; the overall structure is a twisted flux tube whose magnetic

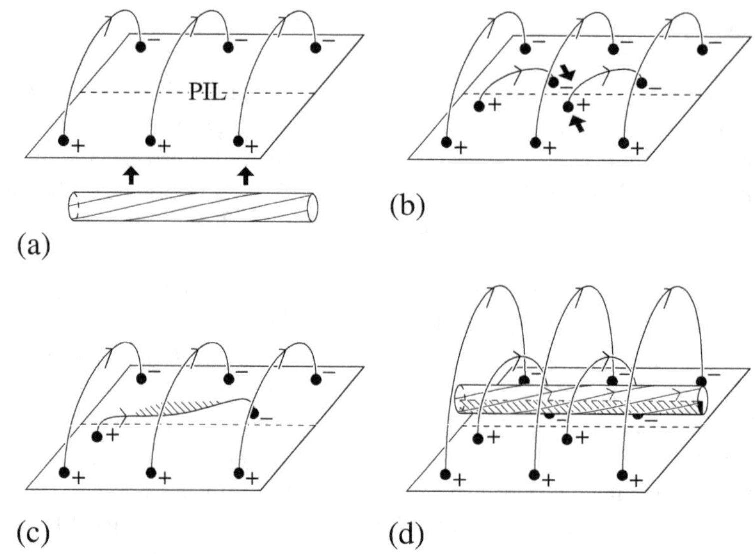

Figure 11.7. The Dextral and Sinistral Model (here illustrated for the dextral case): (a) a left-twisted flux tube is formed below a northern hemisphere PIL; (b) strands of the flux tube emerge as a series of small bipoles along the broad PIL; (c) some magnetic fragments are swept to reconnect with the surrounding network, where they add shear to the arcade; others are carried in and reconnect near the PIL to create long field lines along it and to lift cool plasma into the corona; (d) a fully formed flux tube after many reconnections.

helicity is created below the solar surface; the flux emerges in fragments and is rearranged by reconnection to form first a filament channel with a strong horizontal field along it and later a horizontal twisted coronal flux tube; the small-scale reconnections help to lift chromospheric plasma into the corona.

Mackay and Priest (1997) modelled filament channels, showing how minority polarity plagettes make a filament kink and, if their flux is large enough, they disrupt the filament and break it into two. Also, Mackay et al. (1998) considered a sheared active region that emerges and interacts with a remnant active region by large-scale convergence and flux cancellation to form a filament channel and a filament. They showed how it is the correct sign of magnetic helicity in the active region that leads to the correct filament chirality. Indeed, active regions are known to have mainly negative magnetic helicity ($\alpha < 0$) in the northern hemisphere and the opposite in the southern hemisphere (Pevtsov et al. 1995).

The next step was by van Ballegooijen et al. (1998), who built on previous work (DeVore et al. 1984; Sheeley et al. 1987) to set up a numerical model for the global flux transport of the radial component of the photospheric magnetic field due to differential rotation, meridional flow and magnetic diffusion, starting with an observed magnetogram. They found that with no new flux emergence equal numbers of dextral and sinistral prominences are produced in each hemisphere.

A major advance by van Ballegooijen et al. (2000) was to develop a mean-field model for filament channel formation. They assumed an evolution through nonlinear force-free fields in the corona, driven by the above random and organised photospheric motions as well as flux emergence. Thus, the lowest-order radial field at the photosphere (B_r) evolves according to

$$\frac{\partial B_r}{\partial t} = -\frac{1}{r\sin\theta}\frac{\partial}{\partial\theta}(v_\theta B_r \sin\theta) - \left[\Omega + \frac{f(B_r)}{r^2\sin\theta}\frac{d\beta}{d\theta}\right]\frac{\partial B_r}{\partial\phi} + D\nabla_\perp^2 B_r,$$

where $\nabla_\perp^2 = \nabla^2 - \partial^2/\partial r^2$, v_θ is the imposed meridional flow, $\Omega(\theta)$ is the imposed angular velocity (differential rotation), D is the photospheric diffusion due to random footpoint motions, and the electric field associated with flux emergence is $\mathbf{E} = \beta(\theta) f(B_r)\boldsymbol{\nabla}_\perp B_r$. In the corona the field evolves through approximately nonlinear force-free states according to the nonideal induction equation

$$\frac{\partial \mathbf{B}}{\partial t} = \boldsymbol{\nabla} \times (\mathbf{v} \times \mathbf{B} - \eta_c \mathbf{j}),$$

with coronal diffusion $\eta_c = \eta_0(1 + cj/B)$ that is enhanced in regions of strong current. The velocity is determined by a magnetofrictional equation

$$\mathbf{v} = \frac{\mathbf{j} \times \mathbf{B}}{\nu B^2} + v_0 e^{-(2.5R_\odot - r)/r_w} \hat{\mathbf{r}},$$

with friction ν, where v_0 and r_w are constants and the second term is a radial outflow designed to make the field lines radial at a source surface ($r = 2.5R_\odot$).

Then Mackay and van Ballegooijen (2005) applied the model to a pair of magnetic bipoles in a localised region of the corona. For the first time, they showed how the sheared fields in filament channels and their hemispheric pattern are produced by the effects of the observed dominant helicity and tilt angles on the evolution and interaction of emerging active regions as they are transported poleward. Importantly, they also showed how exceptions to the hemispheric pattern arise in a natural way when bipoles have a large tilt or their helicity has the opposite sign from normal.

This was extended by Yeates et al. (2008) to a much more realistic six-month global simulation of nonlinear force-free evolution using observed magnetograms and inserting bipoles with the correct location and magnetic helicity sign whenever each of 119 active regions emerged (Figure 11.8). Impressively, the chirality of 97 per cent of the 109 filaments below 65 degrees latitude identified on Big Bear Solar Observatory Hα images is predicted correctly by the model, including exceptions to the hemispheric pattern. This proportion increases with the duration of the simulation run. Thus, key aspects of the evolution of

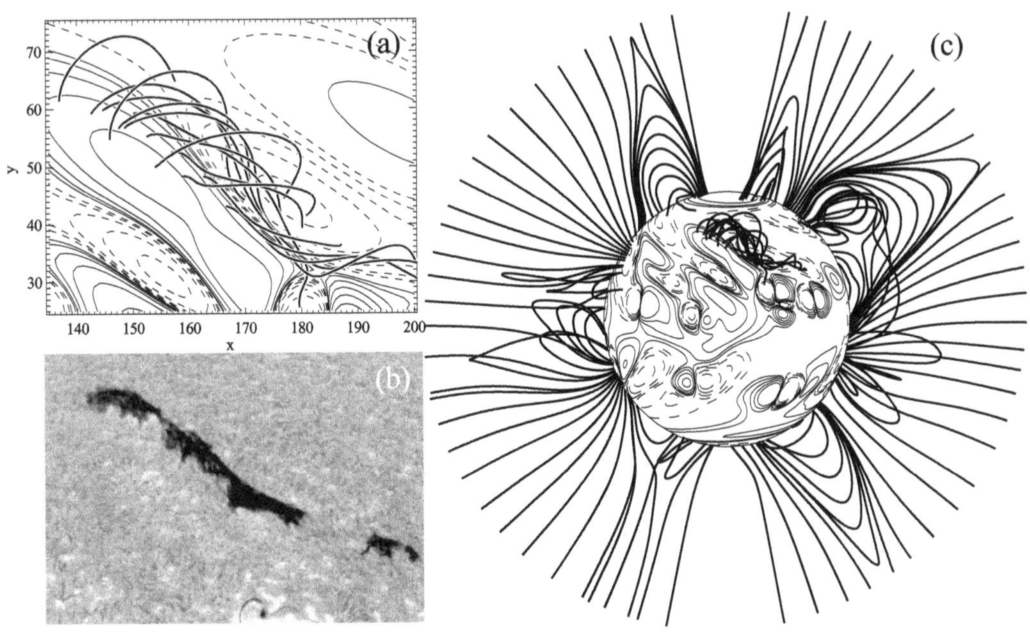

Figure 11.8. (a) Simulated nonlinear force-free structure, with positive and negative radial photospheric magnetic flux shown by solid and dashed contours, respectively, and thicker lines denoting coronal field lines; (b) Big Bear Solar Observatory Hα image of filament at the same location; (c) context of the filament in the global simulation (Yeates et al. 2008).

a nonlinear force-free coronal field are its long-term memory, emergence of new active regions and global transport and diffusion of magnetic helicity.

In future, it will be interesting to extend the study to active region interiors, where higher spatial resolution is required, and to the polar crown, where longer integration times are needed and it is more difficult to determine chirality (Mackay et al. 2011).

11.4 Three-Dimensional Structure: Barbs (or Feet)

The preceding analytical 2.5D flux-rope models have been invaluable in elucidating the basic physical processes involved in prominence magnetic structure and have the advantage of simplicity and of being able to follow easily the effect of changing parameters and assumptions. They have been extended in two complementary ways, namely, (i) the pioneering discovery of a 3D model (Sec. 11.4.1) for a prominence magnetic field as essentially a twisted flux rope perturbed by parasitic polarities, which explains in a natural way the nature of barbs, and (ii) the building of complex numerical models to reproduce individual observed prominence structure (Sec. 11.4.2).

11.4.1 3D Aulanier-Démoulin (1998) Flux-Rope Model

Prominences consist of bridges of cool plasma that periodically connect to the photosphere in so-called *feet*, seen from the side most prominently in hedgerow prominences. Viewed from above, they project to the sides of the main filament spine as *barbs*. The prominences lie above a broad polarity-inversion line (PIL) that is nearly free of magnetic flux except for small parasitic polarities that turn out to be crucial.

The suspension of barb plasma in the corona without falling down at the free-fall speed suggests that it is supported magnetically in a dip. Thus, the aim is to find the location of dips in a 3D linear force-free field as a generalisation of the 2.5D field $(B_x, B_y, B_z) = \{-(l_1/k)\cos kx, -(\alpha/k)\cos kx, \sin kx\}e^{-l_1 z}$. Such a field (Démoulin et al. 1989) that decays with height (z) and has a PIL at $x=0$ is a sum over positive integers $n>0$ and $m \geqslant 0$ of the nm-th harmonics of amplitude $b_{n,m}$, namely,

$$B_x = \frac{b_{n,m}}{k_{nx}^2 + k_{my}^2}\Big(-l_{nm}k_{nx}\cos(k_{nx}x)\cos(k_{my}y) - \alpha k_{my}\sin(k_{nx}x)\sin(k_{my}y)\Big)e^{-l_{nm}z},$$

$$B_y = \frac{b_{n,m}}{k_{nx}^2 + k_{my}^2}\Big(-\alpha k_{nx}\cos(k_{nx}x)\cos(k_{my}y) - l_{nm}k_{my}\sin(k_{nx}x)\sin(k_{my}y)\Big)e^{-l_{nm}z},$$

$$B_z = b_{n,m}\sin(k_{nx}x)\cos(k_{my}y)e^{-l_{nm}z},$$

where $k_{nx} = 2\pi n/L_x$, $k_{my} = 2\pi m/L_y$ and $l_{nm}^2 = k_{nx}^2 + k_{my}^2 - \alpha^2$ in terms of the x- and y-periodicity lengths (L_x, L_y), with α ranging between 0 and $\alpha_{max} = 2\pi/L_x$.

In 2.5D ($m=0$), the different inverse-field topologies may be classified as: OBP if the top of the dip is an O-point and the bottom is a bald patch (Figure 11.9b); OX if the top and bottom are O- and X-points (Figure 11.9c), and OF if the top is an O-point and the bottom is a flat field line with two X-points either side of the PIL (Figure 11.9d). With two harmonics and sufficient shear, the OBP topology is possible with a flux tube touching the photosphere. Aulanier and Démoulin (1998) were surprised to discover, however, that, with three harmonics (n = 1,2,3), as well as the standard OX topology, the new state OF is also present with small parasitic polarity and a flux tube suspended above the photosphere; in addition, it possesses small secondary dips above the lateral X-points. Moreover, for large enough α, OF occupies a much larger region of $b_{2,0}$-$b_{3,0}$ parameter space than OX.

In 3D, adding two harmonics to the fundamental linear force-free arcade solution enabled Aulanier and Démoulin (1998) to build an OBP-OF model with periodic barbs under the main quiescent prominence

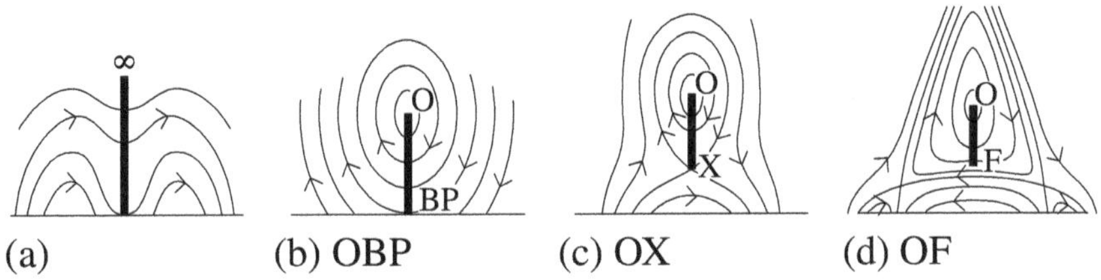

Figure 11.9. Vertical sections across magnetic configurations that possess an inverse-polarity dip either in a potential-field arcade due to (a) parasitic flux or in a force-free flux rope due to: (b) large shear alone, (c) large shear and a third harmonic, and (d) large shear and parasitic flux. In cases (b)–(d) the prominence (dark vertical line) stretches down from an O-point (O) to a bald patch (BP), an X-point (X) or a flat field line (F). Inverse polarity is with respect to the flux that is near to the prominence in (a) and far from it in (d) (after Aulanier and Démoulin 1998).

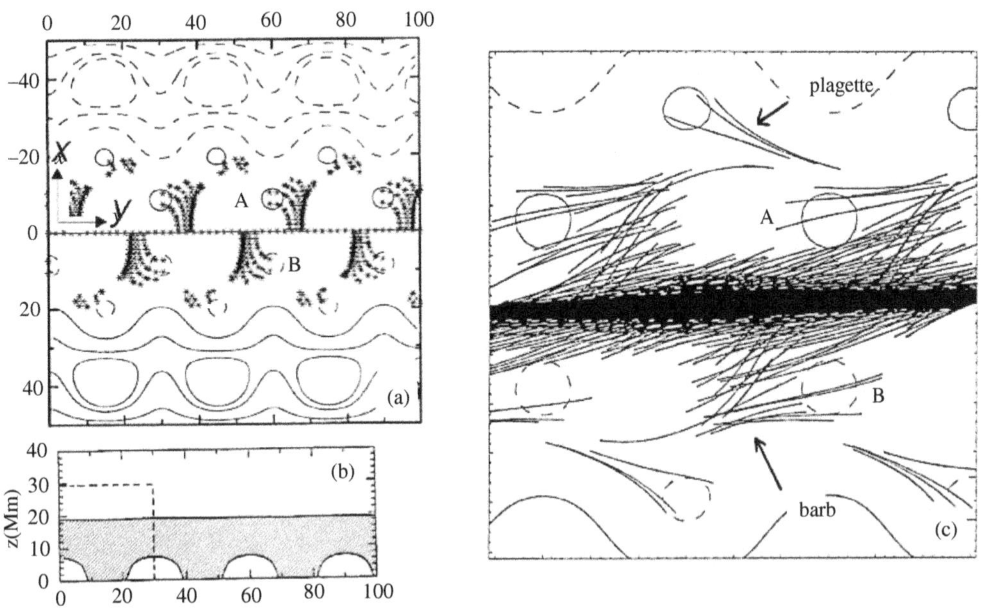

Figure 11.10. A 3D force-free flux-rope model with five harmonics, $L_x = 100$ Mm, $L_y = 30$ Mm and $\alpha/\alpha_{max} = 0.99$. (a) Photospheric flux pattern in the xy-plane viewed from above with $B_z > 0$ (solid), $B_z < 0$ (dashed) and locations of dips (crosses). (b) The prominence (shaded) viewed from the side. (c) Close-up from above near the parasite pair AB of cool plasma that accumulates along field lines at the dips for a sinistral model having axial field direction from right to left (after Aulanier and Démoulin 1998).

body in dips created by the small parasitic polarities. They chose the harmonic amplitudes to give a broad corridor of bipolar photospheric flux around the PIL ($x = 0$), with the parasitic flux distributed along the corridor in a supergranule pattern. In Figure 11.10a, parasites close to the PIL are in pairs at $y = 0, 30, 60, 90$, one such pair being labelled as AB.

The transverse topology oscillates along the PIL between OBP and OF. Viewed from the side (Figure 11.10b), the prominence consists of feet reaching the photosphere in OBP bald patches every 30 Mm at $x = 15, 45, 75$ and of periodic OF arches between them peaking above the parasite pairs at $y = 0, 30, 60, 90$.

The main dips exist on the prominence axis, but secondary lateral OF dips are also suspended above secondary PILs near the parasitic polarities and lying above bald patches. In Figure 11.10a, these secondary dips are just to the right of positive parasite A and to the left of negative parasite B. This is because for this sinistral filament the axial field runs from right to left, and so flux rising up from A bends over to the left as it does so, forming a dip to the right of A. At the same time, flux bending down into B does so from the right and creates a dip to the left of B.

Cool plasma extends along the field line from a dip up to a scale height (i.e., roughly a vertical height of 1 Mm): such extensions of regularly spaced field lines in a dip are indicated in Figure 11.10c and resemble very well the observed barb structures. It can be seen that, in the main prominence body, the dipped field lines are inclined from top right to bottom left, because the axial field is from right to left and the inverse nature of the bald patch gives a field component from top to bottom. Furthermore, the same inclination persists in the field lines from the secondary dips – and it is this which creates the left-bearing nature of the barbs in such a sinistral prominence.

Here, the prominence height is 20 Mm, but it increases with L_x or α. The field strength increases with height in the prominence, as expected for dipped regions. The inclination (θ) of the prominence field to its spine decreases with height to zero at the prominence top, above which it reverses in sign. It is also larger in the barbs.

If the value of α slowly builds up, there is an evolution in which intermittent prominence structures of normal polarity and small vertical extent give way at a critical α to a field aligned with the PIL (a filament channel), followed by the appearance of OF-OBP prominences with the observed chiral structure which grow to produce dips up to heights of 50 Mm when α is close to α_{max}.

Many features are naturally explained by this model, including: the presence of barbs in pairs, periodically produced by and close to bipolar parasitic polarity; prominence chirality, with positive α producing sinistral and left-bearing filaments, while the overlying arcade is right-bearing and the prominence flux rope is a right-handed helix (negative α gives the opposite); so-called *plagettes* due to other dips near other parasitic polarities near the edge of the PIL corridor. Furthermore, the barbs would evolve with a typical supergranular lifetime of a day or so, and may well act as sources of prominence mass supply as the parasites emerge and move.

The model has been extended from a linear to a nonlinear force-free field by Mackay and van Ballegooijen (2009) using van Ballegooijen et al. (2000)'s theory (Sec. 11.3), which couples nonlinear coronal relaxation to driving photospheric motions. They studied the response of a prominence flux tube to the approach of a magnetic bipole. If minority polarity is closest to the filament, right- (or left-) bearing barbs appear for dextral (or sinistral) filaments, since the minority polarity causes the overlying field to bend down. One new feature is that, if the bipole is close enough to the filament and its flux is large enough, it can cause the filament to break into two distinct segments, although the flux tube itself does not break, since field lines develop double dips (see also Dudík et al. 2008). Another new feature is that barbs can also form when the dominant polarity is closest to the filament, since it causes the overlying field to bulge upwards and create a dip nearby. Generalising the model to include gravity when the magnetic field is weak deepens the dips, which increases the tension in the upper part of the prominence and compresses the field near the base (Hillier and van Ballegooijen 2013).

11.4.2 Numerical Flux-Rope Modelling

Aulanier and collaborators pioneered 3D linear force-free models that extrapolate photospheric magnetograms into the corona to compare with observed properties such as chirality (Sec. 11.3) and barbs (Sec. 11.4). The prominence is assumed to be located in dips of helical field lines, which form above *bald patches* (Titov et al. 1993) on parts of a PIL where the field has upwards curvature and inverse polarity.

The location and shape of a filament and its channel have been well represented by a linear force-free field (Mackay et al. 1997; Mackay and Priest 1997). For example, one formed between new and remnant active regions, in a channel of highly sheared, low-lying, flat, dipped field lines that are held down by an

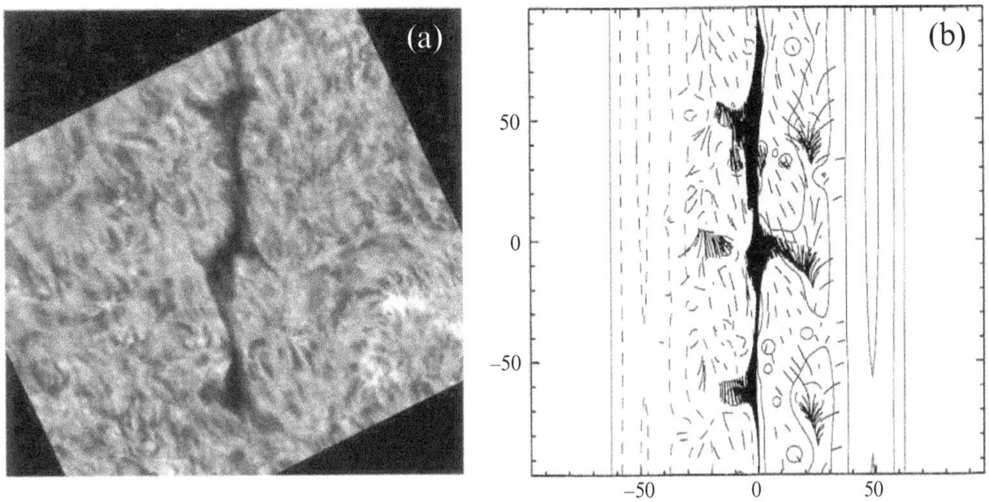

Figure 11.11. (a) A filament observed in Hα with the MDSP instrument on the VTT at Tenerife. (b) Cool plasma accumulating in dips for a linear force-free model based on the observed vertical photospheric magnetic field with $\alpha/\alpha_{max} = 0.995$ (same notation as Figure 11.10) (from Aulanier et al. 1998).

overlying arcade. In other cases, the bending and disruption of a filament is seen as it passes between opposite-polarity fragments (see also Schmieder et al. 2006; Dudík et al. 2008).

Then Aulanier et al. (1998) used a model of a single force-free flux source (Démoulin and Priest 1992) to study its effect on a prominence flux rope and to superpose such sources to obtain a good correlation with the shape of a particular filament and its barbs in Hα (Figure 11.11), or even better using a linear magnetohydrostatic model (Aulanier et al. 1999; Aulanier and Démoulin 2003). Also, the Aulanier-Démoulin model has been applied to observed prominences with great success (Aulanier et al. 1999, 2000; Dudík et al. 2008), with Aulanier and Schmieder (2002) explaining broad EUV filaments as absorption of EUV radiation in cool condensations in low-lying (< 4 Mm altitude) dips above dispersed parasitic polarities in a filament channel.

Vector magnetic fields from THEMIS show that barbs end in bald patches (López Ariste et al. 2006). In one case, the flux rope of an Hα filament consists of two main strands that are twisted around one another and rooted in different locations (Dudík et al. 2008), with an interruption of the filament caused by the splitting of the flux tube due to strong network magnetic flux on the edge of the filament channel.

The main focus is now on flux-rope models, since they naturally form both dips and inverse polarity. However, analytical potential or force-free arcade models of either normal or inverse polarity are also sometimes possible (PROBLEM 11.7) (e.g., Malherbe and Priest 1983; Ridgway et al. 1991; Démoulin and Priest 1993; Cartledge et al. 1996). For example, inverse-polarity prominences can naturally form between two active regions or in the polar crown (Figure 11.12).

An ingenious 3D numerical model (Antiochos et al. 1994; Aulanier et al. 2002) shows how a force-free arcade of inverse polarity can be built up from the field of a subsurface point bipole by strong footpoint shearing along the PIL and confined to a narrow region near it. The inner part of the bipole is dragged into a region of weaker overlying field, producing elongated low-lying field lines that bulge up at their more weakly confined ends and so become dipped near their centres, where they pass under strong unsheared field. Furthermore, the legs of these field lines lean back across the PIL towards weaker field to produce an S-shape seen from above with inverse polarity compared with the original bipole (Figure 11.13). For great enough shear, this field in turn reconnects in the corona to give a partly helical field (DeVore and Antiochos 2000; Aulanier et al. 2002) that is a mixture of inverse and normal.

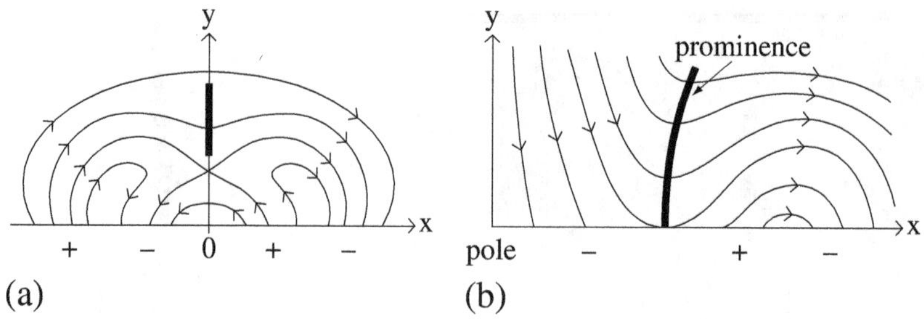

Figure 11.12. Inverse-polarity prominence formation between (a) two active regions or (b) the pole and a bipolar field with opposite leading polarity (after Démoulin and Priest 1993).

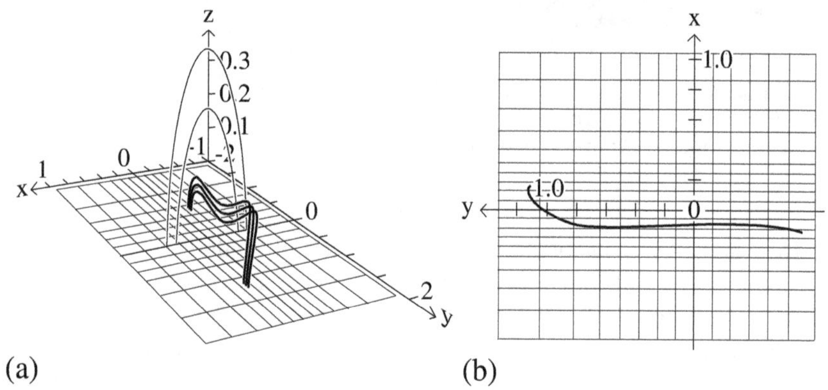

Figure 11.13. An inverse-polarity filament channel produced by shearing a point dipole viewed (a) obliquely and (b) from above (after Antiochos et al. 1994).

Nonlinear force-free modelling is much harder but more realistic than linear modelling (which is reasonable only for local structure). It shows how flux ropes can store energy and magnetic helicity high in the corona (Regnier and Priest 2007). A clever and effective technique for modelling prominences has been developed by van Ballegooijen (2004). He inserts a flux rope along the site of a filament into a potential field based on an observed magnetogram, and then evolves the field to a nonlinear force-free equilibrium by magneto-frictional relaxation. The axial and poloidal fluxes of the rope are varied until a good fit with TRACE and Hα observations is obtained, with field line dips near a magnetic fragment in the filament channel reproducing the shape of a barb (Figure 11.14).

Bobra et al. (2008) presented two more examples of weakly twisted flux ropes just below the stability limit, with the peak α near its edge rather than on its axis. Lionello et al. (2002) used an MHD model for the flux-rope structure and found the thermal structure along each field line. Canou et al. (2009), Canou and Amari (2010) gave examples of nonlinear force-free flux ropes. Guo et al. (2010) computed a nonlinear force-free field to fit the observed vector field below an active-region filament; they found that a narrow part of the filament has a flux-rope field with strength 700 G and twist of 1.4 turns, whereas a broader part of the filament has a quadrupolar structure. Furthermore, the theory (Sec. 11.3) of van Ballegooijen et al. (2000) that couples nonlinear coronal relaxation to driving photospheric motions has been used to show how nonlinear force-free prominence flux ropes may formed and eventually lose equilibrium (Sec. 12.2) (Mackay and van Ballegooijen 2006; Yeates et al. 2008; Yeates and Mackay 2009).

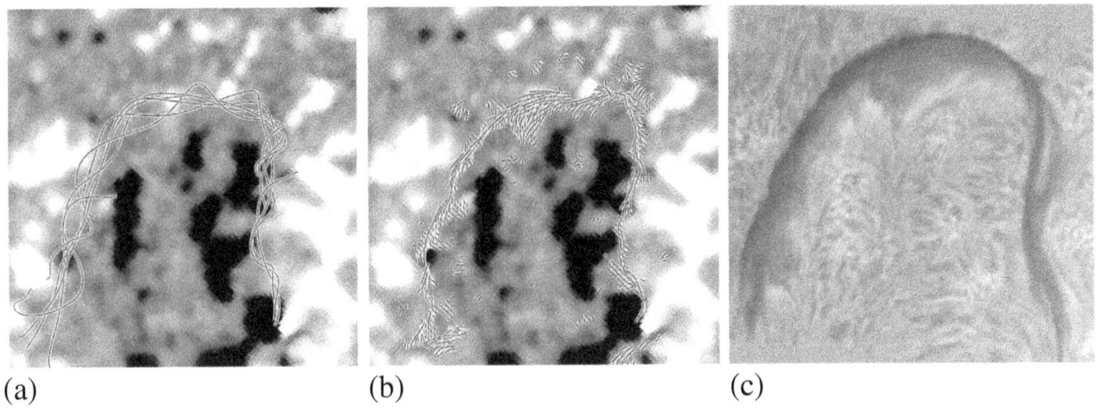

Figure 11.14. Model of a magnetic flux rope supporting a filament: (a) magnetic field lines on a map of $B_z(x,y,0)$ seen from above; (b) locations of dips in the magnetic field; (c) Hα image of the filament from Swedish Solar Telescope (after van Ballegooijen 2004).

11.5 Threads

Prominences consist of many fine threads (width < 100 km) continually forming with lifetimes of only a few mins (Figure 1.35). Some are nearly horizontal and probably aligned with the magnetic field. Others are vertical, suggesting a role for gravity; they are likely to represent condensation in a series of dips stacked one above the other (Sec. 11.4). What produces threads, their widths and lifetimes is not understood at all, but it is likely to be related to the complex process of radiative condensation in a magnetic field in the presence of diffusion (including ambipolar) and gravity.

11.5.1 Local Structure

So far, thread modelling is primitive and has largely focussed on quasi-equilibrium structure. Ballester and Priest (1989) proposed a model for fibril structure in terms of thin flux tubes immersed in a coronal arcade. Then Priest et al. (1991) suggested thread structure is due to filamentation during its formation by radiative instability. They discussed effects of conduction, viscosity, resistivity, gravity and ballooning. These ideas were extended by Hood et al. (1992) and Steele and Priest (1992) to produce models for magnetohydrostatic thread structure, as follows. Magnetic components across (x), along (y) and vertically up (z) the prominence are written

$$\mathbf{B} = \left(-\frac{\partial A}{\partial z}, B_y(A), \frac{\partial A}{\partial x}\right),$$

where the flux function (A) satisfies

$$\nabla^2 A = -\frac{\partial}{\partial A}\left[\mu p(A)\exp\left(-\int \frac{dz}{H(A,z)}\right) + \tfrac{1}{2}B_y(A)^2\right],$$

with scale height $H(A,z) = k_BT/(mg)$. After starting with configurations for which $B_y = 0$, a non-zero B_y is then introduced by a transformation to laminar fields (Low 1982b), for which $B_y/B_x = \alpha$ and $\partial(B_x/B_z)/\partial z = \partial(B_y/B_z)/\partial z = 0$.

Simple models that are independent of z are set up with

$$B_x = B_{x0}, \quad B_z = \sqrt{2\mu p_T}\tanh f(x), \quad p = p_T\ \mathrm{sech}^2 f(x),$$

where $f(x)=\int_0^x \sqrt{2\mu p_T}/(2B_{x0}H)dx$. Fine-scale structure along the magnetic field is modelled by imposing the temperature $[T(x,y)]$ and therefore the scale-height $[H(x,y)]$ to be a function of x alone in order to represent a series of cool slabs separated by hot plasma. Fine-scale structure across the magnetic field is modelled in a similar way by a function of y alone.

Models are also developed with the temperature being a function of x alone, while the magnetic field and pressure are separable in x and z, so that

$$\mathbf{B}=\mathbf{B}(x)e^{-z/(2H_c)}, \quad p=p(x)e^{-z/H_c},$$

in terms of a uniform coronal scale-height (H_c). The magnetic field is then chosen to represent a series of multiple dips supporting cool plasma.

11.5.2 van Ballegooijen–Cranmer Model for Tangled Fields

van Ballegooijen and Cranmer (2010) have proposed an intriguing model for the thread-like nature of a prominence supported in a complex magnetic field that is tangled on a spatial scale of 1 Mm or less and is located within a large-scale vertical current sheet (Figure 11.15a). Here a *tangled field* is not turbulent, since it is assumed to be in equilibrium. They consider the dynamics of plasma along a contorted field line with many dips as the plasma periodically speeds up and slows down. Linear force-free models for tangled fields are developed in which the magnetic pressure of a tangled field of 10 G can support prominence plasma of density 10^{17} m^{-3}.

They first calculate steady adiabatic flow (with $p=k\rho^\gamma$) along a tangled field line, whose height $z(s)$ as a function of distance s along the field is assumed to be a sinusoidal variation of amplitude A superposed on a downward path of slope C:

$$z(s)=A\cos(2\pi s/\Lambda)-Cs,$$

where Λ is the distance between peaks. The equation of motion is simply

$$\rho v\frac{dv}{ds}=-\frac{dp}{ds}-\rho g\frac{dz}{ds},$$

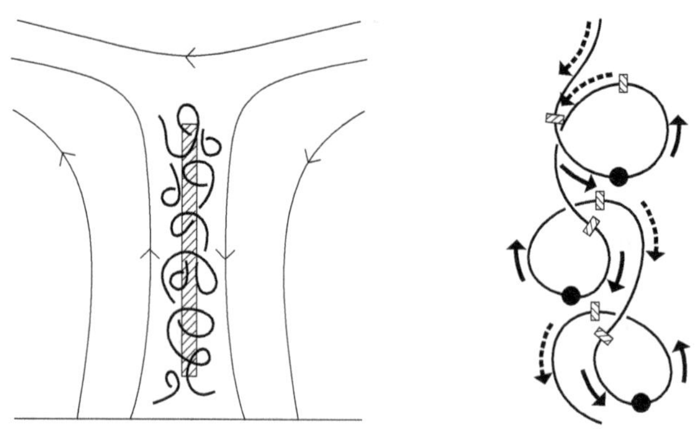

(a) (b)

Figure 11.15. (a) Support of a prominence by a tangled magnetic field. (b) Subsonic (dashed) and supersonic (continuous) flows along a tangled field line, showing dips (large dots) and locations of sonic points (vertical shaded bars at local summits) and shock waves (inclined bars) (after van Ballegooijen and Cranmer 2010).

where mass continuity gives $\rho v = constant$. Thus, eliminating ρ and p, we find

$$\left(v - \frac{c_s^2}{v}\right)\frac{dv}{ds} = -g\frac{dz}{ds},$$

where $c_s = c_{s0}(v/c_{s0})^{-(\gamma-1)/2}$ is the sound speed in terms of its value c_{s0} at a sonic point (where $v = c_s$).

In this model the sonic points ($v = c_s$) occur at local peaks of the field, where $dz/ds = 0$. As plasma falls from such a peak, it accelerates supersonically until it is suddenly slowed down at a shock wave to subsonic values, before reaching a dip and accelerating up to sonic values at the next peak (Figure 11.15b). The net mean vertical downflow is subsonic and considerably smaller than the supersonic free-fall that would occur in a vertical field, indicating that the tangled field is able to provide some support for prominence plasma against gravity.

They suggest that tangled fields tend to relax to a linear force-free field by reconnection and construct linear force-free models for tangled fields as a superposition of planar modes with randomly distributed phases. The field lines have a random walk behaviour when there are more than 10 modes. To produce tangling on a scale of 1 Mm or less needs a much higher magnetic helicity density ($\alpha > 1$ Mm^{-1}) than in the surrounding corona. Such a tangled field is naturally buoyant. Furthermore, its nearly isotropic magnetic pressure can support the weight of prominence plasma by magnetic coupling between the threads and their surroundings.

11.6 Formation of Thermal Structure

The presence of cool plasma (10,000 K or less) in a prominence is a puzzle, but it is clear that the amount of plasma in a prominence is far too large to be provided by the surrounding corona, and so it must somehow flow up from the chromosphere or photosphere, where a huge reservoir is available. Traditionally, the main mechanism has been thought to be a radiative instability (or non-equilibrium) in the corona, which sucks plasma up from the low atmosphere by a plasma pressure imbalance. In the past twenty years, new aspects of this process have been worked out in detail, but three alternative mechanisms have also been proposed, namely, magnetic injection or levitation of cool plasma from below in association with reconnection (Figure 11.16) and also magnetic Rayleigh-Taylor instability.

Observational features to bear in mind are that: prominence plasma is highly dynamic, continually heating and cooling, with horizontal or vertical motions of 10 to 70 km s^{-1}; cool plasma often appears in situ in the corona (Schrijver 2001), but it is sometimes seen to flow up from the chromosphere (Schwartz et al. 2006) or to drain down as *coronal rain* (Müller et al. 2005); prominences are composed of fine threads and knots of width less than 100 km and lengths of 3 to 20 Mm.

11.6.1 Injection from Below (Surges)

Photospheric flux cancellation is frequent in a filament channel below a filament, especially near a barb, and it often produces upflows and downflows in the range 10 to 70 km s^{-1} (typically 30 km s^{-1})

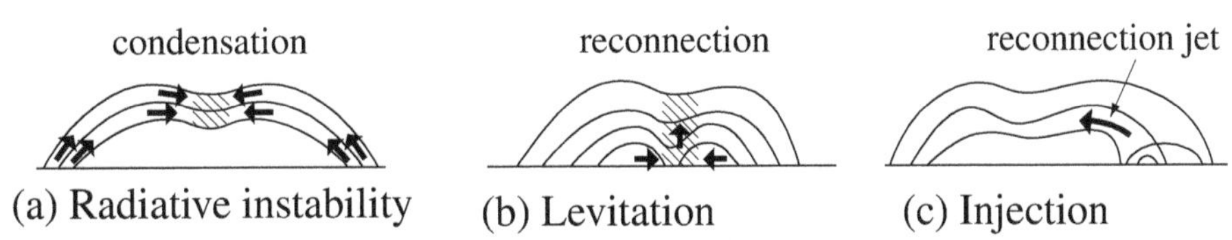

Figure 11.16. The source of cool prominence plasma as (a) radiative instability or non-equilibrium, (b) magnetic levitation, or (c) plasma injection.

(Zirker et al. 1998; Wang 1999). Thus, it has been suggested that prominence plasma may simply be injected ballistically up from below at cool temperatures from a reconnection site.

The ballistic height due to injection at speed u is $h = u^2/(2g)$, and so with g = 274 m s^{-2}, we obtain a height of 2 Mm for u = 30 km s^{-1} or 20 Mm for u = 100 km s^{-1}. We therefore conclude that simple injection may explain some low-lying barbs or active-region prominences, where flows are commonly observed, but another mechanism is required to explain the presence of cool plasma at much greater heights in quiescent prominences. Furthermore, surges do not seem to represent simple ballistic injection, since normal surges of speed 20 to 60 km s^{-1} reach 20 to 35 Mm, whereas large ones at 100 to 200 km s^{-1} reach 100 Mm.

Numerical simulations in 2D of jets due to reconnection between emerging flux (Sec. 9.3.3) and an inclined magnetic field explain surges as hot and cool jets accelerated side by side (Yokoyama and Shibata 1996), although 3D simulations suggest a more complex picture (Sec. 9.3.3). In the 2D simulations, the ejection is not simply ballistic motion from a reconnection point. Flux emergence lifts cool plasma to the transition region (2 to 5 Mm), where coronal pressure halts emergence and reconnection accelerates hot X-ray jets (Sec. 1.6.2) directly to Alfvén speeds over an extended region by slow shocks (Figure 11.17). Nearby cool jets (surges) are created by two effects and accelerated by a further 2 Mm to, say, 6 Mm: the first effect is that some of the cool plasma that has been lifted up is accelerated by the tension in reconnected field lines; the second is that these field lines whiplash to the side and compress nearby open field, squeezing cool plasma upwards by enhanced plasma pressure at the sound speed of say 10 km s^{-1}. In this 2D simulation, the coronal parameters were unrealistic, but it was repeated by Nishizuka et al. (2008) with a realistic temperature (10^6 K), density (10^{16} m^{-3}) and Alfvén speed; they also presented Hinode observations of a hot (5 MK) jet accelerated to 14 Mm, followed after 1 to 2 minutes by a nearby cool (10^4 K) jet moving at 90 km s^{-1}. For other 3D emerging flux simulations, see Section 9.3.3.

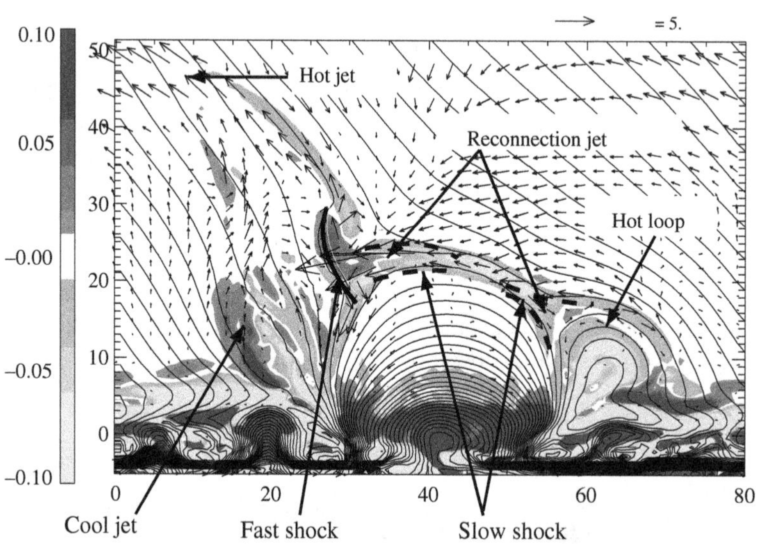

Figure 11.17. Snapshot of a 2D numerical experiment on reconnection in an inclined magnetic field that accelerates both hot and cold jets, showing magnetic field lines, plasma velocity arrows and current density contours. The reconnection jet velocity is 85 km s^{-1}. Units of distance and velocity are 310 km and 12 km s^{-1} (courtesy Takaaki Yokoyama).

11.6.2 Magnetic Levitation of Cool Plasma

Levitation refers to the proposal that cool plasma is lifted up in rising magnetic fields with upward curvature. This may occur when a large-scale twisted flux rope rises directly through the photosphere (Rust and Kumar 1994; Deng et al. 2000), although flux emergence simulations show the difficulty of emerging U-loops containing photospheric plasma breaking through the solar surface (e.g., Archontis 2008).

More likely therefore is the rise of magnetic field lines that have been reconnected in flux cancellation events (von Rekowski and Hood 2008; Litvinenko 2010). The location of the reconnection site during flux cancellation is uncertain: an attractive possibility is the temperature minimum region, where the magnetic diffusivity is lowest and ambipolar diffusion is at work due to partial ionization (Litvinenko and Martin 1999; Arber et al. 2007), but the frequent appearance of X-ray bright points above cancellation sites suggests it is often located in the chromosphere or low corona (Priest 1987). Clearly, more realistic simulations are required in order to determine the likelihood of transport of cool plasma from the low atmosphere by either injection or levitation, which at present seems more feasible for low-lying active-region filaments than huge hedgerow prominences.

11.6.3 Radiative Instability or Non-equilibrium

The most likely mechanism for creating the cool structure of a prominence remains the process of condensation by radiative instability or non-equilibrium (Field 1965).

11.6.3.1 Radiative Instability

Physically, the form of the optically thin radiative loss term in the corona (Figure 2.2) implies that, if the temperature decreases locally, the radiation increases and so it cools even further, leading to a runaway lowering of the temperature until some new effect can balance the radiation or its form changes.

Mathematically, consider hot coronal plasma with temperature T_0 and density ρ_0, in uniform thermal equilibrium under a balance between radiation ($\tilde{Q}\rho^2 = Qn_e^2$, say) and heating $H\rho$ per unit volume. In general, Q is a function of T (Figure 2.2), but we shall for simplicity first assume it is constant and that equilibrium is given by

$$0 = \tilde{Q}\rho_0^2 - H\rho_0.$$

Consider perturbations from this equilibrium at constant pressure, which holds if the time-scale is longer than the sound travel-time, so that pressure equilibrium can be maintained. The perturbations are given by

$$c_p \frac{\partial T}{\partial t} = H - \tilde{Q}\rho + \frac{\kappa_\parallel}{\rho}\frac{\partial^2 T}{\partial s^2},$$

where the last term represents the effect of heat conduction at a distance s along the magnetic field, and the perfect gas law gives

$$\rho = \frac{mp_0}{k_B T}.$$

By putting $T = T_0 + T_1$ and $\rho = \rho_0 + \rho_1$, linearising and eliminating ρ_1, these give an equation for the temperature perturbation, namely,

$$c_p \frac{\partial T_1}{\partial t} = -\frac{\tilde{Q}\rho_0}{T_0}T_1 + \frac{\kappa_\parallel}{\rho_0}\frac{\partial^2 T_1}{\partial s^2}.$$

If the perturbation vanishes at the ends of a coronal loop of length L, T_1 is proportional to $\exp(\omega t + 2\pi i s/L)$, and so this equation gives a growth-rate of

$$\omega = \frac{\tilde{Q}\rho_0}{c_p T_0} - \frac{\kappa_\parallel}{c_p \rho_0}\frac{4\pi^2}{L^2}.$$

Thus, in the absence of conduction, $\omega > 0$ and the plasma is thermally unstable. Furthermore, conduction has a stabilising effect and makes the plasma stable if

$$L < L_m \equiv 2\pi \left(\frac{\kappa_\parallel T_0}{\tilde{Q}\rho_0^2}\right)^{1/2}.$$

However, if the loop is longer than L_m, conduction is so weak that the plasma is thermally unstable and cools down towards a new equilibrium with a prominence-like condensation. This estimate for L_m is modified if a temperature dependence is incorporated in the radiative loss (Sec. 7.5.7).

11.6.3.2 Radiative Non-equilibrium

More importantly, the initial equilibrium is in general only uniform (as assumed in the preceding) if conduction dominates. But, near the point of instability, conduction is just as important as radiation, and its effect is to make the equilibrium temperature $T_0(s)$ non-uniform. Then what happens as L slowly increases is that the temperature profile slowly evolves through a series of equilibria until a *thermal catastrophe* or *thermal nonequilibrium* occurs (Sec. 10.2.3.3). A neighbouring equilibrium no longer exists, and the plasma cools down rapidly towards a new equilibrium, with a rate of cooling that may be much higher than the linear instability rate.

This was demonstrated by Hood and Priest (1979a), who solved the energy balance

$$\frac{d}{ds}\left(\kappa_0 T^{5/2}\frac{dT}{ds}\right) = \tilde{Q}\rho^2 - H\rho,$$

with $\rho = mp_0/(k_B T)$ for a uniform-pressure coronal loop, treated as a single field line. If the temperature is imposed at the base and the temperature gradient vanishes at the summit, they find non-equilibrium if the pressure or loop length are too great or the heating too small. Although the theory is potentially of relevance, it should be borne in mind that the subtle nature of the relation between lower atmosphere and corona may not be well approximated by this simple set of boundary conditions.

This analysis was first applied to a coronal flux tube modelled as a cylinder in magnetostatic equilibrium (Hood and Priest 1979a). After deducing the pressure $[p(r)]$ of the cylinder for a given axial field and twist, the above energy equation was solved to determine $T(s)$ and $n(s)$ along the field lines at each radius. As the twist increases, so the axial pressure increases and enhanced radiative losses lower the temperature, ultimately to a point where thermal non-equilibrium sets in on the tube axis. If the twisting continues, the cool core of the loop broadens (Figure 11.18a). The theory may be relevant to the formation of an *active-region prominence* along a flux tube that has been stretched or twisted so much that thermal conduction can no longer counter the cooling process.

Next, the analysis was repeated (Priest and Smith 1979) for an almost force-free coronal arcade, whose plasma is in thermal and hydrostatic equilibrium given by

$$\frac{d}{ds}\left(\kappa_0 T^{5/2}\frac{dT}{ds}\right) - \frac{\kappa_0 T^{5/2}}{B}\frac{dB}{ds}\frac{dT}{ds} = \tilde{Q}\rho^2 - H\rho, \qquad \frac{dp}{dz} = -\rho g,$$

with $\rho = mp_0/(k_B T)$. These are solved along each field line of a linear force-free arcade subject to $n = n_0$, $T = T_0$ at the base and $dT/ds = 0$ at the summit. When the width, shear or pressure of the arcade become too great, the equilibrium ceases to exist and the plasma cools to form a quiescent prominence. Priest and Smith (1979) therefore suggested that a prominence is a dynamic structure, with prominence plasma

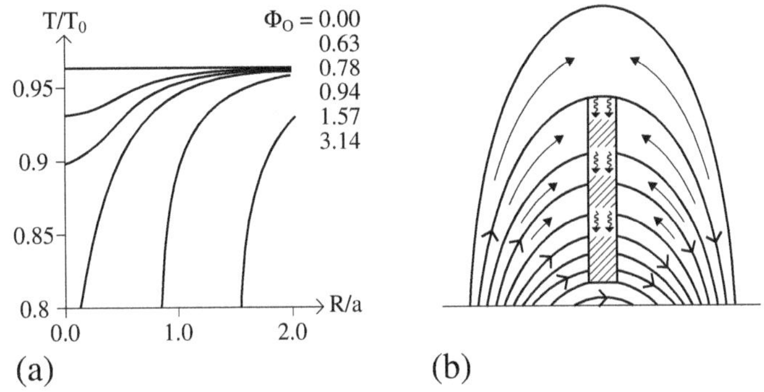

Figure 11.18. (a) Prominence formation by thermal nonequilibrium in a flux rope, showing the effect of increasing the twist $[\Phi_0/(1+R^2/a^2)]$ on the temperature structure $[T(R)]$ across the rope (from Hood and Priest 1979a). (b) A dynamic model for prominence formation in a coronal arcade, in which coronal plasma enters the prominence from the sides and then slowly dribbles through the magnetic field (from Priest and Smith 1979).

continually draining downwards, while new plasma is sucked up along field lines of the ambient arcade into the region of nonequilibrium, where it cools to form new prominence material (Figure 11.18b). The reason why shearing leads to condensation is that it raises and lengthens the field lines, thus reducing the stabilising effect of thermal conduction.

Luna et al. (2012) have simulated formation of a multi-threaded prominence by thermal non-equilibrium. They combine the magnetic structure of a 3D sheared double arcade with 1D simulations of many selected flux tubes. They find that coronal condensations are divided into two populations: threads and blobs. *Threads* are massive condensations that linger in the flux-tube dips. *Blobs* are ubiquitous small condensations throughout the filament that rapidly fall to the chromosphere. The threads provide most of the prominence mass, in agreement with observations for filling factors of order 0.001. The thread motion is horizontal, while blobs move in all directions along the field.

11.6.3.3 Effect of a Magnetic Dip and Footpoint Heating

A problem with condensation near a normal coronal loop summit is that much of the condensing plasma drains away from the summit before it accumulates to form a dip. A cunning way out is to create the dip magnetically rather than by condensation and to focus the heating near chromospheric footpoints of a loop (Antiochos and Klimchuk 1991). Such an enhanced footpoint heating (due to greater complexity) is a natural feature of the coronal tectonics model (Sec. 10.4.5). The effect of slowly increasing the footpoint heating is to raise the density throughout the loop by chromospheric evaporation and initially to raise the summit temperature. However, the increased density causes the radiation to increase near the summit, and eventually it can no longer be balanced by heating and conduction, so that a cool condensation starts to form there by a loss of thermal equilibrium.

Dahlburg et al. (1998) found the new equilibrium after condensation by solving

$$\frac{d}{ds}\left(\kappa_0 T^{5/2}\frac{dT}{ds}\right) = \tilde{Q}\rho^2 - H(s), \qquad \frac{dp}{dz} = -\rho g(s),$$

where $g(s)$ depends on the loop geometry and is the component of gravity along the (assumed symmetric) loop, while $H(s)$ is the spatially dependent heating. Boundary conditions are $T = T_0$ at the base and $dT/ds = 0$ at the base and summit. An adaptive mesh is used to resolve the boundary layers (transition regions) near the feet and at the edges of the summit condensation. Then Antiochos et al. (1999) used a time-dependent code with an adaptive grid and an imposed heating $H(t,s)$ to simulate the growth of a condensation of length a few Mm over 1.5 hr by solving

$$\frac{\partial \rho}{\partial t} + \frac{\partial(\rho v)}{\partial s} = 0, \qquad \frac{\partial(\rho v)}{\partial t} + \frac{\partial}{\partial s}(\rho v^2 + p) = -\rho g_{\|}(s),$$

$$1.5\frac{\partial p}{\partial t} + \frac{\partial}{\partial s}\left(1.5pv - \kappa_0 T^{5/2}\frac{\partial T}{\partial s}\right) = -p\frac{\partial v}{\partial s} + H(t,s) - n^2 Q(T),$$

for the flow (v), density (ρ) and pressure ($p = 2nk_B T$) along the loop. They stress that a prominence is not one continuous structure but is a collection of threads, each of them condensing along separate thin flux tubes.

Later papers clarified various aspects of this non-equilibrium scenario. In particular, quasi-steady asymmetric footpoint heating in long low-lying flux tubes with shallow dips produces a dynamic cycle, in which thread-like condensations repeatedly form, rapidly stream along the field and disappear by draining onto a footpoint (Antiochos et al. 2000; Karpen et al. 2005, 2006). Typical cycle times are from 10 hours to a few days, maximum thread lengths are 25 Mm (with the longest being associated with the flattest field lines), and bulk motions are up to 50 km s^{-1} (lasting several minutes during condensation, coalescence and destruction). Condensations also form when the heating is impulsive in nature and randomly distributed in time, provided the mean time between heating events is less than the coronal radiative time of about 2,000 seconds (Karpen and Antiochos 2008). A particular example of the response to steady heating in a long loop with weaker heating near the right-hand footpoint is shown in Figure 11.19, where no equilibrium is possible: rather, after time t_1, the temperature reaches a maximum at the loop centre, but then the density increases slowly by evaporation; after t_2, a rapid condensation appears at distance $s = 200$ Mm; then the condensation moves uphill at 2 km s^{-1} towards the right and at t_3 it goes over the loop summit at $s = 235$ Mm; it then falls down to the chromosphere, while the high temperature is re-established and the cyclic process repeats.

Thermal non-equilibrium provides a reasonable explanation for many observed properties of active-region and quiescent prominences, as well as barbs, provided the plasma is supported in magnetic dips of field lines longer than a few times the chromospheric heating scale. The observed flows are much slower than

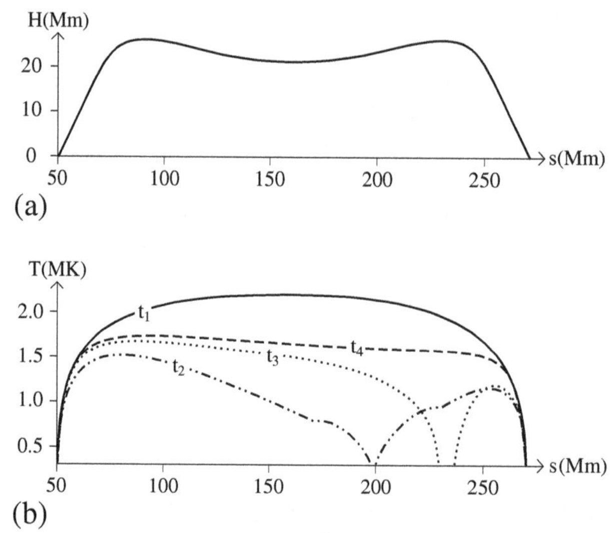

Figure 11.19. Condensation by thermal non-equilibrium, showing: (a) a vertical cross-section through the coronal part of a loop of length 320 Mm and height 25 Mm having a shallow dip that starts 85 Mm from each end; (b) the temperature profile $T(s)$ at several times after heating is switched on ($t_1 = 2.8$, $t_2 = 18.5$, $t_3 = 27.5$, $t_4 = 29.7$ hr) (after Antiochos et al. 2000).

the free-fall or Alfvén speeds, and so are more likely to be driven by pressure gradients (as in condensation models) than by gravity or magnetic forces. Thus, in a flux-rope model (Figures 11.5 and 11.10), the central sheared and twisted dipped field lines can host the main mass of the prominence in dynamic fibril form, while the overlying undipped arcade field lines can host short-lived downflowing condensations (coronal rain).

11.6.4 Magnetic Rayleigh-Taylor Instability

Numerical experiments on a fourth method for forming prominences and their fine structure have recently begun. Hillier et al. (2011, 2012a, 2012b) started with a Kippenhahn-Schlüter model and followed the growth of a buoyant plume in response to the initial insertion of a high-temperature, low-density bubble. The resulting nonlinear interaction between plumes of width 0.5 to 4 Mm produces upflows of 6 km s^{-1} aligned with the magnetic field and downflows of 12 km s^{-1}.

Furthermore, an important step has been to include the effect of partial ionisation in the magnetic Rayleigh-Taylor instability. In the linear regime, an interface between cool plasma lying on top of hot plasma with a uniform horizontal magnetic field is always unstable and ambipolar diffusion lowers the growth rate (Díaz et al. 2012). In the nonlinear regime, a numerical experiment by Khomenko et al. (2013) shows how plumes grow and coalesce, with ambipolar diffusion making small scales grow faster and producing bubbles with larger speeds and higher temperatures.

11.7 Conclusion

The past thirty years have seen a series of major advances in our provisional understanding of the MHD of solar prominences, notably:

- The prominence magnetic field is probably a twisted flux rope that supports the plasma (Priest et al. 1989);
- The twist is due to photospheric reconnection and flux cancellation (van Ballegooijen and Martens 1989);
- The large-scale location on polarity-inversion lines is due to global processes of flux emergence, supergranular diffusion, differential rotation and meridional flow (Mackay and van Ballegooijen 2005; Yeates et al. 2008);
- The barbs represent plasma supported in OBP dips near minority polarity fragments (Aulanier and Démoulin 1998); and
- The dynamic thermal structure is due to condensation by non-equilibrium in magnetic dips with the main heating localised near the chromosphere (Antiochos and Klimchuk 1991; Karpen and Antiochos 2008).

Prominences are, of course, incredibly complex structures that are not yet completely understood. Key open issues include determining:

- observed fine-scale structure of threads and barbs, including their geometry, filling factors, ionization degree and 3D velocity;
- a theoretical explanation for thread structure;
- the nature of the prominence-corona transition-region, including ρ, T and $\mathbf{v}$;
- the details of the energy and mass balances, including the importance of ambipolar diffusion, waves and shocks, and the direction and magnitude of the velocities;
- physical parameters from prominence seismology analysis of the oscillations; and
- the cause of bubbles and plumes and their role (and that of barbs) in mass supply. Perhaps a bubble is the creation of a magnetic arch by the evolution of the field in a filament channel, such as emerging flux (Dudík et al. 2012).

Tackling these issues will need complex observations and theory with ultra-high-resolution (sub-one tenth arc sec) imaging spectroscopy and 3D MHD simulations that couple large-scale and dynamic fine-scale magnetic and thermal structure.

Chapter 12

Solar Flares and Coronal Mass Ejections

12.1 Introduction

Solar flares and coronal mass ejections (CMEs) comprise a huge topic, encompassing most subdisciplines of solar physics. Observations of flares and CMEs have been summarised in Section 1.9, but more comprehensive accounts can be found in Hudson (2010) and Moore et al. (2011).

The present chapter simply focuses on the main MHD aspects of two-ribbon flares and CMEs. These include initiation of the eruptive process responsible for most major events (Sec. 12.2), and the MHD of energy release by reconnection (Sec. 12.3) (see also, e.g., Priest and Forbes 2002; Forbes 2010; Chen 2011; Shibata and Magara 2011). Non-eruptive simple-loop flares (e.g., Priest 1981) and particle acceleration (e.g., Miller et al. 1997; Birn and Priest 2007) are not addressed, even though they are important; indeed, the presence of up to 50 per cent of the energy in some flares in the form of fast particles is at present a major puzzle.

12.1.1 Overview

A prominence may slowly evolve for days or months before reaching a critical point and suddenly erupting into interplanetary space over a few minutes or hours (Figure 1.37). Along with the prominence eruption, a large overlying coronal structure erupts as a CME, often having the appearance of Figures 1.38a and 12.1a with the prominence at its core. This three-part structure is observed in the solar wind as an interplanetary CME consisting of a shock wave, magnetic cloud and prominence material.

When a CME occurs outside an active region, the magnetic field is weak and the whole process can be gentle, with low velocities. However, a CME from inside an active region involves a much stronger magnetic field and so can be much more violent, with higher plasma and particle velocities produced by stronger magnetic and electric forces. In this case, the eruption is accompanied by a *two-ribbon solar flare*, and by the acceleration of high-energy particles. However, the two-ribbon flare, CME and prominence eruption should not be regarded as independent phenomena but as separate manifestations of one magnetic eruptive process. The significance of the prominence is that it is a tracer, indicating that surrounding regions of the corona contain huge amounts of stored magnetic energy with the potential to erupt.

The generally accepted scenario for the events leading up to a two-ribbon flare begins with a magnetic arcade containing a highly sheared (or twisted) central portion that supports a prominence and stores magnetic energy in excess of the potential energy. The central structure responds to the slow photospheric motions of its footpoints by evolving passively through a series of largely force-free equilibria. At some critical point, the configuration erupts outwards, sometimes triggered from outside.

Several theoretical models have been proposed for the cause of the eruption, which is currently a matter of debate (Sec. 12.2). It may occur spontaneously when the arcade shear or height or magnetic helicity pass a threshold such that an MHD eruptive instability or non-equilibrium is initiated. Or it may be close to such a threshold and be triggered by emerging flux or a nearby flare.

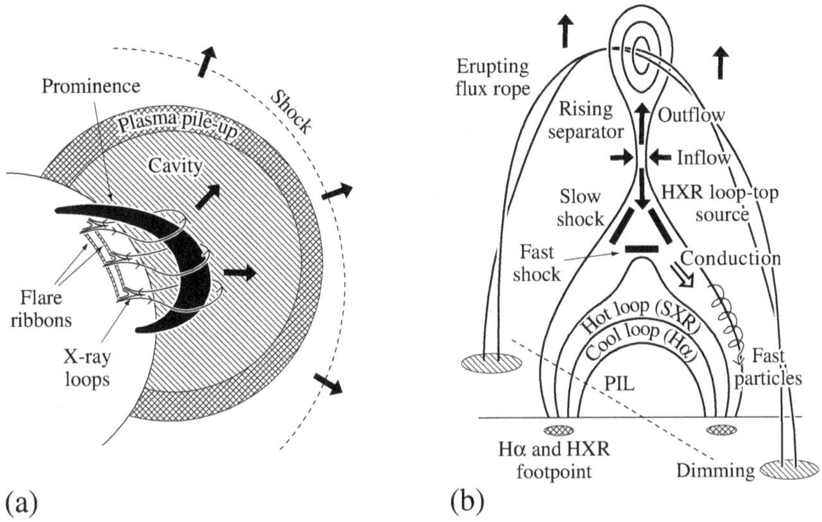

Figure 12.1. The main schematic aspects of an eruptive flare. (a) A global 3D view with underlying flare ribbons and soft X-ray loops (after Forbes 2010). (b) A vertical cross-section of the reconnection process and its effects. Reconnection creates in 2D an erupting plasmoid but in 3D an erupting flux rope (which is in reality very much larger than indicated here). In 2D, reconnection (with its inflow and outflow) occurs at an X-point, but in 3D it occurs at a slowly rising separator or quasi-separator.

Four main ways of initiating or triggering a flare or CME have been proposed:

1. MHD non-equilibrium or catastrophe (Sec. 12.2.1);
2. torus or kink MHD instability (Secs. 12.2.3 to 12.2.6);
3. breakout reconnection between an arcade and and overlying oppositely directed field in a quadrupolar configuration (Sec. 12.2.7); and
4. emerging flux (Sec. 12.2.8).

The source of the flare energy is certainly the magnetic field, and its release is accomplished by magnetic reconnection, in a way that is described in Section 12.3. The onset of reconnection is thought to produce the impulsive phase of a flare, in which the prominence begins to erupt rapidly and open up the field, while bright knots of emission form down below in the chromosphere. The knots then join up to form two ribbons of Hα emission. During the main phase, the ribbons move apart and the opened up magnetic field reconnects (Sec. 12.3) to form a series of rising soft X-ray (SXR) flare loops that join the two ribbons (Figures 1.39, 1.40 and 12.1b). The flare loops were at one time called 'post'-flare loops, but this is misleading, since the long decay in temperature and density during the main phase far exceeds the conduction or radiation time and so requires a continuous energy input (by reconnection).

The flare loops rise rapidly at first (10 to 50 km s^{-1}) and later more slowly (<1 km s^{-1}) for hours or even 1 to 2 days. Cool Hα loops (with strong downflows) are located below the hot SXR loops (which may reach heights of 100 Mm). The summit temperature and density early in the event may be up to 10^7 to 10^8 K and 10^{16} m^{-3}.

Below the reconnection site (Figure 12.1b) is a pair of slow-mode shock waves and sometimes a hard X-ray (HXR) loop-top source. The fast outflow jet is slowed at a fast-mode shock. Energy is transported down from the reconnection site to the lower atmosphere by heat conduction and fast nonthermal particles, which create HXR footpoints, Hα patches and Hα ribbons. The eruption of magnetic flux and plasma decreases the plasma density in the low corona and so creates a dimming region in the lower part of the flux rope. Essentially, the same phenomenon (but less energetic and with fewer high-energy products) occurs when a quiescent filament erupts outside an active region, to give a CME without an Hα flare.

12.2 Eruption by Non-equilibrium, Instability or Breakout

The first question with flares and CMEs from an MHD viewpoint is how to explain the eruptive process, which may either drive reconnection or require reconnection in order to take place. Two approaches to understanding eruption have been tried. The first (and earlier) is to study the linear MHD stability of coronal loops and arcades. The second approach is to seek the existence and multiplicity of equilibrium solutions to the force-free equations in response to the motion of photospheric footpoints. In this case, if the field evolves slowly through a series of equilibria on a time-scale much longer than the Alfvén travel-time, it may erupt dramatically in a few Alfvén times if a neighbouring magnetic equilibrium ceases to exist or if a new equilibrium with lower energy becomes possible.

Demonstrating instability directly is easy only for relatively simple equilibria (such as interfaces between uniform fields). By contrast, the non-equilibrium approach can deal with more complex equilibria. Note that, although a configuration may be linearly unstable, the instability may not necessarily have dramatic consequences, if the timescale is much longer than the Alfvén time, or the wavelength is much shorter than the global length, or if the instability saturates at a low level (Figure 12.2a). Also, in spite of being linearly stable, an instability can be non-linearly unstable (Figure 12.2b). Furthermore, stable states can be triggered into dynamic behaviour by a finite perturbation (e.g., Figure 12.2b or 12.2c).

A full nonlinear stability analysis can reveal several possibilities for the evolution of the magnetic energy $W(t)$ when a critical point is reached (Figure 12.2). In a transcritical or *pitchfork bifurcation*, the equilibrium becomes linearly unstable but non-linearly stable due to the presence of a new nearby stable equilibrium along which the evolution proceeds with no energy release (Figure 12.2a). Secondly, a subcritical bifurcation involves nonlinear instability or metastability, so that the system jumps down to a new equilibrium with significant energy release (Figure 12.2b). Thirdly, a state of non-equilibrium may appear with no neighbouring equilibrium, such that a catastrophe occurs with energy release (Figure 12.2c). [As an

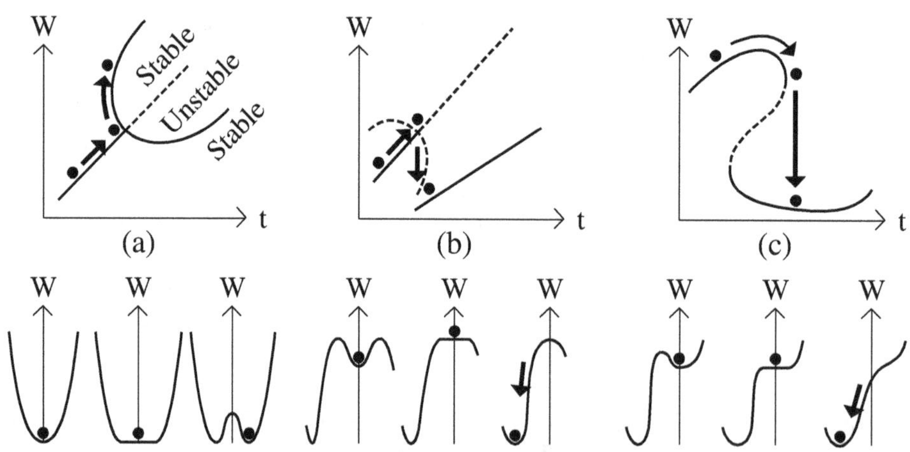

Figure 12.2. Energy diagrams (W) for: (a) a pitch-fork bifurcation, with linear instability but no dynamic behaviour; (b) a subcritical bifurcation, with linear instability and a dynamic transition to another state; and (c) non-equilibrium. Proving the existence of non-equilibrium is a more reliable guide to dynamic behaviour than proving linear instability (see also Figure 7.1).

example, Démoulin and Aulanier (2010) point out the similarity in global balance and imbalance of forces for torus instability and nonequilibrium.] However, proving linear instability alone is a poor guide as to whether or not there will be significant energy release, whereas seeking nonequilibrium or a subcritical bifurcation is a better strategy, since, when it is present, it will always lead to a significant change in state.

Solar coronal loops are observed to be remarkably stable structures; only occasionally do they lose equilibrium or stability and erupt or produce a flare. The dominant stabilising effect is anchoring, or *line-tying*, of coronal magnetic footpoints in the dense photosphere. Due to the inertia of the photospheric plasma, any perturbation that is initiated in the corona must vanish at the loop footpoints.

The *Aly-Sturrock hypothesis* suggests that an initially closed field cannot become fully opened to interplanetary space by an eruptive process. It states that, for an initially closed and simply connected field, the corresponding fully open field with the same footpoints has higher magnetic energy than the closed force-free field (Aly 1991; Sturrock 1991). Thus, a purely MHD process would not appear to be able to open up a force-free field completely. Aly also showed that, for a simply connected force-free field, the magnetic energy is always less than twice the energy of the corresponding potential field. There are several ways around the Aly-Sturrock hypothesis: an eruption is still possible in principle if not all the field lines are opened (Rachmeler et al. 2009), or if some field lines become completely disconnected from the surface, or if some field lines are knotted rather than being simply connected.

Tether cutting (Moore and Roumeliotis 1992; Moore et al. 2011) is a process of reconnection between two sets of short field lines, such as the sheared legs of two parts of a sigmoid (Sec. 1.9.2.3) or prominence field, or between emerging flux and an overlying arcade. It loosens the restraining effect of line-tying by creating longer field lines, which are much more likely to erupt (Figure 12.3).

The causes of eruptions have not yet been identified. A likely possibility is catastrophic non-equilibrium. Such models in both 2D and 3D are described in the following, with the 2D versions illustrating clearly the physical effects involved (Secs. 12.2.1 and 12.2.3). Ideal instabilities, either kink or torus, have also been proposed (Secs. 12.2.3 and 12.2.5). Both non-equilibrium and such instabilities have the effect of driving reconnection after the eruption has been initiated. The alternative is that reconnection is itself responsible for the eruption, by a resistive instability such as the resistive kink (Sec. 12.2.6) or by a change from slow to fast reconnection due to the kinetic microphysics in a current sheet. Furthermore, when tether-cutting or breakout reconnection (Sec. 12.2.7) is present, it is likely to help trigger the eruption of a state that is already close to a threshold for non-equilibrium or instability.

What is required in future in order to identify the initiation mechanisms are more realistic models based on observationally motivated magnetic configurations and detailed comparison with observations, in which all the possibilities are genuinely compared, rather than using a researcher's favourite spectacles that over-emphasise his or her preconceptions.

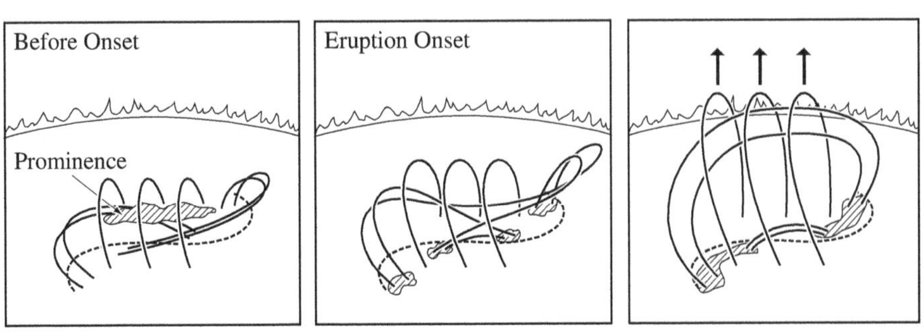

Figure 12.3. Eruption triggered by tether-cutting reconnection (after Moore et al. 2011).

12.2.1 Nonequilibrium or Catastrophe in 2D

When magnetic footpoints on the photospheric boundary slowly evolve, it is important to know whether a coronal equilibrium exists and whether it is stable. An equilibrium that ceases to exist or becomes unstable when the photospheric shear reaches a critical value would allow the release of magnetic energy as the field relaxes back to one of lower energy (such as a potential field). In some circumstances, such non-equilibria can produce large-scale eruptions and drive the formation of the current sheet that is often thought to exist below a rising flux rope.

Priest and Forbes (1990) proposed a simple 2D analytical model for the equilibrium of a horizontal flux rope surrounded by a coronal arcade under a balance between magnetic pressure and tension forces. In response to footpoint motions, the magnetic field evolves slowly through a series of equilibria until, when the flux rope height is too great, force balance is no longer possible. The whole structure erupts and forms a current sheet at which reconnection is driven below the erupting flux rope. This built on earlier ideas of van Tend and Kuperus (1978), who had suggested modelling a flux rope as a horizontal line current with mass, and Démoulin and Priest (1988), Martens and Kuin (1989), van Ballegooijen and Martens (1989) and Amari and Aly (1989), who found non-equilibrium for a flux rope in a linear force-free field.

First of all, a prominence flux rope is modelled at height h above the photosphere (say, $y=0$) as a line current of strength $2\pi I/\mu$, whose magnetic field components (B_x, B_y) may be written concisely as $B_y + iB_x = I/(Z - ih)$, where $Z = x + iy$ is the complex variable. Photospheric line-tying keeps B_y constant at the photosphere as the flux rope moves and may be modelled by adding an image line current $-2\pi I/\mu$ at a distance h below the photosphere. The resulting total field is (Figure 12.4a)

$$B_y + iB_x = \frac{I}{Z - ih} - \frac{I}{Z + ih} = \frac{2ihI}{Z^2 + h^2}.$$

However, a repulsion between these two currents provides an upward force, due to an excess magnetic pressure below the flux rope. This field is therefore not in equilibrium, and will accelerate up from rest to a speed that depends on whether the current, radius or twist is held constant (PROBLEM 12.1: for all problems and solutions, see web page www.cambridge.org/9780521854719).

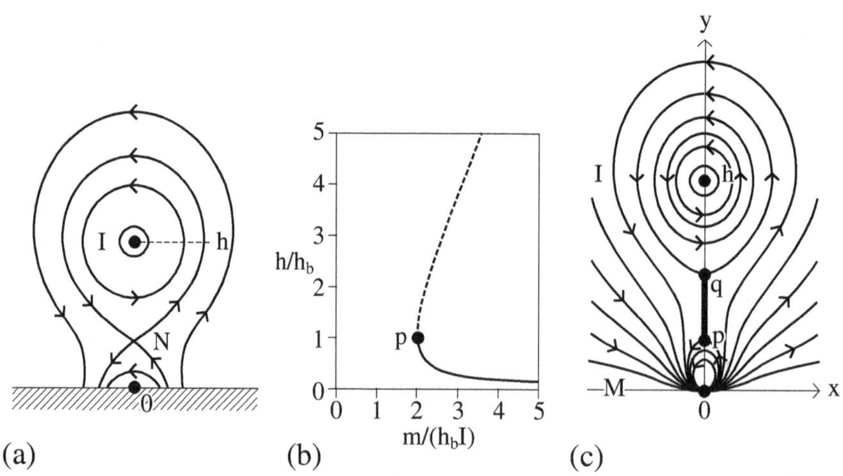

Figure 12.4. (a) The line-tied magnetic field due to a prominence flux rope (current I, height h) in the field of a dipole (of moment m) situated at depth h_b below the photosphere. (b) The corresponding equilibrium flux-rope heights h/h_b as a function of $m/(h_b I)$, with the stable equilibria solid and the unstable ones dashed. (c) The magnetic field that results after non-equilibrium makes the flux rope rise and drive the formation of a current sheet stretching from height p to q on the y-axis (from Priest and Forbes 1990).

Adding next a background field, such as that of a line dipole of moment m at a distance h_b below the photosphere (Figure 12.4a), modifies the total field to

$$B_y + iB_x = \frac{I}{Z - ih} - \frac{I}{Z + ih} + \frac{im}{(Z + ih_b)^2}.$$

If the prominence has mass M, its equation of vertical motion is then (due to the attraction of the prominence current I by its image and repulsion from the dipole)

$$M\frac{d^2h}{dt^2} = \frac{2\pi I}{\mu}\left(\frac{I}{2h} - \frac{m}{(h + h_b)^2}\right). \tag{12.1}$$

Setting the right-hand side of Eq. (12.1) equal to zero gives

$$\frac{h}{h_b} = \frac{m}{h_b I} - 1 \pm \left[\frac{m}{h_b I}\left(\frac{m}{h_b I} - 2\right)\right]^{1/2},$$

namely, two possible equilibrium prominence locations (h) when $m/(h_b I) > 2$, as shown in Figure 12.4b. The lower of these two solutions is stable and the upper unstable (PROBLEM 12.2). The prominence will slowly evolve through a series of stable equilibria with increasing heights as the parameter $m/(h_b I)$ slowly changes, either by weakening the background field (through m or h_b) or increasing the flux-rope twist (through the current I). Eventually, when $m/(h_b I)$ becomes smaller than 2, the flux rope reaches a point of non-equilibrium: it is no longer in equilibrium and the imbalance in forces is upward, since the upward magnetic pressure exceeds the downward magnetic tension, and so it erupts.

As the flux rope erupts, if no reconnection is allowed at the X-point below the prominence or if reconnection proceeds at a certain rate, a current sheet will be formed there, stretching from, say, $y = p$ to $y = q$ on the y-axis. The magnetic field (for the case $h_b = 0$) may be written elegantly by regarding the current sheet as a branch cut in the complex Z-plane (with $Z = x + iy$) as

$$B_y + iB_x = \frac{iA\sqrt{(Z^2 + p^2)(Z^2 + q^2)}}{Z^2(Z^2 + h^2)}, \tag{12.2}$$

where the constants A, p, q and $I(h)$ follow for a given h from flux conservation and the imposed behaviour near the singularities. The rise speed follows from the equation of motion and is found to increase indefinitely, with the electric field at the X-point increasing rapidly to a maximum and then declining towards zero.

Forbes and Isenberg (1991) and Isenberg et al. (1993) adopted a different photospheric boundary condition, namely, the field due to a line dipole or quadrupole at depth d. They used a conformal mapping and a Green's function method to find the field with a current sheet stretching from (0,0) to (0,q). Without reconnection this gave a weak eruption (PROBLEM 12.3). Then Forbes et al. (1994) found the base boundary conditions that maximise the energy release, namely, two line sources for the overlying coronal arcade at, say, $Z = \pm\lambda$, for which the catastrophe releases 21 per cent of the stored energy. This was used by Forbes and Priest (1995) and Lin and Forbes (2000) to model an eruption triggered by converging motion of the sources (i.e., decreasing λ). The field of Eq. (12.2) was generalised to

$$B_y + iB_x = \frac{2iA_0\lambda(h^2 + \lambda^2)\sqrt{(Z^2 + p^2)(Z^2 + q^2)}}{\pi(Z^2 - \lambda^2)(Z^2 + h^2)\sqrt{(\lambda^2 + p^2)(\lambda^2 + q^2)}}, \tag{12.3}$$

where $Z = x + iy$, A_0 is the flux function at the origin, p and q are the heights of the tips of the current sheet, and h is the flux-rope height.

The magnetic field inside the flux rope is assumed to be force-free with vanishing axial field at its surface and a uniform current (I). I can be determined if the slow time-evolution of the magnetic flux between

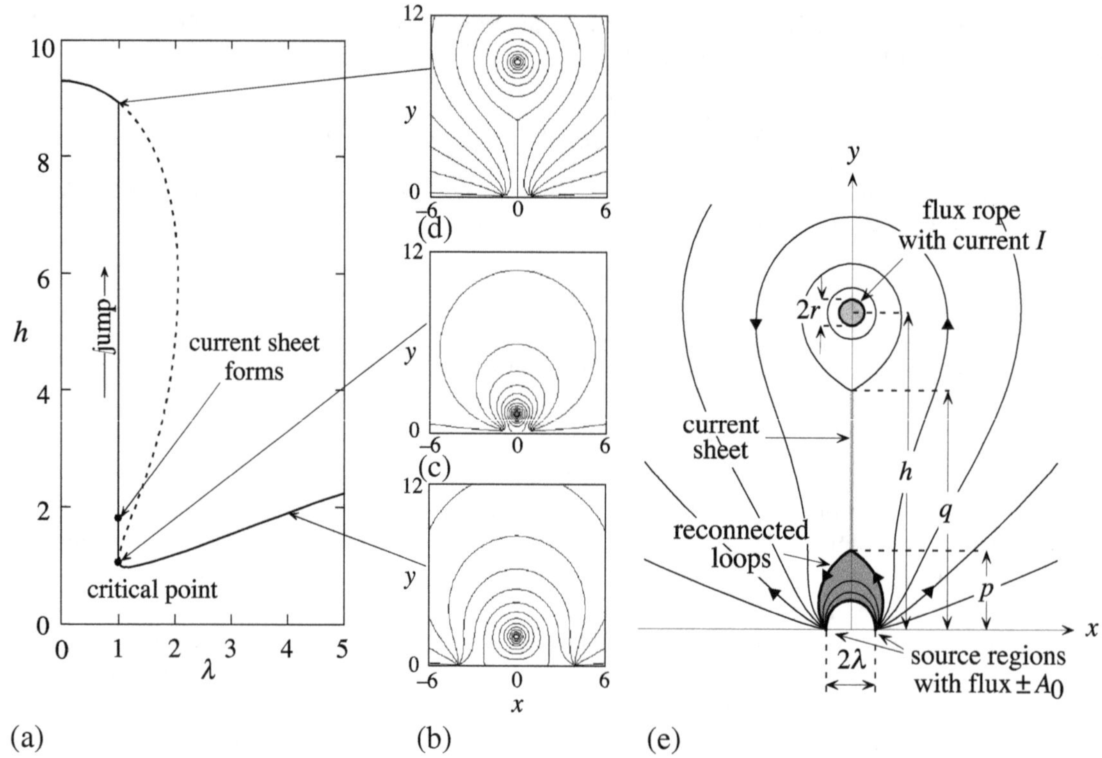

Figure 12.5. Magnetic catastrophe of a flux rope, as shown by (a) the height (h) of the rope as a function of the separation (2λ) of the sources of an overlying arcade, together with (c)–(d) the configurations for three separate locations on the equilibrium curve and (e) the notation (from Forbes and Priest 1995).

the flux rope and the solar surface is known. For instance, if it remains constant in time, the current is

$$I = \frac{c\lambda A_0}{2\pi h}\sqrt{\frac{(h^2-p^2)(h^2-q^2)}{(\lambda^2+p^2)(\lambda^2+q^2)}}.$$

The resulting equilibrium height (h) for the flux rope as a function of the source separation (λ) is shown in Figure 12.5a. For example, when $\lambda = 2.5$, the force-free equilibrium in Figure 12.5 is stable, since, if the flux rope were moved downwards, compression of the underlying field would create an upward restoring magnetic pressure force, whereas, if it were moved upwards, stretching of the overlying field would create a downward restoring magnetic tension force.

As λ is slowly decreased, there is an evolution through a series of equilibria with h decreasing until it is smaller than λ. Then there is no nearby equilibrium state, and the imbalance of forces is upward. The flux rope rapidly moves upwards towards a new equilibrium that contains a current sheet, as shown in Figure 12.5d. If no reconnection is allowed, the amount of energy released is quite small, about 5 per cent, and the flux rope cannot escape. However, if reconnection is fast enough (such that the inflow Alfvén Mach number $M_A > 0.005$), the flux rope will continue rising and escape. Both observations and theory suggest $M_A \approx 0.01 - 0.1$, and so escape is expected. For example, in Figure 12.6, the flux rope expands and moves up at nearly the Alfvén speed due to the magnetic forces. The current sheet's upper tip moves up at half this speed, while its lower tip is controlled by reconnection and so is much slower.

In the above model, non-equilibrium is produced by moving the photospheric sources closer, but it could arise more generally by other changes in photospheric boundary conditions that alter the relative size

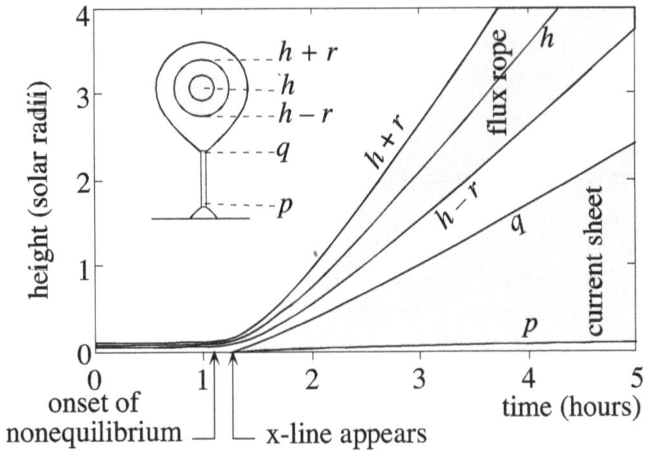

Figure 12.6. Flux-rope and current-sheet trajectories for a catastrophe model of eruption when the reconnection rate (M_A) is 0.1, where h and r are the height and radius of the flux rope, while p and q are the lower and upper height of the current sheet (from Forbes 2010).

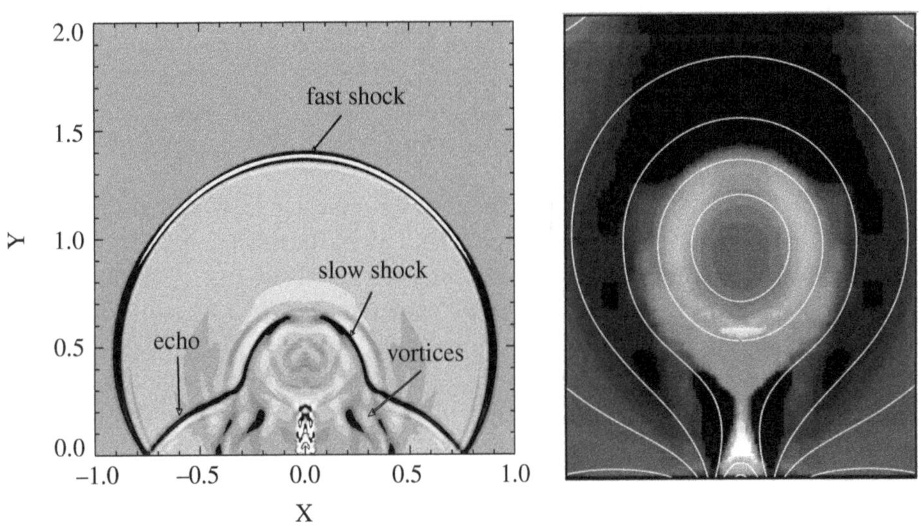

Figure 12.7. Numerical experiments on 2D non-equilibrium, showing a 2D rising flux tube with (a) the shock structure at an early stage (Wang et al. 2009) and (b) temperature variations and field lines later on (Forbes 2010).

of the repulsive and attractive forces acting on the flux rope. Thus, Lin et al. (2001) showed how the emergence of new magnetic flux (PROBLEM 12.3) can cause a loss of equilibrium in a way that depends on the properties of the emergence. For example, the new flux can allow reconnection underneath the flux rope by *tether cutting* (Moore et al. 2011). More realistic numerical MHD simulations produce similar results, with an eruption driving the formation and reconnection of a current sheet below a rising flux rope (Isenberg et al. 1993; Roussev et al. 2003; Linker et al. 2003). They also show how the sudden upward motion launches a fast-mode wave and creates a low-density cavity behind the erupting front, in agreement with CME observations.

Several numerical experiments in 2D have been undertaken to study the onset and consequences of non-equilibrium. For example, Terry Forbes and co-workers considered in detail the effects as a horizontal flux tube erupts and drives reconnection underneath it (Forbes 1990; Lin and Forbes 2000; Wang et al. 2009) (Figure 12.7).

12.2.2 The Hoop Force of a Slender Toroidal Flux Rope

If a field component out of the plane is added to the 2D picture, the magnetic island becomes a straight twisted flux tube, held down by the tension of the overlying field lines acting against the upward magnetic pressure of field lines below the O-point. Now, in 3D in reality a flux rope will curve downwards with its ends anchored in the solar surface, but it is not in equilibrium without an external field since it experiences an extra radially outward *hoop force* (Shafranov 1966). In this section, we consider the magnetic field of a simple isolated toroidal flux rope of major radius R_0, minor radius a and net toroidal current I (Figure 12.8a). Then in Section 12.2.3 we generalise the analysis of Section 12.2.1 to 3D by considering the non-equilibrium and instability of a curved flux rope in 3D. The magnitude of the hoop force is

$$F_{hoop} = \frac{\mu I^2}{4\pi R_0}\left(\log_e \frac{8R_0}{a} - \frac{3}{2} + \beta_p + \frac{l_i}{2}\right), \tag{12.4}$$

when $a \ll R_0$ [first invoked in the corona by Anzer (1978) and later by Chen (1989)]. Here the constants β_p and l_i depend on the internal structure of the flux tube. They are the mean pressure (p) and $B_{\hat{\theta}}^2$ over the volume of the flux rope and are given by

$$\beta_p = \frac{4\mu}{a^2 B_0^2}\int_0^a p\,\hat{r}\,d\hat{r}, \qquad l_i = \frac{\langle B_{\hat{\theta}}\rangle^2}{B_0^2} = \frac{1}{\pi a^2 B_0^2}\int B_{\hat{\theta}}^2\, dV = \frac{2}{a^2 B_0^2}\int_0^a B_{\hat{\theta}}^2\,\hat{r}\,d\hat{r}, \tag{12.5}$$

where $B_0 = B_{\hat{\theta}}(a)$ is the (zeroth-order) field at the flux-rope surface ($\hat{r} = a$) and local polar coordinates $(\hat{r}, \hat{\theta})$ have been taken with respect to a point T on the major axis of the flux rope. As well as the coordinate system $(\hat{r}, \hat{\theta})$ for the location of any point in a vertical plane, we shall also use (R, z), where $R = R_0 + \hat{r}\cos\hat{\theta}$, $z = \hat{r}\sin\hat{\theta}$ (Figure 12.8b). For a thick torus, the above expressions (Eqs. 12.4 and 12.5) remain quite good up to $a/R_0 = 0.1$ (Zic et al. 2007).

In order to prove Eq. (12.4) for the hoop force, consider a flux rope in equilibrium under a balance between several forces and calculate the contribution of the curvature to that balance. (There would be no point attempting a calculation for a flux rope that is not in equilibrium, since the result would depend on

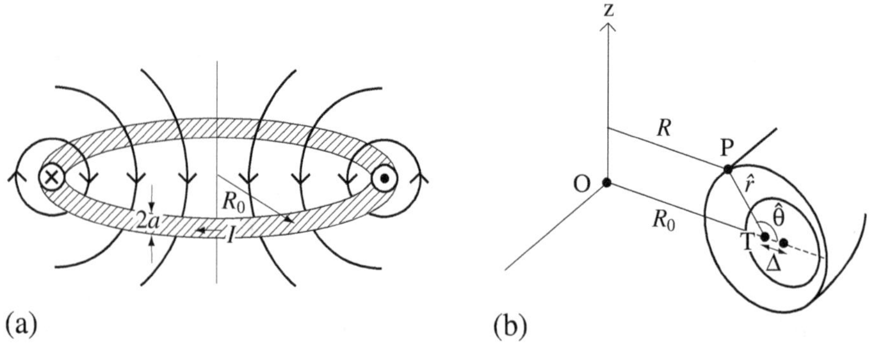

Figure 12.8. (a) An isolated toroidal flux rope of major radius R_0 and minor radius a. (b) A close-up of the two coordinate systems (R, z) and $(\hat{r}, \hat{\theta})$ for the location of P in a $\phi =$ *constant* plane, relative to the origins O and T, respectively. The distance OT $= R_0$ and flux surfaces lie on circles whose centres are distances Δ from T.

the assumed structure.) In particular, suppose the magnetic field of the flux rope plus an external field is in axisymmetric magnetostatic equilibrium (independent of ϕ) with components that are written in terms of a poloidal flux function ($\tilde{A} = RA_\phi$) as

$$(B_R, B_\phi, B_z) = \frac{1}{R}\left(-\frac{\partial \tilde{A}}{\partial z},\ b_\phi(\tilde{A}),\ \frac{\partial \tilde{A}}{\partial R}\right).$$

This satisfies $\nabla \cdot \mathbf{B} = 0$ automatically, with $\tilde{A}$ determined by the Grad-Shafranov equation (cf. Eq. 3.62), namely,

$$\frac{\partial^2 \tilde{A}}{\partial R^2} - \frac{1}{R}\frac{\partial \tilde{A}}{\partial R} + \frac{\partial^2 \tilde{A}}{\partial z^2} = -\mu R^2 \frac{dp}{d\tilde{A}} - \frac{d}{d\tilde{A}}(\tfrac{1}{2}b_\phi^2).$$

The poloidal flux and current are then $2\pi\tilde{A}$ and $2\pi b_\phi/\mu$, respectively. In terms of coordinates $(\hat{r}, \hat{\theta})$, the above two equations become

$$(B_{\hat{r}}, B_\phi, B_{\hat{\theta}}) = \frac{1}{R_0 + \hat{r}\cos\hat{\theta}}\left(-\frac{\partial \tilde{A}}{\hat{r}\partial\hat{\theta}},\ b_\phi(\tilde{A}),\ \frac{\partial \tilde{A}}{\partial \hat{r}}\right),$$

and

$$\left[\frac{1}{\hat{r}}\frac{\partial}{\partial \hat{r}}\left(\hat{r}\frac{\partial \tilde{A}}{\partial \hat{r}}\right) + \frac{1}{\hat{r}^2}\frac{\partial^2 \tilde{A}}{\partial \hat{\theta}^2}\right] - \frac{1}{R_0 + \hat{r}\cos\hat{\theta}}\left(\cos\hat{\theta}\frac{\partial}{\partial \hat{r}} - \frac{\sin\hat{\theta}}{\hat{r}}\frac{\partial}{\partial \hat{\theta}}\right)\tilde{A}$$
$$= -\mu(R_0 + \hat{r}\cos\hat{\theta})^2\frac{dp}{d\tilde{A}} - \frac{d}{d\tilde{A}}\left(\frac{b_\phi^2}{2}\right). \tag{12.6}$$

The proof of Eq. (12.4) involves two parts, as follows. Just as for a 2D line current, we evaluate the force on the toroidal flux rope as the product of its current (I) and the external vertical field (B_v) that exists at the location of the flux rope. The first part (Sec. 12.2.2.1) is to find the flux function (Eq. 12.8) of the field due to a toroidal current alone at a small distance $\hat{r}$ from it. The second part (Sec. 12.2.2.2) is to find B_v by calculating the flux function (Eq. 12.12) of the field just outside a toroidal flux rope in equilibrium and decomposing it into the fields of the toroidal current itself and of the vertical field needed to balance the hoop force.

12.2.2.1 Field of a Toroidal Current Ring

If the torus is treated as a ring of current of radius R_0, then the vector potential $[A_\phi(R,\phi)\hat{\boldsymbol{\phi}}]$ at a point (R,ϕ) in polar coordinates in the plane of the ring relative to the ring's centre may be calculated as follows. In general, the vector potential ($\mathbf{A}$) is such that $\mathbf{B} = \nabla \times \mathbf{A}$ and satisfies Poisson's equation $\nabla^2 \mathbf{A} = -\mu \mathbf{j}$, which has solution $\mathbf{A} = [\mu/(4\pi)] \int \mathbf{j}(\mathbf{r}')/|\mathbf{r} - \mathbf{r}'| dV'$. For our ring of arc length ds', say, $\mathbf{j}\ dV' = I\ \mathbf{ds}' = IR_0\ d\phi'\ \hat{\boldsymbol{\phi}}$, and at any point in the plane of the ring the only magnetic component is $B_z(R)$. Thus, the only component of $\mathbf{A}$ is

$$A_\phi = \frac{\mu I}{4\pi}\int \frac{ds'}{S} = \frac{\mu I R_0}{2\pi}\int_0^\pi \frac{\cos\phi'\ d\phi'}{(R^2 + R_0^2 - 2R_0 R\cos\phi')^{1/2}},$$

where $S = [(R - R_0\cos\phi')^2 + R_0^2\sin^2\phi']^{1/2}$ is the distance between points $(R,0)$ on the x-axis and $(R_0\cos\phi', R_0\sin\phi')$ on the ring.

A_ϕ may be rewritten (PROBLEM 12.5a) in terms of complete elliptic integrals of the 1st and 2nd kind with $k = 2(R/R_0)^{1/2}/(1 + R/R_0)$ as

$$A_\phi = \frac{\mu I}{4\pi}\frac{R_0 + R}{R}[(2 - k^2)K(k) - 2E(k)], \tag{12.7}$$

where

$$K(k) = \int_0^{\pi/2} [1 - k^2 \sin^2\alpha]^{-1/2} d\alpha \qquad \text{and} \qquad E(k) = \int_0^{\pi/2} [1 - k^2 \sin^2\alpha]^{1/2} d\alpha.$$

By expanding in powers of $\hat{r}/R_0$ to first order, it can be shown (PROBLEM 12.5b) that, near the inside of the flux rope, this gives a flux function ($\tilde{A} = RA_\phi$) of

$$\tilde{A} = \frac{\mu I}{4\pi}\left\{2R_0\left(\log_e \frac{8R_0}{\hat{r}} - 2\right) - \hat{r}\left(\log_e \frac{8R_0}{\hat{r}} - 1\right)\right\}. \tag{12.8}$$

12.2.2.2 Field near an Equilibrium Toroidal Flux Rope

The equilibrium field both within and around the flux rope satisfies Eq. (12.6), which we expand in powers of the inverse aspect ratio ($\epsilon = a/R_0 \ll 1$), assuming $B_\phi = B_{\phi 0}(R_0/R)[1 + O(\epsilon^2)]$ and $B_{\hat{\theta}} \sim \epsilon B_{\phi 0}$, where $B_{\phi 0}$ is the potential toroidal field at major radius R_0. If we write $\tilde{A} = \tilde{A}_0(\hat{r}) + \tilde{A}_1(\hat{r},\hat{\theta})$, the zeroth and first-order contributions from Eq. (12.6) are

$$\frac{1}{\hat{r}}\frac{d}{d\hat{r}}\left(\hat{r}\frac{d\tilde{A}_0}{d\hat{r}}\right) = -\mu R_0^2 \frac{dp}{d\tilde{A}_0} - b_\phi(\tilde{A}_0)\frac{db_\phi}{d\tilde{A}_0}, \tag{12.9}$$

$$\frac{1}{\hat{r}}\frac{\partial}{\partial \hat{r}}\left(\hat{r}\frac{\partial \tilde{A}_1}{\partial \hat{r}}\right) + \frac{1}{\hat{r}^2}\frac{\partial^2 \tilde{A}_1}{\partial \hat{\theta}^2} - \frac{\cos\hat{\theta}}{R_0}\frac{d\tilde{A}_0}{d\hat{r}}$$
$$= -\frac{d}{d\hat{r}}\left(\mu R_0^2 \frac{dp}{d\tilde{A}_0} + b_\phi(\tilde{A}_0)\frac{db_\phi}{d\tilde{A}_0}\right)\frac{d\hat{r}}{d\tilde{A}_0}\tilde{A}_1 - 2\mu R_0 \hat{r}\cos\hat{\theta}\frac{dp}{d\tilde{A}_0}. \tag{12.10}$$

The next step is to seek a separable solution of Eq. (12.10) in the form

$$\tilde{A}_1(\hat{r},\hat{\theta}) = -\Delta(\hat{r})B_{\hat{\theta}0}(\hat{r})\cos\hat{\theta},$$

where $B_{\hat{\theta}0}(\hat{r}) = (1/R_0)d\tilde{A}_0/d\hat{r}$. The resulting flux surfaces have cross-sections that are circles whose axes are displaced by a distance $\Delta(\hat{r})$ (the *Shafranov shift*), as in Figure 12.8b (PROBLEM 12.5c).

Within the flux rope, Eq. (12.10) may be solved (PROBLEM 12.5d) to give the field at the inner surface of the flux rope to zeroth plus first order as

$$B_{\hat{\theta}}(a) = B_0\left[1 + \frac{a}{R_0}(\beta_p + \tfrac{1}{2}l_i - 1)\cos\hat{\theta}\right], \tag{12.11}$$

with $B_0 = B_{\hat{\theta}0}(a)$ the zeroth-order $B_{\hat{\theta}}$ at $\hat{r} = a$ and β_p and l_i given by Eq. (12.5).

Outside the flux rope, the field is assumed potential and so the right-hand sides of Eqs. (12.9) and (12.10) vanish. The resulting flux function for $v \ll R_0$ for the field external to the flux rope is then to zeroth plus first order (PROBLEM 12.5e)

$$\tilde{A} = \frac{\mu I}{4\pi}\left\{\left[2R_0\left(\log_e \frac{8R_0}{\hat{r}} - 2\right) + \hat{r}\left(\log_e \frac{8R_0}{\hat{r}} - 1\right)\cos\hat{\theta}\right] + \hat{r}\left(\frac{c_1}{\hat{r}^2} + c_2\right)\cos\hat{\theta}\right\}. \tag{12.12}$$

Calculating $B_{\hat{r}}$ and $B_{\hat{\theta}}$ from this and setting $B_{\hat{r}}(a) = 0$ and $B_{\hat{\theta}}(a)$ on the outer surface of the flux rope equal to the value (12.11) on the inner surface then determines the constants (PROBLEM 12.5f) as

$$c_1 = a^2(l_i - 1)/2, \qquad c_2 = -\log_e(8R_0/a) + 3/2 - \beta_p - l_i/2.$$

Finally, we note that the term in square brackets in Eq. (12.12), when evaluated at $\hat{\theta} = \pi$, is identical to the approximation in Eq. (12.8) for the field of a toroidal current when $\hat{r} \ll R_0$. At large $\hat{r}$, by Eq. (12.7) the field of this current vanishes, and so we are left from Eq. (12.12) with $\tilde{A} \approx (\mu I)/(4\pi)c_2\hat{r}\cos\hat{\theta}$. This corresponds to a vertical field at R_0 of $B_v = (\mu I c_2)/(4\pi R_0)$, which therefore produces an outward hoop force of $F_{hoop} = IB_v$, namely, Eq. (12.4), as required.

12.2.3 Non-equilibrium and Torus Instability in 3D

The 2D non-equilibrium analysis for a straight flux rope under an arcade (Sec. 12.2.1) can be extended to 3D simply by replacing the straight flux rope with a curved one (as in Figure 12.12 of Sec. 12.2.4, where the overlying arcade field is produced by a pair of magnetic point sources $\pm q$). However, let us follow Forbes (2010) by building up to that configuration in three steps, starting with the magnetic field of a simple isolated toroidal flux rope (Figure 12.8). Before doing that, we calculate the constant l_i (Eq. 12.5) and comment on the nature of the hoop force.

12.2.3.1 *Hoop Force of a Force-free Toroidal Flux Rope*

If the flux rope has a uniform toroidal current density $j_\phi = (\mu\hat{r})^{-1}\partial/\partial\hat{r}(\hat{r}B_{\hat{\theta}})$ (with $\mathbf{j}\times\mathbf{B} = \boldsymbol{\nabla}p$), we may put $B_{\hat{\theta}} = \hat{r}B_0/a$, and so $l_i = \frac{1}{2}$. However, more appropriate assumptions for the corona are a zero pressure gradient and a linear force-free (i.e., Lundquist) field inside the flux rope of the form

$$B_{\hat{r}} = 0, \qquad B_{\hat{\theta}} = \frac{B_0 J_1(\alpha\hat{r})}{J_1(\alpha a)}, \qquad B_\phi = \frac{B_0 J_0(\alpha\hat{r})}{J_1(\alpha a)},$$

where J_0 and J_1 are Bessel functions, and the minor radius $\hat{r} = a$ (where $B_{\hat{\theta}} = B_0$) is located at the first zero of J_0, namely, $\alpha a \approx 2.405$, so that the axial field vanishes there. Then, if we change the variable from $\hat{r}$ to $\bar{r} = \hat{r}/a$,

$$l_i = \frac{2}{a^2}\int_0^a \frac{J_1^2(\alpha\hat{r})}{J_1^2(\alpha a)}\hat{r}\,d\hat{r} = \frac{2}{J_1^2(\alpha a)}\int_0^1 J_1^2(\alpha a\bar{r})\bar{r}\,d\bar{r} = 1,$$

where we have used the result that $\int_0^1 xJ_1^2(\alpha x)dx = \frac{1}{2}J_1^2(\alpha)$. With $l_i = 1$ and $\beta_p = 0$ (since the pressure gradient is negligible), the hoop force (Eq. 12.4) therefore becomes

$$F_{hoop} = \frac{\mu I^2}{4\pi R_0}\left(\log_e\frac{8R_0}{a} - 1\right). \tag{12.13}$$

A physical explanation for the hoop force is that the magnetic field on the inside of a torus is stronger than on the outside and so produces an outward magnetic pressure force. An equivalent explanation is that the toroidal currents at two points on opposite sides of the torus are flowing in opposite directions and so repel one another. The repulsive force between two straight wires with oppositely directed currents a distance $2R_0$ apart would be $\mu I^2/(4\pi R_0)$ and so the effect of having a circular ring is simply to produce an extra factor of $[\log_e(8R_0/a) - 1]$. If, e.g., $R_0/a = 10$, say, then this factor is 3.4.

Now, the total magnetic flux inside the ring is given by the value of RA_ϕ at its inner surface, but, if we neglect the flux-rope width and set $R = R_0$, then $k = 1$ in Eq. (12.7) for A_ϕ and the elliptic integral $K(k)$ there becomes infinite. Taking account instead of the flux-rope width and evaluating RA_ϕ at the inner surface $R = R_0 - a$ of the ring makes $k \approx 1 - a^2/(8R_0^2)$, $K(k) \approx \log_e(8R_0/a)$ and $E(k) \approx 1$, so that the flux inside the ring is $\approx \mu R_0 I[\log_e(8R_0/a) - 2]$ from Eq. (12.7) [preferable to using the expression $A \approx \bar{L}I$ in terms of self-inductance ($\bar{L}$), which is only approximate (Jackson 1998)]. The condition that this be conserved (frozen flux) as the major radius changes during an eruption thus implies that the current behaves like

$$I \sim \frac{1}{R_0\ \log_e(R_0/a)}, \tag{12.14}$$

when $R_0 \gg a$, as detailed in PROBLEM 12.6. If the torus is allowed to expand under the effect of the hoop force, then R_0 will increase indefinitely, but I will decrease to zero as the initial magnetic energy of the torus is converted to kinetic energy.

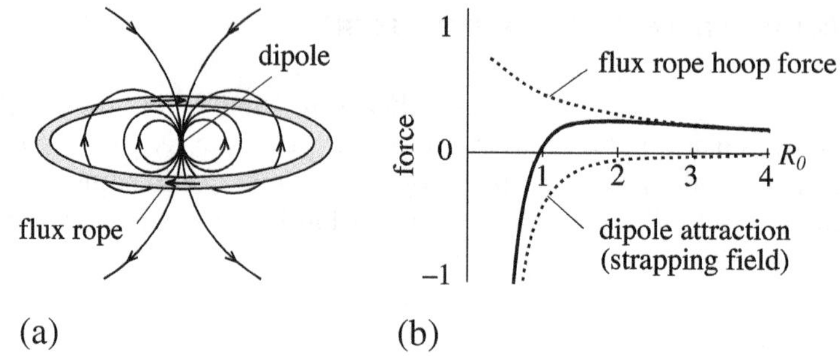

Figure 12.9. (a) A toroidal flux rope in equilibrium in a dipole field and subject to torus instability. (b) The net force (solid) on the flux rope as a function of its radius R_0 in units of $(m/I)^{1/2}$ (after Forbes 2010).

12.2.3.2 Torus Instability of a Force-free Flux Rope

Consider next the addition of a dipole of strength m at the centre of the torus to produce an extra *strapping field* that creates an equilibrium by balancing the hoop force (Figure 12.9a). The net force is then

$$F = \frac{\mu}{4\pi}\left[\frac{I^2}{R_0}\log_e\left(\frac{8R_0}{a}-1\right) - \frac{I\,m}{R_0^3}\right], \tag{12.15}$$

which is sketched as a function of R_0 in Figure 12.9b. Both the hoop force and strapping force decrease with R_0 and balance at a value of about $R_0 = 1$ in the units of this figure. Such an equilibrium is, however, unstable because increasing the tube radius R_0 makes the hoop force dominate, which acts to move the tube further out, whereas decreasing R_0 makes the strapping force dominate and moves the tube further in. Such an instability in the laboratory is called the horizontal tokamak or *torus instability*, which is essentially a *lateral kink instability* (Bateman 1978).

Kliem and Török (2006) revived an interest in the possibility of the torus instability as a cause of CMEs. They followed Bateman (1978) in considering the more general effect of an external strapping field $B_{ext} = \hat{B}R_0^{-n}$ that falls off with distance locally like R_0^{-n} but approaches a constant as $R_0 \to 0$. Then the global equation of motion of the flux rope (of mass M) is

$$M\frac{d^2R_0}{dt^2} = \frac{\mu I^2 c}{4\pi R_0} - I\hat{B}R_0^{-n}, \tag{12.16}$$

where $c = \log_e(8R_0/a) - 1$ is regarded as constant during the eruption of the flux rope: this is exact for a self-similar expansion and otherwise the logarithmic behaviour gives only a small correction. The flux-rope current (I) varies with R_0 in a way that is determined by flux conservation, as follows.

The magnetic flux (F_m) in the configuration consists of two parts

$$F_m = F_{m(I)} + F_{m(ext)} = \mu R_0 I(c-1) - 2\pi\int_0^{R_0} B_{ext}R\,dR, \tag{12.17}$$

due to an isolated flux rope and the external field. There is some debate as to what to hold constant here, bearing in mind that one should include variations in the external field during a perturbation, and often in numerical simulations reconnection occurs below the flux rope, letting it eat its way out through the external field. Suppose for simplicity that we hold $F_{m(I)}$ constant. Then Eq. (12.17) implies (with stars indicating initial equilibrium values) that I decreases with R_0 like

$$I = \frac{R_0^* I^*}{R_0}, \quad \text{where from Eq. (12.16) in equilibrium} \quad I^* = \frac{4\pi\hat{B}(R_0^*)^{1-n}}{\mu c}.$$

This may be substituted into the equation of motion (12.16) to give

$$\frac{d^2\bar{R}_0}{d\bar{t}^2} = \frac{1}{\bar{R}_0^2}\left(\frac{1}{\bar{R}_0} - \bar{R}_0^{1-n}\right),$$

in terms of dimensionless variables $\bar{R}_0 = R_0/R_0^*$ and $\bar{t} = t/t_0$, with the time-scale (t_0) given by $t_0^2 = 4\pi(R_0^*)^2M/[\mu c(I^*)^2]$. A linear perturbation from equilibrium ($\bar{R}_0 = 1$) in the form $\bar{R}_0 = 1 + r_0$ then implies

$$\frac{d^2 r_0}{d\bar{t}^2} = (n-2)r_0,$$

which gives torus instability with r_0 growing exponentially if $n > 2$.

If instead the total flux ($F_m = F_{m(I)} + F_{m(ext)}$) is held constant while both $F_{m(I)}$ and $F_{m(ext)}$ vary with R_0, the condition for instability becomes $n > 3/2 - 1/(2c)$ (PROBLEM 12.7), which reduces to Bateman's original condition ($n > 3/2$) in the large aspect ratio limit ($R_0 \gg a$), for which $c \gg 1$.

12.2.3.3 3D Non-equilibrium of a Toroidal Flux Rope

Torus instability may be stabilised by adding photospheric line-tying. This was accomplished in a global axisymmetric model by Lin et al. (1998) for a toroidal flux rope of radius R_0 suspended in a background dipole field that is tied to a spherical surface of radius $R_\odot$. They created a new equilibrium which is stable, with line-tying of the dipole field producing an extra restoring force (Figure 12.10), although the flux rope is not itself line-tied, since it does not reach the solar surface. However, if the flux-rope current becomes too strong, or the background dipole too weak, a state of catastrophic non-equilibrium is reached, when an equilibrium no longer exists.

Lin et al. (1998) worked in spherical polar coordinates (r,θ,ϕ) with respect to the centre of the Sun. They found the magnetic field due to a circular line current I of radius R_0 suspended in a magnetic field that is dipolar at the solar surface (Figure 12.11a), giving a ϕ-component for the vector potential of

$$A_\phi = \frac{\mu m \sin\theta}{4\pi R_\odot} \quad \text{at} \quad r = R_\odot. \tag{12.18}$$

The current density is therefore $j_\phi = (I/R_0)\delta(\cos\theta)\delta(r - R_0)$, while the form of A_ϕ throughout the corona satisfies the Poisson equation

$$\nabla^2 A_\phi = \mu j_\phi,$$

which is solved for A_ϕ in terms of a Green's function with boundary condition (12.18).

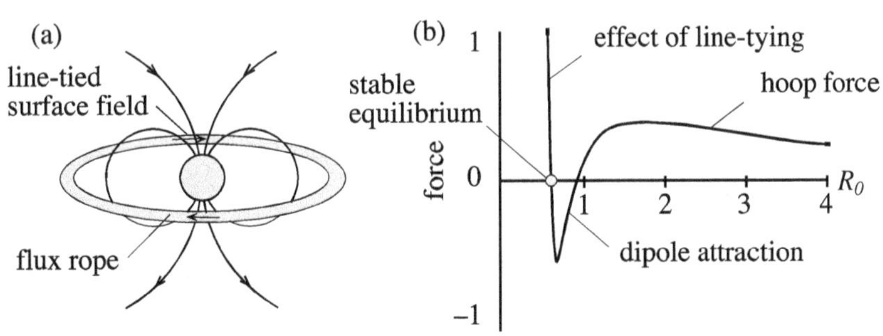

Figure 12.10. (a) A toroidal flux rope in stable equilibrium in a line-tied dipole field. (b) The net force on the flux rope as a function of its radius R_0 in units of $(m/I)^{1/2}$ (Forbes 2010).

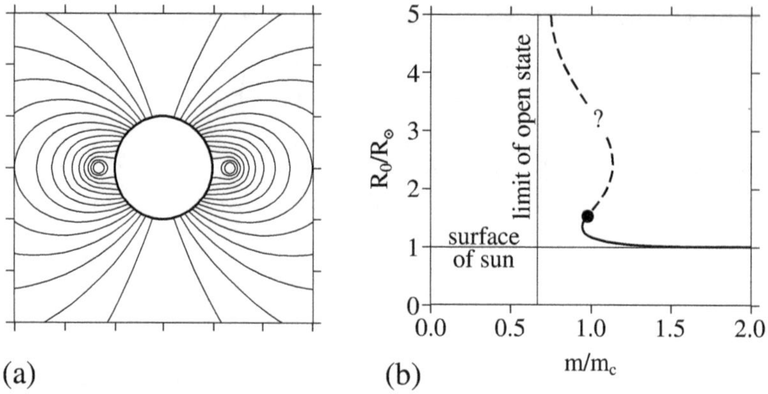

Figure 12.11. (a) Magnetic field lines in a meridional plane for a toroidal flux rope of major radius R_0 and minor radius $a_0 = 0.01 R_\odot$ at the maximum current, suspended in a magnetic field that is dipolar at the solar surface. (b) The major radius R_0 of the flux rope in equilibrium (Eq. 12.21) as a function of the background dipole moment m. The solid circle indicates the position ($m = m_c$, say) where a current sheet forms, while the dashed curve indicates the likely location of equilibria with current sheets (after Lin et al. 1998).

At the location $r = R_0$ of the current ring, the resulting external north-south magnetic field turns out to be

$$B_\theta = \frac{\mu}{4\pi}\left[\frac{4I}{R_\odot}\left(K(\bar{k}) - \frac{E(\bar{k})}{1-\bar{k}^2}\right) + \frac{m}{R_0^3}\right],$$

where $\bar{k} = R_\odot^2/R_0^2$, while $K(\bar{k})$ and $E(\bar{k})$ are complete elliptic integrals of the 1st and 2nd kind, respectively. The interaction between the loop current and this magnetic field produces an inward force IB_θ to counter the outward hoop force [Eq. (12.13)], so that the equation of motion of the current loop under a force F^* becomes

$$M\frac{d^2R_0}{dt^2} = F^* \equiv \frac{\mu}{4\pi}\left[\frac{I^2}{R_0}\log_e\left(\frac{8R_0}{a} - 1\right) - \frac{4I^2}{R_\odot}\left(K(\bar{k}) - \frac{E(\bar{k})}{1-\bar{k}^2}\right) - \frac{Im}{R_0^3}\right]. \tag{12.19}$$

This is a much more accurate version of Eq. (12.15), in which the (new) second term on the right represents the effect of line-tying due to an image current below the solar surface. Force-balance then gives a relation of the form $F^*(I, R_0, a, m) = 0$ between the flux-rope current (I), its location (R_0), its minor radius (a) and the dipole moment (m) of the background field.

However, a is determined by the condition $aI =$ *constant*, which holds when the toroidal flux rope has a linear force-free field for its internal equilibrium (PROBLEM 12.8). Also, a second relation between I, R_0, a and m arises from flux conservation, namely, that the magnetic flux $F_m(I, R_0, a, m)$ between the solar surface and the toroidal flux rope remain constant, that is,

$$F_m \equiv \mu R_0\left\{I\log_e\left(\frac{8R_0}{a} - 2\right) - 2I + \frac{2IR_0}{R_\odot}[E(\bar{k}) - K(\bar{k})] + \frac{m}{2R_0^2}\right\} = constant. \tag{12.20}$$

This is a more accurate version of Eq. (12.17), with the first and last two terms due to the flux rope and background dipole, and the other terms due to the image current.

Then Lin et al. (1998) eliminated I between the force balance $F^*(I,R_0,a,m)=0$ [Eq. (12.19)] and Eq. (12.20) to give the equilibrium radius (R_0) as a function of the background dipole strength (m) from

$$\frac{m}{R_0}\left\{2\log_e\left(\frac{8R_0}{a}-2\right)-4+\log_e\left(\frac{8R_0}{a}-1\right)+\frac{4R_0}{R_\odot}\left[\frac{2-\bar{k}^2}{1-\bar{k}^2}E(\bar{k})-2K(\bar{k})\right]\right\}$$
$$= constant \times \left[\log_e\left(\frac{8R_0}{a}-1\right)+\frac{4R_0}{R_\odot}\left(\frac{E(\bar{k})}{1-\bar{k}^2}-K(\bar{k})\right)\right], \tag{12.21}$$

where $a \sim I^{-1}$ and I is given by Eq. (12.20). The resulting graph of $R_0 = R_0(m)$ (Figure 12.11b) shows that, when the background field is too small, the curve bends back on itself and there is a catastrophic loss of equilibrium. The outward hoop and magnetic pressure forces can no longer be balanced by the tension of the overlying arcade holding it down, and so the flux rope erupts outwards.

When the minor radius (a) of the flux rope is less than $2\times10^{-3}R_\odot$, an X-point is present below the flux rope and so the eruption forms a current sheet there. When a is larger, there is initially no such X-point, but the eruption itself decreases the field below the flux rope until an X-point forms. If no reconnection were allowed at the current sheet, its growth would eventually stop the eruption, but in practice reconnection is likely to occur so that the eruption continues unhindered.

The resulting maximum storage of magnetic energy before equilibrium is lost is 1.51 times the energy of the potential field, which is less than the figure of 1.66 for a fully opened field, so this eruption cannot open all the field lines. The energy storage depends on the flux rope radius (a) and has its maximum value when $a \approx 0.03R_\odot$. For $a < 0.03R_\odot$, the ratio of stored to potential energy increases with a because the curvature force of the flux rope decreases, and so the flux-rope current needed for equilibrium increases, with an accompanying increase in stored energy. Thus, thicker flux ropes tend to produce more energetic eruptions.

12.2.4 Titov-Démoulin Model: An Active-Region Toroidal Flux Rope

After considering the axisymmetric flux rope of Figure 12.11a that is suspended above the photosphere, the next step is to model an active-region flux rope of current I, whose ends are located in the solar surface (although without complete line-tying) (Titov and Démoulin 1999). The rope is toroidal and has major radius R_0 and minor radius a, with an axisymmetric magnetic field comprising three parts ($\mathbf{B}_T+\mathbf{B}_q+\mathbf{B}_L$), which are defined as follows. $\mathbf{B}_T$ is the field of a constant-current, force-free flux rope (with a toroidal field B_e at its surface), in terms of a local polar-coordinate system $(\hat{r},\hat{\theta},\phi)$ relative to the major axis (as in Sec. 12.2.2). It has components

$$B_{\hat{r}}=0, \quad B_{\hat{\theta}}=\frac{\mu I\hat{r}}{2\pi a^2}, \quad B_\phi^2=B_e^2+\left(1-\frac{\hat{r}^2}{a^2}\right)\frac{\mu^2 I^2}{2\pi a^2}. \tag{12.22}$$

The flux rope lies in a vertical plane, with its axis of symmetry a distance d below the solar surface ($z=0$) (Figure 12.12).

The second magnetic component is due to an overlying coronal arcade created by a pair of magnetic charges $\pm q$ representing leading and following sunspots that are located along the flux-rope axis at depth d, namely,

$$\mathbf{B}_q=q\left(\frac{\mathbf{r}_+}{|\mathbf{r}_+|^3}-\frac{\mathbf{r}_-}{|\mathbf{r}_-|^3}\right), \tag{12.23}$$

where $\mathbf{r}_\pm=(x\mp L,y,z+d)$. (Note that, although $\nabla\cdot\mathbf{B}=0$ implies that isolated magnetic charges do not exist within a volume, it is acceptable to consider a field created by fictional magnetic charges on or outside the boundary of a volume. For example, placing such a charge on a boundary is a standard way

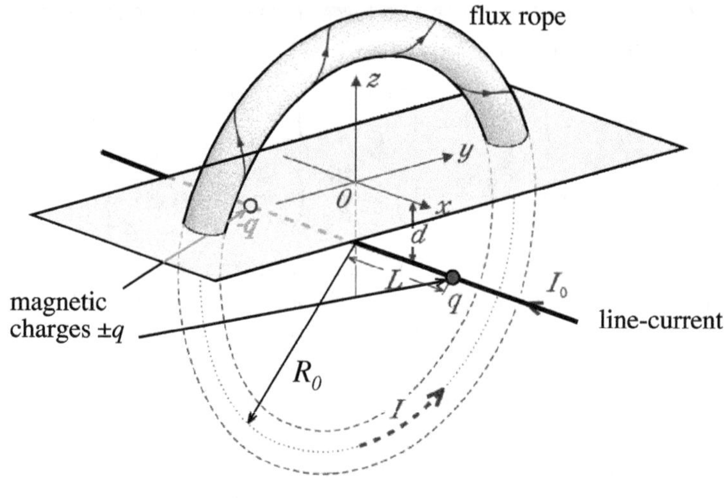

Figure 12.12. A 3D flux-rope model created by a circular flux rope of major radius R_0, a pair of magnetic charges $\pm q$ and a straight line current I_0 (Titov and Démoulin 1999).

of modelling the effect of a thin flux tube that protrudes through the boundary and spreads out into the volume.)

The final contribution ($\mathbf{B}_L$) is a purely toroidal magnetic field parallel to the flux rope axis and created by a line current I_0 along the flux-rope axis at depth d, namely,

$$B_\phi = \frac{\mu I_0}{2\pi[y^2 + (z+d)^2]^{1/2}}, \tag{12.24}$$

which determines the expression for $B_e = \mu I_0/(2\pi R_0)$ at the radius R_0 in Eq. (12.22). As the value of I_0 is increased, the configuration changes from a highly twisted flux rope to a weakly twisted one. For the topology of this configuration [which can include a bald patch (Sec. 6.9.2)], see Titov and Démoulin (1999).

Equilibrium is created by a balance between the outward hoop force of the torus and the inward force of the arcade due to the magnetic charges holding the flux rope down and acting on the toroidal current, namely, the following two terms

$$\frac{\mu I^2}{4\pi R_0}\log_e\left(\frac{8R_0}{a} - \frac{3}{2} + \frac{l_i}{2}\right) - \frac{2qLI}{(R_0^2 + L^2)^{3/2}} = 0, \tag{12.25}$$

where $l_i = 0.5$. The field of the submerged line current exerts no force, since its field is parallel to the toroidal current; the internal flux-rope component (B_θ) does not participate in this global equilibrium equation but contributes only to the internal equilibrium of the flux rope.

Titov and Démoulin (1999) considered the effect of gradual emergence of the toroidal flux rope through the solar surface by supposing that it passes through a series of equilibria as the major radius R_0 slowly increases. The resulting flux-rope current (I) is given from Eq. (12.25) as

$$I = \frac{8\pi qLR_0(R_0^2 + L^2)^{-3/2}}{\mu[\log_e(8R_0/a) - 5/4]}, \tag{12.26}$$

where the minor radius a is determined from a condition that the number of turns (N_t) in the full torus (including the submerged part) remain constant as the flux rope emerges. Since the poloidal field at the flux-rope surface is $B_\theta = \mu I/(2\pi a)$ from Eq. (12.22) and the toroidal field there is $B_\phi = \mu I_0/(2\pi R_0)$ from Eq. (12.24), this condition becomes $N_t = (2\pi R_0 B_\theta)/(2\pi a B_\phi) = IR_0^2/(I_0 a^2) = \textit{constant}$, where the current

I is given by Eq. (12.26). The resulting graph of I as a function of R_0 possesses a maximum at about $R_0 \approx L/\sqrt{2}$, and torus instability sets in at $R_0 \approx \sqrt{2}L$ (PROBLEM 12.9).

A linear instability may or may not lead to dramatic effects, depending on whether it saturates at a low amplitude or grows non-linearly. Instead, the emphasis is now on the possibility of non-equilibrium when the imbalance of forces suggests that dramatic evolution is more likely. In one particular case, namely, when the flux is conserved during the slow evolution through equilibria, the onset of torus instability and torus non-equilibrium occurs at the same threshold (Démoulin and Aulanier 2010). However, when one considers a general track in a general phase space formed by varying all the relevant parameters, a stable equilibrium disappears generically by a loss of equilibrium, with a loss of stability being much rarer (Poston and Stewart 1996).

The Titov-Démoulin model has been improved by completely line-tying the field in a 3D analytical model that is slowly evolved in time (Isenberg and Forbes 2007). Also, some Titov-Démoulin configurations have been shown to be unstable to helical kink instability when the twist exceeds 3.5π (Török et al. 2004) and others to torus instability when the field fall-off with height is too fast (Kliem and Török 2006).

12.2.5 Helical Kink Instability

The above analyses assume the internal forces in a flux rope are in equilibrium. Here we consider the extra possibility of a helical perturbation of the flux-rope axis. A straight force-free flux tube with no line-tying and a non-zero twist is always kink unstable (Anzer 1968). However, the effect of line-tying is to stabilise the flux tube when its twist is small and to enable it to become unstable to the helical kink instability when its twist is large enough (Hood and Priest 1979a, 1981). Kink instability is a possible explanation for the eruptive process in CMEs and many two-ribbon flares, as well as in compact confined flares and subflares (Rachmeler et al. 2010).

Consider a line-tied cylindrical flux rope (of length L and width a, aligned along the z-axis) with pressure $p(R)$ and magnetic field components $[0,\ B_\phi(R),\ B_z(R)]$ in cylindrical polars (R,ϕ,z). The critical twist $[\Phi = (LB_\phi)/(RB_z)$, Eq. (2.79)] for instability depends on the transverse magnetic structure of the flux rope, as well as its plasma beta and aspect ratio (L/a), but it lies typically between 2π and 6π.

The energy method for determining instability (Sec. 7.4.1) may be generalised to include line-tying as follows. Suppose the flux rope is in force balance

$$0 = \frac{dp}{dR} + \frac{d}{dR}\left(\frac{B_\phi^2 + B_z^2}{2\mu}\right) + \frac{B_\phi^2}{\mu R}, \tag{12.27}$$

and that the plasma has a helical displacement of amount

$$\xi = \left[\xi^R(R), -i\frac{B_z}{B}\xi^0(R),\ i\frac{B_\phi}{B}\xi^0(R)\right]\cos\frac{\pi z}{L}e^{i(m\phi+kz)}, \tag{12.28}$$

which vanishes at the ends $(z = \pm\frac{1}{2}L)$ of the loop.

The change in energy (δW) may be minimised with respect to ξ^0 to give

$$\delta W = \frac{1}{\mu}\int_0^\infty F\left(\frac{d\xi^R}{dR}\right) - G\xi_R^2\, dR. \tag{12.29}$$

Here F and G are functions of R, p, B_ϕ, B_z, k and m that generalise Eq. (7.38). In order to minimise Eq. (12.29) with respect to the radial perturbation (ξ^R) one solves the *Euler-Lagrange equation*

$$\frac{d}{dR}\left(F\frac{d\xi^R}{dR}\right) + G\xi^R = 0, \tag{12.30}$$

subject to either $\xi^R = 1, d\xi^R/dR = 0$ at $R = 0$ for $m = 1$ or $\xi^R = 0, d\xi^R/dR = 1$ at $R = 0$ for $m \neq 1$. The boundary between stability and instability for the perturbation (12.28) is found by varying the parameters until $\xi^R(R)$ vanishes somewhere.

The detailed results depend on the forms that are adopted for $p(R), B_\phi(R), B_z(R)$. The *Kruskal-Shafranov limit* [Eq. (7.45)] may be written

$$\Phi(R) = -Lk$$

at radius R: it gives the critical twist for local instability in the absence of plasma pressure gradients and line-tying. When they are included, at large k the loop is less stable than the Kruskal-Shafranov limit (because of the destabilising effect of the pressure gradient), whereas at small k it is more stable (because of line-tying). Hood and Priest (1981) conducted a full stability analysis to all possible line-tied perturbations by solving the partial differential equations. For example, a force-free field of uniform twist becomes unstable at a twist of 2.5π (i.e., 1.25 turns).

During their nonlinear development, numerical simulations of straight and toroidal ropes with line-tying show that the field can become highly contorted locally to form current sheets at *resonant* (or *mode-rational*) surfaces (Sec. 6.8.2) where reconnection occurs (Gerrard and Hood 2004; Gerrard et al. 2004; Hood et al. 2009).

12.2.6 Resistive Kink Instability

The effect of resistive diffusion on the ideal kink mode of the previous section gives a so-called *resistive kink instability*. It is likely that a loop reaches such a resistive instability (Sec. 7.5.5) before the threshold for the onset of ideal instability.

Usually, when linearising about an equilibrium ($\mathbf{B}_0$) by putting $\mathbf{B} = \mathbf{B}_0 + \mathbf{B}_1$ with $B_1 \ll B_0$, the diffusion term in the induction equation

$$\frac{\partial \mathbf{B}_1}{\partial t} = \nabla \times (\mathbf{v}_1 \times \mathbf{B}_0) - \nabla \times (\eta \nabla \times \mathbf{B}_1)$$

is negligible. However, it does become important in singular layers where

$$\nabla \times (\mathbf{v}_1 \times \mathbf{B}_0) \equiv (\mathbf{B}_0 \cdot \nabla)\mathbf{v}_1 - \mathbf{B}_0(\nabla \cdot \mathbf{v}_1) = \mathbf{0},$$

so that $\partial \mathbf{B}_1/\partial t$ can no longer be balanced by $\nabla \times (\mathbf{v}_1 \times \mathbf{B}_0)$. In particular, for an incompressible wave-like perturbation ($\mathbf{v}_1 \sim e^{i\mathbf{k}\cdot\mathbf{r}}$) such that $\nabla \cdot \mathbf{v}_1 = 0$, the locations of the singular layers are given by $\mathbf{k} \cdot \mathbf{B}_0 = 0$. For example, for a straight flux tube in a cylindrical geometry where $\mathbf{v}_1 \sim e^{i(m\phi + kz)}$, the locations ($R = R_s$) of resonant surfaces (Sec. 6.8.2) for each k and m are determined by

$$kB_{0z}(R_s) + \frac{m}{R_s} B_{0\phi}(R_s) = 0.$$

The simplest force-free equilibrium in cylindrical geometry is the constant-α field

$$B_z = B_0 J_0(\alpha R), \qquad B_\phi = B_0 J_1(\alpha R).$$

Its stability may be investigated (following Sec. 6.8.1) by seeking solutions of the form

$$\mathbf{B}_1 = \mathbf{B}_1(R)\exp[\omega t + i(m\phi + kz)]$$

to the linearised resistive MHD equations (Coppi et al. 1976). Assuming ideal kink stability and no line-tying, it is found that a flux tube is unstable to the so-called *resistive internal kink mode* (or *cylindrical tearing mode*). The fastest growing perturbations have long wavelengths ($kR_s \ll 1$) and $m = 1$, with a growth rate

$$\omega \sim \tau_d^{-1/3} \tau_A^{-2/3},$$

in terms of the resistive diffusion and Alfvén travel-times ($\tau_d = R_s^2/\eta$, $\tau_A = R_s/v_A$). Thus, for $R_m = \tau_d/\tau_A \gg 1$, the growth-rate lies between τ_d^{-1} and τ_A^{-1}, and is faster than the value $(\tau_d^{-1/2}\tau_A^{-1/2})$ for a planar (Cartesian) field (Sec. 6.8.1).

The nonlinear evolution of resistive modes is complex and depends on the parameter regime (Sec. 6.8.3, Wesson 1997). Sometimes they *saturate* with an equipartition between kinetic and magnetic energy, both in the *tokamak regime* ($B_z \gg B_\phi$ and $\beta \ll 1$), and in the *reversed-field pinch regime* ($B_z \lesssim B_\phi$ and $\beta \sim 1$). At other times, there is an *overlapping of resonances*, when magnetic islands from neighbouring layers interact. Other nonlinear effects are *mode coupling* of tearing modes of different helicities or *island coalescence* of small islands into longer ones.

An important limitation of the linear and nonlinear analyses of resistive kink instability is that line-tying has not been incorporated. In order of magnitude, it would imply $2\pi/k \leqslant 2L$, so that the wavelength $(2\pi/k)$ can fit into a loop of length $2L$. But k satisfies $\mathbf{k}\cdot\mathbf{B}_0 = 0$, and so this condition reduces to $\Phi \geqslant 2\pi$ in terms of the twist $[\Phi = 2LB_\phi/(RB_z)]$. However, a more precise derivation of the threshold for resistive instability is needed to see whether it is lower or higher than the threshold for the much faster growing ideal kink mode. Furthermore, the simple analysis of resistive instability in a 1D configuration needs to be redone in 2D and 3D magnetic structures that are more realistic for the solar corona.

12.2.7 Breakout

Antiochos et al. (1999) proposed a 2.5D computational *breakout model* that may well be appropriate for some CMEs when the overlying magnetic flux is small enough. It has an axisymmetric quadrupolar geometry, reminiscent of the original Sweet-Parker model for solar flares (Figure 6.6). One difference is that it is regarded as a much more global-scale magnetic field and another is that reconnection is not driven by the approach of the two dipoles (1) and (3) in Figure 12.13a. Instead, the shearing of the smaller equatorial arcade (2) beneath the X-type null point creates a highly sheared force-free magnetic field typical of a prominence (Figure 12.13b). Since the field is axisymmetric, the X-point is really an X-type null-line, located typically between 1.25 $R_\odot$ and 1.75 $R_\odot$ for the maximum energy storage (DeVore and Antiochos 2005). Such a configuration is non-generic and topologically unstable, in the sense that a small perturbation would break up the X-line to create a separator joining two null points. The process may also occur in a delta sunspot (Antiochos 1998).

The initial quadrupolar state is potential with no free energy. As the central arcade [(2) in Figure 12.13a] is slowly sheared in the azimuthal direction, it stores energy and pushes up against the X-line, creating a curved horizontal current sheet at high altitude, called a *breakout current sheet* (Figure 12.13b). Eventually, in these simulations numerical dissipation allows reconnection to start in the breakout current sheet, which permits the equatorial arcade to expand outwards and erupt and its flux to grow. As in the standard non-equilibrium picture of an eruption, this in turn drives the formation of a reconnecting flare current sheet below the erupting flux rope (Figure 12.13c). The fact that not all the field lines are opened enables the Aly-Sturrock hypothesis to be avoided.

Plots of magnetic and kinetic energy (Figure 12.13d) reveal the stages of evolution. First, the magnetic energy slowly builds up as footpoints of the central arcade are sheared and cause a breakout current sheet to form and rise at about 30 km s^{-1}. Then a slow eruption phase (at 60 km s^{-1}) begins when reconnection starts in the breakout current sheet. (The magnetic energy reaches a maximum and starts to decline at the start of this phase, with the kinetic energy growing slowly as the flare current sheet forms.) Next, fast eruption (at 450 km s^{-1}) begins with the start of reconnection in the flare current sheet, a rise in kinetic energy and a fall in magnetic energy. The flare reconnection accelerates the flux rope directly and removes tethers of overlying field lines that would otherwise act to hold down the arcade.

The nature of the mechanism that initiates the slow eruption is unclear, bearing in mind the model's reliance on numerical experiments. Is it because an onset of non-equilibrium drives the formation and

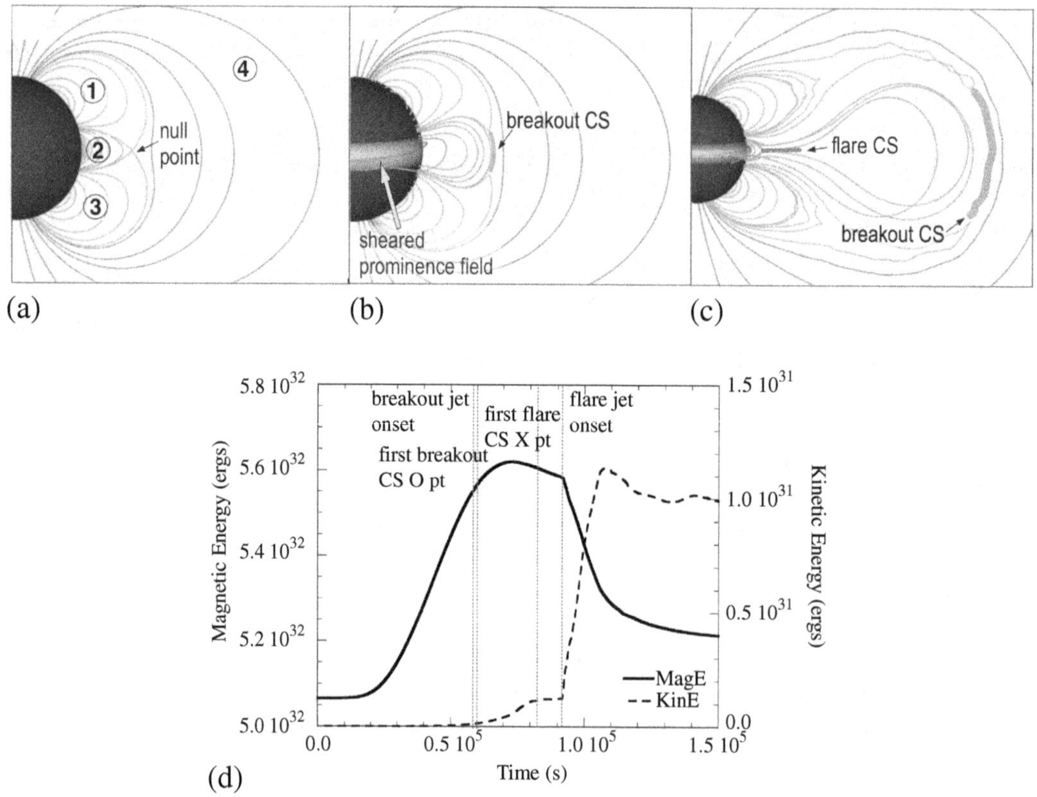

Figure 12.13. The breakout model with an ideal MHD code, showing magnetic flux surface contours at three times: (a) $t = 0$, (b) $t = 0.07 \times 10^5$ sec (before flare onset) and (c) $t = 1.025 \times 10^5$ sec (during the rapid rise phase), as well as (d) the free magnetic and kinetic energy as functions of time (Karpen et al. 2012).

reconnection of the breakout current sheet, or is there some kinetic process that leads to an onset of reconnection when the current sheet is large enough, since in resistive MHD a current sheet tends to reconnect as it forms? Furthermore, what would happen if the arcade were sheared at a very much slower rate than the current unphysical value of $0.1\ v_A$?

Consideration of 3D aspects has begun. Maclean et al. (2005) undertook a topological analysis of 3D breakout by analysing evolution through sequences of potential and linear force-free equilibria, with open breakout field lines created by a global topological bifurcation. They modelled a delta sunspot as a central, positive source surrounded by three negative and two positive sources, simulating the emergence of a new area of positive flux into a pre-existing sunspot configuration to form a delta spot. Altering either the strength or location of the central source (or the parameter α of a force-free field) produces topological breakout by two different mechanisms, namely, *global spine-fan* or *global separator bifurcation* (Brown and Priest 1999).

DeVore and Antiochos (2008) and Lynch et al. (2008) described a 3D computation for breakout from a bipolar active region immersed in a large-scale dipole field. Above the active region lies a coronal null from which a fan surface spreads down to the solar surface as a dome. Photospheric shearing can produce breakout in a qualitatively similar manner to the 2.5D model. Roussev et al. (2007) and Lugaz et al. (2011) have generated eruptions by superimposing a Titov-Démoulin flux rope on fields that are based on magnetograms for actual events. Also, Hood et al. (2009) and Archontis and Hood (2010) and others have constructed computational models for flux emergence into bipolar and quadrupolar fields.

In the future, two developments will be important, namely, to incorporate the kinetic physics of the reconnection and more realistic 3D aspects, beginning with a topology containing a separator rather than an X-line.

12.2.8 Emerging (or Evolving) Flux Model

Heyvaerts et al. (1977) outlined a model inspired by the key fact that flares often follow the emergence of new flux from below the photosphere (e.g., Schrijver et al. 2005; Schrijver 2009). They proposed that the type of flare depends on the magnetic environment into which the flux emerges. Emergence of an ephemeral region with a small magnetic flux in the quiet Sun produces a subflare or an *X-ray bright point* with *X-ray jets* (Sections 1.6.2 and 9.3.3). Emergence of more flux near a unipolar sunspot or into a unipolar area near the edge of an active region may give rise to a *simple-loop* or small *compact flare.* If instead the new flux appears near the sheared field around an active-region prominence, then a *two-ribbon flare* may result, with the emerging flux triggering the release of stored energy in the much more extensive overlying force-free field. In each case, the flux emergence can accelerate an X-ray jet (Shibata et al. 1992; Shibata et al. 2007).

The model suggests that a compact flare is triggered during the interaction of emerging flux with the overlying magnetic field when the current sheet at the interface between new and old flux reaches a critical height (Figure 12.14). As the new flux eats its way into the ambient field, so the current sheet moves up in the atmosphere. At the critical height ($h_{\rm crit}$), which depends on emergence speed and field strength, the current density exceeds a threshold for the onset of microturbulence, and the flare is triggered. In terms of the current-sheet temperature (T_c), this condition is $T_c^2 > T_{\rm turb}^2 \equiv 1.8 \times 10^{16} B_i/v_i$, where B_i (in tesla) is the inflow magnetic field and v_i (in m s^{-1}) is the inflow speed.

T_c is found as a function of height (h) above the photosphere by solving the energy-balance equation inside the sheet (Milne and Priest 1981). Low down in the atmosphere, there is a unique solution to the energy equation with $T_c < T_{\rm turb}$. Eventually, a critical height is reached beyond which there is no thermal equilibrium with such a balance. The sheet heats up rapidly, and, if $T_{\rm turb}$ is exceeded, the flare is triggered. The current sheet broadens rapidly due to the onset of a turbulent resistivity, so that strong electric fields are generated, which accelerate particles to high energies and produce the impulsive phase.

The model also applies to a situation of *evolving flux* rather than emerging flux. When distinct flux systems evolve, current sheets form at the interfaces between the systems; thus, horizontal motions of photospheric flux will produce current sheets in just the same way as vertical motions.

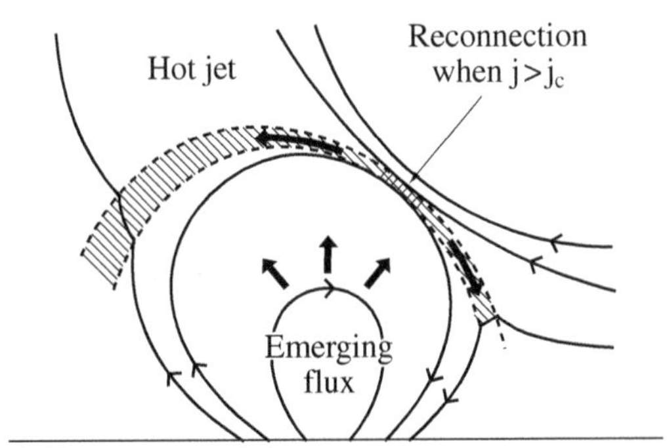

Figure 12.14. An emerging flux model for a flare when the current density exceeds a critical value (j_c) (Heyvaerts et al. 1977).

The process of flux emergence has been modelled in detail numerically starting in 2D and leading on to 3D (Sec. 9.3.3) and its role in initiating eruptions (12.2.9). However, the detailed local plasma physics behaviour of the current sheet in MHD flux emergence has not yet been adequately considered.

12.2.9 Other Numerical Experiments on Initiation of Eruption

Several numerical experiments on 3D non-equilibrium and its consequences have been undertaken. In some cases, a highly sheared or twisted structure (containing energy in excess of potential) emerges from below the photosphere and erupts soon afterwards, whereas in other cases such a structure is slowly built up over several days or weeks from a relatively unsheared or potential field by photospheric motions.

Amari et al. (2000), Amari et al. (2007) and Amari et al. (2010) showed how a flux rope with an overlying arcade can be formed by flux shearing and cancellation (i.e., reconnection beneath it), and how it later erupts due to non-equilibrium with reconnection driven below it (Figure 12.15). Riley et al. (2008) suggested from a simulated eruption that the dark cavity seen in coronal mass ejections (Figure 12.1a) is the flux rope, while the bright front is formed mainly from the overlying coronal streamer within which the flux rope erupts. Aulanier et al. (2010) showed how flux cancellation, shearing and reconnection at a coronal quasi-separator form a coronal flux rope from an initially bipolar field, and how an eruption subsequently occurs by torus instability. Kusano et al. (2004) showed how emerging flux of opposite polarity or reversed shear can trigger eruption and reconnection.

Fan and Gibson (2004, 2007), Fan (2010) and Archontis and Hood (2008, 2010) have investigated the emergence of a flux rope through the photosphere and the subsequent evolution and observational consequences, such as sigmoids, cavities and partial ejection (Gibson et al. 2004; Gibson and Fan 2006, 2008). The mechanism for eruption (non-equilibrium, torus instability, or kink instability) depends on both the amount of twist and also the strength, orientation, and rate of decline with height of the overlying arcade.

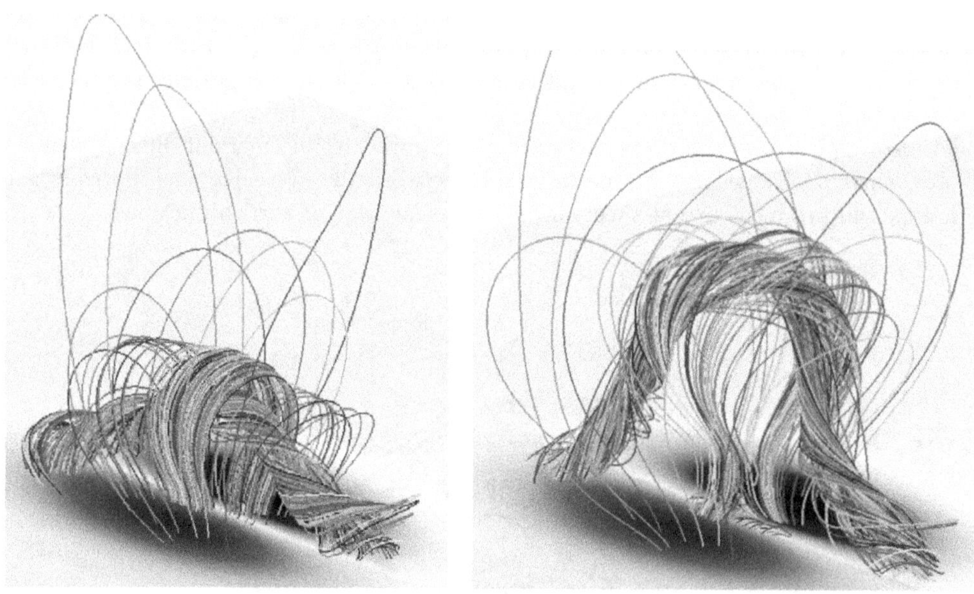

(a) (b)

Figure 12.15. Numerical experiment on 3D nonequilibrium, showing the magnetic field lines (a) of the pre-eruptive configuration and (b) during the eruption (courtesy Tahar Amari).

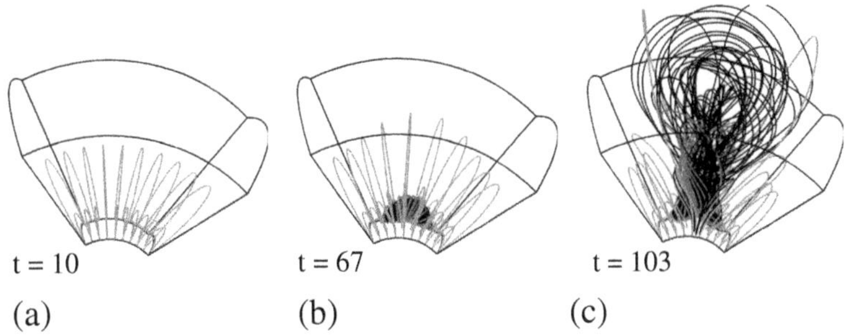

Figure 12.16. Numerical experiment of an eruption due to kink instability of an emerging flux tube: (a) coronal magnetic arcade at $t = 10t_A$, where t_A is the Alfvén travel-time; (b) stable quasi-equilibrium at $t = 57t_A$ after some flux has emerged; (c) a dynamic eruption at $t = 94t_A$ after further slow flux emergence (from Fan 2005).

An impressive series of numerical computations based on observed magnetograms has modelled the global evolution of the solar corona through a series of nonlinear force-free states (Mackay and van Ballegooijen 2006; Yeates and Mackay 2009; Yeates et al. 2010). They show how force-free flux ropes can be built up by flux emergence and photospheric cancellation at sites where prominences form, and how such flux ropes can erupt by nonequilibrium. Indeed, they are able to predict the sites of most coronal mass ejections outside active regions. Reconnection was found to occur both above the flux rope (stripping away the overlying flux and allowing ejection) and also below the flux rope at a quasi-separator or separator.

12.3 Reconnection and the Creation of Flare Loops

After the eruption of a twisted flux rope containing a filament within a CME, the main phase of an eruptive two-ribbon flare is characterised by a rising arcade of flare loops (Sec. 1.9.2). The standard picture for the main phase has been developed by many authors and is as follows (Figure 12.17, Sec. 12.3.1). Magnetic energy is liberated by reconnection in the corona; the energy is carried down to the chromosphere by non-thermal electrons and conduction fronts; most of the energy is radiated by the chromosphere at the loop feet in optical and UV and a little in hard X-rays; the heated chromosphere expands or *evaporates* to create hot dense coronal loops in soft X-rays; as the reconnection location rises, new larger loops are formed, while the earlier loops cool and much of their plasma drains back down to the chromosphere.

The reconnection plays several roles: it allows the eruption to take place or speed up by severing the connection of overlying field lines that would otherwise hold the flux tube down; it also converts magnetic energy into heat, kinetic energy and fast particle energy. Particle acceleration is extremely important in the impulsive phase, but is outside the scope of this MHD treatment (see, e.g., Birn and Priest 2007).

12.3.1 Standard (CSHKP) 2D Reconnection Model

The overall feature of a flare's main phase is that Hα ribbons separate and soft X-rays loops rise, with hot loops lying above cool loops. The details of the way reconnection works in 2D and 3D have been described in Chapter 6, and these have been used in a series of progressively more realistic models of energy conversion in flare loops (Figure 12.18). This began with a series of qualitative ideas by the early authors that gave the CSHKP model its name. Carmichael (1964), Sturrock (1968) and Hirayama (1974) recognised the importance of reconnection in a current sheet lying above closed loops, while Kopp and Pneuman (1976) suggested that the response to reconnection of a pair of flux tubes is to launch a pair of gasdynamic shocks down both legs of the newly closed flux tube.

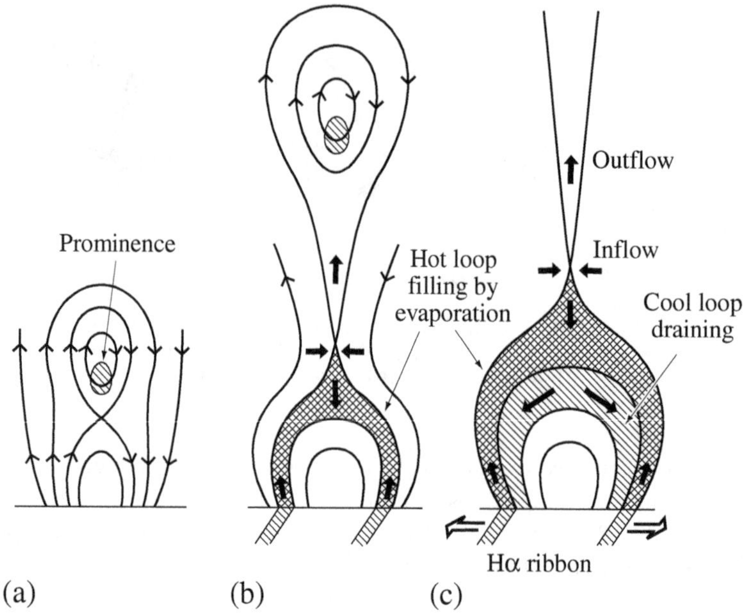

Figure 12.17. Qualitative picture for the creation of new hot loops by the rise of the reconnection location as underlying loops cool down: (a) a vertical section through the pre-eruption equilibrium containing a prominence, (b) eruption driving reconnection below the prominence and filling a hot loop by evaporation, (c) continuing reconnection creating new hot loops and allowing the earlier loops to cool and drain.

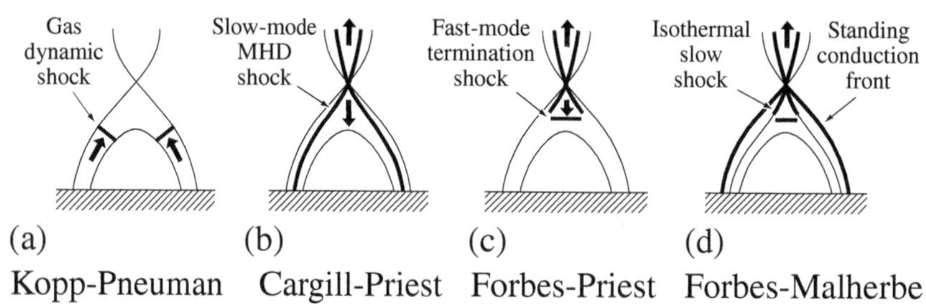

Figure 12.18. Developments of models for flare loops.

Later, more quantitative developments used insights from MHD and numerical experiments. Cargill and Priest (1982) realised that fast reconnection implies that the gasdynamic shocks need to be replaced by standing slow-mode shocks and showed that they can heat the plasma to 30 MK or more (Figure 12.18b). They also suggested the loops should be cusp-shaped, which was confirmed observationally much later by Yohkoh (Tsuneta et al. 1992). Then, numerical experiments of Forbes and Priest (1983) revealed the presence of a fast-mode termination shock, where the downflowing reconnection outflow jet is slowed and diverted sideways as it meets the obstacle created by the underlying closed loops (Figure 12.18c).

The non-perpendicular part of this fast-mode shock is a possible candidate for accelerating high-energy particles in a solar flare, since it is a much better accelerator than the slow-mode shocks (Isenberg 1986). The shock is present when the magnetic field is larger than a few gauss, so that the downflowing jet is super-fast-magnetosonic. However, most of the particle acceleration in the impulsive phase is likely to take place in the current sheet by direct electric field or turbulent acceleration. The shocks are much better at heating and compressing the plasma.

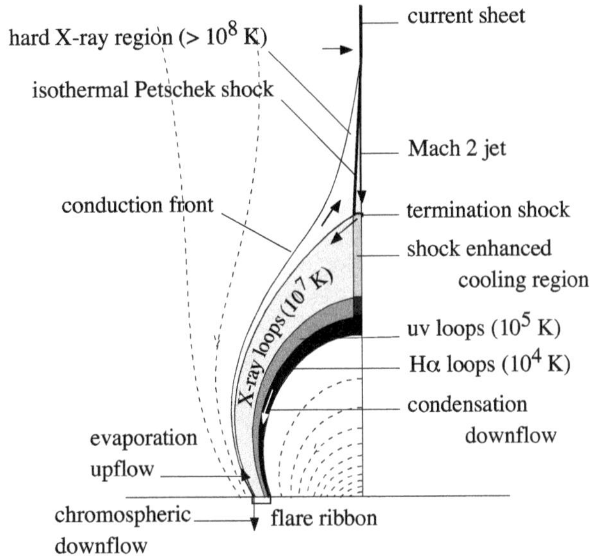

Figure 12.19. Standard picture for reconnection effects in a two-ribbon flare (Forbes 2010).

Then Forbes and Malherbe (1986), Yokoyama and Shibata (1997) and Shiota et al. (2005) introduced thermal conduction into the model, which makes each slow-mode shock bifurcate into an isothermal slow shock (at which the plasma is compressed) and a standing conduction front (at which it is heated) (Figures 12.18d and 12.19). The conduction fronts travel down to the chromosphere and drive evaporation from the outer edges of Hα ribbons. The soft X-rays therefore come from the hot loops that are heated by the conduction fronts, and the slow and fast shocks. The hard X-rays come from the footpoints, where the fast particles and conduction fronts reach the chromosphere, as well as from the downflow jets heated by the slow shocks, and the denser plasma heated and compressed by the fast shock (see also Sec. 12.3.2.3).

Flare-loop temperatures range from 10^4 K to 30 MK or more (with the cooler loops formed below the hot ones when hot loops cool by radiation). The outermost edges of the hot loops map to the outer edges of the chromospheric Hα ribbons, while the inner edges of the cool loops map to the inner edges of the ribbons (Schmieder et al. 1996). Plasma drains downwards in the cool loops at 100 to 500 km s^{-1}.

A simple expression for the rate of reconnection in a 2D configuration was discovered by Forbes and Priest (1984b) in terms of the motion of Hα ribbons. By uncurling (i.e., integrating) Faraday's equation ($\boldsymbol{\nabla} \times \mathbf{E} = -\partial \mathbf{B}/\partial t$) and writing $\mathbf{B} = \boldsymbol{\nabla} \times \mathbf{A}$ in a 2D xy-configuration (for which $\mathbf{E}$ and $\mathbf{A}$ have only z-components), they found that the electric field ($E_0\hat{\mathbf{z}}$) at the reconnection X-line is

$$E_0\hat{\mathbf{z}} = \frac{\partial A_0}{\partial t}\hat{\mathbf{z}} - \boldsymbol{\nabla}\Phi = \frac{\partial A_0}{\partial X_R}\frac{dX_R}{dt}\hat{\mathbf{z}} - \boldsymbol{\nabla}\Phi,$$

where $A_0\hat{\mathbf{z}}$ is the flux function at the separatrix (passing through the X-line) and X_R is the separatrix footpoint location (Figure 12.20). Since the configuration is 2D, $\boldsymbol{\nabla}\Phi$ is constant, and, if the field lines are line-tied to a high-density (stationary) photosphere, the constant vanishes. Thus, this equation reduces to

$$E_0 = B_y(X_R, 0)v_R, \tag{12.31}$$

where $B_y(X_R, 0)$ is the vertical field at $x = X_R$, $y = 0$, and $v_R = dX_R/dt$ is the velocity of the outer edge of the Hα ribbons, assumed to equal the apparent velocity of the separatrix footpoints. This represents the rate of reconnection, namely, the rate at which magnetic flux enters the reconnection region, and therefore

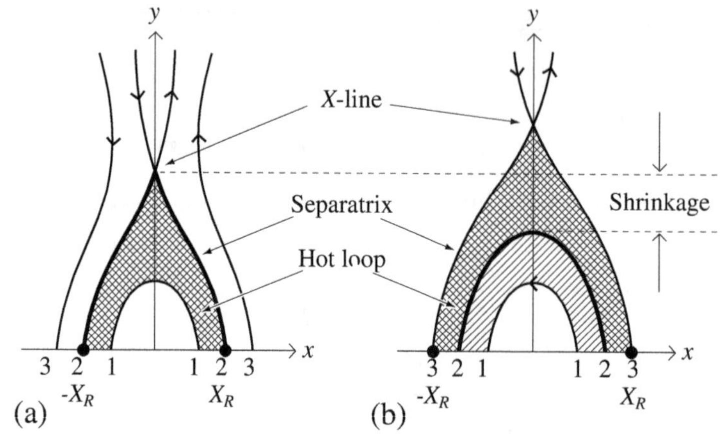

Figure 12.20. The relation between the locations of the reconnection X-line, the separatrices, the hot loops and the outer edges (X_R, shown by large dots) of the Hα-ribbons at (a) one time and (b) a later time (after Forbes 1996).

the rate at which line-tied flux is sampled at the footpoints by field lines that link along the separatrices to the reconnection location.

By measuring the ribbon velocity and photospheric magnetic field, this formula has been used to deduce the electric field in the corona. Typically, the electric field rises over a few minutes to peak values of 200 V m^{-1} and then slowly declines to much lower values during the main phase (Poletto and Kopp 1986; Qiu et al. 2002). Such peak electric fields are enormous – orders of magnitude larger than the Dreicer field (e.g., Wesson 1997) – and so their effect in accelerating particles needs a kinetic plasma treatment. Indeed, even higher values, such as $E = v_i B = 10^4$ V m^{-1}, are possible, arising from a reconnection inflow of, say, $v_i = 0.1 v_A$, a magnetic field of $B = 500$ G (5×10^{-2} T) and an Alfvén speed of $v_A = 2\times10^3$ km s^{-1}.

The ribbon width (W) is given in terms of the ribbon velocity (v_R) by

$$W = v_R \tau_c,$$

where τ_c is the cooling time due to radiation and conduction, the latter dominating at high temperatures and low densities (Cargill et al. 1995). Observed widths have been used to deduce the cooling times by Svestka et al. (1982), who found τ_c = 15 mins early in a flare and τ_c = 240 mins after 10 hours.

An interesting feature of observations and numerical experiments is *shrinkage* of flare loops (Figure 12.20), as field lines shrink or spring back by about 20 to 30 per cent after reconnection (Svestka et al. 1987; Forbes and Acton 1996). It is the upward motion of the X-line that causes the shrinkage. When a field line is connected to the X-line, it is cusp-shaped, but, as the X-line moves up, the field line becomes more rounded.

Another important feature of the numerical experiments (Forbes and Malherbe 1986; Yokoyama and Shibata 1997) is that the reconnection tends to be impulsive and bursty (Priest and Forbes 2000; Shen et al. 2011) as the current sheet splits into a series of current filaments and goes unstable to secondary tearing with a highly time-dependent release of energy. Pairs of null points form and are annihilated in a turbulent manner and a series of plasmoids are ejected (Figure 12.21).

12.3.2 3D Modifications to the Standard Model

The 2D standard model explains some aspects of flare energy release so well that they are likely eventually to be incorporated in generalised ways in a full 3D model. However, there is now a need to go beyond the confines of 2D in order to address other aspects of flare energy release, namely, the

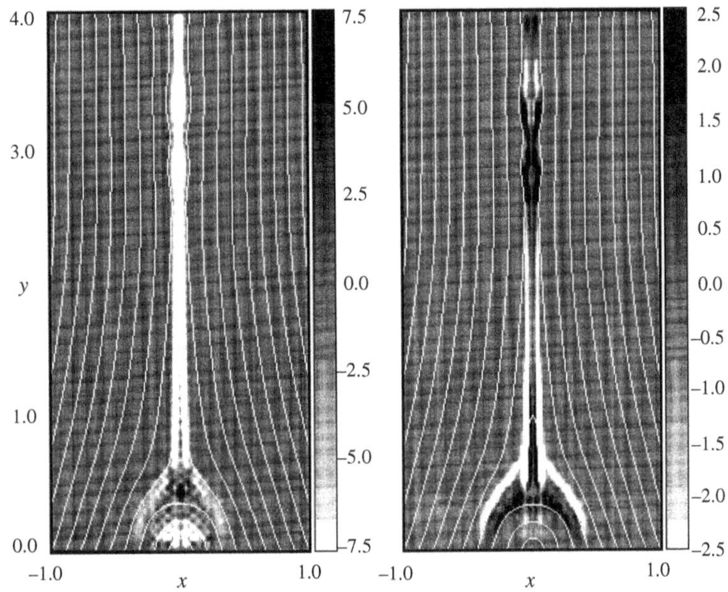

(a) current density (b) vorticity

Figure 12.21. 2D line-tied reconnection, in which the slow-mode shocks and fragmentation of the current sheet show up in plots of (a) current density and (b) vorticity, overlaid on the magnetic field lines (in white) (Forbes et al. 1989).

global 3D nature of the energy release and the fragmentation of the reconnecting current sheet. Also, the Forbes-Priest formula (Eq. 12.31) for using the flux (F_m) swept up by the flare ribbons as a measure of the reconnected flux can be generalised to give

$$\frac{\partial A}{\partial t} = B_n v_R \quad \text{or} \quad F_m = \int B_n v_R \, ds \, dt, \tag{12.32}$$

where $B_n(s,t)$ and $v_R(s,t)$ are the vertical field component and ribbon velocity at time t and location s along a flare ribbon. This has been applied to several flares by Poletto and Kopp (1986), Fletcher et al. (2001), Asai et al. (2002), Qiu et al. (2002) and Qiu et al. (2004) to give peak electric fields of 10–10^4 V m^{-1} and flux transfer rates of 0.5 to 20 GV (where 10 GV corresponds to a field of 100 V m^{-1} over a distance 100 Mm). Such high electric fields could well accelerate particles along separators.

Furthermore, importantly, Qiu et al. (2007) found that the poloidal (azimuthal) flux in interplanetary flux ropes (magnetic clouds) is much larger than their toroidal flux and is comparable with the flux reconnected in the corresponding flare. This suggests that, as a flux rope erupts, most of its twist and helical structure is created by reconnection (as mutual magnetic helicity is transferred to self-helicity) rather than being pre-existing. In addition, Longcope et al. (2007) found, from estimates of flux and magnetic helicity in an active region, that the typical twist in a pre-flaring example was only one-third of a turn, suggesting that the eruption is much more likely to be due to non-equilibrium or torus instability than kink instability.

The indications are, however, that the energy release in 3D is filamentary and patchy, since many small elements are seen to move parallel to a flare ribbon, while hard X-ray footpoints occur at discrete points and tend to move along a ribbon (Fletcher et al. 2004). Furthermore, hard X-ray and ultraviolet (UV) brightenings tend to move mainly parallel to the polarity inversion line during the rise phase, and perpendicular to it in the main phase (Qiu et al. 2010).

In 2D, reconnection can occur only at an X-point, but in 3D it can take place at a *separator*, near a null point, at a *quasi-separator* [the intersection of two quasi-separatrix layers (QSLs), sometimes called a *hyperbolic flux tube*], or indeed at any location where magnetic gradients, and therefore electric currents, build up to much larger values than normal (Chapter 6). So the question arises: whereabouts in a 3D coronal configuration is magnetic energy converted during a solar flare?

Several answers arise from comparing flare-ribbon locations with overlying magnetic structure just before or during the flare. In some cases reconnection is occurring at a separator (Sec. 12.3.2.1) and in others at a quasi-separator (Sec. 12.3.2.2). In others, spine-fan reconnection (Priest and Pontin 2009) is focussed in a current sheet near a coronal null with a separatrix dome and produces field-line flipping (Aulanier et al. 2000; Fletcher et al. 2001; Masson et al. 2009). Occasionally, it occurs at a *bald patch*, where a separatrix touches the solar surface (Titov et al. 1993; Aulanier et al. 1998). Sometimes the basic field structure is relatively simple, such as a quadrupolar field or a field with a single coronal null above an isolated flux patch, and sometimes it is highly complex with many separators (Longcope et al. 2007).

12.3.2.1 3D Fields with Separators [Longcope Minimum-Current Corona (MCC) Model]

A 3D version of the 2D CSHKP model (Figure 12.17) is shown in Figure 12.22: in 2D, reconnection at an X-point creates a closed field line (plasmoid), whereas in 3D it occurs at a series of locations and creates a series of closed loops and a twisted flux rope that is still connected to the Sun. The role of reconnection is then to add twist to the flux rope as it erupts: the flux rope begins as the field around a prominence and is later observed in interplanetary space as a *magnetic cloud* (Burlaga et al. 1981; Gosling et al. 1995). The first quantitative model for such a process was developed by Longcope et al. (2007). They measure the reconnected flux from the photospheric flux swept up by the flare and predict the energy release as well as the flux and magnetic helicity produced in the flux rope, as follows.

When low-β coronal fields evolve in an ideal manner (without reconnection) through a series of equilibria in response to footpoint motions, the equilibria will in general consist of a series of nonlinear force-free fields, with current sheets along any separators that are present. Calculating such fields is a difficult computational exercise, but Longcope (1996, 2001) developed a *Minimum-Current Corona (MCC) model*

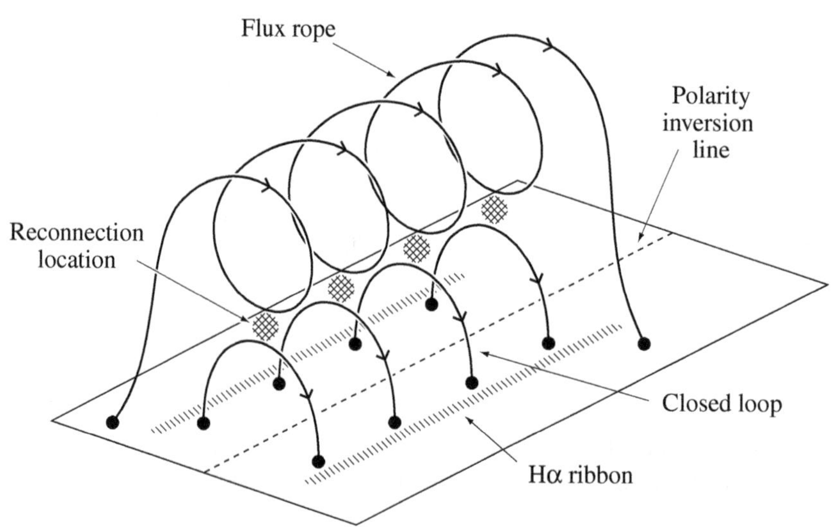

Figure 12.22. A cartoon for a 3D flare model in which reconnection at several sites above the polarity inversion line creates a series of closed loops and a twisted flux rope (after Longcope et al. 2007).

that gives analytical expressions for the net coronal current and a lower bound on the free magnetic energy stored by ideal coronal evolution.

The first assumption in MCC modelling is that the photospheric field consists of a series of discrete isolated unipolar sources or flux patches. This enables every coronal field line to be assigned to a *flux domain* that connects a particular pair of positive (P_i) and negative (N_j) unipolar sources. The total flux joining P_i to N_j is the *domain flux* $F_{m(ij)}$. The interfaces between domains are separatrices which intersect in separators and form the skeleton of the field (Priest et al. 1997).

The second assumption is that the corona evolves through stages (called *flux-constrained equilibria*) with the minimum energy that preserves the domain fluxes. Such fields are current-free except for the presence of current sheets along the separators. As the coronal field evolves, the domain fluxes are held fixed even though the positions of the flux patches change. In general, N_S source regions interconnected in N_D ways will produce $N_C = N_D - N_S + 1$ constraints on N_C separators.

All the free energy in an MCC model is released by reconnection at the separators, which transfers flux from two domains to two other domains and reduces the field to a purely potential one. One limitation of MCC models is that reconnection of a more accurate nonlinear force-free field would conserve the total magnetic helicity and reduce the field to a linear force-free field; another is that the field should be line-tied to the photosphere rather than free to move around within each domain.

Longcope et al. (2007) applied the model to a two-ribbon flare. They partitioned an active region into 28 regions. Observed motions implied that -1.4×10^{43} Mx^2 of magnetic helicity (mostly due to braiding between domains) was injected into the active region during the 40-hour flare build-up. They replaced each flux patch by a positive (P_i) or negative (N_j) source and calculated the skeleton of the initial potential field, as indicated in Figure 12.23a. The photosphere contains spine curves (solid) and intersections (dashed) of separatrix fan surfaces that link positive and negative nulls.

They calculated the domain fluxes ($F_{m(ij)}$) and their expected changes in time if the field were to remain potential. By assuming the domain fluxes instead remain constant during the preflare evolution, the differences ($\Delta F_{m(ij)}$) in domain flux through the separators before and after reconnection were estimated. In total, they find that 5×10^{21} Mx of flux needs to be reconnected during the flare to release the stored energy. This is similar to the flux that is found to be swept up by the flare ribbons from the sum of fluxes in all pixels that brightened in TRACE 1600 Å.

Superposing the 1,600 Å image on the skeleton (Figure 12.23b) shows that the flare ribbons follow a series of spines connecting sets of nulls. The affected domains are shaded dark in Figure 12.23a if they have more flux than the potential field and light if they have less. The whole region has 29 null points and 32 separators, but only 8 of the separators connect null points found on the flare ribbons.

The MCC model uses the flux discrepancies ($\Delta F_{m(ij)}$) to deduce the currents (I_i) along the separators of length L by solving the following equation for I_i

$$\Delta F_{m(ij)} = \frac{I_i\ L}{4\pi} \ln \frac{e I_i^*}{|I_i|}, \tag{12.33}$$

where $I_i^* = (256 L^2 \bar{B}' \delta)/(e^4 \pi)$ is the value of I_i that maximises $\Delta F_{m(ij)}(I_i)$. Here $\bar{B}'$ is the mean shear along the separator and δ is a constant that depends on the loop geometry and equals unity for a circular loop (Longcope and Magara 2004).

Next, the energy (ΔW_{mi}) stored above potential due to the presence of each separator current is found from

$$\Delta W_{mi} = \frac{1}{4\pi} \int_0^{\Delta F_{m(ij)}} I\ d(\Delta F_{m(ij)}) = \frac{L I_i^2}{32\pi^2} \ln \left(\frac{\sqrt{e} I_i^*}{|I_i|} \right). \tag{12.34}$$

Summing over the eight separators gives a lower bound of 8×10^{24} J (8×10^{31} erg) on the energy stored by active-region evolution. This is only 6 per cent of the total active-region energy (1.5×10^{33} erg).

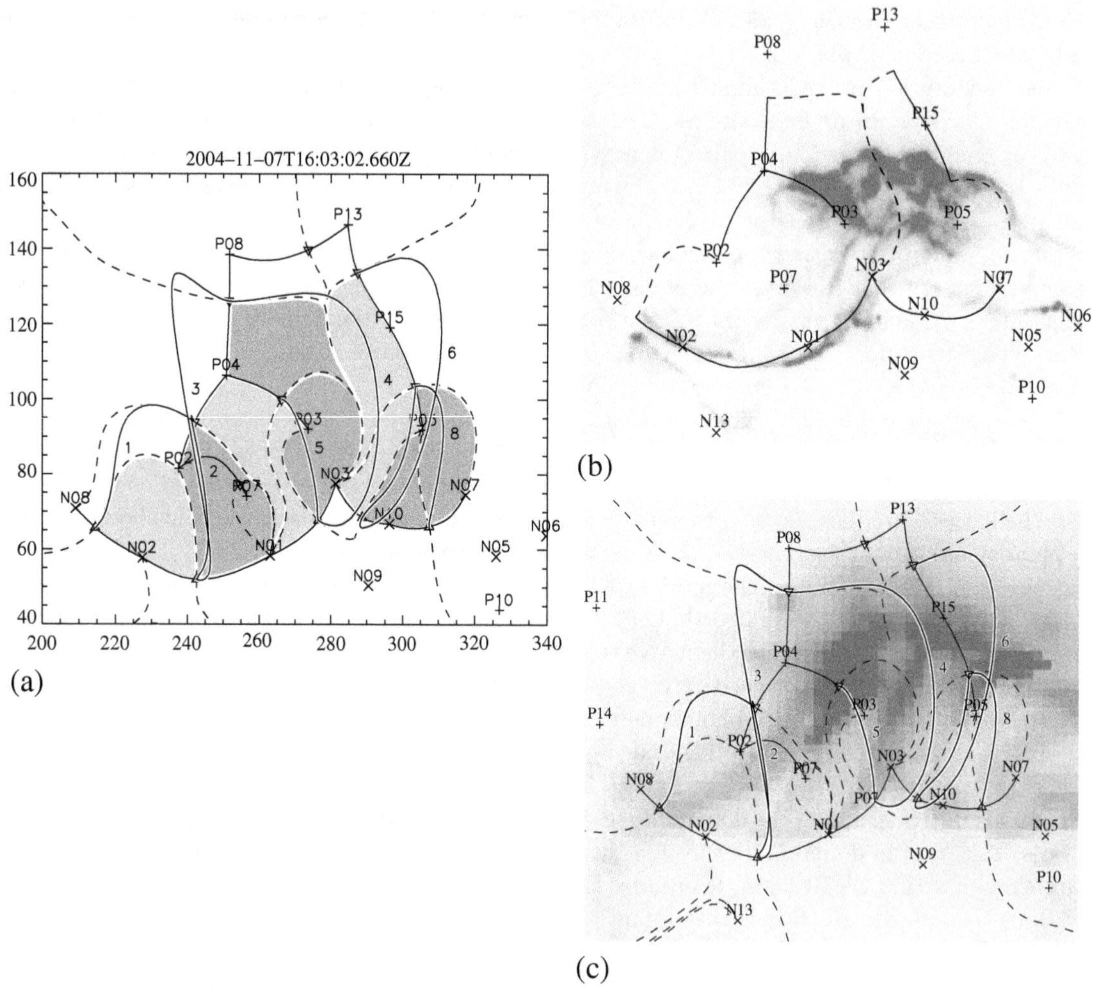

Figure 12.23. (a) The skeleton footprint of an active region, showing positive (+) and negative (×) sources, positive (∇) and negative (Δ) null points, spine curves (solid) and footprints of fan surfaces (dashed), together with separators (thick curves, numbered 1 to 8) and domains that gain (dark) or lose (light) flux. (b) and (c) show 1,600 and 195 Å images from SoHO/EIT superposed on the footprint (after Longcope et al. 2007).

Reconnection transfers flux out of domains such as P02–N02 and into P01–N01, with the field becoming more potential and releasing stored energy. Some domains are adjacent to several separators and so can be donors or recipients of flux. Reconnection in the model therefore occurs in a specific sequence, starting with separators having only one direction of flux transfer and enhancing domains with a deficit.

Such multiple reconnection creates a twisted flux rope by converting helicity associated with separator currents into self helicity and adding half a turn each time. In particular, flux passes from domains 5/3 and 15/10 across separator 7 to 15/3 and then across separator 6 to 13/3 and across 11 to a long overlying domain 8/1. The resulting flux rope joins P08 to N01 and possesses flux 10^{21} Mx with about two turns and ejects a magnetic helicity of 3×10^{42} Mx^2. These are smaller than the values measured in the magnetic cloud later observed in interplanetary space.

By modelling the complex topology of an active region, Longcope thus explains how reconnection spreads through the whole region sequentially at a series of separators. Titov et al. (2012) have shown in a similar way how a series of coronal eruptions (erupting prominences and CMEs) one after another can be explained as a sequence of separator reconnections (Sec. 6.13.3). In his case, each of the erupting prominences was located below a *pseudo-streamer* (Secs. 1.6.1 and 13.9.3), and the separators formed a chain, with separator reconnection at one end triggering the sequence of eruptions.

12.3.2.2 3D Fields with Quasi-separators

The Meudon group of Démoulin and co-workers have made a huge contribution to the MHD of solar flares. In early studies, they found examples of flare ribbons located along separatrices (Mandrini et al. 1991; Démoulin et al. 1993; van Driel-Gesztelyi et al. 1994), while Démoulin et al. (1994) found that coronal null points are present in half of the flares they studied.

Later, Priest and Démoulin (1995) put forward the idea of *quasi-separatrix layers* (or QSLs), which are the remnants of separatrices when the mapping gradients of field-line footpoints are not singular but have steep gradients (Secs. 6.9.5 and 6.13.4), best identified by the *squashing factor,* Q (Eq. 6.51). QSLs in practice behave very similarly to separatrices, in that current sheets tend to form at them, and so 3D reconnection tends to take place (Démoulin et al. 1996; Titov et al. 2003; Aulanier et al. 2005). A method was developed by Démoulin et al. (1996) and applied to flaring active regions by Mandrini et al. (1997) and Démoulin et al. (1997). The basic theory of QSLs was refined by Titov et al. (2002), Titov (2007) and Titov et al. (2009).

A common configuration for a flaring active region is a quadrupolar region with one bipole inside another and inclined to it (e.g., Sweet 1958b; Baum and Bratenahl 1980; Gorbachev and Somov 1988). This was first modelled as the field due to four point sources and produced two separatrix domes intersecting in a separator. Later, a continuous field distribution was assumed on the solar surface produced by point sources below the surface, for which Démoulin et al. (1996) calculated the QSL locations for different source depths (Figure 12.24). When the point sources are a distance 0.1 below the solar surface (in units of the distance between the outer sources), the flux contours on the solar surface are fairly diffuse (Figure 12.24a) and the intersections of the QSLs with the photosphere are fairly short. When the bipole depth is only 0.05 (Figure 12.24b), the sources on the photosphere are more concentrated and the QSLs more extensive. When the point surfaces are located on the solar surface (Figure 12.24c), the QSLs are even more extensive and form closed curves on the photosphere. In fact, they become separatrices, since two null points (N_1 and N_2, which had been present previously below the photosphere) appear on the solar surface and the mapping gradients become singular. The fans (F_1 and F_2) (Sec. 6.9) of these nulls extend outwards and arch over to form the separatrix surfaces in the corona. They end at the solar surface in spine curves, so that the photospheric intersections of the fan surfaces F_1 and F_2 are the spines S_2 and S_1, respectively.

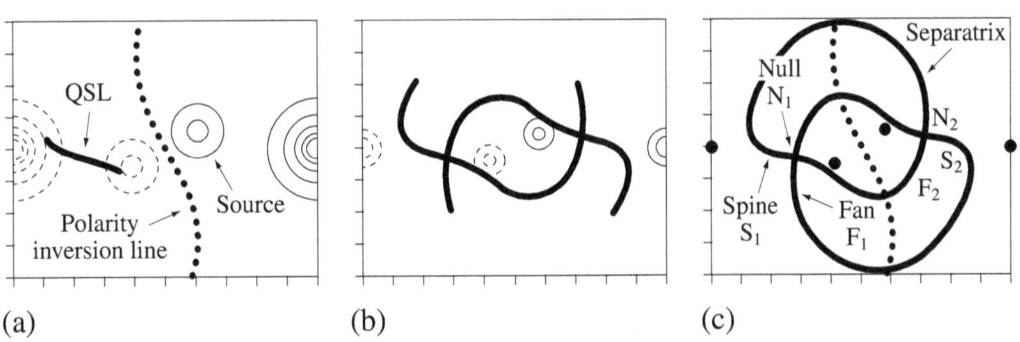

Figure 12.24. The relation between QSLs and separatrices for a quadrupolar field in the photospheric plane, showing flux contours for sources of one polarity (solid) and another (dashed), the polarity inversion line (dotted) and the intersections of the QSLs with the photosphere (thick solid curves) when the four flux sources are (a) diffuse, (b) more concentrated and (c) point-sources (after Démoulin et al. 1996).

The relationship between QSLs and separatrices has been clarified (Restante et al. 2009). Generally, a continuous flux distribution on the solar surface produces null points in the corona and photosphere (and so some separatrices), but also QSLs. However, if you replace the continuous photospheric sources by discrete or point sources, extra null points appear in the photosphere. So, how are the resulting extra coronal separatrices related to the original QSLs? The answer is that the strongest parts of the QSLs are located above the null points that would exist below the photosphere (or would exist if the flux were produced by discrete sources below the photosphere), and the QSLs spread out in a direction parallel to the spines. This is because it is near the nulls, and especially near the spines, that the field lines are separating fastest and so give rise to QSLs. Thus, a QSL starts out near one polarity as an elongated strip over the remnant spine and, as it arches into the corona, it is transformed first into two elongated strips forming an X-shape and then into another elongated strip over the remnant spine at the other polarity.

Reconnection occurs at separators and quasi-separators in a similar way, especially when a QSL thickness is smaller than the dissipation width. Thus, if you want to determine reconnection locations, it often makes little difference whether you model the observed photospheric flux distribution by a continuous flux distribution and find both the separators and quasi-separators or by a series of discrete sources and find the separators. However, in the first case the threshold for defining a QSL affects how far it extends along the remnant separatrix and in the second case the number of discrete sources chosen affects the number of separators.

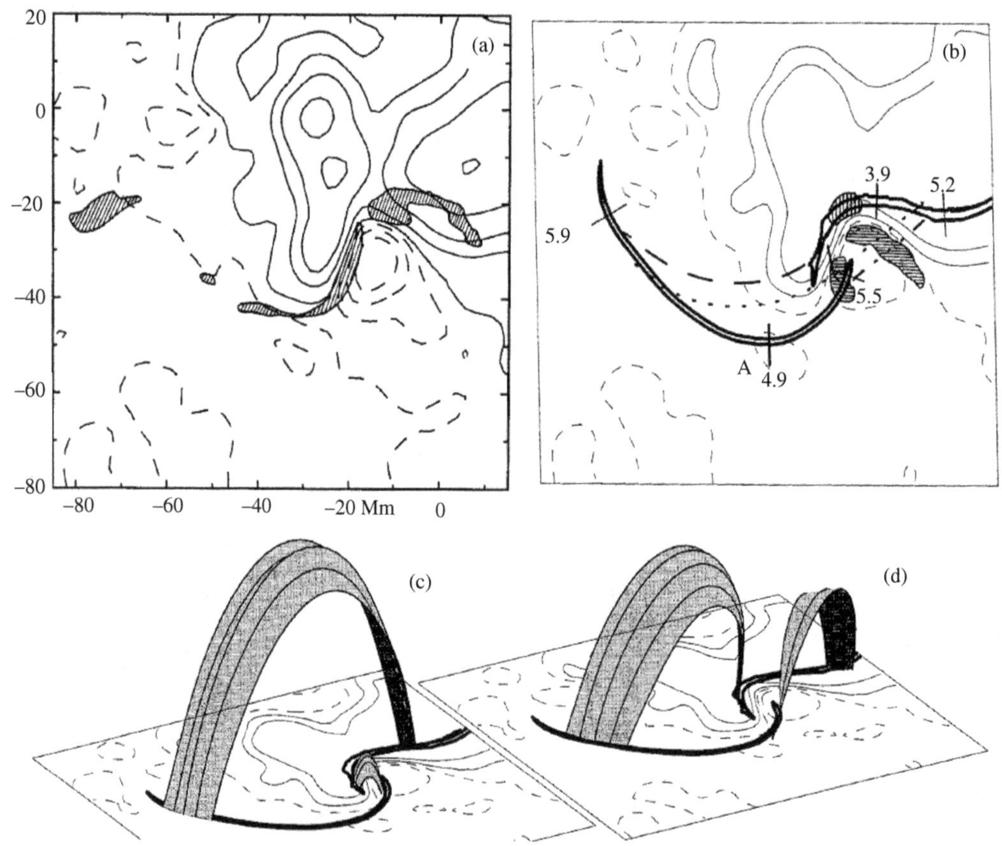

Figure 12.25. For a confined flare: (a) Hα ribbons and contours of longitudinal magnetic field, (b) intersections of QSLs with the photosphere for a linear force-free extrapolation of the longitudinal field, (c) perspective view of magnetic field lines drawn from one side of the QSLs and (d) from the other side (from Démoulin et al. 1997).

Démoulin et al. (1997) considered a typical confined solar flare in an active region with strong magnetic shear along an S-shaped polarity inversion line. They found from a linear force-free extrapolation that the Hα kernels are situated on small regions of the QSLs where the field exceeds 100 G and the kernels are joined by magnetic field lines (Figure 12.25a,b). They also showed how field lines whose footpoints are located a small distance apart on the two sides of a QSL have their opposite feet widely separated, as expected for a QSL.

Quasi-separator reconnection is called *slip-running reconnection* by Aulanier et al. (2006) and Pariat et al. (2006). This refers to the *magnetic flipping* process, which is a common feature of 3D reconnection (discovered by Priest and Forbes 1992b) and which emphasises that, when the field lines are followed from one side of the diffusion region and the QSL is thin enough, their flux velocity on the other side can be super-Alfvénic as they flip through the plasma (see Sec. 6.12.4, Priest et al. 2003; Pontin et al. 2005). They suggested that such rapid changes of field-line connection along a QSL may account for the rapid motion of hard X-ray sources along flare ribbons and of the locations of soft X-ray loops (Aulanier et al. 2007).

12.3.2.3 Effects of Patchy Reconnection

Many observational aspects of flares (such as the appearance of UV ribbons as small kernels, Fletcher et al. 2004) suggest the reconnection is impulsive in time and patchy in nature, and so Longcope and colleagues have explored the process of patchy reconnection, assuming that it is initiated by a small region of enhanced resistivity in a current sheet formed below an erupting prominence.

Linton and Longcope (2006) were aware that a localised transient patch of reconnection in a 2D current sheet (Sec. 6.6.2) produces two tear-shaped plasmoids, enveloped in slow-mode shocks that propagate away from the reconnection site (Semenov et al. 1983) (Figure 12.26a). They therefore suggested that a 3D version would provide a possible explanation for a Super-Arcade Downflow (Sec. 1.9.2.3). They analysed the 3D response to a transient sphere of enhanced resistivity (Figure 12.26b) by using a thin flux-tube model (Spruit 1981) and a full numerical experiment (see also Ugai 2009). Reconnection creates two bent flux tubes, one arching down (Figure 12.26c) and one up. In the post-reconnection evolution of the lower

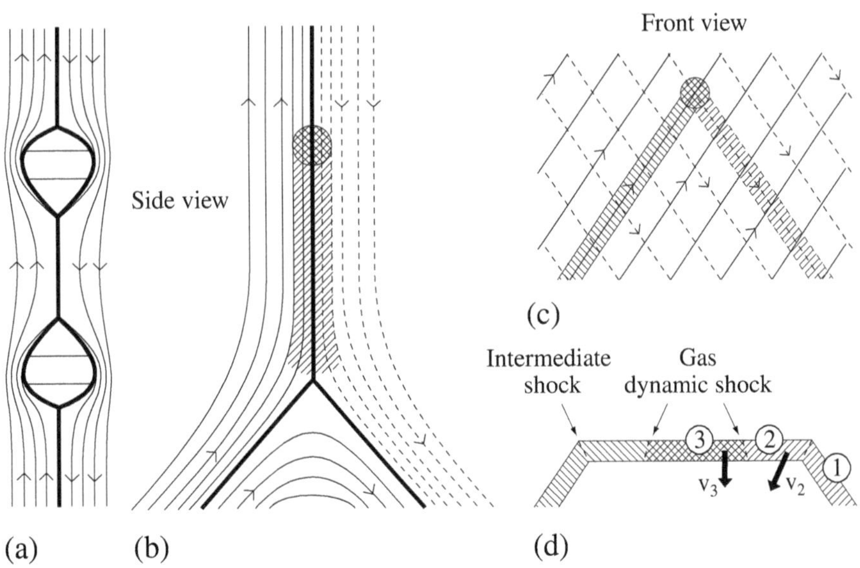

Figure 12.26. (a) A burst of reconnection in a 2D current sheet produces two tear-shaped plasmoids enveloped by slow-mode shocks (thick curves). (b) The basic geometry for a flare current sheet with reconnection initiated by a sphere of enhanced resistivity. (c) A Λ-shaped flux tube (shaded) at its moment of creation. (d) Retraction of the tube due to propagation of intermediate and gas-dynamic shocks (after Longcope et al. 2009).

flux tube, a pair of slow-mode shocks propagates along the tube creating bends that accelerate the tube downwards (as generalisations of Petschek shocks). The retraction speed is the Alfvén speed in thin flux-tube theory and less than this in the numerical experiment due to an extra drag from the underlying field and plasma.

Later, Longcope et al. (2009) analysed the dynamics of plasma along each retracting tube, which shortens at the Alfvén speed and so drives flows along the tube at the Alfvén speed. This is highly supersonic when $\beta \ll 1$, so that the flows collide in a pair of strong gas-dynamic shocks (Figure 12.26d). In other words, the slow-mode shocks split into a sum of intermediate and gas-dynamic shocks, with a two-step energy conversion from magnetic to kinetic and then to thermal.

Super-hot loop-top sources are present in some large flares as compact hard X-ray sources with thermal electron populations > 30 MK and very high densities. They typically appear after the impulsive phase and persist for 10 minutes or more, which is longer than the conductive cooling time. Also, they are independent of a second thermal population at about 20 MK with much higher emission measure, which closely matches the soft X-ray light curve and is probably caused by chromospheric evaporation. Longcope et al. (2010) and Longcope and Guidoni (2011) analysed the thermal effects of patchy reconnection in more detail and suggested that the super-hot sources may be produced by slow-mode shocks generated by an ensemble of flux tubes retracting from a series of sites. Steady shocks of this type can enhance the density for skewed fields by a factor of at most 4, but the transient effects of heat conduction cool the post-shock plasma and increase the density enhancement by 10 to 100, way beyond the normal Rankine-Hugoniot value. Density enhancements of 100 occur only for strong fields ($\geqslant 500$ G) and high shear ($\geqslant 120$ degrees). For one event they produced temperatures of 40 MK with 30 plugs of separate super-hot plasma in different flux tubes, each lasting for about 8 seconds.

The energy release rate by patchy reconnection has been measured in two ways. First of all, Longcope et al. (2010) used the ribbon motion to estimate the reconnection flux [Eq. (12.32)]. Secondly, Qiu et al. (2012) used spatially resolved UV light curves from SDO to construct the heating functions for a few thousand individual flare loops. The light curve of an individual UV pixel reflects the energy release in the thin loop that has the pixel as a footpoint. It typically shows a rapid increase for a few minutes (reflecting the duration and magnitude of impulsive energy release) followed by a slower decline (caused by evaporation and cooling). An EBTEL model (Sec. 10.2.4.2) is solved for the coronal response in each loop to the observationally determined heating function, and the resulting coronal radiation compares well with observed soft X-ray and EUV fluxes from GOES and SDO/AIA.

12.4 Concluding Comment

The two major MHD components of flares and CMEs, namely, the eruption and the reconnection still possess puzzling features. As far as the eruption is concerned, a magnetic flux rope is initially in equilibrium under a balance between magnetic pressure and magnetic tension forces, and so in principle an eruption could occur if the magnetic pressure were increased from below or if the overlying tension force were decreased by reconnection from below, or from above or during flux emergence. This could lead to either an unstable or a non-equilibrium state, but which of these possibilities is occurring in a particular flare has not been determined. Again, as far as the reconnection is concerned, the nature of 3D reconnection is still in its infancy. Thus, much remains to be learned, both from well-designed computational experiments – that is, ones designed to answer physical questions – and also from sophisticated observations that are likewise planned to understand better the nature of the eruption and the 3D reconnection process.

Chapter 13

The Solar Wind

13.1 Introduction

Variations in the Earth's magnetic field were first observed in the nineteenth century as a sudden increase (by about 10^{-3} G) followed by a slow decrease. These *geomagnetic storms* were sometimes found to occur one or two days after large solar flares (Carrington 1859), and further evidence for a link between the two phenomena came with the discovery of an 11-year periodicity in both flare and geomagnetic activity.

In the early 1900s Kristian Birkeland performed experiments to model an aurora produced by charged particles that he thought are ejected continuously from the Sun, but this idea was ignored by the scientific community. Then, Sydney Chapman (1929) suggested that geomagnetic storms are caused by streams of particles occasionally ejected from solar flares, and travelling at 1,000 km s^{-1} through the vacuum that was presumed to exist between the Sun and the Earth. Later, in order to explain the pointing of comet tails away from the Sun, Ludwig Biermann (1951) proposed, in agreement with Birkeland, that the Sun is emitting 'solar corpuscles' continuously rather than sporadically.

The first *theory* for an extended solar corona was by Chapman and Zirin (1957), who considered a static, spherically symmetric atmosphere and deduced that the corona extends to the Earth and beyond (Sec. 13.2). However, the difficulty with the model is that at large distances the pressure far exceeds any reasonable inter-stellar pressure and the density becomes indefinitely large.

Gene Parker (1958) resolved the inconsistency in a classic paper by suggesting that the corona is continuously expanding outwards. In other words, in the absence of a strong pressure at infinity to hold in the corona, it must stream outwards as the 'solar wind'. In his customary elegant manner, Parker modelled a steady expansion and found solutions for which the pressure tends to zero at large distances. However, the ideas were controversial and not widely accepted. The controversy was partly resolved in 1959, when a flux of ions in some unknown direction was directly observed by the satellite Lunik II. It was completely resolved in 1962, when the properties of the solar wind began to be studied in detail by the Venus probe, Mariner II.

The main observed properties of the solar wind at 1 AU (Figure 1.26) are as follows:

- The outflow is continuous, variable (200 to 900 km s^{-1}) and supersonic, with average speed 400 to 500 km s^{-1}; it consists of *fast solar wind* (700 to 750 km s^{-1}) and *slow solar wind* (300 to 400 km s^{-1});
- At solar minimum, *fast solar wind* of low density is emitted from polar coronal holes, while *slow solar wind* of high density originates from magnetically closed regions near the equator in a way that is not understood (Figure 1.27);
- At solar maximum, the solar wind is much more irregular and time-dependent, with narrow regions of fast and slow solar wind at all latitudes (Figure 1.27);

- The proton temperature is higher in the fast solar wind, while the electron temperature remains roughly constant; the mass flux is, surprisingly, the same (10^{12} m^{-2} s^{-1}) to within a factor of two for all wind conditions;
- The plasma consists mainly of electrons, protons and α-particles (1 to 5 per cent);
- The magnetic field forms a spiral pattern due to solar rotation; and
- The solar wind is highly time-dependent, with magnetic fluctuations due to outwardly propagating Alfvénic waves, tangential and rotational discontinuities and magnetoacoustic shocks, generated by coronal mass ejections or complex interactions between streams of different speeds.

For the purposes of numerical estimates in this chapter, we shall adopt the following values: $\tilde{\mu} = 0.6$, $G = 6.67 \times 10^{-11}$ m^{3}s^{-2}kg^{-1}, $M_\odot = 1.99 \times 10^{30}$ kg, $R_\odot = 6.96 \times 10^8$ m. These in turn give isothermal and adiabatic sound speeds [$v_c = (p/\rho)^{1/2}$ and $c_s = (\gamma p/\rho)^{1/2}$] of 120 km s^{-1} and 150 km s^{-1}, respectively, for $T = 1$ MK, but 170 km s^{-1} and 210 km s^{-1} for $T = 2$ MK. By comparison, the escape velocity [$v_{esc} = (2GM_\odot/R_\odot)^{1/2}$] is 620 km s^{-1}. In particular, the expansion of the corona is at first quite astonishing in view of the escape speed being considerably higher than the sound speed.

Observations of the solar wind are outlined in Section 1.6.4, and more details can be found in books by Parker (1963a), Hundhausen (1972), Zirker (1976), Schwenn and Marsch (1990, 1991), Tu and Marsch (1995) and Meyer-Vernet (2007), as well as living review articles on kinetic models (Marsch 2006) and coronal holes (Cranmer 2009). Here the aim is to describe the single-fluid theory of a steady, spherically symmetric fluid expansion (Secs. 13.3 to 13.5), effects of rotation and magnetic field (Sec. 13.6), MHD models of fast and slow wind by wave-turbulence and reconnection (Secs. 13.7 and 13.8), two-dimensional and three-dimensional models for streamers and coronal holes (Sec. 13.9) and large- and small-scale time-dependence (Sec. 13.10). Finally, Section 13.11 discusses limitations of single-fluid theory and briefly outlines two-fluid and kinetic effects.

13.2 Chapman's (1957) Static Corona

Chapman considered a steady, spherically symmetric, static atmosphere with pressure [$p(r)$], density [$\rho(r)$] and temperature [$T(r)$] depending on r alone. Assuming an inverse-square law fall-off of gravity ($g = GM_\odot/r^2$), where $M_\odot$ is the Sun's mass, the equation of hydrostatic equilibrium is

$$\frac{dp}{dr} = -\rho g = -\frac{GM_\odot \rho}{r^2}.$$

The density is here given by the gas law [$\rho = \tilde{\mu} p/(\tilde{R}T)$], where $\tilde{\mu}$ is the mean molecular weight, and so this equation may be integrated to give the pressure as

$$p(r) = p_0 \exp\left(-\int_{r_0}^{r} \frac{2r_c T_0}{r^2 T(r)} dr\right), \tag{13.1}$$

for any form of $T(r)$, where $p(r_0) = p_0$, $T(r_0) = T_0$ and $r_c = GM_\odot \tilde{\mu}/(2\tilde{R}T_0)$.

The simplest model is for an isothermal corona, for which the pressure is

$$p = p_0 \exp[-2r_c(1/r_0 - 1/r)], \tag{13.2}$$

(PROBLEM 13.1: for all problems and solutions, see web page www.cambridge.org/9780521854719), but an alternative is to assume energy transfer by conduction alone, for which the heat flux across any sphere of radius r is constant, so that

$$4\pi r^2 \kappa \frac{dT}{dr} = constant,$$

where $4\pi r^2$ is the surface area, $\kappa dT/dr$ is the heat-flux density, and the coefficient of thermal conduction is $\kappa = \kappa_0 T^{5/2}$. This equation may be integrated to give

$$T = T_0 \left(\frac{r_0}{r}\right)^{2/7}, \tag{13.3}$$

after imposing boundary conditions that $T = T_0$ at $r = r_0$ (the coronal base) and that $T \to 0$ as $r \to \infty$. For $T_0 = 10^6$ K, Eq. (13.3) implies a temperature at 1 AU of about 10^5 K. This suggests that thermal conduction is important in the corona and that the corona extends far from the Sun, so that the Earth is enveloped by an extremely hot plasma. If the heat is assumed to be deposited at $r = r_0$, the temperature profile by conduction below that level may also be deduced (PROBLEM 13.2).

Furthermore, substituting for $T(r)$ from Eq. (13.3) in Eq. (13.1) gives

$$p = p_0 \exp\left\{\frac{14 r_c}{5 r_0}\left[\left(\frac{r_0}{r}\right)^{5/7} - 1\right]\right\}.$$

As $r \to \infty$, so p tends to a constant, namely, $p_\infty \equiv p_0 \exp[-14 r_c/(5 r_0)]$. If we adopt, e.g., $r_0 = 1.2 R_\odot$ and $p_0 = 10^{-3}$ Pa (where a pascal (Pa) is the mks unit of pressure), then we find $p_\infty = 10^{-8}$ Pa, which is 7 orders of magnitude larger than the pressure (10^{-15} Pa) of the inter-stellar medium. At the same time, as $r \to \infty$, so $T \to 0$ and $\rho \sim p/T \to \infty$, which is even more damning for the theory.

13.3 Parker's Isothermal Solution

The equations of mass continuity and motion for the steady, spherically symmetric flow of an isothermal plasma (having density ρ and speed v) are

$$4\pi r^2 \rho v = constant, \tag{13.4}$$

$$\rho v \frac{dv}{dr} = -\frac{dp}{dr} - \frac{GM_\odot \rho}{r^2}, \tag{13.5}$$

where $p = \tilde{R}\rho T/\tilde{\mu}$. Thus, the outflowing plasma is assumed to be acted on by a pressure gradient (pushing it away from the Sun) and the force of gravity.

Since T is uniform, ρ may be eliminated between Eqs. (13.4) and (13.5) to give

$$\left(v - \frac{v_c^2}{v}\right)\frac{dv}{dr} = \frac{2v_c^2}{r^2}(r - r_c), \tag{13.6}$$

where $v_c = (\tilde{R}T/\tilde{\mu})^{1/2}$ is the *isothermal sound speed*, which is somewhat less than the (adiabatic) sound speed, and $r_c = GM_\odot/(2v_c^2) = GM_\odot\tilde{\mu}/(2\tilde{R}T)$. Eq. (13.6) possesses a *critical point* (A in Figure 13.1), where dv/dr is undefined. It occurs when $v = v_c$, $r = r_c$ so that both the coefficient of dv/dr and the right-hand side of the equation vanish.

Equation (13.6) is simple enough to be integrated analytically to give $v(r)$ implicitly from

$$\left(\frac{v}{v_c}\right)^2 - \log_e\left(\frac{v}{v_c}\right)^2 = 4\log_e\frac{r}{r_c} + \frac{4r_c}{r} + C, \tag{13.7}$$

where C is part of the constant of integration and the other part has been incorporated into the log terms for convenience. The solutions for different values of C have the form shown schematically in Figure 13.1, with dv/dr becoming infinite at $v = v_c$ and vanishing at $r = r_c$. The critical point A is a saddle point.

Several types of solution are present in Figure 13.1, depending on the value of C. Types I and II are unacceptable, since they are double-valued and do not connect the solar surface ($r \ll r_c$) to locations far from the Sun ($r \gg r_c$). Type III possesses supersonic speeds ($v > v_c$) at the Sun, which are not observed. Type V solutions (called *breezes*) are everywhere subsonic and have the pressure approaching a constant value at large distances. Thus, the only allowable solution is of Type IV.

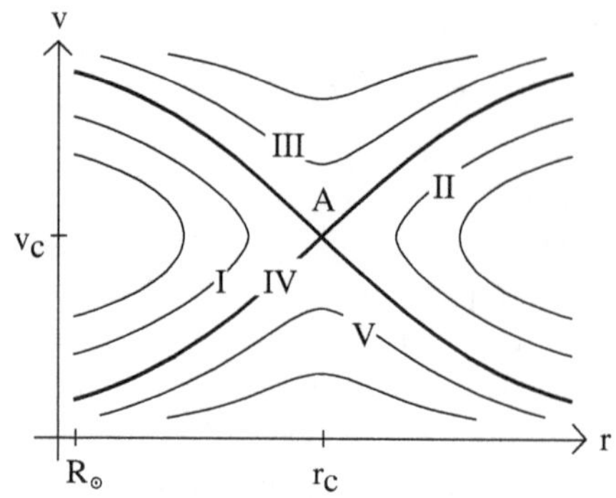

Figure 13.1. A sketch of Parker's isothermal solutions, showing the different classes (I, II, III, IV and V). Type IV is the solar-wind solution, which passes through the critical point (A), where $v = v_c$ and $r = r_c$. Type V represents subsonic solar breeze solutions.

For the solar wind several other properties can be established (PROBLEM 13.3), namely, that, in the low corona, it has a flow speed of 1 to 10 km s^{-1}, depending on the coronal temperature; below the critical point, the density variation is very similar to a static atmosphere; and the mass-loss rate is of the order of $10^{-14} M_\odot$ per year.

The *solar-wind* solution (type IV) starts out subsonically ($v < v_c$) when $r < r_c$ close to the Sun; it passes through the sonic critical point (A) when $(v, r) = (v_c, r_c)$ and becomes supersonic ($v > v_c$) when $r > r_c$. It corresponds to the value $C = -3$, obtained by putting $(v, r) = (v_c, r_c)$ in Eq. (13.7). At large distances (where $v \gg v_c$), the velocity behaves like $v \sim 2v_c(\log_e r/r_c)^{1/2}$, for which the first and third terms are dominant in Eq. (13.7), while the density falls off like $\rho \sim r^{-2}(\log_e r)^{-1/2}$, so that $p \to 0$ as $r \to \infty$, as required. When these solutions were proposed, it was unclear which was the most relevant to the Sun, but the question was resolved when Mariner II found speeds of several 100 km s^{-1} near the Earth and so confirmed the solar-wind solution as the appropriate one.

The solar-wind solution can be written explicitly in terms of the *Lambert W function* $W(x)$ (Cranmer 2004). This is the solution of $We^W = x$, which has two branches ($W_+ > -1$ and $W_- < -1$) when $-1/e < x < 0$. The solution is

$$v^2 = -v_c^2 W[-D(r)], \qquad \text{with} \qquad D(r) = \left(\frac{r_c}{r}\right)^4 \exp\left(3 - \frac{4r_c}{r}\right),$$

where $W = W_+$ when $r \leq r_c$ and $W = W_-$ when $r \geq r_c$

Let us estimate solar-wind values for the Sun. For $T = 10^6$ K, the isothermal sound speed is $v_c =$ 120 km s^{-1} and the critical-point location is $r_c = 7\ R_\odot$, while the predicted flow speed at 1 AU is about 300 km s^{-1} compared with the observed fast-wind speed of 750 km s^{-1}. If instead $T = 1.5 \times 10^6$ K, then $v_c = 140$ km s^{-1} and the critical-point location is r_c=4.5 $R_\odot$, while the flow speed at 1 AU is about 550 km s^{-1}. For a coronal density ρ_0 of, say, $10^{14} m_p$ kg m^{-3}, this gives a mass-loss rate ($4\pi r_0^2 \rho_0 v_0$) of 1.6×10^9 kg s^{-1}. Remarkably for such a simple model, these values are within 50 per cent of the observed values.

The solar-wind solution is analogous to the flow in a *de Laval nozzle*, which accelerates subsonically as the nozzle cross-section decreases, up to the sound speed at the throat. Beyond the throat, the flow continues to accelerate supersonically as the cross-section increases. The solar-wind solution starts out in the low corona with a flow speed much smaller than the sound speed of typically 200 km s^{-1}. Equation (13.6) shows that, when $v < v_c$ and $r < r_c$, the flow accelerates subsonically ($dv/dr > 0$), as if in a converging nozzle, up to the sonic point $v = v_c$ at $r = r_c$. Beyond this point, Equation (13.6) implies that, when $v > v_c$ and $r > r_c$, the flow continues to accelerate supersonically, as if in a diverging nozzle.

The effect on the model of increasing the temperature is to enhance the isothermal sound speed (v_c) and to lower the critical-point radius (r_c) until, when the temperature exceeds 4×10^6 K, r_c falls below the solar surface. Thus, when the temperature is too large, the critical point lies below the coronal base and there is no wind solution (PROBLEM 13.4).

The main difficulty with the isothermal solar-wind solution is that the flow speed is infinite at infinity and requires an infinite source of energy. Also, the predicted density at the Earth is too high, so that the mass flux is too high. More general, spherically symmetric models (Sec. 13.5.2) use an energy equation to determine the temperature; they give a lower mass flux and values for the flow speed (v_0) at the Sun of 1 to 2 km s^{-1}, whereas models for coronal holes (Sec. 13.6.3) whose base areas cover typically $\frac{1}{4}$ of the solar surface have base speeds of 10 to 20 km s^{-1}.

13.4 Breeze, Wind or Accretion?

Solutions of Type V (Figure 13.1) are everywhere subsonic and called *solar* (or *stellar*) *breezes.* At large distances where $v \ll v_c$, the second and third terms balance in Eq. (13.7), and so the flow speed falls off to zero like $v \sim r^{-2}$. At the same time, the density and pressure approach constant values, which may seem to imply these solutions are irrelevant for the Sun, except for the fact that, when a sufficiently rapid fall-off in temperature is incorporated, the pressure does go to zero (PROBLEM 13.5b). The predicted speed at the Earth's orbit is typically only 10 km s^{-1}.

Although the solar wind would now seem to be the obvious solution, there are important and subtle questions which, with great physical insight, Velli (2001, 2009) answered more than 50 years after the original Parker paper. Are the breeze solutions stable? For a given pressure at large distances, how does the Sun know whether to choose plasma ejection rather than accretion inflow, since the steady equation (13.6) is invariant under a change of v into $-v$? What would happen if the Sun entered a denser or thinner part of the Galaxy, so that the external pressure changed?

Note that there are other solutions of interest in Figure 13.1 in addition to those mentioned in Section 13.3. One possibility is a transonic accretion solution (Bondi 1952), starting out with small inflow at large distances and increasing through the critical point A to supersonic inflow at small distances from the Sun. Others are that a shock transition (a sudden jump in v) may join a transonic outflow (IV) at $r > r_c$ to a subsonic outflow (II) lying just above the critical solution (Figure 13.2b) or a transonic inflow at $r < r_c$ to a subsonic inflow (Figure 13.2d).

13.4.1 Stability of Breeze Solutions

On a breeze solution, the pressure (p_∞) at infinity may be determined from $p_\infty = p_0\rho_\infty/\rho_0$, $\rho_\infty v_\infty r^2 = \rho_0 v_0 r_0^2$ and Eq. (13.7) with $v_0 \ll v_c$ and $v_\infty \ll v_c$ as

$$p_\infty = p_0 \exp\left(-\frac{2r_c}{r_0} + \frac{v_0^2}{2v_c^2}\right).$$

Thus, the pressure at large distances is greater than that of a hydrostatic atmosphere [Eq. (13.2)], namely,

$$p_{hs} \equiv p_0 \exp(-2r_c/r_0),$$

by a factor $\exp[v_0^2/(2v_c^2)]$. The breeze solutions depend on the value of v_0. Their pressure at infinity lies between p_{hs} when $v_0 = 0$ and

$$p_{\infty c} \equiv p_{hs} \exp[v_{0c}^2/(2v_c^2)]$$

when $v_0 = v_{0c}$, corresponding to the fastest allowable breeze.

Physically, the breeze solutions are unstable in view of the surprising feature that p_∞ increases with v_∞, since decreasing p_∞ would increase the outward pressure gradient and accelerate the flow; this increases p_∞

and creates a mismatch, which acts in such a way as to reverse the flow, turning the breeze into an inflow. Mathematically, Velli (2001) linearised the equation of motion for acoustic perturbations around a steady state and introduced variables $y^{\pm} = \bar{M} \pm \bar{p}$ in terms of dimensionless Mach number ($\bar{M}$) and pressure ($\bar{p}$) fluctuations and distance ($\bar{r}$). For a time-dependence $y^{\pm} = y^{\pm}(\bar{r})\exp(-i\omega + \gamma)t$ the equations become

$$(M \pm 1)\frac{dy^{\pm}}{d\bar{r}} - i(\omega + i\gamma)y^{\pm} + \tfrac{1}{2}(y^{\pm} + y^{\mp})\frac{dM}{d\bar{r}}\frac{M \mp 1}{M} = 0.$$

In a stationary medium, the wave-energy flux is conserved, but in a medium with flow it is the *wave-action flux* that is conserved. Its evolution equation is

$$\frac{d}{d\bar{r}}\left[\frac{(M+1)^2}{M}|y^+|^2 - \frac{(M-1)^2}{M}|y^-|^2\right] + \frac{2\gamma}{M}\left[(M+1)|y^+|^2 - (M-1)|y^-|^2\right] = 0,$$

where the quantity in the first square brackets is proportional to the wave action. Integrating this equation between $\bar{r} = 1$ and infinity (and assuming $y^+ = y^-$ at both ends and $y^+ = 0$ at infinity) gives an expression for the growth-rate

$$\gamma = \frac{2|y_0^+|^2}{\int_1^{\infty} d\bar{r}[(M+1)|y^+|^2 - (M-1)|y^-|^2]/M},$$

where $|y_0^+|$ is the fluctuation amplitude at the base. Thus, for $0 < M < 1$, the denominator is positive and so $\gamma > 0$ and the breeze outflow solutions are unstable. However, the breeze inflow accretion solutions ($-1 < M < 0$) obtained by reversing the direction of v are stable.

13.4.2 Dependence of Solutions on p_∞ and History

Several types of solution are possible and stable, depending on the value of the pressure (p_∞) at infinity (Figure 13.2). First of all, when p_∞ vanishes, there is a classical Parker solar wind (Figure 13.2a). However, when $0 < p_\infty < p_{hs}$, a *shocked wind* is created, in which the solar-wind solution holds out to a location ($r > r_c$) where a terminal shock wave is set up; this produces a sudden decrease of speed to a subsonic value that continues along part of the downward branch of one of the double-valued solutions beyond r_c (Figure 13.2b).

When $p_{hs} < p_\infty < p_{\infty c}$, the shocked wind continues to exist (Figure 13.2b), but a second stable possibility also arises, namely, an *accretion breeze*, in which the flow direction for a normal breeze solution is reversed to give a subsonic inflow (Figure 13.2f).

When $p_\infty = p_{\infty c}$, there is a *critical breeze*, in which the flow follows the critical solution up to the critical point A, at which there is a jump in velocity derivative, and beyond which the flow speed decreases along the other branch of the critical solution (Figure 13.2c). However, there is also a second stable solution, namely, a *critical accretion*, in which the direction of the flow is reversed to give an inflow (Figure 13.2e).

Finally, when $p_\infty > p_{\infty c}$, the only stable possibility is a *shocked accretion*, in which the flow increases in speed as it comes in up to a location ($r < r_c$), where a shock wave suddenly decreases the speed, and thereafter the inflow continues along part of the subsonic branch of the double-valued solutions below r_c (Figure 13.2d).

Thus, in most ranges of p_∞ there is a unique stable solution, but when $p_{hs} < p_\infty \leqslant p_{\infty c}$ there are two stable possibilities. Which one occurs in practice depends on the history of the development of the flow. For example, suppose that initially there is a solar-wind solution with $p_\infty = 0$ (Figure 13.2a) and its value is gradually increased. First of all, a terminal shock appears and gives a shocked wind (Figure 13.2b). The location of the shock depends on the particular value of p_∞ and is roughly where the ram pressure (ρv^2) of the wind balances the interstellar pressure. As p_∞ increases in value, the shock location moves closer to the Sun, while the shock amplitude (and therefore the jump in v) decreases. Eventually, when $p_\infty = p_{\infty c}$, the shock has disappeared at the critical-point radius r_c to give a *critical breeze*, with the discontinuity in

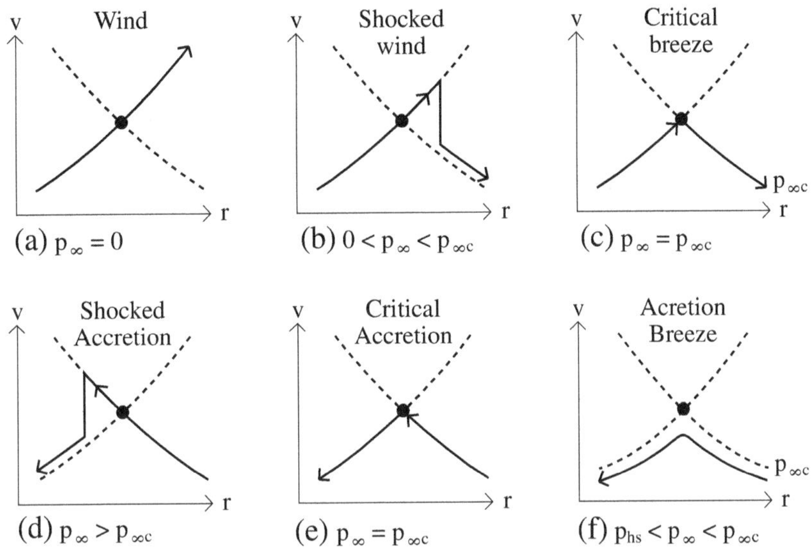

Figure 13.2. Types of solutions (winds, breezes or accretions) that are possible, depending on the pressure (p_∞) at large distances relative to the critical breeze pressure ($p_{\infty c}$) and the hydrostatic pressure (p_{hs}) at infinity.

speed being replaced by a jump in derivative at the critical point A (Figure 13.2c). Then, if $p_\infty > p_{\infty c}$, the critical breeze becomes an inward accretion flow with a shock that moves inwards as p_∞ increases still further (Figure 13.2d).

What happens if the change in external pressure is applied in reverse, starting with a shocked accretion (Figure 13.2d) and decreasing the value of p_∞? When p_∞ reaches $p_{\infty c}$, the nearby stable solution is a critical accretion (Figure 13.2e), and, as p_∞ reduces in value in the range $p_{hs} < p_\infty < p_{\infty c}$, the state passes through a series of stable accretion breezes (Figure 13.2f). When p_∞ becomes less than p_{hs}, however, the flow turns into the only possible solution, namely, a transonic shocked wind (Figure 13.2b).

Thus, in the pressure range $p_{hs} < p_\infty \leqslant p_{\infty c}$, where two stable states are possible, the state depends on its history (either a shocked wind or an accretion breeze). Indeed, the behaviour outlined here has been confirmed by numerical experiments.

13.5 More General Models for a Spherical Expansion

Parker's elegant analytical solution was important in delineating the qualitative features of solar-wind flow. It provides a classic example of how a physical idea may be modelled simply by incorporating only the dominant effects. Subsequently, as the reader may well imagine, the basic model was developed in numerous ways by incorporating many extra effects in an attempt to make the model more realistic. One of the technical difficulties in these often-numerical treatments is to find the solution which passes through the (sometimes-unknown) critical point (or points). We outline here some extensions for a steady, spherically expanding wind, and then discuss in Section 13.9 the departures from spherical symmetry.

13.5.1 Polytropic Solar Wind

The simplest way of allowing spatial variations in temperature is to replace the isothermal assumption by a polytropic energy equation

$$\frac{p}{\rho^\alpha} = K,$$

where K and $\alpha > 1$ are constants, such that $c_{s\alpha}^2(r) = \alpha p/\rho = \alpha\tilde{R}T(r)/\tilde{\mu}$ and the temperature $T = p\tilde{\mu}/(\tilde{R}\rho) = (K\tilde{\mu}/\tilde{R})\rho^{\alpha-1}$ decreases with r as the density decreases. Section 13.3 proved that an isothermal solution ($\alpha = 1$) implies an infinite energy addition. A value of α between 1 and the adiabatic value ($\alpha = \gamma = 5/3$) gives a finite energy injection, and the value $\alpha = 1.1$ works pretty well in reproducing some of the thermal properties of the solar wind, since it corresponds to the required energy addition from coronal heating.

The resulting differential equation for velocity [$v(r)$] is the same as Eq. (13.6), except that v_c is replaced by the polytropic sound speed $c_{s\alpha}(r)$, which varies with r. The equation of motion may be integrated in general to give a Bernoulli equation,

$$\tfrac{1}{2}v^2 + \int \frac{dp}{\rho} - \frac{GM_\odot}{r} = constant.$$

For polytropic flow, it becomes in place of Eq. (13.7) (see PROBLEM 13.5a)

$$\tfrac{1}{2}v^2(r) + \frac{c_{s\alpha}^2(r)}{\alpha - 1} - \frac{GM_\odot}{r} = constant, \tag{13.8}$$

representing kinetic energy, enthalpy and work done against gravity, such that enthalpy is transferred to kinetic energy as the corona expands. Evaluating this at the coronal base (r_0) (where $v \ll v_\infty$ and so the enthalpy exceeds gravitational energy) and at infinity (where $T \to 0$ and so the kinetic energy dominates) gives

$$\frac{\alpha\tilde{R}T_0/\tilde{\mu}}{(\alpha - 1)} - \frac{GM_\odot}{r_0} = \tfrac{1}{2}v_\infty^2.$$

Thus, a hot enough corona will have $v_\infty^2 > 0$ and so will expand, but an adiabatic corona possesses too little enthalpy to drive a solar wind (PROBLEM 13.6).

Several properties may be proved about the nature of the critical point and of the flow through it (PROBLEM 13.7): firstly, if $1 < \alpha < 5/3$, the critical point is a saddle point; secondly, if $1 < \alpha < 3/2$, there is a solution whose speed increases with r from subsonic to supersonic as it passes through the critical point; and thirdly, such a solar-wind solution has a pressure force that always dominates gravity.

13.5.2 Energy Equation

A better way of modelling spatial variations in temperature than a polytropic assumption is to adopt an integrated energy equation, such as

$$\rho v A\left(\tfrac{1}{2}v^2 + \frac{5p}{2\rho} - \frac{GM_\odot}{r}\right) = AQ + E_\infty, \tag{13.9}$$

where $Q = -\kappa dT/dr$ is the heat flux, $\kappa = \kappa_0 T^{5/2}$, $A = 4\pi r^2$ and E_∞ is the energy at infinity. The three terms on the left represent fluxes of kinetic energy, enthalpy and gravitational energy across a sphere of radius r, while the first term on the right is the conductive heat flux (with classical electron conduction). Its differential form [after substituting for $GM_\odot/r$ from Eq. (13.5)] is

$$\rho v\left(\frac{3\tilde{R}}{2\tilde{\mu}}\frac{dT}{dr} - \frac{\tilde{R}T}{\tilde{\mu}\rho}\frac{d\rho}{dr}\right) = \frac{1}{A}\frac{d}{dr}\left(A\kappa\frac{dT}{dr}\right),$$

or, after using $\rho v A = constant$ and $Q = -\kappa dT/dr$,

$$\frac{\tilde{R}}{\tilde{\mu}}\left[\frac{T^{5/2}}{\rho}\frac{d}{dr}\left(\frac{\rho}{T^{3/2}}\right)\right] = \frac{d}{dr}\left(\frac{Q}{\rho v}\right). \tag{13.10}$$

The solution curve topology is similar to Figure 13.1.

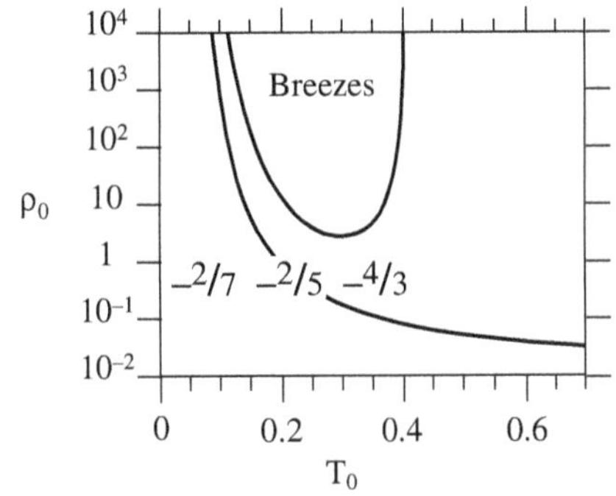

Figure 13.3. Dependence of asymptotic temperature behaviour on surface density (ρ_0) and temperature (T_0) in units of $\kappa_0(GM_\odot r_0 T_0^5)^{1/2}\tilde{R}^{-1}$ and $GM_\odot/(\tilde{R}r_0)$, respectively. Wind solutions with $T \sim r^{-2/7}$ are separated from those with $T \sim r^{-4/3}$ by a curve on which $T \sim r^{-2/5}$. The breeze solutions have zero total energy (after Roberts and Soward 1972).

Several solutions have been discovered that traverse the critical point and have a vanishing temperature at infinity. They possess the following asymptotic behaviour: $T \sim r^{-2/7}$ (Parker 1964), $T \sim r^{-2/5}$ (Whang and Chang 1965) and $T \sim r^{-4/3}$ (Durney 1971). The first solution arises when the conductive flux is constant and dominates Eq. (13.9) at large distances (as in Chapman's analysis, Sec. 13.2). It corresponds closely to Parker's isothermal solution, whereas the second solution occurs when the kinetic energy flux is the largest term. For the third type of solution conduction is negligible, and so the flow is essentially adiabatic. In other words, $p/\rho^{5/3}$ is constant while $\rho \sim r^{-2}$, and so the temperature behaves like $p/\rho \sim \rho^{2/3} \sim r^{-4/3}$.

In a paper relevant for *stellar* wind theory, Roberts and Soward (1972) showed how the type of solution depends on the surface values of density (ρ_0) and temperature (T_0). The Whang-Chang solution (Figure 13.3) is unlikely to occur in practice, since it separates conductively dominated solutions ($T \sim r^{-2/7}$) in cool, low-density winds from adiabatic solutions ($T \sim r^{-4/3}$) in hot, high-density winds. Since the solar wind is closer to conductive than adiabatic, the values of ρ_0 and T_0 for the Sun probably lie just below the Whang-Chang solution.

Let us now evaluate the energy equation (13.9) at the coronal base (r_0), where flow speed is negligible, and at large distances, where enthalpy, heat flux and gravitational energy are negligible, to give

$$\tfrac{1}{2}v_\infty^2 = \frac{5\tilde{R}T_0}{2\tilde{\mu}} - \frac{GM_\odot}{r_0} + \left[\frac{Q}{\rho v}\right]_0 .$$

If we adopt values of $T_0 = 10^6$ K, $r_0 = R_\odot$ and $v_\infty = 800$ km s^{-1}, the kinetic energy at infinity is 3×10^{11} J kg^{-1}, while on the right-hand side of the above equation the enthalpy is 0.3×10^{11} J kg^{-1} and the gravitational energy is -1.9×10^{11} J kg^{-1}, so that the last term (the heat flux per particle) is 4.6×10^{11} J kg^{-1}. However, for a mass-loss rate of $dM_0/dt = 10^9$ kg s^{-1} over the whole surface ($4\pi r^2$), the mass-loss rate per unit area is $\rho v_0 = 1.6\times10^{-10}$ kg m^{-2} s^{-1}. This implies that the heat flux required at the base of the corona to accelerate the fast solar wind is

$$Q_0 \geqslant 80 \text{ W m}^{-2}$$

if the expansion is radial; for the slow solar wind one finds the same heat flux.

However, if a realistic flux-tube expansion in excess of radial is taken into account, the required heating is 3000 W m^{-2} for the slow solar wind and 800 W m^{-2} for the fast solar wind (Cranmer 2012). Thermal conduction is crucial as a kind of 'thermostat' to keep temperatures of coronal loops within a narrow range around 1 to 3 MK, and it is equally important in regulating the solar wind (Owocki 2004). Indeed, if the model is started low down in the atmosphere where the conductive flux vanishes, conduction can be viewed not as a source but as a means of transferring heat from one location to another.

Where does such a fast-wind heat flux come from? A heat flux from collisions would be, for a temperature $T_e \sim r^{-2/7}$ and values $T_0 = 10^6$ K, $r_0 = R_\odot$, $\kappa_0 = 10^{-11}$,

$$Q = -\kappa_0 T_e^{5/2} \frac{dT_e}{dr} \approx \frac{2\kappa_0 T_e^{7/2}}{7r_0} \approx 4 \text{ W m}^{-2}.$$

The slow solar wind (slower velocity but higher mass-loss rate) has a similar required heat flux. Thus, a collisional heat flux is far too small to power the solar wind.

One way out of this conundrum is to recognise that the corona is not sufficiently collisional for the above heat-flux estimate to be correct and so to go beyond a single-fluid picture (Sec. 13.11). Another is to seek extra momentum and energy sources – indeed, the required energy is similar to what is needed to heat the corona outside active regions. If an extra outward force is added to the right-hand side of Eq. (13.5) below the critical point (r_c), it moves r_c inwards and so increases the density at r_c and therefore the mass-loss rate. If an extra source of heat energy is added to the right-hand side of Eq. (13.9) in the subsonic region, it increases the solar-wind mass-flux but does not change v_e, whereas addition in the supersonic region does not affect the mass flux but it can significantly increase v_e.

13.6 Effect of Rotation and a Magnetic Field

If rotation and/or a magnetic field are introduced, the assumption of spherical symmetry, namely, that all variables are functions of r alone in spherical polar coordinates (r,θ,ϕ), is broken. This is because, unlike the force of gravity, the centrifugal and Lorentz forces are no longer central forces – that is, dependent on r alone. Nevertheless, simple analytical progress can be made by restricting our attention to the equatorial plane ($\theta = \pi/2$) and assuming azimuthal symmetry ($\partial/\partial\phi = 0$), so that we may set $\partial/\partial\theta = 0$ and $v_\theta = B_\theta = 0$ and consider only the flow components $[v_r(r), v_\phi(r)]$ and magnetic components $[B_r(r), B_\phi(r)]$ in the equatorial plane as functions of r. We first consider the role of rotation alone (Sec. 13.6.1) and explore the effect of a rotating magnetic field on the wind (Sec. 13.6.2); then we discuss how the magnetic field can produce expansion in a coronal hole that is faster than the radial rate of spherically symmetric expansion (Sec. 13.6.3).

13.6.1 Rotating Wind

Consider a steady-state, thermally driven wind, rotating with angular speed $\Omega_\odot$, but possessing no magnetic field. Its flow components $[v_r(r), v_\phi(r)]$ in the equatorial plane have the following properties (derived in PROBLEM 13.8).

The azimuthal component of the equation of motion is

$$v_r \frac{dv_\phi}{dr} + \frac{v_r v_\phi}{r} = 0, \tag{13.11}$$

which may be integrated to give $r v_\phi = L$ or

$$v_\phi = \frac{L}{r} = \frac{\Omega_\odot r_0^2}{r},$$

where L is a constant, namely, the angular momentum per unit mass.

Then, for an isothermal plasma with $p = v_c^2 \rho$, where $v_c^2 = \tilde{R}T/\tilde{\mu}$, the density may be eliminated between the equations of mass conservation and radial momentum to give in place of Eq. (13.6)

$$\left(v_r - \frac{v_c^2}{v_r}\right)\frac{dv_r}{dr} = \frac{2v_c^2}{r^3}\left(r^2 - r_{c0} r + \tfrac{1}{2}\tau^2 r_0^2\right),$$

where $r_{c0} = GM_\odot/v_c^2$ is the critical-point radius in the absence of rotation, and $\tau = \Omega_\odot r_0/v_c = v_\phi(r_0)/v_c$ measures the importance of rotation.

For $\Omega_\odot = 2.8 \times 10^{-6}$ radians per sec, $v_\phi(r_0) = 2$ km s^{-1}, which is much smaller than typical solar values of v_c, so that rotation has a small effect on the density and outflow speed.

For solar values of rotation, the above equation possesses one critical point, but in general it can be shown that there are larger values (τ_1, τ_2) of τ such that there is one critical point when $0 < \tau < \tau_1$, two when $\tau_1 < \tau < \tau_2$ and none when $\tau > \tau_2$.

13.6.2 Magnetic Field

The magnetic field has several important effects on the properties of the solar wind. It causes differences in some parameters parallel and perpendicular to the field, such as the temperature. It also supports waves and produces micro-instabilities, but the only effect we shall consider in this section is the way it causes plasma close to the Sun to co-rotate with the Sun, almost like a solid body.

Parker (1958) considered qualitatively how a frozen-in magnetic field is dragged out by the solar wind, assuming a small magnetic energy compared with kinetic energy, so that the magnetic field just acts as a tracer and does not affect the wind speed. For a radial flow, the rotation of the Sun makes the magnetic field twist up into a spiral (Figure 13.4), just like a rotating gramophone record with its grooves corresponding to magnetic field lines and the motion of its needle corresponding to the solar-wind outflow (provided the record is rotating backwards).

Suppose the magnetic field is inclined at ψ to the (radial) plasma velocity. Then the component of $\mathbf{v}$ normal to $\mathbf{B}$ (namely $v \sin\psi$) must equal the speed of the (frozen-in) field line in that direction. But the field line rotates with the Sun, and so its speed normal to the radius vector is $\Omega_\odot(r - R_\odot)$ relative to the solar surface. Thus

$$v \sin\psi = \Omega_\odot (r - R_\odot) \cos\psi,$$

or

$$\tan\psi = \frac{\Omega_\odot (r - R_\odot)}{v}, \qquad (13.12)$$

which gives $\psi \approx \frac{1}{4}\pi$ at the Earth's orbit by putting $r = R_E = 1.5 \times 10^{11}$ m $= 215 R_\odot$, $\Omega_\odot = 2.8 \times 10^{-6}$ rad s^{-1} and $v = 400$ km s^{-1}, say. Note from Eq. (13.12) that $\psi = 0$ at $r = R_\odot$, so that the magnetic field has been assumed normal to the solar surface.

In practice, the plasma does have sufficient energy to overcome the field, but it does not completely dominate it, so that the radial and azimuthal flow components are affected by the field. An important parameter is the *Alfvén radius* ($r = r_A$), at which the flow speed becomes Alfvénic. Within that radius, the magnetic field is so strong that it tends to keep the wind rotating with the Sun, and so increases the plasma's angular momentum as it moves out. Beyond the Alfvén radius, the magnetic field has little effect on the solar wind, which therefore conserves its angular momentum.

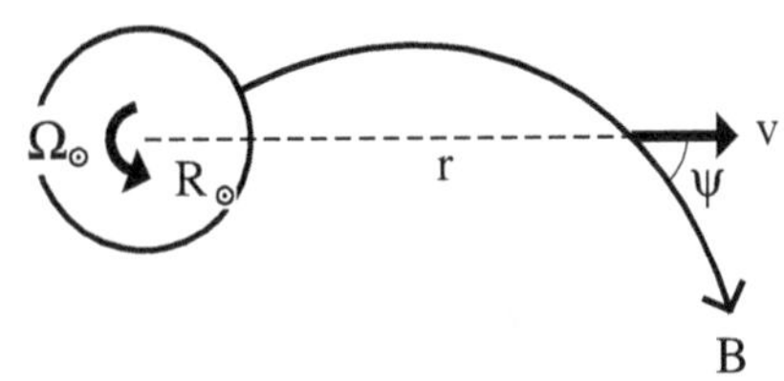

Figure 13.4. A spiral field line attached to a Sun that rotates with angular speed $\Omega_\odot$ in the equatorial plane. The solar wind is here assumed to dominate the magnetic field energetically. It moves radially with speed v inclined at ψ to the magnetic field.

Weber and Davis (1967) modelled this interaction between the solar wind and a magnetic field that is radial and uniform at the solar surface. In spherical polars, the field and flow components in the equatorial plane are assumed to be $(B_r, 0, B_\phi)$ and $(v_r, 0, v_\phi)$ and to depend on r alone. The equation $\nabla \cdot \mathbf{B} = 0$ then implies that

$$B_r = \frac{B_0 R_\odot^2}{r^2}, \tag{13.13}$$

where B_0 is the field strength at the solar surface. Furthermore, the equations of motion, continuity, energy, state and induction for steady, polytropic flow are

$$\rho(\mathbf{v} \cdot \nabla)\mathbf{v} = -\nabla\left(p + \frac{B^2}{2\mu}\right) + (\mathbf{B} \cdot \nabla)\frac{\mathbf{B}}{\mu} - \frac{GM_\odot \rho}{r^2}\hat{\mathbf{r}}, \tag{13.14}$$

$$\rho v_r r^2 = constant, \quad \text{(a)} \qquad \frac{p}{\rho^\alpha} = constant, \quad \text{(b)} \tag{13.15}$$

$$p = \tilde{R}\rho T/\tilde{\mu}, \quad \text{(a)} \qquad \mathbf{0} = \nabla \times (\mathbf{v} \times \mathbf{B}). \quad \text{(b)} \tag{13.16}$$

The polytropic law (13.15b) is adopted as an energy equation on grounds of simplicity, but the new feature is the presence of the Lorentz force in Eq. (13.14).

The induction equation (13.16b) may be integrated to yield

$$v_r B_\phi - v_\phi B_r = \frac{\mathrm{C}}{r}, \tag{13.17}$$

where C is constant and the term on the right is simply the θ-component of the electric field. Assuming $B_\phi = 0$ and $v_\phi = \Omega_\odot R_\odot$ at $r = R_\odot$, where $\Omega_\odot$ is the solar angular speed, implies from Eq. (13.17) that $\mathrm{C} = -\Omega_\odot R_\odot^2 B_0$, or, using Eq. (13.13), $\mathrm{C} = -\Omega_\odot r^2 B_r$. Then Eq. (13.17) determines the azimuthal field as

$$B_\phi = \frac{v_\phi - \Omega_\odot r}{v_r} B_r. \tag{13.18}$$

Near the orbit of Earth, the outflow is almost radial ($v_\phi \ll \Omega_\odot r$), and so, if the radial flow is roughly constant ($v_r \approx V_0$, say), Eq. (13.18) reduces to $B_r/B_\phi = -V_0/(\Omega_\odot r)$. The equation of a field line in the equatorial plane therefore becomes

$$\frac{1}{r}\frac{dr}{d\phi} = \frac{B_r}{B_\phi} = \frac{-V_0}{(\Omega_\odot r)},$$

whose solution is an Archimedean spiral, namely,

$$r = r_0 - \frac{V_0}{\Omega_\odot}(\phi - \phi_0).$$

In general, however, the ϕ-component of Eq. (13.14) becomes [generalising Eq. (13.11)]

$$\rho v_r \frac{d}{dr}(r v_\phi) = \frac{B_r}{\mu}\frac{d}{dr}(r B_\phi),$$

where $\rho v_r / B_r$ is constant by Eqs. (13.13) and (13.15a). Thus, after integrating,

$$r v_\phi - \frac{B_r}{\mu \rho v_r} r B_\phi = L, \tag{13.19}$$

where the constant L is the *total angular momentum* per unit mass carried in both the plasma motion and the magnetic stresses. As the wind carries angular momentum away from the Sun, the Lorentz force transmits a torque to the solar surface, which has slowed the Sun down significantly during its lifetime.

Equation (13.18) may be used to eliminate B_ϕ from Eq. (13.19) to give

$$v_\phi = \Omega_\odot r \frac{M_A^2 L/(\Omega_\odot r^2) - 1}{M_A^2 - 1}, \tag{13.20}$$

in terms of the radial *Alfvén Mach number*, based on the radial Alfvén speed (v_{Ar}),

$$M_A = \frac{v_r}{v_{Ar}} \equiv \frac{v_r}{B_r/(\mu\rho)^{1/2}}.$$

In the low corona, the flow speed and Alfvén speed are typically 1 to 10 km s^{-1} and 100 to 1000 km s^{-1}, respectively, so the flow is sub-Alfvénic, whereas at the Earth's orbit the flow speed is 300 to 800 km s^{-1} and the Alfvén speed is only 50 to 60 km s^{-1}, so the flow is super-Alfvénic. The *Alfvén critical point* ($r = r_A$), where $M_A = 1$ and the solar wind changes from sub- to super-Alfvénic, is believed (from Helios spacecraft measurements) to be located at about $r = 12\ R_\odot$. In order for v_ϕ to remain finite there, it can be seen from Eq. (13.20) that

$$L = \Omega_\odot r_A^2. \tag{13.21}$$

In other words, the total angular-momentum density (L) is just what you would obtain from a solid-body rotation out to r_A. The radius r_A therefore represents the effective lever-arm that the wind uses to brake the Sun's rotation.

Equation (13.20) may be rewritten, after using Eq. (13.21) for L, as

$$v_\phi = \frac{\Omega_\odot r}{M_A^2 - 1}\left(\frac{M_A^2}{(r/r_A)^2} - 1\right),$$

implying that close to the Sun ($M_A \ll 1$) there is rigid rotation with $v_\phi \approx \Omega_\odot r$, while far from it ($M_A \gg 1$) angular momentum is conserved with $v_\phi \approx \Omega_\odot r_A^2/r$.

Finally, the radial component of the equation of motion (13.14) is

$$\rho v_r \frac{dv_r}{dr} - \frac{\rho v_\phi^2}{r} = -\frac{dp}{dr} - \frac{B_\phi}{\mu r}\frac{d}{dr}(rB_\phi) - \frac{GM_\odot \rho}{r^2},$$

which is solved with v_ϕ, B_ϕ, ρ and p given by Eqs. (13.20), (13.18), (13.15a) and (13.15b), respectively. It may be integrated to give a Bernoulli equation [c.f. Eq. (13.8)]:

$$\tfrac{1}{2}v_r^2 + \tfrac{1}{2}(v_\phi - \Omega_\odot r)^2 + \frac{c_{s\alpha}^2(r)}{\alpha - 1} - \frac{GM_\odot}{r} - \tfrac{1}{2}\Omega_\odot^2 r^2 = constant.$$

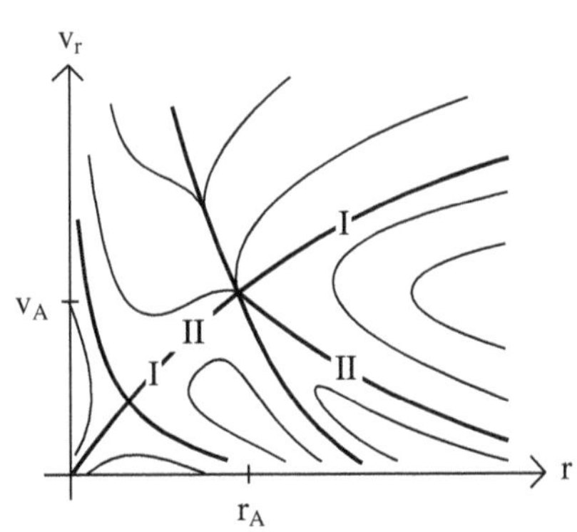

Figure 13.5. A sketch of the solution topology found by Weber and Davis (1967) for a spherical expansion from a rotating magnetised Sun.

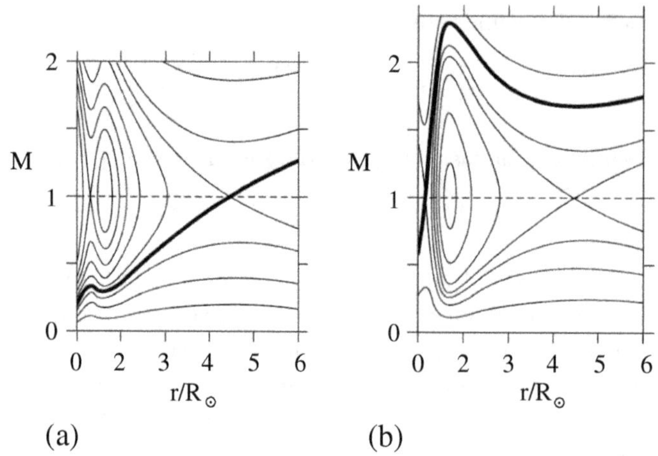

Figure 13.6. The Mach number ($M = v/c_s$) as a function of radial distance (r) from the solar centre for polytropic flow in a coronal hole whose net non-radial divergence (f_{max}) is (a) 3 and (b) 12 (after Kopp and Holzer 1976).

The v_r-r phase-plane (Figure 13.5) is complicated by the presence of three critical points, where the flow speed attains characteristic values for slow, Alfvén and fast-mode waves. Since $p \ll B^2/(2\mu)$ at the Alfvén critical point, the Alfvén and fast-mode speeds are close together, and so are the sound and slow-mode speeds.

Only the two solutions I and II which pass through all three critical points have a vanishing pressure at infinity. Type II must be ruled out, because it predicts much too small a flow speed at the Earth (9 km s^{-1}). Type I gives radial and azimuthal speeds there of 425 km s^{-1} and 1 km s^{-1}. Brandt (1970) repeated the analysis with a full energy equation and found $n = 6 \times 10^6$ m^{-3}, v_r = 315 km s^{-1}, $T = 3.2 \times 10^5$ K, $v_\phi = 2.5$ km s^{-1} and $\psi = 55$ degrees at 1 AU, which agree reasonably with observations.

The mass-loss rate is of the order of $10^{-14} M_\odot$ per year (PROBLEM 13.3), which is much too small to produce significant changes in the solar interior over the lifetime ($\sim 10^{10}$ yr) of the Sun. However, the time-scale for angular-momentum loss using the current angular velocity and magnetic field strength is comparable with the Sun's lifetime (PROBLEM 13.9). Indeed, since the Sun probably rotated much more rapidly and had a stronger magnetic field in the past, angular-momentum loss would certainly have been important in slowing down the Sun's rotation.

13.6.3 Early 1D Coronal-Hole Models

A simple coronal-hole model may be set up by solving the solar-wind equation in a prescribed magnetic geometry and neglecting the back-reaction on the magnetic field. Observationally, one finds that the coronal-hole density is lower than that of the average corona by at least a factor of three, and coronal holes have been identified as the source of high-speed streams (with a typical particle flux of 3×10^{12} m^{-2}s^{-1}).

Kopp and Holzer (1976) considered polytropic flow in an open field of area $A(r)$ by solving the equations of continuity, momentum and energy in the form

$$\rho v A = constant,$$

$$\rho v \frac{dv}{dr} = -\frac{dp}{dr} - \frac{GM_\odot \rho}{r^2}, \qquad \frac{p}{\rho^\alpha} = constant,$$

where $p = \tilde{R}\rho T/\tilde{\mu}$ and α is a constant (taken as 1.1). These may be combined to give an equation for the Mach number ($M = v/c_{\alpha s}$, where $c_{\alpha s}^2 = \alpha \tilde{R}T/\tilde{\mu}$) in the form

$$\frac{M^2-1}{2M^2}\frac{dM^2}{dr} = \tfrac{1}{2}\left(\frac{\alpha+1}{\alpha-1}\right)\left[1+\left(\frac{\alpha-1}{2}\right)M^2\right]\frac{1}{g}\frac{dg}{dr}, \tag{13.22}$$

where $g(r) = A(r)^{2(\alpha-1)/(\alpha+1)}(E + GM_\odot/r)$. Also,

$$E = \tfrac{1}{2}v^2 + \frac{\alpha}{\alpha-1}\frac{p}{\rho} - \frac{GM_\odot}{r}$$

is the total energy. [For the simpler case of an isothermal corona ($\alpha = 1$) with an area function $A(r) = ar^n$, see PROBLEM 13.10.]

Just as for Parker's solar-wind solutions, Eq. (13.22) may be integrated analytically to give $v(r)$ implicitly, and it possesses a critical point where $M^2 = 1$ and $dg/dr = 0$. However, unlike the case of spherical expansion [$A(r) \sim r^2$], there may be more than one critical point, and they may be either X-type or O-type in nature, depending on whether $g(r)$ possesses a minimum or a maximum, respectively. This raises the possibility of multiple wind states with shock transitions involving different critical points (Habbal and Tsinganos 1983).

The area function is written as $A(r) = A_\odot(r/r_0)^2 f(r)$, where $f(r)$ is a function that increases monotonically from 1 at $r = r_0$ to f_{max} (>1) at large distances, most of the increase occurring between 1.4 r_0 and 1.6 r_0, say.

A Parker-type critical point is always present at 4.5 r_0, but when $f_{max} > 2$ an extra pair of critical points appears between 1 r_0 and 2 r_0. When $2 < f_{max} < 7.5$, the flow is close to spherical and the only solar-wind flow (with a continuous connection of subsonic values at the Sun to supersonic values at large distances) is similar to Parker's solution (Figure 13.6a). However, when the coronal hole expands rapidly, ($f_{max} > 7.5$), the solar-wind flow is rapidly accelerated to supersonic speeds at the inner critical point low down in the corona (1.15 r_0 for $f_{max} = 12$ in Figure 13.6b). The flow speed then attains a local maximum and thereafter remains supersonic.

The main point of the Kopp-Holzer paper is that it explains qualitatively the low densities that are observed near the base of coronal holes. However, the rapid field divergence does not account for the high speeds that are observed at Earth, since all of their solar-wind solutions, regardless of the value of f_{max}, possess the same asymptotic flow speed at large distances. Any such polytropic model, even with a spherical expansion, can provide a high enough speed at 1 AU if the base temperature is taken high enough (2×10^6 K for the Kopp-Holzer model), but when a more general energy equation is employed such high speeds are no longer produced.

The other contribution of Kopp and Holzer was to elucidate the topology of the solutions, with the two basic types shown in Figure 13.6. Observations of a coronal hole (Munro and Jackson 1977) gave high speeds low down in the coronal hole (200 km s^{-1} at 2.5 $R_\odot$), which suggests that a flow similar to Figure 13.6b (with rapid acceleration through the lower critical point) is the relevant type.

For a conductive model with an energy equation of the form (13.9), Holzer and Leer (1980) found that the solar-wind speed (v_e) at 1 AU cannot be increased to high-speed values by either a rapidly diverging geometry or collisionless inhibition of heat conduction. However, high-speed solar-wind streams may be produced in three possible ways (see Sec. 13.7):

(i) subsonic momentum loss by, for instance, the magnetic field;

(ii) transfer of energy by waves from the subsonic region to the supersonic region; and

(iii) propagation of energy (as Alfvén or fast-mode waves) from the coronal base and its deposition in the supersonic region. (For a 10 G base field, a wave amplitude of 20 km s^{-1} would make v_e = 700–800 km s^{-1}.)

13.7 Fast Solar Wind: Self-Consistent Wave-Turbulence Models

Two classes of model have been proposed for driving the solar wind. Wave-turbulence models deposit energy and momentum from magnetic fluctuations in an open flux tube and are most natural for fast wind coming from large coronal holes, but they may also work in slow wind when it originates from small coronal holes or near the boundaries of large coronal holes (Sec. 1.6.4). On the other hand, reconnection models (Sec. 13.8) provide energy and momentum from reconnections between two flux tubes (one open and the other open or closed) and are likely to be operating in most of the highly variable slow solar wind but possibly also in fast wind. Wave-turbulence models are relatively straightforward, since they involve flow along single flux tubes and so are highly developed. Reconnection models, being inherently 3D, are much more complex and so at present relatively primitive.

Five ingredients are essential for a self-consistent fast wind (Cranmer 2012):

(i) A physically motivated explicit form for the coronal heating. This may be deposited at large scales and invoke a transfer or cascade to dissipation scales, or it may describe the microphysics of the dissipation;
(ii) Extra momentum deposition, since an observed mean coronal-hole temperature $[\frac{1}{2}(T_p+T_e)]$ is not sufficient to create wind speeds of 700 to 800 km s^{-1} at 1 AU;
(iii) A self-regulating mass flux determined by a balance in the transition region between downward heat conduction, radiation and upward enthalpy flux;
(iv) Extended collisionless heating and electron conduction to 1 AU to produce much shallower radial gradients of T_p and T_e than adiabatic (Cranmer et al. 2009); and
(v) Funnel-type magnetic expansion from the chromospheric supergranular network, where plasma outflow appears to originate (Dowdy et al. 1986; Cranmer and van Ballegooijen 2005; Marsch et al. 2008; Pucci et al. 2010).

Note that, when extra heat is added below the critical point, it increases the temperature and therefore the scale-height, so that the wind becomes more massive and so slower. On the other hand, when heat is added above the critical point, the subsonic region is relatively unaffected, and so the local energy increase naturally goes into kinetic energy and produces a faster, less-dense wind.

Convective motions in the solar surface generate Alfvénic (mainly fast-mode) oscillations at supergranule boundaries that propagate up into the extended corona. They partially reflect back down towards the Sun and develop into MHD turbulence that gradually dissipates. Indeed, waves and turbulent motions are known to be present from the low solar atmosphere out into the solar wind at 1 AU and beyond (Tu and Marsch 1995; Cranmer and van Ballegooijen 2005). Models based on this process can produce realistic fast (and slow) wind conditions with wave amplitudes similar to those observed in the corona and solar wind (Hollweg 1974, 1986; Velli et al. 1991; Ofman 2010; Matsumoto and Shibata 2010; Matthaeus and Velli 2011).

For example, Cranmer et al. (2007) set up a single-fluid model (called ZEPHYR) along a 1D flux tube. They solve the conservation equations of mass, momentum and energy from the photosphere out into the solar wind, together with transport equations for Alfvén and acoustic energy fluxes. Acoustic waves steepen into shocks, while outgoing Alfvén waves are dissipated by MHD turbulence. The wind is accelerated by gradients of plasma pressure and wave pressure. The parameters of the model are the magnetic field as a function of distance and base forms of the acoustic and Alfvénic spectra in the 5 to 10 minute range, the latter being deduced from motions of G-band bright points. The adopted form of heating rate is

$$Q \approx \rho \frac{z_-^2 z_+ + z_+^2 z_-}{4L_\perp},$$

where $z_\pm$ are the magnitudes of the *Elsässer variables* ($\mathbf{z}_\pm = \mathbf{v}_1 \pm \mathbf{B}_1/\sqrt{\mu\rho}$) calculated from wave-transport equations (Sec. 2.4.3), $L_\perp$ is a perpendicular correlation length of the turbulence (taken proportional to flux-tube width). This is reminiscent of the earlier Kolmogorov heating rate ($\rho v^3/L$), originally derived via dimensional analysis and first applied to the solar wind by Hollweg (1986).

The Alfvén waves dissipate after being partially reflected by Alfvén-speed gradients, so that the counter-propagating waves interact and generate a nonlinear turbulent cascade (Kraichnan 1965). This transfers energy from large to small scales and terminates in dissipation and heating that extends from the low corona out into the heliosphere (Matthaeus et al. 1999; Chandran and Hollweg 2009; Cranmer 2010). The resulting turbulent heating rate depends on the radial gradient of the Alfvén speed (which drives the non-WKB reflection) and the radial growth of the turbulent correlation length. In fact, at the base of the corona, these and other effects combine to make the heating rate roughly proportional to the magnetic field, in agreement with observations (Pevtsov et al. 2003; Schwadron and McComas 2008).

The location of the critical point depends on the rate of flux-tube expansion (Sec. 13.6.3, Figure 13.6). Thus, in varying the expansion from rapid to slow, the critical point shifts abruptly from a location near the Sun to one much farther out. At the same time in the ZEPHYR models, the nature of the flow changes from one characteristic of fast solar wind (low density, rapid speed, heating in the supersonic outer corona) to one typical of slow solar wind (high density, slow speed, heating in the subsonic inner corona). As well as turbulent heating, the wave-pressure forces in the ZEPHYR models are crucial in accelerating the fast wind; for slow winds they play a smaller role since the Alfvén-wave amplitudes do not grow as large.

Cranmer et al. (2007) applied this to a 2D stretched-dipole model for solar minimum (Figure 13.7). At large distances, they found fast winds at high latitudes and slow winds below a latitude of 20 degrees (cf. Ulysses observations of Goldstein et al. 1996). Although the model is in a sense self-consistent, it does invoke a number of assumptions and is not universally accepted. For example, Roberts (2010) suggests that, although waves (turbulent or not) are likely to play a role in differential heating and acceleration of minor ions, magnetic spectra from Helios and Ulysses imply that their amplitude is insufficient to power the fast solar wind. Later, Cranmer and van Ballegooijen (2012) calculated collisionless heating of protons, electrons and heavy ions by MHD turbulence, while van Ballegooijen et al. (2011) extended the model to coronal loops (Sec. 10.5.1).

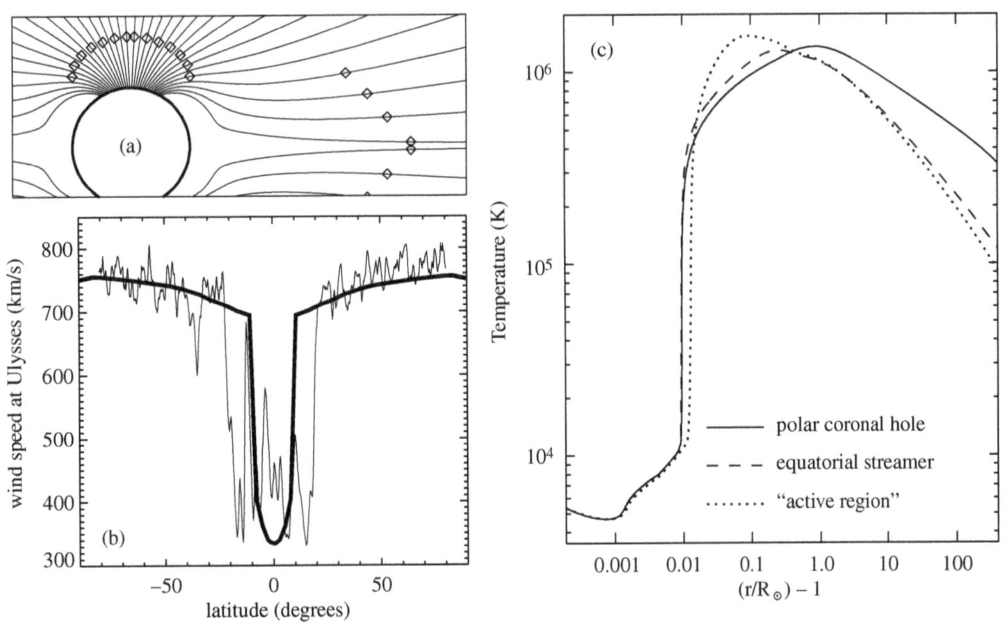

Figure 13.7. ZEPHYR wave-turbulence solar-wind models for solar-minimum conditions, showing: (a) the magnetic geometry with critical-point locations (diamonds); (b) wind speed as a function of latitude (thick curve) compared with Ulysses observations (thin curve); (c) T as a function of r in a polar hole, near a streamer and in open field from an active region (courtesy Steve Cranmer).

13.8 Slow Solar Wind: Reconnection Models

The source and cause of the slow solar wind are unclear (Sec. 1.6.4). It may be the same mechanism as the fast wind but originate in small short-lived coronal holes (such as occur near active regions or after coronal mass ejections) or in the boundary near the edge of a large hole. Another possibility is diffusion across closed regions due to small-scale instabilities such as ballooning or interchange (Sec. 7.5.1). Two other possibilities are large-scale interchange reconnection (between open and closed structures) and the continual reconnection of the topologically complex corona that is also likely to be heating the corona in response to photospheric motions.

The magnetic field of the low solar atmosphere is highly complex and comes through the solar surface in myriads of small flux tubes of both polarities that are continually moving about and interacting at granulation and supergranulation boundaries. In this *magnetic carpet* (Schrijver et al. 1997; Title and Schrijver 1998; Parnell 2001), the magnetic field is reconnecting so efficiently that all the field lines reconnect in a time of order an hour or smaller (Close et al. 2005). The resulting reconnection transfers mass, momentum and energy onto open field lines from closed and other open field lines and so can act as a dynamic source for the solar wind.

Such reconnection-driven energy exchange can occur on a variety of scales, ranging from large-scale streamers (Einaudi et al. 1999; Antiochos et al. 2011) to small supergranular scales (Axford and McKenzie 1992; Fisk et al. 1999; Fisk 2003; Schwadron and McComas 2003, 2008) or even smaller granular scales. Indeed, observed correlations between loop lengths, coronal temperature and wind speed at 1 AU suggest a transfer of Poynting flux from loops to open-field regions (Feldman et al. 1999; Gloeckler et al. 2003). Also, on large scales, reconnection at the edges of coronal holes may explain their observed rigid rotation (Lionello et al. 2006).

In order to assess properly the efficiency of this process, one needs time-dependent 3D models of the magnetic carpet. Questions that need to be answered include: what is the rate at which open fields are being created and how much of the energy is used to drive a wind? Many small jets have been observed by SoHO and Hinode (Wang et al. 1998; Shibata et al. 2007), but what fraction of the solar wind is created by such jets? Furthermore, how is the energy release in reconnection partitioned into kinetic energy, thermal energy, wave energy and particle energy, each of which may help accelerate a wind? Alfvénic waves are observed low down in the atmosphere as spicules waving to and fro with a period of 5 minutes (Tomczyk et al. 2007; De Pontieu et al. 2007), but how much of that energy ends up accelerating a wind and feeding into the much longer-period (hours) waves observed in the solar wind?

Fisk (2003, 2005) suggested so-called *interchange reconnection* between large-scale open magnetic field lines and coronal loops as a qualitative mechanism for heating coronal plasma and accelerating the solar wind [see also Schwadron and McComas (2003)]. Such reconnection makes the footpoints of open flux undergo a random walk and diffuse across the solar surface, but it is unclear how it is converted into an outward mass flux. Note that the discovery of pseudo-streamer topology (Sec. 13.9.3) implies that only very large loops (of size about 1.5 $R_{\odot}$) near open separators are involved in interchange reconnection (Titov et al. 2011).

A preliminary Monte Carlo model of the photospheric magnetic carpet (Cranmer and van Ballegooijen 2010) imposes flux balance for the quiet Sun and strong imbalance for coronal holes. It computes the rate at which closed fields open up above a 200-Mm square, assuming an evolution through potential-field states. For quiet regions, the energy flux is too small to accelerate the solar wind but for coronal holes it may be large enough. However, it is possible that an MHD model with supergranular flows and a larger range of fluxes could increase the energy flux.

MHD simulations of the corona have revealed a highly complex structure with many separatrices (and quasi-separatrix layers). For example, Linker et al. (2011) introduced two bipoles into a coronal hole and advected them towards its boundary. In one case, open flux closed down and created disconnected flux in

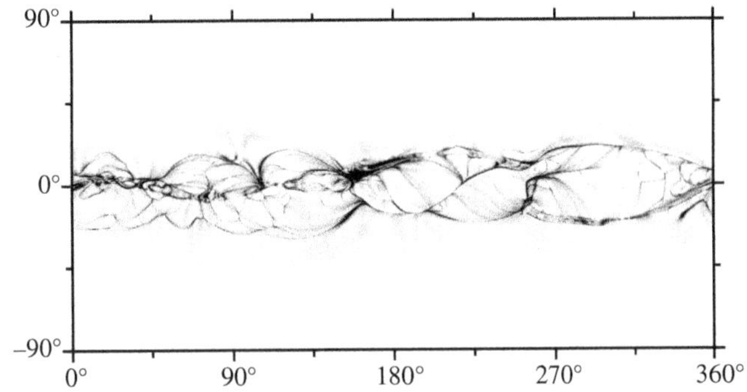

Figure 13.8. The intersection with the sphere $r = 3\ R_\odot$ of the web of quasi-separatrix layers and separatrices from high values of Q (Sec. 6.9.5) in the streamer belt of a solar-wind model for the 2008 eclipse near solar minimum (after Titov et al. 2011).

the heliosphere, while in another an apparently detached coronal hole was in fact linked to the main hole by a separatrix surface (Sec. 6.9.2). Then an analytical model clarified the topology and showed how a single parasitic polarity can produce multiple null points in the corona joined by separators (Titov et al. 2011; Antiochos et al. 2011). It also led to a detailed description of the complex topology of pseudo-streamers (Sec. 13.9.3). This was used to show how the presence of parasitic polarity can cause low-latitude coronal holes to be disconnected from polar coronal holes, in the sense that no finite flux joins them, which disproves a previous hypothesis that narrow open-flux corridors will always join coronal holes (Antiochos et al. 2007).

These authors proposed an *S-Web* model for the slow solar wind (Figure 13.8), whose advantages over other models include natural explanations for the slow wind's high variability, composition (that of the closed-field corona) and large angular width (up to 60 degrees). In the model, the most likely source of the solar wind is 3D reconnection at a complex network of separators (Secs. 6.13.3 and 10.4.5) in the coronal streamer belt (Titov et al. 2012); these are locations where a web of separatrix curtains drapes down and intersects separatrix domes and streamer separatrices as pseudo-streamers lying above parasitic polarities (see Sec. 13.9.3). Quasi-separators that originate in narrow open-flux corridors at the photosphere may also play a role. As the photospheric boundary conditions change, plasma flows are driven along the separatrix curtains by separator reconnection, which is required to change the locations of parasitic polarities and particularly the number and locations of null points.

13.9 2D and 3D Models: Streamers and Coronal Holes

The assumption that the solar-wind expansion is spherically symmetric is a poor approximation close to the Sun where most of the acceleration takes place. On both eclipse and soft X-ray (or EUV) photographs (e.g., Figures 1.3d, 1.21c and 1.25a), one sees evidence of some regions with a predominantly open magnetic field and others with closed fields. The latter consist of coronal loops (Sec. 1.6.2) and the former are coronal holes, from which the fast solar wind is escaping (Sec. 1.6.3).

The large-scale closed structures surmounted by open fields (and known as *coronal helmet streamers*) contain a high-pressure, high-density plasma because the closed magnetic field is strong enough to hold the plasma down. By contrast, the density of an open field falls off rapidly with height because the plasma can easily escape. Also, the heating is likely to be more efficient in the closed regions and so make the temperature higher than in coronal holes. Since the total pressure (plasma plus magnetic) must remain continuous across the interface between closed and open regions, the fall-off of pressure with height in the open region means that there is a jump in magnetic field strength at the interface. The interface is therefore a curved vortex-current sheet, across which the tangential flow speed is discontinuous. Pneuman and Kopp

(1971) solved the isothermal MHD equations self-consistently (Sec. 13.9.1), thus going beyond the simple 1D coronal-hole models (Sec. 13.6.3).

13.9.1 2D Helmet-Streamer Model (Pneuman-Kopp)

Pneuman and Kopp (1971) solved numerically the MHD equations for steady, axisymmetric, coronal expansion from a sphere whose surface magnetic field is dipolar. The usual steady-state equations of motion, induction and continuity are

$$\rho(\mathbf{v}\cdot\boldsymbol{\nabla})\mathbf{v} = -\boldsymbol{\nabla} p - \frac{GM_{\odot}\rho}{r^2}\hat{\mathbf{r}} + \mathbf{j}\times\mathbf{B}, \qquad \mathbf{0} = \boldsymbol{\nabla}\times(\mathbf{v}\times\mathbf{B}), \tag{13.23}$$

$$\boldsymbol{\nabla}\cdot(\rho\mathbf{v}) = 0, \qquad \mathbf{j} = \boldsymbol{\nabla}\times\mathbf{B}/\mu, \qquad \boldsymbol{\nabla}\cdot\mathbf{B} = 0, \tag{13.24}$$

where T is uniform and $p = \tilde{R}\rho T/\tilde{\mu}$. All variables are functions of r and θ alone (in spherical polars). As boundary conditions, they adopted $\rho = \rho_0 =$ constant and $B_r = B_0 \cos\theta$ at the solar surface. A reasonable value for the mean base field strength (and so B_0) is 8 G, obtained from a field of 4×10^{-5} G at 1 AU when coronal holes occupy 20 per cent of the Sun's surface area.

The resulting magnetic structure (Figure 13.9) has an appearance similar to a large helmet streamer. The field lines are dragged out by the solar wind to form a current sheet (dashed) at the boundary of closed and open flux and above a cusp at 2.5 $R_{\odot}$. The model is axisymmetric with a disc-like *heliospheric current sheet* extending out from 2.5 $R_{\odot}$, and it represents qualitatively the structure of the solar wind near sunspot minimum, when polar coronal holes dominate. However, in reality, the current sheet is warped like the rim of a sombrero due to departures from axisymmetry in the surface magnetic field (Figure 1.25b). This means that the low-latitude solar wind flow past the Earth is much more varied than it is at higher latitudes.

The effect of solar rotation on the Pneuman-Kopp model has been evaluated (Priest and Pneuman 1974) by extending the spherically symmetric analysis (Sec. 13.6.2). Departures from spherical symmetry reduce the angular-momentum loss-rate due to the presence of closed magnetic field regions with no loss and also the latitudinal variation in Alfvén-point location (r_A). Since the Alfvén radius is closer to the Sun for field lines near the equator, the angular-momentum density L (Eq. 13.21) is smaller there. The net result is that the angular-momentum loss-rate is only 15 per cent of that for a monopole field. A model for steady flow in the current sheet extending from the cusp point has also been developed (Pneuman 1972; Priest and Smith 1972). The current-sheet interior possesses a transverse magnetic field component, so that the field

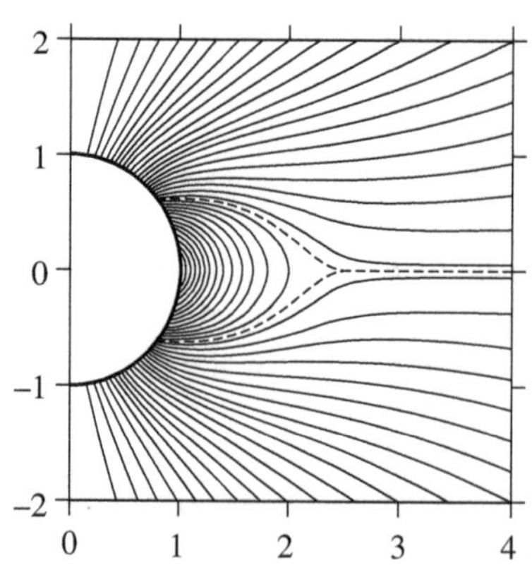

Figure 13.9. Magnetic field lines for a model of isothermal axisymmetric expansion from a dipolar surface field. A heliospheric current sheet beyond 2.5 $R_{\odot}$ is shown dashed (after Pneuman and Kopp 1971).

lines form stationary loops connected to the solar surface. The sheet is thin enough that the plasma is able to slip outwards across the magnetic loops by magnetic diffusion. Further physical insight into the nature of helmet streamers has been given in analytical models (Suess and Nerney 2002; Nerney and Suess 2005).

13.9.2 Two-Dimensional Separable Solution for a Rotating Wind

Analytical progress can be made by seeking separable solutions of the steady-state ideal MHD equations. Thus, the Weber-Davis model may be extended out of the ecliptic plane by setting $v_\theta = B_\theta = 0$ in spherical polar coordinates (r,θ,ϕ) and assuming that the remaining components (v_r, v_ϕ), (B_r, B_ϕ) are products of functions of r and θ alone, so that the field lines and streamlines are helicoidal on cones (Lima and Priest 1993; Lima et al. 2001). Then the MHD equations become

$$\frac{\partial}{\partial r}(B_r r^2) = 0, \qquad \frac{\partial}{\partial r}(\rho v_r r^2) = 0, \qquad \frac{\partial}{\partial r}(r v_\phi B_r - r v_r B_\phi) = 0,$$

$$\rho v_r \frac{\partial v_r}{\partial r} - \rho \frac{v_\phi^2}{r} = -\frac{\partial p}{\partial r} - \frac{B_\phi}{\mu}\frac{\partial B_\phi}{\partial r} - \frac{B_\phi^2}{\mu r} - \frac{\rho G M_\odot}{r^2},$$

$$\rho v_\phi^2 \frac{\cos\theta}{\sin\theta} = \frac{\partial p}{\partial \theta} + \frac{B_\phi^2}{\mu}\frac{\cos\theta}{\sin\theta} + \frac{\partial}{\partial\theta}\left(\frac{B_r^2 + B_\phi^2}{2\mu}\right),$$

$$\rho v_r \frac{\partial v_\phi}{\partial r} + \rho \frac{v_r v_\phi}{r} = \frac{B_r}{\mu}\frac{\partial B_\phi}{\partial r} + \frac{B_r B_\phi}{\mu r}.$$

The first two equations imply that $B_r \sim b_r(\theta)/r^2$ and $\rho v_r \sim f_r(\theta)/r^2$ and then, after eliminating p between the r- and θ-components of the equation of motion, we are able to deduce the forms of the solutions, such as

$$v_r(r,\theta) = v_0 f(r) \sqrt{\frac{1 + \alpha \sin^{2\epsilon}\theta}{1 + \delta \sin^{2\epsilon}\theta}},$$

and similar expressions for the other variables in terms of the function $f(r)$, which is obtained numerically, and the parameters α, δ, which depend on the ratios of the density and flow speed at the equator to those at the pole, and ϵ, which controls the profile of the speed. The solutions possess a total angular momentum $L(\theta)$ (Eq. 13.19) that now varies with latitude; they have two critical points, one at the Alfvénic point and the other where the radial speed equals the fast-mode wave speed in the radial direction. When $\epsilon = 0$, we recover the usual non-rotating spherically symmetric wind and, as ϵ increases, the amount of anisotropy increases. More general separable solutions that are important for astrophysical outflows and jets have been reviewed by Heyvaerts and Norman (2003) and Tsinganos (2010).

13.9.3 3D Topology of Pseudo Coronal Streamers and Coronal Holes

Several geometrical features are of particular importance in determining the topological skeleton of coronal streamers and of the coronal magnetic field in general (Sec. 6.9.4). Separatrix surfaces (or separatrices) separate the volume into topologically distinct regions. They begin either at 3D null points as the fan of field lines emanating from or converging on the null point (Sec. 6.9.1) or at bald patches (Secs. 6.9.2 and 6.9.4), where field lines touch the solar surface from above (Figure 13.10b). In the corona, separatrices either form *separatrix domes* as they curve down towards the solar surface from a coronal null or towards a bald patch and enclose photospheric flux of parasitic polarity, or they form *separatrix curtains* (Titov et al. 2011), which are open flux surfaces that arch high into the corona, or they form *streamer separatrices* that separate the closed field of a bipolar region from open field lines. Separatrix

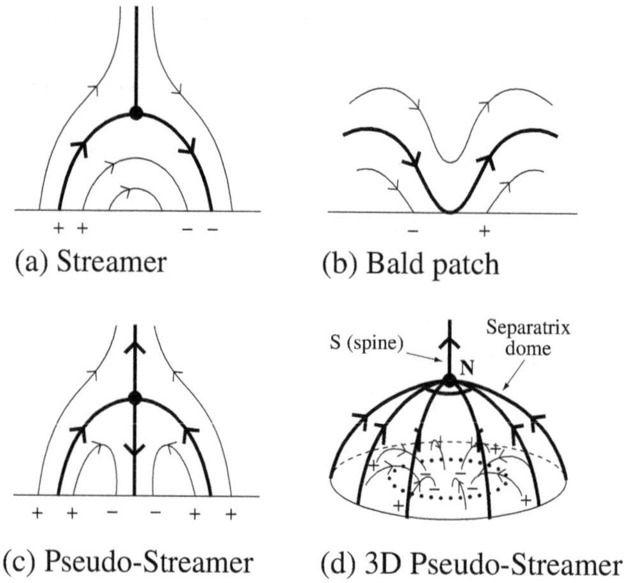

Figure 13.10. Separatrix curves (thick) and field lines (thin curves) for (a) a simple 2D coronal streamer configuration (a bipolar streamer), (b) a field with a bald patch, (c) a simple 2D pseudo-streamer (or unipolar streamer) with one null point and (d) a simple 3D pseudo-streamer with a single null point (N) and a separatrix dome above parasitic polarity.

curtains that are bounded by spines are called *open separatrices* and intersect a spherical (e.g., source) surface in non-closed curves (Figure 6.28; Sec. 6.9.2). Separatrix curtains that intersect a spherical surface in a closed curve are *closed separatrices*

In 2D, a *simple coronal streamer* has a topology indicated in Figure 13.10a, with a T-point, Y-point or cusp-point above bipolar closed flux and separating coronal fields of opposite polarity, surmounted by a heliospheric current sheet. The type of null point depends on whether the plasma pressure in the closed region is smaller (a T), the same (a Y) or larger (a cusp) than in the open region. For a purely potential field model, where the plasma pressure is negligible, there is a Y-point, but, in a more realistic model with an enhanced plasma pressure in the closed region, there is a cusp-point. In an axisymmetric model, the null point becomes a circular null line (i.e., a line of nulls). For example, for a potential-field source-surface model the null line lies on the source surface. In a fully 3D situation, the heliospheric current sheet becomes warped and distorted (Figure 1.25), and the nature of the generalisation of the null point depends on the model: for a potential-field source-surface model, it becomes a curved null line on the source surface; however, for a more realistic static or MHD model, the null line is topologically unstable and so is expected to become a curved separator linking at least two 3D null points.

More common than a simple streamer is a *simple pseudo-streamer* (Sec. 1.6.1), which possesses a single X-point above a tripolar field, separating coronal holes of the same polarity (Figure 13.10c). In 3D, the X-point becomes a 3D null point (N), with a spine (S) projecting upwards and the fan (dark circle) becomes a separatrix dome that encloses all the flux from the parasitic polarity (Figure 13.10d). 3D reconnection can occur at the 3D null, with a non-zero transfer of flux across the dome through the current sheet that forms about the null (Sec. 6.13.2). Note that this 3D null reconnection at such a simple separatrix dome topology is a likely candidate for the occurrence of coronal hole or anemone jets (Shibata et al. 2007) above parasitic polarity (Sec. 1.9.2.3) and also for some solar flares (Secs. 12.2.7 and 12.3.2).

Titov et al. (2011) realised that the topology of a pseudo-streamer in practice is much more complex than the above simple topology, since usually there are several coronal null points on the separatrix dome.

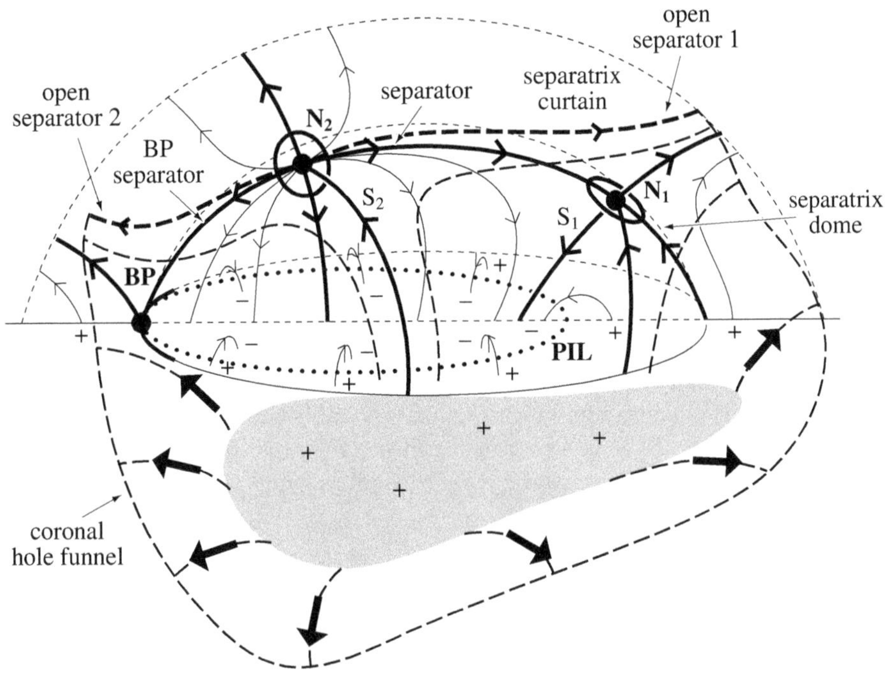

Figure 13.11. The topological skeleton of a realistic (i.e., complex) coronal pseudo-streamer next to a coronal hole (shaded) with two null points (N_1 and N_2) and a bald patch (BP). N_1 possesses a spine S_1 and a fan (dark circle) that extends out as a separatrix dome, while N_2 possesses a spine S_2 and a fan that extends upwards as a separatrix curtain. The dome and curtain intersect in a separator that joins the nulls and the bald patch. The photospheric polarity inversion line is a dotted curve. The neighbouring coronal hole is bounded by a separatrix funnel of open magnetic field lines (dashed). The funnel intersects the separatrix curtain in the open separators and touches the separatrix dome at the spine S_2, which belongs to both the funnel and the dome.

Figure 13.11 shows a potential-field example with two nulls (N_1 and N_2) and a bald patch. The blade of the streamer is a separatrix curtain, whose base is bounded by a separatrix dome, one part of which originates at N_1 and the other at a bald-patch (BP). N_1 is similar to N in Figure 13.10d, with its fan forming part of a separatrix dome. However, the fan of N_2 is orthogonal to it, so that its spine (S_2) lies in the separatrix dome, while its fan forms a separatrix curtain that cuts across the dome and arches high into the corona. The half-domes and the curtain intersect in two separators. To the left of N_2 and its spine, the separator appears as a bald-patch separator (N_2-BP) and the dome continues as a bald-patch separatrix surface of field lines that arch down and touch the photosphere at the bald patch. The polarity inversion line that surrounds the unipolar region in the photosphere touches the separatrix dome at the bald patch. Indeed, in most pseudo-streamers in practice, part of the separatrix dome is a BP separatrix.

The separatrix curtain also intersects the boundaries of coronal holes that are separated by the dome of the pseudo-streamer. This intersection lies along two open separators both connected to the same null N2 (Figure 13.11). The boundaries of the coronal holes have the shape of funnels that bend down towards the dome and join each other along the open separators, so that N2 is actually a saddle point for such a connection. Near the photosphere, the region between the funnel and the dome/curtain is filled with a corridor of closed field.

As the flux sources evolve, reconnection can occur at the separators, and the topology can change due to bifurcations. For example, the bald patch can become a third null (N_3) and move up into the corona along the separator, and later N_2 and N_3 can combine to form a higher-order null that then disappears, reducing the topology into that of a simple separatrix dome with one null (Figure 13.10d), surrounded on all sides by a closed-field region. Similar transformations accompany the displacement of the parasitic polarity into a simple dome structure inside an open-field region.

Titov et al. (2012) have analysed in detail the topological structure of three pseudo-streamers in a potential-field source-surface configuration computed from SoHO/MDI magnetograms, which confirms the results of their previous analytical model. They also found that, due to greater complexity in the surrounding field, the parasitic polarity in pseudo-streamers can be covered by several part-domes, which intersect a separatrix curtain along several closed separators. Low-latitude coronal holes that are disconnected at the photosphere from polar coronal holes lie above pseudo-streamers, which are themselves bounded by separatrix domes, bald-patch separatrices and separatrix funnels. Thus, a separatrix curtain can intersect several separatrix domes in a chain of closed separators and the coronal-hole separatrix funnel (streamer separatrix) in two open separators.

In response to continual photospheric footpoint motion, reconnection can occur at both closed and open separators (Sec. 6.13.3). The way in which this may possibly drive the slow solar wind by delivering plasma to the open corona (Sec. 13.8) is a matter for future research; perhaps it starts with a cascade of reconnections at low-lying closed separators and ends with interchange reconnection at open separators (V Titov, private correspondence).

13.9.4 Global Three-Dimensional Modelling

Great advances have been made in modelling numerically the global structure of the solar corona (Mackay 2012). The most common technique has been to find the potential coronal field subject to the observed radial photospheric component together with an outer boundary condition that the field is radial ($B_\theta = B_\phi = 0$) at a *source surface* at $r = 2.5\ R_\odot$, which simulates the opening out of the field at large distances by the solar wind (Schatten et al. 1969). An improvement to this *Potential-Field Source-Surface (PFSS) Model* is a *Current-Sheet Surface Model* (Zhao and Hoeksema 1995), which possesses both an inner potential region out to a cusp surface and also an extra outer region containing currents which redistribute the coronal field, so that it becomes both radial and uniform at an outer boundary (in agreement with Ulysses observations). A difficulty with the models is that the photospheric magnetic field is poorly known near the poles and on the back side of the Sun. However, they are easy to use and give reasonable first approximations to the global structure (Riley et al. 2006).

A more realistic nonlinear force-free model for the corona (van Ballegooijen et al. 2000; Mackay and van Ballegooijen 2006) uses a data-driven flux-transport model to produce a continuous evolution of the photospheric magnetic field that provides the lower boundary conditions for a coronal nonlinear force-free code. The flux-transport model evolves the radial magnetic field at the photosphere in response to observed flux emergence as well as diffusion, differential rotation and meridional flow; it has been highly successful in reproducing many observed properties of the photospheric field (Sheeley 2005). The resulting coronal field includes highly twisted flux ropes, slightly sheared coronal arcades and almost-potential open field lines.

The Sun's open magnetic flux varies by a factor of 2 through the solar cycle while the photospheric magnetic flux varies by a factor of 5. The open flux peaks 1 to 2 years after cycle maximum and gives the location of coronal holes and of the fast solar wind. An intriguing result from Ulysses is that at 1 AU the open flux is distributed uniformly, since the radial interplanetary magnetic field (IMF) is independent of latitude. The observed open flux agrees well with predictions from the nonlinear force-free model and possesses three contributions: a background flux due to the location of flux sources, an enhancement due to an inflation of the field from currents and a sporadic component due to flux-rope ejections in CMEs (Yeates et al. 2010).

The nonlinear force-free model has been used to predict with remarkable reliability the locations of prominences as sites of large-scale low-lying twisted flux ropes lying above a global polarity-inversion line

(Sec. 11.3). It also predicts the chirality of prominences with 96 per cent accuracy and the locations of erupting prominences (Yeates et al. 2008, 2010), indicating that the Sun has a long-term memory about the build-up and transport of magnetic helicity across the Sun.

Another important complementary line of approach has been to go beyond force-free modelling and develop 3D MHD models of the corona, so far mainly for static photospheric magnetogram boundary conditions and a steady-state corona (Mikić et al. 1999; Roussev et al. 2003). Such simulations have replaced earlier polytropic equations by more realistic energy equations and have been compared with PFSS models and observed eclipse, EUV and X-ray images (Mikić et al. 2007; Nakamizo et al. 2009; Lionello et al. 2009; Downs et al. 2010; Riley et al. 2011).

13.10 Large- and Small-Scale Time-Dependence

Two major types of large-scale time-variation occur in the solar wind. The first is due to interactions between slow and fast streams, when fast wind collides with slow wind and creates a *corotating interaction region* that rotates with the Sun. This includes compressions that steepen into collisionless shocks beyond a few AU.

The second type is caused by coronal mass ejections (CMEs) that account for roughly 10 per cent of the solar-wind mass-flux, ejecting 10^{12} to 10^{13} kg of mass at up to a few $\times 10^3$ km s^{-1}, roughly twice per day at solar minimum and eight times per day at solar maximum. They interact with the ambient solar wind, either being slowed down or accelerated by the wind, and create shock waves when the difference in speed is too large. Both co-rotation interaction regions and CMEs can act as barriers to cosmic rays, accelerate particles and interact with the Earth's magnetosphere.

On small scales at 1 AU, turbulent fluctuations are continually observed in the solar wind (Bruno and Carbone 2005). In homogeneous isotropic fluid turbulence, the energy is fed in at large scales and is transferred down to smaller scales by nonlinear interactions in a so-called *inertial range*, until eventually it reaches very small scales at which viscosity can dissipate the energy. In the inertial range, the energy has a *Kolmogorov spectrum*, behaving as a function of wavenumber (k) like $E(k) \sim \epsilon^{2/3} k^{-5/3}$, where ϵ is the energy dissipation rate. However, when the magnetic field dominates, there is instead the possibility of a *Kraichnan spectrum* $[E(k) \sim (\epsilon v_A)^{1/2} k^{-3/2}]$ and of an inverse cascade of magnetic helicity (Matthaeus et al. 1996).

In the fast solar wind, the fluctuations are, however, strongly intermittent and anisotropic and they evolve with distance from the Sun. There is a break in the spectral slope from -1 at low frequencies to roughly -1.6 at higher frequencies. This latter inertial range lies between 10^{-4} and 10^{-1} Hz at 1 AU in the ecliptic plane, but it is not possible to distinguish between a Kolmogorov and Kraichnan spectrum (Tu and Marsch 1995). The frequency break decreases with distance from 6×10^{-3} Hz at 0.3 AU to 10^{-5} Hz at 10 AU.

Fluctuations of magnetic field and velocity are highly correlated, such that $\delta \mathbf{v} \sim \pm \delta \mathbf{B}/\sqrt{\mu\rho}$ and represent large-amplitude Alfvén waves, in which there is little change in density or magnetic field magnitude but the magnetic field direction rotates. Evaluating the *cross-helicity* ($\frac{1}{2}\langle \mathbf{v} \cdot \mathbf{B}_1 \rangle$) or the energy in the Elsässer variables (Sec. 2.4.3) shows that the Alfvén waves are propagating mainly away from the Sun. Presumably, any magnetoacoustic modes generated at the Sun are dissipated very close to it. Out-of-the-ecliptic Ulysses observations at high latitude have shown that the Alfvénic turbulence evolves in the same way as in the ecliptic plane but more slowly (e.g., Goldstein et al. 1995; Horbury et al. 1995).

13.11 Going Beyond a Single-Fluid Picture

Since the solar wind is far from adiabatic (Sec. 13.5.1), fluid models require an assumption about heat conduction, but its classical expression is invalid, since heat is carried mainly by the electrons, which experience virtually no collisions. Furthermore, a fluid description requires the plasma to be close to thermodynamic equilibrium, which does not hold in the outer solar atmosphere. The particle mean-free path increases from about 1 km in the chromosphere (when $T = 10^3$ K, $n = 10^{16}$ m^{-3}) to 1 Mm in the corona ($T = 1$ MK, $n = 10^{13}$ m^{-3}) and to 10^4 Mm at 1 AU ($T = 10^5$ K, $n = 10^7$ m^{-3}). Ways of going beyond a simple fluid picture have been reviewed by Marsch (2006) and Meyer-Vernet (2007), whom we follow here.

All the simple fluid solar-wind models have faults. An isothermal model (Sec. 13.3) is unphysical, since the velocity tends to infinity at large distances and requires an infinite amount of energy. An adiabatic corona (Sec. 13.5.1) has changes that are too fast for heat conduction to act, but the enthalpy is too small to overcome gravity and produce a solar wind (PROBLEM 13.6). A more general polytropic model does overcome the above difficulties when α is close to unity, but is unsatisfactory, since it hides a lot of physics in the parameter α.

A two-fluid picture helps explain the observations (Sec. 13.11.1), but ultimately we need a much more complex kinetic picture (starting with an exospheric model, Sec. 13.11.2). This treats particles and non-equilibrium plasmas and enables us to calculate the particle distributions and the heat flux rather than impose their forms in an ad-hoc way. Indeed, the observed particle distributions are far from Maxwellian and exhibit the presence of particle beams and temperature anisotropies.

13.11.1 Two-Fluid Model

Since the interaction between electrons and protons is too weak to equalise their temperatures, it is natural to set up a two-fluid model, in which the electrons and protons are pictured as separate fluids that possess the same number density (by charge neutrality) and the same velocity (since the Sun is electrically neutral), but different temperatures (T_p, T_e) and heat fluxes (Q_p, Q_e).

Proton and electron masses differ by a factor $m_p/m_e \approx 1837$, so that electrons have a greater thermal speed by a factor $(m_p/m_e)^{1/2} \approx 43$. The consequences are:

- Protons are attracted much more strongly by the Sun's gravitational pull; electrons hardly notice gravity, since their thermal speed ($\approx 5.5\times10^6$ m s^{-1} for $T_e = 10^6$ K) is an order of magnitude larger than the escape speed;
- an electric field is set up between electrons and protons to preserve quasi-neutrality, since they experience forces of different sizes;
- the electrons carry heat much faster than the protons; and
- electron and proton temperatures differ, since they are weakly collisional; their energy exchange-rate is a factor m_p/m_e slower than momentum exchange.

The equations of mass continuity and proton and electron momentum are then (assuming $m_e \ll m_p$, so that inertia and gravity are negligible for the electrons)

$$\begin{aligned} 4\pi r^2 nv &= constant, \\ nm_p v\frac{dv}{dr} &= -\frac{d}{dr}(nk_BT_p) + neE - \frac{GM_\odot nm_p}{r^2}, \\ 0 &= -\frac{d}{dr}(nk_BT_e) - neE, \end{aligned}$$

where v and n are the common flow speed and number density and $nk_BT = p$.

If collisional energy-transfer between the particles is negligible, energy equations of the form (13.10) apply separately for protons and electrons, and the proton heat flux is much smaller than the electron heat

Table 13.1. *Mean density and temperature of protons and electrons in the low corona and in the fast wind at 1 AU, with polytropic indices* (α)

Fast Wind	n_0 (corona)	T_0 (corona)	n (1 AU)	T (1 AU)	α (1 AU)
Protons	10^{14} m^{-3}	2×10^6 K	3×10^6 m^{-3}	3×10^5 K	1.5–1.7
Electrons	10^{14} m^{-3}	10^6 K	3×10^6 m^{-3}	10^5 K	1.2

flux, so that

$$\frac{d}{dr}\left(\frac{n}{T_p^{3/2}}\right)=0 \tag{13.25}$$

and

$$\frac{k_B T_e^{5/2}}{n}\frac{d}{dr}\left(\frac{n}{T_e^{3/2}}\right)=\frac{d}{dr}\left(\frac{Q_e}{nv}\right). \tag{13.26}$$

The first equation implies that the proton fluid is adiabatic, with $T_p \sim n^{2/3}$, and so at large distances $v\sim$ constant, $n\sim r^{-2}$ and

$$T_p \sim r^{-4/3}.$$

The second equation then implies that, if the electron heat flux (Q_e) falls off more rapidly than nv, T_e will fall off less rapidly than T_p.

This may be generalised to include the proton heat-flux and terms $\sim \nu(T_e - T_p)$ on the right of Eqs. (13.25) and (13.26) to simulate energy exchange by Coulomb collisions, which is important only near the Sun. Then a numerical integration gives the temperature profiles shown in Figure 13.12. A single-fluid model (the limit as $\nu\to\infty$) gives a temperature of 2.1×10^5 K at 1 AU, whereas the two-fluid model gives a somewhat higher T_e (3.5×10^5 K) and a much lower T_p (0.04×10^5 K). Because $\kappa_e \gg \kappa_p$, the electron temperature decreases with distance less rapidly than the proton temperature. The protons cool almost adiabatically, while the electrons remain hot, because they are no longer forced to share their energy with the protons.

Measuring the temperatures in the solar wind is very difficult because the velocity distributions are far from Maxwellian, the pressures are not isotropic and each species is made up of several populations. Table 13.1 (after Meyer-Vernet 2007) shows results for the fast solar wind. The slow wind is more problematic, since it is variable in space and time and comprises a mixture of states: it is typically three times

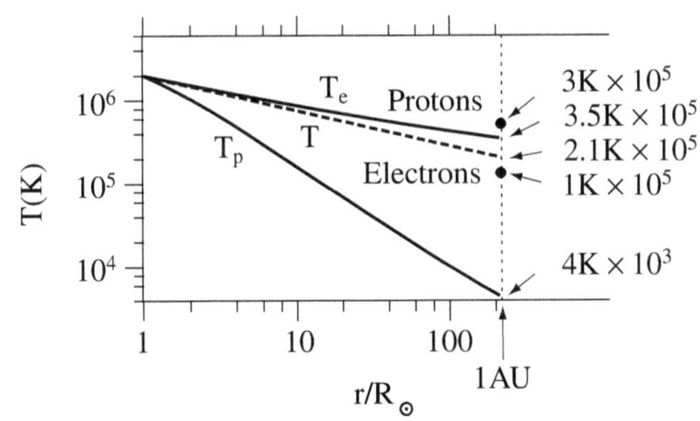

Figure 13.12. Electron (T_e) and proton (T_p) temperatures as functions of distance for the two-fluid solar-wind model of Hartle and Sturrock (1968). The single-fluid profile is shown dashed and the observed values of high-speed streams are shown by dots.

denser than the fast wind, and its T_e is twice as hot in the corona and 50 per cent hotter at 1 AU with a larger α, whereas T_p is six times cooler and closer to adiabatic.

A key question is: what creates the heat flux responsible for the strong deviation of electrons from adiabaticity in Table 13.1? The classical expression for heat flux is reasonable in the low corona, where the mean-free path (λ_{mfp}) is much smaller than the scale height (H), but it is certainly inapplicable in the solar wind beyond a height of about 1 $R_\odot$, where $\lambda_{mfp} > H$. Some authors would place this height much lower (say, 0.01 $R_\odot$) on the grounds that the Spitzer-Harm flux (and therefore the classical transport coefficients) cease to be valid when $\lambda_{mfp} > 0.001\,H$ (Scudder and Olbert 1983; Shoub 1983).

If an empirical heat flux is adopted in Eqs. (13.25) and (13.26) of the form $Q_e = K_e n v k_B T_e$, then Eq. (13.26) implies a polytropic law $p \sim n^{\alpha_e}$ with

$$\alpha_e = \frac{5+2K_e}{3+2K_e}, \qquad \text{so that} \qquad K_e = \frac{5-3\alpha_e}{2(\alpha_e-1)}.$$

At large distances, where $n \sim r^{-2}$, the electron temperature then falls off as

$$T_e \sim r^{-2(\alpha_e-1)}.$$

Observed values of $\alpha_e = 1.2$, n and T_e (Table 13.1) imply that $K_e = 3.5$ and $Q_e = 10^{-5}$ W m^{-2} at 1 AU. This value for the heat flux and its variation ($Q_e \sim r^{-2.4}$) with r at 1 AU are consistent with observations.

A two-fluid model (in which ions carry momentum and electrons carry heat) is a useful attempt to allow for $T_e \neq T_p$, but it does not agree well with fast solar-wind observations, since T_e is too high, T_p too low, v too low and the heat flux is unexplained. However, including an extra energy source for the protons above the critical point (such as collisionless damping of MHD waves) can enhance the flow speed and proton temperature significantly (van Der Holst et al. 2010). Also, a multi-fluid treatment with anisotropic temperatures (i.e., different parallel and perpendicular to the magnetic field) has been undertaken by allowing for nearly bi-Maxwellian distribution functions (Hu et al. 2000), while a 16-moment model has been used to model heavy ions (Lie-Svendsen and Esser 2005).

13.11.2 Kinetic Effects

Above an altitude of about 1 $R_\odot$, the corona ceases to be strongly collisional, and so the classical Spitzer transport coefficients (such as heat conduction) become invalid. *Exospheric models* aim to determine collisionless electron and proton motions in electrostatic, gravitational and magnetic fields. In reality, the solar wind consists mainly of massive protons (which tend to be trapped by gravity and carry most of the momentum) and much lighter electrons (which escape and carry most of the heat). The escaping electrons tend to leave the ions behind, but this produces an outward electric field that outweighs gravity beyond some distance, and so accelerates the protons to form a supersonic wind. Initially, the size of the electric field was underestimated and thought to be strong enough only for a breeze (Chamberlain 1960), but later it was corrected (Lemaire and Scherer 1969, 1971; Hollweg 1970; Jockers 1970).

For purely radial fields, the particle potential energies are

$$\Psi_e = -e\Phi_e, \qquad \Psi_p = e\Phi_e - m_p M_\odot G/r,$$

in terms of the electrostatic and gravitational potentials (Φ_e and $-M_\odot G/r$). For the electrons, gravity is negligible and the potential is monotonic (Figure 13.13b) so that the electric field always attracts them towards the Sun. Their energy ($m_e v^2/2 - e\Phi_e$) and magnetic moment ($v^2 \sin^2\theta/B$) are conserved, where θ is the pitch angle. They therefore have three types of orbit (Figure 13.13a): escaping electrons have a large enough speed to leave the potential well; trapped electrons (which comprise most of the electrons) bounce up and down along a field line between a magnetic mirror-point and an electric reflection-point; and ballistic electrons have too low a speed and fall back to the low corona. For the protons, the electric

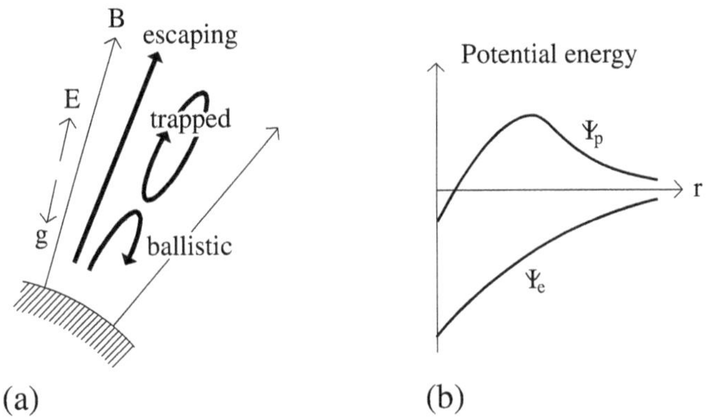

Figure 13.13. An exospheric model showing (a) the different types of electron motion in radial magnetic, electric and gravitational fields and (b) the total potential energies of electrons and protons as functions of r.

field dominates at large distances and repels them towards infinity, but close to the Sun gravity dominates and attracts them to the Sun (Figure 13.13b).

Thus, in an exospheric model, the solar wind is the evaporation of a hot corona, in which the thermal speed for electrons is larger than their speed of escape, while for the much heavier protons it is smaller. The escape of electrons in the high-speed tail of a distribution that is far from Maxwellian sets up an electric field that drags the protons out too. The electric field adjusts itself to equalise the electron and proton fluxes. However, even though this picture gives useful physical insights, it does not account in detail for the observed velocity distributions.

What is needed for a full *kinetic model* is to include weak particle collisions and plasma waves produced by various microinstabilities (e.g., Pierrard et al. 2001). Such a description deals with equations of motion for particles whose distribution functions $[f(\mathbf{r},t,\mathbf{v})]$ vary in space and time and are far from Maxwellian. (For a Maxwellian, $f \sim \exp -(v^2/v_{th}^2)$, where $v_{th}^2 = 2k_BT/m$.) Then the macroscopic quantities are calculated as moments of $f(\mathbf{v})$. Thus, the number density (n) and mean velocity ($\mathbf{v}_w$) are given by the first two moments

$$n = \int f(\mathbf{v})\ d^3v, \qquad n\mathbf{v}_w = \int \mathbf{v}f(\mathbf{v})\ d^3v,$$

while the temperature is the width of the distribution

$$T = \frac{m}{3nk_B} \int |\mathbf{v}-\mathbf{v}_w|^2 f(\mathbf{v})\ d^3v$$

and the heat flux is its skewness

$$\mathbf{Q} = \tfrac{1}{2} m \int (\mathbf{v}-\mathbf{v}_w)|\mathbf{v}-\mathbf{v}_w|^2 f(\mathbf{v})\ d^3v.$$

A kinetic theory would solve the Vlasov-Boltzmann equation for the distribution function of each species, namely,

$$\left[\frac{\partial}{\partial t} + \mathbf{v}\cdot\frac{\partial}{\partial \mathbf{r}} + \left(\mathbf{g} + \frac{e}{m}(\mathbf{E}+\mathbf{v}\times\mathbf{B})\right)\cdot\frac{\partial}{\partial \mathbf{v}}\right] f = \left[\frac{df}{dt}\right]_{c,w},$$

where the right-hand side represents Coulomb collisions (by, e.g., a Fokker-Planck integral) and wave-particle interactions (by, e.g., a quasi-linear diffusion operator). Various microinstabilities (e.g., ion-acoustic, firehose or mirror) and wave-particle interactions will modify the distribution functions and transfer energy between electrons and ions and between parallel and perpendicular temperatures.

The theory would proceed to provide expressions for the collisionless transport coefficients (viscosity, magnetic diffusivity and thermal conduction). One may at first sight imagine that in principle these could be incorporated in a new MHD theory. However, this may be over-optimistic, since in a weakly collisional plasma the transport coefficients do not depend on local derivatives of the moments and so are non-local. Multi-moment models have generally been abandoned, since they lack convergence. Thus, one is left with a more pragmatic approach of adopting a series of different models that offer different insights and complement one another. For example, one can either consider simplified versions of the kinetic equations or take moments of the Vlasov-Boltzmann equation to yield the continuity and momentum equations and adopt an approximate closure of the moment equations, so that the heat flux and other transport expressions can be related to n, T and $\mathbf{v}$. Useful simplified kinetic models include: a gyrotropic approach in which the distribution function depends on only two velocity components $(v_\parallel, v_\perp)$ (Olsen and Leer 1999); a semi-kinetic hybrid model that reduces the distribution function even further by integrating over $v_\perp$ (Vocks and Marsch 2002) and a hybrid model that treats the ions as particles and the electrons as a fluid (e.g., Ofman and Viñas 2007).

Observed proton and electron velocity-distributions in the solar wind deviate strongly from Maxwellian and therefore from local thermodynamic equilibrium. Protons have different temperatures parallel and perpendicular to the magnetic field $(T_\perp > T_\parallel)$ and there is preferential heating of heavy ions $(T_{ion} \gg T_p > T_e)$. Protons consist roughly of the sum of two bi-Maxellians $[f(\mathbf{v}) \sim \exp -(v_\parallel^2/v_{th\parallel}^2 + v_\perp^2/v_{th\perp}^2)]$. Electrons are close to bi-Maxwellian at small speeds $(v^2 \ll v_{the}^2)$, but at larger speeds their distribution declines with v much more slowly (like a power law rather than an exponential), and it also possesses an excess of particles moving parallel to the magnetic field. A combination of Maxwellian at small v and power law at large v can be approximated by a *kappa distribution* $f_\kappa \sim [1 + v^2/(\kappa v_{th}^2)]^{-(\kappa+1)}$.

The observed preferential ion heating $(T_\perp > T_\parallel)$ is highly suggestive of acceleration by a resonance at 10^2 to 10^4 Hz between their gyrofrequency as ions spiral around the magnetic field and the frequency of *ion-cyclotron waves.* It is not clear how such waves are excited in the corona, but Cranmer et al. (2009) have reviewed some possibilities. The first is that small-scale reconnection events in the low corona of the network may generate the waves directly, which then propagate up to a height where they become resonant with the ions (Axford and McKenzie 1992; Tu and Marsch 1997). Alternately, such microflaring may generate fast shocks that accelerate the ions or bursts of electron beams that later produce ion-cyclotron waves. A second general possibility is MHD turbulence in the extended corona, which cascades energy from low-frequency (0.01 Hz) MHD waves, either to low-frequency kinetic Alfvén waves $(k_\perp \gg k_\parallel)$ and then to ion-cyclotron waves or directly to ion-cyclotron waves via nonlinear coupling. Thirdly, low-frequency Alfvén waves may undergo polarization drift or stochastic heating that accelerates ions.

13.12 Conclusion

Solar-wind theory has progressed enormously since the early days of Parker's theory. It is an immense subject with many effects that are not covered here, including interactions of the solar wind with planets and its slowing down to subsonic speeds at a *termination shock* at about 100 AU before reaching the *heliopause.*

Particularly problematical is the slow solar wind, whose origin is unclear: it may start near the boundaries of helmet streamers or in transient coronal holes or be a response to continual reconfiguration of the complex coronal field driven by photospheric motions. Moreover, the relative importance of wave/turbulence and reconnection models for accelerating the solar wind needs to be assessed quantitatively. In addition, at least in the high corona, a unified mechanism of coronal heating and wind acceleration is likely to be at work and so the two problems should not be treated separately.

Underlying issues here are to study the physics of MHD wave generation, propagation and dissipation in more realistic 3D configurations, as well as the effect of myriads of reconnection events in heating and accelerating coronal plasma. Also, the role of jets and polar plumes in contributing to the fast solar wind is not understood.

In future, as in many areas of solar MHD, a complementary approach is required, including observations, data analysis, 3D computational modelling, and simple analytical models that incorporate the key physics. For the solar wind and outer corona, however, an extra feature is to go beyond a single-fluid picture and develop multi-fluid and kinetic models that explain the observed non-Maxwellian distribution functions and produce the magnitude, scaling and form of the transport coefficients. An in-depth plasma-physics analysis is needed to understand in detail the mechanisms for accelerating the fast and slow solar wind. The same is true for thin collisionless dissipative regions or current sheets in solar flares and in coronal heating locations.

Appendix 1

Units

The question of what units to adopt has often complicated the subject of electromagnetism. It clearly would be advantageous if a single system could be accepted by all, even though the transition to standard units may be a painful one. I myself was reared on the gaussian system, and some years ago I suffered the trauma of changing to rationalised mks, which is part of the now internationally adopted SI system.

In the present book I have employed mks units, and have commonly quoted magnetic fields (**B**) in gauss (G) and magnetic flux in maxwells (Mx), which is an acceptable option in SI. In formulae, however, the tesla (T) and the weber (Wb) have been used unless otherwise stated, such that

$$1\ \text{T} = 10^4\ \text{G}, \qquad 1\ \text{Wb} = 10^8\ \text{Mx}.$$

Also, in the text, lengths have often been quoted in megametres (Mm), such that

$$1\ \text{Mm} = 10^3\ \text{km} = 10^6\ \text{m},$$

although lengths in formulae are measured in m unless otherwise stated.

At one time, it appeared that each book on electromagnetism used a different set of units. The reasons for confusion are as follows.

(i) A set of electromagnetic quantities may be measured in either electrostatic units (esu) or electromagnetic units (emu) or a combination of the two.
(ii) The magnetic permeability (μ) and dielectric constant (ϵ) may be either dimensionless or dimensional. In the former case, mass, length and time are fundamental, whereas in the latter case some extra fundamental unit is introduced.
(iii) Mass and length may be measured in either centimetres and grams (cgs) or metres and kilograms (mks), respectively.
(iv) A system may be 'rationalised' to remove the factor 4π from the equations.

The two main sets of units are *gaussian cgs* (for which the defining equations are simpler, so it is a 'natural' system) or *rationalised mks* (for which Maxwell's equations are simpler, so it is easier to use).

(a) Gaussian CGS Units

Here, **E**, **D** and e are measured in esu, while **j**, **B** and **H** are in emu. Also, ϵ and μ are dimensionless. Length, mass and time are fundamental in this set of units, and then the other units are defined in terms of them by appropriate equations. For instance: Newton's Second Law of Motion gives force; the force of repulsion between charges e_1 and e_2, say, separated by a distance r gives charge; the force per unit length between currents I_1 and I_2, say, separated by r gives current; and the magnetic induction at a distance r from a straight wire carrying a current I gives magnetic induction.

Quantity	Defining Equation	Unit
Length	–	1 cm
Mass	–	1 g
Time	–	1 s
Force	$F = m\dfrac{d^2x}{dt^2}$	1 dyne = 1 g cm s^{-2}
Charge	$F = \dfrac{e_1e_2}{r^2}$	1 statcoulomb = 1 cm dyne$^{1/2}$
Electric field	$E = \dfrac{e}{r^2}$	1 statvolt cm^{-1} = 1 dyne$^{1/2}$ cm^{-1}
Electric displacement	$D = \epsilon E$	1 statvolt cm^{-1}
Current	$\dfrac{F}{l} = \dfrac{2I_1I_2}{r}$	1 abamp = 1 dyne$^{1/2}$
Magnetic induction	$B = \dfrac{2I}{r}$	1 gauss (G) = 1 abamp cm^{-1}
Magnetic field	$H = \dfrac{B}{\mu}$	1 oersted = 1 gauss (G)
Magnetic flux	$F_M = \int_S \mathbf{B}\cdot d\mathbf{S}$	1 maxwell (Mx) = 1 G cm^2

If the current is measured in esu rather than emu, it is defined as the time rate of change of charge, and so has units of 1 statamp = 1 statcoulomb s^{-1}. The ratio of the values of I measured in the two units is

$$\frac{I_{\text{in esu}}}{I_{\text{in emu}}} = c, \text{ say.}$$

The constant of proportionality (c) has units of cm s^{-1} and is the speed of light in a vacuum; it has an approximate value 3×10^{10} cm s^{-1}.

Maxwell's equations in this system are

$$\nabla\times\mathbf{E} = -\frac{\partial\mathbf{B}}{\partial t}, \quad \nabla\cdot\mathbf{D} = 4\pi\rho, \quad \nabla\times\mathbf{H} = \frac{1}{c}\frac{\partial\mathbf{D}}{\partial t} + 4\pi\mathbf{j}, \quad \nabla\cdot\mathbf{B} = 0.$$

If $\mathbf{j}$ is measured in esu instead of emu, $4\pi\mathbf{j}$ is replaced by $4\pi\mathbf{j}/c$

(b) Rationalised MKS Units

Here, length, mass, time and current (amp) are fundamental, and the other units are defined in terms of them by the appropriate equations. μ_0 is defined to be $4\pi \times 10^{-7}$ henry m^{-1} and is the vacuum value of μ, where 1 henry (H) = 1 NA^{-2} m is the mks unit of inductance. The unit of current is defined such that two wires carrying currents of 1 amp and separated by 1 m feel a force per unit length of 2×10^{-7} N m^{-1}. ϵ_0 is the vacuum value of ϵ and has units of farad m^{-1}, where 1 farad (F) = 1 A^2 s^2 N^{-1} m^{-1} is the mks unit of electrical capacitance. Its value is determined (as approximately 8.854×10^{-12}) from Coulomb's law $[F = e_1e_2/(4\pi\epsilon_0r^2)]$ such that two charges of 1 C separated by 1 m feel a repulsive force of $(4\pi\epsilon_0)^{-1}$N.

Quantity	Defining Equation	Unit
Length	–	1 m
Mass	–	1 kg
Time	–	1 s
Force	$F = m\dfrac{d^2x}{dt^2}$	1 newton (N) = 1 kg m s^{-2}
Current	$\dfrac{F}{l} = \dfrac{\mu_0}{4\pi}\dfrac{2I_1I_2}{r}$	1 amp (A)
Magnetic induction	$B = \dfrac{\mu_0}{4\pi}\dfrac{2I}{r}$	1 tesla (T) = 1 N A^{-1} m^{-1}
Magnetic field	$H = \dfrac{B}{\mu}$	1 A m^{-1}
Magnetic flux	$F_M = \int_S \mathbf{B}\cdot d\mathbf{S}$	1 weber (Wb) = 1 T m^2
Charge	$I = \dfrac{dq}{dt}$	1 coulomb (C) = 1 A s
Electric field	$E = \dfrac{1}{4\pi\epsilon_0}\dfrac{e}{r^2}$	1 V m^{-1} = 1 N C^{-1}
Electric displacement	$D = \epsilon E$	1 C m^{-2}

Maxwell's equations in this system are

$$\nabla \times \mathbf{E} = -\frac{\partial \mathbf{B}}{\partial t}, \quad \nabla \cdot \mathbf{D} = \rho, \quad \nabla \times \mathbf{H} = \frac{\partial \mathbf{D}}{\partial t} + \mathbf{j}, \quad \nabla \cdot \mathbf{B} = 0.$$

The wave equation in a vacuum becomes

$$\left(\nabla^2 - \epsilon_0\mu_0\frac{\partial^2}{\partial t^2}\right)\mathbf{E} = \mathbf{0},$$

and so $c^2 = (\epsilon_0\mu_0)^{-1}$.

(c) Transformation of Symbols in Equations

A comparison of, for instance, Coulomb's Law in the two sets of units shows that, in going from cgs to mks, e becomes $e(4\pi\epsilon_0)^{-1/2}$. The transformation for the other symbols is as follows.

Quantity	Gaussian	Rationalised MKS
Speed of light	c	$(\epsilon_0\mu_0)^{-1/2}$
Charge	e	$e(4\pi\epsilon_0)^{-1/2}$
Current	j	$j(4\pi\epsilon_0)^{-1/2}$
Electric field	E	$E(4\pi\epsilon_0)^{1/2}$
Magnetic induction	B	$B(4\pi/\mu_0)^{1/2}$
Electric displacement	D	$D(4\pi/\epsilon_0)^{1/2}$
Magnetic field	H	$H(4\pi\mu_0)^{1/2}$
Dielectric constant	ϵ	ϵ/ϵ_0
Magnetic permeability	μ	μ/μ_0
Electrical conductivity	σ	$\sigma(4\pi\epsilon_0)^{-1}$

(d) Relations Between Units

Quantity	Gaussian	Rationalised MKS
Length	1 cm	$= 10^{-2}$ m
Mass	1 g	$= 10^{-3}$ kg
Time	1 s	$= 1$ s
Force	1 dyne	$= 10^{-5}$ newton (N)
Pressure	1 dyne cm^{-2}	$= 0.1$ N m^{-2}
Energy	1 erg	$= 10^{-7}$ joule (J)
	1 calorie	$= 4.185$ J
Power	1 erg s^{-1}	$= 10^{-7}$ watt (W)
Energy flux	1 erg cm^{-2} s^{-1}	$= 10^{-3}$ W m^{-2}
Charge	1 statcoulomb	$= \frac{1}{3} \times 10^{-9}$ coulomb (C)
Electric field	1 statvolt cm^{-1}	$= 3 \times 10^{4}$ V m^{-1}
Electric displacement	1 statvolt cm^{-1}	$= \dfrac{1}{12\pi} \times 10^{-5}$ C m^{-2}
Current	1 statamp	$= \frac{1}{3} \times 10^{-9}$ A
Current density	1 statamp cm^{-2}	$= \frac{1}{3} \times 10^{-5}$ A m^{-2}
Electrical conductivity	1 s^{-1}	$= \frac{1}{9} \times 10^{-9}$ mho m^{-1}
Magnetic induction	1 G	$= 10^{-4}$ tesla (T)
Magnetic flux	1 maxwell (Mx)	$= 10^{-8}$ weber (Wb)
Magnetic field	1 oersted	$= \dfrac{1}{4\pi} \times 10^{3}$ A m^{-1}

Appendix 2

Useful Values and Expressions

Physical Constants

Speed of light	$c = 2.998 \times 10^8$ m s^{-1}
Electron charge	$e = 1.602 \times 10^{-19}$ C
Electron mass	$m_e = 9.109 \times 10^{-31}$ kg
Proton mass	$m_p = 1.673 \times 10^{-27}$ kg
Mass ratio	$m_p/m_e = 1837$
Electron volt	1 eV $= 1.602 \times 10^{-19}$ J $= 11{,}605$ K
Boltzmann constant	$k_B = 1.381 \times 10^{-23}$ J deg^{-1}
Gravitational constant	$G = 6.672 \times 10^{-11}$ N m^2 kg^{-2}
Gas constant	$\tilde{R} = 8.3 \times 10^3$ m^2 s^{-2} deg^{-1}
Permeability of free space	$\mu_0 = 4\pi \times 10^{-7} = 1.257 \times 10^{-6}$ henry m^{-1}
Permittivity of free space	$\epsilon_0 = 8.854 \times 10^{-12}$ farad m^{-1}

Plasma Properties (n_e in m^{-3}, B in G, T in ^{o}K, v in m s^{-1}, l in m)

Sound speed	$c_s = 152\ T^{1/2}$ m s^{-1}
Alfvén speed	$v_A = 2.8 \times 10^{12}\ B\ n^{-1/2}$ m s^{-1}
Plasma beta	$\beta = 3.5 \times 10^{-21}\ n\ T\ B^{-2}$
Magnetic Reynolds number	$R_m = 2 \times 10^{-9}\ l\ v\ T^{3/2}$
Scale-height	$H = 50\ T$ m for $\tilde{\mu} = 0.6$
Electrical conductivity	$\sigma = 10^{-3}\ T^{3/2}$ mho m^{-1}
Magnetic diffusivity	$\eta = 10^9\ T^{-3/2}$ m^2 s^{-1}
Thermal conductivity	$\kappa_\parallel = 10^{-11}\ T^{5/2}$ W m^{-1} deg^{-1}
Electron plasma frequency	$\omega_{pe} = 56.4\ n_e^{1/2}$ rad s^{-1}
Electron gyro-frequency	$\Omega_e = 1.76\ B \times 10^7$ rad s^{-1}
Ion (proton) gyro-frequency	$\Omega_i = 9.58\ B \times 10^3$ rad s^{-1}
Debye length	$\lambda_D = 69\ (T_e/n_e)^{1/2}$ m
Electron gyro-radius	$r_e = 2.38 \times 10^{-2}\ (T_e/11{,}605)^{1/2}\ B^{-1}$ m
Ion (proton) gyro-radius	$r_i = 1.02\ (T_i/11{,}605)^{1/2}\ B^{-1}$ m
Electron thermal speed	$v_{Te} = (k_B T_e/m_e)^{1/2} = \Omega_e r_e = 4.19 \times 10^5 (T_e/11{,}605)^{1/2}$ m s^{-1}
Ion (proton) thermal speed	$v_{Ti} = (k_B T_i/m_p)^{1/2} = \Omega_i r_i = 9.79 \times 10^3 (T_i/11{,}605)^{1/2}$ m s^{-1}

Solar Properties

Age		$= 4.5 \times 10^9$ yr
Mass	$M_\odot$	$= 1.99 \times 10^{30}$ kg
Radius	$R_\odot$	$= 696$ Mm $= 6.96 \times 10^8$ m $= 109$ Earth radii
Surface gravity	$g_\odot$	$= 274$ m s^{-2}
Escape velocity		$= 618$ km s^{-1}
Luminosity	$L_\odot$	$= 3.86 \times 10^{26}$ W
Equatorial rotation speed		$= 26$ days
Angular momentum		$= 1.7 \times 10^{41}$ kg m^2 s^{-1}
Mass-loss rate		$= 10^9$ kg s^{-1}
Surface distances	1 arcsec	$= 726$ km
Distance to Earth	1 AU	$= 1.50 \times 10^{11}$ m $= 215\ R_\odot$

Typical Sizes (1 Mm = 1000 km = 10^6 m)

Depth of convection zone	= 200 Mm
Depth of photosphere	= 500 km
Height of transition region	= 1–10 Mm
Width of granule	= 1 Mm
Width of supergranule	= 30 Mm
Width of giant cell	= 200 Mm
Width of sunspot	= 30 Mm

Temperature

Effective	= 5785 K
Solar centre	$= 1.5 \times 10^7$ K
Base of photosphere	= 6000 K
Temperature minimum	= 4300 K
Chromosphere	$= 10^4$ K
Transition region	$= 10^5$ K
Corona	$= 1\text{–}5 \times 10^6$ K
At 1 AU	$= 10^5$ K

Density

Solar centre	$= 10^{32}$ m^{-3}
Photosphere	$= 10^{23}$ m^{-3}
Transition region	$= 10^{15}$ m^{-3}
Corona	$= 1\text{–}50 \times 10^{14}$ m^{-3}
At 1 AU	$= 10^7$ m^{-3}

Cylindrical Polar Coordinates (R, ϕ, z)

$$\nabla A = \frac{\partial A}{\partial R}\hat{\boldsymbol{R}} + \frac{1}{R}\frac{\partial A}{\partial \phi}\hat{\boldsymbol{\phi}} + \frac{\partial A}{\partial z}\hat{\boldsymbol{z}}.$$

$$\nabla \cdot \mathbf{B} = \frac{1}{R}\frac{\partial}{\partial R}(RB_R) + \frac{1}{R}\frac{\partial B_\phi}{\partial \phi} + \frac{\partial B_z}{\partial z}.$$

$$\nabla \times \mathbf{B} = \left(\frac{1}{R}\frac{\partial B_z}{\partial \phi} - \frac{\partial B_\phi}{\partial z}\right)\hat{\boldsymbol{R}} + \left(\frac{\partial B_R}{\partial z} - \frac{\partial B_z}{\partial R}\right)\hat{\boldsymbol{\phi}} + \left(\frac{1}{R}\frac{\partial}{\partial R}(RB_\phi) - \frac{1}{R}\frac{\partial B_R}{\partial \phi}\right)\hat{\boldsymbol{z}}.$$

$$\nabla^2 A = \frac{1}{R}\frac{\partial}{\partial R}\left(R\frac{\partial A}{\partial R}\right) + \frac{1}{R^2}\frac{\partial^2 A}{\partial \phi^2} + \frac{\partial^2 A}{\partial z^2}.$$

$$(\mathbf{B}\cdot\nabla)\mathbf{B} = \left(B_R\frac{\partial B_R}{\partial R} + \frac{B_\phi}{R}\frac{\partial B_R}{\partial \phi} - \frac{B_\phi^2}{R} + B_z\frac{\partial B_R}{\partial z}\right)\hat{\boldsymbol{R}}$$
$$+ \left(\frac{B_R}{R}\frac{\partial}{\partial R}(RB_\phi) + \frac{B_\phi}{R}\frac{\partial B_\phi}{\partial \phi} + B_z\frac{\partial B_\phi}{\partial z}\right)\hat{\boldsymbol{\phi}} + \left(B_z\frac{\partial B_z}{\partial z} + B_R\frac{\partial B_z}{\partial R} + \frac{B_\phi}{R}\frac{\partial B_z}{\partial \phi}\right)\hat{\boldsymbol{z}}.$$

Spherical Polar Coordinates (r, θ, ϕ)

$$\nabla A = \frac{\partial A}{\partial r}\hat{\boldsymbol{r}} + \frac{1}{r}\frac{\partial A}{\partial \theta}\hat{\boldsymbol{\theta}} + \frac{1}{r\sin\theta}\frac{\partial A}{\partial \phi}\hat{\boldsymbol{\phi}}.$$

$$\nabla \cdot \mathbf{B} = \frac{1}{r^2}\frac{\partial}{\partial r}(r^2 B_r) + \frac{1}{r\sin\theta}\frac{\partial}{\partial \theta}(\sin\theta B_\theta) + \frac{1}{r\sin\theta}\frac{\partial B_\phi}{\partial \phi}.$$

$$\nabla \times \mathbf{B} = \frac{1}{r\sin\theta}\left(\frac{\partial}{\partial \theta}(\sin\theta B_\phi) - \frac{\partial B_\theta}{\partial \phi}\right)\hat{\boldsymbol{r}} + \left(\frac{1}{r\sin\theta}\frac{\partial B_r}{\partial \phi} - \frac{1}{r}\frac{\partial}{\partial r}(rB_\phi)\right)\hat{\boldsymbol{\theta}} + \left(\frac{1}{r}\frac{\partial}{\partial r}(rB_\theta) - \frac{1}{r}\frac{\partial B_r}{\partial \theta}\right)\hat{\boldsymbol{\phi}}.$$

$$\nabla^2 A = \frac{1}{r^2}\frac{\partial}{\partial r}\left(r^2\frac{\partial A}{\partial r}\right) + \frac{1}{r^2\sin\theta}\frac{\partial}{\partial \theta}\left(\sin\theta\frac{\partial A}{\partial \theta}\right) + \frac{1}{r^2\sin^2\theta}\frac{\partial^2 A}{\partial \phi^2}.$$

$$(\mathbf{B}\cdot\nabla)\mathbf{B} = \left(B_r\frac{\partial B_r}{\partial r} + \frac{B_\theta}{r}\frac{\partial B_r}{\partial \theta} - \frac{B_\theta^2 + B_\phi^2}{r} + \frac{B_\phi}{r\sin\theta}\frac{\partial B_r}{\partial \phi}\right)\hat{\boldsymbol{r}}$$
$$+ \left[B_r\frac{\partial B_\theta}{\partial r} + \frac{B_\theta}{r}\frac{\partial B_\theta}{\partial \theta} + \frac{B_r B_\theta}{r} + \frac{B_\phi}{r\sin\theta}\left(\frac{\partial B_\theta}{\partial \phi} - \cos\theta B_\phi\right)\right]\hat{\boldsymbol{\theta}}$$
$$+ \left(B_r\frac{\partial B_\phi}{\partial r} + \frac{B_r B_\phi}{r} + \frac{B_\phi}{r\sin\theta}\frac{\partial B_\phi}{\partial \phi} + \frac{B_\theta}{r\sin\theta}\frac{\partial}{\partial \theta}(B_\phi\sin\theta)\right)\hat{\boldsymbol{\phi}}.$$

Orthogonal Curvilinear Coordinates (u_1, u_2, u_3)

$$\mathbf{B} = \mathbf{B}_1\mathbf{e}_1 + \mathbf{B}_2\mathbf{e}_2 + \mathbf{B}_3\mathbf{e}_3.$$

$$\nabla A = \frac{1}{h_1}\frac{\partial A}{\partial u_1}\mathbf{e}_1 + \frac{1}{h_2}\frac{\partial A}{\partial u_2}\mathbf{e}_2 + \frac{1}{h_3}\frac{\partial A}{\partial u_3}\mathbf{e}_3.$$

$$\nabla \cdot \mathbf{B} = \frac{1}{h_1 h_2 h_3}\left[\frac{\partial}{\partial u_1}(h_2 h_3 B_1) + \frac{\partial}{\partial u_2}(h_3 h_1 B_2) + \frac{\partial}{\partial u_3}(h_1 h_2 B_3)\right].$$

$$\nabla \times \mathbf{B} = \frac{1}{h_2 h_3}\left[\frac{\partial}{\partial u_2}(h_3 B_3) - \frac{\partial}{\partial u_3}(h_2 B_2)\right]\mathbf{e}_1 + \frac{1}{h_3 h_1}\left[\frac{\partial}{\partial u_3}(h_1 B_1) - \frac{\partial}{\partial u_1}(h_3 B_3)\right]\mathbf{e}_2 + \frac{1}{h_1 h_2}\left[\frac{\partial}{\partial u_1}(h_2 B_2) - \frac{\partial}{\partial u_2}(h_1 B_1)\right]\mathbf{e}_3.$$

$$\nabla^2 A = \frac{1}{h_1 h_2 h_3}\left[\frac{\partial}{\partial u_1}\left(\frac{h_2 h_3}{h_1}\frac{\partial A}{\partial u_1}\right) + \frac{\partial}{\partial u_2}\left(\frac{h_3 h_1}{h_2}\frac{\partial A}{\partial u_2}\right) + \frac{\partial}{\partial u_3}\left(\frac{h_1 h_2}{h_3}\frac{\partial A}{\partial u_3}\right)\right].$$

For instance, to transfer to toroidal coordinates (u, v, ϕ) we write cartesians as

$$x = \frac{a\sinh v \cos\phi}{\cosh v - \cos u}, \quad y = \frac{a \sinh v \sin\phi}{\cosh v - \cos u} \quad z = \frac{a \sin u}{\cosh v - \cos u},$$

and put

$$h_1^2 = h_2^2 = \frac{a^2}{(\cosh v - \cos u)^2}, \qquad h_3^2 = \frac{a^2 \sinh^2 v}{(\cosh v - \cos u)^2}.$$

Special Functions

Gamma function

$$\Gamma(x) = \int_0^\infty e^{-u}\, u^{x-1}\, du.$$

Beta function

$$B(x_1, x_2) = \int_0^1 u^{x_1 - 1}\,(1-u)^{x_2 - 1}\, du.$$

Bessel functions $J_\nu(x)$ and $J_{-\nu}(x)$ satisfy

$$x^2\frac{d^2 y}{dx^2} + x\frac{dy}{dx} + (x^2 - \nu^2)y = 0.$$

Hypergeometric function $F(a, b, c, x)$ satisfies

$$(x^2 - x)\frac{d^2 y}{dx^2} + [(1 + a + b)x - c]\frac{dy}{dx} + aby = 0.$$

Error function

$$\operatorname{erf}(x) = \frac{2}{\sqrt{\pi}} \int_0^x e^{-u^2}\ du.$$

Legendre functions $P_\nu(x)$ and $Q_\nu(x)$ satisfy

$$(1-x^2)\frac{d^2y}{dx^2} - 2x\frac{dy}{dx} + \nu(\nu+1)y = 0.$$

Dawson integral function

$$\operatorname{Daw}(x) = e^{-x^2} \int_0^x e^{X^2}\ dX.$$

References

Abbett, W. P. and Fisher, G. H. (2003). A coupled model for the emergence of active region magnetic flux into the solar corona. *Astrophys. J.* **582**, 475–485.

Acheson, D. J. (1978). On the instability of toroidal magnetic fields and differential rotation in stars. *Phil. Trans. Roy. Soc. Lond. A* **289**, 459–500.

Acheson, D. J. (1979). Instability by magnetic buoyancy. *Solar Phys.* **62**, 23–50.

Alexander, D. and Fletcher, L. (1999). High-resolution observations of plasma jets in the solar corona. *Solar Phys.* **190**, 167–184.

Altschuler, M. D. and Newkirk, G. (1969). Magnetic fields and the structure of the solar corona. I. Methods of calculating coronal fields. *Solar Phys.* **9**, 131–149.

Aly, J. J. (1991). How much energy can be stored in a three-dimensional force-free magnetic field? *Astrophys. J. Letts.* **375**, L61–L64.

Amari, T. and Aly, J. J. (1989). Interaction between a line current and a two-dimensional constant-alpha force-free field – an analytical model for quiescent prominences. *Astron. Astrophys.* **208**, 261–270.

Amari, T. and Aly, J. J. (1990). Current sheets in two-dimensional potential magnetic fields. II. Asymptotic limits of indefinitely sheared force-free fields. *Astron. Astrophys.* **227**, 628–633.

Amari, T., Aly, J. J., Luciani, J. F., Boulmezaoud, T. Z., and Mikic, Z. (1997). Reconstructing the solar coronal magnetic field as a force-free magnetic field. *Solar Phys.* **174**, 129–149.

Amari, T., Aly, J. J., Mikic, Z., and Linker, J. (2007). Coronal mass ejection initiation and complex topology configurations in the flux cancellation and breakout models. *Astrophys. J. Letts.* **671**, L189–L192.

Amari, T., Aly, J. J., Mikic, Z., and Linker, J. (2010). Coronal mass ejection initiation: on the nature of the flux cancellation model. *Astrophys. J. Letts.* **717**, L26–L30.

Amari, T., Boulmezaoud, T. Z., and Aly, J. J. (2006). Well-posed reconstruction of the solar coronal magnetic field. *Astron. Astrophys.* **446**, 691–705.

Amari, T., Boulmezaoud, T. Z., and Maday, Y. (1998). A regularization method for the extrapolation of the photospheric solar magnetic field. I. Linear force-free field. *Astron. Astrophys.* **339**, 252–260.

Amari, T., Boulmezaoud, T. Z., and Mikic, Z. (1999). An iterative method for the reconstruction of the solar coronal magnetic field. I. Method for regular solutions. *Astron. Astrophys.* **350**, 1051–1059.

Amari, T. and Luciani, J. F. (2000). Helicity redistribution during relaxation of astrophysical plasmas. *Phys. Rev. Letts.* **84**, 1196–1199.

Amari, T., Luciani, J. F., and Aly, J. J. (2005). Non-current-free coronal closure of subphotospheric MHD models. *Astrophys. J. Letts.* **629**, L37–L40.

Amari, T., Luciani, J. F., Mikic, Z., and Linker, J. (2000). A twisted flux rope model for coronal mass ejections and two-ribbon flares. *Astrophys. J.* **529**, L49–L52.

Anderson, J. E. (1963). *MHD Shock Waves.* (MIT Press, Cambridge, Mass.).

Antiochos, S. K. (1998). The magnetic topology of solar eruptions. *Astrophys. J. Letts.* **502**, L181–L184.

Antiochos, S. K., Dahlburg, R., and Klimchuk, J. A. (1994). The magnetic field of solar prominences. *Astrophys. J.* **420**, L41–L44.

Antiochos, S. K., DeVore, C. R., Karpen, J. T., and Mikić, Z. (2007). Structure and dynamics of the Sun's open magnetic field. *Astrophys. J.* **671**, 936–946.

Antiochos, S. K., DeVore, C. R., and Klimchuk, J. A. (1999). A model for solar coronal mass ejections. *Astrophys. J.* **510**, 485–493.

Antiochos, S. K., Karpen, J. T., DeLuca, E. E., Golub, L., and Hamilton, P. (2003). Constraints on active region coronal heating. *Astrophys. J.* **590**, 547–553.

Antiochos, S. K. and Klimchuk, J. A. (1991). A model for the formation of solar prominences. *Astrophys. J.* **378**, 372–377.

Antiochos, S. K., MacNeice, P. J., and Spicer, D. S. (2000). The thermal nonequilibrium of prominences. *Astrophys. J.* **536**, 494–499.

Antiochos, S. K., MacNeice, P. J., Spicer, D. S., and Klimchuk, J. A. (1999). The dynamic formation of prominence condensations. *Astrophys. J.* **512**, 985–991.

Antiochos, S. K., Mikić, Z., Titov, V. S., Lionello, R., and Linker, J. A. (2011). A model for the sources of the slow solar wind. *Astrophys. J.* **731**, 112.

Antiochos, S. K. and Sturrock, P. A. (1978). Evaporative cooling of flare plasma. *Astrophys. J.* **220**, 1137–1143.

Antolin, P., Shibata, K., and Vissers, G. (2010). Coronal rain as a marker for coronal heating mechanisms. *Astrophys. J.* **716**, 154–166.

Antolin, P., Vissers, G., and Rouppe van der Voort, L. (2012). On-disk coronal rain. *Solar Phys.* **280**, 457–474.

Anzer, U. (1968). The stability of force-free magnetic fields with cylindrical symmetry in the context of solar flares. *Solar Phys.* **3**, 298–315.

Anzer, U. (1972). A method to calculate electric currents in quiescent prominences. *Solar Phys.* **24**, 324–335.

Anzer, U. (1978). Can coronal loop transients be driven magnetically? *Solar Phys.* **57**, 111–118.

Arber, T. D., Botha, G. J. J., and Brady, C. S. (2009). Effect of solar chromospheric neutrals on equilibrium field structures. *Astrophys. J.* **705**, 1183–1188.

Arber, T. D., Haynes, M., and Leake, J. E. (2007). Emergence of a flux tube through a partially ionized solar atmosphere. *Astrophys. J.* **666**, 541–546.

Archontis, V. (2008). Magnetic flux emergence in the Sun. *J. Geophys. Res. 113,* 12, 3–15.

Archontis, V. and Hood, A. W. (2008). A flux emergence model for solar eruptions. *Astrophys. J. Letts.* **674**, L113–L116.

Archontis, V. and Hood, A. W. (2010). Flux emergence and coronal eruption. *Astron. Astrophys.* **514**, A56.

Archontis, V. and Hood, A. W. (2012). Magnetic flux emergence: a precursor of solar plasma expulsion. *Astron. Astrophys.* **537**, A62.

Arendt, U. and Schindler, K. (1988). On the existence of three-dimensional magnetohydrostatic equilibria. *Astron. Astrophys.* **204**, 229–234.

Arnol'd, V. (1974). The asymptotic Hopf invariant and its applications. *Sel. Math. Sov.* **5**, 327–345.

Arregui, I., Oliver, R., and Ballester, J. L. (2012). Prominence oscillations. *Liv. Rev. Solar Phys.* **9**, 2.

Asai, A., Masuda, S., Yokoyama, T., Shimojo, M., Isobe, H., Kurokawa, H., and Shibata, K. (2002). Difference between spatial distributions of the Hα kernels and hard X-ray sources in a solar flare. *Astrophys. J. Letts.* **578**, L91–L94.

Asai, A., Shimojo, M., Isobe, H., Morimoto, T., Yokoyama, T., Shibasaki, K., and Nakajima, H. (2001). Periodic acceleration of electrons in the 1998 November 10 solar flare. *Astrophys. J. Letts.* **562**, L103–L106.

Asai, A., Yokoyama, T., Shimojo, M., and Shibata, K. (2004). Downflow motions associated with impulsive nonthermal emissions observed in the 2002 July 23 solar flare. *Astrophys. J. Letts.* **605**, L77–L80.

Aschwanden, M. J. (2001). An evaluation of coronal heating models for active regions based on Yohkoh, SOHO, and TRACE observations. *Astrophys. J.* **560**, 1035–1044.

Aschwanden, M. J. and Benz, A. O. (1997). Electron densities in solar flare loops, chromospheric evaporation upflows, and acceleration sites. *Astrophys. J.* **480**, 825–839.

Aschwanden, M. J., Fletcher, L., Schrijver, C. J., and Alexander, D. (1999). Coronal loop oscillations observed with the Transition Region and Coronal Explorer. *Astrophys. J.* **520**, 880–894.

Aschwanden, M. J., Schrijver, C. J., and Alexander, D. (2001). Modeling of coronal EUV loops observed with TRACE. I. Hydrostatic solutions with nonuniform heating. *Astrophys. J.* **550**, 1036–1050.

Aschwanden, M. J., Stern, R. A., and Güdel, M. (2008). Scaling laws of solar and stellar flares. *Astrophys. J.* **672**, 659–673.

Aschwanden, M. J., Tarbell, T. D., Nightingale, R. W., Schrijver, C. J., Title, A., Kankelborg, C. C., Martens, P., and Warren, H. P. (2000). Time variability of the "Quiet" Sun observed with TRACE. II. Physical parameters, temperature evolution, and energetics of extreme-ultraviolet nanoflares. *Astrophys. J.* **535**, 1047–1065.

Asplund, M., Grevesse, N., and Sauval, A. J. (2005). The solar chemical composition. *Astron. Soc. Pac. Conf. Ser.* **336**, 25–38.

Atanasiu, C. V., Günter, S., Lackner, K., and Miron, I. G. (2004). Analytical solutions to the Grad-Shafranov equation. *Phys. Plasmas* **11**, 3510–3518.

Athay, R. G. and Moreton, G. E. (1961). Impulsive phenomena of the solar atmosphere. I. Some optical events associated with flares showing explosive phase. *Astrophys. J.* **133**, 935–945.

Aulanier, G., DeLuca, E. E., Antiochos, S. K., McMullen, R. A., and Golub, L. (2000). The topology and evolution of the Bastille Day flare. *Astrophys. J.* **540**, 1126–1142.

Aulanier, G. and Démoulin, P. (1998). 3D magnetic configurations supporting prominences. I. The natural presence of lateral feet. *Astron. Astrophys.* **329**, 1125–1137.

Aulanier, G. and Démoulin, P. (2003). Amplitude and orientation of prominence magnetic fields from constant-alpha magnetohydrostatic models. *Astron. Astrophys.* **402**, 769–780.

Aulanier, G., Démoulin, P., Mein, N., van Driel-Gesztelyi, L., Mein, P., and Schmieder, B. (1999). 3D magnetic configurations supporting prominences. III. Evolution of fine structures observed in a filament channel. *Astron. Astrophys.* **342**, 867–880.

Aulanier, G., Démoulin, P., Schmieder, B., Fang, C., and Tang, Y. H. (1998). Magnetohydrostatic model of a bald-patch flare. *Solar Phys.* **183**, 369–388.

Aulanier, G., Démoulin, P., van Driel-Gesztelyi, L., Mein, P., and Deforest, C. (1998). 3D magnetic configurations supporting prominences. II. The lateral feet as a perturbation of a twisted flux-tube. *Astron. Astrophys.* **335**, 309–322.

Aulanier, G., DeVore, C., and Antiochos, S. K. (2002). Prominence magnetic dips in three-dimensional sheared arcades. *Astrophys. J. Letts.* **567**, L97–L101.

Aulanier, G., Golub, L., DeLuca, E. E., Cirtain, J. W., Kano, R., Lundquist, L. L., Narukage, N., Sakao, T., and Weber, M. A. (2007). Slipping magnetic reconnection in coronal loops. *Science* **318**, 1588–1591.

Aulanier, G., Pariat, E., and Démoulin, P. (2005). Current-sheet formation in quasi-separatrix layers and hyperbolic flux tubes. *Astron. Astrophys.* **444**, 961–976.

Aulanier, G., Pariat, E., Démoulin, P., and DeVore, C. R. (2006). Slip-running reconnection in quasi-separatrix layers. *Solar Phys.* **238**, 347–376.

Aulanier, G. and Schmieder, B. (2002). The magnetic nature of wide EUV filament channels and their role in the mass loading of CMEs. *Astron. Astrophys.* **386**, 1106–1122.

Aulanier, G., Srivastava, N., and Martin, S. F. (2000). Model prediction for an observed filament. *Astrophys. J.* **543**, 447–456.

Aulanier, G., Török, T., Démoulin, P., and DeLuca, E. E. (2010). Formation of torus-unstable flux ropes and electric currents in erupting sigmoids. *Astrophys. J.* **708**, 314–333.

Aurass, H., Vršnak, B., and Mann, G. (2002). Shock-excited radio burst from reconnection outflow jet? *Astron. Astrophys.* **384**, 273–281.

Avrett, E. H. and Loeser, R. (2008). Models of the solar chromosphere and transition region from SUMER and HRTS observations: formation of the extreme-ultraviolet spectrum of Hydrogen, Carbon, and Oxygen. *Astrophys. J. Suppl.* **175**, 229–276.

Axford, W. and McKenzie, J. (1992). The origin of high-speed solar-wind streams. In *Solar Wind Seven*, E. Marsch, and R. Schwenn, eds. COSPAR Colloquia Series, vol. 3. (Pergamon Press Ltd, Oxford), 1–5.

Babcock, H. W. (1961). The topology of the Sun's magnetic field and the 22-year cycle. *Astrophys. J.* **133**, 572–587.

Babcock, H. W. and Babcock, H. D. (1952). Mapping the magnetic fields of the Sun. *Pub. Astron. Soc. Pac.* **64**, 282–287.

Babcock, H. W. and Babcock, H. D. (1955). The Sun's magnetic field, 1952–1954. *Astrophys. J.* **121**, 349–366.

Bąk-Stęślicka, U., Gibson, S. E., Fan, Y. E., Bethge, C. W., Forland, B., and Rachmeler, L. A. (2013). The magnetic structure of solar prominence cavities: New observational signature revealed by coronal magnetometry. *Astrophys. J. Letts.,* **770**, L28.

Bagenal, F. and Gibson, S. (1991). Modeling the large-scale structure of the solar corona. *J. Geophys. Res.* **96**, 17663–17674.

Balbus, S. A. and Hawley, J. F. (1991). A powerful local shear instability in weakly magnetized disks. I. Linear analysis. *Astrophys. J.* **376**, 214–222.

Baldwin, P. and Roberts, P. H. (1972). On resistive instabilities. *Phil. Trans. Roy. Soc. Lond.* **272**, 303–330.

Baliunas, S. L., Donahue, R. A., Soon, W. H., Horne, J. H., Frazer, J., Woodard-Eklund, L., Bradford, M., Rao, L. M., Wilson, O. C., Zhang, Q., Bennett, W., Briggs, J., Carroll, S. M., Duncan, D. K., Figueroa, D., Lanning, H. H., Misch, T., Mueller, J., Noyes, R. W., Poppe, D., Porter, A. C., Robinson, C. R., Russell, J., Shelton, J. C., Soyumer, T., Vaughan, A. H., and Whitney, J. H. (1995). Chromospheric variations in main-sequence stars. *Astrophys. J.* **438**, 269–287.

Ballester, J. L. (2006). Seismology of prominence fine structures: observations and theory. *Space Sci. Rev.* **122**, 129–135.

Ballester, J. L. and Priest, E. R. (1987). A 2D model for a solar prominence. *Solar Phys.* **109**, 335–350.

Ballester, J. L. and Priest, E. R. (1989). Model for the fibril structure of solar prominences. *Astron. Astrophys.* **225**, 213–221.

Banerjee, D., Erdélyi, R., Oliver, R., and O'Shea, E. (2007). Present and future observing trends in atmospheric magnetoseismology. *Solar Phys.* **246**, 3–29.

Banerjee, D., Pérez-Suárez, D., and Doyle, J. G. (2009). Signatures of Alfvén waves in the polar coronal holes as seen by EIS/Hinode. *Astron. Astrophys.* **501**, L15–L18.

Bareford, M. R., Browning, P. K., and van der Linden, R. A. M. (2011). The flare-energy distributions generated by kink-unstable ensembles of zero-net-current coronal loops. *Solar Phys.* **273**, 93–115.

Bárta, M., Karlický, M., and Žemlička, R. (2008). Plasmoid dynamics in flare reconnection and the frequency drift of the drifting pulsating structure. *Solar Phys.* **253**, 173–189.

Basu, S. and Antia, H. M. (2003). Changes in solar dynamics from 1995 to 2002. *Astrophys. J.* **585**, 553–565.

Batchelor, G. K. (1967). *An Introduction to Fluid Dynamics.* (Cambridge University Press, Cambridge, UK).

Bateman, G. (1978). *MHD Instabilities.* (MIT Press, Cambridge, USA).

Baty, H., Priest, E. R., and Forbes, T. G. (2009a). Effect of nonuniform resistivity in Petschek reconnection. *Phys. Plasmas* **13**, 022312/1–7.

Baty, H., Priest, E. R., and Forbes, T. G. (2009b). Petschek reconnection with a nonlocalized resistivity. *Phys. Plasmas* **16**, 012102/1–6.

Baum, P. and Bratenahl, A. (1980). Flux linkages of bipolar sunspot groups – a computer study. *Solar Phys.* **67**, 245–258.

Bazer, J. and Ericson, W. B. (1959). Hydromagnetic shocks. *Astrophys. J.* **129**, 758–785.

Beck, J. G. and Schou, J. (2000). Supergranulation rotation. *Solar Phys.* **193**, 333–343.
Beckers, J. M. (1972). Solar spicules. *Ann. Rev. Astron. Astrophys.* **10**, 73–100.
Beckers, J. M. and Schultz, R. B. (1972). Oscillatory motions in sunspots. *Solar Phys.* **27**, 61–70.
Beckers, J. M. and Tallant, P. (1969). Chromospheric inhomogeneities in sunspot umbrae. *Solar Phys.* **7**, 351–365.
Beer, J. (2000). Long-term indirect indices of solar variability. *Space Sci. Rev.* **94**, 53–66.
Beer, J., Tobias, S., and Weiss, N. O. (1998). An active Sun throughout the Maunder minimum. *Solar Phys.* **181**, 237–249.
Beliën, A. J. C., Martens, P. C. H., and Keppens, R. (1999). Coronal heating by resonant absorption: the effects of chromospheric coupling. *Astrophys. J.* **526**, 478–493.
Bellot Rubio, L. R., Balthasar, H., Collados, M., and Schlichenmaier, R. (2003). Field-aligned Evershed flows in the photosphere of a sunspot penumbra. *Astron. Astrophys.* **403**, L47–L50.
Bender, C. M. and Orszag, S. A. (1999). *Advanced Mathematical Methods for Scientists and Engineers.* (Springer-Verlag, Berlin).
Benz, A. O. and Krucker, S. (2002). Energy distribution of microevents in the quiet solar corona. *Astrophys. J.* **568**, 413–421.
Berdyugina, S. V. (2005). Starspots: a key to the stellar dynamo. *Liv. Rev. in Solar Phys. 2,* 8, 1–46.
Berger, M. A. (1984). Rigorous new limits on magnetic helicity dissipation in the solar corona. *Geophys. Astrophys. Fluid Dyn.* **30**, 79–104.
Berger, M. A. (1988). An energy formula for nonlinear force-free magnetic fields. *Astron. Astrophys.* **201**, 355–361.
Berger, M. A. (1999). Magnetic helicity in space physics. In *Magnetic Helicity in Space and Laboratory Plasmas*, M. R. Brown, R. C. Canfield, and A. A. Pevtsov, eds. (Washington, DC), 1–11.
Berger, M. A. and Field, G. (1984). The topological properties of magnetic helicity. *J. Fluid Mech.* **147**, 133–148.
Berger, T. E., De Pontieu, B., Fletcher, L., Schrijver, C. J., Tarbell, T. D., and Title, A. M. (1999). What is moss? *Solar Phys.* **190**, 409–418.
Berger, T. E., Liu, W., and Low, B. C. (2012). SDO/AIA detection of solar prominence formation within a coronal cavity. *Astrophys. J. Letts.* **758**, L37.
Berger, T. E., Rouppe van der Voort, L., and Löfdahl, M. (2007). Contrast analysis of solar faculae and magnetic bright points. *Astrophys. J.* **661**, 1272–1288.
Berger, T. E., Rouppe van der Voort, L., Löfdahl, M. G., Carlsson, M., Fossum, A., Hansteen, V. H., Marthinussen, E., Title, A., and Scharmer, G. (2004). Solar magnetic elements at 0.1 arcsec resolution. General appearance and magnetic structure. *Astron. Astrophys.* **428**, 613–628.
Berger, T. E., Shine, R. A., Slater, G. L., Tarbell, T. D., Title, A. M., Okamoto, T. J., Ichimoto, K., Katsukawa, Y., Suematsu, Y., Tsuneta, S., Lites, B. W., and Shimizu, T. (2008). Hinode SOT observations of solar quiescent prominence dynamics. *Astrophys. J. Letts.* **676**, L89–L92.
Berger, T. E., Slater, G., Hurlburt, N., Shine, R., Tarbell, T., Title, A., Lites, B. W., Okamoto, T. J., Ichimoto, K., Katsukawa, Y., Magara, T., Suematsu, Y., and Shimizu, T. (2010). Quiescent prominence dynamics observed with the Hinode Solar Optical Telescope. I. Turbulent upflow plumes. *Astrophys. J.* **716**, 1288–1307.
Berger, T. E., Testa, P., Hillier, A., Boerner, P., Low, B. C., Shibata, K., Schrijver, C., Tarbell, T., and Title, A. (2011). Magneto-thermal convection in solar prominences. *Nature* **472**, 197–200.
Berghmans, D. and Clette, F. (1999). Active region EUV transient brightenings – first Results by EIT of SOHO JOP80. *Solar Phys.* **186**, 207–229.
Bernstein, I., Frieman, E., Kruskal, M. D., and Kulsrud, R. M. (1958). An energy principle for hydromagnetic stability problems. *Proc. Roy. Soc. Lond.* **A244**, 17–40.
Beveridge, C. and Longcope, D. W. (2006). A hierarchical application of the minimum current corona. *Astrophys. J.* **636**, 453–461.
Bewsher, D., Innes, D. E., Parnell, C. E., and Brown, D. S. (2005). Comparison of blinkers and explosive events: a case study. *Astron. Astrophys.* **432**, 307–317.
Bhattacharjee, A., Huang, Y. M., Yang, H., and Rogers, B. (2009). Fast reconnection in high-Lundquist-number plasmas due to the plasmoid instability. *Phys. Plasmas 16,* 11 (Nov.), 112102.
Biermann, L. (1941). Der gegenwrtige Stand der Theorie konvektiver Sonnenmodelle. *Vierteljahresschr. Astron. Ges.* **76**, 194–200.
Biermann, L. (1946). Zur Deutung der chromosphärischen Turbulenz und des Exzesses der UV-Strahlung der Sonne. *Naturwissenschaften* **33**, 118–119.
Biermann, L. (1951). Kometenschweife und solare Korpuskularstrahlung. *Zeitschrift fr Astrophysik* **29**, 274–286.
Biernat, H. K. (1987). Unsteady Petschek reconnection. *J. Geophys. Res.* **92**, 3392–3396.
Bineau, M. (1972). Existence of force-free magnetic fields. *Comm. on Pure and Applied Mathematics* **25**, 77.
Bingert, S. and Peter, H. (2011). Intermittent heating in the solar corona employing a 3D MHD model. *Astron. Astrophys.* **530**, A112.
Birn, J., Drake, J. F., Shay, M. A., Rogers, B. N., Denton, R. E., Hesse, M., Kuznetsova, M., Ma, Z. W., Bhattacharjee, A., Otto, A., and Pritchett, P. L. (2001). Geospace environmental modeling (GEM) magnetic reconnection challenge. *J. Geophys. Res.* **106**, 3715–3720.

Birn, J., Goldstein, H., and Schindler, K. (1978). A theory of the onset of solar eruptive processes. *Solar Phys.* **57**, 81–101.

Birn, J., Hesse, M., and Schindler, K. (1989). Filamentary structure of a three-dimensional plasmoid. *J. Geophys. Res.* **94**, 241–251.

Birn, J. and Priest, E. R. (2007). *Reconnection of Magnetic Fields: MHD and Collisionless Theory and Observations.* (Cambridge University Press, Cambridge, UK).

Biskamp, D. (1982). Effect of secondary tearing instability on the coalescence of magnetic islands. *Phys. Lett* **87A**, 357–360.

Biskamp, D. (1986). Magnetic reconnection via current sheets. *Phys. Fluids* **29**, 1520–1531.

Biskamp, D. (1993). *Nonlinear Magnetohydrodynamics.* (Cambridge University Press, Cambridge, UK).

Biskamp, D. (1994). Magnetic reconnection. *Physics Reports 237,* 4, 179–247.

Bloomfield, D. S., Lagg, A., and Solanki, S. K. (2007). The nature of running penumbral waves revealed. *Astrophys. J.* **671**, 1005–1012.

Bobra, M. G., van Ballegooijen, A. A., and DeLuca, E. E. (2008). Modeling nonpotential magnetic fields in solar active regions. *Astrophys. J.* **672**, 1209–1220.

Bobrova, N. A. and Syrovatsky, S. I. (1979). Singular lines of one-dimensional force-free magnetic fields. *Solar Phys.* **61**, 379–387.

Bogdan, T. J., Carlsson, M., Hansteen, V. H., McMurry, A., Rosenthal, C. S., Johnson, M., Petty-Powell, S., Zita, E. J., Stein, R. F., McIntosh, S. W., and Nordlund, Å. (2003). Waves in the magnetized solar atmosphere. II. Waves from localized sources in magnetic flux concentrations. *Astrophys. J.* **599**, 626–660.

Bogdan, T. J. and Judge, P. G. (2006). Observational aspects of sunspot oscillations. *Phil. Trans. Roy. Soc. Lond. Series A* **364**, 313–331.

Bogdan, T. J. and Knölker, M. (1991). Scattering of acoustic waves from a magnetic flux tube embedded in a radiating fluid. *Astrophys. J.* **369**, 219–2236.

Bogdan, T. J. and Low, B. C. (1986). The three-dimensional structure of magnetostatic atmospheres. II – Modeling the large-scale corona. *Astrophys. J.* **306**, 271–283.

Böhm-Vitense, Z. (1958). Über die Wasserstoffkonvektionszone in Sternen verschiedener Effektivtemperaturen und Leuchtkräfte. *Zeitschrift fur Astrophysik* **46**, 108–143.

Bommier, V. and Leroy, J. L. (1998). Global pattern of the magnetic field vectors above neutral lines from 1974 to 1982: Pic-du-Midi observations of prominences. *Astron. Soc. Pac. Conf. Ser.* **150**, 434–438.

Bommier, V. and Stein, R. (2013). The internetwork magnetic flux tube diameter inferred from THEMIS and HINODE observations compared to simulations. *Solar Phys.*, in press.

Bondi, H. (1952). On spherically symmetrical accretion. *Mon. Not. Roy. Astron. Soc.* **112**, 195–204.

Bonet, J. A., Márquez, I., Sánchez Almeida, J., Cabello, I., and Domingo, V. (2008). Convectively driven vortex flows in the Sun. *Astrophys. J. Letts.* **687**, L131–L134.

Bonet, J. A., Márquez, I., Sánchez Almeida, J., Palacios, J., Martínez Pillet, V., Solanki, S. K., del Toro Iniesta, J. C., Domingo, V., Berkefeld, T., Schmidt, W., Gandorfer, A., Barthol, P., and Knölker, M. (2010). SUNRISE/IMaX observations of convectively driven vortex flows in the Sun. *Astrophys. J. Letts.* **723**, L139–L143.

Bonnet, R. M. and Dupree, A. K. (1981). *Solar Phenomena in Stars and Stellar Systems.* (D. Reidel, Dordrecht).

Boström, R. (1973). Models of force-free magnetic fields in a passive medium. *Astrophys. Space Sci.* **22**, 353–380.

Botha, G. J. J., Arber, T. D., Nakariakov, V. M., and Zhugzhda, Y. D. (2011). Chromospheric resonances above sunspot umbrae. *Astrophys. J.* **728**, 84.

Boulmezaoud, T. Z. and Amari, T. (2000). On the existence of non-linear force-free fields in three-dimensional domains. *Zeitschrift Angewandte Mathematik und Physik* **51**, 942–967.

Boyd, T. J. M. and Sanderson, J. J. (2003). *The Physics of Plasmas.* (Cambridge University Press, Cambridge, UK).

Bradshaw, S. J. and Cargill, P. J. (2006). Explosive heating of low-density coronal plasma. *Astron. Astrophys.* **458**, 987–995.

Bradshaw, S. J. and Cargill, P. J. (2010). A new enthalpy-based approach to the transition region in an impulsively heated corona. *Astrophys. J. Letts.* **710**, L39–L43.

Bradshaw, S. J., Klimchuk, J. A., and Reep, J. W. (2012). Diagnosing the Time-dependence of Active Region Core Heating from the Emission Measure. I. Low-frequency Nanoflares. *Astrophys. J.* **758**, 53.

Braginsky, S. I. (1965a). Self-excitation of a magnetic field during the motion of a highly conducting fluid. *Sov. Phys. JETP* **20**, 726–735.

Braginsky, S. I. (1965b). Transport processes in a plasma. In *Reviews of Plasma Physics*, M. A. Leontovich, ed. (Consultants Bureau, New York), 205–311.

Braithwaite, J. (2006). A differential rotation driven dynamo in a stably stratified star. *Astron. Astrophys.* **449**, 451–460.

Brandenburg, A. and Schmidt, D. (1998). Simulations of an alpha-effect due to magnetic buoyancy. *Astron. Astrophys.* **338**, L55–L58.

Brandt, J. C. (1970). *Introduction to the Solar Wind.* (Freeman, San Francisco).

Brizard, A. J. and Hahm, T. S. (2007). Foundations of nonlinear gyrokinetic theory. *Rev. Mod. Phys.* **79**, 421–468.

Brooks, D. H. and Warren, H. P. (2011). Establishing a connection between active region outflows and the solar wind: abundance measurements with EIS/Hinode. *Astrophys. J. Letts.* **727**, L13.

Brown, D. S. and Priest, E. R. (1999). Topological bifurcations in 3D magnetic fields. *Proc. Roy. Soc. Lond.* **A455**, 3931–3951.

Brown, D. S. and Priest, E. R. (2001). The topological behaviour of 3D null points in the Sun's corona. *Astron. Astrophys.* **367**, 339–346.

Brown, J. C. (1971). The deduction of energy spectra of non-thermal electrons in flares from the observed dynamic spectra of hard X-ray bursts. *Solar Phys.* **18**, 489–502.

Brown, T. M., Christensen-Dalsgaard, J., Dziembowski, W. A., Goode, P. R., Gough, D. O., and Morrow, C. A. (1989). Inferring the Sun's angular velocity from observed p-mode frequency splittings. *Astrophys. J.* **343**, 526–546.

Browning, M. K., Brun, A. S., Miesch, M. S., and Toomre, J. (2007). Dynamo action in simulations of penetrative solar convection with an imposed tachocline. *Astron. Nachr.* **328**, 1100–1103.

Browning, P. K., Gerrard, C., Hood, A. W., Kevis, R., and van der Linden, R. A. M. (2008). Heating the corona by nanoflares: simulations of energy release triggered by a kink instability. *Astron. Astrophys.* **485**, 837–848.

Browning, P. K. and Priest, E. R. (1982). The structure of untwisted magnetic flux tubes. *Geophys. Astrophys. Fluid Dyn.* **21**, 237–263.

Browning, P. K. and Priest, E. R. (1984). Kelvin-Helmholtz instability of a phased-mixed Alfvén wave. *Astron. Astrophys.* **131**, 283–290.

Browning, P. K., Sakurai, T., and Priest, E. R. (1986). Coronal heating in closely-packed flux tubes – a Taylor-Heyvaerts relaxation theory. *Astron. Astrophys.* **158**, 217–227.

Brueckner, G. and Bartoe, J.-D. F. (1983). Observations of high-energy jets in the corona above the quiet Sun, the heating of the corona, and the acceleration of the solar wind. *Astrophys. J.* **272**, 329–348.

Brummell, N. H., Hurlburt, N. E., and Toomre, J. (1996). Turbulent compressible convection with rotation. I. Flow structure and evolution. *Astrophys. J.* **473**, 494–513.

Brummell, N. H., Hurlburt, N. E., and Toomre, J. (1998). Turbulent compressible convection with rotation. II. Mean flows and differential rotation. *Astrophys. J.* **493**, 955–969.

Brun, A. S. (2004). On the interaction between differential rotation and magnetic fields in the Sun. *Solar Phys.* **220**, 333–345.

Brun, A. S., Miesch, M. S., and Toomre, J. (2004). Global-scale turbulent convection and magnetic dynamo action in the solar envelope. *Astrophys. J.* **614**, 1073–1098.

Brun, A. S., Miesch, M. S., and Toomre, J. (2011). Modeling the dynamical coupling of solar convection with the radiative interior. *Astrophys. J.* **742**, 79.

Brun, A. S., Turcke-Chièze, S., and Zahn, J.-P. (1999). Standard solar models in the light of new helioseismic constraints. II. Mixing below the convective zone. *Astrophys. J.* **525**, 1032–1041.

Brun, A. S. and Zahn, J.-P. (2006). Magnetic confinement of the solar tachocline. *Astron. Astrophys.* **457**, 665–674.

Bruno, R. and Carbone, V. (2005). The solar wind as a turbulence laboratory. *Liv. Rev. in Solar Phys.* **2**, 4.

Büchner, J. (2006). Locating current sheets in the solar corona. *Space Sci. Rev.* **122**, 149–160.

Bühler, D., Lagg, A., and Solanki, S. K. (2013). Quiet Sun magnetic fields observed by Hinode: support for a local dynamo. *Astron. Astrophys.,* **555**, A33.

Bulanov, S. V., Sakai, D. I., and Syrovatsky, S. I. (1979). Tearing instability in quasi-stationary magnetohydrodynamic configurations. *Soviet J. Plasma Phys.* **5**, 280–290.

Bulanov, S. V. and Sasorov, P. V. (1978). Tearing of a current sheet and reconnection of magnetic lines of force. *Soviet J. Plasma Phys.* **4**, 746–757.

Bulanov, S. V., Shasharaina, S., and Pegoraro, F. (1990). MHD modes near the X-line of a magnetic configuration. *Plasma Phys. and Control. Fusion* **32**, 377–389.

Bungey, T. N. and Priest, E. R. (1995). Current sheet configurations in potential and force-free fields. *Astron. Astrophys.* **293**, 215–224.

Burgess, D. (1995). Collisionless shocks. In *Introduction to Space Physics*, M. G. Kivelson and C. T. Russell, eds. (Cambridge University Press, Cambridge, UK), 129–163.

Burlaga, L., Sittler, E., Mariani, F., and Schwenn, R. (1981). Magnetic loop behind an interplanetary shock – Voyager, Helios, and IMP 8 observations. *J. Geophys. Res.* **86**, 6673–6684.

Bushby, P. J., Houghton, S. M., Proctor, M. R. E., and Weiss, N. O. (2008). Convective intensification of magnetic fields in the quiet Sun. *Mon. Not. Roy. Astron. Soc.* **387**, 698–706.

Caligari, P., Moreno-Insertis, F., and Schüssler, M. (1995). Emerging flux tubes in the solar convection zone. I. Asymmetry, tilt, and emergence latitude. *Astrophys. J.* **441**, 886–902.

Cally, P. S. and Bogdan, T. J. (1997). Simulation of f- and p-mode interactions with a stratified magnetic field concentration. *Astrophys. J. Letts.* **486**, L67–L70.

Cally, P. S., Crouch, A. D., and Braun, D. C. (2003). Probing sunspot magnetic fields with p-mode absorption and phase shift data. *Mon. Not. Roy. Astron. Soc.* **346**, 381–389.

Cally, P. S. and Goossens, M. (2008). Three-dimensional MHD wave propagation and conversion to Alfvén waves near the solar surface. I. Direct numerical solution. *Solar Phys.* **251**, 251–265.

Cameron, A. C. (2007). Differential rotation on rapidly rotating stars. *Astron. Nach.* **328**, 1030–1033.

Cameron, R. H., Jiang, J., Schmitt, D., and Schüssler, M. (2010). Surface Flux Transport Modeling for Solar Cycles 15-21: Effects of Cycle-Dependent Tilt Angles of Sunspot Groups. *Astrophys. J.* **719**, 264–270.

Cameron, R. H. and Schüssler, M. (2012). Are the strengths of solar cycles determined by converging flows towards the activity belts? *Astron. Astrophys.* **548**, A57.

Canfield, R. C., Cheng, C.-C., Dere, K. P., Dulk, G. A., McLean, D. J., Schmahl, E. J., Robinson, R. D., and Schoolman, S. A. (1980). Radiative energy output of the 5 September 1973 flare. In *Skylab Solar Workshop II*, P. A. Sturrock, ed. (Colorado Ass. Univ. Press, Boulder, Colorado), 451–469.

Canfield, R. C., Hudson, H. S., and McKenzie, D. E. (1999). Sigmoidal morphology and eruptive solar activity. *Geophys. Res. Lett.* **26**, 627–630.

Canfield, R. C., Reardon, K. P., Leka, K. D., Shibata, K., Yokoyama, T., and Shimojo, M. (1996). H-alpha surges and X-ray jets in AR 7260. *Astrophys. J.* **464**, 1016–1029.

Canou, A. and Amari, T. (2010). A twisted flux rope as the magnetic structure of a filament in active region 10953 observed by Hinode. *Astrophys. J.* **715**, 1566–1574.

Canou, A., Amari, T., Bommier, V., Schmieder, B., Aulanier, G., and Li, H. (2009). Evidence for a pre-eruptive twisted flux rope using the Themis vector magnetograph. *Astrophys. J. Letts.* **693**, L27–L30.

Canuto, V. and Mazzitelli, I. (1991). Stellar turbulent convection – a new model and applications. *Astrophys. J.* **370**, 295–311.

Carcedo, L., Brown, D. S., Hood, A. W., Neukirch, T., and Wiegelmann, T. (2003). A quantitative method to optimise magnetic field line fitting of observed coronal loops. *Solar Phys.* **218**, 29–40.

Cargill, P. J. (1994). Some implications of the nanoflare concept. *Astrophys. J.* **422**, 381–393.

Cargill, P. J., Bradshaw, S. J., and Klimchuk, J. A. (2012a). Enthalpy-based thermal evolution of loops. II. Improvements to the model. *Astrophys. J.* **752**, 161.

Cargill, P. J., Bradshaw, S. J., and Klimchuk, J. A. (2012b). Enthalpy-based thermal evolution of loops. III. Comparison of zero-dimensional models. *Astrophys. J.* **758**, 5.

Cargill, P. J. and De Moortel, I. (2011). Solar physics: waves galore. *Nature* **475**, 463–464.

Cargill, P. J. and Klimchuk, J. A. (1997). A nanoflare explanation for the heating of coronal loops observed by Yohkoh. *Astrophys. J.* **478**, 799–806.

Cargill, P. J. and Klimchuk, J. A. (2004). Nanoflare heating of the corona revisited. *Astrophys. J.* **605**, 911–920.

Cargill, P. J., Mariska, J. T., and Antiochos, S. K. (1995). Cooling of solar flares plasmas. I. Theoretical considerations. *Astrophys. J.* **439**, 1034–1043.

Cargill, P. J. and Priest, E. R. (1980). Siphon flows in coronal loops. *Solar Phys.* **65**, 251–269.

Cargill, P. J. and Priest, E. R. (1982). Slow shock heating and the Kopp-Pneuman model for post-flare loops. *Solar Phys.* **76**, 357–375.

Carlsson, M. and Hansteen, V. (2005). Chromospheric waves. In *Chromospheric and Coronal Magnetic Fields*, D. E. Innes, A. Lagg, and S. A. Solanki, eds. (ESA SP-596, Noordwijk), 39.1–39.11.

Carlsson, M., Hansteen, V. H., De Pontieu, B., McIntosh, S., Tarbell, T. D., Shine, D., Tsuneta, S., Katsukawa, Y., Ichimoto, K., Suematsu, Y., Shimizu, T., and Nagata, S. (2007). Can high frequency acoustic waves heat the quiet Sun chromosphere? *Publ. Astron. Soc. Japan* **59**, 663–668.

Carlsson, M., Hansteen, V. H., and Gudiksen, B. V. (2010). Chromospheric heating and structure as determined from high resolution 3D simulations. *Mem. della Soc. Astron. Ital.* **81**, 582–587.

Carlsson, M. and Stein, R. F. (1995). Does a nonmagnetic solar chromosphere exist? *Astrophys. J. Letts.* **440**, L29–L32.

Carlsson, M. and Stein, R. F. (1997). Formation of solar calcium H and K bright grains. *Astrophys. J.* **481**, 500–514.

Carlsson, M., Stein, R. F., Nordlund, Å., and Scharmer, G. B. (2004). Observational manifestations of solar magnetoconvection: center-to-limb variation. *Astrophys. J. Letts.* **610**, L137–L140.

Carmichael, H. (1964). A process for flares. In *AAS-NASA Symposium on Physics of Solar Flares.* (NASA, SP-50, Washington), 451–456.

Cartledge, N., Titov, V. S., and Priest, E. R. (1996). A 2-D model for the support of a polar-crown solar prominence. *Solar Phys.* **166**, 287–310.

Casini, R., Bevilacqua, R., and López Ariste, A. (2005). Principal component analysis of the He I D_3 polarization profiles from solar prominences. *Astrophys. J.* **622**, 1265–1274.

Cattaneo, F. (1999). On the origin of magnetic fields in the quiet photosphere. *Astrophys. J.* **515**, L39–L42.

Cattaneo, F., Emonet, T., and Weiss, N. O. (2003). On the interaction between convection and magnetic fields. *Astrophys. J.* **588**, 1183–1198.

Cattaneo, F. and Hughes, D. W. (1988). The nonlinear breakup of a magnetic layer: instability to interchange modes. *J. Fluid Mech.* **196**, 323–344.

Cattaneo, F. and Hughes, D. W. (1996). Nonlinear saturation of the turbulent α-effect. *Phs. Rev. E* **54**, 4532–4535.

Cattaneo, F. and Hughes, D. W. (2006). Dynamo action in a rotating convective layer. *J. Fluid Mech.* **553**, 401–418.

Cattaneo, F. and Hughes, D. W. (2009). Problems with kinematic mean-field electrodynamics at high magnetic Reynolds numbers. *Mon. Not. Roy. Astron. Soc.* **395**, L48–L51.

Centeno, R., Socas-Navarro, H., Lites, B., Kubo, M., Frank, Z., Shine, R., Tarbell, T., Title, A., Ichimoto, K., Tsuneta, S., Katsukawa, Y., Suematsu, Y., Shimizu, T., and Nagata, S. (2007). Emergence of small-scale magnetic loops in the quiet-Sun internetwork. *Astrophys. J. Letts.* **666**, L137–L140.

Chamberlain, J. W. (1960). Interplanetary gas. II. Expansion of a model solar corona. *Astrophys. J.* **131**, 47–58.

Chandran, B. D. G. and Hollweg, J. V. (2009). Alfvén wave reflection and turbulent heating in the solar wind from 1 solar radius to 1 AU: an analytical treatment. *Astrophys. J.* **707**, 1659–1667.

Chandrasekhar, S. (1961). *Hydrodynamic and Hydromagnetic Stability.* (Cambridge University Press, Cambridge, UK).

Chaplin, W. J. (2006). *Music of the Sun.* (Oneworld Publications, Oxford).

Chaplin, W. J., Christensen-Dalsgaard, J., Elsworth, Y., Howe, R., Isaak, G. R., Larsen, R. M., New, R., Schou, J., Thompson, M. J., and Tomczyk, S. (1999). Rotation of the solar core from BiSON and LOWL frequency observations. *Mon. Not. Roy. Astron. Soc.* **308**, 405–414.

Chaplin, W. J., Elsworth, Y., Howe, R., Isaak, G. R., McLeod, C. P., Miller, B. A., van der Raay, H. B., Wheeler, S. J., and New, R. (1996). BiSON performance. *Solar Phys.* **168**, 1–18.

Chapman, S. (1929). Solar streams of corpuscles: their geometry, absorption of light, and penetration. *Mon. Not. Roy. Astron. Soc.* **89**, 456–470.

Chapman, S. and Zirin, H. (1957). Notes on the solar corona and the terrestrial ionosphere. *Smithsonian Contributions to Astrophysics* **2**, 1–11.

Charbonneau, P. (2010). Dynamo models of the solar cycle. *Liv. Rev. in Solar Phys.* **7**, 3.

Charbonneau, P. (2013). *Solar and Stellar Dynamos.* Vol. 39. (Springer-Verlag, Berlin).

Charbonneau, P., Beaubien, G., and St-Jean, C. (2007). Fluctuations in Babcock-Leighton dynamos. II. Revisiting the Gnevyshev-Ohl rule. *Astrophys. J.* **658**, 657–662.

Charbonneau, P. and MacGregor, K. B. (1997). Solar interface dynamos: II linear, kinematic models in spherical geometry. *Astrophys. J.* **486**, 502–520.

Chatterjee, P., Nandy, D., and Choudhuri, A. R. (2004). Full-sphere simulations of a circulation-dominated solar dynamo: exploring the parity issue. *Astron. Astrophys.* **427**, 1019–1030.

Chemin, J.-Y., Gallagher, I., Desjardins, B., and Grenier, E. (2006). *Mathematical Geophysics.* (Oxford University Press, Oxford).

Chen, C. J. and Lykoudis, P. S. (1972). Velocity oscillations in solar plage regions. *Solar Phys.* **25**, 380–401.

Chen, J. (1989). Effects of toroidal forces in current loops embedded in a background plasma. *Astrophys. J.* **338**, 453–470.

Chen, L. and Hasegawa, A. (1974). Plasma heating by spatial resonance of Alfvén wave. *Physics of Fluids* **17**, 1399–1403.

Chen, P. F. (2011). Coronal mass ejections: models and their observational basis. *Liv. Rev. in Solar Phys.* **8**, 1.

Chen, P. F., Wu, S. T., Shibata, K., and Fang, C. (2002). Evidence of EIT and Moreton waves in numerical simulations. *Astrophys. J. Letts.* **572**, L99–L102.

Chen, Y., Li, X., Song, H. Q., Shi, Q. Q., Feng, S. W., and Xia, L. D. (2009). Intrinsic instability of coronal streamers. *Astrophys. J.* **691**, 1936–1942.

Cheung, M. C. M., Moreno-Insertis, F., and Schüssler, M. (2006). Moving magnetic tubes: fragmentation, vortex sheets and the limit of the approximation of thin flux tubes. *Astron. Astrophys.* **451**, 303–317.

Cheung, M. C. M., Rempel, M., Title, A. M., and Schüssler, M. (2010). Simulation of the formation of a solar active region. *Astrophys. J.* **720**, 233–244.

Cheung, M. C. M., Schüssler, M., and Moreno-Insertis, F. (2007). Magnetic flux emergence in granular convection: radiative MHD simulations and observational signatures. *Astron. Astrophys.* **467**, 703–719.

Chiu, Y. T. and Hilton, H. H. (1977). Exact Green's function method of solar force-free magnetic field computations with constant alpha. I. Theory and basic test cases. *Astrophys. J.* **212**, 873–885.

Chou, Y. P. and Low, B. C. (1994). A class of force-free magnetic fields for modeling pre-flare coronal magnetic configurations. *Solar Phys.* **153**, 255–285.

Choudhuri, A. R. (1990). On the possibility of an $\alpha^2\omega$-type dynamo in a thin layer inside the Sun. *Astrophys. J.* **355**, 733–744.

Choudhuri, A. R. (1998). *Physics of Fluids and Plasmas.* (Cambridge University Press, Cambridge, UK).

Choudhuri, A. R. (1999). The solar dynamo. *Current Science* **77**, 1475–1485.

Choudhuri, A. R. (2003). On the connection between mean-field dynamo theory and flux tubes. *Solar Phys.* **215**, 31–55.

Choudhuri, A. R., Chatterjee, P., and Jiang, J. (2007). Predicting solar cycle 24 with a solar dynamo model. *Phys. Rev. Letts.* **98**, 131103–131107.

Choudhuri, A. R., Chatterjee, P., and Nandy, D. (2004). Helicity of solar active regions from a dynamo model. *Astrophys. J. Letts.* **615**, L57–L60.

Choudhuri, A. R. and Dikpati, M. (1999). On the large-scale diffuse magnetic field of the Sun – II. The contribution of active regions. *Solar Phys.* **184**, 61–76.

Choudhuri, A. R. and Gilman, P. A. (1987). The influence of the Coriolis force on flux tubes rising through the solar convection zone. *Astrophys. J.* **316**, 788–800.

Choudhuri, A. R., Schüssler, M., and Dikpati, M. (1995). The solar dynamo with meridional circulation. *Astron. Astrophys.* **303**, L29–L32.

Christensen-Dalsgaard, J., Dappen, W., Ajukov, S. V., Anderson, E. R., Antia, H. M., Basu, S., Baturin, V. A., Berthomieu, G., Chaboyer, B., Chitre, S. M., Cox, A. N., Demarque, P., Donatowicz, J., Dziembowski, W. A., Gabriel, M., Gough, D. O., Guenther, D. B., Guzik, J. A., Harvey, J. W., Hill, F., Houdek, G., Iglesias, C. A., Kosovichev, A. G., Leibacher, J. W., Morel, P., Proffitt, C. R., Provost, J., Reiter, J., Rhodes, Jr., E. J., Rogers, F. J., Roxburgh, I. W., Thompson, M. J., and Ulrich, R. K. (1996). The current state of solar modeling. *Science* **272**, 1286–1292.

Christensen-Dalsgaard, J., Duvall, T. L., Gough, D. O., Harvey, J. W., and Rhodes, E. J. (1985). Speed of sound in the solar interior. *Nature* **315**, 378–382.

Christensen-Dalsgaard, J. and Gough, D. O. (1976). Towards a heliological inverse problem. *Nature* **259**, 89–92.

Christensen-Dalsgaard, J., Gough, D. O., and Thompson, M. J. (1991). The depth of the solar convection zone. *Astrophys. J.* **378**, 431–437.

Ciaravella, A. and Raymond, J. C. (2008). The current sheet associated with the 2003 November 4 coronal mass ejection: density, temperature, thickness, and line width. *Astrophys. J.* **686**, 1372–1382.

Cirtain, J. W., Golub, L., Lundquist, L., van Ballegooijen, A. A., Savcheva, A., Shimojo, M., DeLuca, E., Tsuneta, S., Sakao, T., Reeves, K., Weber, M., Kano, R., Narukage, N., and Shibasaki, K. (2007). Evidence for Alfvén waves in solar X-ray jets. *Science* **318**, 1580–1582.

Claverie, A., Isaak, G. R., McLeod, C. P., van der Raay, H. B., and Cortes, T. R. (1979). Solar structure from global studies of the 5-minute oscillation. *Nature* **282**, 591–594.

Close, R. M., Parnell, C., Mackay, D. H., and Priest, E. R. (2004). Statistical flux tube properties of 3D magnetic carpet fields. *Solar Phys.* **212**, 251–275.

Close, R. M., Parnell, C. E., Longcope, D. W., and Priest, E. R. (2005). Coronal flux recycling times. *Solar Phys.* **231**, 45–70.

Close, R. M., Parnell, C. E., and Priest, E. R. (2004). Recycling of the solar corona's magnetic field. *Astrophys. J.* **612**, L81–L84.

Coppi, B., Pellat, R., Rosenbluth, M., Rutherford, P., and Galvao, R. (1976). Resistive internal kink modes. *Soviet J. Plasma Phys.* **2**, 961–966.

Courant, R. and Hilbert, D. (1963). *Methods of Mathematical Physics, Vol. 2.* (Interscience, New York). p. 367.

Courvoisier, A., Hughes, D. W., and Tobias, S. M. (2006). α-effect in a family of chaotic flows. *Phys. Rev. Letts. 96,* 034503, 1–4.

Cowling, T. G. (1934). The magnetic field of sunspots. *Mon. Not. Roy. Astron. Soc.* **94**, 39–48.

Cowling, T. G. (1953). Solar electrodynamics. In *The Sun*, G. Kuiper, ed. (Univ. Chicago Press, Chicago), 532–591.

Cowling, T. G. (1958). *Magnetohydrodynamics.* (Interscience, New York).

Cowling, T. G. (1976). *Magnetohydrodynamics.* (Adam Hilger, Bristol, UK).

Cox, D. P. and Tucker, W. H. (1969). Ionization equilibrium and radiative cooling of a low-density plasma. *Astrophys. J.* **157**, 1157–1167.

Craig, I. J. D. and Fabling, R. B. (1996). Exact solutions for steady-state, spine, and fan magnetic reconnection. *Astrophys. J.* **462**, 969–976.

Craig, I. J. D., Fabling, R. B., Henton, S. M., and Rickard, G. J. (1995). An exact solution for steady-state magnetic reconnection in three dimensions. *Astrophys. J. Letts.* **455**, L197–L199.

Craig, I. J. D. and Henton, S. M. (1995). Exact solutions for steady-state incompressible magnetic reconnection. *Astrophys. J.* **450**, 280–288.

Craig, I. J. D. and McClymont, A. N. (1991). Dynamic magnetic reconnection at an X-type neutral point. *Astrophys. J.* **371**, L41–L44.

Craig, I. J. D. and McClymont, A. N. (1993). Linear theory of fast reconnection at an X-type neutral point. *Astrophys. J.* **405**, 207–215.

Craig, I. J. D., McClymont, A. N., and Underwood, J. H. (1978). The temperature and density structure of active region coronal loops. *Astron. Astrophys.* **70**, 1–11.

Craig, I. J. D. and Sneyd, A. D. (1986). A dynamic relaxation technique for determining the structure and stability of coronal magnetic fields. *Astrophys. J.* **311**, 451–459.

Craig, I. J. D. and Sneyd, A. D. (2005). The Parker problem and the theory of coronal heating. *Solar Phys.* **232**, 41–62.

Cranmer, S. R. (2004). New views of the solar wind with the Lambert W function. *American Journal of Physics* **72**, 1397–1403.

Cranmer, S. R. (2009). Coronal holes. *Liv. Rev. in Solar Phys.* **6**, 3.

Cranmer, S. R. (2010). An efficient approximation of the coronal heating rate for use in global Sun-heliosphere simulations. *Astrophys. J.* **710**, 676–688.

Cranmer, S. R. (2012). Self-consistent models of the solar wind. *Space Sci. Rev.* **172**, 145–156.

Cranmer, S. R., Matthaeus, W. H., Breech, B. A., and Kasper, J. C. (2009). Empirical constraints on proton and electron heating in the fast solar wind. *Astrophys. J.* **702**, 1604–1614.

Cranmer, S. R. and van Ballegooijen, A. A. (2005). On the generation, propagation, and reflection of Alfvén waves from the solar photosphere to the distant heliosphere. *Astrophys. J. Suppl.* **156**, 265–293.

Cranmer, S. R. and van Ballegooijen, A. A. (2010). Can the solar wind be driven by magnetic reconnection in the Sun's magnetic carpet? *Astrophys. J.* **720**, 824–847.

Cranmer, S. R. and van Ballegooijen, A. A. (2012). Proton, electron, and ion heating in the fast solar wind from nonlinear coupling between Alfvénic and fast-mode turbulence. *Astrophys. J.* **754**, 92.

Cranmer, S. R., van Ballegooijen, A. A., and Edgar, R. J. (2007). Self-consistent coronal heating and solar wind acceleration from anisotropic magnetohydrodynamic turbulence. *Astrophys. J. Supp.* **171**, 520–551.

Culhane, L., Harra, L. K., Baker, D., van Driel-Gesztelyi, L., Sun, J., Doschek, G. A., Brooks, D. H., Lundquist, L. L., Kamio, S., Young, P. R., and Hansteen, V. H. (2007). Hinode EUV study of jets in the Sun's south polar corona. *Publ. Astron. Soc. Japan* **59**, 751–756.

Curdt, W., Tian, H., and Kamio, S. (2012). Explosive events: Swirling transition region jets. *Solar Phys.* **280**, 417–424.

Czaykowska, A., De Pontieu, B., Alexander, D., and Rank, G. (1999). Evidence for chromospheric evaporation in the late gradual flare phase from SOHO/CDS observations. *Astrophys. J. Letts.* **521**, L75–L78.

Démoulin, P., Cuperman, S., and Semel, M. (1992). Determination of force-free magnetic fields above the photosphere using three-component boundary conditions. II. Analysis and minimization of scale-related growing modes and of computational induced singularities. *Astron. Astrophys.* **263**, 351–360.

Dahlburg, R. B., Antiochos, S. K., and Klimchuk, J. A. (1998). Prominence formation by localized heating. *Astrophys. J.* **495**, 485–490.

Dahlburg, R. B., Klimchuk, J. A., and Antiochos, S. K. (2005). An explanation for the "switch-on" nature of magnetic energy release and its application to coronal heating. *Astrophys. J.* **622**, 1191–1201.

Dalda, A. S. and Martínez Pillet, V. M. (2008). In-situ flux losses in active regions. *Astron. Soc. Pac. Conf. Ser.* **383**, 115–122.

Danielson, R. E. (1961). The structure of sunspot penumbras. II. Theoretical. *Astrophys. J.* **134**, 289–311.

Danielson, R. E. (1964). The structure of sunspot umbras. I. Observations. *Astrophys. J.* **139**, 45–47.

Danilovic, S., Beeck, B., Pietarila, A., Schüssler, M., Solanki, S. K., Martínez Pillet, V., Bonet, J. A., del Toro Iniesta, J. C., Domingo, V., Barthol, P., Berkefeld, T., Gandorfer, A., Knölker, M., Schmidt, W., and Title, A. M. (2010). Transverse component of the magnetic field in the solar photosphere observed by SUNRISE. *Astrophys. J. Letts.* **723**, L149–L153.

Danilovic, S., Schüssler, M., and Solanki, S. K. (2010). Magnetic field intensification: comparison of 3D MHD simulations with Hinode/SP results. *Astron. Astrophys.* **509**, A76.

Davis, Jr., R. and Evans, J. C. (1978). Experimental limits on extraterrestrial sources of neutrinos. *Proceedings 13th International Cosmic Ray Conference* **3**, 2001–2006.

De Groof, A. and Goossens, M. (2002). Fast and Alfvén waves driven by azimuthal footpoint motions. II. Random driver. *Astron. Astrophys.* **386**, 691–698.

De Moortel, I. and Galsgaard, K. (2006). Numerical modelling of 3D reconnection. II. Comparison between rotational and spinning footpoint motions. *Astron. Astrophys.* **459**, 627–639.

De Moortel, I., Hood, A. W., Ireland, J., and Arber, T. D. (1999). Phase mixing of Alfvén waves in a stratified and open atmosphere. *Astron. Astrophys.* **346**, 641–651.

De Moortel, I., Hood, A. W., Ireland, J., and Walsh, R. (2002). Longitudinal intensity oscillations in coronal loops observed with TRACE. I. Overview of measured parameters. *Solar Phys.* **209**, 61–88.

De Moortel, I., Ireland, J., and Walsh, R. W. (2000). Observation of oscillations in coronal loops. *Astron. Astrophys.* **355**, L23–L26.

De Moortel, I. and Nakariakov, V. M. (2012). Magnetohydrodynamic waves and coronal seismology: an overview of recent results. *Phil. Trans. Roy. Soc. Lond. A* **370**, 3193–3216.

De Moortel, I. and Pascoe, D. J. (2012). The effects of line-of-sight integration on multistrand coronal loop oscillations. *Astrophys. J.* **746**, 31.

De Pontieu, B. (2002). High-resolution observations of small-scale emerging flux in the photosphere. *Astrophys. J.* **569**, 474–486.

De Pontieu, B., Carlsson, M., Rouppe van der Voort, L., Rutten, R. J., Hansteen, V. H., and Watanabe, H. (2012). Ubiquitous torsional motions in type II spicules. *Astrophys. J. Letts.* **752**, L12.

De Pontieu, B., Erdélyi, R., and James, S. P. (2004). Solar chromospheric spicules from the leakage of photospheric oscillations and flows. *Nature* **430**, 536–539.

De Pontieu, B. and McIntosh, S. W. (2010). Quasi-periodic propagating signals in the solar corona: the signature of magnetoacoustic waves or high-velocity upflows? *Astrophys. J.* **722**, 1013–1029.

De Pontieu, B., McIntosh, S. W., Carlsson, M., Hansteen, V. H., Tarbell, T. D., Boerner, P., Martinez-Sykora, J., Schrijver, C. J., and Title, A. M. (2011). The origins of hot plasma in the solar corona. *Science* **331**, 55–58.

De Pontieu, B., McIntosh, S. W., Carlsson, M., Hansteen, V. H., Tarbell, T. D., Schrijver, C. J., Title, A. M., Shine, R. A., Tsuneta, S., Katsukawa, Y., Ichimoto, K., Suematsu, Y., Shimizu, T., and Nagata, S. (2007). Chromospheric Alfvénic waves strong enough to power the solar wind. *Science* **318**, 1574–1577.

De Pontieu, B., McIntosh, S. W., Hansteen, V. H., Carlsson, M., Schrijver, C. J., Tarbell, T. D., Title, A. M., Shine, R. A., Suematsu, Y., Tsuneta, S., Katsukawa, Y., Ichimoto, K., Shimizu, T., and Nagata, S. (2007). A tale of two spicules: the impact of spicules on the magnetic chromosphere. *Publ. Astron. Soc. Japan* **59**, 655–662.

De Pontieu, B., McIntosh, S. W., Hansteen, V. H., and Schrijver, C. J. (2009). Observing the roots of solar coronal heating in the chromosphere. *Astrophys. J. Letts.* **701**, L1–L6.

De Rosa, M. L. and Toomre, J. (2004). Evolution of solar supergranulation. *Astrophys. J.* **616**, 1242–1260.

de Wijn, A. G., Stenflo, J. O., Solanki, S. K., and Tsuneta, S. (2009). Small-scale solar magnetic fields. *Space Sci. Rev.* **144**, 275–315.

DeBenedetti, M. and Pegoraro, F. (1995). Resistive modes at a magnetic X-point. *Plasma Physics and Controlled Fusion* **37**, 103–116.

DeForest, C. E. (2007). On the size of structures in the solar corona. *Astrophys. J.* **661**, 532–542.

DeForest, C. E., Lamy, P. L., and Llebaria, A. (2001). Solar polar plume lifetime and coronal hole expansion: determination from long-term observations. *Astrophys. J.* **560**, 490–498.

Degenhardt, D. and Wiehr, E. (1991). Spatial variation of the magnetic field inclination in a sunspot penumbra. *Astron. Astrophys.* **252**, 821–826.

Deinzer, W. (1965). On the magneto-hydrostatic theory of sunspots. *Astrophys. J.* **141**, 548–563.

Del Zanna, G. (2008). Flows in active region loops observed by Hinode EIS. *Astron. Astrophys.* **481**, L49–L52.

Delaboudinière, J.-P., Artzner, G. E., Brunaud, J., Gabriel, A. H., Hochedez, J. F., Millier, F., Song, X. Y., Au, B., Dere, K. P., Howard, R. A., Kreplin, R., Michels, D. J., Moses, J. D., Defise, J. M., Jamar, C., Rochus, P., Chauvineau, J. P., Marioge, J. P., Catura, R. C., Lemen, J. R., Shing, L., Stern, R. A., Gurman, J. B., Neupert, W. M., Maucherat, A., Clette, F., Cugnon, P., and van Dessel, E. L. (1995). EIT: Extreme-Ultraviolet Imaging Telescope for the SOHO mission. *Solar Phys.* **162**, 291–312.

Delannée, C., Török, T., Aulanier, G., and Hochedez, J.-F. (2008). A new model for propagating parts of EIT waves: a current shell in a CME. *Solar Phys.* **247**, 123–150.

DeLuca, E. E. and Gilman, P. A. (1986). Dynamo theory for the interface between the convection zone and the radiative interior of a star: I. Model equations and exact solutions. *Geophys. Astrophys. Fluid Dyn.* **37**, 85–127.

Démoulin, P. and Aulanier, G. (2010). Criteria for flux rope eruption: non-equilibrium versus torus instability. *Astrophys. J.* **718**, 1388–1399.

Démoulin, P., Bagala, L. G., Mandrini, C. H., Hénoux, J. C., and Rovira, M. G. (1997). Quasi-separatrix layers in solar flares. II. Observed magnetic configurations. *Astron. Astrophys.* **325**, 305–317.

Démoulin, P., Henoux, J., and Mandrini, C. H. (1994). Are magnetic null points important in solar flares? *Astron. Astrophys.* **285**, 1023–1037.

Démoulin, P., Henoux, J. C., Mandrini, C. H., and Priest, E. R. (1997). Can we extrapolate a magnetic field when its topology is complex? *Solar Phys.* **174**, 73–89.

Démoulin, P., Henoux, J. C., Priest, E. R., and Mandrini, C. H. (1996). Quasi-separatrix layers in solar flares. I. Method. *Astron. Astrophys.* **308**, 643–655.

Démoulin, P. and Priest, E. R. (1988). Instability of a prominence supported in a linear force-free field. *Astron. Astrophys.* **206**, 336–347.

Démoulin, P. and Priest, E. R. (1989). A twisted flux model for solar prominences. II. Formation of a dip in a magnetic structure before the formation of a solar prominence. *Astron. Astrophys.* **214**, 360–368.

Démoulin, P. and Priest, E. R. (1992). The properties of sources and sinks of a linear force-free field. *Astron. Astrophys.* **258**, 535–541.

Démoulin, P. and Priest, E. R. (1993). A model for an inverse-polarity prominence supported in a dip of a quadrupolar region. *Solar Phys.* **144**, 283–305.

Démoulin, P., Priest, E. R., and Anzer, U. (1989). A three-dimensional model for solar prominences. *Astron. Astrophys.* **221**, 326–337.

Démoulin, P., van Driel-Gesztelyi, L., Schmieder, B., Henoux, J., Csepura, G., and Hagyard, M. (1993). Evidence for magnetic reconnection in solar flares. *Astron. Astrophys.* **271**, 292–307.

Deng, Y. Y., Schmieder, B., Engvold, O., DeLuca, E., and Golub, L. (2000). Emergence of sheared magnetic flux tubes in an active region observed with the SVST and TRACE. *Solar Phys.* **195**, 347–366.

Dere, K. P., Barto, J.-D. F., and Brueckner, G. E. (1989). Explosive events in the solar transition zone. *Solar Phys.* **123**, 41–68.

Dere, K. P., Brueckner, G. E., Howard, R. A., Michels, D. J., and Delaboudiniere, J. P. (1999). LASCO and EIT observations of helical structure in coronal mass ejections. *Astrophys. J.* **516**, 465–474.

Dere, K. P., Landi, E., Young, P. R., Del Zanna, G., Landini, M., and Mason, H. E. (2009). CHIANTI – an atomic database for emission lines. IX. Ionization rates, recombination rates, ionization equilibria for the elements hydrogen through zinc and updated atomic data. *Astron. Astrophys.* **498**, 915–929.

Deubner, F. L. (1975). Observations of low-wavenumber nonradial eigenmodes of the Sun. *Astron. Astrophys.* **44**, 371–375.

DeVore, C. R. and Antiochos, S. K. (2000). Dynamical formation and stability of helical prominence magnetic fields. *Astrophys. J.* **539**, 954–963.

DeVore, C. R. and Antiochos, S. K. (2005). Magnetic free energies of breakout coronal mass ejections. *Astrophys. J.* **628**, 1031–1045.

DeVore, C. R. and Antiochos, S. K. (2008). Homologous confined filament eruptions via magnetic breakout. *Astrophys. J.* **680**, 740–756.

DeVore, C. R., Boris, J. P., and Sheeley, N. R. (1984). The concentration of the large-scale solar magnetic field by a meridional surface flow. *Solar Phys.* **92**, 1–14.

Diamond, P. H., Hazeltine, R. D., An, Z. G., Carreras, B. A., and Hicks, H. R. (1984). Theory of anomalous tearing mode growth and the major tokamak disruption. *Phys. Fluids* **27**, 1449–1462.

Díaz, A. J., Soler, R., and Ballester, J. L. (2012). Rayleigh-Taylor instability in partially ionized compressible plasmas. *Astrophys. J.* **754**, 41.

Dikpati, M. and Choudhuri, A. R. (1994). The evolution of the Sun's poloidal field. *Astron. Astrophys.* **291**, 975–989.

Dikpati, M. and Gilman, P. A. (2006). Simulating and predicting solar cycles using a flux-transport dynamo. *Astrophys. J.* **649**, 498–514.

Dikpati, M. and Gilman, P. A. (2007). Global solar dynamo models: simulations and predictions of cyclic photospheric fields and long-term non-reversing interior fields. *New Journal of Physics* **9**, 297–321.

Dixon, A., Browning, P., and Priest, E. R. (1988). Coronal heating by relaxation in a sunspot magnetic field. *Geophys. Astrophys. Fluid Dyn.* **40**, 294–327.

Dobrowolny, M., Veltri, P., and Mangeney, A. (1983). Dissipative instabilities of magnetic neutral layers with velocity shear. *J. Plasma Phys.* **29**, 393–407.

Donati, J.-F. (1999). Magnetic cycles of HR 1099 and LQ Hydrae. *Mon. Not. Roy. Astron. Soc.* **302**, 457–481.

Donati, J.-F., Cameron, A. C., Semel, M., Hussain, G. A. J., Petit, P., Carter, B. D., Marsden, S. C., Mengel, M., López Ariste, A., Jeffers, S. V., and Rees, D. E. (2003). Dynamo processes and activity cycles of the active stars AB Doradus, LQ Hydrae and HR 1099. *Mon. Not. Roy. Astron. Soc.* **345**, 1145–1186.

Donati, J.-F., Forveille, T., Cameron, A. C., Barnes, J. R., Delfosse, X., Jardine, M. M., and Valenti, J. A. (2006). The large-scale axisymmetric magnetic topology of a very-low-mass fully convective star. *Science* **311**, 633–635.

Donati, J.-F. and Semel, M. (1990). Zeeman-Doppler imaging – A new option for magnetic field study of AP and solar-type stars. *Solar Phys.* **128**, 227–242.

Dorch, S. B. F. and Nordlund, Å. (2001). On the transport of magnetic fields by solar-like stratified convection. *Astron. Astrophys.* **365**, 562–570.

Dormy, E. and Soward, A. M. (2007). *Mathematical Aspects of Natural Dynamos.* (CRC Press, London).

Doschek, G. A. (2012). The dynamics and heating of active-region loops. *Astrophys. J.* **754**, 153.

Doschek, G. A., Warren, H. P., Mariska, J. T., Muglach, K., Culhane, J. L., Hara, H., and Watanabe, T. (2008). Flows and nonthermal velocities in solar active regions observed with the EUV Imaging Spectrometer on Hinode: a tracer of active region sources of heliospheric magnetic fields? *Astrophys. J.* **686**, 1362–1371.

Dove, J. B., Gibson, S. E., Rachmeler, L. A., Tomczyk, S., and Judge, P. (2011). A ring of polarized light: evidence for twisted coronal magnetism in cavities. *Astrophys. J. Letts.* **731**, L1.

Dowdy, J. F., Rabin, D., and Moore, R. L. (1986). On the magnetic structure of the quiet transition region. *Solar Phys.* **105**, 35–45.

Downs, C., Roussev, I. I., van der Holst, B., Lugaz, N., Sokolov, I. V., and Gombosi, T. I. (2010). Toward a realistic thermodynamic magnetohydrodynamic model of the global solar corona. *Astrophys. J.* **712**, 1219–1231.

Drake, J. F., Cassak, P. A., Shay, M. A., Swisdak, M., and Quataert, E. (2009). A magnetic reconnection mechanism for ion acceleration and abundance enhancements in impulsive flares. *Astrophys. J. Letts.* **700**, L16–L20.

Drazin, P. G. and Reid, W. H. (1981). *Hydrodynamic Stability.* (Cambridge University Press, Cambridge, UK).

D'Silva, S. and Choudhuri, A. R. (1993). A theoretical model for tilts of bipolar magnetic regions. *Astron. Astrophys.* **272**, 621–633.

Dudík, J., Aulanier, G., Schmieder, B., Bommier, V., and Roudier, T. (2008). Topological departures from translational invariance along a filament observed by THEMIS. *Solar Phys.* **248**, 29–50.

Dudík, J., Aulanier, G., Schmieder, B., Zapiór, M., and Heinzel, P. (2012). Magnetic topology of bubbles in quiescent prominences. *Astrophys. J.* **761**, 9.

Dungey, J. W. (1953). Conditions for the occurrence of electrical discharges in astrophysical systems. *Phil. Mag* **44**, 725–738.

Dunn, R. B. and Zirker, J. B. (1973). The solar filigree. *Solar Phys.* **33**, 281–304.

Durney, B. (1971). A new type of supersonic solution for the inviscid equations of the solar wind. *Astrophys. J.* **166**, 669–673.

Durney, B. R. (1995). On a Babcock-Leighton dynamo model with a deep-seated generating layer for the toroidal magnetic field. *Solar Phys.* **160**, 213–235.

Durney, B. R. (1997). On a Babcock-Leighton solar dynamo model with a deep-seated generating layer for the toroidal magnetic field. IV. *Astrophys. J.* **486**, 1065–1077.

Duvall, T. L. (1979). Large-scale solar velocity fields. *Solar Phys.* **63**, 3–15.

Duvall, T. L. (1982). A dispersion law for solar oscillations. *Nature* **300**, 242–243.

Duvall, T. L., Dziembowski, W. A., Goode, P. R., Gough, D. O., Harvey, J. W., and Leibacher, J. W. (1984). Internal rotation of the Sun. *Nature* **310**, 22–25.

Duvall, T. L. and Harvey, J. W. (1983). Observations of solar oscillations of low and intermediate degree. *Nature* **302**, 24–27.

Duvall, T. L. and Harvey, J. W. (1984). Rotational frequency splitting of solar oscillations. *Nature* **310**, 19–22.

Duvall, T. L., Jefferies, S. M., Harvey, J. W., Osaki, Y., and Pomerantz, M. A. (1993). Asymmetries of solar oscillation line profiles. *Astrophys. J.* **410**, 829–836.

Duvall, T. L., Jefferies, S. M., Harvey, J. W., and Pomerantz, M. A. (1993). Time-distance helioseismology. *Nature* **362**, 430–432.

Duvall, T. L., Kosovichev, A. G., Scherrer, P. H., Bogart, R. S., Bush, R. I., DeForest, C., Hoeksema, J. T., Schou, J., Saba, J. L. R., Tarbell, T. D., Title, A. M., Wolfson, C. J., and Milford, P. N. (1997). Time-distance helioseismology with the MDI instrument: initial results. *Solar Phys.* **170**, 63–73.

Dymova, M. V. and Ruderman, M. S. (2005). Non-axisymmetric oscillations of thin prominence fibrils. *Solar Phys.* **229**, 79–94.

Eddy, J. A. (1976). The Maunder minimum. *Science* **192**, 1189–1202.

Edenstrasser, J. W. (1980). The only three classes of symmetric MHD equilibria. *J. Plasma Phys.* **24**, 515–518.

Edlén, B. (1943). Die Deutung der Emissionslinien im Spektrum der Sonnenkorona. Mit 6 Abbildungen. *Zeitschrift fur Astrophysik* **22**, 30–64.

Edwin, P. M. and Roberts, B. (1982). Wave propagation in a magnetically structured atmosphere. III. The slab in a magnetic environment. *Solar Phys.* **76**, 239–259.

Edwin, P. M. and Roberts, B. (1983). Wave propagation in a magnetic cylinder. *Solar Phys.* **88**, 179–191.

Egbers, C. and Pfister, G. (2000). *Physics of Rotating Fluids.* (Springer Verlag, Berlin).

Einaudi, G., Boncinelli, P., Dahlburg, R. B., and Karpen, J. T. (1999). Formation of the slow solar wind in a coronal streamer. *J. Geophys. Res.* **104**, 521–534.

Elsässer, W. M. (1950). The hydromagnetic equations. *Physical Review* **79**, 183–183.

Elsworth, Y., Howe, R., Isaak, G. R., McLeod, C. P., and New, R. (1990). Variation of low-order acoustic solar oscillations over the solar cycle. *Nature* **345**, 322–324.

Emonet, T. and Moreno-Insertis, F. (1998). The physics of twisted magnetic tubes rising in a stratified medium: two-dimensional results. *Astrophys. J.* **492**, 804–821.

Emslie, A. G., Dennis, B. R., Holman, G. D., and Hudson, H. S. (2005). Refinements to flare energy estimates: a followup to "Energy partition in two solar flare/CME events" by A. G. Emslie et al. *J. Geophys. Res.* **110**, A11103.

Evershed, J. (1909). Radial movement in sun-spots. *Mon. Not. Roy. Astron. Soc.* **69**, 454–457.

Fabbian, D., Khomenko, E., Moreno-Insertis, F., and Nordlund, Å. (2010). Solar abundance corrections derived through three-dimensional magnetoconvection simulations. *Astrophys. J.* **724**, 1536–1541.

Falconer, D. A., Moore, R. L., Porter, J. G., and Hathaway, D. H. (1999). Large-scale coronal heating, clustering of coronal bright points, and concentration of magnetic flux. *Space Sci. Rev.* **87**, 181–184.

Fan, Y. (2001). Nonlinear growth of the three-dimensional undular instability of a horizontal magnetic layer and the formation of arching flux tubes. *Astrophys. J.* **546**, 509–527.

Fan, Y. (2005). Coronal mass ejections as loss of confinement of kinked magnetic flux ropes. *Astrophys. J.* **630**, 543–551.

Fan, Y. (2008). The three-dimensional evolution of buoyant magnetic flux tubes in a model solar convective envelope. *Astrophys. J.* **676**, 680–697.

Fan, Y. (2010). On the eruption of coronal flux ropes. *Astrophys. J.* **719**, 728–736.

Fan, Y. and Gibson, S. E. (2004). Numerical simulations of three-dimensional coronal magnetic fields resulting from the emergence of twisted magnetic flux tubes. *Astrophys. J.* **609**, 1123–1133.

Fan, Y. and Gibson, S. E. (2007). Onset of coronal mass ejections due to loss of confinement of coronal flux ropes. *Astrophys. J.* **668**, 1232–1245.

Faurobert-Scholl, M. (1991). Hanle effect with partial frequency redistribution. I. Numerical methods and first applications. *Astron. Astrophys.* **246**, 469–480.

Fedun, V., Shelyag, S., and Erdélyi, R. (2011). Numerical modeling of footpoint-driven magneto-acoustic wave propagation in a localized solar flux tube. *Astrophys. J.* **727**, 17.

Feldman, U., Widing, K. G., and Warren, H. P. (1999). Morphology of the quiet solar upper atmosphere in the $4 \times 10^4 < T_e < 1.4 \times 10^6$ K temperature regime. *Astrophys. J.* **522**, 1133–1147.

Felipe, T., Khomenko, E., and Collados, M. (2010). Magneto-acoustic waves in sunspots: first results from a new three-dimensional nonlinear magnetohydrodynamic code. *Astrophys. J.* **719**, 357–377.

Ferraro, V. C. A. and Plumpton, C. (1961). *An Introduction to Magnetofluid Mechanics.* (Oxford University Press, Oxford).

Ferriz-Mas, A., Schmitt, D., and Schüssler, M. (1994). A dynamo effect due to instability of magnetic flux tubes. *Astron. Astrophys.* **289**, 949–956.

Ferriz-Mas, A. and Schüssler, M. (1994). Instabilities of magnetic flux tubes in a stellar convection zone. I. Equatorial flux rings in differentially rotating stars. *Geophys. Astrophys. Fluid Dyn.* **72**, 209–247.

Ferriz-Mas, A. and Schüssler, M. (1995). Instabilities of magnetic flux tubes in a stellar convection zone. II. Flux rings outside the equatorial plane. *Geophys. Astrophys. Fluid Dyn.* **81**, 233–265.

Field, G. B. (1965). Thermal instability. *Astrophys. J.* **142**, 531–567.

Finn, J. M. and Kaw, P. K. (1977). Coalescence instability of magnetic islands. *Phys. Fluids* **20**, 72–78.

Fisher, G. H., Canfield, R. C., and McClymont, A. N. (1985a). Flare loop radiative hydrodynamics – part seven – dynamics of the thick target heated chromosphere. *Astrophys. J.* **289**, 434–441.

Fisher, G. H., Canfield, R. C., and McClymont, A. N. (1985b). Flare loop radiative hydrodynamics – part six – chromospheric evaporation due to heating by nonthermal electrons. *Astrophys. J.* **289**, 425–433.

Fisher, G. H., Fan, Y., Longcope, D. W., Linton, M. G., and Pevtsov, A. A. (2000). The solar dynamo and emerging flux (invited review). *Solar Phys.* **192**, 119–139.

Fisher, G. H., Longcope, D. W., Metcalf, T. R., and Pevtsov, A. A. (1998). Coronal heating in active regions as a function of global magnetic variables. *Astrophys. J.* **508**, 885–898.

Fisk, L. A. (2003). Acceleration of the solar wind as a result of the reconnection of open magnetic flux with coronal loops. *J. Geophys. Res.* **108**, 1157.

Fisk, L. A. (2005). The open magnetic flux of the Sun. I. Transport by reconnections with coronal loops. *Astrophys. J.* **626**, 563–573.

Fisk, L. A., Schwadron, N. A., and Zurbuchen, T. H. (1999). Acceleration of the fast solar wind by the emergence of new magnetic flux. *J. Geophys. Res.* **104**, 19765–19772.

Fletcher, L. and De Pontieu, B. (1999). Plasma diagnostics of transition region "moss" using SOHO/CDS and TRACE. *Astrophys. J. Letts.* **520**, L135–L138.

Fletcher, L., Metcalf, T. R., Alexander, D., Brown, D. S., and Ryder, L. A. (2001). Evidence for the flare trigger site and three-dimensional reconnection in multiwavelength observations of a solar flare. *Astrophys. J.* **554**, 451–463.

Fletcher, L., Pollock, J. A., and Potts, H. E. (2004). Tracking of TRACE ultraviolet flare footpoints. *Solar Phys.* **222**, 279–298.

Fontenla, J. M., Curdt, W., Haberreiter, M., Harder, J., and Tian, H. (2009). Semiempirical models of the solar atmosphere. III. Set of non-LTE models for far-ultraviolet/extreme-ultraviolet irradiance computation. *Astrophys. J.* **707**, 482–502.

Forbes, T. G. (1990). Numerical simulation of a catastrophe model for coronal mass ejections. *J. Geophys. Res.* **95**, 11919–11931.

Forbes, T. G. (1996). Reconnection theory for flares. *Astron. Soc. Pac. Conf. Ser.* **111**, 259–267.

Forbes, T. G. (2000). A review on the genesis of coronal mass ejections. *J. Geophys. Res.* **105**, 23153–23166.

Forbes, T. G. (2010). Models of coronal mass ejections and flares. In *Heliophysics: Space Storms and Radiation: Causes and Effects*, C. J. Schrijver, and G. L. Siscoe, eds. (Cambridge University Press, Cambridge, UK), 159–194.

Forbes, T. G. and Acton, L. W. (1996). Reconnection and field line shrinkage in solar flares. *Astrophys. J.* **459**, 330–341.

Forbes, T. G. and Isenberg, P. (1991). A catastrophe mechanism for coronal mass ejections. *Astrophys. J.* **373**, 294–307.

Forbes, T. G. and Malherbe, J. M. (1986). A shock-condensation mechanism for loop prominences. *Astrophys. J.* **302**, L67–L70.

Forbes, T. G., Malherbe, J. M., and Priest, E. R. (1989). The formation flare loops by magnetic reconnection and chromospheric ablation. *Solar Phys.* **120**, 285–307.

Forbes, T. G. and Priest, E. R. (1983). A numerical experiment relevant to line-tied reconnection in two-ribbon flares. *Solar Phys.* **84**, 169–188.

Forbes, T. G. and Priest, E. R. (1984a). Numerical simulation of reconnection in an emerging magnetic flux region. *Solar Phys.* **94**, 315–340.

Forbes, T. G. and Priest, E. R. (1984b). Reconnection in solar flares. In *Solar Terrestrial Phys.: Present and Future*, D. Butler and K. Papadopoulos, eds. (NASA RP-1120, Washington, DC), 1–35.

Forbes, T. G. and Priest, E. R. (1987). A comparison of analytical and numerical models for steadily-driven magnetic reconnection. *Rev. Geophys.* **25**, 1583–1607.

Forbes, T. G. and Priest, E. R. (1995). Photospheric magnetic field evolution and eruptive flares. *Astrophys. J.* **446**, 377–389.

Forbes, T. G., Priest, E. R., and Isenberg, P. (1994). On the maximum energy release in flux-rope models of eruptive flares. *Solar Phys.* **150**, 245–266.

Forbes, T. G., Priest, E. R., Seaton, D. B., and Litvinenko, Y. E. (2013). Indeterminacy and instability in Petschek reconnection. *Phys. Plasmas* **20**, 052902.

Fossat, E. (1991). The IRIS network for full disk helioseismology – present status of the programme. *Solar Phys.* **133**, 1–12.

Fossum, A. and Carlsson, M. (2006). Determination of the acoustic wave flux in the lower solar chromosphere. *Astrophys. J.* **646**, 579–592.

Franz, M. and Schlichenmaier, R. (2009). The velocity field of sunspot penumbrae. I. A global view. *Astron. Astrophys.* **508**, 1453–1460.

Frazier, E. N. (1968a). An observational study of the hydrodynamics of the lower solar photosphere. *Astrophys. J.* **152**, 557–575.

Frazier, E. N. (1968b). A spatio-temporal analysis of velocity fields in the solar photosphere. *Zeit. f. Astrophys.* **68**, 345–356.

Friedberg, J. P. (1987). *Ideal Magnetohydrodynamics.* (Plenum Press, New York).

Frieman, F. and Rotenberg, M. (1960). On hydromagnetic stability of stationary equilibria. *Rev. Mod. Phys.* **32**, 898–902.

Frisch, U., Pouquet, A., Leorat, J., and Mazure, A. (1975). Possibility of an inverse cascade of magnetic helicity in magnetohydrodynamic turbulence. *J. Fluid Mech.* **68**, 769–778.

Fuentes-Fernández, J., Parnell, C. E., Hood, A. W., Priest, E. R., and Longcope, D. W. (2012). Consequences of spontaneous reconnection at a two-dimensional non-force-free current layer. *Phys. Plasmas 19,* 2 (Feb.), 022901.

Fuller, J. and Gibson, S. E. (2009). A survey of coronal cavity density profiles. *Astrophys. J.* **700**, 1205–1215.

Furth, H. P., Killeen, J., and Rosenbluth, M. N. (1963). Finite-resistivity instabilities of a sheet pinch. *Phys. Fluids* **6**, 459–484.

Furth, H. P., Rutherford, P. H., and Selberg, H. (1973). Tearing mode in the cylindrical tokamak. *Phys. Fluids* **16**, 1054–1063.

Gabriel, A. H. (1976). A magnetic model of the solar transition region. *Phil. Trans. Roy. Soc. Lond. A* **281**, 339–352.

Gaizauskas, V. (2008). Development of flux imbalances in solar activity nests and the evolution of filament channels. *Astrophys. J.* **686**, 1432–1446.

Gaizauskas, V., Zirker, J. B., Sweetland, C., and Kovacs, A. (1997). Formation of a solar filament channel. *Astrophys. J.* **479**, 448–457.

Galloway, D. J. and Proctor, M. R. E. (1983). The kinematics of hexagonal magnetoconvection. *Geophys. Astrophys. Fluid Dyn.* **24**, 109–136.

Galloway, D. J., Proctor, M. R. E., and Weiss, N. O. (1978). Magnetic flux ropes and convection. *J. Fluid Mech.* **87**, 243–261.

Galloway, D. J. and Weiss, N. O. (1981). Convection and magnetic fields in stars. *Astrophys. J.* **243**, 945–953.

Galsgaard, K. (2002). Flux braiding in a stratified atmosphere. In *ESA SP-505: SOLMAG 2002. Proceedings of the Magnetic Coupling of the Solar Atmosphere Euroconference.* Noordwijk), 269–271.

Galsgaard, K., Mackay, D., Priest, E. R., and Nordlund, Å. (1999). On the location of energy release and temperature profiles along coronal loops. *Solar Phys.* **189**, 95–108.

Galsgaard, K. and Nordlund, Å. (1996a). The heating and activity of the solar corona: I. Boundary shearing of an initially homogeneous magnetic field. *J. Geophys. Res.* **101**, 13445–13460.

Galsgaard, K. and Nordlund, Å. (1996b). Heating and activity of the solar corona. III. Dynamics of a low-beta plasma with 3D null points. *J. Geophys. Res.* **102**, 231–248.

Galsgaard, K., Priest, E. R., and Titov, V. S. (2003). Numerical experiments on wave propagation towards a 3D null point due to rotational motions. *J. Geophys. Res. 108,* A1, SSH 10.1–10.9.

Garaud, P. (2007). Magnetic confinement of the solar tachocline. In *The Solar Tachocline*, D. W. Hughes, R. Rosner, and N. O. Weiss, eds. (Cambridge University Press, Cambridge, UK), 147–181.

Garaud, P. and Garaud, J. (2008). Dynamics of the solar tachocline – II. The stratified case. *Mon. Not. Roy. Astron. Soc.* **391**, 1239–1258.

Gebhardt, U. and Kiessling, M. (1992). The structure of ideal magnetohydrodynamics with incompressible steady flow. *Phys. Fluids B* **4**, 1689–1701.

Gelly, B., Fierry-Fraillon, D., Fossat, E., Palle, P., Cacciani, A., Ehgamberdiev, S., Grec, G., Hoeksema, J. T., Khalikov, S., Lazrek, M., Loudagh, S., Pantel, A., Regulo, C., and Schmider, F. X. (1997). Solar p-mode frequencies from the IRIS network. *Astron. Astrophys.* **323**, 235–242.

Gerrard, C. L. and Hood, A. W. (2004). Current build-up as a result of the kink instability in a loop. *Solar Phys.* **223**, 143–154.

Gerrard, C. L., Hood, A. W., and Brown, D. S. (2004). Effect of photospheric twisting motions on a confined toroidal loop. *Solar Phys.* **222**, 79–94.

Ghizaru, M., Charbonneau, P., and Smolarkiewicz, P. K. (2010). Magnetic cycles in global large-eddy simulations of solar convection. *Astrophys. J. Letts.* **715**, L133–L137.

Gibson, S. E. and Bagenal, F. (1995). Large-scale magnetic field and density distribution in the solar minimum corona. *J. Geophys. Res.* **100**, 19865–19880.

Gibson, S. E. and Fan, Y. (2006). Coronal prominence structure and dynamics: a magnetic flux rope interpretation. *J. Geophys. Res.* **111**, A12103.

Gibson, S. E. and Fan, Y. (2008). Partially ejected flux ropes: implications for interplanetary coronal mass ejections. *J. Geophys. Res.* **113**, A09103.

Gibson, S. E., Fan, Y., Mandrini, C., Fisher, G., and Démoulin, P. (2004). Observational consequences of a magnetic flux rope emerging into the corona. *Astrophys. J.* **617**, 600–613.

Gibson, S. E., Foster, D., Burkepile, J., de Toma, G., and Stanger, A. (2006). The calm before the storm: the link between quiescent cavities and coronal mass ejections. *Astrophys. J.* **641**, 590–605.

Gibson, S. E., Kucera, T. A., Rastawicki, D., Dove, J., de Toma, G., Hao, J., Hill, S., Hudson, H. S., Marqué, C., McIntosh, P. S., Rachmeler, L., Reeves, K. K., Schmieder, B., Schmit, D. J., Seaton, D. B., Sterling, A. C., Tripathi, D., Williams, D. R., and Zhang, M. (2010). Three-dimensional morphology of a coronal prominence cavity. *Astrophys. J.* **724**, 1133–1146.

Gilman, P. A. (1970). Instability of magnetohydrostatic stellar interiors from magnetic buoyancy. I. *Astrophys. J.* **162**, 1019–1029.

Gilman, P. A. (1977). Nonlinear dynamics of Boussinesq convection in a deep rotating spherical shell. I. *Geophys. Astrophys. Fluid Dyn.* **8**, 93–135.

Gilman, P. A. (1983). Dynamically consistent nonlinear dynamos driven by convection in a rotating spherical shell. II – Dynamos with cycles and strong feedbacks. *Astrophys. J. Suppl.* **53**, 243–268.

Gilman, P. A. and Glatzmaier, G. A. (1981). Compressible convection in a rotating spherical shell. I – III. *Astrophys. J. Suppl.* **45**, 335–388.

Gilman, P. A. and Miller, J. (1981). Dynamically consistent nonlinear dynamos driven by convection in a rotating spherical shell. *Astrophys. J. Suppl.* **246**, 211–238.

Giman, P. A. (2007). Global MHD instabilities of the tachocline. In *The Solar Tachocline*, D. W. Hughes, R. Rosner, and N. O. Weiss, eds. (Cambridge University Press, Cambridge, UK), 243–274.

Gingerich, O., Noyes, R. W., Kalkofen, W., and Cuny, Y. (1971). The Harvard-Smithsonian Reference Atmosphere. *Solar Phys.* **18**, 347–365.

Giovanelli, R. G. (1972). Oscillations and waves in a sunspot. *Solar Phys.* **27**, 71–79.

Gizon, L., Birch, A. C., and Spruit, H. C. (2010). Local helioseismology: three-dimensional imaging of the solar interior. *Ann. Rev. Astron. Astrophys.* **48**, 289–338.

Glatzmaier, G. A. (1985). Numerical simulations of stellar convective dynamos. II. Field propagation in the convection zone. *Astrophys. J.* **291**, 300–307.

Glatzmaier, G. A. and Gilman, P. A. (1981). Compressible convection in a rotating spherical shell. IV. Effects of viscosity, conductivity, boundary conditions and zone depth. *Astrophys. J. Suppl.* **47**, 103–115.

Glatzmaier, G. A. and Gilman, P. A. (1982). Compressible convection in a rotating spherical shell. V. Induced differential rotation and meridional circulation. *Astrophys. J.* **256**, 316–330.

Gloeckler, G., Zurbuchen, T. H., and Geiss, J. (2003). Implications of the observed anticorrelation between solar wind speed and coronal electron temperature. *J. Geophys. Res.* **108**, 1158.

Goedbloed, J. P. and Halberstadt, G. (1994). Magnetohydrodynamic waves in coronal flux tubes. *Astron. Astrophys.* **281**, 265–301.

Goedbloed, J. P., Keppens, R., and Poedts, S. (2004). *Principles of Magnetohydrodynamics.* (Cambridge University Press, Cambridge, UK).

Goldstein, B. E., Neugebauer, M., Phillips, J. L., Bame, S., Gosling, J. T., McComas, D., Wang, Y.-M., Sheeley, N. R., and Suess, S. T. (1996). ULYSSES plasma parameters: latitudinal, radial, and temporal variations. *Astron. Astrophys.* **316**, 296–303.

Goldstein, B. E., Smith, E. J., Balogh, A., Horbury, T. S., Goldstein, M. L., and Roberts, D. A. (1995). Properties of magnetohydrodynamic turbulence in the solar wind as observed by Ulysses at high heliographic latitudes. *Geophys. Res. Lett.* **22**, 3393–3396.

Golub, L., Davis, J. M., and Krieger, A. S. (1979). Anticorrelation of X-ray bright points with sunspot number, 1970–1978. *Astrophys. J.* **229**, L145–L150.

Golub, L., Deluca, E., Austin, G., Bookbinder, J., Caldwell, D., Cheimets, P., Cirtain, J., Cosmo, M., Reid, P., Sette, A., Weber, M., Sakao, T., Kano, R., Shibasaki, K., Hara, H., Tsuneta, S., Kumagai, K., Tamura, T., Shimojo, M., McCracken, J., Carpenter, J., Haight, H., Siler, R., Wright, E., Tucker, J., Rutledge, H., Barbera, M., Peres, G., and Varisco, S. (2007). The X-Ray Telescope (XRT) for the Hinode mission. *Solar Phys.* **243**, 63–86.

Golub, L., Krieger, A., Silk, J., Timothy, A., and Vaiana, G. (1974). Solar X-ray bright points. *Astrophys. J.* **189**, L93–L97.

Golub, L. and Pasachoff, J. M. (1997). *The Solar Corona.* (Cambridge University Press, Cambridge, UK).

Gomez, D. O. and Ferro Fontan, C. (1988). Coronal heating by selective decay of MHD turbulence. *Solar Phys.* **116**, 33–44.

Gomez, D. O., Martens, P. C. H., and Golub, L. (1993). Normal Incidence X-Ray Telescope power spectra of X-ray emission from solar active regions. II. Theory. *Astrophys. J.* **405**, 773–781.

Gömöry, P., Beck, C., Balthasar, H., Rybák, J., Kučera, A., Koza, J., and Wöhl, H. (2010). Magnetic loop emergence within a granule. *Astron. Astrophys.* **511**, A14.

Goossens, M. (2003). *An Introduction to Plasma Astrophysics and Magnetohydrodynamics.* (Kluwer Academic, London).

Goossens, M. (2008). Seismology of kink oscillations in coronal loops: two decades of resonant damping. *IAU Symposium* **247**, 228–242.

Goossens, M., Erdélyi, R., and Ruderman, M. S. (2011). Resonant MHD waves in the solar atmosphere. *Space Sci. Rev.* **158**, 289–338.

Goossens, M., Ruderman, M. S., and Hollweg, J. V. (1995). Dissipative MHD solutions for resonant Alfvén waves in 1-dimensional magnetic flux tubes. *Solar Phys.* **157**, 75–102.

Gorbachev, V. and Somov, B. (1988). Photospheric vortex flows as a cause for two-ribbon flares – a topological model. *Solar Phys.* **117**, 77–88.

Gosling, J. T., Birn, J., and Hesse, M. (1995). Three-dimensional magnetic reconnection and the magnetic topology of coronal mass ejection events. *Geophys. Res. Lett.* **22**, 869–872.

Gough, D. O. (1969). The anelastic approximation for thermal convection. *J. Atmos. Sci.* **26**, 448–456.

Gough, D. O. (1977). Random remarks on solar hydrodnamics. In *IAU Colloquium No. 36*, R. M. Bonnet and P. Delache, eds. (G. de Bussac, Clairmont-Ferrand), 3–36.

Gough, D. O. and McIntyre, M. E. (1998). Inevitability of a magnetic field in the Sun's radiative interior. *Nature* **394**, 755–757.

Gough, D. O. and Scherrer, P. H. (2001). The solar interior. In *The Century of Space Science*, J. Bleeker, J. Geiss, and M. Huber, eds. (Kluwer, Dordrecht), 1–29.

Gough, D. O. and Thompson, M. J. (1991). The inversion problem. In *Solar Interior and Atmosphere*, A. N. Cox, W. C. Livingston, and M. S. Matthews, eds. (University of Arizona Press, Tucson, AZ), 519–561.

Gourgouliatos, K. N. (2008). Self-similar magnetic arcades. *Mon. Not. Roy. Astron. Soc.* **385**, 875–882.

Grad, H. and Rubin, H. (1958). Hydromagnetic equilibria and force-free fields. *Proc. 2nd Internat. Conf. on Peaceful Uses of Atomic Energy* **31**, 190–197.

Grec, G., Fossat, E., and Pomerantz, M. (1980). Solar oscillations – full disk observations from the geographic South Pole. *Nature* **288**, 541–544.

Green, L. M. and Kliem, B. (2009). Flux rope formation preceding coronal mass ejection onset. *Astrophys. J. Letts.* **700**, L83–L87.

Green, L. M., Kliem, B., and Wallace, A. J. (2011). Photospheric flux cancellation and associated flux rope formation and eruption. *Astron. Astrophys.* **526**, A2.

Green, R. M. (1965). Modes of annihilation and reconnection in magnetic fields. In *Stellar and Solar Magnetic Fields. Proc. IAU Symp. 22*, R. Lust, ed. (North-Holland, Amsterdam), 398–404.

Grotrian, W. (1939). Zur Frage der Deutung der Linien in Spektrum der Sonnenkorona. *Naturwissenschaften* **27**, 214.

Guarrasi, M., Reale, F., and Peres, G. (2010). Coronal fuzziness modeled with pulse-heated multi-stranded loop systems. *Astrophys. J.* **719**, 576–582.

Güdel, M. (2004). X-ray astronomy of stellar coronae. *Astron. and Astrophys. Rev.* **12**, 71–237.

Gudiksen, B. V., Carlsson, M., Hansteen, V. H., Hayek, W., Leenaarts, J., and Martínez-Sykora, J. (2011). The stellar atmosphere simulation code Bifrost. Code description and validation. *Astron. Astrophys.* **531**, A154.

Gudiksen, B., V. and Nordlund, Å. (2005). An ab initio approach to the solar coronal heating problem. *Astrophys. J.* **618**, 1020–1030.

Guo, Y., Schmieder, B., Démoulin, P., Wiegelmann, T., Aulanier, G., Török, T., and Bommier, V. (2010). Coexisting flux rope and dipped arcade sections along one solar filament. *Astrophys. J.* **714**, 343–354.

Gurnett, D. A. and Bhattacharjee, A. (2005). *Introduction to Plasma Physics.* (Cambridge University Press, Cambridge, UK).

Habbal, S. R. and Tsinganos, K. (1983). Multiple transonic solutions with a new class of shock transitions in steady isothermal solar and stellar winds. *J. Geophys. Res.* **88**, 1965–1975.

Hagenaar, H. J. (2001). Ephemeral regions on a sequence of full-disk Michelson Doppler Imager magnetograms. *Astrophys. J.* **555**, 448–461.

Hagenaar, H. J., De Rosa, M. L., and Schrijver, C. J. (2008). The dependence of ephemeral region emergence on local flux imbalance. *Astrophys. J.* **678**, 541–548.

Hagenaar, H. J., Schrijver, C. J., and Title, A. M. (1997). The distribution of cell sizes of the solar chromospheric network. *Astrophys. J.* **481**, 988–995.

Hagenaar, H. J., Schrijver, C. J., and Title, A. M. (2003). The properties of small magnetic regions on the solar surface and the implications for the solar dynamo(s). *Astrophys. J.* **584**, 1107–1119.

Haigh, J. D. (2007). The Sun and the Earth's climate. *Liv. Rev. in Solar Phys.* **4**, 2.

Handy, B. N., Acton, L. W., Kankelborg, C. C., Wolfson, C. J., Akin, D. J., Bruner, M. E., Caravalho, R., Catura, R. C., Chevalier, R., Duncan, D. W., Edwards, C. G., Feinstein, C. N., Freel and , S. L., Friedlaender, F. M., Hoffmann, C. H., Hurlburt, N. E., Jurcevich, B. K., Katz, N. L., Kelly, G. A., Lemen, J. R., Levay, M., Lindgren, R. W., Mathur, D. P., Meyer, S. B., Morrison, S. J., Morrison, M. D., Nightingale, R. W., Pope, T. P., Rehse, R. A., Schrijver, C. J., Shine, R. A., Shing, L., Strong, K. T., Tarbell, T. D., Title, A. M., Torgerson, D. D., Golub, L., Bookbinder, J. A., Caldwell, D., Cheimets, P. N., Davis, W. N., Deluca, E. E., McMullen, R. A., Warren, H. P., Amato, D., Fisher, R., Maldonado, H., and Parkinson, C. (1999). The Transition Region and Coronal Explorer. *Solar Phys.* **187**, 229–260.

Hannah, I. G., Hudson, H. S., Battaglia, M., Christe, S., Kašparová, J., Krucker, S., Kundu, M. R., and Veronig, A. (2011). Microflares and the statistics of X-ray flares. *Space Sci. Rev.* **159**, 263–300.

Hansteen, V. H., De Pontieu, B., Rouppe van der Voort, L., van Noort, M., and Carlsson, M. (2006). Dynamic fibrils are driven by magnetoacoustic shocks. *Astrophys. J. Letts.* **647**, L73–L76.

Hansteen, V. H., Hara, H., De Pontieu, B., and Carlsson, M. (2010). On redshifts and blueshifts in the transition region and corona. *Astrophys. J.* **718**, 1070–1078.

Hansteen, V. H. and Leer, E. (1995). Coronal heating, densities, and temperatures and solar wind acceleration. *J. Geophys. Res.* **100**, 21577–21594.

Hara, H., Watanabe, T., Harra, L. K., Culhane, J. L., and Young, P. R. (2011). Plasma motions and heating by magnetic reconnection in a 2007 May 19 flare. *Astrophys. J.* **741**, 107.

Hara, H., Watanabe, T., Harra, L. K., Culhane, J. L., Young, P. R., Mariska, J. T., and Doschek, G. A. (2008). Coronal plasma motions near footpoints of active region loops revealed from spectroscopic observations with Hinode EIS. *Astrophys. J. Letts.* **678**, L67–L71.

Harra, L. K., Hara, H., Imada, S., Young, P. R., Williams, D. R., Sterling, A. C., Korendyke, C., and Attrill, G. D. R. (2007). Coronal dimming observed with Hinode: outflows related to a coronal mass ejection. *Pub. Astron. Soc. Japan* **59**,

801–806.

Harrison, R. A. (1986). Solar coronal mass ejections and flares. *Astron. Astrophys.* **162**, 283–291.

Harrison, R. A. (1996). Coronal magnetic storms: a new perspective on flares and the 'Solar Flare Myth' debate. *Solar Phys.* **166**, 441–444.

Harrison, R. A. (1997). EUV blinkers: the significance of variations in the extreme ultraviolet quiet Sun. *Solar Phys.* **175**, 467–485.

Harrison, R. A., Davies, J. A., Möstl, C., Liu, Y., Temmer, M., Bisi, M. M., Eastwood, J. P., de Koning, C. A., Nitta, N., Rollett, T., Farrugia, C. J., Forsyth, R. J., Jackson, B. V., Jensen, E. A., Kilpua, E. K. J., Odstrcil, D., and Webb, D. F. (2012). An analysis of the origin and propagation of the multiple coronal mass ejections of 2010 August 1. *Astrophys. J.* **750**, 45.

Harrison, R. A., Davies, J. A., Rouillard, A. P., Davis, C. J., Eyles, C. J., Bewsher, D., Crothers, S. R., Howard, R. A., Sheeley, N. R., Vourlidas, A., Webb, D. F., Brown, D. S., and Dorrian, G. D. (2009). Two years of the STEREO Heliospheric Imagers. Invited review. *Solar Phys.* **256**, 219–237.

Hart, A. B. (1954). Motions in the Sun at the photospheric level. IV. The equatorial rotation and possible velocity fields in the photosphere. *Mon. Not. Roy. Astron. Soc.* **114**, 17–38.

Hartle, R. E. and Sturrock, P. A. (1968). Two-fluid model of the solar wind. *Astrophys. J.* **151**, 1155–1170.

Harvey, J. W. (1969). Magnetic fields associated with solar active-region prominences. Ph.D. thesis, University of Colorado at Boulder.

Harvey, K. L. (1996). Observations of X-ray bright points. *Astron. Soc. of Pacif. Conf. Series* **111**, 9–18.

Harvey, K. L. and Harvey, J. W. (1973). Observations of moving magnetic features near sunspots. *Solar Phys.* **28**, 61–71.

Harvey, K. L. and Martin, S. F. (1973). Ephemeral active regions. *Solar Phys.* **32**, 389–402.

Hassler, D. M., Dammasch, I. E., Lemaire, P., Brekke, P., Curdt, W., Mason, H. E., Vial, J.-C., and Wilhelm, K. (1999). Solar wind outflow and the chromospheric magnetic network. *Science* **283**, 810–813.

Hathaway, D. H. (1996). Doppler measurements of the Sun's meridional flow. *Astrophys. J.* **460**, 1027–1033.

Hathaway, D. H. (2012a). Supergranules as probes of solar convection zone dynamics. *Astrophys. J. Letts.* **749**, L13.

Hathaway, D. H. (2012b). Supergranules as probes of the Sun's meridional circulation. *Astrophys. J.* **760**, 84.

Hathaway, D. H., Beck, J. G., Bogart, R. S., Bachmann, K. T., Khatri, G., Petitto, J. M., Han, S., and Raymond, J. (2000). The photospheric convection spectrum. *Solar Phys.* **193**, 299–312.

Hathaway, D. H., Beck, J. G., Han, S., and Raymond, J. (2002). Radial flows in supergranules. *Solar Phys.* **205**, 25–38.

Hathaway, D. H. and Rightmire, L. (2010). Variations in the Sun's meridional flow over a solar cycle. *Science* **327**, 1350–1352.

Haynes, A. L. and Parnell, C. E. (2010). A method for finding three-dimensional magnetic skeletons. *Phys. Plasmas 17,* 9, 092903.

Haynes, A. L., Parnell, C. E., Galsgaard, K., and Priest, E. R. (2007). Magnetohydrodynamic evolution of magnetic skeletons. *Proc. Roy. Soc. Lond.* **463**, 1097–1115.

Heasley, J. N. and Mihalas, D. (1976). Structure and spectrum of quiescent prominences: energy balance and hydrogen spectrum. *Astrophys. J.* **205**, 273–285.

Heinemann, T., Nordlund, Å., Scharmer, G. B., and Spruit, H. C. (2007). MHD simulations of penumbra fine structure. *Astrophys. J.* **669**, 1390–1394.

Heinzel, P., Schmieder, B., and Tziotziou, K. (2001). Why are solar filaments more extended in extreme-ultraviolet lines than in Hα? *Astrophys. J. Letts.* **561**, L223–L227.

Herzenberg, A. (1958). Geomagnetic dynamos. *Phil. Trans. Roy. Soc. Lond. A* **250**, 543–583.

Hesse, M. and Birn, J. (1990). Parallel electric fields in a simulation of magnetotail reconnection and plasmoid evolution. In *Phys. of Magnetic Flux Ropes.* (A.G.U., Washington), 679–685.

Hesse, M., Forbes, T. G., and Birn, J. (2005). On the relation between reconnected magnetic flux and parallel electric fields in the solar corona. *Astrophys. J.* **631**, 1227–1238.

Hesse, M. and Schindler, K. (1988). A theoretical foundation of General Magnetic Reconnection. *J. Geophys. Res.* **93**, 5539–5567.

Heyn, M. F. (1996). Rapid reconnection in compressible plasma. *Phys. Plasmas* **3**, 2725–2741.

Heyvaerts, J. (1974). The thermal instability in a magnetohydrodynamic medium. *Astron. Astrophys.* **37**, 65–73.

Heyvaerts, J. and Norman, C. (2003). Global asymptotic solutions for nonrelativistic magnetohydrodynamic jets and winds. *Astrophys. J.* **596**, 1270–1294.

Heyvaerts, J. and Priest, E. R. (1983). Coronal heating by phase-mixed shear Alfvén waves. *Astron. Astrophys.* **117**, 220–234.

Heyvaerts, J. and Priest, E. R. (1984). Coronal heating by reconnection in DC current systems – a theory based on Taylor's hypothesis. *Astron. Astrophys.* **137**, 63–78.

Heyvaerts, J. and Priest, E. R. (1992). A self-consistent model for solar coronal heating. *Astrophys. J.* **390**, 297–308.

Heyvaerts, J., Priest, E. R., and Rust, D. M. (1977). An emerging flux model for the solar flare phenomenon. *Astrophys. J.* **216**, 123–137.

Hickey, J. R., Stowe, L. L., Jacobowitz, H., Pellegrino, P., Maschhoff, R. H., House, F., and Vonder Haar, T. H. (1980). Initial solar irradiance determinations from Nimbus 7 cavity radiometer measurements. *Science* **208**, 281–283.

Hildner, E. (1974). The formation of solar quiescent prominences by condensation. *Solar Phys.* **35**, 123–136.

Hill, F. (1989). Solar oscillation ring diagrams and large-scale flows. *Astrophys. J. Letts.* **343**, L69–L71.

Hillier, A., Berger, T., Isobe, H., and Shibata, K. (2012). Numerical simulations of the magnetic Rayleigh-Taylor instability in the Kippenhahn-Schlüter prominence model. I. Formation of upflows. *Astrophys. J.* **746**, 120.

Hillier, A., Isobe, H., Shibata, K., and Berger, T. (2011). Numerical simulations of the magnetic Rayleigh-Taylor instability in the Kippenhahn-Schlüter prominence model. *Astrophys. J. Letts.* **736**, L1.

Hillier, A., Isobe, H., Shibata, K., and Berger, T. (2012). Numerical simulations of the magnetic Rayleigh-Taylor instability in the Kippenhahn-Schlüter Prominence model. II. Reconnection-triggered downflows. *Astrophys. J.* **756**, 110.

Hillier, A. and van Ballegooijen, A. (2013). On the support of solar prominence material by the dips of a coronal flux tube. *Astrophys. J.* **766**, 126.

Hirayama, T. (1974). Theoretical model of flares and prominences. I. Evaporating flare model. *Solar Phys.* **34**, 323–338.

Hirzberger, J., Feller, A., Riethmüller, T. L., Schüssler, M., Borrero, J. M., Afram, N., Unruh, Y. C., Berdyugina, S. V., Gandorfer, A., Solanki, S. K., Barthol, P., Bonet, J. A., Martínez Pillet, V., Berkefeld, T., Knölker, M., Schmidt, W., and Title, A. M. (2010). Quiet-Sun intensity contrasts in the near-ultraviolet as measured from SUNRISE. *Astrophys. J. Letts.* **723**, L154–L158.

Hirzberger, J., Gizon, L., Solanki, S. K., and Duvall, T. L. (2008). Structure and evolution of supergranulation from local helioseismology. *Solar Phys.* **251**, 417–437.

Hollweg, J. V. (1970). Collisionless solar wind. 1. Constant electron temperature. *J. Geophys. Res.* **75**, 2403–2418.

Hollweg, J. V. (1974). Transverse Alfvén waves in the solar wind: arbitrary k, v_0, B_0, and $|\delta B|$. *J. Geophys. Res.* **79**, 1539–1541.

Hollweg, J. V. (1981). Alfvén waves in the solar atmosphere. II – Open and closed magnetic flux tubes. *Solar Phys.* **70**, 25–66.

Hollweg, J. V. (1982). On the origin of solar spicules. *Astrophys. J.* **257**, 345–353.

Hollweg, J. V. (1986). Transition region, corona, and solar wind in coronal holes. *J. Geophys. Res.* **91**, 4111–4125.

Hollweg, J. V. (1986). Viscosity and the Chew-Goldberger-Low equations in the solar corona. *Astrophys. J.* **306**, 730–739.

Hollweg, J. V., Jackson, S., and Galloway, D. (1982). Alfvén waves in the solar atmosphere. III – Nonlinear waves on open flux tubes. *Solar Phys.* **75**, 35–61.

Holman, G. D., Sui, L., Schwartz, R. A., and Emslie, A. G. (2003). Electron bremsstrahlung, hard X-ray spectra, electron distributions, and energetics in the 2002 July 23 solar flare. *Astrophys. J. Letts.* **595**, L97–L101.

Holzer, T. E. and Leer, E. (1980). Conductive solar wind models in rapidly diverging flow geometries. *J. Geophys. Res.* **85**, 4665–4679.

Hood, A. W. (1985). MHD instabilities. In *Solar System Magnetic Fields*, E. R. Priest, ed. (D Reidel, Dordrecht), 80–120.

Hood, A. W. and Anzer, U. (1990). A model for quiescent solar prominences with normal polarity. *Solar Phys.* **126**, 117–133.

Hood, A. W., Archontis, V., Galsgaard, K., and Moreno-Insertis, F. (2009). The emergence of toroidal flux tubes from beneath the solar photosphere. *Astron. Astrophys.* **503**, 999–1011.

Hood, A. W., Brooks, S. J., and Wright, A. N. (2002). Coronal heating by the phase mixing of individual pulses propagating in coronal holes. *Proc. Roy. Soc. Lond.* **458**, 2307–2325.

Hood, A. W., Brooks, S. J., and Wright, A. N. (2005). Phase mixing of Alfvén pulses and wavetrains propagating in coronal holes. *Proc. Roy. Soc. Lond.* **461**, 237–252.

Hood, A. W., Browning, P. K., and van der Linden, R. A. M. (2009). Coronal heating by magnetic reconnection in loops with zero net current. *Astron. Astrophys.* **506**, 913–925.

Hood, A. W. and Priest, E. R. (1979a). The equilibrium of solar coronal magnetic loops. *Astron. Astrophys.* **77**, 233–251.

Hood, A. W. and Priest, E. R. (1979b). Kink instability of solar coronal loops as the cause of solar flares. *Solar Phys.* **64**, 303–321.

Hood, A. W. and Priest, E. R. (1981). Critical conditions for magnetic instabilities in force-free coronal loops. *Geophys. Astrophys. Fluid Dyn.* **17**, 297–318.

Hood, A. W., Priest, E. R., and Anzer, U. (1992). The fibril structure of prominences. *Solar Phys.* **138**, 331–351.

Horbury, T. S., Balogh, A., Forsyth, R. J., and Smith, E. J. (1995). Observations of evolving turbulence in the polar solar wind. *Geophys. Res. Lett.* **22**, 3401–3404.

Hornig, G. (2001). The geometry of reconnection. In *An Introduction to the Geometry and Topology of Fluid Flows*, R. L. Ricca, ed. (Kluwer, Dordrecht), 295–313.

Hornig, G. and Priest, E. R. (2003). Evolution of magnetic flux in an isolated reconnection process. *Phys. Plasmas* **10**, 2712–2721.

Hornig, G. and Schindler, K. (1996). Magnetic topology and the problem of its invariant definition. *Phys. Plasmas* **3**, 781–791.

Howard, R. F. and Labonte, B. J. (1980). The Sun is observed to be a torsional oscillator with a period of 11 years. *Astrophys. J. Letts.* **239**, L33–L36.

Howe, R., Christensen-Dalsgaard, J., Hill, F., Komm, R., Schou, J., and Thompson, M. J. (2005). Solar convection-zone dynamics, 1995–2004. *Astrophys. J.* **634**, 1405–1415.

Howe, R., Christensen-Dalsgaard, J., Hill, F., Komm, R. W., Larsen, R. M., Schou, J., Thompson, M. J., and Toomre, J. (2000). Dynamic variations at the base of the solar convection zone. *Science* **287**, 2456–2460.

Hoyt, D. V. and Schatten, K. H. (1996). How well was the Sun observed during the Maunder minimum? *Solar Phys.* **165**, 181–192.

Hu, Y. Q., Esser, R., and Habbal, S. R. (2000). A four-fluid turbulence-driven solar wind model for preferential acceleration and heating of heavy ions. *J. Geophys. Res.* **105**, 5093–5112.

Huba, J. D. (2003). A tutorial on Hall MHD. In *Space Simulations*, M. Scholer, C. Dum, and J. Büchner, eds. (Springer, New York), 170–197.

Huba, J. D. and Rudakov, L. I. (2002). Three-dimensional Hall magnetic reconnection. *Phys. Plasmas* **9**, 4435–4438.

Hudson, H. S. (1991). Solar flares, microflares, nanoflares, and coronal heating. *Solar Phys.* **133**, 357–369.

Hudson, H. S. (2010). Observations of solar and stellar eruptions, flares and jets. In *Heliophysics: Space Storms and Radiation: Causes and Effects*, Schrijver, C. J. & Siscoe, G. L., ed. (Cambridge University Press, Cambridge), 123–157.

Hughes, D. W. (1985). Magnetic buoyancy instabilities incorporating rotation. *Geophys. Astrophys. Fluid Dyn.* **34**, 99–142.

Hughes, D. W. (2007). Magnetic buoyancy instabilities in the tachocline. In *The Solar Tachocline*, D. W. Hughes, R. Rosner, and N. O. Weiss, eds. (Cambridge University Press, Cambridge, UK), 275–298.

Hughes, D. W., Rosner, R., and Weiss, N. O. (2007). *The Solar Tachocline.* (Cambridge University Press, Cambridge, UK).

Hundhausen, A. J. (1972). *Coronal Expansion and the Solar Wind.* (Springer-Verlag, New York).

Hundhausen, A. J. (1977). An interplanetary view of coronal holes. In *Coronal Holes and High Speed Wind Streams*, J. B. Zirker, ed. (Colo. Ass. Univ. Press, Boulder, Colorado), 225–329.

Hundhausen, A. J. (1999). Coronal mass ejections. In *The Many Faces of the Sun: a Summary of the Results from NASA's Solar Maximum Mission.*, K. T. Strong, J. L. R. Saba, B. M. Haisch, and J. T. Schmelz, ed. (Springer, New York), 143.

Hundhausen, J. R., Hundhausen, A. J., and Zweibel, E. G. (1981). Magnetostatic atmospheres in a spherical geometry and their application to the solar corona. *J. Geophys. Res.* **86**, 1117–1126.

Hunter, J. H. (1970). Generalized thermal instability and its application to the interstellar gas. *Astrophys. J.* **161**, 451–455.

Hurlburt, N. E., Proctor, M. R. E., Weiss, N. O., and Brownjohn, D. P. (1989). Nonlinear compressible magnetoconvection. I – Travelling waves and oscillations. *J. Fluid Mech.* **207**, 587–628.

Hurlburt, N. E. and Rucklidge, A. M. (2000). Development of structure in pores and sunspots: flows around axisymmetric magnetic flux tubes. *Mon. Not. Roy. Astron. Soc.* **314**, 793–806.

Hurlburt, N. E. and Toomre, J. (1988). Magnetic fields interacting with nonlinear compressible convection. *Astrophys. J.* **327**, 920–932.

Ichimoto, K., Suematsu, Y., Tsuneta, S., Katsukawa, Y., Shimizu, T., Shine, R. A., Tarbell, T. D., Title, A. M., Lites, B. W., Kubo, M., and Nagata, S. (2007). Twisting motions of sunspot penumbral filaments. *Science* **318**, 1597–1599.

Imshennik, V. S. and Syrovatsky, S. I. (1967). Two-dimensional flow of an ideally conducting gas in the vicinity of the zero line of a magnetic field. *Sov. Phys. JETP* **25**, 656–664.

Inhester, B. and Wiegelmann, T. (2006). Nonlinear force-free magnetic field extrapolations: comparison of the Grad-Rubin and Wheatland-Sturrock-Roumeliotis algorithms. *Solar Phys.* **235**, 201–221.

Innes, D. E. (2001). Coordinated observations of the quiet Sun transition region using SUMER spectra, TRACE images and MDI magnetograms. *Astron. Astrophys.* **378**, 1067–1077.

Innes, D. E., Cameron, R. H., and Solanki, S. K. (2011). EUV jets, type III radio bursts and sunspot waves investigated using SDO/AIA observations. *Astron. Astrophys.* **531**, L13.

Innes, D. E., Inhester, B., Axford, W. I., and Wilhelm, K. (1997). Bi-directional plasma jets produced by magnetic reconnection on the Sun. *Nature* **386**, 811–813.

Innes, D. E., McKenzie, D. E., and Wang, T. (2003). Observations of 1000 km s^{-1} Doppler shifts in 10^7 K solar flare supra-arcade. *Solar Phys.* **217**, 267–279.

Inverarity, G. W. and Priest, E. R. (1995a). Turbulent coronal heating. II. Twisted flux tube. *Astron. Astrophys.* **296**, 395–404.

Inverarity, G. W. and Priest, E. R. (1995b). Turbulent coronal heating. III. Wave heating in coronal loops. *Astron. Astrophys.* **302**, 567–577.

Isenberg, P. A. (1986). On a difficulty with accelerating particles at slow-mode shocks. *J. Geophys. Res.* **91**, 1699.

Isenberg, P. A. and Forbes, T. G. (2007). A three-dimensional line-tied magnetic field model for solar eruptions. *Astrophys. J.* **670**, 1453–1466.

Isenberg, P. A., Forbes, T. G., and Démoulin, P. (1993). Catastrophic evolution of a force-free flux rope: a model for eruptive flares. *Astrophys. J.* **417**, 368–386.

Ishikawa, R. and Tsuneta, S. (2009). Comparison of transient horizontal magnetic fields in a plage region and in the quiet Sun. *Astron. Astrophys.* **495**, 607–612.

Ishikawa, R. and Tsuneta, S. (2010). Spatial and temporal distributions of transient horizontal magnetic fields with deep exposure. *Astrophys. J. Letts.* **718**, L171–L175.

Ishikawa, R. and Tsuneta, S. (2011). The relationship between vertical and horizontal magnetic fields in the quiet Sun. *Astrophys. J.* **735**, 74.

Isliker, H., Anastasiadis, A., and Vlahos, L. (2000). MHD consistent cellular automata (CA) models. I. Basic features. *Astron. Astrophys.* **363**, 1134–1144.

Isliker, H., Anastasiadis, A., and Vlahos, L. (2001). MHD consistent cellular automata (CA) models. II. Applications to solar flares. *Astron. Astrophys.* **377**, 1068–1080.

Isliker, H. and Vlahos, L. (2003). Random walk through fractal environments. *Phys. Rev. 67,* 2, 026413.

Isobe, H., Miyagoshi, T., Shibata, K., and Yokoyama, T. (2005). Filamentary structure on the Sun from the magnetic Rayleigh-Taylor instability. *Nature* **434**, 478–481.

Isobe, H. and Shibata, K. (2009). Reconnection in solar flares: outstanding questions. *J. of Astrophys. Astron.* **30**, 79–85.

Ito, H., Tsuneta, S., Shiota, D., Tokumaru, M., and Fujiki, K. (2010). Is the polar region different from the quiet region of the Sun? *Astrophys. J.* **719**, 131–142.

Jackson, J. D. (1998). *Classical Electrodynamics, 3rd Edition.* (John Wiley and Sons, Hoboken).

Jahn, K. (1989). Current sheet as a diagnostic for the subphotospheric structure of a spot. *Astron. Astrophys.* **222**, 264–292.

Janse, Å. M. and Low, B. C. (2009). The topological changes of solar coronal magnetic fields. I. Spontaneous current sheets in three-dimensional fields. *Astrophys. J.* **690**, 1089–1104.

Javadi, S., Büchner, J., Otto, A., and Santos, J. C. (2011). About the relative importance of compressional heating and current dissipation for the formation of coronal X-ray bright points. *Astron. Astrophys.* **529**, A114.

Jeffrey, A. and Taniuti, T. (1964). *Non-Linear Wave Propagation.* (Academic Press, New York).

Jeffrey, A. and Taniuti, T. (1966). *Magnetohydrodynamic Stability and Thermonuclear Containment.* (Academic Press, New York).

Jennings, R. L. and Weiss, N. O. (1991). Symmetry breaking in stellar dynamos. *Mon. Not. Roy. Astron. Soc.* **252**, 249–260.

Jess, D. B., Mathioudakis, M., Crockett, P. J., and Keenan, F. P. (2008). Do all flares have white-light emission? *Astrophys. J. Letts.* **688**, L119–L122.

Jess, D. B., Mathioudakis, M., Erdélyi, R., Crockett, P. J., Keenan, F. P., and Christian, D. J. (2009). Alfvén waves in the lower solar atmosphere. *Science* **323**, 1582–1585.

Ji, H., Cao, W., and Goode, P. R. (2012). Observation of ultrafine channels of solar corona heating. *Astrophys. J. Letts.* **750**, L25.

Jiang, J., Chatterjee, P., and Choudhuri, A. R. (2007). Solar activity forecast with a dynamo model. *Mon. Not. Roy. Astron. Soc.* **381**, 1527–1542.

Jockers, K. (1970). Solar wind models based on exospheric theory. *Astron. Astrophys.* **6**, 219–239.

Jouve, L., Brun, A. S., Arlt, R., Brandenburg, A., Dikpati, M., Bonanno, A., Käpylä, P. J., Moss, D., Rempel, M., Gilman, P., Korpi, M. J., and Kosovichev, A. G. (2008). A solar mean-field dynamo benchmark. *Astron. Astrophys.* **483**, 949–960.

Jouve, L., Brun, A. S., and Aulanier, G. (2013). Global dynamics of subsurface solar active regions. *Astrophys. J.* **762**, 4.

Kadomtsev, B. B. (1975). Disruptive instability in tokamaks. *Soviet J. Plasma Phys.* **1**, 389–391.

Kano, R., Shimizu, T., and Tarbell, T. D. (2010). Hinode observation of photospheric magnetic activities triggering X-ray microflares around a well-developed sunspot. *Astrophys. J.* **720**, 1136–1145.

Kano, R. and Tsuneta, S. (1995). Scaling law of solar coronal loops obtained with Yohkoh. *Astrophys. J.* **454**, 934–944.

Kantrowitz, A. R. and Petschek, H. E. (1966). MHD characteristics and shock waves. In *Plasma Physics in Theory and Applications*, W. B. Kunkel, ed. (McGraw-Hill, New York).

Kappraff, J. M. and Tataronis, J. A. (1977). Resistive effects on Alfvén wave heating. *J. Plasma Phys.* **18**, 209–226.

Karpen, J. T. and Antiochos, S. K. (2008). Condensation formation by impulsive heating in prominences. *Astrophys. J.* **676**, 658–671.

Karpen, J. T., Antiochos, S. K., and DeVore, C. R. (2012). The mechanisms for the onset and explosive eruption of coronal mass ejections and eruptive flares. *Astrophys. J.* **760**, 81.

Karpen, J. T., Antiochos, S. K., and Klimchuk, J. A. (2006). The origin of high-speed motions and threads in prominences. *Astrophys. J.* **637**, 531–540.

Karpen, J. T., Antiochos, S. K., Picone, J. M., and Dahlburg, R. B. (1989). Nonlinear thermal instability in magnetized solar plasmas. *Astrophys. J.* **338**, 493–508.

Karpen, J. T., Tanner, S. E. M., Antiochos, S. K., and DeVore, C. R. (2005). Prominence formation by thermal nonequilibrium in the sheared-arcade model. *Astrophys. J.* **635**, 1319–1328.

Katsukawa, Y., Berger, T. E., Ichimoto, K., Lites, B. W., Nagata, S., Shimizu, T., Shine, R. A., Suematsu, Y., Tarbell, T. D., Title, A. M., and Tsuneta, S. (2007). Small-scale jetlike features in penumbral chromospheres. *Science* **318**, 1594–1597.

Katsukawa, Y. and Tsuneta, A. (2002). Small fluctuations of coronal X-ray intensity: a signature of nanoflares. In *Multi-Wavelength Observations of Coronal Structure and Dynamics*, P. C. H. Martens and D. Cauffman, eds. (Elsevier Science, Maryland Heights), 61.

Kazachenko, M. D., Canfield, R. C., Longcope, D. W., and Qiu, J. (2010). Sunspot rotation, flare energetics, and flux rope helicity: the Halloween flare on 2003 October 28. *Astrophys. J.* **722**, 1539–1546.

Kazachenko, M. D., Canfield, R. C., Longcope, D. W., and Qiu, J. (2012). Predictions of energy and helicity in four major eruptive solar flares. *Solar Phys.* **277**, 165–183.

Keller, C. U. (1995). Properties of solar magnetic fields from speckle polarimetry. (Ludwig Biermann Award Lecture 1994). *Rev. Mod. Astron.* **8**, 27–60.

Keller, C. U., Schüssler, M., Vögler, A., and Zakharov, V. (2004). On the origin of solar faculae. *Astrophys. J. Letts.* **607**, L59–L62.

Khater, A. H., Callebaut, D. K., and Kamel, E. S. (1998). Painlevé analysis and nonlinear periodic solutions for isothermal magnetostatic atmospheres. *Solar Phys.* **178**, 285–315.

Khater, A. H., Callebaut, D. K., Malfliet, W., and Kamel, E. S. (2002). Bäcklund transformations and exact solutions for some nonlinear evolution equations in solar magnetostatic models. *Journal of Computational and Applied Mathematics* **140**, 435–467.

Khater, A. H., Ibrahim, R. S., Shamardan, A. B., and Callebaut, D. K. (1997). Bäcklund transformations and Painlevé analysis: exact solutions for a Grad-Shafranov-type MHD equilibrium. *IMA Journal of Applied Mathematics* **58**, 51–69.

Khomenko, E. and Collados Vera, M. (2012a). Heating of the Magnetized Solar Chromosphere by Partial Ionization Effects. *Astrophys. J.* **747**, 87.

Khomenko, E. and Collados Vera, M. (2012b). Simulations of chromospheric heating by ambipolar diffusion. *Astron. Soc. Pac. Conf. Ser.* **463**, 1–8.

Khomenko, E., Diaz, A., de Vicente, A., Luna, M., and Collados Vera, M. (2013). Rayleigh-Taylor instability in prominences from numerical simulations including partial ionization effects. *IAU Symp.* **300**.

Khomenko, E. V., Collados, M., Solanki, S. K., Lagg, A., and Trujillo Bueno, J. (2003). Quiet-Sun inter-network magnetic fields observed in the infrared. *Astron. Astrophys.* **408**, 1115–1135.

Kiepenheuer, K. O. (1953). Solar activity. In *The Sun*, Kuiper, G. P., ed. (Univ. Chicago Press, Chicago), 322–465.

Kippenhahn, R. and Schlüter, A. (1957). Eine Theorie der Solaren Filamente. *Zs. Ap.* **43**, 36–62.

Kitchatinov, L. L. and Rüdiger, G. (1993). Λ-effect and differential rotation in stellar convection zones. *Astron. Astrophys.* **276**, 96–102.

Kitchatinov, L. L. and Rüdiger, G. (1996). Transition to rigid rotation in the radiative zone of the Sun: the effect of a fossil magnetic field? *Astron. Letts.* **22**, 279–284.

Kitchatinov, L. L. and Rüdiger, G. (2005). Differential rotation and meridional flow in the solar convection zone and beneath. *Astron. Nach.* **326**, 379–385.

Kliem, B., Karlický, M., and Benz, A. O. (2000). Solar flare radio pulsations as a signature of dynamic magnetic reconnection. *Astron. Astrophys.* **360**, 715–728.

Kliem, B. and Török, T. (2006). Torus instability. *Phys. Rev. Letts. 96,* 25, 255002.

Klimchuk, J. A. (2000). Cross-sectional properties of coronal loops. *Solar Phys.* **193**, 53–75.

Klimchuk, J. A. (2002). Scaling laws for solar and stellar coronae. In *Stellar Coronae in the Chandra and XMM-NEWTON Era*, F. Favata and J. J. Drake, eds. Astron. Soc. Pacific Conf. Ser., vol. 277. (San Francisco), 321–332.

Klimchuk, J. A. (2006). On solving the coronal heating problem. *Solar Phys.* **234**, 41–77.

Klimchuk, J. A. (2009). Coronal loop models and those annoying observations! (keynote). *Astron. Soc. of Pac. Conf. Series* **415**, 221–233.

Klimchuk, J. A. (2012). The role of type II spicules in the upper solar atmosphere. *J. Geophys. Res.* **117**, 12102.

Klimchuk, J. A., Patsourakos, S., and Cargill, P. J. (2008). Highly efficient modeling of dynamic coronal loops. *Astrophys. J.* **682**, 1351–1362.

Klimchuk, J. A., Sturrock, P. A., and Yang, W.-H. (1988). Coronal magnetic fields produced by photospheric shear. *Astrophys. J.* **335**, 456–467.

Knobloch, E., Tobias, S. M., and Weiss, N. O. (1998). Modulation and symmetry changes in stellar dynamos. *Mon. Not. Roy. Astron. Soc.* **297**, 1123–1138.

Kobel, P., Solanki, S. K., and Borrero, J. M. (2012). The continuum intensity as a function of magnetic field. II. Local magnetic flux and convective flows. *Astron. Astrophys.* **542**, A96.

Kohl, J. L., Noci, G., Cranmer, S. R., and Raymond, J. C. (2006). Ultraviolet spectroscopy of the extended solar corona. *Astron. Astroph. Rev.* **13**, 31–157.

Komm, R. W., Howard, R. F., and Harvey, J. W. (1993). Meridional flow of small photospheric magnetic features. *Solar Phys.* **147**, 207–223.

Kopp, R. A. and Holzer, T. E. (1976). Dynamics of coronal hole regions. I – Steady polytropic flows with multiple critical points. *Solar Phys.* **49**, 43–56.

Kopp, R. A. and Pneuman, G. W. (1976). Magnetic reconnection in the corona and the loop prominence phenomenon. *Solar Phys.* **50**, 85–98.

Kosovichev, A. G. (1996). Helioseismic constraints on the gradient of angular velocity at base of the solar convection zone. *Astrophys. J.* **469**, L61–L64.

Kosovichev, A. G., Schou, J., Scherrer, P. H., Bogart, R. S., Bush, R. I., Hoeksema, J. T., Aloise, J., Bacon, L., Burnette, A., DeForest, C., Giles, P. M., Leibrand, K., Nigam, R., Rubin, M., Scott, K., Williams, S. D., Basu, S., Christensen-Dalsgaard, J., Dappen, W., Rhodes, E. J., Duvall, T. L., Howe, R., Thompson, M. J., Gough, D. O., Sekii,

T., Toomre, J., Tarbell, T. D., Title, A. M., Mathur, D., Morrison, M., Saba, J. L. R., Wolfson, C. J., Zayer, I., and Milford, P. N. (1997). Structure and rotation of the solar interior: initial results from the MDI medium-l program. *Solar Phys.* **170**, 43–61.

Kosovichev, A. G. and Zharkova, V. V. (1998). X-ray flare sparks quake inside Sun. *Nature* **393**, 317–318.

Kosugi, T., Sakao, T., Masuda, S., Makishima, K., Inda, M., Murakami, T., Ogawara, Y., Yaji, K., and Matsushita, K. (1992). The Hard X-ray Telescope (HXT) onboard Yohkoh – its performance and some initial results. *Publ. Astron. Soc. Japan* **44**, L45–L49.

Kraichnan, R. H. (1965). Inertial-range spectrum of hydromagnetic turbulence. *Phys. Fluids* **8**, 1385–1387.

Krause, F. and Rädler, K. H. (1980). *Mean-Field Magnetohydrodynamics and Dynamo Theory.* (Pergamon Press, Oxford).

Kretzschmar, M., de Wit, T. D., Schmutz, W., Mekaoui, S., Hochedez, J.-F., and Dewitte, S. (2010). The effect of flares on total solar irradiance. *Nature Physics* **6**, 690–692.

Krucker, S., Battaglia, M., Cargill, P. J., Fletcher, L., Hudson, H. S., MacKinnon, A. L., Masuda, S., Sui, L., Tomczak, M., Veronig, A. L., Vlahos, L., and White, S. M. (2008). Hard X-ray emission from the solar corona. *Astron. Astrophys. Rev.* **16**, 155–208.

Krucker, S. and Benz, A. O. (1998). Energy distribution of heating processes in the quiet solar corona. *Astrophys. J. Letts.* **501**, L213–L216.

Krucker, S., White, S. M., and Lin, R. P. (2007). Solar flare hard X-ray emission from the high corona. *Astrophys. J. Letts.* **669**, L49–L52.

Kruskal, M. D. and Schwarzschild, M. (1954). Some instabilities of a completely ionized plasma. *Proc. Roy. Soc. Lond.* **A223**, 348–360.

Kubo, M., Lites, B. W., Ichimoto, K., Shimizu, T., Suematsu, Y., Katsukawa, Y., Tarbell, T. D., Shine, R. A., Title, A. M., Nagata, S., and Tsuneta, S. (2008). Disintegration of magnetic flux in decaying sunspots as observed with the Hinode SOT. *Astrophys. J.* **681**, 1677–1687.

Kubo, M., Lites, B. W., Shimizu, T., and Ichimoto, K. (2008). Magnetic flux loss and flux transport in a decaying active region. *Astrophys. J.* **686**, 1447–1453.

Kubo, M., Shimizu, T., and Tsuneta, S. (2007). Vector magnetic fields of moving magnetic features and flux removal from a sunspot. *Astrophys. J.* **659**, 812–828.

Kucera, T. A., Gibson, S. E., Schmit, D. J., Landi, E., and Tripathi, D. (2012). Temperature and extreme-ultraviolet intensity in a coronal prominence cavity and streamer. *Astrophys. J.* **757**, 73.

Kuckein, C., Centeno, R., Martínez Pillet, V., Casini, R., Manso Sainz, R., and Shimizu, T. (2009). Magnetic field strength of active-region filaments. *Astron. Astrophys.* **501**, 1113–1121.

Kuckein, C., Martínez Pillet, V., and Centeno, R. (2012a). An active-region filament studied simultaneously in the chromosphere and photosphere. I. Magnetic structure. *Astron. Astrophys.* **539**, A131.

Kuckein, C., Martínez Pillet, V., and Centeno, R. (2012b). An active-region filament studied simultaneously in the chromosphere and photosphere. II. Doppler velocities. *Astron. Astrophys.* **542**, A112.

Kulidzanishvili, V. I. and Zhugzhda, I. D. (1983). On the problem of spicular oscillations. *Solar Phys.* **88**, 35–41.

Kulsrud, R. M. and Hahm, T. S. (1982). Forced magnetic reconnection. *Physica Scripta Volume T* **2**, 525–528.

Kumar, N. and Roberts, B. (2003). Ion-neutral collisions effect on MHD surface waves. *Solar Phys.* **214**, 241–266.

Kuperus, M. and Raadu, M. A. (1974). The support of prominences formed in neutral sheets. *Astron. Astrophys.* **31**, 189–193.

Kusano, K., Maeshiro, T., Yokoyama, T., and Sakurai, T. (2004). The trigger mechanism of solar flares in a coronal arcade with reversed magnetic shear. *The Astrophysical Journal* **610**, 537–549.

Labrosse, N., Heinzel, P., Vial, J., Kucera, T., Parenti, S., Gunár, S., Schmieder, B., and Kilper, G. (2010). Physics of solar prominences: I. Spectral diagnostics and non-LTE modelling. *Space Science Reviews* **151**, 243–332.

Lagg, A., Solanki, S. K., Riethmüller, T. L., Martínez Pillet, V., Schüssler, M., Hirzberger, J., Feller, A., Borrero, J. M., Schmidt, W., del Toro Iniesta, J. C., Bonet, J. A., Barthol, P., Berkefeld, T., Domingo, V., Gandorfer, A., Knölker, M., and Title, A. M. (2010). Fully resolved quiet-Sun magnetic flux tube observed with the SUNRISE/IMAX instrument. *Astrophys. J. Letts.* **723**, L164–L168.

Landi, E. and Cranmer, S. R. (2009). Ion temperatures in the low solar corona: polar coronal holes at solar minimum. *Astrophys. J.* **691**, 794–805.

Lang, K. R. (2009). *The Sun From Space.* (Springer-Verlag, Berlin).

Langhans, K., Scharmer, G. B., Kiselman, D., Löfdahl, M. G., and Berger, T. E. (2005). Inclination of magnetic fields and flows in sunspot penumbrae. *Astron. Astrophys.* **436**, 1087–1101.

Larmor, J. (1919). How could a rotating body such as the Sun become a magnet? *Rep. Brit. Ass. Adv. Sci.* **1919**, 159–160.

Lau, Y.-T. and Finn, J. M. (1990). Three-dimensional kinematic reconnection in the presence of field nulls and closed field lines. *Astrophys. J.* **350**, 672–691.

Laval, G., Mercier, C., and Pellat, R. (1965). Necessity of the energy principles for magnetostatic stability. *Nuclear Fusion* **5**, 156–158.

Leake, J. E. and Arber, T. D. (2006). The emergence of magnetic flux through a partially ionised solar atmosphere. *Astron. Astrophys.* **450**, 805–818.

Lee, L. C. and Fu, Z. (1986a). Multiple X-line reconnection. 1. A criterion for the transition from a single X-line to a multiple X-line reconnection. *J. Geophys. Res.* **91**, 6807–6815.

Lee, L. C. and Fu, Z. (1986b). A simulation study of magnetic reconnection: transition from a fast mode to a slow mode expansion. *J. Geophys. Res.* **91**, 4551–4556.

Lee, M. A. and Roberts, B. (1986). On the behavior of hydromagnetic surface waves. *Astrophys. J.* **301**, 430–439.

Leer, E., Holzer, T. E., and Fla, T. (1982). Acceleration of the solar wind. *Space. Sci. Rev.* **33**, 161–200.

Leibacher, J. W. and Stein, R. F. (1971). A new description of the solar five-minute oscillation. *Astrophys. J. Letts.* **7**, 191–192.

Leighton, R. B. (1960). Preview on granulation: observational studies. *IAU Symp* **12**, 321–327.

Leighton, R. B. (1969). A magneto-kinematic model of the solar cycle. *Astrophys. J.* **156**, 1–26.

Leighton, R. B., Noyes, R. W., and Simon, G. W. (1962). Velocity fields in the solar atmosphere: I. Preliminary report. *Astrophys. J.* **135**, 474–499.

Leka, K. D., Canfield, R. C., McClymont, A. N., and van Driel-Gesztelyi, L. (1996). Evidence for current-carrying emerging flux. *Astrophys. J.* **462**, 547–560.

Lemaire, J. and Scherer, M. (1969). Le champ électrique de polarisation dans l'exosphère ionique polaire. *C.R. Acad. Sci. Paris Ser. B* **269**, 666–669.

Lemaire, J. and Scherer, M. (1971). Kinetic models of the solar wind. *J. Geophys. Res.* **76**, 7479–7490.

Lemen, J. R., Title, A. M., Akin, D. J., Boerner, P. F., Chou, C., Drake, J. F., Duncan, D. W., Edwards, C. G., Friedlaender, F. M., Heyman, G. F., Hurlburt, N. E., Katz, N. L., Kushner, G. D., Levay, M., Lindgren, R. W., Mathur, D. P., McFeaters, E. L., Mitchell, S., Rehse, R. A., Schrijver, C. J., Springer, L. A., Stern, R. A., Tarbell, T. D., Wuelser, J.-P., Wolfson, C. J., Yanari, C., Bookbinder, J. A., Cheimets, P. N., Caldwell, D., Deluca, E. E., Gates, R., Golub, L., Park, S., Podgorski, W. A., Bush, R. I., Scherrer, P. H., Gummin, M. A., Smith, P., Auker, G., Jerram, P., Pool, P., Soufli, R., Windt, D. L., Beardsley, S., Clapp, M., Lang, J., and Waltham, N. (2012). The Atmospheric Imaging Assembly (AIA) on the Solar Dynamics Observatory (SDO). *Solar Phys.* **275**, 17–40.

Leroy, J. L. (1989). Observation of prominence magnetic fields. In *Dynamics and Structure of Quiescent Solar Prominences*, E. R. Priest, ed. (Kluwer Academic, Dordrecht), 77–113.

Leroy, J. L., Bommier, V., and Sahal-Brechot, S. (1983). The magnetic field in prominences of the polar crown. *Solar Phys.* **83**, 135–142.

Levine, R. H. (1974). A new theory of coronal heating. *Astrophys. J.* **190**, 457–466.

Li, X., Morgan, H., Leonard, D., and Jeska, L. (2012). A solar tornado observed by AIA/SDO: Rotational flow and evolution of magnetic helicity in a prominence and cavity. *Astrophys. J. Letts.* **752**, L22.

Lie-Svendsen, Ø. and Esser, R. (2005). Modeling the energy budget of solar wind minor ions: implications for temperatures and abundances. *Astrophys. J.* **618**, 1057–1073.

Lighthill, M. J. (1953). On sound generated aerodynamically. I. General theory. *Proc. Roy. Soc. Lond.* **A211**, 564–587.

Lima, J. and Priest, E. R. (1993). Two-dimensional models for solar and stellar winds – hydrodynamic effects. *Astron. Astrophys.* **268**, 641–649.

Lima, J., Priest, E. R., and Tsinganos, K. (2001). An analytical MHD wind model with latitudinal dependences obtained using separation of the variables. *Astron. Astrophys.* **371**, 240–249.

Lin, H., Penn, M. J., and Tomczyk, S. (2000). A new precise measurement of the coronal magnetic field strength. *Astrophys. J. Letts.* **541**, L83–L86.

Lin, J. and Forbes, T. G. (2000). Effects of reconnection on the coronal mass ejection process. *J. Geophys. Res.* **105**, 2375–2392.

Lin, J., Forbes, T. G., and Isenberg, P. A. (2001). Prominence eruptions and coronal mass ejections triggered by newly emerging flux. *J. Geophys. Res.* **106**, 25053–25074.

Lin, J., Forbes, T. G., Isenberg, P. A., and Démoulin, P. (1998). The effect of curvature on flux-rope models of coronal mass ejections. *Astrophys. J.* **504**, 1006–1019.

Lin, J., Ko, Y.-K., Sui, L., Raymond, J. C., Stenborg, G. A., Jiang, Y., Zhao, S., and Mancuso, S. (2005). Direct observations of the magnetic reconnection site of an eruption on 2003 November 18. *Astrophys. J.* **622**, 1251–1264.

Lin, R. P., Dennis, B. R., Hurford, G. J., Smith, D. M., Zehnder, A., Harvey, P. R., Curtis, D. W., Pankow, D., Turin, P., Bester, M., Csillaghy, A., Lewis, M., Madden, N., van Beek, H. F., Appleby, M., Raudorf, T., McTiernan, J., Ramaty, R., Schmahl, E., Schwartz, R., Krucker, S., Abiad, R., Quinn, T., Berg, P., Hashii, M., Sterling, R., Jackson, R., Pratt, R., Campbell, R. D., Malone, D., Landis, D., Barrington-Leigh, C. P., Slassi-Sennou, S., Cork, C., Clark, D., Amato, D., Orwig, L., Boyle, R., Banks, I. S., Shirey, K., Tolbert, A. K., Zarro, D., Snow, F., Thomsen, K., Henneck, R., McHedlishvili, A., Ming, P., Fivian, M., Jordan, J., Wanner, R., Crubb, J., Preble, J., Matranga, M., Benz, A., Hudson, H., Canfield, R. C., Holman, G. D., Crannell, C., Kosugi, T., Emslie, A. G., Vilmer, N., Brown, J. C., Johns-Krull, C., Aschwanden, M., Metcalf, T., and Conway, A. (2002). The Reuven Ramaty High-Energy Solar Spectroscopic Imager (RHESSI). *Solar Phys.* **210**, 3–32.

Lin, R. P., Schwartz, R. A., Kane, S. R., Pelling, R. M., and Hurley, K. C. (1984). Solar hard X-ray microflares. *Astrophys. J.* **283**, 421–425.

Lin, Y., Engvold, O., Rouppe van der Voort, L., Wiik, J. E., and Berger, T. E. (2005). Thin threads of solar filaments. *Solar Phys.* **226**, 239–254.

Lin, Y., Martin, S. F., and Engvold, O. (2008). Filament substructures and their interrelation. *Astron. Soc. Pac. Conf. Ser.* **383**, 235–242.

Linardatos, D. (1993). Magnetostatic equilibria and analogous Euler flows. *J. Fluid Mech.* **246**, 569–591.

Lindsey, C. and Braun, D. C. (2000). Helioseismic imaging of sunspots at their antipodes. *Solar Phys.* **126**, 101–115.

Linker, J. A., Lionello, R., Mikić, Z., Titov, V. S., and Antiochos, S. K. (2011). The evolution of open magnetic flux driven by photospheric dynamics. *Astrophys. J.* **731**, 110.

Linker, J. A., Mikić, Z., Lionello, R., Riley, P., Amari, T., and Odstrcil, D. (2003). Flux cancellation and coronal mass ejections. *Phys. Plasmas* **10**, 1971–1978.

Linton, M. G. and Longcope, D. W. (2006). A model for patchy reconnection in three dimensions. *Astrophys. J.* **642**, 1177–1192.

Lionello, R., Linker, J. A., and Mikić, Z. (2009). Multispectral emission of the Sun during the first whole Sun month: magnetohydrodynamic simulations. *Astrophys. J.* **690**, 902–912.

Lionello, R., Linker, J. A., Mikić, Z., and Riley, P. (2006). The latitudinal excursion of coronal magnetic field lines in response to differential rotation: MHD simulations. *Astrophys. J. Letts.* **642**, L69–L72.

Lionello, R., Mikić, Z., Linker, J. A., and Amari, T. (2002). Magnetic field topology in prominences. *Astrophys. J.* **581**, 718–725.

Lites, B. W. (2002). Characterization of magnetic flux in the quiet Sun. *Astrophys. J.* **573**, 431–444.

Lites, B. W., Elmore, D. F., Seagraves, P., and Skumanich, A. P. (1993). Stokes profile analysis and vector magnetic fields. VI. Fine scale structure of a sunspot. *Astrophys. J.* **418**, 928.

Lites, B. W., Kubo, M., Socas-Navarro, H., Berger, T., Frank, Z., Shine, R., Tarbell, T., Title, A., Ichimoto, K., Katsukawa, Y., Tsuneta, S., Suematsu, Y., Shimizu, T., and Nagata, S. (2008). The horizontal magnetic flux of the quiet-Sun internetwork as observed with the Hinode Spectro-Polarimeter. *Astrophys. J.* **672**, 1237–1253.

Lites, B. W., Leka, K. D., Skumanich, A., Martinez Pillet, V., and Shimizu, T. (1996). Small-scale horizontal magnetic fields in the solar photosphere. *Astrophys. J.* **460**, 1019–1026.

Lites, B. W., Scharmer, G. B., Berger, T. E., and Title, A. M. (2004). Three-dimensional structure of the active-region photosphere as revealed by high angular resolution. *Solar Phys.* **221**, 65–84.

Littlejohn, R. G. (1979). A guiding center Hamiltonian: A new approach. *J. Math. Phys.* **20**, 2445–2458.

Litvinenko, Y. E. (2010). Evolution of the axial magnetic field in solar filament channels. *Astrophys. J.* **720**, 948–952.

Litvinenko, Y. E. and Martin, S. F. (1999). Magnetic reconnection as the cause of a photospheric canceling feature and mass flows in a filament. *Solar Phys.* **190**, 45–58.

Liu, C., Deng, N., Liu, R., Ugarte-Urra, I., Wang, S., and Wang, H. (2011). A standard-to-blowout jet. *Astrophys. J. Letts.* **735**, L18.

Liu, R., Lee, J., Wang, T., Stenborg, G., Liu, C., and Wang, H. (2010). A reconnecting current sheet imaged in a solar flare. *Astrophys. J. Letts.* **723**, L28–L33.

Liu, W., Ofman, L., Nitta, N. V., Aschwanden, M. J., Schrijver, C. J., Title, A. M., and Tarbell, T. D. (2012). Quasi-periodic fast-mode wave trains within a global EUV wave and sequential transverse oscillations detected by SDO/AIA. *Astrophys. J.* **753**, 52.

Liu, Y. and Lin, H. (2008). Observational test of coronal magnetic field models. I. Comparison with potential field model. *Astrophys. J.* **680**, 1496–1507.

Livingston, W. and Harvey, J. W. (1971). The Kitt Peak Magnetograph. IV: 40-channel probe and the detection of weak photospheric fields. In *Solar Magnetic Fields*, R. Howard, ed. IAU Symposium, vol. 43. Dordrecht), 51–61.

Livingston, W., Harvey, J. W., Malanushenko, O. V., and Webster, L. (2006). Sunspots with the strongest magnetic fields. *Solar Phys.* **239**, 41–68.

Lockwood, G. W., Skiff, B. A., Henry, G. W., Henry, S., Radick, R. R., Baliunas, S. L., Donahue, R. A., and Soon, W. (2007). Patterns of photometric and chromospheric variation among Sun-like stars: a 20-year perspective. *Astrophys. J. Suppl.* **171**, 260–303.

Lockyer, N. (1868). Spectroscopic observation of the Sun, No. II. *Proc. Roy. Soc.* **17**, 131–132.

Longcope, D. W. (1996). Topology and current ribbons: a model for current, reconnection and flaring in a complex, evolving corona. *Solar Phys.* **169**, 91–121.

Longcope, D. W. (1998). A model for current sheets and reconnection in X-ray bright points. *Astrophys. J.* **507**, 433–442.

Longcope, D. W. (2001). Separator current sheets: generic features in minimum-energy magnetic fields subject to flux constraints. *Phys. Plasmas* **8**, 5277–5289.

Longcope, D. W. (2005). Topological methods for the analysis of solar magnetic fields. *Liv. Rev. in Solar Phys.* **2**, 1–72.

Longcope, D. W. and Beveridge, C. (2007). A quantitative, topological model of reconnection and flux rope formation in a two-ribbon flare. *Astrophys. J.* **669**, 621–635.

Longcope, D. W., Beveridge, C., Qiu, J., Ravindra, B., Barnes, G., and Dasso, S. (2007). Modeling and measuring the flux reconnected and ejected by the two-ribbon flare/CME event on 7 November 2004. *Solar Phys.* **244**, 45–73.

Longcope, D. W. and Cowley, S. (1996). Current sheet formation along three-dimensional magnetic separators. *Phys. Plasmas* **3**, 2885–2897.

Longcope, D. W., Des Jardins, A. C., Carranza-Fulmer, T., and Qiu, J. (2010). A quantitative model of energy release and heating by time-dependent, localized reconnection in a flare with thermal loop-top X-ray source. *Solar Phys.* **267**, 107–139.

Longcope, D. W., Fisher, G. H., and Pevtsov, A. A. (1998). Flux tube twist resulting from helical turbulence: the sigma-effect. *Astrophys. J.* **507**, 417–432.

Longcope, D. W. and Guidoni, S. E. (2011). A model for the origin of high density in looptop X-ray sources. *Astrophys. J.* **740**, 73.

Longcope, D. W., Guidoni, S. E., and Linton, M. G. (2009). Gas-dynamic shock heating of post-flare loops due to retraction following localized, impulsive reconnection. *Astrophys. J. Letts.* **690**, L18–L22.

Longcope, D. W. and Magara, T. (2004). A comparison of the minimum current corona to a magnetohydrodynamic simulation of quasi-static coronal evolution. *Astrophys. J.* **608**, 1106–1123.

Longcope, D. W., McKenzie, D., Cirtain, J., and Scott, J. (2005). Observations of separator reconnection to an emerging active region. *Astrophys. J.* **630**, 596–614.

Longcope, D. W. and Priest, E. R. (2007). Fast magnetosonic waves launched by transient, current sheet reconnection. *Phys. Plasmas* **14**, 122905.

Longcope, D. W. and Silva, A. V. R. (1998). A current ribbon model for energy storage and release with application to the flare 1992 Jan. 7. *Solar Phys.* **179**, 349–377.

Longcope, D. W. and Strauss, H. R. (1993). The coalescence instability and the development of current sheets in two-dimensional magnetohydrodynamics. *Physics of Fluids B* **5**, 2858–2869.

López Ariste, A., Aulanier, G., Schmieder, B., and Sainz Dalda, A. (2006). First observation of bald patches in a filament channel and at a barb endpoint. *Astron. Astrophys.* **456**, 725–735.

López Ariste, A. and Sainz Dalda, A. (2012). Scales of the magnetic fields in the quiet Sun. *Astron. Astrophys.* **540**, A66.

López Fuentes, M. C., Klimchuk, J. A., and Mandrini, C. H. (2007). The temporal evolution of coronal loops observed by GOES SXI. *Astrophys. J.* **657**, 1127–1136.

Lorenz, E. N. (1963). Deterministic nonperiodic flow. *J. Atmos. Sci.* **20**, 131–141.

Loureiro, N. F., Schekochihin, A. A., and Cowley, S. C. (2007). Instability of current sheets and formation of plasmoid chains. *Phys. Plasmas 14,* 10, 100703.

Low, B. C. (1973). Resistive diffusion of force-free magnetic fields in a passive medium. *Astrophys. J.* **181**, 209–226.

Low, B. C. (1974). Resistive diffusion of force-free magnetic fields in a passive medium. IV. The dynamical theory. *Astrophys. J.* **193**, 243–252.

Low, B. C. (1975). Nonisothermal magnetostatic equilibria in a uniform gravity field. II. Sheet models of quiescent prominences. *Astrophys. J.* **198**, 211–217.

Low, B. C. (1977). Evolving force-free magnetic fields. I. The development of the preflare stage. *Astrophys. J.* **212**, 234–242.

Low, B. C. (1982a). Magnetic field configurations associated with polarity intrusion in a solar active region. I. The force-free fields. *Solar Phys.* **77**, 43–61.

Low, B. C. (1982b). Magnetostatic atmospheres with variations in three dimensions. *Astrophys. J.* **263**, 952–969.

Low, B. C. (1984). Three-dimensional magnetostatic atmospheres – Magnetic field with vertically oriented tension force. *Astrophys. J.* **277**, 415–421.

Low, B. C. (1985). Three-dimensional structures of magnetostatic atmospheres. I. Theory. *Astrophys. J.* **293**, 31–43.

Low, B. C. (1988). On the hydromagnetic stability of a class of laminated force-free magnetic fields. *Astrophys. J.* **330**, 992–996.

Low, B. C. (1991). Three-dimensional structures of magnetostatic atmospheres. III. A general formulation. *Astrophys. J.* **370**, 427–434.

Low, B. C. (1992). Three-dimensional structures of magnetostatic atmospheres. IV. Magnetic structures over a solar active region. *Astrophys. J.* **399**, 300–312.

Low, B. C. (1993a). Three-dimensional structures of magnetostatic atmospheres. V. Coupled electric current systems. *Astrophys. J.* **408**, 689–692.

Low, B. C. (1993b). Three-dimensional structures of magnetostatic atmospheres. VI. Examples of coupled electric current systems. *Astrophys. J.* **408**, 693–706.

Low, B. C. (2005). Three-dimensional structures of magnetostatic atmospheres. VII. Magnetic flux surfaces and boundary conditions. *Astrophys. J.* **625**, 451–462.

Low, B. C. (2010). The Parker magnetostatic theorem. *Astrophys. J.* **718**, 717–723.

Low, B. C., Berger, T., Casini, R., and Liu, W. (2012). The hydromagnetic interior of a solar quiescent prominence. I. Coupling between force balance and steady energy transport. *Astrophys. J.* **755**, 34.

Low, B. C., Liu, W., Berger, T., and Casini, R. (2012). The hydromagnetic interior of a solar quiescent prominence. II. Magnetic discontinuities and cross-field mass transport. *Astrophys. J.* **757**, 21.

Low, B. C. and Lou, Y. Q. (1990). Modeling solar force-free magnetic fields. *Astrophys. J.* **352**, 343–352.

Low, B. C. and Petrie, G. J. D. (2005). The internal structures and dynamics of solar quiescent prominences. *Astrophys. J.* **626**, 551–562.

Low, B. C. and Wolfson, R. (1988). Spontaneous formation of electric current sheets and the origin of solar flares. *Astrophys. J.* **324**, 574–581.

Lu, E. T. and Hamilton, R. J. (1991). Avalanches and the distribution of solar flares. *Astrophys. J. Letts.* **380**, L89–L92.

Lugaz, N., Downs, C., Shibata, K., Roussev, I. I., Asai, A., and Gombosi, T. I. (2011). Numerical investigation of a coronal mass ejection from an anemone active region: reconnection and deflection of the 2005 August 22 eruption. *Astrophys. J.* **738**, 127–141.

Luna, M., Karpen, J. T., and DeVore, C. R. (2012). Formation and evolution of a multi-threaded solar prominence. *Astrophys. J.* **746**, 30.

Lundquist, S. (1951). On the stability of magneto-hydrostatic fields. *Phys. Rev.* **83**, 307–311.

Lüst, R. and Schlüter, A. (1954). Kraftfreie Magnetfelder. *Z. Astrophys.* **34**, 263–282.

Lynch, B. J., Antiochos, S. K., DeVore, C. R., Luhmann, J. G., and Zurbuchen, T. H. (2008). Topological evolution of a fast magnetic breakout CME in three dimensions. *Astrophys. J.* **683**, 1192–1206.

Lynden-Bell, D. and Boily, C. (1994). Self-similar solutions up to flashpoint in highly wound magnetostatics. *Mon. Not. Roy. Astron. Soc.* **267**, 146–152.

Mackay, D. H. (2012). The Sun's global corona. *Phil. Trans. Roy. Soc.* **370**, 3151–3168.

Mackay, D. H., Gaizauskas, V., Rickard, G. J., and Priest, E. R. (1997). Force-free and potential models of a filament channel in which a filament forms. *Astrophys. J.* **486**, 534–549.

Mackay, D. H., Gaizauskas, V., and Yeates, A. R. (2008). Where do solar filaments form?: consequences for theoretical models. *Solar Phys.* **248**, 51–65.

Mackay, D. H., Green, L. M., and van Ballegooijen, A. A. (2011). Modeling the dispersal of an active region: quantifying energy input into the corona. *Astrophys. J.* **729**, 97.

Mackay, D. H., Karpen, J. T., Ballester, J. L., Schmieder, B., and Aulanier, G. (2010). Physics of solar prominences: II. Magnetic structure and dynamics. *Space Science Review* **151**, 333–399.

Mackay, D. H. and Priest, E. R. (1997). Basic magnetic field configurations for filament channels and filaments. *Astronomical and Astrophysical Transactions* **13**, 111–120.

Mackay, D. H., Priest, E. R., Gaizauskas, V., and van Ballegooijen, A. A. (1998). Role of helicity in the formation of intermediate filaments. *Solar Phys.* **180**, 299–312.

Mackay, D. H. and van Ballegooijen, A. A. (2005). New results in modeling the hemispheric pattern of solar filaments. *Astrophys. J. Letts.* **621**, L77–L80.

Mackay, D. H. and van Ballegooijen, A. A. (2006). Models of the large-scale corona. I. Formation, evolution, and liftoff of magnetic flux ropes. *Astrophys. J.* **641**, 577–589.

Mackay, D. H. and van Ballegooijen, A. A. (2006). Models of the large-scale corona. II. Magnetic connectivity and open flux variation. *Astrophys. J.* **642**, 1193–1204.

Mackay, D. H. and van Ballegooijen, A. A. (2009). A non-linear force-free field model for the evolving magnetic structure of solar filaments. *Solar Phys.* **260**, 321–346.

Maclean, R. C., Beveridge, C., Longcope, D. W., Brown, D. S., and Priest, E. R. (2005). A topological analysis of the magnetic breakout model for an eruptive solar flare. *Proc. Roy. Soc. Lond.* **A461**, 2099–2120.

Maclean, R. C. and Priest, E. R. (2007). Topological aspects of global magnetic field reversal in the solar corona. *Solar Phys.* **243**, 171–191.

Madjarska, M. S. (2011). Dynamics and plasma properties of an X-ray jet from SUMER, EIS, XRT, and EUVI A and B simultaneous observations. *Astron. Astrophys.* **526**, A19.

Madjarska, M. S., Doyle, J. G., Hochedez, J.-F., and Theissen, A. (2006). Macrospicules and blinkers as seen in shutterless EIT 304 Å. *Astron. Astrophys.* **452**, L11–L14.

Magara, T. and Longcope, D. W. (2003). Injection of magnetic energy and magnetic helicity into the solar atmosphere by an emerging magnetic flux tube. *Astrophys. J.* **586**, 630–649.

Malherbe, J. M. and Priest, E. R. (1983). Current sheet models for solar prominences. I. Magnetohydrostatics of support and evolution through quasi-static models. *Astron. Astrophys.* **123**, 80–88.

Malkus, W. V. R. and Proctor, M. R. E. (1975). The macrodynamics of alpha-effect dynamos in rotating fluids. *J. Fluid Mech.* **67**, 417–443.

Mandrini, C. H., Démoulin, P., Bagala, L. G., van Driel-Gesztelyi, L., Henoux, J. C., Schmieder, B., and Rovira, M. G. (1997). Evidence of magnetic reconnection from Hα, soft X-ray and photospheric magnetic field observations. *Solar Phys.* **174**, 229–240.

Mandrini, C. H., Démoulin, P., Henoux, J., and Machado, M. (1991). Evidence for the interaction of large-scale magnetic structures in solar flares. *Astron. Astrophys.* **250**, 541–547.

Mandrini, C. H., Démoulin, P., and Klimchuk, J. A. (2000). Magnetic field and plasma scaling laws: their implications for coronal heating models. *Astrophys. J.* **530**, 999–1015.

Manheimer, W. M. and Lashmore-Davies, C. N. (1989). *MHD and Microinstabilities in Confined Plasmas.* (Adam Hilger, Bristol).

Marsch, E. (2006). Kinetic physics of the solar corona and solar wind. *Liv. Rev. in Solar Phys.* **3**, 1.

Marsch, E., Tian, H., Sun, J., Curdt, W., and Wiegelmann, T. (2008). Plasma flows guided by strong magnetic fields in the solar corona. *Astrophys. J.* **685**, 1262–1269.

Martens, P. C. H. (2010). Scaling laws and temperature profiles for solar and stellar coronal loops with non-uniform heating. *Astrophys. J.* **714**, 1290–1304.

Martens, P. C. H., Kankelborg, C. C., and Berger, T. E. (2000). On the nature of the moss observed by TRACE. *Astrophys. J.* **537**, 471–480.

Martens, P. C. H. and Kuin, N. P. M. (1989). A circuit model for filament eruptions and two-ribbon flares. *Solar Phys.* **122**, 263–302.

Martens, P. C. H. and Zwaan, C. (2001). Origin and evolution of filament-prominence systems. *Astrophys. J.* **558**, 872–887.

Martin, S. F. (1986). Recent observations of the formation of filaments. In *Coronal and Prominence Plasmas*, A. Poland, ed. (NASA CP 2442, Washington, DC), 73–80.

Martin, S. F. (1998). Conditions for the formation and maintenance of filaments (Invited Review). *Solar Phys.* **182**, 107–137.

Martin, S. F., Bilimoria, R., and Tracadas, P. (1994). Magnetic field configurations basic to filament channels. In *Solar Surface Magnetism*, R. Rutten and C. Schrijver, eds. (Springer-Verlag, New York), 303.

Martin, S. F., Livi, S., and Wang, J. (1985). The cancellation of magnetic flux. II. In a decaying active region. *Astrophys. J.* **38**, 929–959.

Martin, S. F., Marquette, W. H., and Bilimoria, R. (1992). The solar cycle pattern in the direction of the magnetic field along the long axes of polar filaments. *Astron. Soc. Pac. Conf. Ser.* **27**, 53–66.

Martínez González, M. J., Asensio Ramos, A., Manso Sainz, R., Khomenko, E., Martínez Pillet, V., Solanki, S. K., López Ariste, A., Schmidt, W., Barthol, P., and Gandorfer, A. (2011). Unnoticed magnetic field oscillations in the very quiet sun revealed by SUNRISE/IMaX. *Astrophys. J. Letts.* **730**, L37.

Martínez González, M. J. and Bellot Rubio, L. R. (2009). Emergence of small-scale magnetic loops through the quiet solar atmosphere. *Astrophys. J.* **700**, 1391–1403.

Martínez González, M. J., Collados, M., Ruiz Cobo, B., and Beck, C. (2008). Internetwork magnetic field distribution from simultaneous 1.56 μm and 630 nm observations. *Astron. Astrophys.* **477**, 953–965.

Martínez González, M. J., Collados, M., Ruiz Cobo, B., and Solanki, S. K. (2007). Low-lying magnetic loops in the solar internetwork. *Astron. Astrophys.* **469**, L39–L42.

Martínez González, M. J., Manso Sainz, R., Asensio Ramos, A., and Bellot Rubio, L. R. (2010). Small magnetic loops connecting the quiet surface and the hot outer atmosphere of the Sun. *Astrophys. J. Letts.* **714**, L94–L97.

Martínez González, M. J., Manso Sainz, R., Asensio Ramos, A., and Hijano, E. (2012). Dead calm areas in the very quiet Sun. *Astrophys. J.* **755**, 175.

Martínez Pillet, V. (2002). Decay of sunspots. *Astron. Nach.* **323**, 342–348.

Martínez-Sykora, J., De Pontieu, B., and Hansteen, V. (2012). Two-dimensional radiative magnetohydrodynamic simulations of the importance of partial ionization in the chromosphere. *Astrophys. J.* **753**, 161.

Martínez-Sykora, J., Hansteen, V., De Pontieu, B., and Carlsson, M. (2009). Spicule-like structures observed in three-dimensional realistic magnetohydrodynamic simulations. *Astrophys. J.* **701**, 1569–1581.

Martínez-Sykora, J., Hansteen, V., and Moreno-Insertis, F. (2011). On the origin of the type II spicules: dynamic three-dimensional MHD simulations. *Astrophys. J.* **736**, 9.

Mason, J., Hughes, D. W., and Tobias, S. M. (2002). The competition in the solar dynamo between surface and deep-seated α-effects. *Astrophys. J.* **580**, L89–L92.

Masson, S., Aulanier, G., Pariat, E., and Klein, K.-L. (2012). Interchange slip-running reconnection and sweeping SEP beams. *Solar Phys.* **276**, 199–217.

Masson, S., Pariat, E., Aulanier, G., and Schrijver, C. J. (2009). The nature of flare ribbons in coronal null-point topology. *Astrophys. J.* **700**, 559–578.

Masuda, S., Kosugi, T., Hara, H., Tsuneta, S., and Ogawara, Y. (1994). A loop-top hard X-ray source in a compact solar flare as evidence for magnetic reconnection. *Nature* **371**, 495–497.

Mathew, S. K., Martínez Pillet, V., Solanki, S. K., and Krivova, N. A. (2007). Properties of sunspots in cycle 23. I. Dependence of brightness on sunspot size and cycle phase. *Astron. Astrophys.* **465**, 291–304.

Matsumoto, T. and Shibata, K. (2010). Nonlinear propagation of Alfvén waves driven by observed photospheric motions: application to the coronal heating and spicule formation. *Astrophys. J.* **710**, 1857–1867.

Matthaeus, W. H., Ghosh, S., Oughton, S., and Roberts, D. (1996). Anisotropic three-dimensional MHD turbulence. *J. Geophys. Res.* **101**, 7619–7630.

Matthaeus, W. H. and Velli, M. (2011). Who needs turbulence? A review of turbulence effects in the heliosphere and on the fundamental process of reconnection. *Space Sci. Rev.* **160**, 145–168.

Matthaeus, W. H., Zank, G. P., Oughton, S., Mullan, D. J., and Dmitruk, P. (1999). Coronal heating by magnetohydrodynamic turbulence driven by reflected low-frequency waves. *Astrophys. J. Letts.* **523**, L93–L96.

Matthews, P. C., Hughes, D. W., and Proctor, M. R. E. (1995). Magnetic buoyancy, vorticity, and three-dimensional flux-tube formation. *Astrophys. J.* **448**, 938–941.

McAllister, A. H., Mackay, D. H., and Martin, S. F. (2002). The skew of high-latitude X-ray arcades in the declining phase of cycle 22. *Solar Phys.* **211**, 155–163.

McClymont, A. N. and Craig, I. J. D. (1996). Dynamical finite-amplitude magnetic reconnection at an X-type neutral point. *Astrophys. J.* **466**, 487–495.

McIntosh, S. W. (2012). Recent observations of plasma and Alfvénic wave energy injection at the base of the fast solar wind. *Space Sci. Rev.* **172**, 69–87.

McIntosh, S. W., De Pontieu, B., Carlsson, M., Hansteen, V., Boerner, P., and Goossens, M. (2011). Alfvénic waves with sufficient energy to power the quiet solar corona and fast solar wind. *Nature* **475**, 477–480.

McIntosh, S. W., Leamon, R. J., and De Pontieu, B. (2011). The spectroscopic footprint of the fast solar wind. *Astrophys. J.* **727**, 7.

McIntosh, S. W., Tian, H., Sechler, M., and De Pontieu, B. (2012). On the Doppler velocity of emission line profiles formed in the "coronal contraflow" that is the chromosphere-corona mass cycle. *Astrophys. J.* **749**, 60.

McKenzie, D. E. and Canfield, R. C. (2008). Hinode XRT observations of a long-lasting coronal sigmoid. *Astron. Astrophys.* **481**, L65–L68.

McKenzie, D. E. and Hudson, H. S. (1999). X-ray observations of motions and structure above a solar flare arcade. *Astrophys. J. Letts.* **519**, L93–L96.

McKenzie, D. E. and Savage, S. L. (2009). Quantitative examination of supra-arcade downflows in eruptive solar flares. *Astrophys. J.* **697**, 1569–1577.

McLaughlin, J. A., De Moortel, I., and Hood, A. W. (2011). Phase mixing of nonlinear visco-resistive Alfvén waves. *Astron. Astrophys.* **527**, A149.

Mestel, L. (1999). *Stellar Magnetism.* (Oxford University Press, Oxford).

Mestel, L. (2012). *Stellar Magnetism: Second Edition.* (Oxford University Press, Oxford).

Mestel, L. and Spruit, H. C. (1987). On magnetic braking of late-type stars. *Mon. Not. Roy. Astron. Soc.* **226**, 57–66.

Mestel, L. and Weiss, N. O. (1987). Magnetic fields and nonuniform rotation in stellar radiative zones. *Mon. Not. Roy. Astron. Soc.* **226**, 123–135.

Metcalf, T. R., De Rosa, M. L., Schrijver, C. J., Barnes, G., van Ballegooijen, A. A., Wiegelmann, T., Wheatland, M. S., Valori, G., and McTtiernan, J. M. (2008). Nonlinear force-free modeling of coronal magnetic fields. II. Modeling a filament arcade and simulated chromospheric and photospheric vector fields. *Solar Phys.* **247**, 269–299.

Meunier, N., Roudier, T., and Rieutord, M. (2008). Supergranules over the solar cycle. *Astron. Astrophys.* **488**, 1109–1115.

Meyer, F. and Schmidt, H. (1967). Die erzeugung von schwingungen in der sonnenatmosphäre durch einzelne granula. *Z. Astrophys.* **65**, 274–298.

Meyer, F., Schmidt, H. U., and Weiss, N. O. (1977). The stability of sunspots. *Mon. Not. Roy. Astron. Soc.* **179**, 741–761.

Meyer, F., Schmidt, H. U., Wilson, P. R., and Weiss, N. O. (1974). The growth and decay of sunspots. *Mon. Not. Roy. Astron. Soc.* **169**, 35–57.

Meyer, K. A., Mackay, D. H., and van Ballegooijen, A. A. (2012). Solar magnetic carpet. II. Coronal interactions of small-scale magnetic fields. *Solar Phys.* **278**, 149–175.

Meyer, K. A., Mackay, D. H., van Ballegooijen, A. A., and Parnell, C. E. (2011). Solar magnetic carpet. I. Simulation of synthetic magnetograms. *Solar Phys.* **272**, 29–58.

Meyer-Vernet, N. (2007). *Basics of the Solar Wind.* (Cambridge University Press, Cambridge, UK).

Miesch, M. S. (2005). Large-scale dynamics of the convection zone and tachocline. *Liv. Rev. in Solar Phys. 2*, 1, 1–56.

Miesch, M. S., Brun, A. S., DeRosa, M. L., and Toomre, J. (2008). Structure and evolution of giant cells in global models of solar convection. *Astrophys. J.* **673**, 557–575.

Miesch, M. S., Elliott, J. R., Toomre, J., Clune, T. L., Glatzmaier, G. A., and Gilman, P. A. (2000). Three-dimensional spherical simulations of solar convection. I – Differential rotation and pattern evolution achieved with laminar and turbulent states. *Astrophys. J.* **532**, 593–615.

Miesch, M. S. and Toomre, J. (2009). Turbulence, magnetism, and shear in stellar interiors. *Annual Review of Fluid Mechanics* **41**, 317–345.

Mikhailovsky, A. B. (1974). *Theory of Plasma Instabilities, Vols. 1 and 2.* (Consultants Bureau, New York).

Mikić, Z., Linker, J. A., Lionello, R., Riley, P., and Titov, V. S. (2007). Predicting the structure of the solar sorona for the total solar eclipse of March 29, 2006. *Astron. Soc. Pac. Conf. Ser.* **370**, 299–307.

Mikić, Z., Linker, J. A., Schnack, D. D., Lionello, R., and Tarditi, A. (1999). Magnetohydrodynamic modeling of the global solar corona. *Phys. Plasmas* **6**, 2217–2224.

Mikić, Z. and McClymont, A. N. (1994). Deducing coronal magnetic fields from vector magnetograms. *Astron. Soc. Pac. Conf. Ser.* **68**, 225–232.

Mikić, Z., Schnack, D. D., and Van Hoven, G. (1990). Dynamical evolution of twisted magnetic flux tubes. I – Equilibrium and linear stability. *Astrophys. J.* **361**, 690–700.

Miller, J. A., Cargill, P. J., Emslie, A. G., Holman, G. D., Dennis, B. R., LaRosa, T. N., Winglee, R. M., Benka, S., and Tsuneta, S. (1997). Critical issues for understanding particle acceleration in impulsive solar flares. *J. Geophys. Res.* **102**, 14631–14659.

Milne, A. M. and Priest, E. R. (1981). Internal structure of reconnecting current sheets and the emerging flux model for solar flares. *Solar Phys.* **73**, 157–181.

Milne, A. M., Priest, E. R., and Roberts, B. (1979). A model for quiescent solar prominences. *Astrophys. J.* **232**, 304–317.

Mitalas, R. and Sills, K. R. (1992). On the photon diffusion time scale for the Sun. *Astrophys. J.* **401**, 759–760.

Mizerski, K. A. and Tobias, S. M. (2011). The effect of stratification and compressibility on anelastic convection in a rotating plane layer. *Geophys. Astrophys. Fluid Dyn.* **105**, 566–585.

Mocanu, G., Marcu, A., Ballai, I., and Orza, B. (2008). The problem of phase mixed shear Alfvén waves in the solar corona revisited. *Astron. Nach.* **329**, 780.

Moffatt, H. K. (1978). *Magnetic Field Generation in Electrically Conducting Fluids.* (Cambridge University Press, Cambridge, UK).

Moffatt, H. K. (1985). Magnetostatic equilibria and analogous Euler flows of arbitrarily complex topology. I. Fundamentals. *J. Fluid Mech.* **159**, 359–378.

Moffatt, H. K. (1990). Structure and stability of solutions of the Euler equations: a Lagrangian approach. *Phil. Trans. R. Soc. Lond. A.* **333**, 321–342.

Molodensky, M. M. (1974). Equilibrium and stability of force-free magnetic fields. *Solar Phys.* **39**, 393–404.

Molodensky, M. M. (1975). Equilibrium and stability of force-free magnetic fields. II. Stability. *Solar Phys.* **43**, 311–316.

Molodensky, M. M. (1976). Equilibrium and stability of force-free magnetic fields. III. *Solar Phys.* **49**, 279–282.

Montgomery, D. (1983). Theory of hydromagnetic turbulence. *NASA Conf. Pub.* **228**, 107–130.

Moore, R. L., Cirtain, J. W., Sterling, A. C., and Falconer, D. A. (2010). Dichotomy of solar coronal jets: Standard jets and blowout jets. *Astrophys. J.* **720**, 757–770.

Moore, R. L. and Roumeliotis, G. (1992). Triggering of eruptive flares – destabilization of the preflare magnetic field configuration. In *IAU Colloq. 133: Eruptive Solar Flares*, Z. Svestka, B. V. Jackson, & M. E. Machado, ed. Lecture Notes in Physics, vol. 399. (Springer Verlag, Berlin), 69–78.

Moore, R. L., Sterling, A. C., Gary, G. A., Cirtain, J. W., and Falconer, D. A. (2011). Observed aspects of reconnection in solar eruptions. *Space Sci. Rev.* **160**, 73–94.

Moreno Insertis, F. (1983). Rise times of horizontal magnetic flux tubes in the convection zone of the sun. *Astron. Astrophys.* **122**, 241–250.

Moreno Insertis, F. (1992). The motion of magnetic flux tubes in the convection zone and the subsurface origin of active regions. In *NATO ASIC Proc. 375: Sunspots. Theory and Observations*, J. H. Thomas and N. O. Weiss, ed. Dordrecht), 385–410.

Moreno Insertis, F. (2012). Sources and removal of magnetic flux in the solar atmosphere. *Astron. Soc. Pac. Conf. Ser.* **455**, 91.

Moreno- Insertis, F. and Galsgaard, K. (2013). Plasma jets and eruptions in solar coronal holes: a 3D flux emergence experiment. *Astrophys. J.* **771**, 20.

Moreno Insertis, F., Galsgaard, K., and Ugarte Urra, I. (2008). Jets in coronal holes: Hinode observations and three-dimensional computer modeling. *Astrophys. J. Letts.* **673**, L211–L214.

Morin, J., Donati, J.-F., Forveille, T., Delfosse, X., Dobler, W., Petit, P., Jardine, M. M., Cameron, A. C., Albert, L., Manset, N., Dintrans, B., Chabrier, G., and Valenti, J. A. (2008). The stable magnetic field of the fully convective star V374 Peg. *Mon. Not. Roy. Astron. Soc.* **384**, 77–86.

Morse, P. M. and Feshbach, H. (1953). *Methods in Theoretical Physics.* (McGraw-Hill, New York).

Müller, D. A. N., De Groof, A., Hansteen, V. H., and Peter, H. (2005). High-speed coronal rain. *Astron. Astrophys.* **436**, 1067–1074.

Munro, R. H. and Jackson, B. V. (1977). Physical properties of a polar coronal hole from 2 to 5 solar radii. *Astrophys. J.* **213**, 874–886.

Murawski, K., Zaqarashvili, T. V., and Nakariakov, V. M. (2011). Entropy mode at a magnetic null point as a possible tool for indirect observation of nanoflares in the solar corona. *Astron. Astrophys.* **533**, A18.

Nagai, F. (1980). A model of hot loops associated with solar flares. I – Gasdynamics in the loops. *Solar Phys.* **68**, 351–379.

Nagata, S., Tsuneta, S., Suematsu, Y., Ichimoto, K., Katsukawa, Y., Shimizu, T., Yokoyama, T., Tarbell, T. D., Lites, B. W., Shine, R. A., Berger, T. E., Title, A. M., Bellot Rubio, L. R., and Orozco Suárez, D. (2008). Formation of solar magnetic flux tubes with kilogauss field strength induced by convective instability. *Astrophys. J. Letts.* **677**, L145–L147.

Nakagawa, Y. (1974). Dynamics of the solar magnetic field. I. Method of examination of force-free magnetic fields. *Astrophys. J.* **190**, 437–440.

Nakagawa, Y. and Raadu, M. A. (1972). On the practical representation of magnetic field. *Solar Phys.* **25**, 127–135.

Nakamizo, A., Tanaka, T., Kubo, Y., Kamei, S., Shimazu, H., and Shinagawa, H. (2009). Development of the 3D MHD model of the solar corona-solar wind combining system. *J. Geophys. Res.* **114**, A07109.

Nakariakov, V. M. and Melnikov, V. F. (2009). Quasi-periodic pulsations in solar flares. *Space Sci. Rev.* **149**, 119–151.

Nakariakov, V. M., Melnikov, V. F., and Reznikova, V. E. (2003). Global sausage modes of coronal loops. *Astron. Astrophys.* **412**, L7–L10.

Nakariakov, V. M., Ofman, L., and Arber, T. D. (2000). Nonlinear dissipative spherical Alfvén waves in solar coronal holes. *Astron. Astrophys.* **353**, 741–748.

Nakariakov, V. M., Ofman, L., DeLuca, E. E., Roberts, B., and Davila, J. M. (1999). TRACE observation of damped coronal loop oscillations: implications for coronal heating. *Science* **285**, 862–864.

Nakariakov, V. M. and Verwichte, E. (2005). Coronal waves and oscillations. *Liv. Rev. in Solar Phys.* **2**, 1–65.

Nakariakov, V. M. and Zimovets, I. V. (2011). Slow magnetoacoustic waves in two-ribbon flares. *Astrophys. J. Letts.* **730**, L27.

Nandy, D. (2006). Magnetic helicity and flux tube dynamics in the solar convection zone: comparison between observation and theory. *J. Geophys. Res.* **111**, A12S01.

Nandy, D. and Choudhuri, A. R. (2002). Explaining the latitudinal distribution of sunspots with a deep meridional flow. *Science* **296**, 1671–1673.

Nandy, D., Muñoz-Jaramillo, A., and Martens, P. C. H. (2011). The unusual minimum of sunspot cycle 23 caused by meridional plasma flow variations. *Nature* **471**, 80–82.

Narain, U. and Ulmschneider, P. (1996). Chromospheric and Coronal Heating Mechanisms. II. *Space Sci. Rev.* **75**, 453–509.

Narukage, N. and Shibata, K. (2006). Statistical analysis of reconnection inflows in solar flares observed with SOHO EIT. *Astrophys. J.* **637**, 1122–1134.

Nelson, N. J., Brown, B. P., Sacha Brun, A., Miesch, M. S., and Toomre, J. (2013). Buoyant magnetic loops generated by global convective dynamo action. *Solar Phys.,* in press.

Nerney, S. and Suess, S. T. (2005). Stagnation flow in thin streamer boundaries. *Astrophys. J.* **624**, 378–391.

Neukirch, T. (2005). Magnetic field extrapolation. In *Chromospheric and Coronal Magnetic Fields*, D. E. Innes, A. Lagg, and S. A. Solanki, eds. (ESA SP-596, Dordrecht, Noordwijk), 12.1–12.13.

Neukirch, T. and Rastätter, L. (1999). A new method for calculating a special class of self-consistent 3D magnetohydrostatic equilibria. *Astron. Astrophys.* **348**, 1000–1004.

Newcomb, W. A. (1958). Motion of magnetic lines of force. *Ann. Phys.* **3**, 347–385.

Newcomb, W. A. (1960). Hydromagnetic stability of a diffuse linear pinch. *Ann. Phys.* **10**, 232–267.

Ng, C. S. and Bhattacharjee, A. (2008). A constrained tectonics model for coronal heating. *Astrophys. J.* **675**, 899–905.

Nickeler, D. H. and Wiegelmann, T. (2010). Thin current sheets caused by plasma flow gradients in space and astrophysical plasma. *Annales Geophys.* **28**, 1523–1532.

Nickeler, D. H. and Wiegelmann, T. (2012). Relation between current sheets and vortex sheets in stationary incompressible MHD. *Annales Geophys.* **30**, 545–555.

Nikolsky, G. M. and Platova, A. G. (1971). Motions of Hα-spicules along the solar limb. *Solar Phys.* **18**, 403–409.

Nishio, M., Yaji, K., Kosugi, T., Nakajima, H., and Sakurai, T. (1997). Magnetic field configuration in impulsive solar flares inferred from coaligned microwave/X-ray images. *Astrophys. J.* **489**, 976–991.

Nishizuka, N., Shimizu, M., Nakamura, T., Otsuji, K., Okamoto, T. J., Katsukawa, Y., and Shibata, K. (2008). Giant chromospheric anemone jet observed with Hinode and comparison with magnetohydrodynamic simulations: evidence of propagating Alfvén waves and magnetic reconnection. *Astrophys. J. Letts.* **683**, L83–L86.

Nordlund, Å. (1985). Solar convection. *Solar Phys.* **100**, 209–235.

Nordlund, Å. and Scharmer, G. B. (2010). Convection and the origin of Evershed flows. In *Magnetic Coupling between the Interior and Atmosphere of the Sun*, S. S. Hasan and R. J. Rutten, ed. (Springer-Verlag, Heidelberg), 243–254.

Nordlund, Å. and Stein, R. F. (1990). Solar magnetoconvection. *IAU Symp.* **138**, 191–211.

Nordlund, Å., Stein, R. F., and Asplund, M. (2009). Solar surface convection. *Liv. Rev. in Solar Phys. 6,* 2, 1–116.

November, L. J., Toomre, J., Gebbi, K. B., and Simon, G. W. (1981). The detection of mesogranulation on the Sun. *Astrophys. J.* **245**, L123–L126.

Noyes, R. W. (1982). *The Sun Our Star.* (Harvard University Press, Cambridge, MA, USA).

Noyes, R. W., Baliunas, S. L., and Guinan, E. F. (1991). *What can other stars tell us about the Sun?* (University of Arizona Press, Tucson, AZ, USA), 1161–1185.

Ofman, L. (2010). Wave modeling of the solar wind. *Liv. Rev. in Solar Phys.* **7**, 4.

Ofman, L. and Davila, J. M. (1995). Nonlinear resonant absorption of Alfvén waves in three dimensions, scaling laws and coronal heating. *J. Geophys. Res.* **100**, 23427–23441.

Ofman, L., Klimchuk, J. A., and Davila, J. M. (1998). A self-consistent model for resonant heating of coronal loops: the effects of coupling with the chromosphere. *Astrophys. J.* **493**, 474–479.

Ofman, L., Morrison, P. J., and Steinolfson, R. S. (1993). Nonlinear evolution of resistive tearing mode instability with shear flow and viscosity. *Physics of Fluids B* **5**, 376–387.

Ofman, L., Romoli, M., Poletto, G., Noci, G., and Kohl, J. L. (1997). Ultraviolet Coronagraph Spectrometer observations of density fluctuations in the solar wind. *Astrophys. J. Letts.* **491**, L111–L114.

Ofman, L. and Viñas, A. F. (2007). Two-dimensional hybrid model of wave and beam heating of multi-ion solar wind plasma. *J. Geophys. Res. 112,* A11, A06104.

Ogilvie, G. I. (2007). Instabilities, angular momentum transport and magnetohydrodynamic turbulence. In *The Solar Tachocline*, D. W. Hughes, R. Rosner, and N. O. Weiss, eds. (Cambridge University Press, Cambridge, UK), 299–315.

Ogura, Y. and Phillips, N. A. (1962). Scale analysis of deep and shallow convection in the atmosphere. *J. Atmos. Sci.* **19**, 173–179.

Okamoto, T. J. and De Pontieu, B. (2011). Propagating waves along spicules. *Astrophys. J. Letts.* **736**, L24.

Okamoto, T. J., Tsuneta, S., Berger, T. E., Ichimoto, K., Katsukawa, . Y., Lites, B. W., Nagata, S., Shibata, K., Shimizu, T., Shine, R. A., Suematsu, Y., Tarbell, T. D., and Title, A. M. (2007). Coronal transverse magnetohydrodynamic waves in a solar prominence. *Science* **318**, 1577–1580.

Oliver, R. (2009). Prominence seismology using small amplitude oscillations. *Space Sci. Rev.* **149**, 175–197.

Olsen, E. L. and Leer, E. (1999). A study of solar wind acceleration based on gyrotropic transport equations. *J. Geophys. Res.* **104**, 9963–9974.

Orozco Suárez, D., Asensio Ramos, A., and Trujillo Bueno, J. (2012). Evidence for rotational motions in the feet of a quiescent solar prominence. *Astrophys. J. Letts.* **761**, L25.

Osherovich, V. A. (1982). A new magneto-hydrostatic theory of sunspots. *Solar Phys.* **77**, 63–68.

Ossendriver, M., Stix, M., and Brandenburg, A. (2001). Magnetoconvection and dynamo coefficients: dependence of the alpha-effect on rotation and magnetic field. *Astron. Astrophys.* **376**, 713–726.

Osterbrock, D. E. (1961). The heating of the solar chromosphere, plages, and corona by magnetohydrodynamic waves. *Astrophys. J.* **134**, 347–388.

Otsuji, K., Kitai, R., Ichimoto, K., and Shibata, K. (2011). Statistical study on the nature of solar-flux emergence. *Publ. Astron. Soc. Japan* **63**, 1047–1057.

Owens, M. J. (2008). Combining remote and in situ observations of coronal mass ejections to better constrain magnetic cloud reconstruction. *J. Geophys. Res.* **113**, 12102.

Owens, M. J., Crooker, N. U., and Horbury, T. S. (2009). The expected imprint of flux rope geometry on suprathermal electrons in magnetic clouds. *Annales Geophysicae* **27**, 4057–4067.

Owens, M. J., Schwadron, N. A., Crooker, N. U., Hughes, W. J., and Spence, H. E. (2007). Role of coronal mass ejections in the heliospheric Hale cycle. *Geophys. Res. Lett.* **34**, 6104.

Owocki, S. (2004). Stellar wind mechanisms and instabilities. In *EAS Publications Series*, M. Heydari-Malayeri, P. Stee, and J.-P. Zahn, eds. Vol. 13. (Paris), 163–250.

Pallavacini, R., Serio, S., and Vaiana, G. S. (1977). A survey of soft X-ray limb flare images: the relation between their structure in the corona and other physical parameters. *Astrophys. J.* **216**, 108–222.

Parenti, S., Buchlin, E., Cargill, P. J., Galtier, S., and Vial, J.-C. (2006). Modeling the radiative signatures of turbulent heating in coronal loops. *Astrophys. J.* **651**, 1219–1228.

Parenti, S. and Vial, J. (2007). Prominence and quiet-Sun plasma parameters derived from FUV spectral emission. *Astron. Astrophys.* **469**, 1109–1115.

Parenti, S. and Young, P. R. (2008). On the ultraviolet signatures of small scale heating in coronal loops. *Astron. Astrophys.* **492**, 857–862.

Pariat, E., Aulanier, G., and Démoulin, P. (2006). A new concept for magnetic reconnection: slip-running reconnection. In *SF2A-2006: Semaine de l'Astrophysique Francaise*, D. Barret, F. Casoli, G. Lagache, A. Lecavelier, and L. Pagani, eds. (Paris), 559–562.

Park, S. and Chae, J. (2012). Ca II transient brightenings associated with canceling magnetic features. *Solar Phys.* **280**, 103–110.

Park, W., Monticello, D. A., and White, R. B. (1984). Reconnection rates of magnetic fields including the effects of viscosity. *Phys. Fluids* **27**, 137–149.

Parker, E. N. (1953). Instability of thermal fields. *Astrophys. J.* **117**, 431–436.

Parker, E. N. (1955a). The formation of sunspots from the solar toroidal field. *Astrophys. J.* **121**, 491–507.

Parker, E. N. (1955b). Hydromagnetic dynamo models. *Astrophys. J.* **122**, 293–314.

Parker, E. N. (1957). Sweet's mechanism for merging magnetic fields in conducting fluids. *J. Geophys. Res.* **62**, 509–520.

Parker, E. N. (1958). Dynamics of the interplanetary gas and magnetic fields. *Astrophys. J.* **128**, 664–676.

Parker, E. N. (1963a). *Interplanetary Dynamical Processes.* (Interscience, New York).

Parker, E. N. (1963b). The solar flare phenomenon and the theory of reconnection and annihilation of magnetic fields. *Astrophys. J.* **8**, 177–212.

Parker, E. N. (1964). Dynamical properties of stellar coronas and stellar winds. II. Integration of the heat-flow equation. *Astrophys. J.* **139**, 93–122.

Parker, E. N. (1966). The dynamical state of the interstellar gas and field. *Astrophys. J.* **145**, 811–833.

Parker, E. N. (1972). Topological dissipation and the small-scale fields in turbulent gases. *Astrophys. J.* **174**, 499–510.

Parker, E. N. (1973). Comments on the reconnection rate of magnetic fields. *J. Plasma Phys.* **9**, 49–63.

Parker, E. N. (1974). The dynamical properties of twisted ropes of magnetic field and the vigor of new active regions on the Sun. *Astrophys. J.* **191**, 245–254.

Parker, E. N. (1975). Generation of magnetic fields in astrophysical bodies: X – Magnetic buoyancy and the solar dynamo. *Astrophys. J.* **198**, 205–209.

Parker, E. N. (1977). The origin of solar activity. *Ann. Rev. Astron. Astrophys.* **15**, 45–68.

Parker, E. N. (1979a). *Cosmical Magnetic Fields.* (Oxford University Press, Oxford).

Parker, E. N. (1979b). Sunspots and the physics of magnetic flux tubes. I – The general nature of the sunspot. II - Aerodynamic drag. *Astrophys. J.* **230**, 905–923.

Parker, E. N. (1984). Magnetic buoyancy and the escape of magnetic fields from stars. *Astrophys. J.* **281**, 839–845.

Parker, E. N. (1988). Nanoflares and the solar X-ray corona. *Astrophys. J.* **330**, 474–479.

Parker, E. N. (1993). A solar dynamo surface wave at the interface between convection and nonuniform rotation. *Astrophys. J.* **408**, 707–719.

Parker, E. N. (1994). *Spontaneous Current Sheets in Magnetic Fields.* (Oxford University Press, New York, NY).

Parnell, C. E. (2001). A model of the solar magnetic carpet. *Solar Phys.* **200**, 23–45.

Parnell, C. E. (2004). The role of dynamic brightenings in coronal heating. In *SOHO 15 Coronal Heating*, R. W. Walsh, J. Ireland, D. Danesy, and B. Fleck, eds. ESA Special Publication, vol. 575. (Noordwijk), 227–234.

Parnell, C. E., Bewsher, D., and Harrison, R. A. (2002). Transition-region blinkers. II. Active-region properties. *Solar Phys.* **206**, 249–271.

Parnell, C. E. and De Moortel, I. (2012). A contemporary view of coronal heating. *Phil. Trans. Roy. Soc. Lond. A* **370**, 3217–3240.

Parnell, C. E., DeForest, C. E., Hagenaar, H. J., Johnston, B. A., Lamb, D. A., and Welsch, B. T. (2009). A power-law distribution of solar magnetic fields over more than five decades in flux. *Astrophys. J.* **698**, 75–82.

Parnell, C. E. and Galsgaard, K. (2004). Elementary heating events – interaction between two flux sources. II. Rates of flux reconnection. *Astron. Astrophys.* **428**, 595–612.

Parnell, C. E., Haynes, A. L., and Galsgaard, K. (2008). Recursive reconnection and magnetic skeletons. *Astrophys. J.* **675**, 1656–1667.

Parnell, C. E., Haynes, A. L., and Galsgaard, K. (2010). Structure of magnetic separators and separator reconnection. *J. Geophys. Res. 115,* A14, 2102.

Parnell, C. E. and Jupp, P. E. (2000). Statistical analysis of the energy distribution of nanoflares in the quiet Sun. *Astrophys. J.* **529**, 554–569.

Parnell, C. E., Maclean, R. C., and Haynes, A. L. (2010). The detection of numerous magnetic separators in a three-dimensional magnetohydrodynamic model of solar emerging flux. *Astrophys. J. Letts.* **725**, L214–L218.

Parnell, C. E. and Priest, E. R. (1995). A converging flux model for the formation of an X-ray bright point above a supergranule cell. *Geophys. Astrophys. Fluid Dyn.* **80**, 255–276.

Parnell, C. E., Priest, E. R., and Golub, L. (1994). The three-dimensional structures of X-ray bright points. *Solar Phys.* **151**, 57–74.

Parnell, C. E., Smith, J. M., Neukirch, T., and Priest, E. R. (1996). The structure of 3D magnetic neutral points. *Phys. Plasmas* **3**, 759–770.

Pascoe, D. J., Wright, A. N., and De Moortel, I. (2011). Propagating coupled Alfvén and kink oscillations in an arbitrary inhomogeneous corona. *Astrophys. J.* **731**, 1–9.

Patsourakos, S., Pariat, E., Vourlidas, A., Antiochos, S. K., and Wuelser, J. P. (2008). STEREO SECCHI stereoscopic observations constraining the initiation of polar coronal jets. *Astrophys. J. Letts.* **680**, L73–L76.

Peres, G. (2000). Models of dynamic coronal loops. *Solar Phys.* **193**, 33–52.

Perez, A., Mussack, K., Däppen, W., and Mao, D. (2009). The solar-interior equation of state with the path-integral formalism. I. Domain of validity. *Astron. Astrophys.* **505**, 735–742.

Peter, H. and Judge, P. G. (1999). On the Doppler shifts of solar ultraviolet emission lines. *Astrophys. J.* **522**, 1148–1166.

Petrie, G. J. D. and Neukirch, T. (1999). Self-consistent three-dimensional steady state solutions of the MHD equations with field-aligned incompressible flow. *Geophys. Astrophys. Fluid Dyn.* **91**, 269–302.

Petrie, G. J. D. and Neukirch, T. (2000). The Green's function method for a special class of linear 3D magnetohydrostatic equilibria. *Astron. Astrophys.* **356**, 735–746.

Petrie, G. J. D. and Sudol, J. J. (2010). Abrupt longitudinal magnetic field changes in flaring active regions. *Astrophys. J.* **724**, 1218–1237.

Petschek, H. E. (1964). Magnetic field annihilation. In *AAS-NASA Symposium on the Physics of Solar Flares.* (NASA Spec. Publ. SP-50, Washington), 425–439.

Pevtsov, A. A., Canfield, R. C., and Latushko, S. M. (2001). Hemispheric helicity trend for solar cycle 23. *Astrophys. J.* **549**, L261–L263.

Pevtsov, A. A., Canfield, R. C., and Metcalf, T. R. (1995). Latitudinal variation of helicity of photospheric magnetic fields. *Astrophys. J.* **440**, L109–L112.

Pevtsov, A. A., Canfield, R. C., Sakurai, T., and Hagino, M. (2008). On the solar-cycle variation of the hemispheric helicity rule. *Astrophys. J.* **677**, 719–722.

Pevtsov, A. A., Fisher, G. H., Acton, L. W., Longcope, D. W., Johns-Krull, C. M., Kankelborg, C. C., and Metcalf, T. R. (2003). The relationship between X-ray radiance and magnetic flux. *Astrophys. J.* **598**, 1387–1391.

Pevtsov, A. A., Maleev, V., and Longcope, D. W. (2003). Helicity evolution in emerging active regions. *Astrophys. J.* **593**, 1217–1225.

Piddington, J. H. (1977). Solar magnetic fields and convection. IX – A primordial magnetic field. *Astrophys. and Space Science* **47**, 319–340.

Pierrard, V., Maksimovic, M., and Lemaire, J. (2001). Self-consistent model of solar wind electrons. *J. Geophys. Res.* **106**, 29305–29312.

Pike, C. D. and Mason, H. E. (1998). Rotating features observed with the SOHO Coronal Diagnostic Spectrometer. *Solar Phys.* **182**, 333–348.

Pneuman, G. W. (1972). On neutral sheets in the solar wind. *Solar Phys.* **23**, 223–237.

Pneuman, G. W. (1981). Two-ribbon flares: post-flare loops. In *Solar Flare Magnetohydrodynamics.* (Gordon and Breach Science Publishers, London), 379–428.

Pneuman, G. W. and Kopp, R. A. (1971). Gas-magnetic field interactions in the solar corona. *Solar Phys.* **18**, 258–270.

Poedts, S. and Goedbloed, J. P. (1997). Nonlinear wave heating of solar coronal loops. *Astron. Astrophys.* **321**, 935–944.

Poletto, G. and Kopp, R. (1986). Macroscopic electric fields during two-ribbon flares. In *The Lower Atmosphere of Solar Flares; Proceedings of the Solar Maximum Mission Symposium*, D. Neidig, ed. (National Solar Observatory, Sunspot, New Mexico), 453–465.

Pontin, D. I., Bhattacharjee, A., and Galsgaard, K. (2007a). Current-sheet formation and nonideal behavior at three-dimensional magnetic null points. *Phys. Plasmas* **14**, 052106.

Pontin, D. I., Bhattacharjee, A., and Galsgaard, K. (2007b). Current sheets at three-dimensional magnetic nulls: effect of compressibility. *Phys. Plasmas* **14**, 052109.

Pontin, D. I. and Craig, I. J. D. (2005). Current singularities at finitely compressible three-dimensional magnetic null points. *Phys. Plasmas* **12**, 072112.

Pontin, D. I. and Galsgaard, K. (2007). Current amplification and magnetic reconnection at a three-dimensional null point: physical characteristics. *J. Geophys. Res.* **112**, A03103.

Pontin, D. I., Galsgaard, K., Hornig, G., and Priest, E. R. (2005). A fully magnetohydrodynamic simulation of 3D non-null reconnection. *Phys. Plasmas* **12**, 052307.

Pontin, D. I., Hornig, G., and Priest, E. R. (2004). Kinematic reconnection at a magnetic null point: spine-aligned current. *Geophys. Astrophys. Fluid Dyn.* **98**, 407–428.

Pontin, D. I., Hornig, G., and Priest, E. R. (2005). Kinematic reconnection at a magnetic null point: fan-aligned current. *Geophys. Astrophys. Fluid Dyn.* **99**, 77–93.

Pontin, D. I. and Huang, Y.-M. (2012). On the formation of current sheets in response to the compression or expansion of a potential magnetic field. *Astrophys. J.* **756**, 7.

Pontin, D. I., Wilmot-Smith, A. L., Hornig, G., and Galsgaard, K. (2011). Dynamics of braided coronal loops. II. Cascade to multiple small-scale reconnection events. *Astron. Astrophys.* **525**, A57.

Poston, T. and Stewart, I. (1996). *Catastrophe Theory and Its Applications.* (Dover Publications, Mineola, New York).

Prichett, P. L. and Wu, C. C. (1979). Coalescence of magnetic islands. *Phys. Fluids* **22**, 2140–2146.

Priest, E. R. (1981). *Solar Flare Magnetohydrodynamics.* (Gordon and Breach Science Publishers, London).

Priest, E. R. (1986). Magnetic reconnection on the Sun. *Mit. Astron. Ges.* **65**, 41–51.

Priest, E. R. (1987). Appearance and disappearance of magnetic flux at the solar surface. In *The Role of Fine-Scale Magnetic Fields on the Structure of the Solar Atmosphere*, A. W. E. Schroter, M. Vazquez, ed. (Cambridge University Press., Cambridge, UK), 297–316.

Priest, E. R. (1989). *Structure and Dynamics of Quiescent Solar Prominences.* (Kluwer, Dordrecht).

Priest, E. R. (1992). Basic magnetic configuration and energy supply processes for an Interacting Flux Model of eruptive solar flares. In *IAU Colloq. 133 on Eruptive solar flares*, B. Z.Svestka, B. Jackson, and M. Machado, eds. (Springer-Verlag, Berlin), 15–32.

Priest, E. R., Bungey, T. N., and Titov, V. S. (1997). The 3D topology and interaction of complex magnetic flux systems. *Geophys. Astrophys. Fluid Dyn.* **84**, 127–163.

Priest, E. R. and Démoulin, P. (1995). Three-dimensional magnetic reconnection without null points. 1. Basic theory of magnetic flipping. *J. Geophys. Res.* **100**, 23,443–23,463.

Priest, E. R., Foley, C. R., Heyvaerts, J., Arber, T. D., Culhane, J. L., and Acton, L. W. (1998). The nature of the heating mechanism for the diffuse solar corona. *Nature* **393**, 545–547.

Priest, E. R., Foley, C. R., Heyvaerts, J., Arber, T. D., Mackay, D., Culhane, J. L., and Acton, L. W. (2000). A method to determine the heating mechanisms of the solar corona. *Astrophys. J.* **539**, 1002–1022.

Priest, E. R. and Forbes, T. G. (1986). New models for fast, steady-state reconnection. *J. Geophys. Res.* **91**, 5579–5588.

Priest, E. R. and Forbes, T. G. (1989). Steady magnetic reconnection in three dimensions. *Solar Phys.* **119**, 211–214.

Priest, E. R. and Forbes, T. G. (1990). Magnetic field evolution during prominence eruptions and two-ribbon flares. *Solar Phys.* **126**, 319–350.

Priest, E. R. and Forbes, T. G. (1992a). Does fast magnetic reconnection exist? *J. Geophys. Res.* **97**, 16757–16772.

Priest, E. R. and Forbes, T. G. (1992b). Magnetic flipping – reconnection in three dimensions without null points. *J. Geophys. Res.* **97**, 1521–1531.

Priest, E. R. and Forbes, T. G. (2000). *Magnetic Reconnection: MHD Theory and Applications.* (Cambridge University Press, Cambridge, UK).

Priest, E. R. and Forbes, T. G. (2002). The magnetic nature of solar flares. *Astron. and Astrophys. Rev.* **10**, 313–377.

Priest, E. R., Heyvaerts, J., and Title, A. (2002). A Flux Tube Tectonics model for solar coronal heating driven by the magnetic carpet. *Astrophys. J.* **576**, 533–551.

Priest, E. R., Hood, A. W., and Anzer, U. (1989). A twisted flux-tube model for solar prominences. I. General properties. *Astrophys. J.* **344**, 1010–1025.

Priest, E. R., Hood, A. W., and Anzer, U. (1991). The fibril structure of prominences. *Solar Phys.* **132**, 199–202.

Priest, E. R., Hood, A. W., and Bewsher, D. (2002). The nature of blinkers and the solar transition region. *Solar Phys.* **205**, 249–264.

Priest, E. R., Hornig, G., and Pontin, D. I. (2003). On the nature of three-dimensional magnetic reconnection. *J. Geophys. Res. 108,* A7, SSH 6.1–6.8.

Priest, E. R., Longcope, D., and Heyvaerts, J. (2005). Coronal heating at separators and separatrices. *Astrophys. J.* **624**, 1057–1071.

Priest, E. R., Longcope, D. W., and Titov, V. S. (2003). Binary reconnection and the heating of the solar corona. *Astrophys. J.* **598**, 667–677.

Priest, E. R., Lonie, D. P., and Titov, V. S. (1996). Bifurcations of magnetic topology by the creation or annihilation of null points. *J. Plasma Phys.* **56**, 507–530.

Priest, E. R. and Milne, A. M. (1979). Force-free magnetic arcades relevant to two-ribbon solar flares. *Solar Phys.* **65**, 315–346.

Priest, E. R. and Parnell, C. E. (2014). MHD reconnection theory. *Liv. Rev. in Solar Phys.,* in press.

Priest, E. R., Parnell, C. E., and Martin, S. F. (1994). A converging flux model of an X-ray bright point and an associated canceling magnetic feature. *Astrophys. J.* **427**, 459–474.

Priest, E. R. and Pneuman, G. W. (1974). The influence of non-uniform solar wind expansion on the angular momentum loss from the Sun. *Solar Phys.* **34**, 231–241.

Priest, E. R. and Pontin, D. I. (2009). Three-dimensional null point regimes. *Phys. Plasmas* **16**, 122101.

Priest, E. R. and Raadu, M. A. (1975). Preflare current sheets in the solar atmosphere. *Solar Phys.* **43**, 177–188.

Priest, E. R. and Smith, D. F. (1972). Current sheets in coronal streamers. *Astrophys. Letters* **12**, 25–29.

Priest, E. R. and Smith, E. A. (1979). The structure of coronal arcades and the formation of solar prominences. *Solar Phys.* **64**, 267–286.

Priest, E. R. and Titov, V. S. (1996). Magnetic reconnection at three-dimensional null points. *Phil. Trans. Roy. Soc. Lond.* **355**, 2951–2992.

Priest, E. R., Titov, V. S., and Rickard, G. K. (1994). The formation of magnetic singularities by nonlinear time-dependent collapse of an X-type magnetic field. *Phil. Trans. Roy. Soc. Lond.* **A351**, 1–37.

Priest, E. R., van Ballegooijen, A. A., and Mackay, D. H. (1996). A model for dextral and sinistral prominences. *Astrophys. J.* **460**, 530–543.

Pritchett, P. L., , Lee, Y. C., and Drake, J. F. (1980). Linear analysis of the double tearing mode. *Phys. Fluids* **23**, 1368–1374.

Proctor, M. R. E. (1994). Convection and magnetoconvection in a rapidly rotating sphere. In *Lectures on Solar and Planetary Dynamos*, M. R. E. Proctor and A. D. Gilbert, eds. (Cambridge University Press, Cambridge, UK), 97–115.

Proctor, M. R. E. (2005). Magnetoconvection. In *Fluid Dynamics and Dynamos in Astrophysics and Geophysics*, A. M. Soward, C. A. Jones, D. W. Hughes, and N. O. Weiss, ed. (CRC Press, Boca Raton, FL), 235–276.

Proctor, M. R. E. and Gilbert, A. D. (1994). *Lectures on Solar and Planetary Dynamos.* (Cambridge University Press, Cambridge, UK).

Pucci, S., Lie-Svendsen, Ø., and Esser, R. (2010). Elemental abundances in the fast solar wind emanating from chromospheric funnels. *Astrophys. J.* **709**, 993–1002.

Qiu, J., Hu, Q., Howard, T. A., and Yurchyshyn, V. B. (2007). On the magnetic flux budget in low-corona magnetic reconnection and interplanetary coronal mass ejections. *Astrophys. J.* **659**, 758–772.

Qiu, J., Lee, J., Gary, D. E., and Wang, H. (2002). Motion of flare footpoint emission and inferred electric field in reconnecting current sheets. *Astrophys. J.* **565**, 1335–1347.

Qiu, J., Liu, W.-J., Hill, N., and Kazachenko, M. (2010). Reconnection and energetics in two-ribbon flares: a revisit of the Bastille-day flare. *Astrophys. J.* **725**, 319–330.

Qiu, J., Liu, W.-J., and Longcope, D. W. (2012). Heating of flare loops with observationally constrained heating functions. *Astrophys. J.* **752**, 124.

Qiu, J., Wang, H., Cheng, C. Z., and Gary, D. E. (2004). Magnetic reconnection and mass acceleration in flare-coronal mass ejection events. *Astrophys. J.* **604**, 900–905.

Raadu, M. A. (1972). Suppression of the kink instability for magnetic flux ropes in the chromosphere. *Solar Phys.* **22**, 425–433.

Rachmeler, L. A., DeForest, C. E., and Kankelborg, C. C. (2009). Reconnectionless CME eruption: putting the Aly-Sturrock conjecture to rest. *Astrophys. J.* **693**, 1431–1436.

Rachmeler, L. A., Gibson, S. E., Dove, J., DeVore, C., and Fan, Y. (2013). Polarimetric properties of flux ropes and sheared arcades in coronal prominence cavities. *Solar Phys.,* in press.

Rachmeler, L. A., Pariat, E., DeForest, C. E., Antiochos, S., and Török, T. (2010). Symmetric coronal jets: a reconnection-controlled study. *Astrophys. J.* **715**, 1556–1565.

Radziszewski, K., Rudawy, P., and Phillips, K. J. H. (2011). High time resolution observations of solar Hα flares. II. Search for signatures of electron beam heating. *Astron. Astrophys.* **535**, A123.

Rae, I. C. and Roberts, B. (1981). Surface waves and the heating of the corona. *Geophys. Astrophys. Fluid Dyn.* **18**, 197–226.

Raouafi, N.-E. and Solanki, S. K. (2006). Sensitivity of solar off-limb line profiles to electron density stratification and the velocity distribution anisotropy. *Astron. Astrophys.* **445**, 735–745.

Rappazzo, A. F. and Velli, M. (2011). Magnetohydrodynamic turbulent cascade of coronal loop magnetic fields. *Phys. Rev. E 83,* 6 (June), 065401.

Rappazzo, A. F., Velli, M., and Einaudi, G. (2010). Shear photospheric forcing and the origin of turbulence in coronal loops. *Astrophys. J.* **722**, 65–78.

Rappazzo, A. F., Velli, M., Einaudi, G., and Dahlburg, R. B. (2008). Nonlinear dynamics of the Parker scenario for coronal heating. *Astrophys. J.* **677**, 1348–1366.

Rayleigh, L. (1916). On convection currents in a horizontal layer of fluid, when the higher temperature is on the under side. *Phil. Mag.* **32**, 529–546.

Raymond, J. C., Ciaravella, A., Dobrzycka, D., Strachan, L., Ko, Y.-K., Uzzo, M., and Raouafi, N.-E. (2003). Far-ultraviolet spectra of fast coronal mass ejections associated with X-class flares. *Astrophys. J.* **597**, 1106–1117.

Reale, F. (2010). Coronal loops: observations and modeling of confined plasma. *Liv. Rev. in Solar Phys.* **7**, 5.

Reale, F., Guarrasi, M., Testa, P., DeLuca, E. E., Peres, G., and Golub, L. (2011). Solar Dynamics Observatory discovers thin high temperature strands in coronal active regions. *Astrophys. J. Letts.* **736**, L16.

Reale, F., Parenti, S., Reeves, K. K., Weber, M., Bobra, M. G., Barbera, M., Kano, R., Narukage, N., Shimojo, M., Sakao, T., Peres, G., and Golub, L. (2007). Fine thermal structure of a coronal active region. *Science* **318**, 1582–1585.

Reale, F., Peres, G., Serio, S., Betta, R. M., DeLuca, E. E., and Golub, L. (2000). A brightening coronal loop observed by TRACE. II. Loop modeling and constraints on heating. *Astrophys. J.* **535**, 423–437.

Reale, F., Testa, P., Klimchuk, J. A., and Parenti, S. (2009). Evidence of widespread hot plasma in a nonflaring coronal active region from Hinode/X-Ray Telescope. *Astrophys. J.* **698**, 756–765.

Rechester, A. B. and Stix, T. H. (1976). Magnetic braiding due to weak asymmetry. *Phys. Rev. Lett.* **40**, 38–41.

Régnier, S., Amari, T., and Kersalé, E. (2002). 3D coronal magnetic field from vector magnetograms: non-constant-alpha force-free configuration of the active region NOAA 8151. *Astron. Astrophys.* **392**, 1119–1127.

Regnier, S. and Priest, E. R. (2007). Two active regions with very different structure. *Astron. Astrophys.* **468**, 701–709.

Reid, J. and Laing, E. W. (1979). The resistive evolution of force-free magnetic fields. I – Slab geometry. *J. Plasma Phys.* **21**, 501–510.

Rempel, M. (2005). Solar differential rotation and meridional flow: the role of a subadiabatic tachocline for the Taylor-Proudman balance. *Astrophys. J.* **622**, 1320–1332.

Rempel, M. (2011). Penumbral fine structure and driving mechanisms of large-scale flows in simulated sunspots. *Astrophys. J.* **729**, 5.

Rempel, M. (2012). Numerical sunspot models: robustness of photospheric velocity and magnetic field structure. *Astrophys. J.* **750**, 62.

Rempel, M., Schüssler, M., Cameron, R. H., and Knölker, M. (2009). Penumbral structure and outflows in simulated sunspots. *Science* **325**, 171–174.

Rempel, M., Schüssler, M., and Knölker, M. (2009). Radiative magnetohydrodynamic simulation of sunspot structure. *Astrophys. J.* **691**, 640–649.

Restante, A. L., Aulanier, G., and Parnell, C. E. (2009). How skeletons turn into quasi-separatrix layers in source models. *Astron. Astrophys.* **508**, 433–443.

Rickard, G. J. and Titov, V. S. (1996). Current accumulation at a three-dimensional magnetic null. *Astrophys. J.* **472**, 840–852.

Ridgway, C., Priest, E. R., and Amari, T. (1991). Prominence sheets supported by constant-current force-free fields. I – Imposition of normal magnetic field components at the current sheet and the photosphere. *Astrophys. J.* **378**, 773–784.

Ridgway, C., Priest, E. R., and Amari, T. (1991). A twisted flux tube model for solar prominences. III. Magnetic support. *Astrophys. J.* **367**, 321–332.

Riethmüller, T. L., Solanki, S. K., and Lagg, A. (2008). Stratification of sunspot umbral dots from inversion of Stokes profiles recorded by Hinode. *Astrophys. J. Letts.* **678**, L157–L160.

Rieutord, M., Meunier, N., Roudier, T., Rondi, S., Beigbeder, F., and Parès, L. (2008). Solar supergranulation revealed by granule tracking. *Astron. Astrophys.* **479**, L17–L20.

Rieutord, M. and Rincon, F. (2010). The Sun's supergranulation. *Liv. Rev. in Solar Phys.* **7**, 2.

Rieutord, M., Roudier, T., Rincon, F., Malherbe, J.-M., Meunier, N., Berger, T., and Frank, Z. (2010). On the power spectrum of solar surface flows. *Astron. Astrophys.* **512**, A4.

Rijinbeek, R. P. and Semenov, V. S. (1993). Features of a Petschek-type reconnection model. *Trends in Geophys. Res.* **2**, 247–268.

Riley, P., Linker, J. A., and Mikic, Z. (2013). On the application of ensemble modelling techniques to improve ambient solar wind models. *J. Geophys. Res.* **118**, 600–607.

Riley, P., Linker, J. A., Mikić, Z., Lionello, R., Ledvina, S. A., and Luhmann, J. G. (2006). A comparison between global solar magnetohydrodynamic and potential field source surface model results. *Astrophys. J.* **653**, 1510–1516.

Riley, P., Lionello, R., Linker, J. A., Mikic, Z., Luhmann, J., and Wijaya, J. (2011). Global MHD modeling of the solar corona and inner heliosphere for the whole heliosphere interval. *Solar Phys.* **274**, 361–377.

Riley, P., Lionello, R., Mikić, Z., and Linker, J. (2008). Using global simulations to relate the three-part structure of coronal mass ejections to in situ signatures. *Astrophys. J.* **672**, 1221–1227.

Robbrecht, E., Berghmans, D., and van der Linden, R. A. M. (2009). Automated LASCO CME catalog for solar cycle 23: are CMEs scale invariant? *Astrophys. J.* **691**, 1222–1234.

Roberts, B. (2000). Waves and oscillations in the corona (invited review). *Solar Phys.* **193**, 39–56.

Roberts, B., Edwin, P. M., and Benz, A. O. (1984). On coronal oscillations. *Astrophys. J.* **279**, 857–865.

Roberts, B. and Webb, A. R. (1979). Vertical motions in an intense magnetic flux tube. III – On the slender flux tube approximation. *Solar Phys.* **64**, 77–92.

Roberts, D. A. (2010). Demonstrations that the solar wind is not accelerated by waves or turbulence. *Astrophys. J.* **711**, 1044–1050.

Roberts, G. O. (1972a). Dynamo action of fluid motions with two-dimensional periodicity. *Phil. Trans. Roy. Soc. Lond. A* **271**, 411–454.

Roberts, P. H. (1967). *An Introduction to Magnetohydrodynamics.* (Longmans, London).

Roberts, P. H. (1972b). Kinematic Dynamo Models. *Phil. Trans. Roy. Soc.A* **272**, 663–698.

Roberts, P. H. and Soward, A. M. (1972). Stellar winds and breezes. *Proc. Roy. Soc. Lond. A* **328**, 185–215.

Robinson, R. D., Worden, S. P., and Harvey, J. W. (1980). Observations of magnetic fields on two late-type dwarf stars. *Astrophys. J.* **236**, L155–L158.

Rosenthal, C. S., Bogdan, T. J., Carlsson, M., Dorch, S. B. F., Hansteen, V., McIntosh, S. W., McMurry, A., Nordlund, Å., and Stein, R. F. (2002). Waves in the magnetized solar atmosphere. I. Basic processes and internetwork oscillations. *Astrophys. J.* **564**, 508–524.

Rosner, R. and Knobloch, E. (1982). On perturbations of magnetic field configurations. *Astrophys. J.* **262**, 349–357.

Rosner, R., Tucker, W. H., and Vaiana, G. S. (1978). Dynamics of the quiescent solar corona. *Astrophys. J.* **220**, 643–645.

Roudier, T. and Muller, R. (1986). Structure of the solar granulation. *Solar Phys.* **107**, 11–26.

Roumeliotis, G. (1993). A new class of exact, nonlinear solutions to the Grad-Shafranov equation. *Astrophys. J.* **404**, 781–787.

Roumeliotis, G. (1996). The "stress-and-relax" method for reconstructing the coronal magnetic field from vector magnetograph data. *Astrophys. J.* **473**, 1095–1103.

Rouppe van der Voort, L., Hansteen, V. H., Carlsson, M., Fossum, A., Marthinussen, E., van Noort, M. J., and Berger, T. E. (2005). Solar magnetic elements at 0.1 arcsec resolution. II. Dynamical evolution. *Astron. Astrophys.* **435**, 327–337.

Rouppe van der Voort, L., Leenaarts, J., De Pontieu, B., Carlsson, M., and Vissers, G. (2009). On-disk counterparts of type II spicules in the Ca II 854.2 nm and Hα lines. *Astrophys. J.* **705**, 272–284.

Rouppe van der Voort, L., Löfdahl, M. G., Kiselman, D., and Scharmer, G. B. (2004). Penumbral structure at 0.1 arcsec resolution. I. General appearance and power spectra. *Astron. Astrophys.* **414**, 717–726.

Roussev, I. I., Gombosi, T. I., Sokolov, I. V., Velli, M., Manchester, IV, W., DeZeeuw, D. L., Liewer, P., Tóth, G., and Luhmann, J. (2003). A three-dimensional model of the solar wind incorporating solar magnetogram observations. *Astrophys. J. Letts.* **595**, L57–L61.

Roussev, I. I., Lugaz, N., and Sokolov, I. V. (2007). New physical insight on the changes in magnetic topology during coronal mass ejections: case studies for the 2002 April 21 and August 24 Events. *Astrophys. J. Letts.* **668**, L87–L90.

Rucklidge, A. M. (1994). Chaos in magnetoconvection. *Nonlinearity* **7**, 1565–1591.

Rucklidge, A. M., Weiss, N. O., Brownjohn, D. P., Matthews, P. C., and Proctor, M. R. E. (2000). Compressible magnetoconvection in three dimensions: pattern formation in a strongly stratified layer. *J. Fluid Mech.* **419**, 283–323.

Ruderman, M. S. and Erdélyi, R. (2009). Transverse oscillations of coronal loops. *Space Sci. Rev.* **149**, 199–228.

Ruderman, M. S., Goossens, M., Ballester, J. L., and Oliver, R. (1997). Resonant Alfvén waves in coronal arcades driven by footpoint motions. *Astron. Astrophys.* **328**, 361–370.

Ruderman, M. S., Nakariakov, V. M., and Roberts, B. (1998). Alfvén wave phase mixing in two-dimensional open magnetic configurations. *Astron. Astrophys.* **338**, 1118–1124.

Ruderman, M. S. and Roberts, B. (2002). The damping of coronal loop oscillations. *Astrophys. J.* **577**, 475–486.

Rüdiger, G. (1989). *Differential Rotation and Stellar Convection.* (Akademie-Verlag, Berlin).

Rüdiger, G. and Kitchatinov, L. L. (1997). The slender solar tachocline: a magnetic model. *Astron. Nach.* **318**, 273–279.

Rüdiger, G. and Küker, M. (2002). Theory of meridional flow and the advection-dominated solar dynamo. In *Solar Variability: From Core to Outer Frontiers*, A. Wilson, ed. ESA Special Publication, vol. 506. (Noordwijk), 811–814.

Rust, D. M. (1967). Magnetic fields in quiescent solar prominences. I. Observations. *Astrophys. J.* **150**, 313–321.

Rust, D. M. and Kumar, A. (1994). Helical magnetic fields in filaments. *Solar Phys.* **155**, 69–97.

Rust, D. M. and Kumar, A. (1996). Evidence for helically kinked magnetic flux ropes in solar eruptions. *Astrophys. J.* **464**, L199–202.

Rutherford, P. H. (1973). Nonlinear growth of the tearing mode. *Phys. Fluids* **16**, 1903–1908.

Rutten, R. J. (2006). On the nature of the solar chromosphere. *Astron. Soc. Pac. Conf. Ser.* **354**, 276–283.

Sakai, J. I. (1983). Forced reconnexion by fast magnetosonic waves in a current sheet with stagnation-point flows. *J. Plasma Phys.* **30**, 109–124.

Sakai, J. I. and Smith, P. D. (2009). Two-fluid simulations of coalescing penumbra filaments driven by neutral-hydrogen flows. *Astrophys. J. Letts.* **691**, L45–L48.

Sakao, T., Kano, R., Narukage, N., Kotoku, J., Bando, T., DeLuca, E. E., Lundquist, L. L., Tsuneta, S., Harra, L. K., Katsukawa, Y., Kubo, M., Hara, H., Matsuzaki, K., Shimojo, M., Bookbinder, J. A., Golub, L., Korreck, K. E., Su, Y., Shibasaki, K., Shimizu, T., and Nakatani, I. (2007). Continuous plasma outflows from the edge of a solar active region as a possible source of solar wind. *Science* **318**, 1585–1588.

Sakurai, T. (1979). A new approach to the force-free field and its application to the magnetic field of solar active regions. *Publ, Astron. Soc. Japan* **31**, 209–230.

Sakurai, T. (1981). Calculation of force-free magnetic field with non-constant alpha. *Solar Phys.* **69**, 343–359.

Sakurai, T. (1982). Green's function method for potential magnetic fields. *Solar Phys.* **76**, 301–322.

Sakurai, T. (1989). Computational modeling of magnetic fields in solar active regions. *Space Sci. Rev.* **51**, 11–48.

Sakurai, T., Goossens, M., and Hollweg, J. V. (1991). Resonant behaviour of MHD waves on magnetic flux tubes. I – Connection formulae at the resonant surfaces. *Solar Phys.* **133**, 227–245.

Sakurai, T. and Levine, R. H. (1981). Generation of coronal electric currents due to convective motions on the photosphere. *Astrophys. J.* **248**, 817–829.

Sánchez Almeida, J., Márquez, I., Bonet, J. A., Domínguez Cerdeña, I., and Muller, R. (2004). Bright points in the internetwork quiet Sun. *Astrophys. J. Letts.* **609**, L91–L94.

Sato, T. (1979). Strong plasma acceleration by slow shocks resulting from magnetic reconnection. *J. Geophys. Res.* **84**, 7177–7190.

Savage, S. L. and McKenzie, D. E. (2011). Quantitative examination of a large sample of supra-arcade downflows in eruptive solar flares. *Astrophys. J.* **730**, 98.

Savage, S. L., McKenzie, D. E., and Reeves, K. K. (2012). Re-interpretation of supra-arcade downflows in solar flares. *Astrophys. J. Letts.* **747**, L40.

Savage, S. L., McKenzie, D. E., Reeves, K. K., Forbes, T. G., and Longcope, D. W. (2010). Reconnection outflows and current sheet observed with Hinode/XRT in the 2008 April 9 "Cartwheel CME" flare. *Astrophys. J.* **722**, 329–342.

Savcheva, A., Cirtain, J., Deluca, E. E., Lundquist, L. L., Golub, L., Weber, M., Shimojo, M., Shibasaki, K., Sakao, T., Narukage, N., Tsuneta, S., and Kano, R. (2007). A study of polar jet parameters based on Hinode XRT observations. *Publ. Astron. Soc. Japan* **59**, 771.

Scharmer, G. B., Gudiksen, B. V., Kiselman, D., Löfdahl, M. G., and Rouppe van der Voort, L. H. M. (2002). Dark cores in sunspot penumbral filaments. *Nature* **420**, 151–153.

Schatten, K. H., Wilcox, J. M., and Ness, N. F. (1969). A model of interplanetary and coronal magnetic fields. *Solar Phys.* **6**, 442–455.

Schatzman, E. (1965). Model of force-free fields. *IAU Symp.* **22**, 337.

Schekochihin, A. A., Cowley, S. C., Dorland, W., Hammett, G. W., Howes, G. G., Quataert, E., and Tatsuno, T. (2009). Astrophysical gyrokinetics: Kinetic and fluid turbulent cascades in magnetized weakly collisional plasmas. *Astrophys. J. Suppl.* **182**, 310–377.

Schindler, K., Hesse, M., and Birn, J. (1988). General Magnetic Reconnection, parallel electric fields, and helicity. *J. Geophys. Res.* **93**, 5547–5557.

Schindler, K., Hesse, M., and Birn, J. (1991). Magnetic field-aligned electric potentials in nonideal plasma flows. *Astrophys. J.* **380**, 293–301.

Schlichenmaier, R., Rezaei, R., Bello González, N., and Waldmann, T. A. (2010). The formation of a sunspot penumbra. *Astron. Astrophys.* **512**, L1.

Schlüter, A. and Temesváry, S. (1958). The internal constitution of sunspots. *IAU Symp.* **6**, 263–274.

Schmelz, J. T., Jenkins, B. S., Worley, B. T., Anderson, D. J., Pathak, S., and Kimble, J. A. (2011). Isothermal and multithermal analysis of coronal loops observed with AIA. *Astrophys. J.* **731**, 49.

Schmelz, J. T., Rightmire, L. A., Saar, S. H., Kimble, J. A., Worley, B. T., and Pathak, S. (2011). Warm and fuzzy: temperature and density analysis of an Fe XV EUV Imaging Spectrometer loop. *Astrophys. J.* **738**, 146.

Schmelz, J. T., Saar, S. H., DeLuca, E. E., Golub, L., Kashyap, V. L., Weber, M. A., and Klimchuk, J. A. (2009). Hinode X-Ray Telescope detection of hot emission from quiescent active regions: a nanoflare signature? *Astrophys. J. Letts.* **693**, L131–L135.

Schmidt, H. U. (1964). On the observable effects of magnetic energy storage and release connected with solar flares. In *NASA Symp. on Phys. of Solar Flares*, W. Hess, ed. (NASA SP-50, Washington, DC), 107–114.

Schmidt, H. U. (1991). Sunspots. *Geophys. Astrophys. Fluid Dyn.* **62**, 249–270.

Schmieder, B., Aulanier, G., Démoulin, P., van Driel-Gesztelyi, L., Roudier, T., Nitta, N., and Cauzzi, G. (1997). Magnetic reconnection driven by emergence of sheared magnetic field. *Astron. Astrophys.* **325**, 1213–1225.

Schmieder, B., Aulanier, G., Mein, P., and López Ariste, A. (2006). Evolving photospheric flux concentrations and filament dynamic changes. *Solar Phys.* **238**, 245–259.

Schmieder, B., Chandra, R., Berlicki, A., and Mein, P. (2010). Velocity vectors of a quiescent prominence observed by Hinode/SOT and the MSDP (Meudon). *Astron. Astrophys.* **514**, A68.

Schmieder, B., Démoulin, P., Aulanier, G., and Golub, L. (1996). Differential magnetic field shear in an active region. *Astrophys. J.* **467**, 881.

Schmieder, B., Fontenla, J., and Tandberg-Hanssen, E. (1991). A microflare-related activation of a filament observed in Hα and C IV lines. *Astron. Astrophys.* **252**, 343–352.

Schmieder, B., Heinzel, P., van Driel-Gesztelyi, L., and Lemen, J. R. (1996). Post-flare loops of 26 June 1992, II. *Solar Phys.* **165**, 303–328.

Schmieder, B., Heinzel, P., Wiik, J. E., Lemen, J., Anwar, B., Kotrc, P., and Hiei, E. (1995). Relation between cool and hot post-flare loops of 26 June 1992 derived from optical and X-ray (SXT-Yohkoh) observations. *Solar Phys.* **156**, 337–361.

Schmieder, B., Malherbe, J. M., Mein, P., and Tanberg-Haassen, E. (1984). Dynamics of solar filaments. III. Analysis of steady flows in H-alpha and CIV lines. *Astron. Astrophys.* **136**, 81–88.

Schmieder, B., Mein, N., Deng, Y., Dumitrache, C., Malherbe, J., Staiger, J., and Deluca, E. E. (2004). Magnetic changes observed in the formation of two filaments in a complex active region: TRACE and MSDP observations. *Solar Phys.* **223**, 119–141.

Schmieder, B., Shibata, K., van Driel-Gesztelyi, L., and Freeland, S. (1995). H-alpha surges and associated soft X-ray loops. *Solar Phys.* **156**, 245–264.

Schmit, D. J. and Gibson, S. E. (2011). Forward modeling cavity density: a multi-instrument diagnostic. *Astrophys. J.* **733**, 1.

Schmitt, J. H. M. M. and Rosner, R. (1983). Doubly diffusive magnetic buoyancy instability in the solar interior. *Astrophys. J.* **265**, 901–924.

Scholer, M. (1989). Undriven magnetic reconnection in an isolated current sheet. *J. Geophys. Res.* **94**, 8805–8812.

Schou, J. (1999). Migration of zonal flows detected using Michelson Doppler Imager f-mode frequency splittings. *Astrophys. J.* **523**, L181–L184.

Schou, J., Antia, H. M., Basu, S., Bogart, R. S., Bush, R. I., Chitre, S. M., Christensen-Dalsgaard, J., di Mauro, M. P., Dziembowski, W. A., Eff-Darwich, A., Gough, D. O., Haber, D. A., Hoeksema, J. T., Howe, R., Korzennik, S. G., Kosovichev, A. G., Larsen, R. M., Pijpers, F. P., Scherrer, P. H., Sekii, T., Tarbell, T. D., Title, A. M., Thompson, M. J., and Toomre, J. (1998). Helioseismic studies of differential rotation in the solar envelope by the solar oscillations investigation using the Michelson Doppler Imager. *Astrophys. J.* **505**, 390–417.

Schrijver, C. J. (2001). Catastrophic cooling and high-speed downflow in quiescent solar coronal loops observed with TRACE. *Solar Phys.* **198**, 325–345.

Schrijver, C. J. (2009). Driving major solar flares and eruptions: a review. *Advances in Space Research* **43**, 739–755.

Schrijver, C. J., Aschwanden, M. J., and Title, A. M. (2002). Transverse oscillations in coronal loops observed with TRACE. I. An overview of events, movies and a discussion of common properties and required conditions. *Solar Phys.* **206**, 69–98.

Schrijver, C. J., Aulanier, G., Title, A. M., Pariat, E., and Delannée, C. (2011). The 2011 February 15 X2 flare, ribbons, coronal front, and mass ejection: interpreting the three-dimensional views from the Solar Dynamics Observatory and STEREO guided by magnetohydrodynamic flux-rope modeling. *Astrophys. J.* **738**, 167.

Schrijver, C. J., De Rosa, M. L., Metcalf, T., Barnes, G., Lites, B., Tarbell, T., McTiernan, J., Valori, G., Wiegelmann, T., Wheatland, M. S., Amari, T., Aulanier, G., Démoulin, P., Fuhrmann, M., Kusano, K., Régnier, S., and Thalmann, J. K. (2008). Nonlinear force-free field modeling of a solar active region around the time of a major flare and coronal mass ejection. *Astrophys. J.* **675**, 1637–1644.

Schrijver, C. J., De Rosa, M. L., Title, A. M., and Metcalf, T. R. (2005). The nonpotentiality of active-region coronae and the dynamics of the photospheric magnetic field. *Astrophys. J.* **628**, 501–513.

Schrijver, C. J., Derosa, M. L., Metcalf, T. R., Liu, Y., McTiernan, J., Régnier, S., Valori, G., Wheatland, M. S., and Wiegelmann, T. (2006). Nonlinear force-free modeling of coronal magnetic fields part I: a quantitative comparison of methods. *Solar Phys.* **235**, 161–190.

Schrijver, C. J., Sandman, A. W., Aschwanden, M. J., and De Rosa, M. L. (2004). The coronal heating mechanism as identified by full-Sun visualizations. *Astrophys. J.* **615**, 512–525.

Schrijver, C. J. and Title, A. M. (2003). The magnetic connection between the solar photosphere and the corona. *Astrophys. J. Letts.* **597**, L165–L168.

Schrijver, C. J., Title, A. M., Harvey, K. L., Sheeley, N. R., Wang, Y.-M., van den Oord, G. H. J., Shine, R. A., Tarbell, T. D., and Hurlburt, N. E. (1998). Large-scale coronal heating by the small-scale magnetic field of the Sun. *Nature* **394**, 152–154.

Schrijver, C. J., Title, A. M., van Ballegooijen, A. A., Hagenaar, H. J., and Shine, R. A. (1997). Sustaining the quiet photospheric network: the balance of flux emergence, fragmentation, merging and cancellation. *Astrophys. J.* **487**, 424–436.

Schrijver, C. J. and Zwaan, C. (2000). *Solar and Stellar Magnetic Activity.* (Cambridge University Press, Cambridge, UK).

Schunker, H. and Cally, P. S. (2006). Magnetic field inclination and atmospheric oscillations above solar active regions. *Mon. Not. Roy. Astron. Soc.* **372**, 551–564.

Schüssler, M. (1987). Magnetic fields and the rotation of the solar convection zone. In *The Internal Solar Angular Velocity*, B. R. Durney and S. Sofia, eds. (Reidel, Dordrecht), 303–320.
Schüssler, M. (2001). Numerical simulation of solar magneto-convection. *Astron. Soc. Pac. Conf. Ser.* **236**, 343–354.
Schüssler, M. (2005). Flux tubes, surface magnetism, and the solar dynamo: constraints and open problems. *Astron. Nach.* **326**, 194–204.
Schüssler, M., Caligari, P., Ferriz-Mas, A., and Moreno-Insertis, F. (1994). Instability and eruption of magnetic flux tubes in the solar convection zone. *Astron. Astrophys.* **281**, L69–L72.
Schüssler, M. and Rempel, M. (2005). The dynamical disconnection of sunspots from their magnetic roots. *Astron. Astrophys.* **441**, 337–346.
Schüssler, M. and Vögler, A. (2006). Magnetoconvection in a sunspot umbra. *Astrophys. J. Letts.* **641**, L73–L76.
Schwadron, N. A. and McComas, D. J. (2003). Solar wind scaling law. *Astrophys. J.* **599**, 1395–1403.
Schwadron, N. A. and McComas, D. J. (2008). The solar wind power from magnetic flux. *Astrophys. J. Letts.* **686**, L33–L36.
Schwartz, P., Heinzel, P., Schmieder, B., and Anzer, U. (2006). Study of an extended EUV filament using SoHO/SUMER observations of the hydrogen Lyman lines. *Astron. Astrophys.* **459**, 651–661.
Schwarzschild, M. (1948). On noise arising from the solar granulation. *Astrophys. J.* **107**, 1–5.
Schwenn, R. (2006). Solar wind sources and their variations over the solar cycle. *Space Sci. Rev.* **124**, 51–76.
Schwenn, R. and Marsch, E. (1990). *Physics of the Inner Heliosphere I. Large-Scale Phenomena.* (Springer-Verlag, Berlin).
Schwenn, R. and Marsch, E. (1991). *Physics of the Inner Heliosphere II. Particles, Waves and Turbulence.* (Springer-Verlag, Berlin).
Scudder, J. D. and Olbert, S. (1983). The collapse of the local, Spitzer-Haerm formulation and a global-local generalization for heat flow in an inhomogeneous, fully ionized plasma. In *NASA Conference Publication.* NASA Conference Publication, vol. 2280. (Washington, DC), 163–182.
Secchi, A. (1877a). *Le Soleil, Vol 1.* (Gauthier-Villars, Paris).
Secchi, A. (1877b). *Le Soleil, Vol 2.* (Gauthier-Villars, Paris).
Sedlacek, Z. (1971). Electrostatic oscillations in cold inhomogeneous plasma. I. Differential equation approach. *J. Plasma Phys.* **5**, 239–263.
Sekse, D. H., Rouppe van der Voort, L., and De Pontieu, B. (2012). Statistical properties of the disk counterparts of type II spicules from simultaneous observations of rapid blueshifted excursions in Ca II 8542 and Hα. *Astrophys. J.* **752**, 108.
Semenov, V. S., Kubyshkin, I. V., and Heyn, M. F. (1983). Asymptotic solution for field-line reconnection. Compressible case of Petschek's model. *J. Plasma Phys.* **30**, 303–320.
Serio, S., Peres, G., Vaiana, G. S., Golub, L., and Rosner, R. (1981). Closed coronal structures. II – Generalized hydrostatic model. *Astrophys. J.* **243**, 288–300.
Shafranov, V. D. (1966). Plasma equilibrium in a magnetic field. *Rev. Plasma Phys.* **2**, 103–151.
Shay, M. A. and Drake, J. F. (1998). The role of electron dissipation on the rate of collisionless magnetic reconnection. *Geophys. Res. Lett.* **25**, 3759–3762.
Shay, M. A., Drake, J. F., and Swisdak, M. (2007). Two-scale structure of the electron dissipation region during collisionless magnetic reconnection. *Phys. Rev. Lett. 99,* 15 (Oct.), 155002.
Sheeley, N. R. (1969). The evolution of the photospheric network. *Solar Phys.* **9**, 347–357.
Sheeley, N. R. (2005). Surface evolution of the Sun's magnetic field: a historical review of the flux-transport mechanism. *Liv. Rev. in Solar Phys.* **2**, 5.
Sheeley, N. R., DeVore, C. R., and Boris, J. P. (1985). Simulations of the mean solar magnetic field during sunspot cycle-21. *Solar Phys.* **98**, 219–239.
Sheeley, N. R., Nash, A. G., and Wang, Y. (1987). The origin of rigidly rotating magnetic field patterns on the Sun. *Astrophys. J.* **319**, 481–502.
Sheeley, N. R., Wang, Y.-M., Hawley, S. H., Brueckner, G. E., Dere, K. P., Howard, R. A., Koomen, M. J., Korendyke, C. M., Michels, D. J., Paswaters, S. E., Socker, D. G., St. Cyr, O. C., Wang, D., Lamy, P. L., Llebaria, A., Schwenn, R., Simnett, G. M., Plunkett, S., and Biesecker, D. A. (1997). Measurements of flow speeds in the corona between 2 and 30 $R_\odot$. *Astrophys. J.* **484**, 472–478.
Shen, C., Lin, J., and Murphy, N. A. (2011). Numerical experiments on fine structure within reconnecting current sheets in solar flares. *Astrophys. J.* **737**, 14.
Shibata, K., Ishido, Y., Acton, L. W., Strong, K., Hirayama, T., Uchida, Y., McAllister, A., Matsumoto, R., Tsuneta, S., Shimizu, T., Hara, H., Sakurai, T., Ichimoto, K., Nishino, Y., and Ogawara, Y. (1992). Observations of X-ray jets with the Yohkoh Soft X-ray Telescope. *Publ. Astron. Soc. Japan* **44**, L173–L179.
Shibata, K. and Magara, T. (2011). Solar flares: magnetohydrodynamic processes. *Liv. Rev. in Solar Phys.* **8**, 6.
Shibata, K., Masuda, S., Shimojo, M., Hara, H., Yokoyama, T., Tsuneta, S., Kosugi, T., and Ogawara, Y. (1995). Hot-plasma ejections associated with compact-loop solar flares. *Astrophys. J. Letts.* **451**, L83–L85.
Shibata, K., Nakamura, T., Matsumoto, T., Otsuji, K., Okamoto, T. J., Nishizuka, N., Kawate, T., Watanabe, H., Nagata, S., UeNo, S., Kitai, R., Nozawa, S., Tsuneta, S., Suematsu, Y., Ichimoto, K., Shimizu, T., Katsukawa, Y., Tarbell, T. D., Berger, T. E., Lites, B. W., Shine, R. A., and Title, A. M. (2007). Chromospheric anemone jets as evidence of ubiquitous reconnection. *Science* **318**, 1591–1594.

Shibata, K., Nishikawa, T., Kitai, R., and Suematsu, Y. (1982). Numerical hydrodynamics of the jet phenomena in the solar atmosphere. II. Surges. *Solar Phys.* **77**, 121–151.

Shibata, K., Tajima, T., Steinolfson, R. S., and Matsumoto, R. (1989). Two-dimensional magnetohydrodynamic model of emerging magnetic flux in the solar atmosphere. *Astrophys. J.* **345**, 584–596.

Shimizu, T. (1995). Energetics and occurrence rate of active-region transient brightenings and implications for the heating of the active-region corona. *Publ. Astron. Soc. Japan* **47**, 251–263.

Shimizu, T., Lites, B. W., Katsukawa, Y., Ichimoto, K., Suematsu, Y., Tsuneta, S., Nagata, S., Kubo, M., Shine, R. A., and Tarbell, T. D. (2008). Frequent occurrence of high-speed local mass downflows on the solar surface. *Astrophys. J.* **680**, 1467–1476.

Shimizu, T., Tsuneta, S., Acton, L. W., Lemen, J., and Uchida, Y. (1992). Transient brightenings in active regions observed by the Soft X-Ray Telescope on Yokhoh. *Publ. Astron. Soc. Japan* **44**, L147–L153.

Shimojo, M., Hashimoto, S., Shibata, K., Hirayama, T., Hudson, H. S., and Acton, L. W. (1996). Statistical study of solar X-ray jets observed with the Yohkoh Soft X-ray Telescope. *Publ. Astron. Soc. Japan* **48**, 123–136.

Shimojo, M., Narukage, N., Kano, R., Sakao, T., Tsuneta, S., Shibasaki, K., Cirtain, J. W., Lundquist, L. L., Reeves, K. K., and Savcheva, A. (2007). Fine structures of solar X-ray jets observed with the X-Ray Telescope aboard Hinode. *Publ. Astron. Soc. Japan* **59**, 745–750.

Shimojo, M. and Shibata, K. (2000). Physical parameters of solar X-ray jets. *Astrophys. J.* **542**, 1100–1108.

Shine, R. A., Simon, G. W., and Hurlburt, N. E. (2000). Supergranule and mesogranule evolution. *Solar Phys.* **193**, 313–331.

Shiota, D., Isobe, H., Chen, P. F., Yamamoto, T. T., Sakajiri, T., and Shibata, K. (2005). Self-consistent magnetohydrodynamic modeling of a coronal mass ejection, coronal dimming, and a giant cusp-shaped arcade formation. *Astrophys. J.* **634**, 663–678.

Shoub, E. C. (1983). Invalidity of local thermodynamic equilibrium for electrons in the solar transition region. I – Fokker-Planck results. *Astrophys. J.* **266**, 339–369.

Simon, G. W. and Leighton, R. (1964). Velocity fields in the solar atmosphere. III. Large-scale motions, the chromospheric network, and magnetic fields. *Astrophys. J.* **140**, 1120–1147.

Simon, G. W., Title, A. M., and Weiss, N. O. (2001). Sustaining the Sun's magnetic network with emerging bipoles. *Astrophys. J.* **561**, 427–434.

Simon, G. W. and Weiss, N. O. (1970). On the magnetic field in pores. *Solar Phys.* **13**, 85–103.

Simon, G. W., Weiss, N. O., and Nye, A. H. (1983). Simple models for magnetic flux tubes. *Solar Phys.* **87**, 65–75.

Smith, P. D., Tsiklauri, D., and Ruderman, M. S. (2007). Enhanced phase mixing of Alfvén waves propagating in stratified and divergent coronal structures. *Astron. Astrophys.* **475**, 1111–1123.

Socas-Navarro, H., Martínez Pillet, V., Sobotka, M., and Vázquez, M. (2004). The thermal and magnetic structure of umbral dots from the inversion of high-resolution full Stokes observations. *Astrophys. J.* **614**, 448–456.

Solanki, S. K. (2003). Sunspots: an overview. *Astron. Astrophys. Rev.* **11**, 153–286.

Solanki, S. K., Barthol, P., Danilovic, S., Feller, A., Gandorfer, A., Hirzberger, J., Lagg, A., Riethmüller, T. L., Schüssler, M., Wiegelmann, T., Bonet, J. A., Pillet, V. M., Khomenko, E., del Toro Iniesta, J. C., Domingo, V., Palacios, J., Knölker, M., González, N. B., Borrero, J. M., Berkefeld, T., Franz, M., Roth, M., Schmidt, W., Steiner, O., and Title, A. M. (2011). The Sun at high resolution: first results from the Sunrise mission. *IAU Symp.* **273**, 226–232.

Solanki, S. K., Barthol, P., Danilovic, S., Feller, A., Gandorfer, A., Hirzberger, J., Riethmüller, T. L., Schüssler, M., Bonet, J. A., Martínez Pillet, V., del Toro Iniesta, J. C., Domingo, V., Palacios, J., Knölker, M., Bello González, N., Berkefeld, T., Franz, M., Schmidt, W., and Title, A. M. (2010). SUNRISE: instrument, mission, data, and first results. *Astrophys. J. Letts.* **723**, L127–L133.

Solanki, S. K. and Fligge, M. (2002). Solar irradiance variations and climate. *Journal of Atmospheric and Solar-Terrestrial Physics* **64**, 677–685.

Solanki, S. K., Inhester, B., and Schüssler, M. (2006). The solar magnetic field. *Rep. Prog. in Phys.* **69**, 563–668.

Solanki, S. K. and Montavon, C. A. P. (1993). Uncombed fields as the source of the broad-band circular polarization of sunspots. *Astron. Astrophys.* **275**, 283–292.

Soler, R., Oliver, R., and Ballester, J. L. (2010). Time damping of non-adiabatic magnetohydrodynamic waves in a partially ionized prominence plasma: effect of Helium. *Astron. Astrophys.* **512**, A28.

Solov'ev, L. S. (1967). Symmetric magnetohydrodynamic flow and helical waves in a circular plasma cylinder. In *Reviews of Plasma Physics*, M. A. Leontovich, ed. (Consultants Bureau, New York), 277–325.

Solov'ev, L. S. (1968). The theory of hydromagnetic stability of toroidal plasma configurations. *Sov. Phys. JETP* **26**, 400–407.

Somov, B. and Syrovatsky, S. I. (1976). Hydrodynamic plasma flow in a strong magnetic field. In *Lebedev Phys. Inst. 74*. Moscow, 13–72.

Somov, B. and Vernata, A. I. (1994). Tearing instability of reconnecting current sheets in space plasmas. *Space Sci. Rev.* **65**, 253–288.

Sonnerup, B. and Priest, E. R. (1975). Resistive MHD stagnation-point flows at a current sheet. *J. Plasma Phys.* **14**, 283–294.

Soward, A. M. (1972). A kinematic theory of large magnetic Reynolds number dynamos. *Phil. Trans. Roy. Soc. A* **272**, 431–462.

Soward, A. M. and Priest, E. R. (1977). Fast magnetic field-line reconnection. *Phil. Trans. Roy. Soc.* **A284**, 369–417.

Spanier, J. and Oldham, B. (1987). *An Atlas of Functions.* (Hemisphere Publishing Corp., New York).

Spiegel, E. A. and Weiss, N. O. (1980). Magnetic activity and variations in solar luminosity. *Nature* **287**, 616–617.

Spiegel, E. A. and Zahn, J.-P. (1992). The solar tachocline. *Astron. Astrophys.* **265**, 106–114.

Spies, G. O. (1974). Nonlinear magnetohydrodynamic stability and plasma containment with closed field lines. *Phys. Fluids* **17**, 1188–1197.

Spitzer, L. (1962). *Physics of Fully Ionized Gases.* (Interscience, New York).

Spruit, H. C. (1974). A model of the solar convection zone. *Solar Phys.* **34**, 277–290.

Spruit, H. C. (1981). Motion of magnetic flux tubes in the solar convection zone and chromosphere. *Astron. Astrophys.* **98**, 155–160.

Spruit, H. C. (2002). Dynamo action by differential rotation in a stably stratified interior. *Astron. Astrophys.* **381**, 923–932.

Spruit, H. C. and Bogdan, T. J. (1992). The conversion of p-modes to slow modes and the absorption of acoustic waves by sunspots. *Astrophys. J. Letts.* **391**, L109–L112.

Spruit, H. C. and Scharmer, G. B. (2006). Fine structure, magnetic field and heating of sunspot penumbrae. *Astron. Astrophys.* **447**, 343–354.

Spruit, H. C. and van Ballegooijen, A. A. (1982). Stability of toroidal flux tubes in stars. *Astron. Astrophys.* **106**, 58–66.

St. Cyr, O. C., Plunkett, S. P., Michels, D. J., Paswaters, S. E., Koomen, M. J., Simnett, G. M., Thompson, B. J., Gurman, J. B., Schwenn, R., Webb, D. F., Hildner, E., and Lamy, P. L. (2000). Properties of coronal mass ejections: SOHO LASCO observations from January 1996 to June 1998. *J. Geophys. Res.* **105**, 18169–18186.

Steele, C. D. C. and Priest, E. R. (1992). A model for the fibril structure of normal-polarity solar prominences. *Solar Phys.* **140**, 289–306.

Steenbeck, M. and Krause, F. (1969). On the dynamo theory of stellar and planetary magnetic fields. I. AC dynamos of solar type. *Astron. Nach.* **291**, 49–84.

Steenbeck, M., Krause, F., and Rädler, K. H. (1966). Berechnung der mittleren Lorentz-Feldstärke $\overline{\mathbf{v}\times\mathbf{B}}$ für ein elektrisch leitendendes Medium in turbulenter, durch Coriolis-Kräfte beeinfluβter Bewegung. *Z. Naturforsch* **21a**, 369–376.

Stein, R. F., Benson, D., Georgobiani, D., and Nordlund, Å. (2006). Supergranule-scale convection simulations. In *Proc. SOHO 18/ GONG 2006/ HELIAS I*, M. Thompson, ed. (ESA SP-624, Noordwijk), 79–82.

Stein, R. F., Lagerfjärd, A., Nordlund, Å., and Georgobiani, D. (2011). Solar flux emergence simulations. *Solar Phys.* **268**, 271–282.

Stein, R. F. and Nordlund, Å. (1989). Topology of convection beneath the solar surface. *Astrophys. J. Letts.* **342**, L95–L98.

Stein, R. F. and Nordlund, Å. (2006). Solar small-scale magnetoconvection. *Astrophys. J.* **642**, 1246–1255.

Steiner, O., Franz, M., Bello González, N., Nutto, C., Rezaei, R., Martínez Pillet, V., Bonet Navarro, J. A., del Toro Iniesta, J. C., Domingo, V., Solanki, S. K., Knölker, M., Schmidt, W., Barthol, P., and Gandorfer, A. (2010). Detection of vortex tubes in solar granulation from observations with SUNRISE. *Astrophys. J. Letts.* **723**, L180–L184.

Steiner, O., Grossmann-Doerth, U., Knölker, M., and Schüssler, M. (1998). Dynamical interaction of solar magnetic elements and granular convection: results of a numerical simulation. *Astrophys. J.* **495**, 468–484.

Steinhilber, F., Abreu, J. A., and Beer, J. (2008). Solar modulation during the Holocene. *Astrophys and Space Sci. Trans.* **4**, 1–6.

Stenflo, J. O. (1973). Magnetic-field structure of the photospheric network. *Solar Phys.* **32**, 41–63.

Stenflo, J. O. (1982). The Hanle effect and the diagnostics of turbulent magnetic fields in the solar atmosphere. *Solar Phys.* **80**, 209–226.

Stenflo, J. O. and Kosovichev, A. G. (2012). Bipolar magnetic regions on the Sun: global analysis of the SOHO/MDI data set. *Astrophys. J.* **745**, 129.

Sterling, A. C. (2000). Solar spicules: a review of recent models and targets for future observations (invited review). *Solar Phys.* **196**, 79–111.

Sterling, A. C. and Hollweg, J. V. (1989). A rebound shock mechanism for solar fibrils. *Astrophys. J.* **343**, 985–993.

Stix, M. (1972). Non-linear dynamo waves. *Astron. Astrophys.* **20**, 9–12.

Stix, M. (1991). The solar dynamo. *Geophys. Astrophys. Fluid Dyn.* **62**, 211–228.

Stix, M. (2002). *The Sun, 2nd Edition.* (Springer-Verlag, Berlin).

Strachan, N. R. and Priest, E. R. (1994). A general family of non-uniform reconnection models with separatrix jets. *Geophys. Astrophys. Fluid Dyn.* **74**, 245–274.

Strauss, H. R. (1981). Resistive ballooning modes. *Phys. Fluids* **24**, 2004–2009.

Stucki, K., Solanki, S. K., Schühle, U., Rüedi, I., Wilhelm, K., Stenflo, J. O., Brković, A., and Huber, M. C. E. (2000). Comparison of far-ultraviolet emission lines formed in coronal holes and the quiet Sun. *Astron. Astrophys.* **363**, 1145–1154.

Sturrock, P. A. (1968). A model of solar flares. *IAU Symp.* **35**, 471–479.

Sturrock, P. A. (1991). Maximum energy of semi-infinite magnetic field configurations. *Astrophys. J.* **380**, 655–659.

Sturrock, P. A. and Uchida, Y. (1981). Coronal heating by stochastic magnetic pumping. *Astrophys. J.* **246**, 331–336.

Sturrock, P. A. and Woodbury, E. T. (1967). Force-free magnetic fields and solar filaments. In *Plasma Astrophysics*, P. Sturrock, ed. (Academic Press, London), 155–167.

Su, Y., Wang, T., Veronig, A., Temmer, M., and Gan, W. (2012). Solar magnetized 'tornadoes': relation to filaments. *Astrophys. J. Letts.* **756**, L41.

Subramanian, S., Madjarska, M. S., and Doyle, J. G. (2010). Coronal hole boundaries evolution at small scales. II. XRT view. Can small-scale outflows at CHBs be a source of the slow solar wind? *Astron. Astrophys.* **516**, A50.

Subramanian, S., Madjarska, M. S., Doyle, J. G., and Bewsher, D. (2012). What is the true nature of blinkers? *Astron. Astrophys.* **538**, A50.

Sudol, J. J. and Harvey, J. W. (2005). Longitudinal magnetic field changes accompanying solar flares. *Astrophys. J.* **635**, 647–658.

Suematsu, Y. (1990). Influence of photospheric 5-minute oscillations on the formation of chromospheric fine structures. In *Progress of Seismology of the Sun and Stars*, Y. Osaki and H. Shibahashi, eds. (Springer Verlag, Berlin), 211–214.

Suess, S. T. and Nerney, S. F. (2002). Stagnation flow in streamer boundaries. *Astrophys. J.* **565**, 1275–1288.

Sui, L., Holman, G. D., and Dennis, B. R. (2004). Evidence for magnetic reconnection in three homologous solar flares observed by RHESSI. *Astrophys. J.* **612**, 546–556.

Suydam, B. R. (1959). Stability of a linear pinch. *Proc. 2nd UN Int. Conf. on Peaceful Uses of Atomic Energy* **31**, 157–159.

Svestka, Z., Dodson-Prince, H. W., Mohler, O., Martin, S. F., Moore, R. L., Nolte, J., and Petrasso, R. (1982). Study of the post-flare loops on 29 July 1973. IV – Revision of T and n_e values and comparison with the flare of 21 May 1980. *Solar Phys.* **78**, 271–285.

Švestka, Z., Fárník, F., Hudson, H. S., and Hick, P. (1998). Large-scale active coronal phenomena in Yohkoh SXT images. IV. Solar wind streams from flaring active regions. *Solar Phys.* **182**, 179–193.

Svestka, Z. F., Fontenla, J. M., Machado, M. E., Martin, S. F., and Neidig, D. F. (1987). Multi-thermal observations of newly formed loops in a dynamic flare. *Solar Phys.* **108**, 237–250.

Sweet, P. A. (1958a). The neutral point theory of solar flares. In *Electromagnetic Phenomena in Cosmical Physics*, B. Lehnert, ed. (Cambridge University Press, Cambridge, UK), 123–134.

Sweet, P. A. (1958b). The production of high-energy particles in solar flares. *Nuovo Cimento Suppl.* **8**, 188–196.

Syrovatsky, S. I. (1966). Dynamic dissipation of a magnetic field and particle acceleration. *Soviet Astronomy* **10**, 270–280.

Tandberg-Hanssen, E. (1995). *The Nature of Solar Prominences.* (Kluwer Academic Publishers, Dordrecht, Holland).

Tandberg-Hanssen, E. and Emslie, A. (1988). *The Physics of Solar Flares.* (Cambridge University Press, Cambridge, UK).

Tarr, L. and Longcope, D. (2012). Calculating energy storage due to topological changes in emerging active region NOAA AR 11112. *Astrophys. J.* **749**, 64.

Tataronis, J. and Grossmann, W. (1973). Decay of MHD waves by phase mixing. *Zeit. fur Phys.* **261**, 203–216.

Taylor, J. B. (1974). Relaxation of toroidal plasma and generation of reverse magnetic fields. *Phys. Rev. Lett.* **33**, 1139–1141.

Taylor, J. B. (1976). Relaxation of toroidal discharges. In *Pulsed High Beta Plasmas*, D. Evans, ed. (Pergamon Press, Oxford), 59–67.

Taylor, J. B. and Hastie, R. J. (1968). Stability of general plasma equilibria – I. Formal theory. *Plasma Phys.* **10**, 479–494.

Terradas, J., Andries, J., and Verwichte, E. (2011). Linear coupling between fast and slow MHD waves due to line-tying effects. *Astron. Astrophys.* **527**, A132.

Thiessen, G. (1950). The structure of the sunspot umbra. *The Observatory* **70**, 234–235.

Thomas, J. H. and Montesinos, B. (1993). A siphon-flow model of the photospheric Evershed flow in a sunspot. *Astrophys. J.* **407**, 398–401.

Thomas, J. H. and Weiss, N. O. (2008). *Sunspots and Starspots.* (Cambridge University Press, Cambridge, UK).

Thomas, J. H., Weiss, N. O., Tobias, S. M., and Brummel, N. H. (2002). Downward pumping of magnetic flux as the cause of filamentary structures in sunspot penumbrae. *Nature* **420**, 390–393.

Thompson, B. J., Cliver, E. W., Nitta, N., Delannée, C., and Delaboudinière, J.-P. (2000). Coronal dimmings and energetic CMEs in April-May 1998. *Geophys. Res. Lett.* **27**, 1431–1434.

Thompson, B. J., Plunkett, S. P., Gurman, J. B., Newmark, J. S., St. Cyr, O. C., and Michels, D. J. (1998). SOHO/EIT observations of an Earth-directed coronal mass ejection on May 12, 1997. *Geophys. Res. Lett.* **25**, 2465–2468.

Thompson, M. J., Christensen-Dalsgaard, J., Miesch, M. S., and Toomre, J. (2003). The internal rotation of the Sun. *Ann. Rev. Astron. Astrophys.* **41**, 599–643.

Thornton, L. M. and Parnell, C. E. (2010). Small-scale flux emergence observed using Hinode/SOT. *Solar Phys.* **269**, 13–40.

Threlfall, J., McClements, K. G., and De Moortel, I. (2011). Alfvén wave phase-mixing and damping in the ion cyclotron range of frequencies. *Astron. Astrophys.* **525**, A155.

Ting, A. C., Montgomery, D., and Matthaeus, W. H. (1986). Turbulent relaxation processes in magnetohydrodynamics. *Phys. Fluids* **29**, 3261–3274.

Title, A. M., Frank, Z. A., Shine, R. A., Tarbell, T. D., Topka, K. P., Scharmer, G., and Schmidt, W. (1993). On the magnetic and velocity field geometry of simple sunspots. *Astrophys. J.* **403**, 780–796.

Title, A. M. and Schrijver, C. J. (1998). The Sun's magnetic carpet. *Astron. Soc. Pac. Conf. Ser.* **154**, 345–360.

Title, A. M., Tarbell, T. D., Topka, K. P., Ferguson, S. H., Shine, R. A., and SOUP Team. (1989). Statistical properties of solar granulation derived from the SOUP instrument on Spacelab 2. *Astrophys. J.* **336**, 475–494.

Titov, V. S. (1992). On the method of calculating two-dimensional potential magnetic fields with current sheets. *Solar Phys.* **139**, 401–404.

Titov, V. S. (2007). Generalized squashing factors for covariant description of magnetic connectivity in the solar corona. *Astrophys. J.* **660**, 863–873.

Titov, V. S. and Démoulin, P. (1999). Basic topology of twisted magnetic configurations in solar flares. *Astron. Astrophys.* **351**, 707–720.

Titov, V. S., Forbes, T. G., Priest, E. R., Mikić, Z., and Linker, J. A. (2009). Slip-squashing factors as a measure of three-dimensional magnetic reconnection. *Astrophys. J.* **693**, 1029–1044.

Titov, V. S., Galsgaard, K., and Neukirch, T. (2003). Magnetic pinching of hyperbolic flux tubes. I. Basic estimations. *Astrophys. J.* **582**, 1172–1189.

Titov, V. S., Hornig, G., and Démoulin, P. (2002). Theory of magnetic connectivity in the solar corona. *J. Geophys. Res.* **107**, SSH 3–1.

Titov, V. S., Mikić, Z., Linker, J. A., Lionello, R., and Antiochos, S. K. (2011). Magnetic topology of coronal hole linkages. *Astrophys. J.* **731**, 111.

Titov, V. S., Mikic, Z., Török, T., Linker, J. A., and Panasenco, O. (2012). 2010 August 1-2 sympathetic eruptions. I. Magnetic topology of the source-surface background field. *Astrophys. J.* **759**, 70.

Titov, V. S. and Priest, E. R. (1993). The collapse of an X-type neutral point to form a reconnecting time-dependent current sheet. *Geophys. Astrophys. Fluid Dyn.* **72**, 249–276.

Titov, V. S., Priest, E. R., and Démoulin, P. (1993). Conditions for the appearance of bald patches at the solar surface. *Astron. Astrophys.* **276**, 564–570.

Tiwari, S. K., van Noort, M., Lagg, A., and Solanki, S. K. (2013). The structure of sunspot penumbral filaments, a remarkable uniformity of properties. *Astron. Astrophys.,* **557**, A25.

Tobias, S. M. (2002). Modulation of solar and stellar dynamos. *Astron. Nach.* **323**, 417–423.

Tobias, S. M., Brummell, N. H., Clune, T. L., and Toomre, J. (1998). Pumping of magnetic fields by turbulent penetrative convection. *Astrophys. J.* **502**, L177–L180.

Tobias, S. M., Brummell, N. H., Clune, T. L., and Toomre, J. (2001). Transport and storage of magnetic fields by overshooting turbulent compressible convection. *Astrophys. J.* **549**, 1183–1203.

Tobias, S. M. and Hughes, D. W. (2004). The influence of velocity shear on magnetic buoyancy instability in the solar tachocline. *Astrophys. J.* **603**, 785–802.

Tobias, S. M. and Weiss, N. O. (2007). The solar dynamo and the tachocline. In *The Solar Tachocline*, D. W. Hughes, R. Rosner, and N. O. Weiss, eds. (Cambridge University Press, Cambridge, UK), 319–350.

Tobias, S. M., Weiss, N. O., and Kirk, K. (1995). Chaotically modulated stellar dynamos. *Mon. Not. Roy. Astron. Soc.* **273**, 1150–1166.

Tomczyk, S. and McIntosh, S. W. (2009). Time-distance seismology of the solar corona with CoMP. *Astrophys. J.* **697**, 1384–1391.

Tomczyk, S., McIntosh, S. W., Keil, S. L., Judge, P. G., Schad, T., Seeley, D. H., and Edmondson, J. (2007). Alfvén waves in the solar corona. *Science* **317**, 1192–1196.

Török, T., Kliem, B., and Titov, V. S. (2004). Ideal kink instability of a magnetic loop equilibrium. *Astron. Astrophys.* **413**, L27–L30.

Tousey, R. (1973). The solar corona. In *Space Research*, M. J. Rycroft and S. K. Runcorn, eds. (Akademie-Verlag, Berlin), 713–730.

Tripathi, D., Gibson, S. E., Qiu, J., Fletcher, L., Liu, R., Gilbert, H., and Mason, H. E. (2009). Partially-erupting prominences: a comparison between observations and model-predicted observables. *Astron. Astrophys.* **498**, 295–305.

Tripathi, D., Isobe, H., and Jain, R. (2009). Large-amplitude oscillations in prominences. *Space Sci. Rev.* **149**, 283–298.

Tripathi, D., Mason, H. E., Dwivedi, B. N., Del Zanna, G., and Young, P. R. (2009). Active-region loops: Hinode/Extreme-Ultraviolet Imaging Spectrometer observations. *Astrophys. J.* **694**, 1256–1265.

Tripathi, D., Mason, H. E., and Klimchuk, J. A. (2012). Active region moss: Doppler shifts from Hinode/Extreme-Ultraviolet Imaging Spectrometer observations. *Astrophys. J.* **753**, 37.

Tritschler, A., Schlichenmaier, R., Bellot Rubio, L. R., the KAOS Team, Berkefeld, T., and Schelenz, T. (2004). Two-dimensional spectroscopy of a sunspot. I. Properties of the penumbral fine structure. *Astron. Astrophys.* **415**, 717–729.

Trujillo Bueno, J., Shchukina, N., and Asensio Ramos, A. (2004). A substantial amount of hidden magnetic energy in the quiet Sun. *Nature* **430**, 326–329.

Trujillo Bueno, J., Štěpán, J., and Casini, R. (2011). The Hanle effect of the hydrogen Lyα line for probing the magnetism of the solar transition region. *Astrophys. J. Letts.* **738**, L11.

Tsinganos, K. (2010). MHD theory of astrophysical jets. *Mem. della Soc. Astron. Ital.* **15**, 102–113.

Tsinganos, K. C., Rosner, R., and Distler, J. (1984). On the topological stability of magnetostatic equilibria. *Astrophys. J.* **278**, 409–419.

Tsuneta, S. (1996). Structure and dynamics of magnetic reconnection in a solar flare. *Astrophys. J.* **456**, 840–849.

Tsuneta, S., Acton, L., Bruner, M., Lemen, J., Brown, W., Caravalho, R., Catura, R., Freel and , S., Jurcevich, B., and Owens, J. (1991). The Soft X-ray Telescope for the SOLAR-A mission. *Solar Phys.* **136**, 37–67.

Tsuneta, S., Hara, H., Shimizu, T., Acton, L. W., Strong, K. T., Hudson, H. S., and Ogawara, Y. (1992). Observation of a solar flare at the limb with the Yohkoh Soft X-ray Telescope. *Publ. Astron. Soc. Japan* **44**, L63–L69.

Tsuneta, S., Ichimoto, K., Katsukawa, Y., Lites, B. W., Matsuzaki, K., Nagata, S., Orozco Suárez, D., Shimizu, T., Shimojo, M., Shine, R. A., Suematsu, Y., Suzuki, T. K., Tarbell, T. D., and Title, A. M. (2008). The magnetic landscape of the Sun's polar region. *Astrophys. J.* **688**, 1374–1381.

Tsuneta, S., Ichimoto, K., Katsukawa, Y., Nagata, S., Otsubo, M., Shimizu, T., Suematsu, Y., Nakagiri, M., Noguchi, M., Tarbell, T., Title, A., Shine, R., Rosenberg, W., Hoffmann, C., Jurcevich, B., Kushner, G., Levay, M., Lites, B., Elmore, D., Matsushita, T., Kawaguchi, N., Saito, H., Mikami, I., Hill, L. D., and Owens, J. K. (2008). The Solar Optical Telescope for the Hinode Mission: an overview. *Solar Phys.* **249**, 167–196.

Tsuneta, S., Masuda, S., Kosugi, T., and Sato, J. (1997). Hot and superhot plasmas above an impulsive flare loop. *Astrophys. J.* **478**, 787–798.

Tu, C.-Y. and Marsch, E. (1995). MHD structures, waves and turbulence in the solar wind: observations and theories. *Space Sci. Rev.* **73**, 1–210.

Tu, C.-Y. and Marsch, E. (1997). Two-fluid model for heating of the solar corona and acceleration of the solar wind by high-frequency Alfvén waves. *Solar Phys.* **171**, 363–391.

Tucker, W. H. (1973). Heating of solar active regions by magnetic energy dissipation: the steady-state case. *Astrophys. J.* **186**, 285–290.

Tur, T. J. (1977). *Aspects of Current Sheet Theory.* (PhD thesis, St Andrews University, St Andrews).

Tur, T. J. and Priest, E. R. (1976). The formation of current sheets during the emergence of new magnetic flux from below the photosphere. *Solar Phys.* **48**, 89–100.

Uberoi, C. (1972). Alfvén waves in inhomogeneous magnetic fields. *Phys. Fluids* **15**, 1673–1675.

Uchida, Y. (1969). A mechanism for the acceleration of solar chromospheric spicules. *Publ. Astron. Soc. Japan* **21**, 128–140.

Uchida, Y. (1970). Diagnosis of coronal magnetic structure by flare-associated hydromagnetic disturbances. *Publ. Astron. Soc. Japan* **22**, 341–164.

Uchida, Y. and Low, B. C. (1981). Equilibrium configurations of the magnetosphere of a star loaded with accreted magnetized mass. *J. Astrophys. Astron.* **2**, 405–419.

Ugai, M. (2009). Fast reconnection evolution in an arcade-like magnetic loop structure. *Phys. Plasmas 16,* 6 (June), 062312.

Ugai, M. and Tsuda, T. (1977). Magnetic field-line reconnection by localised enhancement of resistivity. I. Evolution in a compressible MHD fluid. *J. Plasma Phys.* **17**, 337–356.

Ulrich, R. K. (1970). The five-minute oscillations on the solar surface. *Astrophys. J.* **162**, 993–1002.

Vaiana, G. S., Davis, J. M., Giacconi, R., Krieger, A. S., Silk, J. K., Timothy, A. F., and Zombeck, M. (1973). X-ray observations of characteristic structures and time variations from the solar corona: preliminary results from SKYLAB. *Astrophys. J. Letts.* **185**, L47–L51.

van Ballegooijen, A. A. (1982). The overshoot layer at the base of the solar convective zone and the problem of magnetic flux storage. *Astron. Astrophys.* **113**, 99–112.

van Ballegooijen, A. A. (1985). Electric currents in the solar corona and the existence of magnetostatic equilibrium. *Astrophys. J.* **298**, 421–430.

van Ballegooijen, A. A. (2004). Observations and modeling of a filament on the Sun. *Astrophys. J.* **612**, 519–529.

van Ballegooijen, A. A., Asgari-Targhi, M., Cranmer, S. R., and DeLuca, E. E. (2011). Heating of the solar chromosphere and corona by Alfvén wave turbulence. *Astrophys. J.* **736**, 3.

van Ballegooijen, A. A., Cartledge, N. P., and Priest, E. R. (1998). Magnetic flux transport and the formation of filament channels on the Sun. *Astrophys. J.* **501**, 866–881.

van Ballegooijen, A. A. and Cranmer, S. R. (2008). Hyperdiffusion as a mechanism for solar coronal heating. *Astrophys. J.* **682**, 644–653.

van Ballegooijen, A. A. and Cranmer, S. R. (2010). Tangled magnetic fields in solar prominences. *Astrophys. J.* **711**, 164–178.

van Ballegooijen, A. A. and Martens, P. C. H. (1989). Formation and eruption of solar prominences. *Astrophys. J.* **343**, 971–984.

van Ballegooijen, A. A. and Martens, P. C. H. (1990). Magnetic fields in quiescent prominences. *Astrophys. J.* **361**, 283–289.

van Ballegooijen, A. A., Priest, E. R., and Mackay, D. H. (2000). Mean field model for the formation of filament channels on the Sun. *Astrophys. J.* **539**, 983–994.

van Der Holst, B., Manchester, W., Frazin, R., Vasquez, A., Toth, G., and Gombosi, T. (2010). A data-driven, two-temperature solar wind model with Alfvén waves. *Astrophys. J.* **725**, 1373–1383.

van Doorsselaere, T., Brady, C. S., Verwichte, E., and Nakariakov, V. M. (2008). Seismological demonstration of perpendicular density structuring in the solar corona. *Astron. Astrophys.* **491**, L9–L12.

van Driel-Gesztelyi, L. (2009). Magnetic reconnection and energy release on the Sun and solar-like stars. *IAU Symp.* **259**, 191–200.

van Driel-Gesztelyi, L., Hofmann, A., Démoulin, P., Schmieder, B., and Csepura, G. (1994). Relationship between electric currents, photospheric motions, chromospheric activity, and magnetic field topology. *Solar Phys.* **149**, 309–330.

van Driel-Gesztelyi, L., Schmieder, B., Cauzzi, G., Mein, N., Hofmann, A., Nitta, N., Kurokawa, H., Mein, P., and Staiger, J. (1996). X-ray bright point flares due to magnetic reconnection. *Solar Phys.* **163**, 145–170.

van Driel-Gesztelyi, L., Wiik, J. E., Schmieder, B., Tarbell, T., Kitai, R., Funakoshi, Y., and Anwar, B. (1997). Post-flare loops of 26 June 1992. IV. Formation and expansion of hot and cool loops. *Solar Phys.* **174**, 151–162.

Van Hoven, G., Hendrix, D. L., and Schnack, D. D. (1995). Diagnosis of general magnetic reconnection. *J. Geophys. Res.* **100**, 19819–19828.

van Tend, W. and Kuperus, M. (1978). The development of coronal electric current systems in active regions and their relation to filaments and flares. *Solar Phys.* **59**, 115–127.

Vaquero, J. M., Gallego, M. C., Usoskin, I. G., and Kovaltsov, G. A. (2011). Revisited sunspot data: a new scenario for the onset of the Maunder minimum. *Astrophys. J. Letts.* **731**, L24.

Vargas Dominguez, S., Rouppe van der Voort, L., Bonet, J. A., Martínez Pilet, V., van Noort, M., and Katsukawa, Y. (2008). Moat flow in the vicinity of sunspots for various penumbral configurations. *Astrophys. J.* **679**, 900–909.

Vasheghani Farahani, S., Nakariakov, V. M., Verwichte, E., and van Doorsselaere, T. (2012). Nonlinear evolution of torsional Alfvén waves. *Astron. Astrophys.* **544**, A127.

Vekstein, G. E. and Bian, N. (2005). Forced magnetic reconnection at a neutral X-point. *Astrophys. J. Letts.* **632**, L151–L154.

Vekstein, G. E. and Priest, E. R. (1992). Magnetohydrodynamic equilibria and cusp formation at an X-type neutral line by footpoint shearing. *Astrophys. J.* **384**, 333–340.

Vekstein, G. E. and Priest, E. R. (1993). Magnetostatic equilibria and current sheets in a sheared magnetic field with an X-point. *Solar Phys.* **146**, 119–125.

Vekstein, G. E., Priest, E. R., and Amari, T. (1990). Formation of current sheets in force-free magnetic fields. *Astron. Astrophys.* **243**, 492–500.

Vekstein, G. E., Priest, E. R., and Steele, C. D. C. (1991). Magnetic reconnection and energy release in the solar corona by Taylor relaxation. *Solar Phys.* **131**, 297–318.

Vekstein, G. E., Priest, E. R., and Steele, C. D. C. (1993). On the problem of magnetic coronal heating by turbulent relaxation. *Astrophys. J.* **417**, 781–789.

Velli, M. (2001). Hydrodynamics of the solar wind expansion. *Astrophys. and Space Sci.* **277**, 157–167.

Velli, M. (2009). Physical processes in the solar wind. In *Turbulence in Space Plasmas*, P. Cargill and L. Vlahos, eds. Lecture Notes in Physics, vol. 778. (Springer Verlag, Berlin), 247–268.

Velli, M., Grappin, R., and Mangeney, A. (1991). Waves from the Sun. *Geophys. Astrophys. Fluid Dyn.* **62**, 101–121.

Verdini, A., Grappin, R., and Velli, M. (2012). Coronal heating in coupled photosphere-chromosphere-coronal systems: turbulence and leakage. *Astron. Astrophys.* **538**, A70.

Vernazza, J. E., Avrett, E. H., and Loeser, R. (1981). Structure of the solar chromosphere. III. Models of the EUV brightness components of the quiet Sun. *Astrophys. J. Suppl.* **45**, 635–725.

Veronig, A. M. and Brown, J. C. (2004). A coronal thick-target interpretation of two hard X-ray loop events. *Astrophys. J. Letts.* **603**, L117–L120.

Verth, G., Erdélyi, R., and Goossens, M. (2010). Magnetoseismology: eigenmodes of torsional Alfvén waves in stratified solar waveguides. *Astrophys. J.* **714**, 1637–1648.

Verth, G., Terradas, J., and Goossens, M. (2010). Observational evidence of resonantly damped propagating kink waves in the solar corona. *Astrophys. J. Letts.* **718**, L102–L105.

Verwichte, E., Aschwanden, M. J., van Doorsselaere, T., Foullon, C., and Nakariakov, V. M. (2009). Seismology of a large solar coronal loop from EUVI/STEREO observations of its transverse oscillation. *Astrophys. J.* **698**, 397–404.

Verwichte, E., Marsh, M., Foullon, C., van Doorsselaere, T., De Moortel, I., Hood, A. W., and Nakariakov, V. M. (2010). Periodic spectral line asymmetries in solar coronal structures from slow magnetoacoustic waves. *Astrophys. J. Letts.* **724**, L194–L198.

Verwichte, E., van Doorsselaere, T., Foullon, C., and White, R. S. (2013). Coronal Alfvén speed determination: consistency between seismology using AIA/SDO transverse loop oscillations and magnetic extrapolation. *Astrophys. J.* **767**, 16.

Vlahos, L. (2009). Solar and stellar active regions. In *Chaos in Astronomy*, G. Contopoulos, and P. A. Patsis, eds. (Springer-Verlag, Berlin), 423–433.

Vlahos, L., Fragos, T., Isliker, H., and Georgoulis, M. (2002). Statistical properties of the energy release in emerging and evolving active regions. *Astrophys. J. Letts.* **575**, L87–L90.

Vlahos, L., Georgoulis, M., Kluiving, R., and Paschos, P. (1995). The statistical flare. *Astron. Astrophys.* **299**, 897–911.

Vlahos, L., Isliker, H., and Lepreti, F. (2004). Particle acceleration in an evolving network of unstable current sheets. *Astrophys. J.* **608**, 540–553.

Vocks, C. and Marsch, E. (2002). Kinetic results for ions in the solar corona with wave-particle interactions and Coulomb collisions. *Astrophys. J.* **568**, 1030–1042.

Vögler, A. and Schüssler, M. (2007). A solar surface dynamo. *Astron. Astrophys.* **465**, L43–L46.

Vögler, A., Shelyag, S., Schüssler, M., Cattaneo, F., Emonet, T., and Linde, T. (2005). Simulations of magneto-convection in the solar photosphere. Equations, methods, and results of the MURaM code. *Astron. Astrophys.* **429**, 335–351.

von Rekowski, B. and Hood, A. W. (2008). Photospheric cancelling magnetic features and associated phenomena in a stratified solar atmosphere. *Mon. Not. Roy. Astron. Soc.* **385**, 1792–1812.

Vonmoos, M., Beer, J., and Muscheler, R. (2006). Large variation in Holocene solar activity: constraints from ^{10}Be in the Greenland Ice Core Project. *J. Geophys. Res.* **111**, 10105.

Vranjes, J., Poedts, S., Pandey, B. P., and De Pontieu, B. (2008). Energy flux of Alfvén waves in weakly ionized plasma. *Astron. Astrophys.* **478**, 553–558.

Vrsnak, B., Ruzdjak, V., and Rompolt, B. (1991). Stability of prominences exposing helical-like patterns. *Solar Phys.* **136**, 151–167.

Waddell, B. V., Rosenbluth, M. N., Monticello, D. A., and White, R. B. (1976). Non-linear growth of the $m = 1$ tearing mode. *Nuclear Fusion* **16**, 528–532.

Wagner, G., Beer, J., Masarik, J., Muscheler, R., Kubik, P. W., Mende, W., Laj, C., Raisbeck, G. M., and Yiou, F. (2001). Presence of the solar De Vries cycle ($\sim$ 205 years) during the last ice age. *Geophys. Res. Letts.* **28**, 303–306.

Waldmeier, M. (1956). Synoptische Karten der Sonnenkorona. Mit 16 Textabbildungen. *Z. Astrophys.* **38**, 219–236.

Wang, H., Shen, C., and Lin, J. (2009). Numerical experiments of wave-like phenomena caused by the disruption of an unstable magnetic configuration. *Astrophys. J.* **700**, 1716–1731.

Wang, T., Sui, L., and Qiu, J. (2007). Direct observation of high-speed plasma outflows produced by magnetic reconnection in solar impulsive events. *Astrophys. J. Letts.* **661**, L207–L210.

Wang, T. J. (2003). Slow-mode standing waves observed by SUMER in hot coronal loops. *Astron. Astrophys.* **402**, L17–L20.

Wang, T. J. (2011). Standing slow-mode waves in hot coronal loops: observations, modelling and coronal seismology. *Space Sci. Rev.* **158**, 397–419.

Wang, T. J. and Solanki, S. K. (2004). Vertical oscillations of a coronal loop observed by TRACE. *Astron. Astrophys.* **421**, L33–L36.

Wang, Y. M. (1999). The jetlike nature of He II λ304 prominences. *Astrophys. J. Letts.* **520**, L71–L74.

Wang, Y. M., Biersteker, J. B., Sheeley, N. R., Koutchmy, S., Mouette, J., and Druckmüller, M. (2007). The solar eclipse of 2006 and the origin of raylike features in the white-light corona. *Astrophys. J.* **660**, 882–892.

Wang, Y. M. and Muglach, K. (2007). On the formation of filament channels. *Astrophys. J.* **666**, 1284–1295.

Wang, Y. M., Nash, A. G., and Sheeley, N. R. (1989). Magnetic flux transport on the Sun. *Science* **245**, 712–718.

Wang, Y. M. and Sheeley, N. R. (1994). The rotation of photospheric magnetic fields: a random walk transport model. *Astrophys. J.* **430**, 399–412.

Wang, Y. M., Sheeley, N. R., and Nash, A. G. (1991). A new solar-cycle model including meridional circulation. *Astrophys. J.* **383**, 431–442.

Wang, Y. M., Sheeley, N. R., Socker, D. G., Howard, R. A., Brueckner, G. E., Michels, D. J., Moses, D., St. Cyr, O. C., Llebaria, A., and Delaboudinière, J.-P. (1998). Observations of correlated white-light and extreme-ultraviolet jets from polar coronal holes. *Astrophys. J.* **508**, 899–907.

Warmuth, A. (2010). Large-scale waves in the solar corona: the continuing debate. *Adv. in Space Res.* **45**, 527–536.

Warren, H. P. (2005). A solar minimum irradiance spectrum for Wavelengths below 1200 Å. *Astrophys. J. Supp.* **157**, 147–173.

Warren, H. P., Brooks, D. H., and Winebarger, A. R. (2011). Constraints on the heating of high-temperature active region loops: observations from Hinode and the Solar Dynamics Observatory. *Astrophys. J.* **734**, 90.

Warren, H. P., O'Brien, C. M., and Sheeley, N. R. (2011). Observations of reconnecting flare loops with the Atmospheric Imaging Assembly. *Astrophys. J.* **742**, 92.

Warren, H. P., Ugarte-Urra, I., Brooks, D. H., Cirtain, J. W., Williams, D. R., and Hara, H. (2007). Observations of transient active-region heating with Hinode. *Publ. Astron. Soc. Japan* **59**, 675–681.

Warren, H. P., Ugarte-Urra, I., Doschek, G. A., Brooks, D. H., and Williams, D. R. (2008). Observations of active region loops with the EUV Imaging Spectrometer on Hinode. *Astrophys. J. Letts.* **686**, L131–L134.

Warren, H. P., Winebarger, A. R., and Brooks, D. H. (2010). Evidence for steady heating: observations of an active region core with Hinode and TRACE. *Astrophys. J.* **711**, 228–238.

Warren, H. P., Winebarger, A. R., and Mariska, J. T. (2003). Evolving active region loops observed with the Transition Region and Coronal Explorer. II. Time-dependent hydrodynamic simulations. *Astrophys. J.* **593**, 1174–1186.

Watanabe, H., Kitai, R., and Ichimoto, K. (2009). Characteristic dependence of umbral dots on their magnetic structure. *Astrophys. J.* **702**, 1048–1057.

Webb, D. F., Burkepile, J., Forbes, T. G., and Riley, P. (2003). Observational evidence of new current sheets trailing coronal mass ejections. *J. Geophys. Res.* **108**, 1440.

Webb, D. F. and Howard, T. A. (2012). Coronal mass ejections: observations. *Liv. Rev. in Solar Phys.* **9**, 3.

Webb, D. F., Kahler, S. W., McIntosh, P. S., and Klimchuck, J. A. (1997). Large-scale structures and multiple neutral lines associated with coronal mass ejections. *J. Geophys. Res.* **102**, 24161–24174.

Weber, E. and Davis, L. (1967). The angular momentum of the solar wind. *Astrophys. J.* **148**, 217–227.

Wedemeyer, S., Freytag, B., Steffen, M., Ludwig, H.-G., and Holweger, H. (2004). Numerical simulation of the three-dimensional structure and dynamics of the non-magnetic solar chromosphere. *Astron. Astrophys.* **414**, 1121–1137.

Wedemeyer-Böhm, S. and Rouppe van der Voort, L. (2009). Small-scale swirl events in the quiet-Sun chromosphere. *Astron. Astrophys.* **507**, L9–L12.

Wedemeyer-Böhm, S., Scullion, E., Rouppe van der Voort, L., Bosnjak, A., and Antolin, P. (2013). Are giant tornadoes the legs of solar prominences? *Astrophys. J.* **774**, 123.

Wedemeyer-Böhm, S., Scullion, E., Steiner, O., Rouppe van der Voort, L., de la Cruz Rodriguez, J., Fedun, V., and Erdélyi, R. (2012). Magnetic tornadoes as energy channels into the solar corona. *Nature* **486**, 505–508.

Weiss, N. O. (1966). The expulsion of magnetic flux by eddies. *Proc. Roy. Soc. A* **293**, 310–328.

Weiss, N. O. (1981). Convection in an imposed magnetic field. Part 2. The dynamical regime. *J. Fluid Mech.* **108**, 273–289.

Weiss, N. O. (1991). Magnetoconvection. *Geophys. Astrophys. Fluid Dyn.* **62**, 229–248.

Weiss, N. O. (2003). Modelling solar and stellar magnetoconvection. In *Stellar Astrophysical Fluid Dynamics*, M.J. Thompson and J. Christensen-Dalsgaard, eds. (Cambridge University Press, Cambridge, UK), 329–343.

Weiss, N. O. (2005). Linear and nonlinear dynamos. *Astron. Nach.* **326**, 157–165.

Weiss, N. O. (2011). Chaotic behaviour in low-order models of planetary and stellar dynamos. *Geophys. Astrophys. Fluid Dyn.* **105**, 256–272.

Weiss, N. O., Brownjohn, D. P., Hurlburt, N. E., and Proctor, M. R. E. (1990). Oscillatory convection in sunspot umbrae. *Mon. Not. Roy. Astron. Soc.* **245**, 434–452.

Weiss, N. O., Brownjohn, D. P., Matthews, P. C., and Proctor, M. R. E. (1996). Photospheric convection in strong magnetic fields. *Mon. Not. Roy. Astron. Soc.* **283**, 1153–1164.

Weiss, N. O., Cattaneo, F., and Jones, C. A. (1984). Periodic and aperiodic dynamo waves. *Geophys. Astrophys. Fluid Dyn.* **30**, 305–341.

Weiss, N. O., Proctor, M. R. E., and Brownjohn, D. P. (2002). Magnetic flux separation in photospheric convection. *Mon. Not. Roy. Astron. Soc.* **337**, 293–304.

Weiss, N. O., Thomas, J. H., Tobias, S. M., and Brummel, N. H. (2004). The origin of penumbral structure in sunspots: downward pumping of magnetic flux. *Astrophys. J.* **600**, 1073–1090.

Weiss, N. O. and Thompson, M. J. (2009). The solar dynamo. *Space Sci. Revs.* **144**, 53–66.

Wentzel, D. G. and Seiden, P. E. (1992). Solar active regions as a percolation phenomenon. *Astrophys. J.* **390**, 280–289.

Wesson, J. (1997). *Tokamaks.* (Clarendon Press, Oxford).

Westendorp Plaza, C., del Toro Iniesta, J. C., Ruiz Cobo, B., Martínez Pillet, V., Lites, B. W., and Skumanich, A. (1997). Evidence for a downward mass flux in the penumbral region of a sunspot. *Nature* **389**, 47–49.

Whang, Y. C. and Chang, C. C. (1965). An inviscid model of the solar wind. *J. Geophys. Res.* **70**, 4175–4180.

Wheatland, M. S. (1999). A better linear force-free field. *Astrophys. J.* **518**, 948–953.

Wheatland, M. S. (2004). Parallel construction of nonlinear force-free fields. *Solar Phys.* **222**, 247–264.

Wheatland, M. S., Sturrock, P. A., and Roumeliotis, G. (2000). An optimization approach to reconstructing force-free fields. *Astrophys. J.* **540**, 1150–1155.

White, R. B., Monticello, D. A., Rosenbluth, M. N., and Waddell, B. V. (1977). Saturation of the tearing mode. *Phys. Fluids* **20**, 800–805.

White, R. S. and Verwichte, E. (2012). Transverse coronal loop oscillations seen in unprecedented detail by AIA/SDO. *Astron. Astrophys.* **537**, 1–10.

Wiegelmann, T. (2004). Optimization code with weighting function for the reconstruction of coronal magnetic fields. *Solar Phys.* **219**, 87–108.

Wiegelmann, T. (2008). Nonlinear force-free modeling of the solar coronal magnetic field. *J. Geophys. Res.* **113**, 3.

Wiegelmann, T. and Inhester, B. (2010). How to deal with measurement errors and lacking data in nonlinear force-free coronal magnetic field modelling? *Astron. Astrophys.* **516**, A107.

Wiegelmann, T., Inhester, B., and Sakurai, T. (2006). Preprocessing of vector magnetograph data for a nonlinear force-free magnetic field reconstruction. *Solar Phys.* **233**, 215–232.

Wiegelmann, T. and Neukirch, T. (2002). Including stereoscopic information in the reconstruction of coronal magnetic fields. *Solar Phys.* **208**, 233–251.

Wiegelmann, T. and Neukirch, T. (2003). Computing nonlinear force-free coronal magnetic fields. *Nonlin. Proc. in Geophys.* **10**, 313–322.

Wiegelmann, T., Solanki, S. K., Borrero, J. M., Martínez Pillet, V., del Toro Iniesta, J. C., Domingo, V., Bonet, J. A., Barthol, P., Gandorfer, A., Knölker, M., Schmidt, W., and Title, A. M. (2010). Magnetic loops in the quiet Sun. *Astrophys. J. Letts.* **723**, L185–L189.

Wiegelmann, T., Thalmann, J. K., Schrijver, C. J., De Rosa, M. L., and Metcalf, T. R. (2008). Can we improve the preprocessing of photospheric vector magnetograms by the inclusion of chromospheric observations? *Solar Phys.* **247**, 249–267.

Wilhelm, K. (2000). Solar spicules and macrospicules observed by SUMER. *Astron. Astrophys.* **360**, 351–362.

Wilhelm, K. (2006). Solar coronal-hole plasma densities and temperatures. *Astron. Astrophys.* **455**, 697–708.

Wilhelm, K., Dammasch, I. E., Marsch, E., and Hassler, D. M. (2000). On the source regions of the fast solar wind in polar coronal holes. *Astron. Astrophys.* **353**, 749–756.

Wilhelm, K., Marsch, E., Dwivedi, B. N., Hassler, D. M., Lemaire, P., Gabriel, A. H., and Huber, M. C. E. (1998). The solar corona above polar coronal holes as seen by SUMER on SOHO. *Astrophys. J.* **500**, 1023–1038.

Williams, D. R., Mathioudakis, M., Gallagher, P. T., Phillips, K. J. H., McAteer, R., Keenan, F. P., Rudawy, P., and Katsiyannis, A. (2002). An observational study of a magneto-acoustic wave in the solar corona. *Mon. Not. Roy. Astron. Soc.* **336**, 747–752.

Wilmot-Smith, A. L., Hornig, G., and Pontin, D. I. (2009a). Magnetic braiding and parallel electric fields. *Astrophys. J.* **696**, 1339–1347.

Wilmot-Smith, A. L., Hornig, G., and Pontin, D. I. (2009b). Magnetic braiding and quasi-separatrix layers. *Astrophys. J.* **704**, 1288–1295.

Wilmot-Smith, A. L., Martens, P. C. H., Nandy, D., Priest, E. R., and Tobias, S. M. (2005). Low-order stellar dynamo models. *Mon. Not. Roy. Astron. Soc.* **363**, 1167–1172.

Wilmot-Smith, A. L., Nandy, D., Hornig, G., and Martens, P. C. H. (2006). A time-delay model for solar and stellar dynamos. *Astrophys. J.* **652**, 696–708.

Wilmot-Smith, A. L., Pontin, D. I., and Hornig, G. (2010). Dynamics of braided coronal loops. I. Onset of magnetic reconnection. *Astron. Astrophys.* **516**, A5.

Wilmot-Smith, A. L., Pontin, D. I., Yeates, A. R., and Hornig, G. (2011). Heating of braided coronal loops. *Astron. Astrophys.* **536**, A67.

Wilmot-Smith, A. L. and Priest, E. R. (2007). Flux tube disconnection – an example of 3D reconnection. *Phys. Plasmas* **14**, 102903 (1–9).

Wilmot-Smith, A. L., Priest, E. R., and Hornig, G. (2005). Magnetic diffusion and the motion of magnetic field lines. *Geophys. Astrophys. Fluid Dyn.* **99**, 177–197.

Wilson, O. C. (1978). Chromospheric variations in main-sequence stars. *Astrophys. J.* **226**, 379–396.

Winebarger, A. R., Warren, H. P., and Falconer, D. A. (2008). Modeling X-ray loops and EUV "moss" in an active region core. *Astrophys. J.* **676**, 672–679.

Winebarger, A. R., Warren, H. P., and Mariska, J. T. (2003). TRACE and SXT active region loop observations: comparisons with static solutions of the hydrodynamic equations. *Astrophys. J.* **587**, 439–449.

Withbroe, G. and Noyes, R. W. (1977). Mass and energy flow in the solar chromosphere and corona. *Ann. Rev. Astron. Astrophys.* **15**, 363–387.

Woltjer, L. (1958). A theorem on force-free fields. *Proc. Nat. Acad. Sci. USA* **44**, 489–491.

Wragg, M. A. and Priest, E. R. (1981). The temperature-density structure of coronal loops in hydrostatic equilibrium. *Solar Phys.* **70**, 293–313.

Wragg, M. A. and Priest, E. R. (1982). Thermally isolated coronal loops in hydrostatic equilibrium. *Solar Phys.* **80**, 309–312.

Yan, M., Lee, L. C., and Priest, E. R. (1992). Fast magnetic reconnection with small shock angles. *J. Geophys. Res.* **97**, 8277–8293.

Yan, M., Lee, L. C., and Priest, E. R. (1993). Magnetic reconnection with large separatrix angles. *J. Geophys. Res.* **98**, 7593–7602.

Yan, Y. (1995). The 3D boundary element formulation of linear force-free magnetic fields with finite energy content in semi-infinite space. *Solar Phys.* **159**, 97–113.

Yan, Y. (2003). On the application of the boundary element method in coronal magnetic field reconstruction. *Space Sci. Rev.* **107**, 119–138.

Yan, Y. and Sakurai, T. (2000). New boundary integral equation representation for finite energy force-free magnetic fields in open space above the Sun. *Solar Phys.* **195**, 89–109.

Yashiro, S., Akiyama, S., Gopalswamy, N., and Howard, R. A. (2006). Different power-law indices in the frequency distributions of flares with and without coronal mass ejections. *Astrophys. J. Letts.* **650**, L143–L146.

Yashiro, S., Gopalswamy, N., Michalek, G., St. Cyr, O. C., Plunkett, S. P., Rich, N. B., and Howard, R. A. (2004). A catalog of white-light coronal mass ejections observed by the SOHO spacecraft. *J. Geophys. Res.* **109**, A07105.

Yeates, A. R., Attrill, G. D. R., Nandy, D., Mackay, D. H., Martens, P. C. H., and van Ballegooijen, A. A. (2010). Comparison of a global magnetic evolution model with observations of coronal mass ejections. *Astrophys. J.* **709**, 1238–1248.

Yeates, A. R. and Hornig, G. (2011). Dynamical constraints from field line topology in magnetic flux tubes. *J. of Phys. A 44,* 26 (July), 265501.

Yeates, A. R. and Hornig, G. (2013). Unique topological characterization of braided magnetic fields. *Phys. Plasmas 20,* 1, 012102.

Yeates, A. R., Hornig, G., and Wilmot-Smith, A. L. (2010). Topological constraints on magnetic relaxation. *Phys. Rev. Lett. 105,* 8 (Aug.), 085002.

Yeates, A. R. and Mackay, D. H. (2009). Initiation of coronal mass ejections in a global evolution model. *Astrophys. J.* **699**, 1024–1037.

Yeates, A. R., Mackay, D. H., and van Ballegooijen, A. A. (2008). Modelling the global solar corona. II. Coronal evolution and filament chirality comparison. *Solar Phys.* **247**, 103–121.

Yeates, A. R., Mackay, D. H., van Ballegooijen, A. A., and Constable, J. A. (2010). A nonpotential model for the Sun's open magnetic flux. *J. Geophys. Res.* **115**, 9112.

Yelles Chaouche, L., Moreno-Insertis, F., Martínez Pillet, V., Wiegelmann, T., Bonet, J. A., Knölker, M., Bellot Rubio, L. R., del Toro Iniesta, J. C., Barthol, P., Gandorfer, A., Schmidt, W., and Solanki, S. K. (2011). Mesogranulation and the solar surface magnetic field distribution. *Astrophys. J. Letts.* **727**, L30.

Yokoyama, T., Akita, K., Morimoto, T., Inoue, K., and Newmark, J. (2001). Clear evidence of reconnection inflow of a solar flare. *Astrophys. J. Letts.* **546**, L69–L72.

Yokoyama, T. and Shibata, K. (1995). Magnetic reconnection as the origin of X-ray jets and Hα surges on the Sun. *Nature* **375**, 42–44.

Yokoyama, T. and Shibata, K. (1996). Numerical simulation of solar coronal X-ray jets based on the magnetic reconnection model. *Pub. Astron. Soc Japan* **48**, 353–376.

Yokoyama, T. and Shibata, K. (1997). Magnetic reconnection coupled with heat conduction. *Astrophys. J. Letts.* **474**, L61–L64.

Yoshida, T. and Tsuneta, S. (1996). Temperature structure of solar active regions. *Astrophys. J.* **459**, 342–346.

Zacharias, P., Peter, H., and Bingert, S. (2011). Ejection of cool plasma into the hot corona. *Astron. Astrophys.* **532**, A112.

Zagrodziński, J. (1976). The solutions of the two-dimensional sine-Gordon equation. *Phys. Lett. A* **57**, 213–214.

Zaitsev, V. V. and Stepanov, A. V. (1975). On the origin of pulsations of type IV solar radio emission. Plasma cylinder oscillations (I). *Issledovaniia Geomagnetizmu Aeronomii i Fizike Solntsa* **37**, 3–10.

Zaitsev, V. V. and Stepanov, A. V. (2002). On the origin of the hard X-ray pulsations during solar flares. *Sov. Astron. Lett.* **8**, 132–134.

Zaqarashvili, T. V., Khodachenko, M. L., and Rucker, H. O. (2011). Magnetohydrodynamic waves in solar partially ionized plasmas: two-fluid approach. *Astron. Astrophys.* **529**, A82.

Zel'dovich, Y. B. (1957). The magnetic field in the two-dimensional motion of a conducting turbulent fluid. *Sov. Phys. JETP* **4**, 460–462.

Zhang, J., Solanki, S. K., and Wang, J. (2003). On the nature of moving magnetic feature pairs around sunspots. *Astron. Astrophys.* **399**, 755–761.

Zhang, Y. Z., Shibata, K., Wang, J. X., Mao, X. J., Matsumoto, T., Liu, Y., and Su, J. T. (2012). Revision of solar spicule classification. *Astrophys. J.* **750**, 16.

Zhao, M., Wang, J.-X., Jin, C.-L., and Zhou, G.-P. (2009). Magnetic non-potentiality on the quiet Sun and the filigree. *Res. in Astron. and Astrophys.* **9**, 933–944.

Zhao, X. and Hoeksema, J. T. (1994). A coronal magnetic field model with horizontal volume and sheet currents. *Solar Phys.* **151**, 91–105.

Zhao, X. and Hoeksema, J. T. (1995). Prediction of the interplanetary magnetic field strength. *J. Geophys. Res.* **100**, 19–33.

Zhao, X. P. and Webb, D. F. (2003). Source regions and storm effectiveness of frontside full halo coronal mass ejections. *J. Geophys. Res.* **108**, 1234.

Zhugzhda, Y. D. and Nakariakov, V. M. (1999). Linear and nonlinear magnetohydrodynamic waves in twisted magnetic flux tubes. *Phys. Lett. A* **252**, 222–232.

Zic, T., Vrsnak, B., and Skender, M. (2007). The magnetic flux and self-inductivity of a thick toroidal current. *J. Plasma Phys.* **73**, 741–756.

Zirin, H. and Stein, A. (1972). Observations of running penumbral waves. *Astrophys. J.* **178**, L85–L87.

Zirker, J. B. (1976). *Coronal Holes and High-Speed Wind Streams.* (Colorado Associated University Press, Boulder, Colorado).

Zirker, J. B. (2002). *Journey from the Center of the Sun.* (Princeton University Press, Princeton, NJ).

Zirker, J. B., Engvold, O., and Martin, S. F. (1998). Counter-streaming gas flows in solar prominences as evidence for vertical magnetic fields. *Nature* **396**, 440–441.

Zweibel, E. G. and Hundhausen, A. J. (1982). Magnetostatic atmospheres – a family of isothermal solutions. *Solar Phys.* **76**, 261–299.

Zweibel, E. G. and Li, H.-S. (1987). The formation of current sheets in the solar atmosphere. *Astrophys. J.* **312**, 423–430.

Zweibel, E. G. and Yamada, M. (2009). Magnetic reconnection in astrophysical and laboratory plasmas. *Ann. Rev. Astron. Astrophys.* **47**, 291–332.

Index

α-effect, 283–304
ω-effect, 283–289

acceleration
  CME, 64
  gravitational, 11, 81, 99
accretion, 455–457
acoustic gravity wave, 144, 157
acoustic travel-time, 6–20
acoustic wave, *see* sound wave
active region, 8, 29–54, 104
  bipolar, 45
  complex, 39, 45, 49, 66, 69
  coronal loops, 37, 46
  development, 45
  filament, 399, 406, 411
  flux rope, 431–433
  heating (numerical experiments), 355–356
  model, 130–132
  motions, 46
  prominence, 57–69, 391–412
  remnant, 25–30, 45–46, 62, 392–404
  structure, 45–46
  transient brightenings, 46
active-region transient brightening, 336, 369
activity nest, 46
adiabatic equilibrium, 153
adiabatic perturbation, 83, 84, 145
adiabatic temperature gradient, 13
age of Sun, 4
AIA (Atmospheric Imaging Assembly), 9, 23, 63, 175, 340–342, 369
Alfvén continuum, 163–170
Alfvén Mach number, 73, 90
Alfvén radius, 7, 44, 461
Alfvén speed, 41, 73, 173, 487
Alfvén theorem, 92, 98, 237, 240
Alfvén turbulence, 389
Alfvén wave, 7, 41, 44, 144, 148–176, 357–358, 360
  finite-amplitude (nonlinear), 151
  frequency, 154, 160, 167–170
  kinetic, 336
  phase mixing, 361
  resonant absorption, 359
  surface, 164
  torsional, 382
  travel-time, 6
Alfvénic wave, 35, 44, 165–174, 357–364
Alfvén Mach number, 215
almost-uniform reconnection, 190, 212–217
Aly-Sturrock hypothesis, 419, 435
ambipolar (Cowling) conductivity, 77–79
ambipolar diffusion, 78–79, 321, 395, 407, 411, 415
Ampère's law, 78
anelastic approximation, 89
angular momentum, 460–463, 470, 471
  of Sun, 5
  rate of loss from Sun, 464, 470
anomalous conductivity, 80
anti-dynamo theorem, 3, 287–289
arch-filament system, 45
arcsecond, 5, 488
astronomical unit, 4
Aulanier-Démoulin prominence model, 402–404
avalanche, 389–390

bald patch, 58, 59, 202, 226–231, 391, 402–405, 444
  separatrix, 226–231
ballooning (or flute) mode, 166
barb (prominence), 23, 59, 60, 391, 402–415
barb magnetic field, 405
barb tornado, 23, 59
Beltrami field, 107
Bennett's relation, 113
Bernoulli equation, 458, 463
Bernstein's energy principle, 264–266, 270, 324
Bessel function, 490
beta (plasma), 90, 336, 395
  prominence, 395
beta function, 347, 490
bifurcation, 229, 302, 303, 309, 310, 312
  double-separator, 228
  global, 229, 230, 436
  global separator, 436
  global spine-fan, 436
  local, 229, 230
  pitchfork, 418
  subcritical, 418
  transcritical, 418
Big Bear Solar Observatory, 401
bipolar active region (or sunspot group), 45, 48, 52, 69
BiSON (Birmingham Solar Oscillations Network), 17–19
blinker, 34, 336, 368
blowout jet, 38
body wave, 160–166
Boltzmann constant, 81, 487
boundary-element method, 134
Boussinesq approximation, 89, 301, 308, 325
braid (pigtail or Borromean), 372

braiding, 203
  Parker's hypothesis, 367–375
breakout, 435–437
bright points
  photospheric, 23–27
  X-ray, 3, 26–45
Brunt-Väisälä frequency, 152–157, 316, 319
buckling instability, 274
buoyancy force, 152–158, 307
butterfly diagram, 54–56, 282, 285, 301
  magnetic, 55, 56

cancelling magnetic flux (CMF), 27–53, 62, 72, 337, 369, 392–415, 438, 439
canopy over supergranule, 339
Carrington, 2, 47, 66
  flare, 66
  period, 14
cascade (turbulent), 355, 362, 371, 387, 388
catastrophe, *see* nonequilibrium
cavity (coronal), 61, 64
centrifugal force, 145
Chandrasekhar number, 90, 308
charge neutrality, 76, 82
chirality, 59–62, 399–402, 404
Chiu-Hilton model, 142
Christopherson cell, 278
chromosphere, 7, 31–34
  energy losses, 7, 8
  evaporation, 32, 46, 68, 342–368, 390
  heating, 334–390
    energy required, 334
    flux-tube tectonics, 334, 337, 353, 367, 374–386
    phase mixing, 361–363
    resonant absorption, 358–361
  loop, 337–356
    wave heating, 363–364
  magnetic field, 337–339
  network, 23, 31, 44
  recycling time, 337
Clebsch potential, *see* Euler potential
closed separator, 231, 474
closed separatrix, 227–231, 472
CME (coronal mass ejection), 4, 41–73, 475
  acceleration, 64
  halo, 64, 65
  magnetic field, 65
CNO (Carbon-Nitrogen-Oxygen) cycle, 10
coalescence instability, 220, 224
coefficient of radiative conductivity, 84
coefficient of thermal expansion, 307
coefficient of viscosity, 81
collective mode, *see* quasi-mode
collision time, 80
  anomalous, 80
  electron, 83
  electron-ion, 77
  electron-neutral, 77
  ion-ion, 83
  turbulent, 80
collisionless plasma, 74, 336
collisionless reconnection, 190, 336, 339, 390
collisionless shock, 181
collisionless solar wind, 475–480
CoMP (Coronal Multi-Channel Polarimeter), 44, 176, 364
compact solar flare, 66
complex of activity, 46
complex variables, 196–202, 219
compressional Alfvén wave, 144–151, 159–161
condensation, 392–415
  in a loop (draining), 342, 347, 349, 350, 363, 390
conduction [heat (or thermal)], 83–87, 334–353, 417, 441, 450, 453, 458, 459, 475
  front, 439, 441
conduction time, 351
conductivity
  anomalous, 80
  coefficient of radiative, 84
  Cowling (ambipolar), 77–79
  direct (Pedersen), 78
  electrical, 76–80
  Hall, 78
  turbulent, 80
conformal mapping, 200
connectivity map, 229
conservation
  electromagnetic flux, 191, 192, 237–240
  energy, 82–86
  magnetic field line, 93–95, 191, 192, 237–240, 246
  magnetic flux, 93–95, 101, 191, 199, 209–211, 233–246, 421, 428, 430
  mass, 80
  momentum, 80–81
  torque, 116
conservation (or jump) relations
  shock wave, 178–186
constant-α force-free field, 121–143, 396–408
constant-psi approximation, 222, 224
contact (or entropy) discontinuity, 187
continuous spectrum, 159, 161, 167–171, 360
continuum
  Alfvén, 163–170
  MHD, 167–172
  slow (or cusp), 163, 168
convection, 104
  overturning, 307–333
  penumbral, 333
  Taylor, 304
  umbral, 325
convection zone, 5–6, 12–15, 306–318, 320–321
convective collapse, 26, 53, 312–316
convective instability, 11–13, 153, 278–279, 309, 324
converging flux model for X-ray bright point, 369–370
coordinates
  curvilinear, 490
  cylindrical polar, 489
  spherical polar, 489
Coriolis force, 15, 23, 54, 144–155, 285–301, 320, 399
corona, 7–8, 34–44
  3D global model, 474–475
  cavity, 61, 64
  Chapman global hydrostatic model, 452–453
  dimming, 39, 64
  energy losses, 7, 8, 344
  heating, 176, 334–390
    avalanche, 389–390
    energy required, 334
    flux-tube tectonics, 337, 353, 367, 374–386
    functional form, 344
    MHD turbulence, 367, 386–390
    MHD waves, 356–364
    nanoflare, 341–390

phase mixing, 354, 361–363
reconnection, 336, 364–386
resonant absorption, 358–361
separator, 379–380
intensity, 34, 56
magnetic field, 337–339
roughly potential?, 119
magnetohydrostatic model, 141
minimum-current (MCC), 380
potential-field model, 120
recycling time, 337
coronagraph, 3, 36, 38, 57, 64
coronal loop
active-region, 37
coronal arcade, 110, 124, 125, 130, 137–139, 143
coronal hole, 38–40
1D model, 464–465
3D topology, 471–474
formation, 39
intensity, 39
polar, 28–41
transient, 39, 64
coronal jet, *see* X-ray jet
coronal loop, 37–38, 337–356
active-region, 37, 46, 341
analytical model, 346–347
condensation (or draining), 342, 347, 349, 350, 363, 390
cross-section, 341, 344, 345
cusp-shaped, 40, 72, 73
dynamic model, 349
EBTEL model, 342, 350–352, 450
effects of magnetic field, 339
evolution, 350
flare, 66–68, 70–73, 417, 439–450
shrinkage, 442
form of heating, 352–355
hot, 340–343
instability, 267
interconnecting, 37
model, 132–142, 339, 343–344
static, 344–348
multi-thermal, 340–342
numerical experiments, 355–356
numerical model, 347–348
oscillation, 71, 173–175
overdense, 342, 343, 348
post-flare, 68
properties, 341
quiet-region, 37
scaling law (RTV), 345–353
siphon flow, 341, 349–350
space observations, 340–343
strand, 341–372
thermally isolated, 345–348
time-dependent flows, 350–352
underdense, 342
warm, 340–342
wave heating, 363–364
coronal mass ejection (CME), 4, 41, 45, 63–66, 68–70, 72, 73, 416–450, 475
interplanetary, 65
coronal rain, 32, 46, 348–351, 409, 415
coronal ray (or polar plume), 39–44, 56
coronal seismology, 172–176, 357
coronal streamer, 36–44, 56, 63
bipolar, 36
unipolar, 37
coronal tectonics model, *see* tectonics
coronium, 2
corotating interaction region, 475
correlation time, 292, 304
Couette flow, 274
Coulomb logarithm, 79
Cowling (ambipolar) conductivity, 77–79
Cowling's anti-dynamo theorem, 99, 287–289
cross-helicity, 386, 387, 475
current
displacement, 75
Hall, 78
current sheet, 35–73
annular, 199
behaviour, 104–106
below erupting flux rope, 420–450
breakout, 435, 436
current, 377–378
curved, 199–202, 322
dimensions, 377–378
filamentation, 322
formation, 106, 196–203, 366–367
Parker braiding, 370–374
tectonics, 366–367
heliospheric, 41–44, 470–472
neutral, 105
prominence, 393–399
properties, 106
separator, 377–380
separatrix, 379
sudden dissipation, 383–386
tectonics, 366–367, 383–386
current-free (potential) magnetic field, 117–120
general solutions, 118
curvilinear coordinates, 490
cusp (or slow or tube) speed, 155
cusp (or slow) frequency, 160
cusp-point, 200, 203, 472
cusp-shaped loop, 40, 72, 73, 440, 442
cut-off frequency, 157, 166
cycle (sunspot), 2, 30, 37–69
cyclotron (or ion-gyration) frequency, 156
cylindrical polar coordinates, 489
cylindrical tearing-mode (or resistive kink) instability, 223

Dalton minimum, 53
Dawson integral function, 206, 491
de Laval nozzle, 454
Debye length, 76, 87, 487
delta prime, 222–224
delta sunspot, 48
density
corona, 34, 36, 37, 46, 67
coronal hole, 39
mean of Sun, 4–5
prominence, 57, 59
solar wind, 41–44
dextral and sinistral prominence model, 399, 400
dextral prominence or filament, 59–61, 393–404
diagnostic diagram, 157, 158, 172
differential rotation, 14–19, 47, 56, 283–305, 399, 400, 415
diffusion
ambipolar, 78–79, 321, 407, 411, 415
current sheet, 204–205
equation, 91, 204

diffusion (*cont.*)
  global solar magnetic field, 282
  hyper-, 388
  magnetic, 134–135, 192–208, 240–241
  viscous, 90
diffusion region
  1D, 207, 208
  2D, 189, 190, 193, 208–220, 224, 242, 244
  3D, 242–252
  ion, 190
diffusivity
  Cowling (ambipolar), 77–79
  eddy, 80, 91, 282–294, 330
  Hall, 77–79
  magnetic, 77–80, 91
  non-uniform magnetic, 220
  ratio, 90, 308
  thermal, 84
  turbulent, 190, 215, 219, 285–304
dimensionless equations, 87
dimensionless parameters, 89–90
dimming (coronal), 39, 64
dip in prominence magnetic field, 391–415
direct (Pedersen) conductivity, 78
disconnection (flux tube), 245, 247
discontinuity
  contact (or entropy), 187
  rotational, 41, 188
  tangential, 41, 105, 187
discrete spectrum, 159, 168–170
disparition brusque, 63
dispersion relation, 145
dispersive wave, 147
displacement current, 75
dissipation
  Ohmic (or Joule), 83, 86, 98, 106
  viscous, 85, 86
double separator, 227, 228
  bifurcation, 228
double tearing-mode instability, 223
double-spiral null point, 251
driven reconnection, 215, 244
dynamic fibril, 33
dynamo, 281–305
  α-ω, 283–304
  α^2, 284, 286, 293
  Babcock-Leighton, 294
  fast, 283
  flux-transport, 30, 56, 284–287, 294–297
  global computation, 301–302, 305
  history, 284–287
  interface, 298–300
  kinematic, 283–296
  large-scale, 284, 286, 287
  low-order, 302–303
  nonlinear, 283–304
  overshoot, 286, 298–302
  Parker, 285, 289–291
  small-scale, 284, 287, 304
  stellar, 282, 303–304
  tachocline, 286, 298–301
  turbulent, 289–294
dynamo number, 290, 293, 303
dynamo wave, 285–299
EBTEL (Enthalpy-based Thermal Evolution of Loops) model, 342, 350–352, 450
eclipse (solar), 1–9, 36–42, 56–57, 335, 339
eddy diffusivity, 80, 91
effective temperature of Sun, 5
EIS (EUV Imaging Spectrometer), 44
EIT (Extreme ultraviolet Imaging Telescope), 43, 340–368
EIT wave, 70
Ekman number, 90
electric current
  laminated, 139–141
  uniform, 112
electric field (reconnection), 421, 441–443
electric field (total), 77, 92
electrical conductivity, 76–80, 487
electrical neutrality, 76, 82
electromagnetic energy equation, 211
electromagnetic flux conservation, 191, 192, 237–240
electron charge, 487
electron collision-time, 77, 79, 83
electron gyro-frequency, 77, 487
electron gyro-radius, 487
electron mass, 487
electron plasma frequency, 487
electron thermal speed, 487
electron volt, 487
Ellerman bomb, 46, 49
Elsässer variables, 88, 389, 466, 475
emerging (or evolving) flux model, 369, 437–438
emerging magnetic flux, 26–46, 71, 199, 234, 236, 286, 296, 297, 302, 304, 315, 320–322, 336, 337, 350, 367, 369, 380, 389, 416, 417, 423, 432, 436, 438, 439
energy
  conservation, 82–86
  electrical, 86
  equation, 82–86
    solar wind, 458–460
  internal, 82–86, 100
  loss
    chromospheric, 7, 8
    coronal, 7, 8
    function, 82–84
  spectrum function, 292
enthalpy, 82, 86
  flux, 342–344, 351
  time, 351
entropy, 82, 83, 85, 182
  condition, 178–181
  discontinuity, 187
  wave, 156
ephemeral region, 25–38, 45
equation
  diffusion, 91
  electromagnetic energy, 211
  energy, 82–86
  generalised Hain-Lüst, 165
  Grad-Shafranov, 129–130, 137, 143, 202
  heat, 82–83
  Helmholtz, 121, 126, 141, 143
  induction, 78–79, 91–95, 204
  kinematic, 247
  Laplace, 70, 71, 117, 131, 216
  Liouville, 130, 143
  mechanical energy, 86, 211
  motion, 80–81
  Poisson, 216

sine-Poisson, 130
state, 81
vorticity, 91
equations
dimensionless, 87
incompressible, 81, 88, 91, 261, 265, 270
kinematic, 89
Maxwell, 74–76
MHD, 106
equilibrium
adiabatic, 153
flux-constrained, 380, 445
hydrostatic, 6, 10, 12, 338, 344
magnetohydrostatic, 107–110, 135–143
magnetostatic, 135–137
thermal, 88, 343, 411
equipartition magnetic field, 26, 311, 313
ergodic field line (or flux surface), 121, 135
error function, 205, 491
erupting prominence, 23, 62–63, 66, 416–450
eruptive flare, 45, 63, 66, 68
escape velocity of Sun, 4
Euler potential, 108–143, 246
existence, 108
Euler-Lagrange equation, 267–269, 433
Eulerian coordinates, 259
evanescent disturbance, 157–172
evaporation (chromospheric), 32, 46, 68, 342–368, 390, 439–441, 450
Evershed flow, 2, 47, 51–52, 143, 327–333, 349–350
evolutionary condition for shock wave, 180
examples, xviii
expansion
fast-mode, 215–217
slow-mode, 216
wave, 178
exploding granules, 21, 22
explosive event, 34, 336, 341, 365, 367, 368
explosive instability, 257
external bipolar-region filaments (EBR), 58

F-corona, 34
f-mode, 16
f-spot, 45, 71
faculae, 23–26, 45–46, 56
polar, 24
facular points, 23
fan, 225
reconnection, 208, 247–250
separatrix, 230, 252
fast body mode, 163
fast kink mode, 163–175
fast magnetoacoustic wave, 151–173
fast sausage mode, 166, 173–175
fast solar wind, 36–44
fast surface wave, 166
fast-mode expansion, 215–217
fast-mode shock, 70–72, 179, 186–188, 440–441
fast-mode wave, *see* fast magnetoacoustic wave
feet of prominence, 402–406
Ferraro's isorotation law, 99
fibril, 33–59, 71, 317
dynamic, 33, 317
prominence, 407, 415
superpenumbral, 33, 46–52
field-aligned flow, 143
filament, 45, 57–63
active-region, 399, 406, 411
channel, 7, 59–66, 399–402
formation, 62, 399–402
sinistral, 61
dextral, 393–404
formation, 62
penumbral, 47, 51, 52, 327–333
sinistral, 393–404
spine, 59–60
filament channel, 391–409
filigree, 24, 26, 46
FIP (first ionisation potential), 39–41, 44
five-minute oscillation, 3, 15–20, 50, 144, 172, 328
flare, *see* solar flare
flare loop, *see* solar flare loop
flute (or ballooning) mode, 166
flute instability, 263, 264, 270–272
flux cancellation, 27–53, 62, 72, 337, 369, 438, 439
flux concentration, 310–314, 316
flux conservation, 93, 101, 421, 428, 430
flux emergence, 199, 234, 236
flux expulsion, 310–312
flux function, 103, 106, 196, 202
magnetic, 108
topological, 234, 374
flux pile-up reconnection, 216, 217, 224
flux pumping, 327–332
flux rope, 100, 103, 391–415
prominence, 393–399, 420
toroidal, 424–427
flux surface, 103–104
ergodic, 121, 136
rational, 121, 135, 136
flux tube, 100
behaviour, 100–104
convective collapse, 315–316
cross-sectional area, 84
disconnection, 245, 247
equilibrium, 111–117
force-free, 113
intense, 23, 25, 26, 29, 40, 44, 104, 313–317, 337–375
width, 314
isolated, 100
oscillation, 150, 151
pitch, 102
strength, 100, 101
thin (or slender), 312–317, 321, 325, 424–426
convective collapse, 315–316
dynamics, 315
equations, 315
equilibrium, 314–315
width, 314
toroidal, 424–426
twist, 102, 111–116
twisted, *see* flux rope
uniform-current, 112
uniform-twist, 114–116, 268, 269
variable-twist, 116
wave, 103, 164–167
flux velocity, 92, 95, 192, 237–243
dual, 192, 240, 243
failure in three dimensional reconnection, 241–242
theorem, 241
flux-constrained equilibrium, 380, 445
flux-rope prominence model, 396–398, 402–406

flux-transport dynamo, 30, 56
flux-tube tectonics, *see* tectonics
  current-sheet formation, 366–367
fly-by, 380–383
force
  buoyancy, 152–158
  Coriolis, 144–155, 285–301
  gravitational, 81
  Lorentz, 96–98
  magnetic buoyancy, 104
  magnetic pressure, 96
  magnetic tension, 96
  restoring, 144
  viscous, 80, 81, 90
force-free field, 107, 120–135, 200–202
  alpha-best method, 132
  boundary-element method, 134
  constant-α, 121–143
  constant-alpha, 396–408
  constant-current, 200
  diffusion, 134–135
  existence, 133
  Grad-Rubin method, 133
  laminated, 131
  linear, 121–143, 396–408
    general solutions, 126–129
    inadequacies, 126
  Low-Lou, 132
  Lundquist, 113, 124, 427
  minimum-energy theorem, 98, 121
  nonlinear, 113–135, 356, 400, 401, 404, 406, 474
  numerical methods, 132–134
  one-dimensional, 113
  optimization method, 133
  preprocessing, 134
  relaxation method, 133
  theorems, 121
  toroidal flux rope, 427–429
  uniform-twist, 114–116, 268, 269
  uniqueness, 126–130, 132, 133
  vertical integration method, 133
forced reconnection, 201–202
Fourier series, 119, 126, 127, 132, 141
free-fall speed, 107
frequency
  Alfvén, 154, 160, 167–170
  Brunt-Väisälä, 152–157, 316, 319
  cusp (or slow), 160
  cut-off, 157, 166
  cyclotron (or ion-gyration), 156
  gyration, 77
  inertial, 154
  ion gyro-, 83
  quasi-mode, 360
  slow (or cusp), 167
frozen flux
  theorem, 92
full ionisation, 77–84
function
  Bessel, 490
  beta, 347, 490
  Dawson integral, 206, 491
  energy spectrum, 292
  error, 205, 491
  flux, 196, 202
  gamma, 490
  Green's, 119–142, 168, 169, 171, 205, 292
  hypergeometric, 490
  Legendre, 491

g-modes, 20
gamma function, 490
gamma-ray flare, 70
gas constant, 487
Gaussian cgs units, 483–484
general magnetic reconnection, 190, 244, 246
giant cells, 14
Gleissberg cycle, 54
global invariant (for turbulence), 386
GONG (Global Oscillation Network Group), 18
Grad-Rubin method, 133
Grad-Shafranov equation, 129–130, 137, 143, 202, 425
granular buffeting, 316, 317
granular magnetic loops, 22, 26, 27
granulation, 2, 14, 21–24, 32
  abnormal, 312
  intensity, 21
  umbral, 49
gravitational acceleration, 4–5, 11, 81, 99
gravitational constant, 81, 487
gravitational potential, 99
gravitational resistive mode, 220
gravitational time, 351
gravity wave, *see* internal gravity wave
Green's function, 119–142, 168, 169, 171, 205, 292, 421, 429
group velocity, 147, 156
gyro-frequency, 77, 83, 487
gyro-radius, 487
gyrokinetic theory, 74

Hα
  jet, 31, 33, 46
  knot or kernel, 67, 68, 71
  ribbon, 63–71, 417, 439, 441–445, 447–449
    velocity, 441–443, 450
    width, 442
  surge, 46, 49
Hain-Lüst equation
  generalised, 165
hairy ball, 119, 120
Hale cycle, 54
Hale's law, 30, 54, 282
Hall conductivity and diffusivity, 77–79
Hall MHD, 78, 189, 190, 363
halo CME, 64, 65
Hanle effect, 26, 57
hard X-rays, 67, 439–450
Hartmann number, 90, 279
heat (Ohmic or Joule), 83
heat (or thermal) conduction, 83–87, 334–353, 417, 441, 450, 453, 458, 459, 475
heat equation, 82–83
heat flux, 83, 342, 343, 349, 350, 386, 452, 458–460, 476, 478, 479
heating
  Ohmic (or Joule), 86, 98, 106
hedgerow prominence, 57, 402, 411
helical kink instability, 102
helical symmetry, 130

helicity, 154
cross, 386, 387, 475
magnetic, 123–124, 233–236, 368–388, 392, 393, 400, 401, 406, 409
relative, 386
mutual, 233, 236, 247
self, 233–247
helioseismic wave, 20, 71
helioseismology, 15–20
local, 20
heliosphere, 7
current sheet, 41–44, 470–472
Helium, 2–3
helmet streamer, 36–43, 61
2D model, 470–471
Helmholtz equation, 121, 126, 141, 143
heteroclinic separatrix, 229
Heyvaerts-Priest relaxation, 234
Hi-C (High-resolution Coronal Imager), 37, 340, 341
high-FIP element, 39
Hinode, 4
EIS (EUV Imaging Spectrometer), 44
SOT (Solar Optical Telescope), 27–61
XRT (X-ray Telescope), 44, 340, 341, 368
history, 1–4, 34, 47
dynamo theory, 284–287
magnetic reconnection, 190–191
HMI (Helioseismic and Magnetic Imager), 9, 25
homoclinic separatrix, 229
homologous solar flare, 69
hoop force, 424–428
hydromagnetic inertial wave, 155
hydrostatic equilibrium, 6, 10, 12, 338, 344
hyperbolic flux tube, *see* quasi-separator
hyperdiffusion, 388
hypergeometric function, 490

ideal MHD, 92–94, 237
impulsive bursty reconnection, 65, 217–224
impulsive phase of flare, 66–70
Imshennik-Syrovatsky solution, 194–196
incompressible flow, 81, 88, 91, 149, 261, 265, 270
induction equation, 78–79, 91–95, 204
inertial wave, 144, 153–155
initial-value problem, 168–170
instability, 256–280
breakout, 435–437
buckling, 274
coalescence, 220, 224
convective, 11–13, 153, 278–279, 309, 324
cylindrical tearing-mode (or resistive kink), 223
double tearing-mode, 223
explosive, 257
flute, 263, 264, 270–272
gravitational resistive mode, 220
helical kink, 102, 267, 273, 274
interchange (or exchange), 223, 263, 264, 270–272, 275, 325
Kelvin-Helmholtz, 271, 275
kink, 223, 245, 266–280, 367, 388, 433–434
lateral kink, 266, 428
leak, 279, 307, 330
magnetic buoyancy, 287, 299–301, 318–320
magnetic Rayleigh-Taylor (or Kruskal-Schwarzschild), 263, 272, 319, 415
micro- (current-induced), 384
nonlinear, 256–257, 270, 280, 418
nonlinear tearing-mode, 223–224
normal-mode method, 256–264
radiative, 407–415
Rayleigh-Taylor, 260–264, 271–272, 322
resistive, 219–224, 275–278
ballooning mode, 223
gravitational mode, 276
interchange (or resistive-g), 223
kink, 223, 434–435
rippling mode, 220, 277
tearing mode, 277, 278
sausage, 263, 271, 273
structural (or topological), 203, 218, 229
tearing-mode, 190, 197, 220–223, 363, 367, 383
thermal, 258, 279–280
torus, 427–431
variational (or energy) method, 257, 264–270, 433
X-point, 98
intense flux tube, 23, 25, 26, 29, 40, 44, 104, 313–317, 337–375
equations, 315
width, 314
intensity
corona, 34, 56
corona-hole, 39
granule, 21
penumbra, 51
solar-flare, 66, 67
spectral line, 8
sunspot, 48–49
umbral-dot, 49
inter- (or inner) network, 31, 32
element, 27
magnetic field, 25, 27–28
interaction distance (for photospheric flux), 366, 370
interchange (or exchange) instability, 223, 263, 264, 270–272, 275, 325
interchange reconnection, 468, 474
interconnecting coronal loops, 37
interface boundary conditions, 262, 263
interface wave, 161–162
intergranular lane, 21
interior (solar), 5–6, 10–20
intermediate (or Alfvén) wave, 155, 179–188
intermediate prominence, 57
internal energy, 82–86, 100
internal gravity wave, 144, 152–159, 172
inverse-polarity prominence, 391–405
ion gyro- (or cyclotron) frequency, 77, 83, 156, 487
ion gyro-radius, 487
ion thermal speed, 487
ion-cyclotron resonance, 44
ion-ion collision-time, 83
ionisation degree (or fraction), 77
IRIS (International Research on the Interior of the Sun), 17, 18
irradiance (solar), 3, 56, 69

jet
blowout, 38
coronal, *see* X-ray jet
Hα, 31, 33, 46
penumbral, 51
reconnection, 32, 73

jet (*cont.*)
  separatrix, 219
  X-ray, 38, 67, 321, 322, 366, 410, 437
Joule (or Ohmic) dissipation, 83, 86, 98, 106
Joy's law, 54, 282
jump (or conservation) relations
  shock wave, 178–186

K-corona, 34
Kelvin-Helmholtz instability, 271, 275
Kelvin-Helmholtz time, 5
kernel (Hα), 67
kinematic model, 89, 310, 312
  reconnection, 247–249
kinetic theory, 74, 478–480
kink instability, 223, 245, 266–280, 367, 388, 433–434
  helical, 267, 273, 274
  lateral, 266, 273, 428
kink speed, 166, 173, 175
kink wave, 162–167, 357, 363
Kippenhahn-Schlüter model, 139, 392–396
knot (Hα), 67, 68, 71
Kolmogorov spectrum, 387, 466, 475
Kraichnan spectrum, 387, 467, 475
Kruskal-Shafranov criterion, 273, 434
Kuperus-Raadu model, 396

Lagrange multiplier, 267
Lagrangian coordinates, 259
Lambda effect, 15
Lambert function (W), 454
laminated electric current, 139–141
laminated magnetic field, 131, 138–139
Laplace equation, 117, 131, 216
Laplace transform, 168, 172
LASCO (Large Angle and Spectrometric Coronagraph), 43, 64
lateral kink instability, 428
law
  Ampère's, 78
  Ferraro's isorotation, 99
  generalised Ohm's, 76–78
  Hale's, 30, 54
  Joy's, 54
  Newton's, 84
  Ohm's, 76, 92, 94
  perfect gas, 10, 81
  prominence-chirality, 59
  Spörer's, 2, 54
  Stefan-Boltzmann, 11
  sunspot-polarity, 54
  waveguide, 17
leak instability, 279, 307, 330
leaky mode, 161–173
Legendre function, 16, 491
levitation (prominence plasma), 409, 411
Lie-derivative theorem, 240
light bridge in sunspot, 49, 53
Lighthill mechanism, 172
line current, 420, 432
  circular, 425–426, 429
line-tying (photospheric), 69, 194, 223–224, 249, 419–434
linear force-free field, 121–143, 396–408
  general solutions, 126–129
linear pinch, 113, 272, 273
linking number, 236
Liouville equation, 130, 143
Liouville theorem, 246
lithium problem, 301
little ice age, 54
local helioseismology, 20
loop
  coronal, *see* coronal loop
  sunspot, 49
loop-top source, 67–73, 417, 450
Lorentz force, 96–98
low-FIP element, 39–41, 44
Low-Lou force-free field, 132
luminosity of Sun, 5
Lundquist force-free field, 113, 124, 427

Mach number, 90
macrospicule, 33
  tornado, 23–33, 60
magnetic annihilation, 204–208
  reconnective, 207–208
magnetic buoyancy, 285–302, 315, 317–322
  force, 104
  instability, 287, 299–301, 318–320
magnetic carpet, 26, 28–29, 284, 337, 356, 367, 374, 377, 468
magnetic charge, 431
magnetic cloud, 63, 444
magnetic diffusion, 134–135, 192–208, 240–241
  current sheet, 204–205
  time-scale, 6
magnetic diffusivity, 77–80, 91, 487
  non-uniform, 220
  turbulent, 190, 215, 219
magnetic element, 22–29, 52
magnetic energy, 121–123
  stored, 65
magnetic evolution (different classes), 191, 192
magnetic field
  barb, 405
  Beltrami, 107
  CME, 65
  current-free (or potential or vacuum), 117–120, 311, 324–325
  equipartition, 26, 311, 313
  force-free, *see* force-free field
  Fourier series, 119, 126, 127, 132, 141
  inter- (or inner) network, 25, 27–28
  laminated, 131, 138–139
  magnetohydrostatic, 107–110, 135–143
  magnetostatic, 116, 135–137
  network, 4, 22–32, 39–40
  penumbral, 51
  photospheric, 9, 25–30, 55
  poloidal, 283–304
  potential (or current-free or vacuum), 117–120, 311, 324–325
  prominence, 57–59, 392, 405, 408, 415
  prominence barb, 405
  quadrupolar, 202, 203, 253, 417, 421, 435, 436, 444, 447
  sheared, 201
  solar, 281–282
  solar flare, 69
  solar wind, 41, 43
  sunspot, 48, 131
  sunspot dot, 49
  tangled, 408–409
  toroidal, 223–224, 274, 283–305

vacuum (or potential or current-free), 117, 120, 311, 324–325
magnetic field line, 100
conservation, 93–95, 191, 192, 237–240, 246
ergodic and stochastic, 121, 135
pitch, 111
singular, 232
magnetic flipping, 232, 243–248, 444, 449
magnetic flux
cancelling, 27, 53, 62, 72, 337, 369, 392–415, 438, 439
conservation, 93–95, 101, 191, 199, 209–211, 233–246, 421, 428, 430
emerging, 26–46, 71, 199, 234, 236, 286, 296, 297, 302, 304, 315, 320–322, 336, 337, 350, 367, 369, 380, 389, 416, 417, 423, 432, 436–439
function, 103, 106, 108, 196, 202
parasitic, 398–405
surface, 103–104
ergodic, 121, 136
existence, 121
rational, 121
velocity, *see* flux velocity
magnetic flux rope, *see* flux rope
magnetic flux tube, *see* flux tube
magnetic helicity, 123–124, 233–236, 368–388, 392, 393, 400, 401, 406, 409
coronal arcade, 236
flux tube, 236
in 3D reconnection, 247
inverse cascade, 284
relative, 235, 386
time-variation, 235–236
magnetic levitation, 409, 411
magnetic monopole, 75
magnetic nonequilibrium, 367
magnetic null point, *see* null point
magnetic permeability, 75
magnetic Prandtl number, 90
magnetic pressure force, 96
magnetic reconnection, *see* reconnection
magnetic relaxation, 201, 234, 373, 374, 381, 386–389
magnetic Reynolds number, 89, 91, 190, 487
turbulent, 292, 304
magnetic scalar potential, 108, 117
magnetic shearing, *see* sheared magnetic field
magnetic slippage, 192, 240–251
magnetic tension force, 96
magnetic topology, *see* topology
magnetic vector potential, 108
magneto-frictional method, 133
magneto-gravity wave, 159
magnetoacoustic gravity (or magneto-atmospheric) wave, 144, 157–159
magnetoacoustic wave, 144, 155–159
surface, 161
magnetoconvection, 306–333
magnetofrictional relaxation, 401, 406
magnetogram, 22–28, 50
magnetograph, 3
magnetohydrostatic equilibrium, 79, 89, 107–110, 135–143, 322–324
magnetostatic equilibrium, 116, 135–137
Malkus-Proctor effect, 286, 301
marginal stability, 309
Martens model (coronal loop), 346–347
Martin's law (prominence chirality), 59
mass conservation (or continuity), 80
mass of Sun, 4–5
mass-loss rate of Sun, 5, 454, 459, 464
Masuda flare, 67
Maunder minimum, 2, 53, 54, 281, 282, 302
Maxwell equations, 74–76
MDI (Michelson Doppler Imager), 18–22, 337, 338, 355, 356
mean atomic weight, 81
mean particle mass, 81
mean-field theory, 285, 286, 290–294, 304–305
mechanical energy equation, 86, 211
meridional flow, 14–20, 55, 56, 284–287, 294–305
mesogranulation, 14, 21–24
metastability, 256–257, 418
method of characteristics, 310
MHD
assumptions, 87–88
continuum, 167–172
equations, 106
linearised, 258–260
reduced form, 88–89
going beyond, 475–480
Hall, 78, 189, 190
ideal, 92–94, 237
nonideal, 238–239
reduced, 380, 389, 390
relaxation, 133
shock wave, 179–181
three-fluid, 76
turbulence
coronal heating, 367, 386–390
relaxation, 386–389
solar wind, 466–467, 475, 480
two-fluid, 74, 476–478
MHD wave, *see* wave
micro-bright point, 370
microflare, 336, 341, 364–370, 380, 389
microinstability (current-induced), 384
minimum-current corona (MCC) model, 380, 444, 445
minimum-energy theorem
force-free, 121
potential, 117
Woltjer, 123
mixing-length, 11–14
moat, 20, 52–53, 325–333
mode coupling, 160
monopole, 75
Moreton wave, 70, 144
moss, 341
mottle, 33
moving magnetic feature (MMF), 52, 53
MSDP (Multichannel Subtractive Double Pass spectrograph), 61
mutual helicity, 233, 236, 247

nanoflare, 32, 341–390
flux-tube tectonics, 341, 374–386
network, 3, 31–34
chromospheric, 23, 31, 44
element, 26, 29
magnetic field, 4, 22–32, 39–40
photospheric, 22–32, 40
neutral (or marginal) stability, 256–258, 279
neutral point, *see* null point
neutrino, 10
Newton's law of cooling, 84

non-uniform reconnection, 219
nonequilibrium
  2D, 420–424
  3D, 427–431, 438
  magnetic, 367
  radiative, 391, 409–415
  thermal, 347
nonideal MHD, 94–95, 238–239
nonlinear force-free field, 113–135, 400, 401, 404, 406, 474
nonlinear instability, 256–257, 270, 280
nonlinear stability, 256–257
null line, 229
null point
  cluster, 229
  creation, 227, 228
  cusp-type, 200, 203, 472
  degenerate, 229
  double-spiral, 251
  improper radial, 226
  linear, 225
  O-type, 193, 246, 250–253, 402, 403
  oblique, 226
  proper radial, 226
  second-order, 229
  spiral, 226
  T-type, 200, 472
  three-dimensional, 225–226
    collapse, 249
  three-dimensional reconnection, 247–252
  two-dimensional, 193–196
    collapse, 193–196, 202
  X-type, 190, 193–203, 215, 226–232, 242–244, 253–255, 402, 403
  Y-type, 197–203, 472
null sheet, 229
number
  Alfvén Mach, 90, 209, 215
  Chandrasekhar, 90, 308
  dynamo, 290, 293–303
  Ekman, 90
  Hartmann, 90, 279
  linking, 236
  Mach, 90
  magnetic Prandtl, 90
  magnetic Reynolds, 89, 91, 190
  Nusselt, 90
  Prandtl, 90, 278–279, 308
  Rayleigh, 90, 278–279, 307–308
  Reynolds, 89
  Richardson, 275
  Rossby, 90, 303
  Taylor, 90, 278, 301
  Wolf, 53
numerical experiment
  active region heating, 355–356
  braiding, 367, 371
  coronal heating, 388
  dynamo, 287, 301–302, 304, 305
  emerging flux, 295, 304, 321
  flare or CME eruption, 438–439
  flux expulsion, 310
  force-free field, 132–134
  magnetic buoyancy, 320
  magnetoconvection, 311
  reconnection, 190, 217–218, 224, 228, 230, 247, 249–250, 253
  separator reconnection, 380–382
  sunspot, 330–333
Nusselt number, 90

O-point, 193, 246, 250–253, 402, 403
Ohm's law, 76, 92, 94
  generalised, 76–78
Ohmic (or Joule) dissipation, 83, 86, 98, 106, 180
opacity (mass absorption coefficient), 84
open separator, 231, 468, 473–474
open separatrix, 227–231, 472
optical depth, 7
optically thick or thin plasma, 84
optimization method, 133
oscillation
  coronal loop, 71, 173–175
  five-minute, 3, 15–20, 50, 144, 172, 328
  flux-tube, 150, 151
  prominence, 62, 172, 393, 396, 415
  solar-flare, 175–176
  sunspot, 328–329
  three-minute, 50–52, 175, 328
  torsional, 19, 56, 301
  umbral, 50
overstability, 258, 260, 270, 279, 307–309, 313, 320, 324, 330

p-modes, 16, 17, 50, 328
  high-degree, 16
  low-degree, 16, 18, 19
p-spot, 45, 71
parasitic flux, 49, 398–405
Parker braiding, 370–374
  hypothesis, 367–375
Parker dynamo, 285, 289–291
partial ionisation, 14, 77, 79, 82, 411
Pedersen (direct) conductivity, 78
penumbra
  convection, 333
  filament, 47, 51, 52, 327, 333
  formation, 52
  grain, 51
  intensity, 51
  jet, 51
  magnetic field, 51
  model, 327–331
  spine, 51
percolation model, 389
perfect gas law, 10, 81
permeability, 487
permittivity, 75, 487
perpendicular shock, 180, 182–184
Petschek mechanism, 190, 211–218, 224
  instability, 190, 218
  non-steady, 214
phase mixing, 167, 168, 170, 354, 361–363
  in space, 361–362
  in time, 362–363
phase velocity, 147, 156
photon travel-time, 6
photosphere, 6–7, 21–30
  bright points, 23–27
  line-tying, 69
  magnetic field, 9, 25–30, 55
    interaction distance, 366, 370
  network, 22–32, 40
pigtail or Borromean braid, 372

PIL (polarity-inversion line), 14, 45, 57–72, 391, 392, 399, 400, 402–405
pinch
  linear, 113
  reversed-field, 124
pitch of flux tube, 102, 111
plage, 31–34, 44–46
plagette, 400, 404
plane-wave solution, 146, 150
plasma
  beta, 90, 336, 395, 487
    prominence, 395
  definition of, 76, 87
  fully ionised, 77–84
  partially ionised, 77, 79, 82
plasmoid, 417, 442, 444, 449
plume
  polar, 39–44, 56
  prominence, 61
  sunspot, 49
Poiseuille flow, 274
Poisson equation, 216, 425, 429
polar coronal hole, 28–41
polar crown, 62
  prominence, 57, 72, 392, 399, 402, 405
polar diagram, 149, 156
polar faculae, 24
polar jet, 39
polar plume (or coronal ray), 39–44, 56
polarity-inversion line (PIL), 14, 45, 57–72, 231, 391, 392, 399, 400, 402–405
  formation, 62
polytropic solar wind, 457–458
pore, 44–52, 322–325
post-flare coronal loop, 68
potential
  energy, 256–258
  Euler, 108–143
  gravitational, 99
  magnetic scalar, 108, 117
  magnetic vector, 108
  minimum-energy theorem, 117
potential magnetic field, 117–120
  coronal, 119
  general solutions, 118
  minimum energy theorem, 98
  source-surface (PFSS) model, 119, 472, 474
  uniqueness, 117
potential-field source-surface (PFSS) model, 119
Poynting flux, 86, 211, 335–336, 355–356, 365, 383–388
PP (proton-proton) chain, 10
Prandtl number, 90, 278–279, 308
pre-flare phase of flare, 64–72
preprocessing of boundary conditions, 134
pressure
  scale-height, 108–110, 137, 394, 404, 407, 408
  total, 110–112, 137–138
problems, xviii
prominence, 2–3, 45, 57–63, 232, 391–415
  activated, 62
  active-region, 57–69, 391–412
  barb, 23, 59, 60, 391, 402–415
  barb magnetic field, 405
  blob, 413
  bubble, 61
  chirality, 59–62, 399–402, 404, 475
  chirality law, 59
  classification, 58
  current sheet, 393–399
  development, 62
  dextral, 59–61, 393–404
  equilibrium, 393, 395, 406, 408
  erupting, 23, 62–63, 66, 416–450
  feet, 402–406
  fibril, 407, 415
  flux rope, 393–399, 420
  formation, 57–62, 391, 406, 409–415
  hedgerow, 57, 402, 411
  intermediate, 57
  inverse-polarity, 58, 391–405
  leg, 59
  magnetic dip, 391–415
  magnetic field, 57–59, 392, 405, 408, 415
  mass, 395–397, 413, 415
  model, 138
  normal-polarity, 391, 392, 395, 404, 405
  oscillations, 62, 172, 393, 396, 415
  plume, 61
  polar crown, 392, 399, 402
  polar-crown, 57, 62, 72, 405
  quiescent, 57–63, 391–415
    properties, 391–393
  seismology, 393, 415
  sinistral, 59, 393–404
  spine, 59–60
  stability, 395
  structure, 59–61
  thread, 57–61, 391, 393, 407–409, 413–415
  tornado, 23, 61, 62
  winking, 62
prominence model
  dextral and sinistral, 399, 400
  flux-rope, 396–398, 402–406
  Kippenhahn-Schlüter, 139, 392–396
  Kuperus-Raadu, 396
  van Ballegooijen-Cranmer, 408
  van Ballegooijen-Martens, 396
proton mass, 487
pseudo-streamer, 37, 231, 468–474
  3D topology, 471–474

QPP (quasi-periodic pulsations), 176
QSL (quasi-separatrix layer), 71
quadrupolar magnetic field, 202, 203, 253
quasi- (or collective) mode, 168, 170–172, 360
quasi-separator (or hyperbolic flux tube), 191, 193, 224–232, 245, 254–255, 417, 438, 439, 444, 447–449, 469
  reconnection, 232, 254–255
quasi-separatrix layer (QSL), 193, 224, 232–233, 254, 373–380, 468, 469
quasi-skeleton, 224, 232–233
quasi-topology, 229, 232
quiescent prominence, 57–63, 391–415
  properties, 391–393
quiet Sun, 4

radiation, 84–85
  time, 351
radiative equilibrium, 11
radiative flux, 84
radiative instability, 407–415
radiative loss function, 84, 85

radiative nonequilibrium, 391, 409–415
radiative relaxation time-scale, 84
radiative zone, 5, 282, 300
radio burst
  type II, 65, 70, 72, 177
  type III, 70–72
  type IV, 70
radius of Sun, 4–5
rain (coronal), 32, 46, 409, 415
Rankine-Hugoniot relations, 181
rapid blue-shift event, 33
ratio of specific heats, 82
rational flux surface, 121, 135, 223
rationalised mks units, 484–485
ray path, 358
Rayleigh criterion, 274
Rayleigh number, 90, 278–279, 307–308
Rayleigh-Taylor instability, 260–264, 271–272, 322
  magnetic (or Kruskal-Schwarzschild), 263, 272, 319, 415
reconnection, 106, 189–255, 439–450
  2D, 241–242
    almost-uniform, 190, 212–217
    fast, 211, 212, 214, 215, 217, 218, 224
    flux pile-up, 216, 217, 224
    impulsive bursty, 217–224
    non-uniform, 214, 215, 218–219
    Petschek, 190, 211–214, 216–218, 224
    Petschek (nonsteady), 214
    properties, 242
    standard CSHKP flare model, 439–442
    Sweet-Parker, 190, 208–212
    Sweet-Parker (compressible), 210
    Sweet-Parker (energetics), 211
    unsteady, 219–224
  3D, 191, 192, 224–255
    classification, 243–247
    concepts, 240–247
    conditions, 241
    definition, 243–247
    differences from 2D, 192, 242–243
    fan, 208, 247–250
    finite-B (or non-null), 246–247
    flare model, 442–444
    flux velocity, 241–242
    general, 190, 244, 246
    magnetic helicity, 247
    null point, 247–252
    patchy, 449–450
    properties, 242
    quasi-separator (or QSL), 232, 254–255
    quasi-separator flare model, 447–449
    regimes, 193, 247–255
    separator, 245–253
    separator flare model, 444–447
    slip-running (or quasi-separator or QSL), 232, 254–255, 449
    spine, 208, 247–250
    spine-fan, 251–252
    torsional fan, 251
    torsional spine, 250–251
  binary, 380–383
  collisionless, 190
  concepts, 191–193
  conditions for existence, 192
  coronal heating, 336, 364–386
  creation of flare loop, 439–450
  driven, 215, 244
  electric field, 378–379, 421, 441–443
  energy release, 378–379
  evidence from space, 71–73, 367–369
  fly-by, 380–383
  forced, 201–202
  history, 190–191
  impulsive bursty, 65, 442
  instability, 219–224
  interchange, 468, 474
  jet, 32, 73
  numerical experiment, 218, 249–250
  rate, 208
    2D, 195, 196
    3D, 191, 244–252
    collisionless, 190
    external, 211–217
    fast, 216–218
    in flare (Forbes-Priest formula), 441, 443
    Petschek, 211, 216
    Sweet-Parker, 190, 208–210
  recursive (or multiple), 253, 374, 382
  separator, 380, 469, 473
  singular, 246
  spicules, 316–317
  spontaneous, 215, 244, 247
  stick-slip, 379, 380
  switch-on, 383
  tether cutting, 419, 423, 435
reconnective annihilation, 207–208
recursive (or multiple) reconnection, 253, 374, 382
recycling time (coronal and chromospheric), 337
reduced MHD, 380, 389, 390
relative magnetic helicity, 235
relaxation
  intermediate, 386
  magnetic, 133, 201, 234, 373, 374, 381, 386–389
  magnetofrictional, 401, 406
  MHD turbulence, 386–389
  partial, 386
remnant active region, 25–30, 45–46, 62, 392–404
resistive instability, 219–224, 275–278
  ballooning mode, 223
  gravitational mode, 276
  interchange (or resistive-g), 223
  kink (or cylindrical tearing-mode), 223, 434–435
  rippling mode, 277
resistivity (electrical), 78
resonant (or mode-rational) surface, 201, 223, 247, 434
resonant absorption, 358–361
  footpoint driving, 360–361
  sideways driving, 359–360
restoring force, 144
reversed-field pinch, 124, 234
Reynolds number, 89
Reynolds stress, 15, 300–302
ribbon (Hα), 417, 442, 448
ribbon (Hα), 63–71, 417, 439, 441, 443–445, 447, 449
  velocity, 441–443, 450
  width, 442
Richardson number, 275
rippling resistive mode, 220
Rossby number, 90, 303
Rosseland mean opacity, 11
rotating reference frame (eqn. of motion), 81
rotation period of Sun, 5

rotational discontinuity, 41, 188
rotational slippage, 250–251
RTV scaling law (coronal loops), 345–353
running penumbral wave, 52, 144, 329
Rutherford scaling, 224

S-Web model for solar wind, 469
SAD (supra-arcade downflow), 71–73, 220, 449
SADL (supra-arcade downflowing loop), 73
safety factor, 223
satellite sunspot, 49, 71
sausage instability, 263, 271, 273
sausage wave mode, 163–167
scalar potential (magnetic), 108, 117
scale-height, 108–110, 137, 157, 394, 404, 407, 408, 487
  modified, 110
scattering, 34
  Thomson, 64
Schwarzschild criterion, 13, 153
SDO (Solar Dynamics Observatory), 4, 25
  AIA (Atmospheric Imaging Assembly), 9, 23, 63, 175, 340–342, 369
  HMI (Helioseismic and Magnetic Imager), 9, 25
sector boundary, 42
seismology
  coronal, 172–176, 357
  prominence, 393, 415
self-helicity, 233–247
self-inductance, 427
self-organised critical (SOC) state, 389, 390
self-similar (or similarity) solution, 190, 194–196, 202, 323, 330, 428
separator, 71, 225–228, 252, 374–386, 417, 435–447, 469, 473
  closed, 231, 474
  coronal heating, 379–380
  current sheet, 377–380
  double, 227, 228
  intercluster and intracluster, 229
  open, 231, 468, 473–474
  reconnection, 245–253, 380, 469, 473
  single, 227
separatrix, 71, 191, 200–203, 226–228
  bald patch, 226–231
  closed, 227–231, 472
  current sheet, 379
  curtain, 230–231, 469–474
  dome, 228, 230, 231, 469, 471–474
  fan, 226, 230, 252
  heteroclinic, 229
  homoclinic, 229
  in flare, 442
  jet, 219
  open, 227–231, 472
  streamer, 231, 469–474
Shafranov shift, 426
shear Alfvén wave, 149–151, 167, 175
sheared magnetic field, 201–203
shock rebound model, 316
shock wave, 105, 177–188, 423
  collisionless, 181
  conservation (or jump) relations, 178–186
  fast-mode, 70–72, 179, 186–188
  fast-mode termination, 417–441
  formation, 177–181
  hydrodynamic, 181–182
  magnetic, 179–181
  oblique, 180, 184–188
  perpendicular, 180, 182–184
  slow-mode, 70, 72, 179, 186–188, 211–219, 417, 440–450
  strong, 177
  switch-off, 180, 187–188, 212–213
  switch-on, 180, 187–188
  weak, 177
  width, 178–180
sidereal period, 14
sigmoid, 61, 72, 419, 438
sine-Poisson equation, 130
single separator, 227
singular field line, 232
  reconnection, 246
singular layer, 222–224
singularity
  current sheet, 249
  current, 195–198
  flux velocity, 192, 241–243
  in sheared field, 201
  Jacobian, 229
  velocity, 192, 241, 248
sinistral filament channel, 61
sinistral prominence or filament, 59, 393–404
siphon flow, 49, 104, 341, 349–350
skeleton, 224, 226, 229–231, 245, 337, 338, 380–383
  quasi-, 232–233
  structural, 224
slab wave, 162–164
slender (or thin) flux tube, 313–317
  equations, 315
  width, 314
slip-running (or quasi-separator or QSL) reconnection, 232, 254–255, 449
slippage
  magnetic, 192, 240–251
  rotational, 250–251
  velocity, 238, 241, 244
slow (or cusp or tube) speed, 155, 161
slow (or cusp) continuum, 163, 168
slow (or cusp) frequency, 160, 167
slow magneto-acoustic-gravity (MAG) wave, 328, 329
slow magnetoacoustic wave, 155–160, 168
  body mode, 163, 166
  kink mode, 163, 174
  sausage mode, 163, 164, 174
slow solar wind, 40–43
slow surface wave, 167
slow-mode
  compression, 216
  expansion, 216
  shock wave, 70, 72, 179, 186–188, 211–219, 417, 440–450
  wave, *see* slow magnetoacoustic wave
soft X-rays, 67, 71, 417–450
  coronal, 3, 34–37, 42, 56
soft-hard-soft evolution, 67
SoHO (Solar and Heliospheric Observatory), 4, 17, 19
  EIT (Extreme ultraviolet Imaging Telescope), 43, 340–368
  LASCO (Large Angle and Spectrometric Coronagraph), 43, 64
  MDI (Michelson Doppler Imager), 18–22, 337, 338, 355, 356
  SUMER (Solar Ultraviolet Measurements of Emitted Radiation), 40, 44, 365, 368
  UVCS (Ultraviolet Coronagraph Spectrometer), 40, 44
solar atmosphere, 6–7

solar constant, 3
solar core, 10
solar cycle, 2, 44–45, 53–57, 281, 305
solar eclipse, 1–9, 36–42, 56–57, 335, 339
solar flare, 2–4, 63, 65–73, 219, 416–450
  Masuda, 67
  Carrington, 2, 66
  classification, 66
  compact (or confined), 66, 433, 437, 448, 449
  electron-rich, 70
  emerging (or evolving) flux model, 369, 437–438
  energy release, 68–71
  eruptive, 45, 63, 66, 68
  flash phase, 66
  gamma-ray, 70
  gradual, 66, 67
  hard X-rays, 66–70, 439–450
  homologous, 69
  impulsive, 66, 67
  impulsive phase, 66–70
  intensity, 66, 67
  loop, 66–68, 70–73, 417, 439–450
    shrinkage, 442
  loop-top source, 67–73
  magnetic field, 69
  main phase, 66–72, 417, 439
  oscillation, 175–176
  phases, 66–69
  pre-flare phase, 64–72
  reconnection, 71–73, 439–450
    3D model, 442–444
    3D patchy, 449–450
    3D quasi-separator model, 447–449
    3D separator model, 444–447
    standard CSHKP model, 439–442
  rise phase, 66, 68, 70
  soft X-rays, 66–71, 417–450
  soft-hard-soft evolution, 67
  sympathetic, 69
  thick-target model, 67, 70
  trigger, 417
  two-ribbon, 45–66, 416–445
solar interior, 5–6, 10–20
solar irradiance, 3, 56, 69
solar magnetic field, 25–30
solar maximum, 35–58
solar minimum, 35–56
solar neutrino problem, 10, 17
solar wind, 7, 40–44, 451–481
  2D Pneuman-Kopp model, 470–471
  2D separable solution, 471
  breeze solution, 454–457
    stability, 455–456
  coronal hole model, 464–465
  energy equation, 458–460
  exospheric model, 478, 479
  fast, 36–44, 451, 452, 459, 469, 474, 475, 477
    wave-turbulence model, 466–467
  kinetic model, 478–480
  magnetic field, 41, 43, 460–464
  mass-loss rate, 454, 459, 460, 464
  non-MHD effects, 475–480
  Parker isothermal, 453–455
  polytropic, 457–458
  properties, 451, 452, 477
  rotating, 460–464, 471
  S-Web, 469
  slow, 40–43, 451, 459, 460, 466, 467, 477
    reconnection model, 468–469
  temperature, 452
  time-dependence, 475
  turbulence, 466–467, 475, 480
  two-fluid model, 476–478
  velocity, 451
Sonnerup-Priest model, 205–207
SOT (Solar Optical Telescope), 27–61
sound speed, 173, 487
sound wave, 144, 147, 156, 177
Spörer's law, 2, 54, 282
space observations
  reconnection, 71–73, 367–369
  solar flares, 71–73
space weather, 63
specific heat, 82
spectral line, 2–3, 7–8
spectroheliograph, 2
spectrum
  continuous, 159, 161, 167–171
  discrete, 159, 168–170
  Kolmogorov, 387, 466, 475
  Kraichnan, 387, 467, 475
speed
  Alfvén, 173
  cusp (or slow or tube), 155
  free-fall, 107
  kink, 166, 173, 175
  light, 75, 487
  slow (or cusp or tube), 155, 161
  sound, 173
  tube (or cusp or slow), 155, 161, 315, 316
spherical polar coordinates, 489
spicule, 2–3, 23, 31, 32, 38, 44, 349, 358, 363, 364, 390
  generation, 316–317
  tornado, 23, 31
  twisting, 317
  type I, 31–33, 355
  type II, 31, 33, 176, 317, 337, 355, 357, 363
spin-down time of Sun, 6
spine, 225
  filament, 59–60
  penumbral, 51
  prominence, 59–60
  reconnection, 208, 247–250
spine-fan reconnection, 251–252
spontaneous reconnection, 215, 244, 247
squashing factor (Q), 229, 233, 254, 373, 447
SST (Swedish Solar Telescope), 23–33, 48–50, 60, 407
stability
  meta-, 256–257
  neutral (or marginal), 256–258, 279
  nonlinear, 256–257
stagnation-point flow, 205–207, 310, 314
standard solar model, 10–12
Stefan-Boltzmann constant, 11, 84
Stefan-Boltzmann law, 11
stellar wind, 459
STEREO (Solar TErrestrial RElations Observatory), 65
stochastic field line, 121, 135
strand of coronal loop, 341–372
strapping field, 428
straw (chromospheric), 33

streamer
  3D topology, 471–474
  bipolar, 36, 472
  coronal, 36–44, 56, 63
  helmet, 36–43, 61
    2D model, 470–471
  pseudo-, 37
  separatrix, 231
  unipolar, 37, 472
structural (or topological) instability, 203, 218, 229
structural skeleton, 224
SUMER (Solar Ultraviolet Measurements of Emitted Radiation), 40, 44, 365, 368
Sun
  age, 4, 488
  angular momentum, 5, 488
  density, 4–5
  distance to Earth, 1–2, 4, 488
  effective temperature, 5
  equatorial rotation speed, 488
  escape velocity, 4, 488
  luminosity, 5, 488
  mass, 4–5, 488
  mass-loss rate, 5, 488
  radius, 4–5, 488
  rotational period, 5
  surface gravity, 4–5, 488
Sunrise, 22–28
sunspot, 1–3, 44–45, 47–53, 104, 306–333
  bands, 25, 30
  computational model, 330–333
  cooling, 312–313
  cycle, 2, 30, 37–69
  decay, 329–330
  delta, 48
  Evershed flow, 2, 47, 51, 143, 349–350
  evolution, 52–53
  flows, 52
  flux tube model, 323, 324
  following, 45, 71
  formation, 329
  group, 48
    bipolar, 48, 52, 69
    complex, 48
  intensity, 48–49
  leader, 45
  light bridge, 49, 53
  loop, 49
  magnetic field, 131
  maximum, 35–58
  minimum, 35–56
  moat, 20, 52–53, 325–333
  moving magnetic feature (MMF), 52, 53
  oscillation, 328–329
  penumbra, 47–49, 51–52, 326–328
  plume, 44, 49
  polarity laws, 54
  preceding, 45, 71
  proper motion, 52
  satellite, 49, 71
  seismology, 357
  spaghetti model, 324
  stability, 324–325
  structure, 322–325
  umbra, 47–51, 325–326
  umbral flash, 51
supergranule, 3, 14–46
  overlying magnetic field, 338–339
superpenumbral fibril, 33, 46–52
supra-arcade downflow (SAD), 71–73, 220
supra-arcade downflowing loop (SADL), 73
surface wave, 160–164, 170, 171
  Alfvén, 164
  gravity, 153
surge, 46, 49, 349–350, 409–410
Suydam's criterion, 274
Swedish Solar Telescope (SST), 23–33, 48–50, 60, 317, 407
Sweet-Parker model, 190, 208–212
  compressibility, 210
  energetics, 211
switch-off shock, 180, 187–188, 212–213
switch-on shock, 180, 187–188
sympathetic solar flare, 69
synodic period, 5, 14

T-point, 200, 472
tachocline, 5, 15–19, 56, 286–287, 294–304
tangential discontinuity, 41, 105, 187
tangled magnetic field, 408–409
Taylor convection, 304
Taylor hypothesis, 123–125
Taylor number, 90, 278, 301
Taylor relaxation, *see* relaxation magnetic
Taylor-Proudman theorem, 18, 99
tearing-mode instability, 190, 197, 220–223, 277, 278, 363, 367, 383
  nonlinear, 223–224
tectonics, 334–390
  chromosphere, 334, 337, 353, 367, 374–386
  coronal heating, 337, 353, 367, 374–386
  current sheet
    dissipation, 383–386
    formation, 366–367
temperature
  active region, 46
  coronal, 37, 39, 46
  coronal hole, 39, 40
  flare loop, 71
  minimum, 30
  penumbra, 51
  prominence, 57
  solar flare, 68
  solar interior, 10
  solar wind, 41
  spicule, 31
  umbral dot, 49
  VAL model, 31
tensor
  thermal (or heat) conduction, 83
  viscous stress, 81
termination shock, 417–441
theorem
  Alfvén's frozen flux, 92, 98, 237, 240
  Cowling's anti-dynamo, 3, 99, 287–289
  flux velocity, 241
  force-free fields, 121
  force-free minimum-energy, 98, 121
  frozen-flux, 92
  Lie-derivative, 240
  Liouville's, 246
  potential field minimum-energy, 98, 117

theorem (*cont.*)
  Taylor-Proudman, 18, 99
  virial, 99–100, 121
  vorticity, 93
  Woltjer minimum-energy, 98, 123
thermal conduction, 83–87, 334–353, 417, 441, 450, 453, 458, 459, 475
thermal conductivity, 487
thermal diffusion time, 6
thermal diffusivity, 84
thermal equilibration time, 6
thermal equilibrium, 88, 343, 411
thermal expansion coefficient, 307
thermal instability, 258, 279–280
thermal nonequilibrium, 347, 391, 409–415
thermal relaxation time, 6
thermal speed, 487
thermal structure
  formation, 409–415
thick-target model of solar flare, 67, 70
Thomson scattering, 64
thread
  formation, 407, 413
  prominence, 57–61, 391, 393, 407–409, 413–415
three-minute oscillation, 50–52, 175
time
  anomalous collision, 80
  collision, 80
  correlation, 292, 304
  electron collision, 79, 83
  electron-ion collision, 77
  ion-ion collision, 83
  turbulent collision, 80
  turnover, 21, 23, 311
time-scale
  acoustic travel, 6–20
  Alfvén travel, 6
  conduction, 351
  enthalpy, 351
  gravitational, 351
  Kelvin-Helmholtz, 5
  magnetic diffusion, 6
  photon travel, 6
  radiation, 351
  radiative relaxation, 84
  spin-down, 6
  thermal diffusion, 6
  thermal equilibration, 6
  thermal relaxation, 6
  viscous diffusion, 6
Titov-Démoulin model, 431–433
tokamak, 124, 136
topological (or structural) instability, 203, 218, 229
topology, 92, 191, 203
  building blocks, 231
  conservation, 191, 200, 237
  conservation laws, 240
  degree, 234, 374
  entropy, 234, 374
  flux function, 234, 374
  global invariant, 233
  instability, 229, 244
  linking number, 236
  magnetic charge, 230
  null pair, 227, 228
  skeleton, 224, 226, 229–231, 245, 337, 338, 380–383
  structures, 224–233
  three-source, 230
tornado, 336, 368
  barb, 23, 59
  macrospicule, 23–33, 60
  prominence, 23, 61, 62
  spicule, 23, 31
toroidal current ring (magnetic field), 425–426
toroidal flux rope
  hoop force, 424–427
  instability, 428–429
  nonequilibrium, 429–431
  Titov-Démoulin equilibrium, 431–433
toroidal magnetic field, 223–224, 274
toroidal-poloidal decomposition, 108, 126
torque on plasma, 116
torsional Alfvén wave, 149–151, 163–175
torsional fan reconnection, 251
torsional oscillation, 19, 56, 301
torsional spine reconnection, 250–251
torus instability, 427–431
total electric field, 77, 92
total pressure, 110–112, 137–138
TRACE (Transition Region and Coronal Explorer), 150, 151, 166–168, 174, 340–375
transient brightenings
  active-region, 46
transient coronal hole, 39, 64
transition region, 7–8, 31–34, 39, 57, 334–351, 355
  model, 342–351
  pressure, 7, 349
transverse wave, 147, 149
tube (or cusp or slow) speed, 155, 161, 315, 316
tube wave, 163
turbulence
  Alfvén, 389
  cascade, 355, 362, 371, 387, 388
  collision-time, 80
  convective, 299, 302
  coronal heating, 367, 386–390
  gyrotropic, 293
  helical, 286, 289, 296, 297
  isotropic, 292
  magnetic diffusivity, 190, 215, 219
  MHD, 284, 291, 304–305
  pseudo-isotropic (non-mirror symmetric), 291, 292
  pumping, 298
  solar wind, 466–467, 475, 480
turnover-time, 21, 23, 311
twist, 111–116
twisted flux tube, 391–415
two-ribbon flare, 45–66, 416–445
type II radio burst, 65, 70, 72, 177
type III radio burst, 70–72
type IV radio burst, 70

Ulysses, 40, 42–44, 467, 474, 475
umbral convection, 325
umbral dot, 49, 312, 313, 325, 326, 331, 333
  central, 49
  intensity, 49
  magnetic field, 49
  peripheral, 49, 51
umbral flash, 51
umbral granulation, 49

umbral magnetic field, 48
umbral oscillation, 50
uniform-twist field, 114–116, 268, 269
unipolar region, 14, 25–39, 62
unipolar streamer, 37
units, 483–486
  Gaussian cgs, 483–484
  rationalised mks, 484–485
UVCS (Ultraviolet Coronagraph Spectrometer), 40, 44

vacuum (or potential or current-free) magnetic field, 117–120, 311, 324–325
VAL (Vernazza Avrett Loeser) model, 7, 30, 77, 79
van Ballegooijen-Cranmer prominence model, 408
van Ballegooijen-Martens filament-channel model, 400
van Ballegooijen-Martens prominence model, 396, 415
variable-twist field, 116
variational method, 133, 134
vector potential
  magnetic, 108
  uniqueness, 108
velocity
  dual magnetic flux, 192, 240, 243
  group, 147, 156
  magnetic field line, 237, 239
  magnetic flux, 92, 95, 237–243
  phase, 147, 156
  slippage, 238, 241, 244
vertical integration method, 133
virial theorem, 99–100, 121
virtual flow, 243
viscous coefficient, 81
viscous diffusion, 6, 90
viscous dissipation, 85, 86
viscous force, 80, 81, 90
viscous stress tensor, 81
Vlasov-Boltzmann equation, 479
vortex tubes, 23, 28, 59, 61
vorticity, 81
  analogy with magnetic field, 91, 93
  equation, 91
  theorem, 93

wave, 144–176
  slow sausage, 163
  slow surface, 167
  acoustic gravity, 144, 157
  Alfvén, 144, 148–176, 357–358, 360
    kinetic, 336
    phase mixing, 361
    resonant absorption, 359
    torsional, 382
  Alfvénic, 165–174, 357–364
  body, 160–166
  circularly polarised, 150, 151, 154
  compression, 178
  compressional Alfvén, 144–151, 159–161
  coronal heating, 356–364
  damping length, 358
  dispersive, 147
  dynamo, 285–299
  EIT, 70
  energy flux, 358, 363, 364
  entropy, 156
  expansion, 178
  fast body, 163
  fast kink, 163–175
  fast magnetoacoustic, 151–173
  fast sausage, 166, 173–175
  fast surface, 166
  fast-mode, *see* fast magnetoacoustic wave
  flute (or ballooning), 166
  flux-tube, 164–167
  helioseismic, 20, 71
  hydromagnetic inertial, 155
  in non-uniform medium, 159–167
    continuous planar field, 160–161
    continuous twisted flux tube, 164–165
    magnetic interface, 161–162
    magnetic slab, 162–164
    uniform untwisted flux tube, 165–167
  inertial, 144, 153–155
  interface, 161–162
  intermediate (or Alfvén), 155, 179–188
  internal gravity, 144, 152–159, 172
  kink, 162–167, 357, 363
  longitudinal, 147
  magneto-gravity, 159
  magnetoacoustic, 144, 155–159
  magnetoacoustic gravity (or magneto-atmospheric), 144, 157–159
  magnetoacoustic surface, 161
  Moreton, 70, 144
  propagation, 357–358
  running penumbral, 52, 144, 329
  sausage, 163–167
  shear Alfvén, 149–151, 167, 175
  shock, 177–188
  slab, 162–164
  slow body, 163, 166
  slow kink, 163, 174
  slow magneto-acoustic-gravity (MAG), 328, 329
  slow magnetoacoustic, 155–160, 168
  slow sausage, 164, 174
  slow-mode, *see* slow magnetoacoustic wave
  sound, 144, 147, 156, 177
  surface, 160–164, 170, 171
  surface gravity, 153
  torsional Alfvén, 149–151, 163–175
  transverse, 147, 149
  tube, 163
wave equation (generalised), 146
wave frequency, 146
waveguide law, 17
wavenumber, 146
Wilson effect, 47, 323–324, 331
Wind spacecraft, 43
winking prominence, 62
WKB approximation, 146, 159
Wolf number, 53
Woltjer minimum-energy theorem, 98, 123
writhing number, 236

X-point, 97, 98, 106, 190, 193–203, 215, 226–232, 242–244, 253–255, 402, 403
  instability, 98, 194
X-ray bright point, 3, 26–45, 199, 370
  converging flux model, 369–370
X-ray jet, 38, 67, 321, 322, 366, 410

X-rays
  hard, 67, 439–450
  soft, 67, 71, 417–450
  soft coronal, 3, 34–37, 42, 56
X-type null point, *see* X-point
XRT (X-ray Telescope), 44, 340, 341, 368

Y-point, 197–203, 472
Yohkoh, 3, 35, 36, 38, 70, 72, 73, 340, 352–354, 368

Zeeman effect, 26, 36, 47, 57
Zweibel-Hundhausen solution, 137, 142

For EU product safety concerns, contact us at Calle de José Abascal, 56–1°,
28003 Madrid, Spain or eugpsr@cambridge.org.

www.ingramcontent.com/pod-product-compliance
Lightning Source LLC
LaVergne TN
LVHW062126200926
848484LV00007B/1770
* 9 7 8 0 5 2 1 8 5 4 7 1 9 *